Lecture Notes in Computer Science 3514

Commenced Publication in 1973
Founding and Former Series Editors:
Gerhard Goos, Juris Hartmanis, and Jan van Leeuwen

Editorial Board

David Hutchison
   Lancaster University, UK
Takeo Kanade
   Carnegie Mellon University, Pittsburgh, PA, USA
Josef Kittler
   University of Surrey, Guildford, UK
Jon M. Kleinberg
   Cornell University, Ithaca, NY, USA
Friedemann Mattern
   ETH Zurich, Switzerland
John C. Mitchell
   Stanford University, CA, USA
Moni Naor
   Weizmann Institute of Science, Rehovot, Israel
Oscar Nierstrasz
   University of Bern, Switzerland
C. Pandu Rangan
   Indian Institute of Technology, Madras, India
Bernhard Steffen
   University of Dortmund, Germany
Madhu Sudan
   Massachusetts Institute of Technology, MA, USA
Demetri Terzopoulos
   New York University, NY, USA
Doug Tygar
   University of California, Berkeley, CA, USA
Moshe Y. Vardi
   Rice University, Houston, TX, USA
Gerhard Weikum
   Max-Planck Institute of Computer Science, Saarbruecken, Germany

Vaidy S. Sunderam   Geert Dick van Albada
Peter M.A. Sloot   Jack J. Dongarra (Eds.)

Computational Science – ICCS 2005

5th International Conference
Atlanta, GA, USA, May 22-25, 2005
Proceedings, Part I

 Springer

Volume Editors

Vaidy S. Sunderam
Emory University
Dept. of Math and Computer Science
400 Dowman Dr W430, Atlanta, GA 30322, USA
E-mail: vss@mathcs.emory.edu

Geert Dick van Albada
University of Amsterdam
Department of Mathematics and Computer Science
Kruislaan 403, 1098 SJ Amsterdam, The Netherlands
E-mail: dick@science.uva.nl

Peter M.A. Sloot
University of Amsterdam
Department of Mathematics and Computer Science
Kruislaan 403, 1098 SJ Amsterdam, The Netherlands
E-mail: sloot@science.uva.nl

Jack J. Dongarra
University of Tennessee
Computer Science Department
1122 Volunteer Blvd., Knoxville, TN 37996-3450, USA
E-mail: dongarra@cs.utk.edu

Library of Congress Control Number: 2005925759

CR Subject Classification (1998): D, F, G, H, I, J, C.2-3

ISSN   0302-9743
ISBN-10   3-540-26032-3 Springer Berlin Heidelberg New York
ISBN-13   978-3-540-26032-5 Springer Berlin Heidelberg New York

This work is subject to copyright. All rights are reserved, whether the whole or part of the material is concerned, specifically the rights of translation, reprinting, re-use of illustrations, recitation, broadcasting, reproduction on microfilms or in any other way, and storage in data banks. Duplication of this publication or parts thereof is permitted only under the provisions of the German Copyright Law of September 9, 1965, in its current version, and permission for use must always be obtained from Springer. Violations are liable to prosecution under the German Copyright Law.

Springer is a part of Springer Science+Business Media

springeronline.com

© Springer-Verlag Berlin Heidelberg 2005
Printed in Germany

Typesetting: Camera-ready by author, data conversion by Scientific Publishing Services, Chennai, India
Printed on acid-free paper   SPIN: 11428831   06/3142   5 4 3 2 1 0

Preface

The Fifth International Conference on Computational Science (ICCS 2005) held in Atlanta, Georgia, USA, May 22–25, 2005, continued in the tradition of previous conferences in the series: ICCS 2004 in Krakow, Poland; ICCS 2003 held simultaneously at two locations, in Melbourne, Australia and St. Petersburg, Russia; ICCS 2002 in Amsterdam, The Netherlands; and ICCS 2001 in San Francisco, California, USA.

Computational science is rapidly maturing as a mainstream discipline. It is central to an ever-expanding variety of fields in which computational methods and tools enable new discoveries with greater accuracy and speed. ICCS 2005 was organized as a forum for scientists from the core disciplines of computational science and numerous application areas to discuss and exchange ideas, results, and future directions. ICCS participants included researchers from many application domains, including those interested in advanced computational methods for physics, chemistry, life sciences, engineering, economics and finance, arts and humanities, as well as computer system vendors and software developers. The primary objectives of this conference were to discuss problems and solutions in all areas, to identify new issues, to shape future directions of research, and to help users apply various advanced computational techniques. The event highlighted recent developments in algorithms, computational kernels, next generation computing systems, tools, advanced numerical methods, data-driven systems, and emerging application fields, such as complex systems, finance, bioinformatics, computational aspects of wireless and mobile networks, graphics, and hybrid computation. Keynote lectures were delivered by John Drake – High End Simulation of the Climate and Development of Earth System Models; Marian Bubak – Towards Knowledge – Based Computing: Personal Views on Recent Developments in Computational Science and the CrossGrid Project; Alok Choudhary – Large Scale Scientific Data Management; and David Keyes – Large Scale Scientific Discovery through Advanced Computing.

In addition, four invited presentations were delivered by representatives of industry: David Barkai from Intel Corporation, Mladen Karcic from IBM, Steve Modica from SGI and Dan Fay from Microsoft. Seven tutorials preceded the main technical program of the conference: Tools for Program Analysis in Computational Science by Dieter Kranzlmüller and Andreas Knüpfer; Computer Graphics and Geometric Modeling by Andrés Iglesias; Component Software for High Performance Computing Using the CCA by David Bernholdt; Computational Domains for Explorations in Nanoscience and Technology, by Jun Ni, Deepak Srivastava, Shaoping Xiao and M. Meyyappan; Wireless and Mobile Communications by Tae-Jin Lee and Hyunseung Choo; Biomedical Literature Mining and Its Applications in Bioinformatics by Tony Hu; and Alternative Approaches to

Grids and Metacomputing by Gunther Stuer. We would like to thank all keynote, invited and tutorial speakers for their interesting and inspiring talks.

Aside from the plenary lectures, the conference included 10 parallel oral sessions and 3 poster sessions. Ever since the first meeting in San Francisco, ICCS has attracted an increasing number of researchers involved in the challenging field of computational science. For ICCS 2005, we received 464 contributions for the main track and over 370 contributions for 24 originally-proposed workshops. Of these submissions, 134 were accepted as full papers accompanied by oral presentations, and 89 for posters in the main track, while 241 papers were accepted for presentations at 21 workshops. This selection was possible thanks to the hard work of the 88-member Program Committee and 362 reviewers. The author index contains 1395 names, and over 500 participants from 41 countries and all continents attended the conference. The ICCS 2005 proceedings consists of three volumes. The first volume, LNCS 3514, contains the full papers from the main track of the conference, while volumes 3515 and 3516 contain the papers accepted for the workshops and short papers. The papers cover a wide range of topics in computational science, ranging from numerical methods, algorithms, and computational kernels to programming environments, grids, networking and tools. These contributions, which address foundational and computer science aspects are complemented by papers discussing computational applications in a variety of domains. ICCS continues its tradition of printed proceedings, augmented by CD-ROM versions for the conference participants. We would like to thank Springer for their cooperation and partnership. We hope that the ICCS 2005 proceedings will serve as a major intellectual resource for computational science researchers for many years to come. During the conference the best papers from the main track and workshops as well as the best posters were nominated and commended on the ICCS 2005 Website. A number of papers will also be published in special issues of selected journals.

We owe thanks to all workshop organizers and members of the Program Committee for their diligent work, which led to the very high quality of the event. We would like to express our gratitude to Emory University and Emory College in general, and the Department of Mathematics and Computer Science in particular, for their wholehearted support of ICCS 2005. We are indebted to all the members of the Local Organizing Committee for their enthusiastic work towards the success of ICCS 2005, and to numerous colleagues from various Emory University units for their help in different aspects of organization. We very much appreciate the help of Emory University students during the conference. We owe special thanks to our corporate sponsors: Intel, IBM, Microsoft Research, SGI, and Springer; and to ICIS, Math & Computer Science, Emory College, the Provost's Office, and the Graduate School at Emory University for their generous support. ICCS 2005 was organized by the Distributed Computing Laboratory at the Department of Mathematics and Computer Science at Emory University, with support from the Innovative Computing Laboratory at the University of Tennessee and the Computational Science Section at the University of Amsterdam, in cooperation with the Society for Industrial and Applied Mathematics (SIAM). We invite you to visit the ICCS 2005 Website (http://www.iccs-meeting.org/ICCS2005/) to recount the events leading up to the conference, to view the technical program, and to recall memories of three and a half days of engagement in the interest of fostering and advancing computational science.

June 2005
Vaidy Sunderam
on behalf of
G. Dick van Albada
Jack J. Dongarra
Peter M.A. Sloot

Organization

ICCS 2005 was organized by the Distributed Computing Laboratory, Department of Mathematics and Computer Science, Emory University, Atlanta, GA, USA, in cooperation with Emory College, Emory University (USA), the University of Tennessee (USA), the University of Amsterdam (The Netherlands), and the Society for Industrial and Applied Mathematics (SIAM). The conference took place on the campus of Emory University, in Atlanta, Georgia, USA.

Conference Chairs

Scientific Chair – Vaidy Sunderam (Emory University, USA)
Workshops Chair – Dick van Albada (University of Amsterdam, The Netherlands)
ICCS Series Overall Chair – Peter M.A. Sloot (University of Amsterdam, The Netherlands)
ICCS Series Overall Co-Chair – Jack Dongarra (University of Tennessee, USA)

Local Organizing Committee

Dawid Kurzyniec (Chair)
Piotr Wendykier
Jeri Sandlin
Erin Nagle
Ann Dasher
Sherry Ebrahimi

Sponsoring Institutions

Intel Corporation
IBM Corporation
Microsoft Research
SGI Silicon Graphics Inc.
Emory University, Department of Mathematics and Computer Science
Emory University, Institute for Comparative and International Studies
Emory University, Emory College
Emory University, Office of the Provost
Emory University, Graduate School of Arts and Sciences
Springer

Program Committee

Jemal Abawajy, Deakin University, Australia
David Abramson, Monash University, Australia
Dick van Albada, University of Amsterdam, The Netherlands
Vassil Alexandrov, University of Reading, UK
Srinivas Aluru, Iowa State University, USA
Brian d'Auriol, University of Texas at El Paso, USA
David A. Bader, University of New Mexico, USA
Saeid Belkasim, Georgia State University, USA
Anne Benoit, University of Edinburgh, UK
Michele Benzi, Emory University, USA
Rod Blais, University of Calgary, Canada
Alexander Bogdanov, Institute for High Performance Computing and Information Systems, Russia
Anu Bourgeois, Georgia State University, USA
Jan Broeckhove, University of Antwerp, Belgium
Marian Bubak, Institute of Computer Science and ACC Cyfronet – AGH, Poland
Rajkumar Buyya, University of Melbourne, Australia
Tiziana Calamoneri, University of Rome "La Sapienza", Italy
Serge Chaumette, University of Bordeaux, France
Toni Cortes, Universitat Politecnica de Catalunya, Spain
Yiannis Cotronis, University of Athens, Greece
Jose C. Cunha, New University of Lisbon, Portugal
Pawel Czarnul, Gdansk University of Technology, Poland
Frederic Desprez, INRIA, France
Tom Dhaene, University of Antwerp, Belgium
Hassan Diab, American University of Beirut, Lebanon
Beniamino Di Martino, Second University of Naples, Italy
Jack Dongarra, University of Tennessee, USA
Craig Douglas, University of Kentucky, USA
Edgar Gabriel, University of Stuttgart, Germany
Marina Gavrilova, University of Calgary, Canada
Michael Gerndt, Technical University of Munich, Germany
Yuriy Gorbachev, Institute for High Performance Computing and Information Systems, Russia
Andrzej Goscinski, Deakin University, Australia
Eldad Haber, Emory University, USA
Ladislav Hluchy, Slovak Academy of Science, Slovakia
Alfons Hoekstra, University of Amsterdam, The Netherlands
Yunqing Huang, Xiangtan University, China
Andrés Iglesias, University of Cantabria, Spain
Hai Jin, Huazhong University of Science and Technology, China
Peter Kacsuk, MTA SZTAKI Research Institute, Hungary
Jacek Kitowski, AGH University of Science and Technology, Poland

Dieter Kranzlmüller, Johannes Kepler University Linz, Austria
Valeria Krzhizhanovskaya, University of Amsterdam, The Netherlands
Dawid Kurzyniec, Emory University, USA
Domenico Laforenza, Italian National Research Council, Italy
Antonio Lagana, University of Perugia, Italy
Francis Lau, The University of Hong Kong, China
Laurent Lefevre, INRIA, France
Bogdan Lesyng, ICM Warszawa, Poland
Thomas Ludwig, University of Heidelberg, Germany
Emilio Luque, Universitat Autònoma de Barcelona, Spain
Piyush Maheshwari, University of New South Wales, Australia
Maciej Malawski, Institute of Computer Science AGH, Poland
Michael Mascagni, Florida State University, USA
Taneli Mielikäinen, University of Helsinki, Finland
Edward Moreno, Euripides Foundation of Marilia, Brazil
Wolfgang Nagel, Dresden University of Technology, Germany
Genri Norman, Russian Academy of Sciences, Russia
Stephan Olariu, Old Dominion University, USA
Salvatore Orlando, University of Venice, Italy
Robert M. Panoff, Shodor Education Foundation, Inc., USA
Marcin Paprzycki, Oklahoma State University, USA
Ron Perrott, Queen's University of Belfast, UK
Richard Ramaroson, ONERA, France
Rosemary Renaut, Arizona State University, USA
Alistair Rendell, Australian National University, Australia
Paul Roe, Queensland University of Technology, Australia
Dale Shires, U.S. Army Research Laboratory, USA
Charles Shoniregun, University of East London, UK
Magda Slawinska, Gdansk University of Technology, Poland
Peter Sloot, University of Amsterdam, The Netherlands
Gunther Stuer, University of Antwerp, Belgium
Boleslaw Szymanski, Rensselaer Polytechnic Institute, USA
Ryszard Tadeusiewicz, AGH University of Science and Technology, Poland
Pavel Tvrdik, Czech Technical University, Czech Republic
Putchong Uthayopas, Kasetsart University, Thailand
Jesus Vigo-Aguiar, University of Salamanca, Spain
Jerzy Waśniewski, Technical University of Denmark, Denmark
Greg Watson, Los Alamos National Laboratory, USA
Peter H. Welch, University of Kent, UK
Piotr Wendykier, Emory University, USA
Roland Wismüller, University of Siegen, Germany
Baowen Xu, Southeast University Nanjing, China
Yong Xue, Chinese Academy of Sciences, China
Xiaodong Zhang, College of William and Mary, USA
Alexander Zhmakin, SoftImpact Ltd., Russia

Krzysztof Zielinski, ICS UST / CYFRONET, Poland
Zahari Zlatev, National Environmental Research Institute, Denmark
Elena Zudilova-Seinstra, University of Amsterdam, The Netherlands

Reviewers

Adrian Kacso
Adrian Sandu
Akshaye Dhawan
Alberto
Sanchez-Campos
Alex Tiskin
Alexander Bogdanov
Alexander Zhmakin
Alexandre Dupuis
Alexandre Tiskin
Alexandros Gerbessiotis
Alexey S. Rodionov
Alfons Hoekstra
Alfredo Tirado-Ramos
Ali Haleeb
Alistair Rendell
Ana Ripoll
A. Kalyanaraman
Andre Merzky
Andreas Hoffmann
Andrés Iglesias
Andrew Adamatzky
Andrzej Czygrinow
Andrzej Gościński
Aneta Karaivanova
Anna Morajko
Anne Benoit
Antonio Lagana
Anu G. Bourgeois
Ari Rantanen
Armelle Merlin
Arndt Bode
B. Frankovic
Bahman Javadi
Baowen Xu
Barbara Głut
Bartosz Baliś
Bas van Vlijmen

Bastien Chopard
Behrooz Shirazi
Ben Jackson
Beniamino Di Martino
Benjamin N. Jackson
Benny Cheung
Biju Sayed
Bogdan Lesyng
Bogdan Smolka
Boleslaw Szymanski
Breanndan O'Nuallain
Brian d'Auriol
Brice Goglin
Bruce Boghosian
Casiano Rodrguez León
Charles Shoniregun
Charles Stewart
Chen Lihua
Chris Homescu
Chris R. Kleijn
Christian Glasner
Christian Perez
C. Schaubschlaeger
Christoph Anthes
Clemens Grelck
Colin Enticott
Corrado Zoccolo
Craig C. Douglas
Craig Lee
Cristina Negoita
Dacian Daescu
Daewon W. Byun
Dale Shires
Danica Janglova
Daniel Pressel
Dave Roberts
David Abramson
David A. Bader

David Green
David Lowenthal
David Roberts
Dawid Kurzyniec
Dick van Albada
Diego Javier Mostaccio
Dieter Kranzlmüller
Dirk Deschrijver
Dirk Roekaerts
Domenico Laforenza
Donny Kurniawan
Eddy Caron
Edgar Gabriel
Edith Spiegl
Edward Moreno
Eldad Haber
Elena Zudilova-Seinstra
Elisa Heymann
Emanouil Atanassov
Emilio Luque
Eunjoo Lee
Eunjung Cho
Evarestov
Evghenii Gaburov
Fabrizio Silvestri
Feng Tan
Fethi A. Rabhi
Floros Evangelos
Francesco Moscato
Francis Lau
Francisco J. Rosales
Franck Cappello
Frank Dehne
Frank Dopatka
Frank J. Seinstra
Frantisek Capkovic
Frederic Desprez
Frederic Hancke

Frédéric Gava
Frédéric Loulergue
Frederick T. Sheldon
Gang Kou
Genri Norman
George Athanasopoulos
Greg Watson
Gunther Stuer
Haewon Nam
Hai Jin
Hassan Diab
He Jing
Holger Bischof
Holly Dail
Hongbin Guo
Hongquan Zhu
Hong-Seok Lee
Hui Liu
Hyoung-Key Choi
Hyung-Min Lee
Hyunseung Choo
I.M. Navon
Igor Mokris
Igor Schagaev
Irina Schweigert
Irina Shoshmina
Isabelle Guérin-Lassous
Ivan Dimov
Ivana Budinska
J. Kroc
J.G. Verwer
Jacek Kitowski
Jack Dongarra
Jan Broeckhove
Jan Glasa
Jan Humble
Jean-Luc Falcone
Jean-Yves L'Excellent
Jemal Abawajy
Jens Gustedt
Jens Volkert
Jerzy Waśniewski
Jesus Vigo-Aguiar
Jianping Li
Jing He

Jinling Yang
John Copeland
John Michopoulos
Jonas Latt
Jongpil Jeong
Jose L. Bosque
Jose C. Cunha
Jose Alberto Fernandez
Josep Jorba Esteve
Jun Wu
Jürgen Jähnert
Katarzyna Rycerz
Kawther Rekabi
Ken Nguyen
Ken C.K. Tsang
K.N. Plataniotis
Krzysztof Boryczko
Krzysztof Grzda
Krzysztof Zieliński
Kurt Vanmechelen
Ladislav Hluchy
Laurence T. Yang
Laurent Lefevre
Laurent Philippe
Lean Yu
Leigh Little
Liang Cheng
Lihua Chen
Lijuan Zhu
Luis M. Portela
Luoding Zhu
M. Mat Deris
Maciej Malawski
Magda Sławińska
Marcin Paprzycki
Marcin Radecki
Marcin Smtek
Marco Aldinucci
Marek Gajcki
Maria S. Pérez
Marian Bubak
Marina Gavrilova
Marios Dikaiakos
Martin Polak
Martin Quinson

Massiomo Coppola
Mathilde Romberg
Mathura Gopalan
Matthew Sottile
Matthias Kawski
Matthias Müller
Mauro Iacono
Michał Malafiejski
Michael Gerndt
Michael Mascagni
Michael Navon
Michael Scarpa
Michele Benzi
Mikhail Zatevakhin
Miroslav Dobrucky
Mohammed Yousoof
Moonseong Kim
Moshe Sipper
Nageswara S. V. Rao
Narayana Jayaram
NianYan
Nicola Tonellotto
Nicolas Wicker
Nikolai Simonov
Nisar Hundewale
Osni Marques
Pang Ko
Paul Albuquerque
Paul Evangelista
Paul Gray
Paul Heinzlreiter
Paul Roe
Paula Fritzsche
Paulo Afonso Lopes
Pavel Tvrdik
Paweł Czarnul
Paweł Kaczmarek
Peggy Lindner
Peter Brezany
Peter Hellinckx
Peter Kacsuk
Peter Sloot
Peter H. Welch
Philip Chan
Phillip A. Laplante

Pierre Fraigniaud
Pilar Herrero
Piotr Bala
Piotr Wendykier
Piyush Maheshwari
Porfidio Hernandez
Praveen Madiraju
Putchong Uthayopas
Qiang-Sheng Hua
R. Vollmar
Rafał Wcisło
Rafik Ouared
Rainer Keller
Rajkumar Buyya
Rastislav Lukac
Renata Słota
Rene Kobler
Richard Mason
Richard Ramaroson
Rob H. Bisseling
Robert M. Panoff
Robert Schaefer
Robin Wolff
Rocco Aversa
Rod Blais
Roeland Merks
Roland Wismüller
Rolf Rabenseifner
Rolf Sander
Ron Perrott
Rosemary Renaut
Ryszard Tadeusiewicz
S. Lakshmivarahan
Saeid Belkasim
Salvatore Orlando
Salvatore Venticinque
Sam G. Lambrakos

Samira El Yacoubi
Sang-Hun Cho
Sarah M. Orley
Satoyuki Kawano
Savio Tse
Scott Emrich
Scott Lathrop
Seong-Moo Yoo
Serge Chaumette
Sergei Gorlatch
Seungchan Kim
Shahaan Ayyub
Shanyu Tang
Sibel Adali
Siegfried Benkner
Sridhar Radharkrishnan
Srinivas Aluru
Srinivas Vadrevu
Stefan Marconi
Stefania Bandini
Stefano Marrone
Stephan Olariu
Stephen Gilmore
Steve Chiu
Sudip K. Seal
Sung Y. Shin
Takashi Matsuhisa
Taneli Mielikäinen
Thilo Kielmann
Thomas Ludwig
Thomas Richter
Thomas Worsch
Tianfeng Chai
Timothy Jones
Tiziana Calamoneri
Todor Gurov
Tom Dhaene

Tomasz Gubała
Tomasz Szepieniec
Toni Cortes
Ulrich Brandt-Pollmann
V. Vshivkov
Vaidy Sunderam
Valentina Casola
V. Krzhizhanovskaya
Vassil Alexandrov
Victor Malyshkin
Viet D. Tran
Vladimir K. Popkov
V.V. Shakhov
Włodzimierz Funika
Wai-Kwong Wing
Wei Yin
Wenyuan Liao
Witold Alda
Witold Dzwinel
Wojtek Gościński
Wolfgang E. Nagel
Wouter Hendrickx
Xiaodong Zhang
Yannis Cotronis
Yi Peng
Yong Fang
Yong Shi
Yong Xue
Yumi Choi
Yunqing Huang
Yuriy Gorbachev
Zahari Zlatev
Zaid Zabanoot
Zhenjiang Hu
Zhiming Zhao
Zoltan Juhasz
Zsolt Nemeth

Workshops Organizers

High Performance Computing in Academia: Systems and Applications

Denis Donnelly – Siena College, USA
Ulrich Rüde – Universität Erlangen-Nürnberg

Tools for Program Development and Analysis in Computational Science

Dieter Kranzlmüller – GUP, Joh. Kepler University Linz, Austria
Arndt Bode – Technical University Munich, Germany
Jens Volkert – GUP, Joh. Kepler University Linz, Austria
Roland Wismüller – University of Siegen, Germany

Practical Aspects of High-Level Parallel Programming (PAPP)

Frédéric Loulergue – Université Paris Val de Marne, France

2005 International Workshop on Bioinformatics Research and Applications

Yi Pan – Georgia State University, USA
Alex Zelikovsky – Georgia State University, USA

Computer Graphics and Geometric Modeling, CGGM 2005

Andrés Iglesias – University of Cantabria, Spain

Computer Algebra Systems and Applications, CASA 2005

Andrés Iglesias – University of Cantabria, Spain
Akemi Galvez – University of Cantabria, Spain

Wireless and Mobile Systems

Hyunseung Choo – Sungkyunkwan University, Korea
Eui-Nam Huh Seoul – Womens University, Korea
Hyoung-Kee Choi – Sungkyunkwan University, Korea
Youngsong Mun – Soongsil University, Korea

Intelligent Agents in Computing Systems -The Agent Days 2005 in Atlanta

Krzysztof Cetnarowicz – Academy of Science and Technology AGH, Krakow, Poland
Robert Schaefer – Jagiellonian University, Krakow, Poland

Programming Grids and Metacomputing Systems - PGaMS2005

Maciej Malawski – Institute of Computer Science, Academy of Science and Technology AGH, Krakow, Poland
Gunther Stuer – Universiteit Antwerpen, Belgium

Autonomic Distributed Data and Storage Systems Management – ADSM2005

Jemal H. Abawajy – Deakin University, Australia
M. Mat Deris – College University Tun Hussein Onn, Malaysia

GeoComputation

Yong Xue – London Metropolitan University, UK

Computational Economics and Finance

Yong Shi – University of Nebraska, Omaha, USA
Xiaotie Deng – University of Nebraska, Omaha, USA
Shouyang Wang – University of Nebraska, Omaha, USA

Simulation of Multiphysics Multiscale Systems

Valeria Krzhizhanovskaya – University of Amsterdam, The Netherlands
Bastien Chopard – University of Geneva, Switzerland
Yuriy Gorbachev – Institute for High Performance Computing & Data Bases, Russia

Dynamic Data Driven Application Systems

Frederica Darema – National Science Foundation, USA

2nd International Workshop on Active and Programmable Grids Architectures and Components (APGAC2005)

Alex Galis – University College London, UK

Parallel Monte Carlo Algorithms for Diverse Applications in a Distributed Setting

Vassil Alexandrov – University of Reading, UK
Aneta Karaivanova – Institute for Parallel Processing, Bulgarian Academy of Sciences
Ivan Dimov – Institute for Parallel Processing, Bulgarian Academy of Sciences

Grid Computing Security and Resource Management

Maria Pérez – Universidad Politécnica de Madrid, Spain
Jemal Abawajy – Deakin University, Australia

Modelling of Complex Systems by Cellular Automata

Jiri Kroc – Helsinki School of Economics, Finland
S. El Yacoubi – University of Perpignan, France
M. Sipper – Ben-Gurion University, Israel
R. Vollmar – University of Karlsruhe, Germany

International Workshop on Computational Nanoscience and Technology

Jun Ni – The University of Iowa, USA
Shaoping Xiao – The University of Iowa, USA

New Computational Tools for Advancing Atmospheric and Oceanic Sciences

Adrian Sandu – Virginia Tech, USA

Collaborative and Cooperative Environments

Vassil Alexandrov – University of Reading, UK
Christoph Anthes – GUP, Joh. Kepler University Linz, Austria
David Roberts – University of Salford, UK
Dieter Kranzlmüller – GUP, Joh. Kepler University Linz, Austria
Jens Volkert – GUP, Joh. Kepler University Linz, Austria

Table of Contents – Part I

Numerical Methods

Computing for Eigenpairs on Globally Convergent Iterative Method for Hermitian Matrices
Ran Baik, Karabi Datta, Yoopyo Hong 1

2D FE Quad Mesh Smoothing via Angle-Based Optimization
Hongtao Xu, Timothy S. Newman 9

Numerical Experiments on the Solution of the Inverse Additive Singular Value Problem
G. Flores-Becerra, Victor M. Garcia, Antonio M. Vidal 17

Computing Orthogonal Decompositions of Block Tridiagonal or Banded Matrices
Wilfried N. Gansterer 25

Adaptive Model Trust Region Methods for Generalized Eigenvalue Problems
P.-A. Absil, C.G. Baker, K.A. Gallivan, A. Sameh 33

On Stable Integration of Stiff Ordinary Differential Equations with Global Error Control
Gennady Yur'evich Kulikov, Sergey Konstantinovich Shindin 42

Bifurcation Analysis of Large Equilibrium Systems in MATLAB
David S. Bindel, James W. Demmel, Mark J. Friedman, Willy J.F. Govaerts, Yuri A. Kuznetsov 50

Sliced-Time Computations with Re-scaling for Blowing-Up Solutions to Initial Value Differential Equations
Nabil R. Nassif, Dolly Fayyad, Maria Cortas 58

Application of the Pseudo-Transient Technique to a Real-World Unsaturated Flow Groundwater Problem
Fred T. Tracy, Barbara P. Donnell, Stacy E. Howington, Jeffrey L. Hensley 66

Optimization of Spherical Harmonic Transform Computations
J.A.R. Blais, D.A. Provins, M.A. Soofi 74

Predictor-Corrector Preconditioned Newton-Krylov Method for Cavity Flow
Jianwei Ju, Giovanni Lapenta 82

Algorithms and Computational Kernels

A High-Order Recursive Quadratic Learning Algorithm
Qi Zhu, Shaohua Tan, Ying Qiao 90

Vectorized Sparse Matrix Multiply for Compressed Row Storage Format
Eduardo F. D'Azevedo, Mark R. Fahey, Richard T. Mills 99

A Multipole Based Treecode Using Spherical Harmonics for Potentials of the Form $r^{-\lambda}$
Kasthuri Srinivasan, Hemant Mahawar, Vivek Sarin 107

Numerically Stable Real Number Codes Based on Random Matrices
Zizhong Chen, Jack Dongarra 115

On Iterated Numerical Integration
Shujun Li, Elise de Doncker, Karlis Kaugars 123

Semi-Lagrangian Implicit-Explicit Two-Time-Level Scheme for Numerical Weather Prediction
Andrei Bourchtein 131

Occlusion Activity Detection Algorithm Using Kalman Filter for Detecting Occluded Multiple Objects
Heungkyu Lee, Hanseok Ko 139

A New Computer Algorithm Approach to Identification of Continuous-Time Batch Bioreactor Model Parameters
Suna Ertunc, Bulent Akay, Hale Hapoglu, Mustafa Alpbaz 147

Automated Operation Minimization of Tensor Contraction Expressions in Electronic Structure Calculations
Albert Hartono, Alexander Sibiryakov, Marcel Nooijen, Gerald Baumgartner, David E. Bernholdt, So Hirata, Chi-Chung Lam, Russell M. Pitzer, J. Ramanujam, P. Sadayappan 155

Regularization and Extrapolation Methods for Infrared Divergent Loop Integrals
Elise de Doncker, Shujun Li, Yoshimitsu Shimizu, Junpei Fujimoto, Fukuko Yuasa 165

Use of a Least Squares Finite Element Lattice Boltzmann Method to
Study Fluid Flow and Mass Transfer Processes
    Yusong Li, Eugene J. LeBoeuf, P.K. Basu 172

Nonnumerical Algorithms

On the Empirical Efficiency of the Vertex Contraction Algorithm for
Detecting Negative Cost Cycles in Networks
    K. Subramani, D. Desovski 180

Minimal Load Constrained Vehicle Routing Problems
    İmdat Kara, Tolga Bektaş 188

Multilevel Static Real-Time Scheduling Algorithms Using Graph
Partitioning
    Kayhan Erciyes, Zehra Soysert 196

A Multi-level Approach for Document Clustering
    Suely Oliveira, Sang-Cheol Seok 204

A Logarithmic Time Method for Two's Complementation
    Jung-Yup Kang, Jean-Luc Gaudiot 212

Parallel Algorithms

The Symmetric–Toeplitz Linear System Problem in Parallel
    Pedro Alonso, Antonio Manuel Vidal 220

Parallel Resolution with Newton Algorithms of the Inverse
Non-symmetric Eigenvalue Problem
    Pedro V. Alberti, Victor M. García, Antonio M. Vidal 229

Computational Challenges in Vector Functional Coefficient
Autoregressive Models
    *Ioana Banicescu, Ricolindo L. Cariño, Jane L. Harvill,
    John Patrick Lestrade* 237

Multi-pass Mapping Schemes for Parallel Sparse Matrix Computations
    Konrad Malkowski, Padma Raghavan 245

High-Order Finite Element Methods for Parallel Atmospheric Modeling
    Amik St.-Cyr, Stephen J. Thomas 256

Environments and Libraries

Continuation of Homoclinic Orbits in MATLAB
  M. Friedman, W. Govaerts, Yu.A. Kuznetsov, B. Sautois 263

A Numerical Tool for Transmission Lines
  Hervé Bolvin, André Chambarel, Philippe Neveux 271

The COOLFluiD Framework: Design Solutions for High Performance Object Oriented Scientific Computing Software
  Andrea Lani, Tiago Quintino, Dries Kimpe, Herman Deconinck, Stefan Vandewalle, Stefaan Poedts 279

A Problem Solving Environment for Image-Based Computational Hemodynamics
  Lilit Abrahamyan, Jorrit A. Schaap, Alfons G. Hoekstra, Denis Shamonin, Frieke M.A. Box, Rob J. van der Geest, Johan H.C. Reiber, Peter M.A. Sloot 287

MPL: A Multiprecision MATLAB-Like Environment
  Walter Schreppers, Franky Backeljauw, Annie Cuyt 295

Performance and Scalability

Performance and Scalability Analysis of Cray X1 Vectorization and Multistreaming Optimization
  Sadaf Alam, Jeffrey Vetter 304

Super-Scalable Algorithms for Computing on 100,000 Processors
  Christian Engelmann, Al Geist 313

"gRpas", a Tool for Performance Testing and Analysis
  Laurentiu Cucos, Elise de Doncker 322

Statistical Methods for Automatic Performance Bottleneck Detection in MPI Based Programs
  Michael Kluge, Andreas Knüpfer, Wolfgang E. Nagel 330

Programming Techniques

Source Templates for the Automatic Generation of Adjoint Code Through Static Call Graph Reversal
  Uwe Naumann, Jean Utke 338

A Case Study in Application Family Development by Automated
Component Composition: h-p Adaptive Finite Element Codes
  Nasim Mahmood, Yusheng Feng, James C. Browne 347

Determining Consistent States of Distributed Objects Participating in
a Remote Method Call
  Magdalena Sławińska, Bogdan Wiszniewski 355

Storage Formats for Sparse Matrices in Java
  *Mikel Luján, Anila Usman, Patrick Hardie, T.L. Freeman,
  John R. Gurd* 364

Coupled Fusion Simulation Using the Common Component Architecture
  *Wael R. Elwasif, Donald B. Batchelor, David E. Bernholdt,
  Lee A. Berry, Ed F. D'Azevedo, Wayne A. Houlberg, E.F. Jaeger,
  James A. Kohl, Shuhui Li* 372

Networks and Distributed Algorithms

A Case Study in Distributed Locking Protocol on Linux Clusters
  Sang-Jun Hwang, Jaechun No, Sung Soon Park 380

Implementation of a Cluster Based Routing Protocol for Mobile
Networks
  Geoffrey Marshall, Kayhan Erciyes 388

A Bandwidth Sensitive Distributed Continuous Media File System
Using the Fibre Channel Network
  Cuneyt Akinlar, Sarit Mukherjee 396

A Distributed Spatial Index for Time-Efficient Aggregation Query
Processing in Sensor Networks
  Soon-Young Park, Hae-Young Bae 405

Fast Concurrency Control for Distributed Inverted Files
  Mauricio Marín 411

An All-Reduce Operation in Star Networks Using All-to-All Broadcast
Communication Pattern
  Eunseuk Oh, Hongsik Choi, David Primeaux 419

Parallel and Distributed Computing

S^2F^2M - Statistical System for Forest Fire Management
  Germán Bianchini, Ana Cortés, Tomàs Margalef, Emilio Luque 427

Concurrent Execution of Multiple NAS Parallel Programs on a Cluster
  Adam K.L. Wong, Andrzej M. Goscinski 435

Model-Based Statistical Testing of a Cluster Utility
  W. Thomas Swain, Stephen L. Scott 443

Accelerating Protein Structure Recovery Using Graphics Processing Units
  Bryson R. Payne, G. Scott Owen, Irene Weber 451

A Parallel Software Development for Watershed Simulations
  Jing-Ru C. Cheng, Robert M. Hunter, Hwai-Ping Cheng, David R. Richards 460

Grid Computing

Design and Implementation of Services for a Synthetic Seismogram Calculation Tool on the Grid
  Choonhan Youn, Tim Kaiser, Cindy Santini, Dogan Seber 469

Toward GT3 and OGSI.NET Interoperability: GRAM Support on OGSI.NET
  James V.S. Watson, Sang-Min Park, Marty Humphrey 477

GEDAS: A Data Management System for Data Grid Environments
  Jaechun No, Hyoungwoo Park 485

SPURport: Grid Portal for Earthquake Engineering Simulations
  Tomasz Haupt, Anand Kalyanasundaram, Nisreen Ammari, Krishnendu Chandra, Kamakhya Das, Shravan Durvasula 493

Extending Existing Campus Trust Relationships to the Grid Through the Integration of Pubcookie and MyProxy
  Jonathan Martin, Jim Basney, Marty Humphrey 501

Generating Parallel Algorithms for Cluster and Grid Computing
  Ulisses Kendi Hayashida, Kunio Okuda, Jairo Panetta, Siand Wun Song 509

Relationship Networks as a Survivable and Adaptive Mechanism for Grid Resource Location
  Lei Gao, Yongsheng Ding 517

Deployment-Based Security for Grid Applications
  Isabelle Attali, Denis Caromel, Arnaud Contes 526

Grid Resource Selection by Application Benchmarking for
Computational Haemodynamics Applications
  *Alfredo Tirado-Ramos, George Tsouloupas, Marios Dikaiakos,
  Peter Sloot* 534

AGARM: An Adaptive Grid Application and Resource Monitor
Framework
  Wenju Zhang, Shudong Chen, Liang Zhang, Shui Yu, Fanyuan Ma .. 544

Failure Handling

Reducing Transaction Abort Rate of Epidemic Algorithm in Replicated
Databases
  Huaizhong Lin, Zengwei Zheng, Chun Chen 552

Snap-Stabilizing k-Wave Synchronizer
  Doina Bein, Ajoy K. Datta, Mehmet H. Karaata, Safaa Zaman 560

A Service Oriented Implementation of Distributed Status Monitoring
and Fault Diagnosis Systems
  Lei Wang, Peiyu Li, Zhaohui Wu, Shangjian Chen 568

Adaptive Fault Monitoring in Fault Tolerant CORBA
  Soo Myoung Lee, Hee Yong Youn, We Duke Cho 576

Optimization

Simulated Annealing Based-GA Using Injective Contrast Functions for
BSS
  J.M. Górriz, C.G. Puntonet, J.D. Morales, J.J. delaRosa 585

A DNA Coding Scheme for Searching Stable Solutions
  Intaek Kim, HeSong Lian, Hwan Il Kang 593

Study on Asymmetric Two-Lane Traffic Model Based on Cellular
Automata
  Xianchuang Su, Xiaogang Jin, Yong Min, Bo Peng 599

Simulation of Parasitic Interconnect Capacitance for Present and
Future ICs
  Grzegorz Tosik, Zbigniew Lisik, Malgorzata Langer, Janusz Wozny .. 607

Self-optimization of Large Scale Wildfire Simulations
  Jingmei Yang, Huoping Chen, Salim Hariri, Manish Parashar 615

Modeling and Simulation

Description of Turbulent Events Through the Analysis of POD Modes in Numerically Simulated Turbulent Channel Flow
Giancarlo Alfonsi, Leonardo Primavera 623

Computational Modeling of Human Head Conductivity
Adnan Salman, Sergei Turovets, Allen Malony, Jeff Eriksen, Don Tucker 631

Modeling of Electromagnetic Waves in Media with Dirac Distribution of Electric Properties
André Chambarel, Hervé Bolvin 639

Simulation of Transient Mechanical Wave Propagation in Heterogeneous Soils
Arnaud Mesgouez, Gaëlle Lefeuve-Mesgouez, André Chambarel 647

Practical Modelling for Generating Self-similar VBR Video Traffic
Jong-Suk R. Lee, Hae-Duck J. Jeong 655

Image Analysis and Processing

A Pattern Search Method for Image Registration
Hong Zhou, Benjamin Ray Seyfarth 664

Water Droplet Morphing Combining Rigid Transformation
Lanfen Lin, Shenghui Liao, RuoFeng Tong, JinXiang Dong 671

A Cost-Effective Private-Key Cryptosystem for Color Image Encryption
Rastislav Lukac, Konstantinos N. Plataniotis 679

On a Generalized Demosaicking Procedure: A Taxonomy of Single-Sensor Imaging Solutions
Rastislav Lukac, Konstantinos N. Plataniotis 687

Tile Classification Using the CIELAB Color Model
Christos-Nikolaos Anagnostopoulos, Athanassios Koutsonas, Ioannis Anagnostopoulos, Vassily Loumos, Eleftherios Kayafas 695

Graphics and Visualization

A Movie Is Worth More Than a Million Data Points
Hans-Peter Bischof, Jonathan Coles 703

A Layout Algorithm for Signal Transduction Pathways as
Two-Dimensional Drawings with Spline Curves
  Donghoon Lee, Byoung-Hyon Ju, Kyungsook Han 711

Interactive Fluid Animation and Its Applications
  Jeongjin Lee, Helen Hong, Yeong Gil Shin 719

ATDV: An Image Transforming System
  *Paula Farago, Ligia Barros, Gerson Cunha, Luiz Landau,
  Rosa Maria Costa* 727

An Adaptive Collision Detection and Resolution for Deformable
Objects Using Spherical Implicit Surface
  Sunhwa Jung, Min Hong, Min-Hyung Choi 735

Computation as a Scientific Paradigm

Automatic Categorization of Traditional Chinese Painting Images with
Statistical Gabor Feature and Color Feature
  Xiaohui Guan, Gang Pan, Zhaohui Wu 743

Nonlinear Finite Element Analysis of Structures Strengthened with
Carbon Fibre Reinforced Polymer: A Comparison Study
  X.S. Yang, J.M. Lees, C.T. Morley 751

Machine Efficient Adaptive Image Matching Based on the
Nonparametric Transformations
  Bogusław Cyganek 757

Non-gradient, Sequential Algorithm for Simulation of Nascent
Polypeptide Folding
  Lech Znamirowski 766

Hybrid Computational Methods

Time Delay Dynamic Fuzzy Networks for Time Series Prediction
  Yusuf Oysal 775

A Hybrid Heuristic Algorithm for the Rectangular Packing Problem
  Defu Zhang, Ansheng Deng, Yan Kang 783

Genetically Dynamic Optimization Based Fuzzy Polynomial Neural
Networks
  Ho-Sung Park, Sung-Kwun Oh, Witold Pedrycz, Yongkab Kim 792

Genetically Optimized Hybrid Fuzzy Neural Networks Based on
Simplified Fuzzy Inference Rules and Polynomial Neurons
  Sung-Kwun Oh, Byoung-Jun Park, Witold Pedrycz, Tae-Chon Ahn . . 798

Modelling and Constraint Hardness Characterisation of the
Unique-Path OSPF Weight Setting Problem
  Changyong Zhang, Robert Rodosek 804

Complex Systems

Application of Four-Dimension Assignment Algorithm of Data
Association in Distributed Passive-Sensor System
  Li Zhou, You He, Xiao-jing Wang 812

Using Rewriting Techniques in the Simulation of Dynamical Systems:
Application to the Modeling of Sperm Crawling
  Antoine Spicher, Olivier Michel 820

Specifying Complex Systems with Bayesian Programming. An Alife
Application
  Fidel Aznar, Mar Pujol, Ramón Rizo 828

Optimization Embedded in Simulation on Models Type System
Dynamics – Some Case Study
  Elżbieta Kasperska, Damian Słota 837

A High-Level Petri Net Based Decision Support System for Real-Time
Scheduling and Control of Flexible Manufacturing Systems: An
Object-Oriented Approach
  Gonca Tuncel, Gunhan Mirac Bayhan 843

Applications

Mesoscopic Simulation for Self-organization in Surface Processes
  David J. Horntrop 852

Computer Simulation of the Anisotropy of Fluorescence in Ring
Molecular Systems
  Pavel Heřman, Ivan Barvík 860

The Deflation Accelerated Schwarz Method for CFD
  J. Verkaik, C. Vuik, B.D. Paarhuis, A. Twerda 868

The Numerical Approach to Analysis of Microchannel Cooling Systems
  *Ewa Raj, Zbigniew Lisik, Malgorzata Langer, Grzegorz Tosik,
  Janusz Wozny* 876

Simulation of Nonlinear Thermomechanical Waves with an Empirical
Low Dimensional Model
  Linxiang Wang, Roderick V.N. Melnik 884

A Computational Risk Assessment Model for Breakwaters
  Can Elmar Balas 892

Wavelets and Wavelet Packets Applied to Termite Detection
  *Juan-José González de-la-Rosa, Carlos García Puntonet,
  Isidro Lloret Galiana, Juan Manuel Górriz* 900

Algorithms for the Estimation of the Concentrations of Chlorophyll A
and Carotenoids in Rice Leaves from Airborne Hyperspectral Data
  Yanning Guan, Shan Guo, Jiangui Liu, Xia Zhang 908

Multiresolution Reconstruction of Pipe-Shaped Objects from Contours
  Kyungha Min, In-Kwon Lee 916

Biomedical Applications

Multi-resolution LOD Volume Rendering in Medicine
  Kai Xie, Jie Yang, Yue Min Zhu 925

Automatic Hepatic Tumor Segmentation Using Statistical Optimal
Threshold
  Seung-Jin Park, Kyung-Sik Seo, Jong-An Park 934

Spatio-Temporal Patterns in the Depth EEG During the Epileptic
Seizure
  *Jung Ae Kim, Sunyoung Cho, Sang Kun Lee, Hyunwoo Nam,
  Seung Kee Han* 941

Prediction of Ribosomal Frameshift Signals of User-Defined Models
  Yanga Byun, Sanghoon Moon, Kyungsook Han 948

Effectiveness of Vaccination Strategies for Infectious Diseases According
to Human Contact Networks
  Fumihiko Takeuchi, Kenji Yamamoto 956

Data Mining and Computation

A Shape Constraints Based Method to Recognize Ship Objects from High Spatial Resolution Remote Sensed Imagery
Min Wang, Jiancheng Luo, Chenghu Zhou, Dongping Ming 963

Statistical Inference Method of User Preference on Broadcasting Content
Sanggil Kang, Jeongyeon Lim, Munchurl Kim 971

Density-Based Spatial Outliers Detecting
Tianqiang Huang, Xiaolin Qin, Chongcheng Chen, Qinmin Wang ... 979

The Design and Implementation of Extensible Information Services
Guiyi Wei, Guangming Wang, Yao Zheng, Wei Wang 987

Approximate B-Spline Surface Based on RBF Neural Networks
Xumin Liu, Houkuan Huang, Weixiang Xu 995

Efficient Parallelization of Spatial Approximation Trees
Mauricio Marín, Nora Reyes 1003

Education in Computational Science

The Visualization of Linear Algebra Algorithms in Apt Apprentice
Christopher Andrews, Rodney Cooper, Ghislain Deslongchamps, Olivier Spet 1011

A Visual Interactive Framework for Formal Derivation
Paul Agron, Leo Bachmair, Frank Nielsen 1019

ECVlab: A Web-Based Virtual Laboratory System for Electronic Circuit Simulation
Ouyang Yang, Dong Yabo, Zhu Miaoliang, Huang Yuewei, Mao Song, Mao Yunjie 1027

MTES: Visual Programming Environment for Teaching and Research in Image Processing
JeongHeon Lee, YoungTak Cho, Hoon Heo, OkSam Chae 1035

Emerging Trends

Advancing Scientific Computation by Improving Scientific Code Development: Symbolic Execution and Semantic Analysis
Mark Stewart 1043

Scale-Free Networks: A Discrete Event Simulation Approach
  Rex K. Kincaid, Natalia Alexandrov 1051

Impediments to Future Use of Petaflop Class Computers for Large-Scale Scientific/Engineering Applications in U.S. Private Industry
  Myron Ginsberg 1059

The SCore Cluster Enabled OpenMP Environment: Performance Prospects for Computational Science
  H'sien. J. Wong, Alistair P. Rendell 1067

Author Index 1077

Table of Contents – Part II

Workshop On "High Performance Computing in Academia: Systems and Applications"

Teaching High-Performance Computing on a High-Performance Cluster
Martin Bernreuther, Markus Brenk, Hans-Joachim Bungartz, Ralf-Peter Mundani, Ioan Lucian Muntean 1

Teaching High Performance Computing Parallelizing a Real Computational Science Application
Giovanni Aloisio, Massimo Cafaro, Italo Epicoco, Gianvito Quarta .. 10

Introducing Design Patterns, Graphical User Interfaces and Threads Within the Context of a High Performance Computing Application
James Roper, Alistair P. Rendell 18

High Performance Computing Education for Students in Computational Engineering
Uwe Fabricius, Christoph Freundl, Harald Köstler, Ulrich Rüde 27

Integrating Teaching and Research in HPC: Experiences and Opportunities
M. Berzins, R.M. Kirby, C.R. Johnson 36

Education and Research Challenges in Parallel Computing
L. Ridgway Scott, Terry Clark, Babak Bagheri 44

Academic Challenges in Large-Scale Multiphysics Simulations
Michael T. Heath, Xiangmin Jiao 52

Balancing Computational Science and Computer Science Research on a Terascale Computing Facility
Calvin J. Ribbens, Srinidhi Varadarjan, Malar Chinnusamy, Gautam Swaminathan 60

Computational Options for Bioinformatics Research in Evolutionary Biology
Michael A. Thomas, Mitch D. Day, Luobin Yang 68

Financial Computations on Clusters Using Web Services
Shirish Chinchalkar, Thomas F. Coleman, Peter Mansfield 76

"Plug-and-Play" Cluster Computing: HPC Designed for the
Mainstream Scientist
    Dean E. Dauger, Viktor K. Decyk 84

Building an HPC Watering Hole for Boulder Area Computational
Science
    E.R. Jessup, H.M. Tufo, M.S. Woitaszek 91

The Dartmouth Green Grid
    *James E. Dobson, Jeffrey B. Woodward, Susan A. Schwarz,
John C. Marchesini, Hany Farid, Sean W. Smith* 99

Resource-Aware Parallel Adaptive Computation for Clusters
    James D. Teresco, Laura Effinger-Dean, Arjun Sharma 107

Workshop on "Tools for Program Development and Analysis in Computational Science"

New Algorithms for Performance Trace Analysis Based on Compressed
Complete Call Graphs
    Andreas Knüpfer and Wolfgang E. Nagel 116

PARADIS: Analysis of Transaction-Based Applications in Distributed
Environments
    Christian Glasner, Edith Spiegl, Jens Volkert 124

Automatic Tuning of Data Distribution Using Factoring in
Master/Worker Applications
    Anna Morajko, Paola Caymes, Tomàs Margalef, Emilio Luque 132

DynTG: A Tool for Interactive, Dynamic Instrumentation
    Martin Schulz, John May, John Gyllenhaal 140

Rapid Development of Application-Specific Network Performance Tests
    Scott Pakin 149

Providing Interoperability for Java-Oriented Monitoring Tools with
JINEXT
    Włodzimierz Funika, Arkadiusz Janik 158

RDVIS: A Tool That Visualizes the Causes of Low Locality and Hints
Program Optimizations
    Kristof Beyls, Erik H. D'Hollander, Frederik Vandeputte 166

CacheIn: A Toolset for Comprehensive Cache Inspection
  Jie Tao, Wolfgang Karl 174

Optimization-Oriented Visualization of Cache Access Behavior
  Jie Tao, Wolfgang Karl 182

Collecting and Exploiting Cache-Reuse Metrics
  Josef Weidendorfer, Carsten Trinitis 191

Workshop on "Computer Graphics and Geometric Modeling, CGGM 2005"

Modelling and Animating Hand Wrinkles
  X.S. Yang, Jian J. Zhang 199

Simulating Wrinkles in Facial Expressions on an Anatomy-Based Face
  Yu Zhang, Terence Sim, Chew Lim Tan 207

A Multiresolutional Approach for Facial Motion Retargetting Using Subdivision Wavelets
  Kyungha Min, Moon-Ryul Jung 216

New 3D Graphics Rendering Engine Architecture for Direct Tessellation of Spline Surfaces
  Adrian Sfarti, Brian A. Barsky, Todd J. Kosloff, Egon Pasztor, Alex Kozlowski, Eric Roman, Alex Perelman 224

Fast Water Animation Using the Wave Equation with Damping
  Y. Nishidate, G.P. Nikishkov 232

A Comparative Study of Acceleration Techniques for Geometric Visualization
  Pascual Castelló, José Francisco Ramos, Miguel Chover 240

Building Chinese Ancient Architectures in Seconds
  Hua Liu, Qing Wang, Wei Hua, Dong Zhou, Hujun Bao 248

Accelerated 2D Image Processing on GPUs
  Bryson R. Payne, Saeid O. Belkasim, G. Scott Owen, Michael C. Weeks, Ying Zhu 256

Consistent Spherical Parameterization
  Arul Asirvatham, Emil Praun, Hugues Hoppe 265

Mesh Smoothing via Adaptive Bilateral Filtering
Qibin Hou, Li Bai, Yangsheng Wang 273

Towards a Bayesian Approach to Robust Finding Correspondences in Multiple View Geometry Environments
Cristian Canton-Ferrer, Josep R. Casas, Montse Pardàs 281

Managing Deformable Objects in Cluster Rendering
Thomas Convard, Patrick Bourdot, Jean-Marc Vézien 290

Revolute Quadric Decomposition of Canal Surfaces and Its Applications
Jinyuan Jia, Ajay Joneja, Kai Tang 298

Adaptive Surface Modeling Using a Quadtree of Quadratic Finite Elements
G. P. Nikishkov 306

MC Slicing for Volume Rendering Applications
A. Benassarou, E. Bittar, N. W. John, L. Lucas 314

Modelling and Sampling Ramified Objects with Substructure-Based Method
Weiwei Yin, Marc Jaeger, Jun Teng, Bao-Gang Hu 322

Integration of Multiple Segmentation Based Environment Models
SeungTaek Ryoo, CheungWoon Jho 327

On the Impulse Method for Cloth Animation
Juntao Ye, Robert E. Webber, Irene Gargantini 331

Remeshing Triangle Meshes with Boundaries
Yong Wu, Yuanjun He, Hongming Cai 335

SACARI: An Immersive Remote Driving Interface for Autonomous Vehicles
Antoine Tarault, Patrick Bourdot, Jean-Marc Vézien 339

A 3D Model Retrieval Method Using 2D Freehand Sketches
Jiantao Pu, Karthik Ramani 343

A 3D User Interface for Visualizing Neuron Location in Invertebrate Ganglia
*Jason A. Pamplin, Ying Zhu, Paul S. Katz,
Rajshekhar Sunderraman* 347

Workshop on "Modelling of Complex Systems by Cellular Automata"

The Dynamics of General Fuzzy Cellular Automata
Angelo B. Mingarelli 351

A Cellular Automaton SIS Epidemiological Model with Spatially Clustered Recoveries
David Hiebeler 360

Simulating Market Dynamics with CD++
Qi Liu, Gabriel Wainer 368

A Model of Virus Spreading Using Cell-DEVS
Hui Shang, Gabriel Wainer 373

A Cellular Automata Model of Competition in Technology Markets with Network Externalities
Judy Frels, Debra Heisler, James Reggia, Hans-Joachim Schuetze ... 378

Self-organizing Dynamics for Optimization
Stefan Boettcher 386

Constructibility of Signal-Crossing Solutions in von Neumann 29-State Cellular Automata
William R. Buckley, Amar Mukherjee 395

Evolutionary Discovery of Arbitrary Self-replicating Structures
Zhijian Pan, James Reggia 404

Modelling Ant Brood Tending Behavior with Cellular Automata
Daniel Merkle, Martin Middendorf, Alexander Scheidler 412

A Realistic Cellular Automata Model to Simulate Traffic Flow at Urban Roundabouts
Ruili Wang, Mingzhe Liu 420

Probing the Eddies of Dancing Emergence: Complexity and Abstract Painting
Tara Krause 428

Workshop on "Wireless and Mobile Systems"

Enhanced TCP with End-to-End Bandwidth and Loss Differentiation Estimate over Heterogeneous Networks
Le Tuan Anh, Choong Seon Hong 436

Content-Aware Automatic QoS Provisioning for UPnP AV-Based
Multimedia Services over Wireless LANs
  Yeali S. Sun, Chang-Ching Yan, Meng Chang Chen 444

Simulation Framework for Wireless Internet Access Networks
  Hyoung-Kee Choi, Jitae Shin 453

WDM: An Energy-Efficient Multi-hop Routing Algorithm for Wireless
Sensor Networks
  Zengwei Zheng, Zhaohui Wu, Huaizhong Lin, Kougen Zheng 461

Forwarding Scheme Extension for Fast and Secure Handoff in
Hierarchical MIPv6
  *Hoseong Jeon, Jungmuk Lim, Hyunseung Choo,
  Gyung-Leen Park* 468

Back-Up Chord: Chord Ring Recovery Protocol for P2P File Sharing
over MANETs
  *Hong-Jong Jeong, Dongkyun Kim, Jeomki Song, Byung-yeub Kim,
  Jeong-Su Park* 477

PATM: Priority-Based Adaptive Topology Management for Efficient
Routing in Ad Hoc Networks
  Haixia Tan, Weilin Zeng, Lichun Bao 485

Practical and Provably-Secure Multicasting over High-Delay Networks
  *Junghyun Nam, Hyunjue Kim, Seungjoo Kim, Dongho Won,
  Hyungkyu Yang* 493

A Novel IDS Agent Distributing Protocol for MANETs
  Jin Xin, Zhang Yao-Xue, Zhou Yue-Zhi, Wei Yaya 502

ID-Based Secure Session Key Exchange Scheme to Reduce Registration
Delay with AAA in Mobile IP Networks
  Kwang Cheol Jeong, Hyunseung Choo, Sang Yong Ha 510

An Efficient Wireless Resource Allocation Based on a Data Compressor
Predictor
  Min Zhang, Xiaolong Yang, Hong Jiang 519

A Seamless Handover Mechanism for IEEE 802.16e Broadband Wireless
Access
  Kyung-ah Kim, Chong-Kwon Kim, Tongsok Kim 527

Fault Tolerant Coverage Model for Sensor Networks
  Doina Bein, Wolfgang W. Bein, Srilaxmi Malladi 535

Detection Algorithms Based on Chip-Level Processing for DS/CDMA
Code Acquisition in Fast Fading Channels
  Seokho Yoon, Jee-Hyong Lee, Sun Yong Kim 543

Clustering-Based Distributed Precomputation for Quality-of-Service
Routing
  Yong Cui, Jianping Wu... 551

Traffic Grooming Algorithm Using Shortest EDPs Table in WDM Mesh
Networks
  Seungsoo Lee, Tae-Jin Lee, Min Young Chung, Hyunseung Choo 559

Efficient Indexing of Moving Objects Using Time-Based Partitioning
with R-Tree
  Youn Chul Jung, Hee Yong Youn, Ungmo Kim 568

Publish/Subscribe Systems on Node and Link Error Prone Mobile
Environments
  *Sangyoon Oh, Sangmi Lee Pallickara, Sunghoon Ko, Jai-Hoon Kim,
  Geoffrey Fox* 576

A Power Efficient Routing Protocol in Wireless Sensor Networks
  Hyunsook Kim, Jungpil Ryu, Kijun Han 585

Applying Mobile Agent to Intrusion Response for Ad Hoc Networks
  Ping Yi, Yiping Zhong, Shiyong Zhang 593

A Vertical Handoff Decision Process and Algorithm Based on Context
Information in CDMA-WLAN Interworking
  Jang-Sub Kim, Min-Young Chung, Dong-Ryeol Shin............... 601

Workshop on "Dynamic Data Driven Application Systems"

Dynamic Data Driven Applications Systems: New Capabilities for
Application Simulations and Measurements
  Frederica Darema 610

Dynamic Data Driven Methodologies for Multiphysics System Modeling
and Simulation
  *J. Michopoulos, C. Farhat, E. Houstis, P. Tsompanopoulou,
  H. Zhang, T. Gullaud* 616

Towards Dynamically Adaptive Weather Analysis and Forecasting in
LEAD
   Beth Plale, Dennis Gannon, Dan Reed, Sara Graves,
   Kelvin Droegemeier, Bob Wilhelmson, Mohan Ramamurthy 624

Towards a Dynamic Data Driven Application System for Wildfire
Simulation
   Jan Mandel, Lynn S. Bennethum, Mingshi Chen, Janice L. Coen,
   Craig C. Douglas, Leopoldo P. Franca, Craig J. Johns,
   Minjeong Kim, Andrew V. Knyazev, Robert Kremens,
   Vaibhav Kulkarni, Guan Qin, Anthony Vodacek, Jianjia Wu,
   Wei Zhao, Adam Zornes 632

Multiscale Interpolation, Backward in Time Error Analysis for
Data-Driven Contaminant Simulation
   Craig C. Douglas, Yalchin Efendiev, Richard Ewing, Victor Ginting,
   Raytcho Lazarov, Martin J. Cole, Greg Jones, Chris R. Johnson 640

Ensemble–Based Data Assimilation for Atmospheric Chemical
Transport Models
   Adrian Sandu, Emil M. Constantinescu, Wenyuan Liao,
   Gregory R. Carmichael, Tianfeng Chai, John H. Seinfeld,
   Dacian Dăescu 648

Towards Dynamic Data-Driven Optimization of Oil Well Placement
   Manish Parashar, Vincent Matossian, Wolfgang Bangerth,
   Hector Klie, Benjamin Rutt, Tahsin Kurc, Umit Catalyurek, Joel
   Saltz, Mary F. Wheeler 656

High-Fidelity Simulation of Large-Scale Structures
   Christoph Hoffmann, Ahmed Sameh, Ananth Grama 664

A Dynamic Data Driven Grid System for Intra-operative Image Guided
Neurosurgery
   Amit Majumdar, Adam Birnbaum, Dong Ju Choi, Abhishek Trivedi,
   Simon K. Warfield, Kim Baldridge, Petr Krysl 672

Structure-Based Integrative Computational and Experimental
Approach for the Optimization of Drug Design
   Dimitrios Morikis, Christodoulos A. Floudas, John D. Lambris 680

Simulation and Visualization of Air Flow Around Bat Wings During
Flight
   I.V. Pivkin, E. Hueso, R. Weinstein, D.H. Laidlaw, S. Swartz,
   G.E. Karniadakis 689

Integrating Fire, Structure and Agent Models
 *A.R. Chaturvedi, S.A. Filatyev, J.P. Gore, A. Hanna, J. Means,
 A.K. Mellema* 695

A Dynamic, Data-Driven, Decision Support System for Emergency
Medical Services
 Mark Gaynor, Margo Seltzer, Steve Moulton, Jim Freedman 703

Dynamic Data Driven Coupling of Continuous and Discrete Methods
for 3D Tracking
 Dimitris Metaxas, Gabriel Tsechpenakis 712

Semi-automated Simulation Transformation for DDDAS
 *David Brogan, Paul Reynolds, Robert Bartholet, Joseph Carnahan,
 Yannick Loitière* 721

The Development of Dependable and Survivable Grids
 *Andrew Grimshaw, Marty Humphrey, John C. Knight,
 Anh Nguyen-Tuong, Jonathan Rowanhill, Glenn Wasson,
 Jim Basney* 729

On the Fundamental Tautology of Validating Data-Driven Models and
Simulations
 John Michopoulos, Sam Lambrakos 738

Workshop on "Practical Aspects of High-Level Parallel Programming (PAPP)"

Managing Heterogeneity in a Grid Parallel Haskell
 A. Al Zain, P.W. Trinder, H-W. Loidl, G.J. Michaelson 746

An Efficient Equi-semi-join Algorithm for Distributed Architectures
 M. Bamha, G. Hains 755

Two Fundamental Concepts in Skeletal Parallel Programming
 Anne Benoit, Murray Cole 764

A Formal Framework for Orthogonal Data and Control Parallelism
Handling
 Sonia Campa 772

Empirical Parallel Performance Prediction from Semantics-Based
Profiling
 Norman Scaife, Greg Michaelson, Susumu Horiguchi 781

Dynamic Memory Management in the *Loci* Framework
  Yang Zhang, Edward A. Luke 790

Workshop on "New Computational Tools for Advancing Atmospheric and Oceanic Sciences"

On Adaptive Mesh Refinement for Atmospheric Pollution Models
  Emil M. Constantinescu, Adrian Sandu 798

Total Energy Singular Vectors for Atmospheric Chemical Transport Models
  Wenyuan Liao, Adrian Sandu 806

Application of Static Adaptive Grid Techniques for Regional-Urban Multiscale Air Quality Modeling
  Daewon Byun, Peter Percell, Tanmay Basak 814

On the Accuracy of High-Order Finite Elements in Curvilinear Coordinates
  Stephen J. Thomas, Amik St.-Cyr 821

Analysis of Discrete Adjoints for Upwind Numerical Schemes
  Zheng Liu and Adrian Sandu 829

The Impact of Background Error on Incomplete Observations for 4D-Var Data Assimilation with the FSU GSM
  I. Michael Navon, Dacian N. Daescu, Zhuo Liu 837

2005 International Workshop on Bioinformatics Research and Applications

Disjoint Segments with Maximum Density
  Yen Hung Chen, Hsueh-I Lu, Chuan Yi Tang 845

Wiener Indices of Balanced Binary Trees
  Sergey Bereg, Hao Wang 851

What Makes the Arc-Preserving Subsequence Problem Hard?
  Guillaume Blin, Guillaume Fertin, Romeo Rizzi, Stéphane Vialette .. 860

An Efficient Dynamic Programming Algorithm and Implementation for RNA Secondary Structure Prediction
  Guangming Tan, Xinchun Liu, Ninghui Sun 869

Performance Evaluation of Protein Sequence Clustering Tools
Haifeng Liu, Loo-Nin Teow 877

A Data-Adaptive Approach to cDNA Microarray Image Enhancement
*Rastislav Lukac, Konstantinos N. Plataniotis, Bogdan Smolka,,
Anastasios N. Venetsanopoulos* 886

String Kernels of Imperfect Matches for Off-target Detection in RNA Interference
Shibin Qiu, Terran Lane 894

A New Kernel Based on High-Scored Pairs of Tri-peptides and Its Application in Prediction of Protein Subcellular Localization
Zhengdeng Lei, Yang Dai 903

Reconstructing Phylogenetic Trees of Prokaryote Genomes by Randomly Sampling Oligopeptides
Osamu Maruyama, Akiko Matsuda, Satoru Kuhara 911

Phylogenetic Networks, Trees, and Clusters
Luay Nakhleh, Li-San Wang 919

SWAT: A New Spliced Alignment Tool Tailored for Handling More Sequencing Errors
Yifeng Li, Hesham H. Ali 927

Simultaneous Alignment and Structure Prediction of RNAs Are Three Input Sequences Better Than Two?
Beeta Masoumi, Marcel Turcotte 936

Clustering Using Adaptive Self-organizing Maps (ASOM) and Applications
Yong Wang, Chengyong Yang, Kalai Mathee, Giri Narasimhan 944

Experimental Analysis of a New Algorithm for Partial Haplotype Completion
Paola Bonizzoni, Gianluca Della Vedova, Riccardo Dondi, Lorenzo Mariani 952

Improving the Sensitivity and Specificity of Protein Homology Search by Incorporating Predicted Secondary Structures
Bin Ma, Lieyu Wu, Kaizhong Zhang 960

Profiling and Searching for RNA Pseudoknot Structures in Genomes
Chunmei Liu, Yinglei Song, Russell L. Malmberg, Liming Cai 968

Integrating Text Chunking with Mixture Hidden Markov Models for Effective Biomedical Information Extraction
Min Song, Il-Yeol Song, Xiaohua Hu, Robert B. Allen 976

k-Recombination Haplotype Inference in Pedigrees
Francis Y.L. Chin, Qiangfeng Zhang, Hong Shen 985

Improved Tag Set Design and Multiplexing Algorithms for Universal Arrays
Ion I. Măndoiu, Claudia Prăjescu, Dragoş Trincă 994

A Parallel Implementation for Determining Genomic Distances Under Deletion and Insertion
Vijaya Smitha Kolli, Hui Liu, Michelle Hong Pan, Yi Pan 1003

Phasing and Missing Data Recovery in Family Trios
Dumitru Brinza, Jingwu He, Weidong Mao, Alexander Zelikovsky .. 1011

Highly Scalable Algorithms for Robust String Barcoding
B. DasGupta, K.M. Konwar, I.I. Măndoiu, A.A. Shvartsman 1020

Optimal Group Testing Strategies with Interval Queries and Their Application to Splice Site Detection
Ferdinando Cicalese, Peter Damaschke, Ugo Vaccaro 1029

Virtual Gene: A Gene Selection Algorithm for Sample Classification on Microarray Datasets
Xian Xu, Aidong Zhang 1038

Workshop on "Programming Grids and Metacomputing Systems – PGaMS2005"

Bulk Synchronous Parallel ML: Modular Implementation and Performance Prediction
Frédéric Loulergue, Frédéric Gava, David Billiet 1046

Fast Expression Templates
Jochen Härdtlein, Alexander Linke, Christoph Pflaum 1055

Solving Coupled Geoscience Problems on High Performance Computing Platforms
Dany Kemmler, Panagiotis Adamidis, Wenqing Wang, Sebastian Bauer, Olaf Kolditz 1064

H2O Metacomputing - Jini Lookup and Discovery
  *Dirk Gorissen, Gunther Stuer, Kurt Vanmechelen,
  Jan Broeckhove* 1072

User Experiences with Nuclear Physics Calculations on a H2O
Metacomputing System and on the BEgrid
  P. Hellinckx, K. Vanmechelen, G. Stuer, F. Arickx, J. Broeckhove .. 1080

Author Index 1089

Table of Contents – Part III

Workshop on "Simulation of Multiphysics Multiscale Systems"

Multiscale Finite Element Modeling of the Coupled Nonlinear Dynamics of Magnetostrictive Composite Thin Film
Debiprosad Roy Mahapatra, Debi Prasad Ghosh, Gopalakrishnan Srinivasan 1

Large-Scale Fluctuations of Pressure in Fluid Flow Through Porous Medium with Multiscale Log-Stable Permeability
Olga Soboleva 9

A Computational Model of Micro-vascular Growth
Dominik Szczerba, Gábor Székely 17

A Dynamic Model for Phase Transformations in 3D Samples of Shape Memory Alloys
D.R. Mahapatra, R.V.N. Melnik 25

3D Finite Element Modeling of Free-Surface Flows with Efficient $k-\epsilon$ Turbulence Model and Non-hydrostatic Pressure
Célestin Leupi, Mustafa Siddik Altinakar 33

Cluster Computing for Transient Simulations of the Linear Boltzmann Equation on Irregular Three-Dimensional Domains
Matthias K. Gobbert, Mark L. Breitenbach, Timothy S. Cale 41

The Use of Conformal Voxels for Consistent Extractions from Multiple Level-Set Fields
Max O. Bloomfield, David F. Richards, Timothy S. Cale 49

Nonlinear OIFS for a Hybrid Galerkin Atmospheric Model
Amik St.-Cyr, Stephen J. Thomas 57

Flamelet Analysis of Turbulent Combustion
R.J.M. Bastiaans, S.M. Martin, H. Pitsch, J.A. van Oijen, L.P.H. de Goey 64

Entropic Lattice Boltzmann Method on Non-uniform Grids
C. Shyam Sunder, V. Babu 72

A Data-Driven Multi-field Analysis of Nanocomposites for Hydrogen Storage
  John Michopoulos, Nick Tran, Sam Lambrakos 80

Plug and Play Approach to Validation of Particle-Based Algorithms
  Giovanni Lapenta, Stefano Markidis 88

Multiscale Angiogenesis Modeling
  Shuyu Sun, Mary F. Wheeler, Mandri Obeyesekere, Charles Patrick Jr 96

The Simulation of a PEMFC with an Interdigitated Flow Field Design
  S.M. Guo 104

Multiscale Modelling of Bubbly Systems Using Wavelet-Based Mesh Adaptation
  Tom Liu, Phil Schwarz 112

Computational Study on the Effect of Turbulence Intensity and Pulse Frequency in Soot Concentration in an Acetylene Diffusion Flame
  Fernando Lopez-Parra, Ali Turan 120

Application Benefits of Advanced Equation-Based Multiphysics Modeling
  Lars Langemyr, Nils Malm 129

Large Eddy Simulation of Spanwise Rotating Turbulent Channel and Duct Flows by a Finite Volume Code at Low Reynolds Numbers
  Kursad Melih Guleren, Ali Turan 130

Modelling Dynamics of Genetic Networks as a Multiscale Process
  Xilin Wei, Roderick V.N. Melnik, Gabriel Moreno-Hagelsieb 134

Mathematical Model of Environmental Pollution by Motorcar in an Urban Area
  Valeriy Perminov 139

The Monte Carlo and Molecular Dynamics Simulation of Gas-Surface Interaction
  Sergey Borisov, Oleg Sazhin, Olesya Gerasimova 143

Workshop on "Grid Computing Security and Resource Management"

GIVS: Integrity Validation for Grid Security
  Giuliano Casale, Stefano Zanero 147

On the Impact of Reservations from the Grid on Planning-Based
Resource Management
  Felix Heine, Matthias Hovestadt, Odej Kao, Achim Streit 155

Genius: Peer-to-Peer Location-Aware Gossip Using Network
Coordinates
  *Ning Ning, Dongsheng Wang, Yongquan Ma, Jinfeng Hu, Jing Sun,
  Chongnan Gao, Weiming Zheng* 163

DCP-Grid, a Framework for Conversational Distributed Transactions
on Grid Environments
  Manuel Salvadores, Pilar Herrero, María S. Pérez, Víctor Robles ... 171

Dynamic and Fine-Grained Authentication and Authorization
Architecture for Grid Computing
  *Hyunjoon Jung, Hyuck Han, Hyungsoo Jung,
  Heon Y. Yeom* 179

GridSec: Trusted Grid Computing with Security Binding and
Self-defense Against Network Worms and DDoS Attacks
  *Kai Hwang, Yu-Kwong Kwok, Shanshan Song, Min Cai Yu Chen,
  Ying Chen, Runfang Zhou, Xiaosong Lou* 187

Design and Implementation of DAG-Based Co-scheduling of RPC in
the Grid
  *JiHyun Choi, DongWoo Lee, R.S. Ramakrishna, Michael Thomas,
  Harvey Newman* 196

Performance Analysis of Interconnection Networks for Multi-cluster
Systems
  Bahman Javadi, J.H. Abawajy, Mohammad K. Akbari 205

Autonomic Job Scheduling Policy for Grid Computing
  J.H. Abawajy 213

A New Trust Framework for Resource-Sharing in the Grid Environment
  Hualiang Hu, Deren Chen, Changqin Huang 221

An Intrusion-Resilient Authorization and Authentication Framework
for Grid Computing Infrastructure
  Yuanbo Guo, Jianfeng Ma, Yadi Wang 229

2nd International Workshop on Active and Programmable Grids Architectures and Components (APGAC2005)

An Active Platform as Middleware for Services and Communities Discovery
 Sylvain Martin, Guy Leduc 237

p2pCM: A Structured Peer-to-Peer Grid Component Model
 *Carles Pairot, Pedro García, Rubén Mondéjar,
 Antonio F. Gómez Skarmeta* 246

Resource Partitioning Algorithms in a Programmable Service Grid Architecture
 *Pieter Thysebaert, Bruno Volckaert, Marc De Leenheer,
 Filip De Turck, Bart Dhoedt, Piet Demeester* 250

Triggering Network Services Through Context-Tagged Flows
 Roel Ocampo, Alex Galis, Chris Todd 259

Dependable Execution of Workflow Activities on a Virtual Private Grid Middleware
 A. Machì, F. Collura, S. Lombardo 267

Cost Model and Adaptive Scheme for Publish/Subscribe Systems on Mobile Grid Environments
 *Sangyoon Oh, Sangmi Lee Pallickara, Sunghoon Ko, Jai-Hoon Kim,
 Geoffrey Fox* 275

Near-Optimal Algorithm for Self-configuration of Ad-hoc Wireless Networks
 Sung-Eok Jeon, Chuanyi Ji 279

International Workshop on Computational Nano-Science and Technology

The Applications of Meshfree Particle Methods at the Nanoscale
 Weixuan Yang, Shaoping Xiao 284

Numerical Simulation of Self-heating InGaP/GaAs Heterojunction Bipolar Transistors
 Yiming Li, Kuen-Yu Huang 292

Adaptive Finite Volume Simulation of Electrical Characteristics of Organic Light Emitting Diodes
 Yiming Li, Pu Chen 300

Characterization of a Solid State DNA Nanopore Sequencer Using Multi-scale (Nano-to-Device) Modeling
 Jerry Jenkins, Debasis Sengupta, Shankar Sundaram 309

Comparison of Nonlinear Conjugate-Gradient Methods for Computing the Electronic Properties of Nanostructure Architectures
 Stanimire Tomov, Julien Langou, Andrew Canning, Lin-Wang Wang, Jack Dongarra................................. 317

A Grid-Based Bridging Domain Multiple-Scale Method for Computational Nanotechnology
 Shaowen Wang, Shaoping Xiao, Jun Ni 326

Signal Cascades Analysis in Nanoprocesses with Distributed Database System
 Dariusz Mrozek, Bożena Małysiak, Jacek Fraczek, Paweł Kasprowski.. 334

Workshop on "Collaborative and Cooperative Environments"

Virtual States and Transitions, Virtual Sessions and Collaboration
 Dimitri Bourilkov 342

A Secure Peer-to-Peer Group Collaboration Scheme for Healthcare System
 Byong-In Lim, Kee-Hyun Choi, Dong-Ryeol Shin 346

Tools for Collaborative VR Application Development
 Adrian Haffegee, Ronan Jamieson, Christoph Anthes, Vassil Alexandrov... 350

Multicast Application Sharing Tool – Facilitating the eMinerals Virtual Organisation
 Gareth J. Lewis, S. Mehmood Hasan, Vassil N. Alexandrov, Martin T. Dove, Mark Calleja................................ 359

The Collaborative P-GRADE Grid Portal
 Gareth J. Lewis, Gergely Sipos, Florian Urmetzer, Vassil N. Alexandrov, Peter Kacsuk 367

An Approach for Collaboration and Annotation in Video Post-production
 Karsten Morisse, Thomas Sempf 375

A Toolbox Supporting Collaboration in Networked Virtual Environments
  Christoph Anthes, Jens Volkert 383

A Peer-to-Peer Approach to Content Dissemination and Search in Collaborative Networks
  Ismail Bhana, David Johnson 391

Workshop on "Autonomic Distributed Data and Storage Systems Management – ADSM2005"

TH-VSS: An Asymmetric Storage Virtualization System for the SAN Environment
  Da Xiao, Jiwu Shu, Wei Xue, Weimin Zheng 399

Design and Implementation of the Home-Based Cooperative Cache for PVFS
  In-Chul Hwang, Hanjo Jung, Seung-Ryoul Maeng, Jung-Wan Cho .. 407

Improving the Data Placement Algorithm of Randomization in SAN
  Nianmin Yao, Jiwu Shu, Weimin Zheng 415

Safety of a Server-Based Version Vector Protocol Implementing Session Guarantees
  Jerzy Brzeziński, Cezary Sobaniec, Dariusz Wawrzyniak 423

Scalable Hybrid Search on Distributed Databases
  Jungkee Kim, Geoffrey Fox 431

Storage QoS Control with Adaptive I/O Deadline Assignment and Slack-Stealing EDF
  Young Jin Nam, Chanik Park 439

High Reliability Replication Technique for Web-Server Cluster Systems
  M. Mat Deris, J.H. Abawajy, M. Zarina, R. Mamat 447

An Efficient Replicated Data Management Approach for Peer-to-Peer Systems
  J.H. Abawajy 457

Workshop on "GeoComputation"

Explore Disease Mapping of Hepatitis B Using Geostatistical Analysis Techniques
  Shaobo Zhong, Yong Xue, Chunxiang Cao, Wuchun Cao,
  Xiaowen Li, Jianping Guo, Liqun Fang 464

eMicrob: A Grid-Based Spatial Epidemiology Application
  *Jianping Guo, Yong Xue, Chunxiang Cao, Wuchun Cao,
  Xiaowen Li, Jianqin Wang, Liqun Fang* 472

Self-organizing Maps as Substitutes for K-Means Clustering
  Fernando Bação, Victor Lobo, Marco Painho 476

Key Technologies Research on Building a Cluster-Based Parallel
Computing System for Remote Sensing
  Guoqing Li, Dingsheng Liu 484

Grid Research on Desktop Type Software for Spatial Information
Processing
  Guoqing Li, Dingsheng Liu, Yi Sun 492

Java-Based Grid Service Spread and Implementation in Remote Sensing
Applications
  *Yanguang Wang, Yong Xue, Jianqin Wang, Chaolin Wu,
  Yincui Hu, Ying Luo, Shaobo Zhong, Jiakui Tang, Guoyin Cai* 496

Modern Computational Techniques for Environmental Data;
Application to the Global Ozone Layer
  Costas Varotsos 504

PK+ Tree: An Improved Spatial Index Structure of PK Tree
  Xiaolin Wang, Yingwei Luo, Lishan Yu, Zhuoqun Xu 511

Design Hierarchical Component-Based WebGIS
  Yingwei Luo, Xiaolin Wang, Guomin Xiong, Zhuoqun Xu 515

Workshop on "Computational Economics and Finance"

Adaptive Smoothing Neural Networks in Foreign Exchange Rate
Forecasting
  Lean Yu, Shouyang Wang, Kin Keung Lai 523

Credit Scoring via PCALWM
  Jianping Li, Weixuan Xu, Yong Shi 531

Optimization of Bandwidth Allocation in Communication Networks
with Penalty Cost
  Jun Wu, Wuyi Yue, Shouyang Wang 539

Improving Clustering Analysis for Credit Card Accounts Classification
  Yi Peng, Gang Kou, Yong Shi, Zhengxin Chen 548

A Fuzzy Index Tracking Portfolio Selection Model
  Yong Fang, Shou-Yang Wang 554

Application of Activity-Based Costing in a Manufacturing Company:
A Comparison with Traditional Costing
  *Gonca Tuncel, Derya Eren Akyol, Gunhan Mirac Bayhan,
  Utku Koker* 562

Welfare for Economy Under Awareness
  Ken Horie, Takashi Matsuhisa 570

On-line Multi-attributes Procurement Combinatorial Auctions Bidding
Strategies
  Jian Chen, He Huang 578

Workshop on "Computer Algebra Systems and Applications, CASA 2005"

An Algebraic Method for Analyzing Open-Loop Dynamic Systems
  W. Zhou, D.J. Jeffrey, G.J. Reid 586

Pointwise and Uniform Power Series Convergence
  C. D'Apice, G. Gargiulo, R. Manzo 594

Development of SyNRAC
  Hitoshi Yanami, Hirokazu Anai 602

A LiE Subroutine for Computing Prehomogeneous Spaces
Associated with Complex Nilpotent Orbits
  Steven Glenn Jackson, Alfred G. Noël 611

Computing Valuation Popov Forms
  Mark Giesbrecht, George Labahn, Yang Zhang 619

Modeling and Simulation of High-Speed Machining Processes Based on
Matlab/Simulink
  Rodolfo E. Haber, J.R. Alique, S. Ros, R.H. Haber 627

Remote Access to a Symbolic Computation System for Algebraic
Topology: A Client-Server Approach
  Mirian Andrés, Vico Pascual, Ana Romero, Julio Rubio 635

Symbolic Calculation of the Generalized Inertia Matrix of Robots with
a Large Number of Joints
*Ramutis Bansevičius, Algimantas Čepulkauskas, Regina Kulvietienė,
Genadijus Kulvietis* 643

Revisiting Some Control Schemes for Chaotic Synchronization with
Mathematica
Andrés Iglesias, Akemi Galvez 651

Three Brick Method of the Partial Fraction Decomposition of Some
Type of Rational Expression
Damian Słota, Roman Wituła 659

Non Binary Codes and "Mathematica" Calculations: Reed-Solomon
Codes Over GF (2^n)
Igor Gashkov 663

Stokes-Flow Problem Solved Using Maple
Pratibha, D.J. Jeffrey 667

Workshop on "Intelligent Agents in Computing Systems" – The Agent Days 2005 in Atlanta

Grounding a Descriptive Language in Cognitive Agents Using
Consensus Methods
Agnieszka Pieczynska-Kuchtiak 671

Fault-Tolerant and Scalable Protocols for Replicated Services in Mobile
Agent Systems
JinHo Ahn, Sung-Gi Min 679

Multi-agent System Architectures for Wireless Sensor Networks
Richard Tynan, G.M.P. O'Hare, David Marsh, Donal O'Kane 687

ACCESS: An Agent Based Architecture for the Rapid Prototyping of
Location Aware Services
*Robin Strahan, Gregory O'Hare, Conor Muldoon, Donnacha Phelan,
Rem Collier* 695

Immune-Based Optimization of Predicting Neural Networks
Aleksander Byrski, Marek Kisiel-Dorohinicki 703

Algorithm of Behavior Evaluation in Multi-agent System
Gabriel Rojek, Renata Cięciwa, Krzysztof Cetnarowicz 711

Formal Specification of Holonic Multi-agent Systems Framework
Sebastian Rodriguez, Vincent Hilaire, Abder Koukam 719

The Dynamics of Computing Agent Systems
M. Smołka, P. Uhruski, R. Schaefer, M. Grochowski 727

Workshop on "Parallel Monte Carlo Algorithms for Diverse Applications in a Distributed Setting"

A Superconvergent Monte Carlo Method for Multiple Integrals on the Grid
Sofiya Ivanovska, Emanouil Atanassov, Aneta Karaivanova 735

A Sparse Parallel Hybrid Monte Carlo Algorithm for Matrix Computations
Simon Branford, Christian Weihrauch, Vassil Alexandrov 743

Parallel Hybrid Monte Carlo Algorithms for Matrix Computations
V. Alexandrov, E. Atanassov, I. Dimov, S. Branford, A. Thandavan, C. Weihrauch 752

An Efficient Monte Carlo Approach for Solving Linear Problems in Biomolecular Electrostatics
Charles Fleming, Michael Mascagni, Nikolai Simonov 760

Finding the Smallest Eigenvalue by the Inverse Monte Carlo Method with Refinement
Vassil Alexandrov, Aneta Karaivanova 766

On the Scrambled Soboĺ Sequence
Hongmei Chi, Peter Beerli, Deidre W. Evans, Micheal Mascagni 775

Poster Session I

Reconstruction Algorithm of Signals from Special Samples in Spline Spaces
Jun Xian, Degao Li 783

Fast In-place Integer Radix Sorting
Fouad El-Aker 788

Dimension Reduction for Clustering Time Series Using Global Characteristics
Xiaozhe Wang, Kate A. Smith, Rob J. Hyndman 792

On Algorithm for Estimation of Selecting Core
*Youngjin Ahn, Moonseong Kim, Young-Cheol Bang,
Hyunseung Choo* 796

A Hybrid Mining Model Based on Neural Network and Kernel
Smoothing Technique
Defu Zhang, Qingshan Jiang, Xin Li 801

An Efficient User-Oriented Clustering of Web Search Results
Keke Cai, Jiajun Bu, Chun Chen 806

Artificial Immune System for Medical Data Classification
Wiesław Wajs, Piotr Wais, Mariusz Święcicki, Hubert Wojtowicz ... 810

EFoX: A Scalable Method for Extracting Frequent Subtrees
Juryon Paik, Dong Ryeol Shin, Ungmo Kim 813

An Efficient Real-Time Frequent Pattern Mining Technique Using
Diff-Sets
Rajanish Dass, Ambuj Mahanti 818

Improved Fully Automatic Liver Segmentation Using Histogram Tail
Threshold Algorithms
Kyung-Sik Seo 822

Directly Rasterizing Straight Line by Calculating the Intersection Point
Hua Zhang, Changqian Zhu, Qiang Zhao, Hao Shen 826

PrefixUnion: Mining Traversal Patterns Efficiently in Virtual
Environments
Shao-Shin Hung, Ting-Chia Kuo, Damon Shing-Min Liu 830

Efficient Interactive Pre-integrated Volume Rendering
Heewon Kye, Helen Hong, Yeong Gil Shin 834

Ncvtk: A Program for Visualizing Planetary Data
Alexander Pletzer, Remik Ziemlinski, Jared Cohen 838

Efficient Multimodality Volume Fusion Using Graphics Hardware
Helen Hong, Juhee Bae, Heewon Kye, Yeong Gil Shin 842

G^1 Continuity Triangular Patches Interpolation Based on PN Triangles
Zhihong Mao, Lizhuang Ma, Mingxi Zhao 846

Estimating 3D Object Coordinates from Markerless Scenes
Ki Woon Kwon, Sung Wook Baik, Seong-Whan Lee 850

Stochastic Fluid Model Analysis for Campus Grid Storage Service
  Xiaofeng Shi, Huifeng Xue, Zhiqun Deng 854

Grid Computing Environment Using Ontology Based Service
  Ana Marilza Pernas, Mario Dantas 858

Distributed Object-Oriented Wargame Simulation on Access Grid
  Joong-Ho Lim, Tae-Dong Lee, Chang-Sung Jeong 862

RTI Execution Environment Using Open Grid Service Architecture
  Ki-Young Choi, Tae-Dong Lee, Chang-Sung Jeong 866

Heterogeneous Grid Computing: Issues and Early Benchmarks
  *Eamonn Kenny, Brian Coghlan, George Tsouloupas,
  Marios Dikaiakos, John Walsh, Stephen Childs,
  David O'Callaghan, Geoff Quigley* 870

GRAMS: Grid Resource Analysis and Monitoring System
  Hongning Dai, Minglu Li, Linpeng Huang, Yi Wang, Feng Hong 875

Transaction Oriented Computing (Hive Computing) Using GRAM-Soft
  Kaviraju Ramanna Dyapur, Kiran Kumar Patnaik 879

Data-Parallel Method for Georeferencing of MODIS Level 1B Data Using Grid Computing
  Yincui Hu, Yong Xue, Jiakui Tang, Shaobo Zhong, Guoyin Cai 883

An Engineering Computation Oriented Grid Project: Design and Implementation
  Xianqing Wang, Qinhuai Zeng, Dingwu Feng, Changqin Huang 887

Iterative and Parallel Algorithm Design from High Level Language Traces
  Daniel E. Cooke, J. Nelson Rushton 891

An Application of the Adomian Decomposition Method for Inverse Stefan Problem with Neumann's Boundary Condition
  Radosław Grzymkowski, Damian Słota 895

Group Homotopy Algorithm with a Parameterized Newton Iteration for Symmetric Eigen Problems
  Ran Baik, Karabi Datta, Yoopyo Hong 899

Numerical Simulation of Three-Dimensional Vertically Aligned Quantum Dot Array
  Weichung Wang, Tsung-Min Hwang 908

Semi-systolic Architecture for Modular Multiplication over $GF(2^m)$
  Hyun-Sung Kim, Il-Soo Jeon 912

Poster Session II

Meta Services: Abstract a Workflow in Computational Grid Environments
  Sangkeon Lee, Jaeyoung Choi 916

CEGA: A Workflow PSE for Computational Applications
  Yoonhee Kim 920

A Meta-heuristic Applied for a Topologic Pickup and Delivery Problem with Time Windows Constraints
  Jesús Fabián López Pérez 924

Three Classifiers for Acute Abdominal Pain Diagnosis − Comparative Study
  Michal Wozniak 929

Grid-Technology for Chemical Reactions Calculation
  Gabriel Balint-Kurti, Alexander Bogdanov, Ashot Gevorkyan, Yuriy Gorbachev, Tigran Hakobyan, Gunnar Nyman, Irina Shoshmina, Elena Stankova 933

A Fair Bulk Data Transmission Protocol in Grid Environments
  Fanjun Su, Xuezeng Pan, Yong lv, Lingdi Ping 937

A Neural Network Model for Classification of Facial Expressions Based on Dimension Model
  Young-Suk Shin 941

A Method for Local Tuning of Fuzzy Membership Functions
  Ahmet Çinar 945

QoS-Enabled Service Discovery Using Agent Platform
  Kee-Hyun Choi, Ho-Jin Shin, Dong-Ryeol Shin 950

A Quick Generation Method of Sequence Pair for Block Placement
  Mingxu Huo, Koubao Ding 954

A Space-Efficient Algorithm for Pre-distributing Pairwise Keys in Sensor Networks
  Taekyun Kim, Sangjin Kim, Heekuck Oh 958

An Architecture for Lightweight Service Discovery Protocol in MANET
  Byong-In Lim, Kee-Hyun Choi, Dong-Ryeol Shin 963

An Enhanced Location Management Scheme for Hierarchical Mobile
IPv6 Networks
  Myung-Kyu Yi 967

A Genetic Machine Learning Algorithm for Load Balancing in Cluster
Configurations
  M.A.R. Dantas, A.R. Pinto 971

A Parallel Algorithm for Computing Shortest Paths in Large-Scale
Networks
  Guozhen Tan, Xiaohui Ping 975

Exploiting Parallelization for RNA Secondary Structure Prediction in
Cluster
  Guangming Tan, Shengzhong Feng, Ninghui Sun 979

Improving Performance of Distributed Haskell in Mosix Clusters
  Lori Collins, Murray Gross, P.A. Whitlock 983

Investigation of Cache Coherence Strategies in a Mobile Client/Server
Environment
  C.D.M. Berkenbrock, M.A.R. Dantas 987

Parallel Files Distribution
  Laurentiu Cucos, Elise de Doncker 991

Dynamic Dominant Index Set for Mobile Peer-to-Peer Networks
  Wei Shi, Shanping Li, Gang Peng, Xin Lin 995

Task Mapping Algorithm for Heterogeneous Computing System
Allowing High Throughput and Load Balancing
  Sung Chune Choi, Hee Yong Youn 1000

An Approach for Eye Detection Using Parallel Genetic Algorithm
  A. Cagatay Talay 1004

Graph Representation of Nested Software Structure
  Leszek Kotulski 1008

Transaction Routing in Real-Time Shared Disks Clusters
  Kyungoh Ohn, Sangho Lee, Haengrae Cho 1012

Implementation of a Distributed Data Mining System
  Ju Cho, Sung Baik, Jerzy Bala 1016

Hierarchical Infrastructure for Large-Scale Distributed
Privacy-Preserving Data Mining
  Jinlong Wang, Congfu Xu, Huifeng Shen, Yunhe Pan 1020

Poster Session III

Prediction of Protein Interactions by the Domain and Sub-cellular
Localization Information
  Jinsun Hong, Kyungsook Han 1024

Online Prediction of Interacting Proteins with a User-Specified Protein
  Byungkyu Park, Kyungsook Han 1028

An Abstract Model for Service Compositions Based on Agents
  Jinkui Xie, Linpeng Huang 1032

An Approach of Nonlinear Model Multi-step-ahead Predictive Control
Based on SVM
  Weimin Zhong, Daoying Pi, Youxian Sun 1036

Simulation Embedded in Optimization – A Key for the Effective
Learning Process in (about) Complex, Dynamical Systems
  Elżbieta Kasperska, Elwira Mateja-Losa 1040

Analysis of the Chaotic Phenomena in Securities Business of China
  Chong Fu, Su-Ju Li, Hai Yu, Wei-Yong Zhu 1044

Pulsating Flow and Platelet Aggregation
  Xin-She Yang 1048

Context Adaptive Self-configuration System
  Seunghwa Lee, Eunseok Lee 1052

Modeling of Communication Delays Aiming at the Design of Networked
Supervisory and Control Systems. A First Approach
  Karina Cantillo, Rodolfo E. Haber, Angel Alique, Ramón Galán 1056

Architecture Modeling and Simulation for Supporting Multimedia
Services in Broadband Wireless Networks
  Do-Hyeon Kim, Beongku An 1060

Visualization for Genetic Evolution of Target Movement in Battle Fields
  S. Baik, J. Bala, A. Hadjarian, P. Pachowicz, J. Cho, S. Moon 1064

Comfortable Driver Behavior Modeling for Car Following of Pervasive
Computing Environment
  Yanfei Liu, Zhaohui Wu 1068

A Courseware Development Methodology for Establishing
Practice-Based Network Course
  Jahwan Koo, Seongjin Ahn 1072

Solving Anisotropic Transport Equation on Misaligned Grids
  J. Chen, S.C. Jardin, H.R. Strauss 1076

The Design of Fuzzy Controller by Means of Evolutionary Computing
and Neurofuzzy Networks
  Sung-Kwun Oh, Seok-Beom Roh 1080

Boundary Effects in Stokes' Problem with Melting
  Arup Mukherjee, John G. Stevens 1084

A Software Debugging Method Based on Pairwise Testing
  Liang Shi, Changhai Nie, Baowen Xu 1088

Heuristic Algorithm for Anycast Flow Assignment in
Connection-Oriented Networks
  Krzysztof Walkowiak 1092

Isotropic Vector Matrix Grid and Face-Centered Cubic Lattice Data
Structures
  J.F. Nystrom, Carryn Bellomo 1096

Design of Evolutionally Optimized Rule-Based Fuzzy Neural Networks
Based on Fuzzy Relation and Evolutionary Optimization
  Byoung-Jun Park, Sung-Kwun Oh, Witold Pedrycz, Hyun-Ki Kim .. 1100

Uniformly Convergent Computational Technique for Singularly
Perturbed Self-adjoint Mixed Boundary-Value Problems
  Rajesh K. Bawa, S. Natesan 1104

Fuzzy System Analysis of Beach Litter Components
  Can Elmar Balas 1108

Exotic Option Prices Simulated by Monte Carlo Method on Market
Driven by Diffusion with Poisson Jumps and Stochastic Volatility
  Magdalena Broszkiewicz, Aleksander Janicki 1112

Computational Complexity and Distributed Execution in Water
Quality Management
   Maria Chtepen, Filip Claeys, Bart Dhoedt, Peter Vanrolleghem,
   Piet Demeester 1116

Traffic Grooming Based on Shortest Path in Optical WDM Mesh
Networks
   Yeo-Ran Yoon, Tae-Jin Lee, Min Young Chung, Hyunseung Choo ... 1120

Prompt Detection of Changepoint in the Operation of Networked
Systems
   Hyunsoo Kim, Hee Yong Youn 1125

Author Index 1131

Computing for Eigenpairs on Globally Convergent Iterative Method for Hermitian Matrices

Ran Baik[1], Karabi Datta[2], and Yoopyo Hong[2]

[1] Department of Computer Engineering, Honam University,
Gwangju 506-090, Korea
[2] Department of Mathematical Sciences, Northern Illinois University,
DeKalb, IL 60115, USA
baik@honam.ac.kr, {dattak, hong}@math.niu.edu

Abstract. Let $A = A^* \in M_n$ and $\mathcal{L} = \{(U_k, \lambda_k) | \ U_k \in \mathbb{C}^n, ||U_k|| = 1 \text{ and } \lambda_k \in \mathbb{R}\}$ for $k = 1, \cdots, n$ be the set of eigenpairs of A. In this paper we develop a modified Newton method that converges to a point in $\mathcal{L}$ starting from any point in a compact subset $\mathcal{D} \subseteq \mathbb{C}^{n+1}, \mathcal{L} \subseteq \mathcal{D}$.

1 Introduction

We denote by M_n the space of n-by-n complex matrices. We denote by $\sigma(A)$ the set of eigenvalues of $A \in M_n$. Let $A \in M_n$ be Hermitian. Then there is a unitary

$$U = [U_1 \cdots U_n] \in M_n \text{ such that } A = U \begin{bmatrix} \lambda_1 & & 0 \\ & \ddots & \\ 0 & & \lambda_n \end{bmatrix} U^*, \ \lambda_k \in \mathbb{R}.$$

We assume that the eigenvalues λ_k of A are arranged in decreasing order, i.e., $\lambda_1 \geq \cdots \geq \lambda_n$ [3, chapter 4].
Let $\mathcal{L} = \left\{ \begin{bmatrix} U_k \\ \lambda_k \end{bmatrix} | \ U_k \in \mathbb{C}^n, ||U_k|| = 1, \text{ and } \lambda_k \in \mathbb{R} \right\}_{k=1,\ldots,n}$ be the set of eigenpairs of A, and suppose $\mathcal{D} = \left\{ \begin{bmatrix} X \\ \alpha \end{bmatrix} | \ X \in \mathbb{C}^n, \ ||X||_2 = 1, \text{ and } \alpha \in [a, b] \right\}$ be a compact subset of $\mathbb{C}^n \times \mathbb{R}$ such that $\mathcal{L} \subseteq \mathcal{D}$. The purpose of this paper is to compute on globally convergent iteration method which converges to an eigenpair of A, i.e., an element of $\mathcal{L}$, starting from any arbitrary point $\begin{bmatrix} X \\ \alpha \end{bmatrix} \in \mathcal{D}$. The following is the usual Newton method for obtaining an eigenpair of a hermitian matrix $A \in M_n$[4] : Consider $G : \mathbb{C}^n \times \mathbb{R} \to \mathbb{C}^n \times \mathbb{R}$ such that

$$G\left(\begin{bmatrix} X \\ \alpha \end{bmatrix}\right) = \begin{bmatrix} (\alpha I - A)X \\ X^*X - 1 \end{bmatrix}. \qquad (1)$$

Then $\mathcal{L}$ is the set of solutions for $G\left(\begin{bmatrix} X \\ \alpha \end{bmatrix}\right) = 0$. Assuming the matrix $\begin{bmatrix} \alpha I - A & X \\ 2X^* & 0 \end{bmatrix}$
$\in M_{n+1}$ is invertible, then the usual newton iteration is

$$\begin{bmatrix} X' \\ \alpha' \end{bmatrix} = \begin{bmatrix} X \\ \alpha \end{bmatrix} - \begin{bmatrix} \alpha I - A & X \\ 2X^* & 0 \end{bmatrix}^{-1} \begin{bmatrix} (\alpha I - A)X \\ X^*X - 1 \end{bmatrix} \qquad (2)$$

It is well known that the newton's method has a local quadratic convergence rate [2], that is there is a small neighborhood N_{ϵ_k} for each eigenpair $\begin{bmatrix} U_k \\ \lambda_k \end{bmatrix}$ such that if $\begin{bmatrix} X \\ \alpha \end{bmatrix} \equiv \begin{bmatrix} X^{(0)} \\ \alpha^{(0)} \end{bmatrix} \in N_{\epsilon_k}$ then $\left\| \begin{bmatrix} X^{(i+1)} \\ \alpha^{(i+1)} \end{bmatrix} - \begin{bmatrix} U_k \\ \lambda_k \end{bmatrix} \right\|_2 \leq C \left\| \begin{bmatrix} X^{(i)} \\ \alpha^{(i)} \end{bmatrix} - \begin{bmatrix} U_k \\ \lambda_k \end{bmatrix} \right\|_2^2$
for all $i = 0, 1, \cdots$, where $C < \infty$ is a positive constant.

We call N_{ϵ_k} the quadratic convergence neighborhood of the eigenpair $\begin{bmatrix} U_k \\ \lambda_k \end{bmatrix}$.
Although the specific determination of each N_{ϵ_k} is an extremely difficult task, if the method converges to a point in $\mathcal{L}$ then we know the rate of convergence will eventually be quadratic. It can be shown easily that the newton's method is not global. We provide an example:

Example 1. Let $A = \begin{bmatrix} 1.1 + \epsilon & 0 \\ 0 & 0.9 \end{bmatrix}$, $\epsilon > 0$ be the objective matrix with eigenpairs $\begin{bmatrix} U_1 \\ \lambda_1 \end{bmatrix} = \begin{bmatrix} 1 \\ 0 \\ 1.1 + \epsilon \end{bmatrix}$ and $\begin{bmatrix} U_2 \\ \lambda_2 \end{bmatrix} = \begin{bmatrix} 0 \\ 1 \\ 0.9 \end{bmatrix}$. Suppose the initial points are

$\begin{bmatrix} X_1 \\ \alpha_1 \end{bmatrix} = \begin{bmatrix} \frac{1}{\sqrt{2}} \\ \frac{1}{\sqrt{2}} \\ 1 \end{bmatrix}$ and $\begin{bmatrix} X_2 \\ \alpha_2 \end{bmatrix} = \begin{bmatrix} \frac{-1}{\sqrt{2}} \\ \frac{1}{\sqrt{2}} \\ 1 \end{bmatrix}$.

Then the newton iteration (2) becomes

$$\begin{bmatrix} X_1' \\ \alpha_1' \end{bmatrix} = \begin{bmatrix} X_1 \\ \alpha_1 \end{bmatrix} - \begin{bmatrix} \alpha_1 I - A & X_1 \\ 2X_1^* & 0 \end{bmatrix}^{-1} \begin{bmatrix} (\alpha_1 I - A)X_1 \\ X_1^*X_1 - 1 \end{bmatrix}$$

$$= \begin{bmatrix} \frac{1}{\sqrt{2}} \\ \frac{1}{\sqrt{2}} \\ 1 \end{bmatrix} - \begin{bmatrix} -0.1-\epsilon & 0 & \frac{1}{\sqrt{2}} \\ 0 & 0.1 & \frac{1}{\sqrt{2}} \\ \frac{2}{\sqrt{2}} & \frac{2}{\sqrt{2}} & 0 \end{bmatrix}^{-1} \begin{bmatrix} \frac{-(0.1+\epsilon)}{\sqrt{2}} \\ \frac{0.1}{\sqrt{2}} \\ 0 \end{bmatrix}$$

$$= \begin{bmatrix} \frac{1}{\sqrt{2}} \\ \frac{1}{\sqrt{2}} \\ 1 \end{bmatrix} - 1/\epsilon \begin{bmatrix} -1 & 1 & \frac{-0.1}{\sqrt{2}} \\ 1 & -1 & \frac{0.1+\epsilon}{\sqrt{2}} \\ \frac{-0.2}{\sqrt{2}} & \frac{2(0.1+\epsilon)}{\sqrt{2}} & -0.1(0.1+\epsilon) \end{bmatrix} \begin{bmatrix} \frac{-(0.1+\epsilon)}{\sqrt{2}} \\ \frac{0.1}{\sqrt{2}} \\ 0 \end{bmatrix}$$

$$= \begin{bmatrix} \frac{1}{\sqrt{2}} \\ \frac{1}{\sqrt{2}} \\ 1 \end{bmatrix} - 1/\epsilon \begin{bmatrix} \frac{(0.2+\epsilon)}{\sqrt{2}} \\ \frac{-(0.2+\epsilon)}{\sqrt{2}} \\ 0.02 + 0.2\epsilon \end{bmatrix}$$

Thus if ϵ goes to zero, the iteration diverges. Similarly, for the initial eigenpair $\begin{bmatrix} X_2 \\ \alpha_2 \end{bmatrix}$. We modify the newton method in order to have a global convergence. There are several considerations to give for the modification. First, under the modification we desire the pair $\begin{bmatrix} X^{(i)} \\ \alpha^{(i)} \end{bmatrix}$ gets closer to an eigenpair at each step of the iteration, i.e., $d_\mathcal{L}\left(\begin{bmatrix} X^{(i+1)} \\ \alpha^{(i+1)} \end{bmatrix}\right) \leq d_\mathcal{L}\left(\begin{bmatrix} X^{(i)} \\ \alpha^{(i)} \end{bmatrix}\right)$ where $d_\mathcal{L}$ is a suitable distance measure from a point to $\mathcal{L}$. It will ensure the points under the iteration remain in $\mathcal{D}$. Second, we want to modify the method the least amount as possible in order to preserve the original properties of the newton's Method, for example, local quadratic convergence. Third, the modified method should be simple to implement, requires almost the same procedures as the original newton iteration.

2   Modification for Global Newton Iteration

Consider the newton iteration (2): $\begin{bmatrix} X' \\ \alpha' \end{bmatrix} = \begin{bmatrix} X \\ \alpha \end{bmatrix} - \begin{bmatrix} \alpha I - A & X \\ 2X^* & 0 \end{bmatrix}^{-1} \begin{bmatrix} (\alpha I - A)X \\ X^*X - 1 \end{bmatrix}$

Then, assuming $\alpha \neq 0$ and $\alpha \notin \sigma(A)$

$$\begin{bmatrix} X' \\ \alpha' \end{bmatrix} = \left(I - \begin{bmatrix} \alpha I - A & X \\ 2X^* & 0 \end{bmatrix}^{-1} \begin{bmatrix} \alpha I - A & 0 \\ X^* & -1/\alpha \end{bmatrix}\right) \begin{bmatrix} X \\ \alpha \end{bmatrix}$$

$$= \begin{bmatrix} \alpha I - A & X \\ 2X^* & 0 \end{bmatrix}^{-1} \begin{bmatrix} 0 & X \\ X^* & 1/\alpha \end{bmatrix} \begin{bmatrix} X \\ \alpha \end{bmatrix}, \text{ or } \begin{bmatrix} \alpha I - A & X \\ 2X^* & 0 \end{bmatrix} \begin{bmatrix} X' \\ \alpha' \end{bmatrix} = \begin{bmatrix} 0 & X \\ X^* & 1/\alpha \end{bmatrix} \begin{bmatrix} X \\ \alpha \end{bmatrix}.$$

Choose a parameter $t > 0$ so that the method takes the form [1]

$$\begin{bmatrix} \alpha I - A & X \\ 2X^* & 0 \end{bmatrix} \begin{bmatrix} X' \\ \alpha' \end{bmatrix} = \begin{bmatrix} I & 0 \\ 0 & t \end{bmatrix} \begin{bmatrix} 0 & X \\ X^* & 1/\alpha \end{bmatrix} \begin{bmatrix} X \\ \alpha \end{bmatrix}. \tag{3}$$

Then

$$(\alpha I - A)X' + \alpha' X = \alpha X \tag{4}$$

and

$$2X^*X' = t(X^*X + 1) \tag{5}$$

From (4), we have $X' = (\alpha - \alpha')(\alpha I - A)^{-1}X$, and hence from (5) $X^*X' = \frac{1}{2}t(X^*X + 1) = (\alpha - \alpha')X^*(\alpha I - A)^{-1}X$. Set $\beta \equiv X^*(\alpha I - A)^{-1}X$. Then $\frac{t}{2}(X^*X + 1) = \beta(\alpha - \alpha')$, or $(\alpha - \alpha') = \frac{t}{2}(X^*X + 1)\frac{1}{\beta}$. Thus, $X' = \frac{t}{2}(X^*X + 1)\frac{1}{\beta}(\alpha I - A)^{-1}X$, and $\alpha' = \alpha - \frac{t}{2}(X^*X + 1)\frac{1}{\beta}$. If we normalize the vector $X \in \mathbb{C}^n$ in each step of the iteration, $\frac{1}{2}(X^*X + 1) = 1$.

Thus we have $X' = \frac{1}{\|\frac{t}{\beta}(\alpha I - A)^{-1}X\|_2} \frac{t}{\beta}(\alpha I - A)^{-1}X$, and $\alpha' = \alpha - \frac{t}{\beta}$. Set $\hat{\beta} \equiv \|(\alpha I - A)^{-1}X\|_2 = (X^*(\alpha I - A)^{-2}X)^{\frac{1}{2}}$.

Then we have $X' = \frac{1}{\hat{\beta}} \frac{|\beta|}{\beta}(\alpha I - A)^{-1}X$, and $\alpha' = \alpha - \frac{t}{\beta}$. Since $\frac{|\beta|}{\beta} = \pm 1$, we ignore the sign. Then the parameterized newton method takes a form:

$$X' = \frac{1}{\hat{\beta}}(\alpha I - A)^{-1}X, \text{ and} \tag{6}$$

$$\alpha' = \alpha - \frac{t}{\beta}. \tag{7}$$

Now, suppose $\begin{bmatrix} X \\ \alpha \end{bmatrix} \in \mathcal{D}$ and Let $\mathcal{L} = \left\{ \begin{bmatrix} U_k \\ \lambda_k \end{bmatrix} \mid U_k \in \mathbb{C}^n, \|U_k\|_2 = 1, \text{ and } \lambda_k \in \mathbb{R} \right\}$ be the set of all eigenpairs of A. Define a distance measure from a point $\begin{bmatrix} X \\ \alpha \end{bmatrix}$ to $\mathcal{L}$ by

$$d_{\mathcal{L}}\left(\begin{bmatrix} X \\ \alpha \end{bmatrix}\right) \equiv \|(\alpha I - A)X\|_2. \tag{8}$$

Clearly, $d_{\mathcal{L}}\left(\begin{bmatrix} X \\ \alpha \end{bmatrix}\right) \geq 0$, $d_{\mathcal{L}}\left(\begin{bmatrix} X \\ \alpha \end{bmatrix}\right) = 0$ implies $\begin{bmatrix} X \\ \alpha \end{bmatrix} \in \mathcal{L}$, and $d_{\mathcal{L}} : \mathcal{D} \to \mathbb{R}^+$ is continuous (since $\mathcal{D}$ is compact, $d_{\mathcal{L}}$ is actually uniformly continuous)[5].

We have the following.

Lemma 1. *Let $A \in M_n$ be Hermitian. Consider the parameterized newton's method $X' = \frac{1}{\hat{\beta}}(\alpha I - A)^{-1}X$ and $\alpha' = \alpha - \frac{t}{\beta}$, where $\beta = X^*(\alpha I - A)^{-1}X$ and $\hat{\beta} = \|(\alpha I - A)^{-1}X\|_2 = (X^*(\alpha I - A)^{-2}X)^{1/2}$. Then $d_{\mathcal{L}}\left(\begin{bmatrix} X' \\ \alpha' \end{bmatrix}\right)$ is minimized at*

$$t = \left(\frac{\beta}{\hat{\beta}}\right)^2 \text{ with } \min_t d_{\mathcal{L}}\left(\begin{bmatrix} X' \\ \alpha' \end{bmatrix}\right) = \frac{1}{\hat{\beta}}\left(1 - \left(\frac{\beta}{\hat{\beta}}\right)^2\right)^{\frac{1}{2}}.$$

Proof: Suppose $X' = \frac{1}{\hat{\beta}}(\alpha I - A)^{-1}X$ and $\alpha' = \alpha - \frac{t}{\beta}$. Then

$$d_{\mathcal{L}}^2\left(\begin{bmatrix} X' \\ \alpha' \end{bmatrix}\right) = \|((\alpha - \frac{t}{\beta})I - A)\frac{1}{\hat{\beta}}(\alpha I - A)^{-1}X\|_2^2$$

$$= \frac{1}{\hat{\beta}^2}X^*(I - 2\frac{t}{\beta}(\alpha I - A)^{-1} + (\frac{t}{\beta})^2(\alpha I - A)^{-2})X$$

$$= \frac{1}{\hat{\beta}^2}(1 - 2\frac{t}{\beta} \cdot \beta + t^2\left(\frac{\hat{\beta}}{\beta}\right)^2).$$

Thus, $d_{\mathcal{L}}^2\left(\begin{bmatrix} X' \\ \alpha' \end{bmatrix}\right)$ is minimized at $t = \left(\frac{\beta}{\hat{\beta}}\right)^2$ with $\min_t d_{\mathcal{L}}\left(\begin{bmatrix} X' \\ \alpha' \end{bmatrix}\right) = \frac{1}{\hat{\beta}}\left(1 - \left(\frac{\beta}{\hat{\beta}}\right)^2\right)^{\frac{1}{2}}.$ □

Therefore, we have the following modification of the newton's method:

$$X' = \frac{1}{\overset{\wedge}{\beta}}(\alpha I - A)^{-1}X. \quad (9)$$

$$\alpha' = \alpha - \frac{\beta}{\overset{\wedge}{\beta}^2}. \quad (10)$$

The following result shows that the modified iteration (9), (10) is bounded.

Lemma 2. *Let $A \in M_n$ be a Hermitian such that $\sigma(A) = \{\lambda_1 \geq \cdots \geq \lambda_n, \lambda_k \in \mathbb{R}\}$. Let $\mathcal{D} = \left\{ \begin{bmatrix} X \\ \alpha \end{bmatrix} \mid X \in \mathbb{C}^n, \|X\|_2 = 1, \text{ and } \alpha \in [a,b] \right\}$ be such that $a \leq \lambda_n$ and $b \geq \lambda_1$. Suppose $\mathcal{D}' = \left\{ \begin{bmatrix} X^{(i)} \\ \alpha^{(i)} \end{bmatrix} \right\}_{)=1,\infty,\cdots}$ is the sequence of iterates of (9) and (10), i.e., $X^{(i+1)} = \frac{1}{\overset{\wedge}{\beta}^{(i)}}(\alpha^{(i)} I - A)^{-1} X^{(i)}$, and $\alpha^{(i+1)} = \alpha^{(i)} - \frac{\beta^{(i)}}{\left(\overset{\wedge}{\beta}^{(i)} \right)^2}$.*

Then $\mathcal{D}' \subset \mathcal{D}$ whenever $\begin{bmatrix} X^{(0)} \\ \alpha^{(0)} \end{bmatrix} \in \mathcal{D}$.

Theorem 3. *Suppose $\mathcal{D}' = \left\{ \begin{bmatrix} X^{(i)} \\ \alpha^{(i)} \end{bmatrix} \right\}_{)=1,\infty,\cdots}$ is the sequence of iterates of (9) and (10). Then the sequence $\left\{ d_{\mathcal{L}} \left(\begin{bmatrix} X^{(i)} \\ \alpha^{(i)} \end{bmatrix} \right) \right\}_{i=0,1,\cdots}$ is convergent.*

Note from Theorem 3 that since $\left\{ d_{\mathcal{L}} \left(\begin{bmatrix} X^{(i)} \\ \alpha^{(i)} \end{bmatrix} \right) \right\}$ is a monotone decreasing sequence that is bounded below by zero, The sequence $\left\{ d_{\mathcal{L}} \left(\begin{bmatrix} X^{(i)} \\ \alpha^{(i)} \end{bmatrix} \right) \right\}$ converges to either (i) zero, or (ii) a positive constant L. Suppose $\lim_{i \to \infty} d_{\mathcal{L}} \left(\begin{bmatrix} X^{(i)} \\ \alpha^{(i)} \end{bmatrix} \right) = 0$. Then clearly, $\left\{ d_{\mathcal{L}} \left(\begin{bmatrix} X^{(i)} \\ \alpha^{(i)} \end{bmatrix} \right) \right\}$ converges to an eigenpair of A. In the following section we discuss the case $\lim_{i \to \infty} d_{\mathcal{L}} \left(\begin{bmatrix} X^{(i)} \\ \alpha^{(i)} \end{bmatrix} \right) = L > 0$ which requires some detailed analysis. We summarize the results. Under the modified newton iteration, the sequence $\{d_{\mathcal{L}}\}$ converges to either zero or $L > 0$. If $\{d_{\mathcal{L}}\}$ converges to $L > 0$, then iterates $\left\{ \begin{bmatrix} X^{(i)} \\ \alpha^{(i)} \end{bmatrix} \right\}$ has an accumulation point $\begin{bmatrix} X \\ \alpha \end{bmatrix}$ where $d_{\mathcal{L}} \left(\begin{bmatrix} X \\ \alpha \end{bmatrix} \right) = L > 0$ such that the point $\alpha \in \mathbb{R}$ lies exactly at the midpoint of two distinct eigenvalues (each eigenvalue may have the algebraic multiplicity more than one) such that corresponding components of the vector U^*X have equal weights that are $\frac{1}{2}$ each, see Figure 1.

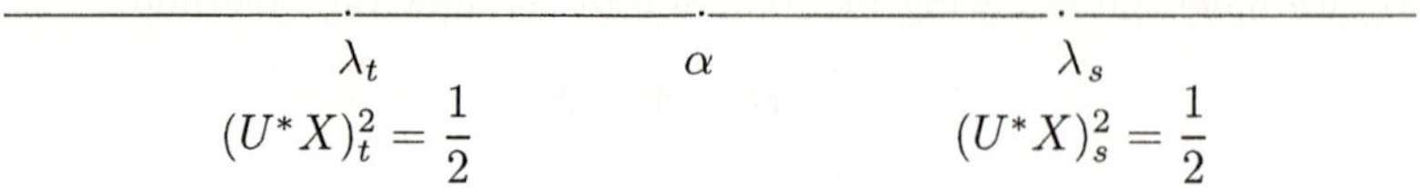

$$
\begin{array}{ccc}
\lambda_t & \alpha & \lambda_s \\
(U^*X)_t^2 = \dfrac{1}{2} & & (U^*X)_s^2 = \dfrac{1}{2}
\end{array}
$$

Fig. 1.

Therefore, $d_{\mathcal{L}}^2 = \dfrac{1}{\hat{\beta}}^2 = \sum_{k=1}^{n}(\alpha - \lambda_k)^2|y_k|^2 = (\alpha - \lambda_k)^2 = \dfrac{(\lambda_s - \lambda_t)^2}{2}$, $\lambda_s > \lambda_t$.

We conclude this section with the following consequence of above results and an example.

Theorem 4. *Suppose* $d_{\mathcal{L}}\left(\begin{bmatrix} X \\ \alpha \end{bmatrix}\right) = L$. *Then both* $\alpha + \dfrac{1}{\hat{\beta}}$ *and* $\alpha - \dfrac{1}{\hat{\beta}}$ *are the eigenvalues of A and* $\dfrac{1}{\hat{\beta}'}[(\alpha + \dfrac{1}{\hat{\beta}})I - A]^{-1}X$ *and* $\dfrac{1}{\hat{\beta}''}[(\alpha - \dfrac{1}{\hat{\beta}})I - A]^{-1}X$ *are the corresponding eigenvector of A, where* $\hat{\beta}' = \left\| \left[(\alpha + \dfrac{1}{\hat{\beta}})I - A\right]^{-1} X \right\|_2$, *and*

$\hat{\beta}'' = \left\| \left[(\alpha - \dfrac{1}{\hat{\beta}})I - A\right]^{-1} X \right\|_2$.

Example 2. Consider the Example 1. Let $A = \begin{bmatrix} 1.1 + \epsilon & 0 \\ 0 & 0.9 \end{bmatrix}$, $\epsilon > 0$. Suppose we start with initial eigenpair $\begin{bmatrix} X_1^{(0)} \\ \alpha_1^{(0)} \end{bmatrix} = \begin{bmatrix} \frac{1}{\sqrt{2}} \\ \frac{1}{\sqrt{2}} \\ 1 \end{bmatrix}$. Then

$$\beta = X^*(\alpha I - A)^{-1}X = [\tfrac{1}{\sqrt{2}} \ \tfrac{1}{\sqrt{2}}] \begin{bmatrix} -(0.1-\epsilon) & 0 \\ 0 & 0.1 \end{bmatrix}^{-1} \begin{bmatrix} \tfrac{1}{\sqrt{2}} \\ \tfrac{1}{\sqrt{2}} \end{bmatrix}$$

$$= [\tfrac{1}{\sqrt{2}} \ \tfrac{1}{\sqrt{2}}] \begin{bmatrix} -1/(0.1+\epsilon) & 0 \\ 0 & 10 \end{bmatrix} \begin{bmatrix} \tfrac{1}{\sqrt{2}} \\ \tfrac{1}{\sqrt{2}} \end{bmatrix}$$

$$= \dfrac{-1}{2(0.1+\epsilon)} + 5 = 5 - \dfrac{1}{0.2 + 2\epsilon}, \text{ and}$$

$\hat{\beta} = \|(\alpha I - A)^{-1} X\|_2 = \left\| \begin{bmatrix} -1/(0.1+\epsilon) & 0 \\ 0 & 10 \end{bmatrix} \begin{bmatrix} \tfrac{1}{\sqrt{2}} \\ \tfrac{1}{\sqrt{2}} \end{bmatrix} \right\|_2$

$$= \left(\dfrac{1}{2}\left(\dfrac{-1}{0.1+\epsilon}\right)^2 + 50\right)^{1/2} = \left(\dfrac{1}{0.02 + 0.4\epsilon + \epsilon^2} + 50\right)^{1/2}.$$

If ϵ goes to zero, then $\beta \to 0$ and $\hat{\beta} \to 10$. Notice that for $\epsilon \cong 0$

$$d_\mathcal{L}\left(\begin{bmatrix} X^{(0)} \\ \alpha^{(0)} \end{bmatrix}\right) = d_\mathcal{L}\left(\begin{bmatrix} X^{(m)} \\ \alpha^{(m)} \end{bmatrix}\right) = \ldots = 0.1.$$

Therefore by Theorem 4, we have $d_\mathcal{L}\left(\begin{bmatrix} X^{(n)} \\ \alpha^{(n)} \end{bmatrix}\right) = \frac{1}{\hat{\beta}}$.

Hence $\lambda_1(A) = 1 + 1/\hat{\beta} = 1 + .1 = 1.1$ and $\lambda_2(A) = 1 - 1/\hat{\beta} = 1 - .1 = .9$.
We obtain $X_1 = \begin{bmatrix} 1 \\ 0 \end{bmatrix}$ and $X_2 = \begin{bmatrix} 0 \\ 1 \end{bmatrix}$ by solving $(A - \lambda_1 I)X_1 = 0$ and $(A - \lambda_2 I)X_2 = 0$.

3 Examples with Modified Newton's Iterations

Example 3. Let $H = \begin{bmatrix} 1 & \frac{1}{2} & \frac{1}{3} & \cdots & \frac{1}{n} \\ \frac{1}{2} & \frac{1}{3} & & & \vdots \\ \vdots & & \ddots & & \\ \vdots & & & \ddots & \\ \frac{1}{n} & \frac{1}{n+1} & \cdots & & \frac{1}{2n-1} \end{bmatrix}$ be an n by n Hilbert matrix H.

The Hilbert matrix is a well-known example of an ill-conditioned positive definite matrix. Because the smallest eigenvalue λ_{12} of $H \in M_{12}$ is so near zero, many conventional algorithms produce $\lambda_{12} = 0$. Our method gives the following expermental results: Set the convergence criterion, $\epsilon = 2 \times 10^{-16}$, i.e., $\|(H - \alpha_k^{(i)} I)X_k^{(i)}\|_2 < \epsilon$.

Suppose $\mathcal{D} = \left\{ \begin{bmatrix} X \\ \alpha \end{bmatrix} \middle| X \in \{e_1, \cdots, e_{12}\}, \text{ and } \alpha \in \{h_{1,1}, \cdots, h_{12,12}\} \right\}$ is the initial set of points where e_i is the ith column of the identity matrix and $h_{i,i}$ is the ith diagonal entry of H.

Table 1. Eigenvalues of H_{12} by Modified Newton Iteration

Eigenvalues	Eigenvalues of H	$\|(H - \alpha_k^{(i)} I)X_k^{(i)}\|_2$
1st	1.7953720595620	4.5163365159057D-17
2nd	0.38027524595504	9.5107769421299D-17
3rd	4.4738548752181D-02	9.3288777118150D-17
4th	3.7223122378912D-03	9.5107769421299D-17
5th	2.3308908902177D-04	6.5092594645258D-17
6th	1.1163357483237D-05	6.6374428417771D-17
7th	4.0823761104312D-07	1.9236667674542D-16
8th	1.1228610666749D-08	4.9553614188006D-17
9th	2.2519644461451D-10	6.0015952254039D-17
10th	3.1113405079204D-12	6.5125904614112D-17
11th	2.6487505785549D-14	1.0932505712948D-16
12th	1.1161909467844D-16	6.0015952254039D-17

Example 4. Let $H_3 = \begin{bmatrix} 10^{40} & 10^{19} & 10^{19} \\ 10^{19} & 10^{20} & 10^{9} \\ 10^{19} & 10^{9} & 1 \end{bmatrix}$ be a 3 by 3 graded matrix. Suppose $\mathcal{D} = \left\{ \begin{bmatrix} X \\ \alpha \end{bmatrix} | X \in \{e_1, e_2, e_3\}, \text{ and } \alpha \in \{h_{1,1}, h_{2,2}, h_{3,3}\} \right\}$ is the initial set of points where e_i is the ith column of the identity matrix and $h_{i,i}$ is the ith diagonal entry of H. The graded matrix is a well-known example of an ill-conditioned symmetric matrix. The following is the result obtained by MATLAB and our method.

Table 2. Eigenvalues of H_3 by Modified Newton Iteration

	Modified Newton Method		MATLAB
Eigenvalues	Eigenvalues of H	Iteration	Eigenvalues of H
λ_1	1.0000000000000D+40	2	9.999999999999999e+39
λ_2	1.0000000000000D+20	1	0
λ_3	0.98000000000020	2	-1.000000936789517+20

References

1. Karabi Datta and Yoopyo Hong and Ran Baik Lee : Parameterized Newton's iteration for computing an Eigenpairs of a Real Symmetric Matrix in an Interval, Computational Methods in Applied Mathemaics, **3**.(2003) 517–535
2. R. A. Horn and C. R. Johnson,: Matrix Analysis, Cambridge University Press, (1985).
3. J. M. Ortega: Numerical Analysis, A Second Course. SIAM Series in Classical in Applied Mathematics, Philadelphia, SIAM Publications, (1990).
4. J. M. Ortega and W. C. Rheinboldt: Iterative Solution Of Nonlinear Equations In Several Variables, Academic Press, New York and London, (1970).
5. H.L. Royden: Real Analysis, Macmillan Publishing Company, New York,(1968).

2D FE Quad Mesh Smoothing via Angle-Based Optimization

Hongtao Xu and Timothy S. Newman

Department of Computer Science, University of Alabama in Huntsville,
Huntsville, Alabama 35899
{hxu, tnewman}@cs.uah.edu

Abstract. A new mesh smoothing algorithm that can improve quadrilateral mesh quality is presented. Poor quality meshes can produce inaccurate finite element analysis; their improvement is important. The algorithm improves mesh quality by adjusting the position of the mesh's internal nodes based on optimization of a torsion spring system using a Gauss-Newton-based approach. The approach obtains a reasonably optimal location of each internal node by optimizing the spring system's objective function. The improvement offered by applying the algorithm to real meshes is also exhibited and objectively evaluated using suitable metrics.

1 Introduction

Finite element (FE) analysis acts on mesh elements that are usually generated by applying mesh generation algorithms. Accurate FE analysis results depend on the mesh being valid (i.e., having valid elements), having no slivers, and conforming to the given domain's boundary. It is also best if the mesh's density varies smoothly.

Meshes generated by mesh generation algorithms can often be optimized with a mesh smoothing algorithm. A smoothing algorithm relocates nodes so that the mesh will be of a higher quality. Mesh smoothing is usually done in an iterative process that does not change element connectivity. One popular 2D mesh smoothing algorithm is the Laplacian Smoothing Algorithm [1]. Laplacian smoothing often produces satisfactory smoothing results. However, it can sometimes generate meshes with sliver-like elements or with invalid elements.

In this paper, we consider an alternate smoothing algorithm, Zhou and Shimada's [2] torsion spring-based algorithm, that is better at avoiding generation of slivers and invalid elements. The Zhou and Shimada algorithm can be implemented easily, but it does not optimally utilize the torsion spring system it is based on. The new smoothing approach presented here smooths a mesh of quadrilateral elements in a manner that is provably optimal for the torsion spring formulation proposed by Zhou and Shimada.

This paper is organized as follows. Section 2 discusses related work. Section 3 introduces the new algorithm. Section 4 presents results and an evaluation of the algorithm. Section 5 presents the conclusion.

2 Previous Work

A number of mesh smoothing methods for producing acceptable quality meshes have been presented previously (e.g., [1, 3, 4, 5, 6, 7, 8]), including approaches that minimize a distortion metric (e.g., as in [3]), that disconnect invalid elements from the remaining mesh (e.g., as in [4]), that solve a generalized linear programming problem (e.g., as in [5]), that combine constrained Laplacian smoothing together with an optimization-based smoothing algorithm (e.g., as in [1, 6]), that use parallel algorithms (e.g., as in [7]), and that generalize Laplacian smoothing (e.g., as in [8]). In this section, we describe the Laplacian smoothing and Zhou and Shimada's smoothing, which are relevant to our work.

The Laplacian Smoothing Algorithm [9] is an iterative method. It is widely used due to its relative efficiency and simplicity in application, although it may generate slivers or invalid elements in some cases. In 2D meshes, the Laplacian Smoothing Algorithm attempts to improve mesh quality by moving internal mesh nodes in one or more smoothing passes. In each pass, it moves each internal node to the centroid $(\bar{x}, \bar{y})$ of the polygon about the internal node. By polygon about the internal node, we mean the polygon whose vertices are the nodes connected to the internal node.

Zhou and Shimada [2] have presented a physically-based mesh smoothing algorithm. It accomplishes smoothing by moving each internal node to a better location based on modeling the edges that connect nodes as a torsion spring system. For a set of nodes connected to an internal node, if the edges that connect these nodes to the internal node are viewed as a torsion spring system, then this torsion spring system's energy can be expressed as:

$$E = \sum_{i=0}^{2(n-1)} \frac{1}{2}k\theta_i^2, \qquad (1)$$

where n is the number of nodes that are connected to the internal node by an edge, k is a constant, and θ_i is the angle between a polygon edge and the line from the internal node to the i-th connected node.

Zhou and Shimada's algorithm minimizes the energy of the torsion spring system (i.e., that was shown in Eqn. 1) and then uses a heuristic to relocate the interior nodes of the mesh. More detail can be found in [2].

3 Angle-Based Optimization Smoothing

As described earlier, the Laplacian smoothing can generate slivers and invalid elements. Other mesh smoothing methods, such as optimization-based methods including the Zhou and Shimada algorithm, can often give better quality meshes than Laplacian smoothing, especially for meshes that contain badly shaped finite elements. However, the Zhou and Shimada algorithm uses a non-optimal heuristic. It is possible to instead use an optimization approach to optimally relocate each internal node. In this section, our optimization approach that accurately minimizes the energy of the torsion spring system to produce a well-smoothed quadrilateral mesh is presented. The approach better-optimizes the new locations of all internal nodes. The approach is an extension of our earlier work on triangular meshes [10].

3.1 Objective Function for Optimal Angle

The energy of a torsion spring system model on the edges that connect nodes is provably minimal when the angle at each vertex is divided by a bisecting line. However, the bisectional lines for a polygon about an internal node seldom intersect at the same point. Zhou and Shimada's heuristic estimates an internal node's new location using averaging, which is not optimal. To find the optimal new location of nodes, our approach uses the following objective function s:

$$s = \sum_{i=0}^{n-1}[distance(\hat{D}', L_i)]^2, \quad (2)$$

where L_i is the bisectional line of the internal angle at node D_i, D_i is one node of the polygon about an internal node $\hat{D}$, and where $\hat{D}'$ is the new (i.e., optimized) position of the internal node.

We can define $t = (p, q)$ as the coordinate of an internal node. Since the function $distance(\hat{D}', L_i)$ is a function of the coordinates of an internal node, the distance can be expressed as a function f_i of the coordinate t, allowing the objective function s in Eqn. 2 to be re-written as:

$$s = \sum_{i=0}^{n-1} f_i(t)^2. \quad (3)$$

The least squares formulation for an objective function s is:

$$\min \quad s(t) = \frac{1}{2} f(t)^T f(t), \quad (4)$$

where $t = (t_1, t_2, ..., t_n)$ is a vector with n components, $f(t)$ is a column vector of m functions $f_i(t)$ and $f(t)^T$ is the row vector $f(t)^T = (f_1(t), ..., f_m(t))$.

The formulation of Eqn. 4 can be solved by any optimization that minimizes the objective function $s(t)$. When $m = n$, this problem becomes a set of non-linear equations in the form $f(t) = 0$.

3.2 Derivation of Linear Equation System

By taking the derivative of s in Eqn. 4, the following equation can be obtained:

$$P = \frac{\partial s}{\partial t} = \sum_{i=1}^{n} f_i \frac{\partial f_i}{\partial t} = \frac{\partial f^T}{\partial t} f = J^T f, \quad (5)$$

using $J^T = \frac{\partial f^T}{\partial t}$, where J^T is the transpose of the Jacobi matrix, and using $P = \left(\frac{\partial s}{\partial t_1} \frac{\partial s}{\partial t_2} \cdots \frac{\partial s}{\partial t_n}\right)^T$. Thus, the Jacobi matrix can be expanded as:

$$J = \begin{pmatrix} J_{11} & \cdots & J_{1n} \\ \vdots & \cdots & \vdots \\ J_{m1} & \cdots & J_{mn} \end{pmatrix}, \qquad (6)$$

where $J_{ij} = \frac{\partial f_i}{\partial t_j}$.

To minimize the objective function in Eqn. 4, we should have $P = 0$. By expanding P at point $t = t^n$ with a Taylor series that deletes derivatives of second and above order, we obtain

$$P(t) = P(t^n) + \left.\frac{\partial P}{\partial t}\right|_{t^n}(t - t^n). \qquad (7)$$

Then, substituting $P = 0$ into Eqn. 7, we can obtain

$$t = t^n - \left(\left.\frac{\partial P}{\partial t}\right|_{t^n}\right)^{-1} P(t^n). \qquad (8)$$

Clearly,

$$\left.\frac{\partial P}{\partial t}\right|_{t^n} = \left.\frac{\partial (J^T f)}{\partial t}\right|_{t^n}. \qquad (9)$$

Deleting the derivatives of second and above order, as is done in Eqn. 7, we obtain

$$\left.\frac{\partial P}{\partial t}\right|_{t^n} = J_n^T J_n, \qquad (10)$$

where J_n is the value of J at $t = t^n$.

By substituting Eqns. 10 and 5 into Eqn. 8, and using t^{n+1} as the next step value, the following equation can be obtained:

$$t^{n+1} = t^n - (J_n^T J_n)^{-1} J_n^T f_n. \qquad (11)$$

Next, we define

$$d^n = -(J_n^T J_n)^{-1} J_n^T f_n, \qquad (12)$$

which allows Eqn. 11 to be simplified to

$$t^{n+1} = t^n + d^n. \qquad (13)$$

In Eqn. 13, it is possible to prove that d^n is always negative. Thus, t^n decreases with increasing n [11]. Although d^n can be very small, there are still chances that a better point exists in the interval $[0, d^n]$. To further optimize Eqn. 13 in the interval, a scalar λ^n with domain $[0, 1]$ can be introduced into Eqn. 13, leading to the following equation:

$$t^{n+1} = t^n + \lambda^n d^n. \qquad (14)$$

Iterative search for a solution t is then used until a specified precision is reached. In each iteration, a one dimensional search of λ is made. The solution t can be substituted into Eqn. 4 to find an optimal s, solving the problem.

3.3 Optimization Algorithm

Next, we describe our approach's use of Gauss-Newton optimization to optimize the objective function s. Optimization of s requires finding vector t. To solve Eqn. 3, it is necessary to find the optimal λ^n in Eqn. 14. Therefore, the following problem, where only λ^n is unknown, needs to be solved:

$$\min s(t^n + \lambda^n d^n). \qquad (15)$$

To solve this problem, the following steps are used. These steps require the solution's precision ϵ to be pre-specified. That is $\left| \frac{s(t^n + d^n) - s(t^n)}{s(t^n)} \right| < \epsilon$.

Step 1. Calculate $s(t^n + d^n)$ and $s(t^n)$.
Step 2. If solution's ϵ has been reached or it is iteration $j_{\max}$, go to Step 8.
Step 3. Set $\lambda^n = 1$.
Step 4. If $s(t^n + d^n) < s(t^n)$, set $t^n = t^n + d^n$ and go to Step 1.
Step 5. Assume $s(t^n + \lambda^n d^n)$ with respect to λ^n is quadratic and find the coefficients of the quadratic polynomial using the values of s at $\lambda^n = 0$, $\lambda^n = 1$, and the derivative of s at $\lambda^n = 0$.
Step 6. Find minimum value of $s(t^n + \lambda^n d^n)$ for $0 \le \lambda^n \le 1$.
Step 7. a. If $s(t^n + d^n) < s(t^n)$, go to Step 1.
    b. Set $\lambda^n = \lambda^n / 2$.
    c. Set $t^n = t^n + \lambda^n d^n$.
    d. Go to Step 7a.
Step 8. Stop.

In practice, we have used 10 iterations ($j_{max} = 10$) which has led to reasonable solutions.

4 Results and Discussion

In this section, the qualitative characteristics and computational results of the new smoothing algorithm are presented. The algorithm's characteristics and performance are also compared with the Laplacian smoothing algorithm.

The comparison uses mesh angle and length ratios. An element's angle ratio is the ratio of the minimum angle (of the four angles of an element) to the maximum angle (of the four angles of the same element). The length ratio is the ratio of an element's minimum side length to its maximum side length.

Metrics that are derived from the angle and length ratios are used to evaluate mesh quality. The basic derived metric is the *metric ratio*. The metric ratio is the product of the angle ratio and the length ratio.

The *element area* is also used to determine quality. The metrics derived from this measure and used in evaluation are the maximum and minimum element areas for the whole mesh.

Next, we report experiments that test mesh quality for three scenarios, which we call Cases 1, 2, and 3.

Shape Improvement. The meshes for the Case 1, 2, and 3 scenarios are shown in Figure 1 (a), (b), and (c), respectively. Figure 1(d), (e), and (f) show the meshes created by applying Laplacian smoothing to the original mesh. Figure 1(g), (h), and (i) show the meshes created by applying the new smoothing algorithm to the original mesh. In all three cases, the meshes produced by both Laplacian smoothing and the new smoothing appear to be more uniform in shape than the original mesh. The mesh quality metrics for the scenarios are shown in Table 1. In this table, the worst metric ratios of the original mesh, the mesh generated by Laplacian smoothing, and the mesh generated by the new smoothing algorithm are shown. The metric values for Laplacian smoothing are all greater than the metric values for the original mesh, which means that the overall mesh uniformity in shape is improved by Laplacian smoothing. In particular, the larger worst metric value means that the worst elements are better in the mesh smoothed by the Laplacian algorithm than they are in the original mesh. The new smoothing algorithm's worst metric values are greater than the metric values for Laplacian smoothing in all three cases, which means that the new algorithm produced a mesh with a less extreme worst element.

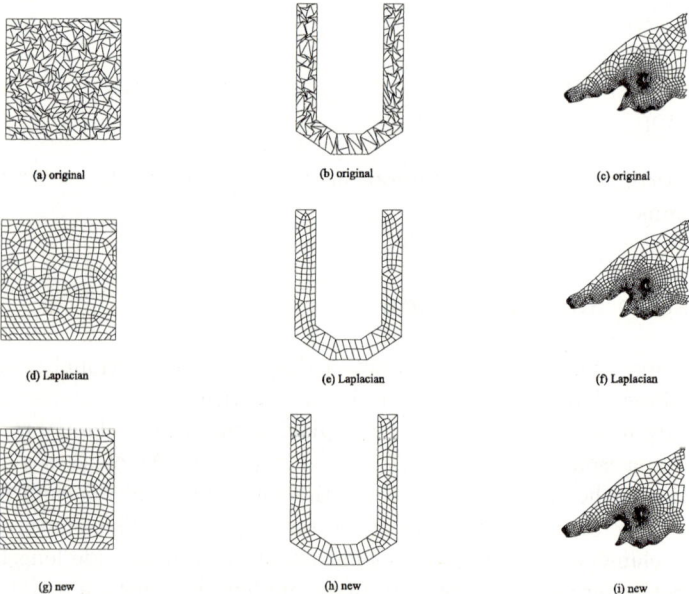

Fig. 1. Case 1, 2, 3 meshes and results of smoothings on them

Table 1. Mesh Quality Metrics for Case 1, 2 and 3 Scenario

	Metric Name	Original Mesh	Laplacian	New Smoothing
Case 1	Worst metric ratio	0.0017	0.0586	0.0591
Case 2	Worst metric ratio	0.0011	0.0793	0.0912
Case 3	Worst metric ratio	0.0019	0.0111	0.0127

In summary, the new algorithm appears to produce reasonable meshes whose worst elements are of a much higher quality than in the Laplacian smoothing; the new algorithm's worst elements are less sliver-like.

Size Improvement. Minimum and maximum element areas are used to measure the size uniformity for the quadrilateral meshes produced by Laplacian smoothing and the new smoothing. Table 2 reports these metrics for the original mesh, the mesh generated with Laplacian smoothing and the mesh generated with the new smoothing for the three cases discussed here. With the exception of the Case 2 scenario minimum area, the new smoothing's meshs' minimum areas are all greater than the minimum areas from Laplacian smoothing. The maximum areas for the new smoothing are also all less than the maximum areas from Laplacian smoothing. More importantly, the variation in extreme size is lowest for the new algorithm. Thus, the meshes generated by the new smoothing algorithm tend to have less extreme element sizes. While our examples demonstrate that Laplacian smoothing can improve mesh quality, it can be further seen that the new smoothing algorithm can generate a mesh with less extreme variation in element size than does Laplacian smoothing.

Table 2. Mesh Quality Metrics for 3 Scenarios

		Original Mesh	Laplacian	New Smoothing
Case 1	Min. area	0.0251	0.0516	0.0676
	Max. area	0.6954	0.4405	0.4125
Case 2	Min. area	2.8810	6.9078	6.8607
	Max. area	94.979	50.102	44.036
Case 3	Min. area	0.0141	0.0285	0.0326
	Max. area	14.960	14.199	12.967

5 Conclusion

In this paper, a new smoothing algorithm for quadrilateral mesh smoothing has been presented. The new smoothing algorithm uses an angle-based optimization method to optimize a torsion spring system. The formulation is set up to optimize the locations of all internal nodes. The solution is found based on the Gauss-Newton optimization.

Here, the new mesh smoothing algorithm's performance in reducing sliver elements was also reported. The testing results lead to the following conclusions. First, the new smoothing algorithm produces better quadrilateral element shapes than does the Laplacian Smoothing Algorithm. Furthermore, the new algorithm gives better mesh uniformity than that generated with Laplacian smoothing.

References

1. L. Freitag: On Combining Laplacian and Optimization-Based Mesh Smoothing Techniques. In: Proc., 6th Int'l Mesh. Roundtable, AMD-Vol. 220, London, 1997, pp. 375-390.
2. T. Zhou and K. Shimada: An Angle-Based Approach to Two-Dimensional Mesh Smoothing. In: Proc., 9th Int'l Mesh. Roundtable, New Orleans, 2000, pp. 373-384.
3. S. Canann, M. Stephenson, and T. Blacker: Optismoothing: An optimization-driven approach to mesh smoothing. Finite Elements in Analysis and Design **13** (1993), 185-190.
4. T. Li, S. Wong, Y. Hon, C. Armstrong, and R. McKeag: Smoothing by optimisation for a quadrilateral mesh with invalid element. Finite Elements in Analysis and Design **34** (2000), 37-60.
5. N. Amenta, M. Bern, and D. Eppstein: Optimal point placement for mesh smoothing. In: Proc., 8th ACM-SIAM Symp. on Disc. Alg., New Orleans, 1997, pp. 528-537.
6. S. Canann, J. Tristano, and M. Staten: An Approach to Combined Laplacian and Optimization-Based Smoothing for Triangular, Quadrilateral, and Quad-Dominant Meshes. In: Proc., 7th Int'l Mesh. Roundtable, Dearborn, Mich., 1998, pp. 479-494.
7. J. Freitag, M. Jones, and P. Plassmann: a Parallel Algorithm for Mesh Smoothing. SIAM J. on Scientific Computing **20** (1999), 2023-2040.
8. P. Hansbo: Generalized Laplacian Smoothing of Unstructured Grids. Communications in Numerical Methods in Engineering **11** (1995), 455-464.
9. D. Field: Laplacian Smoothing and Delaunay Triangulations. Comm. in Applied Numerical Methods **4** (1988), 709-712.
10. H. Xu: An Optimization Approach for 2D Finite Element Mesh Smoothing, M. S. Thesis, Dept. of Comp. Sci., Univ. of Ala. in Huntsville, Huntsville, 2003.
11. J. Nocedal and S. Wright: Numerical Optimization. Springer-Verlag, New York, 1999.

Numerical Experiments on the Solution of the Inverse Additive Singular Value Problem

G. Flores-Becerra[1,2], Victor M. Garcia[1], and Antonio M. Vidal[1]

[1] Departamento de Sistemas Informáticos y Computación,
Universidad Politécnica de Valencia,
Camino de Vera s/n, 46022 Valencia, España,
{gflores, vmgarcia, avidal}@dsic.upv.es
[2] Departamento de Sistemas y Computación,
Instituto Tecnológico de Puebla,
Av. Tecnológico 420, Col. Maravillas, C.P. 72220, Puebla, México

Abstract. The work presented here is an experimental study of four iterative algorithms for solving the Inverse Additive Singular Value Problem (IASVP). The algorithms are analyzed and evaluated with respect to different points of view: memory requirements, convergence, accuracy and execution time, in order to observe their behaviour with different problem sizes and to identify those capable to solve the problem efficiently.

1 Introduction

Inverse problems are of interest for different applications in Science and Engineering, such as Geophysics, Computerized Tomography, Simulation of Mechanical Systems, and many more [4] [9] [11] [12]. Two specific inverse problems are the Inverse Eigenvalue Problem (IEP) and the Inverse Singular Value Problem (ISVP). The goal of these problems is to build a matrix with some structure features, and with eigenvalues (or singular values) previously fixed. In this paper we study a particular case of the ISVP, the Inverse Additive Singular Value Problem (IASVP), which can be defined as:

Given a set of matrices $A_0, A_1, ..., A_n \in \Re^{m \times n}$ ($m \geq n$) and a set of real numbers $S^* = \{S_1^*, S_2^*, ..., S_n^*\}$, where $S_1^* > S_2^* > ... > S_n^*$, find a vector $c = [c_1, c_2, ..., c_n]^t \in \Re^n$, such that S^* are the singular values of

$$A(c) = A_0 + c_1 A_1 + ... + c_n A_n. \qquad (1)$$

There are several well known methods for the IEP. Friedland et al. [8] propose several methods for the IEP based in Newton's method, named Method I, Method II, Method III and Method IV. Chen and Chu [2] proposed the Lift&Project method, where the IEP is solved as a series of minimum squares problems.

The methods for the IASVP are usually derived from those for the IEP. In [3], Chu proposes two methods for the IASVP, one of them (The "discrete" method) is obtained by adapting the Method III for the IEP to the IASVP; we shall name this method as MIII. The convergence of MIII was proved in [5].

The MI [7] and LP [6] methods for the IASVP are other methods derived of Method I and Lift&Project, respectively. On the other hand, the simplest way to set up a Newton iteration for resolution of the IASVP is to write directly the problem as a system of nonlinear equations; we call this a Brute Force (BF) approach.

Some experimental results of IASVP resolution were given for size of problem smaller than 6 (by example, $m = 5$ and $n = 4$ in [3] and [5]); so, it is necesary to study the behaviour of the algorithms for greater sizes.

The goal of this work has been to make a fair comparison among some of these methods with larger problem sizes than those used in previous studies. To do so, we have implemented FB, MI, MIII and LP methods. The performances of these algorithms have been analyzed and evaluated, for different values of m and n, regarding memory requirements, convergence, solution accuracy and execution time, through an experimental study, to determine those of better characteristics to solve the IASVP.

In section 2 the algorithms are briefly described; the performance of the algorithms is analyzed in section 3, and, finally, the conclusions are given in section 4.

2 Resolution Methods for IASVP

2.1 Brute Force (BF)

Let c^* be a solution of the IASVP; then, the singular value decomposition of $A(c^*)$, must be

$$A(c^*) = P diag(S^*) Q^t \qquad (2)$$

with $P \in \Re^{m \times n}$ and $Q \in \Re^{n \times n}$, orthogonals.

A system of nonlinear equations of the form $F(z) = 0$, can be built by using (2) and the orthogonality of P and Q. In this system, the unknown vector would be $z = [Q_{1,1}, Q_{1,2}, ..., Q_{n,n}, P_{1,1}, P_{1,2}, ..., P_{m,n}, c_1, c_2, ..., c_n,]^t$, n unknowns of c, mn unknowns of P [1] and n^2 unknowns of Q, then $F(z) = 0$ is $F(Q, P, c) = 0$, where

$$F_{(i-1)n+j}(Q, P, c) = (A_0 + c_1 A_1 + \cdots + c_n A_n - PS^*Q^t)_{i,j}; \quad i = 1:m; j = 1:n; \qquad (3)$$

$$F_{mn+(i-1)n+j-i+1}(Q, P, c) = (P^t P - I_m)_{i,j}; \quad i = 1:n; j = i:n; \qquad (4)$$

$$F_{mn+n\frac{n+1}{2}+(i-1)n+j-i+1}(Q, P, c) = (Q^t Q - I_n)_{i,j}; \quad i = 1:n; j = i:n; \qquad (5)$$

A nonlinear system with $mn + n^2 + n$ equations and $mn + n^2 + n$ unknowns has been defined from (3), (4) and (5); its solution can be approximated through Newton's method, computing a succession of $(Q^{(0)}, P^{(0)}, c^{(0)}), (Q^{(1)}, P^{(1)}, c^{(1)})$,

---
[1] Only mn unknowns of P because we need only $min\{m,n\}$ singular values.

..., $(Q^{(k)}, P^{(k)}, c^{(k)})$ that approximates to the solution of $F(Q, P, c) = 0$. Then, if $z^{(k)}$ is the k-th element of this succession, the $(k+1)$-th element is given by the expression [10]:

$$z^{(k+1)} = z^{(k)} - J(z^{(k)})^{-1} F(z^{(k)}) \qquad (6)$$

where $J(z^{(k)})$ is the Jacobian matrix of $F(z)$ evaluated at $z^{(k)}$. The Jacobian matrix of $F(Q, P, c)$ is

$$J(Q, P, c) = \left[\frac{\partial F_r(z)}{\partial z_t} \right]_{r=1:mn+n^2+n\ ;\ t=1:mn+n^2+n} = \begin{bmatrix} J_{1,1} & 0 & 0 \\ 0 & J_{2,2} & 0 \\ J_{3,1} & J_{3,2} & J_{3,3} \end{bmatrix}$$

where $J_{1,1}, J_{2,2}, J_{3,1}, J_{3,2}$ y $J_{3,3}$ are blocks of size $n\frac{n+1}{2} \times n^2$, $n\frac{n+1}{2} \times mn$, $mn \times n^2$, $mn \times mn$ and $mn \times n$, respectively, such that

for $i = 0 : n-1;\ j = 1 : n;\ row = 1 + \sum_{k=0}^{i-1}(n-k);\ col = i + (j-1)n + 1$

$$(J_{1,1})_{row,col} = 2Q_{j,i+1}$$

$$(J_{1,1})_{row+1:row+n-i-1,col} = Q_{j,i+2:n}^t$$

$$(J_{1,1})_{row+a,col+a} = Q_{j,i+1} \qquad a = 1 : n-i-1;$$

for $i = 0 : n-1;\ j = 1 : m;\ row = 1 + \sum_{k=0}^{i-1}(n-k);\ col = i + (j-1)n + 1$

$$(J_{2,2})_{row,col} = 2P_{j,i+1}$$

$$(J_{2,2})_{row+1:row+n-i-1,col} = P_{j,i+2:n}^t$$

$$(J_{2,2})_{row+a,col+a} = P_{j,i+1} \qquad a = 1 : n-i-1;$$

and for $i = 1 : m;\ j = 1 : n;\ t = 1 : n$

$$(J_{3,1})_{(i-1)n+j,(j-1)n+t} = S_t^* P_{i,t}$$

$$(J_{3,2})_{(i-1)n+j,(i-1)n+t} = S_t^* Q_{j,t}$$

$$(J_{3,3})_{(i-1)n+j,t} = (A_t)_{i,j}$$

This iterative method converges quadratically to the solution of $F(Q, P, c)$ if the initial guess $Q^{(0)}$, $P^{(0)}$ and $c^{(0)}$ is close enough to the solution [10].

2.2 MI

As mentioned before, the MI method for the IASVP follows the ideas from Method I for IEP [8]. First, for any c we can obtain the singular value decomposition of (1) as $A(c) = P(c)S(c)Q(c)^t$. Then, the IASVP can be stated as finding the solution of the nonlinear system in c: $F(c) = [S_i(c) - S_i^*]_{i=1,n} = 0$. If Newton's method is applied to solve this nonlinear system, the Jacobian matrix is needed; it can be obtained as in [7]: $J = [p_i^t A_j q_i]_{i,j=1,n}$, and the Newton's iteration (6) is given by the expression [7]

$$J^{(k)} c^{(k+1)} = b^{(k)} \qquad (7)$$

where $b^{(k)} = S^* - \left[p_i^{(k)t} A_0 q_i^{(k)}\right]_{i=1,n}$. MI converges quadratically to c^* if $c^{(0)}$ is close enough to c^* [10].

2.3 LP

LP is developed in a similar way to Lift&Project in [2]. Let us define $\Gamma(S^*)$, the set of matrices in $\Re^{m\times n}$, $(m \geq n)$, which can be written in the form PS^*Q^t, where $P \in \Re^{m\times n}$ and $Q \in \Re^{n\times n}$ are orthogonal matrices; and let $\Lambda(c)$ be the set of matrices that can be expressed as in (1). The goal is to find the intersection of both sets, using distance minimization techniques. The distance between two matrices U and V is defined as $d(U,V) = \|U - V\|_F$.

LP is an iterative algorithm, with two stages for each iteration:

1) The Lift stage, which consists in, given $c^{(k)}$ (given $A(c^{(k)}) \in \Lambda(c)$) find $X^{(k)} \in \Gamma(S^*)$ such that $d(A(c^{(k)}), X^{(k)}) = d(A(c^{(k)}), \Gamma(S^*))$. This is achieved by computing the singular value decomposition of $A(c^{(k)})$, $P^{(k)}S^{(k)}Q^{(k)t}$, and then computing $X^{(k)} = P^{(k)}S^*Q^{(k)t}$, which turns out to be the element of $\Gamma(S^*)$ closest to $A(c^{(k)})$ [7].

2) The Projection stage consist in, given $X^{(k)} \in \Gamma(S^*)$, find $c^{(k+1)}$ (find $A(c^{(k+1)}) \in \Lambda(c)$) such that $d(X^{(k)}, A(c^{(k+1)})) = d(X^{(k)}, \Lambda(c^{(k+1)}))$. This is achieved by finding $c^{(k+1)}$ as the solution of the nonlinear least squares problem $\min_{c^{(k+1)}} \|A^{(k+1)} - P^{(k)}S^*Q^{(k)t}\|_F^2$. This problem can be solved by equating the gradient of $\|A^{(k+1)} - P^{(k)}S^*Q^{(k)t}\|_F^2$ to zero and solving the linear system resulting $A_{tr} c^{(k+1)} = b_{tr}^{(k)}$, where [7] $A_{tr} = [\mathrm{tr}(A_i^t A_r)]_{r,i=1,l}$ and $b_{tr} = [\mathrm{tr}(A_r^t(X^{(k)} - A_0^t))]_{r=1,l}$. This LP algorithm converges to a stationary point in the sense that [7] $\|A^{(k+1)} - X^{(k+1)}\|_F \leq \|A^{(k)} - X^{(k)}\|_F$.

2.4 MIII

The method MIII described here is the method presented in [3] by Chu. This method finds the intersection of $\Gamma(S^*)$ and $\Lambda(c)$, defined in (2.3), using an iterative Newton-like method. In the iteration k, given $X^{(k)} \in \Gamma(S^*)$, there exist matrices $P^{(k)}$ and $Q^{(k)}$ such that $X^{(k)} = P^{(k)}S^*Q^{(k)t}$ and the tangent vector to $\Gamma(S^*)$ which starts from the point $X^{(k)}$ and crosses $A(c^{(k+1)})$, can be expressed as

$$X^{(k)} + X^{(k)}L^{(k)} - H^{(k)}X^{(k)} = A(c^{(k+1)}) \qquad (8)$$

where $L^{(k)} \in \Re^{n\times n}$ and $H^{(k)} \in \Re^{m\times m}$ are skew-symmetric matrices. Because $X^{(k)} = P^{(k)}S^*Q^{(k)t}$, (8) can be expressed as

$$S^* + S^*\tilde{L}^{(k)} - \tilde{H}^{(k)}S^* = W^{(k)}, \qquad (9)$$

where $\tilde{L}^{(k)} = Q^{(k)t} L^{(k)} Q^{(k)}$, $\tilde{H}^{(k)} = P^{(k)t} H^{(k)} P^{(k)}$, $W^{(k)} = P^{(k)t} A(c^{(k+1)}) Q^{(k)}$. Equating the diagonal elements of (9), we obtain the linear system (7), that calculate $c^{(k+1)}$ ($A(c^{(k+1)})$ and $W^{(k)}$). Equating the off-diagonal elementos of (9), we calculate $\tilde{H}^{(k)}$ and $\tilde{L}^{(k)}$ [3].

In order to calculate $X^{(k+1)}$ from $A(c^{(k+1)})$, e.g. $P^{(k+1)}$ and $Q^{(k+1)}$, matrix $A(c^{(k+1)})$ must be lifted to a point in $\Gamma(S^*)$. Then $X^{(k+1)}$ is defined as $X^{(k+1)} = P^{(k+1)} S^* Q^{(k+1)t}$ and $P^{(k+1)}$ and $Q^{(k+1)}$ are orthogonal matrices which can be approximated by $P^{(k+1)} \approx P^{(k)} R$ and $Q^{(k+1)} \approx Q^{(k)} T$, being R and T the Cayley transforms: $R = \left(I + \frac{1}{2} H^{(k)}\right) \left(I - \frac{1}{2} H^{(k)}\right)^{-1}$ and $T = \left(I + \frac{1}{2} L^{(k)}\right) \left(I - \frac{1}{2} L^{(k)}\right)^{-1}$. See [1] and [3] for details.

3 Numerical Experiments

In order to observe the behaviour of the algorithms when $m, n > 5$, they are analyzed experimentally, evaluating memory requirements, convergence, accuracy and efficiency. By each analyzed aspect, we compare the algorithms to determine those most suitable to solve the IASVP.

The numerical experiments have been carried out taking matrices sizes of $m = n = \{5, 10, 15, 20, 25, 30, 50\}$, using random values for matrices and vectors of the IASVP and taking different initial guesses $c^{(0)}$.

The algorithms have been implemented in Matlab and executed in a 2.2 GHz Intel Xeon biprocessor with 4 GBytes of RAM and operating system Linux Red Hat 8.

Memory Requirements. The storage required for $n+1$ matrices of size $m \times n$ (A_i, $i = 1 : n$), two vectors of size n (S^*, c) and the vectors and matrices required by each algorithm (singular values and vectors, Jacobians matrices, skew-symmetric matrices, ...), can be seen in Table 1. The FB method has the greatest memory needs, thereby is hardly feasible to implement for $m > 50$. The best algorithms from this point of view are MI and LP.

Table 1. Memory Requirements in KBytes(KB), MBytes(MB), GBytes(GB) and TBytes(TB)

$m = n$	4	10	30	50	100	500	1000
BF	12 KB	366 KB	28 MB	205 MB	4 GB	2 TB	32 TB
MI	1.2 KB	12 KB	250 KB	1.1 MB	8.3 MB	1 GB	8 GB
LP	1.2 KB	12 KB	250 KB	1.1 MB	8.3 MB	1 GB	8 GB
MIII	1.7 KB	15 KB	270 KB	1.2 MB	8.6 MB	1 GB	8.1 GB

Convergence. The convergence of the algorithms BF, MI and MIII is really sensitive to the initial guess. When the initial guess is taken as $c_i^{(0)} = c_i^* + \delta$ ($i = 1 : n$), for small δ such as $\delta = 0.1$ all of them converge; but when $\delta = 1.0$

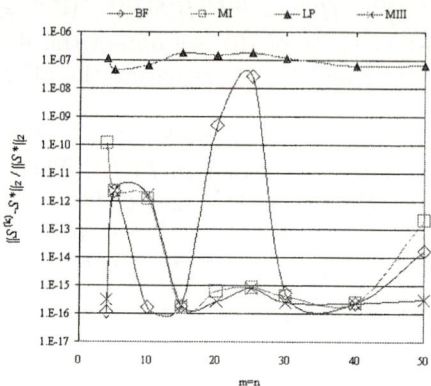

Fig. 1. Accuracy of Algorithms for different values of m, with $c_i^{(0)} = c_i^* + 0.1$ $(i = 1:n)$

they only converge for sizes $m < 50$ and when $\delta = 10.0$ they only converge for $m < 10$. Then, as, the initial guess is farther from the solution, these algorithms converge only for smaller problem sizes. This phenomenon is not suffered by LP, since it has the property $\|A^{(k+1)} - X^{(k+1)}\|_F \leq \|A^{(k)} - X^{(k)}\|_F$ [7].

Solution Accuracy. The accuracy of the solution is measured by obtaining the relative errors of the obtained singular values with respect to the correct ones (Figure 1), with $\delta = 0.1$. MIII gives the approximations S^*, while LP gives the worst; however, the LP errors are smaller than $1e - 6$.

Time Complexity estimates and Execution Times. The time complexity estimates are a function of the problem size (m,n) and of the number of iterations to convergence (K); this estimates can be split in two, the cost of the start-up phase $(T(m,n)_{startUp})$ and the cost of each iteration inside the main

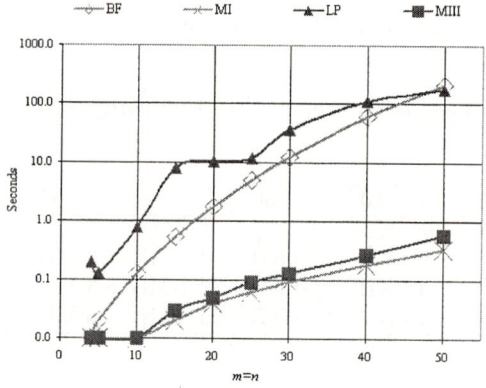

Fig. 2. Execution times (seconds) required to reach convergence, with different values of m and $\delta = 0.1$

Table 2. Number of iterations required for convergence with $\delta = 0.1$

$m = n$	4	5	10	15	20	25	30	40	50
BF	3	2	3	3	3	3	3	3	3
MI	2	2	2	3	3	3	3	3	3
LP	365	206	561	3398	2629	1998	4048	6304	5303
MIII	3	2	2	3	3	3	3	3	4

loop ($T(m,n)_{loop}$) so that the time complexity can be written as: $T(m,n,K) = T(m,n)_{startUp} + KT(m,n)_{loop}$. Then, the estimates for each method (for $m = n$) can be written as:

$$T(m,K)_{BF} \approx (62/3)m^3 + O(m^2) + K\{(16/3)m^6 + O(m^5)\}$$

$$T(m,K)_{MI} \approx (44/3)m^3 + O(m) + K\{2m^4 + (m^3)\}$$

$$T(m,K)_{LP} \approx m^4 + O(m^3) + K\{(56/3)m^3 + O(m^2)\}$$

$$T(m,K)_{MIII} \approx (56/3)m^3 + O(m^2) + K\{2m^4 + O(m^3)\}$$

The time estimate of BF shows that it is not an acceptable algorithm. On the other hand, LP has the smallest time estimate in the iterative part, but it needs many more iterations for convergence (and execution time) than the other methods. This can be checked in Table 2, where the number of iterations K is shown for some experiments, and in Figure 2, where the execution times are shown.

4 Conclusions

We have implemented and analyzed a set of algorithms for resolution of the IASVP and we have analized their behaviour for different values of m ($m > 5$).

The study carried out shows that the FB approach must be discarded as a practical algorithm given its high costs, both in memory ($O(m^4)$) and execution time ($O(m^6)$).

From the memory point of view, MI and LP are those of smaller requirements of memory; from the execution time point of view, MI is most efficient. The highest accuracy of the solutions, in most of the tested cases, is reached with MIII.

However, MI and MIII, like BF, have the drawback of being too sensitive to the quality of the initial guess. This sensitivity becomes worse as the size of the problem increases. In contrast, LP does not suffer this problem.

Since none of the algorithms possess properties good enough on its own, it seems desirable to combine several algorithms in a single one. A possible follow-up of this work is the development of this kind of algorithms; some approaches can be found in [2] and [7].

The problem of the high costs of storage and execution times for problems of larger size ($m > 50$) can be tackled by parallelizing the algorithms, so that any processor involved stores only a part of the data structures and executes part of the calculations.

Acknowledgement

This work has been supported by Spanish MCYT and FEDER under Grant TIC2003-08238-C02-02 and DGIT-SUPERA-ANUIES (México).

References

1. Chan, R., Bai, Z., Morini, B.: On the Convergence Rate of a Newton-Like Method for Inverse Eigenvalue and Inverse Singular Value Problems. Int. J. Appl. Math., Vol. 13 (2003) 59-69
2. Chen, X., Chu, M.T.: On the Least Squares Solution of Inverse Eigenvalue Problems. SIAM, Journal on Numerical Analysis, Vol. 33, No. 6 (1996) 2417-2430
3. Chu, M.T.: Numerical Methods for Inverse Singular Value Problems. SIAM, Journal Numerical Analysis, Vol. 29, No. 3 (1992) 885-903
4. Chu, M.T.: Inverse Eigenvalue Problems. SIAM, Review, Vol. 40 (1998)
5. Bai, Z., Morini, B., Xu, S.: On the Local Convergence of an Iterative Approach for Inverse Singular Value Problems. Submitted
6. Flores G., Vidal A.: Paralelización del Método de Elevación y Proyección para la Resolución del Problema Inverso de Valores Singulares. Primer Congreso Internacional de Computación Paralela, Distribuida y Aplicaciones, (2003).
7. Flores, G., Vidal A.: Parallel Global and Local Convergent Algorithms for Solving the Inverse Additive Singular Value Problem. Submitted
8. Friedland, S., Nocedal, J., Overton, M.L.: The Formulation and Analysis of Numerical Methods for Inverse Eigenvalue Problems. SIAM, Journal on Numerical Analysis, Vol. 24, No. 3 (1987) 634-667
9. Groetsch, C.W.: Inverse Problems. Activities for Undergraduates. The mathematical association of America (1999)
10. Kelley, C.: Iterative Methods for Linear and Nonlinear Equations. SIAM (1995)
11. Neittaanmki, P., Rudnicki, M., Savini, A.: Inverse Problems and Optimal Design in Electricity and Magnetism. Oxford: Clarendon Press (1996)
12. Sun, N.: Inverse Problems in Groundwater Modeling. Kluwer Academic (1994)

Computing Orthogonal Decompositions of Block Tridiagonal or Banded Matrices

Wilfried N. Gansterer

Institute of Distributed and Multimedia Systems,
University of Vienna, Austria

Abstract. A method for computing orthogonal URV/ULV decompositions of block tridiagonal (or banded) matrices is presented. The method discussed transforms the matrix into structured triangular form and has several attractive properties: The block tridiagonal structure is fully exploited; high data locality is achieved, which is important for high efficiency on modern computer systems; very little fill-in occurs, which leads to no or very low memory overhead; and in most practical situations observed the transformed matrix has very favorable numerical properties. Two variants of this method are introduced and compared.

1 Introduction

In this paper, we propose a method for computing an orthogonal decomposition of an irreducible symmetric block tridiagonal matrix

$$M_p := \begin{pmatrix} B_1 & C_1 & & & \\ A_1 & B_2 & C_2 & & \\ & A_2 & B_3 & \ddots & \\ & & \ddots & \ddots & C_{p-1} \\ & & & A_{p-1} & B_p \end{pmatrix} \in \mathbb{R}^{n \times n} \quad (1)$$

with $p > 1$. The blocks $B_i \in \mathbb{R}^{k_i \times k_i}$ ($i = 1, 2, \ldots, p$) along the diagonal are quadratic (but not necessarily symmetric), the off-diagonal blocks $A_i \in \mathbb{R}^{k_{i+1} \times k_i}$ and $C_i \in \mathbb{R}^{k_i \times k_{i+1}}$ ($i = 1, 2, \ldots, p-1$) are arbitrary. The block sizes k_i satisfy $1 \leq k_i < n$ and $\sum_{i=1}^{p} k_i = n$, but are otherwise arbitrary.

We emphasize that the class of matrices of the form (1) comprises *banded* symmetric matrices, in which case the C_i are upper triangular and $C_i = A_i^\top$. Alternatively, given a banded matrix with upper and lower bandwidth b, a block tridiagonal structure is determined by properly selecting the block sizes k_i.

Motivation. Banded matrices arise in numerous applications. We also want to highlight a few situations where block tridiagonal matrices occur. One example from acoustics is the modelling of vibrations in soils and liquids with different layers using finite elements [1]. Every finite element is only linked to two other elements (a deformation at the bottom of one element influences the deformation at the top of the next element). Consequently, when arranging the local finite element matrices into a global system matrix, there is an overlap between these local matrices resulting in a block tridiagonal global matrix. Another example is the block tridiagonalization procedure introduced by Bai et al. [2]. Given a symmetric matrix, this procedure determines a symmetric block tridiagonal matrix whose eigenvalues differ at most by a user defined accuracy tolerance from the ones of the original matrix.

In this paper, we present two orthogonal factorizations of M_p—a URV and a ULV factorization, where U and V are orthogonal matrices, and R and L have special upper and lower triangular structure, respectively. These decompositions are important tools when block tridiagonal or banded linear systems have to be solved. This aspect is discussed in more detail in Section 3.

Although many research activities on ULV/URV decompositions have been documented in the literature (see, for example, [3, 4, 5]), we are not aware of work specifically targeted towards block tridiagonal matrices. A distinctive feature of the approach described in this paper is that it fully exploits the special structure defined in Eqn. (1).

Synopsis. The algorithm for computing the two orthogonal decompositions and their properties are described in Section 2, important applications are summarized in Section 3, and concluding remarks are given in Section 4.

2 Factorization of Block Tridiagonal Matrices

In this section, we summarize two closely related methods for computing an orthogonal factorization of the block tridiagonal matrix M_p defined by Eqn. (1).

The basic idea behind the algorithms is to eliminate off-diagonal blocks one after the other by computing singular value decompositions of submatrices and performing the corresponding update. Depending on how the submatrices are chosen, there are two basic variants of this factorization, a URV and a ULV factorization. In the first one, M_p is transformed into upper triangular structure, whereas in the second one, M_p is transformed into lower triangular structure. A comparison between the two variants for specific applications is given in Section 3.2.

2.1 URV Decomposition

Based on the SVD of the block comprising the first diagonal and the first subdiagonal block of M_p,

$$\begin{pmatrix} B_1 \\ A_1 \end{pmatrix} = U_1 \begin{pmatrix} \Sigma_1 \\ \mathbf{0} \end{pmatrix} V_1^\top = \begin{pmatrix} U_1^1 & U_1^3 \\ U_1^2 & U_1^4 \end{pmatrix} \begin{pmatrix} \Sigma_1 \\ \mathbf{0} \end{pmatrix} V_1^\top$$

we can transform M_p into

$$\begin{pmatrix} \Sigma_1 & \tilde{C}_1 & \tilde{F}_1 & & & \\ 0 & \tilde{B}_2 & \tilde{C}_2 & & & \\ & A_2 & B_3 & \ddots & & \\ & & \ddots & \ddots & C_{p-1} & \\ & & & A_{p-1} & B_p & \end{pmatrix}$$

by updating with $U_1^\top$ from the left and with V_1 from the right. The existing blocks C_1, B_2, and C_2 are modified in this process (indicated by a tilde),

$$\tilde{C}_1 = {U_1^1}^\top C_1 + {U_1^2}^\top B_2 \tag{2}$$

$$\tilde{B}_2 = {U_1^3}^\top C_1 + {U_1^4}^\top B_2 \tag{3}$$

$$\tilde{C}_2 = {U_1^4}^\top C_2, \tag{4}$$

and one block in the second upper diagonal is filled in:

$$\tilde{F}_1 = {U_1^2}^\top C_2. \tag{5}$$

Next, we continue with the SVD of the subblock comprising the updated diagonal block and the subdiagonal block in the second block column:

$$\begin{pmatrix} \tilde{B}_2 \\ A_2 \end{pmatrix} = U_2 \begin{pmatrix} \Sigma_2 \\ 0 \end{pmatrix} V_2^\top.$$

Again, we perform the corresponding updates on $\tilde{C}_2$, B_3, C_3, the fill-in of $\tilde{F}_2$ with $U_2^\top$ from the left, and the update of the second block column with V_2 from the right. Then, we compute the SVD of

$$\begin{pmatrix} \tilde{B}_3 \\ A_3 \end{pmatrix},$$

update with $U_3^\top$ and with V_3, and so on.

After $p-1$ such steps consisting of SVD and update operations, followed by the SVD of the diagonal block in the bottom right corner

$$\tilde{B}_p = U_p \Sigma V_p^\top$$

we get a factorization

$$M_p = URV^\top \tag{6}$$

with upper triangular R (see Fig. 1). More specifically,

$$R = \begin{pmatrix} \Sigma_1 & \tilde{C}_1 & \tilde{F}_1 & & & & \\ & \Sigma_2 & \tilde{C}_2 & \tilde{F}_2 & & & \\ & & \ddots & \ddots & \ddots & & \\ & & & \Sigma_{p-2} & \tilde{C}_{p-2} & \tilde{F}_{p-2} & \\ & & & & \Sigma_{p-1} & \tilde{C}_{p-1} \\ & & & & & \Sigma_p \end{pmatrix}, \tag{7}$$

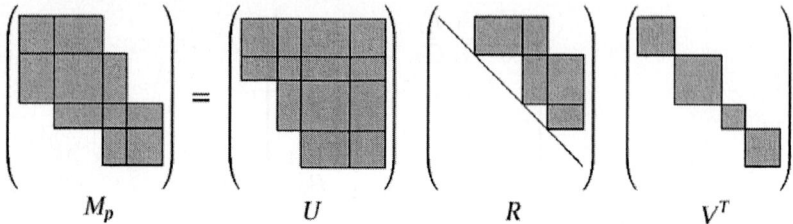

Fig. 1. URV decomposition of M_p

U is orthogonal and represented as the product

$$U = \begin{pmatrix} \begin{matrix} U_1^1 & U_1^3 \\ U_1^2 & U_1^4 \end{matrix} & & & & \\ & I & & & \\ & & I & & \\ & & & \ddots & \\ & & & & I \end{pmatrix} \times \begin{pmatrix} I & & & & \\ & \begin{matrix} U_2^1 & U_2^3 \\ U_2^2 & U_2^4 \end{matrix} & & & \\ & & I & & \\ & & & \ddots & \\ & & & & I \end{pmatrix} \times \cdots$$

$$\cdots \times \begin{pmatrix} I & & & & \\ & I & & & \\ & & \ddots & & \\ & & & I & \\ & & & & \begin{matrix} U_{p-1}^1 & U_{p-1}^3 \\ U_{p-1}^2 & U_{p-1}^4 \end{matrix} \end{pmatrix} \times \begin{pmatrix} I & & & & \\ & \ddots & & & \\ & & I & & \\ & & & I & \\ & & & & U_p \end{pmatrix},$$

and V is also orthogonal and block diagonal:

$$V = \text{block-diag}(V_1, V_2, \ldots, V_p).$$

The correctness of this decomposition algorithm can be verified directly by multiplying out (6) and taking into account the respective relationships corresponding to Eqns. (2, 3, 4, 5).

2.2 ULV Decomposition

We may as well start the process described in the previous section with the SVD of the block comprising the first diagonal and the first *super*diagonal block of M_p,

$$(B_1 \; C_1) = U_1 \, (\Sigma_1 \; 0) \, V_1^T = U_1 \, (\Sigma_1 \; 0) \begin{pmatrix} V_1^1 & V_1^3 \\ V_1^2 & V_1^4 \end{pmatrix}^T.$$

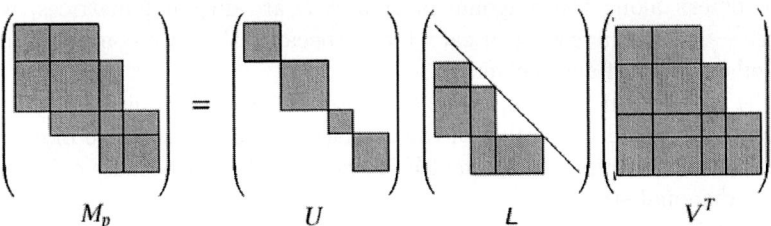

Fig. 2. ULV decomposition

The corresponding updates with $U_1^\top$ from the left and with V_1 from the right will transform M_p into

$$\begin{pmatrix} \Sigma_1 & 0 & & & & \\ \tilde{A}_1 & \tilde{B}_2 & C_2 & & & \\ \tilde{F}_1 & A_2 & B_3 & \ddots & & \\ & \ddots & \ddots & \ddots & C_{p-1} & \\ & & & & A_{p-1} & B_p \end{pmatrix}.$$

Next, we compute the SVD

$$\begin{pmatrix} \tilde{B}_2 & C_2 \end{pmatrix} = U_2 \begin{pmatrix} \Sigma_2 & 0 \end{pmatrix} V_2^\top,$$

update again, compute the next SVD in the next block row, etc.

After $p-1$ such steps, again consisting of SVD and update operations, followed by the SVD of the diagonal block in the bottom right corner

$$\tilde{B}_p = U_p \Sigma V_p^\top,$$

this time we get a factorization

$$M_p = ULV^\top \qquad (8)$$

with *lower* triangular L and orthogonal U and V (see Fig. 2).

2.3 Properties

The decompositions produced by the algorithms described in the previous section have the following important properties.

- $p-1$ lower or upper off-diagonal blocks are "pushed" onto the other side of the diagonal blocks, filling in the $p-2$ blocks F_i in the second upper or lower block diagonal. This implies that there is no increase in memory requirements, except when the transformation matrices U or V need to be stored.

- The blocks along the diagonal of R and L are diagonal matrices, whose entries are non-increasing in each block (because they are computed as the singular values of another matrix).
- In the URV decomposition, explicit representation of U requires considerable computational effort because of the overlap between consecutive block rows, whereas explicit representation of V is available directly because of its simple block diagonal structure.
- In the ULV decomposition, we have the opposite situation: Analogously to the previous argument, explicit representation of V requires considerable computational effort, whereas explicit representation of U is available directly.
- The factorizations (6) and (8) are not guaranteed to be rank-revealing (cf. [3]). However, our experience and current work shows that in practice it turns out to be rank-revealing in most cases and that it is very useful even if it turns out *not* to be rank-revealing [6].
- The arithmetic complexity of the algorithms described in Sections 2.1 and 2.2 is roughly $c_1 n \max_i k_i^2 + c_2 \max_i k_i^3$ with moderate constants c_1 and c_2. Consequently, the factorizations (6) and (8) can be computed highly efficiently, especially so for small block sizes k_i ("narrow" M_p).

3 Applications

The investigation of various applications of the orthogonal decompositions described in this paper is work in progress. In the following, we summarize the basic directions of this approach.

Clearly, one idea is to utilize the decompositions described here in the context of solving linear systems. Compared to standard methods for solving block tridiagonal (in particular, banded) linear systems, the approaches proposed in this paper do not involve pivoting. Consequently, they are expected to have important advantages in terms of efficiency in most practical situations. Some rare numerically "pathological" cases, which may require slightly more expensive techniques, are currently subject of intense research.

Our motivation comes from very specific linear systems arising in the context of the *block divide-and-conquer* eigensolver.

3.1 Block Divide-and-Conquer Eigensolver

The block divide-and-conquer (*BD&C*) eigensolver [7, 8] was developed in recent years for approximating the eigenpairs of *symmetric* block tridiagonal matrices M_p (B_i symmetric and $C_i = A_i^\top$ in Eqn. (1)).

This approach proved to be highly attractive for several reasons. For example, it does not require tridiagonalization of M_p, it allows to approximate eigenpairs to arbitrary accuracy, and it tends to be very efficient if big parts of the spectrum need to be approximated to medium or low accuracy. However, in some situations eigenvector accumulation may become the bottleneck of BD&C—when accuracy requirements are not low and/or when not the entire spectrum, but only small parts of it have to be approximated.

With this in mind, we are currently developing a new approach for computing approximate eigenvectors of a symmetric block tridiagonal matrix as an alternative to eigenvector accumulation [6]. This new approach has the potential of significantly improving arithmetic complexity, proportionally reducing the computational effort if only a (small) part of the spectrum but not the full spectrum needs to be computed, and being highly parallelizable and scalable for large processor numbers.

3.2 Eigenvector Computation

We know that the eigenvector x corresponding to a eigenvalue λ of M_p is the solution of

$$(M_p - \lambda I) x = \mathbf{0}. \tag{9}$$

Since the shift operation preserves block tridiagonal structure, we can use one of the algorithms presented in this paper compute an orthogonal decomposition of the matrix $S := M_p - \lambda I$ and exploit it for solving the linear system (9) efficiently.

URV vs. ULV. In this specific context, there is a clear preference for one of the variants discussed in Section 2, as outlined in the following.

If we want to solve Eqn. (9) based on one of the decompositions described in this paper, we need to first solve $Ry = \mathbf{0}$ or $Ly = \mathbf{0}$ for the URV or the ULV decomposition, respectively. Then, we need to form $x = Vy$ to find a solution of Eqn. (9).

As discussed in Section 2.3, in the URV decomposition explicit representation of V is available directly and it has simple block diagonal structure, whereas multiplication with V from the ULV decomposition requires considerably higher computational effort. This clearly shows that the URV decomposition is preferable for this application context.

Backsubstitution. The rich structure of the matrix R resulting from the orthogonal decomposition described in this paper obviously allows for a very efficient backsubstitution process when solving a system $Ry = b$.

In the special context of eigenvector computations, many specific numerical aspects have to be taken into account in order to achieve certain requirements in terms of residuals and orthogonality properties (cf. [9] and several publications resulting from this work).

4 Conclusion

We have presented two related algorithms for computing orthogonal URV or ULV decompositions of general block tridiagonal matrices. We have shown that the matrices R and L can not only be computed efficiently, but that they also have very attractive structural properties.

Because of these properties the methods introduced in this paper are expected to have numerous applications. In particular, we expect them to be integral part of a new approach for computing approximate eigenvectors of symmetric block tridiagonal matrices.

Current and Future Work

Our current work focusses on reliable and efficient methods exploiting the decompositions presented in this paper for approximating eigenvectors in the context of the block divide-and-conquer eigensolver for symmetric block tridiagonal matrices. This goal requires careful investigation of several important numerical aspects of the decompositions described here, especially in the backsubstitution process. Our findings will be summarized in [6].

Moreover, we are also investigating other application contexts for the decompositions presented here.

Acknowlegdements

We would like to thank Robert C. Ward for many inspiring discussions when developing the concepts presented here and the anonymous referees for their comments.

References

1. Acoustics Research Institute of the Austrian Academy of Sciences: Vibrations in soils and liquids – transform method for layers with random properties. FWF-Project P16224-N07 (2004)
2. Bai, Y., Gansterer, W.N., Ward, R.C.: Block tridiagonalization of "effectively" sparse symmetric matrices. ACM Trans. Math. Softw. **30** (2004) 326–352
3. Stewart, G.W.: Updating a rank-revealing ULV decomposition. SIAM J. Matrix Anal. Appl. **14** (1993) 494–499
4. Park, H., Elden, L.: Downdating the rank-revealing URV decomposition. SIAM J. Matrix Anal. Appl. **16** (1995) 138–155
5. Stewart, M., Van Dooren, P.: Updating a generalized URV decomposition. SIAM J. Matrix Anal. Appl. **22** (2000) 479–500
6. Gansterer, W.N., Kreuzer, W., Ward, R.C.: Computing eigenvectors of block tridiagonal matrices. (2005) *in preparation*.
7. Gansterer, W.N., Ward, R.C., Muller, R.P.: An extension of the divide-and-conquer method for a class of symmetric block-tridiagonal eigenproblems. ACM Trans. Math. Softw. **28** (2002) 45–58
8. Gansterer, W.N., Ward, R.C., Muller, R.P., Goddard, III, W.A.: Computing approximate eigenpairs of symmetric block tridiagonal matrices. SIAM J. Sci. Comput. **25** (2003) 65–85
9. Dhillon, I.S.: A New $O(n^2)$ Algorithm for the Symmetric Tridiagonal Eigenvalue/Eigenvector Problem. PhD thesis, Computer Science Division (EECS), University of California at Berkeley (1997)

Adaptive Model Trust Region Methods for Generalized Eigenvalue Problems[*]

P.-A. Absil[1], C.G. Baker[1], K.A. Gallivan[1], and A. Sameh[2]

[1] School of Computational Science,
Florida State University, Tallahassee,
FL 32306-4120, USA
{absil, cbaker, gallivan}@csit.fsu.edu,
http://www.csit.fsu.edu/{~absil,~cbaker,~gallivan}
[2] Department of Computer Sciences,
Purdue University, West Lafayette,
IN 47907-2066, USA

Abstract. Computing a few eigenpairs of large-scale matrices is a significant problem in science and engineering applications and a very active area of research. In this paper, two methods that compute extreme eigenpairs of positive-definite matrix pencils are combined into a hybrid scheme that inherits the advantages of both constituents. The hybrid algorithm is developed and analyzed in the framework of model-based methods for trace minimization.

1 Introduction

We consider the computation of a few smallest eigenvalues and the corresponding eigenvectors of the generalized eigenvalue problem

$$Ax = \lambda Bx, \qquad (1)$$

where A and B are $n \times n$ symmetric positive-definite matrices. Positive definiteness of A and B guarantees that all the eigenvalues are real and positive. This eigenvalue problem appears in particular in the computation of the lower modes of vibration of a mechanical structure, assuming that there are no rigid-body modes and that all the degrees of freedom are mass-supplied: A is the stiffness matrix, B is the mass matrix, x is a mode of vibration and λ is the square of the circular frequency associated with the mode x.

Inverse iteration (INVIT) and Rayleigh quotient iteration (RQI) are the conceptually simplest methods for the generalized eigenvalue problem [1–§15-9]. Interestingly, they have complementary properties: unshifted INVIT converges to the smallest (or *leftmost*) eigenpair for almost all initial conditions but with linear local convergence only; RQI has cubic local convergence but it can converge to different eigenpairs depending on the initial condition.

---

[*] This work was supported by NSF Grants ACI0324944 and CCR9912415, and by the School of Computational Science of Florida State University.

It is therefore quite natural to try to combine these two methods and obtain a hybrid method that enjoys strong global convergence and fast local convergence; see in particular Szyld [2] for the problem of finding an eigenvalue in a given interval. For the computation of the leftmost eigenpair, such a hybrid method would use INVIT until reaching the basin of attraction of the leftmost eigenvector under RQI; then the method would switch to RQI in order to exploit its superlinear convergence. However, to our knowledge, a practical and reliable switching criterion between the two methods has yet to be found that guarantees global convergence of the hybrid method: if the switch is made too early, RQI may not converge to the leftmost eigenspace. A second drawback of *exact* INVIT and RQI is their possibly high computational cost, since large-scale linear systems with system matrix $(A - \sigma B)$ have to be solved exactly.

In this paper, we propose a remedy to these two difficulties. In Phase I, in order to reduce the computational cost, we replace INVIT with the Basic Tracemin algorithm of Sameh et al. [3,4]. Basic Tracemin proceeds by successive unconstrained approximate minimization of inexact quadratic local models of a generalized Rayleigh quotient cost function, using the stiffness matrix A as the model Hessian. This method is closely related to INVIT: if the model minimizations are carried out exactly (which happens in particular when an exact factorization of A is used for preconditioning), then Basic Tracemin is mathematically equivalent to (block) INVIT.

Extending an observation made by Edelman et al. [5–§4.4], we point out that the stiffness matrix used as the model Hessian is quite different from the true Hessian of the generalized Rayleigh quotient—this is why the model is called "inexact". However, we emphasize that the choice of an inexact Hessian in Phase I does not conflict with the findings in [5]: using the exact Hessian is important only when the iteration gets close to the solution, in order to achieve superlinear convergence. Therefore, using an inexact Hessian in Phase I is not necessarily a liability. On the contrary, the stiffness matrix as the model Hessian offers a useful property: as shown in [3, 4], any decrease in the inexact model induces a decrease in the cost function (i.e., the Rayleigh quotient). Therefore, in the presence of an inexact preconditioner, Basic Tracemin can be thought of as an inexact INVIT, that reduces computational cost per step while preserving the global convergence property. Due to its link with INVIT, it quickly purges the eigenvectors whose eigenvalues are well separated from the leftmost eigenvalues. This is particularly true when a good preconditioner is available, as is often the case in the very sparse problems encountered in structural mechanics. Consequently, the Basic Tracemin iteration is efficient when the iterates are still far away from the solution.

On the other hand, close to the solution, Basic Tracemin suffers from the linear rate of convergence (due to the use of an inexact Hessian), especially when the leftmost eigenvalues are not well separated from the immediately higher ones. This is why a superlinear method, like RQI, is required in Phase II. However, we want to remedy both drawbacks of convergence failure and high computational cost mentioned above. This motivates the use of the recently proposed Riemannian trust-region (RTR) algorithm [6, 7, 8]. Unlike Basic Tracemin, this method uses the *true* Hessian as the model Hessian (for superlinear convergence), along with a trust-region safeguard that prevents convergence to non-leftmost eigenpairs that could otherwise occur if switching takes place too early. Moreover, the computational cost is reduced by using a truncated conjugate-gradient algorithm to solve the trust-region problems inexactly while preserving the global and locally superlinear convergence.

However, in spite of its convergence properties (which make it an excellent choice for Phase II), there is a reason not to use the trust-region method in Phase I: far from the solution, the trust-region confinement often makes it difficult to exploit the full power of a good preconditioner—efficient preconditioned steps are likely to be rejected for falling outside of the trust region. A possible remedy, which we are currently investigating, is to relax the trust-region requirement so as to accept efficient preconditioned steps. Another remedy, which we study in this paper, is to use a Basic Tracemin / RTR hybrid.

In summary, we use Basic Tracemin in Phase I (far away from the solution) and the RTR algorithm with exact Hessian in Phase II (close to the solution). In this work, we develop and analyze the hybrid scheme by unifying the two constituents and their combination using the framework of adaptive model-based algorithms for the minimization of the generalized Rayleigh quotient.

2   Model-Based Scheme for Trace Minimization

We want to compute the p leftmost eigenpairs of the generalized eigenproblem (1), where A and B are positive definite $n \times n$ matrices. We denote the eigenpairs by $(\lambda_1, v_1), \ldots, (\lambda_n, v_n)$ with $0 < \lambda_1 \leq \ldots \leq \lambda_n$ and take $v_1, \ldots, v_n$ B-orthonormal; see, e.g., [9] for details. The methods presented here aim at computing the leftmost p-dimensional eigenspace $\mathcal{V}$ of (A, B), namely, $\mathcal{V} = \mathrm{span}(v_1, \ldots, v_p)$. To ensure uniqueness of $\mathcal{V}$, we assume that $\lambda_p < \lambda_{p+1}$. When p is small, which is inherent in most applications, it is computationally inexpensive to recover the eigenvectors $v_1, \ldots, v_p$ from $\mathcal{V}$ by solving a reduced-order generalized eigenvalue problem.

It is well known (see for example [4]) that the leftmost eigenspace $\mathcal{V}$ of (A, B) is the column space of any minimizer of the Rayleigh cost function

$$f : \mathbb{R}_*^{n \times p} \to \mathbb{R} : Y \mapsto \mathrm{trace}((Y^T B Y)^{-1}(Y^T A Y)), \tag{2}$$

where $\mathbb{R}_*^{n \times p}$ denotes the set of full-rank $n \times p$ matrices. It is readily checked that the right-hand side only depends on $\mathrm{colsp}(Y)$. Therefore, f induces a well-defined real-valued function on the set of p-dimensional subspaces of $\mathbb{R}^n$.

The proposed methods iteratively compute the minimizer of f by (approximately) minimizing successive models of f. The minimization of the models themselves is done via an iterative process, which is referred to as *inner iteration*, to distinguish it with the principal, *outer iteration*. We present here the process in a way that does not require a background in differential geometry; we refer to [7] for the mathematical foundations of the technique.

Let Y, a full-rank $n \times p$ matrix, be the current iterate. The task of the inner iteration is to produce a correction S of Y such that $f(Y + S) < f(Y)$. A difficulty is that corrections of Y that do not modify its column space do not affect the value of the cost function. This situation leads to unpleasant degeneracy if it is not addressed. Therefore, we require S to satisfy some complementarity condition with respect to the space $\mathcal{V}_Y := \{YM : M\ p \times p \text{ invertible}\}$. Here, in order to simplify later developments, we impose complementarity via B-orthogonality, namely $S \in \mathcal{H}_Y$ where

$$\mathcal{H}_Y := \{Z \in \mathbb{R}^{n \times p} : Y^T B Z = 0\}.$$

Consequently, the inner iteration aims at minimizing the function

$$\widehat{f}_Y(S) := \operatorname{trace}\left(\left((Y+S)^T B(Y+S)\right)^{-1}\left((Y+S)^T A(Y+S)\right)\right), \quad S \in \mathcal{H}_Y. \tag{3}$$

A Taylor expansion of $\widehat{f}_Y$ around $S = 0$ yields the "exact" quadratic model

$$m_Y^{\text{exact}}(S) = \operatorname{trace}((Y^T BY)^{-1}(Y^T AY)) + \operatorname{trace}\left((Y^T BY)^{-1} S^T\, 2AY\right)$$
$$+ \frac{1}{2}\operatorname{trace}\left((Y^T BY)^{-1} S^T\, 2\left(AS - BS(Y^T BY)^{-1} Y^T AY\right)\right), \quad S \in \mathcal{H}_Y. \tag{4}$$

Throughout this paper, we let $P := I - BY(Y^T B^2 Y)^{-1} Y^T B$ denote the orthogonal projector onto $\mathcal{H}_Y$, where Y is the current iterate. From (4), and using the inner product

$$\langle Z_1, Z_2 \rangle := \operatorname{trace}\left((Y^T BY)^{-1} Z_1^T Z_2\right), \quad Z_1, Z_2 \in \mathcal{H}_Y, \tag{5}$$

we identify $2PAY$ to be the gradient of $\widehat{f}_Y$ at $S = 0$ and the operator $S \mapsto 2P\left(AS - BS(Y^T BY)^{-1} Y^T AY\right)$ to be the Hessian of $\widehat{f}_Y$ at $S = 0$; see [7] for details. In the sequel, we will use the more general form

$$m_Y(S) = \operatorname{trace}((Y^T BY)^{-1}(Y^T AY)) + \operatorname{trace}\left((Y^T BY)^{-1} S^T\, 2AY\right)$$
$$+ \frac{1}{2}\operatorname{trace}\left((Y^T BY)^{-1} S^T\, H_Y[S]\right), \quad S \in \mathcal{H}_Y \tag{6}$$

to allow for the use of inexact quadratic expansions.

The proposed general algorithm is a model trust-region scheme defined as follows.

Algorithm 1 (outer iteration)
Data: symmetric positive-definite $n \times n$ matrices A and B.
Parameters: $\bar{\Delta} > 0$, $\Delta_0 \in (0, \bar{\Delta}]$, and $\rho' \in [0, \frac{1}{4})$.

Input: initial iterate Y_0 (full-rank $n \times p$ matrix).
Output: sequence of iterates $\{Y_k\}$.
for $k = 0, 1, 2, \ldots$ until an outer stopping criterion is satisfied:

- Using Algorithm 2, obtain S_k that (approximately) solves the trust-region subproblem

$$\min_{S \in \mathcal{H}_{Y_k}} m_{Y_k}(S) \quad \text{s.t.} \quad \|S\|_M^2 := \operatorname{trace}\left((Y_k^T B Y_k)^{-1} S^T M S\right) \leq \Delta_k^2, \quad (7)$$

  where m is defined in (6) and M is a preconditioner.
- Evaluate

$$\rho_k := \frac{\widehat{f}_{Y_k}(0) - \widehat{f}_{Y_k}(S_k)}{m_{Y_k}(0) - m_{Y_k}(S_k)} \quad (8)$$

  where $\widehat{f}$ is defined in (3).
- Update the trust-region radius:
  if $\rho_k < \frac{1}{4}$
    $\Delta_{k+1} = \frac{1}{4}\Delta_k$
  else if $\rho_k > \frac{3}{4}$ and $\|S_k\| = \Delta_k$
    $\Delta_{k+1} = \min(2\Delta_k, \bar{\Delta})$
  else
    $\Delta_{k+1} = \Delta_k$;
- Update the iterate:
  if $\rho_k > \rho'$,
    $Y_{k+1} = \operatorname{orth}(Y_k + S_k)$, where orth denotes an orthonormalization process which prevents loss of rank;
  else
    $Y_{k+1} = Y_k$;

end (for).

The inner iteration (Algorithm 2) employed to generate S_k is a preconditioned truncated conjugate gradient method, directly inspired from the work of Steihaug [10] and Toint [11]. The notation $(PMP)^\dagger R$ below denotes the solution $\tilde{R}$ of the system $PMP\tilde{R} = R$, $P\tilde{R} = \tilde{R}$, $PR = R$, which is given by the Olsen formula $\tilde{R} = M^{-1}R - M^{-1}BY(Y^T BM^{-1}BY)^{-1}Y^T BM^{-1}R$.

Algorithm 2 (inner iteration)
Set $S_0 = 0$, $R_0 = PAY_k = AY_k - BY_k(Y_k^T B^2 Y_k)^{-1} Y_k^T BAY_k$, $\tilde{R}_0 = (PMP)^\dagger R_0$, $\delta_0 = -\tilde{R}_0$;
for $j = 0, 1, 2, \ldots$ until an inner stopping criterion is satisfied, perform the following operations, where $\langle , \rangle$ denotes the inner product (5) and H_{Y_k} denotes model Hessian in (6).
  if $\langle \delta_j, H_{Y_k} \delta_j \rangle \leq 0$
    Compute τ such that $S = S_j + \tau \delta_j$ minimizes $m(S)$ in (6) and satisfies $\|S\|_M = \Delta$;
    return S;

Set $\alpha_j = \langle R_j, \tilde{R}_j \rangle / \langle \delta_j, H_{Y_k} \delta_j \rangle$; Set $S_{j+1} = S_j + \alpha_j \delta_j$;
if $\|S_{j+1}\|_M \geq \Delta$
   Compute $\tau \geq 0$ such that $S = S_j + \tau \delta_j$ satisfies $\|S\|_M = \Delta$;
   return S;
Set $R_{j+1} = R_j + \alpha H_{Y_k} \delta_j$; Set $\tilde{R}_{j+1} = (PMP)^\dagger R_{j+1}$;
Set $\beta_{j+1} = \langle R_{j+1}, \tilde{R}_{j+1} \rangle / \langle R_j, \tilde{R}_j \rangle$; Set $\delta_{j+1} = -\tilde{R}_{j+1} + \beta_{j+1} \delta_j$;
end (for).

We refer to [7] for a convergence analysis of Algorithm 1.

The two-phase scheme outlined in Section 1 can now be formalized.

Algorithm 3 *Phase I: Iterate Basic Tracemin—or, formally, Algorithm 1 with $H[S] := PAPS$, $\Delta_0 := +\infty$ and $\rho' := 0$—until some switching criterion is satisfied.*
Phase II: Continue using Algorithm 1 with $H[S] := P(AS - BS(Y^T B Y)^{-1} Y^T A Y)$, some initial Δ and some $\rho' \in (0, \frac{1}{4})$.

In the algorithms above, stopping and switching criteria, as well as the choices of some parameters and preconditioner, were left unspecified: depending on the information available about the structure of the problem and on the accuracy/speed requirements, various choices may be appropriate. We refer to the next section for examples of specific choices.

3 Numerical Experiments

In order to illustrate the practical relevance of the scheme proposed in Algorithm 3, we conducted experiments on the Calgary Olympic Saddledome arena matrices ($n = 3562$) available on Matrix Market: $A = $ BCSSTK24 and $B = $ BCSSTM24. The task was to compute the $p = 5$ leftmost eigenpairs of (A, B), which correspond to the lower modes of vibration.

With a view to achieving superlinear convergence when the exact model (4) is utilized, we used an inner stopping criterion of the form

$$\|R_j\| \leq \|R_0\| \min(\|R_0\|^\theta, \kappa), \qquad (9)$$

for some $\theta > 0$ and $\kappa > 0$. We chose $\theta = 1$ (striving for quadratic local convergence) and $\kappa = .5$. Several other choices are possible, notably based on the discussion in [3] in the case of Phase I.

In the first step of Phase II, we chose $\Delta := \|S_-\|_M$, where S_- is the last S computed in Phase I, and we chose $\rho' = .1$. The outer stopping criterion was a threshold on the norm of PAY_k. Transition between Phase I and Phase II was forced after various prescribed numbers of outer iterations, to illustrate the effect of different switching points on the behaviour of the algorithm. An initial iterate Y_0 was selected from a normal distribution. In all experiments, the preconditioner was kept constant throughout the iteration.

In the first set of experiments (Figure 1–left), the preconditioner M was chosen as an incomplete Cholesky factorization of A with relative drop tolerance set to 10^{-6}. Basic Tracemin converges linearly and (relatively) slowly. RTR (curve '0') is eventually faster that Basic Tracemin but it is initially slower, due to the trust-region constraint. Curves '5' and '10' show that there is a rather large "sweet-spot" for efficient switching from Basic Tracemin to RTR such that the hybrid method performs better than both of its pure components. If switching is done too late (curve '20'), then superlinear convergence is still observed and the problem is solved, but with less efficiency than the properly switched versions. Preliminary results suggest that the evolution of the trace function (2) gives valuable information on when switching should take place. Note that all that is at stake in the choice of the switching criterion between Basic Tracemin and RTR is efficiency, and neither success nor accuracy.

In a second set of experiments (Figure 1–right), we initially applied a permutation on A and B as returned by the the Matlab function `symamd(A)`. The preconditioner was then defined to be the exact Cholesky factorization of A, which is very sparse due to the approximate minimum degree permutation. Consequently, all algorithms converged (much) faster. Note that Phase I is mathematically equivalent to INVIT in this case. We observe a similar sweet-spot for the switching and the superiority of the global superlinear convergence of the trust-region-based method.

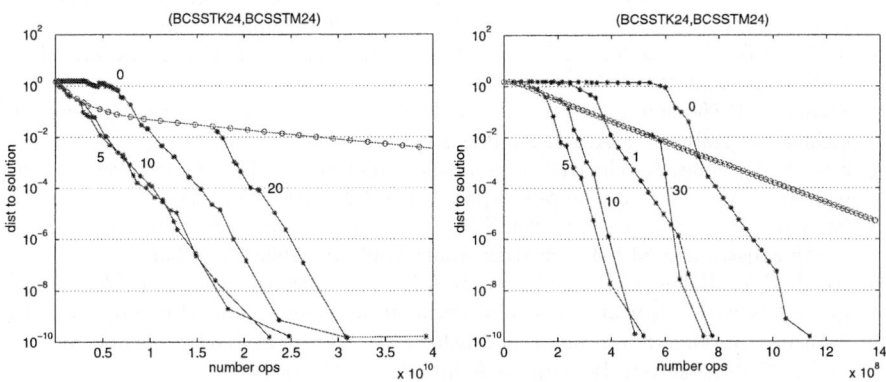

Fig. 1. Left: Experiments on Algorithm 3 with an incomplete Cholesky factorization of A as preconditioner. The distance to the solution is measured as the largest principal angle between the column space of Y and the leftmost p-dimensional eigenspace. Since the numerical cost of inner iterations differs in both phases, the distance is plotted versus an estimation of the number of operations. Circles and stars correspond to outer iterations of Basic Tracemin and RTR, respectively. The numbers on the curves indicate after how many steps of Basic Tracemin switching occurred to RTR. Right: Same, with exact preconditioner after approximate minimum degree permutation. Note that the formation of the preconditioner does not appear in the operation count

Finally, we conducted experiments where RTR was replaced by a block RQI.[1] Because RQI does not have global convergence to the leftmost eigenspace, convergence to a non-leftmost eigenspace may occur if switching is done too early. This was observed in the experiments. To our knowledge, there is no switching criterion that guarantees convergence to the leftmost eigenspace when RQI is used. This is a major reason for using RTR instead of RQI in Phase II (another reason is to reduce the computational cost with the truncated CG inner iteration).

4 Conclusion

We have shown that an appropriate combination of the Basic Tracemin algorithm of [3, 4] and the Riemannian trust-region algorithm of [8] yields an efficient method for high precision computation of the smallest eigenpairs of positive-definite generalized eigenproblems. In future work, we will further investigate the choice of the inner stopping criterion and the switching criterion between the two algorithms. For the latter, there is evidence that the decrease of the trace function (2) provides an useful guideline.

References

1. Parlett, B.N.: The Symmetric Eigenvalue Problem. Prentice-Hall, Inc., Englewood Cliffs, N.J. 07632 (1980) republished by SIAM, Philadelphia, 1998.
2. Szyld, D.B.: Criteria for combining inverse and Rayleigh quotient iteration. SIAM J. Numer. Anal. **25** (1988) 1369–1375
3. Sameh, A.H., Wisniewski, J.A.: A trace minimization algorithm for the generalized eigenvalue problem. SIAM J. Numer. Anal. **19** (1982) 1243–1259
4. Sameh, A., Tong, Z.: The trace minimization method for the symmetric generalized eigenvalue problem. J. Comput. Appl. Math. **123** (2000) 155–175
5. Edelman, A., Arias, T.A., Smith, S.T.: The geometry of algorithms with orthogonality constraints. SIAM J. Matrix Anal. Appl. **20** (1998) 303–353
6. Absil, P.A., Baker, C.G., Gallivan, K.A.: Trust-region methods on Riemannian manifolds with applications in numerical linear algebra. In: Proceedings of the 16th International Symposium on Mathematical Theory of Networks and Systems (MTNS2004), Leuven, Belgium, 5–9 July 2004. (2004)
7. Absil, P.A., Baker, C.G., Gallivan, K.A.: Trust-region methods on Riemannian manifolds. Technical Report FSU-CSIT-04-13, School of Computational Science, Florida State University (2004) http://www.csit.fsu.edu/~absil/Publi/RTR.htm.
8. Absil, P.A., Baker, C.G., Gallivan, K.A.: A truncated-CG style method for symmetric generalized eigenvalue problems. submitted (2004)
9. Stewart, G.W.: Matrix algorithms, Vol II: Eigensystems. Society for Industrial and Applied Mathematics, Philadelphia (2001)

---

[1] We used the NG algorithm mentioned in [12, 13], which reduces to RQI in the $p = 1$ case. This algorithm refines estimates of invariant subspaces, without favoring convergence to the leftmost one.

10. Steihaug, T.: The conjugate gradient method and trust regions in large scale optimization. SIAM J. Numer. Anal. **20** (1983) 626–637
11. Toint, P.L.: Towards an efficient sparsity exploiting Newton method for minimization. In Duff, I.S., ed.: Sparse Matrices and Their Uses. Academic Press, London (1981) 57–88
12. Lundström, E., Eldén, L.: Adaptive eigenvalue computations using Newton's method on the Grassmann manifold. SIAM J. Matrix Anal. Appl. **23** (2002) 819–839
13. Absil, P.A., Sepulchre, R., Van Dooren, P., Mahony, R.: Cubically convergent iterations for invariant subspace computation. SIAM J. Matrix. Anal. Appl. **26** (2004) 70–96

On Stable Integration of Stiff Ordinary Differential Equations with Global Error Control*

Gennady Yur'evich Kulikov and Sergey Konstantinovich Shindin

School of Computational and Applied Mathematics,
University of the Witwatersrand, Private Bag 3,
Wits 2050, Johannesburg, South Africa
{gkulikov, sshindin}@cam.wits.ac.za

Abstract. In the paper we design an adaptive numerical method to solve stiff ordinary differential equations with any reasonable accuracy set by the user. It is a two-step second order method possessing the A-stability property on any nonuniform grid [3]. This method is also implemented with the local-global step size control developed earlier in [8] to construct the appropriate grid automatically. It is shown that we are able to extend our technique for computation of higher derivatives of fixed-coefficient multistep methods to variable-coefficient multistep methods. We test the new algorithm on problems with exact solutions and stiff problems as well, in order to confirm its performance.

1 Introduction

The problem of an automatic global error control for the numerical solution of ordinary differential equations (ODEs) of the form

$$x'(t) = g(t, x(t)), \quad t \in [t_0, t_0 + T], \quad x(t_0) = x^0, \tag{1}$$

where $x(t) \in \mathbf{R}^n$ and $g : D \subset \mathbf{R}^{n+1} \to \mathbf{R}^n$ is a sufficiently smooth function, is one of the challenges of modern computational mathematics. ODE (1) is quite usual in applied research and practical engineering (see, for example, [1], [4], [6], [7]). Often, problem (1) is stiff and requires numerical methods with special properties of stability. A-stable methods are desired in such a situation (see, for example, [1], [4], [7]).

Unfortunately, there are very few algorithms with the property indicated above among linear multistep formulas because of the Dahlquist's second barrier [2]. It says that there exist no A-stable multistep methods of any order higher than two (even on uniform grids). The problem becomes more complicated on nonuniform grids. Thus, the only known now family of multistep formulas which

---

* This work was supported in part by the National Research Foundation of South Africa.

are A-stable on any grid is the one-parameter family of two-step methods derived by Dahlquist et al. [3]. Note there is no sense to consider "longer" methods because, anyway, they will be of order two at most. We concentrate on two particular choices of the parameter. One of them was made in the paper mentioned above (see [3]). Another one is our own choice. We compare both algorithms on numerical examples.

The methods presented in [3] are the good choice for a variable step size implementation. They have no step size restriction with a point of view of stability. Our idea is to supply these methods with the local-global step size control [8] aiming to attain any reasonable accuracy for the numerical solution of problem (1) in automatic mode. We also extend the technique of computation of higher derivatives [9] to variable-coefficient multistep methods. It imposes a weaker condition on the right-hand side of problem (1) for the step size selection to be correct, than in [8], where we differentiated interpolating polynomials.

The paper is organized as follows: Sect. 2 presents the family of A-stable two-step methods on uniform and nonuniform grids. Sect. 3 is devoted to the local and global errors estimation technique for the numerical methods mentioned above. The last section in the paper gives us numerical experiments confirming practical importance of the algorithms under consideration.

2 A-Stable Two-Step Methods

Further, we suppose that ODE (1) possesses a unique solution $x(t)$ on the whole interval $[t_0, t_0 + T]$. To solve problem (1) numerically, we introduce a uniform grid w_τ with step size τ on the interval $[t_0, t_0 + T]$ and apply the A-stable linear two-step method of order 2 in the form

$$\sum_{i=0}^{2} a_i x_{k+1-i} = \tau \sum_{i=0}^{2} b_i g(t_{k+1-i}, x_{k+1-i}), \quad k = 1, 2, \ldots, K-1, \qquad (2)$$

where

$$a_0 = \frac{1}{\gamma+1}, \quad a_1 = \frac{\gamma-1}{\gamma+1}, \quad a_2 = -\frac{\gamma}{\gamma+1},$$

$$b_0 = \frac{3\gamma+1}{2(\gamma+1)^2}, \quad b_1 = \frac{(\gamma-1)^2}{2(\gamma+1)^2}, \quad b_2 = \frac{\gamma(\gamma+3)}{2(\gamma+1)^2}$$

and the free parameter satisfies the condition $0 < \gamma \leq 1$. Note that we have used a slightly different way to present the family of stable two-step methods from [3]. The starting values x_k, $k = 0, 1$, are considered to be known.

We apply the following idea in order to fix the parameter γ. Let us consider the linear test equation $x' = \lambda x$ where λ is a complex number. We want to provide the best stability at infinity for method (2). This property is close to L-stability of Ehle [5] and useful when integrating very stiff ODEs. It means for multistep methods that we need to minimize the spectral radius of the companion matrix of method (2) (see [7]).

The companion matrix of method (2) when Re $\mu \to -\infty$ has the following form:

$$C_\infty(\gamma) \stackrel{\text{def}}{=} \lim_{\text{Re } \mu \to -\infty} \begin{pmatrix} \frac{\mu b_1 - a_1}{a_0 - \mu b_0} & \frac{\mu b_2 - a_2}{a_0 - \mu b_0} \\ 1 & 0 \end{pmatrix} = \begin{pmatrix} -\frac{b_1}{b_0} & -\frac{b_2}{b_0} \\ 1 & 0 \end{pmatrix}$$

where $\mu = \tau\lambda$. Unfortunately, $\rho(C_\infty(\gamma)) > 0$ (i.e., the spectral radius of the matrix $C_\infty(\gamma)$ is greater than zero) for any $0 < \gamma \leq 1$ because both coefficients b_1 and b_2 cannot vanish simultaneously (see (2)). Nevertheless, a simple computation shows that eigenvalues of the matrix $C_\infty(\gamma)$ are

$$\lambda_{1,2} = \frac{-(\gamma-1)^2 \pm (\gamma+1)\sqrt{\gamma^2 - 18\gamma + 1}}{6\gamma + 2}.$$

Then, we easily calculate that the minimum of the expression $\max\{|\lambda_1|, |\lambda_2|\}$ will be achieved when $\gamma = \gamma_1 = 9 - 4\sqrt{5} \approx 0.055$. Thus, we conclude that $\rho(C_\infty(\gamma_1)) = |\lambda_1| = |\lambda_2| = |18 - 8\sqrt{5}|/|3\sqrt{5} - 7| \approx 0.381$.

We remark that Dahlquist et al. [3] suggested another choice for γ. They tried to minimize the error constant of method (2) and preserve good stability properties. Their choice was $\gamma_2 = 1/5$.

Formula (2) implies that the step size τ is fixed. Unfortunately, the latter requirement is too restrictive for many practical problems. Therefore we determine continuous extensions to nonuniform grids for both methods (2) with different γ's and come to the following formulas:

$$\begin{aligned} x_{k+1} + (8 - 4\sqrt{5})x_k + (4\sqrt{5} - 9)x_{k-1} = \tau_k \Bigg(& \frac{\theta_k^2 + (2\theta_k + 1)(9 - 4\sqrt{5})}{2\theta_k(\theta_k + 9 - 4\sqrt{5})} \\ \times g(t_{k+1}, x_{k+1}) + & \frac{\theta_k^2(2\sqrt{5} - 4) + 76 - 34\sqrt{5}}{\theta_k(\theta_k + 9 - 4\sqrt{5})} g(t_k, x_k) \\ + & \frac{(9 - 4\sqrt{5})(\theta_k^2 + 2\theta_k + 9 - 4\sqrt{5})}{2\theta_k(\theta_k + 9 - 4\sqrt{5})} g(t_{k-1}, x_{k-1}) \Bigg), \end{aligned} \quad (3)$$

$$\begin{aligned} x_{k+1} - \frac{4}{5}x_k - \frac{1}{5}x_{k-1} = \tau_k \Bigg(& \frac{5\theta_k^2 + 2\theta_k + 1}{2\theta_k(5\theta_k + 1)} g(t_{k+1}, x_{k+1}) \\ + & \frac{10\theta_k^2 - 2}{5\theta_k(5\theta_k + 1)} g(t_k, x_k) + \frac{5\theta_k^2 + 10\theta_k + 1}{10\theta_k(5\theta_k + 1)} g(t_{k-1}, x_{k-1}) \Bigg) \end{aligned} \quad (4)$$

where τ_k is a current step size of the nonuniform grid w_τ with a diameter τ (i.e., $\tau \stackrel{\text{def}}{=} \max_k\{\tau_k\}$) and $\theta_k \stackrel{\text{def}}{=} \tau_k/\tau_{k-1}$ is a ratio of adjacent step sizes. We have used our choice for γ, i.e. γ_1, in formula (3) and γ_2 to obtain method (4).

3 Local and Global Errors Estimation

We recall that both methods (3) and (4) are A-stable on an arbitrary nonuniform grid. Thus, we control step sizes by the accuracy requirement only. With this idea in mind, we impose the following restriction on the step size change:

$$\tau/\tau_{\min} \leq \Omega < \infty. \tag{5}$$

Formula (5) implies that the ratio of the maximum step size to the minimum one is bounded with the constant Ω. We need the latter formula for the local-global step size control to be correct (see, for example, [10]). On the other hand, any code solving real life problems must be provided with bounds for the maximum step size and for the minimum one, that is equivalent to (5), because of an asymptotic form of the theory of ODE methods and round-off errors. Thus, condition (5) gives us nothing new in practice.

Further, we present the theory of local and global errors computation for methods (3) and (4) together. So, it is convenient to consider the family of numerical methods [3] in the general form

$$\frac{\theta_k}{\theta_k+\gamma}x_{k+1} + \frac{\theta_k(\gamma-1)}{\theta_k+\gamma}x_k - \frac{\theta_k\gamma}{\theta_k+\gamma}x_{k-1}$$
$$= \tau_k\frac{\theta_k^2+(2\theta_k+1)\gamma}{2(\theta_k+\gamma)^2}g(t_{k+1},x_{k+1}) + \tau_k\frac{(1-\gamma)(\theta_k^2-\gamma)}{2(\theta_k+\gamma)^2}g(t_k,x_k) \tag{6}$$
$$+\tau_k\frac{\gamma(\theta_k^2+2\theta_k+\gamma)}{2(\theta_k+\gamma)^2}g(t_{k-1},x_{k-1})$$

where γ is the free parameter and θ_k is the most recent step size ratio.

For method (6), the standard theory in [8] gives

$$\Delta\tilde{x}_{k+1} \approx \frac{-1}{6}\Big(a_0(k)I_n - \tau_k b_0(k)\partial_x g(t_{k+1},\tilde{x}_{k+1})\Big)^{-1}\tilde{x}_{k+1}^{(3)}$$
$$\times \tau_k^3 \sum_{i=1}^{2}\Big(a_i(k)\psi_i^3(\theta_k) + 3b_i(k)\psi_i^2(\theta_k)\Big), \tag{7}$$

$$\Delta x_{k+1} \approx \Big(a_0(k)I_n - \tau_k b_0(k)\partial_x g(t_{k+1},x_{k+1})\Big)^{-1}$$
$$\times \sum_{i=1}^{2}\Big(\tau_k b_i(k)\partial_x g(t_{k+1-i},x_{k+1-i}) - a_i(k)I_n\Big)\Delta x_{k+1-i} + \Delta\tilde{x}_{k+1}, \tag{8}$$

$k = l-1, l, \ldots, K-1$, where $a_i(k)$ and $b_i(k)$, $i = 0, 1, 2$, are the correspondent coefficients of method (6), and functions ψ_i in formula (7) are defined as follows:

$$\psi_i(\theta_k) \stackrel{\text{def}}{=} 1 + \sum_{m=1}^{i-1}\theta_k^{-1}, \quad i = 1, 2.$$

Here, the corrected numerical solution $\tilde{x}_{k+1} \stackrel{\text{def}}{=} x_{k+1} + \Delta x_{k+1}$ is of order 3, $\partial_x g(t_{k+1},x_{k+1})$ denotes a partial derivative of the mapping $g(t_{k+1},x_{k+1})$ with respect to the second variable, I_n is the identity matrix of dimension n. The starting errors Δx_k, $k = 0, 1, \ldots, l-1$, are considered to be zero because the starting values are computed accurately enough (see the starting procedure in [10] or [11]).

Note that formulas (7) and (8) have been derived with errors of $O(\tau^4)$ and $O(\tau^3)$, respectively. We also point out that formula (3) has been given in a slightly different form than it was presented in [8]. Here, we have derived the local error of method (6) with respect to the more recent step size τ_k and the necessary step size ratio θ_k rather than with respect to the step sizes τ_k and τ_{k-1}. We have done that for a convenience of presentation of further results concerning derivative computation.

To calculate the approximate derivative $\tilde{x}_{k+1}^{(3)}$ (at most with an error of $O(\tau)$) one can use a Newton (or Hermite) interpolation formula of sufficiently high degree [8]. On the other hand, it imposes an unnecessarily stiff smoothness requirement. Therefore we show how to adapt the method of derivative computation in [9] to variable-coefficient method (6).

First of all we introduce the matrix

$$V_k(l,s) \stackrel{\text{def}}{=} \begin{pmatrix} 1 & 0 & 0 & \cdots & 0 \\ 1 & \bigl(-\psi_1(\Theta_k)\bigr)^1 & \bigl(-\psi_1(\Theta_k)\bigr)^2 & \cdots & \bigl(-\psi_1(\Theta_k)\bigr)^s \\ 1 & \bigl(-\psi_2(\Theta_k)\bigr)^1 & \bigl(-\psi_2(\Theta_k)\bigr)^2 & \cdots & \bigl(-\psi_2(\Theta_k)\bigr)^s \\ \vdots & \vdots & \vdots & \ddots & \vdots \\ 1 & \bigl(-\psi_l(\Theta_k)\bigr)^1 & \bigl(-\psi_l(\Theta_k)\bigr)^2 & \cdots & \bigl(-\psi_l(\Theta_k)\bigr)^s \\ 0 & 1 & 0 & \cdots & 0 \\ 0 & 1 & 2\bigl(-\psi_1(\Theta_k)\bigr)^1 & \cdots & s\bigl(-\psi_1(\Theta_k)\bigr)^{s-1} \\ 0 & 1 & 2\bigl(-\psi_2(\Theta_k)\bigr)^1 & \cdots & s\bigl(-\psi_2(\Theta_k)\bigr)^{s-1} \\ \vdots & \vdots & \vdots & \ddots & \vdots \\ 0 & 1 & 2\bigl(-\psi_{l-s+1}(\Theta_k)\bigr)^1 & \cdots & s\bigl(-\psi_{l-s+1}(\Theta_k)\bigr)^{s-1} \end{pmatrix} \quad (9)$$

where

$$\psi_i(\Theta_k) \stackrel{\text{def}}{=} \psi_i(\theta_k, \theta_{k-1}, \ldots, \theta_{k-l+1}) \stackrel{\text{def}}{=} 1 + \sum_{m=1}^{i-1} \prod_{j=1}^{m} \theta_{k+1-j}^{-1}, \quad i = 1, 2, \ldots, l, \quad (10)$$

for any l-step method of order s, when computing the $(s+1)$-th derivative of a numerical solution. Formula (9) is a generalization of the extended Vandermonde matrices in [9] to nonuniform grids. The principal point for us is that the matrix $V_k(l,s)$ is nonsingular for any grid. The latter follows from Lemma 2 in the paper mentioned above, formula (10) and the fact that all the step size ratios are positive. Then, the way presented in [9] gives a formula for computation of the necessary derivative $\tilde{x}_{k+1}^{(3)}$ with an error of $O(\tau)$. Thus, formula (7) is transformed to the convenient form

$$\Delta \tilde{x}_{k+1} \approx \bigl(a_0(k)I_n - \tau_k b_0(k)\partial_x g(t_{k+1}, \tilde{x}_{k+1})\bigr)^{-1} \tau_k \sum_{i=1}^{2} c_i(k) g(t_{k+1-i}, \tilde{x}_{k+1-i})$$

where

$$c_0(k) = \frac{-P(\theta_k)}{6\theta_k^2(\theta_k+1)(\theta_k+\gamma)}, \quad c_1(k) = \frac{P(\theta_k)}{6\theta_k^2(\theta_k+\gamma)}, \quad c_2(k) = \frac{-P(\theta_k)}{6\theta_k(\theta_k+1)(\theta_k+\gamma)}$$

and $P(\theta_k) = \theta_k^4 + 4\gamma\theta_k^3 + 6\gamma\theta_k^2 + 4\gamma\theta_k + \gamma^2$ is a polynomial with respect to the parameter γ, being a fixed number, and the most recent step size ratio θ_k.

We further refer to [10] or [11] for the local-global step size control algorithm and for the starting procedure. The step size selection is based on the error estimates presented above.

4 Numerical Experiments

In this section, we give a number of numerical examples confirming the efficiency of the methods presented above for nonstiff and stiff integrations as well. We start with numerical experiments on problems with known solutions. They are nonstiff, and our goal is to check the capacity of both algorithms with the local-global step size control to attain the set accuracy of computation in automatic mode.

The first test problem is taken from [6] and it has the form

$$x_1'(t) = 2tx_2(t)^{\frac{1}{5}} x_4(t), \quad x_2'(t) = 10t\exp\Big(5\big(x_3(t)-1\big)\Big)x_4(t), \quad (11a)$$

$$x_3'(t) = 2tx_4(t), \quad x_4'(t) = -2t\ln\big(x_1(t)\big), \quad t \in [0,3] \quad (11b)$$

with $x(0) = (1,1,1,1)^T$. Problem (11) possesses the exact solution

$$x_1(t) = \exp\big(\sin t^2\big), \; x_2(t) = \exp\big(5\sin t^2\big), \; x_3(t) = \sin t^2 + 1, \; x_4(t) = \cos t^2.$$

Therefore it is convenient to verify how our adaptive methods will reach the required accuracy.

The second problem is quite practical. This is the restricted three body problem (see, for example, [6]):

$$x_1''(t) = x_1(t) + 2x_2'(t) - \mu_1\frac{x_1(t)+\mu_2}{y_1(t)} - \mu_2\frac{x_1(t)-\mu_1}{y_2(t)}, \quad (12a)$$

$$x_2''(t) = x_2(t) - 2x_1'(t) - \mu_1\frac{x_2(t)}{y_1(t)} - \mu_2\frac{x_2(t)}{y_2(t)}, \quad (12b)$$

$$y_1(t) = \Big((x_1(t)+\mu_2)^2 + x_2(t)^2\Big)^{3/2}, \quad y_2(t) = \Big((x_1(t)-\mu_1)^2 + x_2(t)^2\Big)^{3/2}, \quad (12c)$$

where $t \in [0,T]$, $T = 17.065216560157962558891$, $\mu_1 = 1 - \mu_2$ and $\mu_2 = 0.012277471$. The initial values of problem (12) are: $x_1(0) = 0.994$, $x_1'(0) = 0$, $x_2(0) = 0$, $x_2'(0) = -2.00158510637908252240$. It has no analytic solution, but its solution-path is periodic. Thus, we are also capable to observe the work of both methods in practice.

Having fixed the global error bounds and computed the local tolerances by the formula $\epsilon_l = \epsilon_g^{3/2}$, we apply methods (3) and (4) with the local-global step size control to problems (11) and (12) and come to the data collected in Tables 1, 2. We see that both choices of the parameter γ in the family of numerical

Table 1. Global errors obtained for variable-coefficient methods (3) and (4) (with the local-global step size control) applied to problem (11)

Method	required accuracy				
	$\epsilon_g = 10^{-01}$	$\epsilon_g = 10^{-02}$	$\epsilon_g = 10^{-03}$	$\epsilon_g = 10^{-04}$	$\epsilon_g = 10^{-05}$
(3)	8.005×10^{-02}	7.305×10^{-03}	7.281×10^{-04}	9.702×10^{-05}	9.823×10^{-06}
(4)	7.191×10^{-02}	9.041×10^{-03}	8.648×10^{-04}	7.758×10^{-05}	7.566×10^{-06}

Table 2. Global errors obtained for variable-coefficient methods (3) and (4) (with the local-global step size control) applied to problem (12)

Method	required accuracy				
	$\epsilon_g = 10^{-01}$	$\epsilon_g = 10^{-02}$	$\epsilon_g = 10^{-03}$	$\epsilon_g = 10^{-04}$	$\epsilon_g = 10^{-05}$
(3)	$2.083 \times 10^{+00}$	9.445×10^{-03}	7.113×10^{-04}	7.077×10^{-05}	7.081×10^{-06}
(4)	9.092×10^{-02}	9.373×10^{-03}	7.704×10^{-04}	7.703×10^{-05}	7.714×10^{-06}

methods (6) lead to quite nice results. Both methods have computed the numerical solutions with the set accuracy. We only want to point out that our choice (method (3)), when $\gamma = 9 - 4\sqrt{5}$, gives the required numerical solutions faster. The average execution time for method (3) is less by a factor of 1.4 for the first test problem and by a factor of 1.3 for the second test problem compared with method (4).

Now we try methods (3) and (4) on the Van der Pol's equation

$$x'_1(t) = x_2(t), \quad x'_2(x) = \mu^2\Big((1 - x_1(t)^2)x_2(t) - x_1(t)\Big), \quad t \in [0, 2] \quad (13)$$

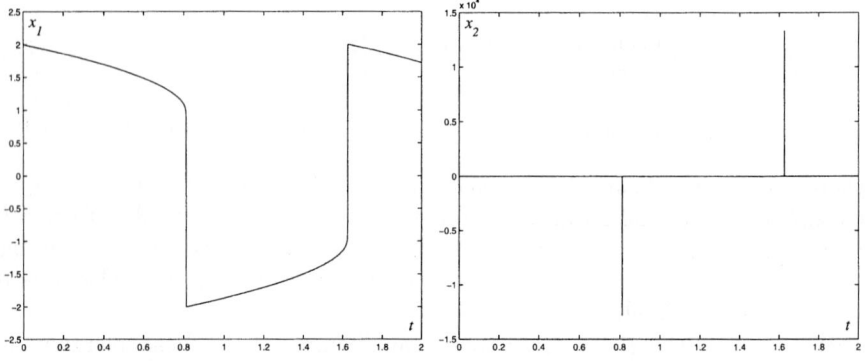

Fig. 1. The components x_1 and x_2 of the Van der Pol's equation calculated by methods (3) and (4) with $\epsilon_g = 10^{-1}$

where $x(0) = (2,0)^T$, and $\mu = 100$. Problem (13) is considered to be very stiff when the parameter μ is a big number. Despite the small order of the methods under consideration the results obtained are quite promising. The components of the numerical solution of problem (13) are given in Figure 1. Both methods have produced the same result (up to an error of 10^{-1}) which completely corresponds to the picture in [7].

The final point to mention is that our choice of γ (method (3)) has again computed the numerical solution of Van der Pol's equation faster (with a factor of 1.4). This is a good reason to implement it in practice.

References

1. Butcher, J.C.: Numerical methods for ordinary differential equations. John Wiley and Son, Chichester, 2003
2. Dahlquist, G.: A special stability problem for linear multistep methods. BIT. **3** (1963) 27–43
3. Dahlquist, G.G., Liniger W., Nevanlinna, O.: Stability of two-step methods for variable integration steps. SIAM J. Numer. Anal. **20** (1983) 1071–1085
4. Dekker, K., Verwer, J.G. Stability of Runge-Kutta methods for stiff nonlinear differential equations. North-Holland, Amsterdam, 1984
5. Ehle B.L.: On Padé approximations to the exponential function and A-stable methods for the numerical solution of initial value problems. Research report CSRR 2010, Dept. AACS, Univ. of Waterloo, Ontario, Canada, 1969
6. Hairer, E., Nørsett, S.P., Wanner, G.: Solving ordinary differential equations I: Nonstiff problems. Springer-Verlag, Berlin, 1987
7. Hairer, E., Wanner, G.: Solving ordinary differential equations II: Stiff and differential-algebraic problems. Springer-Verlag, Berlin, 1996
8. Kulikov, G.Yu., Shindin, S.K.: A technique for controlling the global error in multistep methods. (*in Russian*) Zh. Vychisl. Mat. Mat. Fiz. **40** (2000) No. 9, 1308–1329; *translation in* Comput. Math. Math. Phys. **40** (2000) No. 9, 1255–1275
9. Kulikov, G.Yu., Shindin, S.K.: On effective computation of asymptotically correct estimates of the local and global errors for multistep methods with fixed coefficients. (*in Russian*) Zh. Vychisl. Mat. Mat. Fiz. **44** (2004) No. 5, 847–868; *translation in* Comput. Math. Math. Phys. **44** (2004) No. 5, 794–814
10. Kulikov, G.Yu., Shindin, S.K.: On interpolation-type multistep methods with automatic global error control. (*in Russian*) Zh. Vychisl. Mat. Mat. Fiz. **44** (2004) No. 8, 1400–1421; *translation in* Comput. Math. Math. Phys. **44** (2004) No. 8, 1314–1333
11. Kulikov, G.Yu., Shindin, S.K.: On automatic global error control in multistep methods with polynomial interpolation of numerical solution. In: Antonio Lagana et al (eds.): Computational Science and Its Applications — ICCSA 2004. International Conference, Assisi, Italy, May 2004. Proceedings, Part III. Lecture Notes in Computer Science. **3045** (2004) 345–354

Bifurcation Analysis of Large Equilibrium Systems in MATLAB

David S. Bindel[1], James W. Demmel[2], Mark J. Friedman[3],
Willy J.F. Govaerts[4], and Yuri A. Kuznetsov[5]

[1] Department of Electrical Engineering and Computer Science,
University of California at Berkeley,
Berkeley, CA 94704
dbindel@eecs.berkeley.edu

[2] Department of Electrical Engineering and Computer Science,
Department of Mathematics,
University of California at Berkeley,
Berkeley, CA 94704
demmel@eecs.berkeley.edu

[3] Mathematical Sciences Department,
University of Alabama in Huntsville,
Huntsville, AL 35899
friedman@math.uah.edu

[4] Department of Applied Mathematics and Computer Science,
Ghent University,
Krijgslaan 281-S9, B-9000 Ghent, Belgium
Willy.Govaerts@UGent.be

[5] Mathematisch Instituut, Budapestlaan 6,
3584CD Utrecht, The Netherlands
kuznet@math.uu.nl

Abstract. The Continuation of Invariant Subspaces (*CIS*) algorithm produces a smoothly varying basis for an invariant subspace $\mathcal{R}(s)$ of a parameter-dependent matrix $A(s)$. In the case when $A(s)$ is the Jacobian matrix for a system that comes from a spatial discretization of a partial differential equation, it will typically be large and sparse. CL_MATCONT is a user-friendly MATLAB package for the study of dynamical systems and their bifurcations. We incorporate the CIS algorithm into CL_MATCONT to extend its functionality to large scale bifurcation computations via subspace reduction.

1 Introduction

Parameter-dependent Jacobian matrices provide important information about dynamical systems

$$\frac{du}{dt} = f(u, \alpha), \text{ where } u \in \mathbb{R}^n, \ \alpha \in \mathbb{R}, \ f(u, \alpha) \in \mathbb{R}^n. \tag{1}$$

For example, to analyze stability at branches $(u(s), \alpha(s))$ of steady states

$$f(u, \alpha) = 0, \tag{2}$$

we look at the linearization $f_u(u(s), \alpha(s))$. For general background on dynamical systems theory we refer to the existing literature, in particular [15]. If the system comes from a spatial discretization of a partial differential equation, then f_u will typically be large and sparse. In this case, an invariant subspace $\mathcal{R}(s)$ corresponding to a few eigenvalues near the imaginary axis provides information about stability and bifurcations. We are interested in continuation and bifurcation analysis of large stationary problems (2).

Numerical continuation for large nonlinear systems of this form is an active area of research, and the idea of subspace projection is common in many methods being developed. The continuation algorithms are typically based on Krylov subspaces, or on recursive projection methods which use a time integrator instead of a Jacobian multiplication as a black box to identify the low-dimensional invariant subspace where interesting dynamics take place; see e.g. [17, 5, 11, 4, 6], and references there.

CL_MATCONT [9] and its GUI version MATCONT [8] are MATLAB packages for the study of dynamical systems and their bifurcations for small and moderate size problems. The MATLAB platform is attractive because it makes them user-friendly, portable to all operating systems, and allows a standard handling of data files, graphical output, etc.

Recently, we developed the Continuation of Invariant Subspaces *(CIS) algorithm* for computing a smooth orthonormal basis for an invariant subspace $\mathcal{R}(s)$ of a parameter-dependent matrix $A(s)$ [7, 10, 12, 3]. The CIS algorithm uses projection methods to deal with large problems. See also [12] for similar results.

In this paper we consider integrating the CIS algorithm into CL_MATCONT. Standard bifurcation analysis algorithms, such as those used in CL_MATCONT, involve computing functions of $A(s)$. We adapt these methods to large problems by computing the same functions of a much smaller restriction $C(s) := A(s)|_{\mathcal{R}(s)}$ of $A(s)$ onto $\mathcal{R}(s)$. Note, that the CIS algorithm ensures that only eigenvalues of $C(s)$ can cross the imaginary axis, so that $C(s)$ provides all the relevant information about bifurcations. In addition, the continued subspace is adapted to track behavior relevant to bifurcations.

2 Bifurcations for Large Systems

Let $x(s) = (u(s), \alpha(s)) \in \mathbb{R}^n \times \mathbb{R}$ be a smooth local parameterization of a solution branch of the system (2). We write the Jacobian matrix along this path as $A(s) := f_u(x(s))$. A solution point $x(s_0)$ is a *bifurcation point* if $\text{Re}\,\lambda_i(s_0) = 0$ for at least one eigenvalue $\lambda_i(s_0)$ of $A(s_0)$.

A *test function* $\phi(s) := \psi(x(s))$ is a (typically) smooth scalar function that has a regular zero at a bifurcation point. A bifurcation point between consecutive continuation points $x(s_k)$ and $x(s_{k+1})$ is *detected* when

$$\psi(x(s_k))\,\psi(x(s_{k+1})) < 0. \tag{3}$$

Once a bifurcation point has been detected, it can be *located* by solving the system
$$\begin{cases} f(x) = 0, \\ g(x) = 0, \end{cases} \quad (4)$$
for an appropriate function g.

We consider here the case of a (generic codimension-1 bifurcation) *fold* or *limit point* (LP) and a *branch point* (BP), on a solution branch (2). For both detecting and locating these bifurcations, CL_MATCONT uses the following test functions (see e.g. [15], [13], [9]):

$$\psi_{BP}^M(x(s)) := \det \begin{bmatrix} A & f_\alpha \\ \dot{u}^T & \dot{\alpha} \end{bmatrix}, \quad (5)$$

$$\psi_{LP}^M(x(s)) := \det(A(s)) = \prod_{i=1}^n \lambda_i(s), \quad (6)$$

where $\dot{x} := dx/ds$. The bifurcations are defined by:

$$\text{BP} : \psi_{BP}^M = 0, \qquad \text{LP} : \psi_{LP}^M = 0, \psi_{BP}^M \neq 0. \quad (7)$$

For some $m \ll n$, let
$$\Lambda_1(s) := \{\lambda_i(s)\}_{i=1}^m, \ \operatorname{Re} \lambda_m \leq \ldots \leq \operatorname{Re} \lambda_{m_u+1} < 0 \leq \operatorname{Re} \lambda_{m_u} \leq \ldots \leq \operatorname{Re} \lambda_1, \quad (8)$$

be a small set consisting of rightmost eigenvalues of $A(s)$ and let $Q_1(x(s)) \in \mathbb{R}^{n \times m}$ be an orthonormal basis for the invariant subspace $\mathcal{R}(s)$ corresponding to $\Lambda_1(s)$. Then an application of the CIS algorithm to $A(s)$ produces

$$C(x(s)) := Q_1^T(x(s))A(s)Q_1(x(s)) \in \mathbb{R}^{m \times m}, \quad (9)$$

which is the restriction of $A(s)$ onto $\mathcal{R}(s)$. Moreover, the CIS algorithm ensures that the only eigenvalues of $A(s)$ that can cross the imaginary axis come from $\Lambda_1(s)$, and these are exactly the eigenvalues of $C(x(s))$. We use this result to construct new methods for detecting and locating bifurcations. Note, that $\Lambda_1(s)$ is computed automatically whenever $C(x(s))$ is computed.

2.1 Fold

Detecting Fold. We replace $\psi_{BP}^M(x(s))$ and $\psi_{LP}^M(x(s))$, respectively, by

$$\psi_{BP}(x(s)) := \operatorname{sign}\left(\det \begin{bmatrix} A & f_\alpha \\ \dot{u}^T & \dot{\alpha} \end{bmatrix}\right), \quad (10)$$

$$\psi_{LP}(x(s)) := \prod_{i=1}^m \lambda_i(s). \quad (11)$$

Then LP is detected as:

$$\text{LP} : \psi_{BP}(x(s_k))\psi_{BP}(x(s_{k+1})) > 0 \text{ and } \psi_{LP}(x(s_k))\psi_{LP}(x(s_{k+1})) < 0. \quad (12)$$

Locating Fold. Let $x_0 = x(s_0)$ be a fold point. Then $A(s_0)$ has rank $n-1$. To locate x_0, we use a *minimally augmented system* (see [13], [9]), with A replaced by C, whenever possible. The system consists of $n+1$ scalar equations for $n+1$ components $x = (u, \alpha) \in \mathbb{R}^n \times \mathbb{R}$,

$$\begin{cases} f(x) = 0, \\ g(x) = 0, \end{cases} \tag{13}$$

where $g = g(x)$ is computed as the last component of the solution vector $(v, g) \in \mathbb{R}^m \times \mathbb{R}$ to the $(m+1)$-dimensional *bordered system*:

$$\begin{bmatrix} C(x) & w_{bor} \\ v_{bor}^T & 0 \end{bmatrix} \begin{bmatrix} v \\ g \end{bmatrix} = \begin{bmatrix} 0_{m \times 1} \\ 1 \end{bmatrix}, \tag{14}$$

where $v_{bor} \in \mathbb{R}^m$ is close to a nullvector of $C(x_0)$, and $w_{bor} \in \mathbb{R}^m$ is close to a nullvector of $C^T(x_0)$ (which ensures that the matrix in (14) is nonsingular). For $g = 0$, system (14) implies $Cv = 0$, $v_{bor}^T v = 1$. Thus (13) and (14) hold at $x = x_0$, which is a regular zero of (13).

The system (13) is solved by the Newton method, and its Jacobian matrix is:

$$J := \begin{bmatrix} f_x \\ g_x \end{bmatrix} = \begin{bmatrix} A & f_\alpha \\ g_u & g_\alpha \end{bmatrix} \in \mathbb{R}^{(n+1) \times (n+1)}, \tag{15}$$

where g_x is computed as

$$g_x = -w^T C_x v, \tag{16}$$

with w obtained by solving

$$\begin{bmatrix} C^T(x) & v_{bor} \\ w_{bor}^T & 0 \end{bmatrix} \begin{bmatrix} w \\ g \end{bmatrix} = \begin{bmatrix} 0 \\ 1 \end{bmatrix}. \tag{17}$$

Here $C_x v$ is computed as

$$C_x(x)v \approx Q_1^T \frac{f_x(u + \delta \frac{z}{\|z\|}, \alpha) - f_x(u - \delta \frac{z}{\|z\|}, \alpha)}{2\delta} \|z\|, \quad z := Q_1 v \in \mathbb{R}^n. \tag{18}$$

Finally we note that at each Newton step for solving (13), linear systems with the matrix (15) should be solved by the mixed block elimination (see [13] and references there), since the matrix (15) has the form

$$M := \begin{bmatrix} A & b \\ c^T & d \end{bmatrix}, \tag{19}$$

where $A \in \mathbb{R}^{n \times n}$ is large and sparse, $b, c \in \mathbb{R}^n$, $d \in \mathbb{R}$, and A can be ill conditioned.

Once the fold point $x_0 = (u_0, \alpha_0)$ is computed, the corresponding quadratic normal form coefficient

$$a := \frac{1}{2} \widehat{w}^T f_{uu} [\widehat{v}, \widehat{v}]$$

is computed approximately as

$$a \approx \frac{1}{2\delta^2} \widehat{w}^T [f(u_0 + \delta \widehat{v}, \alpha_0) + f(u_0 - \delta \widehat{v}, \alpha_0)], \quad \widehat{v} \approx \frac{Q_1 v}{\|Q_1 v\|}, \quad \widehat{w} \approx \frac{Q_1 w}{\widehat{v}^T Q_1 w}. \tag{20}$$

Fold Continuation. We use again the system (13) of $n+1$ scalar equations, but for $n+2$ components $x = (u, \alpha) \in \mathbb{R}^n \times \mathbb{R}^2$, in this case. Again g is obtained by solving (14), where g_x is computed using (16), (17), and (18).

There are four generic codimension-2 bifurcations on the fold curve: *Bogdanov-Takens (or double zero) point* (BT), *Zero - Hopf point* (ZH), *Cusp point* (CP), and a *branch point* (BP). These are detected and located by the corresponding modifications of CL_MATCONT test functions. For example, test function to detect ZH is

$$\psi_{ZH}(x(s)) := \prod_{m \geq i > j} (\operatorname{Re} \lambda_i(s) + \operatorname{Re} \lambda_j(s)). \tag{21}$$

2.2 Branch Points

Detecting Branching. The branch point is detected as:

$$BP : \psi_{BP}(x(s_k)) \psi_{BP}(x(s_{k+1})) < 0, \tag{22}$$

where ψ_{BP} is defined by (10).

Locating Branching. Let $x_0 = x(s_0)$ be a branch point such that $f_x^0 = f_x(x_0)$ has rank $n - 1$ and

$$\mathcal{N}(f_x^0) = \operatorname{Span}\{v_1^0, v_2^0\}, \quad \mathcal{N}\left((f_x^0)^T\right) = \operatorname{Span}\{\psi^0\}.$$

We use a minimally augmented system ([14], [2], [1]) of $n+2$ scalar equations for $n+2$ components $(x, \mu) = (u, \alpha, \mu) \in \mathbb{R}^n \times \mathbb{R} \times \mathbb{R}$,

$$\begin{aligned} f(x) + \mu w_{bor} &= 0, \\ g_1(x) &= 0, \\ g_2(x) &= 0, \end{aligned} \tag{23}$$

where μ is an unfolding parameter, $w_{bor} \in \mathbb{R}^n$ is fixed, and $g_1 = g_1(x)$, $g_2 = g_2(x) \in \mathbb{R}$ are computed as the last row of the solution matrix $\begin{bmatrix} v_1 & v_2 \\ g_1 & g_2 \end{bmatrix}$, $v_1, v_2 \in \mathbb{R}^{n+1}$, to the $(n+2)$-dimensional bordered system:

$$\begin{bmatrix} f_x(x) & w_{bor} \\ V_{bor}^T & 0_{2 \times 1} \end{bmatrix} \begin{bmatrix} v_1 & v_2 \\ g_1 & g_2 \end{bmatrix} = \begin{bmatrix} 0_{n \times 2} \\ I_2 \end{bmatrix}, \quad V_{bor} = [v_{1,bor}\ v_{2,bor}], \tag{24}$$

where $v_{1,bor}, v_{2,bor} \in \mathbb{R}^{n+1}$ are close to an orthonormal basis of $\mathcal{N}(f_x^0)$, and w_{bor} is close to the nullvector of $(f_x^0)^T$.

The system (23) is solved by the Newton method [14] with the modifications in [2]. The Newton method is globalized by combining it with the bisection on the solution curve.

3 Examples

All computations are performed on a 3.2 GHz Pentium IV laptop.

Example 1. 1D Brusselator, a well known model system for autocatalytic chemical reactions with diffusion:
$$\frac{d_1}{l^2}u'' - (b+1)u + u^2v + a = 0, \quad \frac{d_2}{l^2}v'' + bu - u^2v = 0, \quad \text{in } \Omega = (0,1),$$
$$u(0) = u(1) = a, \qquad v(0) = v(1) = \frac{b}{a}. \tag{25}$$

This problem exhibits many interesting bifurcations and has been used in the literature as a standard model for bifurcation analysis (see e.g. [16]). We discretize the problem with a standard second-order finite difference approximation for the second derivatives at N mesh points. We write the resulting system, which has dimension $n = 2N$, in the form (2). This discretization of the Brusselator is used in a CL_MATCONT example [9]. In Figure 1 a bifurcation diagram in two parameters (l, b) is shown in the case $n = 2560$. We first continue an equilibrium branch with a continuation parameter l (15 steps, 13.5 secs) and locate LP at $l = 0.060640$. We next continue the LP branch in two parameters l, b (200 steps, 400.8 secs) and locate ZH at $(l, b) = (0.213055, 4.114737)$.

Example 2. Deformation of a 2D arch. We consider the snap-through of an elastic arch, shown in Figure 2. The arch is pinned at both ends, and the y displacement of the center of the arch is controlled as a continuation parameter.

Let $\Omega_0 \subset \mathbb{R}^2$ be the interior of the undeformed arch (Figure 2, top left), and let the boundary $\Gamma = \Gamma_D \cup \Gamma_N$, where Γ_D consists of the two points where the arch is pinned, and Γ_N is the remainder of the boundary, which is free. At equilibrium, material points $X \in \Omega_0$ in the deformed arch move to positions $x = X + u$. Except at the control point X_{center} in the center of the arch, this deformation satisfies the equilibrium force-balance equation [18]

$$\sum_{J=1}^{2} \frac{\partial S_{IJ}}{\partial X_J} = 0, \quad X \in \Omega_0, \quad I = 1, 2. \tag{26}$$

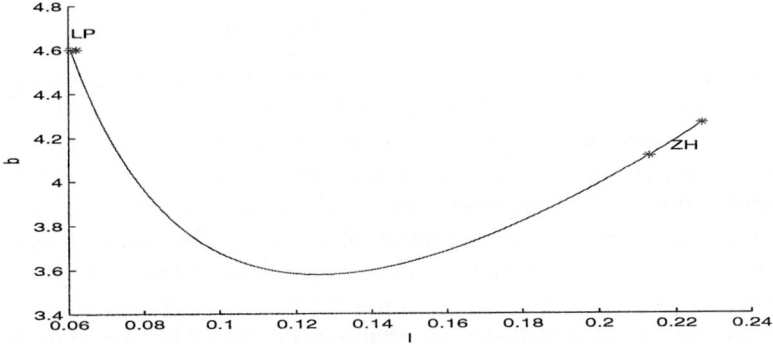

Fig. 1. Bifurcation diagram for a 1D Brusselator

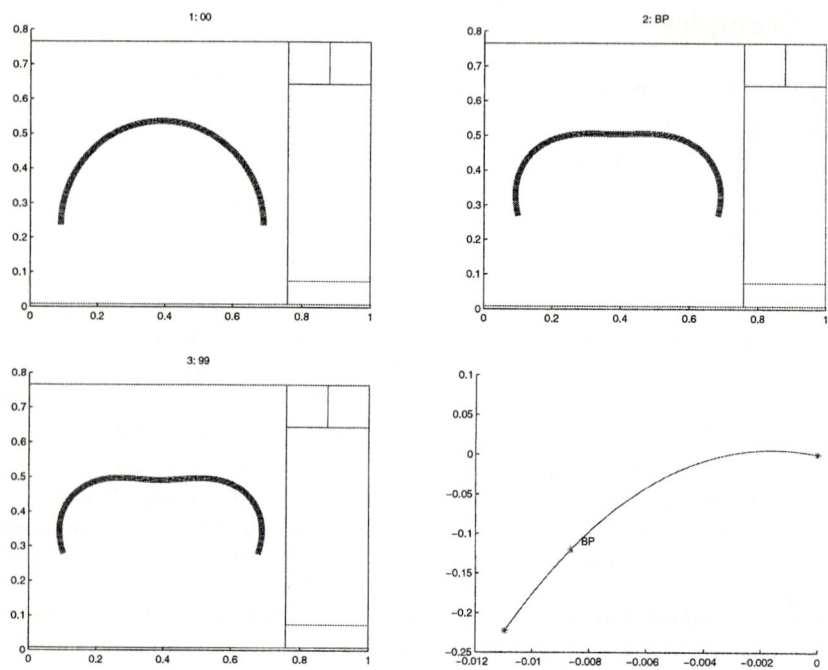

Fig. 2. Top left: the undeformed arch, top right: the arch at the bifurcation point, bottom left: the arch at the end of continuation, bottom right: the bifurcation diagram

where the second Piola-Kirchhoff stress tensor S is a nonlinear function of the Green strain tensor E, where $E := \frac{1}{2}(F^T F - I)$, $F := \frac{\partial u}{\partial X}$. Equation (26) is thus a fully nonlinear second order elliptic system. The boundary and the control point X_{center} are subject to the boundary conditions

$$u = 0 \text{ on } \Gamma_D, \quad SN = 0 \text{ on } \Gamma_N, \text{ where } N \text{ is an outward unit normal}, \quad (27)$$

$$e_2 \cdot u = \alpha, \quad e_1 \cdot (FSN) = 0. \quad (28)$$

The first condition at X_{center} says that the vertical displacement is determined; the second condition says that there is zero force in the horizontal direction.

We discretize (26) with biquadratic isoparametric Lagrangian finite elements. Let m be the number of elements through the arch thickness, and n be the number of elements along the length of the arch; then there are $(2m+1)(2n+1)$ nodes, each with two associated degrees of freedom. The Dirichlet boundary conditions are used to eliminate four unknowns, and one of the unknowns is used as a control parameter, so that the total number of unknowns is $N = 2(2m+1)(2n+1) - 5$. Figure 2 displays the results in the case when the continuation parameter is y displacement of node in middle of arch; $m = 4$, $n = 60$, $N = 2173$ unknowns; 80 steps took 258.4 secs, one BP was found.

Acknowledgment

Mark Friedman was supported in part under under NSF DMS-0209536.

References

1. E. ALLGOWER AND K. GEORG, *Numerical path following*, in Handbook of Numerical Analysis, volume 5, P. G. Ciarlet and J. L. Lions, eds., North-Holland, 1997, pp. 3–207.
2. E. ALLGOWER AND H. SCHWETLICK, *A general view of minimally extended systems for simple bifurcation points*, Z. angew. Math. Mech., 77 (1997), pp. 83–98.
3. D. BINDEL, J. DEMMEL, AND M. FRIEDMAN, *Continuation of invariant subspaces for large bifurcation problems*, in Proceedings of the SIAM Conference on Linear Algebra, Williamsburg, VA, 2003.
4. E. A. BURROUGHS, R. B. LEHOUCQ, L. A. ROMERO, AND A. J. SALINGER, *Linear stability of flow in a differentially heated cavity via large-scale eigenvalue calculations*, Tech. Report SAND2002-3036J, Sandia National Laboratories, 2002.
5. C. S. CHIEN AND M. H. CHEN, *Multiple bifurcations in a reaction-diffusion problem*, Computers Math. Applic., 35 (1998), pp. 15–39.
6. K. A. Cliffe, A. Spence, and S. J. Tavener, *The numerical analysis of bifurcations problems with application to fluid mechanics*, Acta Numerica, (2000), pp. 1–93.
7. J. W. DEMMEL, L. DIECI, AND M. J. FRIEDMAN, *Computing connecting orbits via an improved algorithm for continuing invariant subspaces*, SIAM J. Sci. Comput., 22 (2001), pp. 81–94.
8. A. Dhooge, W. Govaerts, and Yu.A. Kuznetsov, *MATCONT: A MATLAB package for numerical bifurcation analysis of odes.*, ACM TOMS., 29 (2003), pp. 141–164.
9. A. DHOOGE, W. GOVAERTS, YU.A. KUZNETSOV, W. MESTROM, AND A. M. RIET, *MATLAB continuation software package CL_MATCONT*, Jan. 2003. http://www.math.uu.nl/people/kuznet/cm/.
10. L. DIECI AND M. J. FRIEDMAN, *Continuation of invariant subspaces*, Numerical Linear Algebra and Appl., 8 (2001), pp. 317–327.
11. E. J. DOEDEL AND H. SHARIFI, *Collocation methods for continuation problems in nonlinear elliptic PDEs, issue on continuation*, Methods in Fluid Mechanics, Notes on Numer. Fluid. Mech, 74 (2000), pp. 105–118.
12. W-J. BEYN AND W. KLESSAND V. THÜMMLER", *Ergodic Theory, Analysis, and Efficient Simulation of Dynamical Systems" Continuation of low-dimensional Invariant Subspaces in dynamical systems of large dimension*, Springer, 2001, pp. 47– 72.
13. W. GOVAERTS, *Numerical methods for bifurcations of dynamical equilibria*, SIAM, Philadelphia, 2000.
14. A. GRIEWANK AND G. REDDIEN, *Characterization and computation of generalized turning points*, SIAM J. Numer. Anal., 21 (1984), pp. 176–185.
15. YU. A. KUZNETSOV, *Elements of Applied Bifurcation Theory, Third edition*, Springer-Verlag, New York, 2004.
16. D. SCHAEFFER AND M. GOLUBITSKY, *Bifurcation analysis near a double eigenvalue of a model chemical reaction*, Arch. Rational Mech. Anal., 75 (1981), pp. 315–347.
17. G. M. SHROFF AND H. B. KELLER, *Stabilization of unstable procedures: The recursive projection method*, SIAM J. Numer. Anal, 30 (1993), pp. 1099–1120.
18. O. ZIENKIEWICZ AND R.T.TAYLOR, *The Finite Element Method: Volume 2, Solid Mechanics, 5th edition*, Butterworth Heinemann, Oxford, 2000.

Sliced-Time Computations with Re-scaling for Blowing-Up Solutions to Initial Value Differential Equations

Nabil R. Nassif[1], Dolly Fayyad[2], and Maria Cortas[3]

[1] American University of Beirut, Beirut Lebanon
[2] Lebanese University, Faculty of Science II, Fanar Lebanon
[3] Lebanese University, Computer Center, Museum Square, Beirut Lebanon

Abstract. In this paper, we present a new approach to simulate time-dependent initial value differential equations which solutions have a common property of blowing-up in a finite time. For that purpose, we introduce the concept of "sliced-time computations", whereby, a sequence of time intervals (slices) $\{[T_{n-1}, T_n] | n \geq 1\}$ is defined on the basis of a change of variables (re-scaling), allowing the generation of computational models that share symbolically or numerically "similarity" criteria. One of these properties is to impose that the re-scaled solution computed on each slice do not exceed a well-defined cut-off value (or threshold) S. In this work we provide fundamental elements of the method, illustrated on a scalar ordinary differential equation $y' = f(y)$ where $f(y)$ verifies $\int_0^\infty f(y) dy < \infty$. Numerical results on various ordinary and partial differential equations are available in [7], some of which will be presented in this paper.

1 Introduction

Let Ω be an open domain of $\mathbb{R}^n$ with boundary Γ, sufficiently regular. We consider in this work some numerical aspects related to the blow-up in finite time, of the solution of the semi-linear parabolic equation consisting in finding $\{T_b, u\}$ where $0 < T_b < \infty, u : \overline{\Omega} \times [0, T_b)$ such that:

$$u_t - \Delta u = f(u), x \in \Omega, 0 \leq t < T_b, u(x,0) = u_0(x) \geq 0, \forall x \in \Omega,$$
$$u(x,t) = 0, \forall x \in \Gamma, 0 < t < T_b. \quad (1)$$

The function f is such that $f(y) = O(y^p)$ or $f(y) = O(e^{ky})$ as $y \to \infty$ and verifies:

$$f : \mathbb{R} \to \mathbb{R}^+, \text{ with } f \in C^1, f \text{ and } f' \text{ strictly monotone increasing on } \mathbb{R}^+. \quad (2)$$

Under (2), there exists a unique solution to (1) such that, if $u_0 \geq 0$, then $u(.,t) \geq 0$, for all t in the time domain of existence for u. Furthermore, it is easy to verify that under (2.3), then $\int_a^\infty \frac{1}{f(x)} dx < \infty$ for some $a \geq 0$. In fact, it is shown in ([4] pp. 55-56), that this condition is necessary for finite blow-up time to problem (1). Thus, the solution $\{T_b, u\}$ is assumed to satisfy a blowing-up property, specifically the existence of a time T_b, such that $\lim_{t \to T_b^-} ||u(.,t)||_\infty = \infty$. where the L^∞-norm is defined by $||g||_\infty = \max_{x \in \Omega} |g(x)|$, for all functions $g : \mathbb{R} \to \mathbb{R}$. Under (2), the blow-up time T_b, verifies (see [4]), $\int_0^\infty [f(z)]^{-1} dz < T_b < \int_0^\infty [f(z) - \lambda_1 z]^{-1} dz$, with $\delta^* = \lambda_1 \sup\{\frac{u}{f(u)}, u \geq 0\} < 1$, $u_0 =$, λ_1 being the first eigenvalue of the problem $-\Delta \Phi = \lambda \Phi$. As to the points of Ω where the explosion occurs, Friedman et McLeod ([10]) proved that the set E of blow up points of (1), form a compact subset E in Ω. In case the domain Ω satisfies a radial symmetry property around the origin o, and u_0 is a radial function, then E reduces to a single point which is the origin itself. This property is confirmed by [4] who gives also the asymptotic behavior of the solution when $t \to T_b$. Space discretization of (1) leads naturally to very stiff solutions that demonstrate rapid growth in a short range of time, particularly when numerical time integration gets close to the blow-up time. This rapid growth imposes major restrictions on the time step. It is easily verified in such case, that even adaptive numerical procedures (available for example in the various **ode** MATLAB solvers) fail to provide an acceptable approximation to either the solution or to the blow-up time. Thus, our objective in this work is to obtain numerical algorithms for finding an approximation to u and as well an estimate of the blow-up time T_b, using a new **rescaling technique**. In previous papers (see [8], [9]) we have implemented a numerical method to solve (1) based on an ϵ-perturbation that transforms the system into an equivalent one that possesses a "mass conservation property" (see [13]). The solution of this resulting problem is known to be global, thus subject to be solved using standard numerical methods for stiff systems. Several authors have attempted to solve numerically finite blow-up time problems. To our knowledge, initial work can be found in [11],[12], when $f(u) = \lambda e^u$. The authors compute an approximation to the blow-up time, for several values of λ, through a semi-discretization of the space variables, then solve the resulting ordinary differential system of equations using Runge-Kutta methods. More sophisticated algorithms based on rescaling techniques have been considered by Chorin [6], and also Berger and Kohn [5] when $f(u) = u^\gamma$. Such technique can describe accurately the behavior of the solution near the singularity. However, one of its main disadvantages is a varying mesh width and time step at each point in space-time, linked to the rapid growth in the magnitude of the solution. Consequently, this requires simultaneously the adjustment of boundary conditions to avoid the loss of accuracy far from the singularity. In References [1],[2], the authors analyze the blow-up behavior of semidiscretizations of reaction diffusion equations. More recently, Acosta, Duran and Rossi derived in [3] an adaptive time step procedure for a parabolic problem with a non-linear boundary condition that causes blow-up behavior of the solution. They used also an explicit Runge Kutta scheme on the resulting system of ordinary differential equations, obtained from the semi-discretization of the space variable using a standard piecewise linear finite element method followed by lumping the mass matrix. A time rescaling procedure is also introduced with necessary and sufficient conditions for the existence of blow-up for the numerical solution.

Our approach in this paper differs from those just cited in the sense that we attempt to rescale simultaneously the time and space variables, aiming at the generation of a sequence of "slices" of time intervals $\{[T_{n-1}, T_n] : n = 1, 2, ...\}$, such that on each of these subintervals, the computation of the solution can be easily controlled by a preset **threshold (or cut-off)** value S, avoiding thus overflow in the global computations. We illustrate the method on an elementary differential equation $y' = f(y)$ having blow-up under (2). In fact, the technique we are presenting was motivated by an invariance property in the cases when $f(y) = e^y$ or $f(y) = y^p$. The scalar problem consists in finding $\{T_b, y\}$, where $0 < T_b < \infty, y : [0, T_b) \to R$ such that:

$$\frac{dy}{dt} = f(y),\ 0 < t < T_b,\ y(0) = y_0,\ \lim_{t \to T_b} y(t) = \infty. \tag{3}$$

In order to determine the first time subinterval $[0, T_1]$, we start by normalizing (3), by introducing the normalizing parameters $\alpha^{(1)}$ and $\beta^{(1)}$ through the change of variables $y(t) = y_0 + \alpha^{(1)} z(s)$, $t = \beta^{(1)} s$. Thus, the **base Model** corresponding to (3), is given by:

$$\frac{dz}{ds} = g_1(s) = \frac{\beta^{(1)}}{\alpha^{(1)}} f(y_0 + \alpha^{(1)} z), 0 < s \le s_1,\ z(0) = 0,\ z(s_1) = S. \tag{4}$$

where $s_1 = \int_0^S \frac{dz}{g_1(z)}$. Stepping up the solution forward, we define the nth slice by rescaling both variables y and t in the initial value problem

$$\frac{dy}{dt} = f(y),\ T_{n-1} < t,\ y(T_{n-1}) = y_{n-1},$$

using the change of variables:

$$t = T_{n-1} + \beta^{(n)} s,\ y(t) = y_{n-1} + \alpha^{(n)} z(s) \tag{5}$$

Thus, the subinterval $[T_{n-1}, T_n]$ is determined recurrently for $n \ge 2$ using the **rescaled model**:

$$\frac{dz}{ds} = g_n(z),\ 0 < s \le s_n,\ z(0) = 0,\ z(s_n) = S. \tag{6}$$

The function $g_n(.)$ is given by $g_n(z) = [f(y_{n-1}) + \alpha^{(n)} z] \frac{\beta^{(n)}}{\alpha^{(n)}}$, with $s_n = \int_0^S g_n(z) dz$. Note that, the base and rescaled models are respectively defined by the parameters: $\{g_1, s_1, g_1(0), g_1(S)\}$ and $\{g_n, s_n, g_n(0), g_n(S)\}$. When $g_n = g_1$, both the rescaled and base models are identical. Such is the case when $f(y) = e^y$ or $f(y) = y^p$, with respectively $\alpha_n = 1\ \beta_n = e^{-y_{n-1}}$ and $\alpha_n = y_{n-1}\ \beta_n = (y_{n-1})^{1-p}$. In general, the sequences $\{T_n\}$ and $\{y_n = y(T_n)\}$ verify:

$$T_n = T_{n-1} + \beta^{(n)} s_n,\ y_n = y_{n-1} + \alpha^{(n)} S \tag{7}$$

Determining the sequences $\{\alpha^{(n)}\}$ and $\{\beta^{(n)}\}$ appears to be crucial in the implementation of our rescaling method. This will be done in sections 2 and 3 respectively. More precisely, the sequence $\{\alpha^{(n)}\}$ is determined consistently with the explosive behavior of the solution of the initial value problem (3). In section 3, we analyze the sequence $\{\beta^{(n)}\}$ which allows the computation of the sequence $\{T_n\}$, using (5.2), $T_n = T_{n-1} + \beta^{(n)} s_n$. In this view, we start by defining a "similarity" criterion between the base and rescaled models identified by the parameters:

Definition 1. *The rescaled model (6) is said to be similar to the base model (4), if there exists two positive constants c and C independant of n, such that:*

$$c g_1(0) \leq g_n(0) \leq g_n(z) \leq g_n(S) \leq C g_1(S).$$

The sequences $\{\alpha^{(n)}\}$ and $\{\beta^{(n)}\}$ are determined as follows.

1. The main criteria on which we rely to determine the adequate values of the parameters $\{\alpha^{(n)}\}$ is the explosive behavior of the initial problem (3). That is, the sequences $\{T_n\}$ and $\{y_n\}$ generated by (7) should satisfy the blow-up behavior of the solution, $\lim_{n \to \infty} y_n = \infty$, $y_n = y(T_n)$, $\lim_{n \to \infty} T_n = T_b$. On that basis and given the estimate $\frac{\alpha^{(n)} S}{f(y_n)} \leq T_n - T_{n-1} \leq \frac{\alpha^{(n)} S}{f(y_{n-1})}$, the sequence $\{\alpha^n\}$ must verify the necessary conditions: (i) the infinite series $\sum_{n=1}^{\infty} \alpha^{(n)}$ is divergent and (ii) $\lim_{n \to \infty} \left(\frac{\alpha^{(n)}}{f(y_n)} \right) = 0$.

    Note that the 2 cases, $\alpha^{(n)} = 1$ and $\alpha^{(n)} = y_{n-1}$ (corresponding to $g_n = g_1$, when respectively, $f(y) = e^y$ and $f(y) = y^p$) verify such necessary conditions. Moreover, the resulting sequences $\{y_n\}$ reveal to be respectively geometric or arithmetic, which significantly simplifies the implementation of the rescaling method. Although conditions (i) and (ii) provide a wide range of choices for the sequence $\{\alpha^{(n)}\}$ we restrict ourselves to the choices $\alpha^{(n)} = y_{n-1}$ in the case $f(y) = O(y^p)$ and $\alpha^{(n)} = 1$ if $f(y) = O(e^y)$.

2. For the sequence $\{\beta^{(n)}\}$, we rely on the estimate $g_n(0) \leq g_n(z) \leq g_n(S)$ which implies $\frac{S}{g_n(S)} \leq s_n \leq \frac{S}{g_n(0)}$, $\forall n$. Thus, fixing the bounds on s_n can be obtained directly by fixing those of g_n. Since $g_n(0) = \frac{\beta^{(n)}}{\alpha^{(n)}} f(y_{n-1})$ and $g_n(S) = \frac{\beta^{(n)}}{\alpha^{(n)}} f(y_n)$, hence $g_n(z)$ depends directly on the ratio $\frac{\beta^{(n)}}{\alpha^{(n)}}$ and the sequence $\{y_n\}$. Since we cannot fix simultaneously $g_n(0)$ and $g_n(S)$, we let $r_n = \frac{f(y_n)}{f(y_{n-1})} = \frac{g_n(S)}{g_n(0)}$ and choose $g_n(0) = g_1(0)$, implying that $\beta^{(n)} = \alpha^{(n)} \frac{g_1(0)}{f(y_{n-1})} = \alpha^{(n)} \frac{\beta^{(1)}}{\alpha^{(1)}} \frac{f(0)}{f(y_{n-1})}$. Thus $g_1(0) \leq g_n(z) \leq g_1(0) r_n$ implying $\frac{S}{r_n g_1(0)} \leq s_n \leq \frac{S}{g_1(0)}$. If $f(y) = O(y^p)$, $\alpha^{(n)} = y_{n-1}$ or $f(y) = O(e^y)$, $\alpha^{(n)} = 1$, the sequence $\{r_n\}$ is uniformly bounded and one has similarity between the base and rescaled models according to Definition 1. Furthermore, $\frac{T_n - T_{n-1}}{T_1} = \beta^{(n)} s_n \leq \beta^{(n)} \frac{S}{T_1 g_1(0)}$.

Thus, re-scaling Algorithm Main features are described by the following steps:

1. Choice of the cut-off value S, and of the computational tolerance ε_{Tol}
2. Given y^f_{max} the maximum acceptable argument for $f(.)$ in the computational environment, verify **Compatibility Test** to insure no **overflows** would cause process interruption. When $\alpha^{(n)} = 1$, Test is given by: $\lceil \frac{1}{S} f^{-1} \left(\frac{S}{\varepsilon_{Tol}} \right) \rceil + 1 < \lfloor \frac{y^f_{max}}{S} \rfloor$.
3. Using RK4 (or other method) solve the base model to determine $[0, T_1]$ and $y(t)$ on $[0, T_1]$ using cut-off value S.
   (a) **Start the recurrence for** $n > 1$
   (b) **Re-scaling the variables** $(t = T_{n-1} + \beta^{(n)} s \, y(t) = y_{n-1} + \alpha^{(n)} z(s))$
   (c) **Specify and program the function** g_n.
   (d) **Solve the re-scaled model using RK4**: Determine s_n using cut-off value S.
   (e) **Deduce** $T_n = T_{n-1} + \beta^{(n)} s_n$ and $y(t)$ on $[T_{n-1}, T_n]$.
   (f) **End recurrence when** $\frac{T_n - T_{n-1}}{T_1} \leq \varepsilon_{Tol}$. $n_{STOP} := n$
4. Set the **approximation blow-up** as $T_b := T_{n_{STOP}}$
5. Compile the **global solution** $y(t)$ **on** $[0, T_b]$.
6. Visualize the numerical solutions.

2 Application to a Blowing-Up Semi-linear Parabolic System

Applying the method of sliced-time computations on the problem (1) is done through a first step consisting of a semi-discretization in space of this system. This leads to a system of ordinary differential equations. In fact, finding an invariance, or even defining a similarity on (1) seems to be a very complex task. On the other hand, dealing with a a first order semi-discretized system resulting from (1), allows an eventual track to applying the method of sliced-time computations. We introduce some notations.

Definition 2. *If* $V = (V_1, V_2, ..., V_k)^T \in \mathbb{R}^k$, *then* $F(V) = (f(V_1), f(V_2), ..., f(V_k))^T \in \mathbb{R}^k$. *On the other hand,* D_V *is the diagonal matrix whose diagonal elements are the components of the vector* V, *i.e.* $D_V = Diag(V) \in \mathbb{R}^{k \times k}$.

We choose the finite differences method for discretization. Let $h = (h_1, h_2, ..., h_d)^T \in \mathbb{R}^d$, $(|h| = \max_i |h_i|)$, the discretization parameter. $k = k(h)$ is the number of nodes in Ω as well as the dimension of the semi-discretized problem. The solution u of (1) being known on the boundaries Γ, the approached solution $U(t)$ of (8) is only about the nodes in Ω. $k(h) = O(|h|^{-d})$ and $\lim_{|h| \to 0} k(h) = \infty$. Under such conditions, the ith component of $U_i(t)$ is an approximation of the solution u on the node $x_i \in \Omega$ and on the time t. Let $u_i(t) = u(x_i, t)$, $\forall i = 1, ..., k$ and let $u(t) = \{u_i(t) : 1 \leq i \leq k\}$ the resulting vector of $\mathbb{R}^k$. The semi-discretization leads to the following problem:

$$\frac{dU}{dt} + AU = F(U), \; U(0) = U_0, \quad (8)$$

having $U(t) \in \mathbb{R}^k$ and $F : \mathbb{R}^k \to \mathbb{R}^k$, and $A \in \mathbb{R}^{k \times k}$, A is a square matrix representing the standard *finite differences* discretization corresponding to $-\Delta$ on Ω. An nth slice $[T_{n-1}, T_n]$ for (8), starts with the initial value problem $\frac{dU}{dt} = -AU + F(U)$, $T_{n-1} \leq t$, $U(T_{n-1}) = U_{n-1}$. If we consider the change of variables:

$$U_i(t) = U_{i,n-1} + \alpha_i^{(n)} Z_i(s), \ \alpha_i^{(n)} \neq 0, \ \forall i, \ \forall n, \ t = T_{n-1} + \beta^{(n)} s, \ \beta^{(n)} \neq 0. \quad (9)$$

Written in matrix, this is equivalent to:

$$U(t) = U_{n-1} + D_{\alpha^{(n)}} Z(s). \quad (10)$$

which gives the following form for the rescaled model:

$$\frac{dZ}{ds} = -A_n(Z) + G_n(Z), \ Z(0) = 0, \ \|Z(s_n)\|_\infty = S. \quad (11)$$

where

$$G_n(Z) := \beta^{(n)} D_{\alpha^{(n)}}^{-1} [F(U_{n-1} + D_{\alpha^{(n)}} Z)], \quad (12)$$

and $A_n(Z) := \beta^{(n)} D_{\alpha^{(n)}}^{-1} [A(U_{n-1} + D_{\alpha^{(n)}} Z)]$. Using (9) and (10), we have for $s = s_n$, $U_n = U(T_n) = U_{n-1} + D_{\alpha^{(n)}} Z(s_n)$, $T_n = T_{n-1} + \beta^{(n)} s_n$. For $n = 1$, (11) defines the base model. No invariance can be found between the base model and the rescaled model. We will therefore be looking for similarity criteria, with $f(.)$ verifying (2). Since our numerical tests are limited to the case of a single blow-up point, we assume the explosion of the vector $U(t)$ occurs for the component i_0, i.e. we assume the existence of i_0 independent of h such that the solution $U(t)$ of (8) verifies $\|U(t)\|_\infty = |U_{i_0}(t)|$. As for $\{\alpha_i^{(n)}, 1 \leq i \leq k\}$, we discern two cases: (i) $f(.)$ is exponential order leading to $\alpha_i^{(n)} = 1$, $\forall i$. and (ii) $f(.)$ is polynomial order in which case we choose $\alpha_i^{(n)} = U_{i,n-1}$, $\forall i$, ($\alpha_i^{(1)} = 1$, lorsque $U_0 = 0$). In fact, (9) added to the last identity lead respectively to $\|U_n\|_\infty = \|U_0\|_\infty + nS$ or $\|U_n\|_\infty = (1+S)^n \|U_0\|_\infty$. Using (9), we have respectively $|U_{i,n}| \leq |U_{i,0}| + nS$, if $\alpha_i^{(n)} = 1$ and $|U_{i,n}| \leq (1+S)^n |U_{i,0}|$, if $\alpha_i^{(n)} = U_{i,n-1}$. Both relations are consistent with $\lim_{n \to \infty} \|U_n\|_\infty = \infty$. We aim now to define the similarity for (11). In order to obtain $\{\beta^{(n)}\}$, we noted that in (11), $\frac{dZ}{ds}$ is the sum of a diffusion term $-A_n(Z)$, and a reaction term $G_n(Z)$. Since the reaction term rules the behavior of the solution, $\beta^{(n)}$ will be obtained through a lower similarity property of the form $\|G_n(0)\|_\infty = \|G_1(0)\|_\infty$. This leads to: $\beta^{(n)} \|D_{\alpha^{(n)}}^{-1} F(U_{n-1})\|_\infty = \beta^{(1)} \|D_{\alpha^{(1)}}^{-1} F(U_0)\|_\infty$, which can also be written as:

$$\beta^{(n)} = \beta^{(1)} \frac{\|D_{\alpha^{(1)}}^{-1}\|_\infty \|F(U_0)\|_\infty}{\|D_{\alpha^{(n)}}^{-1}\|_\infty \|F(U_{n-1})\|_\infty} = \beta^{(1)} \frac{\|D_{\alpha^{(n)}}\|_\infty \|F(U_0)\|_\infty}{\|D_{\alpha^{(1)}}\|_\infty \|F(U_{n-1})\|_\infty}. \quad (13)$$

Thus, $\beta^{(n)} = \beta^{(1)} \frac{f(U_{i_0,0})}{f(U_{i_0,n-1})}$ if $\alpha_i^{(n)} = 1$ and $\beta^{(n)} = \beta^{(1)} \frac{|U_{i_0,n-1}|}{|U_{i_0,0}|} \frac{f(U_{i_0,0})}{f(U_{i_0,n-1})}$ if $\alpha_i^{(n)} = U_{i,n-1}$. Computing $Z(s)$ on the nth slice $[T_{n-1}, T_n]$ gives $S_n = Z(s_n)$.

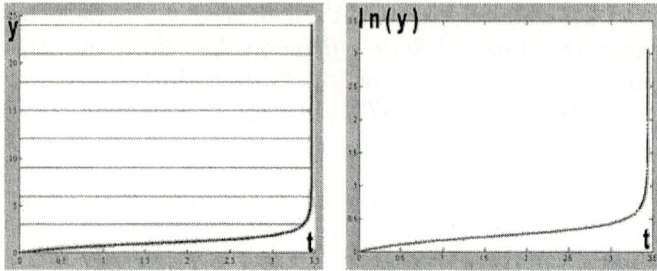

Fig. 1. Variation of the single point of the solution u and of $\ln(u)$ against the time t. Ω is unidimensional, $S = 3$, $h = \frac{1}{4}$ and $\tau = \frac{h^2}{2} = \frac{1}{16}$. Note the equal slices due to $\alpha_i^{(n)} = 1$

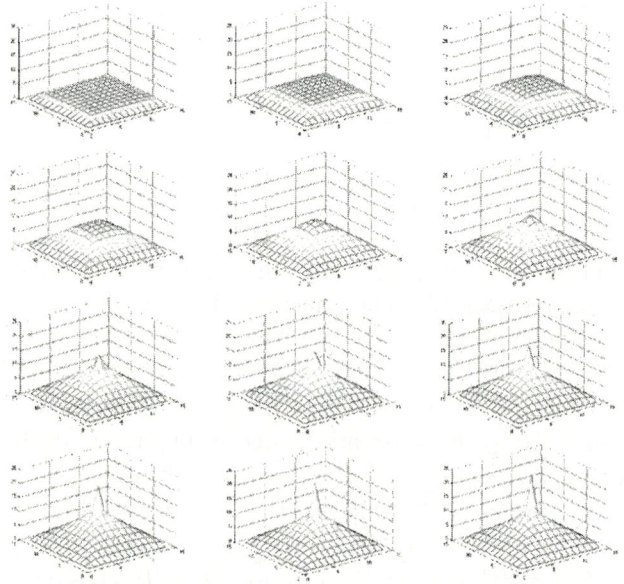

Fig. 2. Variation of the solution u of the problem $u' = \Delta u + e^y$, over the slices having $u(0) = 0$, $S = 2$, Ω is bidimensionnel $-1 \leq x \leq 1$ et $-1 \leq y \leq 1$ having $15 \times 15 = 225$ nodes

Under the validity of (12), we conclude that $G_n(S_n) = \beta^{(n)} D_{\alpha^{(n)}}^{-1} F(U_n)$. Consequently, $\|G_n(S_n)\|_\infty = \beta^{(n)} \|D_{\alpha^{(n)}}^{-1}\|_\infty \|F(U_{n-1})\|_\infty \frac{|f(U_{i_0,n})|}{|f(U_{i_0,n-1})|}$ $= \|G_n(0)\|_\infty \frac{f(U_{i_0,n})}{f(U_{i_0,n-1})}$. Furthermore, since $U_n(t) = U_{n-1} + D_{\alpha^{(n)}} Z(s)$, one has: $\|G_n(Z)\|_\infty = \beta^{(n)} \|D_{\alpha^{(n)}}^{-1}\|_\infty \|F(U_{n-1})\|_\infty \frac{f(U_{i_0}(t))}{f(U_{i_0,n-1})}$ and $\|G_n(Z)\|_\infty = \|G_n(0)\|_\infty \frac{f(U_{i_0}(t))}{f(U_{i_0,n-1})} \leq \|G_n(0)\|_\infty \frac{f(U_{i_0,n})}{f(U_{i_0,n-1})}$. Since $f(U_{i_0}(t)) \leq f(U_{i_0,n})$, one has under the assumptions on $f(.)$ added to (13), $\lim_{n \to \infty} \beta^{(n)} = 0$ and $\|G_n(Z)\|_\infty \leq C \|G_n(0)\|_\infty$ if $\frac{f(U_{i_0,n})}{f(U_{i_0,n-1})} \leq C$, $\forall n$.

A numerical application is conducted on $\frac{\partial u}{\partial t} = \Delta u + e^u$. The rescaled models (11) are given by

$$\frac{d\boldsymbol{Z}}{ds} = \beta^{(n)} \boldsymbol{D}_{\alpha^{(n)}}^{-1}[-(\boldsymbol{A}\boldsymbol{U}_{n-1} + \boldsymbol{D}_{\alpha^{(n)}} \boldsymbol{Z})]$$
$$+ \beta^{(n)} \boldsymbol{D}_{\alpha^{(n)}}^{-1}[e^{(\boldsymbol{U}_{n-1}+\boldsymbol{D}_{\alpha^{(n)}} \boldsymbol{Z})}] = -\boldsymbol{A}_n(\boldsymbol{Z}) + \boldsymbol{G}_n(\boldsymbol{Z}),\ 0 < s \leq s_n,$$

$\boldsymbol{Z}(0) = 0, \|\boldsymbol{Z}(s_n)\|_\infty = S$. We pick $\alpha_i^{(n)} = 1, \forall i, \forall n$, implying $\boldsymbol{D}_{\alpha^{(n)}} = \boldsymbol{I}$, and from (13), $\beta^{(n)} \|\boldsymbol{D}_{\alpha^{(n)}}^{-1} e^{\boldsymbol{U}_{n-1}}\|_\infty = \beta^{(1)} \|\boldsymbol{D}_{\alpha^{(1)}}^{-1} e^{\boldsymbol{U}_0}\|_\infty$, leading to $\beta^{(n)} = \beta^{(1)} \frac{e^{\|\boldsymbol{U}_0\|_\infty}}{e^{\|\boldsymbol{U}_{n-1}\|_\infty}}$. The figures below are few results of numerical experiments conducted on blow-up problems. (See [7]).

References

1. Abia L.M. , Lopez-Marcos J.C. , Martinez J. *Blow-up for semidiscretizations of reaction diffusion equations.* Appl. Numer. Math. 20(1996)145-156.
2. Abia L.M. , Lopz-Marcos J.C. , Martinez J. *On the blow-up time convergence of semidiscretizations of reaction diffusion equations.* Appl. Numer. Math. 26(1998)399-414.
3. Acosta G. , Duran R. , Rossi J. *An adaptive time step procedure for a parabolic problem with blow-up.*
4. Bebernes J., Eberley D. *Mathematical Problems from Combustion Theory.* Springer-Verlag, (1989).
5. Berger M. , Kohn R. *Rescaling algorithm for the numerical calculation of blowing-up solutions.* Commun. Pure Appl. Math. 41(1988)841-863.
6. Chorin A. *Estimates of intermittency, spectra, and blow-up in developed turbulence.* Commun. Pure Appl. Math. 34(1981)853-866.
7. Cortas M. *Calcul par tranches pour les équations différentielles à variable temps à caractère explosif.* Thesis, Université de Reims, January 2005.
8. Fayyad D. , Nassif N. *Approximation of blowing-up solutions to semilinear parabolic equations using "mass-controlled" parabolic systems.* Mathematical Modeling and Methods in the Applied Sciences,9(1999) pp. 1077-1088.
9. Fayyad D. , Nassif N. *On the Numerical Computation of blowing-up solutions for semi-linear parabolic equations.* Mathematical Methods in the Applied Sciences,24(2001)pp. 641-657.
10. Friedman A., McLeod B. *Blowup of positive solutions of semilinear heat equations.* Indiana Univ. Math. J 34 (1985) pp 425-447.
11. Kapila A. K. *Reactive-diffusive system with Arrhenius kinetics: Dynamics of ignition.* SIAM J. Appl. Math. 39(1980)21-36.
12. Kassoy D. , Poland J. *The induction period of a thermal explosion in a gas between infinite parallel plates.* Combustion and Flame50(1983)259-274.
  Crandall M. et al. , Academic Press(1987)159-201.
13. Pierre M. , Schmitt D. *Blow-up in reactio-diffusion systems with dissipation of mass.* Rapport INRIA, 2652(1995).

Application of the Pseudo-Transient Technique to a Real-World Unsaturated Flow Groundwater Problem

Fred T. Tracy[1], Barbara P. Donnell[2], Stacy E. Howington[2], and Jeffrey L. Hensley[1]

[1] Information Technology Laboratory,
Major Shared Resource Center (MSRC)
[2] Coastal and Hydraulics Laboratory,
Engineer Research and Development Center (ERDC),
Vicksburg, MS, USA 39180

Abstract. Modeling unsaturated flow using numerical techniques such as the finite element method can be especially difficult because of the highly nonlinear nature of the governing equations. This problem is even more challenging when a steady-state solution is needed. This paper describes the implementation of a pseudo-transient technique to drive the solution to steady-state and gives results for a real-world problem. The application discussed in this paper does not converge using a traditional Picard nonlinear iteration type finite element solution. Therefore, an alternate technique needed to be developed and tested.

1 Introduction

Modeling unsaturated flow using numerical techniques such as the finite element method can be especially difficult because of the highly nonlinear nature of the governing equations. This challenge is even more exacerbated when (1) a steady-state solution is needed, (2) soil properties such as relative hydraulic conductivity go from almost horizontal to almost vertical (see Fig. 1), (3) an influx of water from rainfall occurs at the top of very dry, low hydraulic conductivity unsaturated soil, (4) injection wells are in the unsaturated zone (see IG-1 in Fig. 2), and (5) some pumped wells become totally above the water table or in the unsaturated zone. An early version of the data set for the application discussed in this paper has all of these traits, and the steady-state solution does not converge using a traditional Picard nonlinear iteration finite element solution. The data set needed to be modified to relieve some of the above problems, but a converged solution was needed to know what changes to make. Therefore, an alternate solution was developed and tested. This paper describes the implementation of a modified version of a pseudo-transient technique described in [3] to drive the solution to steady-state and gives computational results for this real-world application.

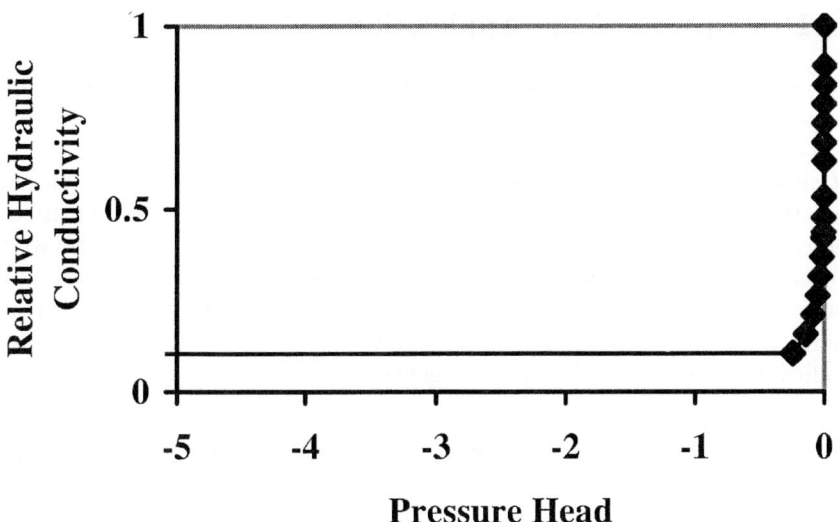

Fig. 1. This figure illustrates the relative hydraulic conductivity versus pressure head curve for soils such as sand where the curve goes from almost horizontal to near vertical rather quickly

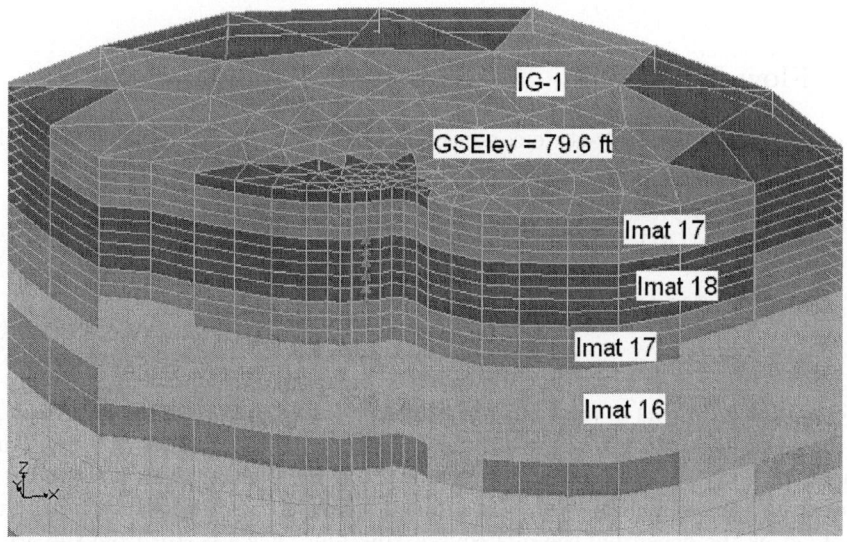

Fig. 2. This figure shows the top twenty layers of the mesh. There are 18 different material types. A large injection well, IG-1, in the unsaturated zone is also shown

The computer model used in this study is the parallel version [7] of FEMWATER [5], which is a groundwater flow and transport code. The finite element mesh is partitioned using METIS [2].

2 Description of the Application

The problem consists of modeling the remediation of a military site using a pump-and-treat system. The finite element mesh uses 402,628 nodes and 779,328 three-dimensional prism finite elements. There are 18 different soil types, 9 injection wells, 1 chimney drain which also injects water, and 38 extraction wells. Once the model is properly calibrated with observed data, it is then used to predict the efficiency of a proposed pump-and-treat system. Table 1 shows the saturated hydraulic conductivities for the three material types at the top of the ground (16-18 in Fig.2). The incredibly small $k_z = 0.0001$ ft/hour where the ground is very dry and water from precipitation is coming into the system adds significantly to the difficulty of convergence.

Table 1. Saturated hydraulic conductivities (ft/hour)

Material Number	k_x	k_y	k_z
16	1.042	1.042	0.008
17	0.08	0.08	0.008
18	0.01	0.01	0.0001

3 Flow Equations

Pressure head for steady-state conditions in FEMWATER is modeled by applying conservation of mass to obtain,

$$\nabla \cdot \left[k_r(h) \mathbf{k}_s \cdot \left(\nabla h + \frac{\rho}{\rho_0} \nabla z \right) \right] + \sum_{m=1}^{N_{ss}} \frac{\rho_m^*}{\rho_0} Q_m \delta(\mathbf{r} - \mathbf{r}_m) = 0 , \qquad (1)$$

$$h = \frac{p}{\rho_0 g} . \qquad (2)$$

where δ is the Dirac delta function, g is the acceleration due to gravity, h is the pressure head, $k_r(h)$ is the highly nonlinear relative hydraulic conductivity, $\mathbf{k}_s$ is the saturated hydraulic conductivity tensor, N_{ss} is the number of source/sink nodes for flow, Q_m is the quantity of flow at the mth source/sink node, p is the pressure, ρ is the density with contaminant, ρ_0 is the density without contaminant, ρ_m^* is the density of the mth source/sink fluid, $\mathbf{r}$ is a vector from the origin to an (x, y, z) point in space, and $\mathbf{r}_m$ is the location of the mth source/sink node.

The Galerkin finite element method is then applied to obtain

$$\mathbf{K}(\mathbf{h}) \mathbf{h} = \mathbf{Q}'(\mathbf{h}) . \qquad (3)$$

Here $\mathbf{K}(\mathbf{h})$ is the stiffness matrix, $\mathbf{h}$ is a vector of pressure heads at the finite element nodes, and $\mathbf{Q}'(\mathbf{h})$ is a collection of flow type terms for the right-hand side. Eq. 3 is the resulting system of nonlinear equations to be solved.

4 Solution of the System of Nonlinear Equations

A Newton iteration [4] to reduce the nonlinear residual,

$$\mathbf{F}(\mathbf{h}) = \mathbf{Q}'(\mathbf{h}) - \mathbf{K}(\mathbf{h})\mathbf{h} . \tag{4}$$

is given by

$$\mathbf{h}_{n+1} = \mathbf{h}_n - \mathbf{F}'(\mathbf{h}_n)^{-1} \mathbf{F}(\mathbf{h}_n) , \tag{5}$$

$$\mathbf{F}'(\mathbf{h})_{ij} = \frac{\partial (\mathbf{F})_i}{\partial (\mathbf{h})_j}(\mathbf{h}) . \tag{6}$$

where $\mathbf{F}'(\mathbf{h})$ is the Jacobian matrix. FEMWATER uses a Picard iteration, which is equivalent to approximating the Jacobian matrix by $-\mathbf{K}$. This produces

$$\mathbf{h}_{n+1} = \mathbf{h}_n + \mathbf{K}_n^{-1}\left(\mathbf{Q}'_n - \mathbf{K}_n \mathbf{h}_n\right) = \mathbf{K}_n^{-1}\mathbf{Q}'_n . \tag{7}$$

The disadvantage of the Picard approximation is the loss of potentially important terms, but the advantage is that the system of simultaneous, linear equations to be solved (Eq. 7) remains symmetric and positive-definite. Thus, a parallel preconditioned conjugate gradient solver works quite well.

4.1 Convergence of the Steady-State Problem

Because of factors such as the severe relative hydraulic conductivity curve shown in Fig. 1, the steady-state solution for this data set did not converge. Convergence is helped somewhat by adding a dynamic type relaxation to the Picard iteration as follows:

$$\mathbf{h}_{n+1} = (1 - \alpha_{n+1})\mathbf{h}_n + \alpha_{n+1}\bar{\mathbf{h}}_{n+1} \quad 0 < \alpha_{n+1} \leq 1 , \tag{8}$$

$$\bar{\mathbf{h}}_{n+1} = \mathbf{K}_n^{-1}\mathbf{Q}'_n . \tag{9}$$

where the relaxation factor α_{n+1} is adjusted based on whether the maximum absolute value of the pressure head change $|\triangle h|_{n+1}^{max}$ between h_n and $\bar{h}_{n+1}$ for all the N finite element nodes,

$$|\triangle h|_{n+1}^{max} = \max_{i=1}^{N}\left|(\bar{\mathbf{h}}_{n+1})_i - (\mathbf{h}_n)_i\right| . \tag{10}$$

decreased or increased from $|\triangle h|_n^{max}$. The adjustment used is

$$\alpha_{n+1} = \min(\alpha_n + \epsilon_\alpha, \alpha_{max}) \quad |\triangle h|_{n+1}^{max} \leq |\triangle h|_n^{max} , \tag{11}$$

$$\alpha_{n+1} = \max(f_\alpha \alpha_n, \alpha_{min}) \quad |\triangle h|_{n+1}^{max} > |\triangle h|_n^{max}, \; 0 < f_\alpha < 1 . \tag{12}$$

where ϵ_α, f_α, α_{min}, and α_{max} are input variables. Values that worked well for this application are $\epsilon_\alpha = 0.005$, $f_\alpha = 0.667$, $\alpha_{min} = 0.01$, and $\alpha_{max} = 0.5$.

4.2 Pseudo Time-Step Implementation

For those problems such as the initial data set used in this study where the steady-state Picard iteration would not converge, an additional term is added to Eq. 3 to produce

$$f^{m+1}\mathbf{M}\left(\mathbf{h}^{m+1}\right)\left(\mathbf{h}^{m+1}-\mathbf{h}^{m}\right)+\mathbf{K}\left(\mathbf{h}^{m+1}\right)\mathbf{h}^{m+1}=\mathbf{Q}^{'}\left(\mathbf{h}^{m+1}\right) . \quad (13)$$

where f^{m+1} is a multiplication factor equivalent to the reciprocal of a time increment for pseudo time-step $m+1$, $\mathbf{h}^m$ is pressure head for pseudo time-step m, and after some experimentation, $\mathbf{M}$ was chosen to be the diagonal of $\mathbf{K}$. Adding this term is acceptable because eventually steady-state will be achieved, and thus $\mathbf{h}^{m+1} \approx \mathbf{h}^m$, causing the additional term to vanish. The Picard iteration for this implicit Euler approximation now becomes,

$$\mathbf{h}_{n+1}^{m+1}=\left(f^{m+1}\mathbf{M}_{n}^{m+1}+\mathbf{K}_{n}^{m+1}\right)^{-1}\left(\mathbf{Q}_{n}^{'m+1}+f^{m+1}\mathbf{M}_{n}^{m+1}\mathbf{h}^{m}\right) . \quad (14)$$

To compute f^{m+1}, the norm of the residual must first be computed. The residual is computed from

$$\mathbf{F}_{1}^{m+1}=\mathbf{Q}_{1}^{'m+1}+f^{m+1}\mathbf{M}_{1}^{m+1}\mathbf{h}^{m}-\left(f^{m+1}\mathbf{M}_{1}^{m+1}+\mathbf{K}_{1}^{m+1}\right)\mathbf{h}_{1}^{m+1} . \quad (15)$$

But since the pressure head at time-step m is the same as that of the first nonlinear iteration of time-step $m+1$,

$$\mathbf{h}_{1}^{m+1}=\mathbf{h}^{m} , \quad (16)$$

Eq. 15 becomes

$$\mathbf{F}^{m}=\mathbf{Q}^{'m}-\mathbf{K}^{m}\mathbf{h}^{m} . \quad (17)$$

Using the discrete l^2 norm on R^N,

$$\|\mathbf{F}^{m}\|=\sqrt{\frac{1}{N}\sum_{i=1}^{N}(\mathbf{F}^{m})_{i}^{2}} , \quad (18)$$

f^{m+1} is computed by using a version of the switched evolution relaxation (SER) algorithm [6] as follows:

$$f^{m+1}=f^{m}\frac{\|\mathbf{F}^{m}\|}{\|\mathbf{F}^{m-1}\|} . \quad (19)$$

In most cases, the nonlinear iteration for a pseudo time-step converged in a few Picard iterations. However, occasionally the same instability that caused the need for the pseudo time-step algorithm in the first place generated a lack of convergence after 50 iterations. If that happened, f^{m+1} was modified by

$$f^{m+1}=f^{m+1}+\frac{1}{2}f^{0} . \quad (20)$$

every 50 iterations until convergence occurred. f^0 is computed from the input value of Δt by

$$f^0 = \frac{1}{\Delta t} \ . \tag{21}$$

Also,

$$f^1 = f^0 \ . \tag{22}$$

The advantage of this approach is that each new pseudo time-step brings the solution closer to steady-state, and the added term in Eq. 13 is simpler than the real transient term that could be used. Doing only one nonlinear iteration per pseudo time-step was also tried, but for this application, the instability so dominated that a full convergence of each pseudo time-step seemed best. For this data set, f^{m+1} always became so small that the inherent instability reemerged such that f^{m+1} would become bigger. However, the residual gradually became smaller for most pseudo time-steps. Initially, a number of traditional Picard nonlinear iterations ($f^{m+1} = 0$) are done, and only if convergence is not achieved the traditional way is the pseudo time-stepping started. This way, only isolated areas in the unsaturated zone need further adjustment.

5 Results

The application data set was run on the ERDC MSRC SGI 3900 with 32 processors using the material properties given in Table 1 with additional data described in Table 2. Time required to do the computation was 19,593 seconds (5.4425 hour). It was not practical to do a true transient solution with the material properties given in Table 1, as convergence was so difficult, even when increasing the pumping rates and precipitation flux gradually. Thus the material properties were modified as shown in Table 3 to an easier problem. The consequence of changing the material data is that a different problem is solved. It will, however, allow the comparison of errors between a true transient solution and a more difficult problem solved by the pseudo time-stepping technique.

To determine accuracy, six wells in the unsaturated zone are examined for error between the input pumping rates and those produced after the solution process. The true transient solution was run for a simulation time of 288 hours

Table 2. Computational data

Number of traditional Picard iterations	1,000
Tolerance for convergence of Picard iterations	0.0005
Number of pseudo time-steps	1,000
Δt for computing f^0	0.05

Table 3. Modified saturated hydraulic conductivities (ft/hour)

Material Number	k_x	k_y	k_z
16	1.042	1.042	0.1042
17	0.04	0.04	0.004
18	0.02	0.02	0.002

Table 4. Errors in well pumping rates

Well Number	Pseudo time-stepping	True transient
1	3.12E-09	1.52E-05
2	-3.22E-07	1.20E-05
3	-4.67E-05	7.39E-06
4	4.15E-09	4.64E-06
5	2.31E-09	-1.86E-04
6	-6.67E-07	2.63E-05

using the relaxed soil properties. Table 4 shows a comparison of the error in the well pumping rates for the two different techniques.

6 Conclusions

The pseudo time-stepping algorithm gives an acceptable alternative to achieving a steady-state solution to the highly nonlinear unsaturated flow groundwater problem when convergence is not obtained by traditional methods. Further, automatic determination of the pseudo time-step size is easily found. Results also show that accuracy can be achieved easier than doing a true transient computation. For this application, the pseudo time-stepping algorithm is currently the only way found thus far to achieve a solution because of the complexities of the true transient term which is also highly nonlinear.

7 Future Work

As described in [1], alternate approaches will next be investigated. One example is to use Newton iterations with line search after a given number of Picard iterations have been completed. Also, the nonlinear Newton-Krylov solvers described in [8] will be investigated.

Acknowledgment

This work was supported in part by a grant of computer time from the Department of Defense High Performance Computing Modernization Program at the ERDC MSRC.

References

1. Farthing, M.W., Kees, C.E., Coffet, T.S., Kelley, C.T., and Miller, C.T.: Efficient Steady-State Solution Techniques for Variably Saturated Groundwater Flow. Adv. in Water Res., 26(2003), 833-849
2. Karypis, G.: METIS (computer program). http://www.users.cs.umn.edu/~karypis/metis/, University of Minnesota, Minneapolis, MN (2004)
3. Kawanagh, K.R. and Kelley, C.T.: Pseudo Transient Continuation for Nonsmooth Nonlinear Equations. Draft paper, North Carolina State University, Raleigh, NC (2003)
4. Kelley, C.T.: Solving Nonlinear Equations with Newton's Method, SIAM, Philadelphia (2003), 2
5. Lin, H.J., Richards, D.R., Talbot, C.A., Yeh, G.T., Cheng, J.R., Cheng, H.P., and Jones, N.L.: FEMWATER: A Three-Dimensional Finite Element Computer Model for Simulating Density-Dependent Flow and Transport in Variably Saturated Media. Technical Report CHL-97-12, U.S. Army Engineer Research and Development Center (ERDC), Vicksburg, MS (1997)
6. Mulder, W. and Lebr, B.V.: Experiments with Typical Upwind Methods for the Euler Equations. J. Comp. Phys., 59(1985), 232-246
7. Tracy, F.T., Talbot, C.A., Holland, J.P., Turnbull, S.J., McGehee, T.L., and Donnell, B.P.: The Application of the Parallelized Groundwater Model FEMWATER to a Deep Mine Project and the Remediation of a Large Military Site. DoD HPC Users Group Conference Proceedings, Monterey, CA (1999)
8. Wheeler, M.F., Kle, H., Aksoylu, B., and Eslinger, O.: Nonlinear Krylov Based Solvers. PET Report, University of Texas at Austin, Austin, TX (2005)

Optimization of Spherical Harmonic Transform Computations

J.A.R. Blais[1], D.A. Provins[2], and M.A. Soofi[2]

[1,2] Department of Geomatics Engineering,
[1] Pacific Institute for the Mathematical Sciences,
University of Calgary, Calgary, AB, T2N 1N4, Canada
{blais, soofi}@ucalgary.ca, provinsd@telusplanet.net
www.ucalgary.ca/~blais

Abstract. Spherical Harmonic Transforms (SHTs) which are essentially Fourier transforms on the sphere are critical in global geopotential and related applications. Discrete SHTs are more complex to optimize computationally than Fourier transforms in the sense of the well-known Fast Fourier Transforms (FFTs). Furthermore, for analysis purposes, discrete SHTs are difficult to formulate for an optimal discretization of the sphere, especially for applications with requirements in terms of near-isometric grids and special considerations in the polar regions. With the enormous global datasets becoming available from satellite systems, very high degrees and orders are required and the implied computational efforts are very challenging. The computational aspects of SHTs and their inverses to very high degrees and orders (over 3600) are discussed with special emphasis on information conservation and numerical stability. Parallel and grid computations are imperative for a number of geodetic, geophysical and related applications, and these are currently under investigation.

1 Introduction

On the spherical Earth and neighboring space, spherical harmonics are among the standard mathematical tools for analysis and synthesis (e.g., global representation of the gravity field and topographic height data). On the celestial sphere, COBE, WMAP and other similar data are also being analyzed using spherical harmonic transforms (SHTs) and extensive efforts have been invested in their computational efficiency and reliability (see e.g. [6], [7] and [17]).

In practice, when given discrete observations on the sphere, quadrature schemes are required to obtain spherical harmonic coefficients for spectral analysis of the global data. Various quadrature strategies are well known, such as equiangular and equiareal. In particular, Gaussian quadratures are known to require the zeros of the associated Legendre functions for orthogonality in the discrete computations. As these zeros are not equispaced in latitude, the Gaussian strategies are often secondary in applications where some regularity in partitioning the sphere is critical. Under such requirements for equiangular grids in latitude and longitude, different optimization schemes are available for the computations which become quite complex and intensive for high degrees and orders. The approach of Driscoll and Healy [5] using Chebychev quadrature with an equiangular grid is advantageous in this context.

2 Continuous and Discrete SHTs

The orthogonal or Fourier expansion of a function $f(\theta, \lambda)$ on the sphere $\mathbf{S}^2$ is given by

$$f(\theta,\lambda) = \sum_{n=0}^{\infty} \sum_{|m| \le n} f_{n,m} Y_n^m(\theta,\lambda) \tag{1}$$

using colatitude θ and longitude λ, where the basis functions $Y_n^m(\theta,\lambda)$ are called the spherical harmonics satisfying the (spherical) Laplace equation $\Delta_{S^2} Y_n^m(\theta,\lambda) = 0$, for all $|m| \le n$ and $n = 0, 1, 2, \ldots$. This is an orthogonal decomposition in the Hilbert space $L^2(\mathbf{S}^2)$ of functions square integrable with respect to the standard rotation invariant measure $d\sigma = \sin\theta\, d\theta\, d\lambda$ on $\mathbf{S}^2$. In particular, the Fourier or spherical harmonic coefficients appearing in the preceding expansion are obtained as inner products

$$\begin{aligned} f_{n,m} &= \int_{S^2} f(\theta,\lambda)\, \overline{Y}_n^m(\theta,\lambda)\, d\sigma \\ &= \sqrt{\frac{(2n+1)(n-m)!}{4\pi(n+m)!}} \int_{S^2} f(\theta,\lambda)\, P_n^m(\cos\theta)\, e^{im\lambda}\, d\sigma \\ &= (-1)^m \sqrt{\frac{(2n+1)(n-m)!}{4\pi(n+m)!}} \int_{S^2} f(\theta,\lambda)\, P_{nm}(\cos\theta)\, e^{im\lambda}\, d\sigma \end{aligned} \tag{2}$$

in terms of the associated Legendre functions $P_{nm}(\cos\theta) = (-1)^m P_n^m(\cos\theta)$, with the overbar denoting the complex conjugate. In most practical applications, the functions $f(\theta,\lambda)$ are band-limited in the sense that only a finite number of those coefficients are nonzero, i.e. $f_{n,m} \equiv 0$ for all $n \ge N$.

The usual geodetic spherical harmonic formulation is slightly different with

$$f(\theta,\lambda) = \sum_{n=0}^{\infty} \sum_{m=0}^{n} [\bar{C}_{nm} \cos m\lambda + \bar{S}_{nm} \sin m\lambda]\, \bar{P}_{nm}(\cos\theta) \tag{3}$$

where

$$\begin{Bmatrix} \bar{C}_{nm} \\ \bar{S}_{nm} \end{Bmatrix} = \frac{1}{4\pi} \int_{S^2} f(\theta,\lambda) \begin{Bmatrix} \cos m\lambda \\ \sin m\lambda \end{Bmatrix} \bar{P}_{nm}(\cos\theta)\, d\sigma \tag{4}$$

and

$$\begin{aligned} \bar{P}_{nm}(\cos\theta) &= \sqrt{\frac{2(2n+1)(n-m)!}{(n+m)!}}\, P_{nm}(\cos\theta) \\ \bar{P}_n(\cos\theta) &= \sqrt{2n+1}\, P_n(\cos\theta) \end{aligned} \tag{5}$$

In this geodetic formulation, the dashed overbars refer to the assumed normalization. Colombo [1981] has discretized this formulation with $\theta_j = j\pi/N$, $j = 1, 2, ..., N$ and $\lambda_k = k\pi/N$, $k = 1, 2, ..., 2N$.

Legendre quadrature is however well known to provide an exact representation of polynomials of degrees up to $2N - 1$ using only N data values at the zeros of the Legendre polynomials. This, for example, was the quadrature employed by Mohlenkamp [12] in his sample implementation of a fast Spherical Harmonic Transform (SHT). Driscoll and Healy [5] have exploited these quadrature ideas in an exact algorithm for a reversible SHT using the following $(2N)^2$ grid data for degree and order $N - 1$:

Given discrete data $f(\theta,\lambda)$ at $\theta_j = \pi j/2N$ and $\lambda_k = \pi k/N$, $j,k = 0,...,2N-1$, the analysis using a discrete SHT gives

$$f_{n,m} = \frac{1}{N} \cdot \sqrt{\frac{\pi}{2}} \sum_{j=0}^{2N-1} \sum_{k=0}^{2N-1} a_j \, f(\theta_j,\lambda_k) \, \bar{Y}_n^m(\theta_j,\lambda_k) \tag{6}$$

with the following explicit expressions for the Chebychev quadrature weights a_j, assuming N to be a power of 2,

$$a_j = \frac{\sqrt{2}}{N} \sin\left(\frac{\pi j}{2N}\right) \sum_{h=0}^{N-1} \frac{1}{2h+1} \sin\left((2h+1)\frac{\pi j}{2N}\right) \tag{7}$$

and for the synthesis using an inverse discrete SHT (or SHT^{-1}),

$$f(\theta_j,\lambda_k) = \sum_{n=0}^{N-1} \sum_{|m|\le n} f_{n,m} \, Y_n^m(\theta_j,\lambda_k). \tag{8}$$

These Chebychev weights a_j are the analytical solution of the following equations

$$\sum_{j=0}^{2N-1} a_j P_k(\cos\frac{\pi j}{2N}) = \sqrt{2}\,\delta_{k0}, \quad k = 0,1,...,2N-1 \tag{9}$$

which are also considered in the so-called (second) Neumann method where the corresponding numerical solutions are evaluated for different choices of distinct parallels θ_j, not necessarily equispaced in latitude [18]. Other related formulations are briefly discussed in [2], [3], and also in [15].

3 Optimization of Discrete SHTs

A number of modifications have been developed for the preceding Driscoll and Healy [5] formulation in view of the intended applications.

First, the geodetic normalization and conventions have been adopted and implemented in the direct and inverse discrete SHTs. Geopotential and gravity models usually follow the geodetic conventions. However, software packages such as SpherePack [1] and SpharmonKit [14] use the mathematical normalization convention. Also, the requirement that N be a power of 2 as stated explicitly by

Driscoll and Healy [5] does not seem to be needed in the included mathematical derivation of the Chebychev weights.

Second, the latitude partition has been modified to avoid polar complications, especially in practical geodetic and geoscience applications. Two options have been experimented with: First, using the previous latitude partition, $\theta_j = \pi j / 2N$, with $j = 1, \ldots, 2N-1$, as for $j=0$, the weight $a_0 = 0$, and hence the North pole ($\theta = 0$) needs not be carried in the analysis and synthesis computations. Second, the latitude partition can be redefined using $\theta_j = \pi (j + \frac{1}{2}) / 2N$ for $j = 0, \ldots, 2N-1$, with the same longitude partition, i.e., $\lambda_k = \pi k / N$, $k = 0, \ldots, 2N-1$. In this case, the Chebychev weights a_j (when N is not necessarily a power of 2) are redefined as d_j, where

$$d_j = \frac{\sqrt{2}}{N} \sin\left(\frac{\pi(j+\frac{1}{2})}{2N}\right) \sum_{h=0}^{N-1} \frac{1}{2h+1} \sin\left((2h+1)\frac{\pi(j+\frac{1}{2})}{2N}\right) \qquad (10)$$

for $j = 0,\ldots,2N-1$, which are symmetric about mid-range. The grids of 2Nx2N nodes with $\Delta\theta = \frac{1}{2}\Delta\lambda$ have also been modified to 2Nx4N nodes with $\Delta\theta = \Delta\lambda$ for the majority of practical applications.

Third, hemispherical symmetries have been implemented. The equatorial symmetry of this latitude partition is also advantageous to exploit the symmetries of the associated Legendre functions, i.e.

$$P_{nm}(\cos(\pi - \theta)) = (-1)^{n+m} P_{nm}(\cos\theta). \qquad (11)$$

This can be verified using the definition of the associated Legendre functions [19] and is important for the efficiency of SHTs computations.

Fourth, the computations along parallels involve functions of longitude only and lend themselves to Discrete Fourier Transforms (DFTs), and hence Fast Fourier Transforms (FFTs) for computational efficiency in practice.

Fifth, to ensure numerical stability for degrees over 2000, quadruple precision computation has been implemented. Also, parallel and grid computations are under development.

4 Numerical Analysis Considerations

For high degrees and orders, the normalized associated Legendre functions $\ddot{P}_{nm}(\cos\theta)$ have to be used. However, as the normalizing factors in

$$\ddot{P}_{nm}(\cos\theta) = \sqrt{\frac{2(2n+1)(n-m)!}{(n+m)!}} P_{nm}(\cos\theta)$$

$$\ddot{P}_n(\cos\theta) = \sqrt{2n+1}\, P_n(\cos\theta) \qquad (12)$$

get quite large for large n, it is important to use the recursive formulas directly in terms of the normalized associated Legendre functions $\ddot{P}_{nm}(\cos\theta)$. Otherwise, significant loss in numerical accuracy is observed for degrees over 60 even with such software as the intrinsic spherical harmonic and Legendre functions in Mathematica©.

The normalized associated Legendre functions $\ddot{P}_{nm}(\cos\theta)$ are computed following Rapp [16] formulation. These recursive formulas have been used in geodesy for all kinds of geopotential field applications for degree and order up to 1800, e.g. [20]. Attempts to exceed this limit using FORTRAN compilers available at the time found that numerical accuracy which ranged from 10^{-11} to 10^{-13} for all colatitudes, degraded substantially beyond that point. Wenzel observed accuracy of only 10^{-3} by degree 2000, which has been confirmed using REAL*8 specifications in Win 32 and AMD 64 environments. Experimentation with REAL*16 in AMD 64 and DEC Alpha environments has demonstrated numerical stability for degrees over 3600. Other recursive formulas are discussed in [15] and other publications, such as [10], [12] and [19].

5 Numerical Experimentation

As indicated earlier, there are several formulations for employing spherical harmonics as an analysis tool. One popular code that is readily available is Spherepack [1] of which a new version has been released recently. Other experimental codes are those of Driscoll and Healy [5] and the follow-ons, such as [8] and [14], plus the example algorithm described by Mohlenkamp [11][12] and offered as a partial sample implementation in [13]. Experimentation with these codes has shown scaling differences from that which is expected in a geodetic context [15].

Using the Driscoll and Healy [5] formulation modified as described in Section 3, extensive experimentation using different grids on several computer platforms in double precision (i.e. REAL*8) and quadruple precision (i.e. REAL*16) has been carried out. The first synthesis started with unit coefficients,

$$a_{nm} = b_{nm} = 1, \text{ except for } b_{n0} = 0, \tag{13}$$

for all degrees n and orders m, which corresponds to white noise. Then, following analysis of the generated spatial grid values, the coefficients are recomputed and root-mean-square (RMS) values are given for this synthesis/analysis. Then after another synthesis using these recomputed coefficients, RMS values of recomputed grid residuals are given for the second synthesis. Hence, starting with arbitrary coefficients $\{c_{nm}\}$, the procedure can be summarized as follows:

$SHT[SHT^{-1}[\{c_{nm}\}]] - [\{c_{nm}\}]$ $\rightarrow$ RMS of first synthesis/analysis,
and
$SHT^{-1}[SHT[SHT^{-1}[\{c_{nm}\}]]] - SHT^{-1}[\{c_{nm}\}]$ $\rightarrow$ RMS of second synthesis.

Notice that the first RMS is in the spectral domain while the second is in the spatial domain. The SHTs and SHT^{-1}s are evaluated and re-evaluated explicitly to study their numerical stability and the computational efficiency. The only simplification implemented is in not recomputing the Fourier transforms in the second synthesis following the inverse Fourier transforms in the second part of the analysis. The above procedure is repeated for coefficients corresponding to $1/\text{degree}^2$, i.e. explicitly,

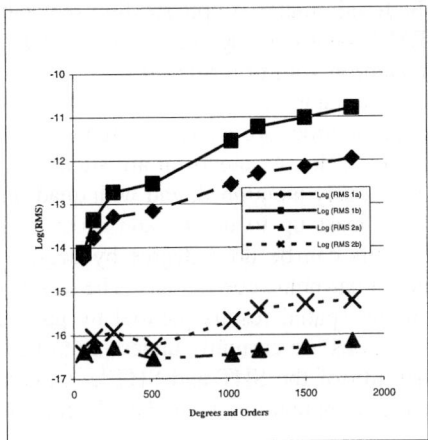
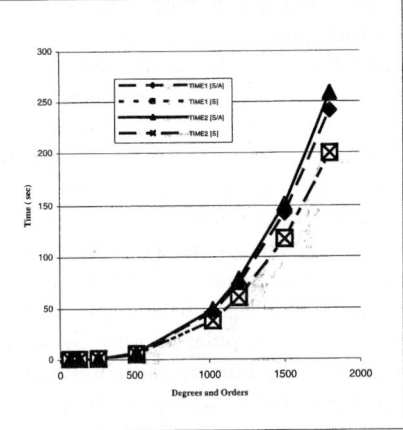

Fig. 1. Results from computation with $\Delta\theta = \frac{1}{2}\Delta\lambda$ in REAL*8 Precision on AMD 64 Athlon FX-53 PC. Left:.SHT RMS Values for Synthesis/Analysis [a] and Synthesis [b] of Simulated Series (unit coefficients [RMS1] and 1/degree^2 coefficients [RMS2]). Right:.SHT Time Values for Synthesis/Analysis [S/A] and Synthesis [S] of Simulated Series (unit coefficients [TIME1] and 1/degree^2 coefficients [TIME2])

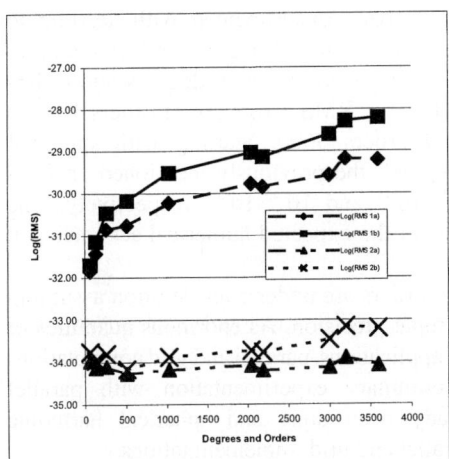
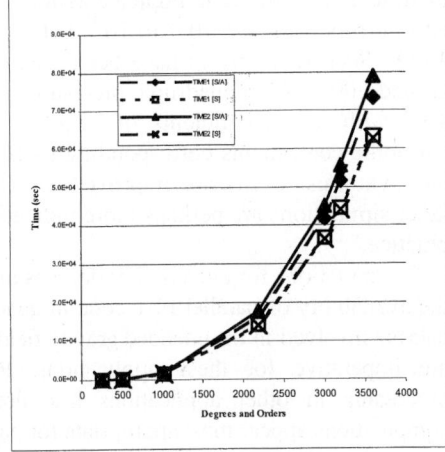

Fig. 2. Results from computation with $\Delta\theta = \Delta\lambda$ in REAL*16 Precision on DEC Alpha Computer. Left: SHT RMS Values for Synthesis/Analysis [a] and Synthesis [b] of Simulated Series (unit coefficients [RMS1] and 1/degree^2 coefficients [RMS2]). Right: SHT Time Values for Synthesis/Analysis [S/A] and Synthesis [S] of Simulated Series (unit coefficients [TIME1] and 1/degree^2 coefficients [TIME2])

$$a_{nm} = 1/(n+1)^2, \tag{14}$$

$$b_{nm} = 0 \text{ for m=0, and } 1/(n+1)^2, \text{ otherwise,} \tag{15}$$

for all degrees n and orders m, which simulate a physically realizable situation. Figures 1 show the plots of the RMS values in logarithmic scale and the computation times for double precision computation with grid $\Delta\theta = \frac{1}{2}\Delta\lambda$. The above procedure is repeated for quadruple precision computations and for the grid $\Delta\theta = \Delta\lambda$. The results of the RMS values and the computation times are plotted in Figures 2.

The analysis and synthesis numerical results in double precision are stable up to approximately degrees and orders 2000, as previously mentioned, and with quadruple precision, the computations are very stable at least up to degree and order 3600. Spectral analysis of the synthesis/analysis results can be done degree by degree to study the characteristics of the estimated spectral harmonic coefficients. The results of the second synthesis also enable a study of the spatial results parallel by parallel, especially for the polar regions. Such investigations are currently underway to better characterize the numerical stability and reliability of the SHT and SHT^{-1}. Ongoing experimentation is attempting to carry out the computations to higher degrees and orders.

6 Concluding Remarks

Considerable work has been done on solving the computational complexities, and enhancing the speed of calculation of spherical harmonic transforms. The approach of Driscoll and Healy [5] is exact for exact arithmetic, and with a number of modifications, different implementations have been experimented with, leading to RMS errors of orders 10^{-16} to 10^{-12} with unit coefficients of degrees and orders up to 1800. Such RMS errors have been seen to increase to 10^{-3} with degrees and orders around 2000. With quadruple precision arithmetic, RMS errors are of orders 10^{-31} to 10^{-29} with unit coefficients of degrees and orders 3600. Starting with spherical harmonic coefficients corresponding to $1/degree^2$, the previously mentioned analysis and synthesis results are improved to 10^{-17}-10^{-16} and 10^{-35}-10^{-34}, respectively. The latter simulations are perhaps more indicative of the expected numerical accuracies in practice.

Computations for even higher degrees and orders are under consideration assuming the availability of parallel FFT code in quadruple precision. As enormous quantities of data are involved in the intended gravity field applications, parallel and grid computations are imperative for these applications. Preliminary experimentation with parallel processing in other applications has already been done and spherical harmonic computations appear most appropriate for parallel and grid implementations.

Acknowledgement

The authors would like to acknowledge the sponsorship of the Natural Science and Engineering Research Council in the form of a Research Grant to the first author on Computational Tools for the Geosciences. Special thanks are hereby expressed to Dr. D. Phillips of Information Technologies, University of Calgary, for helping with the optimization of our code for different computer platforms. Comments and suggestions from a colleague, Dr. N. Sneeuw, are also gratefully acknowledged.

References

1. Adams, J.C. and P.N. Swarztrauber [1997]: SPHEREPACK 2.0: A Model Development Facility. http://www.scd.ucar.edu/softlib/SPHERE.html
2. Blais, J.A.R. and D.A. Provins [2002]: Spherical Harmonic Analysis and Synthesis for Global Multiresolution Applications. Journal of Geodesy, vol.76, no.1, pp.29-35.
3. Blais, J.A.R. and D.A. Provins [2003]: Optimization of Computations in Global Geopotential Field Applications. Computational Science – ICCS 2003, Part II, edited by P.M.A. Sloot, D. Abramson, A.V. Bogdanov, J.J. Dongarra, A.Y. Zomaya and Y.E. Gorbachev. Lecture Notes in Computer Science, vol.2658, pp.610-618. Springer-Verlag.
4. Colombo, O. [1981]: Numerical Methods for Harmonic Analysis on the Sphere. Report no. 310, Department of Geodetic Science and Surveying, The Ohio State University
5. Driscoll, J.R. and D.M. Healy, Jr. [1994]: Computing Fourier Transforms and Convolutions on the 2-Sphere. Advances in Applied Mathematics, 15, pp. 202-250.
6. Gorski, K.M., E. Hivon and B.D. Wandelt [1998]: Analysis Issues for Large CMB Data Sets. Proceedings of Evolution of Large Scale Structure, Garching, Preprint from http://www.tac.dk/~healpix (August 1998).
7. Górski, K.M., B.D. Wandelt, E. Hivon, F.K. Hansen and A.J. Banday [1999]: The HEALPix Primer, http://arxiv.org/abs/astro-ph/9905275 (May 1999).
8. Healy, D., Jr., D. Rockmore, P. Kostelec and S. Moore [1998]: FFTs for the 2-Sphere - Improvements and Variations, To appear in Advances in Applied Mathematics, Preprint from http://www.cs.dartmouth.edu/~geelong/publications (June 1998).
9. Holmes, S.A. and W.E. Featherstone [2002a]: A unified approach to the Clenshaw summation and the recursive computation of very-high degree and order normalised associated Legendre functions. Journal of Geodesy, 76, 5, pp. 279-299.
10. Holmes, S.A. and W.E. Featherstone [2002b]: SHORT NOTE: Extending simplified high-degree synthesis methods to second latitudinal derivatives of geopotential. Journal of Geodesy, 76, 8, pp. 447-450.
11. Mohlenkamp, M.J. [1997]: A Fast Transform for Spherical Harmonics. PhD thesis, Yale University.
12. Mohlenkamp, M.J. [1999]: A Fast Transform for Spherical Harmonics. The Journal of Fourier Analysis and Applications, 5, 2/3, pp. 159-184, Preprint from http://amath.colorado.edu/faculty/mjm.
13. Mohlenkamp, M.J. [2000]: Fast spherical harmonic analysis: sample code. http://amath.colorado.edu/faculty/mjm.
14. Moore, S., D. Healy, Jr., D. Rockmore and P. Kostelec [1998]: SpharmonKit25: Spherical Harmonic Transform Kit 2.5, http://www.cs.dartmouth.edu/~geelong/sphere/.
15. Provins, D.A. [2003]: Earth Synthesis: Determining Earth's Structure from Geopotential Fields, Unpublished PhD thesis, University of Calgary, Calgary.
16. Rapp, R.H. [1982]: A FORTRAN Program for the Computation of Gravimetric Quantities from High Degree Spherical Harmonic Expansions. Report no. 334, Department of Geodetic Science and Surveying, The Ohio State University.
17. Schwarzschild, B. [2003]: WMAP Spacecraft Maps the Entire Cosmic Microwave Sky With Unprecedented Precision. Physics Today, April, pp. 21-24.
18. Sneeuw, N. [1994]: Global Spherical Harmonic Analysis by Least-Squares and Numerical Quadrature Methods in Historical Perspective. Geophys. J. Int. 118, 707-716.
19. Varshalovich, D.A., A.N. Moskalev and V.K. Khersonskij [1988]: Quantum Theory of Angular Momentum. World Scientific Publishing, Singapore.
20. Wenzel, G. [1998]: Ultra High Degree Geopotential Models GPM98A, B and C to Degree 1800. Preprint, Bulletin of International Geoid Service, Milan.

Predictor-Corrector Preconditioned Newton-Krylov Method for Cavity Flow

Jianwei Ju and Giovanni Lapenta

Los Alamos National Laboratory, Los Alamos, NM 87545, USA
{jju, lapenta}@lanl.gov

Abstract. The Newton-Krylov method is used to solve the incompressible Navier-Stokes equations. In the present study, two numerical schemes are considered for the method: employing the predictor-corrector method as preconditioner, and solving the equations without the preconditioner. The standard driven cavity flow is selected as the test problem to demonstrate the efficiency and the reliability of the present preconditioned method. It is found that the Newton-Krylov method becomes more efficient if combined with the preconditioner.

1 Introduction

A classic problem of computational science and engineering is the search for an efficient numerical scheme for solving the incompressible Navier-Stokes equations. Explicit and semi-implicit methods can provide simple solution techniques but are seriously limited by the time step limitations for stability (explicit methods) and accuracy (implicit methods).

Recently, significant progress has been made in the development of a new fully implicit approach for solving nonlinear problems: the inexact Newton method [1, 2, 3]. The method is developed from the Newton iterative method, by applying a linear iterative method to the Jacobian equation for the Newton step and terminating that iteration when the convergence criterion holds [4]. The rapid increases in both speed and memory capacity of computing resources makes it practical to consider inexact Newton methods for incompressible flow problem.

For the solution of the linear Jacobian equation, Krylov methods are often the choice, leading to the Newton-Krylov (NK) approach. However, for most cases, Krylov solvers can be extremely inefficient solvers. The need for good preconditioners techniques becomes a constraining factor in the development of NK solvers.

In the field of incompressible and geophysical flows, recent work based on the multi-grid preconditioners [7, 9] have shown near ideal scaling, resulting in extremely competitive alternatives to classic explicit and semi-implicit methods.

In the present study, we present a new approach to preconditioning: the predictor-corrector (PC) preconditioner. The approach has two novelties. First, it preconditions directly the non-linear equations rather than the linear Jacobian equation for the Newton step. The idea is not new [4], but it is implemented here in a new way that leads to great simplification of the implementation. We note that this simplification is designed also to minimize the effort in refitting existing semi-implicit codes into full fledged implicit codes, representing perhaps a greater advance in software engineering than in computational science. Second, we test new ways of preconditioning the equations by using a combination of predictor-corrector semi-implicit preconditioning.

The fundamental idea is to use a predictor to advance a semi-implicit discretization of the governing equations and use a corrector Newton step to correct for the initial state of the predictor step. In substance, the Newton method iterates for a modification of the actual initial state to find a modified initial state that makes the semi-implicit predictor step give the solution of the fully implicit method.

Two advantages are obvious. First, the initial step is likely to be a better first guess for the modified initial step of the predictor than it is for the final state of the corrector step. Second, by modifying the non-linear function, the PC preconditioner gives the same type of speed-up of the Krylov convergence without requiring to formulate an actual preconditioning of the Krylov solver.

We use the standard driven cavity flow problem as the test problem to demonstrate the efficiency and the reliability of the present preconditioned method.

2 Governing Equations and Numerical Method

In the present study, we selected the standard driven 2-D incompressible cavity flow as the test example. The geometry and the velocity boundary conditions are shown in Fig. 1. The following non-dimensional variables are introduced:

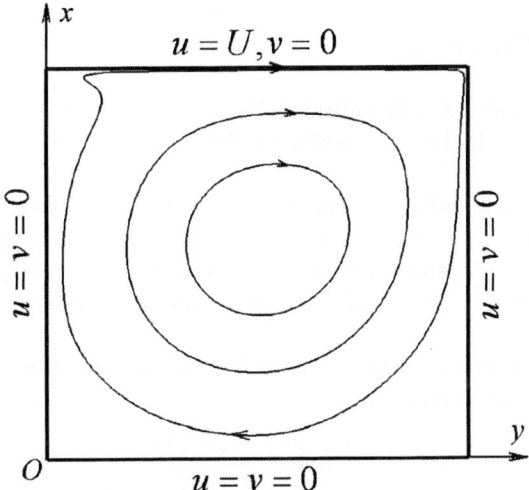

Fig. 1. Standard driven cavity flow and the velocity boundary conditions

$$(u,v) = \frac{(\hat{u}, \hat{v})}{U}, \quad (x,y) = \frac{(\hat{x}, \hat{y})}{L} \qquad (1)$$

where the hatted variables represent the dimensional variables. The scales are the cavity width L and the upper boundary velocity U. Time is normalized according to eq.(1). The governing equation, in term of the vorticity ω, and the stream function ψ, can be expressed as:

$$\frac{\partial^2 \psi}{\partial x^2} + \frac{\partial^2 \psi}{\partial y^2} = -\omega \qquad (2)$$

$$\frac{\partial \omega}{\partial t} + u \frac{\partial \omega}{\partial x} + v \frac{\partial \omega}{\partial y} = \frac{1}{Re} \left(\frac{\partial^2 \omega}{\partial x^2} + \frac{\partial^2 \omega}{\partial y^2} \right) \qquad (3)$$

where $u = \partial \psi / \partial y$, $v = -\partial \psi / \partial x$, $\omega = \partial v / \partial x - \partial u / \partial y$, and $Re = UL/\nu$ is the Reynolds number based on the viscosity ν. Dirichlet boundary conditions are applied for the stream function and the boundary conditions for vorticity is determined by the physical boundary conditions on the velocity [11]. Thus, at the left wall $\omega = \partial v / \partial x = -\partial^2 \psi / \partial y^2$. We can obtain expressions for ω at other walls in an analogous manner.

Eqs. (2) and (3) are discretized using the centered difference scheme in space. Two types of discretization are considered in time, the semi-implicit Euler scheme (where the velocity in eq.(3) is taken at the initial time level of the time step) and the fully implicit Euler scheme.

The resulting set of coupled non-linear difference equations are solved using the Newton-Krylov method. As Krylov solver we use GMRES since the Jacobian system is non-symmetric. The resulting method is completely matrix-free as only matrix-vector products, rather than details of the matrix itself are needed. This circumstance greatly simplifies the application of the method to complex problems. We design the preconditioner to maintain this property.

3 Preconditioners

In the present study, a preconditioner is constructed by using the predictor-corrector method. The key idea lies on modifying the target point of the Newton iteration.

Most discretization schemes can be expressed as a set of difference equations for a set of unknowns **x** representing the unknown fields on a spatial grid. Once the time is discretized, the state vector **x** is computed at a sequence of discrete time levels. We label the initial state of a time step as $\mathbf{x}^0$ and the final time as $\mathbf{x}^1$.

When the time discretization scheme is fully implicit, the most general two-level scheme can be formulated as:

$$\mathbf{x}^1 = \mathbf{x}^0 + f(\mathbf{x}^0, \mathbf{x}^1) \qquad (4)$$

where the vector function f depends both on the initial and the final states. The implicit nature of the scheme resides in the fact that the function f is a function of the new time level, requiring the solution of a set of non-linear (if the function f is non-linear) coupled equations. As noted above this can be accomplished with the NK method [4]. The method is based on solving the Jacobian equation obtained linearizing the difference eq. (4) around the current available estimate $\mathbf{x}_k^1$ of the solution in the Newton iteration:

$$\delta\mathbf{x} + \mathbf{x}_k^1 = \mathbf{x}^0 + f(\mathbf{x}^0, \mathbf{x}_k^1) + J\delta\mathbf{x} \qquad (5)$$

where $J = \partial f/\partial \mathbf{x}$ is the Jacobian matrix and $\delta\mathbf{x}$ is the correction leading to the new estimation by the Newton iteration: $\mathbf{x}_{k+1}^1 = \mathbf{x}_k^1 + \delta\mathbf{x}$.

The solution of eq. (5) is conducted with a Krylov solver, we use here GMRES since the Jacobian matrix is non symmetric. While the pure inexact NK method works in giving a solution, the number of Krylov iterations required for each Newton step to solve eq. (5) can be staggering. In particular, as the grid is refined and the size of the unknown vector $\mathbf{x}^1$ is increased the number of Krylov iterations tends to increase. This is the reason why a preconditioner is needed.

Here we propose to use the predictor-corrector method as a preconditioner. The approach requires to design alongside the fully implicit scheme in eq. (4), a second semi-implicit method. We note that this is typically no hardship as semi-implicit methods were developed and widely used before the implicit methods became tractable. Using the same notation, we can write the most general two-level semi-implicit algorithm as:

$$\mathbf{x}^1 = \mathbf{x}^0 + A\mathbf{x}^1 + f_{SI}(\mathbf{x}^0) \qquad (6)$$

where A is a linear operator (matrix) and the function f_{SI} depends only on the initial state $\mathbf{x}^0$. The semi-implicit nature of the scheme resides on the fact that the difference eq. (6) depends non-linearly on the (known) initial state $\mathbf{x}^0$ but only linearly on the new (unknown) state $\mathbf{x}^1$.

In the classic implementation of preconditioners [8], the equation for the semi-implicit scheme (6) is rewritten in terms of the modification $\delta\mathbf{x}$ in a given Newton iteration, essentially leading to the matrix A of the semi-implicit scheme to be used as a preconditioner for the Jacobian matrix J of eq. (5). The approach has been extremely successful in terms of providing a robust and effective solution scheme. For example in the case of incompressible flows, the number of Krylov iteration has been shown [9, 7] to be reduced drastically and to become nearly independent of the grid size.

However, a substantial modification of existing codes follows from the need to modify the GMRES solver to use the matrix A as a preconditioner, especially when the method is formulated in a matrix-free form where the matrix J and the matrix A are not explicitly computed and stored.

We propose a different approach. We consider the following predictor-corrector algorithm:

$$\begin{cases} (P) \ \mathbf{x}^1 = \mathbf{x}^\star + A\mathbf{x}^1 + f_{SI}(\mathbf{x}^0) \\ (C) \ \mathbf{r} = \mathbf{x}^1 - \mathbf{x}^0 - f(\mathbf{x}^0, \mathbf{x}^1) \end{cases} \qquad (7)$$

The predictor step uses the semi-implicit scheme to predict the new state $\mathbf{x}^1$ starting from a modification of the initial state $\mathbf{x}^\star$. The corrector step computes the residual **r** for the fully implicit scheme when $\mathbf{x}^1$ from the predictor step is used.

We propose to use scheme (7) by using $\mathbf{x}^0$ as the initial guess of $\mathbf{x}^*$ and using the NK method to find the solution for $\mathbf{x}^*$ that makes the residual **r** of the corrector equation vanish. Once $\mathbf{r} = 0$ (within a set tolerance), the fully implicit scheme is solved, but it is solved not iterating directly for $\mathbf{x}^1$ but iterating for the $\mathbf{x}^*$ that makes the predictor step predict the correct solution $\mathbf{x}^1$ of the corrector step.

Two points are worth noting.

First, we have modified the task of the NK iteration changing our unknown variable from $\mathbf{x}^1$ to $\mathbf{x}^*$. This corresponds to change the non-linear residual function that the Newton method needs to solve. To first order in the Taylor series expansion leading to the Jacobian equation this practice is identical to applying the traditional preconditioners directly to the Jacobian equation [4]. However, to higher order this might be a better approach as it reduces the distance between the initial guess ($\mathbf{x}^0$) and the solution for $\mathbf{x}^*$. If the semi-implicit method works properly, $\mathbf{x}^0$ is closer to the converged $\mathbf{x}^*$ than to the final state $\mathbf{x}^1$.

Second, programming the PC preconditioner is easier. The NK solver can be used as a black box, without any need to formally go into it and modify the Jacobian eq. (5) by adding a preconditioner. The semi-implicit method can be used directly on the actual states and not on their variation $\delta\mathbf{x}$ between two subsequent Newton iterates. This latter operation is complex as boundary conditions and source terms in equations need to be treated differently.

The approach described above is ideally suited for refitting an existing semi-implicit code by simply taking an off the shelf NK solver and wrapping it around the semi-implicit method already implemented. The only change being that in the semi-implicit scheme the initial state $\mathbf{x}^0$ is replaced by the guess of $\mathbf{x}^*$ provided by the NK solver. We have indeed proceeded in this fashion by wrapping the standard NK solver provided in the classic textbook by Kelley [5] around our previously written semi-implicit solver for the incompressible flow equations.

4   Results and Discussion

In the example below we present results for a case with a mesh of 129×129 cells. The classic cavity flow solution is computed starting from a stagnant flow and allowing the boundary conditions to drive the cavity to a steady state. The time evolution of the vorticity at the center of the cavity is shown in Fig. 2.

The flow condition at steady state is shown in Fig. 3. The figure is generated using the same contour lines used in the reference benchmark solution presented by Chia et al. [10]. We compared visually our solution with the published reference benchmark obtaining complete agreement.

We have compared the efficiency of the NK solver with and without the PC preconditioner described above. The number of GMRES iterations per Newton iteration for the two simulations are shown in Fig. 4 as a function of time. At the beginning of the simulation when the flow is in a transient state and is

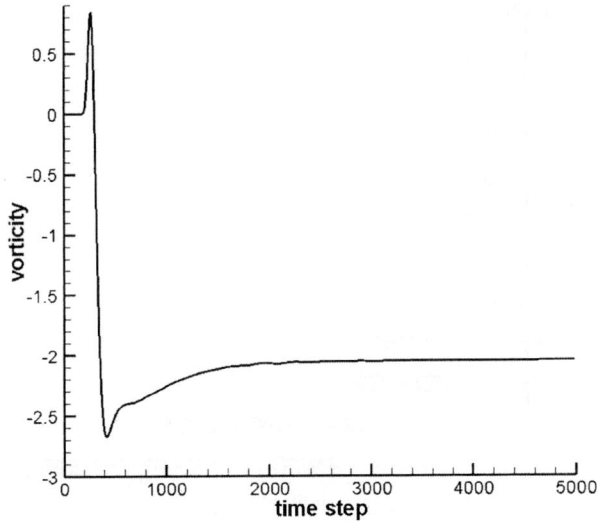

Fig. 2. Evolution of the vorticity at the cavity center as a function of time step

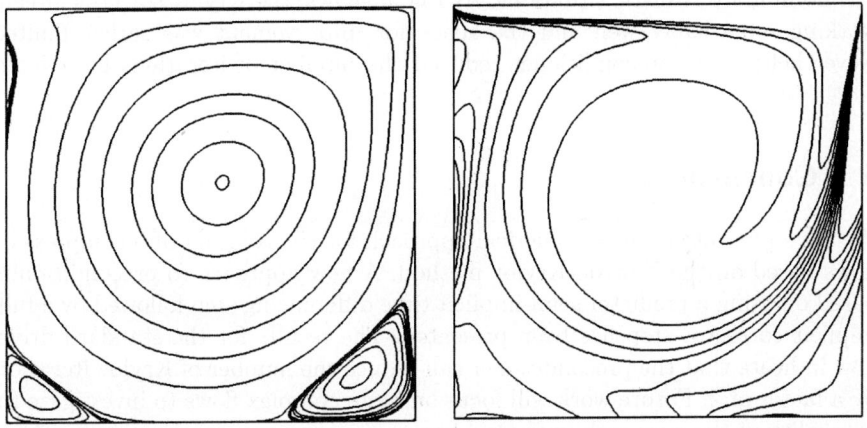

Fig. 3. (a) Contours of stream function, (b) Contours of vorticity. Flow structure at steady state for Re=1000

adjusting to the driving upper boundary condition, the evolution is dynamical. In that condition, the solver needs more Krylov iterations. The average number of iterations before 2000 time steps is 5.3.

At later stages of the simulation, the system reaches a steady state and flow remains unchanged from time step to time step. In this very favorable condition for the NK method, the average number of Krylov iterations after 2000 time steps is 1.3 for the preconditioned simulation and 2.5 for un-preconditioned simulation.

We should point out that the test provides confirmation that the preconditioning approach presented here is promising. But it cannot test how well it

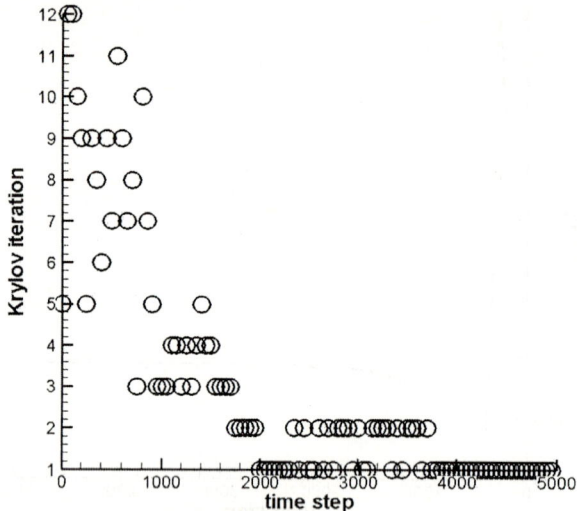

Fig. 4. Number of GMRES iterations per Newton iteration

will do in more realistic situations. As Fig. 4 shows, the Krylov solver is already working remarkably well and the space for improvement was rather limited. Nevertheless, the preconditioner reduced the number of iterations by a factor of 2.

5 Conclusions

We have presented a new numerical approach for the solution of incompressible flows based on the Newton-Krylov method. A new approach to preconditioning based on using a predictor semi-implicit time differencing step followed by a fully implicit corrector step has been presented. The results for the standard driven flow indicate that the preconditioner can reduce the number of Krylov iterations by a factor of 2. Future work will focus on more complex flows to investigate the generality of the approach presented here.

References

1. Dembo, R.S , Eisenstat, S.C , Steihaug, T.: Inexact Newton Method. SIAM J. Numer. Anal.,19 (1982), 400-408.
2. Eisenstat, S.C , Walker, H.F : Globally Convergent Inexact Newton Methods. SIAM J. Optim., 4 (1994), 393-422.
3. Eisenstat, S.C , Walker, H.F : Choosing The Forcing Terms In An Inexact Newton Method. SIAM J. Sci. Comput., 17 (1996), 16-32.
4. Kelley, C.T : Solving Nonlinear Equations with Newton's Method. Society for Industrial and Applied Mathematics (2003).
5. Kelley, C.T : Iterative Methods for Linear and Nonlinear Equations. Society for Industrial and Applied Mathematics (1995).

6. Kim, J.H , Kim, Y.H : A Predictor-Corrector Method For Structural Nonlinear Analysis. Comput. Methods Appl. Mech. Engrg. 191 (2001) 959-974.
7. Knoll, DA, Mousseau, VA: On Newton-Krylov multigrid methods for the incompressible Navier-Stokes equations., J. Comput. Phys. 163 (2000), 262-267.
8. Knoll, DA; Keyes, DE: Jacobian-free Newton-Krylov methods: a survey of approaches and applications, J. Comput. Phys. 193 (2004), 357-397.
9. Pernice, M, Tocci, M.D : A Multigrid-Preconditioned Newton-Krylov Method for the Incompressible Navier-Stokes Equations. SIAM J. Sci. Comput., 23 (2001). 398-418.
10. Chia, U., Chia, K. N., Shin, T., High-Re Solutions for Incompressible Flow Using the Navier-Stokes Equations and a Multigrid Method. J. Comput. Phys. 48 (1982), 387-411.
11. Roache, P.J.: Fundamentals of Computational Fluid Dynamics. Hermosa Publishers (1998).

A High-Order Recursive Quadratic Learning Algorithm

Qi Zhu[1,*], Shaohua Tan[2], and Ying Qiao[3]

[1] University of Houston - Victoria, TX 77901
Tel.: +1-361-570-4312, Fax: +1-361-570-4207
zhuq@uhv.edu
[2] National University of Singapore,
Singapore, 119260
[3] Virginia Polytechnic Institute and State University,
Virginia, 24060

Abstract. A k-order Recursive Quadratic learning algorithm is proposed and its features are described in detail in this paper. Simulations are carried out to illustrate the efficiency and effectiveness of this new algorithm by comparing the results with both the projection algorithm and the conventional least squares algorithm.

1 Introduction

The Least Squares (LS) algorithm [3] and its adaptive version Recursive Least Squares (RLS) algorithm [2] are well-known algorithms widely used in areas such as Data Compression [8], Neural Network [5], Parameter identification [10], Pattern Recognition [4], Graphics [7], and Gene/DNA studies[9]. RLS is often used in linear-in-the-parameter (LIP) models. However, some undesirable features of RLS algorithm are that the regression matrix should be of full rank, convergence slows down considerably at certain low level error, and it persists over a considerable number of iteration steps before dropping eventually below the prescribed error bound. This is due to the fact that at low error level, RLS algorithm takes very small step sizes to ensure the convergence.

This paper develops a *high-order Recursive Quadratic (RQ for short) learning algorithm*, initially proposed in [11], which avoids the problem of RLS for identifying a general class of linear and nonlinear LIP models. RQ algorithm is thoroughly investigated and we reveal its features as a high-order extension of the Projection algorithm with a quadratic cost function. The convergence and accuracy analysis of the algorithm is performed along with the simulations to demonstrate various specialized properties of the algorithm.

---

* Corresponding author.

2 The RQ Algorithm

2.1 Preliminaries

The *Linear-in-the-Parameters (LIP)* is one of the most widely used model structures with the following general form [12]:

$$y(t) = \sum_{l=1}^{m} \xi_l(x(t))\omega_l = \varphi^T(x(t))\theta^\star \qquad (1)$$

where $\{x(t), y(t)\}(t > 0)$ is a set of sample data, $\theta^\star = [\omega_1, ..., \omega_l, ...]^T$ is the desired weight, and $\varphi(x(t)) = [\xi_1(x(t)), ..., \xi_l(x(t)), ...]^T$ is an m-vector of basis functions.

We define a kt-dimensional matrix Λ_t to be the system forgetting factor:

$$\Lambda_t = \begin{bmatrix} 0_{k(t-1)} & 0 \\ 0 & \Lambda(kt, k) \end{bmatrix} \qquad (2)$$

where $\Lambda(kt, k)$ is a k-dimensional diagonal matrix $\Lambda(kt, k) = diag[\lambda_1, ..., \lambda_k]$ and λ_i ($i = 1, 2, ..., k$) are some positive real scalars. The constant k is the order of the algorithm.

2.2 RQ Learning Algorithm for LIP Models

For a particular k, we define the abbreviated notations of input matrix, output vector, and output error vector as follows:

$$\Phi_t = \Phi_t(kt, k) \triangleq [\varphi_t(x(k(t-1)+1)), \varphi_t(x(k(t-1)+2)), ..., \varphi_t(x(k(t-1)+k))]^T \qquad (3)$$

$$Y_t = Y_t(kt, k) \triangleq [y_t(k(t-1)+1), y_t(k(t-1)+2), ..., y_t(k(t-1)+k)]^T \qquad (4)$$

$$E_t = E_t(kt, k) \triangleq Y_t(kt, k) - \Phi_t(kt, k)\hat{\theta}_{t-1}$$
$$= [e_t(k(t-1)+1), e_t(k(t-1)+2), ..., e_t(k(t-1)+k)]^T \qquad (5)$$

where $e_t(kt) = y_t(kt) - \varphi_t^T(x(kt))\hat{\theta}_{t-1}$, subscript t denotes that the parameters are estimated at time t. Introduce J_t to be the quadratic function:

$$J_t = J_t(kt, k) \triangleq \frac{1}{2} E_t^T \Lambda(kt, k) E_t = \frac{1}{2}(Y_t - \Phi_t \hat{\theta}_{t-1})^T \Lambda(kt, k)(Y_t - \Phi_t \hat{\theta}_{t-1})$$
$$= \frac{1}{2}\hat{\theta}_{t-1}^T \Phi_t^T \Lambda(kt, k)\Phi_t \hat{\theta}_{t-1} - \hat{\theta}_{t-1}^T \Phi_t^T \Lambda(kt, k) Y_t + \frac{1}{2} Y_t^T \Lambda(kt, k) Y_t$$
$$= \frac{1}{2}\hat{\theta}_{t-1}^T P_t \hat{\theta}_{t-1} - \hat{\theta}_{t-1}^T Q_t + \frac{1}{2} R_t \qquad (6)$$

where $\Lambda(kt, k)$ is a k-dimensional identity matrix if we select $\lambda_i = 1 (\forall i = 1, 2, ..., k)$, and

$$P_t = P_t(kt, k) \triangleq \Phi_t^T(kt, k)\Lambda(kt, k)\Phi_t(kt, k) = \Phi_t^T \Phi_t$$
$$Q_t = Q_t(kt, k) \triangleq \Phi_t^T(kt, k)\Lambda(kt, k)Y_t(kt, k) = \Phi_t^T Y_t$$
$$R_t = R_t(kt, k) \triangleq Y_t^T(kt, k)\Lambda(kt, k)Y_t(kt, k) = Y_t^T Y_t$$

Using the above notations we can introduce **the k-order RQ algorithm** as follows:

$$\hat{\theta}_t = = \hat{\theta}_{t-1} + \frac{\alpha J_t(Q_t - P_t\hat{\theta}_{t-1})}{\beta + (Q_t - P_t\hat{\theta}_{t-1})^T(Q_t - P_t\hat{\theta}_{t-1})} \quad (7)$$

where $t = 1, 2, ...$; and $\beta > 0$, $0 < \alpha < 4$.

Theorem 1. The algorithm (7) is obtained by solving the following optimization problem: Given $\hat{\theta}_{t-1}$ and Y_t, determine $\hat{\theta}_t$ so that $J_{\hat{\theta}} = \frac{1}{2}\|\hat{\theta}_t - \hat{\theta}_{t-1}\|^2$ is minimized subject to

$$(Y_t - \Phi_t\hat{\theta}_{t-1})^T Y_t = (Y_t - \Phi_t\hat{\theta}_{t-1})^T \Phi_t^T \hat{\theta}_t \quad (8)$$

Proof. Introducing a Lagrange multiplier λ for the constraint (8), we have the augmented function as $J'_{\hat{\theta}} = \frac{1}{2}\|\hat{\theta}_t - \hat{\theta}_{t-1}\|^2 + 2\lambda[Y_t - \Phi_t\hat{\theta}_{t-1}]^T[Y_t - \Phi_t\hat{\theta}_t]$. The necessary conditions for an optimization are $\partial J'_{\hat{\theta}}/\partial \hat{\theta}_t = 0$ and $\partial J'_{\hat{\theta}}/\partial \lambda = 0$, which are

$$\hat{\theta}_t - \hat{\theta}_{t-1} - 2\lambda\Phi_t^T[Y_t - \Phi_t\hat{\theta}_{t-1}] = 0 \quad (9)$$

$$[Y_t - \Phi_t\hat{\theta}_{t-1}]^T[Y_t - \Phi_t\hat{\theta}_t] = 0 \quad (10)$$

From (9) we obtain $\hat{\theta}_t = \hat{\theta}_{t-1} + 2\lambda\Phi_t^T[Y_t - \Phi_t\hat{\theta}_{t-1}]$, substituting into (10) gives

$$\lambda = \frac{\frac{1}{2}[Y_t - \Phi_t\hat{\theta}_{t-1}]^T[Y_t - \Phi_t\hat{\theta}_{t-1}]}{[Y_t - \Phi_t\hat{\theta}_{t-1}]^T\Phi_t\Phi_t^T[Y_t - \Phi_t\hat{\theta}_{t-1}]} \quad (11)$$

And we have $\Phi_t^T(Y_t - \Phi_t\hat{\theta}_{t-1}) = \Phi_t^T E_t = Q_t - P_t\hat{\theta}_{t-1}$, then substituting this as well as (5),(6), and (11) into (9) gives

$$\hat{\theta}_t = \hat{\theta}_{t-1} + \frac{2J_t(Q_t - P_t\hat{\theta}_{t-1})}{(Q_t - P_t\hat{\theta}_{t-1})^T(Q_t - P_t\hat{\theta}_{t-1})}$$

To avoid division by zero, a small constant β is added to the denominator of the above formula. In order to adjust the convergence rate of the algorithm, we multiply a constant α to the numerator of the algorithm that erases the constant 2. This leads to the slightly modified form (7) of the new high-order RQ algorithm. □

Figure 1 illustrates the geometric interpretation of the RQ algorithm with parameter of two dimensions $\hat{\theta} = (\theta_0 \; \theta_1)$ and the order $k = 4$. The parameters θ_0 and θ_1 span a plane if they are linearly independent. The input matrix $\Phi_t = \{\varphi_{t1}, \varphi_{t2}, \varphi_{t3}, \varphi_{t4}\}$ and each vector φ has two dimensions. Through Figure 1, we know that $\hat{\theta}_t$ is convergent to the desired parameter value θ^* using the shortest path, and it eventually reaches θ^*, i.e., $\|\hat{\theta}_t - \hat{\theta}_{t-1}\|$ is minimized by RQ learning algorithm.

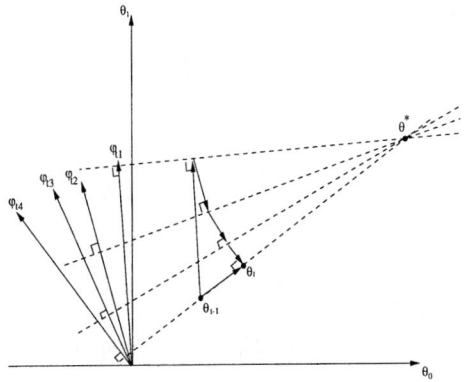

Fig. 1. Geometric interpretation of the Recursive Quadratic learning algorithm

2.3 Properties of RQ Learning Algorithm

Lemma 1 provides the convergent assessment of the RQ algorithm.

Lemma 1. For any given initial value $\hat{\theta}_0$, (7) has the following properties:

(i)
$$\| \hat{\theta}_t - \theta^* \| \leq \| \hat{\theta}_{t-1} - \theta^* \| \leq \cdots \leq \| \hat{\theta}_0 - \theta^* \| \text{ for any } t \geq 1.$$

(ii)
$$\lim_{t \to \infty} J_t = 0$$

(iii)
$$\lim_{t \to \infty} \| \hat{\theta}_t - \hat{\theta}_{t-s} \| = 0 \text{ for any finite positive integer } s.$$

Proof. The complete proof is given in [11].

In Lemma 2, we consider the k-order RQ algorithm in the presence of a Gaussian noise.

Lemma 2. Modify the system model (1) into :

$$y_t = \varphi_t^T \theta^* + w_t \qquad (12)$$

w_t is a sequence of Gaussian noise, so that $\lim_{t \to \infty} E[w_t] = 0$ and $\lim_{t \to \infty} E[w_t^2] = \sigma^2$. Define the k-order noise vector $W_t \triangleq [w_{k(t-1)+1}, w_{k(t-1)+2}, ..., w_{k(t-1)+k}]^T$, then the RQ algorithm in (7) has the following properties:

(i) The output error converges in the mean

$$\lim_{t \to \infty} E[e_t] = 0$$

(ii)
$$\lim_{t\to\infty} E[e_t^T e_t] = \sigma^2$$

Proof. (i)
$$e_t = y_t - \varphi_t^T \hat{\theta}_{t-1} = -\varphi_t^T \tilde{\theta}_{t-1} + w_t$$

Then
$$\lim_{t\to\infty} E[e_t] = \lim_{t\to\infty} E[-\varphi_t \tilde{\theta}_{t-1} + w_t]$$
$$= -\lim_{t\to\infty} E[\varphi_t] \lim_{t\to\infty} E[\tilde{\theta}_{t-1}] + \lim_{t\to\infty} E[w_t]$$

As w_t is white noise, $\lim_{t\to\infty} E[w_t] = 0$, with the parameter $\hat{\theta}_t$ being unbiased and the parameter error converging to zero, i.e. $\lim_{t\to\infty} \tilde{\theta}_{t-1} = \lim_{t\to\infty} (\hat{\theta}_{t-1} - \theta^\star) = 0$, then $\lim_{t\to\infty} E[e_t] = 0$.

To prove (ii), we observe first that
$$\sum_{i=k(t-1)+1}^{kt} e_i^2 = E_t^T E_t = (Y_t - \Phi_t \hat{\theta}_{t-1})^T (Y_t - \Phi_t \hat{\theta}_{t-1})$$
$$= (-\Phi_t \tilde{\theta}_{t-1} + W_t)^T (-\Phi_t \tilde{\theta}_{t-1} + W_t)$$
$$= \tilde{\theta}_{t-1}^T \Phi_t^T \Phi_t \tilde{\theta}_{t-1} - \tilde{\theta}_{t-1}^T \Phi_t^T W_t - W_t^T \Phi_t \tilde{\theta}_{t-1} + W_t^T W_t$$

The covariance estimate of the output error is:
$$\lim_{t\to\infty} E[E_t^T E_t] = \lim_{t\to\infty} E[\tilde{\theta}_{t-1}^T \Phi_t^T \Phi_t \tilde{\theta}_{t-1}] - \lim_{t\to\infty} E[\tilde{\theta}_{t-1}^T \Phi_t^T] \lim_{t\to\infty} E[W_t]$$
$$- \lim_{t\to\infty} E[W_t^T] \lim_{t\to\infty} E[\Phi_t \tilde{\theta}_{t-1}] + \lim_{t\to\infty} E[W_t^T W_t]$$

As the vector W_t is composed of a sequence of white noise w_t, we can conclude that $\lim_{t\to\infty} E[W_t] = 0_k$ is a k-order zero vector, and $\lim_{t\to\infty} E[W_t^T W_t] = k\sigma^2$. Since $\lim_{t\to\infty} \tilde{\theta}_{t-1} = 0$, and $\lim_{t\to\infty} \{\tilde{\theta}_{t-1}^T \Phi_t^T \Phi_t \tilde{\theta}_{t-1}\}$ is a scalar, we have

$$\lim_{t\to\infty} E[\tilde{\theta}_{t-1}^T \Phi_t^T \Phi_t \tilde{\theta}_{t-1}] = \lim_{t\to\infty} E[tr\{\tilde{\theta}_{t-1}^T \Phi_t^T \Phi_t \tilde{\theta}_{t-1}\}]$$
$$= \lim_{t\to\infty} E[tr\{\tilde{\theta}_{t-1}^T \tilde{\theta}_{t-1} \Phi_t^T \Phi_t\}]$$
$$= \lim_{t\to\infty} tr\{E[\tilde{\theta}_{t-1}^T \tilde{\theta}_{t-1}] E[\Phi_t^T \Phi_t]\} = 0$$

Therefore
$$\lim_{t\to\infty} E[E_t^T E_t] = \lim_{t\to\infty} \sum_{i=k(t-1)+1}^{kt} E[e_i^2] = k\sigma^2$$

Finally
$$\lim_{t\to\infty} E[e_t^2] = \sigma^2$$

Lemma 2 allows us to conclude that (7) converges under the white noisy data as well. □

Lemma 3. When $k = 1$, the new 1st-order RQ learning algorithm is equivalent to the Projection algorithm in [1].

$$\hat{\theta}_t = \hat{\theta}_{t-1} + \frac{\alpha \phi_t}{\beta + \phi_t^T \phi_t}[y_t - \phi_t^T \hat{\theta}_{t-1}] \quad (13)$$

Proof. When $k = 1$, the 1st-order RQ learning algorithm becomes:

$$\hat{\theta}_t = \hat{\theta}_{t-1} + \frac{\alpha' e_t^2 \varphi_t}{\beta + e_t^2 \varphi_t^T \varphi_t}(y_t - \varphi_t^T \hat{\theta}_{t-1}) \quad (14)$$

where $t > 0, \beta > 0, 0 < \alpha' = \frac{1}{2}\alpha < 2$, and $e_t = y_t - \varphi_t^T \hat{\theta}_{t-1}$.

Since β can be chosen as any positive number for preventing the denominator to be zero, selecting $\beta' = \beta/e_t^2$, (14) can be interpreted in terms of (13). Thus, the Projection algorithm is a special case of the new RQ learning algorithm when we choose a set of specific parameters. □

3 Simulations

In this section, we present an example [1] to assess the computational features of the RQ algorithm. We provide the performance comparisons among the k-order RQ learning algorithm, the projection algorithm, and the conventional Recursive Least Squares algorithm as well as the data statistics.

Example. The moving average (MA) model

$$x_t = 0.1(t - 1)$$
$$y_t = 5 \sin x_t - 1 \cos x_t e^{\frac{1}{x_t+10}} + 2\ln(x_t + 10) + w_t$$

where $t = 1, 6, 11, ..., 101$ and w_t is a sequence of Gaussian noise with variance 0.01, $m = 3$ is the number of basis functions.

Figure 2 show the parameters convergence using 21 input and output data for identifying the system for $k = 15$, where k is the order of the RQ learning algorithm. With the choice of k, the number of multiplications (NM) is $3m^2 + 4m + km^2 + km + k + 3$ and the number of additions (NA) is $m^2 + m + m^2k + mk + k$.

These results are also compared to the Projection algorithm (13) and the RLS algorithm in [6]. For the Projection algorithm, the NM is $3m + 1$ and NA is $3m - 1$ per iteration step. For RLS algorithm, the NM is $6m^2 + 3m + 2$ and NA is $5m^2 - m$. We choose the same initial parameter value $\hat{\theta}_0 = [0, 0, 0]^T$ and the same error bound (3×10^{-4}) for all the three algorithms.

---

[1] Limit to space, we just show one example here. However, we have done many simulations for all kinds of LIP models

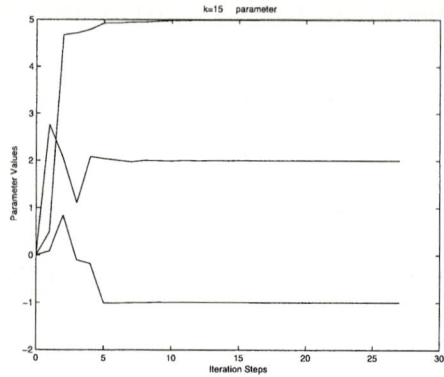

Fig. 2. When $k = 15$, RQ algorithm converges to 3×10^{-4} in 28 steps

The Projection algorithm can reach the error bound with 2339 iteration steps, its convergence rate is the slowest. The RLS algorithm converges faster than the Projection algorithm with 1387 iteration steps. However, after the initial fast convergence, the step length of the algorithm changes to be very small at low error level to avoid the convergence to the wrong parameter values.

The k-order RQ learning algorithm, on the other hand, can reach the error bound in only 28 iterations with $k = 15$. This shows that the speed of convergence of the k-order RQ learning algorithm is much faster than both the RLS algorithm and the Projection algorithm. Counting the total number of multiplications, additions and CPU time needed for the convergence, the RQ algorithm are 6636, 5796 and 1.43 seconds respectively, which are much less than RLS algorithm (90155 for NM, 58254 for NA and 22.57 seconds for CPU time) and Projection algorithm (23390 for NM, 18712 for NA and 71.70 seconds for CPU time).

We also observe that the choice of the order k is very critical. As shown in the Figure 3, if k is chosen to be large, then the convergence is fast but the

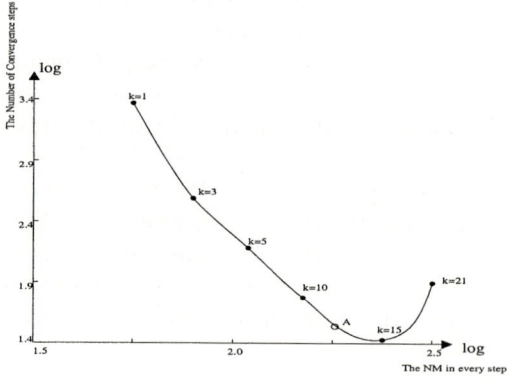

Fig. 3. Relationship between the steps and NM (both axes are scaled logarithmically)

computation at each iteration step is quite intensive. When k is too large, the convergence slows down again, at the same time, the NM and NA are very high. On the other hand, if k is chosen to be too small, then the computation is much simpler at each step, but it has much slower convergence. In this example, k around point A is the optimal choice, both the NM for each step and the total number of convergence steps are low. Currently the choice of k is based largely on intuitions rather than analytical means.

4 Conclusion

In this paper, we have developed a new high-order Recursive Quadratic (RQ) learning algorithm for Linear-in-the-Parameter models. This new RQ learning algorithm is derived as a high-order extension of the Projection algorithm with a new form of quadratic cost function. RQ algorithm is thoroughly described along with complete investigations to reveal its various features. The convergence and accuracy analysis of the algorithm is performed along with the simulations to demonstrate various specialized properties of the RQ algorithm.

We only developed the high-order RQ learning algorithm by choosing a specific Λ_t in this paper. One future research is to explore the learning algorithm by choosing different kinds of matrix Λ_t. In our research, choosing an appropriate order k is very critical. The larger the order, the faster the convergent speed and the more complex the computation at very iteration step. Hence, in future research it is very important that how can we choose a proper order k such that the convergent speed is reasonably fast, and yet the computation at every iteration step is reasonably simple in practice. Moreover, we can also extend the RQ learning algorithm for the purpose of on-line identification.

References

1. K.J. Astrom and B. Wittenmark, "Adaptive Control." Addison-Wesley, Mass., 1989.
2. Y. Boutalis, C. Papaodysseus, E. Koukoutsis, "A New Multichannel Recursive Least Squares Algorithm for Very Robust and Efficient Adaptive Filtering." *Journal of Algorithms*. Vol. 37, No.2, pp. 283-308, 2000.
3. L. E. Ghaoui and H. Lebret, "Robust Least Squares and Applications." *In Proc. CDC*, pp. 249-254, Kobe, Japan, 1996.
4. S. Ghosal, R. Udupa, N. K. Ratha, S. Pankanti, "Hierarchical Partitioned Least Squares Filter-Bank for Fingerprint Enhancement." *In Proc. ICPR*. pp. 3338-3341, 2000.
5. A. P. Grinko, M. M. Karpuk, "Modification of Method of Least Squares for Tutoring Neural Networks." *In Proc. Intelligent Information Systems*. pp. 157-164, 2002.
6. T. C. Hsia, "System Identification: Least-Squares methods." Lexington Books, 1977.
7. H. A. Lensch, J. Kautz, M. Goesele, W. Heidrich, H. Seidel, "Image-based Reconstruction of Spatial Appearance and Geometric Detail." *ACM Transactions on Graphics*. Vol. 22, No. 2, pp: 234 - 257, 2003.

8. B. Meyer, P. E. Tischer, "Glicbawls - Grey Level Image Compression by Adaptive Weighted Least Squares." *Data Compression Conference.* pp. 503, 2001.
9. D. V. Nguyen, D. M. Rocke, "Multi-class Cancer Classification Via Partial Least Squares with Gene Expression Profiles. " *Bioinformatics.* Vol. 18, No.9, pp. 1216-1226, 2002.
10. B. D. Saunders, "Black Box Methods for Least Squares Problems." *In Proc. ISSAC.* pp. 297-302, 2001.
11. S. Tan, X. H. Zhang and Q. Zhu, "On a new kth-order Quadratic Learning Algorithm." *IEEE Trans. Circ. and Syst.* Vol. 44, No. 1, pp. 186-190, 1997 .
12. W. X. Zheng, "Modified Least-squares Identification of Linear Systems with Noisy Input and Output Observations." *In Proc. CDC,* pp. 1067-1068, Kobe, Japan, 1996.

Vectorized Sparse Matrix Multiply for Compressed Row Storage Format[*]

Eduardo F. D'Azevedo[1], Mark R. Fahey[2], and Richard T. Mills[2]

[1] Computer Science and Mathematics Division
[2] Center for Computational Sciences,
Oak Ridge National Laboratory,
Oak Ridge, TN 37831, USA
{dazevedoef, faheymr, rmills}@ornl.gov

Abstract. The innovation of this work is a simple vectorizable algorithm for performing sparse matrix vector multiply in compressed sparse row (CSR) storage format. Unlike the vectorizable jagged diagonal format (JAD), this algorithm requires no data rearrangement and can be easily adapted to a sophisticated library framework such as PETSc. Numerical experiments on the Cray X1 show an order of magnitude improvement over the non-vectorized algorithm.

1 Introduction

There is a revival of vector architecture in high end computing systems. The Earth Simulator[1] consists of 640 NEC SX-6 vector processors and is capable of sustaining over 35 Tflops/s on LINPACK benchmark. It was the fastest machine in the TOP500[2] list in 2002 and 2003. The Cray X1[3] series of vector processors are also serious contenders for a 100 Tflops/s machine in the National Leadership Computing Facility (NLCF) to be built at the Oak Ridge National Laboratory (ORNL). Vector machines have the characteristic that long regular vector operations are required to achieve high performance. The performance gap between vectorized and non-vectorized scalar code may be an order of magnitude or more.

The solution of sparse linear systems using a preconditioned iterative method forms the core computational kernel for many applications. This work of developing a vectorized matrix-vector multiply algorithm was motivated by issues arising

---

[*] This Research sponsored by the Laboratory Directed Research and Development Program of Oak Ridge National Laboratory (ORNL). This research used resources of the Center for Computational Sciences at the Oak Ridge National Laboratory, which is supported by the Office of Science of the U.S. Department of Energy under Contract No. DE-AC05-00OR22725.
[1] See http://www.es.jamstec.go.jp/ for details.
[2] See http://www.top500.org for details.
[3] See http://www.cray.com/products/x1/ for details.

from porting finite element codes that make extensive use of the PETSc [1] library framework for solving linear systems on the Cray X1 vector supercomputer.

Matrix-vector multiply, triangular solves and incomplete LU factorization are common computational kernels of the linear solver. The compressed sparse row storage (CSR) format is used in PETSc and achieves good performance on scalar architectures such as the IBM Power 4. However, it is difficult to achieve high performance on vector architectures using the straight-forward implementation of matrix-vector multiply with CSR. A vectorized iterative linear solver was developed for the Earth Simulator for solving sparse linear equations from finite elements modeling [2]. The algorithm used jagged diagonal storage (JAD) format with multi-coloring of nodes (or rows) to expose independent operations and parallelism. Another approach for sparse matrix multiply on the Cray C90 vector machines was based on fast prefix sum and segment scan [3]. The algorithm was fairly complicated and may require coding in assembly language for best efficiency. Extra storage was also needed to hold the partial prefix sums. The SIAM books by Dongarra [4, 5] are useful references on solving linear equations on vector processors.

The main contribution of this work is the development of a simple vectorized algorithm to perform sparse matrix vector multiply in CSR format *without* data rearrangement. This makes it attractive to implement such a vectorized algorithm in a sophisticated library framework such as PETSc. Numerical experiments on the Cray X1 show an order of magnitude improvement in sparse matrix multiply over the non-vectorized scalar implementation.

The background of vectorizing sparse matrix multiply is contained in Section 2. The new algorithm, compressed sparse row storage with permutation (CSRP), is described in Section 3. Results of numerical experiments on the Cray X1 are described in Section 4.

2 Background

There are many ways to store a general sparse matrix [6, 7]. The commonly used sparse matrix storage format for general nonsymmetric sparse matrices include the compressed row storage (CSR), ELLPACK-ITPACK [8] (ELL) and jagged diagonal (JAD) format.

In CSR, the matrix multiply operation, $y = A * x$, is described in Fig. 1. Here JA contains the column indices, A contains the nonzero entries, and IA points to the beginning of each row. The algorithm is efficient on scalar processors since it has unit stride access for A and JA. Moreover, the variable YI can be held in fast registers. Each visit through the inner loop performs one addition and one multiplication but requires memory fetches for A(J), JA(J), and X(JA(J)). The computational efficiency is usually limited by memory bandwidth and the actual attainable performance is only a small fraction of peak performance for the processor. On a vector machine, the vectorization across the row index J is limited by the number of nonzeros (IEND-ISTART+1) per row, which may be only 7 for a regular finite difference stencil in three dimensions.

```
1   DO I=1,N
2     ISTART = IA(I);IEND = IA(I+1)-1
3     YI = 0.0
4     DO J=ISTART,IEND
5       YI = YI + A(J) * X( JA(J) )
6     ENDDO
7     Y(I) = YI
8   ENDDO
```

Fig. 1. Matrix multiply in CSR format

```
1   Y(1:N) = 0.0
2   DO J=1,NZ
3     Y(1:N) = Y(1:N) +  A(1:N,J)*X( JA(1:N,J) )
4   ENDDO
```

Fig. 2. Matrix multiply in ELLPACK format

If every row of the matrix has approximately equal number of nonzeros, then the ELL format is more efficient. The nonzero entries and column indices are stored in rectangular N × NZ arrays, where N is the number of rows and NZ is the maximum number of nonzeros per row. The computation has more regularity and is easily optimized by vectorizing compilers. One algorithm (see Fig. 2) vectorizes along the rows to give long vectors. However, it will incur more traffic to memory since the vector Y will be repeatedly read in and written out again. Another variant (see Fig. 3) uses a "strip-mining" approach to hold a short array YP in vector registers. This will avoid repeated reading and writing of the Y vector. However, the ELL format is not suitable for matrices with widely different number of nonzeros per row. This would lead to wasted storage and unnecessary computations.

The jagged diagonal format may be consider a more flexible version of ELL (see Fig. 4). The rows are permuted or sorted in increasing number of nonzeros (see Fig. 5) and the data rearranged to form long vectors. Conceptually the vectorization is down along the rows but a column-oriented variant might also be efficient. If a matrix is already available in CSR format, extra storage is required to copy and convert the matrix into JAD format. This may be a significant drawback if the application is already trying to solve the largest problem possible.

3 CSR with Permutation (CSRP)

The vectorizable algorithm proposed here performs the matrix vector multiply operation using the CSR format but with a permutation vector such that rows with the same number of nonzeros are grouped together. Conceptually, this may be considered a variant of the JAD format. The algorithm is also similar to the ELL algorithm where "strip-mining" is employed to reuse vector registers. The

```
1   DO I=1,N,NB
2       IEND = MIN(N,I+NB-1)
3       M = IEND-I+1
4       YP(1:M) = 0.0
5   ! ------------------------------------------------
6   ! Consider YP(1:M) as vector registers
7   ! NB is multiple of the size of vector registers
8   ! ------------------------------------------------
9       DO J=1,NZ
10          YP(1:M) = YP(1:M) + A(I:IEND,J) * X( JA(I:IEND,J) )
11      ENDDO
12      Y(I:IEND) = YP(1:M)
13  ENDDO
```

Fig. 3. Variant of matrix multiply in ELLPACK format

```
1   Y(1:N) = 0.0
2   IP = 1
3   DO J=1,NZ
4       I = ISTART(J)
5       M = N - I + 1
6       Y(I:N) = Y(I:N) + A(IP:(IP+M-1)) * X( JA(IP:(IP+M-1)) )
7       IP = IP + M
8   ENDDO
```

Fig. 4. Matrix multiply in JAD format

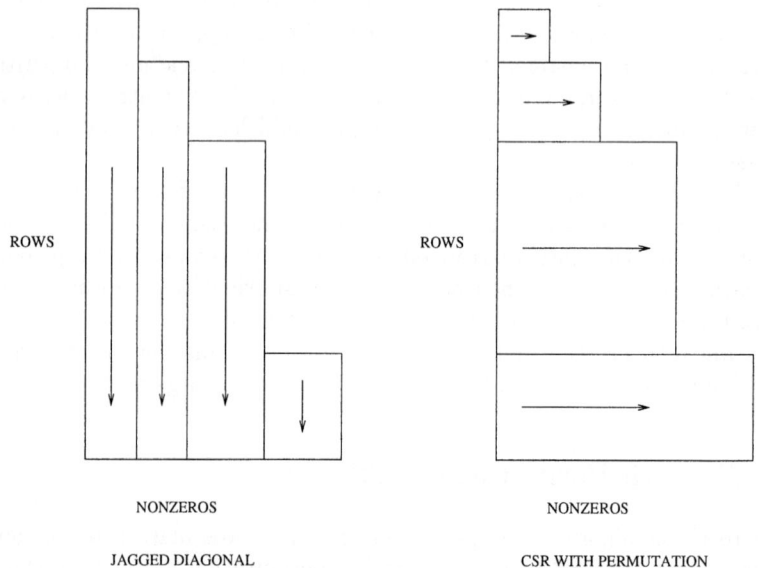

Fig. 5. Matrix profile after rows are permuted in increasing number of nonzeros

key difference is the data remain in place and are accessed indirectly using the permutation vector. The major drawback for leaving the data in place is the irregular access to arrays A and JA. The algorithm is described in Fig. 6. Here

```
1    DO IGROUP=1,NGROUP
2      JSTART = XGROUP(IGROUP)
3      JEND = XGROUP(IGROUP+1)-1
4      NZ = NZGROUP(IGROUP)
5    ! -----------------------------------------------------------
6    ! Rows( IPERM(JSTART:JEND) ) all have same NZ nonzeros per row
7    ! -----------------------------------------------------------
8      DO I=JSTART,JEND,NB
9        IEND = MIN(JEND,I+NB-1)
10       M = IEND - I + 1
11       IP(1:M) = IA( IPERM(I:IEND) )
12       YP(1:M) = 0.0
13   ! ----------------------------------------
14   ! Consider YP(:), IP(:) as vector registers
15   ! ----------------------------------------
16       DO J=1,NZ
17         YP(1:M) = YP(1:M) + A( IP(1:M) ) * X( JA(IP(1:M)) )
18         IP(1:M) = IP(1:M) + 1
19       ENDDO
20     ENDDO
21     Y( IPERM(I:IEND) ) = YP(1:M)
22   ENDDO
```

Fig. 6. Matrix multiply for CSR with permutation

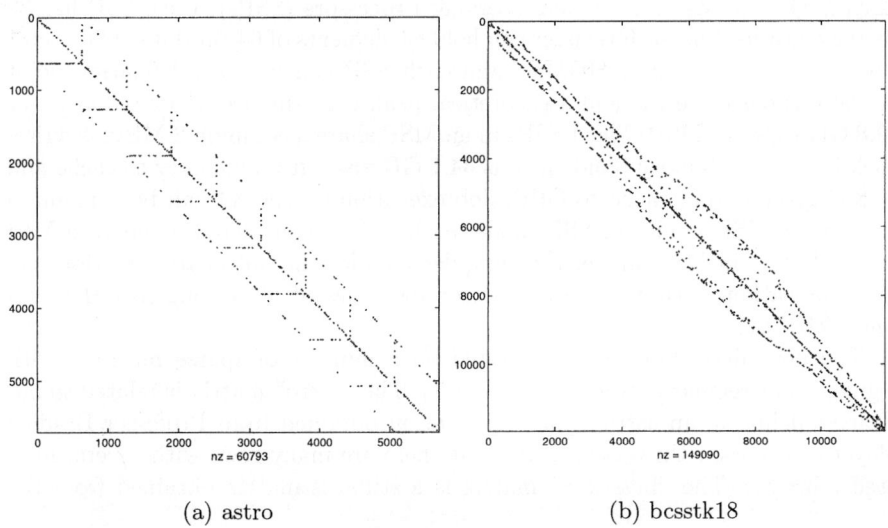

(a) astro   (b) bcsstk18

Fig. 7. Sparsity patterns for test matrices

IPERM is the permutation vector, XGROUP points to beginning indices of groups in IPERM. If extra storage is available, it is feasible to copy and convert each group of rows into ELLPACK format (CSRPELL) to get better performance since access to A and JA would be unit-stride. This would be equivalent to the ITPACK permuted blocks (ITPER) format investigated by Peters [9] for the IBM 3090VF vector processor. For CSRP to be effective, there is an implicit assumption that there are many rows with the same number of nonzeros. This is often the case for sparse matrices arising from mesh based finite element or finite difference codes.

The permutation vector can be easily constructed using bucket sort and two passes over the IA vector in $O(N)$ work (N is the number of rows). The algorithm only requires knowing the number of nonzeros per row and does not need to examine the larger JA array.

4 Numerical Experiments on the Cray X1

The CSRP algorithm for sparse matrix multiply has been implemented in the Mat_SeqAIJ "class"[4] in version 2.2.1 of PETSc. The parallel distributed memory sparse matrix class locally uses Mat_SeqAIJ on each processor. Therefore testing with just the sequential matrix class yields important insights in the effectiveness of vectorization with CSRP. The permutation vector is generated in MatAssemblyEnd_SeqAIJ and destroyed in MatDestroy_SeqAIJ. The procedure MatMult_SeqAIJ is modified to use the CSRP and CSRPELL algorithms. The implementation uses C (with Fortran kernels) to minimize the changes to PETSc in only the files aij.h and aij.c.

This implementation has been tested in sequential mode on the Cray X1 at the Center for Computational Sciences at the Oak Ridge National Laboratory. The processing units on the X1 consist of Multi-Streaming Processors (MSPs). Each MSP consists of 4 Single-Streaming Processors (SSPs). Each SSP has 32 vector registers, and each register can hold 64 elements of 64-bit data. The vector clock on an SSP runs at 800MHz and each SSP can perform 4 floating point operations per cycle to yield a theoretical peak performance of 3.2 Gflops/s, or 12.8 Gflops/s per MSP. The 4 SSPs in an MSP share a common 2 MBytes writeback L2 cache. Memory bandwidth is 34.1 GBytes/s from memory to cache and 76.8 GBytes/s from cache to CPU. Job execution on the X1 can be configured for SSP or MSP mode. In SSP mode, each SSP is considered a separate MPI task; whereas in MSP mode, the compiler handles the automatic creation and synchronization of threads to use vector resources of all 4 coupled SSPs as a single MPI task.

The new algorithm has been tested on a number of sparse matrices with regular and irregular patterns (see Table 1). The "astro" matrix is related to nuclear modeling in an astrophysics application obtained from Professor Bradley Meyer at Clemson University. Note that there are many rows with several hundred nonzeros. The "bcsstk18" matrix is a stiffness matrix obtained from the

---

[4] PETSc is written in C in a disciplined object-oriented manner.

Table 1. Description of matrices

Name	N	Nonzeros	Description
astro	5706	60793	Nuclear Astrophysics problem from Bradley Meyer
bcsstk18	11948	149090	Stiffness matrix from Harwell Boeing Collection
7pt	110592	760320	7 point stencil in 48 × 48 × 48 grid
7ptb	256000	7014400	4 × 4 blocks 7-pt stencil in 40 × 40 × 40 grid

Table 2. Performace (in MFlops/s) of sparse matrix multiply using CSR, CSRP and CSRPELL in PETSc

	SSP			MSP		
Problem	CSR	CSRP	CSRPELL	CSR	CSRP	CSRPELL
astro	26	163	311	14	214	655
bcsstk18	28	315	340	15	535	785
7pt	12	259	295	8	528	800
7ptb	66	331	345	63	918	1085

Harwell-Boeing collection of matrices[5]. The "7pt" matrix is constructed from a 7 point stencil on a 48 × 48 × 48 rectangular mesh. The "7ptb" matrix is similarly constructed using 4 × 4 blocks from a 7 point stencil on a 40 × 40 × 40 grid. Table 2 shows the performance (in Mflops/s) for the original CSR algorithm versus the new vectorizable CSRP and CSRPELL algorithms in SSP and MSP modes. The Megaflop rate is computed by timing over 100 calls to PETSc `MatMult`. The data suggest the CSRP algorithm is an order of magnitude faster than the original non-vectorizable algorithm. The CSRP with ELLPACK format (CSRPELL) algorithm achieves even better performance with unit stride access to data, but requires rearranging the data into another copy of the matrix. However, even with this rearrangement, the CSRPELL algorithm achieves less than 11% of theoretical peak performance of the Cray X1 in SSP and MSP modes.

5 Summary

We have presented a simple vectorizable algorithm for computing sparse matrix vector multiply in CSR storage format. Although the method uses non-unit stride access to the data, it is still an order of magnitude faster than the original scalar algorithm. The method requires no data rearrangement and can be easily incorporated into a sophisticated library framework such as PETSc. Further work is still required in vectorizing the generation of sparse incomplete LU factorization and the forward and backward triangular solves.

---

[5] Available at http://math.nist.gov/MatrixMarket/data/.

References

1. Balay, S., Buschelman, K., Eijkhout, V., Gropp, W.D., Kaushik, D., Knepley, M.G., McInnes, L.C., Smith, B.F., Zhang, H.: PETSc users manual. Technical Report ANL-95/11 - Revision 2.2.0, Argonne National Laboratory (2004) See also http://www.mcs.anl.gov/petsc.
2. Nakajima, K., Okuda, H.: Parallel iterative solvers for finite-element methods using a hybrid programming model on SMP cluster architectures. Technical Report GeoFEM 2003-003, RIST, Tokyo (2003) See also http://geofem.tokyo.rist.or.jp/members/nakajima.
3. Blelloch, G.E., Heroux, M.A., Zagha, M.: Segmented operations for sparse matrix computation on vector multiprocessors. Technical Report CMU-CS-93-173, Department of Computer Science, Carnegie Mellon University (1993) See also http://www-2.cs.cmu.edu/~guyb/publications.html.
4. Dongarra, J.J., Duff, I.S., Sorensen, D.C., van der Vorst, H.A.: Solving Linear Systems on Vector and Shared Memory Computers. SIAM, Philadelphia, PA (1991)
5. Dongarra, J.J., Duff, I.S., Sorensen, D.C., van der Vorst, H.: Numerical Linear Algebra for High-Performance Computers. SIAM, Philadelphia, PA (1998)
6. Duff, I.S., Erisman, A.M., Reid, J.K.: Direct Methods for Sparse Matrices. Clarendon Press, Oxford (1986)
7. Saad, Y.: Iterative Methods for Sparse Linear Systems. Second edn. SIAM, Philadelphia, PA (2003) See also http://www-users.cs.umn.edu/~saad/books.html.
8. Kincaid, D.R., Young, D.M.: The ITPACK project: Past, present and future. In Birkhoff, G., Schoernstadt, A., eds.: Elliptic Problem Solvers II Proc. (1983) 53–64
9. Peters, A.: Sparse matrix vector multiplication techniques on the IBM 3090 VF. Parallel Computing **17** (1991) 1409–1424

A Multipole Based Treecode Using Spherical Harmonics for Potentials of the Form $r^{-\lambda}$*

Kasthuri Srinivasan**, Hemant Mahawar, and Vivek Sarin

Department of Computer Science, Texas A&M University,
College Station, TX, U.S.A.
{kasthuri, mahawarh, sarin}@cs.tamu.edu

Abstract. In this paper we describe an efficient algorithm for computing the potentials of the form $r^{-\lambda}$ where $\lambda \geq 1$. This treecode algorithm uses spherical harmonics to compute multipole coefficients that are used to evaluate these potentials. The key idea in this algorithm is the use of Gegenbauer polynomials to represent $r^{-\lambda}$ in a manner analogous to the use of Legendre polynomials for the expansion of the Coulomb potential r^{-1}. We exploit the relationship between Gegenbauer and Legendre polynomials to come up with a natural generalization of the multipole expansion theorem used in the classical fast multipole algorithm [2]. This theorem is used with a hierarchical scheme to compute the potentials. The resulting algorithm has known error bounds and can be easily implemented with modification to the existing fast multipole algorithm. The complexity of the algorithm is $O(p^3 N \log N)$ and has several advantages over the existing Cartesian coordinates based expansion schemes.

1 Introduction

Computing the potentials of the form $r^{-\lambda}$ where $\lambda \geq 1$ is an important task in many fields like molecular dynamics [10], computational chemistry [9] and fluid mechanics [11]. Potentials such as Lennard-Jones, Van der Wall's forces and H-bonds require evaluation of functions of the form r^{-6}, r^{-10}, etc. The naive brute force algorithm based on particle-particle interaction takes $O(N^2)$, where N is the number of particles. There are various approximation algorithms with lower complexity that have been proposed for the potential evaluation problem. Appel's algorithm [12], Barnes-Hut algorithm [1] etc. reduce the complexity to $O(N \log N)$ by making use of the cluster-particle interactions. As opposed to brute force algorithm that evaluates the potential exactly, approximation algorithms estimate the potential to the desired accuracy that can be controlled by certain parameters. The fast multipole method (FMM), proposed by Greengard and Rokhlin [2], is the fastest contemporary approximation algorithm to

---
* This work has been supported in part by NSF under the grant NSF-CCR0113668, and by the Texas Advanced Technology Program grant 000512-0266-2001.
** Corresponding author.

solve the potential evaluation problem without losing much accuracy. FMM is based on cluster-cluster interaction of the particle system and has a complexity $O(N)$. Although FMM has complexity that is linear in the number of particles in the system, its applicability is limited to operators whose multipole expansions are available analytically. In absence of such expansions treecodes such as Barnes-Hut, Appel's algorithm may be used [3].

For potentials of the form $\Phi(r) = r^{-1}$, FMM exploits the separability of the Greens function kernel using spherical harmonics. However, for kernels of the form $r^{-\lambda}, \lambda \geq 1$, advances have been made by using Cartesian coordinates only. Duan et al. [3] propose a treecode which uses Gegenbauer polynomials and recurrence relations. Chowdhury et al. [5] recent approach generalizes single level fast multipole algorithm for these potentials using addition theorem for Gegenbauer polynomials. They develop necessary operators and error bounds based on these addition theorems. Although, their approach can be used for $r^{-\lambda}$ kernels, it is restricted to single level multipole expansions. The algorithm by Elliott et al. [4] uses multivariate Taylor expansions. Although, these methods are designed to compute such potentials, they strongly make use of Cartesian coordinates and special recurrence relations specific to their schemes.

In this paper we propose a multipole based treecode that uses spherical harmonics to evaluate potentials of the form $r^{-\lambda}$, where $\lambda \geq 1$. This treecode has several advantages over existing Cartesian coordinate based expansion schemes. First, the use of spherical harmonics gives an analytic formula to compute multipole coefficients efficiently. This analytic formula is a natural generalization of the multipole expansion theorem for r^{-1} potential. Second, any previous implementation of FMM can be modified to evaluate $r^{-\lambda}$ potentials using this approach. Third, our method allows precomputing certain vectors that significantly reduce the time to compute multipole coefficients. Finally, there is no need to use special recurrence relations or spherical harmonics other than the ones used in FMM.

The paper is organized as follows: Section 2 discusses the multipole expansion theorem for r^{-1} functions. Section 3 describes the Gegenbauer polynomials and their relationship with Legendre polynomials. Section 4 describes a tree code based on the hierarchical multipole method. Section 5 discusses the complexity and implementation issues. Conclusions are presented in Section 6.

2 Multipole Expansion Theorem for r^{-1} Potential

This section describes the multipole expansion theorem from the classical FMM [2]. Let $P(r, \theta, \phi)$ and $Q(\rho, \alpha, \beta)$ be two points with spherical coordinates. Also let $P - Q = R(r', \theta', \phi')$ and γ be the angle between P and Q taken in anticlockwise direction. Then, if $\rho < r$, the potential $\Phi(P)$ at P due to a charge q at Q is given by

$$\Phi(P) = \frac{q}{r'} = \sum_{n=0}^{\infty} \frac{q \cdot \rho^n}{r^{n+1}} P_n(\cos \gamma), \tag{1}$$

where $P_n(\cos\gamma)$ is the Legendre polynomial of degree n. We also have the addition theorem for Legendre polynomials

$$P_n(\cos\gamma) = \sum_{m=-n}^{n} Y_n^{-m}(\alpha,\beta) Y_n^m(\theta,\phi), \tag{2}$$

where Y_n^m are the spherical harmonics given by

$$Y_n^m(x,y) = \sqrt{\frac{n-|m|}{n+|m|}} P_n^m(\cos x) e^{imy}, \tag{3}$$

in which P_n^m are the associated Legendre functions evaluated at $\cos x$. Using (1) and (2) we have the following theorem.

Theorem 1. *Suppose that a charge of strength q is located at $Q(\rho,\alpha,\beta)$ with $\rho < a$. At any point $P(r,\theta,\phi)$ with $r > a$, the potential $\Phi(P)$ is given by*

$$\Phi(P) = \frac{q}{r'} = \sum_{n=0}^{\infty} \sum_{m=-n}^{n} \frac{M_n^m}{r^{n+1}} Y_n^m(\theta,\phi), \tag{4}$$

where $M_n^m = q\rho^n Y_n^{-m}(\alpha,\beta)$.

If there are several charges $\{q_i : i = 0, 1, \ldots k\}$ around Q with coordinates $\{(\rho_i,\alpha_i,\beta_i) : i = 1, 2, \ldots k\}$ we can superpose them at Q and the resulting multipole moment at Q would be $M_n^m = \sum_{i=0}^{k} q_i \rho_i^n Y_n^{-m}(\alpha_i,\beta_i)$.

3 Gegenbauer Polynomials

Gegenbauer polynomials are *higher dimensional* generalization of Legendre polynomials [7,8]. Gegenbauer polynomials are eigenfunctions of the generalized angular momentum operator just as Legendre polynomials are eigenfunctions of the angular momentum operator in three dimensions. Gegenbauer polynomials allow addition theorem using hyperspherical harmonics similar to Legendre polynomials which allow addition theorem using spherical harmonics. One can refer to [7] for list of similarities between Gegenbauer and Legendre polynomials. Gegenbauer polynomials are also a generalization of Legendre polynomials in terms of the underlying generating function. Thus, if $x, y \in \Re$, then

$$\frac{1}{(1-2xy+y^2)^{\lambda/2}} = \sum_{n=0}^{\infty} C_n^\lambda(x) y^n, \tag{5}$$

where $C_n^\lambda(x)$ is a Gegenbauer polynomial of degree n. This generating function can be used to expand $r^{-\lambda}$ using Gegenbauer polynomials. Let $P(r,\theta,\phi)$ and $Q(\rho,\alpha,\beta)$ be points with spherical coordinates. Using (5) we have,

$$\frac{1}{(r')^\lambda} = \frac{1}{r^\lambda(1-2u\mu+\mu^2)^{\lambda/2}} = \sum_{n=0}^{\infty} \frac{\rho^n}{r^{n+\lambda}} C_n^\lambda(u), \tag{6}$$

where $\mu = \rho/r$ and $u = \cos\gamma$. There are addition theorems for Gegenbauer polynomials using hyperspherical harmonics which allow us to separate $C_n^\lambda(u)$ in terms of the coordinates, but we require hyperspherical harmonics. Instead, we use a relation that exists between Gegenbauer and the Legendre polynomials [6] which allows us to use spherical harmonics in three dimensions. Let $P_n(x)$ and $C_n^\lambda(x)$ be Legendre and Gegenbauer polynomials, of degree n, respectively, Then

$$C_n^\lambda(x) = \sum_{s=0}^{\lfloor n/2 \rfloor} \frac{(\lambda)_{n-s}(\lambda - 1/2)_s}{(3/2)_{n-s} s!}(2n - 4s + 1)P_{n-2s}(x) \tag{7}$$

where $(p)_s$ is the Pochhammer symbol. We define

$$B_{n,s}^\lambda = \frac{(\lambda)_{n-s}(\lambda - 1/2)_s}{(3/2)_{n-s} s!}(2n - 4s + 1). \tag{8}$$

Using (2) and (7), we derive an addition theorem for Gegenbauer polynomials. The following lemma is found in abstract form in many places [7, 8]. A proof is omitted to conserve space.

Lemma 1. (Addition Theorem for Gegenbauer Polynomials) *Let P and Q be points with spherical coordinates (r, θ, ϕ) and (ρ, α, β), respectively, and let γ be the angle subtended by them at the origin. Then*

$$C_n^\lambda(\cos\gamma) = \sum_{m=0}^{\lfloor n/2 \rfloor} B_{n,m}^\lambda \mathbf{Y_{n,m}}(\theta, \phi) \cdot \overline{\mathbf{Y_{n,m}}}(\alpha, \beta) \tag{9}$$

where $\mathbf{Y_{n,m}^T}(x,y) = [Y_{n-2m}^{-(n-2m)}, Y_{n-2m}^{-(n-2m)+1}, \ldots, Y_{n-2m}^{(n-2m)}]$ is a vector of spherical harmonics of degree $n - 2m$.

Once we have an addition theorem for Gegenbauer polynomials we can prove the multipole expansion theorem for $r^{-\lambda}$ potentials.

Theorem 2. (Multipole Expansion) *Suppose that k charges of strengths $\{q_i, i = 1, \ldots k\}$ are located at the points $\{Q_i = (\rho_i, \alpha_i, \beta_i), i = 1, \ldots, k\}$, with $|\rho_i| < a$. Then for any point $P = (r, \theta, \phi)$ with $r > a$, the potential $\Phi(P)$ is given by*

$$\Phi(P) = \sum_{n=0}^{\infty} \sum_{m=0}^{\lfloor n/2 \rfloor} \frac{1}{r^{n+\lambda}} \mathbf{M_n^m} \cdot \mathbf{Y_{n,m}}(\theta, \phi) \tag{10}$$

where

$$\mathbf{M_n^m} = \sum_{i=1}^{k} q_i \rho_i^n B_{n,m}^\lambda \overline{\mathbf{Y_{n,m}}}(\alpha_i, \beta_i) \tag{11}$$

Furthermore, for any $p \geq 1$,

$$\left| \Phi(P) - \sum_{n=0}^{p} \sum_{m=0}^{\lfloor n/2 \rfloor} \frac{\mathbf{M_n^m}}{r^{n+\lambda}} \cdot \mathbf{Y_{n,m}}(\theta, \phi) \right| \leq \frac{AB}{r^{\lambda-1}(r-a)} \left(\frac{a}{r}\right)^{p+1} \tag{12}$$

where $A = \sum_{i=1}^{k} |q_i|$ and $B = \sum_{m=0}^{\lfloor n/2 \rfloor} |B_{n,m}^\lambda|$

Proof. From equation (6) and Lemma (1), for any q_i at Q_i we have

$$\Phi(P) = \sum_{n=0}^{\infty} \frac{q_i \rho_i^n}{r^{n+\lambda}} C_n^\lambda(u)$$

$$= \sum_{n=0}^{\infty} \sum_{m=0}^{\lfloor n/2 \rfloor} \frac{[q_i \rho_i^n B_{n,m}^\lambda \overline{Y_{n,m}}(\alpha_i, \beta_i)]}{r^{n+\lambda}} Y_{n,m}(\theta, \phi)$$

The moments in (11) are obtained by superposition.

We now prove the error bound. Note that for every $u \in \Re$ with $|u| \leq 1$, we have $|P_n(u)| \leq 1$. From (7), $|C_n^\lambda| \leq B$ where $B = \sum_{m=0}^{\lfloor n/2 \rfloor} |B_{n,m}^\lambda|$ (using triangle inequality). Now, for each q_i located at Q_i having $|\rho_i| < a$, we have

$$\left| \Phi(P) - \sum_{n=0}^{p} \frac{q_i \rho_i^n}{r^{n+\lambda}} C_n^\lambda(u) \right| = \left| \sum_{n=p+1}^{\infty} \frac{q_i \rho_i^n}{r^{n+\lambda}} C_n^\lambda(u) \right|$$

$$\leq \frac{B}{r^{\lambda-1}} \frac{q_i}{r-a} \left(\frac{a}{r} \right)^{p+1}$$

The error bound (12) is obtained by superposition of error bounds for all the k charges.

4 A Treecode for $r^{-\lambda}$ Potentials

The treecode can be viewed either as a variant of Barnes-Hut algorithm [1] or FMM [2] that uses only particle-cluster potential evaluations. The method works in two phases: The tree construction phase and the potential computation phase. In tree construction phase, a spatial tree representation of the domain is derived. At each step of this phase, if the domain contains more than s particles, where s is a preset constant, it is recursively divided into eight equal sub-domains. This process continues until each sub-domain has at most s elements. The resulting tree is an unstructured oct-tree. Each internal node in the tree computes and stores multipole series representation of the particles. Since we don't have any translations, at each level of the tree and for every sub-domain in that level we compute the multipole coefficients of all the particles contained in the sub-domain. These coefficients are obtained using the theorem given in the preceding section. Once the tree has been constructed, the potential at any point can be computed as follows: a *multipole acceptance criterion* is applied to the root of the tree to determine if an interaction can be computed; if not, the node is expanded and the process is repeated for each of its eight children. The multipole acceptance criterion computes the ratio of the distance of the point from the center of the box to the dimension of the box. If the ratio is greater than α, a specific constant, an interaction can be computed. In the following pseudo-code for computing the multipole coefficients and potential evaluation, p is the pre-specified multipole degree and s is the minimum number of particles contained in any leaf box and α, a constant.

```
Multipole_Calculation(p)
```
- For(each level in the Oct-tree)
    - For(each node in the level)
        * Find the multipole coefficients of all the particles in the node using Theorem 2 with respect to the box center of the node

```
Potential_Evaluation()
```
- For(each particle)
    - nodes = Alpha_Criteria(particle,root)
    - If(nodes = leaf)
        * Compute potentials directly
    - Else
        * Use the multipole expansion Theorem 2 to find the potential
    - Add direct and the computed potential

```
nodes = Alpha_Criteria(particle,node)
```
- *ratio* = distance of the particle from the box center/box length
- If($ratio > \alpha$)
    - return nodes;
- Else If(node = leaf)
    - return leafnodes;
- Else
    - For(each children nodes)
        * nodes = Alpha_Criteria(particle,node)
        * return nodes;

5 Complexity and Implementation

It can be seen from the multipole expansion theorem that the complexity for computing the multipole coefficients at each level of the tree is $O(p^3 N)$, where N is the number of particles in the system and p is the multipole degree. Since there are $\log N$ levels in the tree, the total cost of computing the multipole coefficients is $O(p^3 N \log N)$. Similarly, it can be seen that the complexity for the potential evaluation phase is $O(p^3 N \log N)$. Thus, the overall complexity for the algorithm is $O(p^3 N \log N)$.

We have experimentally verified the theorem for r^{-6} and r^{-10} potentials. The logarithmic plot (Fig 1 and Fig 2) shows the relative error and the error bound as multipole degree p varies from 1 to 10. For an efficient implementation of this algorithm, here are few pointers.

- An existing FMM code can be appropriately modified to compute the multipole coefficients in this algorithm.
- The Gegenbauer constants can be precomputed and used in the potential evaluation phase to reduce computation time of the multipole coefficients.

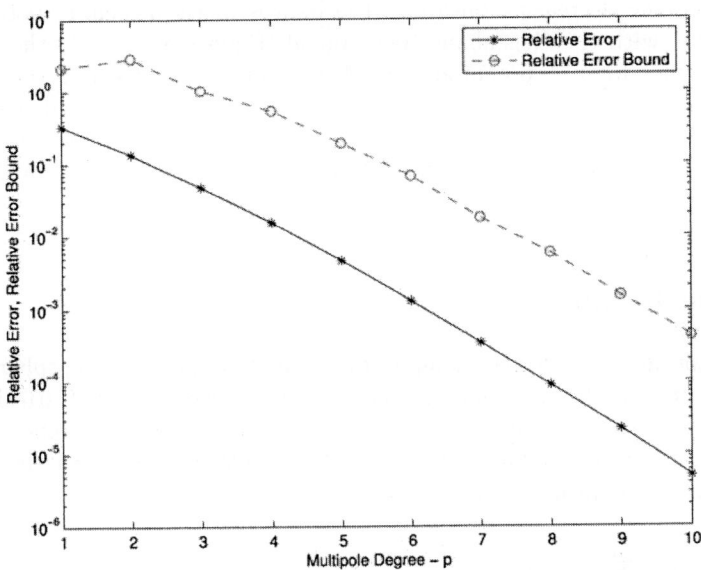

Fig. 1. Relative error and error bound for the potential $\Phi(r) = r^{-6}$

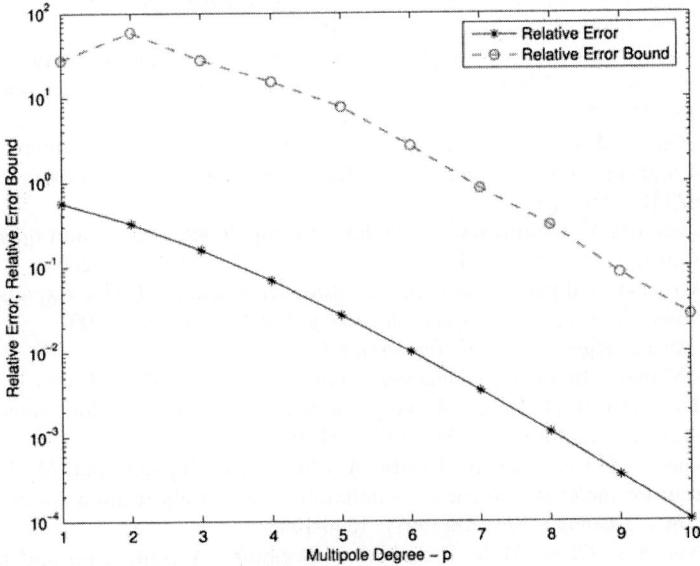

Fig. 2. Relative error and error bound for the potential $\Phi(r) = r^{-10}$

– Due to the decreasing nature of the indices in the Legendre polynomial's relation with the Gegenbauer polynomial (Equation 7), only the spherical harmonics for r^{-1} potential is needed. They can be used to the compute multipole coefficients for the $r^{-\lambda}$ potentials. It should be noted that the set spherical harmonics for $r^{-\lambda}$ potentials are the same as the set of spherical harmonics for r^{-1}, but with the shift $n-2m$ for $m = 0, 1, ... \lfloor n/2 \rfloor$ as Lemma 1 shows.

6 Conclusion

An efficient algorithm for the computation of $r^{-\lambda}$ potentials using spherical harmonics is presented. This algorithm has advantages over existing Cartesian coordinate based expansion schemes. Since, there are no special recurrence relations or spherical harmonics required, an FMM code can be appropriately modified to implement the proposed algorithm.

References

1. J. Barnes and P. Hut. A hierarchical O($n\ log\ n$) force calculation algorithm. *Nature*, Vol. 324:446–449, 1986.
2. L. Greengard. *The Rapid Evaluation of Potential Fields in Particle Systems*. The MIT Press, Cambridge, Massachusetts, 1988
3. Z. Duan, R. Krasny. An adaptive treecode for computing nonbonded potential energy in classical molecular systems. *Journal of Computational Chemistry*, Vol.22, No.2, 184-195 (2001)
4. W.D. Elliott, J.A. Board. Multipole algorithms for molecular dynamics simulation on high performance computers. http://www.ee.duke.edu/ wrankin/SciComp/ Papers/TR95-003.html.
5. I. Chowdhury, V. Jandhyala. Single level multipole expansions and operators for potentials of the form $r^{-\lambda}$. In *UWEE Technical Report*, March 2004.
6. E.J. Weniger. Addition Theorems as Three-Dimensional Taylor Expansions. *International Journal of Quantum Chemistry*, Vol.76, 280-295 (2000)
7. John Avery. *Hyperspherical Harmonics*.
8. Claus Müller. *Analysis of Spherical Symmetries in Euclidean Spaces*.
9. G.H. Kho and W.H. Fink. Rapidly converging lattice sums for nanoelectronic interactions. *J.Comp.Chem*, 23(2001), 447-483
10. Y. Komeiji, M. Uebayasi, R. Takata, A. Shimizu, K. Itsukahi and M. Taiji. Fast and accurate molecular dynamics simulation of a protein using a special purpose computer . *J.Comp.Chem*, 18(1997), 1546-1563
11. X.B. Nei, S.Y. Chen, W.N. E and M.O. Robbins. A continuum and molecular dynamics hybrid method for micro- and nano- fluid flow. *J.Fluid.Mech.*, 500(2004), 55-64
12. A. Appel. An Efficient Program for Many-Body Simulations. *SIAM J Sci Comput*, 1985,6, 85-103

Numerically Stable Real Number Codes Based on Random Matrices*

Zizhong Chen and Jack Dongarra

Computer Science Department, University of Tennessee, Knoxville,
1122 Volunteer Blvd., Knoxville, TN 37996-3450, USA
{zchen, dongarra}@cs.utk.edu

Abstract. Error correction codes defined over real-number field have been studied and recognized as useful in many applications. However, most real-number codes in literature are quite suspect in their numerical stability. In this paper, we introduce a class of real-number codes based on random generator matrices over real-number fields. Codes over complex-number field are also discussed. Experiment results demonstrate our codes are numerically much more stable than existing codes in literature.

1 Introduction

Error correction codes are often defined over finite fields. However, in many applications, error correction codes defined over finite fields do not work. Instead, codes defined over real-number or complex-number fields have to be used to detect and correct errors. For example, in algorithm-based fault tolerance [2, 10, 11, 13] and fault tolerant dynamic systems [8], to provide fault tolerance in computing, data are first encoded using error correction codes and then algorithms are re-designed to operate (using floating point arithmetic) on the encoded data. Due to the impact of the floating-point arithmetic on the binary representation of these encoded data, codes defined over finite fields do not work. But codes defined over real-number and complex-number fields can be used in these applications to correct errors in computing by taking advantage of certain relationships, which are maintained only when real-number (or complex-number) codes are used.

However, most real-number and complex-number codes in literature are quite suspect in their numerical stability. Error correction procedures in most error correction codes involve solving linear system of equations. In computer floating point arithmetic where no computation is exact due to round-off errors, it is

---

* This research was supported in part by the Los Alamos National Laboratory under Contract No. 03891-001-99 49 and the Applied Mathematical Sciences Research Program of the Office of Mathematical, Information, and Computational Sciences, U.S. Department of Energy under contract DE-AC05-00OR22725 with UT-Battelle, LLC.

well known [7] that, in solving a linear system of equations, a condition number of 10^k for the coefficient matrix leads to a loss of accuracy of about k decimal digits in the solution. In the generator matrices of most existing real-number and complex-number codes, there exist ill-conditioned sub-matrices. Therefore, in these codes, when certain error patterns occur, an ill-conditioned linear system of equations has to be solved in the error correction procedure, which can cause the loss of precision of possibly all digits in the recovered numbers.

The numerical issue of the real-number and complex-number codes has been recognized and studied in some literature. In [2], Vandermonde-like matrix for the Chebyshev polynomials was introduced to relieve the numerical instability problem in error correction for algorithm-based fault tolerance. In [5, 6, 9, 12], the numerical properties of the Discrete Fourier Transform codes were analyzed and methods to improve the numerical properties were also proposed. To some extent, these efforts have alleviated the numerical problem of the real-number and complex-number codes. However, how to construct real-number and complex-number codes without numerical problem is still an open problem.

In this paper, we introduce a class of real-number and complex-number codes that are numerically much more stable than existing codes in literature. Our codes are based on random generator matrices over real-number and complex-number fields. The rest of this paper is organized as follow: Section 2 specifies the problem we focus on. In Section 3, we first study the properties of random matrices and then introduce our codes. Section 4 compares our codes with most existing codes in both burst error correction and random error correction. Section 5 concludes the paper and discusses the future work.

2 Problem Specification

Let $x = (x_1, x_2, ..., x_N)^T \in \mathcal{C}^N$ denote the original information, and G denote a M by N real or complex matrix. Let $y = (y_1, y_2, ..., y_M)^T \in \mathcal{C}^M$, where $M = N + K$, denote the encoded information of x with redundancy. The original information x and the encoded information y are related through

$$y = Gx. \qquad (1)$$

Our problem is: how to choose the matrix G such that, after any no more than K erasures in the elements of the encoded information y, a good approximation of the original information x can still be reconstructed from y?

When there are at most K elements of y lost, there are at least N elements of y available. Let J denote the set of indexes of any N available elements of y. Let y_J denote a sub-vector of y consisting of the N available elements of y whose indexes are in J. Let G_J denote a sub-matrix of G consisting of the N rows whose indexes are in J. Then, from (1), we can get the following relationship between x and y_J:

$$y_J = G_J x. \qquad (2)$$

When the matrix G_J is singular, there are infinite number of solutions to (2). But, if the matrix G_J is non-singular, then (2) has one and only one solution, which is the original information vector x.

In computer real-number and complex-number arithmetic where no computation is exact due to round-off errors, it is well known [7] that, in solving a linear system of equations, a condition number of 10^k for the coefficient matrix leads to a loss of accuracy of about k decimal digits in the solution. Therefore, in order to reconstruct a good approximation of the original information x, G_J has to be well-conditioned.

For any N by N sub-matrix G_J of G, there is a erasure pattern of y which requires to solve a linear system with G_J as the coefficient matrix to reconstruct an approximation of the original x. Therefore, to guarantee that a reasonably good approximation of x can be reconstructed after any no more than K erasures in y, the generator matrix G must satisfy: *any N by N sub-matrix of G is well-conditioned*.

3 Real Number Codes Based on Random Matrices

In this section, we will introduce a class of new codes that are able to reconstruct a very good approximation of the original information with high probability regardless of the erasure patterns in the encoded information. Our new codes are based on random matrices over real or complex number fields.

3.1 Condition Number of Random Matrices from Standard Normal Distribution

In this sub-section, we mainly focus on the probability that the condition number of a random matrix is large and the expectation of the logarithm of the condition number. Let $G(m,n)$ be an $m \times n$ real random matrix whose elements are independent and identically distributed standard normal random variables and $\widetilde{G}(m,n)$ be its complex counterpart.

Theorem 1. *Let κ denote the condition number of $G(n,n)$, $n > 2$, and $t \geq 1$, then*

$$\frac{0.13n}{t} < P(\kappa > t) < \frac{5.60n}{t}. \tag{3}$$

Moreover,

$$E(\log(\kappa)) = \log(n) + c + \epsilon_n, \tag{4}$$

where $c \approx 1.537$, $\lim_{n \to \infty} \epsilon_n = 0$,

Proof. The inequality (3) is from Theorem 1 of [1]. The formula (4) can be obtained from Theorem 7.1 of [3]. □

Theorem 2. Let $\widetilde{\kappa}$ denote the condition number of $\widetilde{G}(n,n)$, and $t \geq \sqrt{n}$, then

$$1 - \left(1 - \frac{1}{t^2}\right)^{n^2-1} \leq P(\widetilde{\kappa} > t) \leq 1 - \left(1 - \frac{n}{t^2}\right)^{n^2-1}. \tag{5}$$

Moreover,
$$E(\log(\widetilde{\kappa})) = \log(n) + c + \epsilon_n, \tag{6}$$

where $c \approx 0.982$, $\lim_{n \to \infty} \epsilon_n = 0$,

Proof. Let $\widetilde{\kappa}_D$ denote the scaled condition number (see [4] for definition) of $\widetilde{G}(n,n)$, then
$$P(\frac{\widetilde{\kappa}_D}{\sqrt{n}} > t) \leq P(\widetilde{\kappa} > t) \leq P(\widetilde{\kappa}_D > t). \tag{7}$$

From Corollary 3.2 in [4], we have
$$P(\widetilde{\kappa}_D > t) = 1 - \left(1 - \frac{n}{t^2}\right)^{n^2-1}. \tag{8}$$

Therefore,
$$P(\frac{\widetilde{\kappa}_D}{\sqrt{n}} > t) = P(\widetilde{\kappa}_D > \sqrt{n}t) = 1 - \left(1 - \frac{1}{t^2}\right)^{n^2-1}. \tag{9}$$

The inequality (5) can be obtained from (7), (8) and (9). The formula (6) can be obtained from Theorem 7.2 of [3]. □

In error correction practice, all random numbers used are pseudo random numbers, which have to be generated through a random number generator. Fig.1 shows the empirical probability density functions of the condition numbers of the pseudo random matrix $G(100, 100)$ and $\widetilde{G}(100, 100)$, where $G(100, 100)$ is generated by $randn(100, 100)$ and $\widetilde{G}(100, 100)$ is generated by $randn(100, 100) + \sqrt{-1} * randn(100, 100)$ in MATLAB. From these density functions, we know that most pseudo random matrices also have very small condition numbers. And, for the same matrix size, the tail of the condition number for a complex random matrix is thinner than that of a real one.

We have also tested some other random matrices. Experiments show a lot of other random matrices, for example, uniformly distributed pseudo random matrices, also have small condition numbers with high probability. For random matrices of non-normal distribution, we will report our experiments and some analytical proofs of their condition number properties in a further coming paper.

3.2 Real Number Codes Based on Random Matrices

In this sub-section, we introduce a class of new codes that are able to reconstruct a very good approximation of the original information with very high probability regardless of the erasure patterns in the encoded information.

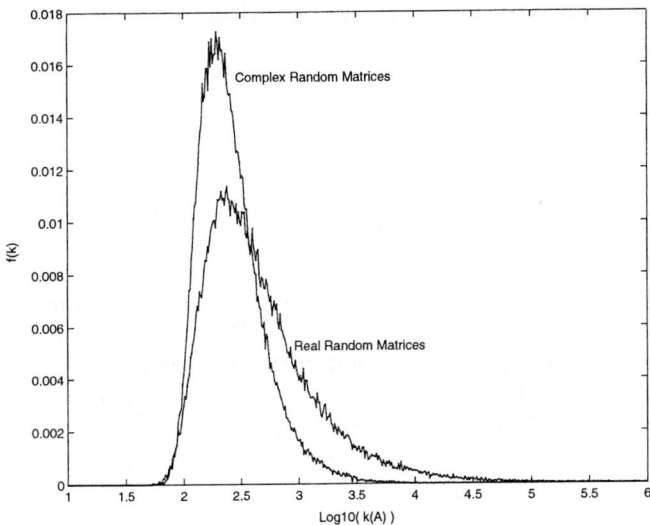

Fig. 1. The density functions of the condition numbers of $G(100,100)$ and $\widetilde{G}(100,100)$

In the real number case, we propose to use $G(M,N)$ or uniformly distributed M by N matrices with mean 0 (denote as $U(M,N)$) as our generator matrices G. In the complex number case, we propose to use $\widetilde{G}(M,N)$ or uniformly distributed M by N complex matrices with mean 0 (denote as $\widetilde{U}(M,N)$) as our generator matrices G.

Take the real-number codes based on random matrix $G(M,N)$ as an example. Since each element of the generator matrix $G(M,N)$ is a random number from the standard normal distribution, so each element of any $N \times N$ sub-matrix $(G_J)_{N \times N}$ of $G(M,N)$ is also a random number from the standard normal distribution. According to the condition number results in Subsection 3.1 , the probability that the condition number of $(G_J)_{N \times N}$ is large is very small. Hence, any N by N sub-matrix $(G_J)_{N \times N}$ of G is well-conditioned with very high probability. Therefore, no mater what erasure patterns occur, the error correction procedure is numerically stable with high probability.

We admit that our real-number and complex-number codes are not perfect. Due to the probability approach we used, the drawback of our codes is that, no matter how small the probability is, there is a probability that a erasure pattern may not be able to be recovered accurately.

However, compared with the existing codes in literature, the probability that our codes fail to recover a good approximation of the original information is negligible (see Section 4 for detail). Moreover, in the error correction practice, we may first generate a set of pseudo random generator matrices and then test each generator matrix until we find a satisfied one.

4 Comparison with Existing Codes

In the existing codes in literature, the generator matrices mainly include: Vandermonde matrix (Vander) [8], Vandermonde-like matrix for the Chebyshev polynomials (Chebvand) [2], Cauchy matrix (Cauchy), Discrete Cosine Transform matrix (DCT), Discrete Fourier Transform matrix (DFT) [6]. These generator matrices all contain ill-conditioned sub-matrices. Therefore, in these codes, when certain error patterns occur, an ill-conditioned linear system has to be solved to reconstruct an approximation of the original information, which can cause the loss of precision of possibly all digits in the recovered numbers. However, in our codes, the generator matrices are random matrices. Any sub-matrix of our generator matrices is still a random matrix, which is well-conditioned with very high probability. Therefore, no mater what erasure patterns occur, the error correction procedure is numerically stable with high probability. In this section, we compare our codes with existing codes in both burst erasure correction and random erasure correction.

4.1 Burst Erasure Correction

We compare our codes with existing codes in burst error correction using the following example.

Example 1. Suppose $x = (1, 1, 1, ..., 1)^T$ and the length of x is $N = 100$. G is a 120 by 100 generator matrix. $y = Gx$ is a vector of length 120. Suppose y_i, where $i = 101, 102, ...120$, are lost. We will use y_j, where $j = 1, 2, ...100$, to reconstruct x through solving (2).

Table 1. The generator matrices of different codes

Name	The generator matrix $G = (g_{mn})_{120 \times 100}$
Vander	$((m+1)^{100-n-1})_{120 \times 100}$
Chebvand	$(T_{m-1}(n))_{120 \times 100}$, where T_{m-1} is the chebyshev polynomial of degree $n-1$
Cauchy	$\left(\frac{1}{m+n}\right)_{120 \times 100}$
DCT	$\left(\sqrt{\frac{i}{120}} \cos \frac{\pi(2n+1)m}{240}\right)_{120 \times 100}$, where if $m = 0, i = 1$, and if $m \neq 0, i = 2$
DFT	$\left(e^{-j\frac{2\pi}{120}mn}\right)_{120 \times 100}$, where $j = \sqrt{-1}$
RandN	randn(120,100) in MATLAB
RandN-C	randn(120,100) + j * randn(120,100) in MATLAB, where $j = \sqrt{-1}$
RandU	rand(120,100) - 0.5 in MATLAB
RandU-C	rand(120,100) - 0.5 + j * (rand(120,100) - 0.5) in MATLAB,

Table 1 shows how the generator matrix of each code is generated. Table 2 reports the accuracy of the recovery for each code. All calculations are done using MATLAB. The machine precision is 16 digits. Table 2 shows our codes are able to reconstruct the original information x with much higher accuracy than the existing codes. The reconstructed x from all existing codes lost all of their 16 effective digits. However, the reconstructed x from the codes we proposed in the last section lost only about 2 effective digits.

Table 2. Burst erasure recovery accuracy of different codes

Name	$\kappa(G_J)$	$\frac{\|x-\tilde{x}\|_2}{\|x\|_2}$	Accurate digits	Number of digits lost
Vander	3.7e+218	2.4e+153	0	16
Chebvand	Inf	1.7e+156	0	16
Cauchy	5.6e+17	1.4e+03	0	16
DCT	1.5e+17	2.5e+02	0	16
DFT	2.0e+16	1.6e+00	0	16
RandN	7.5e+2	3.8e-14	14	2
RandN-C	4.5e+2	6.8e-14	14	2
RandU	8.6e+2	3.7e-14	14	2
RandU-C	5.7e+2	2.6e-14	14	2

4.2 Random Erasure Correction

For any N by N sub-matrix G_J of G, there is a erasure pattern of y which requires to solve a linear system with G_J as the coefficient matrix to reconstruct an approximation of the original x. A random erasure actually results in a randomly picked N by N sub-matrix of G. In Table 3, we compare the proportion of 100 by 100 sub-matrices whose condition number is larger than 10^i, where $i = 4, 6, 8$, and 10, for different kind of generator matrices of size 150 by 100. All generator matrices are defined in Table 1. All results in Table 3 are calculated using MATLAB based on 1,000,000 randomly (uniformly) picked sub-matrices.

From Table 3, we can see, of the 1,000,000 randomly picked sub-matrices from any of our random generator matrices, there are 0.000% sub-matrices whose condition number is larger than 10^8. However, for all existing codes in literature that we have tested, there are at least 21.644% sub-matrices whose condition number is larger than 10^8. Therefore, our codes are much more stable than the existing codes in literature.

Table 3. Percentage of 100 by 100 sub-matrices (of a 150 by 100 generator matrix) whose condition number is larger than 10^i, where $i = 4, 6, 8$, and 10

Name	$\kappa \geq 10^4$	$\kappa \geq 10^6$	$\kappa \geq 10^8$	$\kappa \geq 10^{10}$
Vander	100.000%	100.000%	100.000%	100.000%
Chebvand	100.000%	100.000%	100.000%	100.000%
Cauchy	100.000%	100.000%	100.000%	100.000%
DCT	96.187%	75.837%	48.943%	28.027%
DFT	92.853%	56.913%	21.644%	5.414%
RandN	1.994%	0.023%	0.000%	0.000%
RandN-C	0.033%	0.000%	0.000%	0.000%
RandU	1.990%	0.018%	0.000%	0.000%
RandU-C	0.036%	0.000%	0.000%	0.000%

5 Conclusion and Future Work

In this paper, we have introduced a class of real-number and complex-number codes based on random generator matrices over real-number and complex-number fields. we have compared our codes with existing codes in both burst erasure correction and random erasure correction. Experiment results demonstrate our codes are numerically much more stable than existing codes in literature.

For the future, we will compare real-number codes based on different random matrices with different probability distributions. we would also like to investigate what is the numerically optimal real number codes.

References

1. Azais, J. M. and Wschebor, M.: Upper and lower bounds for the tails of the distribution of the condition number of a gaussian matrix, submitted for publication, 2003
2. Boley, D. L., Brent, R. P., Golub, G. H. and Luk, F. T.: Algorithmic Fault Tolerance Using the Lanczos Method, SIAM Journal on Matrix Analysis and Applications, vol. 13, (1992), pp. 312-332.
3. Edelman, A.: Eigenvalues and Condition Numbers of Random Matrices, Ph.D. thesis, Dept. of Math., M.I.T., 1989.
4. Edelman, A.: On the distribution of a scaled condition number, Mathematics of Computation,vol. 58, (1992), pp. 185-190.
5. Ferreira, P.: Stability issues in error control coding in complex field, interpolation, and frame bounds, IEEE Signal Processing Letters, vol.7 No.3,(2000) pp.57-59.
6. Ferreira, P., Vieira, J.: Stable DFT codes and frames, IEEE Signal Processing Letters, vol.10 No.2,(2003) pp.50-53.
7. Golub, G. H. and Van Loan, C. F.: Matrix Computations, 2nd Ed., The John Hopkins University Press, 1989.
8. Hadjicostis, C. N. and Verghese, G. C.: Coding approaches to fault tolerance in linear dynamic systems, Submitted to IEEE Transactions on Information Theory.
9. Henkel, W.: Multiple error correction with analog codes, Proceedings of AAECC, Springer-Verlag, (1989), pp. 239-249.
10. Huang, H. and Abraham, J. A.: Algorithm-based fault tolerance for matrix operations, IEEE Transactions on Computers, vol. C-39, (1984) pp.300-304.
11. Luk, F. T. and Park, H.: An analysis of algorithm-based fault tolerance techniques, Journal of Parallel and Distributed Computing, vol. 5 (1988), pp. 1434-1438.
12. Marvasti, F., Hasan, M., Echhart, M. and Talebi, S.: Efficient algorithms for burst error recovery using FFT and other transform kernels, IEEE Transactions on Signal Processing, vol.47, No.4, (1999), pp. 1065-1075.
13. Nair, S. S. and Abraham, J. A.: Real-number codes for fault-tolerant matrix operations on processor arrays, IEEE Transactions on Computers, vol. C-39,(1990) pp.300-304.

On Iterated Numerical Integration

Shujun Li, Elise de Doncker, and Karlis Kaugars

Computer Science,
Western Michigan University
{sli, elise, kkaugars}@cs.wmich.edu
http://www.cs.wmich.edu/~parint

Abstract. We revisit the iterated numerical integration method and show that it is extremely efficient in solving certain classes of problems. A multidimensional integral can be approximated by a combination of lower-dimensional or one-dimensional adaptive methods iteratively. When an integrand contains sharp ridges which are not parallel with any axis, iterated methods often outperform adaptive cubature methods in low dimensions. We use examples to support our analysis.

1 Introduction

We will call an integration method *iterated*, if lower-dimensional methods are used for the integration in different coordinate directions [18, 11]. We use an adaptive method from QuadPack [14] to compute the one-dimensional integrals in an iterated method. The work of [5, 6] shows that the iterated method is much more efficient in integrating certain Feynman loop integrals of particle physics. As an example, one of the three-dimensional integrands has a singular behavior located within three narrow adjacent layers of the integration domain. The function has large positive values in the outer two and is negative in the middle layer.

Further study reveals that limitations of the machine precision prevent us from achieving better results for some problems. When a specific integrand parameter becomes small, increasing the function evaluation limit does not lead to a better result. Even when this parameter is not very small, setting the function evaluation limit to a large value results in unnecessary function evaluations and possibly a less accurate result.

Iterated methods deserve a thorough investigation. A prototype implementation is underway for PARINT [16]. After modifying the one-dimensional methods in PARINT to handle discontinuities, the integrals of the Dice functions (see Section 4 below) can be approximated to an accuracy that is limited only by the machine precision.

2 Iterated Method

We will use one-dimensional integration methods iteratively to compute an n-dimensional ($n \geq 2$) integral numerically.

The integral of an n-dimensional scalar function $f(x_1, x_2, ..., x_n)$ over a hyper-rectangular region $\mathcal{D}$ in $\mathcal{R}^n$ is

$$I = \int_{\mathcal{D}} f(x_1, x_2, ..., x_n) \, dx_1 dx_2 ... dx_n, \tag{1}$$

which is the same as

$$I = \int_{x_1^a}^{x_1^b} dx_1 \int_{\mathcal{D}'} f(x_1, x_2, ..., x_n) \, dx_2 ... dx_n. \tag{2}$$

Let

$$F(x_1) = \int_{\mathcal{D}'} f(x_1, x_2, ..., x_n) \, dx_2 ... dx_n, \tag{3}$$

then integral (2) becomes

$$I = \int_{x_1^a}^{x_1^b} F(x_1) \, dx_1, \tag{4}$$

which is a one-dimensional integral. We then start from formula (4), and repeat the process in the remaining coordinate directions.

We found that the iterated method outperforms other methods significantly for a class of problems with steep, narrow ridges for 2D functions, or similar behavior in higher dimensions.

3 Implementation

We can use a one-dimensional integration method to compute an approximation for integral (4). The code in this method will evaluate $F(x_1)$ for a given value of x_1 by calling an integration method. The calls are done recursively in the following pseudo-code. The pseudo-code below uses a C-like syntax.

Iterated Method:

```
n <-- dimension
lower <-- array for lower bounds
upper <-- array for upper bounds
xx <-- temporary array of size n for a point

main() {
    a <-- lower[1]
    b <-- upper[1]
    i <-- 1
    integrate(foo, i, a, b, result, error)
    print(error, result)
}
```

```
foo(n, i, x_i, fcn_value) {
  if i = n then
    integrand(n, x_i, fcn_value)
  else
    xx[i] = x_i
    a <-- lower[i+1]
    b <-- upper[i+1]
    i <-- i + 1
    integrate(foo, i, a, b, result, error)
    fcn_value <-- result;
}

integrand(n, x_i, fcn_value) {
  xx[n] <-- x_i
  fcn_value <-- f(xx)
}
```

The actual code differs from the pseudo-code significantly. It is also more complex. In the pseudo-code n, *lower*, *upper* and *xx* are global variables. *foo* is part of the package. The driver function *main* and the *integrand* function are given by the user. The end users do not need to know how the iterated method is implemented. *integrate* is a one-dimensional adaptive method. Any combination of the directions will be implemented in a future release of PARINT [16]. For example, a one-dimensional method in the x direction can call a two-dimensional method in the y and z directions.

Iterated integration methods were implemented in FORTRAN for the computations of [5, 6]. Other implementations include D01DAF in NAG (a FORTRAN subroutine for two-dimensional iterated numerical integration) [18]. According to its documentation, D01DAF is not well-suited for non-smooth integrands. Two-dimensional iterated numerical integration is also explained in [11] and [14].

For a given total error tolerance, selecting the error tolerances of the inner integrals is non-trivial. Currently we use the same relative error tolerance for all levels. The contribution of the inner and outer integration errors and a heuristic estimation of the total error are outlined in [11], where it is suggested that, for a two-dimensional iterated method, the inner integral be computed about a factor of ten more accurately than the outer. If the total absolute error tolerance is ε_a, then $\varepsilon_a^O = 0.9\varepsilon_a$ and $\varepsilon_{ai}^I = \frac{\varepsilon_a}{10(x_2^b - x_2^a)}$. The total estimate error is given by $error_O + (x_2^b - x_2^a) \max_{x_i} error_I(x_i)$, where $error_O$ is the estimated error of the outer integral, and $error_I(x_i)$ is that of the inner for a given value x_i of x..

Fritsch, Kahaner and Lyness [7] study the error tolerance assignment in a two-dimensional iterated method. The total absolute error tolerance, $\varepsilon_a^T = \varepsilon_a^O + \varepsilon_a^I$, where ε_a^O and ε_a^I are the absolute error tolerances for the outer and the inner integrations, respectively. For m-panel $((m + 1)$-point) closed Newton-Cotes rules, an optimal ratio is $\frac{\varepsilon_a^I}{\varepsilon_a^O} = \frac{m^O+2}{m^I+2}$. The authors discuss the assignment of ε_a^I for the inner integrations. One way is to use a constant error tolerance ε_{ai}^I for all points x_i in the x direction. Another way is to have $\varepsilon_{ai}^I |W_i| = constant$, where W_i is the weight assigned at the first appearance of $F(x_i)$. The former is intended for situations where the function has no peaks or untoward behavior, while the latter assignment will apply if the function is well behaved over most of its domain but has peaks or oscillations in a small portion of the domain.

4 Performance Analysis

Let us address the performance of the iterated method using two sample integrals from [17], which we refer to as DICE1 and DICE2 below.

$$\text{DICE1} \int_0^1 dx \int_0^1 dy \frac{2\varepsilon y}{(x+y-1)^2 + \varepsilon^2}$$

$$\text{DICE2} \int_{-1}^1 dx \int_{-1}^1 dy \frac{\varepsilon y^2 \theta(1 - x^2 - y^2)}{(x^2 + y^2 - b^2)^2 + \varepsilon^2}$$

Here $\theta(t) = 1$ for $t \geq 0$, and 0 otherwise. The DICE1 integrand has a ridge of height $\frac{2y}{\varepsilon}$ along the diagonal $y = 1 - x$. The DICE2 function has a ridge along the circle of radius b centered at the origin, and a discontinuity at the unit circle.

Two-dimensional adaptive cubature methods are not very effective for computing these integrals, which mimic the behavior of certain integration problems arising in high-energy physics computations. Results are given in [4] for $b = 0.8$, $\varepsilon = 10^{-1}, 10^{-2}, \ldots, 10^{-6}$, a relative error tolerance of 10^{-5} and a function evaluation limit of 250 million.

In order to show the behavior of the DICE2 integrand aggregated in the y direction, we graph the inner integral $F(x)$ in Figure 1 *(Left)* and *(Right)* for $\varepsilon = 10^{-1}$ and 10^{-4}, respectively. It emerges that these are rather smooth functions of x. Consequently, the integral in x can be carried out easily and only a moderate number of subdivisions in the x direction is carried out for the iterated integration. This is also true in the y-direction, if the x direction is aggregated first.

Figure 2 *(Left)* and *(Right)* displays visualizations of the function values evaluated by the adaptive cubature and by the iterated method, respectively, for DICE1. We use the (parallel) cubature methods in PARINT, based on the integration rules of Genz and Malik [9, 1].

For the adaptive cubature, subdivision is performed in a 2D surface. For the iterated method, subdivision is mainly done along the y axis. This behavior is also confirmed by the views of $F(x)$ for DICE2 in Figure 1 *(Left)* and *(Right)*. In spite of the ridges on the two-dimensional domains, the functions aggregated to one dimension depict a relatively smooth behavior. Thus the outer integral is easy to calculate.

Let us furthermore examine the function evaluation count. As a rough estimate, the number of function evaluations of the two-dimensional adaptive cubature method is about the square of that of the iterated method for the problems under consideration.

The adaptive cubature method chooses a direction to bisect a region so that the subsequent computation becomes easier on the subregions. If the ridges are

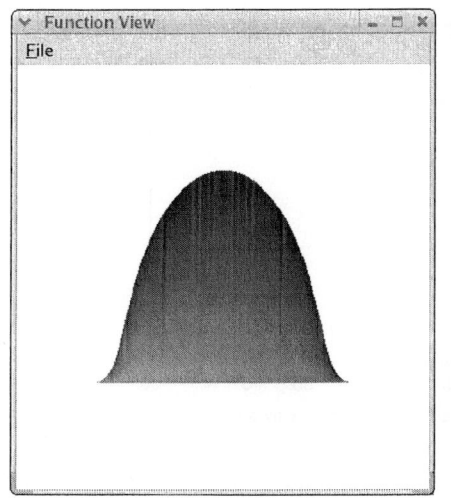 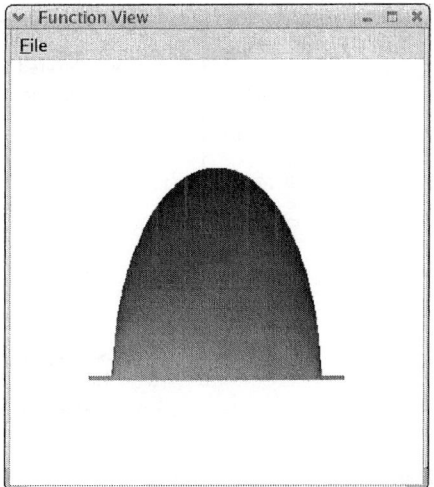

Fig. 1. $F(x)$ of the DICE2 function, where *Left:* $\varepsilon = 0.1$, and *Right:* $\varepsilon = 0.0001$

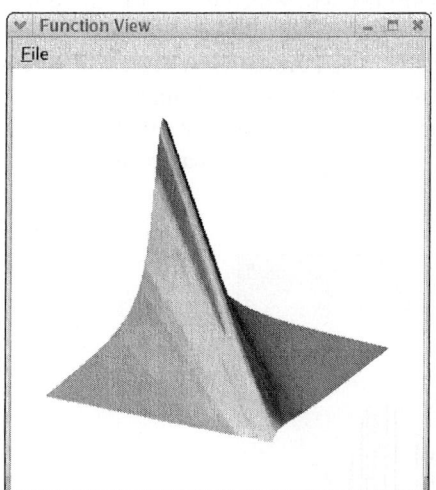 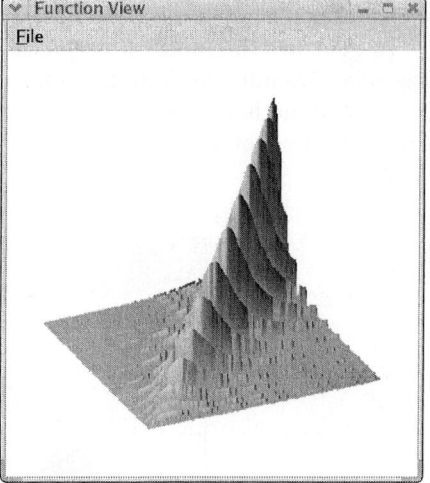

Fig. 2. *Left:* Adaptive cubature method for DICE1, where $b = 0.8$ and $\varepsilon = 0.1$; *Right:* Iterated method for DICE1. This image has been rotated for a better view

parallel to one of the axes, the advantage of the iterated method disappears. We removed y in the numerator of the DICE1 integrand and rotated it with respect to the point $(0.5, 0.5)$. We performed numerical integration for a variety of angles between the ridge and the x axis, with $\varepsilon = 0.01$ and a relative error tolerance of 10^{-10}. Figure 3 illustrates the relationship between the ridge angle and the number of function evaluations performed. When the ridge is parallel to

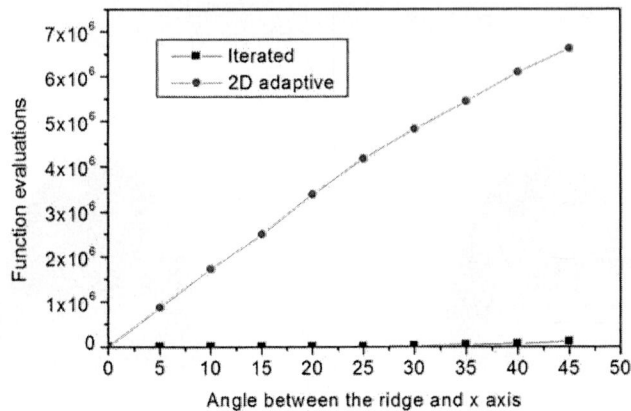

Fig. 3. The relationship between number of function evaluations and ridge orientation

the x axis, the number of function evaluations is 5355 for the adaptive cubature method, which is close to that of the iterated method (6,975). When the ridge is along the diagonal, 6,621,069 evaluations are needed, which is almost the square of the former.

To generate the function visualizations we used AdaptView [12], which utilizes adaptive numerical integration to visualize the integrand function.

Iterated methods are able to produce end-results with a very small estimated error for the problems discussed above. In order to avoid early termination,

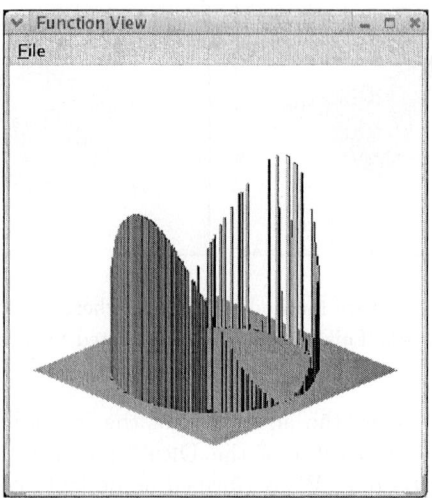

Fig. 4. Visualization of the DICE2 integrand, where $b = 0.8$, $\varepsilon = 10^{-6}$ and the absolute error tolerance is 10^{-6}

especially in cases where the ridges are extremely narrow, the error tolerance should be set small enough, so that the singularities are "discovered" before the computation is trapped in unstable areas or the error estimate has met the requirement. Figure 4 illustrates this effect. For a fixed value of x, the (inner) one-dimensional integration is done in the y variable. When a peak is not sampled by the integration rule, some regions may have an under-estimated error, and will not be evaluated again.

5 Conclusions and Future Work

We demonstrated that the iterated method is extremely fast for an important class of problems with *ridged* integrand behavior. If the integrand function has only a limited number of singular points (i.e., a *localized* singular behavior), or if the singular ridge is parallel to one of the axes, adaptive method are usually effective. We plan to investigate loop integrals of fairly low dimensions (less than 6, for example) [15, 6, 5, 2, 8, 3] using iterated methods. Note that some methods involving multi-dimensional integrals reduce to the computation of many two- and three-dimensional integrals as in [3].

It is feasible to compute many two- and three-dimensional integrals numerically to produce a reasonably small error, because a thousand of function evaluations in one direction can often achieve decent accuracy. Even if the total evaluation count is a billion for, say, a three-dimensional problem, that may not be a real obstacle for even a desktop computer today. We do need to make use of reasonable summation methods [10].

Iterated integration yields a good candidate for parallel/distributed integration methods in view of the large granularity of the inner integral evaluations. We performed preliminary tests on distributed computations with with a Web service based integration system, PI service [13].

Acknowledgment

This work is supported in part by Western Michigan University, and by the National Science Foundation under grants ACI-0000442, ACI-0203776 and EIA-0130857.

References

1. BERNTSEN, J., ESPELID, T. O., AND GENZ, A. Algorithm 698: DCUHRE-an adaptive multidimensional integration routine for a vector of integrals. *ACM Trans. Math. Softw. 17* (1991), 452–456. Available from http://www.sci.wsu.edu/math/faculty/genz/homepage.
2. BINOTH, T., AND HEINRICH, G. An automized algorithm to compute infrared divergent multi-loop integrals. hep-ph/0004013 v2.
3. BINOTH, T., HEINRICH, G., AND KAUER, N. A numerical evaluation of the scalar hexagon integral in the physical region. hep-ph/0210023.

4. DE DONCKER, E., KAUGARS, K., CUCOS, L., AND ZANNY, R. Current status of the ParInt package for parallel multivariate integration. In *Proc. of Computational Particle Physics Symposium (CPP 2001)* (2001), pp. 110–119.
5. DE DONCKER, E., SHIMIZU, Y., FUJIMOTO, J., AND YUASA, F. Computation of loop integrals using extrapolation. *Computer Physics Communications 159* (2004), 145–156.
6. DE DONCKER, E., SHIMIZU, Y., FUJIMOTO, J., YUASA, F., CUCOS, L., AND VAN VOORST, J. Loop integration results using numerical extrapolation for a non-scalar integral. *Nuclear Instruments and Methods in Physics Research Section A 539* (2004), 269–273. hep-ph/0405098.
7. FRITSCH, F. N., KAHANER, D. K., AND LYNESS, J. N. Double integration using one-dimensional adaptive quadrature routines: A software interface problem. *ACM Trans. Math. Softw. 7*, 1 (1981), 46–75.
8. FUJIMOTO, J., SHIMIZU, Y., KATO, K., AND OYANAGI, Y. Numerical approach to one-loop integrals. *Progress of Theoretical Physics 87*, 5 (1992), 1233–1247.
9. GENZ, A. MVNDST Mult. Normal Dist. software, 1998. Available from web page at http://www.sci.wsu.edu/math/faculty/genz/homepage.
10. KAHAN, W. Further remarks on reducing truncation errors. *Comm. ACM 8* (1965), 40.
11. KAHANER, D., MOLER, C., AND NASH, S. *Numerical Methods and Software.* Prentice Hall, 1989.
12. LI, S., KAUGARS, K., AND DE DONCKER, E. Grid-based distributed function visualization.
13. LI, S., KAUGARS, K., AND DE DONCKER, E. Massive scale distributed integration using Web service. In *The Hawaii International Conference on Computer Sciences* (2003). CDROM Proceedings.
14. PIESSENS, R., DE DONCKER, E., ÜBERHUBER, C. W., AND KAHANER, D. K. *QUADPACK, A Subroutine Package for Automatic Integration.* Springer Series in Computational Mathematics. Springer-Verlag, 1983.
15. SON, D. H. *Feynman Loop Integrals and their automatic Computer-aided Evaluation.* PhD dissertation, Johannes Gutenberg-Universität Mainz, June 2003.
16. PARINT GROUP. http://www.cs.wmich.edu/parint, PARINT web site.
17. TOBIMATSU, K., AND KAWABATA, S. Multi-dimensional integration routine DICE. Tech. Rep. 85, Kogakuin University, 1998.
18. ÜBERHUBER, C. *Numerical Computation 2 - Methods, Software, and Analysis.* Springer-Verlag, 1997.

Semi-Lagrangian Implicit-Explicit Two-Time-Level Scheme for Numerical Weather Prediction

Andrei Bourchtein

Mathematics Department, Pelotas State University, Brazil
burstein@terra.com.br

Abstract. Semi-Lagrangian two-time-level finite difference scheme for hydrostatic atmospheric model is considered. Approximation of the gravitational waves in implicit-explicit manner allows to keep balance between extended stability and required accuracy. Both are assured by implicit discretization of the fast principal vertical modes and explicit approximation of the slow secondary gravitational waves. Numerical experiments with actual atmospheric data are carried out to define the most efficient implicit-explicit separation, which produces the accurate forecasts at the less computational cost.

1 Introduction

Semi-Lagrangian (SL) approach has been proved to be an efficient alternative to Eulerian one because it allows to circumvent the Courant-Friedrichs-Lewy (CFL) criterion related to advection velocity, requiring only solution of trajectory equations, which represent the systems of ODE decoupled at each grid point and solved efficiently by iterative algorithm [7,15]. If, additionally, gravitational terms of hydrostatic atmospheric model are discretized with sufficient degree of implicitness, then time step of SL schemes can be chosen on the base of accuracy considerations [7,15]. This is great advantage of such schemes because more straight explicit and implicit discretizations are computationally expensive: explicit approximation requires very small time steps due to fast gravitational waves and implicit approximation requires solution of nonlinear PDEs at each time step.

The choice of the level of implicitness of gravitational terms in SL schemes can be based on considerations of efficiency, including the accuracy of forecasting fields and minimization of computational cost. The most direct approximation is implicit, which is used in the majority of the SL schemes [7,15]. It allows to use a great time steps, but it is not the most efficient way because the 3D linear algebraic systems of the high order $M \times N \times L$ (where M, N and L are the number of points in horizontal and vertical directions) should be solved at each time step. There is a chance to avoid these hard computations by separating the spectrum of gravitational waves. It is well known that highest internal barotropic modes of the vertically decoupled hydrostatic equations contain the gravitational waves with the smallest amplitudes and slowest velocities of propagation [8]. Such waves are secondary and do not impose any relevant restriction on time step and, consequently, they can be discretized in explicit manner on a coarse grid. This way, implicit part can be reduced to K separate 2D linear systems of the order $M \times N$, where K is a small as compared to L. Thus, the computational cost can be reduced if the slow gravitational waves are discretized more explicitly and coarsely.

A similar approach has been used in a three-time-level SL model [3]. The reported results showed strong points of developed scheme for time steps up to 40 min. In the last decade, motivated by results of McDonald [10] and Temperton and Staniforth [16] for shallow water equations, numerical modelers started to substitute tree-time-level SL schemes by two-time-level ones, which allow to achieve the same accuracy with even larger time steps [9,11,17]. In this paper we apply the modified technique of [3] to two-time-level scheme with objective to increase time step up to 60 min with no loss of forecast accuracy.

2 Semi-Lagrangian Implicit-Explicit Time Discretization

Primitive equations in time coordinate t, horizontal Cartesian coordinates x, y and vertical coordinate $\sigma = p/p_s$ can be written as follows [7]:

$$d_t u = f_0 v - G_x + N_u \; , \; d_t v = -f_0 u - G_y + N_v \; , \tag{1}$$

$$G_{\ln \sigma} = -RT \; , d_t P = -D - \dot{\sigma}_\sigma \; , \; c_p d_t T = RT_0 \cdot (d_t P + \dot{\sigma}/\sigma) + c_p N_T \; ; \tag{2}$$

$$N_u = (f - f_0)v - R(T - T_0)P_x \; , \; N_v = -(f - f_0)u - R(T - T_0)P_y \; , \tag{3}$$

$$c_p N_T = -R(T - T_0)(\dot{\sigma}/\sigma - D - \dot{\sigma}_\sigma). \tag{4}$$

Here, u, v and $\dot{\sigma}$ are horizontal and vertical velocity components, $D = u_x + v_y$ is the horizontal divergence, $P = \ln p_s$, p and p_s are the pressure and surface pressure respectively, T is the temperature, $G = gz + RT_0 P$, z is the height. Nonlinear and variable coefficient terms are grouped in N_u, N_v, N_T. Individual 3D derivative is

$$d_t \varphi = \varphi_t + u\varphi_x + v\varphi_y + \dot{\sigma}\varphi_\sigma \; , \; \varphi = u, v, P, T$$

and the following parameters are used: f is the Coriolis parameter with the mean value f_0, g is the gravitational acceleration, R is the gas constant of dry air, c_p is the specific heat at constant pressure, $T_0 = const$ is the reference temperature profile. Hereinafter the subscripts t, x, y, σ denote the partial derivatives with respect to indicated variable.

Let us split solution of (1)-(2) into two steps. The first SL step consists of solution of the advective part of the prognostic equations

$$d_t \mathbf{r} = \mathbf{V} \; , \; \mathbf{r} = (x, y, \sigma) \; , \; \mathbf{V} = (u, v, \dot{\sigma}) \; .$$

These equations are efficiently solved by Robert's iterative algorithm [14,15], which assures the second order of accuracy and converges under limitation on time step expressed in the terms of the wind derivatives [13,15]:

$$\tau \le 2/3 V_d, \quad V_d = \max\left(|u_x|, |u_y|, |u_\sigma|, |v_x|, |v_y|, |v_\sigma|, |\dot\sigma_x|, |\dot\sigma_y|, |\dot\sigma_\sigma|\right),$$

For fine grid with horizontal meshsize about 50km and 20 vertical levels it gives maximum time step about 1 hour.

The other terms, including gravitational waves, are considered on the second step

$$u_t = f_0 v - G_x + N_u, \quad v_t = -f_0 u - G_y + N_v, \tag{5}$$

$$P_t = -D - \dot\sigma_\sigma, \quad c_p T_t = RT_0 \cdot (P_t + \dot\sigma/\sigma) + N_T. \tag{6}$$

Implicit discretization of all the gravitational terms ($G_x, G_y, D, \dot\sigma_\sigma, \dot\sigma/\sigma$) is the most traditional approximation in SL models [7,15,17]. It gives rise to the following time difference equations:

$$\frac{u^{n+1} - u^n}{\tau} = f_0 \frac{v^{n+1} + v^n}{2} - \frac{G_x^{n+1} + G_x^n}{2} + N_u^{n+1/2}, \tag{7}$$

$$\frac{v^{n+1} - v^n}{\tau} = -f_0 \frac{u^{n+1} + u^n}{2} - \frac{G_y^{n+1} + G_y^n}{2} + N_v^{n+1/2} \tag{8}$$

$$\frac{P^{n+1} - P^n}{\tau} = -\frac{D^{n+1} + D^n}{2} - \frac{\dot\sigma_\sigma^{n+1} + \dot\sigma_\sigma^n}{2}, \quad \frac{T^{n+1} - T^n}{\tau} = \frac{RT_0}{c_p}\left(\frac{P^{n+1} - P^n}{\tau} + \frac{\dot\sigma^{n+1} + \dot\sigma^n}{2\sigma}\right) + N_T^{n+1/2}, \tag{9}$$

where τ is the time step, superscript $n+1$ denotes the values at the new time level $t_{n+1} = (n+1)\tau$ and superscript n denotes the values at the current time level $t_n = n\tau$. The nonlinear terms are evaluated by extrapolation to the half way time level $t_{n+1/2}$:

$$N^{n+1/2} = (3N^n - N^{n-1})/2, \quad N = N_u, N_v, N_T.$$

The scheme (7)-(9) is of the second order of accuracy and linear analysis shows that it is absolutely stable. However, it requires the solution of 3D elliptic problem at each time step [11,17].

Another way of approximation of gravitational terms is explicit. For example, forward-backward time differencing gives:

$$\frac{\hat u^{n+1} - u^n}{\tau} = f_0 \frac{\hat v^{n+1} + v^n}{2} - \hat G_x^{n+1} + N_u^{n+1/2}, \tag{10}$$

$$\frac{\hat v^{n+1} - v^n}{\tau} = -f_0 \frac{\hat u^{n+1} + u^n}{2} - \hat G_y^{n+1} + N_v^{n+1/2}, \tag{11}$$

$$\frac{\hat P^{n+1} - P^n}{\tau} = -D^n - \dot\sigma_\sigma^n, \quad \frac{\hat T^{n+1} - T^n}{\tau} = \frac{RT_0}{c_p}\left(\frac{\hat P^{n+1} - P^n}{\tau} + \frac{\dot\sigma^n}{\sigma}\right) + N_T^{n+1/2}. \tag{12}$$

This scheme is of the first order of accuracy and its CFL condition is $\tau \le \sqrt{2} h_g / c_{grav}$, where h_g is meshsize used for approximation of gravitational waves and $c_{grav} \approx 350\,m/s$ is the propagation velocity of the fastest gravitational waves. This condition implies the use of the small time steps about 3 min on a fine horizontal grid with $h_g = 50km$. However, the algorithm of solution at each time step is very simple because this scheme is actually explicit.

Finally, let us consider implicit time differencing for essential fast gravitational waves and explicit one for secondary slow waves. To this end, we should apply vertical transform to separate the different types of the gravitational waves. First we eliminate functions P, T and $\dot{\sigma}$ from (6) to obtain prognostic equation for G:

$$c_p(\sigma G_{t\ln\sigma})_\sigma = R^2 T_0 D - c_p R(\sigma N_T)_\sigma. \tag{13}$$

Then, we use the vertical expansion of the functions

$$\varphi = \sum \varphi_k S_k, \tag{14}$$

where $S_k(\sigma)$ are the first K eigenfunctions of the differential vertical structure equation

$$c_p(\sigma S_{\ln\sigma})_\sigma = -\lambda^{-1} R S$$

or its discrete analogues, that is, the eigenvectors of the difference vertical structure equation on vertical K-level grid. Using (14) we can rewrite equations (5), (13) in the following form for each vertical mode k

$$u_{tk} = f_0 v_k - G_{xk} + N_{uk}, \quad v_{tk} = -f_0 u_k - G_{yk} + N_{vk}, \quad G_{tk} = -c_k^2 D_k + N_{Gk}, \tag{15}$$

where $c_k = \sqrt{RT_0 \lambda_k}$ and $N_G = \lambda_k R((\sigma N_T)_\sigma)_k$. It was shown in [5] for differential vertical structure equation and in [4] for difference equation that all eigenvalues λ_k are positive and have zero limit point as k approaches infinity. Fig.1 shows the values of c_k as function of the mode number k for 20-level Lorenz staggered vertical grid. The results for homogeneous grid and actual grid (with concentration of the levels in boundary layer and higher troposphere) are presented.

Now we can apply different approximation to the fast and slow gravitational waves. The first I principal vertical modes are approximated implicitly with second order of accuracy

$$\frac{u_k^{n+1} - u_k^n}{\tau} = f_0 \frac{v_k^{n+1} + v_k^n}{2} - \frac{G_{xk}^{n+1} + G_{xk}^n}{2} + N_{u\,k}^{n+1/2}, \tag{16}$$

$$\frac{v_k^{n+1} - v_k^n}{\tau} = -f_0 \frac{u_k^{n+1} + u_k^n}{2} - \frac{G_{yk}^{n+1} + G_{yk}^n}{2} + N_{v\,k}^{n+1/2}, \tag{17}$$

$$\frac{G_k^{n+1} - G_k^n}{\tau} = -c_k^2 \frac{D_k^{n+1} + D_k^n}{2} + N_{G\,k}^{n+1/2}, \quad k=1,\dots,I. \tag{18}$$

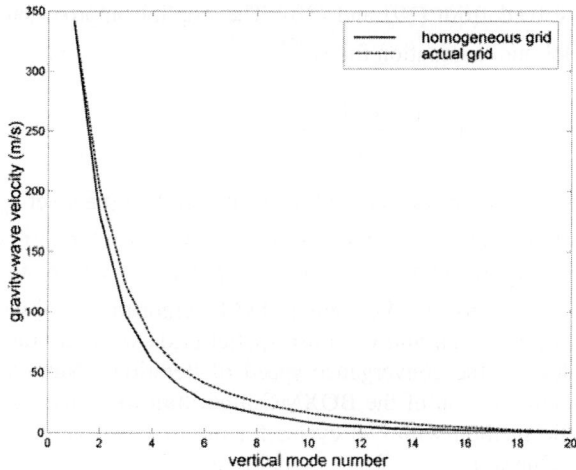

Fig. 1. Gravity-wave velocity as function of the vertical mode number

The remaining secondary modes are approximated explicitly with the first order of accuracy:

$$\frac{u_k^{n+1} - u_k^n}{\tau} = f_0 \frac{v_k^{n+1} + v_k^n}{2} - G_{x\ k}^{n+1} + N_{u\ k}^{n+1/2}, \qquad (19)$$

$$\frac{v_k^{n+1} - v_k^n}{\tau} = -f_0 \frac{u_k^{n+1} + u_k^n}{2} - G_{y\ k}^{n+1} + N_{v\ k}^{n+1/2}, \qquad (20)$$

$$\frac{G_k^{n+1} - G_k^n}{\tau} = -c_k^2 D_k^n + N_{G\ k}^{n+1/2}, \quad k = I+1,...K. \qquad (21)$$

Let us note that although systems (15) are coupled through nonlinear terms, both schemes (16)-(18) and (19)-(21) can be solved separately for each k because nonlinear terms are treated explicitly. This approach generates stability condition in the form $\tau \leq \sqrt{2} h_g / c_{I+1}$, where c_{I+1} is the maximum gravity-wave speed of the modes treated explicitly.

If discrete vertical transform is applied, implicit approximation (16)-(18) of all the vertical modes will result in the scheme (7)-(9) and explicit approximation (19)-(21) of all the modes will give forward-backward scheme (10)-(12).

3 Numerical Experiments

In this section we present the results of the experiments with different configurations of the vertical approximation. At each time step, the explicit approximation (19)-(21) is solved by direct formulas: first G_k^{n+1} are found from (21) and then u_k^{n+1} and v_k^{n+1}

are elementary solved from (19) and (20). The implicit approximation (16)-(18) is reduced to 2D Helmholtz equation for G_k^{n+1}:

$$\nabla^2 G_k^{n+1} - \frac{4+\tau^2 f_0^2}{\tau^2 c_k^2} G_k^{n+1} = F_k^n , \qquad (22)$$

where ritght-hand side is combination of the values at the time level t_n. This equation is solved by multigrid method, which is fast solver for such kind of the problems. Its optimal versions require $O(MN)$ arithmetic operations, where M and N are the number of points in horizontal. We apply BOXMG algorithm [2,6] based on Galerkin type of discretization, which allows to use spatial grids with arbitrary number of the points with no less of the convergence speed of iterations. Numerical experiments showed that optimal version of the BOXMG algorithm for equation (22) consists of using the V-cyclic method with two cycles for the first two vertical modes and one cycle for others. One four-color Gauss-Seidel point relaxation sweep is performed on any grid both before dropping down to the next coarser grid and before interpolation to the previous finer grid. As initial guess for iterations we use G_k^n. After G_k^{n+1} is found, the u_k^{n+1} and v_k^{n+1} are calculated by elementary solution of (16), (17).

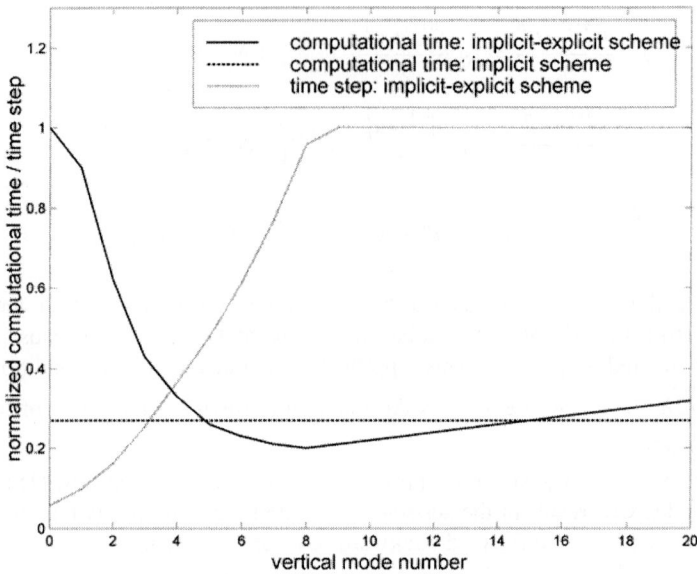

Fig. 2. Time step and computational cost of the implicit-explicit schemes

24-hour integrations of the primitive equations were carried out for different number of the vertical modes approximated implicitly. Fig. 2 shows time step used in these experiments (chosen in accordance with stability criterions) and computational time cost (in percent of the forecast time of the forward-backward scheme (10)-(12))

as functions of the number of the implicitly treated vertical modes. The required computational time for one forecast computed by implicit approximation (7)-(9) is also shown. The last integration is less expensive than that with implicit solution of all vertical modes by formulas (16)-(18). The former requires solution of the 3D elliptic equation but does not need application of vertical tranform, while the latter uses vertical transform and fast MG solvers for a set of 2D elliptic problems. Although multigrid solution of 3D equation is more expensive than fast solution of the decoupled set of 2D equations, the additional computational charge due to vertical transform is too hard. Since both schemes have absolutely stable adjustment step, that is, their maximum time step is about 60 min as defined by advective step, the scheme (7)-(9) required less computational time than (16)-(18) with $I=K$. Nevertheless, some versions of the implicit-explicit algorithm are more computationally efficient than the standard algorithm (7)-(9). For example, implicit treatment of 7-9 vertical modes gives certainly more efficient algorithm.

To evaluate forecasting ability of the above schemes we carried out integrations based on actual atmospheric data. The horizontal domain of 5000x5000 km^2 centered at Porto Alegre city ($30^0 S$, $52^0 W$) was covered by uniform spatial grid C with meshsize $h = 50$ km (we use Arakawa-Mesinger nomenclature of spatial grids [12]). The initial and boundary conditions were obtained from objective analysis and global forecasts of National Centers for Environmental Prediction (NCEP).

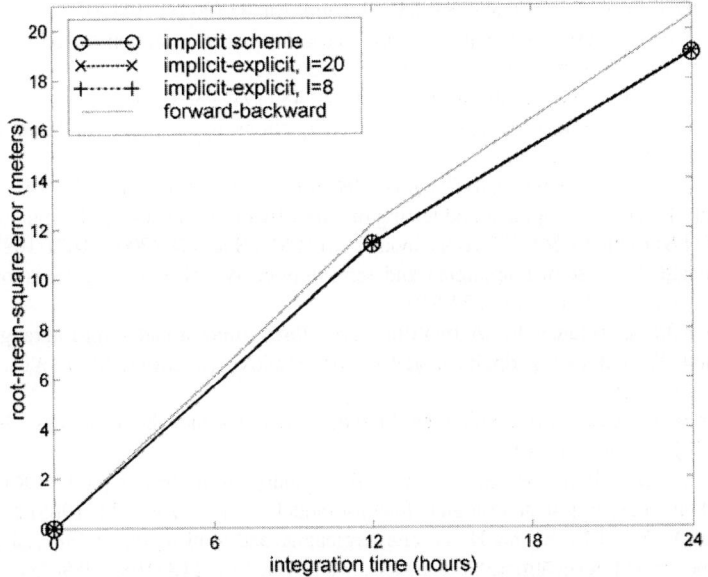

Fig. 3. Root-mean-square error of geopotential forecast at the surface of 500 hPa

It is well expected that after certain period of integration the forecast results will be determined in high degree by boundary conditions. Therefore the standard evaluation time for the regional models is limited to 24 or 36 hours of integration [1]. In Fig. 3

we present the root-mean-square differences in meters between 24-h forecasts and analysis at pressure level 500hPa. This is standard measure of quality of short-range weather forecasts for geopotential fields [1]. Each scheme was run with appropriate time step chosen in accordance with stability criterions. Based on this results we can conclude that the most efficient algorithms are obtained when implicit-explicit algorithm with 7-9 implicit modes is used.

Acknowledgements

This research was supported by brazilian science foundation CNPq under grant 302738/2003-7.

References

1. Anthes R.A., Kuo Y.H., Hsie E.Y., Low-Nam S., Bettge T.W.: Estimation of skill and uncertainty in regional numerical models. Q. J. R. Meteorol. Soc. **115** (1989) 763-806.
2. Bandy V., Sweet R.: A set of three drivers for BOXMG: a black box multigrid solver. Comm. Appl. Num. Methods **8** (1992) 563-571.
3. Bourchtein A.: Semi-Lagrangian semi-implicit space splitting regional baroclinic atmospheric model. Appl. Numer. Math. **41** (2002) 307-326.
4. Bourchtein A., Kadychnikov V.: Well-posedness of the initial value problem for vertically discretized hydrostatic equations. SIAM J. Num. An. **41** (2003) 195-207.
5. Cohn S.E., Dee D.P.: An analysis of the vertical structure equation for arbitrary thermal profiles. Q. J. R. Meteorol. Soc. **115** (1989) 143-171.
6. Dendy J.E.: Black box multigrid. J.Comp.Phys. **48** (1982) 366-386.
7. Durran D.: Numerical Methods for Wave Equations in Geophysical Fluid Dynamics. Springer, New York (1999).
8. Holton J.R.: An Introduction to Dynamic Meteorology. Academic Press, San Diego (1992).
9. Hortal M.: The development and testing of a new two-time-level semi-Lagrangian scheme (SETTLS) in the ECMWF forecast model, Q.J.R.Met.Soc. **128** (2002) 1671-1687.
10. Mcdonald A.: A semi-Lagrangian and semi-implicit two time level integration scheme. Mon. Wea. Rev. **114** (1986) 824-830.
11. McDonald A., Haugen J.: A two-time-level, three-dimensional semi-Lagrangian, semi-implicit, limited-area gridpoint model of the primitive equations. Mon. Wea. Rev. **120** (1992) 2603-2621.
12. Mesinger F., Arakawa A.: Numerical Methods Used in Atmospheric Models. GARP Publ. Ser. 17(I), Geneva (1976).
13. Pudykiewicz J., Benoit R., Staniforth A.: Preliminary results from a partial LRTAP model based on an existing meteorological forecast model. Atmos.-Ocean **23** (1985) 267-303.
14. Robert A., Yee T.L., Ritchie H.: A semi-Lagrangian and semi-implicit numerical integration scheme for multilevel atmospheric models. Mon. Wea. Rev. **113** (1985) 388-394.
15. Staniforth A., Côté J.: Semi-Lagrangian integration schemes for atmospheric models - A review. Mon. Wea. Rev. **119** (1991) 2206-2223.
16. Temperton C., Staniforth A.: An efficient two-time-level semi-Lagrangian semi-implicit integration scheme, Q. J. R. Meteorol..Soc. **113** (1987) 1025-1039.
17. Temperton C., Hortal M., Simmons A.J.: A two-time-level semi-Lagrangian global spectral model. Q. J. R. Meteorol. Soc. **127** (2001) 111-126.

Occlusion Activity Detection Algorithm Using Kalman Filter for Detecting Occluded Multiple Objects

Heungkyu Lee* and Hanseok Ko**

* Dept. of Visual Information Processing
** Dept. of Electronics and Computer Engineering,
Korea University, Seoul, Korea
hklee@ispl.korea.ac.kr, hsko@korea.ac.kr

Abstract. This paper proposes the detection method of occluded moving objects using occlusion activity detection algorithm. When multiple objects are occluded between them, a simultaneous feature based tracking of multiple objects using tracking filters fails. To estimate feature vectors such as location, color, velocity, and acceleration of a target are critical factors that affect the tracking performance and reliability. To resolve this problem, the occlusion activity detection algorithm is addressed. Occlusion activity detection method provides the occlusion status of next state using the Kalman prediction equation. By using this predicted information, the occlusion status is verified once again in its current state. If the occlusion status is enabled, an object association technique using a partial probability model is applied. For an experimental evaluation, the image sequences for a scenario in which three rectangles are moving within the image frames are made and evaluated. Finally, the proposed algorithms are applied to real image sequences. Experimental results in a natural environment demonstrate the usefulness of the proposed method.

1 Introduction

The importance of detection, tracking, and recognition problems has received increased attention since visual tracking began to play an important role in surveillance systems, virtual reality interfaces and a variety of robotic tasks. But many key issues are not solved yet. The tracking of non-rigid objects and classifying their appearance model is a challenging problem in visual surveillance system. Especially, the monitoring and visual surveillance of human activity [1][2][3][4] requires complex tracking algorithms because of the unpredictable situations which occur whenever multiple peoples are moving, stopping, hiding behind obstacles and interacting with each other. Human actions within the field of view have no consistent rules concerning their movement. In addition, when multiple peoples are interacting with each other in a natural scene, a variety of events can occur such as occlusion, partial occlusion or short-time stopping. In such cases, the general tracking filter like a Kalman filter tends to fail because of sudden movements or sudden variation of speed.

Some tracking algorithms have a weakness according to the given specific situation. Feature-based tracking has a weakness in that it's difficult to estimate a centroid or the velocity of moving targets when the targets are occluded from each other. A region-based approach has also same situation. The parametric method needs the calculation of an optimal fitting of the model to pixel the data in a region, and it should be updated continuously to the change of the appearance and the intensity variation while non-rigid objects are moving. However, when multiple peoples are occluded from each other, a parametric estimation of the individual person model is inaccurate. The view-based method to find the best match of a region in a search area with a reference template also has similar weaknesses with a region based approach in occlusion situations. These problems are due to the inaccurate estimate of state information such as centroid, color, velocity and acceleration of targets. Thus, in this paper, we propose the detection method for an accurate estimate of the state information using occlusion activity detection and an object association algorithm as shown in Figure 1. The occlusion activity detection algorithm provides the occlusion status information. By using this information, when the occlusion is activated, the proposed object association algorithm [5] can be applied to estimate the accurate state of information of the occluded multiple objects respectively, and then the general tracking algorithm is applied. On the contrary, if the occlusion is not activated, the general tracking algorithm is applied. This is due to the fact that the tracking algorithm is reliable when the occlusion between the objects has not occurred. Thus, the proposed algorithms can provide the reliable feature vectors for simultaneous multiple tracking algorithms even in the occlusion time.

This paper is organized as follows. In Section 2, we describe the proposed algorithms. It describes the detection method using the occlusion activity detection and object association. In Section 3, we evaluate our proposed algorithms having image sequences. Finally, the conclusive remarks are presented in Section 4.

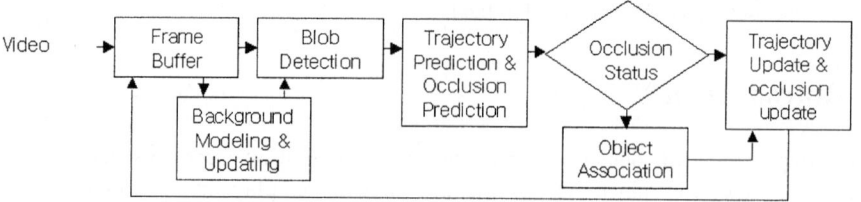

Fig. 1. System block-diagram for detecting occluded multiple objects

2 Detection of Occluded Multiple Objects

2.1 Behavior of Moving Objects

The key issue of multiple targets tracking in the image sequences is the occlusion problem. During the occlusion time, unpredictable events can be activated. Thus, the tracking failure and missed tracking often happened. To cope with this problem, a behavior analysis of multiple targets is required the first time. Figure 2 describes all the possible action flows of targets within the field of view. A specific target enters into the specific field of view, and then it is moving, stopping, interacting with other targets. And finally, it leaves the field of view. We can classify the behaviors of multiple targets according to the action flow as in Figure 2.: (1) A specific target enters into the scene. (2) Multiple targets enter into the scene. (3) A specific target is moving and forms a group with other targets, or just moves beside other targets or obstacles. (4) A specific target within the group leaves a group. (5) A specific target continues to move alone, or stops moving and then starts to move again. (6) Multiple targets in a group continue to move and interact between them, or stop interacting and then start to move again. (7) (8) A specific target or a group leaves a scene. The events of (1), (4), (5), and (7) can be tracked using general tracking algorithms. However, the events of (2), (3), (6) and (8) cannot be tracked reliably. Thus, to resolve this problem, we propose the occlusion reasoning method that detects occlusion activity status using Kalman Filter.

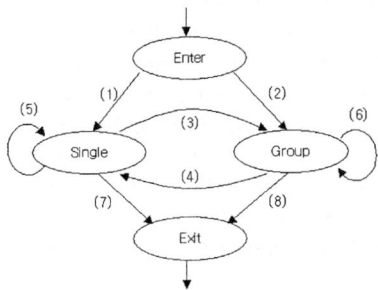

Fig. 2. State Transition Diagram; general action flow of objects within the FOV

2.2 Occlusion Activity Detection

We assume that we found the moving blobs, and then it starts to detect occlusion activity. Occlusion activity detection is an algorithm to provide the current status of occlusion between objects, which are just labeled blobs of a blob detection level. According to the occlusion status, a countermeasure to reliably track can be applied. We assumed that occluded objects from the first time have not appeared, and the objects are non-rigid. The procedure of occlusion activity detection is as follows.

- STEP 1: Occlusion Prediction
As in (a) of Figure 3, this step predicts the next positions (centroids) of blobs employing the usual Kalman prediction model used in JPDAF:

$$\hat{X}(k+1/k) = F(k)\hat{X}(k/k) + u(k) \quad (1)$$

where $X(k+1/k)$ is the state vector at time $k+1$ given cumulative measurements to time k, $F(k)$ is a transition matrix, and $u(k)$ is a sequence of zero-mean, white Gaussian process noise. Using the predicted position computed at equation (1), we can determine the redundancy of objects within the field of view using the intersection measure. The decision of the occlusion is computed by comparing if or not there is an overlapping region between the rectangular blocks MV_i in the predicted center points as follows.

$$F_{oc} \begin{cases} 1 & \text{if } (MV_i \cap MV_j) \neq \phi \\ 0 & \text{otherwise} \end{cases}, \text{ where } i, j = 1,...,m \quad (2)$$

where F_{oc} is an occlusion alarm flag, the subscript i and j are the index of the detected target at the current frame, the block MBR_i that represents the validation region has the fixed range computed at the current frame and m is a number of a target. If a redundant region has occurred at the predicted position, the probability of occlusion occurrence in the next step increases. Therefore, the occlusion alarm flag is set to 1, and current status is maintained.

- STEP 2: Occlusion Status Update

In the current frame, the occlusion status is updated to decide the occlusion occurrence. The first time, the size of the labeled blobs is verified whether they are contained within the validation region or not. If the size of labeled blobs is contained within the validation region, the occlusion status flag is disabled. Then, the Kalman gain is then computed and the measurement equation is updated. If number of the predicted center point is greater than two, then the occlusion alarm flag is set to 1, we can conclude that the occlusion has occurred at the region. Thus, the occlusion status is enabled. At this time, the object association technique is applied to estimate the accurate center points of the respective blobs. From the predicted center points of the previous step including near that region, it is searched. In addition, the process transition mode is changed as in Figure 2.

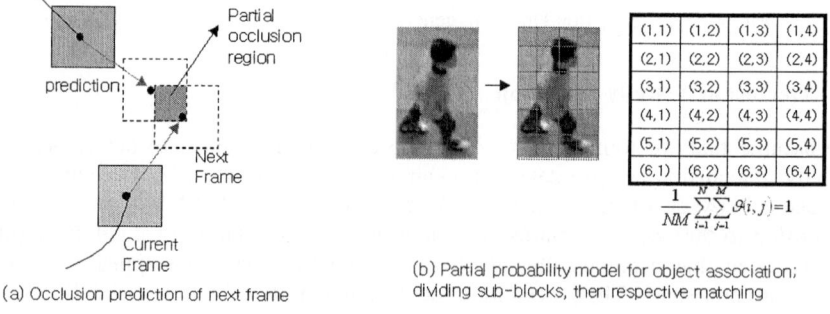

Fig. 3. Occlusion prediction of next frame

2.3 Object Association

For the identity of the occluded blobs, the object association technique can be applied to associate a measured object with a real target when the occlusion status is enabled. That is due to the fact that the labeling procedure measures some targets as one target. In addition, the occlusion status may be maintained during some periods. This case causes a tracking failure due to miss-association if there is no association technique. To resolve this problem, an object association can be applied for not only the position decision in the occlusion state, but also for the decision of the identity of a target between frames. It can be a means for an attribute tracking method, which can be described as the process of combining a position and color information incorporating the data from a prior target model, a target dynamic model, and a feature measurement model through a buffering technique.

To do this, we applied the SEA to verify the target identification for an object association between a priori target model and a feature measurement model. The object pixel data buffered for the prior target model is used. This calculates the matching relationship between the buffered data in a queue and a candidate block. If we assume that the size of a blob is $N \times N$ pixels, the search window is of size $(2N+1) \times (2N+1)$ pixels in a basis of the predicted position. The mean absolute difference (MAD) is used to measure the match between two blocks. The match is performed on the current frame t using a previously stored blob model, a prior target model. The SEA algorithm as in [7] can be computed as

$$R - M(x, y) \leq MAD(x, y) \tag{3}$$

However, matched result of a hidden object behind a specific object may result in a false acceptance. Thus, we divide the reference block into N sub-blocks, then calculate a partial probability of candidate blocks as in Figure 4. It is an alternative evidential reasoning based approach for identity reasoning under the partial probability models. The concept of a typical sequence is defined in terms of a i, j-element partition, P_i, given the true target type i.

$$P_i = \{a_{11}, \ldots, a_{ij} \,/\, \text{target type i}\} \tag{4}$$

We consider it as a target if the sum of probability values of a sub-window is greater than and equal to a given threshold value as follows.

$$p(P_i) = \frac{1}{NM} \sum_{i=1}^{N} \sum_{j=1}^{M} \vartheta(a_{ij}) \geq Th \tag{5}$$

The matching probability of an occluded object is computed using an equation (6) after dividing into $i \times j$ partition window as in (b) of Figure 3.

$$\vartheta(i, j) = \begin{cases} 1 & \text{if } MAD(x, y) \geq R - M(x, y) \\ 0 & \text{oterwise} \end{cases} \tag{6}$$

Thus, we can estimate the occupancy region of the occluded objects. By using this information, the center point of an individual object is calculated again.

3 Experimental Results

The proposed scheme was tested on real image sequences to assess its capabilities for tracking multiple moving targets (two people) in complex road scenes. Two different road scenes with an increasing complexity were considered. Acquired images were sampled at video rate: example 1 (total 180 frames, 15 frames per seconds, and its size is 240×320) and example 2 (total 70 frames, 15 frames per seconds, and its size is 240×320) which is a gray level image. These include the occlusion scenario. In addition, to show the robustness of the proposed methods, the image sequences including the occlusion scenario are depicted.

For the moving blob detection, an adaptive change detection algorithm is performed, and then a binarized algorithm is applied as in Figure 4. If the moving blobs are detected, a labeling process is performed, then the center points of the respective blobs are calculated as feature vectors for tracking.

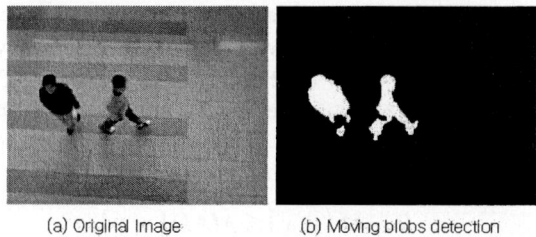

(a) Original Image (b) Moving blobs detection

Fig. 4. Blobs detection experiments

To show the robustness of the proposed algorithm, the test image sequences having three rectangles moving are made as in (a) of Figure 5. The three rectangles are moving randomly with partial occlusion. This scenario file is simply binarized for the detection of moving blobs, and then labeled. Finally, the occlusion activity detection algorithm is applied, and if the occlusion status is enabled, the object association algorithm is applied. The (b) of Figure 5 shows the occlusion status of each rectangle in the image sequences. We can know that the occlusion activity detection can provide the precious information to process the occlusion problem.

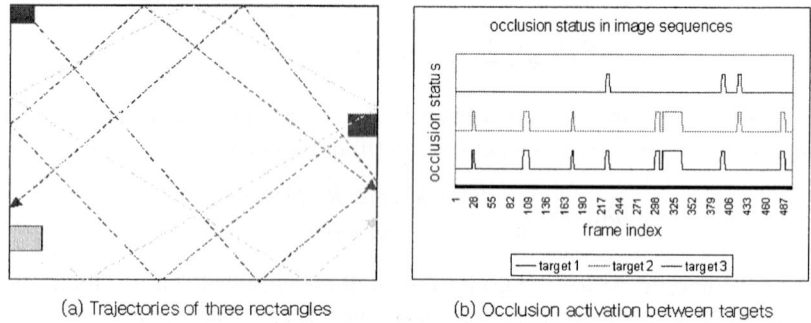

(a) Trajectories of three rectangles (b) Occlusion activation between targets

Fig. 5. Test image sequences and its trajectories

Next, according to the occlusion status, the SEA using a partial probability for the object association is applied to the test image sequences. Table 1 describes the performance of occlusion activity detection rate and the root mean square (RMS) error of an object association algorithm.

The proposed algorithm is applied on the real image sequences having an occlusion scenario. The occlusion activity detection and objection association algorithm is applied according to the occlusion status. When the occlusion status is enabled, the object association algorithm using SEA gave the estimated center points. Using computed center points, the JPDA tracking filter [6] is performed. In the initial value of the JPDA algorithm to track multi-targets, the process noise variance = 10 and the measurement noise variance = 25 are used. The initial position of two people are set to A(17, 60), B(254,147) and A(16,115), B(108,215) in Cartesian coordinates. In example 1, object A moved from left to right and object B moves from top to bottom. Object A is moved from the left-bottom to the right-top, and object B is move from the right-center to the left-center in example 2. An occlusion state is maintained for frames 34 and 24. Figure 6 depicts the tracking results and its trajectories. We know that the targets can be tracked reliably even while the occlusion status is enabled.

Table 1. Simulation results of test image sequences

Analysis of proposed methods	Results
Occlusion time	9
Total occlusion frame number	76
Accuracy of occlusion status	91.566
RMS error in estimated position x,y; object association	1.2

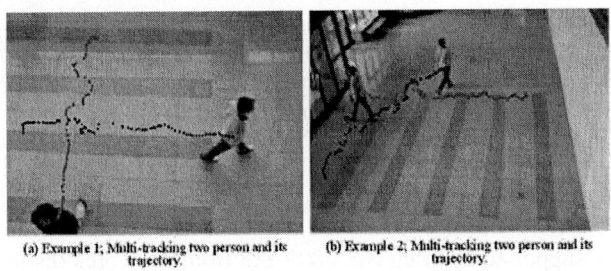

(a) Example 1; Multi-tracking two person and its trajectory.

(b) Example 2; Multi-tracking two person and its trajectory.

Fig. 6. Test image sequences and its trajectories

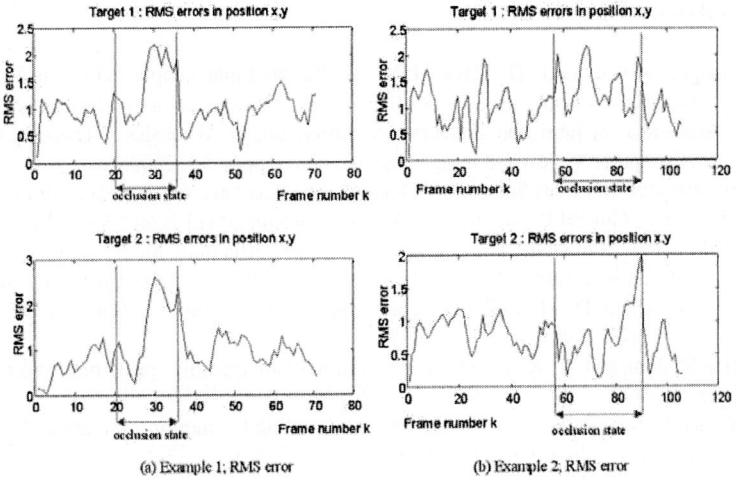

(a) Example 1; RMS error

(b) Example 2; RMS error

Fig. 7. RMS errors of test image sequences

We can evaluate the robustness of the object association algorithm using the RMS error of the computed position values as in Figure 7. In example 1, we know that the RMS error is high. This is due to the fact that the overlapping region is large. Meanwhile, (b) of Figure 7 shows that the RMS error is similar with that of the non-occlusion frames. This is due to that fact that this has a small overlapping region between targets.

4 Conclusions

In the proposed method, the occlusion activity detection gives the rules to cope with decision-making. In addition, the object association algorithm provides the estimated feature vectors even when the occlusion status is enabled. This is due to the fact that we know the target information such as velocity and acceleration. The proposed method reduced the computation time because the extra computation time for occlusion reasoning and association is spent when the occlusion status is enabled.

However, in the case of high variation of illumination change and shadow effects, some missing approximation is computed. Thus, the missing estimation of geometric information and an appearance model can give us a track failure. So, it needs a more accurate approximation and estimation method under a more complex situation and higher illumination variation environments, respectively.

Acknowledgements

This work was supported by grant No. 2003-218 from the Korea Institute of Industrial Technology Evaluation & Planning Foundation.

References

1. Haritaoglu, I, Harwood, D, Davis, L.S, "Hydra: multiple people detection and tracking using silhouettes" Visual Surveillance, Second IEEE Workshop on, pp 6 -13, June 1999.
2. S. J. McKenna, S. jabri and Z. Duric, A. Rosenfeld, H. Wechsler "Tracking Groups of people", Computer Vision and Image Understanding, pp42-56, 2000.
3. Romer Rosales and Stan Sclaroff, "3D Trajectory Recovery for Tracking Multiple Objects and Trajectory Guided Recognition of Actions", In Proc of IEEE on CVPR, June 1999.
4. Huwer, S.and Niemann, H., "Adaptive change detection for real-time surveillance applications", Visual Surveillance, IEEE International Workshop on, pp 37 -46, July 2000.
5. M. J. Swain and D. H. Ballard, "Colour indexing", International journal of Computer Vision, 7(1):11-32, 1991.
6. Y. Bar-Shalom and X. R. Li, Multitarget-multisensor tracking: principles and techniques, YBS Press, 1995.
7. W. Li and E. Salari, "Successive elimination algorithm for motion estimation, "IEEE Trans. Image processing. Vol 4, pp. 105-107, Jan. 1995.

A New Computer Algorithm Approach to Identification of Continuous-Time Batch Bioreactor Model Parameters

Suna Ertunc, Bulent Akay, Hale Hapoglu, and Mustafa Alpbaz

Ankara University, Faculty of Engineering,
Department of Chemical Engineering,
06100 Tandoğan,Ankara,Turkey
ertunc@eng.ankara.edu.tr

Abstract. The performance of a continuous-time Recursive Least Squares (CRLS) and a discrete-time Recursive Least Squares (DRLS) algorithms are examined for the growth medium temperature control of a cooling batch bioreactor in which *Saccharomyces cerevisiae* growth at aerobic condition by using Continuous-time Generalised Predictive Control (CGPC) algorithm. MATLAB programme was utilized for recursive parameter identification algorithms (CRLS and DRLS). The success or otherwise of these algorithms are estimated using parameter norm criterion for the various order of models and several input signals. There is a considerable improvement of identification algorithms with the reduced order of models. It has been shown that the performance of a DRLS algorithm is as successful as the other recursive parameter identification of a continuous-time system model.

1 Introduction

In bioprocesses, an organic compound is converted to a valuable product or products using enzymes or microorganisms called biocatalysts. Since the biocatalysts are very sensitive to changes occuring in their environment it is inevitable to control the operating parameters such as temperature, pH, dissolved oxygen concentration and substrate concentration for maintenance of optimal conditions for product formation in the complex environment in a bioreactor. Temperature is a fundamental parameter regulating microorganism growth, kinetics and overall product yield. For this reason temperature control systems are an integral part of biochemical processes that regulate the quality and the rate at which can be produced. About 40 % to 50 % of the energy stored in a carbon and energy source is converted to biological energy (ATP) during aerobic metabolism, and the rest of the energy is released as heat. Thus heat evolution is directly related to microbial growth [1].

Batch processes are extensively used to produce specialty chemicals, biotechnology, pharmaceutical and agricultural products. *S.cerevisiae* microorganism which is known as Baker's yeast is produced by using batch or fed-batch operation under aerobic conditions [2]. Although setting up, operating and modeling are available in the literature for batch processes, controlling them is quite challenging. In addition, these processes exhibit time variant dynamic behavior and recharacterized by complex, nonlinear physiologial phenomenon that are difficult to model [3,4].

A model may be obtained by examining the internal structure of the system, although it is often the case that a complete picture cannot be achieved due to unknown factor, an element which is not directly measurable or an extremely complicated process [5]. The class and accuracy of a particular model is dependent on its required application. System identification is an effective procedure for the modelling of the systems. The model structures must therefore be defined and evaluations made of the parameters contained within each models. A certain model structure should approximate the system to a chosen degree and contain all the known information about operating conditions. It must also be flexible and lead to fast parameter estimation procedures [6].

Identification of process parameters for control purposes must often be done using discrete-time computation, from samples of input-output observations. On the other hand, the process is usually of continuous-time nature, and modelled in terms of differential equations. In the previously published literatures, several approaches have been developed for the identification of continuous-time model parameters [7,8,9,10].

The major objective of this study is to identify the recursive parameters of certain models in MATLAB for control of the temperature in the growth medium. DRLS and CRLS algorithms are realized for this purpose.

2 Materials and Method

Temperature disturbances range was chosen by taking into account the real temperature change in the bioreactor during the *Saccharomyces cerevisiae* growth at aerobic condition. *Saccharomyces cerevisiae* yeast (NRRL Y-567)utilized in this study was obtained from the ARS culture collection (Northern Regional Research Center, Peoria, IL, U.S.A.). Stock cultures were maintained on agar slants containing (in g/L): Glucose (20), yeast extract (6), K_2HPO_4 (3), $(NH_4)_2SO_4$ (3.35), NaH_2PO_4 (3.76), $MgSO_4.7H_2O$ (0.52), $CaCl_2.4H_2O$ (0.01) and agar (20) (pH 5). The cells growing on the newly prepared slants were inoculated in to the same liquid medium (without agar) and cultivated at 32 °C for 24 h in an incubator-shaker.

The bioreactor given in Fig. 1 was modelled in MATLAB. In this experimental system, bioreactor temperature is measured by a thermocouple. A 2 liters bioreactor has a cooling jacket. Also sensors were placed in this to measure pH and DO in the culture medium. Cooling water was continuously fed into the jacket at changing rates as an input type. Agitation was supplied using a turbine impeller at 600 rpm. An immersed heater for heating the culture medium to the desired operating temperature was also placed in the bioreactor. Air was supplied to the bioreactor by passing through a rotometer and microbiological filter at 1 vvm.

For on-line data acquisition VISIDAQ package programme was utilized. This programming package consists of Task Designer and Display Designer. The on-line computer was used in experimental studies. In the theoretical model identification work was realized by using MATLAB.

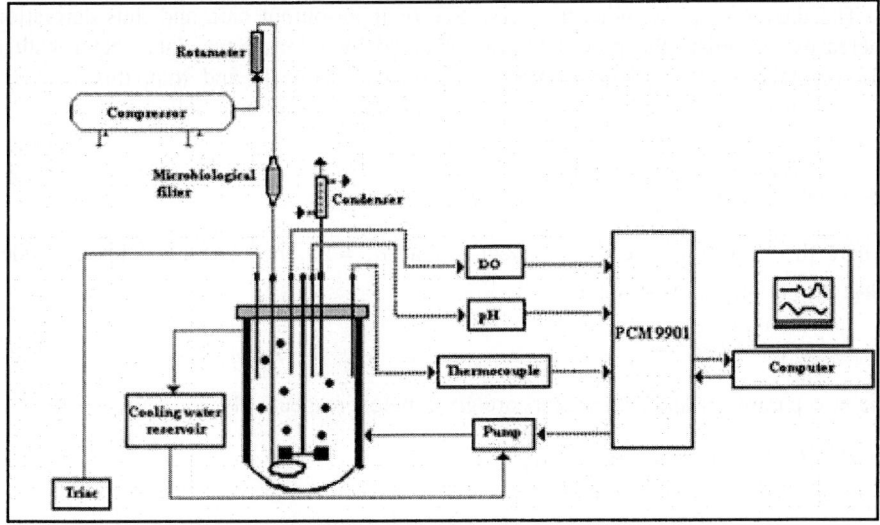

Fig. 1. Experimental system

3 Continuous-Time Recursive Least Squares Estimation (CRLS)

Continuous-time Generalised Predictive Control (CGPC) was based upon a continuous-time system model [11]. To be useful in control applications, the system model parameters should be iteratively estimated. For control purposes, a continuous-time single input-single output system model was utilized to identify the system. The model in the laplace domain is then given by

$$A(s)Y(s) = B(s)U(s) + E(s) \tag{1}$$

$$A(s) = a_0 s^n + a_1 s^{n-1} + ... + a_{n-1} s + a_n$$
$$B(s) = b_0 s^m + b_1 s^{m-1} + ... + b_{m-1} s + b_m \tag{2}$$

Eq. 1 can be rewritten as differential equation;

$$a_0 \frac{d^n y(t)}{dt^n} + a_1 \frac{d^{n-1} y(t)}{dt^{n-1}} + ... + a_n y(t) = b_0 \frac{d^m u(t)}{dt^m} + b_1 \frac{d^{m-1} u(t)}{dt^{m-1}} + ... + b_m u(t) + e(t) \tag{3}$$

The value of the parameter is given as $a_n = 1$, the estimation process can be given as linear in the parameters model;

$$y(t) = \phi^T(t)\theta + e(t) \tag{4}$$

$$\phi^T(t) = [-\frac{d^n y(t)}{dt^n}, -\frac{d^{n-1} y(t)}{dt^{n-1}}, ..., -\frac{dy(t)}{dt}, \frac{d^m u(t)}{dt^m}, \frac{d^{m-1} u(t)}{dt^{m-1}}, ..., \frac{du(t)}{dt}, u(t)] \tag{5}$$

$$\theta = [a_0, a_1, ..., a_{n-1}, b_0, b_1, ..., b_m]^T$$

The data vector includes the derivative of input-output data and thus derivation increases the noise, the new data was obtained by filtering the data vector with a polynomial. For this purpose, Y(s) was added to (Eq. 1) and than this equation divided by T(s) polynomial. The rearrange equation was given as follows:

$$\frac{Y(s)}{T(s)} = \frac{B(s)U(s)}{T(s)} + [1 - A(s)]\frac{Y(s)}{T(s)} + \frac{E(s)}{T(s)} \tag{6}$$

where the conditions are deg(T)≥deg(A) and $a_n = 1$. The rearranged linear model with the parameters is given as

$$Y_f(s) = \phi^T(s)\theta + \varepsilon(s) \tag{7}$$

the new parameter and data vectors are given below respectively;

$$\theta = [b_0, b_1, ..., b_m, -a_0, -a_1, ..., -a_{n-1}]^T \tag{8}$$

$$\phi^T(s) = \frac{1}{T(s)}([s^m, s^{m-1}, ..., 1]U(s) \quad [s^n, s^{n-1}, ..., s]Y(s)) \tag{9}$$

As it is accepted that the parameter estimation vector at time t are given in (Eq. 4), the estimated output and the prediction error are given in (Eq. 10 and 11) respectively.

$$\hat{y}(\tau) = \phi^T(\tau)\hat{\theta}(t) \quad , \tau \leq t \tag{10}$$

$$\varepsilon(t, \tau) = y(\tau) - \hat{y}(\tau) = y(\tau) - \phi^T(t)\hat{\theta}(t) \tag{11}$$

The aim is to choose the current estimate $\hat{\theta}(t)$ such that thus error is minimum over the range $0 \leq \tau \leq t$. Least squares estimation method considers a cost function of the following form to achieve this objective. The cost function for CRLS algorithm is defined as [12];

$$J(\hat{\theta}(t), t) = \frac{1}{2} e^{-\beta_c t} (\hat{\theta}(t) - \hat{\theta}_0)^T S_0 (\hat{\theta}(t) - \hat{\theta}_0) + \frac{1}{2} \int_0^t e^{-\beta_c(t-\tau)} \varepsilon^2(t, \tau) d\tau \tag{12}$$

Where $\beta_c \geq 0$, initial information matrix (S_0) is a possitive definite symmetric matrix, $\hat{\theta}_0(t)$ is the initial estimate of θ. The first term is the cost allow us to include a prior estimate in the algorithm. The second brings in the measured data into the criterion. β_c is the forgetting factor. As time t increases, the effect of old data at time $\tau \langle t$ is discounted exponentially with the elapsed time $t - \tau$.

The main equations for CRLS algorithm are given as;

$$S(t+T) = e^{-\beta_c t} S(t) + \int_t^{t+T} e^{-\beta_c(t+T-\tau)} \phi(\tau) \phi^T(\tau) d\tau \tag{13}$$

$$\hat{\theta}(t+T) = \hat{\theta}(t) + S^{-1}(t+T) \int_t^{t+T} e^{-\beta_c(t+T-\tau)} \phi(\tau)[y(\tau) - \phi^T(\tau)\hat{\theta}(t)]d\tau \qquad (14)$$

In the present work, it is noted that by filtering input-output data with a certain polynomial, the DRLS algorithm can also be used for continuous-time model parameters. The main equation for this method were given in the previously published work [13,14].

4 Results and Discussion

A CRLS algorithm with MATLAB programme was utilized for recursive parameter identification of a continuous-time system model. A DRLS algorithm was also used succesfully for the same purpose. Performance of the both identification algorithm with the estimated models of the system given in Table 1 were investigated theoretically for square wave input signal. These results were given in (Fig. 2-5).

Table 1. Continuous-time models for examining the DRLS and CRLS algorithms

	Model	Poles	Zeros
Model 1	$G(s) = \dfrac{Y(s)}{U(s)} = \dfrac{b_0 s}{a_0 s + a_1} = \dfrac{0.8s}{s + 0.5}$	-0.5	0
Model 2	$G(s) = \dfrac{Y(s)}{U(s)} = \dfrac{b_0 s^2 + b_1 s}{a_0 s^2 + a_1 s + a_2} = \dfrac{0.8s^2 + 1.2s}{s^2 + 1.5s + 0.5}$	-0.5 -1.0	0 -1.5

Affects of model order on performance criteria, parameter error norm ($\|\theta - \hat{\theta}\|/\|\theta\|$), were investigated. It is noted that increasing model order reduces the performance of the both CRLS and DRLS algorithms (Table 2).

The important parameters of the CRLS and DRLS algorithms such as initial parameter vector, covariance matrix, information matrix, forgetting factor and sampling period were determined by using trial and error method. Their values are given 0, 10000, 0.0000001, 0.99, 0.1 respectively.

It is noted that as a new approach, discrete-time identification algorithm can be used acceptably for evaluation of a continuous-time model parameters. When the certain values of the parameter such as forgetting factor, filter polynomial etc., which affect the identification performances was choosen most effectively, the widely used DRLS algorithm in previously published work has been used successfully for continuous-time model identification.

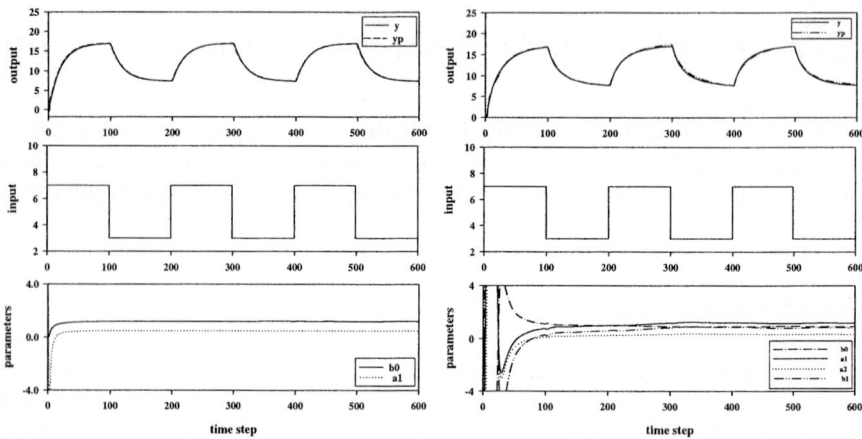

Fig. 2. DRLS identification with Model 1   **Fig. 3.** DRLS identification with Model 2

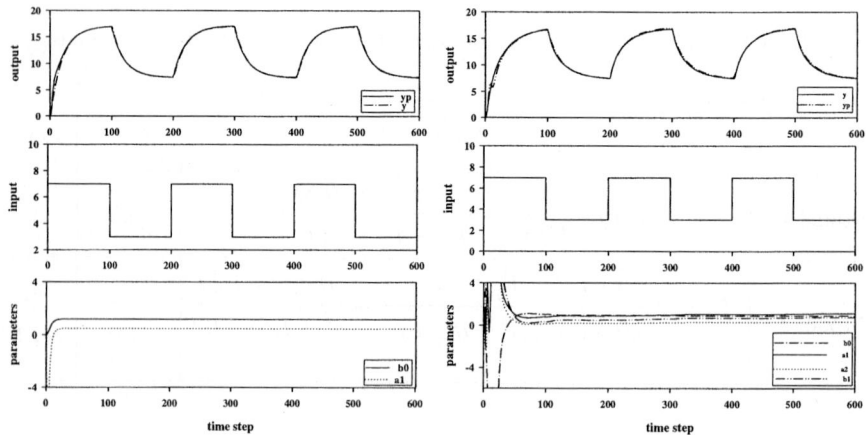

Fig. 4. CRLS identification with Model 1   **Fig. 5.** CRLS identification with Model 2

The results of parameter identification in CRLS and DRLS for Model 1 and several input signals are shown in Fig. 6-9. It is noted that the performance of a CRLS algorithm was more successfull than the performance of a DRLS algorithm (Table 2 and 3).

Table 2. Identification performance of DRLS and CRLS for the square wave input

Algorithm	Model	Parameter error norm
DRLS	Model 1	0.0109
	Model 2	0.4110
CRLS	Model 1	0.0081
	Model 2	0.2896

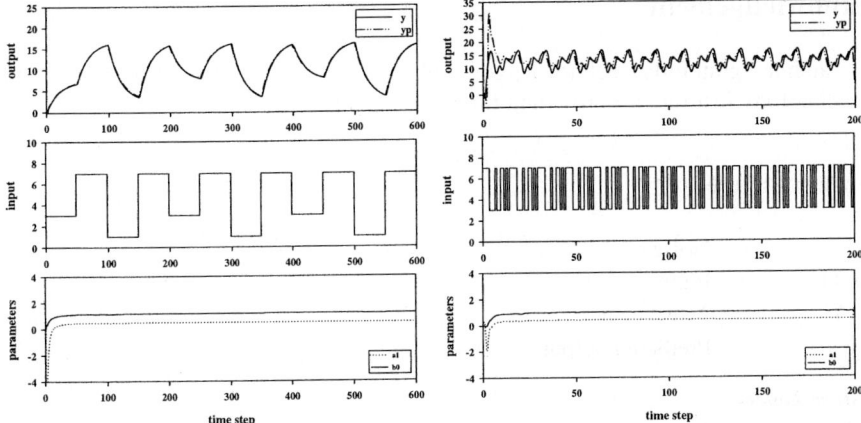

Fig. 6. DRLS identification with Model 1 and ternary input

Fig. 7. DRLS identification with Model 1 and PRBS input

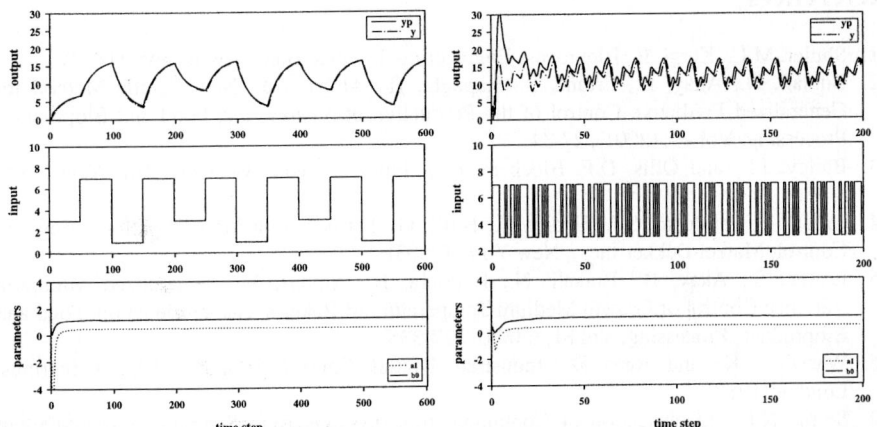

Fig. 8. CRLS identification with Model 1 and ternary input

Fig. 9. CRLS identification with Model 1 and PRBS input

Table 3. Identification performance of DRLS and CRLS for various input type

Algorithm	Model	Input signal	Parameter error norm
DRLS	Model 1	Square wave	0.0109
		PRBS	0.1936
		Ternary	0.0066
CRLS	Model 1	Square wave	0.0081
		PRBS	0.0588
		Ternary	0.0080

Acknowledgement

The author gratefully acknowledge Ankara University Research Fund and Biotechnology Institute for providing financial support.

Nomenclature

e	Noise
u, U	Input
y, Y	Actual output
$\hat{y}$	Predicted output

Greek Letters

θ	Actual parameter vector
$\hat{\theta}$	Estimated parameter vector

References

1. Shuler, M.L., Kargi, F. Bioprocess Engineering, Prentice Hall. New Jersey: (1992).
2. Bursali, N., Akay, B., Ertunc, S., Hapoglu, H., Alpbaz, M., New Tuning Method for Generalized Predictive Control of the Production of *S.cerevisiae*, Food and Bioproducts Processing, Vol 79, (2001), 27-34.
3. Bailey, J.E. and Ollis, D.F. Biochem. Eng. Fund. 2nd edn, Mc Graw Hill New York: (1986).
4. Cinar, A., Parulekar, S.J., Undey, C., Birol, G., Fermentation Modeling Monitoring and Control, Marcel Dekker Inc. ,, New York: (2003).
5. Ertunc, S., Akay, B., Bursali, N., Hapoglu, H., Alpbaz, M., Generalized Minimum Variance Control of Growth Medium Temperature of Baker's Yeast Production, Food and Bioproducts Processing, Vol 81, (2003), 327-335.
6. Warwick, K. and Rees D., Industrial Digital Control Systems, Peter Peregrinus, London:(1988).
7. Sinha, N.K., Identification of Continuous-time Systems from Samples of Input-Output data:An Introduction, Sadhana, Vol 25, part 2, April, (2000) 75-83.
8. Wang, L., Gawthrop, P., On the Estimation of Continuous Time Transfer Functions, International Journal of Control, Vol 74, (2001), 889-904.
9. Kowalczuk, Z. and Kozlowski, J., Continuous-time Approaches to Identification of Continuous-time Systems, Automatica, Vol 36, (2000),1229-1236.
10. Subrahmanyam, A.V.B., Saha, D.C., Rao, G.P., Irreducible Continuous Model Identification via Markov Parameter Estimation, Automatica, Vol 32, (1996), 249-253.
11. Demircioglu, H., Continuous-time Generalised Predictive Control (CGPC): Implementation Issues, Conference on Advances in Model-Based Predictive Control, UK: (1993),145-159.
12. Gawthrop, P.J., Continuous-time Self-Tuning Control, Vol 1-Design, Research Studies Press, Letchworth: (1987).
13. Wellstead, P.E. and Zarrop, M.B., Self-tuning Systems-Control and Signal Processing, John Wiley & Sons, New York: (1991).
14. Rao, M. And Qui, H., Process Control Engineering, Gordon and Breach Science Publishers, Switzerland : (1993).

Automated Operation Minimization of Tensor Contraction Expressions in Electronic Structure Calculations

Albert Hartono[1], Alexander Sibiryakov[1], Marcel Nooijen[3], Gerald Baumgartner[4], David E. Bernholdt[6], So Hirata[7], Chi-Chung Lam[1], Russell M. Pitzer[2], J. Ramanujam[5], and P. Sadayappan[1]

[1] Dept. of Computer Science and Engineering
[2] Dept. of Chemistry, The Ohio State University,
Columbus, OH, 43210 USA
[3] Dept. of Chemistry, University of Waterloo,
Waterloo, Ontario N2L BG1, Canada
[4] Dept. of Computer Science
[5] Dept. of Electrical and Computer Engineering,
Louisiana State University, Baton Rouge, LA 70803 USA
[6] Computer Sci. & Math. Div., Oak Ridge National Laboratory,
Oak Ridge, TN 37831 USA
[7] Quantum Theory Project, University of Florida,
Gainesville, FL 32611 USA

Abstract. Complex tensor contraction expressions arise in accurate electronic structure models in quantum chemistry, such as the Coupled Cluster method. Transformations using algebraic properties of commutativity and associativity can be used to significantly decrease the number of arithmetic operations required for evaluation of these expressions, but the optimization problem is NP-hard. Operation minimization is an important optimization step for the Tensor Contraction Engine, a tool being developed for the automatic transformation of high-level tensor contraction expressions into efficient programs. In this paper, we develop an effective heuristic approach to the operation minimization problem, and demonstrate its effectiveness on tensor contraction expressions for coupled cluster equations.

1 Introduction

Currently, manual development of accurate quantum chemistry models is very tedious and takes an expert several months to years to develop and debug. The Tensor Contraction Engine (TCE) [2, 1] is a tool that is being developed to reduce the development time to hours/days, by having the chemist specify the computation in a high-level form, from which an efficient parallel program is automatically synthesized. This should enable the rapid synthesis of high-performance implementations of sophisticated ab-initio quantum chemistry models, including models that are too tedious for manual development by quantum chemists. An important first step in the synthesis process of the TCE is that of algebraic manipulation of the input tensor contraction expressions, to find an equivalent form with minimized operation count.

We illustrate the operation minimization problem using simple examples. Consider the following tensor contraction expression involving three tensors t, f and s, with a and c representing virtual orbital indices with range V, and i and j representing occupied orbital indices with range O. Computed as a single nested loop computation, the number of multiply-accumulate operations needed would be O^2V^2.

(1) $\quad r_i^a \mathrel{+}= \sum_{c,k} t_i^c f_c^k s_k^a,$ \hfill (cost O^2V^2)

However, by performing a two-step computation with an intermediate I, it is possible to compute the result using $2OV^2$ operations:

(2) $\quad I_c^a = \sum_k f_c^k s_k^a,$ \hfill (cost OV^2)

(3) $\quad r_i^a \mathrel{+}= \sum_c t_i^c I_c^a,$ \hfill (cost OV^2)

Another possibility using O^2V computations, which is more efficient when $V > O$ (as is usually the case), is shown below:

(4) $\quad I_i^k = \sum_c t_i^c f_c^k,$ \hfill (cost O^2V)

(5) $\quad r_i^a \mathrel{+}= \sum_k I_i^k s_k^a,$ \hfill (cost O^2V)

The above example illustrates the problem of single-term optimization, also called strength reduction: find the best sequence of two-tensor contractions to achieve a multi-tensor contraction. Different orders of contraction can result in very different operation costs; for the above example, if the ratio of V/O were 10, there is an order of magnitude difference in the number of arithmetic operations for the two choices.

For more complex expressions with several tensors to be contracted, the number of possible ways of forming intermediates is exponential in the number of tensors. The single-term optimization problem is a generalization of the well known matrix-chain multiplication problem, but while the latter has a simple polynomial time dynamic programming solution, the former problem has been shown to be NP-complete [8].

Let us next consider an expression with two terms:

(6) $\quad r_{ij}^{ab} \mathrel{+}= \sum_{c,d} t_i^c s_j^d v_{cd}^{ab} + \sum_{c,d} u_{ij}^{cd} v_{cd}^{ab},$ \hfill (cost $2O^2V^4$)

If each term were individually optimized via strength reduction, we would have:

(7) $\quad I_{id}^{ab} = \sum_c t_i^c v_{cd}^{ab},$ \hfill (cost OV^4)

(8) $\quad r_{ij}^{ab} \mathrel{+}= \sum_d s_j^d I_{id}^{ab} + \sum_{c,d} u_{ij}^{cd} v_{cd}^{ab},$ \hfill (cost $O^2V^3 + O^2V^4$)

However, a better approach to reducing the overall operation cost would be as follows:

(9) $\quad I_{ij}^{cd} = t_i^c s_j^d + u_{ij}^{cd},$ \hfill (cost O^2V^2)

(10) $\quad r_{ij}^{ab} \mathrel{+}= \sum_{c,d} I_{ij}^{cd} v_{cd}^{ab},$ \hfill (cost O^2V^4)

Thus it can be seen that single-term optimization (strength reduction) is not an optimal strategy. We have to look at the expression in the global context to determine the optimal evaluation. Evaluation of the binary terms in an expression constitute an intrinsic cost to evaluating the tensor product. Little can be done (except for instances of factorization of the form $AC + BC \rightarrow (A+B)C$) to reduce the cost of the binary terms. However, the cost of evaluating terms that are ternary or higher can be greatly reduced by combining them with the binary terms that have to be evaluated anyway. In the present example, the additional cost of evaluating the ternary term is reduced from $OV^4 + O^2V^3$ to O^2V^2. The expensive O^2V^4 multiplication that would be counted in single term optimization disappears as the multiplication has to be done anyway in the binary term.

The goal of operation minimization is to find an optimal or near-optimal factorization of the input tensor contraction expression to evaluate the ternary+ terms, given the presence of the binary terms that carry an intrinsic basic cost. In this paper, we develop algorithms for operation minimization. The solution presented in this paper is an extension of the techniques first reported in [11]. The conceptually simplest approach is to use an exhaustive search algorithm that is guaranteed to determine the optimal factorization. However, its runtime grows exponentially with the number of terms in the tensor contraction expression, making it infeasible for use on the more complex coupled cluster equations. We then develop heuristic search strategies for the operation minimization problem. The best algorithm is found to be a random-descent heuristic, which is then used to explore the generated solutions for a range of values of V/O. The results validate the effectiveness of the use of an automated approach to generating operation-minimal factorizations for large tensor contraction expressions.

2 Related Work

Compilers use common subexpression elimination to reduce the number of arithmetic operations [4]. However they do not consider algebraic properties such as associativity and distributivity. Computer algebra systems typically contain factorization algorithms, e.g., for finding roots of polynomials [3]. Similarly, an algorithm based on factor graphs can be used to factor functions of many variables into products of local functions [6]. However, the emphasis of these approaches is mainly on symbolic manipulation instead of on minimizing operation counts based on index range information. Winograd [13] addressed the general problem of evaluating multiple expressions that share common variables using the minimum number of arithmetic operations. Miller [9] suggested several analytical and numerical techniques for reducing the operation count in computational electromagnetic applications.

The work presented in this paper builds upon methods developed in a recent thesis by Sibiryakov [11]. The problem of strength reduction for arbitrary tensor contraction expressions was addressed in [8, 7]. We are not aware of other work that has addressed in a general manner the operation minimization problem that we consider in this paper. In developing efficient implementations of electronic structure methods such as the coupled cluster methods [10, 12], quantum chemists have used domain heuristics for strength reduction and factorization for specific kinds of tensor contraction expressions, but have not developed approaches to solve the operation minimization problem for arbitrary tensor contraction expressions.

3 Operation Minimization Algorithms

In this section, we outline several algorithms for the operation minimization problem. Due to space limitations, pseudo-code and details are omitted, but may be found in [5].

3.1 Exhaustive Search

We first describe an exhaustive search algorithm that systematically evaluates all possible factorizations of the input tensor contraction expression to determine the form with lowest operation count. This search is implemented recursively using memoization, which is equivalent to a dynamic programming approach implemented in a top-down manner.

Considering a particular tensor as a factor, exhaustive search enumerates all possible factorizations, which grows exponentially. If a factor appears in n terms of the tensor contraction expression, the number of possible factorizations with respect to that factor is $2^n - n$. For instance, all possible factorizations of an expression $AB + AC + AD$ are $AB + AC + AD$, $A(B+C) + AD$, $A(B+D) + AC$, $A(C+D) + AB$, and $A(B+C+D)$.

As in standard dynamic programming, a storage table is maintained with solutions for subexpressions; hence, we need not re-evaluate subexpressions that have been previously considered. Matching two equivalent expressions requires generating canonical forms of both expressions. If a canonical form of a subexpression is found as a key in the storage table, the corresponding entry value, which is the optimal solution of the subexpression, is fetched and replicated. The indices of replica may differ from the original subexpression; thus, renaming indices of the replica is required.

The exhaustive search algorithm is guaranteed to find the operation-minimal factorization of the input expression, but since its time complexity grows exponentially with the number of terms, it may be impractical to use in optimizing expressions with a large number of terms. We therefore also implemented several heuristic search strategies for operation minimization.

3.2 Time-Limited Exhaustive Search with Tier-Based Partitioning

By imposing a time limit, we can avoid an indefinitely long search time that often occurs in exhaustive search. Each time exhaustive search exceeds the specified time limit, we suspend the search and store the result of the partially executed exhaustive search. Afterward the original expression is divided into two smaller groups each with half the original group's terms. These two subsets of terms can be individually factorized using the same time-limited exhaustive search and the partially factored terms can then be recombined. The cost of the combined expression is compared with the result of the timed exhaustive search that was previously interrupted. The result with the minimum operation count is returned. The splitting process is continued till each group of terms is successfully factorized within the time limit.

Prior to each splitting, it is essential to sort the expression terms in a decreasing order of term cost (as determined by single-term optimization), allowing higher-ordered terms to be placed and optimized in the same group after the splitting.

In the worst case, a splitting can occur whenever a time-limited exhaustive search is applied. We can view these recursive splittings as a binary tree in which each node represents one exhaustive search. The maximum number of nodes in such a binary tree will be $\sum_{i=0}^{log_2 N}(\frac{N}{2^i}) \approx 2N$, where N indicates the number of terms in the original expression. Therefore, the time complexity of time-limited exhaustive search is $O(TN)$, where T is the given time limit. For generating experimental results in this paper, the time limit used for exhaustive search was set to ten minutes. In addition, very similar results were also obtained with a shorter time limit of ten seconds, demonstrating the efficacy of the algorithm in finding a reasonable solution quickly.

We evaluated another approach to partitioning based on tier groups. Two terms are placed into the same tier group if they have the same number of tensors. The terms in the input expression are first partitioned into tier groups. Optimizing and combining tier groups is done incrementally. Suppose we have groups at tier 2, tier 3, and tier 4. We first optimize tier 2 with timed exhaustive search. The optimized tier 2 terms are then grouped together with the unoptimized tier 3 terms and then optimized with timed exhaustive search. Then this result is grouped with the unoptimized tier 4 terms and then optimized again with timed exhaustive search.

3.3 Direct Descent Search

Direct descent is a greedy algorithm that chooses the best local factorization at every step. All pairs of terms that can be combined using the distributive property of multiplication over addition are considered, and the transformation that provides the greatest reduction in operation count is chosen. For a particular factor, if the number of factorizable terms in the expression is n the total number of possible two-term factorizations is $n(n-1)/2$. At every step, all possible two-term factorizations are evaluated and the best one is chosen; this process is repeated until no more factorization is possible. Based on the number of factorizations considered at each step, the runtime complexity of direct descent obviously grows polynomially with the degree of terms in the input expression.

3.4 Random Descent Search

Random descent search is a modified version of direct descent search that attempts to avoid some local minima by making random choices for two-term factorization at the initial steps. These random moves are then followed by direct descent moves. Through experimentation, it was found best to make the number of random moves to be one fourth of the total number of terms in the input expressions. Using too many random factorizations at the initial steps was found to give poorer results; too few random factorizations at the initial steps did not contribute a significant improvement. In order to further minimize the operation count, we first execute a direct descent search and store its factorization result; after that, one hundred attempts of random descent are repeated. The best result from these one hundred tries is compared with the result of direct descent that was initially executed. The result with the minimum operation count is returned.

4 Experimental Results

We have implemented the algorithms for searching a formula with minimum number of operations as described previously. They were tested on complex tensor contraction expressions that appear in the "coupled cluster" family of quantum chemical methods. We used the coupled cluster equations including just single and double excitations (CCSD) and also single, double, and triple excitations (CCSDT) as representatives of the many different computational chemistry methods based on tensor contraction expressions. These methods involve coupled equations which determine the single excitation amplitudes (referred to here the "T1" equation), double excitation amplitudes (T2), and in the case of CCSDT, triple excitations (T3).

Table 1 shows the number of terms in each of the equations, along with the number of arithmetic operations of evaluating the equation. The number of arithmetic operations depends upon O and V, which vary depending on the molecule and desired quality of the simulation, but a typical range is $1 \leq V/O \leq 100$. To provide concrete comparisons, we set O to 10 and V to 100, giving the numerical values in Table 1. This is representative of calculations of modest size that could be done on a workstation and will be used throughout this paper unless otherwise noted.

In order to focus on the effects of the optimization algorithms, we eliminate the binary terms from the input equations and consider only the ternary and higher terms, as described in the introduction. The optimal cost of the binary terms of each equation can be seen in the last column of Table 1.

Table 2 illustrates the results obtained by optimizing the five equations with the algorithms described previously. Exhaustive search results are shown only for the T1

Table 1. Characteristics of the fully unfactorized input equations used in this experiment

Equation	Number of terms	Operation count (no optimization)	Operation count (with single term optimization only)	Operation count of optimal binary terms
CCSD T1	14	$1.78 \cdot 10^{10}$	$3.11 \cdot 10^{8}$	$2.24 \cdot 10^{8}$
CCSD T2	31	$4.90 \cdot 10^{13}$	$3.58 \cdot 10^{10}$	$2.27 \cdot 10^{10}$
CCSDT T1	15	$2.08 \cdot 10^{10}$	$2.31 \cdot 10^{9}$	$2.22 \cdot 10^{9}$
CCSDT T2	37	$6.13 \cdot 10^{13}$	$3.00 \cdot 10^{11}$	$2.45 \cdot 10^{11}$
CCSDT T3	47	$9.34 \cdot 10^{16}$	$3.26 \cdot 10^{13}$	$2.26 \cdot 10^{13}$

Table 2. The operation counts of expressions optimized by different algorithms

| Equation | Single Term | Ternary+ operation count | | | |
		Exhaustive	Timed Exhaustive with Tier-Based Partitioning	Direct Descent	Random Descent
CCSD T1	$8.65 \cdot 10^{7}$	$4.73 \cdot 10^{7}$	$4.73 \cdot 10^{7}$	$4.73 \cdot 10^{7}$	$4.73 \cdot 10^{7}$
CCSD T2	$1.31 \cdot 10^{10}$	–	$5.18 \cdot 10^{9}$	$5.33 \cdot 10^{9}$	$5.14 \cdot 10^{9}$
CCSDT T1	$8.65 \cdot 10^{7}$	$4.73 \cdot 10^{7}$	$4.73 \cdot 10^{7}$	$4.73 \cdot 10^{7}$	$4.73 \cdot 10^{7}$
CCSDT T2	$5.52 \cdot 10^{10}$	–	$5.57 \cdot 10^{9}$	$5.57 \cdot 10^{9}$	$5.37 \cdot 10^{9}$
CCSDT T3	$9.94 \cdot 10^{12}$	–	$9.80 \cdot 10^{11}$	$9.17 \cdot 10^{11}$	$9.17 \cdot 10^{11}$

Table 3. Ternary+ operation count for CCSD T2

Optimized for V/O	Actual V/O					
	1	2	3	5	10	100
2	$7.41 \cdot 10^6$	$4.52 \cdot 10^7$	$1.46 \cdot 10^8$	$7.20 \cdot 10^8$	$7.47 \cdot 10^9$	$4.41 \cdot 10^{13}$
5	$1.05 \cdot 10^7$	$5.88 \cdot 10^7$	$1.73 \cdot 10^8$	$7.08 \cdot 10^8$	$5.14 \cdot 10^9$	$4.67 \cdot 10^{12}$
10	$1.07 \cdot 10^7$	$5.95 \cdot 10^7$	$1.74 \cdot 10^8$	$7.13 \cdot 10^8$	$5.15 \cdot 10^9$	$4.67 \cdot 10^{12}$

Opt. V/O	Leading terms of cost function in symbolic form
2	$2O^3V^3 + 2O^4V^2 + \underline{4OV^4} + 10O^2V^3 + 10O^3V^2 + 6O^4V + \ldots$
5	$4O^3V^3 + 4O^4V^2 + 6O^2V^3 + 8O^3V^2 + 6O^4V + \ldots$
10	$4O^3V^3 + 4O^4V^2 + 6O^2V^3 + 8O^3V^2 + 6O^4V + \ldots$

Table 4. Ternary+ operation count for CCSDT T2

Optimized for V/O	Actual V/O					
	1	2	3	5	10	100
2	$7.85 \cdot 10^6$	$4.77 \cdot 10^7$	$1.54 \cdot 10^8$	$7.52 \cdot 10^8$	$7.70 \cdot 10^9$	$4.43 \cdot 10^{13}$
5	$9.88 \cdot 10^6$	$5.89 \cdot 10^7$	$1.78 \cdot 10^8$	$7.55 \cdot 10^8$	$5.63 \cdot 10^9$	$5.28 \cdot 10^{12}$
10	$1.09 \cdot 10^7$	$6.13 \cdot 10^7$	$1.80 \cdot 10^8$	$7.40 \cdot 10^8$	$5.37 \cdot 10^9$	$4.88 \cdot 10^{12}$

Opt. V/O	Leading terms of cost function in symbolic form
2	$2O^3V^3 + 2O^4V^2 + \underline{4OV^4} + 12O^2V^3 + 12O^3V^2 + 6O^4V + \ldots$
5	$4O^3V^3 + 2O^4V^2 + 12O^2V^3 + 16O^3V^2 + 6O^4V + \ldots$
10	$4O^3V^3 + 4O^4V^2 + 8O^2V^3 + 10O^3V^2 + 6O^4V + \ldots$

equations because they are the only ones small enough for this approach to be feasible. All heuristic algorithms find the optimal factorization in small cases (i.e., the T1 equations), and in the other cases produce very similar results, which have less than one percent differences. The direct descent results illustrate its tendency to get stuck in local minima and not find an optimal factorization. Random descent sometimes offers improvement and consistently performs the best of all of the algorithms described.

To examine the behavior of the random descent search in more detail, we examined the effect of varying the V/O ratio. Changing this ratio will change the actual costs of each term and may even change the optimal factorization. Tables 3 and 4 illustrates these effects. Operation counts are shown for V/O ratios ranging from 1 to 10, for factorizations of the CCSD T2 and CCSDT T2 equations that have been optimized explicitly for V/O ratios of 2, 5, and 10. This is representative of how the results of this optimization are likely to be used in practice: code will be automatically generated for selected values of V/O spanning the range of interest, and at runtime, the best available version will be selected based on the actual V and O for the molecule under study. The results clearly illustrate the value of tailoring the factorization to the V/O of interest. For example, using the factorization that is optimal for $V/O = 2$ for a calculation in which $V/O = 100$ yields an operation count that is an order of magnitude larger than in the more optimal case using the algorithm optimized for $V/O = 10$. Conversely, using an algorithm optimized for the large $V/O = 10$ ratio is clearly not optimal if V/O is small, e.g. 1. In Table 3 the symbolic operation count is given for the CCSD T2 equation for the various factorization solutions. For the larger V/O ratios of 5 and 10, no terms are used that scale as V^4 or OV^4, as are present (and underlined) in the factorization scheme optimized for $V/O = 2$. Instead the algorithm prefers to use more terms that scale as O^3V^3 (four instead of two), and there is an interesting tradeoff between terms that formally scale as N^5 vs. N^6, where N indicates O or V.

Similar conclusions can be drawn from Table 4. Comparing the optimized algorithms for $V/O = 2$ and $V/O = 10$ for various values of actual V/O ratios we clearly see that for small V/O values, the $V/O = 10$ solution is about 30% more costly than the $V/O = 2$ solution, while at the other end of the spectrum the $V/O = 10$ solution is about an order of magnitude more efficient. Similar tradeoffs as in case of the CCSD T2 equations are at work to determine the best overall factorization scheme, depending on the actual V and O values.

The present computer optimized factorization can be contrasted with current (handwritten) implementations of coupled cluster methods. In traditional implementations, factorization is considered only at a symbolic level, trying to reduce the V exponent first, then the O exponent, then the factor in front of the cost term; typically, little attention is paid to terms beyond the highest order (in the sum of the O and V exponents). This approach doesn't fully consider the ratio of V/O, and the possibility that terms considered lower order might result in an operation count comparable to higher order terms with a different balance of O and V exponents. The equation parts of Tables 3 and 4 illustrates this idea, with the symbolic costs of the different factorizations found for different V/O ratios. Comparing, for example, the costs of the CCSD T2 equation factored for $V/O = 2$ and $V/O = 5$, we observe that N^6 terms have *higher* coefficients in $V/O = 5$ than in $V/O = 2$, while the OV^4 term has been entirely eliminated from $V/O = 5$. The larger V space means that it is more cost-effective to evaluate more N^6 terms with a lower V exponent than the N^5 term OV^4. A similar cross-over occurs in the CCSDT T2 equations between $V/O = 2$ and $V/O = 5$. Moreover, changes in term coefficients take place in the CCSDT T2 equations where $V/O = 5$ and $V/O = 10$. In the $V/O = 10$ equation the coefficient of the O^4V^2 term is higher than that in the $V/O = 5$ equation, while it is vice versa for the O^2V^3 term. This is an important result because it runs counter to the intuition (and accepted practice) of most quantum chemists. Let us note that it is not only the ratio V/O which determines the optimal factorization, but also their individual sizes as there is a trade-off between overall N^5 and N^6 terms in achieving optimal performance.

Interestingly, the random descent algorithm not necessarily finds the optimal factorization. This can be seen from Table 4 where the algorithm optimized for $V/O = 10$ performs better for the actual $V/O = 5$ ratio than the algorithm that was explicitly optimized for this case. This shows that there is still some room for improvement.

Table 5. Percentage of ternary+ operation count

V/O	1	2	3	5	10	100
%ternary+ (CCSD)	51.72	42.50	36.98	30.47	18.47	2.26
%ternary+ (CCSDT)	18.94	13.57	10.45	7.08	3.87	0.33

To put these results, obtained for the ternary and higher terms only, in proper perspective, Table 5 shows the percentage of the total computational cost due to the ternary and higher terms as a function of the V/O ratio. It is seen that the ternary+ cost is a sizable fraction of the overall calculation for lower V/O ratios, but it can rapidly become a rather small fraction of the overall cost if V/O is large, in particular for CCSDT. In such cases single term optimization might suffice. However it must be noted that these percentages do not tell the entire story. On modern hierarchical memory systems, the binary terms can typically be implemented with significantly greater efficiency than the ternary+ terms, so that a simple operation count underestimates their importance to the overall computation. Further, the present optimization scheme for factorization is quite generally applicable, and the efficiency gains can be expected to be very relevant for computational schemes that are not dominated by the binary terms. The factorization of more complicated, yet efficient theoretical models in quantum chemistry will be explored in our future work.

5 Conclusions

In this paper we presented heuristic and exhaustive search algorithms for operation minimization of complex tensor expressions occurring for example in quantum chemistry. It has been demonstrated that optimal factorization depends on the precise sizes of the index ranges of the tensors involved, and therefore very different computer implementations will be optimal for different size problems. We found that the random descent algorithm works best, although this search algorithm can be expensive for complicated cases. The time-limited exhaustive search with tier-based partitioning algorithm is often a cost-effective alternative; in addition, we believe this can provide a suitable starting point that can be subject to further optimizations.

Acknowledgments. This work has been supported in part by the U.S. National Science Foundation, the Laboratory Directed Research and Development Program of Oak Ridge National Laboratory (ORNL), and by a Discovery grant from the Natural Sciences and Engineering Research Council of Canada. ORNL is managed by UT-Battelle, LLC for the US Dept. of Energy under contract DE-AC-05-00OR22725.

References

1. G. Baumgartner, A. Auer, D. Bernholdt, A. Bibireata, V. Choppella, D. Cociorva, X. Gao, R. Harrison, S. Hirata, S. Krishnamoorthy, S. Krishnan, C. Lam, Q. Lu, M. Nooijen, R. Pitzer, J. Ramanujam, P. Sadayappan, and A. Sibiryakov. Synthesis of high-performance parallel programs for a class of ab initio quantum chemistry models. *Proceedings of the IEEE*, 93(2):276–292, February 2005.
2. G. Baumgartner, D.E. Bernholdt, D. Cociorva, R. Harrison, S. Hirata, C. Lam, M. Nooijen, R. Pitzer, J. Ramanujam, and P. Sadayappan. A high-level approach to synthesis of high-performance codes for quantum chemistry. In *Proc. of Supercomputing 2002*, November 2002.
3. B. Buchberger, G. Collins, and R. Loos, editors. *Computer Algebra: Symbolic and Algebraic Computation*. Springer-Verlag, New York, 1983.

4. C.N. Fischer and R.J. LeBlanc Jr. *Crafting a Compiler*. Benjamin/Cummings, 1991.
5. A. Hartono, A. Sibiryakov, M. Nooijen, G. Baumgartner, D. Bernholdt, S. Hirata, C. Lam, R. Pitzer, J. Ramanujam, and P. Sadayappan. Automated operation minimization of tensor contraction expressions in electronic structure calculations. Technical Report OSU-CISRC-2/05-TR10, Computer Science and Engineering Department, The Ohio State University, 2005.
6. F. Kschischang, B. Frey, and H. Loeliger. Factor graphs and the sum-product algorithm. *IEEE Transactions on Information Theory*, 47(2):498–519, February 2001.
7. C. Lam. *Performance Optimization of a Class of Loops Implementing Multi-Dimensional Integrals*. PhD thesis, The Ohio State University, Columbus, OH, August 1999.
8. C. Lam, P. Sadayappan, and R. Wenger. On optimizing a class of multi-dimensional loops with reductions for parallel execution. *Parallel Processing Letters*, 7(2):157–168, 1997.
9. E.K. Miller. Solving bigger problems by decreasing the operation count and increasing computation bandwidth. *Proceedings of the IEEE*, 79(10):1493–1504, October 1991.
10. G.E. Scuseria, C.L. Janssen, and H.F. Schaefer III. An efficient reformulation of the closed-shell coupled cluster single and double excitation (CCSD) equations. *The Journal of Chemical Physics*, 89(12):7382–7387, 1988.
11. A. Sibiryakov. Operation Optimization of Tensor Contraction Expression. Master's thesis, The Ohio State University, Columbus, OH, August 2004.
12. J.F. Stanton, J. Gauss, J.D. Watts, and R.J. Bartlett. A direct product decomposition approach for symmetry exploitation in many-body methods. I. Energy calculations. *The Journal of Chemical Physics*, 94(6):4334–4345, 1991.
13. S. Winograd. *Arithmetic Complexity of Computations*. Society for Industrial and Applied Mathematics, Philadelphia, PA, 1980.

Regularization and Extrapolation Methods for Infrared Divergent Loop Integrals

Elise de Doncker[1], Shujun Li[1], Yoshimitsu Shimizu[2],
Junpei Fujimoto[2], and Fukuko Yuasa[2]

[1] Western Michigan University,
{elise, sli}@cs.wmich.edu
http://www.cs.wmich.edu/~elise
[2] High Energy Accelerator Research Organization(KEK),
Oho 1-1, Tsukuba, Ibaraki, 305, Japan
{yoshimitsu.shimizu, junpei.fujimoto, fukuko.yuasa}@kek.jp

Abstract. Loop integrals occur in higher order perturbation calculations for the cross section of particle interactions in high energy physics. In previous work we introduced a numerical extrapolation method to handle a class of Feynman loop diagrams where the integrand shows a singular behavior on a hypersurface which may intersect the domain of integration. The integral is considered in the limit as a parameter in the integrand tends to zero. Under certain conditions, the extrapolation process achieves convergence acceleration to the limit. In order to handle massless cases, we apply a dimensional regularization technique to extract infrared divergences from the integral. We illustrate the combined technique using a scalar one-loop sample integral.

1 Introduction

Loop integrals occur in higher order perturbation terms of the scattering amplitude, which is used for cross section computations of particle interactions in high energy physics. The cross section of a particle interaction gives the probability of a given configuration in energy-momentum space (E, p^1, p^2, p^3).

Figure 1 gives an example of a one-loop Feynman diagram [2]. In a Feynman diagram, each line, termed *propagator* is associated with a particle and can be straight or wavy depending on whether the particle is a fermion or boson, respectively. M and m are particle masses. A propagator corresponds to an intermediate state of a particle, where it is not observable. Particles collide at the *vertices* of the diagram, according to a *coupling constant* g which represents the strength of the interaction.

The diagram of Figure 1 is a *one-loop* diagram as it exhibits a single loop. No-loop (*tree diagram*) and multi-loop configurations are possible. The number of vertices N specifies the type of the diagram (e.g., *vertex* for $N = 3$ and *box* for $N = 4$).

The scattering amplitude T is expanded as a (*perturbation*) series in g,

$$T = T_0 + T_1 g + T_2 g^2 + \ldots. \tag{1}$$

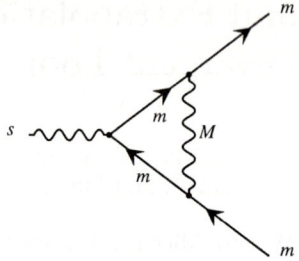

produced by GRACEFIG

Fig. 1. Vertex example

Here, T_0 represents the tree amplitude, and T_k the k-loop amplitude for $k \geq 1$. The cross section is then derived as a series from (1) (see [8]).

A general form of the scalar one-loop N-point integral is given in [3], as a 4-dimensional integral over the loop momenta. In [8], a sample one-loop integral with three propagators is introduced as

$$\mathcal{I} = \frac{1}{(2\pi)^4 i} \int_{-\infty}^{\infty} d^4 l \frac{1}{((p-l)^2 - m^2 + i\varepsilon)((q+l)^2 - m^2 + i\varepsilon)((l^2 - M^2 + i\varepsilon)},$$

where l is the loop momentum, and p and q are given momenta which are assumed to satisfy $p^2 = q^2 = m^2$. Furthermore, $\varepsilon > 0$ is a real constant which is supplied to prevent the integral from diverging. The physical scattering amplitude contains this type of integrals and its value is defined at $\varepsilon = 0$. Using the identity

$$\frac{1}{A_1 A_2 A_3} = \Gamma(3) \int_0^1 dx_1 \int_0^{1-x_1} dx_2 \frac{1}{(x_1 A_1 + x_2 A_2 + (1 - x_1 - x_2) A_3)^3}$$

and integrating over l then results in

$$\mathcal{I} = -\frac{1}{16\pi^2} \int_0^1 dx_1 \int_0^{1-x_1} dx_2 \frac{1}{\mathcal{F}(x_1, x_2) - i\varepsilon}, \tag{2}$$

with

$$\mathcal{F}(x_1, x_2) = -s x_1 x_2 + m^2 (x_1 + x_2)^2 + M^2 (1 - x_1 - x_2) \tag{3}$$

($s = (p+q)^2$ denotes the squared energy). The integral $\mathcal{I}$ of (2) is called a *scalar loop integral* as the numerator of the integrand equals 1. In a *non-scalar* integral, the numerator is generally a polynomial.

A generalization to N particles can be written as

$$I = \int_0^\infty dx_1 dx_2 \ldots dx_N \frac{\delta(1 - \sum_{j=1}^N x_j)}{((\mathcal{F}_N(x_1, x_2, \ldots, x_n) - i\varepsilon)^{N-2}}. \tag{4}$$

While analytical formulations were given for scalar one-loop integrals, e.g., in [9, 10], there is considerable interest in numerical methods which could handle non-scalar cases in a similar way.

In [2] we introduced a numerical extrapolation method to handle Feynman loop diagrams where the integrand shows a singular behavior on a hypersurface which may intersect the domain of integration. The scalar 1-loop vertex integral was considered as a study case.

The extrapolation procedure yields an approximation to the limit as ε tends to 0, under certain conditions on the asymptotic behavior of the integral with respect to ε. The infrared singularity, where $M/m \to 0$, affects the asymptotic behavior. The effect of the singularity (at $x_1 = x_2 = 0$) emerges in the integral (2) above as M is set to 0 in (3).

In this paper we investigate a combination of the extrapolation for $\varepsilon \to 0$ with the dimensional regularization method by Binoth and Heinrich [1]. The latter extracts poles introduced by the infrared singularity (but does not handle the internal singularity). Dimensional regularization is applied in QCD (Quantum Chromodynamics) which deals with the strong interactions between quarks and gluons (both are massless).

2 Regularization Techniques

One method to regulate the infrared divergence is by using the fictitious mass λ of the photon as $M = \lambda$. This results in a singular behavior of the form $\log \lambda$ when (2) is integrated analytically.

Another method uses *dimensional regularization*. For the latter, we re-write the exponent of the denominator in (4) as $N - D/2$ with the dimension $D = 4 - 2\eta$ as $\eta \to 0$. As shown in [1], this results in a Laurent expansion of the integral as a function of η, from which the poles can be extracted.

Both methods introduce non-physical situations: the mass of the photon does not correspond to nature, and the extra dimension from 4 is also non-physical. This implies that the final results should not depend on the fictitious photon mass, or on $1/\eta$. The *Kinoshita-Lee-Nauenberg theorem* [4, 5, 6] implies that, after accumulating the amplitude with respect to one-loop diagrams and emission of the photon, the infrared divergence disappears, i.e., the terms involving the photon mass vanish; or the coefficient of the $1/\eta$ term resulting in the dimensional regularization is zero.

The (linear) extrapolation procedure of [2] gives good extrapolated results for $m = m_e = 0.511 \; 10^{-3}$ GeV, $M = 10^{-5}$ GeV and $s = 100$ GeV^2, using small ε values in the integrand. As an example, the approximated imaginary part gives 0.86805413, in agreement with the analytical expression given in [3], for $\varepsilon = 2^{-32} \; (*2^{-1}) \; 2^{-41}$ and double precision arithmetic. This run takes time of the order of 20-25 seconds on a 2.5 GHz Athlon processor laptop.

The convergence behavior is illustrated for the real part in Table 1. For the ε values listed in the leftmost column, the numerical integration results for (2) (disregarding the constant factor $-1/(16\pi^2)$) are listed in column 2,

Table 1. Extrapolation Results for $m = m_e = 0.511 \; 10^{-3}$ GeV, $M = 10^{-5}$ GeV and $s = 100$ GeV^2

ε	Integration result	Extrapolated result	#Function evals.
2^{-32}	-0.3295240321026314E+01		0.24796950E+07
2^{-33}	-0.3384924822405812E+01	-0.3474609323785311E+01	0.28055550E+07
2^{-34}	-0.3430352719735732E+01	-0.3476171048159099E+01	0.30822750E+07
2^{-35}	-0.3443533393938414E+01	-0.3446671052357742E+01	0.34816650E+07
2^{-36}	-0.3445153187362171E+01	-0.3442193850326485E+01	0.38417550E+07
2^{-37}	-0.3444447779180230E+01	-0.3442652948552781E+01	0.42755850E+07
2^{-38}	-0.3443700367400504E+01	-0.3442689614444009E+01	0.46413150E+07
2^{-39}	-0.3443227592907696E+01	-0.3442688989410037E+01	0.50845950E+07
2^{-40}	-0.3442966496374880E+01	-0.3442688970861624E+01	0.56085150E+07
2^{-41}	-0.3442829784757492E+01	-0.3442688970926735E+01	0.60628650E+07
2^{-42}	-0.3442759890277537E+01	-0.3442688970927178E+01	0.66527850E+07
2^{-43}	-0.3442724558666787E+01	-0.3442688970927342E+01	0.74449050E+07

and the corresponding extrapolated results using linear extrapolation in column 3. The integration was performed iteratively using a one-dimensional code in each direction (DQAGE from Quadpack [7]), with a relative error tolerance of 10^{-12}.

For $M = 10^{-5}$ or 10^{-7} GeV and $s = 100, 10,000$ or $100,000$ GeV^2, results can be obtained to 5 or 6 figure accuracy (starting from, e.g., $\varepsilon = 2^{-42}$, and for a requested integration accuracy of 10^{-7} or 10^{-8} for the integral approximations corresponding to each of the ε values).

It is interesting to note that, even though the procedure breaks down for $M = 10^{-9}$ GeV using double precision arithmetic, we are able to extend it to $M = 10^{-15}$ GeV using quadruple precision and a total number of function evaluations of the order of 10^7 for a requested relative accuracy of 10^{-7}. The method works in cases where the integrals can be obtained efficiently for very small ε values.

3 Dimensional Regularization

In this section we apply a dimensional regularization technique. According to [1] we split the integral (2) into sector integrals. We will omit the term $-i\varepsilon$ in the notation below initially; it is re-introduced later for the computation.
We start from the form

$$I = \int_0^\infty dx_1 \int_0^\infty dx_2 \int_0^\infty dx_3 \frac{\delta(1 - x_1 - x_2 - x_3)}{(-sx_1x_2 + m^2(x_1 + x_2)^2 + M^2 x_3)^{N-D/2}} \quad (5)$$

of dimension $N = 3$ and where $D = 4 - 2\eta$. As we are interested in the infrared singularity, we will consider the case where $M \to 0$ (and m is fixed).

The domain is split into N sectors, which in this case are given by:

$$\{ (x_1, x_2, x_3) \mid 0 \le x_2 \le x_1, \ 0 \le x_3 \le x_1\},$$
$$\{ (x_1, x_2, x_3) \mid 0 \le x_1 \le x_2, \ 0 \le x_3 \le x_2\},$$
$$\{ (x_1, x_2, x_3) \mid 0 \le x_1 \le x_3, \ 0 \le x_2 \le x_3\}.$$

The integral I_1 over the first sector is thus as I but where the integration limits of the x_2 and x_3 ranges are replaced by $(0, x_1)$. Performing the transformation

$$x_1$$
$$x_2 = x_1 t_1$$
$$x_3 = x_1 t_2$$

in (5) yields

$$I_1 = \int_0^\infty dx_1 \int_0^1 dt_1 \int_0^1 dt_2 \, \frac{\delta(1 - x_1(1 + t_1 + t_2))}{x_1^{2N-D-2}(-st_1 + (1+t_1)^2 m^2 + M^2 \frac{t_2}{x_1})^{N-D/2}}.$$

We now use the transformation $x_1 = y_1/(1 + t_1 + t_2)$, so that

$$I_1 = \int_0^\infty dy_1 \int_0^1 dt_1 \int_0^1 dt_2 \, \frac{\delta(1 - y_1)(1 + t_1 + t_2)^{2N-D-3}}{y_1^{2N-D-2}(-st_1 + (1+t_1)^2 m^2 + M^2 \frac{t_2(1+t_1+t_2)}{y_1})^{N-D/2}},$$

and y_1 integrates out in view of the delta function, giving

$$I_1 = \int_0^1 dt_1 \int_0^1 dt_2 \, \frac{(1 + t_1 + t_2)^{2N-D-3}}{(-st_1 + (1+t_1)^2 m^2 + M^2 t_2(1+t_1+t_2))^{N-D/2}}. \quad (6)$$

The integral I_2 over the second sector equals I_1 through symmetry. Similar operations on the third sector integral cast it into the form

$$I_3 = \int_0^1 dt_1 \int_0^1 dt_2 \, \frac{(1 + t_1 + t_2)^{2N-D-3}}{(-st_1 t_2 + (t_1+t_2)^2 m^2 + M^2(1+t_1+t_2))^{N-D/2}}. \quad (7)$$

The extrapolation procedure of [2] applied to the sum of I_1, I_2 and I_3 gives the same results as when applied to I, for example, with $m = 40$ GeV, $M = 93$ GeV, $s = 9000$ GeV^2 and ε (for the extrapolation) ranging over 128 $(*2^{-1})$ 1.

4 Infrared Singularity

Examination of the sector integrals obtained previously as $M \to 0$ indicates that only I_3 in (7) shows the infrared singularity; indeed its denominator

$$(-t_1 t_2 s + (t_1 + t_2)^2 m^2)^{N-D/2} \to 0$$

as both t_1 and t_2 tend to zero, and $N - D/2 = N - 2 + \eta = 1 + \eta \to 1$ as $\eta \to 0$.

We write $I_3 = I_3' + I_3''$ where

$$I_3' = \int_0^1 dt_1 \int_0^{t_1} dt_2 \frac{(1+t_1+t_2)^{2N-D-3}}{(-st_1t_2 + (t_1+t_2)^2 m^2 + M^2(1+t_1+t_2))^{N-D/2}}$$

and

$$I_3'' = \int_0^1 dt_2 \int_0^{t_2} dt_1 \frac{(1+t_1+t_2)^{2N-D-3}}{(-st_1t_2 + (t_1+t_2)^2 m^2 + M^2(1+t_1+t_2))^{N-D/2}}$$

(note that $I_3' = I_3''$ through symmetry). Performing the transformation $t_1 = t_2 t_1'$ in I_3'' and setting $M = 0$ gives

$$I_3'' = \int_0^1 dt_2 \int_0^1 dt_1' \frac{t_2^{D+1-2N}(1+t_2 t_1' + t_2)^{2N-D-3}}{(-st_1' + (t_1'+1)^2 m^2)^{N-D/2}}.$$

Let us write this as

$$I_3'' = \int_0^1 dt_1' \frac{1}{(-st_1' + (t_1'+1)^2 m^2)^{N-D/2}} \int_0^1 dt_2\, t_2^{D+1-2N}(1+t_2 t_1' + t_2)^{2N-D-3},$$

where the exponent of t_2 is $D+1-2N = -1-2\eta$, which tends to -1 as $\eta \to 0$. This corresponds to a logarithmic behavior caused by the infrared singularity.

With respect to the singularity at $t_2 = 0$, we expand the part

$$f(t_2, \eta) = (1+t_2 t_1' + t_2)^{2N-D-3}$$

around $t_2 = 0$. Since we set $D = 4-2\eta$, the exponent of the singular factor in t_2 is $D+1-2N = 4-2\eta+1-6 = -1-2\eta$. Thus we expand

$$f(t_2, \eta) = f^{(0)}(0, \eta) + R(t_2, \eta),$$

with the remainder term

$$R(t_2, \eta) = f(t_2, \eta) - f^{(0)}(0, \eta) = f(t_2, \eta) - 1.$$

Substituting for $f(t_2, \eta)$ in I_3'' gives

$$I_3'' = \int_0^1 dt_1' \frac{1}{(-st_1' + (t_1'+1)^2 m^2)^{N-D/2}} \int_0^1 dt_2\, t_2^{D+1-2N}(f^{(0)}(0,\eta) + R(t_2, \eta)).$$

Thus, with respect to the infrared singularity, the integral is split into two terms, the first one of which accounts for the pole as $\eta \to 0$. The finite part of I_3'' is derived from the term in $R(t_2, \eta)$,

$$\int_0^1 dt_1' \frac{1}{(-st_1' + (t_1'+1)^2 m^2)^{N-D/2}} \int_0^1 dt_2\, t_2^{D+1-2N} R(t_2, \eta).$$

Numerical results can be obtained converging to the finite part by setting $\eta = 0$ and introducing (extrapolating on) the ε parameter to account for the singularity inside the integration region.

5 Conclusions

The dimensional regularization technique of Binoth and Heinrich [1] leads to obtaining a Laurent series expansion as a function of η. Their method does not deal with integrand singularities inside the region of integration. To handle a singularity on a quadratic which intersects the integration region, we introduce a parameter ε in the integrand and perform an extrapolation as ε tends to 0. This technique enables us to evaluate the finite part integral.

Furthermore we consider a regularization with respect to the photon mass parameter, results of which depend on the arithmetic precision used for the computation.

References

1. BINOTH, T., AND HEINRICH, G. An automized algorithm to compute infrared divergent multi-loop integrals. hep-ph/0004013 v2.
2. DE DONCKER, E., SHIMIZU, Y., FUJIMOTO, J., AND YUASA, F. Computation of loop integrals using extrapolation. *Computer Physics Communications 159* (2004), 145–156.
3. FUJIMOTO, J., SHIMIZU, Y., KATO, K., AND OYANAGI, Y. Numerical approach to one-loop integrals. *Progress of Theoretical Physics 87*, 5 (1992), 1233–1247.
4. KINOSHITA, T. *J. Math Phys. 3* (1962), 650.
5. LEE, T. D., AND NAUENBERG, M. *Phys. Rev. 133* (1964), 1549.
6. NAKANISHI, N. *Prog. Theor. Phys. 19* (1958), 150.
7. PIESSENS, R., DE DONCKER, E., ÜBERHUBER, C. W., AND KAHANER, D. K. *QUADPACK, A Subroutine Package for Automatic Integration.* Springer Series in Computational Mathematics. Springer-Verlag, 1983.
8. SHIMIZU, Y. Glossary for perturbative calculations in quantum field theory, August 2002.
9. 'T HOOFT, G., AND VELTMAN, M. *Nucl. Phys. B 153* (1979), 365.
10. VAN OLDENBORGH, G. J., AND VERMASEREN, J. A. M. *Z. Phys. C 46* (1990), 425.

Use of a Least Squares Finite Element Lattice Boltzmann Method to Study Fluid Flow and Mass Transfer Processes

Yusong Li, Eugene J. LeBoeuf, and P.K. Basu

Department of Civil and Environmental Engineering
Vanderbilt University, Nashville,
Tennessee 37325

Abstract. In our previous efforts, a least squares finite element lattice Boltzmann method (LSFE-LBM) was developed and successfully applied to simulate fluid flow in porous media. In this paper, we extend LSFE-LBM to simulate solute transport in bulk fluid and couple it with non-linear sorption/desorption processes at solid particle surfaces. The influences of the Peclet number and sorption non-linearity on solute transport is evaluated. Results of this work demonstrate the capability of using LSFE-LBM to study fluid flow and non-linear mass transfer processes at the pore scale.

1 Introduction

To provide for effective and efficient groundwater contamination prevention and remediation, it is important to possess a clear understanding of the complex mass transfer processes governing solute transport in the subsurface environment. Solute mass transfer in the subsurface includes several processes acting simultaneously: (i) advective-dispersive transport from bulk solution to the boundary layer of a soil or sediment particle; (ii) film diffusion across adsorbed water to the surface of a particle; (iii) sorption/desorption processes at the surface of the soil particle; and (iv) intrasorbent diffusion. Different factors, including transport-related non-equilibrium processes and sorption-related non-equilibrium processes, influence mass transfer in the subsurface [1], leading to non-ideal behaviors; i.e., early breakthrough and tailing breakthrough curves (BTC). Traditional advective-dispersive equations, which employ a local equilibrium assumption (LEA), fail to predict this non-ideal behavior [2]. In the last two decades, many efforts were devoted to better capture both transport-related and sorption-related non-equilibrium processes and elucidate the comparative contributions of different factors [3].

Recently, lattice Boltzmann method has been successfully applied to simulate fluid flow in porous media [4], providing a powerful alternative to model transport-related non-equilibrium processes. In this paper, we use a newly developed least squares finite element lattice Boltzmann method [5] to simulate fluid flow in porous media. Further, we extend LSFE-LBM to simulate solute transport in bulk fluid and couple it with non-linear sorption/desorption processes at particle surfaces.

2 Least Squares Finite Element Lattice Boltzmann Method

Although LBM has been developed as an effective tool to simulate complex fluid flow problems in porous media, one of the challenges with LBM is its inability to allow irregularity in the lattice [6]. We developed a least squares finite element lattice Boltzmann method (LSFE-LBM), which uses a LSFE method [7] in space and Crank-Nicolson method in time to solve the lattice Boltzmann equation. As described in an earlier publication [5], LSFE-LBM was successfully implemented on unstructured mesh to simulate fluid flow in porous media, requiring fewer grid points and consuming significantly less memory than traditional LBM.

2.1 Derivation of LSFE-LBM

Beginning with the basic equations of the LBM with a Bhatnagar-Gross-Krook collision operator:

$$\frac{\partial f_i}{\partial t} + \vec{c}_i \cdot \vec{\nabla} f_i = -\frac{1}{\tau}(f_i - f_i^{eq}) \quad (i = 1,2,...N) \tag{1}$$

where f_i represents particle distribution moving with velocity c_i, τ is the relaxation time, f_i^{eq} is the local equilibrium function, and N is the number of elements per site based on the LB model employed. Discretizing in time with a Crank-Nicholson scheme, a standard form of the governing equation for LSFE-LBM is

$$Lf^{n+1} = p \tag{2}$$

where the differential operator

$$L = c_x \frac{\partial}{\partial x} + c_y \frac{\partial}{\partial y} + A$$

$$A = \frac{2}{\Delta t} + \frac{1}{\tau} \tag{3}$$

$$p_i = (\frac{2}{\Delta t} - \frac{1}{\tau}) f_i^n + \frac{1}{\tau}(f_i^{eq,n+1} + f_i^{eq,n}) - (c_x \frac{\partial f_i^{n+1}}{\partial x} + c_y \frac{\partial f_i^{n+1}}{\partial y})$$

where Δt denotes the time step in the Crank-Nicholson scheme.

In the finite element implementation, the problem domain is first discretized into a set of finite elements, and then an approximate solution, $f_h^{e,n+1}$ in the eth finite element is formulated as:

$$f_h^{e,n+1} = \sum_{j=1}^{n} N_j f_j^{n+1} \tag{4}$$

Here, N_j denotes the element shape function, n represents the number of variables in the element, and f_j is the value of the j-th variable. Introducing this approximation into Eq. (2), the residual error at a point in the element is obtained. Integrating the square of this error over each element and minimizing the integral with respect to the nodal variables of the element, the elemental matrix relationship of the following form is obtained:

$$K_e F_e^{n+1} = P_e \tag{5}$$

where $K_e = \int_{\Omega_e} Q^T Q d\Omega_e$ with the ith element of Q,

$$Q^i = c_x \frac{\partial N_i}{\partial x} + c_y \frac{\partial N_i}{\partial y} + AN_i \qquad (6)$$

F_e^{n+1} is the vector of nodal values at the current time step, and

$$P_e = \int_{\Omega_e} Q^T p_h^e d\Omega_e \qquad (7)$$

Here, the K_e matrix is symmetric and positive definite.

2.2 LSFE-LBM Simulating Fluid Flow

In this study, a two-dimensional, nine-velocity lattice model (D2Q9) [8] is employed to implement LSFE-LBM for fluid flow. Accurate numerical results have been obtained for incompressible Poiseuille flow, Couette flow, and flow past a circular cylinder. Figure 1 is an example application of LSFE-LBM modeling fluid flow in porous media using an unstructured mesh.

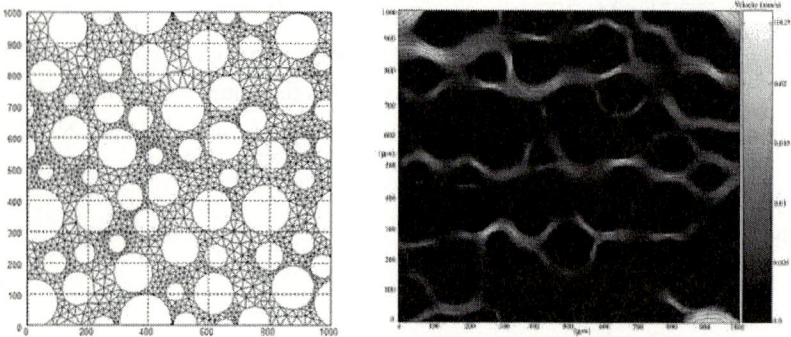

Fig. 1. LSFE-LBM-based unstructured mesh and velocity vectors for fluid flow in porous media

2.3 LSFE-LBM Simulating Solute Transport

In this study, we assume that the solute concentration is sufficiently low that it will not influence solvent flow. In this case, the solute can be described by a separate particle distribution function [9]. To recover the advection-diffusion equation, a simple square lattice with four possible directions is sufficient, which is thus used for implementing LSFE-LBM simulating solute transport.

The validation of LSFE-LBM simulating solute transport is evaluated by a problem describing diffusion between two parallel walls. As illustrated in Figure 2, the two walls are assumed to be porous and a constant normal flow u_a is injected through the lower wall and removed from the upper wall. The concentration of solute at the lower and upper walls is maintained with C_U and C_L, respectively. In this specific problem, C_U is assumed higher than C_L; it follows that solute diffuses counter to the flow of the fluid.

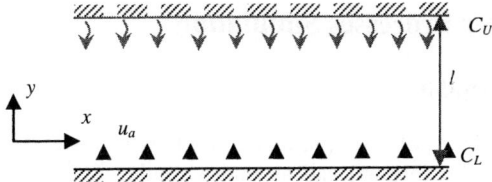

Fig. 2. Diffusion between two parallel walls

The governing equation for this problem is:

$$\frac{\partial \Phi}{\partial t} = D\frac{\partial^2 \Phi}{\partial y^2} + u_a \frac{\partial \Phi}{\partial y} \qquad (8)$$

$$\Phi(y,0) = 0, \quad \Phi(0,t) = 0, \quad \Phi(l,t) = 1 \qquad (9)$$

where, Φ is a normalized concentration defined as: $\Phi = \dfrac{C - C_L}{C_U - C_L}$, and D is the diffusivity of solute. Analytical solutions can be obtained for this problem in two special cases. In Case I, when $u_a = 0$, Eq. (8) will reduce to an unsteady state pure diffusion problem. The analytical solution can be expressed as:

$$\Phi(y,t) = \frac{y}{l} + \frac{2}{\pi}\sum_{n=1}^{\infty}\frac{(-1)^n}{n}\sin\frac{n\pi y}{l}e^{-n^2 t/\lambda}, \text{ where } \lambda = \frac{l^2}{\pi^2 D} \qquad (10)$$

When $u_a \neq 0$ (Case II), analytical solutions are only available for steady-state conditions:

$$\Phi = \frac{Exp(u_a y/D) - 1}{Exp(u_a/D) - 1} \qquad (11)$$

Results presented in Figure 3 illustrate that LSFE-LBM achieves close agreement with the analytical solution for solute transport in both unsteady state and steady state conditions.

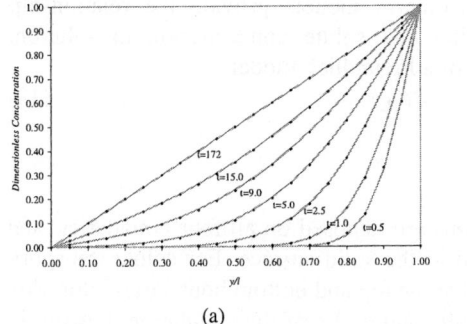
(a)

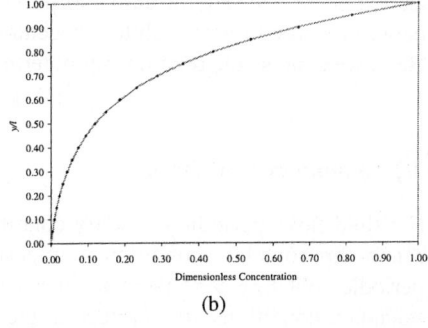
(b)

Fig. 3. Comparison of LSFE-LBM solution (points) and analytical solution (line) for diffusion between two parallel walls. (a) represents unsteady state solutions when water velocity $u_a=0$ and (b) represents steady state solution when water velocity $u_a \neq 0$

3 Mass Transfer Processes Simulation

3.1 Problem Description

To explore the influences of different factors on mass transfer processes, we consider fluid flow and transport through and around a single circular particle, set in a two-dimensional domain with a uniform far-field velocity, as illustrated in Figure 4.

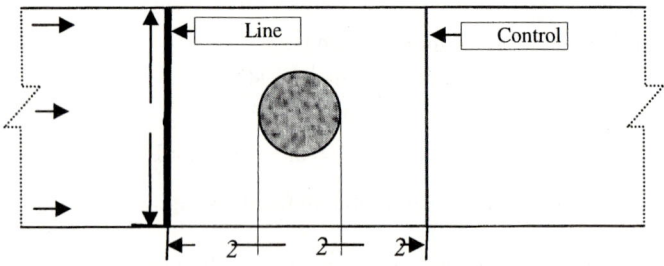

Fig. 4. An illustration of fluid flow and solute transport through and around a circular particle

A constant body force is imposed to drive fluid to flow from the left to right. When fluid flow reaches steady state, an instantaneous slug of solute is injected as a line source, as shown in Figure 4. Solute transport in the bulk fluid is driven by advection and diffusion processes. Several solute particles will diffuse across the water film to the surface of the particle. Sorption/desorption processes will then occur at the particle surface. The sorption rate at the particle surface can be expressed [10] as a function of the concentration difference between the solid and solution phases:

$$\frac{\partial q}{\partial t} = k_a C^n - k_d q \tag{12}$$

where, q is the solute concentration in the solid phase, C is the solute concentration in solution, k_a is a sorption rate coefficient, k_d is a desorption rate coefficient, and m is an exponent. At equilibrium, sorption isotherm models provide the relationship between sorbed-phase solute concentration and solute concentration in solution. Here, we express the relationship in terms of a Freundlich model:

$$q = (k_a / k_d) C^n = k_f C^n \tag{13}$$

3.2 Boundary Conditions

For fluid flow, periodic boundary conditions are imposed on all four boundaries, and a non-slip boundary condition is imposed at the solid surface. For solute transport, periodic boundary conditions are enforced at the top and bottom boundaries. Non-flux boundary conditions are enforced at the inlet and outlet of the simulation domain. In order to provide for a valid non-flux boundary condition, the length of the simulation domain is adjusted such that there is no mass loss at the domain inlet and outlet. At the solid particle surface, the boundary condition can be expressed as:

$$-D \frac{\partial C}{\partial n} = k_a C^n - k_d q \tag{14}$$

where D is the solute diffusion coefficient in the fluid, and n is the direction normal to the interface pointing toward the fluid phase. To implement this boundary condition in LSFE-LBM, a relationship between the concentration gradient and microscale particle parameters is required. For the 4 velocity lattice Boltzmann model, Eq. (15) can be derived by the Chapman-Enskog expansion.

$$\sum_i g_i c_{i\alpha} \cong C u_\alpha - \frac{\tau}{2} \partial_\alpha C \qquad (15)$$

where g_i is the particle distribution function for the solute, and α denotes x and y axis directions for a two-dimensional case. Figure 5 provides an example of this boundary condition at the upper right quadrant of a circular particle.

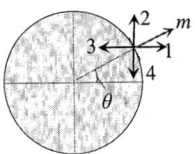

Fig. 5. A boundary node at the upper right part of the particle surface

Using Eq. (15), the following expressions are derived:

$$g_2 - g_4 = -(k_a C^n - k_d q) \sin \theta \qquad (16)$$
$$g_1 - g_3 = -(k_a C^n - k_d q) \cos \theta \qquad (17)$$

Further, we know:
$$g_1 + g_2 + g_3 + g_4 = C \qquad (18)$$

Unknown distribution functions g_1 and g_2 is calculated, by solving this non-linear equation system. At each time step, the solid phase concentration, q, at the particle surface is updated based on Eq. (12).

3.3 Results

The simulation is carried out for a particle with radius $R = 50$ μm in the domain as defined in Figure 4. The evolution of the concentration profile is represented at four selected time steps in Figure 6. This example vividly displays the influence of particle geometry and fluid hydrodynamics on the solute concentration profile, suggesting the need to further evaluate transport-related non-equilibrium processes in more complex systems.

Using the same Reynolds number (Re = 1.0), we examined the influence of the Peclet number on mass transfer processes. (Pe = uL/D, where u is the x-direction specific flow rate, L is the characteristic length of the domain which equals six times the particle diameter, and D is the solute diffusion coefficient in the bulk fluid.) By adjusting the values of D, two different breakthrough curves for Pe=10 and Pe=20 are observed at the control plane (Figure 7(a)). While representing the relative speed of fluid flow and solute diffusion, a lower Pe value denotes further deviation from an ideal symmetric breakthrough, with an earlier breakthrough and longer tailing effect as indicated in Figure 7 (a).

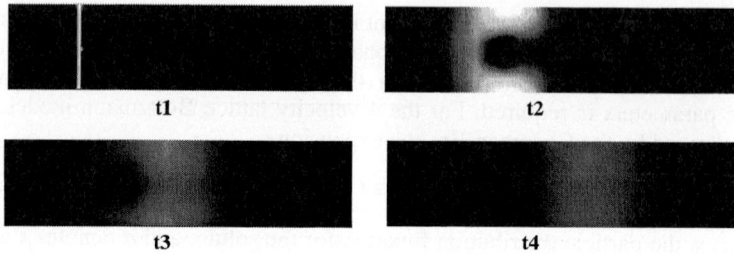

Fig. 6. Concentration profiles of a solute as it passes through and around a circular particle at four time points. Here $t1 < t2 < t3 < t4$

Further, the influence of non-linear sorption is evaluated to explore the influence of sorption-related non-equilibrium on the mass transfer processes. Keeping all the other parameters the same, we simulated two cases with $n = 0.5$ and $n = 1.0$ in the sorption rate equation (Eq. 12). Here, the evolution of solid phase concentration for a point at the surface of the particle is tracked. As expected, the presence of non-linear sorption ($n = 0.5$) leads to a much stronger tailing effect (Figure 7 (b)).

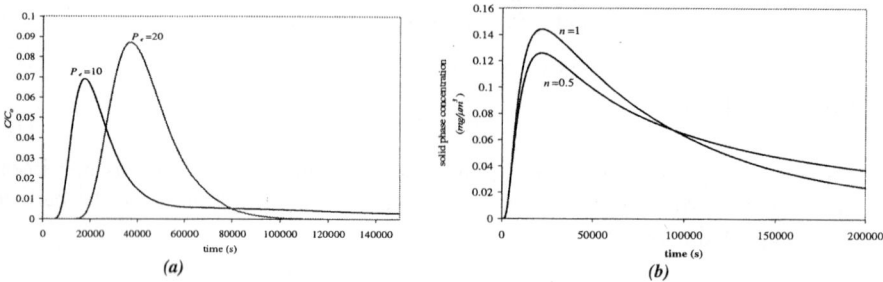

Fig. 7. (a) Breakthrough curves at the control plane for different P_e. (b) Time evolution of the solid phase concentration for a point at the surface of a particle

4 Conclusion

In this paper, we successfully applied our newly developed LSFE-LBM to simulate fluid flow and solute transport. LSEF-LBM was coupled with non-linear sorption/desorption through use of properly developed boundary conditions. The influences of particle geometry, Peclet number, and sorption/desorption non-linearity on solute transport were studied. Results from this work demonstrate the ability of LSFE-LBM to model fluid flow and highly non-linear mass transfer processes at the pore scale. In the future, LSFE-LBM will be applied to more complex systems to explore relative contributions of mass transfer processes with varying degrees of permeability and a variety of sorption/desorption properties.

References

1. M. L. Brusseau and P. S. C. Rao, Critical Reviews in Environmental Control 19, 33 (1989)
2. W. J. Weber and F. A. DiGiano, Process dynamics in environmental systems (John Wiley & Sons Inc., New York, 1996)
3. M. L. Brusseau, R. E. Jessup, and P. S. C. Rao, Water Resources Research 25, 1971 (1989).
4. S. Chen and G. D. Doolen, Annual Review of Fluid Mechanics 1998, 329 (1998)
5. Y. Li, E. J. LeBoeuf, and P. K. Basu, Physical Review E 69, Art. No. 065701 (2004)
6. J. D. Sterling and S. Chen, Journal of Computational Physics 123, 196 (1996)
7. B.-n. Jiang, The least-squares finite element method: theory and applications in computational fluid dynamics and electromagnetics (Springer, New York, 1998)
8. S. Succi, The Lattice Boltzmann equation for fluid dynamics and beyond (New York : Oxford University Press, 2001, 2001)
9. D. A. Wolf-Gladrow, Lattice-gas cellular automata and lattice Boltzmann models : an introduction (Springer, New York, 2000)
10. M. L. Brusseau and P. S. C. Rao, Geoderma 46, 169 (1990)

On the Empirical Efficiency of the Vertex Contraction Algorithm for Detecting Negative Cost Cycles in Networks

K. Subramani and D. Desovski

LDCSEE, West Virginia University,
Morgantown, WV
{ksmani, desovski}@csee.wvu.edu

Abstract. In this paper, we present a comprehensive empirical analysis of the Vertex Contraction (VC) algorithm for the problem of checking whether a directed graph with positive and negative costs on its edges has a negative cost cycle (NCCD). VC is a greedy algorithm, first presented in [SK05], for NCCD and is the only known greedy strategy for this problem. In [SK05] we compared a naive implementation of VC with the "standard" Bellman-Ford (BF) algorithm for the same problem. We observed that our algorithm performed an order of magnitude better than the BF algorithm on a range of randomly generated inputs, thereby conclusively demonstrating the superiority of our approach. This paper continues the study of contrasting greedy and dynamic programming approaches, by comparing VC with a number of sophisticated implementations of the BF algorithm.

1 Introduction

This paper contrasts the performance of the Vertex Contraction (VC) algorithm with existing algorithms for the Negative Cost Cycle detection (NCCD) problem. NCCD is defined as follows: *Given a directed graph* $\mathbf{G} =< \mathbf{V}, \mathbf{E}, \mathbf{c} >$*, where* $\mathbf{V} = \{v_1, v_2, v_3, \ldots, v_n\}$*,* $|\mathbf{V}| = n$*,* $\mathbf{E} = \{e_{ij} : v_i \leadsto v_j\}$*,* $|\mathbf{E}| = m$*, and a cost function* $\mathbf{c} : \mathbf{E} \to \mathbf{Z}$*, is there a negative cost cycle in* $\mathbf{G}$*?* There are no restrictions on the edge costs, i.e., they can be arbitrary integers as opposed to small integers, which is a requirement of scaling algorithms [Gol95]. We note that the problem, as specified, is a decision problem, in that all that is asked of an algorithm is to *detect* the presence of a negative cycle.

Some of the important application areas of NCCD include Image Segmentation, Temporal Constraint Solving, scheduling and System Verification [SK05]. Algorithms for negative cost cycle detection can be broadly classified as comparison based or scaling based. Comparison based algorithms in turn are based on heuristics to efficiently solve the linear programming formulation of the shortest path problem as a min-cost flow problem [CG96, CGR96]. The VC algorithm is a comparison based algorithm which is different from all existing approaches to NCCD in that it is a purely local, greedy approach.

Our work in this paper is motivated primarily by the need for a simple algorithm, with good performance characteristics. Whereas the naive Bellman-Ford approach for NCCD is admittedly simple, it suffers from significant performance drawbacks that are not explained using asymptotic analysis [SK05]. On the other hand, the techniques used to enhance the performance of BF as outlined in [Gol95, CG96] and [AMO93], suffer from the drawbacks of implementational difficulty and application specificity. We are therefore interested in an easy-to-implement comparison-based algorithm which can be modified in simple ways to provide reasonable performance on a wide variety of graphs.

2   The Vertex Contraction Algorithm

The *vertex contraction* procedure consists of eliminating a vertex from the input graph, by merging all its incoming and outgoing edges. Consider a vertex v_i with incoming edge e_{ki} and outgoing edge e_{ij}. When v_i is contracted, e_{ki} and e_{ij} are deleted and a single edge e'_{kj} is added with cost $c_{ki} + c_{ij}$. This process is repeated for each pair of incoming and outgoing edges. Consider the edge e'_{kj} that is created by the contraction; it falls into one of the following categories:

(a) It is the first edge between vertex v_k and v_j. In this case, nothing more is to be done.
(b) An edge e_{kj} already existed between v_k and v_j, prior to the contraction of v_i. In this case, if $c'_{kj} < c_{kj}$, keep the new edge and delete the previously existing edge (since it is redundant); otherwise delete the new edge (since it is redundant).

Algorithm (2.1) is a formal description of our technique. The correctness and analysis of our technique can be found in [SK05].

Function NEGATIVE-COST-CYCLE($\mathbf{G}, n$)
1: **for** $(i = 1$ **to** $n)$ **do**
2:     VERTEX-CONTRACT($\mathbf{G}, v_i$)
3: **end for**
4: **return**(false)

Algorithm 2.1. Negative cost cycle detection

2.1   The Cruel Adversary

The simple Vertex Contraction algorithm always chooses the next vertex to be contracted in a well-defined order; it is well-known that such a selection is susceptible to attack by a malicious adversary. For instance, an adversary could provide the graph in Figure (1) as input.

The above graph is sparse and has exactly $2 \cdot (n-1)$ edges. Observe that if vertex v_n is contracted first, the resultant graph is the complete graph on

```
Function VERTEX-CONTRACT(G, v_i)
 1: for (k = 1 to n) do
 2:   for (j = 1 to n) do
 3:     if (e_{ki} and e_{ij} exist) then
 4:       {Let c_{kj} denote the cost of the existing edge between v_k and v_j; note that
          c_{kj} = ∞ if there does not exist such an edge}
 5:       Create edge e'_{kj} with cost c'_{kj} = c_{ki} + c_{ij}
 6:       if (j = k) then
 7:         {A cycle has been detected}
 8:         if (c'_{jj} < 0) then
 9:           return(true)
10:         else
11:           Delete edge e_{jj}
12:         end if
13:       else
14:         if (c'_{kj} < c_{kj}) then
15:           Replace existing edge e_{kj} with e'_{kj} in G
16:         else
17:           Delete edge e'_{kj}
18:         end if
19:       end if
20:     end if
21:   end for
22:   Delete edges e_{ki} from G.
23: end for
24: for (j = 1 to n) do
25:   Delete edge e_{ij} from G, if it exists.
26: end for
27: Eliminate vertex v_i from G.
```

Algorithm 2.2. Vertex Contraction

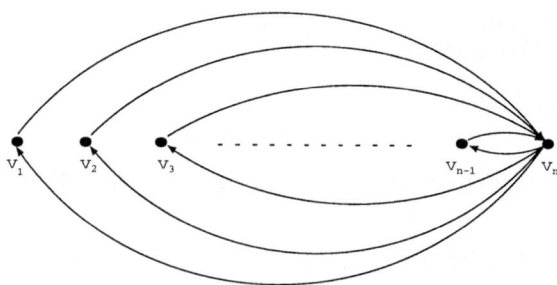

Fig. 1. Sparse graph that becomes dense after vertex contraction

$n - 1$ vertices and therefore dense. We call this graph *the cruel adversary*; in our experiments, we made it a point to test the performance of all algorithms on this input. We could however choose the vertex to be contracted at random,

Function RANDOM-NEGATIVE-COST-CYCLE(G, n)
1: Generate a random permutation Π of the set $\{1, 2, 3 \ldots, n\}$.
2: **for** ($i = 1$ **to** n) **do**
3: $\quad$ VERTEX-CONTRACT($G, v_{\Pi(i)}$)
4: **end for**
5: **return**(false)

Algorithm 2.3. Random negative cost cycle detection algorithm

without affecting the correctness of the algorithm. We have implemented the Vertex Contraction algorithm with three different strategies based on how the vertices to be contracted are chosen:

(i) the vertex to be contracted is chosen in a well-defined order (VC),
(ii) the vertex to be contracted is the vertex with the smallest degree product (the degree product is calculated by multiplying the number of edges coming into a vertex v_i with the number of edges going out of the vertex); this is accomplished by using a heap (HVC),
(iii) the vertex to be contracted is chosen at random (RVC). Algorithm (2.3) is a formal description of the random vertex contraction algorithm.

The experimental results on Cruel Adversary graphs have been relegated to the journal version of this paper. The extended version also contains a detailed description of the related work in the literature.

3 Implementation

Our experiments are classified into various categories, based on the following criteria:

(i) Type of input graph - Sparse with many small negative cycles (Type A), Sparse with a few long negative cycles (Type B), Dense with many small negative cycles (Type C), Dense with a few long negative cycles (Type D), the Cruel Adversary (Type E).
(ii) Type of Algorithm - Simple Vertex Contraction (VC), Vertex Contraction using a heap (HVC), Random Vertex Contraction (RVC), Bellman-Ford using a FIFO queue (BFFI), Bellman-Ford using a predecessor array (BFPR), Bellman-Ford using both a FIFO queue and a predecessor array (BFFP), Bellman-Ford using subtree disassembly (BFCT), and Goldberg-Radzik (GORC).
(iii) Type of Graph Data Structure - Simple Pointer or Advanced Pointers.

3.1 Implementation Details

Two different types of graph data structures were used for the experiments viz., a simple pointer structure and an advanced pointer structure. Both structures require linear space.

The simple pointer structure is also known as the adjacency-list representation [CLR92]. This representation makes use of an array of n lists, one for each of the n vertices of the graph. For each vertex v_i, we store the in-degree d_i^{in} (the number of edges going into the vertex), the out-degree d_i^{out} (the number of edges going out of the vertex), and a singly linked list of edges going out from the vertex along with their weights. The linked lists of each vertex are sorted based on the destination vertex of the edge. Assuming that we are contracting vertex v_i, the vertex contraction operation for the simple pointer structure is performed as follows.

The time required to contract vertex v_i by the simple pointer implementation is: $O(m + d_i^{in} \cdot d_i^{out} + \sum_x^{(v_x,v_i) \in E} d_x^{out})$. In the worst case when the graph is dense $m = O(n^2)$, $d_i^{in} = d_i^{out} = O(n)$, and the time complexity of the vertex contraction operation is $O(n^2)$.

In order to decrease the amount of time taken by the simple pointer structure, we consider the advanced pointer structure. For each vertex v_i, we store the in-degree d_i^{in}, the out-degree d_i^{out}, and two doubly linked lists representing the edges. Each vertex has an *out* list, for the edges going out of the vertex, and an *in* list, for the edges going into the vertex. The *out* lists of each vertex are sorted based on the destination vertex of the edge. Assuming that we are contracting vertex v_i, the vertex contraction operation for advanced pointers is performed as follows. A variation of the advanced pointer structure *in which the in-lists and out-lists are not sorted*, is discussed in [MN99].

Table 1. Time required to perform vertex contraction using simple pointers

Step	Time to Execute
1) Find edges with destination vertex v_i : (v_x, v_i)	$O(m)$
2) For every edge (v_x, v_i) found:	$O(d_i^{in})$
2-1) Remove the edge from the adjacency list of vertex v_x	$O(1)$
2-2) Merge v_i's list with v_x's list by adding those edges which are not present in v_x's list and updating those who are already there	$O(d_i^{out} + d_x^{out})$
2-3) If a negative cost edge (v_x, v_x) is created, the algorithm terminates with the negative cycle being detected	$O(1)$

Table 2. Time required to perform vertex contraction using advanced pointers

Step	Time to Execute
1) For every edge (v_x, v_i) in v_i's *in* list	$O(d_i^{in})$
1-1) Merge v_i's *out* list with v_x's *out* list by adding those edges which are not present in v_x's *out* list and updating those who are already there, and also updating the *in* lists of the vertices appropriately	$O(d_i^{out} + d_x^{out})$
1-2) If a negative cost edge (v_x, v_x) is created, the algorithm terminates with the negative cycle being detected	$O(1)$

The time required to contract vertex v_i by the advanced pointer implementation is: $O(d_i^{in} \cdot d_i^{out} + \sum_x^{(v_x,v_i) \in E} d_x^{out})$. In the event that there exists some constant c such that, for all v_x, $d_x^{out} \leq c \cdot d_i^{out}$, then this time bound simplifies to: $O(d_i^{in} \cdot d_i^{out})$. In the worst case when $d_i^{in} = d_i^{out} = O(n)$ and hence the time complexity of the vertex contraction operation is $O(n^2)$.

3.2 Implementation Remarks

Note that

(i) For each experimental suite, we have provided only the graphical picture in this paper; the numerical tables are provided in the journal version of this paper. Our goal is to show that minor modifications to the VC framework result in significant performance enhancements, making its performance comparable to that of more sophisticated algorithms. Depending on the type of graph under consideration, a particular modification of VC may not be run.

(ii) All performance graphs have been drawn according to the following scale:
  (a) On the x-axis, we represent the logarithm to the base 2, of the number of vertices.
  (b) On the y-axis, we represent the logarithm to the base 2, of the running time in seconds.

3.3 Experimental Setup for Sparse Graphs

Sparse graphs were generated using the generator developed by Andrew Goldberg [CG96], which generates multiple edges between two vertices. Sparse graphs are defined as graphs with $o(n \cdot \log n)$ edges. We generated each graph 5 times using 5 different seeds for the random number generator. The times recorded are the medians over 5 executions of each implementation. (We used the median statistic based on related work in the literature; we have also maintained the worst-case times of each run. The timing profiles of the worst-case times and medians were very similar.)

Graphs of Type A and B were tested, with a number of vertices ranging from 500 to 10,000 in increments of 500.

We define a small negative cycle as one consisting of at most $\frac{n}{100}$ vertices. We define a long negative cycle as one consisting of $\Omega(\frac{n}{2})$ vertices. The number of long negative cycles in the input graphs was set to 4.

3.4 Experimental Results for Sparse Graphs

It is easy to see from Figure (2) and Figure (3) that GORC outperforms all other implementations; this is true for both types of sparse graphs that were tested. BFCT and BFFP are comparable to GORC on most instances. AVC, and AHVC are far superior to BFFI and BFPR; they also outperform HVC and RVC on most instances. Bellman Ford using a FIFO queue (BFFI) and Bellman Ford

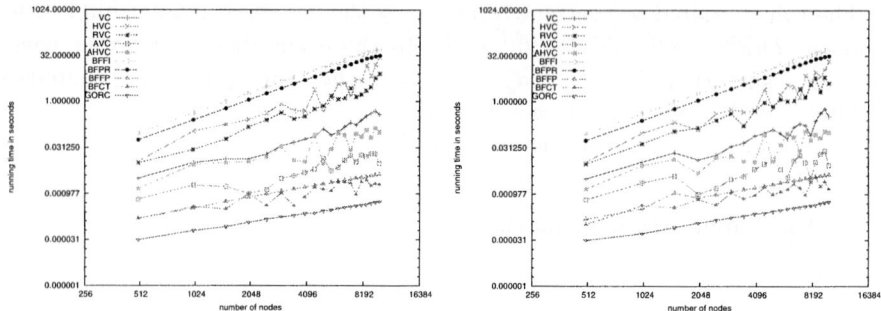

Fig. 2. Implementation execution times required to solve the Negative Cost Cycle detection problem for Type A graphs

Fig. 3. Implementation execution times required to solve the Negative Cost Cycle detection problem for Type B graphs

using a predecessor array (BFPR) have been omitted from the graph, since the times for these two algorithms are much worse than any of the other algorithms tested.

3.5 Experimental Setup for Dense Graphs

Dense graphs were generated using the generator developed by Andrew Goldberg [CG96], which generates multiple edges between two vertices. Dense graphs are defined as those graphs with $\Omega(\frac{n^2}{8})$ edges. We generated each graph 5 times using 5 different seeds for the random number generator. The times recorded are the medians over 5 executions of each implementation.

Graphs of Type C and D were tested, with a number of vertices ranging from 500 to 10,000 in increments of 500, with small negative cycles and long negative cycles defined as in Section §3.3. As mentioned before, HVC and AHVC were

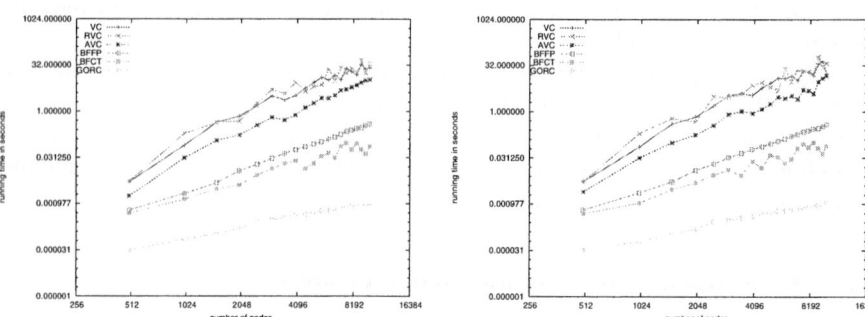

Fig. 4. Implementation execution times required to solve the Negative Cost Cycle detection problem for Type C graphs

Fig. 5. Implementation execution times required to solve the Negative Cost Cycle detection problem for Type D graphs

not run on dense graphs, since we found that there was no gain in terms of performance for the additional complexity in our test cases.

3.6 Experimental Results for Dense Graphs

It is easy to see from Figure (4) and Figure (5), that GORC outperforms all other implementations; this is true for both the types of dense graphs that were tested. BFCT performs slightly better than BFFP, although, both are comparable to GORC. AVC performs the best among the Vertex Contraction algorithms; although AVC, VC, and RVC are all far superior to BFFI and BFPR. Bellman Ford using a FIFO queue (BFFI) and Bellman Ford using a predecessor array (BFPR) have been omitted from the graph, since the times for these two algorithms are much worse than any of the other algorithms tested.

References

[AMO93] R. K. Ahuja, T. L. Magnanti, and J. B. Orlin. *Network Flows: Theory, Algorithms and Applications*. Prentice-Hall, 1993.

[CG96] Boris V. Cherkassky and Andrew V. Goldberg. Negative-cycle detection algorithms. In Josep Díaz and Maria Serna, editors, *Algorithms—ESA '96, Fourth Annual European Symposium*, volume 1136 of *Lecture Notes in Computer Science*, pages 349–363, Barcelona, Spain, 1996. Springer.

[CGR96] Boris V. Cherkassky, Andrew V. Goldberg, and T. Radzik. Shortest paths algorithms: Theory and experimental evaluation. *Mathematical Programming*, 73:129–174, 1996.

[CLR92] T. H. Cormen, C. E. Leiserson, and R. L. Rivest. *Introduction to Algorithms*. MIT Press and McGraw-Hill Book Company, Boston, Massachusetts, 2nd edition, 1992.

[Gol95] Andrew V. Goldberg. Scaling algorithms for the shortest paths problem. *SIAM Journal on Computing*, 24(3):494–504, June 1995.

[MN99] K. Mehlhorn and St. Näher. *The LEDA Platform of Combinatorial and Geometric Computing*. Cambridge University Press, Cambridge, 1999.

[SK05] K. Subramani and L. Kovalchick. A greedy strategy for detecting negative cost cycles in networks. *Future Generation Computer Systems*, 2005. Accepted, In Press.

Minimal Load Constrained Vehicle Routing Problems

İmdat Kara and Tolga Bektaş

Başkent Üniversitesi, Endüstri Mühendisliği Bölümü,
Bağlıca Kampüsü, Eskişehir, Yolu 20. km. 06530 Ankara, Turkey
{ikara, tbektas}@baskent.edu.tr

Abstract. In this paper, the Capacitated Vehicle Routing Problem is extended to the case where each vehicle is restricted to an additional minimal starting or returning load constraint. We refer to this extension as the Minimal Load Constrained Vehicle Routing Problem. We present integer programming formulations for single and multidepot cases. An illustrative example is also provided to show how a decision maker can use the proposed formulation as an aid in distribution planning.

1 Introduction

In this paper, we deal with the well-known Capacitated Vehicle Routing Problem (CVRP), which finds many practical applications in the field of distribution and logistics. The problem is formally defined on a graph $G = (V, A)$ where $V = \{0, 1, 2, ..., n\}$ is the set of nodes (vertices), node 0 is the depot and the remaining nodes are customers. The set $A = \{(i, j) : i, j \in V, i \neq j\}$ is an arc (or edge) set. With each customer $i \in V \setminus \{0\}$ is associated a positive integer demand q_i and with each arc (i, j) is associated a travel cost c_{ij} (which may be symmetric, asymmetric, Euclidean, deterministic, random, etc.). There are m vehicles with identical capacity Q. The classical CVRP consists of determining a set of m vehicle routes satisfying the following conditions simultaneously:

(i) Each route starts and ends at the depot,
(ii) Each customer belongs to exactly one route,
(iii) The total demand of each route does not exceed the vehicle capacity Q,
(iv) The total cost of all routes (m routes) is minimized.

The CVRP has been studied extensively in the literature (for recent publications, see Achuthan et al. [1], Toth and Vigo [2], Ralphs [3]). Most of the existing literature on CVRPs deals with the minimization of the total distance (or time or cost) of the tours, as formulated 45 years ago by Dantzig and Ramser [4]. This definition, although very general, is not sufficient to adequately model situations where the loads of the vehicles are of importance. It is obvious that the cost of a vehicle's tour depends not only on the distance travelled but also on the amount of goods carried. In addition, balancing of the loads is also a concern for the distribution design. For instance, vehicle loads may also be important for drivers, i.e. they may be required to carry similar amounts of goods. Along this line of research, few studies exist that have attempted to extend the definition of the CVRP so as to balance the length of the tours (Jozefowiez et al. [5]) or workload among employees (Lee and Ueng [6]) in addition to minimizing the total distance.

Clearly, the amount of goods carried by a vehicle is directly proportional to the starting or returning loads. Thus, the starting load of a vehicle in the case of delivery, and the returning (final) load of a vehicle in the case of collection may be as important as the capacity of the vehicles. As far as the authors are aware, there exist no formulations imposing minimal starting or returning loads for vehicles. Such a restriction frequently arises in real life applications when distribution of loads across vehicles is important. Thus, there is a need to expand the traditional definition of the problem to be in compliance with practical situations. This is the main motivation and contribution of this study.

This paper extends the CVRP to include additional load constraints, i.e. it imposes a minimal starting load in the case of delivery or a minimal returning load in the case of collection, and proposes integer linear programming formulations (ILPFs). In section 2, we define and present ILPFs for the CVRP and the Minimal Load Constrained VRP with a single depot. Then, an extension to the nonfixed multidepot case is given in section 3. We provide an illustrative example in section 4, showing how the proposed model can be used as an aid in routing decisions. The paper concludes with some suggestions in section 5.

2 Vehicle Routing Problems with Minimal Load Constraints

The first linear programming formulation of the CVRP is due to Dantzig and Ramser [4], in which the authors solved problems having up to 13 nodes (including the depot) by the aid of a heuristic. The first practical solution method was proposed by Clarke and Wright [7]. An integer linear programming formulation of the CVRP with three-index variables was presented by Golden et al. [8]. Subsequent research has mainly focused on two-index variable formulations. We also follow two-index formulations here. Several ILPFs for the CVRP with two-index variables have been proposed in the literature. The first group is based on Dantzig, Fulkerson and Johnson's [9] subtour elimination constraints of the Traveling Salesman Problem (TSP) (see e.g. [1], [10], [2]). The number of inequalities of these types of constraints grows exponentially with the number of nodes. Thus models including such constraints cannot be solved directly by any optimizer. A two-index ILPF for the CVRP with polynomial size was given by Kulkarni and Bhave [11], which was based on Miller-Tucker-Zemlin [12] subtour elimination constraints. Later, Desrochers and Laporte [13] proposed lifted subtour elimination constraints. Recently, Kara et al. [14] suggested a correction for both of these papers and provided a corrected ILPF of the CVRP.

We now formally define the minimal load constrained VRP. Let the following side condition also be imposed on the CVRP in addition to (i)-(iv):

(v) In the case of delivery, the starting load of each vehicle (or in the case of collection, the returning load of each vehicle) cannot be less than a predetermined value.

We refer to the VRP with constraints (i)-(v) as the *Minimal Load Constrained VRP* (MLCVRP). To construct the ILPF of MLCVRP, we further define the following:

- c_{ij}, Q, and q_i are as defined previously
- Q_0 is the minimal starting load (delivery) or the minimal returning load (pick up) of each vehicle
- $x_{ij} = 1$ if the arc (i,j) is on the optimal tour and 0 otherwise
- u_i is the load of a vehicle (amount of goods collected) just after visiting customer i, or is the amount of goods delivered (unloaded space of a vehicle) just after visiting customer i

We now provide a general ILPF for the MLCVRP:

$$\min \sum_{i \in V} \sum_{j \in V} c_{ij} x_{ij} \tag{1}$$

s.t.

$$\sum_{i \in V \setminus \{0\}} x_{0i} \leq m \tag{2}$$

$$\sum_{i \in V} x_{ij} = 1, \quad j \in V \setminus \{0\} \tag{3}$$

$$\sum_{j \in V} x_{ij} = 1, \quad i \in V \setminus \{0\} \tag{4}$$

$$u_i - u_j + Q x_{ij} + (Q - q_i - q_j) x_{ji} \leq Q - q_j, \, i \neq j \in V \setminus \{0\} \tag{5}$$

+Capacity and Minimal Load Constraints

$$u_i \geq 0, \quad i \in V \setminus \{0\} \tag{6}$$

$$x_{ij} \in \{0, 1\}, \quad i, j \in V \tag{7}$$

where x_{ij}'s are defined for $q_i + q_j \leq Q, \forall (i,j) \in A$. In this formulation, constraint (2) is used to allow at most m vehicles, whereas (3)-(4) are the degree constraints. Constraint (5) is a subtour elimination constraint (see [14]) ensuring the solution contains no illegal subtours. Nonnegativity and integrality constraints are given in (6) and (7).

In the following, we define valid inequalities for the MLCVRP that impose capacity and minimal load constraints:

Proposition 1. *Let* $\bar{q}_i = \min_{j \neq i}\{q_j\}$, $Q \geq Q_0$ *and* $Q_0 \geq \bar{q}_i + q_i, \forall i \in V \setminus \{0\}$. *Then,*

$$u_i + (Q - \bar{q}_i - q_i)x_{0i} - \bar{q}_i x_{i0} \leq Q - \bar{q}_i, \qquad i \in V \setminus \{0\} \qquad (8)$$
$$u_i + \bar{q}_i x_{0i} + (q_i + \bar{q}_i - Q_0)x_{i0} \geq q_i + \bar{q}_i, \qquad i \in V \setminus \{0\} \qquad (9)$$
$$x_{0i} + x_{i0} \leq 1, \qquad i \in V \setminus \{0\} \qquad (10)$$

are valid inequalities for the MLCVRP, and are here referred to as bounding constraints.

Proof. Constraints (5) ensure that if $x_{ij} = 1$, then $u_j = u_i + q_j$. Since the inequalities given in (10) do not allow single node visits, we consider three cases: (i) If $x_{0i} = x_{i0} = 0$, then this means that there exist two distinct nodes $k \neq l \neq i \geq 1$ such that $x_{ki} = x_{il} = 1$ by the assignment constraints. In this case, $q_i + q_k \leq u_i \leq Q - q_l$ by (5). Since $q_k \geq min_{j \neq i}\{q_j\}$ and $q_l \geq min_{j \neq i}\{q_j\}$, letting $\bar{q}_i = min_{j \neq i}\{q_j\}$, this inequality can be written as $q_i + \bar{q}_i \leq u_i \leq Q - \bar{q}_i$. This is exactly what constraints (8) and (9) imply when $x_{0i} = x_{i0} = 0$. (ii) If $x_{0i} = 1$ and $x_{i0} = 0$, (8) and (9) imply $u_i = q_i$. (iii) Finally, if $x_{0i} = 0$ and $x_{i0} = 1$, then $Q_0 \leq u_i \leq Q$. □

The proposed model contains $O(n^2)$ binary variables and $O(n^2)$ constraints. Thus, appropriate size real life problems can be solved to optimality by commercial codes. Such an opportunity will also allow us to conduct post-optimality analysis. The following proposition shows that the proposed model can be used for the classical CVRP as well.

Proposition 2. *The constraints*

$$u_i + \bar{q}_i x_{0i} \geq q_i + \bar{q}_i, \qquad i \in V \setminus \{0\} \qquad (11)$$

together with constraints (8) are valid inequalities for the CVRP.

Proof. Similar to the proof of Proposition (1). □

In the case of the classical CVRP, if single customer visits are allowed, then constraints (10) can be dropped.

3 A Multidepot Case

In this section, we give a fairly straightforward extension of the proposed model to the multidepot MLCVRP. The multidepot case is a generalization of the single depot MLCVRP, so as to permit more than one depot and a number of vehicles located at each depot. This problem has immediate applications in supply chain management or distribution planning where such structures generally include more than one facility (such as distribution centers or warehouses) and a set of vehicles that have to be routed from these facilities.

To formally define the problem, let the node set be partitioned such that $V = D \cup V'$, where the first d nodes of V are depot set D, there are m_i vehicles located at depot i initially and the total number of vehicles is m. Also, let $V' = \{d+1, d+2, \ldots, n\}$ be the set of customers. The problem consists of finding tours for all the vehicles such that all customers are visited exactly once, the loads of each vehicle lie within a predetermined interval and the total cost of all the tours is minimized. If the problem is to determine a total of m tours such that each vehicle must return to its original depot, this is referred to as the *fixed destination* case. On the other hand, if the vehicles do not have to return to their original depot but the number of vehicles at each depot after all travel is to be equal to the initial number, we have the *nonfixed destination* case. In this paper, we consider the latter case.

We now provide the relevant ILPF:

$$\min \sum_{i \in V} \sum_{j \in V} c_{ij} x_{ij} \quad (12)$$

s.t.

$$\sum_{j \in V'} x_{ij} \leq m_i, \quad i \in D \quad (13)$$

$$\sum_{i \in V'} x_{ij} \leq m_j, \quad j \in D \quad (14)$$

$$\sum_{i \in V} x_{ij} = 1, \quad j \in V' \quad (15)$$

$$\sum_{j \in V} x_{ij} = 1, \quad i \in V' \quad (16)$$

$$u_i + (Q - \bar{q}_i - q_i) \sum_{k \in D} x_{ki} - \bar{q}_i \sum_{k \in D} x_{ik} \leq Q - \bar{q}_i, \quad i \in V' \quad (17)$$

$$u_i + \bar{q}_i \sum_{k \in D} x_{ki} + (q_i + \bar{q}_i - Q_0) \sum_{k \in D} x_{ik} \geq q_i + \bar{q}_i, \quad i \in V' \quad (18)$$

$$x_{ki} + x_{ik} \leq 1, \quad k \in D, i \in V' \quad (19)$$

$$u_i - u_j + Q x_{ij} + (Q - q_i - q_j) x_{ji} \leq Q - q_j, \quad i \neq j, i, j \in V' \quad (20)$$

$$u_i \geq 0, \quad i \in V' \quad (21)$$

$$x_{ij} \in \{0, 1\}, \quad i, j \in V \quad (22)$$

where the decision variables and parameters are as defined previously. It is easily seen that the subtour elimination constraints of both models (single and multidepot cases) are the same. A careful analysis of the constraints given in (17), (18) and (19) shows that they play the same role as the bounding constraints in the single depot case.

4 An Illustrative Example

In this section, we illustrate how the model proposed above can be used as a decision aid by solving a problem taken from the famous paper of Dantzig and Ramser [4]. The problem is to deliver gasoline to gas stations in such a way that the total distance traveled by four identical trucks, each of which has 6000 unit capacity, is minimized. There are 12 customers (gas stations) denoted by

Table 1. Dantzig and Ramser's solution

Truck number	Tours	Starting Load	Distance
1	P0 P1 P2 P3 P4 P0	5100	54
2	P0 P7 P12 P11 P9 P0	5800	112
3	P0 P6 P10 P8 P0	4900	84
4	P0 P5 P0	1700	44
	Total Distance Travelled		294

P1,..., P12, with P0 as the depot. The symmetric distance matrix and demand of each customer can be found in the original paper [4]. Dantzig and Ramser solved this problem by the aid of a heuristic that they developed, and found the near-optimal solution shown in Table 1.

By using CPLEX 8.0 on a Pentium III computer running at 1400 Mhz, we solved this problem first without a lower bound, and then obtained alternative solutions with various upper and lower bounds using the single depot model with $m = 4$. Our findings are summarized in Table 2.

The solution of the problem given in Table 1 and the first solution given in Table 2 show that there are big variations in the starting loads and distances of the trucks. However, when the decision maker imposes different lower bounds as well as upper bounds on the loads of each vehicle, the solutions change as expected. More specifically, increasing the lower bound decreases the difference between the starting loads of the vehicles. For instance, when a lower bound of $Q_0 = 4500$ is imposed on the original problem where $Q = 6000$, the loads of the vehicles become very close to each other. The table also shows two other

Table 2. Various solutions for Dantzig and Ramser's problem

Q_0	Q	Optimal Value	CPU	Tours	Starting Loads	Distance
None	6000	290	127	P0 P1 P2 P3 P4 P0	5800	54
				P0 P7 P10 P12 P11 P0	5600	112
				P0 P6 P8 P9 P0	5100	80
				P0 P5 P0	1700	44
4500	6000	330	118.5	P0 P1 P8 P6 P0	4500	80
				P0 P4 P3 P2 P0	4600	54
				P0 P5 P10 P7 P0	4500	84
				P0 P9 P11 P12 P0	4600	112
3750	5000	312	360.50	P0 P1 P3 P2 P0	4400	42
				P0 P4 P10 P8 P0	4900	94
				P0 P6 P7 P5 P0	4300	64
				P0 P12 P11 P9 P0	4600	112
5250	7000	244	5374.86	P0 P4 P3 P2 P1 P0	5800	54
				P0 P6 P8 P7 P5 P0	6200	78
				P0 P10 P12 P11 P9 P0	6200	112

solutions with $Q_0 = 3750, Q = 5000$ and $Q_0 = 5250, Q = 7000$. It is interesting to see that the latter solution only requires 3 vehicles to be used, where the total distance travelled is less than that of the former solution which dispatches a total of 4 vehicles.

The results shown in Table 2 indicate that load balancing can indeed be performed by the proposed model. Thus, the decision maker will be able to see alternative solutions to a single problem by imposing varying lower and/or upper bounds on the starting loads of the vehicles. The importance of the difference between the loads of the trucks and the variations in the optimal value of the objective function will enlighten the final decision.

As far as the distances travelled by each vehicle are concerned, the proposed model does not have any effect on the variation in these. This is a topic for further research.

5 Conclusion

In this paper, the classical CVRP is extended to include the case when the starting or returning load of a vehicle is restricted to a predetermined value. Current formulations of the CVRP do not contain such a restriction. Therefore, we define and present new bounding constraints for the single and multiple depot minimal load constrained VRPs. It is further shown that post-optimality analysis may be conducted with respect to various values of minimal and/or maximal loads, so that distributions of loads across vehicles can be considered in the planning phase of the decision process.

The proposed model will allow appropriate sized vehicle routing problems to be directly solved to optimality with the use of commercial software. For larger problems, other solution methodologies either based on this model (such as branch and bound) or model-independent techniques (such as metaheuristics) can be used. In any case, we believe that the problems defined here and the related models will constitute a starting point for such research.

To the best of the authors' knowledge, there is no formulation for a minimal distance constrained VRP (i.e., for a VRP where there is a restriction related to the minimum distance traveled). The modeling approach presented in this paper may be applied to this case, and such an application is currently under consideration.

References

1. Achuthan, N.R., Caccetta, L., Hill, S.P.: A New Subtour Elimination Constraint for the Vehicle Routing Problem. European Journal of Operational Research **91** (1996) 573–586
2. Toth, P., Vigo, D.: The Vehicle Routing Problem. SIAM Monographs on Discrete Mathematics and Applications (2002)
3. Ralphs, T.K.: Parallel Branch and Cut for Capacitated Vehicle Routing. Parallel Computing **29** (2003) 607–629

4. Dantzig, G.B., Ramser, J.H.: The Truck Dispatching Problem. Management Science **6** (1959) 80–91
5. Jozefowiez, N., Semet, F., Talbi, E.: Parallel and Hybrid Models for Multi-Objective Optimization: Application to Vehicle Routing Problem. In: Merelo Guervós, J. J., Adamidis, P., Beyer, H.-G., Fernández-Villacañas, J.-L., Schwefel, H.-P. (eds.): Parallel Problem Solving from Nature - PPSN VII. Lecture Notes in Computer Science, Vol. 2439. Springer-Verlag, Berlin Heidelberg New York (2002) 271–280
6. Lee, T., Ueng, J.: A Study of Vehicle Routing Problems with Load-Balancing. International Journal of Physical Distribution and Management **29** (1999) 646–658
7. Clarke, G., Wright, J.: Scheduling of Vehicles from a Central Depot to a Number of Delivery Points. Operations Research **12** (1964) 568–581
8. Golden, B.L., Magnanti, T.L., Nguyen, H.Q.: Implementing Vehicle Routing Algorithms. Networks **7** (1977) 113–148
9. Dantzig, G.B., Fulkerson, D.R., Johnson, S.M.: Solution of a Large Scale Traveling Salesman Problem. Operations Research **2** (1954) 393–410
10. Laporte, G., Nobert, Y.: A Branch and Bound Algorithm for the Capacitated Vehicle Routing Problem. OR Spektrum **5** (1983) 77-85
11. Kulkarni, R.V., Bhave, P.R.: Integer Programming Formulations of Vehicle Routing Problems. European Journal of Operational Research **20** (1985) 58–67
12. Miller, C.E., Tucker, A.W., Zemlin, R.A.: Integer Programming Formulations and Traveling Salesman Problems. Journal of the Association for Computing Machinery **7** (1960) 326–329
13. Desrochers, M., Laporte, G.: Improvements and Extensions to the Miller-Tucker-Zemlin Subtour Elimination Constraints. Operations Research Letters **10** (1991) 27–36
14. Kara, İ., Laporte, G., Bektaş, T.: A Note on the Lifted Miller-Tucker-Zemlin Subtour Elimination Constraints for the Capacitated Vehicle Routing Problem. European Journal of Operational Research **158** (2004) 793–795

Multilevel Static Real-Time Scheduling Algorithms Using Graph Partitioning

Kayhan Erciyes[1] and Zehra Soysert[2]

[1] Izmir Institute of Technology,
Computer Eng. Dept., Urla, Izmir 35340, Turkey
kayhanerciyes@iyte.edu.tr
[2] Ege University International Computer Institute,
35100 Bornova, Izmir, Turkey
soysert@bornova.ege.edu.tr

Abstract. We propose static task allocation algorithms for the periodic tasks of a distributed real-time system. The cyclic task consists of task threads which may communicate and share resources. A graph partitioning process and a thread sequencing algorithm are applied to these threads to yield local schedules. The exact analysis is then obtained and further refinements are performed if the worst case response time of a task is greater than its deadline.

1 Introduction

Scheduling in real-time systems can be broadly described as static or dynamic. Static scheduling of processes with known release times, deadlines, precedence, and exclusion relations is decribed in [13]. Schedulability tests for *Rate Monotonic* (RM) and *Earliest Deadline First* (EDF) algorithms for cyclic tasks when their deadlines equal their periods are presented in [5]. An exact sheduling test method for RM algorithm [9] is derived in [6]. A method to find the schedulability of a task set when Deadline Monotonic Scheduling is used is described in [1]. The task of scheduling tasks on a multiprocessor/distributed environment is NP-hard [12]. For this reason, various heuristics such as iterative improvement algorithms [7], and the probabilistic optimization as simulated annealing algorithms [10] and genetic algorithms [11] have been proposed. A middleware distributed real-time scheduling method is shown in [2]. The goal of our ongoing work is to investigate the balancing of static load over a distributed real-time system where computation nodes executing real-time kernels are connected by a real-time network that provides bounded mesage delays. Formally, if $T = \{t_1, t_2, ..., t_m\}$ is the set of real-time tasks and $P = \{p_1, p_2, ..., p_n\}$ is the set of processors, we need to derive the mapping function $M : T \rightarrow P$ so that every task meets its deadline. To accomplish this, we consider the periodic tasks of the real-time system consisting of *real-time threads* each with a hard deadline. Threads have interprocess communications and may access shared resources. We sketch the static execution characteristics of these threads which are computation times, interprocess communication patterns and deadlines as a directed graph and use a novel *multilevel graph partitioning* heuristic to divide the graph into n regions. The output of the graph partitioning algorithm are the task thread sets to be executed by each processor. We then sequence these threads for each processor by a *Task Sequencing Algorithm*, providing that the precedence constraints are obeyed and deadlines are met. In the second phase, we apply the exact scheduling analysis to work out the worst case response times R_{ij} for $j = 1..N_i$ of the individual task components taking the blocking times by lower priority threads in the same or other tasks. If for any task component, R_{ij} is larger than deadline, our partitioning is unsuccesful and we go into refinement phase where we try to move threads from one processor to another to provide $R_{ij} < d_{ij}$.

The rest of the paper is organized as follows. Section 2 provides the computation model and the partitioning algorithm output of which is scheduled by the task sequencing algorithm described in Section 3. The exact analysis and the refinement procedure that is invoked if the exact analysis detects missing deadlines of tasks is described in Section 4. An example of operation is shown in Section 5 and future directions are outlined in the Conclusions Section.

2 Task Graph Partitioning Algorithm

For the hard-real time tasks we assume the following computation model:

- Worst case execution time of each task C_i and its deadline D_i are known in advance and for periodic tasks, deadlines equal periods. $(D_i = P_i)$
- Each task T_i consists of a number of threads t_{ik}, each of which has a computation time c_{ik} and a deadline d_{ik} for $k = 1..N_i$ where N_i is the count of threads of t_i.
- Task threads have interprocess communication which determine their precedence. If a task thread t_{ij} has to communicate the result of its computation to task thread t_{ik}, we say $t_{ij} \prec t_{ik}$, that is, the execution of t_{ij} precedes the execution of t_{ik}.
- Tasks threads, of the same or different tasks, may access shared resources, therefore may block, competing for these resources. The maximum blocking time b_i for a task and its threads can be determined priori using the *Priority Ceiling Protocol* analysis [8].

The aim of the our partitioning algorithm is to partition the task thread graph into subgraphs so that each task thread meets its deadline and also to provide a partition such that the load (total execution time of task threads) is averaged over all subgraphs. We use a modified multilevel graph partitioning method of [4] where instead of finding maximal matching, the graph is partitioned around fixed centers as in [3]. The task threads t_{ik} of a periodic real-time task t_i can be constructed using a directed graph $G = (V, E, w)$ where V is the set of task threads, E is the set of edges giving interprocess communication between threads and $w : \Re \rightarrow E$ is the set of weights associated with edges. This method has coarsening, partitioning and uncoarsening phases. During the coarsening phase, a set of smaller graphs are obtained from the initial graph $G_i = (V_i, E_i)$ such that $|V_i| > |V_{i+1}|$. When graph G_{i+1} is to be constructed from graph G_i, a maximal matching $M_i \subseteq E_i$ is found and vertices that are incident on both edge of this matching are collapsed. The rules for collapsing are as follows. If $u, v \in V_i$ are collapsed to form vertex $w \in V_{i+1}$, the total weight of vertices u and v become the weight of w, the edges incident on w is set equal to the union of the edges incident on u and v minus the edge (u, v). If there is an edge that is incident on both u and v, then the weight of this edge is set equal to the sum of the weights of these two edges. Vertices that are not incident on any edge of the matching are simply copied over to G_{i+1} [4].

We have adapted the modified version of the method presented in [3] for the real-time case where a directed graph depicts task threads with precedence relations and strict deadlines. The coarsening phase then needed to be modified as follows. When we want to choose a neighbor to collapse around the fixed centers, we first look for any predecessors not assigned to a group yet. If there is more than one predecessor, we apply a combination heuristic $H = W_1 * EDF + W_2 * HEM$ where HEM is the *Heaviest Edge Matching*. In other words, we consider communication costs as well as the earliest deadlines when collapsing. W_1 and W_2 are the weights that can be adjusted. If there are no predecessors, the same process is repeated for the sucessors until the number of vertices in the coarsened graph equals the number of procesors n. Finally, we uncoarsen the coarsened vertices back to get the original graph. This algorithm is depicted in Fig. 1.

```
1. Procedure Task_Graph_Partition (TTG:Graph(V,E), n:number of processors);
2. Begin
3.    Mark n nodes of the graph as fixed centers;
4.    While there are nodes to be collapsed
5.       Apply H to neighbors of centers;
6.       Add the executon time of neighbors to centers;
7.       Collapse the chosen neighbors to the centers;
8.    Partition the coarsened TTG;
9.    Uncoarsen TTG back to original;
10. End.
```

Fig. 1. Task Graph Partitioning Algorithm (TGPA)

Theorem 1. *TGPA performs partitioning of $G(V, E)$ in $O(\lfloor N/n \rfloor)$ steps for each task where $|V| = N$ is an upperbound on the number of task threads and n is the number of processors. The time complexity of the total collapsing of TGPA is $O(mN)$ where m is the number of tasks.*

Proof. The TGPA collapses n nodes using H at each step for each task for $O(\lfloor N/n \rfloor)$ steps which would be $O(N)$ operations in total for one task. Total number of collapsing for all of the tasks is then $O(mN)$.

3 Thread Sequencing Algorithm

The information from the partitioning phase yields the coarse order of the task threads to be executed in one processor such that if t_{ij} and t_{ik} are assigned to the same processor by the partitioning algorithm and $t_{ij} \prec t_{ik}$, then t_{ij} has to be executed before t_{ik}. If these two threads have no precedence constraints, that is $t_{ij} \parallel t_{ik}$, they can be assigned arbitrarily after their release times. The *Task Thread Sequencing Algorithm (TTSA)* that is employed to provide the final schedule for task threads on the processor is shown in Fig.2 where any thread that does not have any predecessors becomes ready to be scheduled.

```
1.  Procedure Thread_Sequence (TTG:Graph(V,E), n:Number of processors);
2.  Begin
3.    Task_type=READY for all t_{ik} where n_{ik} = 0; {no predecessors}
4.    Insert READY tasks into Sched_list;
5.    While Sched_list is not empty
6.      Get a task thread t_{ik} from the Sched_list;
7.      Switch (Event_type)
8.        case READY :
9.          Event_type = FINISHED;
10.         time = time + Finish_time of t_{ik};
11.         Insert_event into Sched_list;
12.       case FINISHED :
13.         For each immediate successor t_{ip} of t_{ik};
14.           n_{ip} = n_{ip} - 1;
15.           if n_{ip} = 0
16.             Event_type = READY;
17.             time=time+finish_time of t_{ik} ;
18.             Insert event into Sched_list;
19. End.
```

Fig. 2. Thread Sequencing Algorithm (TSA)

4 Exact Analysis and Refinement

In the second phase, we take the partial execution times of tasks per processor as inputs which is the sum of the execution times of all the task threads on the same processor, assuming interprocess communication costs are zero. We then calculate the worst blocking times for these threads as follows. For each thread, we look at the lower priority threads of the same tasks, that is threads that are prior in execution, and all threads of higher priority tasks. From those, we find the semaphores they are using, and choose the semaphores that have a higher or equal ceiling than the task thread under consideration. The maximum critical section time for these semaphores is the blocking time b_{ik} for thread t_{ik}. In this case, a thread cannot be prempted by a lower priority thread of the same task, it can however be preempted by any thread of the higher priority tasks. Based on these assumptions, a worst case response time R_{ij}, that is, the response time of task component t_{ij} on processor j can be calculated as follows:

$$R_{ij}^{n+1} = b_{ij} + c_{ij} + \sum_{\forall l \in hp(i)} (\lceil \frac{R_{ij}^n}{T_l} \rceil * c_{lj})$$

where $hp(i)$ is the set of tasks that have higher priorities than t_i, c_{lj} is the computation time of a task component t_{lj} on processor j. We then find $max(R_{ij})$ for $j = 1..N_i$. If this is greater than deadline d_i of task t_i then our allocation has failed. In this case, we go into refinement phase where we change the weights of the EDF and HEM heuristics to get different partitionings. This procedure is repeated until a feasible schedule is obtained.

5 An Example of Operation

As an example, consider the task set (T_A, T_B, T_C) with the static properties as shown in Table 1. First, a schedulability analysis shows us that this task set is unschedulable in one processor as $U = \Sigma(C_i/P_i) = 1.60$. Our aim is to see whether this set can be scheduled on two processors using our approach.

Table 1. An Example Task Set

T_i	C_i	D_i	P_i
A	16	30	30
B	24	40	40
C	33	60	60

Assume the threads for each task are as shown in Fig 3 with unity communication costs. The graph partitioning algorithm produces the partitions as shown in Table 2 with weights set equal. The initial centers are chosen randomly as 3 and 4 for Task A; 2 and 7 for Task B; and 3 and 6 for Task C. At each iteration, the combination heuristic $H = W_1 * EDF + W_2 * HEM$ is applied around these centers for the collapse operation. We then input these data which represents the rough schedule to the task thread sequencing algorithm (TSA) of Fig. 2 which produces the output depicted in Table 3. We can now calculate the worst case response times for each task component as shown and conclude that this task thread sets are schedulable as $R_{ij} \leq d_i, \forall i$. We have assumed that the blocking times for threads and the network delays are negligible.

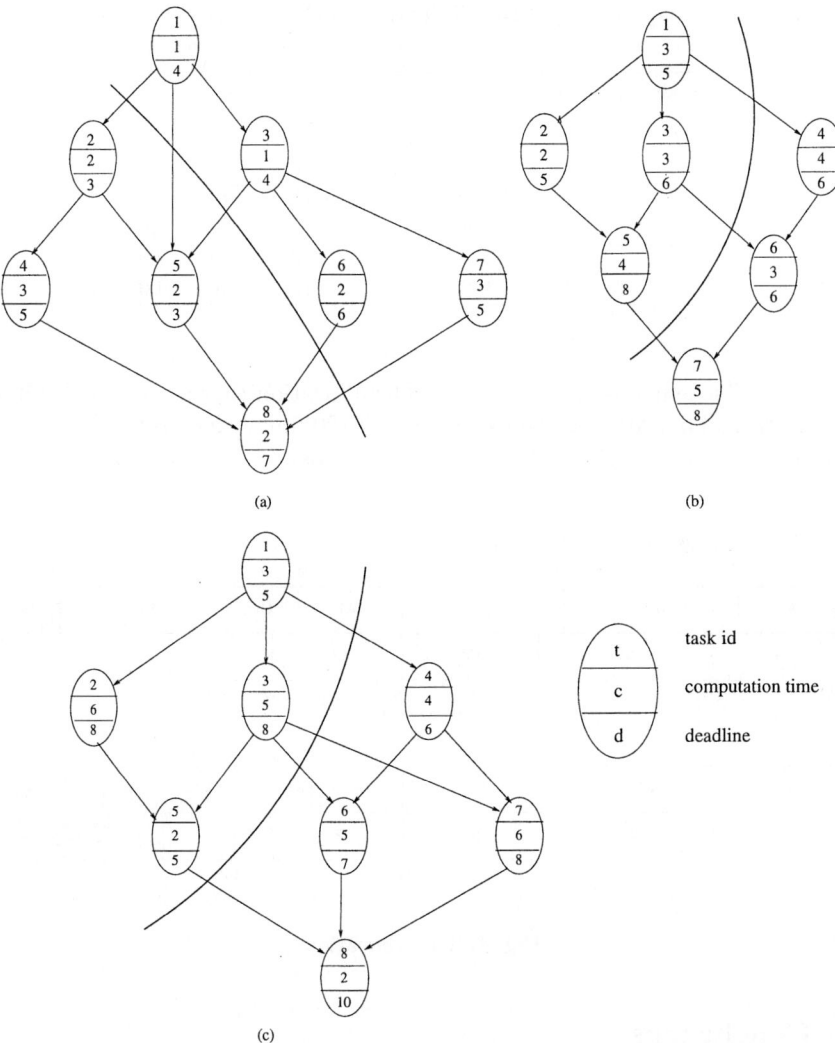

Fig. 3. Tasks: a) Task A b) Task B c) Task C

Table 2. TGPA Iterations

Tasks	Processor	Initial	$iter_1$	$iter_2$	$iter_3$
A	P1	3	3-1	3-1-7	3-1-7-6
	P2	4	4-2	4-2-5	4-2-5-8
B	P1	2	2-1	2-1-3	2-1-3-5
	P2	7	7-6	7-6-4	7-6-4
C	P1	3	3-1	3-1-2	3-1-2-5
	P2	6	6-4	6-4-7	6-4-7-8

Table 3. TSA Output

Tasks	Processor	TSA Output	Σc_{ij}	R_{ij}	P_i
A	P1	1≺3≺6≺7	7	7	30
	P2	2≺5≺4≺8	9	9	30
B	P1	1≺2≺3≺5	12	19	40
	P2	4≺6≺7	12	21	40
C	P1	1≺3≺2≺5	16	42	60
	P2	4≺7≺6≺8	16	46	60

The final schedule is given in Fig. 4. The allocation that meets the deadlines of every thread of every task will repeat every 120 units which is the *hyperperiod* (lowest common multiple of the three task periods).

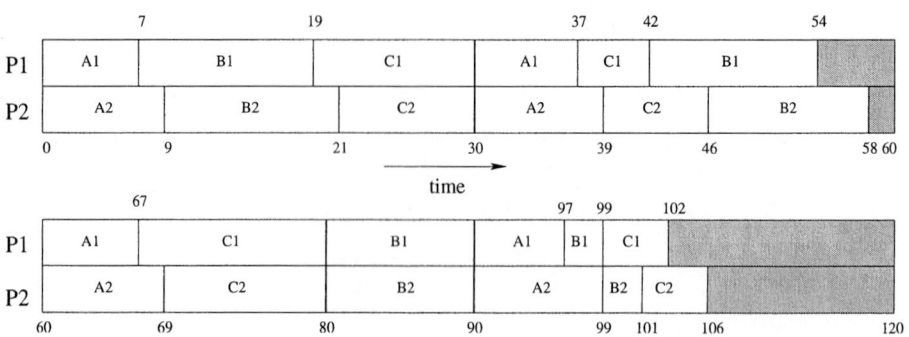

Fig. 4. The Allocation

6 Conclusions

We have presented algorithms based on graph partitioning to yield feasible schedules of real-time tasks to distributed processors. To our knowledge, there is not a significant research effort to accomplish this task using graph partitioning methods. For tasks allocated to different processors, the bounded message delays need to be considered for this model to yield effective results. Also, we need to determine relative weights of the EDF and HEM heuristics to give attainable schedules per processor while partitioning the task graph. Further experimental studies are neccessary to evaluate the proposed algorithms and also different scenarios including threads of different tasks accessing shared resources are needed. It is possible to perform TGPA in parallel for all tasks independently and also for each fixed center collapses within the threads.

References

1. Audsley, N.C., Burns, A., Richardson, M., Wellings, A.: Hard Real-time Scheduling: The Deadline Monotonic Approach, Procs. of IEEE Workshop on Real-Time Operating Systems and Software, (1991), 133-137
2. Dipippo, L. C. et al.: Scheduling and Priority Mapping for Static Real-Time Middleware, Real-time Systems, 20(2), (2001), 155-182
3. Erciyes, K, Marshall, G., A Cluster Based Hierarchical Routing Protocol for Mobile Networks, Springer Verlag, Lecture Notes in Computer Science, ICCSA(3), (2004), 528-537
4. Karypis G., J., Kumar J., Analysis of Multilevel Graph Partitioning, University of Minnesota, Dept. of Computer Science, Tech. Report 95-037
5. Liu J., Layland J.: Scheduling Algorithms for Multiprogramming in a Hard Real-time Environment, Journal of the ACM, 20(1), (1973), 40-61
6. Lehoczky, J.P., Sha, L., Ding, Y.: The Rate Monotonic Scheduling Algorithm: Exact Characterization and Average Case Behaviour, Procs. of IEEE Real-Time Systems Symposium, (1989), 166-171
7. Lin, M., Yang, L.T.: Hybrid Genetic Algorithms for Scheduling Partially Ordered Tasks in a Multi-processor Environment, Procs. of 6th International Conference on Real-Time Computer Systems and Applications. (1999), 382-387
8. Lui, S., Rajkumar R., Lehoczky, J. P.: Priority Inheritance Protocols: An Approach to Real-Time Synchronization, IEEE Transactions on Computers, 39(9), (1990), 1175-1185
9. Lui S., Rajkumar R., Shrish S.: Generalized Rate-Monotonic Scheduling Theory: A Framework for Developing Real-Time Systems, Procs. of the IEEE, 82(1), (1994), 68-82
10. Natale, M.D., Stankovic, J.A.: Scheduling Distributed Real-time Tasks with Minimum Jitter, IEEE Transactions on Computers 49(4), (2000), 303-316
11. Oh, J, Wu, C.: Genetic-algorithm-based real-time task scheduling with multiple goals, Journal of Systems and Software, 71(3), (2004), 245-258
12. Chu, C.: Parallel Machine Scheduling to Minimize Total Tardiness. International Journal of Production Economics, 76(3), (2002), 265-279
13. Xu. J., Parnas, D.: Scheduling Processes with Release Times, Deadlines, Precedence, and Exclusion Relations, IEEE Trans. On Software Engineering, 16(3), (1990), 360-369

A Multi-level Approach for Document Clustering

Suely Oliveira and Sang-Cheol Seok

The University of Iowa, Iowa City IA 52242, USA

Abstract. The divisive MinMaxCut algorithm of Ding et al. [3] produces more accurate clustering results than existing document cluster methods. Multilevel algorithms [4, 1, 5, 7] have been used to boost the speed of graph partitioning algorithms. We combine these two algorithms to construct faster and more accurate algorithm. In this new algorithm, the original graph is coarsened, partitioned by the divisive MinMaxCut algorithm and then decoarsened. A refining algorithm is also applied to improve the accuracy at each level.

1 Introduction

Clustering is the task of classifying a collection of objects, such as documents, into natural categories. Diagnostic tasks in medicine involve classifying a set of symptoms according to the cause and treatment methods. Data mining frequently involves separating documents into different categories so as to provide only relevant information for a user's query. We do not expect that classification or clustering algorithms will ever be 100% accurate, and the related optimization problems are frequently NP-hard. However, we do expect to find algorithms which are highly accurate that are nevertheless very efficient for practical clustering problems.

A variety of algorithms have been proposed and are in widespread use, including the K-means method and its variants [9], the Ratio Cut method [4], and the Normalized Cut method [12]. A recent method that has been proposed is based on the computation of eigenvectors which is called the MinMaxCut algorithm proposed by Ding et al. [3].

The MinMaxCut algorithm proposed by Ding et al. [3] outperformed the existing document clustering methods in accuracy. The algorithm leads to better accuracies because it aims to satisfy both the following desirable properties of a clustering of objects: (P1) nodes in the same cluster are similar, and (P2) nodes in the different clusters are dissimilar. The K-means method mainly achieves (P1), while RatioCut and NormalizedCut mainly achieve (P2). Two-way MinMaxCut is a clustering method which splits the whole cluster into two smaller clusters while divisive MinMaxCut is a K-way MinMaxCut which has as many as K clusters by splitting one of the current clusters repeatedly. A different version of K-way MinMaxCut may have a certain stopping criterion to decide how many clusters it has. The two key algorithms in this divisive method are how to select a cluster to split and how to split the cluster. Two-way MinMaxCut is used to split one cluster into two clusters for divisive MinMaxCut algorithm. Ding et al. [3] presents an efficient way to select one cluster using the sum of similarities of all pairs of each cluster. The two-way MinMaxCut algorithm they use is closely related to the spectral graph partitioning algorithm, which is finding the *Fiedler vector* [11], the second smallest eigenvector of the Graph Laplacian associated with a graph. Finding this eigenvector is, however, expensive and spends the majority of the time required by spectral partitioning techniques.

A class of graph partitioning algorithms [4, 1, 5, 7] coarsen the graph by collapsing vertices and edges, partition the smaller graph, and then decoarsening the partition on the coarsened graph. These methods are called multilevel graph partitioning schemes. These speed up the execution time. Later papers [1] use refining steps to achieve more accurate partitions Hendrickson and Leland [5] presented an improved algorithm for coarsening step by using edge and vertex weights to capture the collapsing of the vertex and edges. The coarsening algorithm of Karypis and Kumar [7] has much smaller coarsened graphs than other multilevel algorithms. They also present a variant of Kernigan-Lin (KL) refinement method which is faster than the original KL algorithm.

In this paper we present a divisive MinMaxCut algorithm to which a multilevel scheme, that is, coarsening, clustering and decoarsening, is applied. Our algorithm take advantage of both fast multilevel algorithm and accurate divisive MinMaxCut algorithm. This method uses edge weights to match 80% to 90% of the nodes by using edges with the highest weights. The other nodes with smaller edge weights, are grouped randomly. The coarsest graphs have less than one hundred nodes. The K-way MinMaxCut method produces a fairly good cut for this coarsest graph. Then during the decoarsening step, the cut for the coarsest graph is decoarsened and refined for all level-ups. The initial cut from the coarsest graph is getting more accurate as the level goes up and finally becomes a very accurate cut for the original graph. The major advantage of multilevel scheme is speed. Furthermore the cut which is decoarsened and refined at last for the original graph is more accurate than that of Ding et al's version of divisive MinMaxCut algorithm.

Our algorithm was tested on newsgroup articles in 20 newsgroups. The accuracy is mostly increasing first and then decreasing and the speed slowly improves as the number of levels increases. With more than 2 levels our algorithm takes less than half the time of Ding et al.'s algorithm. Like other multilevel algorithms, constructing the coarsened graphs takes most of the execution time. So for further improvement, a more efficient coarsening method is highly desirable.

2 K-Way MinMaxCut and Related Issues

In this section we first of all introduce the divisive MinMaxCut algorithm. Then we discuss some issues related to Multilevel approach.

2.1 Two-Way and K-Way MinMaxCut

Divisive MinMaxCut algorithm is based on the two-way MinMaxCut. It starts with a similarity matrix $S = (s_{ij})$, where $s_{ij} \geq 0$ stands for the similarity between node i and j. The more two nodes are similar the bigger s_{ij} is. Similarly, the less two nodes are related the smaller s_{ij} is. It aims to maximize the sum of similarities in a cluster and to minimize the sum of similarities of two nodes in different clusters. We define the sum of similarities between two clusters A and B as $s(A, B) = \sum_{i \in A, j \in B} s_{ij}$, and the sum of similarities of a cluster A as $s(A, A) = \sum_{i \in A, j \in A} s_{ij}$. The two-way MinMaxCut aims to minimize the objective function

$$J_{MMC} = \frac{s(A,B)}{s(A,A)} + \frac{s(A,B)}{s(B,B)} = \frac{s(A,\bar{A})}{s(A,A)} + \frac{s(B,\bar{B})}{s(B,B)} \quad (1)$$

Note that there are many objective functions that satisfy both (P1) and (P2). However, a solution to a continuous relaxation of J_{MMC} can be computed efficiently [2,3]. An indicator vector q is used as the clustering solution. Each component of q gets one of two discrete numbers ($q_i = a$ if $i \in A$, otherwise $q_i = b$). Instead, if we relax the condition that $q_i = a$ or b to $q_i \in \mathbf{R}$, it is well known [3] that the optimal solution of (1) is the eigenvector q_2 associated with the second smallest eigenvalue of the system below with $D = diag(d_1, d_2, \cdots, d_n)$ and $d_i = \sum_j s_{ij}$:

$$(D - S)q = \lambda D q. \quad (2)$$

After searching an optimal dividing point i_{cut}, the final cut is computed

$$A = \{i \mid q_2(i) \leq q_2(i_{cut})\}, \quad B = \{i \mid q_2(i) > q_2(i_{cut})\}. \quad (3)$$

The optimal dividing point i_{cut} is the minimizer of the objective function. The dividing point can be computed in $O(N^2)$ time with a linear search. Note that, however, this does not guarantee that the cut after this linear search has no room to be more accurate. We talk about the refinement method in section 2.3.

A good generalization of the two-way MinMaxCut objective function (1) is:

$$J_{MMC}(C_1, ..., C_K) = \sum_{k=1}^{K} \frac{s(C_k, \bar{C}_k)}{s(C_k, C_k)} = \sum_{1 \leq p \leq q} J_{MMC}(C_p, C_q). \quad (4)$$

where $\bar{C}_i = \bigcup_{j \neq i} C_j$.

2.2 Divisive MinMaxCut

Divisive MinMaxCut algorithm repeatedly performs two main steps. One is selecting a cluster to split and the other is applying the two-way MinMaxCut algorithm. We discussed the two-way MinMaxCut algorithm which is how to split one cluster into two clusters in section 2.1. So we now talk about the way to select the next cluster to split. Ding et al. [3] suggests 5 plans:

Size-priority cluster split: Choose the biggest current cluster.

Average similarity: Select the cluster with smallest $\bar{s}_{kk} := s(C_k, C_k)/|C_k|^2$.

Cluster cohesion: Select the cluster which has the smallest cohesion.
Similarity-cohesion: Select the cluster with the smallest $\bar{s}_{kk} \times$ cohesion.

Greedy: Select the cluster which leads to the minimum objective function value.

The best results are obtained by average similarity cluster selection [3]. For the stopping criterion, we use a user-selected number K; that is, the algorithm selects and splits clusters until there are K clusters. Another criterion is using a threshold on the objective function value which increases monotonically as the number of leaf clusters increase. For details, see [3].

2.3 Refinement

The Kernighan-Lin (KL) refinement [8] method was successfully applied for refining partitions of graphs in a multilevel algorithm [7]. We use the KL algorithm for our refining scheme. KL starts with an initial partition. It iteratively searches for nodes from each cluster of the graph if swapping of a node to one of the other $K - 1$ clusters leads to a better partition. For each node, there would be more than one cluster to give smaller objective function value than the current cut. So the node moves to the cluster that give the biggest improvement. The computation for each node takes only $O(N)$ complexity. So the overall complexity per round does not exceed $O(N^2)$. The iteration terminates when it does not find any node to improve the partition or it finds the predefined number of nodes which lead to a better result. We may apply several iterations of KL to find a better partition.

2.4 Multi-level Approach: Coarsening and Decoarsening

The basic concept is that when we have a big graph $G_0 = (V_0, E_0)$ to cluster, then we construct a smaller graph $G_1 = (V_1, E_1)$ each of whose vertices is a group of several vertices from G_0. We can apply a clustering method to this smaller graph, and transfer this partition to the original graph. This idea is very useful because smaller matrices requires much less time. The process we construct the smaller matrix is called coarsening, and the reverse process is called decoarsening. We can recursively coarsen $G_i = (V_i, E_i)$ to get $G_{i+1} = (V_{i+1}, E_{i+1})$.

The decoarsening step is the way back to the original graph by going through the graphs $G_i, G_{i-1}, \cdots, G_0$. Note that even if the cluster in G_i is a local optimum, the cluster in the next finer level G_{i-1} might not be a local optimum. So we need to check if there is any room to improve the partition in this finer level. Refinement methods are used as the level goes up by one. For more details in the context of graph partitioning, refer to [6, 7].

3 Specific Algorithm and Computational Experiments

3.1 Description

The algorithm consists of three main steps. (1) Coarsening; (2) Clustering the coarsest graph into K subgraphs with the divisive method; (3) Decoarsening and refinement.

The Coarsening and Decoarsening steps are implemented by multiplying special matrices $E_1, \cdots, E_{level}$. We select two nodes to collapse by checking edges from the highest edge weight. Then one column of E is filled with two 1's for the two nodes and 0's for the rest. We collapse less than 90% of nodes by checking the edge weights. The rest of nodes which are not grouped are collapsed randomly. We finally construct $S_0, S_1, \cdots, S_l$ and $E_1, \cdots, E_l$ such as $S_i = E_i' * S_{i-1} * E_i$ and $i = 1, \cdots, l$. The coarsest similarity matrix is used to get the initial partition Cut. During the Decoarsening step the Cut in the current level is multiplied by the proper E for the partition in the next finer level.

We use Divisive MinMaxCut algorithm with average similarity selection scheme. In this algorithm we don't use any specific stopping criterion but predefined number of clusters, let us say K. That is the divisive method stops when we have K clusters in it.

3.2 Source of Data and Preprocessing

The experiments is performed on newsgroup articles in 20 newsgroups (datasets available online [10]). We focus on two sets of 5-clusters cases. The choice of $K = 4, 8$ where the clustering results are less sensitive to cluster section is avoided. The newsgroups chosen are listed in Table 1.

Clusters in M5 overlap at medium level. Meanwhile, clusters in L5 overlap at large. From each set of the newsgroups, we construct two datasets of different sizes: (B) randomly select 100 articles from each newsgroup. (U) randomly select 200, 140, 120, 100, 60 articles from each of the 5 newsgroups, respectively. Dataset(B) has clusters of equal sizes, which is presumably easier to cluster. Dataset(U) has clusters of significantly varying sizes, which is presumably difficult to cluster. Therefore we have 4 newsgroup- cluster size combination categories.

After documents from each category are extracted, we construct a word-document matrix $W = (x_1, \cdots, x_N)$ using standard **tf.idf** scheme. After each

Table 1. 10 newsgroups at different overlapping levels

Dataset M5	Dataset L5
NG2: comp.graphics	NG2: comp.graphics
NG9: rec.motorcycles	NG3: comp.os.ms-windows
NG10: rec.sport.baseball	NG8: rec.autos
NG15: sci.spaces	NG13: sci.electronics
NG18: talk.politics.mideast	NG19: talk.politics.misc

document of W is normalized to 1 using L_2 norm, document-document similarities are calculated as $S = W^T W$.

3.3 Result and Analysis: Accuracy, Time, Objective Function

This section has two main parts. One is comparison of two methods, Divisive MinMaxCut and Multilevel divisive MinMaxCut. The other is some important issues regarding the multi level approach. We constructed 2 different randomly sampled datasets from each category. The average of the time and accuracy are compared with the result from [3]. Their results come from the average of 5 different datasets from each category. For both cases the selection algorithm for the next cluster to split is average similarity selection.

We compare accuracies, time consuming and saturation for Divisive MinMaxCut and Multilevel divisive MinMaxCut in Table2. I/F stands for 'initial' and 'final' accuracies. D and MD mean divisive MinMaxCut and Multilevel divisive MinMaxCut respectively. Initial accuracy of D is the accuracy for the clustering generated by the eigenvector without any refinement. Initial accuracy of MD is the accuracy measured just after clustering the coarsest graph and decoarsening the partition without refinement. The number of levels used for MD is 4. We will see the results with different levels after this. All time consumings are in second and all accuracies are in percent.

Accuracy. We see the initial accuracy of D is mostly better than that of MD but the final accuracy of MD is mostly better than that of M. The reason is that the initial partition of D comes from the whole graph while the initial partition of MD comes from much smaller graph, the coarsest graph. Instead, MD is refining at each level up of decoarsening step. The final result is much different and improved from the initial partition.

Time. MD finishes clustering much faster. MD spends most time for coarsening. For example, it takes 1.63 seconds to go through the first two steps for a dataset of M5B and coarsening takes 1.62 of 1.63 and partitioning takes the rest, 0.01. And 0.19 second is spent for decoarsening and refinement. Note that refinement step for D takes various time consuming depending on the refining scheme and the number of rounds it has.

Saturation comparison: Improvement of accuracy does not exceed some point even though refining step is applied repeatedly. This upper bound is called the

Table 2. Comparison of plain divisive(D) and multidivisive(MD) methods for different categories. Time and accuracy are measured for with/without refinement(I/F) cases. Upper bound of accuracy, saturation(sat), for both cases D and MD are experimented

	D Accuracy I/F	MD Accuracy I/F	D time I/F	MD time I/F	D sat	MD sat
M5B	83.5/91.7	77.2/98.4	2.91/28.53	1.63/1.82	92.5	99.2
M5U	69.3/72.4	70.5/87.5	3.83/55.48	2.99/3.17	91.7	97.42
L5B	88.4/91.7	62.6/81.4	2.01/28.82	1.60/1.71	81.4	95.0
L5U	74.8/74.1	58.5/83.2	3.72/54.94	3.04/3.18	79.0	93.39

saturation of objective function. Saturation of D comes from [3] and that of MD comes from the best result among all levels of each category. This shows MD provides more accurate result than D.

Now we focus on MD itself. We will see how different results MD gives us depending on the various levels. The main focuses here are (a) the relation between time and number of levels and (b) accuracy and the number of levels.

(a) time and level: Table 3 shows how much time is spent for each of three steps. We used two datasets. One is balanced and the other is unbalanced. Time for each of all three steps is measured. As we see the first step, coarsening, takes the most time. Partitioning step takes less time as the number of levels increases. When the number of nodes in the level is less than 100 it takes at most one hundredth second. In both cases the total time consumed decreases slowly as the number of levels increases.

Table 3. Time consuming of all three steps (coarsening/partitioning/decoarsening) of MD for balanced(B) and unbalanced(UB) cases with various levels

level	1	2	3	4
B	1.43/0.33/0.19	1.59/0.05/0.18	1.58/0.00/0.18	1.57/0.01/0.19
UB	2.65/0.53/0.22	2.98/0.08/0.27	3.02/0.01/0.38	2.96/0.00/0.18

(b) accuracy and level: As you see in Table 4, the accuracy and the number of levels curve kind of upside down parabola on average. All 8 dataset are listed, where each 2 datasets are from one of 4 categories. The best accuracy comes from level 3 or 4 where the number of nodes in the coarsest graph is around 50.

We conclude that Multilevel divisive MinMaxCut works very well when compared to the existing cut-based document clustering methods in time and accuracy. More research may be necessary on the relationship between the number of levels and accuracy; that is, how we can decide the optimal number of levels depending the size of the original graph.

Table 4. Relationship between accuracy and level for MD

level	1	2	3	4	5	6
M5Ba	80.4	98.2	98.2	99.2	96.2	68.2
M5Bb	91.8	94.8	95.8	97.6	92.6	60.2
L5Ba	93.4	80.5	95.0	71.2	89.2	82.0
L5Bb	85.6	88.8	54.6	91.6	62.8	65.4
M5Ua	76.0	86.8	79.4	97.3	80.6	91.26
M5Ub	94.3	76.6	97.4	77.7	88.2	78.9
L5Ua	75.5	69.4	82.9	78.9	66.1	54.8
L5Ub	72.3	64.7	93.4	87.6	72.3	59.4

Acknowledgments

We would like to acknowledge the support of the National Science Foundation for this work through grant DMS-02-13305 and the help from Dr. David Eichmann for generating the similarity matrices.

References

1. T. Bui and C. Jones. A heuristic for reducing fill in sparse matrix factorization. In *6th SIAM Conf. Parallel Processing for Scientific Computing*, pages 445–452, 1993.
2. C. Ding, X. He, H. Zha, M. Gu, and H. Simon. A min-max cut algorithm for graph partitioning data clustering. In *ICDM 2001, Proceedings IEEE International Conference on Data Mining, 2001*, pages 107–114. IEEE, 2001.
3. C. Ding, X. He, H. Zha, M. Gu, and H. Simon. A minmaxcut spectral method for data clustering and graph partitioning. Technical Report 54111, LBNL, December 2003.
4. L. Hagen and A. Kahng. Fast spectral methods for ratio cut partitioning and clustering. In *Proceedings of IEEE International Conference on Computer Aided Design*, pages 10–13. IEEE, 1991.
5. B. Hendrickson and R. Leland. A multilevel algorithm for partitioning graphs. Technical Report SAND93-1301, Sandia National Laboratories, 1993.
6. M. Holzrichter and S. Oliveira. A graph based davidson algorithm for the graph partitioning problem. *International Journal of Foundations of Computer Science*, 10:225–246, 1999.
7. G. Karypis and V. Kumar. A fast and high quality multilevel scheme for partitioning irregular graphs. Technical Report 95-035, Department of Computer Science, University of Minnesota, Minneapolis, MN, 1998.
8. B.W. Kernighan and S. Lin. An efficient heuristic procedure for partitioning graphs. *The Bell System Technical Journal*, 1970.
9. J. MacQueen. Some methods for classification and analysis of multivariate observations. *Proc. Fifth Berkeley Symp. Math. Statistics and Probability*, 1:281–296, 1967.
10. A. McCallum. A toolkit for statistical language modeling, text retrieval, classification and clustering., 1996. Available on WWW at URL http://www.cs.cmu.edu/~mccallum/bow.
11. A. Pothen, H. D. Simon, and Kang-Pu K. Liou. Partitioning sparse matrices with eigenvectors of graphs. *SIAM J. Matrix Anal. Appl.*, 11(3):430–452, 1990.
12. J. Shi and J. Malik. Normalized cuts and image segmentation. *IEEE Trans. on Pattern Analysis and Machine Intelligence*, 22(8):888–905, Aug 2000.

A Logarithmic Time Method for Two's Complementation

Jung-Yup Kang[1] and Jean-Luc Gaudiot[2]

[1] Mindspeed Technologies,
Inc., Newport Beach, CA 92660, USA
[2] University of California at Irvine,
Irvine, CA 92612, USA

Abstract. This paper proposes an innovative algorithm to find the two's complement of a binary number. The proposed method works in logarithmic time ($O(logN)$) instead of the worst case linear time ($O(N)$) where a carry has to ripple all the way from LSB to MSB. The proposed method also allows for more regularly structured logic units which can be easily modularized and can be naturally extended to any word size. Our synthesis results show that our method achieves up to 2.8× of performance improvement and up to 7.27× of power savings compared to the conventional method.

1 Introduction

Signed binary numbering representation [5, 7, 11, 12] (based on two's complement numbers) is a nearly universally used numbering representation in the computing world. Thus, in computer systems which are based on this two's complement representation, operations to find the two's complement of a signed binary number are frequently executed. This is indeed true for applications which require to find the absolute value of signed binary numbers. For instance, motion estimation operations of MPEG encodings [9, 10, 15, 17] require to find the absolute value of the difference (of pixel values) for each pixel position for each block comparison. Another example would be multipliers [2, 8, 13] that need to find the two's complement of the multiplicand for the negative encodings of Booth algorithms [2, 3, 13, 14, 18].

Despite the frequent need for finding the two's complement of a signed binary number, two's complementation of a binary number is still carried out using the conventional way of complementing each bit and adding 1 to the complemented number. By doing so, we cannot ignore the possibility of a carry propagating all the way from the LSB (Least Significant Bit) to the MSB (Most Significant Bit). However, we have learned that the speed of finding the two's complement of a binary number is critical to the performance improvement for a group of applications. Our recent study indicates that if the two's complement of the multiplicand of a multiplication was found fast, there can be up to 40% of performance improvement [8]. It has also been reported by Hashemian [6] that in some specific parallel processing applications, it is more effective to expedite the two's complementation by using special-purpose hardware rather than by using an adder and inverters.

Therefore, in this paper, we present an efficient algorithm and architecture to find the two's complement of a binary number in a truly logarithmic time ($O(logN)$) instead of worst case of $O(N)$ time when using an adder. The proposed method also allows for more regularly structured logic units which can be easily modularized and can be naturally extended to any word size. In the next section, the conventional methods to find the two's complement of a binary number and their problems are discussed. Then, our logarithmic method and its possible implementations will be introduced. Finally, the evaluation of the algorithm followed by the module generation techniques for our two's complementation algorithm will be discussed before the summary of this paper.

2 Conventional Methods

As mentioned before, conventionally (and by definition), the two's complement of a binary number is found by complementing each bit and adding 1 to the complemented number. However, by doing so, there is the possibility of a carry propagating all the way from the LSB to the MSB. Therefore, the time complexity of finding the two's complement of a binary number is at least that of one addition (plus the complementation of each bit). However, even this delay is too large for fast multiplier architectures [8] and not efficient for some specific parallel applications [6, 10, 15, 17].

There is another well-known conventional method in which all the bits after the rightmost "1" in the word are complemented and all the other bits are left untouched. For example, the two's complement of the binary number 001010_2 (10_{10}) is 110110_2 (-10_{10}) (Figure 1). For this number, the rightmost "1" happens in bit position 1 (the check mark position in Figure 1). Therefore, values in bit positions 2 to 5 can simply be complemented while values in bit positions 0 and 1 are kept as they were.

Our method is an extension of the latter algorithm. We observed from this algorithm, that two's complementation comes down to finding the conversion signals that are used for selectively complementing some of the input bits. If the conversion signal at any position is "0" (the red crosses in Figure 1), then the value is kept as it is and if the conversion signal is "1" (the green check marks in Figure 1), then the value is complemented. All the conversion signals to the left of the rightmost "1" are 1 and all the conversion signals to the right of the rightmost "1" (and the conversion signal for the rightmost "1") are 0. For example, for data word 00101000_2, the conversion signals would be "11110000_2." Applying these conversion signals to the input (complementing only the most significant 4 bits in this case) would result in the two's complement of the input (11011000_2).

However, this searching for the rightmost "1" could be as time consuming as rippling a carry through to the MSB since the previous bits information must be

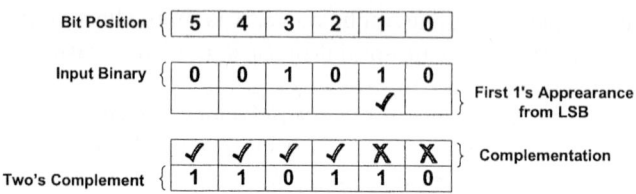

Fig. 1. Two's Complement Conversion Example

transferred to the MSB to determine which one is the rightmost "1." Therefore, we must find a method to expedite this detection of the rightmost "1." As we shall see, this search for the rightmost "1" can be achieved in logarithmic time using our binary search tree-like structure.

One possible way to implement some fast logic which will produce the adequate conversion signals would be to wire each input bit along with the preceding less significant bits to an OR-gate that accepts that many inputs. This allows for each input bit to determine whether there was a "1" in any lower order bit position and to produce its own conversion signal. However, in such cases, although it seems possible to produce the conversion signal in ideal constant time, each input bit must drive a significant number of wires (up to the number of input bits for the LSB). This is not considered to be practical nor efficient (in terms of implementation).

3 Proposed Logarithmic Method

Consequently, in this section, we describe an efficient (in terms of speed and implementation) algorithm which determines the conversion signals needed to perform two's complementation. We first find the conversion signals for a 2-bit group by grouping two consecutive bits (the grouping always starts from the LSB) from the input and find the conversion signals in each group as shown in Figure 2(a). Then we find the conversion signals for a 4-bit group (formed by two consecutive 2-bit groups). Then we find the conversion signals for an 8-bit group (formed by two consecutive 4-bit groups). This divide-and-conquer approach is pursued until the whole input word has been covered.

When grouping two 2^n-bits groups, the leftmost conversion signals from the right group contain the accumulative information of its group about whether a "1" ever appeared in any bit position of its group, so that a conversion signal should force all the conversion signals from the left group all the way to the "1" if it is itself is a "1." For instance, as shown in Figure 2(b), if CS_1 (the leftmost conversion signal from the right group) = "1," the conversion signals from the left group (CS_2 and CS_3) should be forced to a "1," regardless of their previous values. If CS_1 = "0," nothing happens to the conversion signals from the left group. This variable control is shown with a dashed arrow. Likewise, CS_5 may affect conversion signals CS_6 and CS_7. The same goes for CS_3' which may affect the conversion signals (CS_7', CS_6', CS_5', and CS_4').

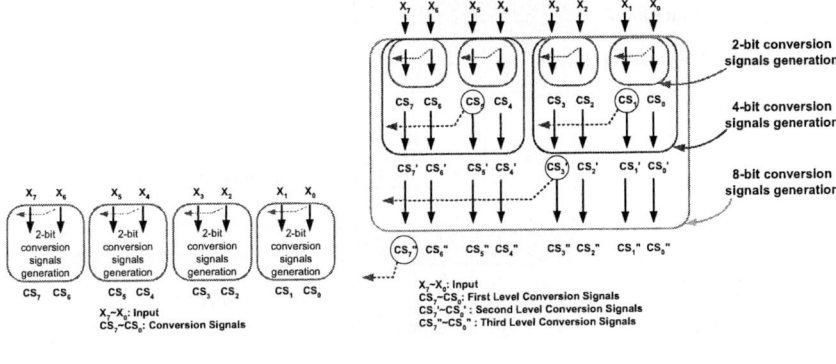

(a) Conversion Signals for Pairs of Bits

(b) Conversion Signals for Eight Bits

Fig. 2. Determining the Conversion Signals

The inputs to the 2-bit group are bits from the original binary number. However, the inputs to the next level groups are conversion signals from the previous level. For instance, the inputs to the 4-bit group are the conversion signals generated from two 2-bit groups. Therefore, from the second level (4-bit grouping) on, the conversion signals are scanned in order to find the rightmost "1."

After determining the conversion signals, two's complementation is a mere complementation of the input binary according to the conversion signals. One possible implementation of our algorithm is shown in Figure 3(a). Figure 3(b) shows another version of the design using NAND, NOR, and inverter gates. Once we have the complete conversion signals, these signals are shifted left 1 bit and EXOR-ed with the input to create the two's complement of the input. In Figure 3(a), X_0 to X_7 represent the input and X_0' to X_7' represent its two's

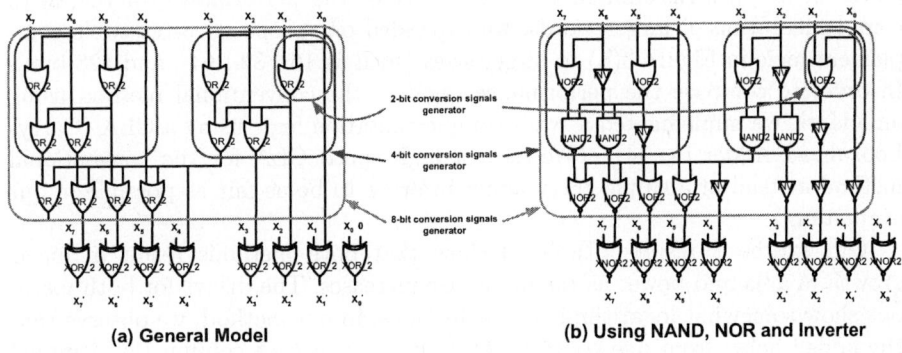

(a) General Model

(b) Using NAND, NOR and Inverter

Fig. 3. A Gate-Level Diagram of 8-bit Two's Complementation Logic Using Our Approach

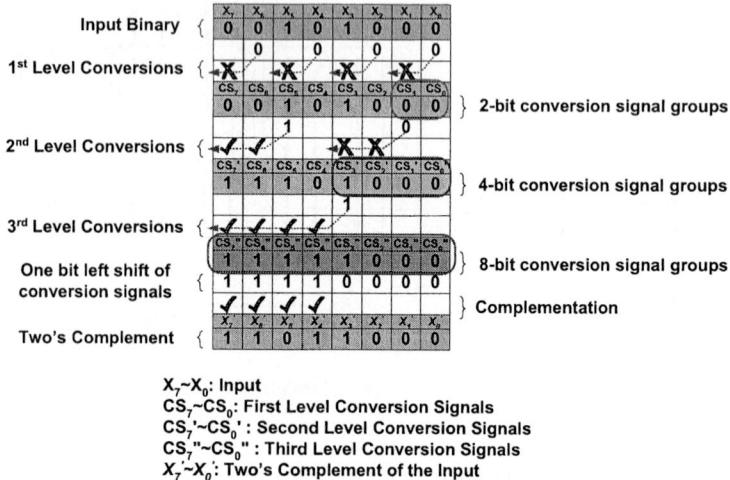

Fig. 4. 8-bit Example of Two's Complementation Using Our Approach

complement. One complete example of two's complementation of "00101000_2" is shown in Figure 4.

4 Performance Evaluation and Analysis

In order to measure the performance of the proposed algorithm and its implementation, we designed our algorithm using Verilog HDL (Hardware Description Language) and synthesized it using Synopsys synthesis tools [16]. Note that we used Artisan TSMC 0.13um 1.2-Volt standard-cell library [1] with "slow corner" operating conditions for our synthesis. We estimated the area, delay, and power of our designs and in order to measure the performance of the two's complementation of larger words, we expanded our proposed 8-bit two's complement logic in Figure 3(b) to larger sizes (such as 16-, 32-, 64-, and 128-bits). In order to compare the performance (against the conventional method using an adder), we implemented a two's complementation logic using a CLA (Carry-Lookahead Adder in [4]). (We used a high speed CLA for the conventional method instead of a ripple carry adder in order to be as fair as possible in our evaluation.)

Our synthesis results (Table 1) show that both methods result in linear growth in area and power as the input size increases. The delays for both methods show somewhat logarithmic characteristics. In our method, we observe that the added delay from one column (2^n-input) to the next column (2^{n+1}-input) is the one additional level of OR-gates, the associated wire delay, and the delay for driving twice the number of OR-gates (note that it would be a perfect logarithmic growth if there were no wire delays or delays due to the high fan-outs required in the last level of OR-gates). This can be made clearer as we observe

Table 1. Synthesis Reports

	Input (n-bit)	8-bit	16-bit	32-bit	64-bit	128-bit
Our Method	Delay (ns)	0.47	0.71	1.06	1.62	2.17
	Area (um²)	201	480	1111	2528	5661
	Power (mW)	0.096	0.18	0.35	0.69	1.37
Conventional (using CLA)	Delay (ns)	1.33	1.64	2.40	2.80	3.35
	Area (um²)	497	997	2122	4274	8614
	Power (mW)	0.51	1.04	2.35	4.74	9.96

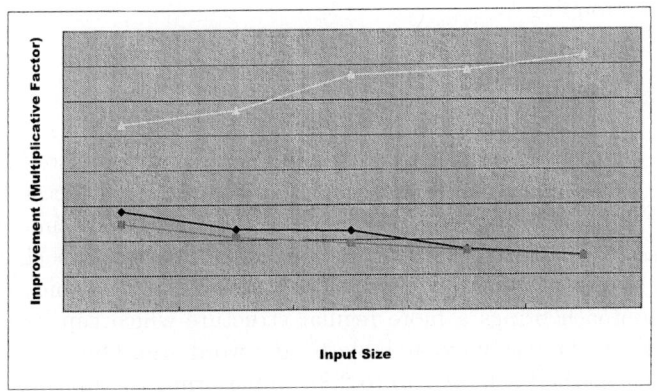

Fig. 5. Improvement (Speed, Area, and Power) of Our Two's Complementation Along with Input Size

the difference from one column to the other as we move right from one column to the other in the table.

When the methods are compared, our approach brings up to 2.8× (when $n = 8$) of performance improvement, up to 2.47× (when $n = 8$) of area savings, and up to 7.27× (when $n = 128$) of power saving when compared to the conventional method. We notice that as we increase the input size, the improvements from delay and area shrink (Figure 5). When $n = 128$, we achieve about 1.54× of performance improvement and about 1.52× of area saving. We believe this phenomenon is due to the fact that, as we increase the size of the operator, the fan-out of the last stage OR-gate is severely impacted which results in greater delays and area penalty. The power savings (in percentage) are about the same across the input sizes.

Related to our method, Hwang [7] and Hashemian [6] have shown similar approaches (finding the conversion signals). However, our method is logarithmic whereas Hwang's method is linear and Hashemian has focused on circuit optimization to improve the performance. Our approach is more general and shows better adaptability to any word size.

5 Module Generation

Our two's complementation algorithm can be easily modularized and expanded to cover binary numbers of any size. First, as shown in Figure 3(a), a 2-bit group can be modularized into a 2-bit conversion signals generator (using one OR-gate). Then, two 2-bit conversion signals generators and two more OR-gates form a 4-bit conversion signals generator. Again, two 4-bit conversion signals generators and four more OR-gates constitute an 8-bit conversion signal generator. In this fashion (two 2^n-bit conversion signals generators and 2^n more OR-gates connected to the left 2^n-bit conversion signals generator), we can continue for any 2^{n+1}-bit grouping.

6 Conclusions

This paper has introduced an innovative and efficient method to find the two's complement of a binary number. When using the proposed method, the two's complement of a binary number can be found in logarithmic time and can be used for cases where faster two's complementation is necessary such as fast multiplications as well as some specific parallel processing applications. At the same time, our approach brings a more regular structure which can be easily modularized and can be easily expandable to any word size. Our synthesis results show that our method achieves up to 2.8× of performance improvement and up to 7.27× of power savings compared to the conventional method.

Acknowledgements

This paper is based upon work supported in part by NSF grants CCR-0234444 and INT-0223647. Any opinions, findings, and conclusions or recommendations expressed in this material are those of the authors and do not necessarily reflect the views of the National Science Foundation.

References

1. Artisan Components. *TSMC 0.13µm Process CL013LV 1.2-Volt SAGE-$X^T M$ Standard Cell Library Databook*. Artisan Components, October 2001.
2. A. D. Booth. A Signed Binary Multiplication Technique. *Quarterly J. Mechanical and Applied Math.*, 4:236–240, 1951.
3. F. Elguibaly. A Fast Parallel Multiplier-Accumulator Using the Modified Booth Algorithm. *IEEE Transactions on Circuits and Systems*, 47(9):902–908, 2000.
4. M. D. Ercegovac and T. Lang. *Digital Arithmetic*. Morgan Kaufmann Publishers, Los Altos, CA 94022, USA, 2003.
5. D. Gajski. Principles of Digital Design. Prentice Hall, 1997.
6. R. Hashemian and C. P. Chen. A New Parallel Technique for Design of Decrement/Increment and Two's Complement Circuits. In *Proceedings of the $34^t h$ Midwest Symposium on Circuits and Systems*, volume 2, pages 887–890, 1991.

7. K. Hwang. *Computer Arithmetic Principles, Architecture and Design*. Wiley, New York, 1979.
8. J.-Y. Kang and J.-L. Gaudiot. A Fast and Well-Structured Multiplier. In *EUROMICRO Symposium on Digital System Design*, pages 508–515, August 2004.
9. J.-Y. Kang, S. Shah, S. Gupta, and J.-L. Gaudiot. An Ecient PIM (Processor-In-Memory) Architecture for Motion Estimation. In *IEEE $14^{t}h$ International Conference on Application-specic Systems, Architectures and Processors*, pages 273–283, June 2003.
10. J. Kneip, S. Bauer, J. Vollmer, B. Schmale, P. Kuhn, and R. Bosch. The MPEG-4 Video Coding Standard - A VLSI Point of View. In *1998 IEEE Workshop on SIGNAL PROCESSING SYSTEMS (SiPS): Design and Implementation*, pages 43–52, Octover 1998.
11. M. Mano and C. Kime. *Logic and Computer Design Fundamentals*. Prentice Hall, 2000.
12. A. Marcovitz. *Introduction to Logic Design*. McGraw Hill, 2002.
13. O. L. Mac Sorley. High Speed Arithmetic in Binary Computers. *IRE Proc.*, 1961.
14. M. R. Santoro and M. Horowitz. SPIM: A Pipelined 64x64-bit Iterative Multiplier. *IEEE Transactions on Circuits and Systems*, 24(2):487–493, 1989.
15. M. Sun and K. Yang. A Flexible VLSI Architecture for Full-search Block-Matching Motion Vector Estimation. In *IEEE Int. Symp. on Circuits and Systems*, pages 179–182, May 1989.
16. Synopsys. Design Compiler User's Guide. *http://www.synopsys.com/*, 2004.
17. K. Yang, M. Sun, and L. Wu. A Family of VLSI Designs for Motion Compensation Block Matching Algorithm. *IEEE Transactions on Circuits and Systems*, 36(10):1317–1325, 1989.
18. W.-C. Yeh and C.-W. Jen. High-Speed Booth Encoded Parallel Multiplier Design. *IEEE Transactions on Computers*, 49(7):692–701, 2000.

The Symmetric–Toeplitz Linear System Problem in Parallel*

Pedro Alonso and Antonio Manuel Vidal

Universidad Politécnica de Valencia, cno. Vera s/n, 46022 Valencia, Spain
{palonso, avidal}@dsic.upv.es

Abstract. Many algorithms exist that exploit the special structure of Toeplitz matrices for solving linear systems. Nevertheless, these algorithms are difficult to parallelize due to its lower computational cost and the great dependency of the operations involved that produces a great communication cost. The foundation of the parallel algorithm presented in this paper consists of transforming the Toeplitz matrix into a another structured matrix called Cauchy–like. The particular properties of Cauchy–like matrices are exploited in order to obtain two levels of parallelism that makes possible to highly reduce the execution time. The experimental results were obtained in a cluster of PC's.

1 Introduction

In this paper, we present a parallel algorithm for the solution of the linear system

$$Tx = b , \qquad (1)$$

where $T \in \mathbb{R}^{n \times n}$ is a symmetric Toeplitz matrix $T = (t_{ij})_{i,j=0}^{n-1} = (t_{|i-j|})_{i,j=0}^{n-1}$ and $b, x \in \mathbb{R}^n$ are the independent and the solution vector, respectively.

It is difficult to obtain efficient parallel versions of *fast* algorithms, because they have a reduced computational cost and they also have many dependencies among fine–grain operations. These dependencies produce many communications, which are a critical factor to obtain efficient parallel algorithms, especially on distributed memory computers. This problem could explain partially the small number of parallel algorithms implemented so far dealing with Toeplitz matrices. For instance, it can be found parallel algorithms to solve Toeplitz systems using systolic arrays [1] or dealing only with positive definite matrices or with symmetric matrices [2]. There also exist parallel algorithms for shared memory computers [3, 4, 5] and, more recently, several parallel algorithms for distributed architectures have been proposed [6].

One of our main goals is to offer efficient parallel algorithms for general purpose architectures, especially, clusters of personal computers. Furthermore, the codes are portable because they are based on standard libraries, both sequential,

---

* Supported by Spanish MCYT and FEDER under Grant TIC 2003-08238-C02-02.

LAPACK [7], and parallel, ScaLAPACK [8]. We are mainly interested in the reduction of parallel runtime because one of the main set of applications requires real time computation of the linear system (1) like digital signal analysis [9].

In the next section, the mathematical background used is summarized. In Sections 3, 4 and 5 the parallel algorithm is described. Finally, the experimental results are shown in the last section.

2 Rank Displacement and Cauchy–Like Matrices

It is said that a matrix of order n is *structured* if its *displacement representation* has a lower rank regarding n. The *displacement representation* of a symmetric Toeplitz matrix T (1) can be defined in several ways depending on the form of the displacement matrices. A useful form for our purposes is

$$\nabla_F T = F T - T F = \mathcal{G} \mathcal{H} \mathcal{G}^T ; \qquad (2)$$

where $F = T(e_1)$, called *displacement matrix*, is a $n \times n$ symmetric Toeplitz matrix with the second column of the identity matrix as the first column; $\mathcal{G} \in \mathbb{R}^{n \times 4}$ is the *generator* matrix and $\mathcal{H} \in \mathbb{R}^{4 \times 4}$ is a skew–symmetric *signature* matrix. The rank of $\nabla_F T$ is 4, that is, lower than n and independent of n.

It is easy to see that the displacement of T with respect to F is a matrix of a considerably sparsity from which it is not difficult to obtain an analytical form of $\mathcal{G}$ and $\mathcal{H}$.

A symmetric *Cauchy–like* matrix C is a structured matrix that can be defined as the unique solution of the displacement equation

$$\nabla_\Lambda C = \Lambda C - C \Lambda = \hat{\mathcal{G}} \mathcal{H} \hat{\mathcal{G}}^T , \qquad (3)$$

being $\Lambda = \text{diag}(\lambda_1, \ldots, \lambda_n)$, where $\text{rank}(\nabla_\Lambda C) \ll n$ and independent of n.

Now, we use the normalized Discrete Sine Transformation (DST) $\mathcal{S}$ as defined in [10]. Since $\mathcal{S}$ is symmetric, orthogonal and $\mathcal{S} F \mathcal{S} = \Lambda$ [11,12], we obtain

$$\mathcal{S}(FT - TF)\mathcal{S} = \mathcal{S}(\mathcal{G}\mathcal{H}\mathcal{G}^T)\mathcal{S} \rightarrow \Lambda C - C\Lambda = \hat{\mathcal{G}}\mathcal{H}\hat{\mathcal{G}}^T ,$$

where $C = \mathcal{S} T \mathcal{S}$ and $\hat{\mathcal{G}} = \mathcal{S}\mathcal{G}$. This shows how it can be transformed (2) into (3).

In this paper, we solve the Cauchy–like linear system $C\hat{x} = \hat{b}$, where $\hat{x} = \mathcal{S}x$ and $\hat{b} = \mathcal{S}b$, by performing the triangular decomposition $C = LDL^T$, being L unit lower triangular and D diagonal. The solution of (1) is obtained by computing $Ly = \hat{b}$, $y \leftarrow D^{-1}y$, $L^T \hat{x} = y$ and $x = \mathcal{S}\hat{x}$.

Solving a symmetric Toeplitz linear system by transforming it into a symmetric Cauchy–like system has an interesting advantage due to the symmetric Cauchy–like matrix has an important sparsity. Matrix C has the form (x only denotes non–zero entries),

$$C = \begin{pmatrix} x & 0 & x & 0 & \ldots \\ 0 & x & 0 & x & \ldots \\ x & 0 & x & 0 & \ldots \\ 0 & x & 0 & x & \ldots \\ \vdots & \vdots & \vdots & \vdots & \ddots \end{pmatrix} .$$

We define the odd–even permutation matrix P_{oe} as the matrix that, after applied to a vector, groups the odd entries in the first positions and the even entries in the last ones, $P_{oe}\begin{pmatrix} x_1 & x_2 & x_3 & x_4 & x_5 & x_6 & \ldots \end{pmatrix}^T = \begin{pmatrix} x_1 & x_3 & x_5 & \ldots & x_2 & x_4 & x_6 & \ldots \end{pmatrix}^T$. Applying transformation $P_{oe}(.)P_{oe}^T$ to a symmetric Cauchy–like matrix C gives

$$P_{oe}CP_{oe}^T = \begin{pmatrix} C_0 & \\ & C_1 \end{pmatrix}, \quad (4)$$

where C_0 and C_1 are symmetric Cauchy–like matrices of order $\lceil n/2 \rceil$ and $\lfloor n/2 \rfloor$, respectively. In addition, it can be shown that matrices C_0 and C_1 have a displacement rank of 2, as opposed to C that has a displacement rank of 4 [5].

The two submatrices arising in (4) have the displacement representation

$$\Lambda_j C_j - C_j \Lambda_j = G_j H_j G_j^T, \quad i = 0, 1, \quad (5)$$

where $\begin{pmatrix} \Lambda_0 & \\ & \Lambda_1 \end{pmatrix} = P_{oe} S \Lambda S P_{oe}^T$ and $H_0 = H_1 = \begin{pmatrix} 0 & 1 \\ -1 & 0 \end{pmatrix}$. As it is shown in [13], given vector $u^T = \begin{pmatrix} 0 & t_2 & t_3 & \cdots & t_{n-2} & t_{n-1} \end{pmatrix}^T$ and the first column of the identity matrix e_0, the generators of (5) can be computed as

$$\begin{pmatrix} G_0 \\ G_1 \end{pmatrix} = \sqrt{2} P_{oe} S \begin{pmatrix} u & e_0 \end{pmatrix}. \quad (6)$$

The odd–even permutation matrix is used to decouple the symmetric Cauchy–like matrix arising from a real symmetric Toeplitz matrix into the following two Cauchy–like systems of linear equations

$$C_j \hat{\tilde{x}}_j = \hat{\tilde{b}}_j, \quad j = 0, 1, \quad (7)$$

where $\hat{\tilde{x}} = \begin{pmatrix} \hat{\tilde{x}}_0^T & \hat{\tilde{x}}_1^T \end{pmatrix}^T = P_{oe} S x$ and $\hat{\tilde{b}} = \begin{pmatrix} \hat{\tilde{b}}_0^T & \hat{\tilde{b}}_1^T \end{pmatrix}^T = P_{oe} S b$.

Each one of both linear systems are of half the size and half the displacement rank so this yields substantial saving over the non–symmetric forms of the displacement equation. Furthermore, it can be exploited in parallel by solving each of the two independent sub–systems into two different processors.

3 The Parallel Algorithm

For the parallel solution we used a two dimensional mesh of $p/2 \times 2$ processors as shown in Fig. 1, where each one of the p processors is denoted by the corresponding row and column index.

We used the ScaLAPACK tools in order to manage data distribution over this logical configuration of the processors. Once the symmetric Toeplitz system has been converted into a symmetric Cauchy–like one, the two subsystems arisen (7) will be solved independently on each "logical column" of the two–dimensional processors mesh. This is what the external loop ($j = 0, 1$) of Algorithm 1 represents, that is, iteration 0 and 1 are concurrently executed by processors column 0 and 1, respectively.

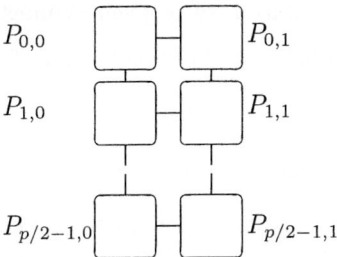

Fig. 1. 2D mesh of processors

Algorithm 1 (Parallel Algorithm for the solution of a symmetric–Toeplitz system with Cauchy–like transformation). *Given $T \in \mathbb{R}^{n \times n}$ a symmetric Toeplitz matrix and $b \in \mathbb{R}^n$ an independent vector, this algorithm returns the solution vector $x \in \mathbb{R}^n$ of the linear system $Tx = b$. For each $P_{i,j}$,*

1. for $j = 0, 1$, do
      for $i = 0, \ldots, p/2 - 1$, do
        1.1. "Previous computations".
        1.2. $C_j = L_j D_j L_j^T$ (4).
        1.3. Solution of $L_j D_j L_j^T \hat{\tilde{x}}_j = \hat{\tilde{b}}_j$ (7).
      end for.
   end for.
2. $P_{0,0}$ computes $x = SP_{oe}^T \left(\hat{\tilde{x}}_0^T \; \hat{\tilde{x}}_1^T \right)^T$.

Steps 1.1 and 1.2 are explained in the next two sections, respectively. Step 1.3 is performed by ScaLAPACK and PBLAS parallel subroutines. This last step can be repeated several times for iterative refinement. Finally, processor $P_{0,0}$ gathers the partial solutions of the two independent Cauchy–like linear systems and computes the solution of (1).

4 Parallel Triangularization of Symmetric Cauchy–Like Matrices

The workload of the step 1.2 of Algorithm 1 falls in the operation with the $n \times 2$ entries of the generators G_0 and G_1 (6). The logical column of processors $P_{i,j}$, $i = 0, \ldots, p/2 - 1$, performs the triangular decomposition of the corresponding matrix $C_j = L_j D_j L_j^T$, $j = 0, 1$, where L_j is unit lower triangular and D_j is diagonal. The parallel algorithm exploits the fact that operations performed by each column of processors can be carried out independently on each row of G_j.

Let

$$\begin{pmatrix} \Lambda_0 & \\ & \Lambda_1 \end{pmatrix} C - C \begin{pmatrix} \Lambda_0 & \\ & \Lambda_1 \end{pmatrix} = \begin{pmatrix} G_0 \\ G_1 \end{pmatrix} H \begin{pmatrix} G_0 \\ G_1 \end{pmatrix}^T, \qquad (8)$$

be the displacement representation of a given symmetric Cauchy–like matrix $C = \begin{pmatrix} C_{00} & C_{01} \\ C_{10} & C_{11} \end{pmatrix}$, then the Schur complement C_{sc} of C_{11} regarding C_{00} is also structured for any partition of C [14] so

$$\Lambda_1 C_{sc} - C_{sc} \Lambda_1 = G'_1 H G'^T_1 . \tag{9}$$

The parallel algorithm uses a sequential algorithm that, given the generator, the displacement matrix and the diagonal of C in equation (8) as entries, returns G'^T_1, the diagonal of C_{sc} of equation (9) and the factorization $C_{00} = L_{00} D_{00} L^T_{00}$.

Let m_j denotes the size of C_j (4), $j = 0, 1$, respectively. Each generator G_j (5) is partitioned in m_j/ν blocks of size $\nu \times 2$ for a given integer ν, $1 \leq \nu \leq (m_j/p)$, and cyclically distributed onto the respective column of processors $P_{k,j}$, for $k = 0, \ldots, p/2 - 1$, in such a way that block $G_{i,j}$, $i = 0, \ldots, m_j/\nu - 1$, belongs to processor $P_{\mathrm{mod}(i,p),j}$. For simplicity in the exposition we will assume in the next that $(m_j \bmod \nu) = 0$. The unit lower triangular factor L_j obtained by the algorithm is partitioned in a two dimensional array of $(m_j/\nu) \times (m_j/\nu)$ square blocks of order ν, where each square block $L^j_{i,l}$, for $l = 0, \ldots, i$, belongs to processor $P_{\mathrm{mod}(i,p),j}$ as the generators blocks. The diagonal matrix D_j is stored in the diagonal entries of L_j since all diagonal entries of L_j are implicitly one. In Fig. 2 it can be seen an example of distribution of both G_j and L_j in the logical column $j = 0, 1$ formed of three processors.

$P_{0,j}$	$G_{0,j}$	$L^j_{0,0}$				
$P_{1,j}$	$G_{1,j}$	$L^j_{1,0}$	$L^j_{1,1}$			
$P_{2,j}$	$G_{2,j}$	$L^j_{2,0}$	$L^j_{2,1}$	$L^j_{2,2}$		
$P_{0,j}$	$G_{3,j}$	$L^j_{3,0}$	$L^j_{3,1}$	$L^j_{3,2}$	$L^j_{3,3}$	
$P_{1,j}$	$G_{4,j}$	$L^j_{4,0}$	$L^j_{4,1}$	$L^j_{4,2}$	$L^j_{4,3}$	$L^j_{4,4}$
$\vdots$	$\vdots$	$\vdots$	$\vdots$	$\vdots$	$\vdots$	$\ddots$

Fig. 2. Example of data distribution in a mesh of 3×2 processors

In each iteration k, $k = 0, \ldots, (m_j/\nu - 1)$, the processor containing block $G_{k,j}$ computes $L^j_{k,k}$ by means of the sequential algorithm described at the beginning of this section and broadcasts the suitable information to the rest of the processors of the column that compute $L^j_{i,k}$ and update the $G_{i,j}$, for $i > k$.

5 Previous Computations

Step 1.1 of Algorithm 1, called "previous computations", involves four different tasks, denoted as Task00, Task10, Task01 and Task11, respectively, in which is divided the computation of G_0 and G_1 (6), the computation of the displacement matrices Λ_0 and Λ_1 (5), the computation of the diagonal of C [5] and the computations of $\hat{\hat{b}}_0$ and $\hat{\hat{b}}_1$ (7).

If there are only two processors ($p = 2$), $P_{0,0}$ computes Task$i$0 while $P_{0,1}$ computes Task$i$1, for $i = 0, 1$. For $p \geq 4$, processor $P_{i,j}$ computes Taskij. After the computation of the tasks, all processors perform two communication steps: 1. A multicast of the results obtained by the tasks within each column; 2. A data interchange between pairs of processors in each row of the processors mesh.

Several a DST's are carried out in step 1.1 of Algorithm 1. There exist algorithms related with the DFT that involves $O(n \log n)$ operations to perform the DST. But the performance of these algorithms highly depends on the size of the greatest prime number of the primes decomposition of $(n+1)$. In our algorithms, we used a method to avoid this dependency by applying several power of 2 of order DFT's using the *Chirp-z* factorization described in [10].

6 Experimental Results

All experimental analysed were carried out in a cluster with 20 nodes connected by a SCI network with a topology of a 4×5 torus. Each node is a two–processor board with two Intel Xeon at 2 GHz. and 1 Gb. of RAM memory per node.

The first analysis concerns the block size ν. At the light of the experiments, it can be concluded that there exist a wide range of values for that minimize the execution time. But, none of these values must be selected close to 1 (this implies too many communications) or close to n/p (this implies not enough concurrency between communications and computations). The best value obtained by experimental tuning is a fixed size of $\nu = 20$ that is only hardware dependent.

The second experimental analysis deals with the weight of each step of Algorithm 1 on its total cost. Table 1 shows the time spent on each step. It can be observed that the time spent in Step 1.1 grows with the problem size. The *Chirp-z* factorization used to perform the DST makes the cost of this step independent of the size of the prime numbers in which $(n + 1)$ is decomposed. The weight of this first step is $\approx 25\%$ of the total computational cost of the algorithm. The most costly step is the second one in which is performed the factorization of one

Table 1. Execution time in seconds of Algorithm 1 in one processor

n	Step 1.1	Step 1.2	Step 1.3	total
4000	0.11	0.24	0.05	0.39
6000	0.25	0.51	0.12	0.87
8000	0.35	0.88	0.20	1.42
10000	0.60	1.35	0.35	2.28
12000	0.75	1.93	0.47	3.13
14000	0.95	2.59	0.63	4.14
16000	1.17	3.36	0.81	5.28
18000	1.71	4.21	1.09	6.96
20000	1.96	5.18	1.30	8.37

of the two Cauchy–like matrices (C_0 or C_1). The weight of this step is $\approx 60\%$ of the total cost of the algorithm. The third step involves $\approx 15\%$ of the total time.

As it was explained in Section 5, the first step is divided in four tasks, each of one is carried out concurrently so it can be obtained a reduction in time in this step using up to 4 processors (Table 2).

Table 2. Execution time in seconds and efficiency of Step 1.1

n	1 processor time	2 processors time	2 processors efficiency	4 processors time	4 processors efficiency
4000	0.11	0.06	92%	0.04	69%
6000	0.25	0.14	89%	0.10	63%
8000	0.35	0.19	92%	0.13	67%
10000	0.60	0.33	91%	0.22	68%
12000	0.75	0.41	91%	0.27	69%
14000	0.95	0.53	90%	0.35	68%
16000	1.17	0.66	89%	0.43	68%
18000	1.71	0.91	94%	0.60	71%
20000	1.96	1.07	92%	0.70	70%

In Table 3 it can be seen the execution time and the efficiency of Step 1.2. The important effort performed in the parallelization of the triangularization process gives a good efficiency even with 10 processors. The low time obtained with the most costly step using several processors lets to obtain a low total time. This result, although it cannot be as efficient as it was desirable, it can be very useful in applications with real time constraints like digital signal processing.

We note that the efficiency obtained with 2 processors is quite good mainly due to the triangular decomposition of the two independent Cauchy–like matrices over two independent processors.

Table 3. Execution time in seconds and efficiency of Step 1.2

n	1 proc. time	2 procs. time	2 procs. effi.	4 procs. time	4 procs. effi.	6 procs. time	6 procs. effi.	8 procs. time	8 procs. effi.	10 procs. time	10 procs. effi.
4000	0.24	0.13	92%	0.09	67%	0.07	57%	0.06	50%	0.05	48%
6000	0.51	0.27	94%	0.16	80%	0.13	65%	0.11	58%	0.10	51%
8000	0.88	0.48	92%	0.28	79%	0.20	73%	0.18	61%	0.16	55%
10000	1.35	0.73	92%	0.39	87%	0.30	75%	0.25	68%	0.23	59%
12000	1.93	1.03	94%	0.54	89%	0.41	78%	0.34	71%	0.30	64%
14000	2.59	1.39	93%	0.76	85%	0.53	81%	0.45	72%	0.39	66%
16000	3.36	1.80	93%	0.91	92%	0.67	84%	0.56	75%	0.49	69%
18000	4.21	2.25	94%	1.21	87%	0.83	85%	0.69	76%	0.60	70%
20000	5.18	2.71	96%	1.48	88%	1.00	86%	0.83	78%	0.71	73%

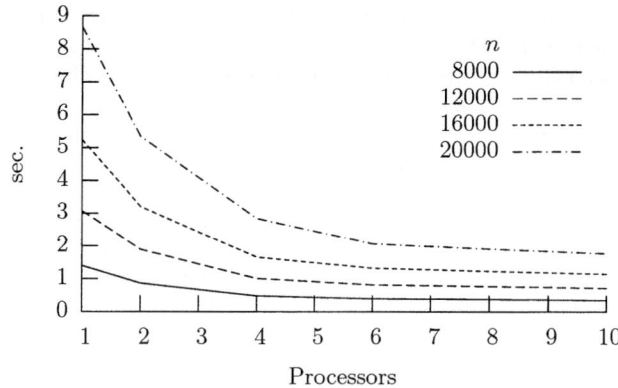

Fig. 3. Time in seconds of the parallel algorithm

Finally, we analyze the execution time of the parallel algorithm. In Fig. 3, it can be seen that the time decreases with the increment of the number of processors. This reduction in time is more significant if the problem size increases. The algorithm also reduces its execution time with more than four processors although Step 1.1 does not exploit more processors in parallel than this quantity. We emphasize the reduction in time regarding the scalability of the parallel algorithm by its utility in applications with real time constraints.

References

1. Sweet, D.R.: The use of linear-time systolic algorithms for the solution of toeplitz problems. Technical Report JCU-CS-91/1, Department of Computer Science, James Cook University (1991) Tue, 23 Apr 1996 15:17:55 GMT.
2. Evans, D.J., Oka, G.: Parallel solution of symmetric positive definite Toeplitz systems. Parallel Algorithms and Applications **12** (1998) 297–303
3. Gohberg, I., Koltracht, I., Averbuch, A., Shoham, B.: Timing analysis of a parallel algorithm for Toeplitz matrices on a MIMD parallel machine. Parallel Computing **17** (1991) 563–577
4. Gallivan, K., Thirumalai, S., Dooren, P.V.: On solving block toeplitz systems using a block schur algorithm. In: Proceedings of the 23rd International Conference on Parallel Processing. Volume 3., Boca Raton, FL, USA, CRC Press (1994) 274–281
5. Thirumalai, S.: High performance algorithms to solve Toeplitz and block Toeplitz systems. Ph.d. th., Grad. College of the U. of Illinois at Urbana–Champaign (1996)
6. Alonso, P., Badía, J.M., Vidal, A.M.: Parallel algorithms for the solution of toeplitz systems of linear equations. Lecture Notes in Computer Science **3019** (2004) 969–976
7. Anderson, E., et al.: LAPACK Users' Guide. SIAM, Philadelphia (1995)
8. Blackford, L., et al.: ScaLAPACK Users' Guide. SIAM, Philadelphia (1997)
9. Alonso, P., Badía, J.M., González, A., Vidal, A.M.: Parallel design of multichannel inverse filters for audio reproduction. In: Parallel and Distributed Computing and Systems, IASTED. Volume II., Marina del Rey, CA, USA (2003) 719–724

10. Loan, C.V.: Computational Frameworks for the Fast Fourier Transform. SIAM Press, Philadelphia (1992)
11. Heinig, G.: Inversion of generalized Cauchy matrices and other classes of structured matrices. Linear Algebra and Signal Proc., IMA **Math. Appl. 69** (1994) 95–114
12. Gohberg, I., Kailath, T., Olshevsky, V.: Fast Gaussian elimination with partial pivoting for matrices with displacement structure. Mathematics of Computation **64** (1995) 1557–1576
13. Alonso, P., Vidal, A.M.: An efficient and stable parallel solution for symmetric toeplitz linear systems. TR DSIC-II/2005, DSIC–Univ. Polit. Valencia (2005)
14. Kailath, T., Sayed, A.H.: Displacement structure: Theory and applications. SIAM Review **37** (1995) 297–386

Parallel Resolution with Newton Algorithms of the Inverse Non-symmetric Eigenvalue Problem

Pedro V. Alberti[1,2], Victor M. García[1], and Antonio M. Vidal[1]

[1] Universidad Politécnica de Valencia,
Camino de Vera s/n, Valencia España
{palberti, vmgarcia, avidal}@dsic.upv.es
[2] Universidad de Magallanes,
Av. Bulnes 01855, Punta Arenas Chile

Abstract. In this work, we describe two sequential algorithms to solve the Inverse Non-Symmetric Eigenvalue Problem. Both approaches have been based on Newton's iteration. The second one has been parallelized by using parallel standard public domain libraries. Good performance has been obtained in a cluster of PC's.

1 Introduction

An Inverse Eigenvalue Problem *(IEP)* is the reconstruction of a matrix from its spectrum, possibly forcing some structure characteristics in the resulting matrix. The goal of the work described in this paper is to solve a kind of inverse problem, the Inverse Additive Non-Symmetric Eigenvalue Problem *(IANSEP)*. This problem appears in several areas of Science and Engineering, such as Seismic Tomography, Pole Assignment Problems, Geophysics and many more [1, 4, 6, 7, 10].

The state of the art of the problem (see, for example, [6]) shows that the vast majority of algorithms proposed are restricted to the symmetric case. Some authors like [6, 8] review the *IANSEP* in a quite general form, but we have not found any proposal for resolution of this problem; the present paper intends to fill this gap.

This paper presents two iterative algorithms based on Newton's method. A parallel version of the second one is also presented.

In section 2 a description of the problem is given, discussing the existence of solutions of the *IANSEP*. In section 3 a sequential resolution algorithm is proposed. It is based on a full statement of the problem as a nonlinear problem, and gives rise to large linear and nonlinear systems; we call it a 'Brute Force' approach. In section 4, another approach is taken, in which the nonlinear system is written only with the differences between the desired eigenvalues and the 'actual' ones. We call it a Restricted Newton method. In that case, the Jacobian matrix is approximated through finite differences. This algorithm is very suitable for parallel processing, and we describe as well the parallel implementation. Finally, some conclusions are given in section 5.

2 The *IANSEP*: Description and Existence of Solution

The problem can be stated as follows:

Given $n+1$ real, $n \times n$ matrices, $A_0, A_1, ... A_n$ and a set of real numbers $\lambda_1^* \leq \lambda_2^* \leq ... \leq \lambda_n^*$, find $d \in \Re^n$ such that the eigenvalues $\lambda_1(d) \leq \lambda_2(d) \leq ... \leq \lambda_n(d)$ of

$$A(d) := A_0 + \sum_{k=1}^{n} d_k A_k \qquad (1)$$

fulfill $\lambda_i(d) = \lambda_i^*$ for $i = 1, ..., n$.

Of course, the main problem that arises if A_i are non-symmetric matrices is that the matrix $A_0 + \sum_{i=1}^{n} d_i A_i$ can have non real eigenvalues. In our work, we shall assume that the matrices A_i, the vector d and the desired eigenvalues λ_i^* are real.

This sets a double existence problem. First, it should be determined when the matrices $A(d)$ have real eigenvalues. Second, it should be determined whether given a vector, $\rho \in \Re^n$ (the desired spectrum), $\exists d \in \Re^n$ such that the spectrum of $A(d)$ ($eig(A(d))$) is ρ.

As far as we know, there are no theorem which can answer to these questions. The study of the 2×2 case can help us to determine under which circumstances will the algorithms work (or fail). A similar analysis can be found in [5] for the symmetric case.

Let us consider $A_0, A_1, ... A_n \in \Re^{2 \times 2}$ and $d \in \Re^2$, and let us try to find out how are the eigenvalues of $A(d) = A_0 + \sum_{i=1}^{n} d_i A_i$.

In this case

$$A_0 + \sum_{i=1}^{n} d_i A_i = \begin{bmatrix} \alpha(d) & \beta(d) \\ \gamma(d) & \delta(d) \end{bmatrix}, \qquad (2)$$

and, therefore, the eigenvalues of $A(d)$ are the solutions of the equation $\lambda^2 - S(d)\lambda + det(d) = 0$, with $S(d) = \alpha(d) + \delta(d)$ and $det(d) = \alpha(d)\delta(d) - \gamma(d)\beta(d)$.

The eigenvalues of $A(d)$ shall be either real or complex depending on the value of the discriminant: $\triangle(d) = (\alpha(d) - \delta(d))^2 + 4\beta(d)\gamma(d)$.

If $\triangle(d) < 0$, $\forall d$, then the matrix $A(d)$ will always have non real eigenvalues, which means that our problem (that is, to find $d \in \Re^2$ such that $eig(A(d)) = \rho \in \Re^2$) does not have 'real' solution. Trivial numerical examples can be formed but we have avoided them for the sake of simplicity.

If $\triangle(d) \geq 0$, $\forall d$, then the matrix $A(d)$ always have real eigenvalues, which means that there can exist a 'real' solution of the problem.

Finally, it can happen that $\triangle(d) < 0$ for some d and $\triangle(d) \geq 0$ for other values of d. This means that the problem can have solution, but during the iterative search the algorithm may enter regions where $\triangle(d) < 0$. This should be avoided or treated with special care.

If we consider now the case in which $\triangle(d) \geq 0$ (and therefore it is possible a real solution) we might ask whether given a $\rho^* \in \Re^2$, there exists always a $d \in \Re^2$ such that $eig(A(d)) = \rho^*$. Clearly, this will not happen always, for example, in the case where $\triangle(d) < 0$.

Let us try to express d as a function of λ^*. Let us set the system of equations:

$$\begin{cases} \alpha(d)\delta(d) = \lambda_1^* + \lambda_2^* \\ \alpha(d)\delta(d) - \gamma(d)\beta(d) = \lambda_1^* \lambda_2^* \end{cases} \quad (3)$$

which is equivalent (in the 2×2 case) to force that $eig(A(d))$ are λ_1^*, λ_2^*, and let us solve it for d:

$$\begin{aligned} d_2 &= \epsilon_1 d_1 + \epsilon_2 \\ \theta_1 d_1^2 + \theta_2 d_1 + \theta_3 &= 0, \end{aligned} \quad (4)$$

being $\theta_1, \theta_2, \theta_3, \epsilon_1, \epsilon_2$ functions of the entries of matrices A_0, A_1, A_2 and of λ_1^*, λ_2^* which can be easily formed. Clearly, if $\theta_2^2 - 4\theta_1\theta_3 < 0$ the problem does not have 'real' solution.

This shows again that the real solution of $IANSEP$ does not always exist.

The results found can be extrapolated to $n \times n$ matrices. Therefore, the situation when $A_i \in \Re^{n \times n}$ and, for a given set of given eigenvalues $\rho^* \in \Re^n$ it is desired to find $d \in \Re^n$ such that $eig(A(d)) = \rho^*$, is the following:

It may happen that for all $d \in \Re^n$ the matrix $A(d)$ has non real eigenvalues. In this case the problem will not have real solution.

It also may happen that for all $d \in \Re^n$ all the eigenvalues of $A(d)$ are real. In this case, it might happen that for some $\rho^* \in \Re^n$ does not exist any $d \in \Re^n$ such that $eig(A(d)) = \rho^*$.

Finally, it can happen that the eigenvalues of $A(d)$, are real for some $d \in \Re^n$ and complex for other values of $d \in \Re^n$.

This leads us to consider the working hypothesis of the algorithms for $IANSEP$, in the same way in which it is considered for the resolution of nonlinear system. Let us suppose that our problem, $IANSEP$, has a solution $d^* \in \Re^n$. From the properties of continuity of eigenvalues it can be stated that there exists a (small enough) region comprising $d^* \in \Re^n$ in which the eigenvalues of any matrix of the form $A_0 + \sum_{i=1}^n d_i A_i$ are also real. If a Newton-like method is used to find d^*, the initial guess should be taken inside this region; since Newton methods are locally convergent, if the region is small enough, quadratic convergence can be achieved.

3 'Brute Force' Method

As a first approach, we will try to rewrite the $IANSEP$ as a system of nonlinear equations.

If the system has real solution $\rho^* = (\lambda_1^*, \lambda_2^* \cdots, \lambda_n^*)$ must satisfy

$$A_0 + \sum_{i=1}^n d_i A_i = Q\tilde{T}Q^T, \quad (5)$$

where the elements of main diagonal of the upper triangular matrix $\tilde{T}$ are ($\lambda_1^* \leq \lambda_2^* \leq ... \leq \lambda_n^*$) and

$$QQ^T = I \tag{6}$$

So, we can set up the nonlinear system $F(v) = 0$, with $n^2 + \frac{n(n+1)}{2}$ equations and unknowns:

$$F(v) = \begin{cases} q_i^T q_j - \delta_{ij} = 0 & \text{with } i = 1, 2, \ldots, n \text{ and } j = i, \ldots, n \\ \left(A_0 + \sum_{k=1}^n d_k A_k\right)_{ij} - q_i^T \tilde{T} q_j = 0 & \text{with } i = 1, 2, \ldots, n \text{ and } j = 1, \ldots, n, \end{cases}$$

whose unknowns are

$$v = [q_{1,1}, q_{2,1}, \cdots, q_{n,1}, \cdots, q_{n,2}, \cdots, q_{n,n},$$
$$t_{12}, t_{1,3}, \cdots, t_{1,n}, t_{2,3} \cdots, t_{2,n} \cdots, t_{(n-1),n} \cdots, d_1, d_2, \cdots, d_n].$$

This nonlinear system can be tackled with Newton's method, through the iteration $v^{(i+1)} = v^{(i)} - J^{-1}(v^{(i)})F(v^{(i)}) = v^{(i)} + s^{(i)}$, with $J(v^{(i)})s^{(i)} = -F(v^{(i)})$. The Jacobian matrix J has the form

$$J = [J_{ij}] = \left[\frac{\partial F_i(v^{(k)})}{\partial v_j^{(k)}}\right] = \begin{bmatrix} J_{1,1} & J_{1,2} & J_{1,3} \\ J_{2,1} & J_{2,2} & J_{2,3} \end{bmatrix}, \tag{7}$$

with:

$J_{1,1} \in \Re^{\frac{n(n+1)}{2} \times n}$ where $\frac{\partial F_i(v^{(k)})}{\partial v_j^{(k)}} = \delta_{i,s} q_{r,j} + \delta_{j,s} q_{r,i}$.

$J_{1,2} \in \Re^{\frac{n(n+1)}{2} \times \frac{n(n-1)}{2}}$ where $\frac{\partial F_i(v^{(k)})}{\partial v_j^{(k)}} = 0$.

$J_{1,3} \in \Re^{\frac{n(n+1)}{2} \times n}$ where $\frac{\partial F_i(v^{(k)})}{\partial v_j^{(k)}} = 0$.

$J_{2,1} \in \Re^{n \times n}$ where
$\frac{\partial F_i(v^{(k)})}{\partial v_j^{(k)}} = -\lambda_s^* q_{j,s} \delta_{i,r} - \lambda_s^* q_{i,s} \delta_{j,r} - \delta_{j,r} \sum_{k=s+1}^n t_{s,k} q_{j,k}$
$\quad - \sum_{z=1}^n q_{i,z} \left(\sum_{k=z+1}^n t_{z,k} \delta_{j,r} \delta_{k,s} \right).$

$J_{2,2} \in \Re^{n \times \frac{n(n-1)}{2}}$ where $\frac{\partial F_i(v^{(k)})}{\partial v_j^{(k)}} = -\sum_{z=1}^n q_{i,z} \left(\sum_{k=z+1}^n q_{j,z} \delta_{r,z} \delta_{s,k} \right)$.

$J_{2,3} \in \Re^{n \times n}$ where $\frac{\partial F_i(v^{(k)})}{\partial v_j^{(k)}} = (A_r)_{i,j}$.

The solutions of the system of equations $J(v^{(i)})s^{(i)} = -F(v^{(i)})$ has computational complexity of $O(n^6)$ which makes this algorithm very expensive. This algorithm, as other Newton-like algorithm in inverse problems, has serious problems of convergence unless the initial guess is close enough to the solution. To alleviate this problem, the algorithm has been implemented including a well-known globalization technique, Armijo's rule [9].

3.1 Experimental Results

The algorithm has been developed in C language, by using BLAS and LAPACK libraries [2]. Experimental results and validations have been carried out in a PC with Intel Xeon processor, at 2GHz. and 1GB. of RAM.

It has been tested using random matrices with known eigenvalues. The initial guess has been chosen by perturbating the solution, $v^{(0)} = v^* + \delta$, with a small $\delta = 0.1$. The execution time of the algorithm can be divided in two parts, the initialization phase and iterative phase; figure 1 shows (for different problem sizes) the time for the initial phase, the iteration time and the number of iterations to reach the solution.

The algorithm needs large memory resources, since it needs to create very large Jacobians matrices. As an example, for a problem of size of 210 × 210 the Jacobian matrix would have a size of 132405 × 132405. As a consequence, it can be applied only to small problems

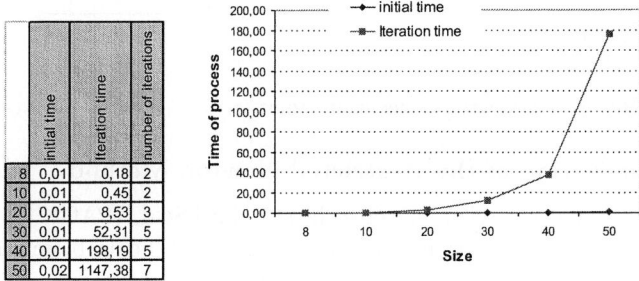

Fig. 1. Time in seconds for 'Brute Force' algorithm

4 Restricted Newton's Method

A different approach to the problem is to solve the nonlinear system

$$F_i(d^{(k)}) = \lambda_i(d^{(k)}) - \lambda_i^* = 0 \quad i = 1, ..., n. \tag{8}$$

Newton's method can be applied as well to this system, through the iteration: $d^{(k+1)} = d^{(k)} - J(d^{(k)})^{-1} F(d^{(k)}) = d^{(k)} + s^{(k)}$ with $J(d^{(k)}) s^{(k)} = -F(d^{(k)})$, where $d \in \Re^n$, $J \in \Re^{n \times n}$ A similar idea is used in [8] for the symmetric case. In this work, we have approximated the Jacobian matrix through Finite Differences:

$$J = \left[\frac{\partial F_i(d)}{\partial d_j}\right] \simeq \left[\frac{F_i(d + e_j h) - F_i(d)}{h}\right] = \left[\frac{eig(A(d) + hA_j)_i - eig(A(d))_i}{h}\right] \tag{9}$$

The main cost of this algorithm is the computation of the Jacobian matrix, since it is needed to compute $(n+1)$ times the eigenvalues of an $n \times n$ matrix, that is, the complexity is $O(n^4)$. The resolution of the linear system $J(d^{(k)}) s^{(k)} = -F(d^{(k)})$ has only a cost of $O(n^3)$. As in the former section, the algorithm was implemented using Armijo's rule.

4.1 Experimental Results: Sequential Implementation

Figure 2 is analogous to figure 1 in the former section, reflecting the costs of the start-up phase, of the iterating phase and the number of iterations.

Matrices and initial guesses were chosen as in the former section. It is quite clear that the execution times of this algorithm are lower than the Brute Force algorithm ones.

However, it can still be improved through parallel computing techniques; this is discussed in the next section.

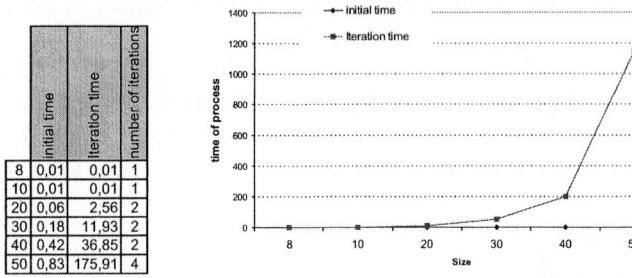

Fig. 2. Restricted Newton: Execution times (s.) and number of iterations

4.2 Parallelization of the Restricted Newton Method

The parallelization was carried out with the standard ScaLAPACK [3] approach, that is, distributing in a 2-D processors mesh the data of the problem; in this case, the matrices $A_0, A_1, A_2, \cdots, A_n$, the initial guess $d^{(0)} = [d_1, d_2, d_3, \cdots, d_n]^T$ and the desired eigenvalues $\rho^* = [\lambda_1^*, \lambda_2^*, \lambda_3^*, \cdots, \lambda_n^*]^T$.

A C language subroutine was written to create the Jacobian matrix:
$J_{ij}^{(k)} = \frac{eig_i(A(d) + hA_j) - eig_i(A(d))}{h} \quad i = 1, 2, \cdots, n; \quad j = 1, 2, , n$.

An obvious parallelization possibility was to assign each column of the Jacobian matrix (solution of a single eigenvalue problem) to each processor. The second option was to perform each different eigenvalue problem by using all the processors, making use of the SCALAPACK 2-D distribution. This second option was chosen to take full advantage of the parallel SCALAPACK routines.

All the other steps in the algorithm correspond to calls to BLACS, PBLAS or ScaLAPACK.

4.3 Numerical Experiments

The code was tested in a cluster of PC's with up to 7 processors. As before, the matrices have random values and have known eigenvalues, and the times of the initial part (T-ini) and of the iterative part (T-iter) have been recorded separately. So, the speed-up can be expressed as a function of the number of iterations k to reach convergence and of the number of processors, p:

$$Sp = \frac{T_{ini\ 1} + kT_{iter\ 1}}{T_{ini\ p} + kT_{iter\ p}}. \qquad (10)$$

In figure 3 it is shown the execution time for several sizes, of both the initial part and the iterative part as a function of the number of processors. The speed-up as a function of the number of processors is shown in figure 4.

It can be noted the influence of the dimensions of the 2-D mesh used in the ScaLAPACK. Thus, for square logical 2-D meshes (i.e. $p = 4$) efficiency is optimum.

It can be seen that the algorithm works quite well for large problem sizes.

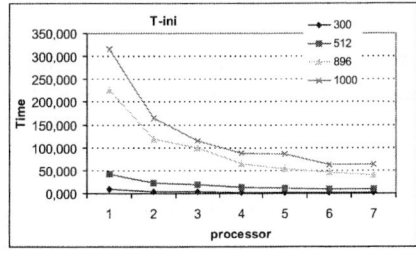 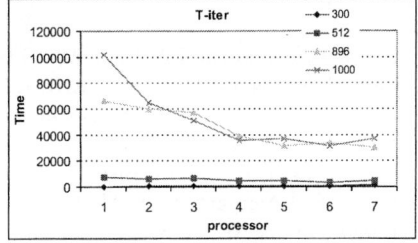

Fig. 3. Execution time of the initial phase (T-ini) and iterative phase (T-iter)

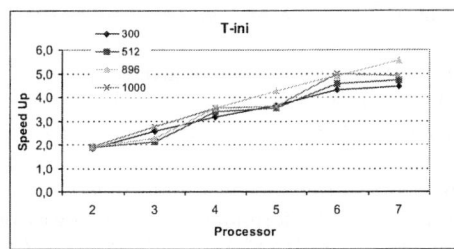 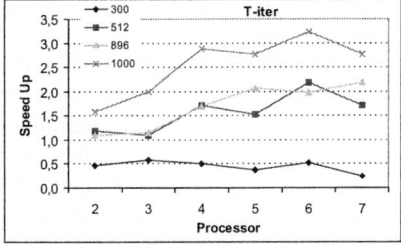

Fig. 4. Experimental Speed Up

5 Conclusions

In this paper, we have discussed the *IANSEP*, analyzed its peculiarities studying the 2×2 case, and proposed two solution methods. One of them has been parallelized, giving good results. The algorithms have been implemented using C language and standard libraries such as LAPACK, ScaLAPACK, so that the resulting codes are efficient and portable.

The study in section 2 shows that the Non-Symmetric problem is far more complex than the symmetric one, and that even if the matrix $A(d)$ has only real eigenvalues, the problem may not have any solution.

The 'Brute Force' approach is an attempt to state the problem directly, that for some problems (and thanks to the power of today's processors) can be successful, but which compares poorly with the other method, the restricted Newton method.

The method in section 3 is a good option for the solution of the *IANSEP*, working quite well both in its sequential and parallel versions.

Acknowledgement

This work has been supported by Spanish MYCT and FEDER under Grant TIC 2003-08238-C02-02.

References

1. Pedro V. Alberti and Antonio M. Vidal. Solución Paralela de Mínimos Cuadrados para el Problema Inverso Aditivo de Valores Propios. Technical Report DSIC-II/03/02, Departamento de Sistemas Informáticos y Computación, Universidad Politécnica de Valencia, Camino de Vera s/n, 46022, Valencia España, 2002.
2. E. Anderson, Z. Bai, C. Bischof, J. Demmel, J. Dongarra, J. Du Croz, A. Greenbaum, S. Hammarling, A. McKenney, S. Ostrouchov, and D. Sorensen. LAPACK User's Guide. *SIAM*, Second edition, 1995.
3. L. S. Blackford, J. Choi, A. Cleary, E. D. Acevedo, J. Demmel, I. Dhillon, J. Dongarra, S. Hammarling, G. Henry, A. Petitet, K. Stanley, D. Walker, and R. C. Whaley. ScaLAPACK User's Guide. *SIAM*, 1997.
4. D. L. Boyle and G. H. Golub. A survey of inverse eigenvalue problems. *Inv. Prob., 595-622*, 1987.
5. Xuzho Chen and Moody T. Chu. On the least squares solution of inverse eigenvalue problems. *SIAM J.Numer. Anal.*, 33(6):2417–2430, 1996.
6. Moody T. Chu. Inverse eigenvalue problem. *SIAM Rev.*, 40(1):1–39, March 1998.
7. Hua Dai. A Numerical Method for Solving Inverse Eigenvalue Problem. Technical Report TR/PA/98/33, Nanjing 210016, People's Republic of China, 1998.
8. S. Friedland, J. Nocedal, and M. L. Overton. The formulation and analysis of numerical methods for inverse eigenvalue problems. *SIAM Anal. Vol 24, june, N3*, 24(5):634–667, June 1987.
9. C. T. Kelley. Iterative methods for linear and nonlinear equations. *SIAM Pub.*, 1995.
10. J. Peinado and A.M. Vidal. A new parrallel approach to the toeplitz inverse eigenproblem using the newton's method. *Lecture Notes in Computer Science. Springer-Verlag*, 1999.

Computational Challenges in Vector Functional Coefficient Autoregressive Models*

Ioana Banicescu[1,2], Ricolindo L. Cariño[2], Jane L. Harvill[3,2], and John Patrick Lestrade[4]

[1] Department of Computer Science and Engineering, PO Box 9637
[2] Center for Computational Sciences ERC, PO Box 9627
[3] Department of Mathematics and Statistics, PO Box MA
[4] Department of Physics and Astronomy, PO Box 5167
Mississippi State University, Mississippi State MS 39762, USA
(ioana@cse,rlc@erc,harvill@math,lestrade@ra).msstate.edu

Abstract. An important research area in statistical computing is found in the literature for vector functional coefficient autoregressive models, a special case of vector nonlinear time series. Methods used are computationally intensive. As a result, analyses and simulations can run into weeks, or even months. Statisticians have been known to base empirical results on a relatively small number of simulation replications, sacrificing precision, accuracy and reliability of results in the interest of time and productivity. The simulations are amenable for parallelization; however, parallel computing technology has not yet been widely used in this specific research area. This paper proposes an approach to the parallelization of statistical simulation codes to address the challenge of long running times, without resorting to extensive code revisions. This approach takes advantage of recent advances in dynamic loop scheduling on workstation clusters to achieve high performance, even with the presence of unpredictable load imbalance factors. Preliminary results of applying this approach in the simulation of normal white noise and threshold autoregressive model obtains efficiencies in the range 95–98% on 8–64 processors.

1 Introduction

A vector time series is a set of observations of multiple related phenomena across time. The mathematical underpinnings of the statistical analysis of time series incorporate the correlation across time and between series – properties that complicate statistical theory. This is especially true for nonlinear models, where mathematical theory may be extremely difficult, even intractable. Consequently, Monte Carlo simulation is often relied upon to give direction, to interpret, and

---

* This work was partially supported by the National Science Foundation Grants: CAREER #9984465, ITR/DMS #0313274, ITR/ACS #0081303, ITR/ACS #0085969, #0132618, and #0082979.

to explain complex analytical results. Gentle [1] gives a thorough discussion of Monte Carlo simulation methods, their usefulness, and an idea of the underlying theory of their application to statistical computing. In general, it can be said that as the number of Monte Carlo replications increases, the result of the simulation approaches the "truth." The more complex the structure, the larger the number of replications needs to be.

On a single processor, to examine statistical properties of methods related to vector functional coefficient autoregressive models via simulations would require execution times of a few weeks or even months. Statisticians have been known to base empirical results on a relatively small number of simulation replications, sacrificing precision, accuracy, and possibly compromising the reliability of results in the interest of time. As a result, techniques which may require thousands of replications for accuracy and reliability often have at most, a few hundred. Fortunately, the replications are amenable for computation in parallel. Thus, parallel processing technology can be exploited to enable the extensive simulation of a variety of models within reasonable running time limits. This paper demonstrates an approach to the simulation of such models which takes advantage of recent advances in dynamic loop scheduling on clusters of workstations, in order to shorten the simulation time.

The remainder of this paper is organized as follows. Section 2 contains a brief overview of vector functional coefficient autoregressive models, along with the computational issues related to the statistical methods that make use of these models. Section 3 describes an approach to facing the challenges in implementing the procedures numerous times, as would be done in a typical Monte Carlo study. Section 4 gives preliminary results which demonstrate the high parallel performance achieved by the proposed approach on general-purpose clusters. Concluding remarks are in Section 5.

2 Vector Functional Coefficient Autoregressive Model

Although complicated in both presentation and theory, vector nonlinear time series are especially useful for describing complicated nonlinear dynamic structures that exists in many time-dependent multivariate series. Let $\boldsymbol{Y}_t = (Y_{1,t}, \ldots, Y_{k,t})'$ denote the vector time series at time $t = 1, 2, \ldots, T$. Then the vector functional coefficient autoregressive model of order p (VFCAR(p)) is defined as

$$\boldsymbol{Y}_t = \boldsymbol{f}^{(0)}(\boldsymbol{Z}_t) + \sum_{j=1}^{p} \boldsymbol{f}^{(j)}(\boldsymbol{Z}_t)\boldsymbol{Y}_{t-j} + \boldsymbol{\varepsilon}_t, \quad t = p+1, \ldots, T, \tag{1}$$

where $\boldsymbol{f}^{(j)}$, $j = 0, \ldots, p$ are $k \times k$ matrices whose elements are real-valued measurable functions that change as a function of the (possible vector-valued) $\boldsymbol{Z}_t$, and which have continuous second-order derivatives. The error terms $\boldsymbol{\varepsilon}_t$ in (1) are such that for each i, the series $\{\varepsilon_{i,t}\}_{t=1}^{T}$ is a white noise sequence, independent of $\{\boldsymbol{Y}_t\}_{t=1}^{T}$. However contemporaneous cross-correlation may exist between $\{\varepsilon_{i,t}\}$ and $\{\varepsilon_{j,t}\}$, $i \neq j$. The primary motivation for studying this model is that specific choices for the elements of the $\boldsymbol{f}^{(j)}$ yield parametric models.

The VFCAR(p) model may be considered a hybrid of parametric and nonparametric models since the autoregressive structure is assumed, but there is little or no information about the form of the elements of the $\boldsymbol{f}^{(j)}$. As such, estimation of the parameters of the VFCAR(p) model is done nonparametrically via local regression. Simultaneous estimation of the elements of the $\boldsymbol{f}^{(j)}$ provides improved statistical efficiency when the error terms have positive cross-correlation. In the process of fitting the model (1), modified multifold cross-validation is used to determine an optimal bandwidth and value for p by finding the pair of values that minimize the accumulated prediction error. This multistage procedure requires an immense number of arithmetic operations on a univariate series. That number increases exponentially for multivariate series.

The VFCAR model in used in statistical tests of model misspecification. However, the null distribution is unknown, and so a procedure that fits the model to a large number of bootstrapped realizations of the series under the null model is used to obtain a numerical p value for the test. Specifics for fitting the model and the testing procedure are found in [2]. Forecasting can be accomplished using the VFCAR model either recursively or via bootstrap as in [3]. The mathematical complexity of the statistical theory of the estimators, testing procedures, and forecasts using the VFCAR model are highly complicated, necessitating the use of simulation to study their statistical properties.

3 Simulation

A high-level outline of the Fortran program for simulation to investigate statistical properties of methods using the VFCAR model is given by Figure 1. Executing the program on a desktop computer with a 1GHz Pentium 4 processor and 256MB RAM, for 1000 replications of a single bivariate model using a sample size 400 takes approximately nine (9) days. An investigation of various models using more error cross correlation values and larger sample sizes or number of replications will run into months on a single machine.

The bulk of the arithmetic (sample generation, parameter estimation, hypothesis testing) is performed by the iterations of the replication loop, in Model testing. These iterations are independent, hence they can be distributed among P processors to shorten the loop completion time. In general, equally distributing the replications among the processors leads to high performance if the replications have the same execution time and the processors are homogeneous. However, this static assignment may result in load imbalance due to heterogeneity of processors, irregular iteration execution times, or unpredictable systemic effects like operating system interference leading to varying effective processor speeds. Load imbalance is typically indicated by highly uneven processor finishing times, and is a major cause of performance degradation in parallel applications. Dynamic loop scheduling is therefore necessary in order to balance processor loads and minimize the loop completion time.

```
{Input model specifications; error cross correlations: no_corrs,
  corrs(1..no_corrs); sample sizes: no_ns, ns(1..no_ns); and
  no. of replications: no_reps}
emp_rr = 0
DO corr_no = 1, no_corrs    ! correlation loop
  {Correlation initializations}
  DO n_no = 1, no_ns        ! sample size loop
    n = ns(n_no)
    DO rep_no = 1, no_reps  ! replication loop
      {Model testing; update emp_rr()}
    END DO
  END DO
END DO
emp_rr = emp_rr/no_reps
Output emp_rr
```

Fig. 1. High-level outline of the serial program

The parallelization of the simulation program was accomplished with minor alterations to the serial code. Essentially, three (3) lines of code are added before and after the original replication loop code, and the loop extents were changed, as illustrated by Figure 2. The routines LS_Finalize, LS_Initialize, etc., are contained in a module specifically designed for quick integration of dynamic loop scheduling into sequential programs containing parallel loops. This module evolved from previously developed code for dynamic loop scheduling in scientific applications that utilize the Message Passing Interface (MPI) library [4]. Prior versions of the module were used to improve the performance of applications such as the profiling of automatic quadrature routines [5,6] and simulation of wave packets by the quantum trajectory method [7].

```
call LS_Initialize (foreman, 1, no_reps, method)
do while ( .not. LS_Terminated() )
  call LS_StartChunk (cStart, cSize)
  DO rep_no = cStart, cStart+cSize-1 ! replication loop
    {Model testing; update emp_rr()}
  END DO
  call LS_FinishChunk()
end do
call LS_Finalize (nIters, workTime)
```

Fig. 2. The modified replication loop for parallelization and dynamic scheduling

The routine LS_Initialize signals the start of dynamic scheduling of loop iterations: foreman specifies the rank of the process that will serve as the scheduler, (1,no_reps) is the loop extent, and method identifies the scheduling technique.

The logical function `LS_Terminated` tests for loop termination. `LS_Finalize` synchronizes the processes, and returns `nIters` and `workTime`, the count and total execution time, respectively, of `Model testing` computations performed by the calling process. These two quantities are useful in analyzing the performance of load balancing; ideally, the work times of the processors should be approximately equal.

The routine `LS_StartChunk` returns the start `cStart` and size `cSize` of a chunk of iterations to be executed by the calling processor. These quantities are determined according the the scheduling technique specified by `method`. After the chunk is finished, the routine `LS_FinishChunk` is invoked. Internally, the calling processor sends performance data pertaining to the recently executed chunk to the scheduler. The scheduler receives the performance data, and if there are remaining iterations, it computes and sends the next (`cStart`, `cSize`) to the processor; otherwise, the scheduler sends a termination message.

In addition to the changes in the replication loop, other modifications to enable parallelization pertain to the use of MPI. The usual calls to MPI_Init(), MPI_Comm_size() and MPI_Comm_rank() are added at the start of the program, and MPI_Finalize() added just before the program ends. Also, each processor computes only a fraction of the total number of replications, giving a partial result in the `emp_rr` array. The full set of results is obtained by performing the summation MPI_Allreduce(..., MPI_SUM, ...) on the `emp_rr` after the correlation loop.

4 Performance Measurements

Performance tests of the Fortran 90+MPI implementation of the model testing program were conducted on a general-purpose Solaris and Linux clusters at the Mississippi State University Engineering Research Center. The Solaris cluster consists of Myrinet-interconnected Sun Microsystems SMPs. Each SMP has four (4) UltraSPARC 400MHz processors, and the cluster has 16 nodes for a total of 64 processors. However, the cluster usage policy allows a job to utilize no more than 32 processors and no more than 48 hours execution time. The Linux cluster has a total of 1038 Pentium III (1.0 GHz and 1.266 GHz) processors, runs the Red Hat Linux operating system, and is connected via fast Ethernet switches with Gigabit Ethernet uplinks. On the Linux cluster, the queuing system (PBS) attempts to assign homogeneous compute nodes to a job, but this is not guaranteed. Other jobs were also running on the clusters along with the experiments, thus network traffic volume may have varied during the experiments, with unpredictable effects on the performance tests.

Table 1 summarizes the tests that were attempted. WNnnn and TARnnn denote a normal white noise and a threshold autoregressive model, respectively, with 'nnn' sample size. The last two columns give the job execution time (in hh:mm format) without loop scheduling (STAT) or with loop scheduling using the adaptive factoring technique (AF) [8, 9, 10]. The AF was chosen since it is the most sophisticated technique, as it addresses all sources of load imbalance. However, the AF has a higher overhead than other techniques, which may cause

Table 1. Performance tests

Cluster	Test Id	Sample size	Replications	P	STAT time	AF time
Sun	TAR30	30	10000	8	11:09	11:25
Solaris	TAR75	75	10000	16	27:47	29:05
	WN150	150	10000	32	31:15	23:28
Intel	WN500	500	10000	64	5:56	4:47
Linux	TAR500	500	10000	64	15:42	15:59

the AF to perform slightly lower than other techniques for problems that exhibit no load imbalance.

The job times (STAT, AF times) demonstrate the dramatic reduction in simulation time achieved by the simple code modification. For instance, running the original code for the TAR500 model on one of the processors of the Linux cluster would take approximately 64*15.7 hours, or 42 days.

Except for WN150 and WN500, the parallelization without loop scheduling (STAT) consumed slightly lesser time than with loop scheduling using the AF technique. This indicates that application-induced load imbalance was not a significant issue. However, the cases of the WN150 and WN500 provide clear evidence for the influence of system-induced load imbalance. Figure 3 for WN150 illustrates this influence. For STAT, where each processor executed (no_reps + $P-1)/P$ replications (the gray bars), the plots of the processor work times (the gray triangles) show unusually longer times for processors 4, 12, 20 and 28. Most likely, these processors belonged to the same SMP node, and that this node had an extraneous process. With the AF technique, system-induced imbalance

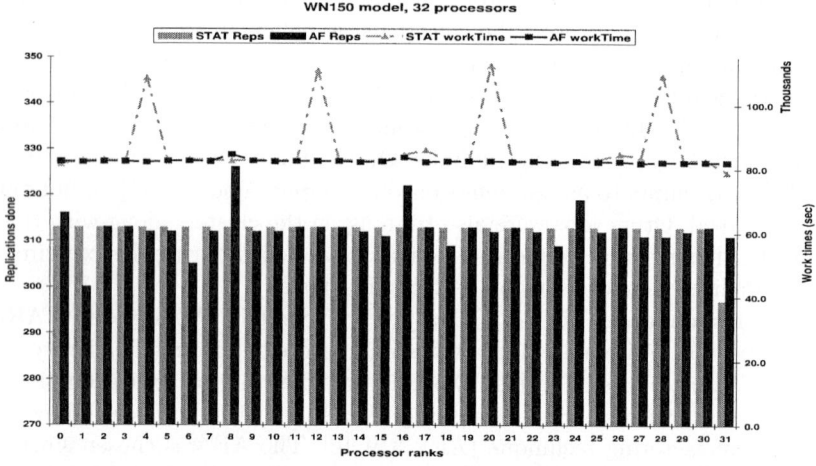

Fig. 3. Loop scheduling summary for WN150 on the Solaris cluster

was also present, as evidenced by the markedly unequal number of replications executed by the processors. The almost flat trend line of the processor work times for the AF technique (dark squares) indicates that the load imbalance was successfully addressed.

The cluster resources allocated to the simulation were very efficiently utilized, as indicated in Table 2 which shows that up to 98% parallel efficiency is achievable. The lowest is 77% for WN150 with STAT, which may be attributed to severe system-induced load imbalance. This load imbalance is successfully addressed by AF obtaining 98% efficiency for the same problem. The speedup (Spd) and efficiency (Eff) metrics were computed as follows. Denote by t_r the time spent by processor r on the modified replication loop, and w_r the workTime. A t_r is the difference of the times taken just before LS_Initialize and right after LS_Finalize. A w_r is the cumulative time spent in Model testing. The parallel time T_P for the replication loop is $T_P = \max t_r$, and an estimate of the serial time T_1 is given by $T_1 = \sum_{\forall r} w_r$. This estimate for T_1 is accurate if the processors are homogeneous. From P, T_P and T_1, Spd=T_1/T_P and Eff=$T_1/(P \times T_P)$.

Table 2. Speedup and efficiency

Test Id	P	No loop scheduling (STAT)				With loop scheduling (AF)			
		T_1	T_P	Spd	Eff	T_1	T_P	Spd	Eff
TAR30	8	315071	40118	7.9	0.98	313965	41121	7.6	0.95
TAR75	16	1589994	100030	15.9	0.99	1594960	104680	15.2	0.95
WN150	32	2757215	112490	24.5	0.77	2636511	84475	31.2	0.98
WN500	64	1066760	21359	49.9	0.78	1067011	17232	61.9	0.97
TAR500	64	3582159	56531	63.4	0.99	3581880	57527	62.3	0.97

5 Conclusion

This paper presents an approach to the parallelization of simulations to investigate statistical properties of methods using VFCAR models. The parallelization allows improved accuracy and precision of statistical results, and is based on novel techniques for improving performance of scientific computing in parallel and distributed environments. The approach is capable of successfully overcoming the challenge of long running times and requires very minimal revisions to an existing sequential code. The revisions generate parallel code which incorporates dynamic load balancing. The parallel code addresses factors that may give rise to load imbalance, such as those that may be inherent in the computations or induced by the computing environment. Tests on a homogeneous Sun Solaris cluster and a heterogeneous Linux cluster indicate that the parallel code achieves very high performance, even with unpredictable system-induced load imbalance, obtaining efficiencies in the range 95–98% for the normal white noise and threshold autoregressive model, with sample sizes up to 500 and 10000 replications, on up to 64 processors. This combination of sample size and replication count, to best knowledge, has not been previously attempted when using simulation to investigate properties of statistical methods using VFCAR models. Future work will include experiments with new models and more error correlation values. Furthermore, this interdisciplinary research work finds application in a variety of other statistical disciplines. The authors plan to to apply these state of-the-art techniques to an important problem in astrophysics, the classification of gamma-ray bursts.

References

1. Gentle, J.: Random Number Generation and Monte Carlo Methods. 2nd edn. Springer, New York (2003)
2. Harvill, J., Ray, B.: Functional coefficient autoregressive models for vector time series. Computational Statistics and Data Analysis (2005) (To appear)
3. Harvill, J., Ray, B.: A note on multi-step forecasting with functional coefficient autoregressive models. International Journal of Forecasting (2005) (To appear)
4. Cariño, R.L., Banicescu, I.: A load balancing tool for distributed parallel loops. In: Proc. Int. Workshop on Challenges of Large Applications in Distributed Environments, IEEE Computer Society Press (2003) 39–46
5. Cariño, R.L., Banicescu, I.: Dynamic scheduling parallel loops with variable iterate execution times. In: Proc. 16th IEEE Int. Parallel and Distributed Processing Symposium - PDSECA, IEEE Computer Society Press (2002) on CDROM
6. Cariño, R.L., Banicescu, I.: Load balancing parallel loops on message-passing systems. In Akl, S., Gonzales, T., eds.: Proc. 14th IASTED Int. Conf. on Parallel and Distributed Computing and Systems, ACTA Press (2002) 362–367
7. Cariño, R.L., Banicescu, I., Vadapalli, R.K., Weatherford, C.A., Zhu, J.: Message-passing parallel adaptive quantum trajectory method. In Yang, L.T., Pan, Y., eds.: High performance Scientific and Engineering Computing: Hardware/Software Support, Kluwer Academic Publishers (2004) 127–139
8. Banicescu, I., Liu, Z.: Adaptive factoring: A dynamic scheduling method tuned to the rate of weight changes. In: Proc. High Performance Computing Symposium. (2000) 122–129
9. Banicescu, I., Velusamy, V.: Load balancing highly irregular computations with the adaptive factoring. In: Proc. 16th IEEE Int. Parallel and Distributed Processing Symposium - HCW, IEEE Computer Society Press (2002) on CDROM
10. Banicescu, I., Velusamy, V., Devaprasad, J.: On the scalability of dynamic scheduling scientific applications with adaptive weighted factoring. Cluster Computing: The Journal of Networks, Software Tools and Applications **6** (2003) 215–226

Multi-pass Mapping Schemes for Parallel Sparse Matrix Computations[*]

Konrad Malkowski and Padma Raghavan

Department of Computer Science and Engineering,
The Pennsylvania State University,
343K IST Building, University Park, PA 16802-6106
{malkowsk, raghavan}@cse.psu.edu

Abstract. Consider the solution of a large sparse linear system $Ax = b$ on multiprocessors. A parallel sparse matrix factorization is required in a direct solver. Alternatively, if Krylov subspace iterative methods are used, then incomplete forms of parallel sparse factorization are required for preconditioning. In such schemes, the underlying parallel computation is tree-structured, utilizing task-parallelism at lower levels of the tree and data-parallelism at higher levels. The *proportional heuristic* has typically been used to map the data and computation to processors. However, for sparse systems from finite-element methods on complex domains, the resulting assignments can exhibit significant load-imbalances. In this paper, we develop a multi-pass mapping scheme to reduce such load imbalances and we demonstrate its effectiveness for a test suite of large sparse matrices. Our scheme can also be used to generate improved mappings for tree-structured applications beyond those considered in this paper.

1 Introduction

Many computational science and engineering applications concern the numeric solution of models based on nonlinear partial-differential-equations on complex domains, which are discretized using finite-element or finite-difference methods. When implicit or semi-implicit schemes are used in the solution process, the total application time can be dominated by the time required for the solution of the underlying sparse linear systems of the form $Ax = b$. Consequently, effective parallel sparse linear system solution is of critical significance in such large-scale applications.

A solution to $Ax = b$, where A is sparse, can be achieved using either direct methods or preconditioned iterative methods. In parallel sparse direct solvers,

---

[*] The work was supported in part by the National Science Foundation through grants NSF ACI-0102537 and NSF CCF-0444345, and by the Director, Office of Science, Division of Mathematical, Information, and Computational Sciences of the U.S. Department of Energy under contract number DE-AC03-76SF00098.

a Cholesky ($A = LL^T$) or an LU ($A = LU$) factorization is first computed and then used for triangular solution [1, 2, 4, 6, 8, 16]. For preconditioning, incomplete counterparts of both types of factorizations can be utilized to compute a sparse approximation to the factors to accelerate the convergence of an iterative method such as Conjugate Gradients or GMRES [7, 10, 17, 18, 19]. Efficient implementations on distributed memory multiprocessors require data and task assignments that can balance the computational load for the factorization step among processors. The computations in the factorization step are tree-structured and are formulated bottom-up on a *supernodal* tree using either effectively dense panels of columns in a left-looking panel scheme [1] or using dense triangular matrices in a multifrontal scheme [8, 9, 13, 16].

In this paper, we focus on mapping tree-structured computations typical of sparse factorizations where assignments generated by the popular proportional mapping [5, 15] scheme often exhibit large imbalances. Our main contribution is the formulation of a new multi-pass refinement scheme that can substantially improve the quality of the assignment and thus the performance of parallel factorization codes. In the next section, we provide a brief review of parallel sparse factorization. In Section 3, we begin with a review of the original proportional mapping scheme [15] and then describe our new multi-pass schemes. In Section 4, we provide an empirical evaluation of the performance of our multi-pass schemes and the original proportional mapping. In Section 5, we summarize our contributions and discuss further extensions and applications.

2   Parallel Tree-Structured Sparse Factorization

Sparse matrix factorization and its incomplete variations typically require a four step process: (1) ordering to compute a fill-reducing numbering, (2) symbolic factorization to determine the nonzero structure of the factor, (3) numeric factorization, and, (4) triangular solution. The first ordering step is also critical for determining the parallelism and the total computational costs over all remaining steps. A well-established practice is to compute orderings, using for example, nested dissection techniques that recursively partition the graph of A using vertex separators [3, 11]. After this step, the parallelism available for the subsequent factorization and triangular solution step can be represented by a tree. This tree can be weighted to represent computation costs and the tree can be mapped to processors to enable load-balanced computation of the factorization step.

We provide a brief overview of parallel sparse Cholesky factorization using a small example to illustrate the main ideas, which are described in greater detail in the survey article by Heath et al. [8]. Figure 1 concerns the sparse matrix A of a five-point 7×7 finite-difference grid, which is widely used as a model problem in this area. The columns of the Cholesky factor L of this matrix, can be grouped into *supernodes* [12]. A supernode is a set of consecutive columns that have nested sparsity structure, and can essentially be treated as a dense block, see for example, the last seven columns in Figure 1. As a consequence of sparsity in L, columns in a supernode need not be updated by columns in all

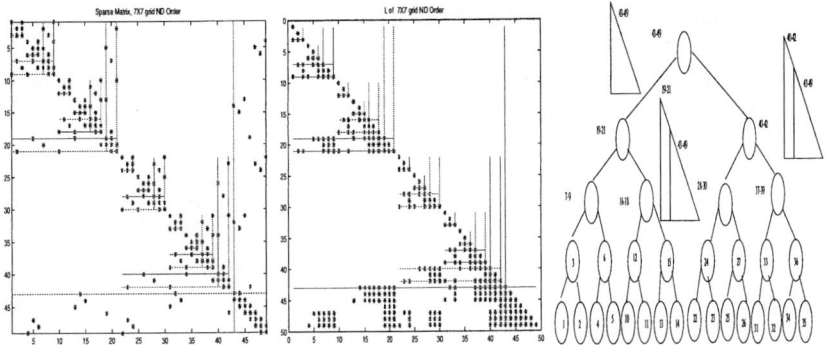

Fig. 1. The structure of a sparse matrix A from a 7×7, 5-point finite-difference grid reordered to reduce fill (left), the structure of L shown with a recursive partition (middle) and a multifrontal scheme on the binary supernodal tree (right)

preceding supernodes in the numeric factorization, instead columns in a supernode v are updated only by columns in supernodes within the subtree rooted at v [12]. The sparsity structure of L can be viewed in terms of effectively dense column-blocks or alternatively, in a recursive manner in terms of submatrices, as shown in Figure 1. The two types of cache-efficient numeric factorizations are a column-block scheme [14] and a multifrontal scheme [2]. The two schemes differ in how they compute and apply updates to columns in a given supernode from columns in earlier supernodes. In a multifrontal scheme, dense triangular matrix operations are used to factor the columns in a supernode and to accumulate and propagate updates from these columns to those at the parent and ancestor supernodes, as shown in Figure 1. Multifrontal schemes typically lead to efficient parallel implementations [6, 16].

Parallel implementations of both left-looking or multifrontal factorization depend on the supernodal tree which can be weighted to represent the corresponding computation and communication costs [5, 6, 16]. For illustrative purposes, view this tree as a complete binary tree with more leaves than the number of processors P and with all nodes having the same computational cost. Now, at some level $l = \log_2 P$, P disjoint nodes can be identified and the subtrees rooted at these nodes can be assigned to distinct processors. These subtrees represent disjoint local computations at processors with ideal task-parallelism. At a level higher than this one, disjoint processor groups of size 2 can cooperate to perform data-parallel computations at the supernode and so on, until all processors participate at the root. This is known as the balanced subtree to processor mapping. The proportional mapping scheme is a generalization of this scheme to derive assignments for practical problems where the supernodal tree can be highly irregular and the computation at a node and total computations in subtrees can vary dramatically.

3 Multi-pass Mapping Schemes

In this section, we begin with an overview of the original proportional mapping scheme [15] and continue with the formulation of our multi-pass assignment schemes. The latter seek to refine and improve an assignment obtained from the proportional mapping, which is used in the first step.

To provide a precise statement of our mapping schemes, we start with a definition of the weighted supernodal tree and a valid mapping, i.e., an assignment of computations represented by the tree to a set of processors. Consider a supernodal tree $T(r) = (V, E, NW, SW)$ with V vertices, E edges, rooted at $r \in V$ with two weighting functions SW and NW representing a suitable measure of computational costs. For each vertex $v \in V$, $NW(v)$ is the nodal weight, corresponding to the cost of computations at the node v. The subtree weight of v, $SW(v)$ is defined as the sum of nodal weights of all vertices in $T(v)$, the subtree rooted at v. A mapping $M = (T, P)$ indicates an assignment of a set of P disjoint subtrees $T(v_0), T(v_1), \cdots T(v_{p-1})$ (including all leaf vertices in V) to processors $0, 1, \cdots (P-1)$. These disjoint subtrees represent local task parallel computations; computations at an interior vertex v are shared equally among all processors assigned subtrees in $T(v)$. However, such computation cannot proceed until all processors can synchronize at the internal node. Consequently, load imbalances among processors along different paths leading to a vertex v, result in some processors remaining idle until all can synchronize and proceed with the computations at v.

The proportional mapping [15] is a recursive scheme with an initial assignment of P processors to the root r and thus $T(r)$. Consider $T(v)$ the subtree at v, which has been assigned p processors. If $p = 1$, the recursion terminates; otherwise for $p > 1$, for each child c of v assign p_c to $T(c)$ where $p_c = p \times \frac{SW(c)}{SW(v) - NW(v)}$. Some rounding scheme must be used to ensure that p_c is an integer number of processors and that $\sum_{c,(c,v) \in E} p_c = p$. We experimented with several rounding schemes to select the one that leads to the best mappings for our test collection. In this scheme, at a vertex v with $T(v)$ assigned p processors, we compute for each child c, $\hat{p}_c = \lfloor p \times \frac{SW(c)}{SW(v) - NW(v)} \rfloor$, and the projected load $\hat{W}(c) = \frac{SW(c)}{\hat{p}_c}$. Next, we compute $\tilde{p} = p - \hat{p}$ where $\hat{p} = \sum_{c,(c,v) \in E} \hat{p}_c$. Let $c_1, c_2, \cdots c_{\tilde{p}}$ be the children vertices of v with the $\tilde{p}$ highest values of the projected load $\hat{W}(c_i)$; for these vertices we set $p_c \leftarrow \hat{p}_c + 1$ while for the others, we set $p_c \leftarrow \hat{p}_c$. Our multi-pass assignment scheme uses this proportional mapping as the first step.

Rounding effects, incurred to ensure an integer number of processors at each node, can be exaggerated by irregular subtree weights resulting in assignments with processor loads that are not well-balanced. Consequently, the highest load at a processor, i.e., the critical path weight, can be substantially higher than the ideal of $SW(r)/P$ as shown in the next section in Figure 2.

Our multi-pass schemes attempt to refine the assignment from the proportional mapping in order to improve the worst load at a processor. We therefore start with a precise definition of this quantity. Consider any mapping $M = (T(r), P)$ of the supernodal tree $T(r) = (V, E, NW, SW)$ with disjoint sub-

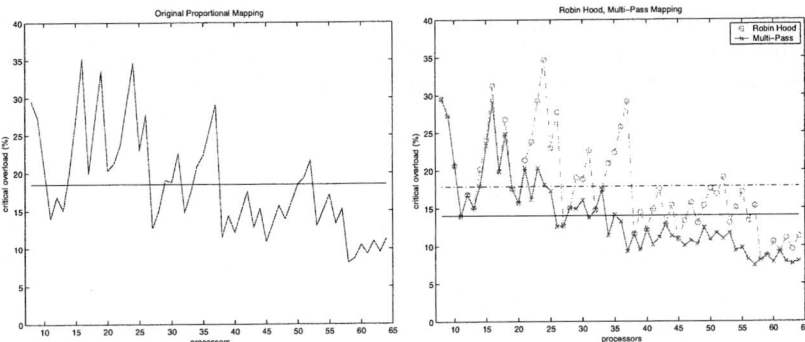

Fig. 2. Critical overload for the original proportional heuristic (left) and and the Robin Hood and multi-pass schemes (right) for augustus7 on 8 – 64 processors

trees $T(v_0), T(v_1), \cdots T(v_{P-1})$ assigned to processors $0, 1, \cdots (P-1)$. Let $\pi(v)$ denote the number of processors assigned to a node v; note that $\pi(r) = P$. Let path(i) denote the intermediate vertices in T from the parent of v_i to the root r. Now the workload of a processor p_i, $0 \leq i \leq (P-1)$ is given $W(p_i) = SW(T(v_i)) + \sum_{v \in \text{path}(v_i)} NW(v)/\pi(v)$. An ideal assignment would lead to the ideal load $I(M) = SW(r)/P$. The heaviest load at a processor, which corresponds to a critical path is given by $H(M) = \max_{p_i, 0 \leq i \leq P-1} \{W(p_i)\}$; likewise, we define the lightest load $L(M) = \min_{p_i, 0 \leq i \leq P-1} \{W(p_i)\}$; Our goal is compute mappings where $H(M)$ is close to $I(M)$ for a given T and P.

We next present two refinement schemes, followed by our final multi-pass scheme which enables their effective combination.

A Robin Hood Refinement Scheme. Assume that an assignment M has been provided either from proportional mapping or the application of one or more refinement schemes. First, determine $H(M)$ and $L(M)$ corresponding to the heaviest and the lightest loads at a processor. Let p_h and p_l denote the corresponding processors assigned to subtrees $T(h)$ and $T(l)$, rooted at vertices h and l respectively. We consider p_h, the overworked processor to be "poor "and the lightly loaded p_m to "rich." Our Robin Hood scheme, removes a processor assigned along the lightest path, path(l) and uses it to refine the mapping of the local subtree $T(h)$ and thus reduce the load along critical path path(h). The Robin Hood scheme is typically applied four times to allow refinement of paths with weights close to the weight of the critical path, and the best assignment is retained.

Iterative Correction with Processors in Reserve. Assume we are seeking an assignment $M = (T(r), P)$ with P processors. We first obtain a mapping with $\tilde{M} = (T(r), \tilde{P})$ with $\tilde{P}$ processors where $\tilde{P} < P$. Let $\hat{P} = P - \tilde{P}$ denote the remaining processors "held in reserve." Our correction algorithm proceeds thus in $\hat{P}$ iterations, starting with a mapping M_1 initialized to $\tilde{M}$ with $P_1 = \tilde{P}$ processors. At iteration i, compute $H(M_i)$ and let p_h be the corresponding processor with the heaviest load. Refine the mapping of $T(h)$ by adding another processor to vertex h to obtain a new mapping M_{i+1} with $P_{i+1} = P_i + 1$ processors. A disadvantage of this scheme is that improvements depend on the initial choice of $\tilde{P}$.

A Multi-pass Mapping Scheme. We now combine the two refinement schemes with the original proportional mapping to present our final multi-pass mapping scheme. The first pass is the proportional mapping scheme. In the second pass, this mapping is refined using the Robin Hood scheme. Let this result in a mapping $\hat{M}$ with the specified number of processors, P. Compute $H(\hat{M})$; if this quantity is more than the ideal load $(SW(r)/P)$, then compute $\tilde{P} = SW(r)/H(\hat{M})$, where $\tilde{P} < P$. Apply the proportional mapping to obtain a mapping $\bar{M}$ with $\tilde{P}$ processors. Refine it using the Robin Hood scheme to obtain a new mapping $\tilde{M}$ with $\tilde{P}$ processors. Finally, refine this mapping using iterative correction with $\hat{P} = P - \tilde{P}$ processors in reserve. This defines our overall multi-pass mapping scheme.

4 Empirical Results

In this section, we empirically evaluate the quality of assignments for performing parallel sparse Cholesky factorization. We report on the improvements observed when our schemes are used to refine the assignments computed by the original proportional scheme. We use a collection of well-known sparse matrices from finite-element analysis of three dimensional structures and shells, and one problem from computational fluid dynamics.

Our test suite of matrices and best observed improvements are shown in Table 1. We consider the factorization of these matrices using 8-64 identical pro-

Table 1. Description of test matrices, relative critical loads (RCL), and numeric factorization times using the original proportional scheme and the multi-pass scheme. Each matrix - processor pair corresponds to the best observed improvements of the RCL metric using our multi-pass scheme. The column labeled "Error" indicates the difference between predicted and observed execution times

Matrix Characteristics				Relative Critical Load and Numeric Factorization Time					
Matrix	Rank (10^3)	$|A|$ (10^4)	$|L|$ (10^4)	Processors	Proportional Map		Multi-pass Map		
					RCL	Time (sec)	RCL	Predicted (sec)	Observed (sec)	Error (%)
bmw7st1	141	374	9,134	24	134	97	114	82.52	80.72	2
bmwcra1	148	539	18,924	14	146	493	129	435.6	454	4
bmw3_2_1	227	575	17,998	24	130	263	122	246	242	2
augustus7	1,060	518	79,995	24	133	1,989	117	1,749	1,653	6
augustus5	134	64	4,598	26	145	32.42	117	26.16	25.97	1
af_shell3	504	904	11,425	26	137	43.65	116	36.96	40.11	9
cfd2	123	160	7465	23	154	75.67	118	57.98	57.84	0

cessors after ordering using a nested dissection scheme. The supernodal trees were weighted to represent computational costs (for floating point operations, and not for communication or for other integer operations) in a parallel multifrontal scheme, for example, such as the scheme in the DSCPACK software [16].

Our results concern the quality of mappings as defined by the heaviest workload at a processor, i.e., the critical path weights. Consider $T(r)$ corresponding to a specific problem. For each mapping M of $T(r)$ using some P processors in the range 8 — 64, we focus on heaviest load at a processor, i.e., the critical path cost, as the main metric indicative of the quality of the mapping. The closer this metric, $H(M)$ (defined in the earlier section) is to the ideal load $(I = SW(r)/P)$, the better is the quality of the mapping. Over the range of processors, problems and mapping the actual value of this metric can vary significantly making direct comparisons difficult. We therefore use the following two scaled forms: (i) the *relative critical load,* (RCL) defined as $\frac{H(M)}{I} \times 100$, and (ii) the *critical overload,* (CO) defined as $\frac{H(M)-I}{I} \times 100$.

We begin with some experiments (reported in the right half of Table 1) to verify that the relative critical load (and the critical overload) metric corresponds well to the actual performance of the numeric factorization step. We used the DSCPACK software [16] with two different mappings, one from the original proportional scheme and the other from our multi-pass scheme. Our experiments were performed on a cluster with 81 dual-processor compute nodes with AMD Athlon MP2200+ processors, with 71 nodes having 1GB of main memory, and 10 nodes having 2GB of main memory, and a 9 × 9 torus Scali interconnect. Table 1 shows the relative critical load (RCL) and CPU time for numeric factorization using the original proportional mapping. We then compute assignments using our multi-pass scheme and their corresponding values of the relative critical load (also reported in Table 1). Using the relative critical load metrics for the original and multi-pass mappings and the CPU time for numeric factorization with the proportional mapping, we can project the estimated numeric factorization time for the new assignment (reported in the column labeled "Predicted" time). We next performed numeric factorization with the new assignments and observed actual CPU time. As shown in Table 1, these observed times are in close agreement with our predicted values, thus indicating that it is valid to use the relative critical load metric and the closely related critical overload metric to evaluate the quality of assignments.

Figure 2 plots the critical overload metric for the original proportional mapping, the Robin Hood scheme and our final multi-pass scheme for **augustus7** on 8 through 64 processors. The plots indicate that Robin Hood scheme can improve the assignments produced by the original proportional mapping. However, as expected, these improvements are not as substantial as the improvements from our final multi-pass scheme. Consequently, in the remainder of this section, we focus on more detailed comparisons between the quality of assignments produced by the original proportional scheme and our final multi-pass scheme.

Figure 3 (left) indicates how our scheme can improve the worst mapping generated by the original scheme. For each matrix, we select the instance with the largest value of the critical overload metric from an assignment by the pro-

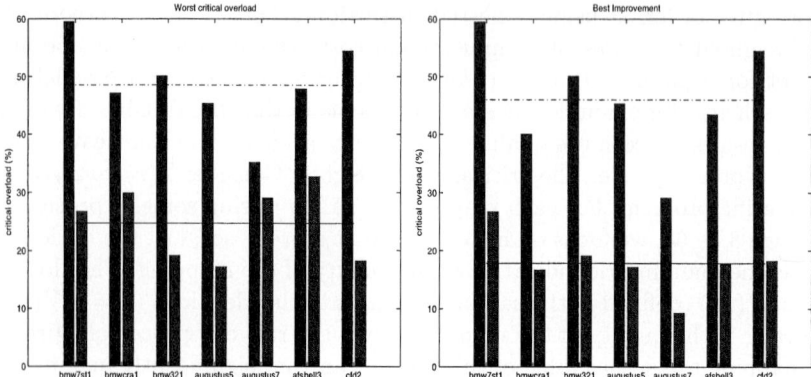

Fig. 3. Quality of assignments for worst load instances (left) and the best observed improvement instances (right); each group of two bars indicates the critical overload from the original proportional scheme and our multi-pass scheme. Average values are indicated by horizontal lines

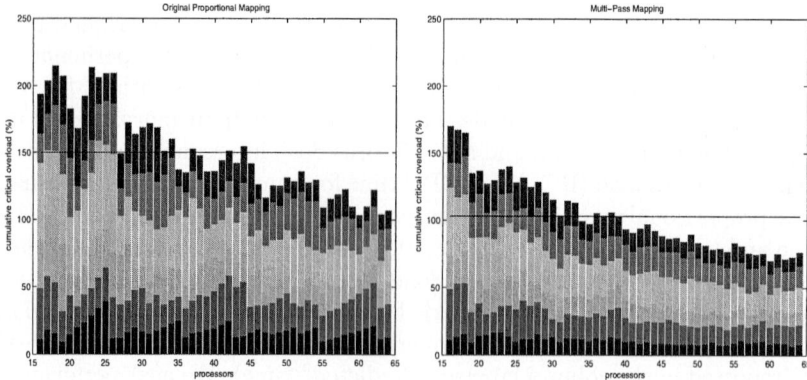

Fig. 4. Cumulative critical overload over all matrices for each processor, from assignments using the original proportional scheme (left) and from our multi-pass scheme (right). Each patch in a stacked bar represents the critical overload for one matrix. Average values are indicated by horizontal lines

portional scheme. For this instance, i.e., problem-processor pair, we show the value of the overload metric when our multi-pass scheme is used. Our multi-pass scheme successfully reduced the worst case overload from almost 60% to 27% for the bmw7st1 matrix. On average, the metric is halved from nearly 50% for the original to 25% for our multi-pass scheme. Figure 3 (right) shows the best observed improvement from our new multi-pass scheme when compared to the original mapping for each matrix in the test set. Assignments from our multi-pass scheme reduced the critical overload for bmw7st1 from approximately 60% to 27% and from 55% to 17% for cfd2. Additionally, on average over these in-

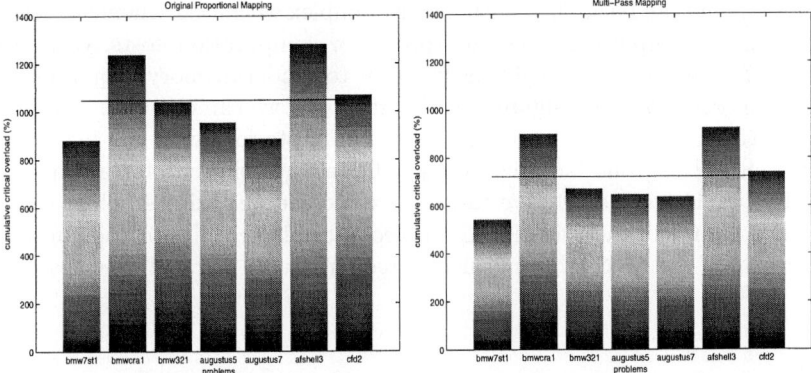

Fig. 5. Cumulative critical overload for 16 – 64 processors for each matrix, from assignments using the original proportional scheme (left) and from our multi-pass scheme (right). Each patch in a stacked bar represents the critical overload for one processor size in the range 16 – 64. Average values are indicated by horizontal lines

stances, the critical overload from the proportional scheme was 46%, whereas it was reduced to 17% from our multi-pass schemes.

We next consider the overall quality of the assignments produced by the original proportional scheme and our multi-pass scheme for the 7 matrices using 16 – 64 processors. We now consider a cumulative form of the critical overload metric shown as a stacked bar in Figures 4 and 5. In Figure 4, each stacked bar represents the critical overload value summed over all 7 matrices for a specific number of processors. Figure 4 clearly indicates that our multi-pass scheme significantly improves the quality of mapping over all problems for the entire range of processors. On average, the cumulative critical overload is reduced from a value of approximately 150 to 100 through the use of our multi-pass scheme. Figure 5 shows the improvements for each matrix cumulatively over the range of processors; each stacked bar represents the critical overload value summed over all 49 processor sizes (from 16 — 64) for a specific matrix. Now the average value from the original scheme is approximately 1050 which is reduced to under 725 by our multi-pass mapping scheme. These results show that our multi-pass schemes are indeed effective in producing assignments that substantially improve the balance of loads among processors.

5 Conclusions

In this paper, we have presented a multi-pass scheme that improves workload distribution among processors in tree-structured computation. Through experiments using trees weighted to represent sparse multifrontal factorizations on distributed memory multiprocessors we show that our multi-pass mapping scheme can significantly improve the quality of assignments.

Our scheme can be applied with a more complex weighting scheme to take into account both computation and interprocessor computation costs. Our scheme can also be used with a weighting function to model memory requirements to produce more balanced assignments. Additionally, we can compute assignments for applications where the triangular solution costs following the factorization are dominant, i.e., one factorization is followed by solutions for a sequence of right-hand-side vectors [20]. We can also extend the weighting schemes to model parallel incomplete factorization for preconditioning [18] and the subsequent application of the preconditioner using tree-structured parallel schemes.

References

1. J. Demmel, S. C. Eisenstat, J. R. Gilbert, X. S. Li, and J. W. H. Liu. A supernodal approach to sparse partial pivoting. Technical Report CSL-94-14, Xerox Palo Alto Research Center, 1995.
2. I.S. Duff. Parallel implementation of multifrontal schemes. *Parallel Computing*, 3:193-204, 1986.
3. A. George and J. W. H. Liu. An automatic nested dissection algorithm for irregular finite element problems. *SIAM J. Numer. Anal.*, 15:1053-1069, 1978.
4. J. A. George and J. W. H. Liu. *Computer Solution of Large Sparse Positive Definite Systems*. Prentice-Hall Inc., Englewood Cliffs, NJ, 1981.
5. L. Grigori and X. S. Li. A new scheduling algorithm for parallel sparse LU factorization with static pivoting. In *Proceedings of the 2002 ACM/IEEE conference on Supercomputing*, pages 1-18. IEEE Computer Society Press, 2002.
6. A. Gupta, F. Gustavson, M. Joshi, G. Karypis, and V. Kumar. PSPASES: An efficient and scalable parallel sparse direct solver, 1999. See http://www-users.cs.umn.edu/$\sim$mjoshi/pspases.
7. A. Gupta, V. Kumar, and A. Sameh. Performance and scalability of preconditioned conjugate gradient methods on the CM-5. In R. F Sincovec, D. E. Keyes, M. R. Leuze, L. R. Petzold, and D. A. Reed, editors, *Proceedings of the Sixth SIAM Conference on Parallel Processing for Scientific Computing*, pages 664-674, Philadephia, PA, 1993. SIAM Publications.
8. M. T. Heath, E. Ng, and B. W. Peyton. Parallel algorithms for sparse linear systems. *SIAM Review*, 33:420-460, 1991.
9. M. T. Heath and P. Raghavan. Performance of a fully parallel sparse solver. *Int. J. Supercomputing Appl.*, 11:49-64, 1997.
10. M. T. Jones and P. E. Plassman. An improved incomplete Cholesky factorization. *ACM Trans. Math. Software*, 21:5-17, 1995.
11. G. Karypis and V. Kumar. METIS: Unstructured graph partitioning and sparse matrix ordering system. Technical report, Department of Computer Science, University of Minnesota, Minneapolis, MN, 1995.
12. J. W. H. Liu. The role of elimination trees in sparse factorization. *SIAM J. Matrix Anal. Appl.*, 11:134-172, 1990.
13. J. W. H. Liu. The multifrontal method for sparse matrix solution: theory and practice. *SIAM Review*, 34:82-109, 1992.
14. E. Ng and B. W. Peyton. A supernodal Cholesky factorization algorithm for shared-memory multiprocessors. *SIAM J. Sci. Comput.*, 14:761-769, 1993.

15. A. Pothen and C. Sun. A mapping algorithm for parallel sparse Cholesky factorization. *SIAM J. Sci. Comput.*, 14(5):1253–1257, 1993.
16. P. Raghavan. DSCPACK: Domain-Separator Codes for the parallel solution of sparse linear systems, 2002. Software for solving sparse linear systems on multiprocessors an d NOWs using C and MPI. Package has a two stage parallel nested dissection, higher-level BLAS, fast repeated solves and an easy t o use parallel interface. See http://www.cse.psu.edu/Dscpack.
17. P. Raghavan, K. Teranishi, and E. Ng. Scalable parallel preconditioning with incomplete factors. In *Proceedings of the Seventh SIAM Conference on Applied Linear Algebra*. 2000.
18. P. Raghavan, K. Teranishi, and E. Ng. A latency tolerant hybrid sparse solver using incomplete Cholesky factorization. *Numerical Linear Algebra*, 10:541–560, 2003.
19. Y. Saad. *Iterative Methods for Sparse Linears Systems*. PWS Publishing Co., Boston, MA, 1996.
20. C. Yang, P. Raghavan, L. Arrowood, B. Sumpter, and D . Noid. Large-scale normal coordinate analysis on distributed memory mu ltiprocessors and nows. *Int. J. Supercomputing Appl.*, 1(4):409–424, 2002.

High-Order Finite Element Methods for Parallel Atmospheric Modeling

Amik St.-Cyr and Stephen J. Thomas

National Center for Atmospheric Research,
1850 Table Mesa Drive, Boulder, 80305 CO, USA
{amik, thomas}@ucar.edu

Abstract. High-order finite element methods for the atmospheric shallow water equations are reviewed. The accuracy and efficiency of nodal continuous and discontinuous Galerkin spectral elements are evaluated using the standard test problems proposed by Williamson et al (1992). The relative merits of strong-stability preserving (SSP) explicit Runge-Kutta and multistep time discretizations are discussed. Distributed memory MPI implementations are compared on the basis of the total computation time required, sustained performance and parallel scalability. Because a discontinuous Galerkin method permits the overlap of computation and communication, higher sustained execution rates are possible at large processor counts.

1 Introduction

High-order finite element methods are well-suited to atmospheric modeling due to their desirable numerical properties and inherent parallelism. A spectral element atmospheric model received an honorable mention in the 2001 Gordon Bell award competition, Loft et al (2001). Discontinuous Galerkin approximations are an extension of low order finite-volume techniques for compressible flows with shocks (Cockburn et al 2000). Either nodal or modal basis functions can be employed in high-order finite elements and the methods are spectrally accurate for smooth solutions. To avoid excessive memory requirements, global assembly of finite element matrices is avoided in the continuous Galerkin method by applying a direct-stiffness summation, Deville et al (2002). Computations within an element are based on tensor-product summations, taking the form of dense matrix-matrix multiplications. These are naturally cache-blocked and can be unrolled to expose instruction level parallelism to processors containing multiple floating point units.

The shallow water equations are a prototype for atmospheric general circulation models. The parallel performance of a 3D model can be estimated by solving identical 2D shallow water problems on multiple layers. To evaluate the efficiency of various time integrators, schemes of equivalent order will be compared on the basis of the total wall-clock time required to solve a given initial value problem (i.e. time to solution). The amount of computation required by a method depends on the Courant number or equivalently the time step size. Moreover, the parallel performance depends on the number of right-hand side evaluations per time step and the associated parallel communication. In the case of the discontinuous Galerkin method, communication of conserved variables can be overlapped with the weak divergence and source term computations. The continuous Galerkin spectral element model developed by Loft et al (2002) will serve as the parallel performance baseline for our simulations.

2 Shallow Water Equations

The shallow water equations contain the essential wave propagation mechanisms found in atmospheric general circulation models. These are the fast-moving gravity waves and nonlinear Rossby waves. The latter are important for correctly capturing nonlinear atmospheric dynamics. The flux form shallow-water equations in curvilinear coordinates are described in Sadourny (1972).

$$\frac{\partial u_1}{\partial t} + \frac{\partial}{\partial x^1}E = \sqrt{G}\,u^2(f+\zeta),$$

$$\frac{\partial u_2}{\partial t} + \frac{\partial}{\partial x^2}E = -\sqrt{G}\,u^1(f+\zeta),$$

$$\frac{\partial}{\partial t}(\sqrt{G}\,\Phi) + \frac{\partial}{\partial x^1}(\sqrt{G}\,u^1\Phi) + \frac{\partial}{\partial x^2}(\sqrt{G}\,u^2\Phi) = 0,$$

where

$$E = \Phi + \frac{1}{2}(u_1 u^1 + u_2 u^2), \quad \zeta = \frac{1}{\sqrt{G}}\left[\frac{\partial u_2}{\partial x^1} - \frac{\partial u_1}{\partial x^2}\right]$$

h is the height above sea level. u^i and u_j are the contravariant and covariant velocities. $\Phi = gh$ the geopotential height. f is the Coriolis parameter. The metric tensor is G_{ij} and $G = \det(G_{ij})$.

3 Space Discretization

The computational domain Ω is partitioned into finite elements Ω_k. An approximate solution u_h belongs to the finite dimensional space $\mathcal{V}_h(\Omega)$. u_h is expanded in terms of a tensor-product of the Lagrange basis functions defined at the Gauss-Lobatto-Legendre points

$$u_h^k = \sum_{i=0}^{N}\sum_{j=0}^{N} u_{ij} h_i(x) h_j(y)$$

A weak Galerkin variational problem is obtained by integrating the equations with respect to a test function $\varphi_h \in \mathcal{V}_h$. In the continuous Galerkin spectral element method, integrals are directly evaluated using Gauss-Lobatto quadrature and continuity is enforced at the element boundaries.

To illustrate the discontinuous Galerkin approach, consider a scalar hyperbolic equation in flux form,

$$u_t + \nabla \cdot \mathcal{F} = S.$$

By applying the Gauss divergence theorem, the weak form becomes

$$\frac{d}{dt}\int_{\Omega_k} \varphi_h u_h \, d\Omega = \int_{\Omega_k} \varphi_h S \, d\Omega + \int_{\Omega_k} \mathcal{F} \cdot \nabla \varphi_h \, d\Omega - \int_{\partial\Omega_k} \varphi_h \mathcal{F} \cdot \hat{n} \, ds$$

The jump discontinuity at an element boundary requires the solution of a Riemann problem where the flux function $\mathcal{F} \cdot \hat{n}$ is approximated by a Lax-Friedrichs numerical flux. The resulting semi-discrete equation is given by

$$\frac{du_h}{dt} = L(u_h).$$

4 Time Discretization

Strong-stability preserving (SSP) time discretization methods were developed for semi-discrete method of lines approximation of hyperbolic PDE's in conservative form, Gottlieb et al (2001). Strong stability is a monotonicity property for the internal stages and the numerical solution. A general m-stage SSP Runge-Kutta method is given by

$$u^{(0)} = u^n$$
$$u^{(i)} = \sum_{k=0}^{i-1} \alpha_{ik} u^{(k)} + \Delta t \beta_{ik} L(u^{(k)}), \quad i = 1, \ldots, m, \tag{1}$$
$$u^{n+1} = u^{(m)}.$$

To compute $u^{(i)}$ for each stage requires up to m evaluations of the right-hand side $L(u^{(k)})$. Thus, higher-order SSP Runge-Kutta methods can be expensive, in terms of the number of floating point operations, memory to store intermediate stages and parallel communication overhead per time step. Linear multistep methods (LMM) substitute time levels for stages.

Higueras (2004) discovered second order SSP Runge-Kutta methods with three stages and having Courant number $C = 2$. Therefore, the most efficient second order explicit integrator for the discontinuous Galerkin approximation would appear to be the three stage SSP Runge-Kutta scheme with an efficiency factor of $C/3 = 2/3$. Indeed, our numerical experiments confirmed this to be the case. In fact it was found in practice that this scheme integrates twice as fast as the two stage method. Moreover, the maximum time step for the three stage SSP RK2-3 matches that of the second order leap frog integrator employed in the continuous Galerkin spectral element model of Loft et al (2001).

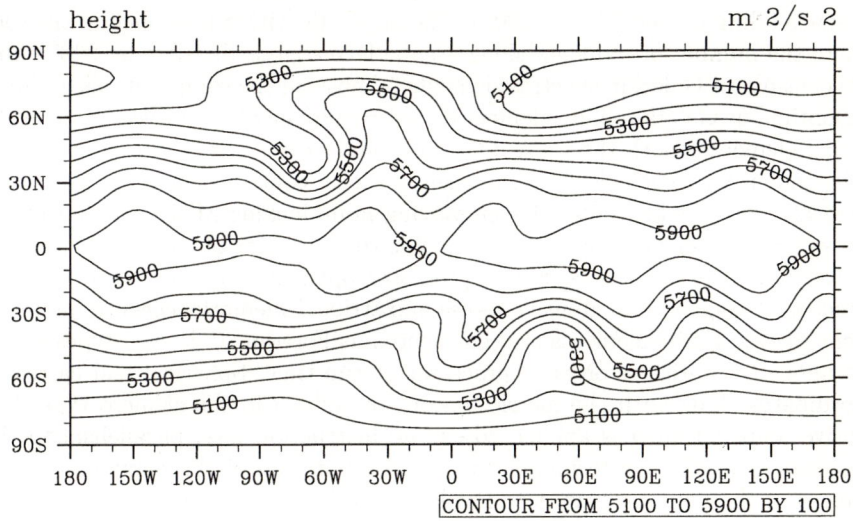

Fig. 1. Shallow water test case 5: Flow impinging on a mountain. 150 spectral elements, 8×8 Gauss-Legendre Lobatto points per element Geopotential height field h at fifteen days produced by discontinuous Galerkin method

5 Numerical Experiments

Our numerical experiments are based on the shallow water test suite of Williamson et al (1992). Test case 5 is a zonal flow impinging on an isolated mountain. The center of the mountain is located at $(3\pi/2, \pi/6)$ with height $h_s = 2000\,(1 - r/R)$ meters, where $R = \pi/9$ and $r^2 = \min[R^2, (\lambda - 3\pi/2)^2 + (\theta - \pi/6)^2]$. The initial wind and height fields take the same form as test case 2 with $\alpha_0 = 0$, $gh_0 = 5960$ m^2/s^2 and $u_0 = 20$ m/s. A total of 150 spectral elements containing 8×8 Gauss-Lobatto-Legendre points are employed. The explicit time step was $\Delta t = 90$ sec. A spatial filter was not applied during this integration. Figure 1 contains a plot of the geopotential height field after 15 days of integration using the discontinuous Galerkin approximation. These results compare favorably with the continuous spectral element model.

6 Parallel Performance Results

Both the continuous and discontinuous Galerkin spectral element models are implemented within a unified software framework. A hybrid MPI/OpenMP programming model is supported where the entire time step is threaded according to an SPMD shared-memory approach. MPI message passing calls are serialized in hybrid mode. The cubed-sphere computational domain is partitioned across the compute nodes of a distributed-memory machine using either Metis or space-filling curves (Dennis 2003). The latter algorithm is applied in the case where the number of elements along a cube face edge is divisible by $2^n 3^m$. Unlike continuous Galerkin spectral elements, almost all the computations within a discontinuous Galerkin finite element do not require any information from neighboring elements. Both the weak divergence and source terms can be computed independently on each element and then the local contribution of the boundary integrals can be added later. By employing non-blocking MPI communication, these computations can be performed while the exchange of conserved variables between elements proceeds. Single processor optimizations are based on data structures designed for extensive re-use and stride-1 memory access to minimize cache misses. Finite elements are represented as Fortran 90 derived types which are allocated statically on each processor at run-time. Spherical and cartesian coordinates along with the metric tensor are defined in the element type. Extensive loop unrolling is applied to expose instruction level parallelism to the processor. An effective technique for fast computation of the maximum eigenvalue α of the flux Jacobian is to invoke vector intrinsics and the use of square roots is minimized.

An experiment was designed to test if overlapping communication and computation in a pure MPI code has any measurable effect on an IBM p690 cluster with a Colony switch. The machine consists of 32-way SMP nodes containing 1.3 Ghz Power4 processors capable of four flops per clock cycle. A node can be configured as either a single 32-way or four 8-way logical partitions. Both the

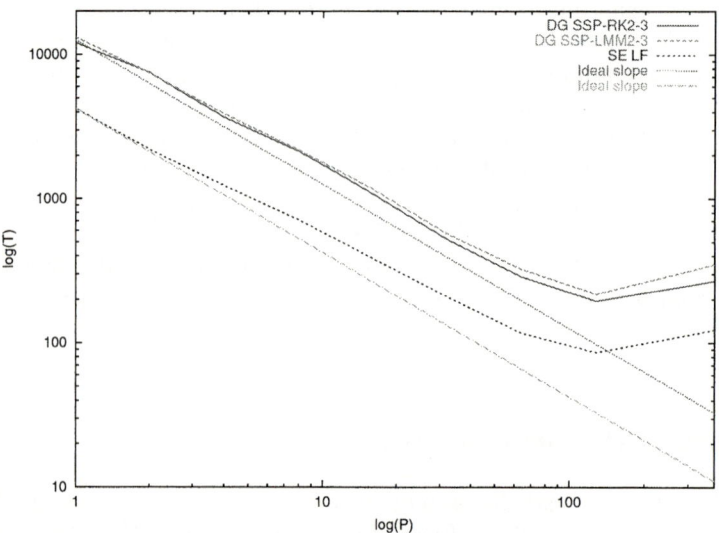

Fig. 2. Shallow water test case 5: Flow impinging on a mountain. 384 spectral elements, 10×10 Gauss-Legendre Lobatto points per element. 40 vertical levels. Integration time in seconds for IBM p690 8-way partitions

continuous and discontinuous Galerkin models were compared using test case 5, but integrated for only five days. For this test, the problem is replicated on 40 independent vertical levels. The discontinuous Galerkin code was integrated using the three step linear multistep method (LMM2-3) with $\Delta t = 15$ sec and three stage SSP Runge-Kutta (RK2-3) scheme with $\Delta t = 60$ sec. The continuous Galerkin model uses a second order leap frog integrator and $\Delta t = 60$ sec. $K = 384$ elements were employed with 10×10 Gauss-Lobatto-Legendre points. The LMM2-3 and RK2-3 single processor execution rates are 667 MFlops and 528 MFlops, respectively. The leap frog code sustains 724 MFlops. The wait time is defined as the average time for the MPI non-blocking communication to complete, summed over all time steps and all processors. We find that the wait time for 8-way is less than for 32-way partitions. Figure 2 is a plot of the integration times using 8-way partitions.

7 Conclusions

The three stage RK2-3 integrator discovered by Higueras (2004) was the most efficient SSP method examined. The discontinuous Galerkin method converges exponentially for smooth solutions and standard error metrics compare favorably with a continuous Galerkin spectral element model. Overlapping communication with computation was suggested by Baggag et al (1999). However, we only communicate u_h and compute fluxes $\mathcal{F}(u_h)$ locally. Both non-overlapping and overlapping implementations were compared. The latter was found to be clearly beneficial on SMP clusters such as the IBM p690 and leads to improved scalability in the strong sense for a fixed problem size. The per performance within a 32-way node is higher but 8-way nodes scale better. Our experience with a nodal Galerkin method indicates that a filter is required for long integrations to stabilize the scheme, thereby extending the results of Nair et al (2004).

Acknowledgements. NCAR is supported by the National Science Foundation. This work was partially supported by an NSF collaborations in mathematics and the geosciences grant (0222282) and the DOE climate change prediction program (CCPP).

References

1. Baggag, A., H. Atkins, D. Keyes, 1999: Parallel implementation of the discontinuous Galerkin method. ICASE Report 99-35, NASA/CR-1999-209546.
2. Cockburn, B., G. E. Karniadakis, and C. W. Shu, 2000: Discontinuous Galerkin Methods. Springer-Verlag, New York, 470 pp.
3. Dennis, J. M., 2003: Partitioning with space-filling curves on the cubed-sphere. Proceedings of Workshop on Massively Parallel Processing at IPDPS'03. Nice, France, April 2003.

4. Deville, M. O., P. F. Fischer, and E. H. Mund, 2002: High-Order Methods for Incompressible Fluid Flow. Cambridge University Press, 499 pp.
5. Fischer, P. F., and J. S. Mullen, 2001: Filter-Based stabilization of spectral element methods. *Comptes Rendus de l'Académie des sciences Paris*, t. 332, Série I - Analyse numérique, 265–270.
6. Giraldo, F. X., J. S. Hesthaven, and T. Warburton, 2003: Nodal high-order discontinuous Galerkin methods for spherical shallow water equations. *J. Comput. Phys.*, **181**, 499-525.
7. Gottlieb, S., C. W. Shu, and E. Tadmor, 2001: Strong stability preserving high-order time discretization methods. *SIAM Review*, **43**, 89–112.
8. Higueras, I., 2004: On strong stability preserving time discretization methods. *J. Sci. Comput.*, **21**, 193-223.
9. Loft, R. L., S. J. Thomas, and J. M. Dennis, 2001: Terascale spectral element dynamical core for atmospheric general circulation models. Proceedings of Supercomputing 01, IEEE/ACM.
10. Nair, R. D., S. J. Thomas, and R. D. Loft, 2004: A discontinuous Galerkin global shallow water model. *Mon. Wea. Rev.*, to appear.
11. Sadourny, R., 1972: Conservative finite-difference approximations of the primitive equations on quasi-uniform spherical grids. *Mon. Wea. Rev.*, **100**, 136–144.
12. Thomas, S. J., and R. D. Loft, 2002: Semi-implicit spectral element atmospheric model. *J. Sci. Comp.*, **17**, 339–350.
13. Williamson, D. L., J. B. Drake, J. J. Hack, R. Jakob, P. N. Swarztrauber, 1992: A standard test set for numerical approximations to the shallow water equations in spherical geometry *J. Comp. Phys.*, **102**, 211–224.

Continuation of Homoclinic Orbits in MATLAB

M. Friedman[2], W. Govaerts[1], Yu.A. Kuznetsov[3], and B. Sautois[1]

[1] Department of Applied Mathematics and Computer Science, Ghent University,
Krijgslaan 281-S9, B-9000 Ghent, Belgium
[2] Mathematical Sciences Department, University of Alabama,
209 Madison Hall, Huntsville, AL 35899, USA
[3] Mathematisch Instituut,
Budapestlaan 6, 3584CD Utrecht, The Netherlands

Abstract. We have added the functionality for continuing homoclinic orbits to CL_MATCONT, a user-friendly MATLAB package for the study of dynamical systems and their bifurcations. It is now possible to continue homoclinic-to-hyperbolic-saddle and homoclinic-to-saddle-node orbits. The implementation is done using the continuation of invariant subspaces, with the Ricatti equations included in the defining system. The continuation can be initiated from a limit cycle with large period or from a Bogdanov-Takens point. All known codimension-two bifurcations are tested for, during continuation. The test functions for inclination-flip bifurcations are implemented in a new and more efficient way.

1 Introduction

A continuous-time dynamical system is usually defined by a set of ordinary differential equations

$$\dot{x} = f(x, \alpha), \qquad (1)$$

where x is a state vector, α is a parameter vector and f is a smooth function. For general background on dynamical systems theory we refer to the existing literature, in particular [1].

CL_MATCONT [2] and its GUI version MATCONT [3] are MATLAB packages for the study of dynamical systems and their bifurcations. Among other things, they support the numerical continuation of equilibria, limit cycles, limit points, Hopf points, fold, flip, and torus bifurcations of cycles. Both packages are freely available at http://allserv.UGent.be/~ajdhooge. CL_MATCONT and MATCONT are successor packages to AUTO [4] and CONTENT[5], which are written in compiled languages (Fortran, C, C++). The MATLAB platform of CL_MATCONT and MATCONT is attractive because it makes them user-friendly, portable to all operating systems, and allows a standard handling of data files, graphical output, etc. On the other hand, it makes the code inevitably slower because MATLAB is not compiled.

Recently we did some successful testing to improve CL_MATCONT ([6]). First, the transparency and readability of the code was improved greatly by organizing it in an object-oriented structure. Eventually, all of CL_MATCONT's total functionality will be transferred into the new structure. Second, a partial inclusion of C-code was used to speed up the computation while preserving the portability of the package. In the case of limit cycle continuations this nearly doubles the speed. This version of CL_MATCONT can be downloaded from http://allserv.UGent.be/~bsautois.

In dynamical systems theory, an orbit corresponding to a solution $\varphi(t)$ is called homoclinic to the equilibrium point x_0 of (1) if $\varphi(t) \to x_0$ as $t \to \pm\infty$. There are two types of homoclinic orbits with codimension 1, namely homoclinic-to-hyperbolic-saddle (HHS), if x_0 is a saddle, and homoclinic-to-saddle-node (HSN), if x_0 is a saddle-node. Codimension 1 means that in generic dynamical systems with two free parameters these orbits exist along curves in the parameter plane. Both types of homoclinic orbits are important in many applications, e.g. as wave solutions in combustion models [7], to model 'bursting' in models of biological cells [8], chemical reactions [9], etc.

In this paper, we describe new functionalities of CL_MATCONT related to homoclinic orbits. We have implemented continuation of both HHS and HSN orbits, starting from a Bogdanov-Takens (BT) point (no other software allows this) or from a limit cycle with high period, and the detection of a large number of codimension 2 bifurcations during the continuation. To compute the relevant eigenspaces of the equilibrium in each step, we use a method to continue invariant subspaces based on [10]. AUTO also has a toolbox for homoclinic continuation, namely HomCont [11]. Important differences with our implementation are that HomCont does not use the continuation of invariant subspaces, and cannot start the continuation of homoclinics from a BT-point. Also, we have implemented test functions for inclination flip bifurcations in a new and more efficient way. Thus, the algorithm combines various ingredients from [10], [11], [12] and [13] but differs from any existing implementation.

2 Extended Defining System for Continuation

2.1 Homoclinic-to-Hyperbolic-Saddle Orbits

To continue HHS orbits in two free parameters, we use an extended defining system that consists of several parts.

First, the infinite time interval is truncated, so that instead of $[-\infty, +\infty]$ we use $[-T, +T]$, which is scaled to $[0, 1]$ and divided into mesh-intervals. The mesh is nonuniform and adaptive. Each mesh interval is further subdivided by equidistant fine mesh points. Also, each mesh interval contains a number of collocation points. (This discretization is the same as that in AUTO for boundary value problems.) The equation

$$\dot{x}(t) - 2T f(x(t), \alpha) = 0, \qquad (2)$$

must be satisfied in each collocation point.

The second part is the equilibrium condition

$$f(x_0, \alpha) = 0. \tag{3}$$

Third, there is a so-called phase condition needed for the homoclinic solution, similar to periodic solutions

$$\int_0^1 \dot{\tilde{x}}^*(t)[x(t) - \tilde{x}(t)]dt = 0. \tag{4}$$

Here $\tilde{x}(t)$ is some initial guess for the solution, typically obtained from the previous continuation step. We note that in the literature another phase condition is also used, see, for example [14]. However, in the present implementation we employ the condition (4).

Fourth, there are the homoclinic-specific constraints to the solution. For these we need access to the stable and unstable eigenspaces of the system in the equilibrium point after each step. It is not efficient to recompute the spaces from scratch in each continuation-step. Instead, we use the algorithm for continuing invariant subspaces, as described in [10]. This method adds two small-sized vectors (Y_S and Y_U) to the system variables, from which the necessary eigenspaces (stable and unstable, respectively) can easily be computed in each step.

If Q_0 is an orthogonal matrix whose first m columns form a basis for the invariant subspace under consideration in the previous step, and $A = f_x(x_0, \alpha)$ is the Jacobian at the new equilibrium point, then we first compute the so-called Ricatti-blocks, T_{ij}, by the formula

$$\begin{bmatrix} T_{11} & T_{12} \\ T_{21} & T_{22} \end{bmatrix} = Q_0^* A \, Q_0. \tag{5}$$

If n is the number of state variables, then T_{11} is of size $m \times m$ and T_{22} is $(n-m) \times (n-m)$. This is done for the stable and unstable eigenspaces separately. Now Y_S and Y_U are obtained from the Ricatti equations

$$\begin{aligned} T_{22U} Y_U - Y_U T_{11U} + T_{21U} - Y_U T_{12U} Y_U &= 0, \\ T_{22S} Y_S - Y_S T_{11S} + T_{21S} - Y_S T_{12S} Y_S &= 0. \end{aligned} \tag{6}$$

Now we can formulate constraints on the behavior of the solution close to the equilibrium x_0. The initial vector of the orbit, $(x(0) - x_0)$, is placed in the unstable eigenspace of the system in the equilibrium. We express that by the requirement that it is orthogonal to the orthogonal complement of the unstable eigenspace. Using Y_U, we can compute the orthogonal complement of the unstable eigenspace. If Q_{0U} is the orthogonal matrix from the previous step, related to the unstable invariant subspace, then a basis for the orthogonal complement in the new step Q_{1U_o} is

$$Q_{1U_o} = Q_{0U} \begin{bmatrix} -Y_U^* \\ I \end{bmatrix}.$$

Note that Q_{1U_o} is not orthogonal. The full orthogonal matrix Q_{1U} needed for the next step, is computed separately after each step. The equations to be added

to the system are (after analogous preparatory computations for the stable eigenspace)
$$Q_{1U}^*(x(0) - x_0) = 0,$$
$$Q_{1S}^*(x(1) - x_0) = 0. \qquad (7)$$

Finally, the distances between $x(0)$ (resp., $x(1)$) and x_0 must be small enough, so that
$$\|x(0) - x_0\| - \epsilon_0 = 0,$$
$$\|x(1) - x_0\| - \epsilon_1 = 0. \qquad (8)$$

A system consisting of all equations (2), (3), (4), (6), (7) and (8), is overdetermined. The basic defining system for the continuation of a HHS orbit in two free parameters consists of (2), (3), (6), (7), and (8) with fixed $\epsilon_{0,1}$, so that the phase condition (4) is not used. The variables in this system are stored in one vector. It contains the values of $x(t)$ in the fine mesh points including $x(0)$ and $x(1)$, the truncation time T, two free system parameters, the coordinates of the saddle x_0, and the elements of the matrices Y_S and Y_U. Alternatively, the phase condition (4) can be added if T is kept fixed but ϵ_0 and ϵ_1 are allowed to vary. It is also possible to fix T and ϵ_0, say, and allow ϵ_1 to vary, again with no phase condition. Other combinations are also possible, in particular, when the homotopy method [10] is used to compute a starting homoclinic solution.

2.2 Homoclinic-to-Saddle-Node Orbits

For a homoclinic orbit to a saddle-node equilibrium, the extended defining system undergoes some small changes. Now $(x(0) - x_0)$ has to be placed in the center-unstable subspace. Analogously, $(x(1) - x_0)$ must be in the center-stable subspace. This again is implemented by requiring that the vector is orthogonal to the orthogonal complement of the corresponding space. So the equations (7) themselves do not really change; the changes happen in the computation of the matrices Q. The defining system now has one equation less than in the HHS case ($n_s + n_u < n$, with n_s the dimension of the stable, and n_u of the unstable eigenspace); the number of equations is restored however, by adding the constraint that the equilibrium must be a saddle-node. For this we use the bordering technique, as described in section 4.2.1 of [15].

3 Starting Strategies

At present, continuation of homoclinic orbits in CL_MATCONT can be started in two ways: either from a Bogdanov-Takens (BT) point or from a limit cycle with large period.

When starting from a limit cycle with large period, we first look for a point on the cycle with smallest $\|f(x, \alpha)\|$. This point is taken as a first approximation to x_0. The mesh points of the limit cycle are kept as mesh points for the homoclinic orbit, except for the mesh interval that contains the current equilibrium approximation. In memory, the stored cycle needs to be rotated, so that the first and last point of the orbit ($x(0)$ and $x(1)$) are stored in the correct locations. To start from a Bogdanov-Takens point, we use the method from [16]. It computes a predictor for the homoclinic orbit, using the coefficients of the normal form at the Bogdanov-Takens point. However, it does take some trial-and-error to set all parameters for the continuation.

4   Bifurcations

Several codimension-two bifurcations can be detected on HHS and HSN curves. In HSN continuation, only one bifurcation is tested for, namely the non-central homoclinic-to-saddle-node orbit or NCHSN. This orbit forms the transition between HHS and HSN curves: a sharp corner, which normally characterizes HHS orbits, appears in the otherwise smooth HSN orbit. The strategy used for detection is taken from HomCont [11].

In HHS continuation, all bifurcations detected in HomCont, are also detected in our implementation. For this, mostly test functions from [11] are used. We refer to that paper for test functions for the following bifurcations:

- Neutral saddle with resonant eigenvalues
- Double real stable leading eigenvalue
- Double real unstable leading eigenvalue
- Neutral saddle, saddle-focus or bi-focus
- Neutrally-divergent saddle-focus (stable)
- Neutrally-divergent saddle-focus (unstable)
- Three leading eigenvalues (stable)
- Three leading eigenvalues (unstable)
- NCHSN
- Shil'nikov-Hopf
- Bogdanov-Takens point
- Orbit-flip with respect to the stable manifold
- Orbit-flip with respect to the unstable manifold

For inclination-flip bifurcations, we also implemented the test functions from [11], but in a more efficient way. We assume that the eigenvalues of $f_x(x_0, \alpha)$ are ordered according to

$$\operatorname{Re} \mu_{n_s} \leq \ldots \leq \operatorname{Re} \mu_1 < 0 < \operatorname{Re} \lambda_1 \leq \ldots \leq \operatorname{Re} \lambda_{n_u},$$

where n_s is the number of eigenvalues with negative real part, and n_u the number of eigenvalues with positive real part. We further assume that μ_1 and λ_1 are real and v_1^S and v_1^U are the eigenvectors of $f_x(x_0, \alpha)$ corresponding to eigenvalues μ_1 and λ_1, respectively. Let $y(t)$ satisfy

$$y'(t) + 2\, T f_x^*(x(t), \alpha)\, y(t) = 0, \tag{9}$$
$$L_{1S}^*\, y(1) = 0, \tag{10}$$
$$L_{1U}^*\, y(0) = 0, \tag{11}$$

where L_{1S} and L_{1U} are matrices whose columns form the orthogonal complements to the unstable and stable eigenspaces of $f^*(x_0, \alpha)$, respectively. These are defined similar to Q_{1S} and Q_{1U} in the previous section. Now introduce

– a test function for the inclination-flip with respect to the stable manifold:
$$\psi_1 = e^{-\mu_1 T} \langle v_1^S, y(0) \rangle = e^{-\mu_1 T} \langle v_1^S, L_{1U} \zeta_2 \rangle;$$
– a test function for the inclination-flip with respect to the unstable manifold:
$$\psi_2 = -e^{-\lambda_1 T} \langle v_1^U, y(1) \rangle = -e^{-\lambda_1 T} \langle v_1^U, L_{1U} \zeta_1 \rangle,$$
where ζ_1 and ζ_2 are found by demanding that
$$[y(1) \; \zeta_1 \; \zeta_2]^* \perp Range\,[(D - 2Tf_x)^* \; (L_{1S}^* \delta_1)^* \; (L_{1U}^* \delta_0)^*]^*.$$

Here D and δ are the differentiation and the evaluation operators, respectively. Our algorithm keeps track of the left and right singular vectors of this matrix; in each newly computed homoclinic orbit the matrix is bordered with the right and left singular vectors of the previously computed homoclinic orbit to compute the left and right singular vectors of the new matrix.

5 Example

The Morris-Lecar model for the barnacle giant muscle fiber [17] is a famous model in computational neuroscience. The equations are:

$$C\dot{V} = I_{ext} - g_L(V - V_L) - g_{Ca}M_\infty(V - V_{Ca}) - g_K N(V - V_K), \quad \dot{N} = \tau_N(N_\infty - N),$$

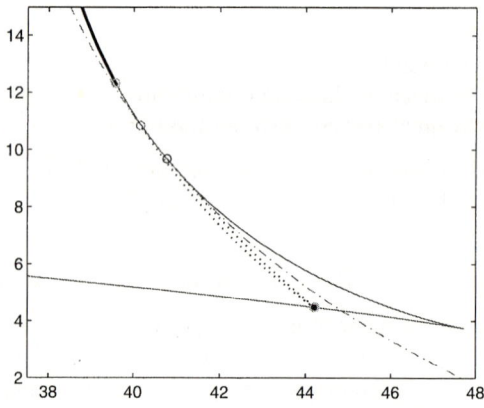

Fig. 1. Bifurcation diagram of the biologically most relevant parameter-range of the Morris-Lecar system. The X-axis shows the value of I, the Y-axis that of V_3. The thin full line is the limit point curve (which clearly shows a cusp point). The dash-dotted line is the Hopf curve. The thick full lines are HSN curves, and the dotted lines are HHS curves. The top and bottom circles indicate the locations of NCHSN points; the bottom circle is actually two circles close together. The second circle from the top is a HNS point, and the third is a BT point

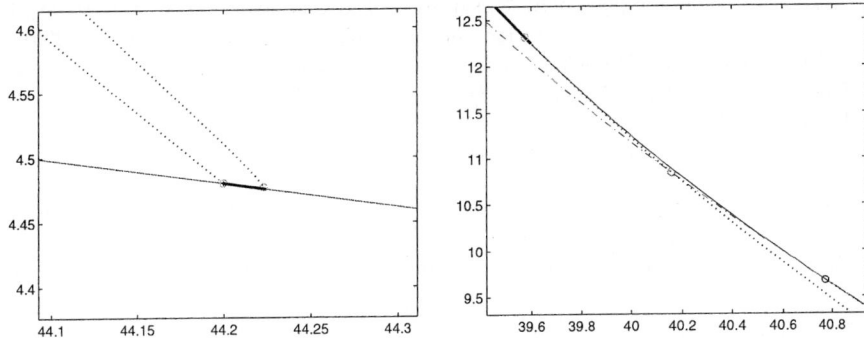

Fig. 2. More detailed views of Figure 1. Left: 2 HHS curves approach the limit point curve. At the intersections, there are NCHSN orbits, and between these, there exists a curve of HSN orbits. Right: Top circle: a NCHSN point forms the transition between HHS and HSN curves. Middle circle: at the intersection of a Hopf curve and a HHS curve (dotted line), a HNS orbit appears. Bottom circle: a BT point exists where a limit point curve (thin solid line) and a Hopf curve (dash-dotted line) touch. From the BT point, a HHS curve arises (which is barely visible here)

where $M_\infty = \frac{1}{2}(1 + tanh((V - V_1)/V_2))$, $N_\infty = \frac{1}{2}(1 + tanh((V - V_3)/V_4))$ and $\tau_N = \phi \; cosh((V - V_3)/2V_4)$.

In our tests, $C = 5$, $g_L = 2$, $V_L = -60$, $g_{Ca} = 4$, $V_{Ca} = 120$, $g_K = 8$, $V_K = -80$, $\phi = \frac{1}{15}$, $V_1 = -1.2$, $V_2 = 18$ and $V_4 = 17.4$ are fixed. I and V_3 are varied. In the biologically most relevant parameter-range for I and V_3, a lot of curves and bifurcations are really close together, among others curves of HHS and HSN orbits. In Figure 1 a bifurcation diagram is shown, and more detailed views in Figure 2. These diagrams were computed completely using CL_MATCONT with our modifications and extensions. A full bifurcation diagram and a more thorough study of the model can be found in [18].

Acknowledgement. Bart Sautois thanks the Fund for Scientific Research Flanders - FWO for funding the research reported in this paper. Mark Friedman was supported in part under NSF DMS-0209536.

References

1. Kuznetsov, Yu.A.: Elements of Applied Bifurcation Theory. 3rd edn. Springer-Verlag, New York, NY (2004).
2. Dhooge, A., Govaerts, W., Kuznetsov, Yu.A., Mestrom, W., Riet, A.M.: A Continuation Toolbox in MATLAB. Manual (2003), http://allserv.UGent.be/~ajdhooge/doc_cl_matcont.zip.
3. Dhooge, A., Govaerts, W., Kuznetsov, Yu.A.: MATCONT: A MATLAB package for numerical bifurcation analysis of ODEs. ACM TOMS 29,2 (2003) 141-164.

4. Doedel, E.J., Champneys, A.R., Fairgrieve, T.F., Kuznetsov, Yu.A., Sandstede, B., Wang, X.J.: AUTO97: Continuation and bifurcation software for ordinary differential equations (with HomCont), http://indy.cs.concordia.ca/auto (Concordia University, Montreal, Canada, 1998)
5. Kuznetsov, Yu.A., Levitin, V.V.: CONTENT: Integrated environment for analysis of dynamical systems, version 1.5, http://ftp.cwi.nl/CONTENT (CWI, Amsterdam, 1998).
6. Govaerts, W., Sautois, B.: Bifurcation software in Matlab with applications in neuronal modeling. Comput. Meth. Prog. Bio. 77,2 (2005) 141-153.
7. Beresticky, H., Nirenberg, L.: Traveling fronts in cylinders. Ann. Inst. Henri Poincaré 9,5 (1992) 497-572.
8. Rinzel, J., Ermentrout, B.: Analysis of neural excitability and oscillations. In Koch, C., Segev, I. (eds.): Methods in neuronal modeling: from synapses to networks. MIT Press, Cambridge, MA (1989) 135-169.
9. Gray, P., Scott, S.K.: Chemical oscillations and instabilities: non-linear chemical kinetics. Oxford University Press, Oxford, UK (1990).
10. Demmel, J.W., Dieci, L., Friedman, M.J.: Computing connecting orbits via an improved algorithm for continuing invariant subspaces. SIAM J. Sci. Comput. 22,1 (2001) 81-94.
11. Champneys, A.R., Kuznetsov, Yu.A., Sandstede, B.: A numerical toolbox for homoclinic bifurcation analysis. Int. J. Bifurcation Chaos 6,5 (1996) 867-887.
12. Champneys, A.R., Kuznetsov, Yu.A.: Numerical detection and continuation of codimension-two homoclinic bifurcations. Int. J. Bifurcation Chaos 4 (1994) 785-822.
13. Bindel, D., Demmel, J., Friedman, M.: Continuation of invariant subspaces for large bifurcation problems. In proceedings, SIAM Conference on Applied Linear Algebra, July 2003, http://www.siam.org/meetings/la03/proceedings/.
14. Doedel, E.J. and Friedman, M.J.: Numerical computation of heteroclinic orbits, J. Comp. Appl. Math. 26 (1989) 155-170.
15. Govaerts, W.: Numerical Methods for Bifurcations of Dynamical Equilibria. SIAM, Philadelphia (2000).
16. Beyn, W.-J., Champneys, A., Doedel, E., Govaerts, W., Kuznetsov, Yu.A., Sandstede, B.: Numerical continuation, and computation of normal forms. In Fiedler, B. (ed.): Handbook of Dynamical Systems, Vol.2, Ch.4. Elsevier Science (2002).
17. Morris, C., Lecar, H.: Voltage oscillations in the barnacle giant muscle fiber. Biophys. J. 35 (1981) 193-213.
18. Govaerts, W., Sautois, B.: The onset and extinction of neural spiking: a numerical bifurcation approach. To appear in J. Comput. Neurosci. (2005).

A Numerical Tool for Transmission Lines

Hervé Bolvin, André Chambarel, and Philippe Neveux

UMR A114 Climate, Soil and Environment
33 rue Louis Pasteur, F-84000 Avignon, France

Abstract. The electric line numerical study is usually made through a harmonic approach with the impedance concept based on the Finite Difference discretization in both the time and space domains. These methods present severe drawbacks when used in impulse working and space dependent parameters. We present here an efficient numerical tool for electric line simulation. The mathematical model is based on the telegrapher's equations. In the present paper, we propose a new approach based on the Finite Element Method associated with an efficient algorithm for the numerical resolution of the telegrapher's equations. In practice the electric line is connected to the entrance and the exit with linear or non linear circuits. The objective of this paper is to provide an efficient numerical tool for electric line simulation in complex configurations.

Keywords: Transmission line, finite element method, differential system, impulse voltage, stiff time problem.

1 Introduction

The electric line model shall be built in order to analyze propagation phenomena [1]. Usually the transmission lines are connected at both ends to other circuits, the supply circuit at the entrance and the utilization circuit at the exit (Fig.1). As a matter of fact, the analysis of the recorded output signal leads to the characterization of the line. Hence, the identification of line properties can be done by coupling this model with an optimization scheme [2][3].

The propagation equations are Partial Differential Equations (PDE) both in time and space. In order to solve the wave propagation problem, various approaches have been developed. Many authors use a harmonic approach sometimes associated with a numerical method for space discretization [4]. This method is generally uneasy to use particularly in impulse working with wave fronts and space dependent electric properties. Moreover it is not available in the case of non linear connected circuits or non linear electric properties. Also, the numerical approach of the problem in the time and space domain is essentially based on the FDTD method [5]. For these reasons we propose a general numerical tool for the simulation of the transmission line for a large class of configurations, based on the Finite Element Method (FEM) for space discretization, and a Modified Backward Difference Method (MBDM) [6] for time discretization. The FEM allows complex boundary conditions to be considered [7], and the MBDM is well recommended for stiff time problems.

2 Statement of the Problem

Let us consider a transmission line with entrance and exit circuits (Fig. 1).

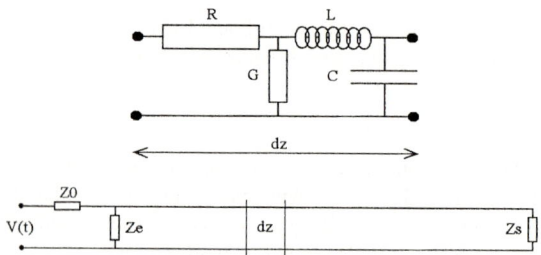

Fig. 1. The electric line pattern

We obtain the following dimensionless model :

$$\frac{\partial i'}{\partial z'} = -C^* \cdot \frac{\partial v'}{\partial t'} - G^* \cdot v'$$
$$\frac{\partial v'}{\partial z'} = -L^* \cdot \frac{\partial i'}{\partial t'} - R^* \cdot i'$$
$$\rightarrow$$
$$\mu_r \cdot \frac{\partial i}{\partial t} = -\frac{\partial v}{\partial z} - a.i$$
$$\varepsilon_r \cdot \frac{\partial v}{\partial t} = -\frac{\partial i}{\partial z} - b.v$$
(1)

We can propose a technique capable of approaching complex boundary conditions:

- At the entrance of the line we can apply a time dependent voltage, and the propagation phenomenon takes place in an open transmission line.
- We can include the electric characteristics of the voltage source with its connection at the entrance and a known impedance at the exit. This circuit possibly contains resistances, capacitances and self-inductions.

3 Finite Element Formulation

We have tested a new approach by the FEM for space discretization. Here the advantage of the FEM concerns the boundary conditions and the space-dependent electric properties. In fact we have natural boundary conditions and essential boundary conditions [8][9]. In accordance with these boundary conditions different Finite Element formulations are possible. So the corresponding boundary conditions can be expressed through Dirichlet's conditions or mixed conditions.

3.1 Weak Formulations

The Galerkin weighting applied to the system (1) gives rise to a weak formulation. In accordance with the boundary conditions of the problem under consideration, one can use one of the following formulations (2) (3) obtained via different integration-by-parts schemes ; we present examples :

$$\begin{cases} \int_0^l \delta i.\mu_r.\frac{\partial i}{\partial t}.dz = -\int_0^l \delta i.\frac{\partial v}{\partial z}.dz - \int_0^l \delta i.a.i.dz \\ \int_0^l \delta v.\varepsilon_r.\frac{\partial v}{\partial t}.dz = -\int_0^l \delta v.\frac{\partial i}{\partial z}.dz - \int_0^l \delta v.b.v.dz \end{cases} \quad (2)$$

$$\begin{cases} \int_0^l \delta i.\mu_r.\frac{\partial i}{\partial t}.dz = -[\delta i.v]_0^l + \int_0^l v.\frac{\partial(\delta i)}{\partial z}.dz - \int_0^l \delta i.a.i.dz \\ \int_0^l \delta v.\varepsilon_r.\frac{\partial v}{\partial t}.dz = -\int_0^l \delta v.\frac{\partial i}{\partial z}.dz - \int_0^l \delta v.b.v.dz \end{cases} \quad (3)$$

These formulations are mathematically equivalent. We can choose the one that best fits the form of the model of boundary conditions.

3.2 Matricial Formulations

For the implementation of the different formulations (2) (3) for users, we must be aware of the details of matricial formulation. For this reason we have to provide the significant example below. Line length is discretized into n_e elements (Ω_e) with $\bigcup_e (\Omega_e) = (\Omega)$. We have used Lagrange's linear polynomial base $<n(z)>$ [7]. We have choosen the first formulation (2) to illustrate the matricial form. The geometrical and analytical discretizations can be written as follows :

$$\sum_{ne} \left(\langle \widetilde{\delta i}_e, \delta v_e \rangle . \int_{(\Omega_e)} \begin{bmatrix} \{n\}\mu_r.\langle n \rangle & 0 \\ 0 & \{n\}\varepsilon_r.\langle n \rangle \end{bmatrix} .dz. \begin{Bmatrix} \frac{\partial i_e}{\partial t} \\ \frac{\partial v_e}{\partial t} \end{Bmatrix} \right) =$$

$$\sum_{ne} \left(\langle \widetilde{\delta i}_e, \delta v_e \rangle . \int_{(\Omega_e)} \begin{bmatrix} -\{n\}.a.\langle n \rangle & -\{n\}.\langle \frac{\partial n}{\partial z} \rangle \\ -\{n\}.\langle \frac{\partial n}{\partial z} \rangle & -\{n\}.b.\langle n \rangle \end{bmatrix} .dz. \begin{Bmatrix} i_e \\ v_e \end{Bmatrix} \right) \quad (4)$$

The general matricial formulation can be summarized under the form:

$$\sum_{ne} \left(\langle \delta u_e \rangle.[m_e]. \left\{ \frac{\partial u_e}{\partial t} \right\} \right) = \sum_{ne} \langle \delta u_e \rangle.(\{f_e\} - [k_e]\{u_e\}) \quad (5)$$

where index e denotes the element number and : $<u_e> = <i_e, v_e>$
After the assemblage process we obtain the following differential system :

$$[M].\frac{d}{dt}\{U\} = \{F\} - [K].\{U\} \quad \text{where} \quad \{U\} = \{i^{(G)}, v^{(G)}\} \quad (6)$$

Index G denotes the global space.

4 Numerical Method

4.1 The Modified Backward Difference Method Technique (MBDM)

For all weak formulations, the general formulation of the differential system is [8]:

$$\frac{d}{dt}\{U\} = \{\Phi(U,t)\} \quad \text{where} \quad \{\Phi(U,t)\} = [M]^{-1}.(\{F\} - [K].\{U\}) \tag{7}$$

For the numerical time-resolution, the MBDM is used. With the proposed formulation one can choose the time order of discretization and the upward scheme [6][10]. In this context, it is possible to apply a *matrix-free technique*. As a matter of fact the mass matrix and the stiffness matrix are never built during the calculation. As a consequence, we obtain a high performance level in terms of CPU time and storage cost. This technique can be summarized in the following algorithm:

$t_n = 0$
while $(t_n \leq t_{max})$

$$\left\{ \begin{array}{c} \{\Delta U_n^i\} = \Delta t_n . \sum_{j=0}^{k-1} \lambda_j . [M_{n-j}^i]^{-1} . \{\Psi_{n-j}(U_{n-j} + \alpha_j . \Delta U_{n-j}^{i-1}, t_n + \alpha_j . \Delta t_n)\} \\ i = 1, 2, \ldots \quad \text{until} \quad \|\Delta U_n^i - \Delta U_n^{i-1}\| \leq tolerence \end{array} \right\} \tag{8}$$

$\{U_{n+1}\} = \{U_n\} + \{\Delta U_n\}$
$t_{n+1} = t_n + \Delta t_n$
end while

4.2 The Software

Fig. 2 shows the general structure of the compact code. It is organized in three classes corresponding to the functional blocks of the FEM's different stages [11].

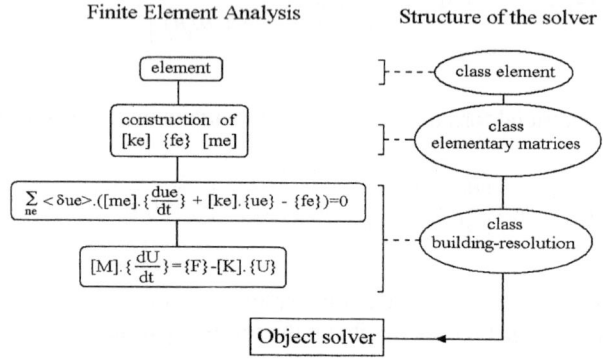

Fig. 2. Object structure of a standard solver

5 Numerical Results

5.1 The Benchmark

Let us consider a transmission line with an open circuit at the exit. So we have Dirichlet conditions at both ends of the line. In this case, weak formulation (2) before the discretization of the mathematical model is the most convenient one.

In order to test the proposed method, a very unfavorable case is considered : line without loss, small space-discretization (1000 elements), first order time-discretization and time discontinuity as initial condition (step of voltage).

The main difficulty of the numerical computation of the wave front propagation is that numerical instabilities may occur. The MBDM technique provides an upward time-parameter α. The numerical instability control has been calculated for different values of parameter α : thus we can determine an optimal value for α.

Fig. 3 shows the time dependent intensity signal at the entrance of the line and Fig. 4, the space dependent intensity signal at $t = 0.5$.

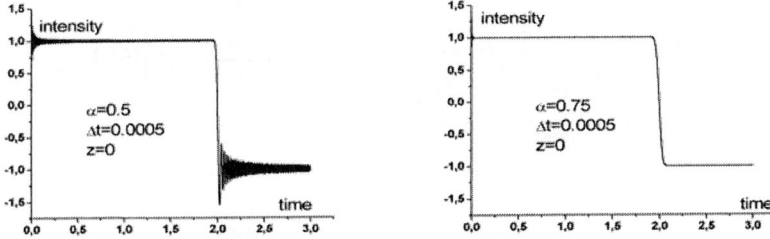

Fig. 3. Time dependent intensity at the entrance of the line for two values of α

One can notice from Fig. 3 and Fig. 4, that if the α-value is lower than 0.5, no stable numerical solution can be obtained with a realistic time step. By using an optimal value of α, we obtain the well-known signal with a rectangular form (Fig.5).

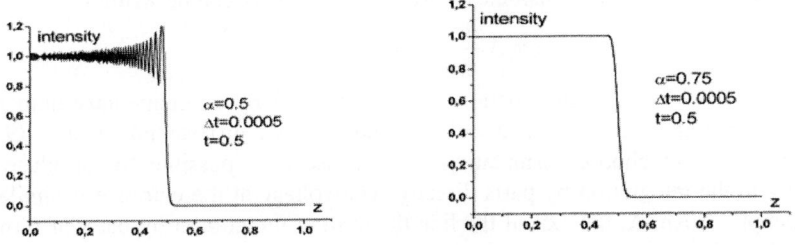

Fig. 4. Space dependent intensity at $t =0.5$ for different values of α

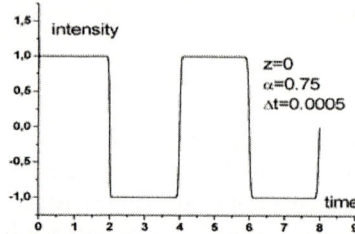

Fig. 5. Time dependent intensity at the entrance ($z=0$) of the line

5.2 Example of a Non Linear Circuit

Let us examine the case of a non linear circuit (a transistor, for example) connected at the end of the line. Fig. 6 presents a bipolar transistor.

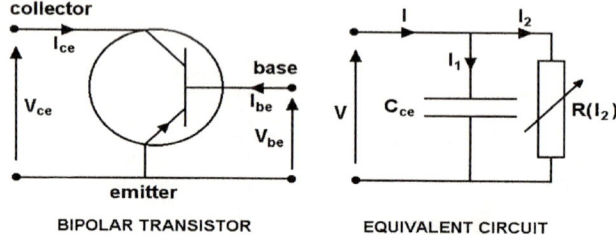

Fig. 6. An example of non linear circuit

We will not give all working details about this classical electronic component. In our context the peculiarity of the problem is the non-linearity of the static current-voltage characteristic of the collector-emitter.

For a given base-emitter voltage, the bipolar transistor can be equivalent to the circuit presented in Fig. 6. At a high frequency the collector-emitter capacitance C_{ce} is not negligible and the static part of the transistor's characteristic is represented by a variable resistance. The dimensionless form of the intensity can be written:

$$i = A.\frac{\partial v}{\partial t} + \varphi(v) \tag{9}$$

The extraction of $\varphi(v)$ in equation (9) is performed by an elementary numerical method. In impulse working this circuit is modelized by a differential equation (9). In that case we also choose formulation (3) because it is possible to introduce the intensity in the integration-by-parts directly. The voltage at the entrance is similar to the example above. At the exit of the line the term of integration-by-parts of formula (3) can be written :

$$[\delta v . i]_0^1 = \langle \delta v_n \rangle . \{n(z)\} \left(A.\langle n(z) \rangle . \left\{ \frac{\partial v_n}{\partial t} \right\} + \varphi(v_n) \right)_{z=1} \tag{10}$$

This matricial formulation is modified because we do not have here a modification of the elementary matrix $[k_e]$ but a modification of elementary electric loading $\{f_e\}$. Function $\varphi(v)$ must be built at each iteration of algorithm (8). Under these conditions and for the last element ($z =1$) the elementary matrices concerned are modified as follows :

$$\begin{Bmatrix} f_i \\ f_v \end{Bmatrix}_{new} = \begin{Bmatrix} f_i \\ f_v \end{Bmatrix}_{old} + \begin{Bmatrix} 0 \\ \{n\}.\varphi(v_n) \end{Bmatrix}$$

$$\begin{bmatrix} m_{ii} & 0 \\ 0 & m_{vv} \end{bmatrix}_{new} = \begin{bmatrix} m_{ii} & 0 \\ 0 & m_{vv} \end{bmatrix}_{old} + \begin{bmatrix} 0 & 0 \\ 0 & \{n\}.A.\langle n \rangle \end{bmatrix} \quad (11)$$

Algorithm (8) is performed with the same numerical characteristics.

In Fig. 7, we notice a transient complex intensity signal at the entrance of the line. If the dimensionless time is lower than 2 a step of intensity propagates along the line, reflects at the end and comes back. After this we can see the influence of the exit circuit.

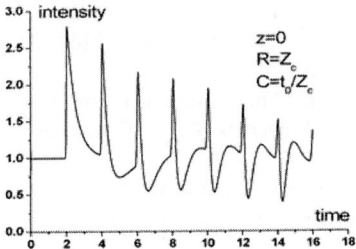

Fig. 7. Time dependent intensity at the entrance of the line in the case of a non-linear circuit connected at the exit

6 Conclusion

For the simulation of the transmission lines, we wish to place at the disposal of electronics specialists, a numerical tool which is effective as much by its simplicity as by its performances. The main applicability of this technique concerns complex cases of line models, these potentialities being associated to a simple data-processing tool that can be used by non specialists. By its various formulations, the method makes it possible to approach the whole of the problems quoted above. Moreover, in this context, we have developed a method of numerical resolution specific to the problems of transmission lines. We have sought significant examples which illustrate the principle of the method as well as concrete application cases.

Acknowledgement. The authors would like to thank Ralph Beisson for his assistance with the English composition of this paper.

References

1. Magnusson P.G., Alexander G.C. and Tripathi V.K.: Transmission Lines and Waves Propagation. 3 rd Edition CRC, Boca Raton
2. Norgren M. and He S.: An Optimization Approach to the Frequency Domain Inverse Problem for a Non-Uniform LCRG Transmission Line. IEEE Transactions on Microwave Theory and Techniques, Vol. 44 (8) (1996) 1503-1507
3. Lundstedt J. and He S.: A Time-Domain Optimization Technique for the Simultaneous Reconstruction of the Characteristic Impedance, Resistance and Conductance of a Transmission Line. Journal of Electromagnetic Waves and Applications, Vol.10 (4) (1996) 581-602
4. Heimovara T.J.: Frequency Domain Analysis of Time Domain Reflectometry Wave Form, Measurement of the Complex Dielectric Permittivity of Soils. Water Resource Research, Vol. 30 (1994) 189-199
5. Hu J.-L., Chan C.H.and Sarkar T. K.: Optimal Simultaneous Interpolation / Extrapolation Algorithm of Electromagnetic Responses in Time and Frequency Domains. IEEE Transactions on Microwave Theory and Techniques, Vol. 49 (10) (2001) 1725-1732
6. Anderson D.A., Tannehill J.C. and Pletcher R.H.: Computational Fluid Mechanics and Heat Transfer. Hemisphere Publishing Corporation Editor,1984.
7. Dhatt G. and Touzot G.: Une Présentation de la Méthode des Eléments Finis (in French). Editions Maloine S.A., Paris (1981)
8. Chambarel A. and Ferry E.: Finite Element Formulation for Maxwell's Equations with Space Dependent Electric Properties. Revue Européenne des Eléments Finis, Vol. 9 (8) (2000) 941-967
9. Assous F., Degond P., Heintze E., Raviard P.A.. and Segre J.: On a Finite Element Method for Solving the Three-Dimensional Maxwell Equations. J. Comp. Phys., Vol.109 (1993) 222-237
10. Sod G.A.: A survey of Several Finite Difference Methods for Systems of Nonlinear Hyperbolic Conservation Laws. J. Comp. Phys., Vol. 27 (1978) 1-31
11. Chambarel A. and Fougère D.. A General Parallel Computing Approach using the Finite Element Method and the objects oriented programming by selected data technique, Lecture Notes in Computer Science, Vol. 2127 (2001) 428-435

Nomenclature

R^*	lineic resistance	$[m_e]$	elementary mass matrix
L^*	lineic inductance	$[k_e]$	elementary electric matrix
C^*	lineic capacitance	$\{f_e\}$	elementary electric loading
ε_r	relative electric permittivity	α	upward time-parameter
μ_r	relative magnetic permeability	$\langle . \rangle$	line matrix
(Ω)	integration domain	$\{.\}$	column matrix
(Ω_e)	element e		

The COOLFluiD Framework: Design Solutions for High Performance Object Oriented Scientific Computing Software

Andrea Lani[1], Tiago Quintino[1,2], Dries Kimpe[2,3], Herman Deconinck[1], Stefan Vandewalle[2], and Stefaan Poedts[3]

[1] Von Karman Institute, Aerospace Dept.,
Chaussee de Waterloo 72,
B-1640 Sint-Genesius-Rode, Belgium
[2] Catholic University Leuven, Computer Science Dept.,
Celestijnenlaan 200A, B-3001 Leuven, Belgium
[3] Catholic University Leuven, Center for Plasma-Astrophysics,
Celestijnenlaan 200B, B-3001 Leuven, Belgium

Abstract. The numerical simulation of complex physical phenomena is a challenging endeavor. Software packages developed for such purpose should combine high performance and extreme flexibility, in order to allow an easy integration of new algorithms, models and functionalities, without penalizing run-time efficiency. COOLFluiD is an object-oriented framework for multi-physics simulations using multiple numerical methods on unstructured grids, aiming at satisfying these needs. To this end, specific design patterns and advanced techniques, combining static and dynamic polymorphism, have been employed to attain modularity and efficiency. Some of the main design and implementation solutions adopted in COOLFluiD are presented in this paper, in particular the Perspective and the Method-Command Patterns, used to implement respectively the physical models and the numerical modules.

1 The COOLFluiD Architecture

COOLFluiD (**C**omputational **O**bject-**O**riented **L**ibrary for **Flui**d **D**ynamics) is a multi-physics and multi-methods platform that combines flexibility and high performance for the simulation of complex fluid dynamical phenomena on unstructured grids. The package is implemented in C++, which, during the last decade, has shown a great potential for scientific applications, offering significant support to develop both flexible and efficient code: *Cogito*, *ELEMD* ([Arge97]), *MOUSE*, *Deal* ([OONum]) are only few examples of available C++ platforms.

An overview of the COOLFluiD framework is sketched in Fig. 1. It consists of a kernel, where *Simulation*, the simulation manager object, and *MeshData*, the basic data-structure object are implemented. Also the abstract interfaces for all the polymorphic objects are defined in the kernel, in particular the ones for the physics description (*PhysicalModel*) and for the numerical

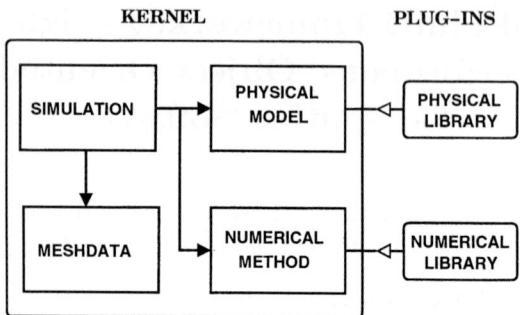

Fig. 1. Simplified overview of the COOLFluiD framework

methods (e.g., *MeshCreator*, *SpaceMethod*, *ConvergenceMethod*, *LinearSystemSolver*). Each concrete numerical method or physical model is enclosed in a separate *plug-in* library. This *plug-in* policy, which provides COOLFluiD with significant modularity and extensibility, relies heavily on two complementary techniques, namely *self-registration* and *self-configuration* of objects, whose basic principles are explained below.

1.1 Self Registration of Objects

The *self registration of objects*, pioneered by [Bev98] in a C++ context, automatizes the creation of polymorphic objects and reduces implementation and compilation dependencies. This is of great help in easing the integrability of new components in the framework, since they can be compiled as external *plug-ins* and loaded dynamically, on demand, into the main core application. A generic polymorphic concrete object (*ConcreteObj*) can be registered by simply instantiating the corresponding *ObjectProvider* in the implementation file:

```
ObjectProvider<BaseObj, ConcreteObj> myProvider("objName");
```

and it can be created by calling the corresponding *Factory*:

```
BaseObj* objPtr = Factory<BaseObj>::getProvider("objName")->create();
```

1.2 Self Configuration of Objects

In COOLFluiD, objects can be self-configurable, meaning that they can create and set their own data. The template configuration function was inspired by the *Yagol* library [Yagol]. An object is made self configurable, by deriving it from a parent class *ConfigObject* and by adding a call to

```
addConfigOption("OptionKey", &configData);
```

in its constructor for each configurable data member *configData*. In particular, *OptionKey* is the configuration key string, used to map the value of *configData*. We consider, for instance, what would appear in a configuration file for the *RK* (Runge-Kutta) time stepper object:

```
ConvergenceMethod = RK
RK.coeff = 0.28 0.61 1.0
```

RK is the self-configuration value for the *ConvergenceMethod*, which is the configuration key for the homonymous polymorphic object. *RK* is also the self-registration key for the Runge-Kutta class, that will be then instantiated and will configure itself with the three given coefficients, whose configuration key is *coeff*.

1.3 Parallel Data-Structure

COOLFluiD uses a parallel layer designed to minimize impact on both users and software developers by exporting a high level platform independent interface. Parallelization is fully transparent and high performance is assured by techniques like parallel IO and remote memory access, when supported by the underlying platform. The intrinsically parallel data-structure is encapsulated in a *Facade* object [Gamma95], *MeshData*, whose main component is *DataStorage*. The latter offers a simple interface to create and handle generic typed arrays of data to be shared among different numerical *Methods* and *Commands*, allowing to treat uniformly both local and distributed data. When running in parallel, all MPI calls are encapsulated in an underlying dynamically growing parallel array, hidden to the clients of *DataStorage*. The following examples show how to create and get local (global) data:

```
getDataStorage()->createData<StorageType>("storageName",
                                          storageSize);
DataHandle<LOCAL, StorageType> myStorage =
  getDataStorage()->getData<StorageType>("storageName");
```

2 Physical Model: Perspective Pattern

The framework is designed to apply different numerical methods for the solution of systems of *Partial Differential Equations*, which typically appear in the form:

$$\frac{\partial \mathbf{U}}{\partial t} + \nabla \cdot \mathbf{F} = \mathbf{S} \qquad (1)$$

where **U** (unknown variables), **F** (convective and viscous fluxes), **S** (source term) depend on the chosen physical model. Such a physical model can be seen as a composition of entities, e.g., coefficients, quantities and thermodynamic properties. Note that one can look at the same physics through different formulations of the equations, involving the use of different sets of variables, transformations, or other adaptations tailored, e.g., towards a particular numerical method.

This logical picture can be translated into an implementation offering multiple views or *perspectives* for the same physics, according to the specific needs. The first advantage is that this design approach is able to break a hypothetical heavy and hard-to-define interface for the physical model object into a limited

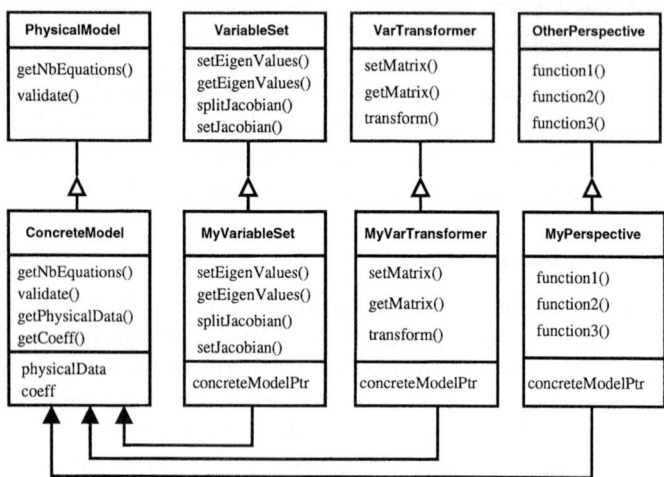

Fig. 2. Perspective pattern applied to a Physical Model module

granularity of independent abstractions. Flexibility and maintainability are positively affected too. If new base *Perspectives* objects are needed in order to provide new points of view for the same physics, they can easily be attached without needing to modify the existing code.

The OMT (Object Modeling Technique) class diagram of the pattern is shown in Fig. 2. It reflects the composition-based *Adapter* described in [Gamma95], but it has a single shared *Adaptee* object (*ConcretePhysicalModel*), and multiple abstract *Targets* (called *Perspectives* here), each one with a number of derived classes (*Adapters*). The *Perspective* pattern is conceptually opposite to the *View Handler* presented in [Busch00], since the former helps to tackle a situation in which the different views (and their interfaces) are not all foreseeable a priori, meaning that it would be impossible to define both an handler object and a single abstract view interface, as required by the latter.

Data and functionalities that are typical of a certain physics, but invariant to all its possible *Perspectives*, should be implemented in the actual *ConcretePhysicalModel* object. The base *PhysicalModel* defines a very general abstract interface. The concrete one implements the virtual methods of the parent class and defines another interface to which the *ConcretePerspective* objects (and only those) are statically bound. As a result, a client makes use of the physical model through an abstract layer, enlargeable if required, given by a number of perspective objects (*VariableSets*, *VariableTransformers*, etc.) while all their collaborations with the concrete physics are completely hidden. As all the other polymorphic objects in COOLFluiD, also *PhysicalModels* and *Perspectives* are self-registrable and, if needed, self-configurable.

3 Method-Command Pattern

Every numerical module is implemented following a common design structure, to which we will refer as the *Method-Command* pattern. It consists of a concrete *Method* object delegating tasks to a number of *Commands* that share a tuple of multiple receivers. A configurable *BaseMethod* object (e.g., *SpaceMethod*, *ConvergenceMethod*, *MeshCreator*) defines the abstract interface for a specific type of Method (Fig. 3). Each *ConcreteMethod* implements the virtual functions of the corresponding parent *BaseMethod* by encapsulating requests for specific actions (setup, unsetup, compute something ...) in ad-hoc *Commands*. *ConcreteMethod* functions can act as *Template Methods* [Gamma95] where the hooks are not virtual functions but polymorphic commands. Each *Command* behaves therefore as a *Strategy* object (see [Gamma95]) in performing a specific task, which can be accomplished in several different ways. All the *Commands* share some common data enclosed in *ConcreteMethodData*, a configurable tuple typically aggregating the multiple receivers. The latter are polymorphic *Strategy* or *Perspective* objects, providing, e.g., the dynamic binding to the physics.

The flexibility yielded by this structural pattern is considerable and does not affect performance, since the fast-path code is wrapped inside the concrete *Commands* or their receivers. The *Method-Command* pattern can be viewed as a sophisticated variant of the *Whole-Part* pattern described in [Busch00], or as a three-layer *Strategy*, where *Methods*, *Commands* and command receivers are completely interchangeable, self registering and self configurable.

Some applications of the pattern within the COOLFluiD framework will now be presented, in order to show the high reusability provided by this approach, but also its suitability to deal with complex numerical problems.

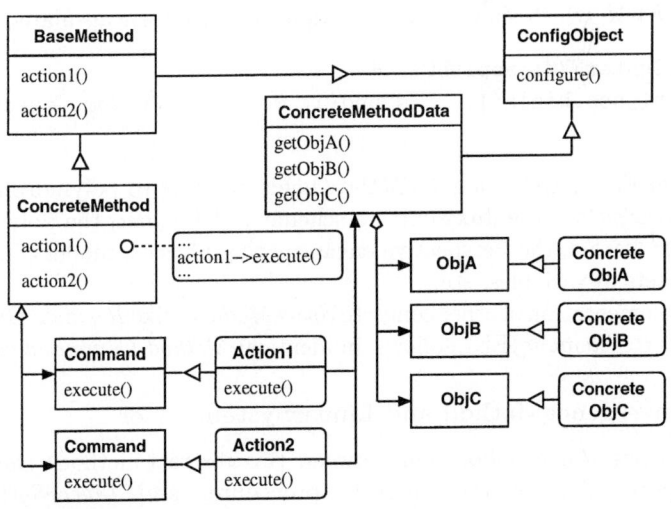

Fig. 3. OMT diagram of a Method-Command pattern

3.1 Space Method

A *SpaceMethod* takes care of the spatial discretization of the given set of partial differential equations, according to a specified numerical scheme on a given mesh. The abstract interface of a *SpaceMethod* is as follows:

```
class SpaceMethod : public ConfigObject {
public:
  // constructor, destructor, accessors, mutators ...
  virtual setup() = 0;     // setup data
  virtual unsetup() = 0;   // unsetup data
  virtual void computeRHS()= 0; // compute residual and jacobian terms
  virtual void applyBC() = 0; // apply boundary conditions
};
```

A possible concrete class is the so-called cell centered *Finite Volume* (FV) method:

```
class CellCenterFVM : public SpaceMethod {
public:
    typedef Command<CellCenterFVMData> FVMCom;
    // overridden parent virtual functions ...
private:
  SharedPtr<CellCenterFVMData> _data; // shared data
  std::auto_ptr<FVMCom> _setup;    // setup command
  std::auto_ptr<FVMCom> _unSetup;  // unsetup command
  std::auto_ptr<FVMCom> _compRHS;  // compute residual command
  std::vector<FVMCom*>  _bcs;      // boundary conditions
};
```

As an example, we present the implementation of the method *applyBC()* in *CellCenterFVM*, which shows how all actions are nicely encapsulated:

```
void CellCenterFVM::applyBC() {
  for_each(_bcs.begin(), _bcs.end(), mem_fun(&FVMCom::execute));
}
```

As shown in Fig. 4, *CellCenterFVMData* holds pointers to polymorphic Strategies like *FluxSplitter*, the flux splitting scheme and *PolyRec*, the polynomial reconstructor; *VarSet*, the Perspective encapsulating physical model traits related to specific sets of variables, etc.

The implementation of other concrete *SpaceMethods*, like *Residual Distribution* (RD) or *Finite Element* (FE), follows an identical *Method-Command* pattern.

3.2 ConvergenceMethod and LinearSystemSolver

Figure 5 shows the collaboration between two abstract methods: *Convergence Method*, responsible of the iterative procedure, and *LinearSystemSolver*. In this case, an implicit convergence method, *BackwardEuler*, delegates polymorphically the solution of the resulting linear system to *PetscLSS*, which interfaces

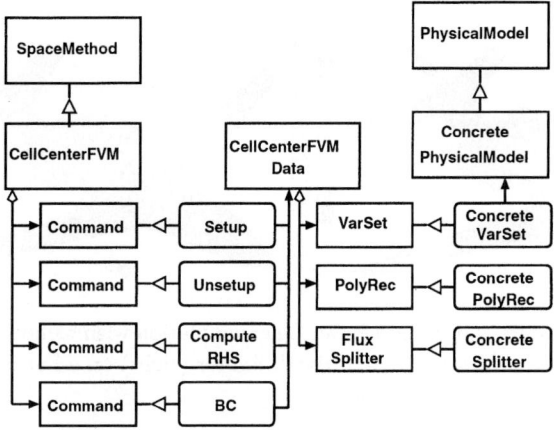

Fig. 4. Finite Volume module

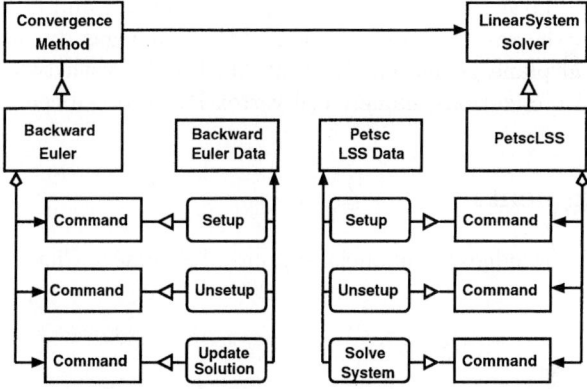

Fig. 5. Backward Euler and Petsc Linear System Solver modules

the *PETSc* library ([Petsc]). *BackwardEuler* makes use of commands for the setup, unsetup and solution update. Also *PetscLSS* delegates tasks to specific commands sharing some data (*PetscLSSData*), such as references to the (parallel) Petsc matrix and the (parallel) Petsc vectors involved in the solution of the linear system.

4 Conclusions

The flexible and reusable design solutions presented in this paper have allowed an easy integration of several components in COOLFluiD: explicit (Runge-Kutta) and implicit (Newton, Crank-Nicholson) time stepping, different spatial discretizations (FV, RD, Space-Time RD, FE), different physical models (Euler, compressible Navier-Stokes, Magneto Hydro Dynamics), a parallel flexible data-

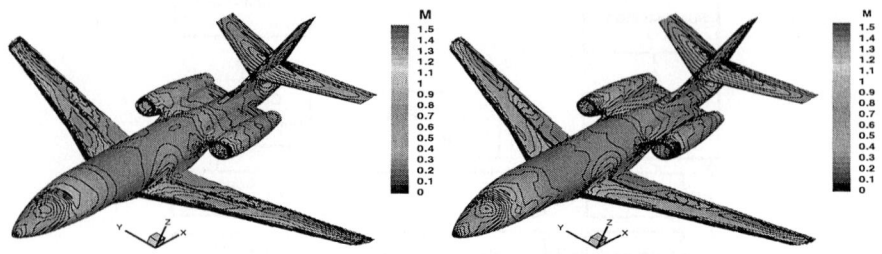

Fig. 6. Mach contours and isolines for an Euler simulation with blended LDA/N scheme (left) and 2nd order Roe scheme with Barth limiter (right) on a Falcon airplane (mesh courtesly provided by Dassault Aviation)

structure supporting the use of hybrid meshes, etc. The implementation of many other functionalities is underway: Aero-Thermo-Chemical models, incompressible Multi-Phase flows, error estimation, mesh movement and adaptation.

Fig. 6 shows the result of a simulation of the Euler gas dynamics equations over a Falcon airplane geometry, flying at Mach 0.85, with two different high order spatial discretizations, namely cell vertex RD and cell centered FV.

Acknowledgments

T. Quintino acknowledges the financial support of Fundaćao Ciencia e Tecnologia.

References

[Bev98] Beveridge, J.: Self-Registering Objects in C++. Dr. Dobbś Journal, 8/1998.
[Gamma95] Gamma, E., Helm, R. Johnson, R., Vlissides, J.: Design Patterns. Elements of Reusable Object-Oriented Software. Addison Wesley, 1995.
[Busch00] Buschmann, F., Meunier, R. Rohnert, H., Sommerlad, P., Stal, M.: Pattern-Oriented Software architecture. A System of Patterns. Wiley, 2000.
[Arge97] Arge, E., Bruaset, A. M. , Langtangen, H. P. eds.: Modern Software Tools for Scientific Computing, Birkhäuser, 1997.
[OONum] Open Systems Laboratory: The Object-Oriented Numerics Page, http://www.oonumerics.org/oon/, 2005.
[Yagol] Pace, J.: Another Getopt Library, http://yagol.sourceforge.net, 2003.
[Petsc] Argonne National Laboratory: PETSc. Portable, Extensible Toolkit for Scientific Computation, http://www-unix.mcs.anl.gov/petsc, 2004.

A Problem Solving Environment for Image-Based Computational Hemodynamics

Lilit Abrahamyan[1,*], Jorrit A. Schaap[2], Alfons G. Hoekstra[1], Denis Shamonin[1], Frieke M.A. Box[2], Rob J. van der Geest[2], Johan H.C. Reiber[2], and Peter M.A. Sloot[1]

[1] Section Computational Science, Laboratory for Computing,
System Architecture and Programming, Faculty of Science,
University of Amsterdam Kruislaan 403,
1098 SJ Amsterdam, The Netherlands
{labraham, alfons, dshamoni, peter}@science.uva.nl
http://www.science.uva.nl/research/scs/

[2] Division of Image Processing, Department of Radiology,
Leiden University Medical Center Albinusdreef 2, 2333 ZA Leiden,
PO Box 9600, 2300 RC Leiden, The Netherlands
{J.A.Schaap, J.H.C.Reiber, R.J.van_der_Geest, F.M.A.Box}@lumc.nl
http://www.lkeb.nl

Abstract. We introduce a complete problem solving environment designed for pulsatile flows in 3D complex geometries, especially arteries. Three-dimensional images from arteries, obtained from e.g. Magnetic Resonance Imaging, are segmented to obtain a geometrical description of the arteries of interest. This segmented artery is prepared for blood flow simulations in a 3D editing tool, allowing to define in- and outlets, to filter and crop part of the artery, to add certain structures (e.g. a by-pass, or stents), and to generate computational meshes as input to the blood flow simulators. Using dedicated fluid flow solvers the time dependent blood flow in the artery during one systole is computed. The resulting flow, pressure and shear stress fields are then analyzed using a number of visualization techniques. The whole environment can be operated from a desktop virtual reality system, and is embedded in a Grid computing environment.

Keywords: Problem Solving Environment, Computational Hemodynamics, blood flow modeling.

1 Introduction

"A problem solving environment (PSE) is a computer system that provides all the computational facilities necessary to solve a target class of problems" [1, 2]. The target class of problems that we chose in our study is associated with cardiovascular diseases, a predominant cause of death [3, 4]. In particular our attention

---

[*] Corresponding author.

is concentrated on vascular disorders caused by atherosclerosis. The goal of our PSE, which we call HemoSolve, is to provide a fully integrated environment for simulation of blood flow in patient specific arteries.

Because of the complex structure of the human vascular system it is not always obvious for surgeons how to solve the problem of bypass and/or stent placement on the deformed part, or to decide on specific treatment alternatives. Having a completely integrated computational hemodynamics environment like HemoSolve can serve as a pre-operational planing tool for surgeons, but also as a useful experimental system for medical students to enlarge their practical skills [5, 6]. It also serves as a environment for biomedical engineers that study e.g. new stent designs.

Moreover, our HemoSolve is merged with Grid technology, thus offering a unified access to different and distant computational and instrumental resources [6]. This is one of the desirable abilities of PSEs in general [2].

We first describe our system for image-based computational hemodynamics, then we provide examples of using it in the case of the abdominal aorta bifurcation and an abdominal aorta aneurysm, and finally the last section represents discussions and conclusions.

2 The Problem Solving Environment - HemoSolve

In ref. [5] Steinman argues that a need exists for robust and user-friendly techniques that can help an operator turn a set of medical images into computational fluid dynamics (CFD) input file in a matter of minutes. HemoSolve not only has this ability but also is a tool which allows to simulate pulsatile (systolic) flows in arteries.

The whole system consists of the following components (See Fig.1):

1. Medical data segmentation to obtain arteries of interest;
2. 3D editing and mesh generation, to prepare for the flow simulation;
3. Flow simulation, computing of blood flow during systole;
4. Analyses of the flow, pressure, and stress fields.

2.1 Medical Data Segmentation

The goal of the segmentation process is to automatically find the lumen border between the blood and non-blood, i.e. the vessel wall, thrombus or calcified plaque. The algorithm consists of three stages: In the first stage a wave front propagation algorithm is used to find an approximation of the centerline of the vessel. In the second stage the volumetric data is resampled into a stack of 2D slices orthogonal to the centerline. Then in each slice a contour delineating the lumen border is detected. Finally in the third stage, the stack of 2D contours is combined to form a 3D surface model, which will serve as input for the 3D editing tool.

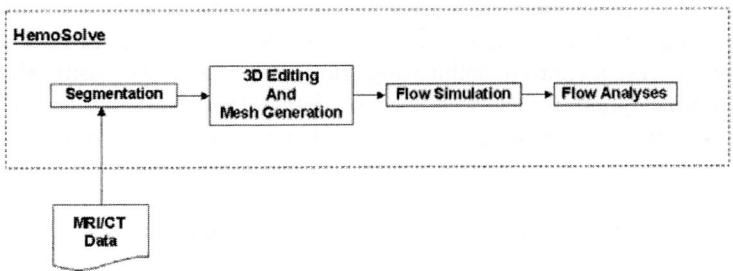

Fig. 1. Functional design of HemoSolve

3D Centerline Extraction: In order to get a first estimate of the centerline of the vessel we use the WaveProp method as described in [7] and [8]. WaveProp is based on the Fast Marching Level Set algorithm introduced by Sethian [9]. The principle is that an image is treated as an inhomogeneous medium through which a wave can propagate. The speed of the wave front is determined by a speed function which maps intensity values to speed values. The propagating wavefront creates a monotonous ascending function with its single minimum at the startpoint; with a steepest decent the shortest path from end point to startpoint is found.

In the 3D computer tomography angiography (CTA) datasets the user indicates with two points the beginning (proximal point) and end (distal point) of the vessel segment of interest. Then, a wave front is initiated at the proximal point and propagated through the vessel until it reaches the distal point marking all visited voxels as lumen (blood pool). This gives us a binary volume of the solid lumen. A distance image [10, 11] is then calculated from the binary lumen volume containing for each voxel its distance to the background. Consequently, there will be a 3D ridge of high values in the distance image, which coincides with the exact path of the centerline of the solid blood pool. This ridge can be tracked as described in [8] using wave front Propagation and Backtracking, resulting in the centerline (Fig. 2).

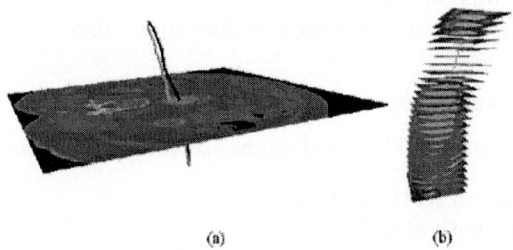

Fig. 2. (a) - is the pathline and the centerline crossing an axial slice, (b) - is a stack of slices resampled along the centerline

Per Slice Contour Detection: Perpendicular to the detected centerline 2D slices are extracted from the original CTA dataset (Fig. 2). In each of these transversal slices, a contour delineating the lumen is detected using the Wave-Prop algorithm similar as it was used to detect the centerline, but now with different speed functions and cost images. Contours delineating objects in an image usually follow the edge of that object. With the proper speed function, a wave propagating through an edge image will flow fast along the object edges and stall in solid parts of the object and in the background. The backtrack method can be used to obtain a contour from one point on the edge to another point on the edge, following that particular edge.

3D Surface Model: The stack of 2D transversal contours can be used to obtain per-slice information, such as diameter, circumference and area of the 2D contour. For the flow simulations however, a 3D surface model is needed. Therefore, the stack of 2D contours is converted into a 3D surface model by connecting each of the contour points of one slice to the closest contour point in the next slice (Fig. 4 (a)).

2.2   3D Editor and Mesh Generator

3D editing is the second component after the segmentation. The 3D stereoscopic image can easily be maintained in this user-friendly editing tool. Here surgeons and students can execute their experimental visualization studies on realistic arterial geometries. They can crop parts of the artery, where important factors in the study of hemodynamics exist, with the help of a clipping instrument. They can add inlet and outlet layers on the end-points of the arterial geometry and can enhance it with structures like bypasses or stents. Also this component allows them to define the geometrical features (e.g. width, length, placement positions) of these structures. Thus, the 3D editing tool allows surgeons and students to mimic the real surgical processes.

The final stage of this component is mesh generation. The prepared arterial geometry, including aneurysms, bifurcations, bypasses and stents, is converted into a computational mesh in several minutes. The mesh could be coarse or fine depending on the wish of user.

The mesh is then ready to be used in flow simulators.

2.3   Hemodynamic Solvers

Two different computational hemodynamic solvers can be used in HemoSolve:

1. Lattice Boltzmann method (LBM)
2. Finite element method (FEM)

In both solvers the flow is time-dependent and after simulation the pressure, velocity, and shear stress fields during one full systolic period are produced and can be visualized. Both solver receive the input geometry mesh from the 3D editing tool.

LBM is a mesoscopic method based on a discretized Boltzmann equation with simplified collision operator [12]. Here the flow is considered Newtonian. In the solver bounce back on links is used as boundary condition on walls and pressure difference boundary condition is applied on inlet/outlets. We have shown that LBM is capable of solving hemodynamic flows in the range of Reynolds and Womersley numbers of interest [13].

To run the simulator, except the input data file from 3D editing tool, one should define several patient-specific free parameters like Reynolds number.

FEM is a general discretization tool for partial differential equations. For blood flow model the incompressible Navier-Stokes equations are used. Input parameters are the velocity profiles of the in- and outflow of the bifurcation and a model which calculates the non-Newtonian properties of the blood in a patient specific manner [14]. The finite element package that was used in this study is called SEPRAN [15].

2.4 Flow Analyses

In order to analyze the blood flow in arteries its velocity, pressure and shear stress profiles need to be examined. Several methods exist for it and among them visualization of the flow is one of the advanced methods that helps to understand the meaning and behavior of flow better. Also visualization techniques are different and can show different features of flow. One of the visualization techniques we use in HemoSolve is based on simulated pathline visualization[16].

3 Examples

As an application example of using HemoSolve we present two case studies complex geometries representing parts of the human vascular system:

1. Aneurysm on upper part of abdominal aorta;
2. Whole abdominal aorta.

3.1 Aneurysm

We consider case of an aneurysm (ballooning out of the artery) in the upper part of the aorta. First the medical data of the upper part of patient's abdominal aorta with aneurysm is segmented by applying the segmentation algorithm (Fig.3 (a,b)). Then the segmented part which includes the aneurysm, is transfered into 3D editing tool where the user crops the structurally interesting part (Fig.3 (c)) and defines inlet and outlet layers (Fig.3 (d)). This placement is easy to control, that is to change the plane of position with simply moving the normal vector in the middle of the layer or to change the size just by movement of corner points of the rectangle. In this example there is one inlet at the top

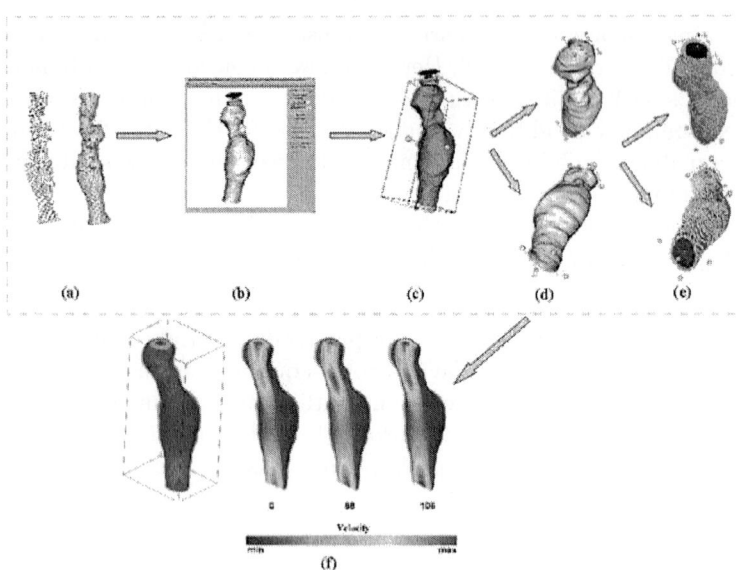

Fig. 3. Aneurysm: main stages of HemoSolve. First segmentation of the raw medical data (a). Then the segmented data (b) is first cropped (c) and inlet/outlet layers are added (d) and the mesh is generated (e). Simulation results of created mesh are presented (f)

and one outlet at the bottom layer. Finally the mesh is generated depending on constructed geometry (Fig.3 (e)). The mesh is then used as an input data for CFD solver. The presented flow is the velocity profile of blood flow simulated by LBM method. The Reynolds number applied to this blood flow is 500. The size of generated mesh is 146x73x55 lattice points and the simulation time is about 20 minutes on 16 processors. As a result three frames during the cardiac systole are captured (Fig.3 (f)).

3.2 Abdominal Aorta

Next we consider the full lower abdominal aorta down to the bifurcation. Again the medical data is segmented by applying the segmentation algorithm (Fig.4 (a)). Then the same steps as in the first example are applied (Fig.4 (b)), except the outlet layers here are six and are laing in different planes (Fig.4 (c)) and the sizes of generated mesh are bigger (Fig. 4 (d)). The ready mesh then used as an input data for CFD solver. The presented flow is the velocity profile of blood flow simulated by LBM method with the same Reynolds number 500. The size of generated mesh is 355x116x64 lattice point and the simulation time is 85 minutes on 16 processors. As a result three frames during cardiac systole are captured (Fig.4 (e)).

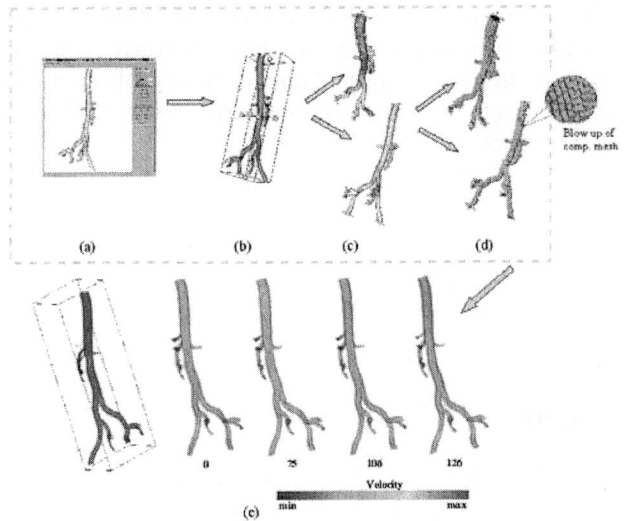

Fig. 4. Abdominal aorta: main stages of HemoSolve. Segmented data (a) is first cropped (b) and inlet/outlet layers are added (c) and the mesh is generated (d). Simulation results of created mesh are presented (e)

4 Discussions and Conclusions

The field of image based hemodynamics needs integrated PSEs, especially to enhance the preparation of computational meshes, and to allow non-specialists to enlarge their practical skills. HemoSolve consists of several stages that make it a complete system. Thus biomedical engineers,surgeons or novice surgeons can take a raw medical data from patients vascular system and after several simple steps within a quite short period can get completed flow fields (velocity, pressure, shear stress) and can analyze them with different visualization tools. One of this tools is personal space station (PSS) which realize 3D visualization and interaction [17]. Moreover with the help of 3D editing tool potential users can add bypasses or stents to vessel and examine the blood flow profile in them. The biomechanists of University of Amsterdam and Leiden University Medical Center have already used Hemosolve in their scientific research.

In order to estimate the efficiency of HemoSolve we compare its main features with the requirements of users from PSEs in general. Those characteristics are [1, 2] :

1. Simple human-computer interaction - It is enhanced by graphical user interface which is easy accessible even for inexperienced users [17].
2. Complete and accurate numerical models - Numerical models used in this PSE are LBM [13] and FEM [14, 15] which are complete and quite accurate for simulation of blood flow in human vascular system.
3. Parallel and distributed computing environment - The solvers in PSE are fully parallelized [18] .

4. Geographically distributed data - This image-based PSE is completely integrated into a Grid environment, which gives huge abilities not only to distribute the data but also to do simulations and diagnoses by grid computing [6].
5. Usefulness for university student - The potential users of this PSE are considered novice surgeons who can first practice their knowledge by doing an operation on PSE and afterward apply their experience on patients.

Thus, we conclude that HemoSolve is a well defined, easy applicable environment for image-based hemodynamics research of time-dependent blood flow in the human vascular system.

Acknowledgements

The work was funded by the Dutch National Science Foundation, NWO, Token 2000 Distributed Interactive Medical Exploratory for 3D Medical Images (DIME) project (634.000.024).

References

1. E.Gallopoulos, E.N. Houstis and J.R. Rice, *IEEE Comp. Sci. Eng. 1* **2**, 11 (1994);
2. E.N. Houstis and J.R. Rice, *Math. and Comp. in Sim.* **54**, 243 (2000);
3. Ch.A. Taylor and M.T. Draney, *Annu. Rev. Fluid. Mech.* **36**, 197 (2004);
4. World Health Organization, 2002, The World Health Report 2002, http://www.who.int/whr/en/ ;
5. D.A. Steinman, *Annals of Biom. Eng.* **30**, 483 (2002);
6. A. Tirado-Ramos, P.M.A. Sloot, A.G. Hoekstra and M. Bubak, *Parallel Computing* **30**, 1037 (2004);
7. J.P. Janssen, G. Koning, P.J.H. de Koning, J.C. Tuinenburg and J.H.C. Reiber, *Int J Cardiovasc Imaging* **18** 317 (2002);
8. P.J.H. de Koning, J.A. Schaap, J.P. Janssen, R.J. van der Geest and J.H.C. Reiber, *Magnetic Resonance in Medicine* **50** 1189 (2003);
9. J.A. Sethian, *Proc Nat Acad Sci USA* **93** 1591 (1996);
10. T. Saito and J.I. Toriwaki, *Pattern Recognition* **27** 1551 (1994);
11. O. Cuisenaire, *PhD thesis, Universit_e catholique de Louvain* (October, 1999);
12. S. Succi, *New York:Oxford* (2001);
13. A.M. Artoli, A.G. Hoekstra and P.M.A. Sloot, *Int. J. of Mod. Phys. B*, **17** 95 (2003);
14. F.M.A. Box, M.C.M. Rutten, M. A. van Buchem, J. Doornbos, R.J. van der Geest, P.J.H. de Koning, J. Schaap, F.N. van de Vosse and J.H.C. Reiber *Computational Science - ICCS 2002*, 255 (April, 2002);
15. G. Segal, *Ingenieursbureau SEPRA, Park Nabij 3, Leidschendam, the Netherlands*
16. D.A. Steinman, *Jour. of Biomech. 33* **5**, 623 (2000);
17. E.V. Zudilova and P.M.A. Sloot *Springer-Verlag Berlin Heildelberg 3345* 184 (2005).
18. D. Kandhai, A.G. Hoekstra, M. Kataja, J. Timonen and P.M.A. Sloot, *Comp. Phys. Commun.*, **111** 14 (1998);

MPL: A Multiprecision MATLAB-Like Environment

Walter Schreppers, Franky Backeljauw, and Annie Cuyt

University of Antwerp (CMI)
Department of Mathematics and Computer Science,
Middelheimlaan 1, B-2020 Antwerpen, Belgium
{walter.schreppers, franky.backeljauw,annie.cuyt}@ua.ac.be

Abstract. A number of generic tools, some developed by the authors, some developed in cooperation with other teams and others available freely, are combined into an environment, called MPL from Multi Precision Lab, which offers a cross-platform variable precision alternative to MATLAB. Among the tools we mention for instance our C/C++ precompiler for type conversion, the GMP arithmetic library complemented with our own IEEE-854 compliant multi-radix multiprecision MpIeee library, the Boost matrix library, our own MATLAB parser, the libraries FFCall and GNU Libtool. The functionality of the well-known MATLAB toolboxes is available through the multiprecision equivalent of one's library of choice, generated using the same tools. We mention, among others, GSL, Numerical Recipes, an automatic differentiation toolkit [1], a hybrid polynomial solver [2] and so on.

1 Introduction

While symbolic computing environments have the tendency to also support variable precision numeric routines besides symbolic and exact arithmetic, popular numeric programming environments such as MATLAB usually do not offer any higher precisions besides the standard hardware precisions.

Since predictions, based on the growth in the size of mathematical models solved as the memory and speed of computers increase, suggest that floating-point arithmetic with unit roundoff of the order of 10^{-32} is needed for some applications on future supercomputers, we want to investigate the possibility to offer high precision floating-point and exact rational arithmetic in a MATLAB-like environment.

Furthermore, numeric code that has passed an experimental stage, is often run in optimized and compiled form and not from within a computing environment. This feature is supported as well.

In the subsequent sections we discuss the building blocks that constitute the MPL environment. The sections 2, 3, 5, 7, 8, and 9 all concern packages developed by ourselves in the past few years.

2 High Precision Arithmetic

In our context the notion high precision varies from more than 64 digits of binary precision to infinite precision rational arithmetic. A finite precision radix β arithmetic implementation ($\beta = 2^i$ or $\beta = 10^j$) is preferably fully IEEE-854 compliant. Such an implementation offers an additional benefit compared to the symbolic computing environments which do not comply with the floating-point standard(s).

Our infinite precision C++ classes `Rational` and `BigInt` for rational and big integer arithmetic are based on the well-known GMP library [3]. For multiprecision floating-point arithmetic, a number of libraries have been developed in the past decade. We refer among others to MpIeee [4], CLN [5], FMLIB [6], MPFR [7] and MPFUN [8].

`MpIeee` is a C++ class that offers fully IEEE-854 compliant, multi-radix and variable precision floating-point aritmetic. Its implementation allows to encapsulate the data structure together with the routines to create, manipulate and destroy this structure, thus offering an easy-to-use interface to the concept it implements. Furthermore, the possibility of operator overloading allows the use of the ordinary mathematical operators (such as $+$, $-$, $/$, $\times$, $\sqrt{}$, but also sin, cos, abs, ...).

Operators have a big impact on the overall runtime performance, as their usage often involves the creation of several temporary objects. These temporaries come from the fact that the operator that is being called, does not know where the result will be stored. Instead, it creates a temporary object to store the result. To actually return the result, yet another (unnamed) temporary object is created, which is nothing more than an implicit copy of the result. It is needed because simply returning a reference to the result would most probably cause memory leakage. It is clearly necessary to try to avoid as many of these temporaries as possible, if not all.

In `MpIeee`, a technique called delayed evaluation is used to avoid the need for temporaries altogether. This technique, as its name indicates, consists in delaying the operation until it knows the existence of a target object to store the result. This is done using some interim object, called a proxy, which simply stores references to the operands and an indication of the operation to which it applies. The actual operation is then performed through a modified assignment operator which takes this interim object as its operand. This way, the assignment operator knows its target object (the left hand side of the operation) as well as which operation needs to be performed on the given operands (the right hand side of the operation). As such, the full computation can be executed without the need for any temporary objects.

This technique can be completely implemented using inline functions, which are almost always resolved at compile time when optimization settings apply. Hence it is the fastest approach we can aim for. In the sequel `MpIeee` is therefore our preferred multiprecision package. Besides being multi-radix and fast, it is also the only one in its class offering full IEEE-854 compliance.

3 Precompiler for Type Conversion

With the exception of MPFUN, none of the high precision libraries comes with a transcription program to automatically convert existing source code, using standard precisions, into code that uses the multiprecision types of the library. This problem is addressed now.

In [9] we describe an easy to use, generic C/C++ transcription program or precompiler for the conversion of C/C++ source code into new code that uses a C++ multiprecision library of choice. The precompiler can convert any type in the input source code to another type in the output source code. The input source can be C or C++, while the output code generated by the precompiler and using the new types, is C++. The type conversion is based on a simple configuration file, provided by the developer of the multiprecision library or by the user of the precompiler.

During the transcription of the code, special care is taken with respect to constants, among others to avoid the default conversion by the C++ compiler of decimal literals to their standard double precision binary representation. Hence constants need to be signaled to the user of the precompiler to make sure that they are provided with sufficient accuracy. This can be guaranteed either by string initialization from the decimal literal or by providing sufficiently accurate representations in different radices for constants such as π, e, $\ln 2$, $\sqrt{2}, \ldots$.

The precompiler can be told to skip the conversion of certain variables, such as the running variables in `for`-loops or even a complete function implementation. Use of the precompiler saves a lot of time and avoids errors that otherwise easily occur in a manual conversion. At the same time, great care has been taken to obtain precompiled code with performance similar to that of manually converted code.

4 High Precision Matrix Library

The basic MATLAB type is a matrix. Several matrix libraries are freely available, among which MTL (Matrix Template Library), TNT (Template Numerical Toolkit) and Boost. While MTL and TNT claim to be fully templated, in reality the code still contains hard coded `float`, `double` and `int` variables. This renders them useless when trying to generate a true multiprecision matrix library by use of the above precompiler. Fortunately, Boost provides templated C++ classes for several types of matrices: dense, sparse, triangular, banded, symmetric, hermitian, etc. The library covers the usual basic linear algebra operations on vectors and matrices and provides BLAS level 1, 2, 3 functionality.

Our matrix library is based on the templated Boost uBLAS routines and uses our high precision data types implemented in the classes `MpIeee`, `Rational` and `BigInt` as its data types. When different types are used in an expression, the arguments are automatically converted to a predefined (larger) type unless the cast operator is used to force conversion to a certain data type. The resulting library is very time and memory efficient by using advanced template techniques similar to the delayed evaluation techniques used by our own high precision classes.

5 MATLAB Parser

MPL (Multi Precision Lab) implements a subset of the MATLAB language but has a superset of multiprecision types. Starting from MATLAB scripts, our lexer and parser construct an abstract syntax tree. This tree can then be interpreted in our environment or compiled into C++ sources. In their turn, these C++ sources can easily be compiled into standalone executables. These executables are faster because no iteration in the abstract syntax tree is required.

Here is a short overview of the implemented subset:

- Block encapsulation: begin, end.
- Loops: while, for.
- Control structures: if, else, elseif, switch, case, otherwise, break.
- Input/Output: disp, print, println, input.
- Functions: function, return.
- Library loader: loadlibrary, calllib, unloadlibrary, libisloaded.
- Matrix creation: zeros, ones, eye.
- Complex variables: i, j.
- Built-in elementary functions: sin, sinh, asin, asinh, cos, cosh, acos, acosh, cotan, cotanh, acotan, acotanh, tan, atan, tanh, atanh, exp, exp2, exp10, log, log2, log10.
- Special constant values: Inf, inf, NaN.
- Built-in functions: transpose, colon (incl. range operations used in for-loops etc.), inv, horzcat, vertcat, help, sqrt, pow, mod, rem.
- Relational operators : <, >, <=, >=, =, , ==, |, &, ||, && with same precedence as MATLAB.
- Arithmetic operations : +,−,*,.*,./,.\,∧,.∧ with same precedence as MAT-LAB.
- Cell array's : basics implemented but not yet complete.

Subsequently, we extend the MATLAB language with functions to alter the current floating-point environment settings:

- mode: with an argument mpieee, rational, complex, double, int or logical to specify the type.
- rounding: with an argument nearest, up, down, zero to specify the rounding.
- exponent: two arguments specifying the minimal exponent L and the maximal exponent U with $L = 1 - U$.
- radix: one argument which sets the value of the radix.
- outputmode: one argument with a value between 1 and 13 for the different output modes and 0 to reset the status flags.

Garbage collection is done using reference counting pointers. This is not as efficient as for instance mark-and-sweep but for our purposes (especially for compiling into readable C++ source code) it is the best option. Also we are able to use the already available shared pointer from the Boost library. Due to the nature of our language we do not have to worry about cycles in object pointers and (inefficient) tracing routines to resolve them.

Strangely, generic memory management libraries do not seem to be freely available. This is a pity because every garbage collecting language (such as Perl, Python, Java, ...) has to implement memory management schemes.

6 Runtime Library Loader

The runtime library loader LibLoader enables the loading of shared libraries at runtime. This way the MATLAB parser can be extended with various algorithms selected from high precision precompiled versions of numerical libraries of choice such as Numerical Recipes [10] and GSL [11]. An overview of the MPL environment and more precisely how LibLoader fits in the picture, is shown in Figure 1.

The implementation of LibLoader consists of various wrapper classes around the free cross-platform libraries FFCall and GNU Libtool. We have to circumvent compile time checking of types by using the void* pointer, since it is the only way to pass class objects to functions that are loaded dynamically. Proper use of the precompiler guarantees that the object types of the loaded library and the types used in our interpreter and compiler are identical.

Another pitfall when loading arbitrary libraries and calling arbitrary routines is the precise knowledge of the arguments and their dimension(s). Without this

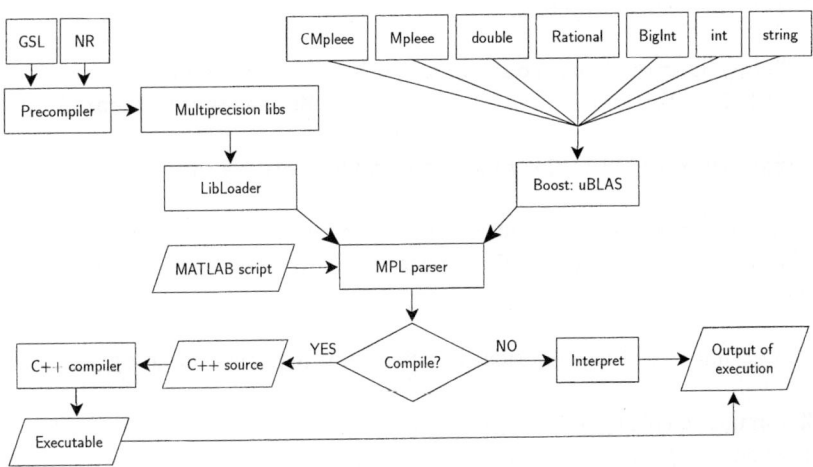

Fig. 1. Overview of MPL environment

information, no successful calls can be made without getting errors, or worse, memory leaks. This is not surprising, since the same caution is required when calling the routines from ordinary C/C++.

7   Interfacing to Numerical Recipes

Let us now describe how a multiprecision version of the Numerical Recipes library can be loaded into MPL. First, the precompiler is used to convert every function file so that it uses a multiprecision type of choice. The precompiled output files are then compiled and linked into a shared library. Unfortunately some files do not link or corrupt the generated multiprecision library because of the use of global or external variables. The shared library can then be loaded as follows:

```
libname = '/home/mpl/scripts/recipes';
loadlibrary( libname );
```

However, loading the library at runtime causes `MpIeee` objects to use a separate environment. This environment has to be synchronized with the environment used inside the parser. Therefore one has to add two functions, called `getEnvironment` and `setEnvironment`, to our shared library. These functions are called before and after every call of a Numerical Recipes routine in order to ensure that the correct internal representation is used. We can automate this by placing these extra calls in an external function file:

```
% External function file callrecipes.m
function result = callrecipes( funcname, varargin )
  libname = 'recipes';

  %copy current environment to library environment
  calllib libname, 'setEnvironment', MpIeeeEnvironment();

  result = calllib libname, funcname, varargin{:};

  %copy library environment back to current environment
  calllib libname, 'getEnvironment', MpIeeeEnvironment();
end
```

The actual call of a Numerical Recipes routine is then done by:

```
% actual calling of recipes routines
libname = '/home/mpl/scripts/recipes';
loadlibrary( libname );
```

```
if libisloaded( libname )
  callrecipes 'rtflsp', 8.8, 9.2, 1e-64
  callrecipes 'rtflsp', 8.8, 9.2, 1e-128  %higher precision
  unloadlibrary( libname );
else
  disp('recipes not loaded');
end
```

8 Interfacing to GSL

To call routines of the GSL library, we need to follow a slightly different approach. We fine tune the conversion of GSL using the precompiler so that all double and float instances in computations are replaced by a multiprecision type. The variables used exclusively for array or matrix indexing are left untouched for better performance. After this conversion we still cannot use the GSL routines directly in the parser like we did with Numerical Recipes. The reason for this is that GSL uses C structs and specific memory allocating calls like malloc for the implementation of vector, matrix and interval types.

To map the types used in GSL to the types of the matrix library used in the parser, we need interfacing wrapper functions. Here is a skeleton of such a wrapper function:

```
#include <iostream>
#include "matrix.h"
#include <gsl/gsl_math.h> //gsl specific includes
#include <gsl/gsl_errno.h>
#include <gsl/gsl_roots.h>

void mexFunction(int nlhs, Matrix *plhs[], int prhs,
                 const Matrix *prhs[]){
  // convert the Matrix arguments in prhs
  // to needed structs defined in gsl.h

  // make the call to the gsl routine, for example:
  // bisection_iterate (void * vstate, gsl_function * f,
  //                    MpIeee * root,
  //                    MpIeee * x_lower,
  //                    MpIeee * x_upper)

  // copy returned value(s) into plhs
}
```

These wrapper functions are very similar to the MATLAB mex functions. One way to automate the construction of these wrapper functions is to use the SWIG tool which is also used successfully by Python to import various libraries.

9 Example

We implement a remarkable example [12, 13] where all hardware precisions, even quadruple precision, go wrong:

```
a = 77617;
b = 33096;
y = 333.75*b*b*b*b*b*b + ...
    a*a * (11*a*a*b*b - b*b*b*b*b*b - 121*b*b*b*b - 2) + ...
    5.5*b*b*b*b*b*b*b*b + a/(2*b)
```

This gives the wrong answer $y = 1.1726$ in MATLAB 6.5 revision 13 on an Intel Pentium 3 based system, as well as in MPL through `mode(double)`. Setting the radix and precision to 10 and 37 in MPL, by adding the commands

```
mode( mpieee );
precision( 37 );
radix( 10 );
```

the computed result for y is

```
y = -8.273960599468213681411650954798162920^-1
```

as it should be. MPL also allows a correct value for y to be computed in rational arithmetic.

This example shows that it is very straightforward to go from hardware to multiprecision using our MPL environment. All the functionality is also available from within a cross-platform GUI: run scripts, set the radix, increase or decrease the precision and exponent range, change the rounding modes and default types.

References

1. Hammer, R., Hocks, M., Kulisch, U., Ratz, D.: C++ Toolbox for Verified Computing. Springer Verlag, Berlin, Heidelberg (1995)
2. Bini, D., Fiorentino, G.: Design, analysis, and implementation of a multiprecision polynomial rootfinder. Numerical Algorithms **23** (2000) 127–173
3. Granlund, T.: GNU MP: The GNU Multiple Precision Arithmetic Library. (2004)
4. Cuyt, A.: http://www.cant.ua.ac.be/arithmos/ (2004)
5. Haible, B.: CLN, a class library for numbers. (1997)
6. Smith, D.M.: Algorithm 693: A FORTRAN package for floating-point multiple-precision arithmetic. ACM Trans. Math. Software **17**(2) (1991) 273–283
7. Zimmermann, P., et al.: MPFR: a library for multiprecision floating-point arithmetic with exact rounding. (2000)
8. Bailey, D.: A FORTRAN 90-based multiprecision system. ACM Trans. Math. Software **21** (1995) 379–387
9. Schreppers, W., Cuyt, A.: A generic C/C++ precompiler. ACM Trans. Math. Software (2004) submitted.
10. Press, W.H., Teukolsky, S.A., Vetterling, W.T., Flannery, B.P.: Numerical recipes in C++. Cambridge University Press, Cambridge (2002)

11. Galassi, M., Davies, J., Theiler, J., Gough, B., Jungman, G., Booth, M., Rossi, F.: GNU Scientific Library Reference Manual. Second edn. Network Theory Ltd. (2003)
12. Cuyt, A., Verdonk, B., Becuwe, S., Kuterna, P.: A remarkable example of catastrophic cancellation unraveled. Computing **66** (2001) 309–320
13. Rump, S.: Algorithms for verified inclusions - theory and practice. In Moore, R., ed.: Reliability in Computing. (1988) 109–126

Performance and Scalability Analysis of Cray X1 Vectorization and Multistreaming Optimization

Sadaf Alam and Jeffrey Vetter

Computer Science and Mathematics Division,
Oak Ridge National Laboratory
{alamsr, vetterjs}@ornl.gov

Abstract. Cray X1 Fortran and C/C++ compilers provide a number of loop transformations, notably vectorization and multistreaming, in order to exploit the multistreaming processor (MSP) hardware resources and its high memory bandwidth. A Cray X1 node is composed of four MSPs, which in turn are composed of four single streaming processors (SSP). Each SSP contains a superscalar processing unit and two vector processing units. Compiler vectorization provides loop level parallelization and uses the vector processing hardware. Multistreaming code generation by the compiler permits execution across the SSPs of an MSP on a block of code. In this paper, we analyze overall impact of loop-level compiler optimization on a scientific application called Parallel Ocean Program (POP). POP has been extensively optimized for X1 by instrumenting the code using X1 compiler directives. We compare and contrast automatic and manual optimization schemes available on X1 and analyze their impact on the code performance and scalability. Our results show that the addition of compiler directives increases the average vector length, thereby improving the single node performance significantly. However, this code scales at a slower rate as the local workload volume decreases and the communication costs increase.

1 Introduction

Modern vector computers like Japan's Earth Simulator and the Cray X1 combine the vector processing architecture and Massively Parallel Processing (MPP) in a single system design [5,9]. They provide a very high memory bandwidth, which is key to realizing a high percentage of theoretical peak performance. The basic building block of the Cray X1 is the 12.8 GFlops multi-streaming processor (MSP) [1]. Each MSP is comprised of four single-streaming processors (SSPs). An SSP has two 32-stage 64-bit floating-point vector units and one 2-way superscalar unit. The SSP uses two clock frequencies, 800 MHz for the vector units and 400 MHz for the scalar unit. Each MSP has a MByte E-cache. Cray X1 systems implement Cray's new vector instruction set architecture, which support decoupled scalar and vector execution for maximum performance. The Cray X1 compiler automatically optimizes a loop for the eight vector units within an MSP [6]. It multistreams a long vectorized loop or unvectorized outer loop.

With aggressive compiler optimization flags, the Cray X1 compiler carries out a number loop transformations, which are presented in this paper.

We conducted experiments on a 512 MSP Cray X1 system at the Oak Ridge National Laboratory using a complete scientific application called Parallel Ocean Program. Several performance critical loops have been manually vectorized in the code [14] using compiler directives and dummy loop indices. We compare and contrast the code optimization strategies, in particular multistreaming and vectorization, and measure their impact on the overall performance of the application code. We also capture, using a hardware performance counter utility from Cray called pat_hwpc [7], detailed execution characteristics for complete application runs.

The layout of the paper is as follows: Section 2 presents the Cray X1 node architecture. Section 3 outlines various loop transformation schemes and compiler directives that exploit the MSP architecture. Section 4 provides a brief description of the scientific application. Section 5 details the experimental setup. Results are presented in section 6 and section 7 concludes the research.

2 Processing Node Architecture

The Cray X1 has a hierarchical processing node, memory and interconnection network design [1]. A processing node is composed of four multistreaming processors (MSPs), 16 GBytes of globally shared local memory, and an inter-MSP communication network. Local memory latency is the same for all processors within a single node and it is accessible by all processors on the node, processors on the other nodes and all I/O devices. Total peak local memory bandwidth for one node is 204.8 GBps, which also supports network traffic and I/O. Interconnect design and programming models for the X1 are described in [5].

Figure 1 shows the design blocks of an MSP. Each of the four SSP contain two vector units and one scalar unit. The X1 provides a large set of registers to reduce the number of memory accesses, reduce register spills, eliminate write-after-read dependencies and hide memory latency. The register set includes thirty-two 64-bit vector registers, 8 vector mask registers, 64 scalar registers, 64 address

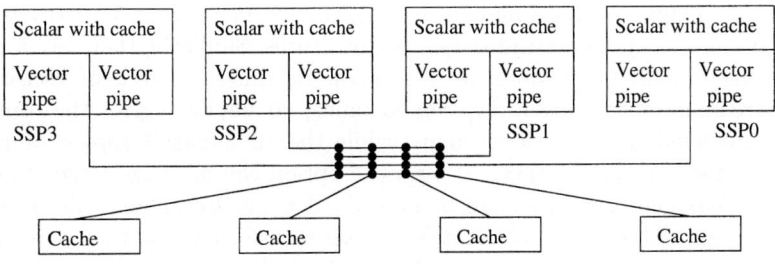

Fig. 1. Cray X1 multi-streaming processor architecture

registers, 8 control registers, a bit-matrix multiply register and a vector carry register [12].

There are three forms of cache in the X1 system. Each SSP has a 16KB scalar data cache and a 16 KB instruction cache, while each MSP has a 2MB instruction and data cache (E-cache) that is shared by the four SSPs within an MSP. Processors cache data from their local node modules only; references to memory on other node modules are not cached locally. In addition, the 32 KB of vector register space effectively functions as the lowest level of the processor data cache, similar to the L1 cache found on the microprocessor-based systems.

3 Vectorization and Multistreaming

Compiler vectorization provides loop-level parallelization of operations and uses the vector processing hardware: execution pipes, load buffers and functional unit groups. At the SSP level, vector instructions allow a large number of Single Instruction Multiple Data (SIMD) operations to execute in a pipeline fashion thereby tolerating memory latency and allowing for high sustained performance. MSP parallelism is achieved by distributing loop iterations across each of the four SSPs, which is referred as multistreaming. Vectorization and multistreaming can be intermixed, and multistreaming often extends outside the loop boundary. Figure 2 shows vectorization of a do loop with multistreaming. Individual iterations of a loop can be distributed over the four SSPs such that four 2-vector units of an SSP work together as a large single vector unit and can achieve an average vector length of 64.

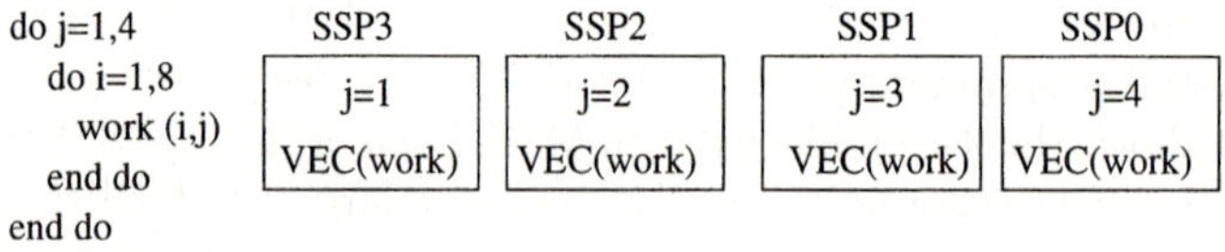

| do j=1,4
  do i=1,8
    work (i,j)
  end do
end do | SSP3
j=1
VEC(work) | SSP2
j=2
VEC(work) | SSP1
j=3
VEC(work) | SSP0
j=4
VEC(work) |

Fig. 2. Multistreaming and vectorization of a nested loop

3.1 Loop Transformations

Presently, compilers offer a number of loop transformations [2, 3], many by default, on almost all modern processor architectures. Similarly, there are a number of flags in the Cray compilers that allow for different levels of vector and multistreaming optimization [11]. For instance, -O vector3 gives the compiler maximum freedom to vectorize loops, while the -O stream3 allows for maximum multistreaming. Further, -O aggress option can optimize large loops. In addition, compiler directives can be embedded manually in the code. Here we briefly describe the Cray X1 compiler's loop transformations that exploit the multistreaming vector architecture. This includes loop interchange, collapsing, unwinding and loop fusion.

Loop Interchange. Compiler attempts to automatically interchange loops to maximize cache and vector register reuse and to reduce memory bandwidth usage. In the example below, the i elements of vector array a can be loaded before the j loop and i elements of vector array a(i) can be stored after the inner loop. After the loop interchange, the completely vectorized inner loop has no array a load and store.

```
before:                         after:
  do j = 1,m                      do i = 1,n
    do i = 1,n                      do j = 1,m
      a(i) = a(i)+b(i,j)              a(i) = a(i)+b(i,j)
    end do                          end do
  end do                          end do
```

Collapsing. The loop collapse schemes result in an increase in vector length and multistreaming and a reduction in loop overheads. Loop overheads are reduced by replacing a nested loop by a single loop.

Loop Fusion. In Fortran array syntax, every statement is considered a loop. Loop fusion minimizes loop overhead and maximizes register use. For example, array initialization a=0; b=0 is fused and vectorized.

3.2 Code Instrumentation

Compiler directives are extensively used in the POP code [14]. Cray compilers offer a number of compiler directives for a range of compiler-directed optimization including vectorization and tasking directives, streaming directives and MSP optimization directives. For instance, a compiler directive can inform the compiler that a given loop can be collapsed (e.g. by inserting !CDIR COLLPASE before the loop). Likewise, other directives can indicate that it is safe to completely unroll, vectorize or inline a loop. Typically, during the code optimization process, a programmer generates the loopmark listing and identifies the optimized and unoptimized loops. A sample loopmark output is listed below:

```
ftn-6204 ftn: VECTOR File = file.f, Line = 625
  A loop starting at line 625 was vectorized.
ftn-6601 ftn: STREAM File = file.f, Line = 625
  A loop starting at line 625 was multi-streamed.
ftn-6004 ftn: SCALAR File = file.f, Line = 627
  A loop starting at line 627 was fused with the loop starting at
  line 625.
ftn-6289 ftn: VECTOR File = file.f, Line = 639
  A loop starting at line 639 was not vectorized because a
  recurrence was found on "TEMP" between lines 641 and 647.
```

Compiler directives can then be used, if due to some ambiguous data dependencies, the compiler is unable to transform or vectorize a loop. Given this information the compiler can generate highly optimized code. In addition to compiler directives, dummy loops can be inserted. This is particularly useful for loop vectorization in Fortran code, which uses a verbose representation for array assignment operations.

4 Application

Cray X1 optimization strategies have been reported for a set of synthetic benchmarks [8]. In this paper, we conduct performance and scaling analysis of a scientific application called Parallel Ocean Program (POP), which was developed at the Los Alamos National Laboratory [13]. This code has been ported to a number of high-end supercomputing systems including two mainstream vector supercomputers: Japan's Earth Simulator and the Cray X1. A comparative performance study of POP on vector computers is presented in [4].

POP code is synchronous, and except for the infrequent I/O operations follows a Single Program Multiple Data (SPMD) programming paradigm. The code executes in a time-step fashion with the number of time steps specified in an input script file. There are two main processes in a POP time-step: baroclinic and barotropic. Baroclinic requires only point-to-point communication and is highly parallelizable. Barotropic contains a conjugate gradient solver, which requires global reduction operations. Moreover, the discretized POP grid is mapped and distributed evenly on a logical, two-dimensional processor grid [14].

Inter-processor communication in POP is performed via the Message Passing Interface (MPI) protocol. The exception is the conjugate gradient solver, which is implemented in Co-array Fortran [10] in order to reduce the communication requirement of this typically expensive component. Co-Array Fortran is a small set of extensions to Fortran 95 for SPMD parallel processing. It offers new set of rules for work distribution and data distribution within a program. POP optimization on Cray X1 are detailed in [14].

5 Experiments

For the performance evaluation experiments, the POP code is compiled with the options listed in table 1.

Additionally, using the -rm flag the compiler generates the loopmark listing, which identifies and explains optimization performed by the compiler on a given loop. The results reported in this paper are gathered from the loopmark files. In most cases a number of techniques identified in the paper have been applied to a single loop. For instance, a single loop can be collapsed, unrolled and vectorized by the compiler. Likewise, a compiler directive can result in a combination of loop transformations.

It is worth noting here that in a parallel code, a compiler hint or a directive should take into account the scaling effect of a given loop or loop indices. In an SPMD program like POP, often loop iterations and memory access operations

Table 1. Loop optimization applied to the POP application

Optimization	Description
default	equivalent to -O2
-O3,aggress	compiler aggressively optimize large loops
-Ovector3,aggress	aggressively vectorize loops
-Ovector3,stream3	maximum freedom to vectorize and multistream
Code instrumentation	as explained in [14]

are divided and distributed according to the size of the logical processor grid i.e. the two-dimensional MPI processor grid. The number of processor in x and y directions are specified at compile time in POP code. The number of simulated time steps in POP is controlled via nstep parameter. Scaling experiments are conducted for a fixed problem size, with a fixed nstep and by varying the number of processors in the x and y directions. Cray's pat_hwpc tool is used for calculating the runtime and average vector length.

6 Results

Table 2 shows the different loop optimization counts using the optimization listed in table 1. The table 2 also lists the runtime, percentage of vector instructions and average vector length, which has been collected by the pat_hwpc tool. The hand-coded version has a large number of single vector iterations and fused loops compared to the code that is generated by the Cray Fortran compiler without the compiler directives. As a result a larger fraction of instructions are executed by the vector processing units. An optimal utilization of the vector resources is translated in an increased average vector length. At the same time, compiler optimized code has a large number of unwounded loops compared to the hand-coded version. There is only a slight variation between different variants of compiler optimization.

Table 2. Number of individual loop transformations with different optimization schemes, pat_hwpc statistics and their impact on a single node performance

Optimization	Run-time (sec)	Average Vector Length	%Vector Instructions	Vector loops	Multi-streamed loops	Single Vector Iterations	Collapsed loops	Un-wounded loops
default (O2)	254.32	37.243	93.06	1291	1243	76	34	192
O3,aggress	255.17	37.273	93.61	1329	1243	76	32	200
vector3,aggress	254.89	37.273	93.61	1329	1222	76	34	200
vector3,stream3	255.14	37.273	93.61	1329	1243	76	34	200
Instrumented	40.34	60.58	99.95	1306	1248	282	70	52

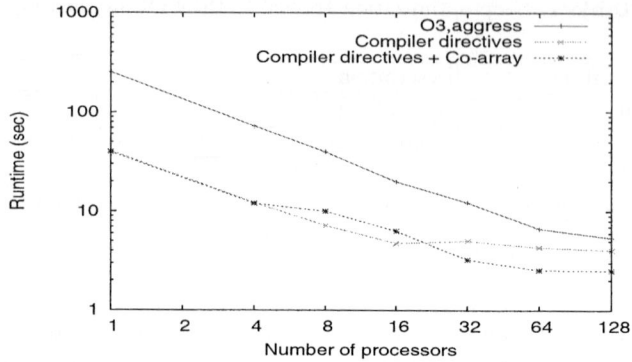

Fig. 3. Parallel efficiency

We also analyzed the scaling behavior of the code. For these fixed problem size experiments, increasing the number of processors did not significantly improve the performance in the instrumented code. Figure 3 shows the runtime values for the POP code generated: (1) with compiler-only optimization, (2) with compiler directives and (3) with compiler directives and co-array Fortran. We also compared the average vector length across different runs (shown in figure 4), which is an indirect measure of how effectively the vector resources are exploited. We note that the vector length first decreases with the decrease in local POP grid dimension and then stabilizes. We attribute this behavior to the vertical dimension k of the POP grid. (i x j) POP grid points scale with the underlying processor grid. The vertical k dimension is mapped across all processors and its value remains constant. That is why we do not observe a gradual drop in vector length by increasing the number of processors. Furthermore, the co-array Fortran optimization, which replaces the frequent MPI communication routines in the POP code result in the efficient execution and scaling of the code.

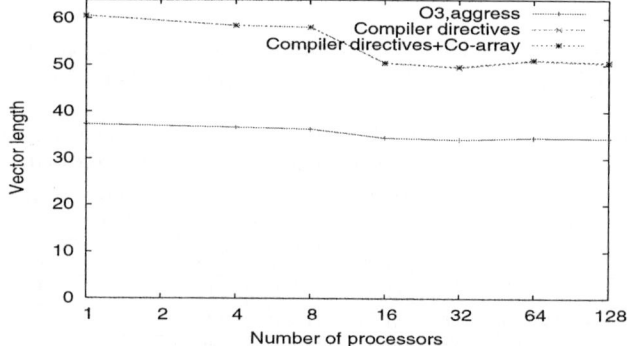

Fig. 4. Average vector length scaling

7 Conclusions

The Cray X1 design embodies advanced vector and superscalar concepts and it is highly optimized for floating-point intensive scientific calculations. We studied the loop vectorization and multistreaming strategies of the Cray X1 supercomputer and quantitatively evaluated their impact on the performance of a scientific application called POP. Furthermore, we analyze the scaling of single node loop optimization strategies, particularly code instrumentation strategies that involve adding compiler directives and dummy loop indices. We demonstrated that the single node performance of the instrumented code far exceeds that of the code which relies exclusively on the compiler for optimization, even when the most aggressive optimization flags are specified. At the same time however, for a fixed problem size, POP scales at a slower rate, as the local workload volume decreases. Hence, for larger problem sizes, the instrumented POP code is likely to more efficient than the code generated with aggressive compiler optimization.

Acknowledgements

This research was sponsored by the Office of Mathematical, Information, and Computational Sciences, Office of Science, U.S. Department of Energy under Contract No. DE-AC05-00OR22725 with UT-Battelle, LLC. Accordingly, the U.S. Government retains a nonexclusive, royalty-free license to publish or reproduce the published form of this contribution, or allow others to do so, for U.S. Government purposes.

References

1. P. K. Agarwal, *et. al. ORNL Cray X1 evaluation status report*, Proceedings of the 46th Cray User Group Conference, 2004.
2. A. V. Aho, M. Hill and J. D. Ullman, *Compilers: Principles, Techniques, and Tools*, Addison-Wesley Longman Publishing Co., Inc. USA, 1986.
3. R. Allen and K. Kennedy, *Optimizing Compilers for Modern Architecture: A Dependence Based Approach*, 1st ed. Morgan Kaufmann Publishers, 2001.
4. T. H. Dunigan, *et. al. Early evaluation of the Cray X1*, Proceedings of the 17th Annual International Conference on Supercomputing, 2003.
5. T. H. Dunigan, *et. al. Performance Evaluation of the Cray X1 Distributed Memory Architecture*, IEEE Micro 25(1), 2005.
6. *Optimizing Applications on the Cray X1 System: Loopmark Listings*. Available at http://www.cray.com
7. *Optimizing Applications on the Cray X1 System: Using CrayPAT Tools*. Available at http://www.cray.com
8. H. Shan and E. Strohmaier, *Performance Characteristics of the Cray X1 and Their Implications for Application Performance Tuning*, Proceedings of the 18th Annual International Conference on Supercomputing, 2004.
9. A. J. van der Steen and J. J. Dongarra, *Overview of Recent Supercomputers*, 2004.

10. R.W. Numrich and J.K. Reid, *Co-Array Fortran for Parallel Programming*, ACM SIGPLAN Fortran Forum, volume 17, no 2, 1998.
11. *Cray Fortran Compiler Commands and Directives Reference Manual*. Available at http://www.cray.com.
12. *Cray X1 System Overview*. Available at http://www.cray.com.
13. The Parallel Ocean Program Homepage, http://climate.lanl.gov/Models/POP
14. P. Worley and J. Levesque, *The Performance Evolution of the Parallel Ocean Program on the Cray X1*, Proceedings of the 46th Cray User Group Conference, 2004.

Super-Scalable Algorithms for Computing on 100,000 Processors[*]

Christian Engelmann and Al Geist

Computer Science and Mathematics Division,
Oak Ridge National Laboratory,
Oak Ridge, TN 37831-6164, USA
{engelmannc, gst}@ornl.gov
http://www.csm.ornl.gov

Abstract. In the next five years, the number of processors in high-end systems for scientific computing is expected to rise to tens and even hundreds of thousands. For example, the IBM BlueGene/L can have up to 128,000 processors and the delivery of the first system is scheduled for 2005. Existing deficiencies in scalability and fault-tolerance of scientific applications need to be addressed soon. If the number of processors grows by a magnitude and efficiency drops by a magnitude, the overall effective computing performance stays the same. Furthermore, the mean time to interrupt of high-end computer systems decreases with scale and complexity. In a 100,000-processor system, failures may occur every couple of minutes and traditional checkpointing may no longer be feasible. With this paper, we summarize our recent research in super-scalable algorithms for computing on 100,000 processors. We introduce the algorithm properties of scale invariance and natural fault tolerance, and discuss how they can be applied to two different classes of algorithms. We also describe a super-scalable diskless checkpointing algorithm for problems that can't be transformed into a super-scalable variant, or where other solutions are more efficient. Finally, a 100,000-processor simulator is presented as a platform for testing and experimentation.

1 Introduction

Today's top supercomputers are able to deliver several tens of TeraFLOPS of sustained performance for computational scientific research in areas like climate modeling, fusion energy and nanotechnology. If the steady increase in computing power stays on the track of Moore's Law, by 2010 the largest supercomputers in the world will be in the PetaFLOPS range. This trend is not solely based on

---

[*] Research sponsored by the Laboratory Directed Research and Development Program of Oak Ridge National Laboratory (ORNL), managed by UT-Battelle, LLC for the U. S. Department of Energy under Contract No. DE-AC05-00OR22725.

improvements of individual processors, but also aided by ever-increasing parallelism. Currently, these systems scale for up to 10,000 processors. In the next five years the number of processors is expected to rise to tens and even hundreds of thousands. For example, the IBM BlueGene/L [1, 2] will have up to 128,000 low-powered processors shipped in densely populated compute nodes. The first BlueGene/L system will be delivered in 2005. A prototype at IBM recently achieved a maximal LINPACK performance of 70 TeraFLOPS and currently holds the top spot in the Top 500 list of supercomputers.

Experiences with existing 10,000-processor machines show that the efficiency of scientific applications can be as low as 1%, which is equal to fully utilizing only 100 processors. Amdahl's Law shows how efficiency can drop off as the number of processors increases. If the number of processors grows by a magnitude and efficiency drops by a magnitude, the overall effective computing performance stays the same. Furthermore, the mean time to interrupt (MTTI) decreases with system scale and complexity. While reliability of individual components, such as network and storage, can be improved by redundancy, the number of system software issues increases due to complexity. Some of today's major supercomputing centers have already scheduled downtimes and unscheduled outages about every 40 hours. In a 100,000-processor machine, such system interrupts may occur as often as every couple of minutes. Network bottlenecks and latencies will make frequent coordinated checkpointing (once every hour) of applications, for fault-tolerance, almost impossible. Even with traditional checkpointing, it does not make sense to restart 99,999 processors because one failed! Finally, at some point the MTTI is going to exceed the time to restart.

In this paper, we summarize our recent research in super-scalable algorithms for high-end scientific computing on extreme-scale systems with 100,000 processors. First, we introduce the algorithmic properties of *scale invariance* and *natural fault tolerance*, and then we discuss how they can be applied to two different classes of algorithms. We also describe a super-scalable diskless checkpointing algorithm for problems that cannot be transformed into a super-scalable variant, or where other solutions are more efficient. We continue with a short description of our efforts in developing a 100,000-processor simulator as a platform for testing and experimentation. Finally, we close with a brief summary of the work and possible future directions.

2 Super-Scalable Algorithms

High-end computing on 100,000-processor systems requires fundamental rethinking of how algorithms can efficiently utilize such an enormous amount of processors. There are two major issues that need to be considered. The first is Amdahl's Law, and the need to reduce the serial fraction to a point where reasonable efficiency can be achieved. The second is the high probability of failures, and the need to survive in a way that does not involve global operations. In order to address these problems, we have established a foundation for a new class of algorithms called *super-scalable algorithms* [3] that have the properties of *scale invariance* and *natural fault tolerance*.

Scale invariance means that the individual tasks in a larger parallel job have a fixed maximum number of other tasks they communicate with, independent of the total number in the application. For example, a finite difference algorithm has a constant number of neighbor tasks defined by its stencil, which is independent of the total number of tasks in the problem. Another example is a binary tree communication infrastructure, where each node is only connected to three other nodes. With scale invariance, individual tasks do not have to be concerned about failures throughout the system unless these failures happen to affect one of their neighbors. Conversely, dynamically adding replacement or additional tasks can be ignored by tasks not communicating with these new tasks.

However, scale invariance alone does not guarantee high efficiency of applications on 100,000-processor computing systems. The serial fraction of a parallel algorithm does not solely depend on the communication footprint, but also on hardware factors, such as I/O latencies and cache misses, that can quickly drive efficiency down even if the best-known algorithms are being used.

Scale invariance does not provide fault tolerance, but it enables isolation of the failure. However, most parallel algorithms designed today will deadlock, or worse, calculate the wrong answer, if one or more tasks fail. Fault tolerance needs to be handled locally by "self-healing" or natural fault tolerance.

Natural fault tolerance is the ability to tolerate failures through the mathematical properties of the algorithm itself, without requiring notification or recovery. It is not that the calculations are taken over by other tasks, but rather that the nature of the algorithm includes natural compensation for the lost information. For example, an iterative algorithm may require more iterations to converge, but it still converges despite lost information [4].

The maximum number of tasks that can fail, yet still obtain the correct answer, is problem dependent and still an open research question. We assume that the actual number of tasks lost during an application run will be a small fraction of the overall number of tasks. We based our research on the assumption that up to 100 out of 100,000 tasks may fail, which is only 0.1%. However, the time-to-solution increases dramatically when using traditional checkpointing or message logging schemes due to the large amount of processors involved and the centralized nature of existing solutions. We discuss a peer-to-peer diskless checkpointing alternative later in this paper.

Scale invariance and natural fault tolerance are rather restrictive requirements on algorithms, and when we began our research it was not clear that anything other than the most trivial applications, using the bag-of-tasks programming paradigm, would be able to meet these definitions. Such applications are (to a certain extend) scale invariant, because each task communicates only to send back its answer. They have fault tolerance, because tasks are farmed out and can be easily replaced. Task farming with on-the-fly fault tolerance by task replacement is a widely used technique today. Examples are SETI@HOME [5] and Condor [6, 7].

In the following sections, we will describe solutions for two different nontrivial classes of super-scalable algorithms. The first is where the problem can be formulated as some function of a local volume, such as for finite difference and finite element applications. The second is where the problem requires global information, like in global minimum or maximum searches, that are often used to determine if an iterative algorithm has converged.

2.1 Local Information

Parallel applications where individual tasks only require information from a local region include finite difference and finite element solutions to differential equations. We combined two ideas, chaotic relaxation [8, 9] and meshless methods [10], to demonstrate that both super-scalable algorithm requirements, scale invariance and natural fault tolerance, can be achieved.

In a meshless finite difference algorithm with chaotic relaxation, each data point in the solution space is assigned to an independent task that asynchronously receives update messages from its neighbors, calculates its own value and sends update messages back to its neighbors. The programming model is similar to active messaging [11], but could be coded using PVM [12] or MPI [13].

We use a coordinate in a virtual space to identify each task. This virtual space may coincide with the solution space, for example in a 2-D Poisson problem. Based on the coordinate, we can form nearest neighbor as well as random peer-to-peer networks to experiment with the algorithm. Each message contains the sender coordinates, so that necessary metrics, such as distance, can be calculated at runtime. Update messages additionally contain the value of the sending task. The update processing routine reflects the mathematical definition of the task and its relation to its neighboring tasks, which in the case of the 2-D Poisson problem is an average of the surrounding values with a distance bias.

Early investigations in the 1970's showed that chaotic relaxation has quite restrictive convergence properties, which is the main reason why it never became popular. However, for 100,000-processor systems it may be time to once again look at this iteration-free method. When failures and failure recovery are factored into the solution time, chaotic relaxation has some attractive recovery properties. The tasks that communicate with a failed task can do recovery independently and locally. Furthermore, the information lost by a failed task does not need to be recovered. The calculations can be formulated to proceed and converge to the solution despite failures.

We experimented with super-scalable finite difference algorithms and observed that simple problems, such as 2-D Poisson, converged despite 100 random failures across the machine. However, multiple failures of neighboring tasks, similar to multi-processor node failures, could cause the error of the solution to be significantly higher. However, this can be avoided if the virtual space is not directly mapped to the physical location of processors. Furthermore, connecting tasks randomly can decrease the overall convergence time.

We also experimented with asynchronous multi-grid variants based on the above ideas and this approach also tolerated failures. However, a master that controlled the "V" and "W" cycles was necessary, since the mathematical model of chaotic relaxation between different levels is not yet well understood at this time.

2.2 Global Information

Parallel algorithms where individual tasks require global information include global maximum searches, such as are often used to determine if an iterative algorithm has converged. First, the global maximum needs to be found among the values of all tasks. Then this value needs to be broadcast to all other tasks.

This is a graph problem that can be solved by creating a logical interconnect topology with the property of high probability message delivery despite failures, and that maintains efficient scale invariance to a low degree.

We conducted experiments with different network architectures, such as nearest neighbor, random, mesh and fully connected. We also implemented a broadcast algorithm. Both algorithms, global maximum search and broadcast, worked very well under various failure conditions.

A serious challenge for the global information algorithms, as well as for the finite difference, is algorithm termination. How does each task know when the complete system is stable and all tasks have the correct answer? Only the observing user knows that there are no messages on the network any more and that the system has converged. A global convergence test can solve this problem, but it needs to be either super-scalable or occur very infrequently.

3 Peer-to-Peer Diskless Checkpointing

Problems that cannot be transformed into a super-scalable variant, or where other existing solutions are more efficient, still need to deal with the expected MTTI of 100,000-processor systems.

To address this, we have developed a super-scalable replication technique based on peer-to-peer diskless checkpointing [14], which equips scientific applications with a self-healing capability for fault-tolerance. We assume that on BlueGene/L like systems local disk storage will no longer be available, due to the associated costs, failure sensitivity and maintenance.

In peer-to-peer diskless checkpointing, every task replicates its own local application state to a set of neighboring tasks using an encoding, such as RAID. The neighbor tasks themselves also replicate their own local application state, each to different sets of neighbor tasks. A scalable peer-to-peer infrastructure of checkpointing tasks is formed with local separation of current application state and multiple redundant backups. The amount of additional information each task needs to hold in its memory is dependent only on the encoding algorithm and on the number of neighbors involved in the replication of the state of one task, i.e. the system-wide degree of fault-tolerance.

The set of neighbor tasks may be derived from the network infrastructure or application algorithm. However, the probability of a failure involving physical neighbors, e.g. multi-processor node failures, may be greater than the probability of a failure involving a set of random or far away neighbors. The physical neighborhood of a task may also change in the case of a restart.

Synchronization of individual checkpoints is not necessary if tasks do not communicate with each other at all or if they do not communicate between synchronizing checkpoints. The traditional global snapshot method, using a barrier, can be used to synchronously checkpoint all tasks at once. Localized asynchronous checkpointing requires additional message logging to make sure that a consistent application state is being saved. We discussed advantages and disadvantages of both approaches in an earlier paper [14].

In the case of a failure, all surviving tasks roll back to their last checkpoint using a locally maintained copy or the remote backup in the neighboring tasks. All failed tasks are replaced using their last checkpoint from their neighboring tasks. An area of future research would be to identify surviving tasks that do not need to roll back if they are not directly dependent on the failed ones. Furthermore, a localized replay of the message log can eliminate the rollback of surviving tasks all together for a certain set of deterministic scientific applications. While centralized and partially localized rollback strategies and message log replay solutions [15, 16] exist, they currently do not scale to 100,000 processors. Initial work [17] has been done recently to address this issue.

Our experience with peer-to-peer diskless checkpointing shows that it can provide super-scalable self-healing capability for algorithms, such as FFT, where every single task holds important information for calculating the correct result. However, checkpointing and recovery scenarios can generally be very complex, especially when using localized asynchronous mechanisms. Furthermore, an application run still fails if the number of simultaneous failures of neighboring tasks is greater than the system-wide degree of fault-tolerance.

4 100,000-Processor Simulator

While the theoretical analysis of super-scalable algorithms gave us some insight into convergence properties and the probability of achieving the right answer, there is a lot of practical analysis data that can only be acquired by testing the algorithms using a variety of different failure situations.

A 100,000-processor machine was not available at the time of this work, the IBM BlueGene/L still under development, and software emulation frameworks, such Charm++ [18], did not reached the necessary scale. Therefore, we developed a simulator (Figure 1) that is able to run hundreds of thousands of tasks and supports rapid prototyping. It is designed to test algorithms at very high scale and provide a platform to develop fault-tolerant applications. It is instrumented to mimic different failure modes, but it does not provide performance estimates or analysis of the applications for a particular machine architecture.

The simulator can handle modules written in multiple languages and runs on different operating systems, e.g. Linux and Windows. It is implemented in Java, but also supports C and Fortran using the Java Native Interface. The

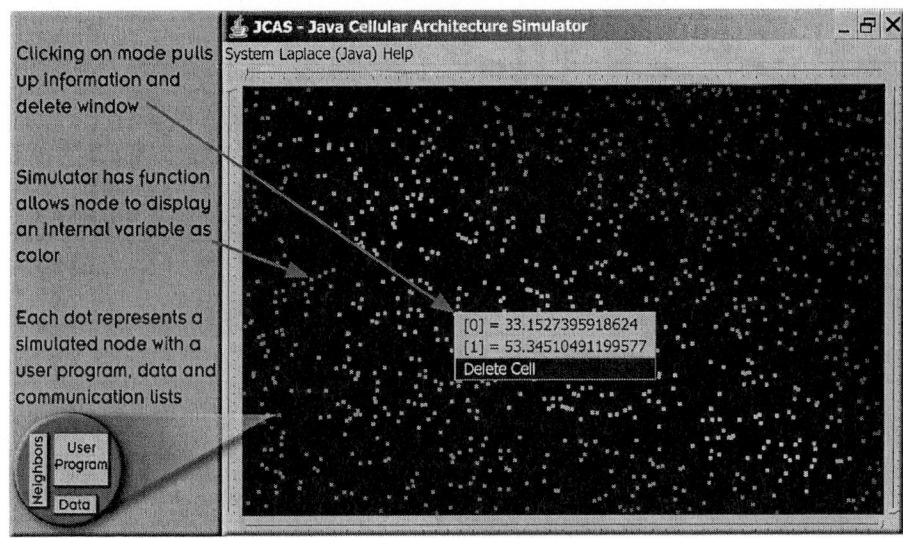

Fig. 1. User Interface to the Simulator

number of nodes that can be simulated depends on the size of the application being simulated and the power of the hardware the simulator is running on. The simulator is itself a parallel application and can run across a Linux cluster. On a 2 GHz Windows laptop we have simulated 10,000 nodes for a small application. Using a 32 processor, large-memory Linux cluster we have simulated half a million nodes running the super-scalable algorithms described earlier in this paper.

The simulated network topology, such as nearest neighbor, mesh, torus and random, can be configured before running an application. The simulator has a number of built in failure modes that the user can specify. It allows the killing of a selected node, block of nodes or a random percentage of nodes in a specified region. The failures are interactively initiated, i.e. the user clicks on a node and kills it, or selects a region and 1% of the nodes in this region die.

5 Conclusions

In this paper we have summarized our recent research at the Oak Ridge National Laboratory in super-scalable algorithms for high-end scientific computing on extreme-scale supercomputer systems with 100,000 processors. We presented the notion of a new class of algorithms called *super-scalable algorithms* that have the properties of *scale invariance* and *natural fault tolerance*. These properties allow scientific algorithms to scale to hundreds of thousands of processors, while maintaining efficiency and fault-tolerance.

We described solutions for two classes of super-scalable algorithms. In the first, the problem can be formulated as some function of a local volume, such as for finite difference applications. In the second, the problem requires global information, like in global maximum searches. We also developed a self-healing FFT based on peer-to-peer diskless checkpointing. Finally, we developed a software simulator that is able to run an enormous number of tasks.

Future research needs to be conducted to further develop appropriate programming models for 100,000-processor machines. Furthermore, scientists will need to rethink the mathematical models used in today's applications to better support the development of super-scalable solutions based on scale invariance and natural fault tolerance.

References

1. Adiga, N.R., et al.: An overview of the BlueGene/L supercomputer. Proceedings of SC, also IBM research report RC22570 (W0209-033) (2002)
2. Lawrence Livermore National Laboratory, Livermore, CA, USA: ASCII BlueGene/L Computing Platform at http://www.llnl.gov/asci/platforms/bluegenel
3. Geist, G.A., Engelmann, C.: Development of naturally fault tolerant algorithms for computing on 100,000 processors. (2002) to be published.
4. Bosilca, G., Chen, Z., Dongarra, J., Langou, J.: Recovery patterns for iterative methods in a parallel unstable environment. Submitted to SIAM Journal on Scientific Computing (2005)
5. Space Sciences Laboratory, University of California Berkeley, USA: SETI@HOME at http://setiathome.ssl.berkeley.edu
6. Basney, J., Livny, M.: Deploying a high throughput computing cluster. In Buyya, R., ed.: High Performance Cluster Computing: Architectures and Systems, Volume 1. Prentice Hall PTR (1999)
7. Computer Sciences Department, University of Wisconsin, USA: Condor at http://www.cs.wisc.edu/condor
8. Chazan, D., Miranker, M.: Chaotic relaxation. Linear Algebra and its Applications **2** (1969) 199–222
9. Baudet, G.M.: Asynchronous iterative methods for multiprocessors. Journal of the ACM **25** (1978) 226–244
10. Liu, G.R.: Mesh Free Methods: Moving beyond the Finite Element Method. CRC Press (2002)
11. von Eicken, T., Culler, D.E., Goldstein, S.C., Schauser, K.E.: Active Messages: A mechanism for integrated communication and computation. In: 19th International Symposium on Computer Architecture, Gold Coast, Australia (1992) 256–266
12. Geist, G.A., Beguelin, A., Dongarra, J.J., Jiang, W., Manchek, R., Sunderam, V.S.: PVM: Parallel Virtual Machine: A Users' Guide and Tutorial for Networked Parallel Computing. MIT Press, Cambridge, MA, USA (1994)
13. Snir, M., Otto, S., Huss-Lederman, S., Walker, D., Dongarra, J.: MPI: The Complete Reference. MIT Press, Cambridge, MA, USA (1996)
14. Engelmann, C., Geist, G.A.: A diskless checkpointing algorithm for super-scale architectures applied to the fast fourier transform. Proceedings of CLADE (2003) 47–52
15. University of Paris South, France: MPICH-V at http://www.lri.fr/ gk/MPICH-V

16. Indiana University, Bloomington, IN, USA: LAM-MPI at http://www.lam-mpi.org
17. Chen, Z., Fagg, G.E., Gabriel, E., Langou, J., Angskun, T., Bosilca, G., Dongarra, J.: Building fault survivable MPI programs with FTMPI using diskless checkpointing. Submitted to PPoPP (2005)
18. Zheng, G., Singla, A.K., Unger, J.M., Kale, L.V.: A parallel-object programming model for petaflops machines and blue gene/cyclops. Proceedings of IPDPS (2002)

"gRpas", a Tool for Performance Testing and Analysis

Laurentiu Cucos and Elise de Doncker

Western Michigan University
{lcucos, elise}@cs.wmich.edu
http://aegis.cs.wmich.edu/~lcucos,
http://www.cs.wmich.edu/~elise

Abstract. This paper presents "gRpas", a tool written in Java and designed to help analyzing test results from scientific computing applications. "gRpas" stands for "gather Results / plot, analyze, and store". As one of its main features, the tool is easy to interface with the user program. Furthermore it provides for one click data filtering and plot generation, effective graphical display of program output, and statistical report generation on algorithm results and performance. gRpas also has built-in functionality for comparison testing between two or more algorithms or algorithm versions. We will present examples of its use with parallel multivariate integration routines. However, its target applications cover a wide class of scientific computing programs.

1 Introduction

Software applications in general and scientific computing programs in particular can be seen as multidimensional functions: $f : R^n \to R^m$. A program can take an n-variate input, and generate an m-variate output. To analyze the program behavior, the investigator selects a set of input values and compares the output with results obtained using a different method. The complexity of evaluating the program arises through a number of factors including: the problem to be solved and its parameters, the algorithm used and its parameters, random factors, total number of processors involved, etc.

For a given problem, some algorithms perform better than others, while for a given algorithm some problems will be solved more efficiently. Efficiency related results are generally affected by the problem parameters and the algorithm parameters.

The random factor is related to the fact that some algorithms require a set of random values. Ideally, the outcome must not depend on the random values; however this is not always the case. By repeating the same test a number of times, the developer is able to see if the output of the algorithm tends to the same value.

The number of processors involved can complicate the analysis. On the one hand, for each processor, the developer must monitor a number of efficiency related parameters; on the other hand, a disturbance of the distributed environment (such as in sharing bandwidth with another program) can possibly affect the outcome, introducing a random factor.

For those problems for which there are no proved solutions (either because they are too hard and require approximations - as for NP complete problems, or because they involve too many extra conditions - as the parallel processing, adaptive numerical integration, etc.), the development cycle often iterates as follows:

a) find a new algorithm (or start from a current version);
b) perform a set of tests;
c) based on results analysis improve the algorithm.

For the later stages, some common questions are: How much testing must be done to obtain significant results? Is the new algorithm better and how much better? Despite the fact that there are no practical limits on the amount of data that can be collected from the tests, it is desired to perform as few tests as possible (to save CPU cycles, or decrease development time).

Tests are usually automated by varying one input parameter while keeping the others constant. If, for the ith parameter, k_i different values are tested, then a problem with n input parameters and m output values generates $m \prod_{1 \leq i \leq n} k_i$ numbers to analyze.

Often, to evaluate improvements, the same set of tests are performed for different versions. In [1], Hooker mentions two types of algorithm comparisons: (i) competitive testing, and (ii) controlled experimentation. Competitive testing is more suitable for development, and tells which algorithm is faster but not why, whereas controlled experimentation is suitable for research and gives insight on how the code behaves under certain conditions. We can relate this classification with the following: (i) compare results from two or more versions. (ii) compare results from the same algorithm version;

For comparing algorithms in (i), it is indicated to examine the difference in result vectors from different algorithms for corresponding input parameters. There is often a trade-off between the amount of work done and the quality of the result. Algorithm A may be pessimistic and B more optimistic, so that requesting less accuracy from A may lead to results equivalent to B's. The performance profiles technique of [4] accounts for this characteristic in the comparison of A and B.

In (ii), the algorithm may be tested with respect to the influence of random factors, either due to the environment or a random component in the algorithm (such as through the use of random numbers). The difference between the (sample) result vectors in m-space is significant to measure performance dependence on the random factor(s). Their distance can be accounted for by using the mean and variance with respect to some or all of the output components.

Performance measures to test algorithm behavior as a function of varying input parameters are problem related. For example, "result" and "estimated error" output parameters can be analyzed as a function of input "tolerated error" by comparing the actual error (if known) and estimated error to the level tolerated. The situation is complicated by the fact that the dependence on different input parameters may be correlated, such as that of "actual error" on "tolerated error" and "allowed number of iterations". As another example, various scalability measures (as a function of the number of processors used) are discussed in [5].

Another issue to take into account is the stability of the algorithm. An unstable algorithm may cause a large change in results for a small change in the input parameters, or, magnify the small errors from earlier computation stages until the result deviates completely from the true answer. Examples can be found in recurrence relations, like in computing the power of $\phi = -\frac{1}{2}(\sqrt{5}+1)$ using the recursive formula $\phi^{n+1} = \phi^{n-1} - \phi^n$. Solving Fredholm equations of the first kind also belongs to this category [6].

In this paper we introduce the gRpas tool, designed to help analyze test results from scientific computing applications. The idea was originally motivated by applications in numerical integration. Section 2 below outlines the numerical integration framework that lead to the development of gRpas. Section 3 presents the tool's main characteristics, and Section 4 concludes the paper.

2 Testing Numerical Integration Algorithms

The two main ingredients of a numerical integration problem are: $I(\alpha)$ the integral, and $M(\beta)$ the integration method, where α and β are various parameters. We denote by I_j a problem $I(\alpha_j)$ and M_k a method $M(\beta_k)$. We consider the integrand function, integration region, requested accuracy and maximum number of function evaluations as part of the integration parameters α. A few examples of integration methods are: adaptive, Monte Carlo and quasi-Monte Carlo methods. Various cubature rules that can be used at the basis of adaptive multivariate integration methods are listed in [2].

A major issue in numerical integration is that often we don't know the characteristics of the integrand function. In general, we don't have answers to questions like: are there any singularities, how steep are they and where are they located?

We denote with M_j^o the best algorithm that solves I_j, that is, the algorithm with the best convergence. Let $C(I, M)$ be the cost of solving I using algorithm M, expressed in an appropriate measurement unit (e.g., number of function evaluations performed).

Given M, it is our goal to develop a set of tests to:

a) find β_k such that on the average over all I_j, $C(I, M)$ is minimum. - Since it is impossible to handle all I_j, we need to determine a suitable test set for which we require that on the average $C(I, M)$ is minimized. Alternatively we may require that $M(\beta_k) = M_j^o$ for I_j in the test set. Or, we may just want to extract as much information as possible about the algorithm behavior.
b) compare M_k with other algorithms for the entire test space.

3 gRpas Tool

gRpas [7] is a small application designed to help analyze results from scientific applications. Its main features are listed bellow.

3.1 Flexible Application Interface

For inter-operability and flexibility, all data is stored in an XML like format. For each input/output parameter, the user specifies a set of attributes: name, type, and tag. These are stored in a type repository and are used by the GUI module and re-used for similar tests. All results, along with input parameters, are stored in a tagged format.

3.2 Option for Filtering or Combining Results

Tests are performed in groups. Usually a set of tests is performed, analyzed and if there are satisfactory results, another group of tests is performed. If the results are not as expected, the code is changed and the same set of tests is repeated. *gRpas* can combine the results from different groups of tests or can present them individually. Figure 1 shows the tree structure: applications - versions - tests. In Figure 2 a Test Set is expanded to show the relation between Individual Tests and input parameters for an application with 4 parameters: p_1, p_2, p_3, p_4, where p_1 gets assigned three values and p_2 two. The Test Set is composed of 6 Individual Tests.

Note that although the version can be seen as an algorithm parameter, from the practical point of view it is better to consider it as a separate entity.

The user can filter certain results across multiple sets of tests for individual analysis, or can combine multiple tests from multiple sources.

3.3 Flexible User Interface

The application is mainly designed for algorithm comparisons, and is mainly intended for scientific software developers. The user interface is simple and is

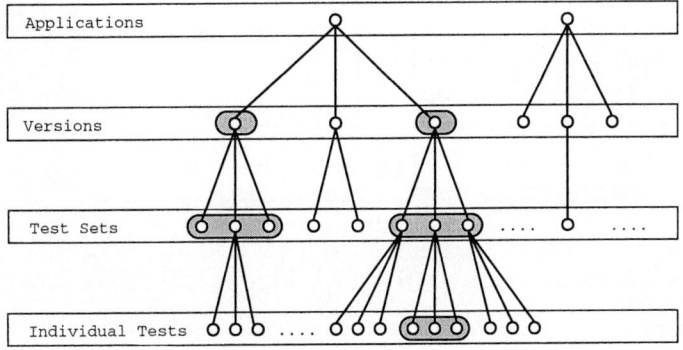

Fig. 1. Tests Tree Structure

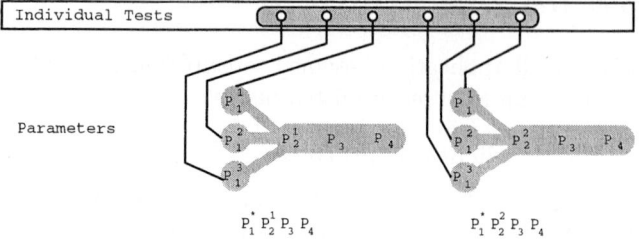

Fig. 2. Individual Tests and Parameters

point-and-click based. This allows for a short learning curve and fast response to the user. After data is collected, extended plots can be generated. Parameters defined in the input files can be selected by their names from drop-down menus. Multiple plots can be drawn in the same window and multiple windows can be displayed at once.

Figure 3 shows the main user interface. The table in the lower part lists all the Test Sets performed for a particular version of code (they correspond to the shaded areas in the *Test Sets* level in Figure 1). In the rightmost column the title shows the name of the code version, and the cells show the time when the

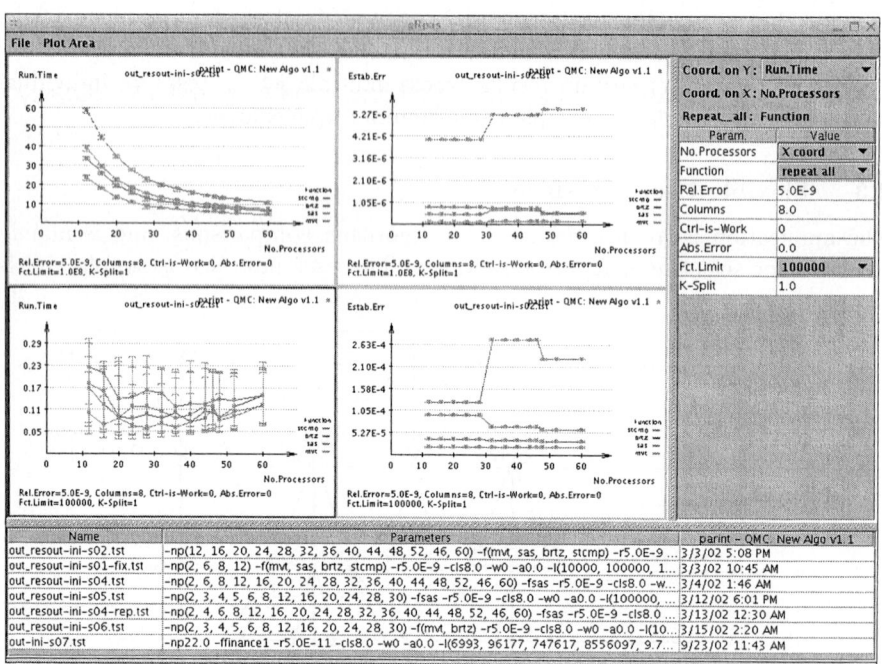

Fig. 3. gRpas: multiple plots, multiple window

Sets were generated. The middle column shows the parameters used in the Set, and the left column shows the associated name for the Set.

Once the user has selected a Test Set, its parameters are shown in the control group situated in the right (they correspond to the type of area that is shaded in the *Individual Tests* level in Figure 1). This section has a double role: to show and to select the parameters to be plotted. In the top part the user selects the Y coordinate from a drop list containing all the output variables, while the X coordinate is selected from the input parameters with multiple values. Once the user has selected a parameter set, the plots are displayed in the center-left area.

The plots in Figure 1 graph results from solving multivariate integration problems using a quasi-Monte Carlo method with up to 60 processors. The variable input parameters are: Integrand Function, Number of Processors, and Function Limit. This particular Set consists of 325 Individual Tests, each repeated a number of times. On the left, the Run Time is plotted versus the Number of Processors for all the integrand functions using a different Function evaluation Limit, whereas on the right the obtained error is plotted. Each plotted point represents the median of multiple values. The data interval is shown as a closed segment.

Figure 4 presents two algorithm versions (note an additional column in the table Set area). The two versions have a number of identical Test Sets: *Test s01*, *Test s02*, *Test s05*, and *Test s06* (which can be identified by the date in the algorithm version column). The plot shows the differences in the result values.

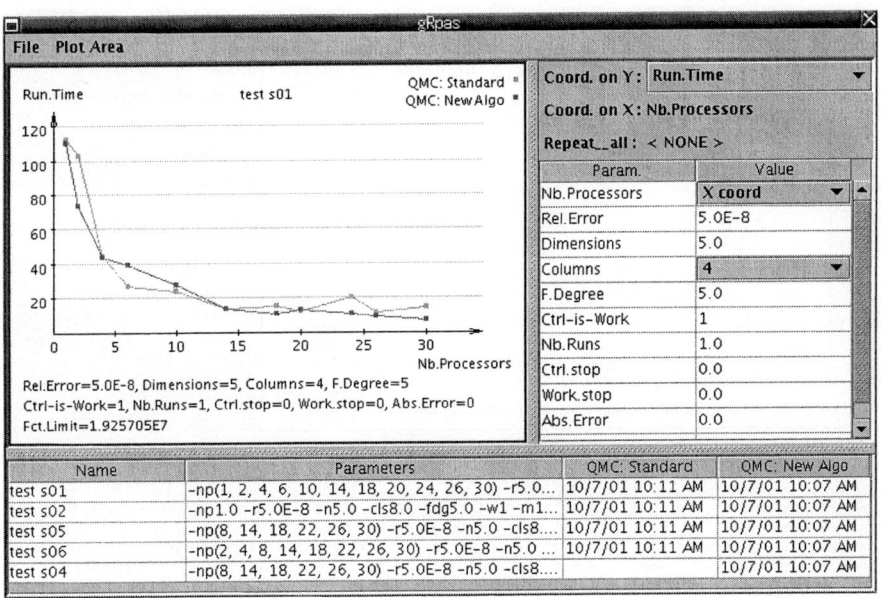

Fig. 4. gRpas: multiple algorithm plot

3.4 User Defined Plugins

Extern modules can be linked to gRpas for additional functionality. The user can pre-process the output before displaying, or can link in a different application. The plugins must be written in java. The extern modules are called using menus. For example, Figure 5 shows the effect of applying the scalability plugin: for each test, the plot shows the inverse of the runtime multiplied with the runtime of the first test (which corresponds to the sequential run).

3.5 Statistical Reports

As mentioned earlier, the developer is interested in two types of comparisons. First, for a given algorithm M, what is the value of β such that $M(\beta)$ gives the best performance for a set of problems I_j. Second, given two or more algorithms, which one performs better for the same set of problems.

Most of the time, a simple inspection of the plots gives enough information to continue analyzing the algorithm or to change it.

To generate more elaborate statistics we have two options, either implement a small set of functions in java as plugins, or export the data to applications like Splus or SAS. It is our goal to implement a set of functions that can perform comparisons at the click of a button.

For further consideration, we envision a new functionality for generating plots from non-common-parameters tests. Instead of performing Test Sets where some parameters are fixed while others vary, is possible to generate indicative plots

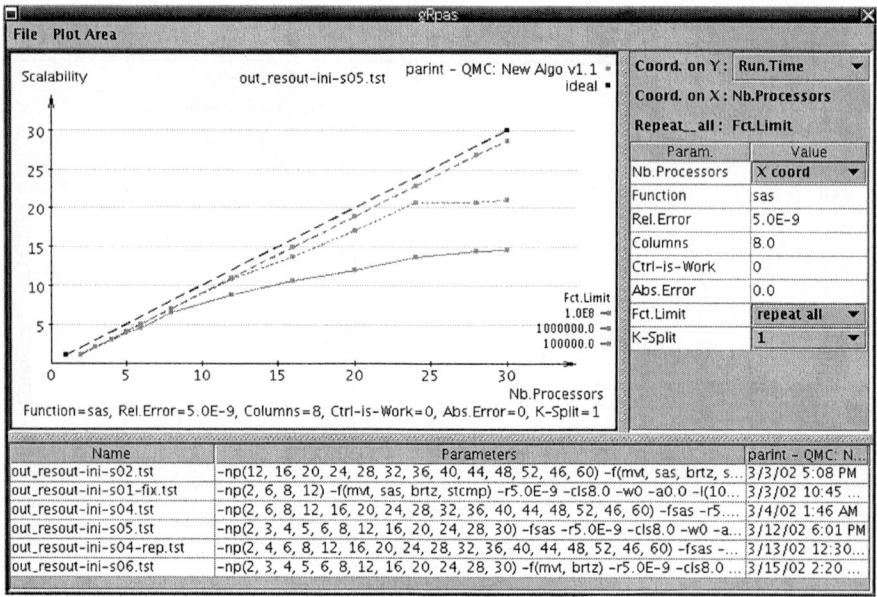

Fig. 5. gRpas: scalability plugin

from Tests that don't share common parameters. In this way it is possible to reduce the number of performed tests, while increasing the problem investigation space.

4 Conclusions

In this paper we presented "gRpas", a tool to help analyze test results for scientific computing applications. Intended to be used for both competitive testing as well as controlled experimentation, the application has been extensively used in our parallel numerical integration research. We included samples of its usage from solving multivariate t-distribution (MVT) problems using a quasi-Monte Carlo method with up to 60 processors. The "gRpas" application and code is available for download at [7]. It can also be executed online as an applet at [7].

Acknowledgements

The authors thank Krishna K, and Brendon Thiede for important programming contributions.

References

1. J. N. Hooker, "Testing heuristics: We have it all wrong", J. Heuristics, vol. 1, no. 1, pp. 33-42, 1995.
2. R. Cools. Monomial cubature rules since "Stroud": a compilation – part 2. J. Comput. Appl. Math., 112(1-2): pp. 21-27, 1999.
3. C. H. Chen, S. D. Wu, L. Dai, "Ordinal Comparisons of Heuristics Algorithms Using Stochastic Optimization", IEEE Transactions on Robotics and Automation, vol. 15, no. 1, pp. 44-56, 1999.
4. J. N. Lyness and J. J. Kaganove, "A Technique for Comparing Automatic Quadrature Routines", The Computer Journal, vol. 20, no. 2, pp. 170-177, 1977.
5. R. Zanny, K. Kaugars and E. de Doncker, "Scalability of Branch-and-Bound and Adaptive Integration", in Proceedings of the Conference on Parallel Distributed Processing Techniques and Applications (PDPTA'01), pp. 674-680, 2001.
6. Press, W. H., Teukolsky, S. A., Vetterling, W. T. and Flannery, B. P, "Numerical Recipes in C - The Art of Scientific Computing", Second Edition, Cambridge University Press",1997.
7. "gRpas", http://aegis.cs.wmich.edu/%7Elcucos/prjs/grpas/doc/gRpas.html

Statistical Methods for Automatic Performance Bottleneck Detection in MPI Based Programs

Michael Kluge, Andreas Knüpfer, and Wolfgang E. Nagel

Technische Universität Dresden, Dresden, Germany
{kluge, knuepfer, nagel}@zhr.tu-dresden.de

Abstract. With increasing number of processors the visualization of trace data for MPI communication primitives becomes harder due to many limitations. An automatic analysis of the recorded data should be able to identify some clues for the optimization of the source code. The approach presented with this paper is able to divide the communication time for each communication relationship into a necessary part and the overhead. We use statistical methods to define a time for each single communication event that we will accept as the 'usual' time for this event. Additional runtime will be considered as overhead. At the end of this article, the benefits from this method are demonstrated on the Sweep3D benchmark.

1 Introduction

2005, the supercomputer IBM 'Blue Gene' will be installed with a processor count of 131072. The first desktop systems with more than a single processor are already available for everyone. To use such systems efficiently parallel programs have to be developed that utilizes all processors. Writing parallel programs today is commonly done by using one (or both) of the two popular programming paradigms, OpenMP [2] and MPI [4]. This work presents a new approach for automatic bottleneck detection in MPI programs.

The first section gives an overview on actual approaches in MPI program analysis. The next sections are dedicated to the motivation and our statistical approach for finding a maximal communication time. Then we will present some tests for the practical relevance of these maximal communication times. At the end of this paper we will show how a profiler using our approach can be used for automatic performance bottleneck detection in the Sweep3D benchmark.

2 State of the Art in MPI Program Analysis

The question how well a program is scaling with an increasing amount of processors has driven the development of many tools (see [11], [10], [3] etc.). Recording all program activities during runtime and an appropriate post-mortem analysis of the recorded data is still state of the art in parallel program analysis. This is due to two main reasons. First of all, in the beginning of an optimization process one does typically not know where to start. What is needed is a trace of all program activities. The outcome of this is at the same time the second reason for the post-mortem analysis, a huge amount of data. The user himself will not be able to do any analysis during runtime. An automated analysis will most probably interfere with the program execution in a way that no meaningful conclusions can be drawn from these data. Actually there exist several approaches for trace file compression or removing of redundant information during the trace process itself ([7],[1]) and during the post-mortem analysis ([8]).

3 Motivation

The aim of this article is to define a value that reflects the maximum execution time for a MPI function call under ideal conditions. Within our approch we try to determine, how the data within a sample generated with a repeated execution of this function are distributed. If those data are distributed like a theoretical distribution (like a Normal/Gaussian or Exponential distribution) it is possible to define for example the time that is associated with 99% of this distribution as the maximum time. First experiments have shown that the data we are looking at are usually not normally distributed. Instead of this we have seen asymetric distributions. As a result of this, a simple approach like calculating the mean μ and variation σ^2 and afterwards using the time associated with $\mu + 3\sigma$ would end up with a maximum execution time without meaning. Because a calculation that $\mu + 3\sigma$ includes 99.73% of the distribution is only true for the assumption of a Gaussian distribution.

4 The Statistical Approach

To be able to divide the time t needed for each MPI communication event into a necessary and a wasted part we will define a time t_{max} as a maximum time we want to accept for an execution without overhead. A communication event is a call of an MPI function of the MPI 1.1 standard that transmits data or completes one of the non-blocking communication calls [4–p. 41]. The time t is the execution time a function call needs during the program execution. To determine t_{max} we take a sample $S = \{t_1, \ldots, t_N\}$ with the sample size N of execution times for a communication event and define the wasted time t_{oh} as follows:
$$t \leq t_{max} \rightarrow t_{oh} = 0$$
$$t > t_{max} \rightarrow t_{oh} = t - t_{max}$$
We assume that a sample taken with a synthetic program represents the typical behavior of the underlying MPI implementation and hardware. The problem that we have to face is that although each time t_i within the sample is taken by executing the same function with the same parameters those times are not

identical. They are distributed in a way that differs from machine to machine. Each time t_i can not be smaller than a minimum t_{min} that is determined by the MPI implementation and the hardware. It is most likely that none of the times t_i will match t_{min} exactly. Instead of that each is superimposed with one or more errors. If we assume an exclusive use of the communication network we can consider that those errors result from the overlapping execution of processes, from the operation system or from the resolution of the timer used for the measurement. If we assume further on that those processes are activated periodically and each activation does only influence one measurement they can be represented by a Poisson distribution. A sum of Poisson distributions is again a Poisson distribution. If we presume t as

$$t = t_{min} + t_{err}, \qquad (1)$$

we can model the distribution of the error t_{err} as a Poisson distribution. During our experiments we have seen that there are other statistical distributions that sometimes fit to the error within a sample better than a Poisson distribution. The upper bound of the error that can be accepted can be fixed by establishing a quantile q_{max} for the adapted theoretical distribution. With this quantile (reasonable values are $0.8 \leq q_{max} \leq 1$) we are now able to calculate a maximum error time $t_{err,max}$. If we assume that $\min(S)$ is close to t_{min} we can define t_{max} as

$$t_{max} = \min(S) + t_{err,max}. \qquad (2)$$

5 Used Models

In this evaluation we have used two theoretical distributions as the base for the models. The first one is the already mentioned Poisson distribution. The probability density function

$$p_k = \frac{\lambda^k}{k!} e^{-\lambda}, \ (k = 0, 1, \ldots) \qquad (3)$$

with the parameter lambda is fitted to the histogram of the error.

The second used model is the discretization of an originally continuous exponential distribution. We have seen in our experiments on different computer architectures that the type of the distribution differs from architecture to architecture. The observed error is sometimes distributed in a way that is similar to an exponential distribution. For the discretization of an exponential we will part the continuous function

$$f(x) = \begin{cases} 0 & : \ x < 0 \\ \lambda e^{-\lambda x} & : \ x \geq 0 \end{cases} \qquad (4)$$

into windows of equal size.

By introducing a window size i in a way that

$$p_k = F((k+1) * i) - F(k * i), \qquad (5)$$

where k is the class number we get

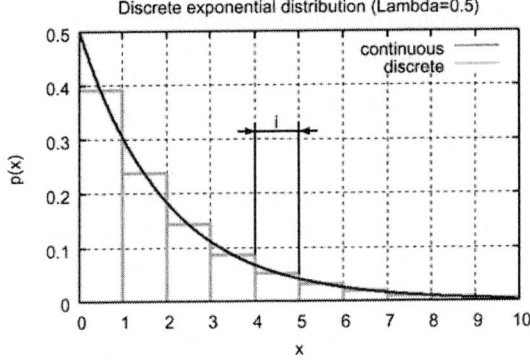

Fig. 1. Relationship between the continuous and discrete exponential distribution

$$p_k = e^{-\lambda k i}(1 - e^{-\lambda i}) \tag{6}$$

as a discrete exponential distribution. Some connections of the original and the discretization can be taken from Figure 1.

6 Fitting the Model to the Data

To actually fit a sample to a model we first create a histogram from the sample. To do this we need to sort the data into windows of a certain size. This window size should not be smaller than the resolution of the timer used to get the sample. Now we are able to use the maximum likelihood method

$$L(x_1, \ldots, x_n; \gamma) = p_1^{f_1}(\gamma) * p_2^{f_2}(\gamma) * \ldots * p_r^{f_r}(\gamma), \tag{7}$$

to fit the model to the data. The result of this process is a set of parameters for a model.

7 Quality of the Model Fitting

To determine how well a model fits to the distribution the χ^2 test is a way to check the quality of the model fitting in the case of discrete distributions. The χ^2 test uses a weighted sum of the square of the difference between the theoretical frequency and the practical frequency for each class.

$$\hat{\chi}^2 = \sum_{i=1}^{k} \frac{(B_i - E_i)^2}{E_i} \tag{8}$$

If the sum $\hat{\chi}^2$ does not exceed an upper limit, we accept the model fitting. This upper limit can be found in tables in adequate books (for example [12]) or can be approximated for those cases that use more than 30 classes. This approximation is defined as

$$\chi_\nu^2 \approx \nu \left(1 - \frac{2}{9\nu} + z_\alpha \sqrt{\frac{2}{9\nu}}\right)^3, \qquad (9)$$

where ν are the degrees of freedom and z_α depends on the selected level of significance and has to be taken from an adequate table. We always try to fit each model to the sample and use the model with the lowest $\hat{\chi}^2$. If all models are rejected, there are two choices: generation of a new sample or building a new model.

To increase the quality of the fitting we can use function (9) as the objective function for an optimization

$$\min_{\lambda \in \mathbb{R}^+} \sum_i \|f_i - p_i(\lambda)\|. \qquad (10)$$

In this case we can use the parameters we got from the maximum likelihood method as the start point for the optimization. The optimization itself will result in slight changes of this parameters and a better fitting of the model in the sense of the χ^2 test.

8 Creating a Complete Profile

One problem that arises while analyzing the MPI related behavior of different applications at different numbers of processors is that there are a lot of varying message sizes. It is not necessary to pay attention to each possible single message size at each possible processor count for each MPI function if the assumptions is made, that the application will always be analyzed with the same environment (hardware, MPI library, environment variables) the model has been built in. If this environment does not change during the model fitting (and the generation of the program traces later) we are able to define a discrete function

$$T_f : p, b \to t_{max} \qquad (11)$$

for each MPI function that uses the number of used processors (p) and the number of bytes (b) to define the time t_{max} for this pair (p, b). In order to be able to evaluate T_f for each possible pair (p, b) we will define T_f for some base points and interpolate all other points from these. To keep the error introduced by this technique as low as possible the initial pairs (p, b) have to be chosen carefully. Within the BenchIt project [5] we have seen that for those communication relations the following heuristic will work well for a start:

- Numbers of processors: each power of two and a number in between. (2^n and $1.5 * 2^n$ with $n = \{1, 2, \ldots\}$)
- Message size: 1 Byte, 10, 20, ..., 100, 200, ..., 1000, and so on up to 4 MB

By using a bilinear interpolation between this base points we are now able to evaluate T_f for each reasonable pair (p, b).

9 Profile for a SGI Origin 3800

As a first example, we present the measurements for a subset of what we have proposed within the last section. We will use just a single processor count and a single message size to create a profile for a MPI_Allreduce() with one integer and eight processors on a SGI Origin 3800. For a first try, we have taken a sample

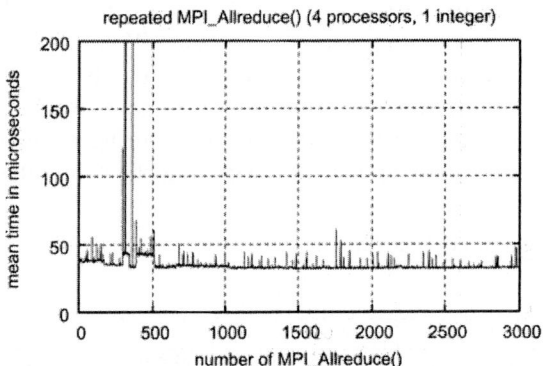

Fig. 2. Runtimes for repeated execution of a MPI_Allreduce() with one integer and eight processors on a SGI Origin 3800

of size 3000. This sample is shown in Figure 2. To actually see how large the sample really needs to be, we have fitted the models to subsets of this sample. The first subset size we used is 50. This size is increased by 10 (up to 3000). Each try produces a t_{max} which is plotted in Figure 3 against the subset size.

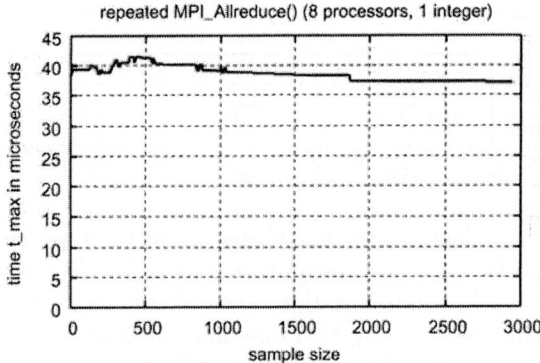

Fig. 3. Result of the model fitting for the model with the Poisson distribution plotted against the sample size

One conclusions from this example and our experiments at all [6] is that sample sizes around 200 are already enough to fit a model to the data. What we have seen during the experiments is that the error $\hat{\chi}^2$ is sensitive to the sample size and has always increased with it even when the result t_{max} is stable.

10 Application to Sweep3D

As a more practical example for the usefulness we have implemented a profiler that is able to trace each call of a MPI function back to a source code location. This source code location is either a source file and the associated line number or the name of the function that called this MPI function. By using the arguments applied to the MPI function (mainly the message size s and the execution time t) and the number of processors participating in a collective MPI function call it is possible to find the (p,b) pair for each call. Now we can evaluate T_f to figure out how long this call should have taken at maximum. By adding up t_{oh} for each source code location it is possible to automatically find the source code locations, where the application is wasting time during communication. These position will be a good entry point for optimizing the communication behavior.

Table 1. Total communication time in seconds for Sweep3D for different numbers of processors

numbers of processors	4	8	12	14	21	24	28
communication time	26.79	55.03	73.78	109.42	129.35	124.40	166.10

As an practical example we have chosen the Sweep3D application [9] from the ASCI benchmark collection. The communication time is mainly claimed by point to point communication. In Table 2 we see the overhead for those two lines in the source code in the source file `msg_stuff.cpp` that need most of the communication time.

Table 2. Overhead in seconds caused by the two source code lines performing the point to point communication within the file msg_stuff.cpp at different number of processors

Line	Function	Numbers of processors						
		4	8	12	14	21	24	28
311	MPI_Send()	5.9508	11.7368	9.3950	12.5326	13.9212	10.0274	12.2072
364	MPI_Recv()	18.0278	36.9847	54.6950	85.4923	98.2872	95.7525	129.9080

With this information and the knowledge about the total time spent during communication (Table 1) we can conclude that there is a late sender/late receiver problem dominating the communication time with the bigger fraction for the late sender. By replacing the blocking MPI function calls with non-blocking calls and one MPI_Waitall() we were able to attenuate this problem.

11 Conclusion and Future Work

We have shown that it is possible to use statistical distributions to model the overhead that occur during a repeated execution of a MPI function. By using a fixed quantile of the distribution we are able to define the time each MPI function call should not exceed. With this information we are able to find the overhead within communication relations. Mapping this overhead back to the source code location gives the user clues about good entry points for program optimization.

Future work will be the an automatic measurement for a complete MPI profile without the heuristics.

References

1. Dong H. Ahn and Jeffrey S. Vetter. Scalable analysis techniques for microprocessor performance counter metrics. In *Proceedings of Supercomputing 2002*, pages 1–16, 2002.
2. OpenMP Architecture Review Board. OpenMP Application Program Interface, Version 2.5. http://www.openmp.org/drupal/mp-documents/draft_spec25.pdf, November 2004.
3. Luiz DeRose and Felix Wolf. CATCH - A Call-Graph Based Automatic Tool for Capture of Hardware Perfromance Metrics for MPI and OpenMP Applications. In *Euro-Par 2002*, 2002.
4. Message Passing Interface Forum. MPI: A Message-Passing Interface Standard. Technical report, University of Tennessee, 1995.
5. Guido Juckeland, Stefan Börner, Michael Kluge, Sebastian Kölling, Wolfgang E. Nagel, Stefan Pflüger, Heike Rödling, Stephan Seidl, Thomas William, and Robert Wloch. BenchIt-Performance Measurement and Comparison for Scientific Applications. In *G. R. Joubert et. al., Eds., Advances in Parallel Computing, Vol. 13*, pages 501–508. Elsevier, 2004.
6. Michael Kluge. Statistische Analyse von Programmspuren für MPI-Programme. Diploma thesis, November 2004.
7. Andreas Knüpfer. A New Data Compression Technique for Event Based Program Traces. In *International Conference on Computational Science 2003*, pages 956–965, 2003.
8. Andreas Knüpfer and Wolfgang E. Nagel. Compressible Memory Data Structures for Event Based Trace Analysis. *Future Generation Computer Systems by Elsevier*, 2004.
9. Lawrence Livermore National Laboratory. The ASCI Sweep3D Benchmark Code. http://www.llnl.gov/asci_benchmarks/asci/limited/sweep3d/asci_sweep3d.html, 1995.
10. Tom LeBlanc and Wagner Meira. Measurement and Prediction of Parallel Program Performance, http://www.cs.rochester.edu/u/leblanc/prediction.html. http://www.cs.rochester.edu/u/leblanc/prediction.html, 1997.
11. Wolfgang E. Nagel, Alfred Arnold, Michael Weber, Hans-Christian Hoppe, and Karl Solchenbach. VAMPIR: Visualization and Analysis of MPI Resources. In *Supercomputer 63, Volume XII, Number 1*, pages 69–80, 1996.
12. Lothar Sachs. *Applied Statistics: A Handbook of Techniques*. Springer, 11 edition, 2002.

Source Templates for the Automatic Generation of Adjoint Code Through Static Call Graph Reversal

Uwe Naumann[1] and Jean Utke[2]

[1] Software and Tools for Computational Engineering,
RWTH Aachen University, D-52056 Aachen, Germany
naumann@stce.rwth-aachen.de,
http://www.stce.rwth-aachen.de
[2] Mathematics and Computer Science Division,
Argonne National Laboratory, 9700 S. Cass Avenue,
Argonne, IL 60439, USA
utke@mcs.anl.gov,
http://www.mcs.anl.gov

Abstract. We present a new approach to the automatic generation of adjoint codes using automatic differentiation by source transformation. Our method relies on static checkpointing techniques applied to an extended version of the program's call graph. A code template is provided to implement a control structure governing the execution of the adjoint and augmented forward versions of each subroutine in the program. These code variants are generated automatically by algorithms that are independent of the programming language of the original code. The major advantage of this new approach is its flexibility with respect to various reversal schemes.

1 Context and Outline

This paper discusses novel algorithmic choices made in the context of the ongoing work on OpenAD (see www.mcs.anl.gov/OpenAD) – a software tool for the automatic generation of adjoint codes. OpenAD is being developed as part of the Adjoint Compiler Technology and Standards (ACTS, see www.autodiff.org/ACTS) project. The main application of this tool within the ACTS project is the MIT General Circulation Model (MITgcm, see mitgcm.org).

The structure of the paper is as follows. In Section 2 we discuss the basics of adjoint code construction and present an example for an adjoint code at the level of single subroutines. Two static call graph reversal modes are explained in Section 3. In Section 4 we consider a new approach to the generation of adjoint code based on code templates. A successful application of the new method is presented in Section 5.

2 Introduction

Inverse methods [1] play an increasingly important role in various application areas of numerical scientific computing. Such methods are of interest, for example, in the context of large-scale gradient-based optimization methods as they eliminate the dependence of the complexity of the gradient computation on the dimension of the parameter space. We assume that we are given a computer program that implements some mathematical model

$$\mathbf{y} = F(\mathbf{x}), \quad \mathbf{x} \in \mathbb{R}^n, \mathbf{y} \in \mathbb{R}^m \quad . \tag{1}$$

Our aim is to derive an adjoint program that computes the product of the transposed Jacobian matrix $(F'(\mathbf{x}))^T$ with an adjoint vector $\bar{\mathbf{y}}$ in the image space $\mathbb{R}^m$. This modification of the program's semantics is to be performed by a source transformation tool for automatic differentiation. In a compiler like fashion the original program for F is parsed into an abstract intermediate representation (ir). The source transformation is performed based on a set of static analyses [2] on ir. The modified internal representation $\bar{ir}$ is unparsed to obtain the adjoint program. This process is illustrated in Figure 1. Automatic differentiation (AD)

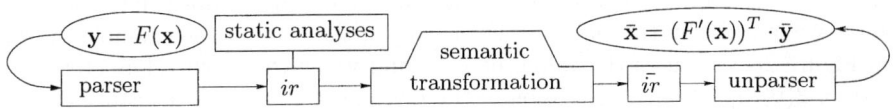

Fig. 1. Source transformation tool for the automatic generation of adjoint codes

[3, 4, 5] is a set of techniques for transforming numerical programs into derivative code that can be used to compute derivatives of vector functions such as Jacobians, Hessians, or higher-order Taylor coefficients. A detailed description of the mathematical foundations of AD is beyond the scope of this paper. Refer to [6] for a discussion of the theory.

The adjoint of a program implementing a vector function as in Equation (1) is obtained by the reverse mode of AD. Given values for $\mathbf{x}$ and the adjoints of the original outputs $\bar{\mathbf{y}}$, the adjoint program computes the transposed Jacobian vector product

$$\bar{\mathbf{x}} = (F')^T \cdot \bar{\mathbf{y}} \quad . \tag{2}$$

This process is best introduced with the help of a simple example that illustrates the basic features. Consider the Fortran implementation of a function $\mathbf{y} = F(\mathbf{x})$ depicted in Figure 2(a), where $\mathbf{x} \in \mathbb{R}^4$, $\mathbf{y} \in \mathbb{R}$. Often the program that implements F consists of a possibly large number of subroutines calling each other. In this paper we consider methods for implementing adjoint codes at the *interprocedural* level. Our approach uses a given solution for the problem of constructing adjoints at the *intraprocedural* level like the one shown in Figure 2(b),(c).

An execution of the *augmented forward code* (as in Figure 2(b)) stores the flow of control [2] in an integer stack (IS) as well as all numerical values (in a

(a)
```
    y=1.
    do i=1,4
      if (i<3) then

        x(i)=sin(x(i))
      else

        x(i)=x(i)*x(i)
      end if

      y=y*x(i)
    end do
```

(b)
```
    y=1.
    do i=1,4
      if (i<3) then
        push(IS,1)
        push(FS,x(i))
        x(i)=sin(x(i))
      else
        push(IS,0)
        push(FS,x(i))
        x(i):=x(i)*x(i)
      end if
      push(FS,y)
      y=y*x(i)
    end do
```

(c)
```
    do i=4,1,-1
      call pop(FS,y)
      x_b(i)=x_b(i)+y*y_b
      y_b=x(i)*y_b
      call pop(IS,branchId)
      if (branchId==1) then
        call pop(FS,x(i))
        x_b(i)=cos(x(i))*x_b(i)
      else
        call pop(FS,x(i))
        x_b(i)=2*x(i)*x_b(i)
      end if
    end do
```

Fig. 2. Intraprocedural adjoint code: original (a), augmented forward (b), adjoint (c)

floating-point stack FS) needed for the adjoint computation during the following reverse sweep. In principle the adjoint code is obtained by applying Equation (2) to each statement $y = \phi(x_1, \ldots, x_k)$ of the original code in reverse order (w.l.o.g. we consider scalar assignments). This application yields $\bar{x}_i = \bar{x}_i + \frac{\partial \phi}{\partial x_i}\bar{y}$ for $i = 1, \ldots, k$. The increment of the current value of $\bar{x}_i$ is due to the fact that x_i can occur on several right-hand sides. Adjoint versions of variables in the original code are marked by the suffix _b. For example, the code in the example shown in Figure 2(c) requires the values of x(i) and y, which are therefore stored in Figure 2(b). A detailed discussion of the reversal of the flow of control inside a subroutine can be found in [7].

Adjoint codes permit the accumulation of the Jacobian at a cost proportional to the number of outputs m. Of particular interest are gradients ($m = 1$) obtained at a small constant multiple of the cost of evaluating the function itself, for example, in the context of data assimilation in the MITgcm. The gradient of some objective with respect to parameters at the grid points of a very fine discretization leads to a number of input variables n on the order of 10^9. Neither forward-mode AD nor approximation by finite difference quotients represents a feasible approach, as either has a complexity of $O(n)$.

3 Call Graph Reversal Modes

In the remainder of this paper we assume that adjoint code is available for all subroutines in the given program generated by a method similar to the one sketched in the previous section. The *call graph* (CG) is usually defined as a graph with nodes representing subroutines and edges representing potential (direct) calls [2]. It is a static entity and generally not acyclic. To be useful for the static construction of adjoint code at the interprocedural level, we need to break the cycles, a process generally possible only for a concrete execution of the program. The result is the *dynamic call tree* (DCT). Edges represent subroutine calls and the order of execution in the context of a depth-first traversal. The vertices represent executions of variants of subroutine code (forward, augmented forward,

```
subroutine 1
    call 2; ... call 4; ... call 2;
end subroutine 1
subroutine 2
    call 3
end subroutine 2
subroutine 4
    call 5
end subroutine 4
```

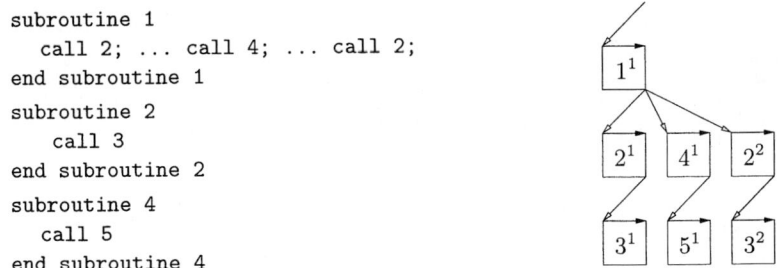

Fig. 3. Dynamic call tree of a simple calling hierarchy

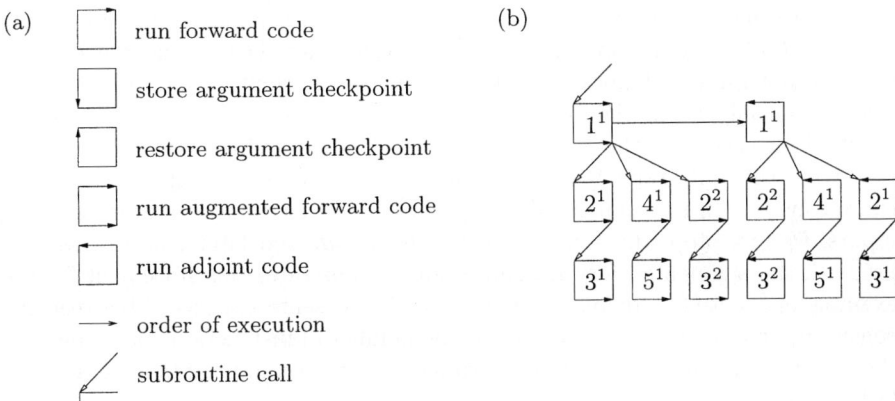

Fig. 4. Symbols for reversal mode graphs (a), DCT of split reversal mode (b)

adjoint). The order of calls to other subroutines shown on the next lower level is implied by the left to right order of edges emanating from a given vertex. An edge pointing to a vertex on the same level indicates the execution of a different code variant of the same subroutine. Hence, the depth-first traversal is well defined. Figure 3 shows the DCT for a simple example. Vertices are labeled with the identifier of the respective subroutine. Superscripts denote the instance of the call, as one and the same subroutine can be called repeatedly. In the example, subroutines 2 and 3 are called twice. The symbols are explained in Figure 4(a). Note that for the purpose of static reversal schemes the DCT is merely a conceptual tool and does not need to be instantiated in practice.

Two basic call graph reversal modes have been proposed in the literature [6]. In *split* reversal mode the adjoint computation is preceded by an execution of the augmented forward code of the entire program. All values that are needed for the correct reversal of the intraprocedural flow of control, as well as those required to evaluate the adjoint assignments correctly, are stored. The split reversal of the example in Figure 3 is shown in Figure 4(b). For computationally challenging problems this approach usually leads to prohibitively large memory requirements.

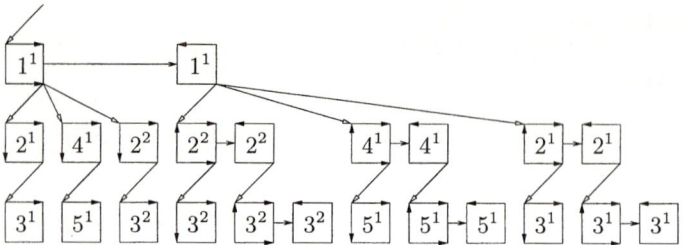

Fig. 5. DCT of adjoint obtained by joint reversal mode

This problem can be mitigated by the *joint* reversal mode, where a trade-off between memory requirement and computational complexity is achieved by *checkpointing* [8] at the level of subroutine arguments. The augmented forward code variant directly followed by the adjoint code variant of the same subroutine is run only when the adjoint values of the inputs of this subroutine are required for executing the adjoint code variant of the calling subroutine. While traversing the call graph backwards, one needs to know the input values to a subroutine to run its augmented forward code. Rather than recomputing these inputs, we first store them as *argument checkpoints* and later restore them for use with the augmented forward run. Figure 5 illustrates this statement for the example in Figure 3. Obviously, joint reversal represents a trade-off between the consumption of memory resources and the number of instructions performed by the adjoint code. Split and joint reversal modes can be combined in a hierarchical fashion as described in [6] and Section 5.

4 Code Templates

Section 3 outlines the principal approach to control a reversal scheme. The basic building blocks are variants S_i of the original subroutine code S_0, each accomplishing one of the tasks shown as a subroutine symbol in Figure 4(a). The S_i are created by transforming the source code contained in the respective subroutine bodies. To integrate the S_i into a particular reversal scheme, we need to be able to make all subroutine calls in the same fashion as in the original code and, at the same time, control which task each subroutine call accomplishes. We replace the original subroutine body with a branch structure in which each branch contains one S_i. The execution of each branch is determined by a global control structure whose members represent the state of execution in the reversal scheme. The branches contain code for pre- and post-state transitions enclosing the respective S_i. This ensures that the transformations producing the S_i do not depend on any particular reversal scheme. In practice we insert the S_i in a subroutine template, shown in Figure 6(a). The template is written in the target language, and the insertion is done by a postprocessing step that identifies specific pragmas in the template code as insertion points for the S_i. Anything related to setting up storage for taping

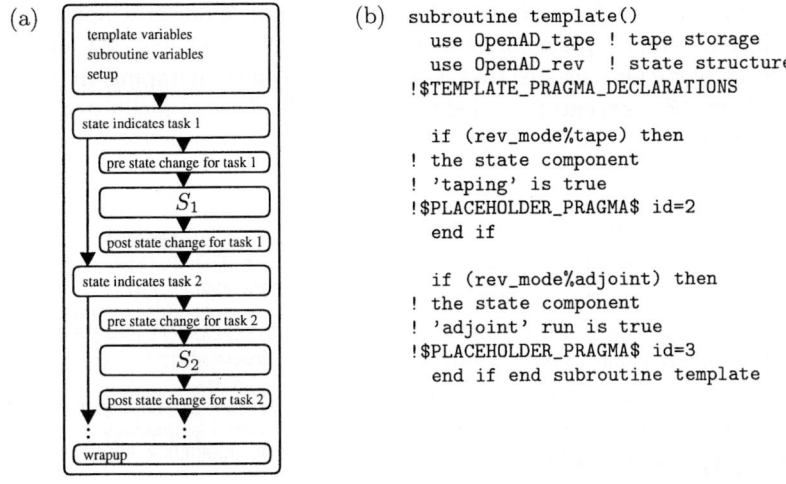

Fig. 6. Subroutine template components (a), split-mode Fortran90 template (b)

or checkpointing, such as declaring data arrays or referring to external modules, can be coded directly into the template. The following paragraphs show examples.

Split Reversal: Split reversal is the simplest static reversal mode. We first execute the entire computation with the augmented forward code (S_2) and then follow with the adjoint (S_3). From the task pattern shown in Figure 4(b) it is apparent that, aside from the top-level routine, there is no change to the state structure within the call tree. Therefore, there is no need for state changes within the template. Since no checkpointing is needed either, we have only two tasks: producing the tape and the adjoint run. Figure 6(b) shows a simplified version of the split-mode template used in OpenAD. The different S_i in the PLACEHOLDER_PRAGMA are identified by their respective $i =$ id. The state is contained in rev_mode, a static Fortran90 variable in module OpenAD_rev of type modeType also defined in this module. In order to perform a split-mode reversal for the entire computation, a driver routine calls the top-level subroutine first in taping mode and then in adjoint mode.

Joint Reversal with Argument Checkpointing: Figure 5 illustrates the task pattern for a joint reversal scheme that requires state changes in the template as well as the insertion of more tasks. Figure 7 shows a simplified version of the template used in OpenAD. The state transitions in the template directly relate to the pattern shown in Figure 5. Each prestate change applies to the callees of the current subroutine. Since the argument store (S_4) and restore (S_6) do not contain any subroutine calls they do not need state changes. Looking at Figure 5, one realizes that the callees of any subroutine executed in plain forward mode (S_1) never store the arguments (only callees of subroutines in taping mode do). This explains lines 18, 25, and 30. Furthermore, all callees of a routine currently in taping mode are not to be taped but instead run in plain forward mode, as reflected in lines 27 and 28. Joint mode in particular means that a subroutine called in taping mode (S_2) has its adjoint (S_3) executed immediately after S_2. This is facilitated by line 33, which makes the condition in line 35 true, and we execute S_3 without leaving the subroutine. Any subroutine executed in adjoint mode has its direct callees called in taping mode, which in turn triggers their respective adjoint run. This is done in lines 37–39. Finally, we have to account for sequence of callees in a subroutine; that is, when we are done with this subroutine, the next subroutine (in reverse order) needs to be adjoined. This process is triggered by calling the subroutine in taping mode, as done in lines 41–43. The respective top-level routine is called by the driver with the state structure having both `tape` and `adjoint` set to `true`.

5 Application

A simplified version of the MITgcm code is the so-called shallow water model, which can be used for problems of variable complexity. A simple split or joint mode implies storage or runtime requirements far beyond what is feasible. Hierarchical checkpointing has traditionally been used to allow for a trade-off between storage and runtime requirements. For the shallow water code we implemented a reversal scheme with a two-level hierarchical checkpointing splitting the main time-stepping loop into an inner loop i and an outer loop o. Wrapping the respective loop bodies into subroutines allows the use of the subroutine-level template approach illustrated in Figure 8. We show two loop iterations for each level, whereas

```
 1: subroutine template()
 2:   use OpenAD_tape
 3:   use OpenAD_rev
 4:   use OpenAD_checkpoints
 5: !$TEMPLATE_PRAGMA_DECLARATIONS
 6:   type(modeType) :: orig_mode
 7:
 8:   if (rev_mode%arg_store) then
 9:   ! store arguments
10: !$PLACEHOLDER_PRAGMA$ id=4
11:   end if
12:   if (rev_mode%arg_restore) then
13:   ! restore arguments
14: !$PLACEHOLDER_PRAGMA$ id=6
15:   end if
16:   if (rev_mode%plain) then
17:     orig_mode=rev_mode
18:     rev_mode%arg_store=.FALSE.
19:   ! run the original code
20: !$PLACEHOLDER_PRAGMA$ id=1
21:     rev_mode=orig_mode
22:   end if
23:   if (rev_mode%tape) then
24:   ! run augmented forward code
25:     rev_mode%arg_store=.TRUE.
26:     rev_mode%arg_restore=.FALSE.
27:     rev_mode%plain=.TRUE.
28:     rev_mode%tape=.FALSE.
29: !$PLACEHOLDER_PRAGMA$ id=2
30:     rev_mode%arg_store=.FALSE.
31:     rev_mode%arg_restore=.FALSE.
32:     rev_mode%plain=.FALSE.
33:     rev_mode%adjoint=.TRUE.
34:   end if
35:   if (rev_mode%adjoint) then
36:   ! run the adjoint code
37:     rev_mode%arg_restore=.TRUE.
38:     rev_mode%tape=.TRUE.
39:     rev_mode%adjoint=.FALSE.
40: !$PLACEHOLDER_PRAGMA$ id=3
41:     rev_mode%plain=.FALSE.
42:     rev_mode%tape=.TRUE.
43:     rev_mode%adjoint=.FALSE.
44:   end if
45: end subroutine template
```

Fig. 7. Joint mode Fortran90 template with argument checkpointing

the real-world problem potentially has thousands of steps at each level. Subroutine calls located outside the outer loop are nonrepetitive allowing for use of the

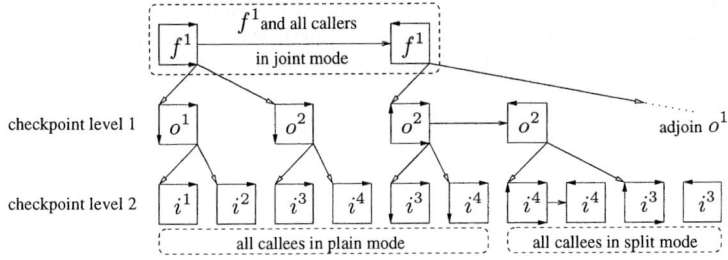

Fig. 8. Two-level hierarchical checkpointing scheme

joint mode because the few checkpoints require little memory. A single execution of the inner-loop body and its callees is short enough to fit all required values on the tape stack, allowing for use of the split mode.

The code transformation uses the templates introduced above for the subroutines handled in either mode respectively. To control the outer and inner level checkpointing, we use two additional templates that are similar to the joint-mode template but have small adjustments in the state transitions to control the checkpointing. The templates and their use in applying OpenAD to the shallow water model code are part of the case studies presented on the OpenAD website.

Acknowledgments

This work was supported by the Mathematical, Information, and Computational Sciences Division subprogram of the Office of Advanced Scientific Computing Research, Office of Science, U.S. Department of Energy, under Contract W-31-109-ENG-38 and by NSF under ITR contract OCE-0205590.

References

1. Wunsch, C.: The Ocean Circulation Inverse Problem. Cambridge U. Press, Cambridge (1996)
2. Aho, A., Sethi, R., Ullman, J.: Compilers. Principles, Techniques, and Tools. Addison-Wesley, Reading, MA (1986)
3. Berz, M., Bischof, C., Corliss, G., Griewank, A., eds.: Computational Differentiation: Techniques, Applications, and Tools. SIAM, Philadelphia (1996)
4. Corliss, G., Faure, C., Griewank, A., Hascoët, L., Naumann, U., eds.: Automatic Differentiation of Algorithms: From Simulation to Optimization. Springer, New York (2002)
5. Griewank, A., Corliss, G., eds.: Automatic Differentiation of Algorithms: Theory, Implementation, and Application. SIAM, Philadelphia (1991)
6. Griewank, A.: Evaluating Derivatives: Principles and Techniques of Algorithmic Differentiation. Number 19 in Frontiers in Appl. Math. SIAM, Philadelphia (2000)

7. Naumann, U., Utke, J., Lyons, A., Fagan, M.: Control flow reversal for adjoint code generation. In: Proceedings of the Fourth IEEE International Workshop on Source Code Analysis and Manipulation (SCAM 2004), Los Alamitos, CA, USA, IEEE Computer Society (2004) 55–64
8. Griewank, A.: Achieving logarithmic growth of temporal and spatial complexity in reverse automatic differentiation. Optimization Methods and Software **1** (1992) 35–54

A Case Study in Application Family Development by Automated Component Composition: h-p Adaptive Finite Element Codes

Nasim Mahmood[1], Yusheng Feng[2], and James C. Browne[1]

[1] Department of Computer Sciences,
University of Texas at Austin, Austin, Texas 78712
{nmtanim, browne}@cs.utexas.edu
[2] Institute for Computational Engineering and Sciences,
University of Texas at Austin, Austin, Texas 78712
feng@ices.utexas.edu

Abstract. This paper reports a case study in automated composition of application families from components. The case study composes multiple instances of an h-p adaptive finite element code. An application family is represented as a structure of components. Each component is encapsulated with an interface giving a semantic specification of the properties and behavior of the component. Instances of the application family can be automatically assembled from a library of components by a compiler and the application instance can be optimized by component replacement during runtime through runtime component selection and binding. The case study demonstrates the benefits of the component composition approach to application family development and shows that execution efficiency is maintained or improved by the componentized development process.

1 Introduction

1.1 Applications as Families of Programs: Problem, Approach and Case Study

Application packages such as h-p adaptive finite element codes can be applied to a wide spectrum of problems in engineering and sciences [1,2]. These packages are usually composed of a large number of parameterized functions including mesh generation, element matrix generation strategy for h-p adaptation, methods for solution of the resulting linear system of equations, etc. These application codes must operate robustly and efficiently on a wide spectrum of problems and on a wide spectrum of execution environments. The properties and behavior of the program may vary widely with the mesh, with the model/equation set, the properties of the material or materials composing the system which is the subject of the computational model, etc.

The current practice in development of h-p adaptive codes is to construct them as an integrated and comprehensive package of functional modules based on common, shared data structures. A package which is robust and offers a wide spectrum of implementations for each of the possible functions may be very complex and very difficult to debug and to maintain and modify and these codes are often sub-optimally efficient on many of the problems to which they are applied and many of the execution environments upon which they may be hosted.

We address these problems through automated component-oriented development. We demonstrate that such automation is feasible, and execution efficiency is maintained or improved by the componentized development process. The semantic parameters of choice in our case study include: model or equation set, problem geometry, mesh structure including mesh generation, strategy for h-p adaptation and methods for solution of resulting linear system of equations. Each functionality is represented as a separate, semantically parameterized, component. There may be several implementations of each component, for example, there may be several different solvers (direct or iterative).

Each component is encapsulated with an interface which specifies its properties. This encapsulation is based on a domain analysis of the problem set the application is intended to address and a domain analysis of applicable computational methods. An application family is developed by specifying properties for each component and invoking a compiler which automatically selects appropriate components through the use of information on the components specified in the interfaces of these components. The compiler can generate a serial or parallel (either multithreaded or MPI) code as desired. The codes may be adapted to evolution of the meshes and approximation functions by monitoring of the code and runtime replacement of components in the computational model.

The approach is illustrated by a case study of its application to a simple but representative h-p adaptive code which contains all of the functions typically found in such codes. The code is reverse engineered to extract components which can be characterized in terms of the domain analysis. Several versions of the application family are realized through the compilation process and the properties and performance of the instances are presented. It is demonstrated that application instances which are efficient for multiple cases of the reachable application family are readily generated.

2 Approach: Component-Oriented Development with Semantic Interfaces

This section sketches the concepts of component-oriented development which form the basis of this case study and describes the implementation technology. A more complete specification can be found in [7].

2.1 The Interface Definition Language

The concepts of the interface definition language which specifies the semantic properties of the components are sketched in the following.

Component: A component is one or more sequential computations, an interface which specifies the information used for selection and matching of components and a state machine which manages the interface, the interactions with other peers and the invocation of the sequential computations.

Associative Interface: An associative interface [7] encapsulates a component. It describes the behavior and functionality of a component. An associative interface consists of an **accepts** specification and a **requires** specification.

Accepts Specification: An accepts interface specifies the set of interactions in which a component is willing to participate. The accepts interface for a component is a set of three-tuples (profile, transaction, protocol). A profile is a set of attribute/value pairs. A transaction is a somewhat extended procedure call and a protocol is a state machine defining a sequence of interactions.

Requires Specification: A requires interface specifies the set of interactions which a component must initiate if it is to complete the interactions it has agreed to accept. The requires interface is a set of three-tuples (selector, transaction, protocol). A selector is a conditional expression over the attributes of all the components in the domain.

Properties of implementations such as degree of parallelism for a given component are also specified in the associative interface as runtime determined parameters.

2.2 Compilation Process

Each attribute name in the selector expression of a component behaves as a variable. The attribute variables in a selector are instantiated with the values defined in the profile of another component. The profile and the selector are said to match when the instantiated conditional expression evaluates to true.

The source program for the compilation process is a component which implements initialization for the program and has a requires interface which specifies the components implementing the first steps of the computation and one or more libraries to search for components. The target language for the compilation process is a generalized data flow graph as defined in [9]. The generated data flow graph is then compiled to a parallel program for a specific architecture by compilation processes implemented in the CODE [9] parallel programming system.

2.3 Runtime Adaptation

The component-composition approach to application family development enables substitution of components implementing different algorithms during execution.

Most operating systems enable runtime linking of components to executable images. The requirement is to identify components which need to be replaced and to specify the properties of the component which is to be substituted for an existing component. Additionally the programs must be instrumented to acquire and analyze the execution properties and behavior of the components of the system.

Composition of a program from components enables and facilitates each of these tasks. Monitoring can be done on a component by component basis; components whose behavior is unlikely to vary need not be monitored. The monitoring code is readily generated by the compiler on a component by component basis. The required analysis and actions can be provided in a separate component or components.

3 Case Study

3.1 Description of the h-p Adaptive Finite Element Code

This case study in application family development is based on an h-p adaptive finite element code structure developed in [3,4,5]. These packages have a common data structure in one-, two-, and three-dimensional space. The major logical components include mesh generation, problem definition, shape function definition, and element routine, linear system of equation solver, error estimation module, and h-p adaptation module. We have used the one-dimensional code in this case study since it has the same structure as the two-D and three-D codes but is of considerably smaller size.

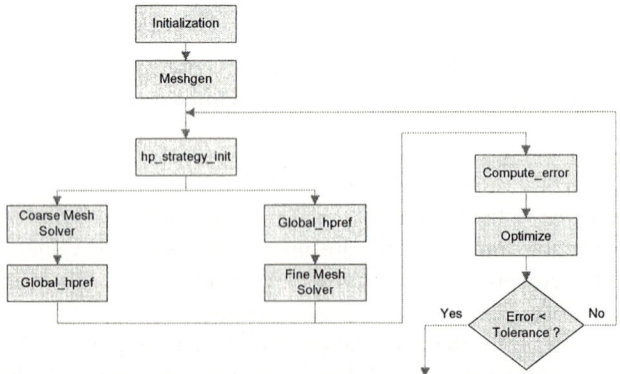

Fig. 1. Workflow Diagram for h-p Adaptive Finite Element Code

3.2 Componentization of the h-p Adaptive Finite Element Code

The set of components is determined by constructing a workflow diagram for the application in which each logical function is identified as a component. Figure 1 is a workflow diagram for a family of codes implementing *h-p* adaptive codes. Figure 1 and the components in Figure 1 were obtained by reverse engineering the one-dimensional code described in the previous section. This componentization does not represent the finest granularity of functional decomposition. The "Coarse Mesh Solver" and the "Fine Mesh Solver" each contain three logical functions, the computational model, the element generator for the stiffness matrix and the solver for the stiffness matrix. Componentization was stopped at the level shown in Figure 1 because component extraction by reverse engineering of the existing code was laborious and because this componentization enables practical experiments in componentization.

3.3 Experiments

The experiments illustrate composition of programs implementing a sequence of models, compile time choice of linear solvers and runtime substitution of the linear solver.

Compile time selection of linear solvers is illustrated by composing application instances first using a direct solver for the coarse mesh and a conjugate gradient solver with a diagonal preconditioner for the fine mesh. Runtime replacement (and optimization) is illustrated by replacement of the direct solver by the conjugate gradient solver after the first cycle of the adaptation demonstrates that the direct solver is not an efficient choice.

Composition of applications based on different computational models for a physical system is illustrated by composing a sequence of applications using successively more accurate models for bioheat transfer. We consider a set of bioheat transfer equations ranging from simple conductivity (Poisson) to incorporation of blood perfusion (Pennes Equation) to incorporation of artery-vein countercurrent (Weinbaum-Jiji Equation [5]).

These models represent progression of complexity and accuracy from the simple Poisson model through the Pennes and Wein-Jiji models. The experiment compares a standard metric resulting from solution of each of the models.

3.4 Illustrations of Automated Composition

3.4.1 Compile Time Selection of Solver and Model

The component library is initialized with two solvers: i) A direct solver that uses LU factorization and back substitution and ii) A Preconditioned Conjugate Gradient (PCG) solver that uses a diagonal preconditioner. Each of the four models sketched in Section 3.3: i) Laplace model ii) Poisson model iii) Pennes model and iv) Weinbaum-Jiji model have been incorporated into a component. The componentization of the h-p adaptive code leaves the model and the solver in the same component although they could readily be separated and would be separated for a production implementation. There are therefore eight implementations of the solver component. Each can be used for the coarse or fine solver so long as the model is the same for both the coarse and fine meshes. These eight implementations were encapsulated using the interface definition language of PCOM[2]. A component that needs a particular combination of solver and model expresses that requirement using the selector interface. The selector of a component that requires a direct solver and Poisson model is shown below (only the attributes part is shown for space limitation).

```
Selector:
    string domain == "application";
    string component == "solver";
    string solver_type == "Direct"
    string model == "Poisson"
```

Similarly a PCG implementation of a solver that uses a Laplace model expresses that information in the profile of that implementation.

```
Profile:
    string domain = "application";
    string component = "solver";
    string solver_type = "PCG"
    string model = "Laplace"
```

The compiler chooses the appropriate component as described in Section 2.2. By changing the selector section of a component the appropriate implementation can be chosen at compile time.

Table 1 compares the solutions obtained from application family instances based on each of Poisson, Pennes and Weinbaum-Jiji computational models. Using Weinbaum-Jiji as a base model, we compared the solution in $H^1(D)$-norm. Table 1 indicates that differences are significant. These quantities in percentage can be used as a criterion for the decision-making in model selection. For example, if the acceptance criterion is set to 20%, then we need to reject both Poisson and Pennes models with respect to more accurate Weinbaum-Jiji model.

Table 1. Properties of Solutions from Multiple models

Model	Poisson	Pennes	Weinaum-Jiji
Solution Norm	0.18787E+06	0.18348E+06	0.14895E+06
Percentage	26%	23%	-

3.4.2 Runtime Optimization by Component Replacement

The PCOM^2 compiler automatically generates performance measures for the execution behavior of each component. This information can be used to determine whether a currently loaded component is performing efficiently and/or robustly. When it is determined that a change of algorithm is needed, the dynamic loading capability of the PCOM^2 runtime system can be used to dynamically replace an implementation of a component. The implementation of the solver component incorporated code to load libraries at runtime depending upon argument values in the transaction specification. Based on the argument (a domain attribute) the implementation can either run the direct solver or load a PCG solver from the library and invoke it. Similarly the PCG solver can be directed to replace itself by a direct solver. In the illustration reported here, during the first iteration the coarse mesh was solved using a direct solver and the fine mesh was solved using a PCG solver. But for large mesh sizes the direct solver component may take a longer time to solver the coarse mesh than the PCG solver takes to solve the fine mesh. After the first iteration, the runtime of the direct solve of the coarse mesh and the PCG solve of the fine mesh are compared component are compared in the *optimize* component, "optimize." If it turns out that the direct solve of the coarse mesh is too slow, an appropriate argument is passed to the coarse mesh solver so that it can load the PCG solver using dynamic loading from the library on the next mesh refinement iteration.

Table 2. Execution Time Improvement with Dynamic Solver Replacement

Iteration	Coarse Mesh Solve	Fine Mesh Solve	Total Solve Time
1	2401x2401Direct 3.162 sec.	5401x5401PCG 1.199 sec.	4.361sec.
2	2404x2404PCG 0.536 sec.	5404x5404PCG 0.972 sec.	1.508 sec.

Table 2 summarizes the results of some experiments with dynamic solver replacement. An appropriate choice of solver cuts the time for solution down by nearly a factor of three.

4 Related Research

For a survey of the research on component-oriented development focusing on generation of parallel and distributed programs, please see [7]

There has been very little research utilizing dynamic replacement of components to enhance performance or robustness of scientific and engineering applications. Adve's PCL [8] system enables runtime adaptation of task graphs at a finer level of granularity (the basic block level).

AspectIX [10] offers the ability to replace an implementation at runtime. The functional and configuration interface in AspectIX is similar to the transaction and attributes of the profile in $PCOM^2$. The transaction provides the syntax of a component and the attributes expresses the semantics in the program domain. AspectIX uses the interface information at runtime. $P-COM^2$ integrates runtime and compile time composition.

5 Conclusions

The feasibility and the potential benefits of automated compiler-based composition of instances of application systems from libraries of components has been demonstrated. The concepts and their applications are quite simple and readily accessible to application developers. Modification and maintenance of families of application systems is simplified. Runtime adaptation at the component level is shown to be straightforwardly implemented and to offer the potential of substantial performance benefits.

Acknowledgements

This work was accomplished through support from ITR NSF grant number 0205181 , "A Computational Infrastructure for Reliable Computer Simulations." NGS NSF grant number 0103725 "Performance-Driven Adaptive Software Design and Control." Additional support came from the Defense Advanced Research Project Agency (DARPA) under Contract NBCH30390004.

References

1. M. Ainsworth and J.T. Oden, A Posteriori Error Estimation in Finite Element Analysis. John Wiley & Sons, New York, (2000).
2. I. Babuska, and T. Strouboulis, Finite Element Method and its Reliability. Oxford Univ. Press (2001)

3. L. Demkowicz and C. W. Kim, 1D hp-Adaptive Finite Element Package. Fortran 90 Implementation (1Dhp90), TICAM Report 99-38, The University of Texas at Austin (1999)
4. L. Demkowicz, 2D hp-Adaptive Finite Element Package (2Dhp90) version 2.0, TICAM Report 02-06, The University of Texas at Austin (2002)
5. L. Demkowicz, D. Pardo, and W. Rachowicz, 3D hp-Adaptive Finite Element Package (3Dhp90) version 2.0: The Ultimate Data Structure for Three Dimensional, Anisotropic hp Refinitement, TICAM Report 02-24, The University of Texas at Austin (2002)
6. J.T. Oden and S. Prudhomme, Estimation of Modeling Error in Computational Mechanics. *J. Comput. Phys.*,182 (2002), 496-515
7. N. Mahmood, G. Deng, and J. C. Browne, Compositional Development of Parallel Programs, *Proceedings of the 16th Workshop on Languages and Compilers for Parallel Computing* (LCPC'03), College Station, TX, 2-4 October 2003.
8. B. Ensink, J. Stanley, and V. Adve, Program Control Language: A Programming Language for Adaptive Distributed Applications, *Journal of Parallel and Distributed Computing* , vol. 63, no. 11, pp. 1082-1104, Nov. 2003
9. P. Newton and J. C. Browne, The CODE 2.0 Graphical Parallel Programming Language, *Proceedings of the ACM International Conference on Supercomputing*, July 1992.
10. F. Hauck, U. Becker, M. Geier, E. Meier, U. Rastofer, and M. Steckermeier, AspectIX an Aspect-Oriented and CORBA-Compliant ORB Architecture, Tech. Report TR-I4-98-08, IMMD IV, Univ. Erlangen-Nürnberg, Sep. 1998.

Determining Consistent States of Distributed Objects Participating in a Remote Method Call

Magdalena Sławińska and Bogdan Wiszniewski

Faculty of Electronics, Telecommunications and Informatics
Gdańsk University of Technology
Narutowicza 11/12, 80-952 Gdańsk, Poland
{magg, bowisz}@eti.pg.gda.pl

Abstract. The article presents the first component of a new approach for testing distributed object-oriented applications called *TestByRep* which is based on the concept of replication of object states. The paper describes key ideas in *TestByRep* like: *hash clocks* for ordering events in distributed object-oriented systems with the unknown number of objects, the *E-path* for representing the execution of the method, construction of the *E-tree* for determining states of objects involved in a remote method call and introduces the *recovery condition* in order to assure that the determined state is consistent.

1 Introduction

Testing non-distributed applications is easier than testing distributed ones since in non-distributed applications it is usually possible to use the *cyclic debugging* technique [1, 2]. The cyclic debugging technique is based on the approach of repeatable execution of a program in order to exercise its code and localise errors. Although it is a very simple technique, it is commonly used by testers and programmers of non-distributed applications. However, adapting the cyclic debugging technique to distributed applications is not straightforward and requires special mechanisms which must be incorporated into the distributed application, the execution platform or the operating system. It is because in distributed systems testers meet difficulties like: determining a global consistent state of a distributed application, physical distribution of application's components, accessibility of application's components, the lack of global clock which makes impossible to reckon about the order of the events in a distributed application [3].

This paper describes the first component of a new methodology called *TestByRep* based on the concept of replication of object states in order to enable the cyclic debugging technique for distributed object-oriented applications (DOA). The idea of replication is widely used in order to increase reliability and accessibility as well as to provide fault-tolerance and support load-balancing [4, 5, 6]. The methodology *TestByRep* shows that the idea of replication can be also applied to test and debug DOA.

TestByRep consists of three main components: (1) the model and key concepts, (2) the infrastructure to capture the relations among objects and object states during the execution of DOA and (3) the *method driven recovery approach* which comprises *reexecution, simulated execution* and *selective execution* of replicated objects [7].

The description of the infrastructure to capture the relations among objects in DOA can be found in [8]. In this paper we present fundamental concepts of the methodology *TestByRep*: *hash clocks, E-paths* and *E-trees*.

2 Hash Clocks

To order the events in DOA, we introduce the concept of *hash clocks*. The *logical clock* and two important conditions: (1) the *clock condition* and (2) the *strong clock condition* were defined in [3]. The logical clock (it is a function) assigns numbers to events in the system. In order to work properly it should satisfy the clock condition. Events in the system have *timestamps* which are concrete values of the logical clock. Timestamps are assigned to events in the system. The clock condition says that earlier events in the system should have lower timestamps [3]. And when it is possible to draw a conclusion based on timestamps about the order of the events in the system, we say that the logical clock satisfies the strong clock condition [3]. Logical clocks can be defined as numbers [3], vectors with the *a priori* known and constant size [9] or matrices also with the *a priori* given constant size [9, 10, 11]. However, sometimes in DOA it is difficult to predict the number of objects in the application, especially if *foreign* objects are considered [8]. A distributed object-oriented application consists of cooperating objects scattered over the network. A tester can have an access to the source code of the object and such an object is called *native* object or s(he) cannot have an access to the source code and such objects are called *foreign* objects [12], e.g., services which are used by native objects.

In this context the assumption about the constant and known sizes of vectors or matrices is irrelevant. In [13] were proposed logical clocks which take into account the dynamic and unknown number of processes in the system. However, they require an additional process which assigns subsequent numbers to new processes or before registering a new process in the system, all other processes must agree on the number which should be assigned to this new process. The presented approach can be appropriate for non-object distributed systems like PVM or MPI (in fact it was proposed for PVM applications), however DOA requires a new approach.

In this paper we propose a new logical clock called a *hash clock*. The construction of the hash clock is based on the *hash table*. The hash table is the set of key-value pairs with effective search of elements [14]. We assume that elements can be dynamically added to the hash table. We also assume that objects in the distributed object-oriented system have unique identifiers. Symbol id_o means the identifier of object o and HC_o denotes the hash clock of object o. Hash clock HC_o is a set of key-value pairs where keys are object identifiers while values are natural numbers representing the logical clock of a given key-object from the point of view of object o. For example: $HC_o = \{(id_o, C_o), (id_{o_1}, C_{o_1}), \ldots, (id_{o_k}, C_{o_k})\}$, where $k = 1, 2, \ldots$, means that from the point of view of object o: object o has the clock value C_o, object o_1 has the clock value C_{o_1}, and so on. Let the function $HC_o()$ give a value of the key argument in hash clock HC_o, i.e., $HC_o(id_{o_1}) = C_{o_1}$. Values of hash clocks will be called *hash timestamps* and denoted by symbol HTS.

Before presenting the algorithm for updating the hash clock, first we should define operations of adding the new value to the hash clock and merging a hash clock with a hash timestamp. The operation of adding the new element to the hash clock is defined as follows:

$$HC_o.add(id_{o_1}, C_1) \equiv HC_o := HC_o \cup \{(id_{o_1}, C_1)\} \tag{1}$$

We assume that if it is required operation $add()$ increases the size of the hash clock and then adds the element to it. The symbol Id_{HC_o} denotes the set of identifiers of the hash clock of object o and Id_{HTS} denotes the set of identifiers of hash timestamp HTS, e.g., $Id_{HC_o} = \{id_o, id_{o_1}, \ldots, id_{o_k}\}$. Function $mergeMax()$ defines how to merge a hash clock and a hash timestamp and is described by Algorithm 1.

Algorithm 1 (Function $mergeMax(HC_o, HTS)$)

1. $\forall_{id \in Id_{HTS}} \ (id \notin Id_{HC_o}) \Rightarrow (HC_o.add(id, HTS(id)))$.
2. $\forall_{id \in Id_{HC_o}} \ (id \in Id_{HTS}) \Rightarrow (HC_o(id) := max(HC_o(id), HTS(id)))$.

Function $mergeMax()$ increases hash clock HC_o (if it is required) and compares the values of logical clocks of relevant identifiers of HC_o and HTS, and then chooses the greater value and assigns it to the relevant object identifier in HC_o.

We assume that objects interact with other objects only by method calls which can be modeled as requests and replies [6,8]. By the *internal* event we understand the event which does not require the communication with another object, e.g., the event of invoking by an object its own method. Otherwise the event is called *external*, i.e., requires the interaction with another object. In the context of requests and replies the method call can be modeled as the event of (1) sending request req by object o_1 to object o_2 denoted by $s(req)$, (2) receiving request req by object o_2 denoted by $r(req)$, (3) performing relevant actions by o_2, (4) sending reply rep by object o_2 to object o_1 denoted by $s(rep)$, and finally (5) receiving reply rep by object o_1 from object o_2 [8]. If it is not relevant to distinguish replies and requests, they will be called *messages* and denoted by symbol m. We assume that messages carry hash timestamps of their senders.

Now, the algorithm of updating the hash clock can be introduced.

Algorithm 2 (Updating the hash clock)

1. $HC_o := \{(id_o, 0)\}$.

2. For each new event e in the system:
   1). $HC_o(id_o) := HC_o(id_o) + 1$, if event e is an internal event
   2). $HC_o := mergeMax(HC_o, HTS(m))$, if $e = r(m)$
   $HC_o.add(id_p, 0)$, if $e = s(m)$ and id_p is a receiver of m and $id_p \notin Id_{HC_o}$
   $HC_o(id_o) := HC_o(id_o) + 1$.

Firstly, the hash clock is initialized and then after each internal event the clock value referring to object o is increased by 1. Then, after each event (external or internal) the hash clock is increased by 1. However, in the case of the receive event the current hash clock is compared with the received hash timestamp of the received message and updated according to Algorithm 1. In the case of the sending event if a receiver is not in the hash clock, the receiver is added to the hash clock and its clock is initialized with 0.

Figure 1 presents Algorithm 2. Let's notice that the size of the hash clock for each object settles at the number of objects in the system if objects in the system interact and information about hash clocks scatters over objects in the system. Let's consider for example the send event $e_{1,2}$ from Figure 1. The hash timestamp assigned to event $e_{1,2}$ is $HC_{o_1}(e_{1,2}) = \{(o_1, 2), (o_3, 0)\}$. Next, this hash timestamp is sent to object o_3. Since HC_{o_3} does not contain the identifier of object o_1, pair $(o_1, 2)$ is added to HC_{o_3} according to Algorithm 1, and the rest elements are updated according to Algorithm 1. Finally, this results in $HC_{o_3}(e_{3,3}) = \{(o_3, 3), (o_2, 1), (o_1, 2)\}$.

Since hash clocks can have different sizes in order to compare them we need to normalize them. Let's define operation *extend()*:

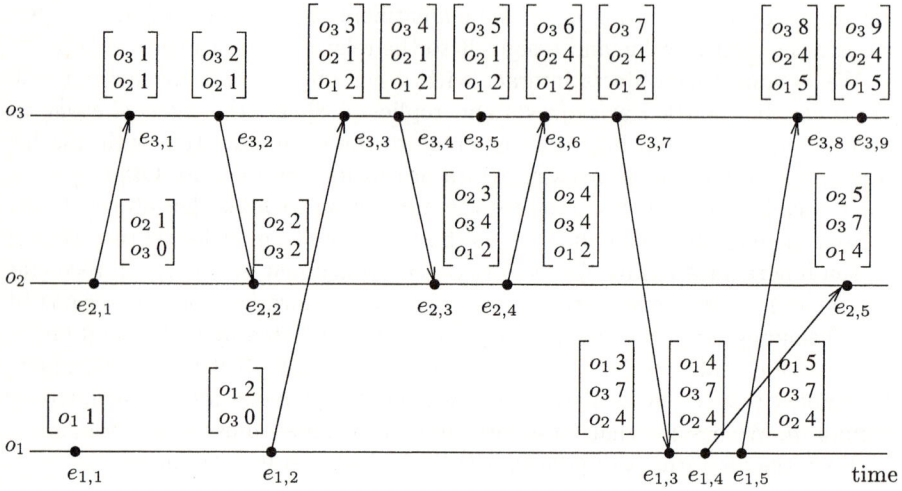

Fig. 1. The hash clock for the dynamic unknown number of objects in the system

Definition 1 (Operation extend(HTS_1, HTS_2)) Let HTS_1 and HTS_2 be hash timestamps. We say that HTS_{ext} is the extended hash timestamp of HTS_1 with respect to HTS_2 if HTS_{ext} was constructed according to the following rules:
1. $HTS_{ext} := HTS_1$.
2. $\forall_{id \in Id_{HTS_2}} (id \notin Id_{HTS_1}) \Rightarrow HTS_{ext}.add(id, 0)$.

Operation $extend()$ returns the hash timestamp which contains all object identifiers and relevant values from its first argument and all object identifiers from its second argument with value 0 if the first argument does not contain them. Let's consider for example $HTS_1 = HTS(e_{2,2}) = \{(o_2, 2), (o_3, 2)\}$ and $HTS_2 = HTS(e_{3,4}) = \{(o_3, 4), (o_2, 1), (o_1, 2)\}$ from Figure 1. Then $HTS_{ext} = extend(HTS_1, HTS_2) = \{(o_1, 0), (o_2, 2), (o_3, 2)\}$.

Considering operation $extend()$ it is possible to define operators $\leqslant, <$ and $\|$ for comparison of hash timestamps [7]. Informally $HTS_1 \leqslant HTS_2$ when for all identifiers in HTS_1 their values are not greater than relevant values in HTS_2. We say that $HTS_1 < HTS_2$ if $HTS_1 \leqslant HTS_2$ and exists at least one identifier id in HTS_1 that $HTS_1(id) < HTS_2(id)$. When $HTS_1 \| HTS_2$, i.e., the hash timestamps are *concurrent*, then $\neg(HTS_1 < HTS_2) \land \neg(HTS_2 < HTS_1)$.

We can prove that hash clocks with updating by Algorithm 2 and operation $extend()$ satisfy the clock condition and strong clock condition [7]. It means that hash clocks can be used to conclude about the order of events in the system and construct consistent state of a set of objects.

3 Consistent States of Distributed Objects

The states of objects registered in corresponding logs during the execution of DOA are called *checkpoints* [10]. They will be denoted as ✕ on diagrams and instead of $e_{i,j}$ we will use symbol $s_{i,j}$. In order to check if a given set of checkpoints is consistent it is sufficient to check if timestamps corresponding to checkpoints are concurrent. If at least one pair of the set of checkpoints is not concurrent the state of the set of objects represented by these checkpoints is not consistent [15].

In order to determine the consistent state of the set of objects we can represent a method call as an *execution path*, shortly called *E-path*.

3.1 E-Path

The *E-path* is the sequence of events beginning from the event of sending a request and ending with the event of receiving the reply corresponding to this request. Let's consider Figure 2.

Since $e_{1,2} = s(req_1)$ and $e_{1,3} = r(rep_1)$, they determine the following *E-path* $= \langle e_{1,2}, e_{2,2}, e_{2,4}, e_{3,2}, e_{3,3}, e_{2,5}, e_{2,7}, e_{3,5}, e_{3,7}, e_{4,4},$
$e_{4,5}, e_{3,8}, e_{3,10}, e_{5,2}, e_{5,3}, e_{3,11}, e_{3,13}, e_{2,8}, e_{2,10}, e_{1,3}\rangle$.

Having the *E-path* of the method call it is possible to determine which objects in the system are involved in the method call. For example in Figure 2 the following objects participate in calling a method in request req_1: $\{o_1, o_2, o_3, o_4, o_5\}$ since all of them belongs to $E\text{-}path(req_1)$.

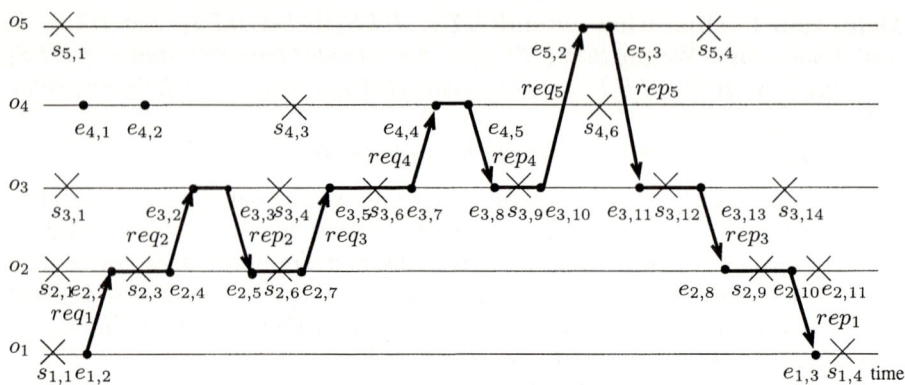

Fig. 2. The E-path

In order to determine the consistent state before the method call, the *E-tree* can be constructed.

3.2 E-Tree

The *E-tree* is the tree of the execution of the method. It is an ordered tree [14] where the root of the tree is the request of the method execution which is sent by a caller to a callee and nodes are requests which belong to *E-path* of the request in the root node. Earlier requests are placed in the most left nodes in the *E-tree*. The *E-tree* for request req_1 from Figure 2 is shown in Figure 3.

Each level of the *E-tree* corresponds to the so-called *method execution interval* of the direct parent in the *E-tree*. The method execution interval denoted by I_m is the period of time determined by the event of receipt of the request by an object and the event of sending the reply to the request. Informally, the method execution interval is the period of time when the callee perform the method requested by the caller. For example in Figure 2: $I_m(req_1) = \langle e_{2,2}, e_{2,10} \rangle$ and $I_m(req_3) = \langle e_{3,5}, e_{3,13} \rangle$. Considering the *E-tree* from Figure 3 we can see that $req_2, req_3 \in I_m(req_1)$ (level $k = 1$) and $req_4, req_5 \in I_m(req_3)$ (level $k = 2$).

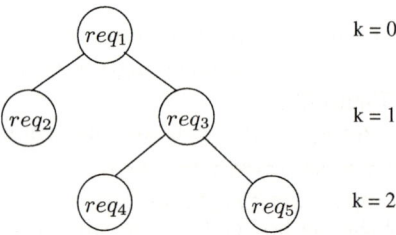

Fig. 3. The *E-tree* for request req_1 from Figure 2

We can construct a consistent state of objects preceding the method call if checkpoints satisfy the *recovery condition* defined as follows:

Definition 2 (The recovery condition.) *Let $e_{a,i}$ be the i-th event in object a. If the state of object a is registered with the following rules the recovery condition is satisfied:*

1. $(e_{a,i+1} = s(req)) \Rightarrow s_{a,i}$
2. $(e_{a,i+1} = r(req)) \Rightarrow s_{a,i}$
3. $(e_{a,i} = s(rep)) \Rightarrow s_{a,i+1}$
4. $(e_{a,i} = r(rep)) \Rightarrow s_{a,i+1}$

Rules 1 and 2 of Definition 2 state that the checkpoint should be registered just before any event of sending and receiving a request. Rules 3 and 4 say that checkpoints should be registered just after sending and receiving replies to requests.

Let's notice that checkpoints in Figure 2 satisfy the recovery condition.

3.3 Consistent State

Determining the consistent state of objects involved in a given remote method call requires that the recovery condition is satisfied and *E-tree* of the given method call is constructed. Then, the consistent state can be determined by examining the *E-tree* beginning from the root level to the lowest level in the following way:

Algorithm 3 (Determining consistent state)

1. Add to set CS the checkpoints directly preceding the event of sending request in the root node in the E-tree and the event of receiving that request. Add to set O the sender and receiver of the request placed in the root node.
2. Initialize variable $k := 0$. It will denote the current level of the E-tree.
3. If k is greater than the depth of the E-tree then stop the algorithm.
4. For each request on level k check if the receiver of the request is in set O and if it is not then do:
   (a) add the receiver of the request to set O and
   (b) add the checkpoint directly preceding the event of the receiving this request to set CS.
5. Increase by 1 variable k, i.e., $k := k + 1$. Do step 3.

Algorithm 3 examines the *E-tree* and adds checkpoints directly preceding the receive events to set CS. However, the checkpoint is added only if its owner does not belong to set O, i.e., this is the first request of a given receiver. For example, for the *E-tree* in Figure 3 Algorithm 3 will generate the following set $O = \{o_1, o_2, o_3, o_4, o_5\}$. It is correct since all objects from set O are involved in the method call requested in request req_1 (compare the *E-path* of request req_1 in Figure 2). The result set $CS = \{s_{1,1}, s_{2,1}, s_{3,1}, s_{4,3}, s_{5,1}\}$. We can prove that the set of checkpoints, generated by Algorithm 3 assuming that the recovery condition is satisfied, is consistent [7] and as a result it can be used to set states of replica objects in order to test a remote method call.

4 Conclusions

The paper presents the first component of the new methodology called *TestByRep*. *TestByRep* allows for adapting the cyclic debugging technique to distributed object-oriented applications. Determining a consistent state is a key element of *TestByRep*. We showed in this article that it can be done with the concept of hash clocks which enable ordering of events in DOA with respect to dynamic and unknown number of objects in DOA, notions of *E-path* and *E-tree* which provide means for representing a remote method call and allow for determining the consistent state of remote objects participating in the method call.

References

1. J. Huselius, "Debugging Parallel Systems: A State of the Art Report," Tech. Rep. 63, Mälardalen Univ., Dept. of Comp. Science and Engineering, Sept. 2002.
2. M. Ronsse, K. D. Bosschere, and J. C. d. Kergommeaux, "Execution replay and debugging," in *the 4th International Workshop on Automated Debugging*, Aug. 2000.
3. L. Lamport, "Time, Clocks, and the Ordering of Events in a Distributed System," *Comm. of the ACM*, vol. 21, pp. 558–565, July 1978.
4. G. Coulouris, J. Dollimore, and T. Kindberg, *Distributed Systems. Concepts and Design*. Addison-Wesley Longman Limited, 1994.
5. S. Maffeis, "Adding Group Communication and Fault-Tolerance to CORBA," in *Proc. of the USENIX Conference on Object-Oriented Technologies*, June 1995.
6. A. S. Tanenbaum, *Distributed Operating Systems*. Prentice-Hall International, Inc., 1995.
7. M. Sławińska, "Efektywna metoda replikacji dynamicznie powiązanych obiektów rozproszonych." Materials for PhD (In Polish), Nov. 2004.
8. M. Sławińska, "Hunting for Bindings in Distributed Object-Oriented Systems," in *Proc. of International Conference on Computational Science – ICCS'2004* (M. Bubak, G. D. van Albada, P. M. A. Sloot, and J. Dongarra, eds.), vol. 3036 of *LNCS*, (Krakow, Poland), Springer, June 2004.
9. M. Raynal and M. Singhal, "Logical Time: A Way to Capture Causality in Distributed Systems," tech. rep., IRISA, Jan. 1995.
10. E. N. M. Elnozahy, L. Alvisi, Y.-M. Wang, and D. B. Johnson, "A survey of rollback-recovery protocols in message-passing systems," *ACM Comput. Surv.*, vol. 34, pp. 375–408, Sept. 2002.
11. M. Raynal and M. Singhal, "Logical Time: Capturing Causality in Distributed Systems," *Computer*, vol. 29, no. 2, pp. 49–56, 1996.
12. M. Sławińska, "Testability for Distributed Objects," in *Parallel Processing and Applied Mathematics, 5th International Conference, PPAM 2003, Czestochowa, Poland, Sept. 7-10, 2003. Revised Papers* (R. Wyrzykowski, J. Dongarra, M. Paprzycki, and J. Wasniewski, eds.), vol. 3019 of *LNCS*, pp. 413–418, Springer, Sept. 2004.
13. M. Neyman, *Metody odtwarzania obliczeń w rozproszonych środowiskach sieciowych*. PhD thesis, Technical Univ. of Gdańsk, ETI, Feb. 2000. (In Polish).

14. T. H. Cormen, C. E. Leiserson, and R. L. Rivest, *Introduction to Algorithms*. The Massachusetts Institute of Technology, 1994.
15. A. P. Goldberg, A. Gopal, A. Lowry, and R. Strom, "Restoring consistent global states of distributed computations," in *Proceedings of the 1991 ACM/ONR workshop on Parallel and distributed debugging*, pp. 144–154, ACM Press, 1991.

Storage Formats for Sparse Matrices in Java

Mikel Luján*, Anila Usman, Patrick Hardie, T.L. Freeman, and John R. Gurd

Centre for Novel Computing, The University of Manchester,
Oxford Road, Manchester M13 9PL, United Kingdom
{mlujan, ausman, hardiep, lfreeman, jgurd}@cs.man.ac.uk

Abstract. Many storage formats (or data structures) have been proposed to represent sparse matrices. This paper presents a performance evaluation in Java comparing eight of the most popular formats plus one recently proposed specifically for Java (by Gundersen and Steihaug [6] – Java Sparse Array) using the matrix-vector multiplication operation.

1 Introduction

Sparse matrices are those matrices which have a substantial minority of nonzero elements – normally less than 10% are nonzero elements. These matrices are pervasive in many computational science and engineering (CS&E) applications. The storage formats for sparse matrices have been proposed to better suit particular CS&E applications or computer architectures. The significant number of different storage formats is the source of a research problem. For example, consider the recently published Basic Linear Algebra Subroutines (BLAS) standard and the part dedicated to sparse matrices (*Sparse BLAS*)[4]. The Sparse BLAS do not state which storage formats must be supported or used. Each specific hardware vendor has the freedom (or problem) to select the storage format (or formats) that perform best for its hardware. In the context of iterative methods [2] and Java, this papers investigates the performance delivered by different storage formats considering a wide variety of sparse matrices.

The structure of the paper is as follows. Section 2 introduces the most commonly used storage formats for sparse matrices. The Java Sparse Array (JSA) storage format was recently proposed by Gundersen and Steihaug [6] to take advantage of Java arrays; Section 3 briefly describes JSA. The performance evaluation (see Section 5) consider a specific kernel from iterative methods, namely matrix-vector multiplication, and compares this operation on two different computational platforms with nine different storage formats. The Java implementation of this matrix operation is described in Section 4. The performance study considers around 200 different sparse matrices representing various CS&E applications as recorded by the Matrix Market repository [1]. To the best of the authors' knowledge, there is no other performance evaluation of storage formats for sparse matrices which consider such a variety of matrices and storage formats. Conclusions and future work are given in Section 6.

---

* ML acknowledges a postdoctoral fellowship from the Basque Government. AU acknowledges a postdoctoral fellowship from the HEC Pakistan.

2 Storage Formats for Sparse Matrices

The objective of storage formats for sparse matrices is to best exploit certain matrix properties by (1) reducing memory space, by storing only nonzero elements of a sparse matrix, and (2) by storing these elements in contiguous memory locations, for more efficient execution of subroutines on the matrix data.

From an implementation point of view, there are two categories of storage formats. *Point entry* is used to categorise storage formats where each entry in the storage format is a single element of the matrix. *Block entry* refers to storage formats where each entry defines a dense block of elements of any two dimensions. For both cases, programming languages provide static and dynamic data structures. However since Fortran 77 has been the dominant language in CS&E and does not support dynamic data structures, the most commonly used storage formats are array-based.

There are many documented versions of different storage formats for sparse matrices. One of the most complete sources is the book by Duff *et al.* [3] (for a historical source see [7]).

2.1 Point Entry Storage Formats

Coordinate Format (COO) — Possibly the most intuitive storage format for a sparse matrix is in terms of coordinates. Instead of storing the matrix densely, a list of the coordinates in terms of row and column numbers is stored, with the associated nonzero values. COO requires no specific structure of the matrix and is a very flexible format. It requires three (unordered) arrays and a single scalar recording the total *number of nonzero elements, nnz*. The combination of the three arrays provides a row i and column j coordinate pair for an element in the matrix along with its value a_{ij}. In general, for a matrix with nnz, COO requires three 1-dimensional arrays of length nnz plus a scalar.

Compressed Sparse Row/Column Storage Formats (CSR/CSC) — As with COO, CSR and CSC storage formats can store any matrix. In CSR, the nonzero values of every row in the matrix are stored, together with their column number, consecutively in two parallel arrays, *Value* and *j*. There is no particular order with respect to the column number, j. The *Size* and *Pointer* for each row define the number of nonzero elements in the row and point to the relative position of the first nonzero element of the next row, respectively. The column based version, CSC, instead stores *Value* and i, in two parallel arrays and *Size* and *Pointer* of each column allows each member of *Value* to be associated with a column as well as the row given in i. The storage requirements are two arrays, each of length the number of rows (or columns), and two further arrays of length nnz, and a scalar to point to the next free location in the arrays i (or j) and *Value*.

The *Diagonal Storage Format* (DIA) and *Skyline Storage Formats* (SKS) are also part of the performance evaluation described in Section 5, but their description is omitted for brevity.

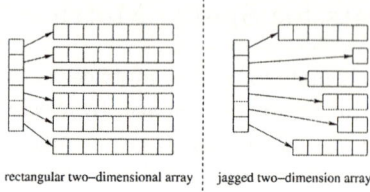

Fig. 1. Examples of two-dimensional arrays in Java

2.2 Block Entry Storage Formats

Block entry storage formats form an extension of certain point entry storage formats based on partitioning matrices into blocks of elements (i.e. sub-matrices). An example of a variable block matrix A is as follows:

$$A = \begin{pmatrix} a_{11} & a_{12} & a_{13} & a_{14} & a_{15} & a_{16} & a_{17} \\ a_{22} & a_{22} & a_{23} & a_{24} & a_{25} & a_{26} & a_{27} \\ a_{31} & a_{32} & a_{33} & a_{34} & a_{35} & a_{36} & a_{37} \\ a_{41} & a_{42} & a_{43} & a_{44} & a_{45} & a_{46} & a_{47} \\ a_{51} & a_{52} & a_{53} & a_{54} & a_{55} & a_{56} & a_{57} \\ a_{61} & a_{62} & a_{63} & a_{64} & a_{65} & a_{66} & a_{67} \end{pmatrix} \text{ or } \begin{pmatrix} A_{11} & A_{12} & A_{13} \\ A_{21} & A_{22} & A_{23} \\ A_{31} & A_{32} & A_{33} \end{pmatrix} \text{ where, for example, } A_{11} = $$

$$\begin{pmatrix} a_{11} & a_{12} \end{pmatrix}, A_{13} = \begin{pmatrix} a_{17} \end{pmatrix}, A_{22} = \begin{pmatrix} a_{23} & a_{24} & a_{25} & a_{26} \\ a_{33} & a_{34} & a_{35} & a_{36} \end{pmatrix} \text{ and } A_{33} = \begin{pmatrix} a_{47} \\ a_{57} \\ a_{67} \end{pmatrix}.$$

In the point entry storage formats, the storage format describes the position in the (*Value*) array of single matrix elements. Block entry storage formats (with length of block lb), instead have a scheme to describe the position of a single block in a $n/lb \times n/lb$ blocked matrix. Each block contains lb^2 elements. In this way, most point entry storage formats can be blocked to generate Block Coordinate storage format (BCO), Block Sparse Row/Column storage format (BSR/BSC) and others where the block does not have constant dimensions (e.g. Variable Block Compressed Sparse Row).

3 Java Sparse Array (JSA)

The storage formats covered so far have been in use for several years. In contrast, a more recent storage format, JSA, has been created to exploit Java's flexible definition of multi-dimensional arrays. In Java, every array is an object storing either primitive types (i.e. float, double, etc.) or other objects. A two-dimensional array is formed as an array of arrays. This definition enables developers to create both rectangular and jagged arrays (see Fig. 1).

JSA is a row oriented storage format similar to CSR. It uses two arrays, each element of which is itself an array (object). One of these arrays, *Value*, stores arrays of the matrix elements – each row in the matrix has its elements in a separate array. All the separate arrays are elements of the *Value* array; that is an array of array objects. The second main array *Index* stores arrays containing the column numbers of the matrix, again one array per row. The memory re-

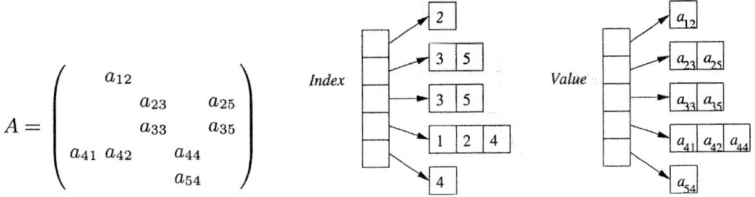

Fig. 2. An example sparse matrix stored using JSA

quirements to store a sparse matrix in JSA are $2nnz + 2n$ array locations. Figure 2 shows an example sparse matrix stored using JSA.

4 Sparse Matrix-Vector Multiplication

The performance evaluation presented in the following section is predicated upon Java implementations of sparse matrix-vector multiplications. These implementations have been developed for this work and follow the structure of the Fortran 95 reference implementation of the Sparse BLAS developed by CERFACS [4]. Object-oriented features are used only for passing parameters to methods (subroutines), but are not used inside the kernels that actually implement the matrix operation. Some simplifications are made compared with the Sparse BLAS reference implementation. The matrix data is assumed to be static once created. This does not modify the implementation of the matrix-vector multiplication, but simplifies the code to create, access and destroy matrices. The implementations concentrate on square and double precision matrices. The Java implementations incorporate external code, such as JSA.

JSA Implementation — Two Java implementations of sparse matrix-vector multiplication are considered for JSA: JSA-GS and JSA2. The JSA-GS implementation is the code made available by Gundersen and Steihaug [6]. This code does not include a specialised case for symmetric matrices[1]. A new subroutine was incorporated into this code (a new method in class JavaSparseArray) to support symmetric matrices. The JSA2 implementation is a reimplementation of JSA following the structure of the Sparse BLAS reference implementation. The main differences compared with JSA-GS is the code for creating, handling and destroying matrices, and one subroutine (or method) which implements the multiplication rather than two as in JSA-GS. When a matrix is created the Sparse BLAS allow users to specify whether the matrix is symmetric. With JSA-GS a program has to check the information provided about the matrix to call either the general subroutine or the symmetric subroutine. With JSA2, a program simply calls the subroutine and the check is performed internally. Otherwise the sets of instructions that implement the multiplication are identical.

---

[1] A matrix A is symmetric when $a_{ij} = a_{ji}$.

5 Performance Evaluation

The aim of this performance evaluation is to analyse the circumstances under which a given storage format performs better than the other storage formats.

Table 1. Legend for the 'storage formats'-axis in Fig. 3

1	COO	6	BCO block size 8	11	BSR block size 16	16	BSC block size 32
2	CSR	7	BCO block size 16	12	BSR block size 32	17	BSC block size 64
3	CSC	8	BCO block size 32	13	BSR block size 64	18	DIA
4	JSA-GS	9	BCO block size 64	14	BSC block size 8	19	SKS
5	JSA2	10	BSR block size 8	15	BSC block size 16		

Experimental Testbed — Matrix test data are (around 200 different matrices) real, symmetric and non-symmetric matrices from Eigenvalue problems and Linear Systems available from the Matrix Market Collection [1]. The test program reads a matrix from file, calculates the multiplication of that matrix with a random vector and records the result as well as the time taken to calculate the product. The test program repeats this computation for the 9 different storage formats (note different block sizes).

The two test machines run the Java Virtual Machines (JVMs) with the minimum and maximum heap sizes of 128 MB and 1536 MB, respectively, and -server flag. System A is an Ultra Sparc 10 at 333 MHz with 256 MB running Solaris 5.8 and Sun Java 2 SDK 1.4.2 Standard Edition (SE). System B is an Intel Pentium 4 at 2.6 GHz with 512 MB running Red Hat 9 kernel 2.4.20-31.9 and Sun Java 2 SDK 1.4.2 SE. The timer accuracy is one millisecond and the time reported is the time spent performing 50 matrix-vector multiplications on System A and 200 on System B. The numbers 50 and 200 are selected so that the times are large enough in relation to the accuracy of the timers.

Performance Results on all Systems and Matrices — Figure 3 presents the average times (out of four runs) for each matrix on both machines. The general pattern is the same on both platforms. The block entry storage formats do not perform significantly better with different block sizes. This suggests that any gain that results from more efficient use of the memory hierarchy is offset by increases in the number of zero elements that need to be processed. Throughout, the point entry storage formats, with the exception of DIA, appear to give the best performance. At best the block entry storage formats get close to the point entry storage formats.

The Fastest Storage Formats — In Fig. 3 the fastest storage formats are COO, CSR, CSC, JSA-GS and JSA2. Figure 4 shows in more detail the execution times for these storage formats. The graphs present the results for JSA-GS, JSA2 and the minimum time taken by the other three storage formats; i.e. min(COO, CSR, CSC) ≡ MCCC. For System A, JSA-GS consistently performs better than JSA2 and JSA2 itself in most cases performs better than or equivalent to MCCC. System B does not offer the same clear conclusion. For the matrices in the region

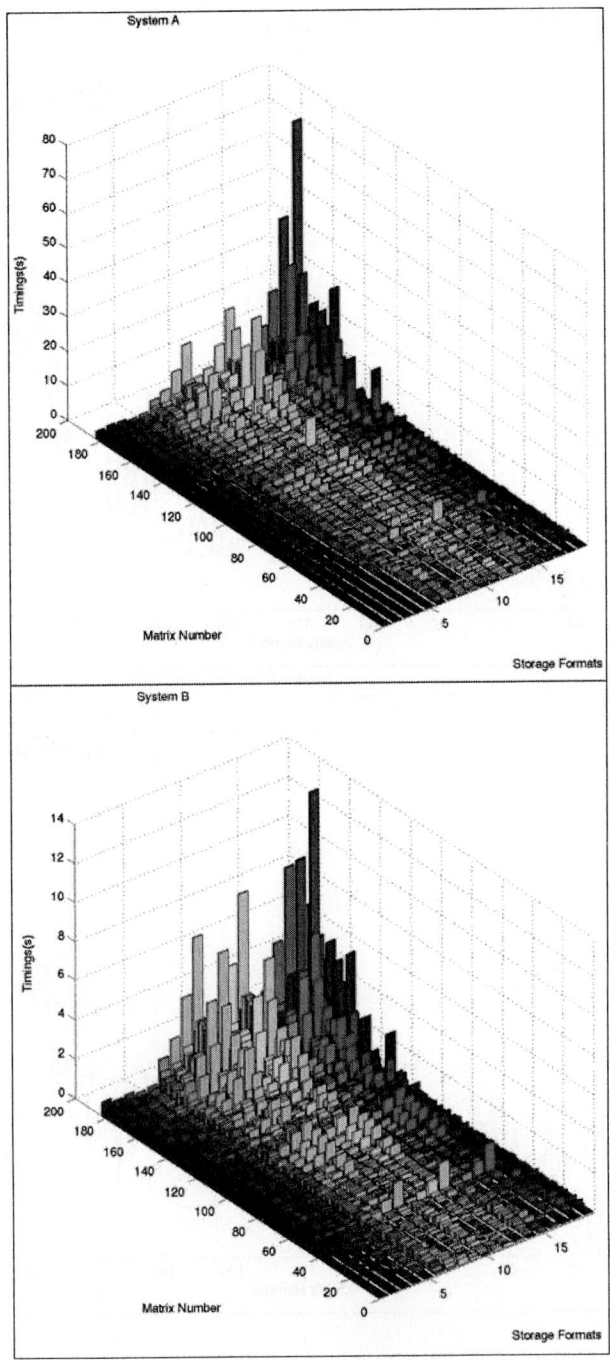

Fig. 3. Time results (seconds) on Systems A and B. The legend for the 'storage formats'-axis is given in Table 1. The 'matrix number'-axis is ordered by nnz

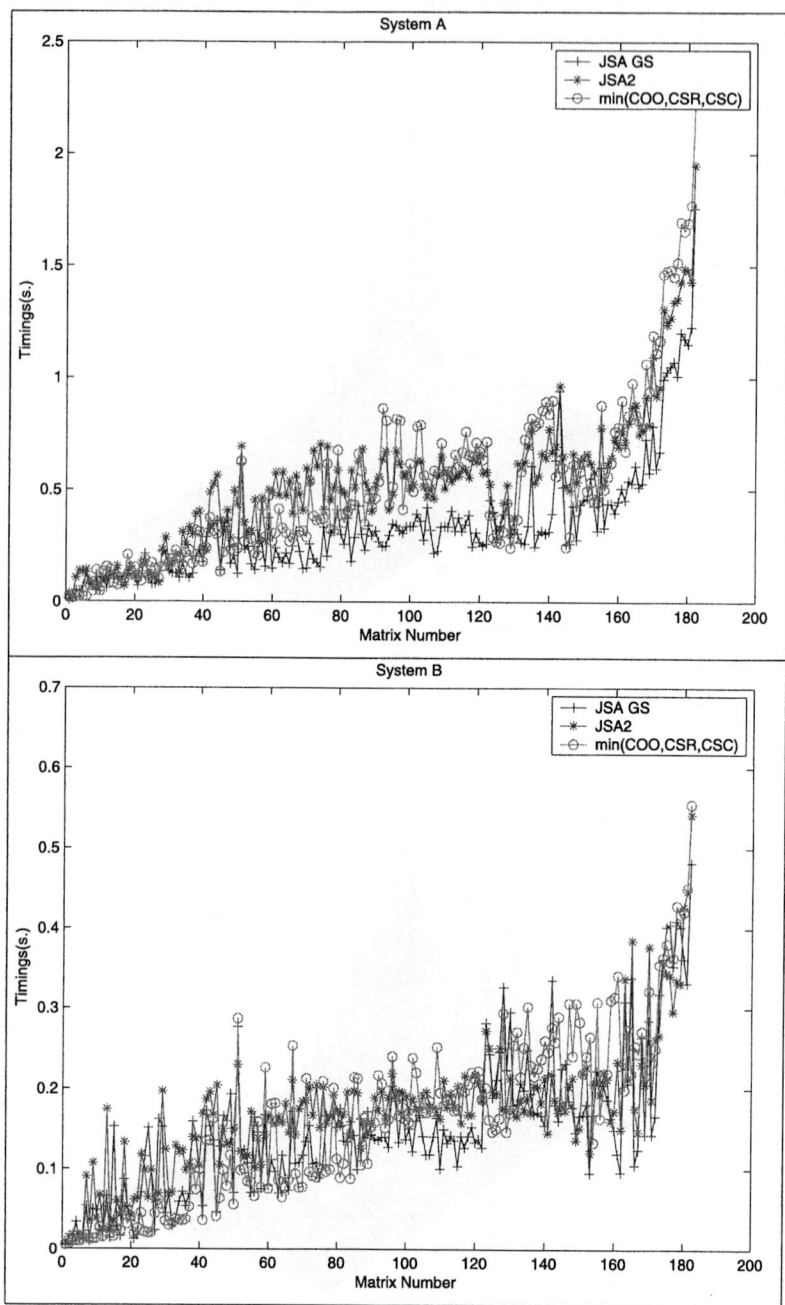

Fig. 4. Time results (seconds) for the fastest storage formats. The matrix number axis is ordered by nnz

1 to 90 (matrices with the fewest nnz), MCCC performs in most cases similarly to JSA-GS. For these matrices JSA2 performs similarly, but slightly worse than JSA-GS. For the rest of the matrices, JSA-GS performs better than JSA2 in most cases and JSA2 performs better than or similarly to MCCC. The exception is in the matrix region around 130 where JSA-GS and JSA2 deliver similar execution times, but both are outperformed by MCCC.

As mentioned in Section 4, the computationally intensive code in JSA-GS and JSA2 is identical. Thus, a reasonably expectation would be that the performance delivered by JSA-GS and JSA2 should be similar. One possible explanation for the observed different performances is that the JVMs find it more amenable to optimise the smaller subroutines (methods) in JSA-GS (symmetric vs. non-symmetric support).

6 Conclusions and Future Work

It would be presumptuous to say that all the storage formats for sparse matrices are covered by this work, especially since there are many minor variations which can create entirely new storage formats. Nonetheless, this paper has presented a comprehensive performance comparison of storage formats for sparse matrices. The results have shown that JSA performes better than the other storage formats for most matrices. The block entry storage formats have not performed as well as the point entry storage formats. Future work underway is to include a similar set of experiments with Fortran implementations and then other operations supported by the Sparse BLAS. The most relevant related work is the Sparsity project [5]. For a given sparse matrix the Sparsity project has developed compile time techniques to optimise automatically several sparse matrix kernels using a specific block entry storage format.

References

1. The matrix market. http://math.nist.gov/MatrixMarket/.
2. R. Barrett *et al. Templates for the Solution of Linear Systems: Building Blocks for Iterative Methods*. SIAM, 1994.
3. I. S. Duff, A. M. Erisman, and J. K. Reid. *Direct Methods for Sparse Matrices*. Oxford University Press, 1986.
4. I. S. Duff, M. A. Heroux, and R. Pozo. An overview of the sparse basic linear algebra subprograms: The new standard from the BLAS technical forum. *ACM Transactions on Mathematical Software*, 28(2):239–267, 2002.
5. Eun-Jin, K. A. Yelick, and R. Vuduc. SPARSITY: An optimization framework for sparse matrix kernels. *International Journal of High Performance Computing Applications*, 18(1):135–158, 2004.
6. G. Gundersen and T. Steihaug. Data structures in Java for matrix computations. *Concurrency and Computation: Practice and Experience*, 16(8):799–815, 2004.
7. U. W. Pooch and A. Nieder. A survey of indexing techniques for sparse matrices. *ACM Computing Surveys*, 5(2):109–133, 1973.

Coupled Fusion Simulation Using the Common Component Architecture*

Wael R. Elwasif[1], Donald B. Batchelor[2], David E. Bernholdt[1], Lee A. Berry[2], Ed F. D'Azevedo[1], Wayne A. Houlberg[2], E.F. Jaeger[2], James A. Kohl[1], and Shuhui Li[3]

[1] Computer Science and Mathematics Division
[2] Fusion Energy Division
Oak Ridge National Laboratory, Oak Ridge, TN 37831 USA
{elwasifwr, batchelordb, bernholdtde, berryla, dazevedoef, houlbergwa, jaegeref, and kohlja}@ornl.gov
[3] Department of Electrical Engineering & Computer Science, Texas A&M University–Kingsville, MSC 192, Kingsville, TX 78363 USA
Shuhui.Li@tamuk.edu

Abstract. The physics of magnetically-confined fusion plasmas involves many different processes with multiple time and length scales that cover many orders of magnitude. As the capability of large parallel computers continues to grow, the goal of an integrated self-consistent simulation of all of the relevant physics draws closer. However, advances in computer science, physics formulations, and algorithms are also needed to achieve this goal. In this paper, we present an overview of an on-going project which is exploring these issues in the context of integrated simulation of radio frequency (RF) heating and transport physics as an initial step toward whole-device modeling. We present our experience in using the common component architecture (CCA) as the underlying framework for the integration of the different physics modules. This work illustrates the viability of using high performance component technology in a complex simulation environment.

1 Introduction

The processes involved in modelling magnetically-confined fusion plasmas include electromagnetic-wave plasma interactions, transport (both collisional and turbulent) particle drift orbit effects, magnetohydrodynamics, atomic physics, and plasma-wall interactions. Research groups have developed codes to model these processes individually, or in a very loosely coupled fashion (generally without self-consistency). However scientific needs for planned systems, such as the International Thermonuclear Experimental Reactor (ITER) project[1], require the capability to model entire devices in an integrated, self-consistent fashion. This coupling requires that the output of one simulation process provides (time varying) boundary conditions or inputs to the other physical

---

* Research sponsored by the Laboratory Directed Research and Development Program of Oak Ridge National Laboratory (ORNL), managed by UT-Battelle, LLC for the U. S. Department of Energy under Contract No. DE-AC05-00OR22725.
[1] http://www.iter.org/

process model simulations. A complete, integrated whole device model (WDM) simulation would encompass time scales ranging from nanoseconds to tens of minutes and length scales from the subatomic to the size of the reactor vessel itself (∼6 meters tall).

The growing capability of large parallel computers continually increases the feasibility of this type of integrated simulation, at least in terms of the raw computational power required. However, significant challenges still must be overcome to make large-scale coupled simulations, such as WDM, a practical reality. Advances must be made in computer science, algorithms and in the physics formulations themselves [1]. For example, while a well-tested physical model will typically provide satisfactory results in its original (uncoupled or loosely coupled) context, the behavior may become more complex and harder to understand when additional physical processes are coupled in because assumptions under which the original model was developed may break down. Likewise, the algorithms and mathematical approaches used in the individual models must often change to accommodate the new coupled interactions. For example, when coupling two simulations where the natural time step differs by a factor of a million, it is often not feasible to actually run a million steps of the fast model between successive steps of the slower one, as a naive integration approach might attempt. Alternative algorithms and solution strategies are required to preserve the essential physics while accelerating the coupled solution to a more practical level. In addition, the integration of codes developed by disparate groups raises software-related issues ranging from reconciling software architectures and facilitating software-based collaboration, to the detailed mechanics of integrating software that may have been developed using different parallel programming models, written in different incompatible languages, and use different data structures.

In order to examine these issues in more detail and lay the groundwork for larger-scale efforts to develop a comprehensive integration fusion simulation capability [1], we have undertaken a modest "prototype" integrated fusion simulation effort. Our project focuses initially on the coupling of just two physical processes: RF heating of the plasma, which occurs on the nanosecond time scale, with transport within the plasma, which occurs on the millisecond time scale. We have developed an "evolutionary preconditioning" approach to address the disparity in time scales between these two models, and have used the Common Component Architecture (CCA)[2] as an efficient and flexible framework in which to integrate the independent simulation codes. In this paper, we present a high-level overview of our ongoing effort, emphasizing the computer science and mathematical aspects of the project.

This paper is organized as follows: in Section 2 we present, briefly, a description of the RF heating and transport processes and how they are simulated; Section 3 describes the evolutionary preconditioning approach used to numerically facilitate the integration; and, Section 4 introduces the Common Component Architecture and its utility for this work. Then in Section 5 we describe our prototype integrated simulation application and share our experiences developing it. We present our Conclusions and plans for Future Work in Section 6.

---

[2] http://www.cca-forum.org

2 Plasma Physics: RF Heating and Transport

Plasma transport, typically particle convection; current and particle diffusion; and thermal conduction, combined with the boundary conditions and sources for the fluxes, control the evolution of macroscopic plasma parameters. The transport coefficients used to model the fluxes can be highly non-linear and, in some cases, non-local in space and time. As a result, the time and space dependence of the temperatures, densities, and magnetic fields that comprise the desired output can be quite complex. Time scales for these processes are in the range of hundreds of milliseconds to tens of seconds for present and planned experiments.

The transport of particles, heat, and current (or equivalently magnetic flux) are each described by a partial differential equation. For example, heat transport for a particular plasma species i is described by an equation of the form

$$\frac{\partial(n_i T_i)}{\partial t} + \vec{\nabla} \cdot K_i \vec{\nabla}(n_i T_i) = S_i \qquad (1)$$

where n_i is the density of species i, T_i is the temperature, K_i is the thermal conductivity, and S_i is the external source/sink of energy. For simplicity, convective terms and the interchange of energy between species due to friction and thermal equilibration have been omitted. Although the plasma is either 2-D or 3-D in space, systems with good confinement have magnetic field lines that form closed, nested surfaces. Because transport along field lines is much faster than that across field likes, we average the transport equations along magnetic field lines and solves for transport across these surfaces. Standard conservative finite different techniques and operator splitting are used for solving the coupled systems. Evolving the transport equations is much less computationally intensive than solving for the interactions of the plasma with RF waves.

Radio frequency (RF) power in the ion cyclotron frequency range (∼50-200 MHz at 10s of MW) is a principle means of plasma heating. Plasma parameters may be assumed constant for the tens of nanoseconds time scale for variation of the RF electromagnetic fields. Thus the waves reach steady state on the electromagnetic time scale before plasma parameters change significantly. Thus wave properties, for example energy deposition, may be calculated with stationary plasma parameters and subsequently used to advance the transport.

For many problems of interest, plasma responses to the applied RF power may be linearized in RF fields. Under these conditions, the All-Orders Spectral Algorithm (AORSA) solves for the wave fields needed for the plasma simulation [2] This code accurately treats the physics of waves that interact with arbitrary harmonics of the ion cyclotron frequency and with perpendicular wavelengths that are smaller than ion gyroradii in the strong equilibrium magnetic field. However, resolving the spatial dependence of the dominant wave-plasma interactions in 2-D requires the solution of from 10,000 to over 100,000 dense linear equations and can use over 10,000 processor hours. As a result, time-dependent plasma simulations were prohibitively expensive as hundreds of AORSA solutions might be required to resolve a few plasma energy confinement times. To address this issue, we have developed an "evolutionary preconditioning" approach to accelerate the coupled simulation.

3 Mathematics: Evolutionary Preconditioning

One of the most time consuming computational kernels in the AORSA code is the solution of a large dense complex linear system by LU factorization in parallel using ScaLAPACK [3]. The matrix is highly non-symmetric and the spectrum has large components in the imaginary plane. This presents a significant challenge to most Krylov iterative solvers such as GMRES [4] and an effective preconditioner is required.

The key to implementing a time dependent simulation was the observation that the plasma parameters (determined by the transport simulation) are changing on a longer time scale, so we need results from a sequence of closely related RF calculations rather than independent solutions to a large number of independent simulations. Thus we use an "evolutionary" approach, using the exact LU decomposition for a set of plasma conditions at one time to precondition the AORSA linear equations that are determined by the parameters for a subsequent time step. Reusing the LU factorization from previous time steps as a preconditioner for an iterative solver offers the possiblity of reducing the $O(N^3)$ cost for factorization to $O(N^2)$ cost in triangular solves and matrix multiplies. When the plasma parameters have evolved sufficiently that the iterative method is no longer effective, a new direct solution will be recomputed. Hopefully, the need for such direct factorization would be limited to a sufficient small number of intermediate points to make the overall simulation practical.

This approach has been tested and proven effective over a reasonable range of plasma parameters. For "sensitive" parameters including electron density and magnetic field strengths, parameters could be varied ±20% and still converge within 100 itera-

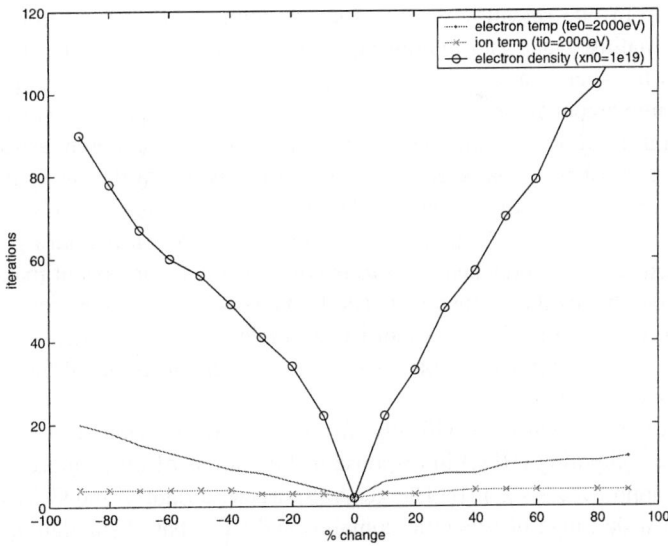

Fig. 1. Sensitivity of convergence of evolutionary preconditioning scheme to changes in various simulation parameters. Tests were done for a 32×32 mode problem, with a reduced matrix size (N) of 1965 complex equations

tions. While for "non-sensitive" parameters such as electron and ion temperatures and toroidal mode number of RF antenna, a variation of ±50% was possible while maintaining good convergence. Fig. 1 illustrates the convergence characteristics of several simulation parameters. These results suggest that a significant number of transport time steps can be taken before a new preconditioner must be generated. We are currently working on incorporating this approach into the integrated simulation.

4 Computer Science: The Common Component Architecture

The computer science and software engineering aspects of building a coupled simulation are often at least as complex as the physics and mathematics involved. Issues range from the software (and social) engineering of getting disparate groups of software developers to collaborate around the integrated simulation to the details of reconciling different software architectures, programming styles, parallel programming models, programming languages, data structures, etc. In this work, we look to ideas of component-based software engineering (CBSE), and in particular the realization of those ideas in the Common Component Architecture (CCA) [5, 6] to provide a software environment in which to integrate the coupled simulation and facilitate the social aspects of collaborative development.

CBSE is, fundamentally, a means to help manage the complexity inherent in modern large-scale software systems. CBSE treats applications as assemblies of software *components* that interact with each other only through well-defined *interfaces* (or *ports* in CCA terminology) within a particular execution environment, or *framework*. Components are units of software functionality, and are a logical means of encapsulating knowledge from one scientific domain (or sub-domain) for use by those in others, thereby facilitating group and multidisciplinary interactions. The glue that binds the components together is a set of common, agreed-upon interfaces, or ports. Multiple component implementations conforming to the same external interface standard should be interoperable, while providing flexibility to accommodate different aspects such as algorithms, performance characteristics, and coding styles. At the same time, the use of common interfaces facilitates the reuse of components across multiple applications. The Common Component Architecture distinguishes between *provides ports*, which are implemented by a component, and *uses ports*, where a component makes calls on a port provided by another component. The framework is responsible for providing an environment in which applications can be assembled from components (by connecting matching *uses* and *provides* ports) while insuring that internals of the components remain opaque to each other.

The CCA is designed specifically for the needs of parallel, scientific high-performance computing (HPC) in response to limitations of other, more widely used component approaches (see [5]). The general-purpose design of the CCA is intended for use in a wide range of scientific domains, and CCA-based simulations are under development in numerous fields, e.g. [7]. The CCA's emphasis on supporting HPC scientific computing has led to a number of important characteristics being incorporated into its design, and into tools that implement the CCA environment.

The CCA's *uses/provides* design pattern allows components in the same process address space to be invoked directly, without intervention by the framework, and with data passed by reference if desired, preserving high performance for components which are local to each other [8]. In parallel computing, interactions among components within the *same* process are handled by the CCA's port-based mechanisms, taking advantage of the "direct connect" approach just mentioned. Interactions across parallel instances of a component are up to the component's developer, and carry no CCA-imposed overheads.

Language interoperability is an important feature of many component models because it allows components to be opaque with respect to implementation *language* as well as other implementation details. In the CCA, the Scientific Interface Definition Language (SIDL) [9] provides a means to express the interfaces between components independent of their implementation language. SIDL works in conjunction with the Babel language interoperability tool[3] [9], which currently supports C, C++, Fortran 77, Fortran 90/95, Python, and Java.

Finally, the CCA employs a minimalist design philosophy to simplify the task of incorporating existing software into the CCA environment. The amount of extra code needed to acclimate a software module to the CCA environment can be as little as one extra (typically short) method, and several additional calls that manage ports provisioning and use. Experience has shown that componentization of existing software in the CCA environment is straightforward when starting from well-organized code [10].

5 Prototype Integrated Fusion Simulation

Our prototype integrated simulation was designed with the idea it should be extensible to integrate other physical processes in addition to the RF heating and transport simulations. In addition, we envisioned a simple, evolutionary migration path from the existing physics software to the componentized integrated simulation. The resulting design includes four types of components: a driver, which orchestrates the coupled simulation; the "physics" components, which encapsulate the original standalone simulation codes, an in-memory "plasma state" component, which holds the data exchanged by the different physics components, and "transformers" which adapt data from the form stored in the plasma state and the forms required by the various physics components. Fig. 2 shows the visual representation of the assembled application provided by the Ccaffeine CCA framework [4].

While the physics components and the transformers are all written in Fortran 90, the driver is written in the Python scripting language to facilitate flexible and rapid experimentation with different driver algorithms and parameters, and the plasma state component is implemented in C++.

The original standalone physics components read and write files directly for their inputs and outputs, and the initial version of the coupled application retains this characteristic (temporarily), using same file formats to transfer data between the plasma state

---

[3] http://www.llnl.gov/CASC/components/babel.html
[4] http://www.cca-forum.org/ccafe/

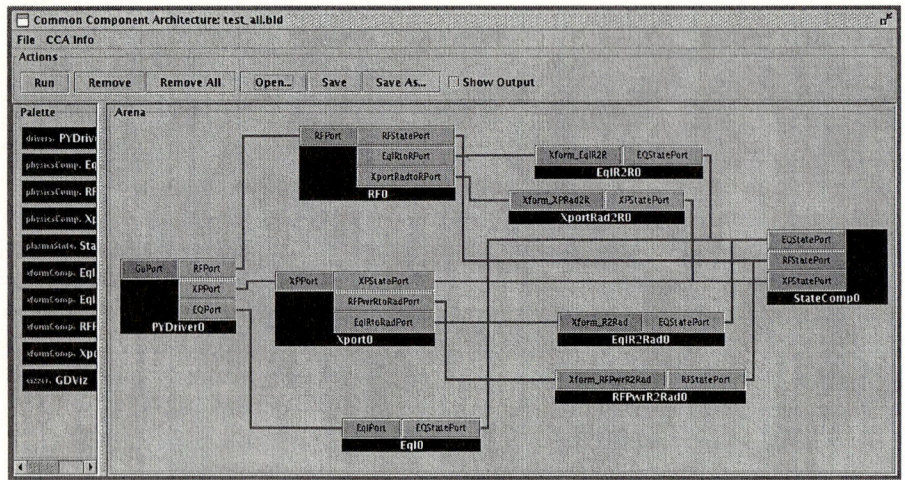

Fig. 2. Plasma simulation components

and the physics components. This approach was taken in order to minimize changes to the physics codes before verifying their correct behavior within the coupled application. We expect to quickly evolve the interfaces to allow for direct data exchange between components as the following step in the evolution of the port interfaces.

Data exchange between different physics module poses another challenge related to the format of the data. The output of one component may need to be re-gridded or transformed into a different coordinate system for another component. Most standalone physics codes in this community have their own set of transformation routines to perform these functions on inputs derived from other codes. In our initial prototype, we decided that the plasma state component should store data in the format in which it was computed. We use transformer components (extracted from the original standalone physics simulations) to implement a "receiver makes right" approach to insure components using the data get it in the format they need. While this approach does not address the n^2 problem of data transformation, we plan on using this prototype to gain a better understanding of the data transform issues in fusion simulations and develop general purpose reusable components that address this issue, drawing on existing efforts such as [11, 12].

The design of the interfaces that define the boundaries between the various components is crucial for the long-term flexibility and extensibility of the model. In our prototype, we opted for an *evolutionary* approach that facilitates rapid code integration, while allowing for future refinement as we gain better understanding of the overall application. A decision was made to adopt an initial design that promoted a *loose component coupling*, where methods that make up port interfaces are kept at a decidedly high level. This decision allows prototyping to proceed with little or no change to the mature physics applications that underly the main physics components. An added benefit of this approach has been the reduction in the amount of testing that had to be performed to validate the correctness of the overall application.

6 Conclusions and Future Work

In this paper, we presented the physical, mathematical, and computer science issues involved in the design and implementation of a coupled fusion simulation using the Common Component Architecture. The prototype we developed couples together several physics modules, along with the necessary data and driver components. Our experience has demonstrated the viability of using CCA as the underlying foundation for such integration efforts. As we continue, our primary focus is on gaining a better understanding of the detailed needs of large-scale coupled fusion simulations and evolving our prototype to meet them. This includes, for example, expanding the capabilities of the prototype by introducing application controlled checkpointing and archival storage into the plasma state component. Another aspect of our future plan involves the refinement of the interfaces, with the ultimate goal of helping to develop interfaces that can be agreed to and used across the larger fusion community; this would be an important step towards coupled simulations integrating more physical processes.

References

1. Dahlburg, J., et al.: Fusion simulation project: Integrated simulation and optimization of fusion systems. J. Fusion Energy **20** (2001) 135–196
2. Jaeger, E., et al.: All-orders spectral calculation of radio-frequency heating in two-dimensional toroidal plasmas. Physics of Plasmas **8** (2001) 1573–1583
3. Blackford, L.S., et al.: ScaLAPACK Users' Guide. SIAM (1997) Online version at http://www.netlib.org/scalapack/slug/index.html.
4. Saad, Y., Schultz, M.: GMRES: A generalized minimal residual algorithm for solving non-symmetric linear systems. SIAM J. Sci. Statit. Comput. **7** (1986) 856–869
5. Bernholdt, D.E., et al.: A component architecture for high-performance scientific computing (2004) submitted to *Intl. J. High-Perf. Computing Appl.*
6. Armstrong, R., et al.: Toward a Common Component Architecture for high-performance scientific computing. In: Proceedings of the Eighth IEEE International Symposium on High Performance Distributed Computing. (1999)
7. McInnes, L.C., et al.: Parallel pde-based simulations using the common component architecture. In Bruaset, A.M., Bjørstad, P., Tveito, A., eds.: Numerical Solution of PDEs on Parallel Computers. Springer-Verlag (2005) invited chapter, accepted.
8. Bernholdt, D.E., Elwasif, W.R., Kohl, J.A., Epperly, T.G.W.: A component architecture for high-performance computing. In: Proceedings of the Workshop on Performance Optimization via High-Level Languages and Libraries (POHLL-02). (2002)
9. Dahlgren, T., Epperly, T., Kumfert, G.: Babel User's Guide. CASC, Lawrence Livermore National Laboratory. (2004) http://www.llnl.gov/CASC/components/babel.html.
10. Norris, B., et al.: Parallel components for PDEs and optimization: Some issues and experiences. Parallel Computing **28** (2002) 1811–1831
11. Bertrand, F., Bramley, R., Damevski, K.B., Kohl, J.A., Bernholdt, D.E., Larson, J.W., Sussman, A.: Data redistribution and remote method invocation in parallel component architectures. In: Proceedings of the 19th International Parallel and Distributed Processing Symposium: IPDPS 2005. (2005) submitted.
12. Glimm, J., Brown, D., Freitag, L.: Terascale Simulation Tools and Technologies (TSTT) Center. http://www.tstt-scidac.org (2001)

A Case Study in Distributed Locking Protocol on Linux Clusters

Sang-Jun Hwang[1], Jaechun No[1], and Sung Soon Park[2]

[1] Dept. of Computer Software,
College of Electronics and Information Engineering,
Sejong University, Seoul, Korea
[2] Dept. of Computer Science & Engineering,
College of Science and Engineering,
Anyang University, Anyang, Korea

Abstract. Today's scientific simulations often generate huge amounts of data for data archival, data analysis, and visualization. These data are stored in high-performance distributed storages that consist of a network-oriented computing environment. In such a computing environment, one of the major issues affecting in achieving substantial I/O performance and scalability is to build an efficient locking protocol. In this paper, we present a distributed locking protocol that enables multiple client nodes to simultaneously write their data to distinct data portions of a file, while providing the consistent view of client cached data, and conclude with an evaluation of the performance of our locking protocol on a Linux cluster.

1 Introduction

Today's scientific simulations often generate huge amounts of data for data archival, data analysis, and visualization, and then these data are stored in high performance distributed storages that consist of a network-oriented computing environment. The common network-oriented storage architecture designates a few number of network-attached servers as a data storage pool and connects the clients to the servers via network, like GigaEthernet or Fibre Channel [1, 2, 3, 5, 6]. In such a storage architecture, a critical issue affecting in achieving high I/O bandwidth and scalability for the scientific simulations is to build an efficient locking protocol which is used for providing the coordinated accesses to remotely stored data and for providing the consistent views of client cached data. The reason is that, in order to produce high I/O bandwidth, many scientific simulations use parallel I/O packages where multiple client nodes simultaneously perform their I/O operations. MPI-IO is among those parallel I/O packages. MPI-IO [8, 10] is specifically designed to enable the optimizations that are critical for generating high-performance I/O. These optimizations include collective I/O and the ability to access noncontiguous data sets. However, in order to achieve high I/O performance using MPI-IO on a network-oriented distributed storage, the distributed file system which is running on top of the storage must provide the ability to lock a file per data section to have multiple concurrent writers. However, many of the locking protocols integrated with distributed file systems are based on a coarse-grained method [1, 2, 3] where only a single client at any given time is allowed to write its data to a file, while the other clients are waiting for the current node to finish its write operation even when the others would write to the different data portions of the same file. This drawback significantly degrades I/O performance in many scientific simulations where supporting parallel write operations happens to be proved generating high I/O bandwidth [4, 7].

In this paper, we present a distributed locking protocol based on multiple reader/single writer semantics for a data portion to be accessed. In this scheme, a single lock is used to synchronize concurrent accesses to a data portion of a file. But, several nodes can simultaneously run on the district data sections in order to support data concurrency. We conclude our paper by discussing the performance evaluation of our locking protocol on a Linux cluster.

2 Design Motivation

Our main objectives in developing a distributed locking protocol were to provide high-performance parallel I/O, to minimize the communication latency occurred during the lock negotiation steps, and to utilize local lock services as much as possible.

- **High-performance I/O** We designed the distributed locking protocol capable of allowing multiple concurrent writers to the same file to achieve high performance I/O. Also, the locking protocol provides data consistency between the data stored in the storage device and the data stored in the client-side cache.

- **Low communication latency** We designed the locking protocol to reduce the network overhead taking place during the lock negotiation steps with Global Lock Manager (GLM). All the lock requests coming from the client nodes are evenly distributed on multiple GLMs. Moreover, in order to minimize the number of callback messages necessary to revoke and release a lock, we grouped all the client nodes into several node groups. If GLM finds the node group where the lock holder belongs to it then sends a lock revocation message to the node group.

- **Use of local lock service** we designed the locking protocol to utilize local lock service to the maximum extents in order not to incur communication overhead with GLM and remote lock holders. By retaining the privileges on data sections even in the absence of active processes on a client, we eliminated the need to communicate with GLM repeatedly for the same data section, and thus can minimize the network latency.

3 Implementation Details

3.1 Overview

Figure 1 illustrates the distributed lock interface that is integrated with distributed file systems. Applications issue I/O requests using local file system interface, on top of VFS layer. Before performing an I/O request, each client should acquire an appropriate distributed lock from GLM in order to maintain data consistency between the cached data on clients and the remote, shared data on servers. The lock request is initiated by calling the lock interface, snq_clm_lock.

As mentioned in section 2, in order to reduce the communication latency occurring in the lock acquire step, we grouped the client nodes into several node groups. In the current implementation, an eight bit integer is used to denote node groups. When a client acquires an appropriate lock to perform I/O operation, the bit corresponding to the node group where the client belongs to is set to 1. Also, if a client requests a lock to GLM, GLM first locates the node group where the lock holder belongs to and then sends a callback message to the nodes of the node group. When the lock holder receives the callback message, it releases the requested lock and sends back an acknowledge to GLM to grant the lock to the requester.

Figure 2 represents a hierarchical overview of the locking construct with two client nodes and one GLM. The lock modes that we provide for are SHARED

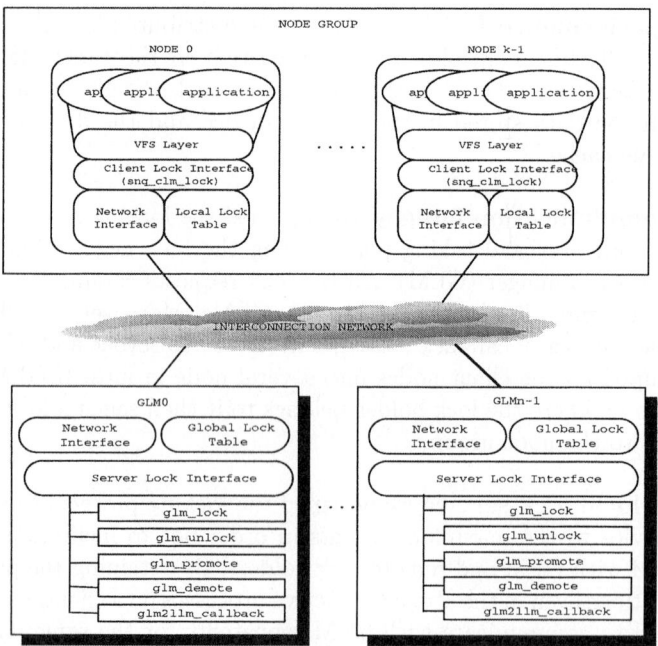

Fig. 1. A distributed lock interface

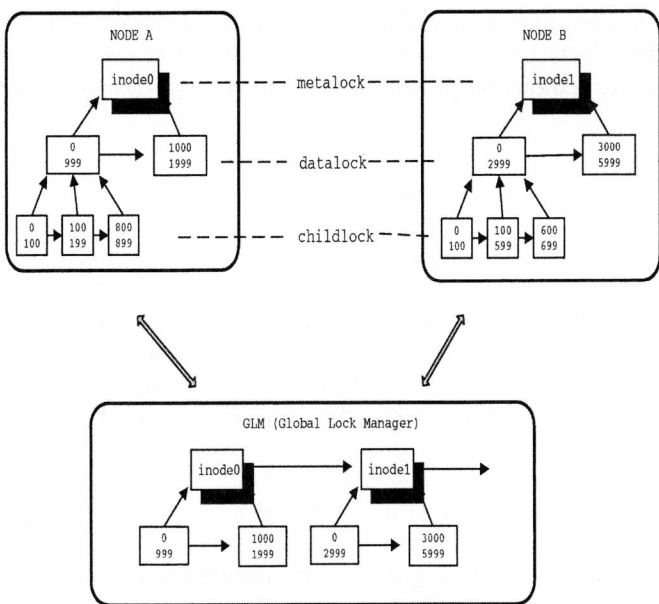

Fig. 2. A hierarchical overview of distributed locking protocol

for multiple read processes and EXCLUSIVE for a single write process. The lock structure consists of three levels: metalock, datalock, and childlock. The metalocks, inode0 on node A and inode1 on node B in Figure 2, synchronize accesses to files and the value of a metalock is an inode number of the corresponding file. Below the metalock is a datalock responsible for coordinating accesses to a data portion. For example, on node A, metalock inode0 is split into two datalocks associated with the data sections 0-999 and 1000-1999 in bytes and, on node B, two datalocks below inode1 are associated with the data sections 0-2999 and 3000-5999 in bytes. In order to grant a datalock, the lock mode of the higher lock (metalock) must be SHARED, meaning that a file is shared between multiple clients.

The lowest level is a childlock that is of a split datalock. As mentioned in section 2, given that a datalock is granted, the datalock can be split further to maximize local lock services as long as the data section to be accessed by a requesting process does not exceed the data section of the datalock held. In other words, in Figure 2, the datalock for the data portion 0-999 is split into three childlocks that control accesses to the data portion 0-100, 100-199, and 800-899, respectively. The childlock is locally granted and therefore the requesting process needs not communicate with GLM to obtain the childlock. However, the childlock is granted only when the lock mode of a childlock is compatible with that of the higher datalock. The datalock and childlock are found by comparing the starting file offset and data length being passed from the local file interface.

GLM contains the global lock information consisting of a list of locks that each GLM is responsible for serving. In Figure 2, GLM contains the metalocks, inode0 and inode1, and the datalocks of the data portions 0-999, 1000-1999, 0-2999, and 3000-5999 held by node A and node B. GLM also contains the node group information indicating those groups where the lock holders belong to.

4 Performance Evaluation

We measured the performance of the distributed locking protocol on the machines that have Pentium3 866MHz CPU, 256 MB of RAM, and 100Mbps of Fast Ethernet. The operating system installed on those machines was RedHat 9.0 with Linux kernel 2.4.20-8. The performance results focused on the time to obtain locks by performing lock revoke, downgrade, and upgrade operations. The time to invalidate client cached data and to write dirty data to disk was not included in the evaluation.

Figures 3 and 4 represent the time to obtain the locks with the exclusive mode in write operations and with the shared mode in read operations, as the number of clients increases from 4 to 16. Also, in Figure 3, one machine was configured as a GLM and, in Figure 4, four machines were configured as GLMs. When four machines were configured as GLMs, each lock request is given to a GLM, according to round robin fashion. All clients read or wrote 1Mbytes of data to the distinct portions of the same file. In this case, the lock requested by each client is newly created on GLM and returned to the requesting client, causing no callback message to be sent to the remote lock holder. Figures 3 and 4 show that if there is no lock revocation occurred with the remote lock holder, then the communication overhead necessary for acquiring a new lock becomes

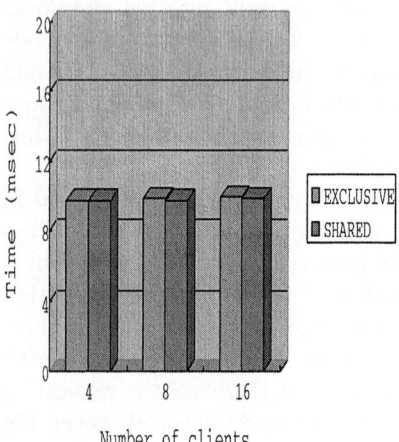
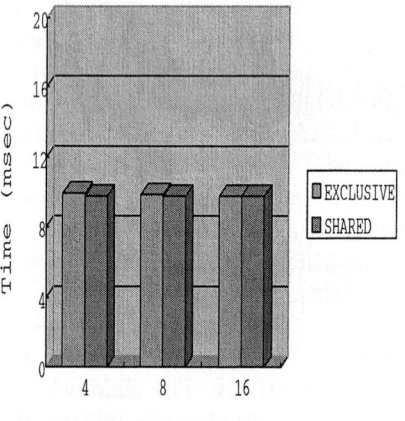

Fig. 3. Time overhead to acquire a distributed lock using one GLM. Each client read or wrote 1Mbytes of data to the distinct section of the same file

Fig. 4. Time overhead to acquire a distributed lock using four GLMs. Each client read or wrote 1Mbytes of data to the distinct section of the same file

small both with the exclusive mode and with the shared mode. Also, changing the number of GLMs from one to four doesn't affect the performance.

Figures 5 and 6 show the time to obtain the locks with the exclusive mode and with the shared mode, while moving each client's data section to access to the one given to the neighbor at the previous step. For example, at the first step, the first client accesses to the first 1Mbytes of data section of a file and the second client accesses to the second 1Mbytes of data section of the same file. At the second step, the first client's data section changes to the second 1Mbytes of data section for which the second client has already acquired the lock at the first step. Therefore, at the second step, the second client should yield the lock held to the first client, while taking a new lock from the third client. Also, Figure 5 shows the time taken by using one GLM and Figure 6 shows the time taken by using four GLMs.

Figures 5 and 6 both illustrate that the overhead of the lock revocation is significant with the exclusive mode because only a single client is allowed to write to a data section at any given time. With the shared mode, there is no need to contact the remote lock holder since a single lock can be shared between multiple nodes. With the shared lock mode, GLM just increases a counter denoting the number of shared lock holders before granting the lock. Finally, the communication overhead is decreased when the number of GLMs is changed from one to four since the lock requests issued by the clients can be distributed on the multiple GLMs.

In order to figure out how much the network latency occurred at the lock negotiation step dominates the performance, we changed the number of clients running on each node, while keeping the total number of clients as 16. If the

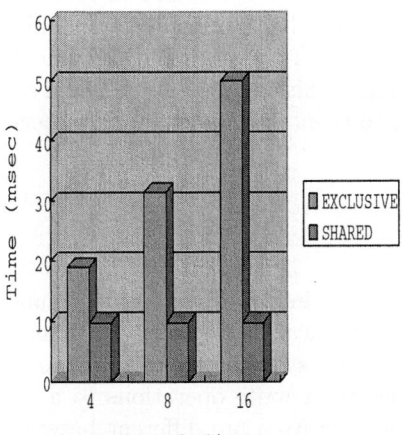
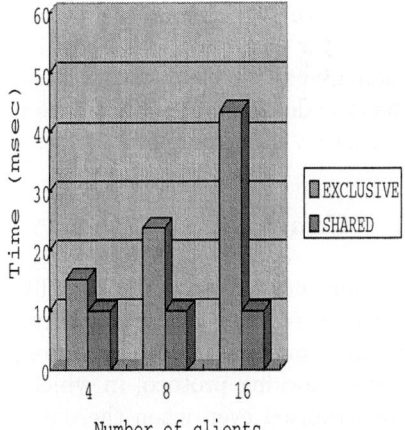

Fig. 5. Time to acquire a distributed lock using one GLM. A client's data section is shifted to the one given to the neighbor at the previous step

Fig. 6. Time to acquire a distributed lock using four GLMs. A client's data section is shifted to the one given to the neighbor at the previous step

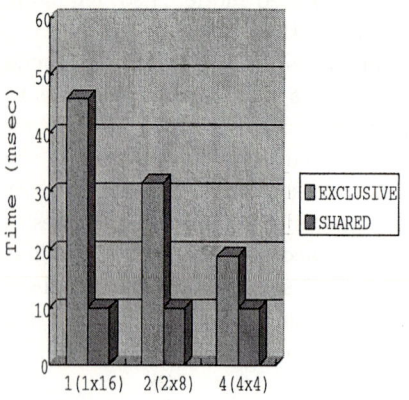 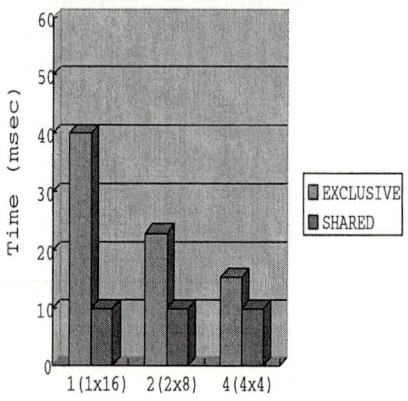

Fig. 7. Time overhead to acquire a distributed lock using one GLM as a function of number of clients running on each node. A client's data access range is shifted right at each step

Fig. 8. Time overhead to acquire a distributed lock using four GLMs as a function of number of clients running on each node. A client's data access range is shifted right at each step

number of clients on a node is one, then every 16 clients runs on the different node. If the number of clients is two, then two clients run on each node and so 8 different nodes are needed to have 16 clients as total. Also, as did in Figures 5 and 6, we changed the data access range of each client to the one given to the neighbor at the previous step. With two clients running on the same node, the callback message is sent to the remote lock holder every two I/O operations to revoke a lock. With four clients, the callback message is sent to the remote lock holder every four I/O operations, resulting in the lock negotiation overhead decrement, compared to two clients on each node. According to this experiment, we could see that the dominating performance factor with 16 clients is the network overhead to contact the remote lock holder.

5 Conclusion

Concurrent accesses to the same file frequently occur in modern scientific simulations where allowing parallel write operations significantly improves I/O bandwidth. However, most distributed client-server file systems support a coarse-grained locking protocol in which all the concurrent write operations to a file are serialized even when the data sections being written are different between writers. In this paper, we presented a distributed locking protocol with which several nodes can simultaneously write to the distinct data portions of a file, while guaranteeing a consistent view of client cached data. The distributed locking protocol has also been designed to exploit locality of lock requests to minimize communication overhead with GLM and remote lock holders.

References

1. Murthy Devarakonda, Bill Kish, and Ajay Mohindra. Recovery in the Calypso file system. ACM Transactions on Computer Systems, 14(3):287–310, August 1996
2. Chandramohan A. Thekkath, Timothy Mann, and Edward K. Lee. Frangipani: A Scalable Distributed File System. In Proceedings of the Symposium on Operating Systems Principles, 1997, pages 224–237
3. Kenneth W. Preslan, Andrew P. Barry, Jonathan E. Brassow, Grant M. Erickson, Erling Nygaard, Christopher J. Sabol, Steven R. Soltis, David C. Teigland, and Matthew T. O'Keefe. A 64-bit Shared Disk File System for Linux. In Proceedings of Sixteenth IEEE Mass Storage Systems Symposium Seventh NASA Goddard Conference on Mass Storage Systems & Technologies, March 15-18, 1999
4. Jean-Pierre Prost, Richard Treumann, Richard Hedges, Bin Jia, Alice Koniges. MPI-IO/GPFS, an Optimized Implementation of MPI-IO on top of GPFS. In Proceedings of Supercomputing, November 2001
5. F. Schmuck and R. Haskin. GPFS: A Shared-Disk File System for Large Computing Clusters. In Proceedings of the First Conference on File and Storage Technologies(FAST), pages 231–244, Jan. 2002
6. P. J. Braam. The Lustre stroage architecture. Technical Report available at - http://www.lustre.org, Lustre, 2002
7. Jaechun No, Rajeev Thakur, and Alok Choudhary. High-Performance Scientific Data Management System. Journal of Parallel and Distributed Computing, (64)4:434-447, April 2003
8. Jean-Pierre Prost. MPI-IO/PIOFS. World-Wide Web page at http://www.research.ibm.com/people/p/ prost/sections/mpiio.html, 1996
9. MacroImpact Inc., SANique CFS. A SAN Based Cluster File System, Version 2.1, Technical Report, August 2002
10. William Gropp and Ewing Lusk and Rajeev Thakur. Using MPI-2: A dvanced Features of the Message-Passing Interface, MIT Press, 1999, Cambridge, MA

Implementation of a Cluster Based Routing Protocol for Mobile Networks

Geoffrey Marshall[1] and Kayhan Erciyes[2]

[1] California State University San Marcos,
Computer Science Dept., 333 S.Twin Oaks Valley Rd.,
San Marcos, CA 92096, U.S.A
[2] Izmir Institute of Technology
Computer Eng. Dept., Urla, Izmir 35430, Turkey
{marsh021, kerciyes}@csusm.edu

Abstract. We show the implementation and the simulation results of a hierarchical, cluster based routing protocol for mobile ad hoc networks using *Parallel Virtual Machine* (PVM). The network represented by a graph is partitioned into clusters by a graph partitioning algorithm and the shortest routes are first calculated locally in each cluster in the first srep. The simplified network which consists only of the nodes that have connections to other clusters called the neighbor nodes is then formed and the shortest routes are calculated for this simple network as the second step. A complete route between the two nodes of different clusters is formed by the union of intra-cluster and inter-cluster routes. We show the implementation results using PVM where a workstation represents a cluster and each node is a PVM process. The results obtained support the theoretical considerations where the efficiency increases by the number of clusters in use.

1 Introduction

Mobile ad hoc networks do not have central administration or fixed infrastructure and consist of mobile wireless nodes that have temporary interconnections to communicate over packet radios. As the topology of a mobile network changes dynamically, routes are needed to be calculated much more frequently than the wired networks. Various methods such as distributed, adaptive and self-stabilizing algorithms are used to perform routing in mobile networks. In *Link reversal routing* algorithms, a node reverses its incident links when it loses routes to the destination. Performance analysis of link reversal algorithms are given in [1] and TORA [9] is an example system that uses link reversal routing. Routing in mobile networks can be performed by clustering, that is, partitioning of the network into smaller subnetworks to limit the amount of routing information stored at individual nodes. In [8], a mobile network is partitioned into clusters of a two level graph. In the *zone routing* proposed in [5] where a zone functions similar to a cluster, the requested routes are first searched within the local zone.

For inter-zone routes, the search is carried by multicast messages to the boundary nodes within the zones. In *k-way clustering*, the mobile network is divided into non-overlapping clusters where two nodes of a cluster are at most k hops away from each other. A k-way clustering method is proposed in [3] where the spanning tree of the network is constructed in the first phase and this tree is partitioned into subtrees with bounded diameters in the second phase.

In this study, we evaluate the performance of a hierarchical, two-level dynamic routing protocol described in [4] using PVM. The protocol consists of three main phases of partitioning the mobile network graph into clusters, calculating local cluster routes and finally calculating the simplified network graph routes. The rest of the paper is organized as follows. The background is given in Section 2, the analysis is discussed in Section 3, the PVM test results are given in Section 4 and the conclusions are outlined in Section 5.

2 Background

2.1 Partitioning of the Mobile Network

Graph partitioning algorithms aim at providing subgraphs such that the number of vertices in each partition is averaged and the number of edges cut between the partitions is minimum with a total minimum cost. An arbitrary network can be constructed as an undirected connected graph $G = (V, E, w)$ where V is the set of routing nodes, E is the set of edges giving the cost of communication between the routing nodes and $w: E \rightarrow \Re$ is the set of weights associated with edges. *Multilevel partitioning* is performed by coarsening, partitioning and uncoarsening phases [6]. During the coarsening phase, a set of smaller graphs are obtained from the initial graph. In the maximal matching, vertices which are not neighbors are searched. In Heaviest edge matching (HEM), the vertices are visited in random order, but the collapsing is performed with the vertex that has the heaviest weight edge with the chosen vertex. In Random Matching (RM) however, vertices are visited in random order and an adjacent vertex is chosen in random. The coarsest graph can then be partitioned and further refinements can be achieved by suitable algorithms like Kernighen and Lin [7]. Finally, the partition of the coarsest graph is iteratively reformed back to the original graph.

We provide a partitioning method called *Fixed Centered Partitioning* (FCP) [4] where several fixed centers are chosen and the graph is then coarsened around these fixed centers by collapsing the heaviest or random edges around them iteratively. Different than [6], FCP does not have a matching phase, therefore iterations are much faster. FCP requires the initial marking of the fixed centers. One possible solution is to choose the fixed centers randomly so that they are all at least some bounded distance from each other. The heuristic for the bound we used is $h = 2d / p$ where d is the diameter of the network and p is the number of partitions (clusters) to be formed. The time complexity of the total collapsing of FCP is $O(n)$. FCP provided much favorable partitions than CM and RM in terms of the average edge cost, time to partition a graph and the quality of the partitions experimentaly [4].

2.2 The Hierarchical Routing Protocol

The routing protocol called the *Neighbor Protocol* for the mobile network is not fully distributed due to the existence of some privileged nodes in the network. The distributed routing architecture consists of hierarchical clusters of routing nodes and each cluster has a controller which is called the *representative*. At the highest level, one of the representatives called the *coordinator*, receives messages to update its view everytime there is an addition or deletion of a node to a cluster. Upon such changes of configuration or periodically gathering of the changes, the coordinator starts a new configuration process by partitioning the network graph into new clusters. The nodes in the cluster that have connections to other clusters are called the *neighbor nodes*. The coordinator chooses one of the neighbor nodes in each cluster as the cluster representative and sends the cluster and neighbor topology information to the representative of such a group. Each representative then distributes the local connectivity information to all of the nodes in its group which concludes the first step of the protocol. In the second step, each node performs All-Pairs Shortest-Paths (APSP) routing within its cluster. At the end of this step, the distances between all pairs of nodes in the cluster including the neighbor nodes are calculated. In the third step, only the neighbor nodes calculate APSP routes for the simplified network graph which consists of neighbor nodes only. Any route is then formed by the union of the route from the source node to its nearest neighbor, the shortest route between the source neighbor and the destination neighbor and the shortest route between the destination neighbor and the destination node.

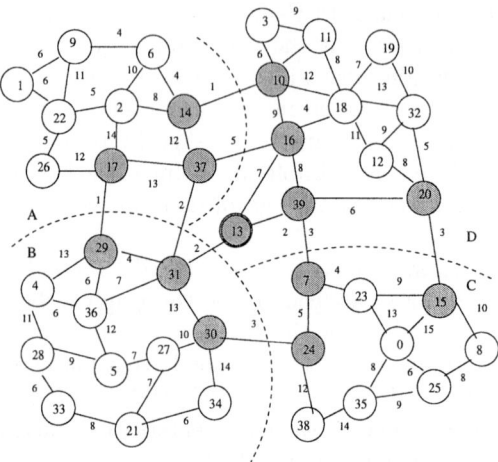

Fig. 1. The Original Network

2.3 An Example Network

An example network is depicted in Fig. 1. The initial centers allocated are 28, 8, 1 and 32. The coordinator and also the representative for cluster D is at 13. The coordinator partitions the graph using FCP as shown. Based on the partitioning information, the representatives chosen from the neighbors as 17, 30 and 15 are informed of their local connection. In the second phase, the representatives transfer this information to local nodes in their clusters in parallel. The ordinary nodes then calculate APSP in parallel, however, the neighbor nodes have to also calculate APSP for the simplified network graph which consists of the neighbor nodes only as shown in Fig. 3. Consider an example where node 26 in cluster A wants to send a message to the node 35 in cluster C. Since destination is not in its own cluster, 26 sends the message to its closest neighbor node, 17. Node 17 sends the message to node 7 which is its closest neighbor node in cluster C over 17-29-31-13-39-7. The neighbor node 7 routes the message to the destination over the shortest path which is 7-23-0-35. The total cost of this path is 49.

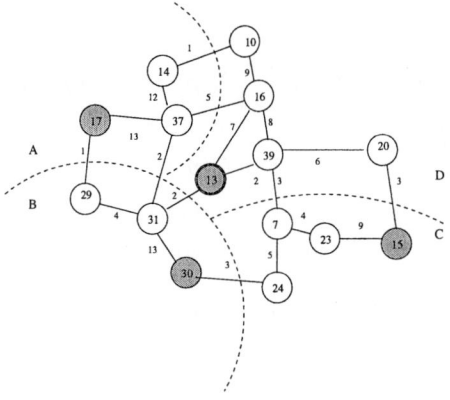

Fig. 2. The Simplified Neighbor Network

3 Analysis

The performance analysis should include the following

1. Partitioning of the network graph by FCP : O_1
2. Distribution of the cluster connectivity messages to the cluster representatives : O_2
3. Distribution of the routing information to the individual nodes by each representative : O_3
4. Intra-cluster route calculation time by the nodes within the cluster : O_4
5. Inter-neighbor route calculation by the neighbors : O_5

It was shown in [4] that the distribution of individual cluster routing information to the nodes (steps 2 and 3 above) take $O_{dist}(m)$ time where m is an upper bound on the number of nodes in a cluster. Also the total time required for intra-cluster and inter-neighbor routing algorithms is $O_{route}(m^3)$. The following theorem showed the Speedup obtained by the proposed protocol[4].

Theorem 1. *The Speedup obtained by the proposed protocol to a pure sequential all-to-all shortest paths protocol is $O(p^3)$ and to the parallel case where each node calculates all of the routes in parallel with others is $O(p^2/m)$.*

Proof. Total time for the protocol (O_{prot}) is :

$$O_{prot} = O_{part}(n) + O_{dist}(m) + O_{route}(m^3) = O(n + m^3) \qquad (1)$$

and assuming a balanced partition, that is, $n = mp$

$$O_{prot} = O(n + m^3) = O(mp + m^3) \qquad (2)$$

Assuming the network has p clusters and m nodes at each cluster, a serial algorithm to compute all routes of this network will take $O_{serial}((p*m)^3)$ operations. The speedup S that can be approximated with respect to pure serial case is :

$$S = O_{serial}/O_{prot} = O((p*m)^3/(mp + m^3)) \qquad (3)$$

and assuming $m \gg p$

$$S = O(p^3) \qquad (4)$$

For the pure parallel case where each node has all of the network connectivity information, $O_{par} = O(p^2 m^2)$ and the speedup now is :

$$S = O(p^2 m^2/m^3) = O(p^2/m) \qquad (5)$$

4 Experimental Results Using PVM

Simulation of the network initialization and routing was performed using PVM. The simulation was performed for cluster sizes of 2, 4, 8, and 12 on a Beowulf cluster of PC's running Linux.

4.1 Initialization

To simulate the Neighbor Protocol, a central coordinator task is started to initialize the configuration of the network as follows:

1. Central coordinator task partitions graph into designated number of clusters.
2. From each cluster, the coordinator selects a neighbor node as representative and spawns a rep task on a new host.
3. Coordinator distributes local (cluster) connectivity information and neighbor connectivity to each rep task.

4. Each rep task spawns a neighbor task for each neighbor (not including itself) in its cluster on its host machine, and distributes local connectivity and neighbor connectivity to the neighbor.
5. The rep task spawns an ordinary task for each remaining non-neighbor node in the cluster and sends only local connectivity.

Every node, in parallel, calculates local routes using APSP. In addition, the neighbor tasks perform APSP for inter-cluster routes. Upon completing routing calculations, each node sends and ACK message to the representative of the cluster. Once the rep has collected ACK messages from all nodes in the cluster it sends ACK to the coordinator. When the coordinator has collected ACK's from all rep tasks, initialization is complete. The results for initialization of the network cluster configuration and routing calculations is shown in Fig. 3 for different size clusters. The times are significantly lower than the times for a non-distributed implementation that does not use the Neighbor Protocol wherein all nodes perform APSP for the entire graph as illustrated below. The measurements for normal APSP are about 5-10 times higher than the NP values and are not shown in graph.

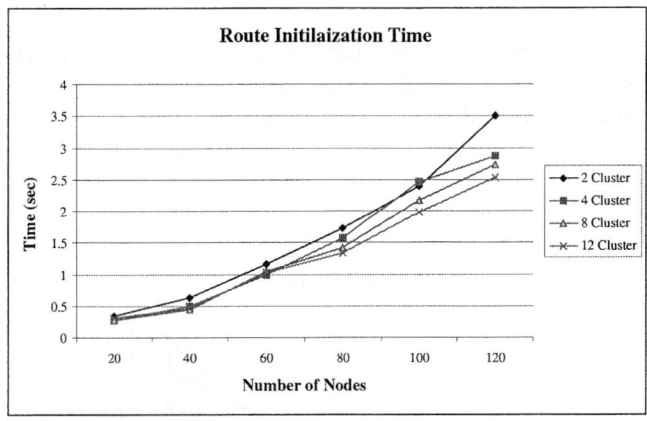

Fig. 3. Initialization Times for Clusters

4.2 Network Communication

Once a node has calculated routes it sends messages to randomly picked nodes in the network. Node u sends a message to node w by first creating a message and then looking up the next node v, in the route to w. For example, if ordinary node in cluster $d1$ wishes to send a message to a node in cluster $d2$, it would send the message to the next node in the shortest path to its closest neighbor, to be routed to cluster $d2$. When an incoming message is received, the node checks the header and either receives it or forwards it to the next node in route to destination. This procedure continues for each node to send one hundred messages and until all messages have been received. We measured communication time required for each node in the Neighbor Protocol to send one hundred messages and have them correctly routed and received. We found that run time is lower for more clusters providing less of a load per cluster as graph size increases.

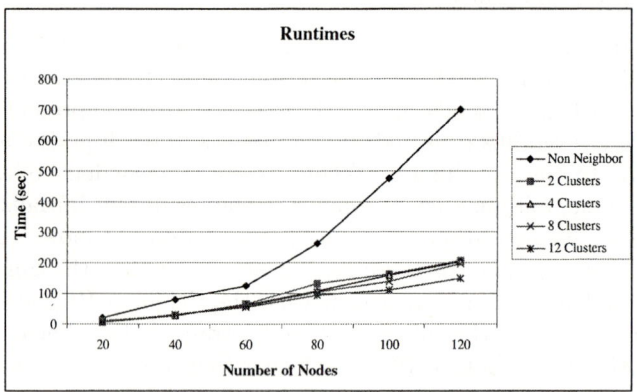

Fig. 4. Runtimes for Clusters

Fig. 4 shows that the execution time is roughly equivalent for smaller graphs of sizes 20 and 40 nodes for all cluster sizes but is improved for the larger graphs of 100 and 120 nodes for larger number of clusters such as 8 and 12. Again, the distributed Neighbor version substantially outperforms the single host implementation as shown.

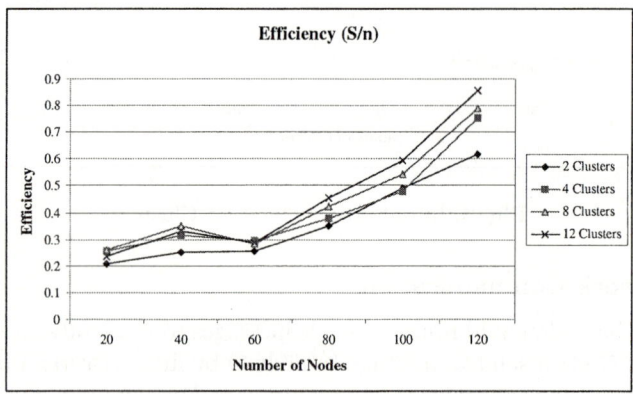

Fig. 5. Efficiency of the Neighbor Protocol

In Fig. 5, the efficiency curves for various cluster and node sizes are plotted where it can be seen that the efficiency rises as the number of clusters increases which is in accordance with the theoretical analysis (Section 3 and [4]).

5 Conclusions

We showed the simulation and further results of a proposed dynamic routing protocol initially described in [4] for a mobile network called the Neighbor Protocol using PVM. The protocol consists of three main steps of by firstly partitioning the mobile network graph, secondly delivery of the connectivity information of each cluster to the representative of the cluster which forwards this to individual nodes which calculate APSP routes within their clusters. In the final step, neighbor nodes calculate APSP routes for the simplified network. We showed that this approach improves performance considerably theoretically and the test results using PVM supported the theoretical analysis that the efficiency of the NP protocol rises as the number of clusters are inreased. The method we propose for routing in mobile networks provides *good* routes which are not necessarily the shortest paths but are comparable to shortest paths as shown by the tests. Further tests that exhibit the dymanic nature of the mobile networks to evaluate the performance of NP in terms of total control traffic against the frequency of route requests and frequency of movement in a mobile network using PVM are required. We are also looking into the fully distributed version of this protocol for mobile ad hoc networks for the case where there is no central coordinator but there are representatives and decisions on the partitioning of the graph and routing are done at the representative level by distributed agreement.

References

1. Busch, C. et al.: Analysis of Link Reversal Routing Algorithms for Mobile Ad hoc Networks. Proc. of the 15th ACM Symp. on Parallel Alg. and Arch. (2003) 210-219
2. E.W. Dijkstra: A Note on Two Problems in Connection with Graphs. Numerische Math., Vol. 1. (1959) 69-271
3. Fernandess,Y., Malki, D.: K-clustering in Wireless Ad hoc Networks, Proc. of the second ACM Int. Workshop on Principles of Mobile Computing, (2002) 31-37
4. Erciyes, K, Marshall, G, : A Cluster-based Hierarchical Routing Protocol for Mobile Networks, SV-LNCS, ICCSA(3) 2004, 528-537.
5. Z.J. Haas, M.R. Pearlman: The Zone Routing Protocol (ZRP) for Ad hoc Networks. Internet Draft, Internet Engineering Task Force. (1997)
6. Karypis, G., Kumar V.: Multilevel k-way Partitioning scheme for Irregular Graphs. Journal of Parallel and Distributed Computing, Vol. 48. (1998) 96-129
7. Kernighan, B., Lin, S.: An Effective Heuristic Procedure for Partitioning Graphs. The Bell System Technical Journal, (1970) 291-308
8. P. Krishna et al.: A Cluster-based Approach for Routing in Dynamic Networks. ACM SIGCOMM Comp. Comm. Rev., Vol. 27(2). (1997) 49 - 64
9. V.D. Park, M.S. Corson: A Highly Adaptive Distributed Routing Algorithm for Mobile Wireless Networks. Proc. IEEE INFOCOM, Vol. 3. (1997) 1405-1413

A Bandwidth Sensitive Distributed Continuous Media File System Using the Fibre Channel Network

Cuneyt Akinlar[1] and Sarit Mukherjee[2]

[1] Computer Eng. Dept., Anadolu University, Eskisehir, Turkey
cakinlar@anadolu.edu.tr
[2] Lucent Technologies, Holmdel NJ, USA
sarit@lucent.com

Abstract. A recent trend in storage systems design is to move disks to a storage area network for direct client access. While several highly scalable file systems have been built using such direct attached disks, they have all been designed for traditional text-based data, and are not well suited for streaming continuous media, i.e., audio and video files, which are characterized by high volumes of data and require strict timing requirements during storage and retrieval. In this paper, we propose a scalable distributed continuous media file system built using Storage Area Network (SAN)-Attached disks, and describe bandwidth and time sensitive read/write procedures for our file system. We present experimental results on the performance of our Linux-based prototype implementation of the file system and show that the file system can provide strict bandwidth guarantees for continuous media streams.

1 Introduction

A new trend in storage systems design is to move disks, that traditionally reside behind a centralized server (e.g., Network File System (NFS) [6], Symphony [7], Continuous Media File System (CMFS) [1]), to a storage area network (SAN) for direct client access. While several highly scalable file systems have been built using such direct attached disks (e.g., Global File System (GFS) [8], File systems for NASD [4], xFS [2] among many others), they have all been designed for traditional text-based data (i.e., file systems consisting of many small files), and are not well suited for streaming continuous media files, which are characterized by large volumes of data and stringent bandwidth requirements.

In this paper we propose a scalable distributed continuous media file system based on Storage Area Network (SAN)-Attached disks that can provide strict bandwidth guarantees to open media streams. We briefly describe the general architecture of the file system and discuss different ways to implement this "real-time" sensitivity. We present experimental results on the performance of our Linux-based prototype implementation of the file system using the Fibre Channel

SAN. Our file system appears as a (Ext2fs) [3] file system to users, and can provide strict bandwidth guarantees for continuous media streams.

2 Architecture of the Distributed File System

The architecture of the Fibre Channel Distributed File System (FCDFS) is shown in Figure 1(a). Main components are a file server (FCDFS-Server) and several Storage Area Network (SAN)-Attached disks, which are directly exposed to the clients. While the SAN is used for data transmission between the disks and the clients, a control network connects the clients and the server and is used for exchange of file system meta-data and control messages. Although the figure shows separate logical networks for the SAN and the Control Network, both can coexist on the same physical network depending on the capabilities of the SAN.

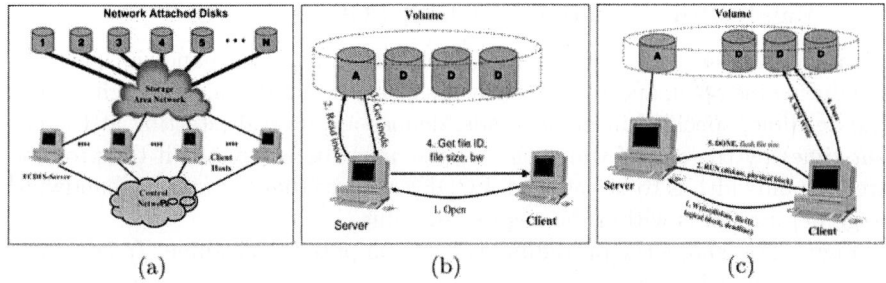

Fig. 1. (a) Architecture of FCDFS: A file server, client and disks connected together, (b) Open, and (c) Read/Write Operations: **A** and **D** denotes attribute and data disks, respectively

FCDFS consists of a client part, which we call the Client-FCDFS, and a FCDFS-Server. While Client-FCDFS is responsible for maintaining open files (streams) and actual reading and writing of the data from/to the disks, the server is responsible for maintaining the meta-data for all volumes, files and directories, and coordinating client access to the data disks. The Client-FCDFS and the server work together in a coordinated fashion to make the operations seamless to the users.

An overview of the file read/write procedure is shown in Figure 1(c). To read or write a block of data, the Client-FCDFS sends a read/write request to the server specifying the logical block of the file to read/write along with the deadline of the request. The server converts the logical block into a (disk, physical block) pair, and schedules the request based on its deadline. Once the client gets a response to its read/write request, it directly accesses the data disk. Therefore, during a read operation, the data directly comes from the disk to the client and during a write operation, the data directly goes from the client to the disk. Excluding the file server from the path of data transfers makes the file system scalable. The server is contacted when the file is opened and closed and for read/write request scheduling and acts like a data disk access coordinator. The actual disk request is carried out by the client.

3 Bandwidth Allocation and Enforcement

To preserve the quality guarantee across all streams, per stream bandwidth allocation and its enforcement is necessary, and involves the following steps:

1. *Bandwidth allocation*: Each stream (user) is allocated a certain bandwidth. FCDFS-Client computes a deadline for each **pull** request using the negotiated bandwidth.
2. *Bandwidth enforcement*: FCDFS-Server employs a time-sensitive scheduling discipline to ensure that no request misses its deadline.

3.1 Bandwidth Allocation by Client-FCDFS

Client-FCDFS keeps per-stream state information, performs bandwidth negotiation during file open, and subsequently assigns a service deadline to each request. The deadline, specified in milliseconds, defines an interval (starting at the current time) by the end of which the request must be completed if the promised stream bandwidth is to be satisfied. FCDFS ensures that the average bandwidth usage per stream is within the negotiated value.

Figure 2(a) shows the procedure used to compute the deadline. To keep track of past bandwidth usage by a stream, the Client-FCDFS keeps a virtual clock [9],

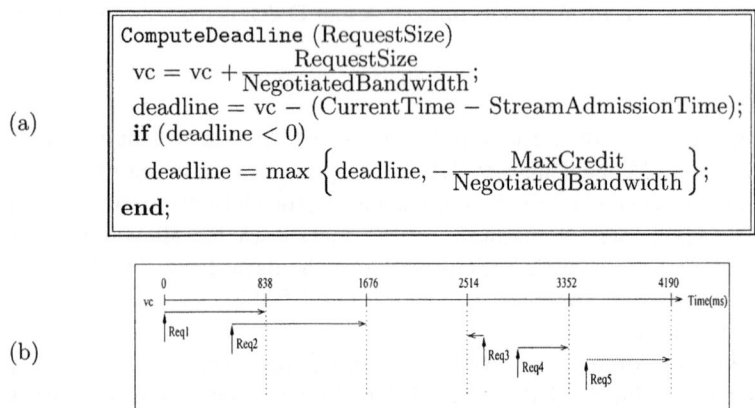

Fig. 2. (a) Algorithm for deadline computation, (b) An example assignment of relative deadlines at the client. Negotiated bandwidth is 10Mbps and requests are in FCDFS-block size of 1MB. Notice that the deadline for the third request is negative, which happens when a stream gets (or asks) less than what it negotiated for a period of time. The negative deadline allows the stream to eventually catch up

vc, for each stream. During the admission of the stream, vc is initialized to 0. It is modified only when there is any I/O request for the stream. vc keeps account of how long it should have taken for a request to complete had the stream obtained the exact negotiated bandwidth. Note that FCDFS allows the deadline to become negative if a stream does not use its allocated bandwidth. However, FCDFS limits the maximum credit (i.e., the parameter MaxCredit in bytes in the following algorithm) that a stream can accumulate during any inactive period (i.e., no I/O). This ensures that the burst size to the system is limited and regulated.

3.2 Bandwidth Enforcement by the FCDFS-Server

The bandwidth enforcement deals with effective and efficient scheduling of the user requests at the server such that a request completes by its deadline. This proceeds in two steps as detailed in the following two sections.

3.2.1 Global Request Deadline Computation

Upon reception of a request, the server converts the relative deadline of the request (recall that each request carries a relative deadline that specifies an interval within which the request must be completed) into an absolute deadline so that the relative order among requests from different client hosts can be constructed and maintained. Computation of the absolute deadline of a request, $s^{request}$, simply is: $s^{request} = t_{cur} + r_{deadline}$, where t_{cur} is the current time at the server, $r_{deadline}$ is the relative deadline specified by the client.

3.2.2 Request Scheduling

To achieve time-sensitive scheduling of client requests, the server keeps an ordered queue of requests (RQ) for each data disk. The requests in RQ are scheduled using the Earliest Deadline First (EDF) algorithm [5] with the absolute deadline, $s^{request}$, as the key.

Upon reception of a new request, the server checks the RQ for the disk. If there is no request accessing the disk, then the server schedules the new request by sending a RUN message to the client. Notice that the server schedules a request for execution, but it is the Client-FCDFS that actually executes the request. Thus the server is not on the path of data flow between the disk and the client host.

If there is already a request accessing the disk when a request arrives, the server has two choices: (1) It can wait until the currently executing request completes and then schedule the next request from among the pending requests or (2) If the new request has a smaller deadline, it can preempt the currently executing request and schedule the new one. These two scheduling disciplines are called **non-preemptive scheduling** and **preemptive scheduling**.

Figure 3(a) shows how 2 requests from different clients are scheduled by the non-preemptive scheduling discipline. First, Client 1 sends a READ request. Because the disk is idle, the server schedules it immediately, and sends a RUN reply. Client 1 starts accessing the disk. While the disk read is in progress,

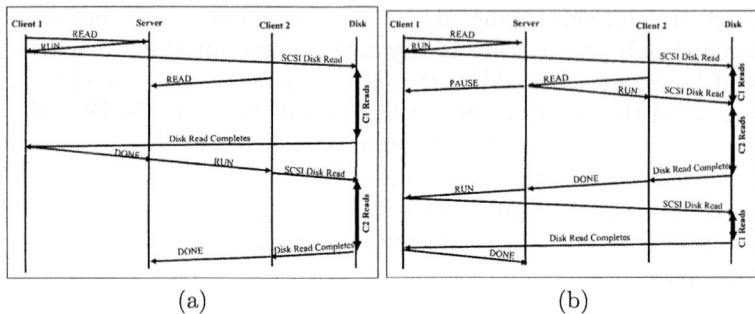

Fig. 3. (a) Sequence of Events in non-preemtive scheduling, (b) Sequence of Events in preemptive scheduling

Client 2 sends a READ request, which waits its turn until the currently executing request is done. When Client 1 is done reading the disk, it sends a DONE message to the server. The server then schedules Client 2's request and sends a RUN reply. Client 2 reads the data from the disk, and sends a DONE message back to the server when done. If Client 2's request had a stricter deadline than Client 1's request, it might miss its deadline because it must wait until the currently executing request is done.

Figure 3(b) shows how 2 requests from different clients are scheduled by the preemptive scheduling discipline. First, Client 1 sends a READ request. Since the disk is idle, the server schedules it immediately, sends a RUN reply. Client 1 starts accessing the disk. While Client 1 is reading from the disk, Client 2 sends a READ request. Because the request from Client 2 has a stricter deadline than the currently executing request from Client 1, the server preempts it by sending a PAUSE to Client 1 and a RUN to Client 2. From this point on, Client 2 starts reading the disk. When Client 2 is DONE, it sends a DONE message back to the server. The server then reschedules Client 1's request by sending a RUN reply. Client 1 then resumes the execution of the request and finishes it. It sends a DONE message to the server when done.

Since a FCDFS-block resides contiguously on a disk, the non-preemptive policy ensures that disk head movement will be minimal during the service. Therefore, the policy can yield high disk utilization. On the other hand, since servicing a 1MB FCDFS-block may take a considerable amount of time (about 100ms for a disk with an average transfer rate of 10MB/s), the higher priority requests may get delayed when the disk is busy, which may cause deadline misses.

3.2.3 Request Execution at the Client-FCDFS

Although requests are scheduled by the server, it is the Client-FCDFS that executes them. To execute disk read/write requests, the Client-FCDFS runs a kernel thread for each data disk, which runs the algorithm shown in figure 4.

```
ExecuteRequest(Request r)
    BatchSize = number of 1KB requests issued to the disk driver at once;
    for (i = 0; i < r.blocks/BatchSize; i++)
        /* each request is for a 1KB disk-block */
        prepare BatchSize disk-block requests;
        give BatchSize requests to the disk driver;
        wait until BatchSize requests are completed;
        if ( r.status == PAUSE ) then
            Save the current execution state; return;
        end-if
    endfor;
    send DONE message to the server;
end-ExecuteRequest
```

Fig. 4. Request execution at the Client-FCDFS

Instead of issuing a single request for an FCDFS block of size 1MB, ExecuteRequest function issues 1024 read requests, each for a 1KB portion of the block. Therefore "r.blocks" in the algorithm is initialized to 1024 and saved properly in the case of a request preemption. ExecuteRequest does not issue each 1KB request separately, rather it prepares groups of requests of size `BatchSize` and gives this group of requests to the disk driver at once. Once `BatchSize` requests are complete, the next group of requests are prepared. This process continues until the whole block is read/written.

When the server employs the preemptive scheduling discipline, it sends a PAUSE message to preempt a request. Upon reception of a PAUSE message, `ExecuteRequest` saves the current state of the request and suspends the execution. The execution resumes when the server reschedules the request by sending a RUN message. Notice that the execution of the current request will be preempted only after `BatchSize` disk requests are complete even if PAUSE message arrives early. Therefore, the value of `BatchSize` determines how quickly (i.e., the frequency) the Client-FCDFS can respond to preemption messages.

4 Numerical Results

To evaluate the bandwidth enforcement policies of our file system, we set up an architecture consisting of 4 client hosts, a server and 8 disks connected to a Fibre Channel Arbitrated Loop (FC_AL) with a maximum transfer rate of 800Mbps. Client hosts and the server are connected with a switched 100Mbps Ethernet network. The block size for the file system was fixed at 1MB [1]. We use a volume with 6 disks, and 18 users equally distributed across 4 client hosts: 2 hosts run 5 and 2 hosts run 4. We then use the following workloads for evaluation:

---

[1] Optimal block size of 1MB was determined by experimentation.

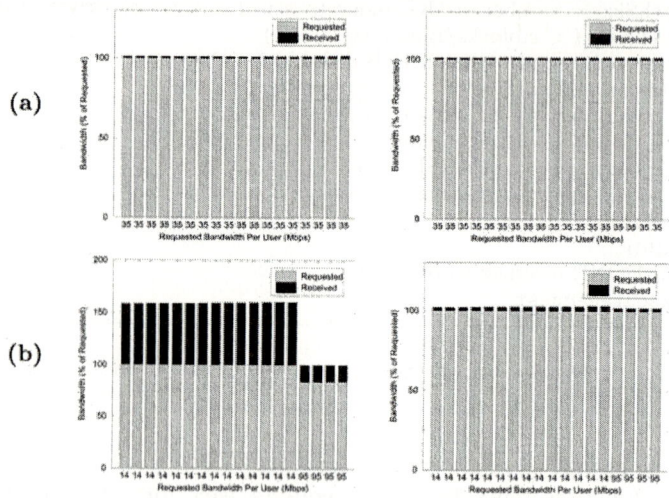

Fig. 5. Workload (a): equal bandwidth distribution, Workload (b): skewed bandwidth distribution. Notice that with non-preemptive scheduling some clients get less than what they negotiated for

(a) *Equal bandwidth distribution:* All 18 users ask for the same bandwidth of 35Mbps for a total of 630Mbps.
(b) *Skewed bandwidth distribution:* 4 very fast users ask for 95Mbps and 14 slow users ask for 14 Mbps for a total of 576Mbps.

Figure 5 shows the results of the experiments. Each user is represented by a vertical bar in the figures. The gray portion of the bar corresponds to the requested bandwidth and is always 100 in the figures. The black portion of the bar shows the percentage of the extra bandwidth received to the requested bandwidth and is simply computed by $\frac{Received - Requested}{Requested} \times 100$. The received bandwidth is overlayed on top of the requested bandwidth. If the client gets at least what it requested, the black portion of the bar will be above the requested bandwidth of 100% and shows the extra bandwidth that the user has received. (This is seen in Figure 5(a)). If however, the user gets less than what it requested, the black portion will be negative and will be below the 100% requested bandwidth overriding the gray portion of the bar (This is seen in Figure 5(b)). In the x-axis we show the actual bandwidth requested by the user in Mbps.

First column of Figure 5(a) and (b) show the results for non-preemptive scheduling: In (a), we see that the requested bandwidths are enforced. Since all users are asking for the same bandwidth, the request deadlines will pretty be similar. So the server will alternate between the requests from the users and non-preemptive EDF scheduling enforces the requested bandwidths for all users.

In (b) however, we observe that negotiated bandwidths are not enforced: The slow users get a lot more than what they asked for, while the fast users get less than what they asked for.

Observing that non-preemptive EDF scheduling does not always enforce bandwidth, we evaluated the effectiveness of preemptive scheduling by conducting the same set of experiments. We have assumed a **BatchSize** of 64 in these experiments [2]. The second column of Figure 5(a) and (b) show the results for each individual user. When all users ask for the same bandwidth, the bandwidths are enforced as in non-preemptive scheduling. When there are a mix of slow and fast users, preemptive scheduling is still able to enforce user bandwidths as shown in Figure 5(b). In this experiment, we observed that 48% of the requests are preempted by the server, which causes total disk bandwidth to go down to 590Mbps instead of 640Mbps with no preemption. That's the price paid to enforce negotiated bandwidths of all users.

5 Concluding Remarks

In this paper we presented the architecture and bandwidth enforcement algorithms of our Continuous Media Fibre Channel Distributed File System (FCDFS). Experimental results obtained from our prototype implementation of the file system in Linux platform are presented to evaluate the effectiveness of the file system in enforcing the real-time bandwidth guarantees. We conclude that the proposed file system is well-suited for emerging continuous media applications.

References

1. D. P. Anderson, Y. Osawa, and R. Govindan. File System for Continuous Media. *ACM Transactions on Computer Systems*, pages 311–337, November 1992.
2. T. E. Anderson, M. D. Dahlin, J. M. Neefe, D. A. Patterson, D. S. Roselli, and R. Y. Wang. Serverless Network File Systems. *ACM Transactions on Computer Systems*, February 1996.
3. M. Beck, H. Bohme, M. Dziadzka, U. Kunitz, R. Magnus, and D. Verworner. *Linux Kernel Internals*. Addison-Wesley, 1998.
4. G. A. Gibson, D. F. Nagle, K. Amiri, J. Butler, F. W. Chang, H. Gobioff, C. Hardin, E. Riedel, D. Rochberg, and J. Zelenka. Filesystems for Network-Attached Secure Disks. Technical Report CMU-CS-97-118, Carnegie Mellon University, July 1997.
5. K. Ramamritham and J. A. Stankovic. Scheduling Algorithms and Operating Systems Support for Real-Time Systems. *Proceedings of the IEEE, 82(1)*, January 1994.
6. R. Sandberg, D. Goldberg, S. Kleiman, D. Walsh, and B. Lyon. Design and Implementation of the Sun Network File System. In *Proceedings of the Summer USENIX Conference*, pages 119–130, 1985.

---

[2] Optimal BatchSize of 64 was determined by experimentation.

7. P. J. Shenoy, P. Goyal, S. S. Rao, and H. M. Vin. Symphony: An Integrated Multimedia File System. In *ACM SIGMETRICS Conference on Modeling and Evaluation of Computer Systems*, 1998.
8. S. R. Soltis, G. M. Erickson, K. W. Preslan, M. T. O'Keefe, and T. M. Ruwart. The Global File System: A File System for Shared Disk Storage. *Submitted to the IEEE Transactions on Parallel and Distributed Systems*, 1997.
9. L. Zhang. Virtual Clock: A New Traffic Control Algorithm for Packet Switching Networks. In *Proceedings of SIGCOMM*, 1990.

A Distributed Spatial Index for Time-Efficient Aggregation Query Processing in Sensor Networks[1]

Soon-Young Park and Hae-Young Bae

Dept. of Computer Science and Information Engineering, Inha University,
Yonghyun-dong, Nam-ku, Inchon, 402-751, Korea
sunny@dblab.inha.ac.kr, hybae@inha.ac.kr

Abstract. Many applications using sensing data require the fast retrieval of aggregated information in sensor networks. In this paper, distributed spatial index structure in sensor networks for time-efficient aggregation query processing is proposed. The main idea is logically to organize sensors in underlying networks into distributed R-Tree structure, named Sensor Tree. Each node of the Sensor Tree has pre-aggregated results which are the collection of the values of aggregated result for sensing data of the same type of sensors within MBR. If a spatial region query is required, the processing of the query searches the location of the target sensor from the root of the Sensor Tree. And then it finally sends the values of pre-aggregated result of that sensor. By the proposed Sensor Tree aggregation query processing on any region can reduce response time and energy consumption since it avoids flooding query to the leaf sensor or non-relevant sensors in the sensor networks.

1 Introduction

In the past few years, smart sensor devices have matured to the point that it is now feasible as large, distributed sensor networks. Sensors are connected to the physical world which they monitor and collect data. Sensors are connected to other sensors in a through a wireless network, and they use multi-hop routing protocol to communicate with sensors that are spatially distance [9, 12]. Sensors of the same type, for example, temperature sensors, light sensors result the sensing data of the same type which has the same schema. Sensor network consists of lots of sensors and provides opportunities for monitoring information about a spatial region of interest. Sensor data might contain noise, and it is often possible to obtain more accurate result by aggregation data from several sensors. Summaries and aggregates of sensing data are thus more useful than individual sensor readings [4].

Sensors in the sensor networks are usually not connected to a fixed infrastructure, they use batteries as their main power supply, and saving of power is the one of the design issues of a sensor network [10]. But sensor networks can be embedded in a variety of environments. Different applications usually have different requirements from accuracy, power consumption to time-efficient processing. In many spatial applications, for example rescue region control system, spatial queries which gather

---

[1] This research was supported by University IT Research Project in Korea (ITRC).

sensing data within a specific region are an essential functionality and require the fast retrieval of aggregated information in sensor networks.

In this paper, distributed spatial index structure in sensor networks is proposed for time-efficient aggregation query processing. The main idea is combining in-network distributed spatial index with the pre-aggregated results, named Sensor Tree. It is to cover the underlying sensors distributed R-Tree using the location of the same type of sensors and to store for each Minimum Bounding Rectangle (MBR), the values of the aggregation function for sensing data of the same type of sensors that are enclosed by the MBR [1, 3]. If a region query is required, the processing of the query is started to search the location of the target sensor from the root of the Sensor Tree and finished by sending the pre-aggregation value of the finding sensor to server which requires the query.

By using pre-aggregation technique of the proposed Sensor Tree region query processing can reduce response time and energy consumption since it avoids flooding query to the leaf sensor or non-relevant sensors in the sensor networks.

The remainder of the paper is structured as follows. Section 2 provides some related work. Section 3 proposes Sensor Tree, which is a distributed index with pre-aggregated results for time-efficient aggregation query processing in sensor networks. Section 4 evaluates the proposed approach by comparison with others. Finally section 5 has concluding remarks of this paper.

2 Related Work

Traditionally, sensors are used as data gathering instruments, which continuously feed a database on server. Each sensor can produce a stream of data about it surroundings. If queries are posed at a powered server, they are flooded to sensors in the networks. Most of all queries over sensor networks are simple and repeatable. In particular, since a message send operation may spend at least 1000 times more battery than a local operation (e.g. sense operation), query processing in sensor network should avoid unnecessary communication as much as possible [6].

There has been much previous work on query processing in sensor networks [7, 8]. Due to the geographical distribution of sensors in a sensor networks, each piece of data generated in the sensor networks has geographic location. And hence to specify the data of the interest over which a query should be answered, each query in a sensor network has a geographical region associated with it [2]. It seems that any kind of index on distributed data requires a hierarchical structure that aggregates information from different region of the networks. Prior work in range query for sensor networks has addressed a number of important issues in constructing such hierarchies. More detailed information can be accessed by top-down traversal of the hierarchy to visit the sensors holding the relevant information [3, 5, 11]. These researches noted the importance of power consumption. They focus on in-network query processing to reduce communication cost and power consumption.

Therefore, there is need for making an efficient access structure on sensor networks in order contact only the relevant sensor nodes for the execution of a query and hence achieve real time response and an accurate result.

3 Sensor Tree: Distributed Index Structure with Pre-aggregated Results in Sensor Networks

In this section, distributed index structure in sensor networks is proposed for time-efficient aggregation query processing. The main idea is to cover the underlying sensors in networks distributed R-Tree with pre-aggregated results, named Sensor Tree. Sensor Tree stores for each Minimum Bounding Rectangle (MBR), the values of the aggregation function for sensing data of the same type of sensors within the MBR. Sensor Tree is built on the sensor of the spatial dimension, therefore its structure is a hierarchical partitioning of the sensor network into rectangle-shaped clusters.

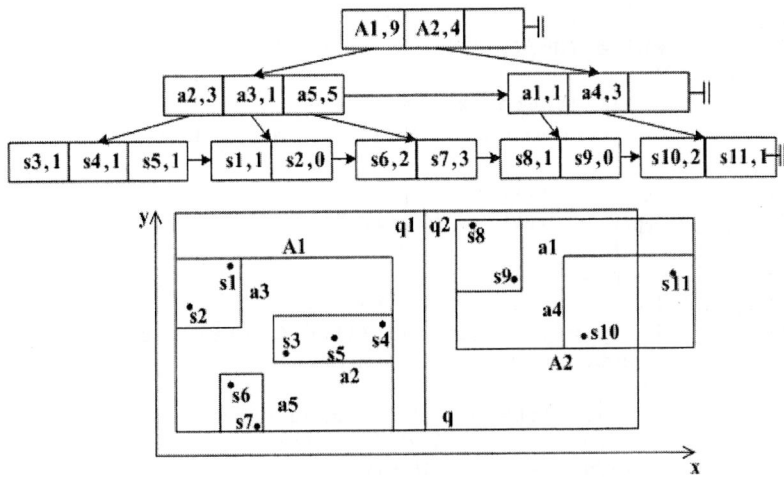

Fig. 1. The Structure of Sensor Tree

It is assumed that every sensor nodes in sensor networks lie in (x, y) coordinate space and have been deployed and arranged themselves into a connected topology, protocols for routing of the messages, and the cost of communication between nodes that are distance d apart takes $O(d)$ amount of energy.

Fig. 1 logically depicts a Sensor Tree which indexes a set of 5 roads $r1, r2, ..., r5$ with MBR is $a1, a2, ..., a5$ respectively when we assume the virtual road is in the rectangle. There are 3 sensors on the road $r2$ and sensing data of sensor $s3, s4, s5$ are 1, 1, 1 respectively. Therefore the total number of sensing data in $r3$ is 3 and there is an entry $(a2, 3)$ in the internal node of the Sensor Tree. Moving one level up, MBR $A1$ contains three roads, $r2, r3, r5$. The total number of sensing data in these roads is 9, and therefore there is an entry $(A1, 9)$ at level one of the Sensor Tree.

Each sensor node has a unique identifier, id. Each node has a field of communication, which it is capable to send/receive messages. All nodes within this boundary are its immediate neighbors by duplex link. For a node v, we denotes v's r-neighborhood, $Nbr(v, r)$, as the set of nodes within a radius r of v. Every nodes v maintain a variable $l.v$, level of v denoting the highest level of the Sensor Tree that v has participated in. Every node in sensor networks cooperates at level 0 of Sensor Tree, only clusterheads of level i cooperate for construction of level $i+1$ of the Sensor Tree. To minimize communication cost, the choice of clusterhead should be tiered into the routing structure. We use a simple clusterhead selection procedure as follows. For every level i, v maintains a variable p, parent node ptr to denote v's clusterhead for level i. Dually, v maintains c, children node ptr, That means, v is the clusterhead of the MBR that contains $c.v(i)$, children of v for level i. Using this doubly linked structure, any node can query the Sensor Tree. Pseudo code of joining cluster for node v at a level i is as follows.

```
00: join_cluster()
01: {
02:     if (l.v = i and p.v(i) = nil)
03:     {
04:         r <- 1;
05:         while (Nbr(v, r) < m or p.v(i) != nil)
06:         {
07:             if (there exists an i such that
                    both u ∈ Nbr(v, r) and l.u = i+ 1)
08:             {
09:                 p.v(i) <- u;
10:                 c.u(i) <- c.u(i) ∪ {v};
11:             }
12:             r <- r + 1;
13:         }
14:         if (p.v(i) = nil and (if u ∈ Nbr(v, r) is true
                than so is p.u(i) = nil))
15:         {
16:             calculate mbr.v(i) using Nbr(v, r);
17:             p.v(i) <- v; l.v <- i + 1;
18:             c.v(i + 1) <- Nbr(v, r);
19:         }
20:     }
21: }
```

A node v with $l.v = i$ executes the join_cluster() if it is not included in the cluster of level $i+1$, that is $p.v(i)$ = nil. In this case v first tries to contact a neighboring node u with $l.u = i + 1$ by searching increasingly larger radii, r, and join u's cluster by setting $p.v(i) = u$. if such u not exists and v encounters n nodes in level i within its r-neighborhood, and become the clusterhead of these n nodes in one step. If v is contacted by another node u with $l.u = i + 1$, v simply joins u's cluster.

During the concurrent executions of join operation by multiple nodes, a clusterhead may end up having more than N children. In the splitting cluster, a clusterhead v with more than N children at level i splits its cluster into clusters with number of children greater than or equal to n but less than N. Split operation is the same in based R-tree. Pseudo code of splitting cluster for node v at a level i is as follows.

```
00: split_cluster()
01: {
02:     if (c.v(i) > N)
03:         split(mbr.v(i));
04: }
```

4 Comparison with Other Approaches

Given a query over a sensor networks, a naive way to run the query will be to simply flood the all of the relevant sensor in sensor network with a query. Each sensor node in the network broadcast the query message exactly once and also remembers the id of the sensor node it receives the query from. If there are n sensors whose sensing regions intersect with the query's region, then using about n message transmissions, a communication routing tree spanning the n sensor could be built within the network. Each node in the built routing tree responds to the query. The responses propagate upwards in the tree towards the root of the query source. This again incurs a cost of n message transmissions, assuming the responses are aggregated at each tree node. Thus, the total communication cost incurred in answering q such queries over the same region is $2qn$ using the naive flooding approach.

Case of in-network query processing to visit only the sensors holding the relevant information, if there are m relevant sensors of the total n sensors, the total communication cost incurred in answering q such queries over the same region is $C + 2qm$, where C, $m \leq C \leq n$, is the communication cost incurred in composing the routing tree by m relevant sensors.

Consider Sensor Tree with pre-aggregated results composed of m sensor. The total cost incurred in executing q queries over the same region will be $D + m$, where D, $m \leq D \leq n$, is the communication cost incurred in composing the Sensor Tree and m is the cost of propagating information from leafs to root of the Sensor Tree in the worst case. If m is substantially less than n, constructing Sensor Tree could result in large saving in communication cost. If there are n sensors, n is 1000, whose sensing regions intersect with the query's region, communication cost incurred in answering query q is as Fig. 2.

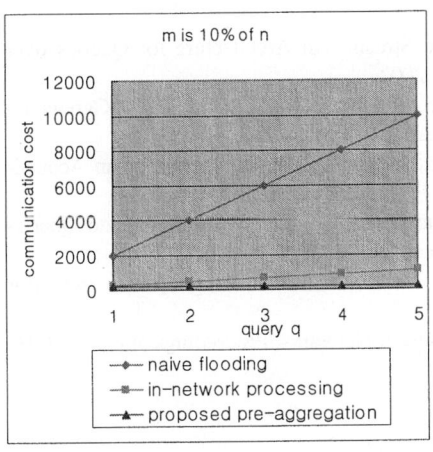 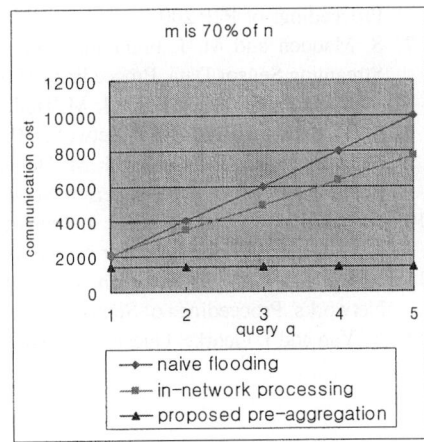

Fig. 2. Comparison of communication cost

5 Conclusion

In this paper, it is focused that efficient execution of region query for aggregation distributed information in sensor networks. It is established that a distributed spatial index structure on sensors of the same type in sensor networks, which is called Sensor Tree. Each node of the Sensor Tree has pre-aggregated result which is the collection of the values of aggregated results for sensing data of the same type of sensors within MBR. Therefore, an aggregation query does not need to access all sensors in networks, since part of the answer is found in the intermediate sensor nodes of the tree.

By using pre-aggregation of the proposed Sensor Tree the query can be answered much more efficiently without processing of aggregation and reduce response time and energy consumption and achieve an accurate response since it avoids flooding query to the leaf sensor or non-relevant sensors in the sensor networks. Our method allows efficient range query in sensor networks.

References

1. A. Guttaman, R-trees: A Dynamic Index Structure for Spatial Searching, Proceedings of SIGMOD 1984.
2. H. Cupta, S. R. Das, and Q. Gu, Connected Sensor Cover: Self-Organization of Sensor Networks for Efficient Query Execution, Proceedings of MobiHoc 2003.
3. I. Lazaridis and S. Mehrotra, Progressive Approximate Aggregate Queries with a Multi-Resolution Tree Structure, Proceedings of SIGMOD 2001.
4. J. Considine, F. Li, G. Kollios, and J. Byers, Approximate Aggregation Techniques for Sensor Databases, Proceedings of ICDE 2004.
5. J. Gao, L. J. Guibas, J. Hershberger, and L. Zhang, Fractionally Cascaded Information in a Sensor Network, Proceedings of IPSN 2004.
6. N. Demirbas and H. Ferhatosmanoglu, Peer-to-Peer Spatial Queries in Sensor Networks, Proceedings of P2P 2003.
7. S. Madden and M. J. Franklin, Fjording the Stream: An Architecture for Queries over Streaming Sensor Data, Proceedings of ICDE 2002.
8. S. Madden, M. J. Franklin, J. M. Hellerstein, and W. Hong, TAG: a Tiny AGgregation Service for Ad-Hoc Sensor Networks, Proceedings of OSDI 2002.
9. S. Madden, M. J. Franklin, J. M. Hellerstein, and W. Hong, The Design of an Acquisitional Query Processor For Sensor Networks, Proceedings of SIDMOD 2003.
10. S. Madden, R. Szewzyk, M. J. Franklin, and D. Culler, Supporting Aggregate Queries Over Ad-Hoc Wireless Sensor Networks, Proceedings of WMCSA 2002.
11. X. Li, Y. J. Kim, R. Govindan, and W. Hong, Multi-dimensional Range Queries in Sensor Networks, Proceedings of SenSys 2003.
12. Y. Yao and J. Gehrke, Query Processing for Sensor Networks, Proceedings of CIDR 2003.

Fast Concurrency Control for Distributed Inverted Files

Mauricio Marín*

Computing Department, University of Magallanes
Casilla 113-D, Punta Arenas, CHILE
mmarin@ona.fi.umag.cl

Abstract. A new method for controlling concurrent read/write operations upon inverted files is proposed and evaluated. Communication and synchronization among processors is effected by ways of the bulk-synchronous parallel model of computing. Thanks to the global synchronization property of this model, a simple but very efficient mechanism for synchronizing read/write operations is feasible at very low overheads in running time. Experimental results using a large text collection show that our method is more efficient than traditional approaches to the synchronization problem.

1 Introduction

The inverted file [2] is a popular data structure that is frequently used as an index for text databases. Its purpose is to speed-up query operations over large text collections. A typical application is in Web search engines in which case the server site must be able to cope efficiently with thousands of query operations per unit time coming from Internet users. This has lead to the consideration of parallel realizations of inverted files [1, 5, 9, 7, 11].

Query operations over parallel search engines are usually read-only requests upon the distributed inverted file. This means that one is not concerned with multiple users attempting to get information from the same text collection. All of them are serviced with no regards for consistency problems since no concurrent updates are performed over the data structure. However, it is becoming relevant to consider mixes of read and write operations. For example, for a large news service we want users to get very fresh texts as answers to their queries. Certainly we cannot stop the server every time we add and index a few news into the text collection. It is more convenient to let writes and reads take place concurrently. Solutions to this problem using traditional approaches from relational databases developments have been proposed for inverted files in [8].

Concurrency control is perfomed by algorithms that are in charge of properly synchronizing simultaneous accesses to the underlying data structure. From the database and parallel discrete-event simulation literature we learn of a number

---

* Partially supported by projects FONDECYT 1030454 and UMAG PRF101IC04.

of synchronization algorithms [3]. They can be divided into conservative and optimistic ones. The two-phases locks and time warp protocols are good examples of conservative and optimistic approaches respectively [4]. In the first case, write operations are performed when it is certain that no reads are to take place whereas in the second one no such restriction is imposed and errors are detected and corrected when necessary.

In this paper we propose a conservative algorithm that departs from previous approaches as it organizes computations in a bulk-synchronous manner as understood in the BSP model of parallel computing [12]. This model is known to be efficient, portable and scalable for a wide range of applications but it has not been widely employed to support distributed indexes for text databases. In BSP, processors are globally synchronized after performing computations on local data and communication actions. Messages are available at their target processors only after the global synchronization.

We take advantage of this fact to synchronize read/write operations in a straightforward manner: queries are timestamped and organized in batches delimited by processor sychronizations. Rules for message availability and processor synchronization ensure that consistency is maintained by just processing queries in timestamp order. We apply the proposed method to a particular realization of Distributed Inverted Files though it can actually applied to any other.

2 The BSP Model and Server Configuration

The bulk-synchronous parallel (BSP) model of computing [12] is a distributed memory model with a well-defined structure that enables the prediction of running time. The practical model of BSP programming is SPMD, which is realized as P program copies running on the P processors, wherein communication and synchronization among copies is performed by ways of libraries such as BSPlib or BSPub. In practice, it is certainly possible to implement BSP programs using the traditional PVM and MPI libraries.

In BSP, the parallel computer is seen as composed of a set of P processor-local-memory components which communicate with each other through messages. The computation is organized as a sequence of *supersteps*. During a superstep, the processors may perform sequential computations on local data and/or send messages to other processors. The messages are available for processing at their destinations by the next superstep, and each superstep is ended with the barrier synchronization of the processors.

We assume a server operating upon a set of P identical machines, each containing its own main and secondary memory (e.g., a cluster of PCs). The text database (documents) is evenly distributed over the P machines.

Clients request service to one broker machine, which in turn distribute them evenly onto the P machines implementing the server. Requests are queries that are solved by using an index data structure distributed on the P processors. We assume that the index is implemented using an inverted file which, as described in the next section, is composed of a vocabulary (set of terms) and a set of identifiers (inverted list) representing all the documents that contain at least one of the words that are members of the vocabulary. The inverted file data structure enables the efficient retrieval of all identifiers for which a given term appears in the respective documents.

We assume that under a situation of heavy traffic the server is able to process batches of $Q = qP$ queries. Every query is composed of one or more vocabulary terms for which it is necessary to retrieve all document identifiers associated with them. Only the identifiers of the K most relevant documents are presented to the user, namely those which more closely match the user information need represented by the query terms. For this, it is necessary to perform a ranking of documents. A widely used strategy for this task is the so-called vector model [2], which provides a measure of how close is a given document to a certain user query.

In order to better exploit the available parallelism we try to minimize the amount of work performed by the *broker* machine. We restrict its functionality to (**a**) receive user requests, (**b**) distribute the queries onto the processors (uniformly at random by means of a hashing function on the terms, i.e., vocabulary words), (**c**) receive the best ranked documents (K in total) from the server, and (**d**) pass them back to the user.

The two most basic operations related to providing answers to user queries are left to the parallel sever. That is, the retrieval of document identifiers and its respective ranking. Both operations are effected in parallel where the broker is responsible for scheduling those in a manner that keeps load balance of processors work as close to the optimal $1/P$ as possible.

For a collection of documents the inverted file strategy can be seen as a vocabulary table in which each entry contains a term (relevant word) found in the collection and a pointer to a list of document's identifiers (inverted list) that contains such term. Thus, for example, a query composed of the logical AND of terms 1 and 2 can be solved by computing the intersection between the inverted-lists associated with the terms 1 and 2. The resulting list of documents can be then ranked so that the user is presented with the most relevant documents first (the technical literature on this kind of topics is large and diverse, e.g., see [2]). Parallelization of this strategy has been tackled using two approaches.

The global index approach is as follows. The whole collection of documents is used to produce a single inverted file index which is identical to the sequential one. Then the T terms that form the global term table (vocabulary) are uniformly distributed onto the P processors along with their respective lists of document identifiers. This is done by ways of the same hashing function employed by the broker. Thus, after the mapping, every processor contains about T/P terms per processor.

In the local index case, each processor contains the same T terms but the length of document identifier lists are closely a fraction $1/P$ of the global index ones. This is the strategy used by most popular search engines such as Google though there has been some discussions in the literature [1,5,9,7,11] about which approach is better (including variations and combinations). This discussion is out of the scope of this paper and we present our results in the context of the global index approach.

The BSP realization of the global index is as follows. Every term is routed to one server processor by the broker. For each term w belonging to a query u the inverted lists associated with terms of u are retrieved in their respective processors. Then these lists are sent to the ranker processor defined for the query u to then proceed in the next superstep like the local inverted lists case. The whole process takes 2 supersteps to complete.

The following pseudo-code shows the major tasks performed by every processor of the server (which is a set of machines supporting the BSP model of computing) for read-only queries,

```
while(true)   // Each BSP processor.
{
 Receive new messages and put them
 in a queue Q.

 Foreach message msg in Q do
 {
  switch( msg.type )
  {
   case BROKER://  term from the broker.
   //  retrieve and rank doc. list
     List=FirstK-ItemsOfList(msg.term);
     subList= preRanking(List);

   // buffer message to be sent to the
   // ranker processor.
     bufferMsg(msg.ranker,RANKING,
               subList);
     break;

   case RANKING:
    if( queueSize(msg.queryId)==
        msg.numTermsQry )
    {
     L=dequeueAll(msg.queryId);
     List=CalculateFinalRanking(L);
     bufferMsg(broker,SERVER,List);
    }
    else//queue up to wait for terms
       enqueue(msg.queryId,msg);
    break;
  } // switch
 } // foreach

 Send all buffered messages to their
 target processors, and globally
 synchronize the processors.
} // while
```

3 Conservative Synchronization Algorithm

Every processor of the BSP machine must execute R/W operations of a large number of queries. They are evenly distributed so that during a superstep all processors maintain about the same number of them.

For each new document to be included in the text collection, a sub-set of the vocabulary terms get their respective inverted-lists modified. Those modifications come in the form of write operations on the inverted file. The parsing process and other calculations on the new document to be included is assumed to be performed by a secondary machine which in turn sends the write operations to the broker machine. Thus the broker send the write operations as they were normal queries. That is, they are routed to the server processors using the hashing function on the vocabulary terms.

A key fact here is that the broker can assign a unique timestamp to each query it sends to the BSP server. The vocabulary terms are those which are assigned timestamps and terms belonging to the same query get identical timestamps. Now, no R/W consistency conflicts can ever take place if server processors process terms with associated R/W operations in increasing timestamps. This is true because at the end of every superstep the processors are barrier synchronized and new messages arriving from other processors are only available by the following superstep. This introduces a global order because batches of queries are send to the server and every processor executes sequentially, one by one, the queries it receives at the begining of each superstep.

Thus the R/W version of the global inverted file algorithm as implemented in the BSP model is as follows,

```
while(true)// Each BSP processor.
{
 Receive new messages and put them                case RANKING:
 in a queue Q.                                     if ( queueSize(msg.queryId) ==
                                                        msg.numTermsQry )
 Sort Q by increasing timestamps                   {
 (ranking messages are not considered).             L= dequeueAll(msg.queryId);
                                                    List= CalculateFinalRanking(L);
 Foreach message msg in Q do                        bufferMsg( broker, SERVER,List);
 {                                                 }
  switch( msg.type )                               else// queue up to wait for terms
  {                                                 enqueue(msg.queryId,msg);
   case BROKER_READ://  R query                   }// switch
    List=FirstK-ItemsOfList(msg.term);           }// foreach
    subList= preRanking(List);
    bufferMsg( msg.ranker,RANKING,                Send all buffered messages to their
              subList);                           target processors,and globally
    break;                                        synchronize the processors.
   case BROKER_WRITE: // W query                 }
    UpdateList( msg.term,
        msg.documentId, msg.info);
    break;
```

As shown in this pseudo-code, the protocol is very simple which is in contrast with previous appraches to synchronization of query operations in inverted files [8], see next section.

4 Experimental Evaluation

We performed experiments using a 2GB sample of the Chilean Web and a query log from www.todocl.cl. This gave us a realistic setting both on the set of terms that compose the text collection and the type of terms that typically are part of user queries. Transactions were generated at random by taking terms from the query log. We started with 60% of the text collection and increased it by including the remaining 40% divided in documents as part of the write transactions generated at random. At the start of this almost real-life system every processor "knows" its set of queries (i.e., we exclude the effect of query traffic and broker operations). We performed our experiments on a high-performance cluster with 16 processors (Pentium IV, 1GB main memory).

We worked with rather large query batches (64 ... 1024) to simulate high query traffic scenarios. For the final ranking of answers to every query, we considered only a small fraction of the involved inverted list since we considered only the top 100 of those answers (Persin's strategy was applied to organize the inverted lists and filtering [10] and parallel priority queue technique proposed in [1]).

We compared our method against the two-phases and time warp protocols. In the **two-phases protocol**, transactions first request locks on the subset of index-terms that are part of a read-only query or the relevant terms found in the document being added to the collection. After all of the locks have been granted, the associated operations are allowed to take place and the locks are released. Deadlocks are avoided by asking locks in lexicographic term order. A direct realization of this protocol on a BSP machine is to request all the required locks in the same superstep, and then wait during one or more supersteps to receive all the pending lock authorisations. If a required lock is being held by another transaction, it is necessary to wait until this transaction releases the lock. Read locks are answered with the data itself to be read. Write locks are requested by sending into the same message the new data to be written. That is, no additional message traffic is necessary for effecting the R/W operations. All messages releasing the granted locks can be sent in the same superstep.

The **Time Warp protocol** is based on the optimistic assumption that no events will probably get into conflict with each other, and if that situation happens to occur a correction procedure is executed by moving backwards the computation, correcting the error, and then moving it forward again but this time taking into consideration the cause of the trouble. The same strategy can be applied to the parallel processing of transactions. That is, they are allowed to perform their R/W operations at will, but each time a data item is read or written a consistency check is executed to detect if it necessary to do a roll-back of all causally related transactions or let them continue forward. A timestamp is assigned to each transaction. This is an increasing integer number. All operations of a given transaction receive the transaction timestamp and the protocol is in charge of ensuring that all operations on records are done in increasing timestamp order. Whenever an operation breaks this rule, all already-executed operations on the involved record that have timestamps greater than the new one are undone and re-executed on the record to obtain the right sequence. Only read operations are allowed to be done in different timestamp order as long as no write operation should have been executed in between.

Every time an operation of a given transaction is undone, it is also necessary to undone all subsequent operations of the same transaction which have already been executed on other records. Note that these records can be located in other processors. Then these operations must be re-executed again since each one in the sequence can depend on the previous one. All this process is call a *roll-back*. Efficiency depends heavily on the amount of roll-backs performed during the computation. Transactions are committed when all their operations become ones with timestamps less than the smallest timestamp of any operation waiting to be executed (this considering all processors). In [6] we propose an efficient BSP algorithm for Time Warp on BSP Computers which can be easily adapted to support this strategy.

In figure 1.a we present speed-ups values for different number of processors obtained in a 16-processors PC cluster running the BSPpub library. The speed-up values where obtained taking the running time achieved by a efficient sequential realization of inverted files, and dividing it by the running time achieved with the method proposed in this paper (CON), the two-phases lock protocol (LOCKS) and the time warp (TW) protocol. The server was assumed to receive batches of 1024 queries. It is observed that the proposed concurrency control method achieves better performance than the other alternative algorithms.

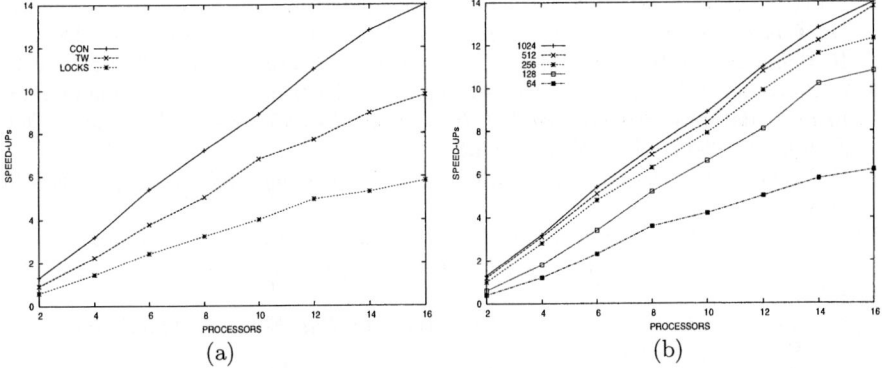

Fig. 1. (a) Speed-ups for different number of processors (b) Speed-ups for different number of queries per batch

To see the effect of batch size on the speed-up we obtained speed-up values for different number of queries per batch. The results are shown in figure 1.b where it can be seen that the proposed method is quite robust to small sized batched. In practice we should expect large size batches as parallelism is justified in situations of heavy query traffic on the server.

5 Conclusions

We have presented a very efficient method for synchronizing concurrent accesses to distributed text databases which are indexed by a parallel realization of inverted files. The method outperforms previous solutions to this problem which are based on the conservative two-phases locks and the optimistic Time Warp approaches. Both have involved implementations and are less efficient than the method proposed in this paper.

The implementation is extremely simple because it only requieres processing the new arriving R/W queries by increasing timestamps. Correctness comes from the following points: (a) The broker send for processing batches of R/W queries, (b) All R/W operations of a given batch are processed during the same superstep (this is possible because the global index approach is employed to implement the inverted file), and (c) At the end of every supersteps are barrier synchronized and new queries only arrive by the begining of the next superstep. Some people could argue that the cost of globally synchronizing the processors after every batch can be too high. We claim that this is not the case because fairly small-sized batches can easily amortize the cost of barrier synchronization of processors. Our empirical results confirm this claim.

References

1. C. Badue, R. Baeza-Yates, B. Ribeiro, and N. Ziviani. Distributed query processing using partitioned inverted files. In *Eighth Symposium on String Processing and Information Retrieval (SPIRE'01)*, pages 10–20. (IEEE CS Press), Nov. 2001.
2. R. Baeza and B. Ribeiro. *Modern Information Retrieval*. Addison-Wesley., 1999.
3. S. Blott and H. Korth. An almost-serial protocol for transaction execution in main-memory database systems. In *28th International Conference on Very Large Data Bases*, Aug. 2002. Hong Kong, China.
4. D.R. Jefferson. Virtual time. *ACM Trans. Prog. Lang. and Syst.*, 7(3):404–425, July 1985.
5. B.S. Jeong and E. Omiecinski. Inverted file partitioning schemes in multiple disk systems. *IEEE Transactions on Parallel and Distributed Systems*, 6(2):142–153, 1995.
6. M. Marín. Time Warp on BSP Computers. In *12th European Simulation Multiconference*, June 1998.
7. M. Marín. Parallel text query processing using Composite Inverted Lists. In *Second International Conference on Hybrid Intelligent Systems (Web Computing Session)*. IO Press, Feb. 2003.
8. M. Marín. Optimistic Concurrency Control for Inverted Files in Text Databases. In *IASTED International Conference on Databases and Applications (DBA2004)*. Acta Press, Feb. 2004.
9. A.A. MacFarlane, J.A. McCann, and S.E. Robertson. Parallel search using partitioned inverted files. In *7th International Symposium on String Processing and Information Retrieval*, pages 209–220. (IEEE CS Press), 2000.
10. M. Persin, J. Zobel, and R. Sacks-Davis. Filtered document retrieval with frequency-sorted indexes. *Journal of the American Society for Information Science*, 47(10):749–764, 1996.
11. B.A. Ribeiro-Neto and R.A. Barbosa. Query performance for tightly coupled distributed digital libraries. In *Third ACM Conference on Digital Libraries*, pages 182–190. (ACM Press), 1998.
12. L.G. Valiant. A bridging model for parallel computation. *Comm. ACM*, 33:103–111, Aug. 1990.

An All-Reduce Operation in Star Networks Using All-to-All Broadcast Communication Pattern

Eunseuk Oh, Hongsik Choi, and David Primeaux

Department of Computer Science, School of Engineering,
Virginia Commonwealth University,
Richmond, VA 23284-3068, USA
{eoh, hchoi, dprimeau}@vcu.edu

Abstract. Most parallel computations require the exchange of data between processing elements. One of important basic communication operations is all-reduce, a variation of the reduction operation. This paper presents an all-reduce communication operation scheme using all-to-all broadcast communication pattern. All-to-all broadcast is the operation in which each processor sends its message to all other processors, and receives messages from all other processors in the system. In this paper, we develop an efficient all-reduce operation scheme in a star network topology with the single-port communication capability. Communication time is compared against known broadcasting schemes to verify the efficiency of the suggested scheme.

Keywords: all-reduce, all-to-all broadcast, distributed memory parallel computing systems, inter-processor communication, star network.

1 Introduction

Due to rapid progress in hardware technology, designing a distributed memory parallel computing system connecting autonomous microprocessors has become feasible. In such a system, high-performance microprocessors communicate by message passing and have no shared memory or global clock. Proper implementation of basic communication operations such as broadcast, reduction, and all-reduce on various parallel computing systems is key to the design of efficient parallel algorithms for distributed memory parallel systems.

One-to-all broadcast is an operation that disseminates information across processors in a multiprocessor system. It is not difficult to see that broadcasting stands as a foundation for many applications on parallel computing systems. To list a few applications that use broadcasting, we mention Fast Fourier Transformation (FFT), parallel matrix algorithms, parallel graph algorithms, and distributed algorithms. The all-reduce operation combines the arriving content in the input buffer of each processor using an associative operator (e.g. sum, maximum), and the result appears in the result buffer of all processors. All-reduce

Table 1. Comparison of topological properties for parallel computer models of similar sizes: n-Cube (hypercube), MCT (mesh connected tree), D_n (de Bruijn network), and $HS_{n,m}$ (Cartesian product of hypercube and star)

Model	Size	Degree	Diameter	Model	Size	Degree	Diameter
S_5	120	4	6	S_6	720	5	7
7-Cube	128	7	7	10-Cube	1024	10	10
$MCT_4(3)$	81	12	16	$MCT_6(3)$	729	18	24
D_7	128	4	7	D_{10}	1024	4	10
$HS_{4,3}$	144	5	7	$HS_{5,3}$	720	6	9

is typically used for barrier synchronization on a distributed memory parallel computing system. Also, all-reduce is one of the most important MPI routines; a case study reveals that more than 40% of the execution time of MPI routines is spent in all-reduce or reduction operations [9].

In this paper, we study an all-reduce communication operation in star networks by using all-to-all broadcast communication pattern. The all-reduce operation is identical to performing an all-to-one reduction which is followed by a one-to-all broadcast of the result. Thus, we will compare the communication time of our scheme with all-reduce operation by using the best known broadcast scheme proposed in [10]. We first design a recursive all-to-all broadcast scheme that can be utilized to perform the all-reduce operation.

The star model has attracted considerable interest in the parallel processing research community [1, 2, 3, 7, 8, 10, 11] due to its numerous desirable properties for building large parallel computer systems. Basic parameters such as size, degree, and diameter for the models whose size is similar to S_n are shown in Table 1.

Broadcasting schemes vary according to the communication capability of the channels or links. With single-port communication capability, every processor can simultaneously send and receive at most one message in one communication step. Also, a channel or link may be bidirectional or unidirectional. Cost measurements for the suggested scheme are provided under the single-port and bidirectional communication capability. Following the terminology used in [4], our scheme is "NODUP" in that there is no duplication of information on messages carried during the communication process.

The remainder of this paper is organized as follows. In Section 2 we introduce the communication model and the assumptions made about that model. In Section 3 we present an all-to-all broadcast scheme. In Section 4 we present an all-reduce operation based on our all-to-all broadcast scheme. In Section 5, we provide concluding remarks.

2 Model and Assumptions

The network model considered here is the star graph model. An n-star graph, S_n, consists of $n!$ nodes labeled with the $n!$ permutations on the symbols $\{1, 2, \ldots, n\}$.

There is a communication link between two processors p_i and p_j in S_n if and only if the permutation label of p_j can be obtained from the permutation label of p_i by exchanging the symbol in the first position in p_i with the symbol in some other position in p_i. If the label of p_j is obtained from the label of p_i by exchanging the first symbol of p_i with the symbol in kth position of p_i, then p_i and p_j are said to be connected along the communication link k.

The pattern of interconnected processors in the star network can be viewed recursively as follows. S_1 is a trivial network with one processor. Suppose that S_{n-1} is defined inductively, then S_n is composed of n graphs, S_{n-1}^i, $i = 1 \ldots n$, where each S_{n-1}^i is an isomorphic copy of S_{n-1} with symbols $\{1, \ldots, n\} - i$, and with symbol i appearing as the nth symbol in each processor in S_{n-1}^i. Connecting the nodes in different copies S_{n-1}^i and S_{n-1}^j is done with respect to the above definition. A node u in S_{n-1}^i is connected to a node v in S_{n-1}^j, $i \neq j$ when the label of v can be obtained from the label of u by exchanging the first symbol with the nth symbol.

The communication model for parallel computers varies depending on the communication hardware and the memory bus bandwidth. Most commercial systems support the single-port model. In the single-port communication model, a processor can send a message on only one of its communication links at a time. The sending and the receiving ports are not necessarily the same. The system model we consider is as follows; (1) The system is completely connected with synchronous communication. (2) A processor sends a message to a connected processor in one communication step. (3) Single-port communication and the communication links are bidirectional.

3 All-to-All Broadcast

All-to-all broadcast is performed recursively. After performing all-to-all broadcast in each S_{n-1}^i, $i = 1, \ldots, n$, all-to-all broadcast in S_n is performed. To avoid sending a message more than once to the same processor in the network, the private memory of each processor will be divided into two parts: the result buffer, and the outgoing message buffer. The partial sum will be stored in the result buffer.

All-to-All broadcast in S_4: The algorithm first calls itself recursively to perform broadcast in each of S_3^1, S_3^2, S_3^3, and S_3^4. The base of the recursion is when the network is an S_2. The algorithm performs broadcast in S_2 by a simple exchange of messages between the two processors. Then each processor in S_2 sends its message along communication link 3, and saves it in its result buffer. After that each processor sends its received message along communication link 2, and broadcast in S_3 is terminated. Now each S_3 in S_4 performs all-to-all broadcast in parallel fashion. At this point every processor computes its message by concatenating the message in its outgoing buffer with the message in its result buffer. Each processor stores the concatenated message in the result buffer, and writes a copy of the concatenated message over the current content of the outgoing buffer. We call this concatenated message the *meta message*. Then, every processor sends its meta message along communication link 4, and saves its message in its result buffer. Once a processor receives the message along communication link 4, it only needs to broadcast within each S_3. It starts this process by sending its message along communication link 3, then 2, meanwhile storing received messages in the result buffer.

All-to-All broadcast in S_n: In general, suppose inductively that all-to-all broadcast has been completed within each S_{n-1}^i, $i = 1 \ldots n$, and let us see how this can be extended to all-to-all broadcast in S_n. Since all-to-all broadcast has been completed in each S_{n-1}^i, each processor in S_{n-1}^i has received the messages from all other processors in S_{n-1}^i, and hence all processors in S_{n-1}^i share the same information. Denote the meta message in S_{n-1}^i by Δ_{n-1}^i. Let $\Delta_n = \bigcup_{i=1}^n \Delta_{n-1}^i$, then all-to-all broadcast in S_n is achieved once every processor in S_n holds Δ_n. All-to-all broadcast is performed as follows. In the first stage every processor p in S_{n-1}^i, $i = 1 \ldots n$, broadcasts its meta message Δ_{n-1}^i along communication link n and saves the meta message in its result buffer. After this stage, each S_{n-1}^i contains all messages in Δ_n among its processors. Thus, the only thing left to be done in the second stage is to propagate the information within each S_{n-1}^i. This step needs to be done with some care so that to avoid sending a message more than once to the same processor. Once p receives the meta message Δ_{n-1}^j, $j \neq i$, along communication link n, it propagates Δ_{n-1}^j across S_{n-1}^i by sending it along communication links $n-1, n-2, \ldots, 2$, respectively. Also, the received message is stored in the result buffer.

Theorem 1. *At the termination of all-to-all broadcast, each processor in S_n holds the meta message Δ_n.*

4 All-Reduce Operation

We perform all-reduce by using the communication pattern of all-to-all broadcast. Throughout this discussion, without loss of generality, we assume that addition is the associative operation performed in the all-reduce. An illustration of the all-reduce operation on S_3 is given in Figure 1. At each node, the final sum is obtained by adding the content in the result buffer and the outgoing buffer.

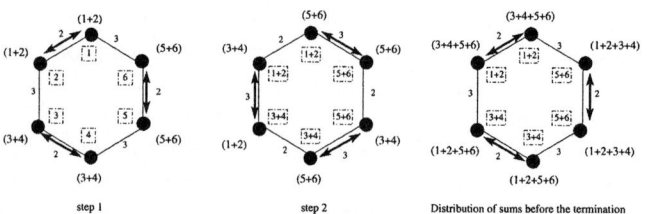

Fig. 1. The all-reduce operation on S_3. At each node, parentheses show the local sum in the outgoing buffer and the contents in the box is the local sum accumulated in the result buffer

All-Reduce

1. each S_{n-1}^i performs All-Reduce recursively;
   /* At this point, all processors in S_{n-1}^i have the sum of corresponding numbers to be added. */
2. each processor in S_{n-1}^i, $i = 1 \ldots n$, sends the local sum along the communication link n and saves the local sum in its result buffer;
3. $d = n - 1$;
4. once each processor in S_{n-1}^i receives the local sum from a processor in S_{n-1}^j, where $j \neq i$, it performs the following
   while $d \geq 2$ **do**
   send the local sum to a neighbor of S_{n-1}^i along the communication link d;
   add the number received from the processor and the content of the result buffer in S_{n-1}^i;
   $d = d - 1$;
5. every processor adds the contents in its result buffer and the outgoing message buffer;

Fig. 2. All-reduce communication operation scheme

Assume that each number in the box, initially in the result buffer, is a number to be added.

An all-reduce operation follows the communication steps of all-to-all broadcast, but adds two numbers instead of concatenating messages. Thus, each message transferred in the all-reduce operation has only one word, where each word hold the partial sum of numbers. At the termination of the all-reduce operation, each node holds the sum $(1 + 2 + \ldots + n!)$. Figure 3 shows all-reduce performed in S_4. The all-reduce scheme **All-Reduce** is shown in Figure 2.

Theorem 2. *At the termination of the all-reduce operation, each processor in S_n holds the sum $\sum_{i=1}^{n!} i$.*

Proof. The statement is vacuously true when $n = 1$. Let $n > 1$, and assume inductively that when **All-reduce** on S_{n-1} terminates, each processor in S_{n-1} contains the sum of corresponding numbers. When **All-reduce** is called on S_n, **All-reduce** calls itself recursively on each S_{n-1}^i, $i = 1 \ldots n$. Since each S_{n-1}^i is a copy of S_{n-1}, by the inductive hypothesis, when each of these recursive calls terminates, each processor in S_{n-1}^i, $i = 1 \ldots n$, holds the sum of corresponding numbers.

From the recursive definition of S_n given in Section 2, each S_{n-1}^i is linked to the other S_{n-1}^j, $j \neq i$ by exactly $(n-2)!$ links along communication link n. Thus, after the execution of step 2 of **All-reduce**, exactly $(n-2)!$ processors in

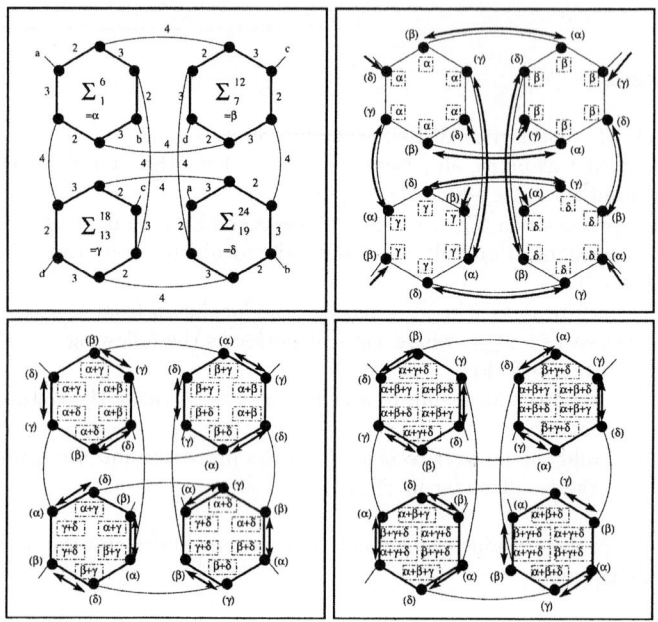

Fig. 3. The all-reduce operation on S_4

S_{n-1}^i holds the sum received from S_{n-1}^j. Also, each processor in S_{n-1}^i has the local sum stored in its result buffer. Now, each S_{n-1}^i holds the partial sums such that adding all partial sums results in $\sum_{i=1}^{n!} i$. Let p_1 be a processor in S_{n-1}^i, and let $\sum_j$ the partial sum received from S_{n-1}^j. From the above discussion, we know that there are exactly $(n-2)!$ processors in S_{n-1}^i that hold the partial sum $\sum_j$. If p_1 is one of these processors then we are done. Suppose this is not the case. Now from step 4 in **All-reduce**, we know that each of these processors will send $\sum_j$ along communication links $n-1, \ldots, 2$, respectively. We first claim that no processor p_1 in S_{n-1}^i is the neighbor of two distinct processors p_2 and p_3 that hold $\sum_j$ at the beginning of step 4. Suppose, for the sake of contradiction, that this were not the case. Let $p_1 = \sigma_1 \ldots \sigma_n$, and suppose that p_1 is the neighbor of p_2 and p_3 along communication links x and $y \in \{1, \ldots, n-1\}$, respectively. Notice first that $x \neq y$ since a processor is connected to exactly one processor along any given communication link. Since p_1 is connected to p_2 along communication link x, $p_2 = \sigma_x, \ldots, \sigma_1, \ldots \sigma_n$. Similarly $p_3 = \sigma_y, \ldots, \sigma_1, \ldots \sigma_n$. Since both p_2 and p_3 have $\sum_j$, both p_2 and p_3 are connected to S_{n-1}^j along communication link n. Now all processors in S_{n-1}^j have the same nth symbol. Since p_2 and p_3 are the neighbors of two processors in S_{n-1}^j along communication link n, it follows that both p_2 and p_3 have the same first symbol and $\sigma_x = \sigma_y$, a contradiction, since σ_x and σ_y are two symbols in the representation of processor p_1, and hence must be distinct.

Table 2. Comparison of all-reduce operations on communication time

Dimension of S_n	Size of S_n	Tseng [11]	Sheu [10]	Ours
3	6	12	6	3
4	24	27	12	6
5	120	48	18	10
6	720	75	26	15
7	5040	108	34	21
8	40320	127	42	28
9	362880	192	50	36
10	3628800	243	60	45

It follows from the above claim that the neighbors of the processors possessing the sum $\sum_j$ at the beginning of step 4 of **All-reduce** are distinct. Now each processor that holds $\sum_j$ broadcasts it to exactly $n-2$ neighbors along communication links $n-1,\ldots,2$. Since all these neighbors are distinct, the number of processors in S_{n-1}^i that receive $\sum_j$ from a processor in S_{n-1}^i at the beginning of step 4 is $(n-2)(n-2)!$. Thus, the total number of processors in S_{n-1}^i that hold $\sum_j$ at the end of **All-reduce** is $(n-2)! + (n-2)(n-2)! = (n-1)!$. It follows that all processors in S_{n-1}^i hold $\sum_j$ and in particular p_1. Since p_1 and j were arbitrarily chosen, every processor in S_{n-1}^i possesses every $\sum_j$ at the end of **All-reduce**, and hence holds the total sum $\sum_{j=1}^{n!} j$. □

Theorem 3. **All-reduce** *performs an all-reduce operation on S_n in time* $n(n-1)/2$.

Proof. The above theorem proves that when the algorithm **All-reduce** terminates, each processor holds the sum $\sum_{i=1}^{n!}$ in the system. Let $T(n)$ be the number of communication steps performed by **All-reduce** on S_n. Each S_{n-1}^i performs an all-reduce operation within itself and then sends a single message along communication link n, and then along communication links $n-1,\ldots,2$. Thus, the number of communication steps performed by each S_{n-1}^i is $T(n-1) + n - 1$. Since all the S_{n-1}^i's do this in parallel, the number of communication steps for S_n is the same as the number of communication steps performed by each S_{n-1}^i. Thus, the total number of communication steps performed by each S_{n-1}^i, and hence, by the whole network is given by the recurrence $T(n) = T(n-1) + n - 1$. It gives $T(n) = n(n-1)/2$. □

5 Concluding Remarks

In this paper we presented an efficient all-reduce communication operation scheme by using the all-to-all broadcast communication pattern. Our scheme performs an all-reduce operation on an n-star network with the single-port capability in $n(n-1)/2$ time steps. If we use an all-to-one reduction followed by a one-to-all broadcast, an all-reduce can be performed in time $2\sum_{i=2}^{n}(\lceil log(i-1)\rceil+1)$ by the broadcast scheme proposed in [10] and in time $3(n-1)^2$ by the scheme proposed in [11]. In terms of the communication time shown in Table 2, our algorithm provides an improvement over the algorithms in [10, 11].

References

1. S. B. AKERS, D. HAREL, AND B. KRISHNAMURTHY, "The Star Graph: An Attractive Alternative to The n-cube," *Proc. Int'l. Conf. of Parallel Processing*, 1987, pp. 393-400.
2. S. B. AKERS AND B. KRISHNAMURTHY, "The Fault Tolerance of Star Graphs," *Proc. 2nd Int'l. Conf. on Supercomputing*, 1987, pp. 270-276.
3. S. G. AKL, K. QIU, AND I. STOJMENOVIC, "Fundamental Algorithms for The Star and Pancake Interconnection Networks With Applications to Computational Geometry," *Networks*, 1993, vol.23, no. 4, pp. 215-225.
4. M. -S. CHEN, P. S. YU, AND K. -L. WU, "Optimal NODUP All-to-All Broadcast Schemes in Distributed Computing Systems," *IEEE Trans. Parallel and Distributed Systems*, 1994, pp. 1275-1284.
5. K. EFE AND A. FERNANDEZ, "Computational Properties of Mesh Connected Trees: Versatile Architecture for Parallel Computation," *Int'l Conference on Parallel Processing*, 1994, pp. 72-76.
6. Intel Corp. iPSC/2 User's Guide, Intel Corp., Mar, 1988.
7. I. M. MKWAWA AND D. D. KOUVATSOS, "An Optimal Neighbourhood Broadcasting Scheme for Star Interconnection Networks," *J. Interconnection Networks*, 2004, vol. 4, pp. 103-112.
8. E. OH AND J. CHEN, "Strong Fault-Tolerance: Parallel Routing in Star Networks with Faults," *J. Interconnection Networks*, 2003, vol.4, pp. 113-126.
9. R. RABENSEIFNER AND P. ADAMIDIS, "Collective Reduction Operation on Cray X1 and Other Platforms," *Proc. the Cray User Group*, 2004
10. J. -P. SHEU, C. -T. WU, AND T. -S. CHEN, "An Optimal Broadcasting Algorithm Without Message Redundancy in Star Graphs," *IEEE Trans. Parallel and Distributed Systems* 1995, vol.6, no. 6, pp. 653-658.
11. Y. -C. TSENG AND J. -P. SHEU, "Toward Optimal Broadcast in A Star Graph Using Multiple Spanning Trees," *IEEE Trans. Computers* 1997, vol. 46, pp. 593-599.

S^2F^2M - Statistical System for Forest Fire Management*

Germán Bianchini, Ana Cortés, Tomàs Margalef, and Emilio Luque

Departament d'Informàtica, E.T.S.E,
Universitat Autònoma de Barcelona,
08193-Bellaterra (Barcelona), Spain

Abstract. One of the most serious problems in wildland fire simulators is the lack of precision for input parameters (moisture content, wind speed, wind direction, etc.). In this paper, a statistical method based on a factorial experiment is presented. This method evaluates a high number of parameter combinations instead of considering a single value for each parameter, in order to obtain a prediction which is closer to reality. The proposed methodology has been implemented in a parallel scheme and tested in a Linux cluster using MPI.

1 Introduction

The main goal of forest fire model developers is to provide models that explain and predict fire behavior. These models can be used to develop simulators and tools for preventing and fighting forest fires [1, 2, 7, 8]. These simulators and tools are integrated into a Decision Support System (DSS). It is possible to define a DSS as "a computer system that helps in the process of making a decision, helping users to form and explore the implications of their judgments, and, therefore to make decisions based on understanding" [14]. Therefore, this type of system should help to form judgments instead of giving general advise as, for example, an information digest does in a database. Nowadays, a DSS has the more ambitious objective of trying to supply accurate information (sometimes in real time) to achieve terrain planning, implementation of preventive rules, efficient monitoring and giving online help while the forest fire is happening.

However, most models are unable to accurately predict the forest fire's behavior. This is due to several reasons but one of the most significant ones is that there are several parameters (i.e. moisture content, wind conditions, etc.) that are difficult to estimate precisely.

It is possible to minimize this input parameter problem by using techniques such as parameter optimization [3], with the aim of determining as precisely

---

* This work has been supported by the Comisión Interministerial de Ciencia y Tecnología (CICYT) under contract TIC2001-2592 and by the European Commission under contract EVG1-CT-2001-00043 SPREAD.

as possible the parameter values that provide the closest prediction of real behavior.

In this paper, although we also focus on processing the parameters, our goal is to develop a methodology based on statistical analysis to determine the most probable behavior of a forest fire and apply this methodology to implement a DSS.

S^2F^2M (Statistical System for Forest Fire Management) does not feed the simulation core with "known" single values, but rather carries out a set of simulations considering a range of possible values for the input parameters that are more uncertain.

This method requires a lot of computations to reach a conclusion because it is necessary to run a large number of simulations. To tackle this problem we have used a parallel scheme (master-worker), applied in a PC cluster. The method has been implemented using MPI as a message pass library and is executed in a Linux cluster. In this paper we analyze the improvements obtained by using the proposed scheme in terms of quality of the prediction and simulation speed-up for burns on experimental fields.

This paper is organized as follows: The factorial experimentation and basic concepts of the system are explained in section 2. The system's implementation is described in section 3. Section 4 includes the results obtained when the method was applied to two forest fires. Finally, the main conclusions are reported in section 5.

2 Factorial Experimentation

The methodology of this work is based on statistics. Statistics deal with collection, presentation, analysis and use of data to make, for example, decisions. There are two possible ways of collecting data about an event. In an **observational study** the researcher only takes notes without interacting in the situation. Data are obtained as they appear.

Another way is through **designed experiments**. In these kinds of experiments it is possible to make deliberate changes in the controlled variables of a system or process. Results are observed and then it is possible to either make an inference or make a decision about variables that are responsible for changes. When there are a lot of significant factors involved (i.e. weather, wind speed, slope, etc.), the best strategy is to use some kind of **factorial experiment**. A factorial experiment is one in which the factors vary at the same time [16](for example, wind conditions, moisture content and vegetation parameters). A **scenario** represents each particular situation that results from a set of values.

For a given time interval, we want to know whether a portion of the terrain (called a cell) will be burnt or not. If n is the total number of scenarios and n_A is the number of scenarios in which the cell was burned, we can calculate the **ignition probability** as:

$$P_{ign}(A) = n_A/n$$

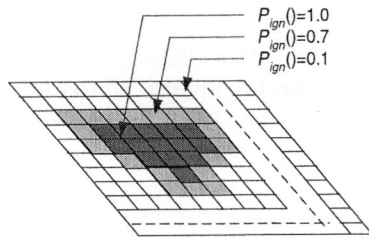

Fig. 1. Generalizing the cells analysis

The next step is to generalize this reasoning and apply it to some cell sets. In this manner we obtain a matrix with values representing the probability of each cell catching fire (Fig. 1).

Hence, we can focus our analysis on the procedure of generating possible scenarios.

2.1 Scenarios Generation

Our system uses a forest fire simulator as a black box which needs to be fed with different parameters in order to work. A particular setting of the set of parameters defines an individual scenario. These parameters correspond to the parameters proposed in the Rothermel [9] model.

For each parameter we define a rank and an increment value, which are used to move throughout the interval. For a given parameter i (which we will refer to as $Parameter\_i$) the associated interval and increment is expressed as:

$$[Inferior\_threshold\_i,\ Superior\_threshold\_i],\ Increment\_i$$

Then, for each parameter i, it is possible to obtain a number C_i (parameter domain cardinality), which is calculated as follows:

$$C_i = ((Superior\_threshold\_i - Inferior\_threshold\_i) + Increment\_i)/Increment\_i$$

Finally, from each parameter's cardinality it is possible to calculate the total number of scenarios obtained from variations of all possible combinations.

$$\#Scenarios = \prod_{i=1}^{n} C_i$$

where n is the number of parameters.

3 S^2F^2M Implementation

The concepts described above has been implemented in an operational system that incorporates a simulation kernel and applies the methodology to evaluate the fitness function. This system has been developed on a PC LINUX cluster using MPI as message passing library.

3.1 The Simulator

S^2F^2M uses as a simulation core the wildland simulator proposed by Collin D. Bevins, which is based on the fireLib library [4]. **fireLib** is a library that encapsulates the BEHAVE fire behavior algorithm [1]. In particular, this simulator uses a cell automata approach to evaluate fire spread. The terrain is divided into square cells and a neighborhood relationship is used to evaluate whether a cell will be burnt and at what time the fire will reach the burnt cells.

As inputs, this simulator accepts maps of the terrain, vegetation characteristics, wind and the initial ignition map.

The output generated by the simulator consists of a map of the terrain in which each cell is labeled with its ignition time.

3.2 The Fitness Function

To evaluate the system's response we defined a fitness function. Since S^2F^2M uses an approximation based on cells, the fitness function is defined as the quotient between the number of cells in the intersection between the simulation results and the real map, and the union of the simulation results and the real situation (Fitness = (cells in the intersection) / (cells in the union)).

Figure 2 shows an example of how to calculate this function for a terrain made up of 5x5 cells. In this case, the fitness function is 7/10 = 0.7.

A fitness value equal to one corresponds to the perfect prediction because it means that the predicted area is equal to the real burned area. On the other hand, a fitness equal to zero indicates the maximum error, because in this case our experiment did not coincide with reality at all.

3.3 Parallelisation

S^2F^2M has to make a large quantity of calculations because it uses a sequential simulator as a kernel [4], and for this reason it needs to make a simulation for each resulting combination of parameters (*#Scenarios*). This high number of simulations requires a lot of time.

To reduce the execution time we used multiple computational resources working in parallel to obtain the desired efficiency. Keeping in mind the nature of the problem that S^2F^2M tries to solve, we believe a master-worker architecture is

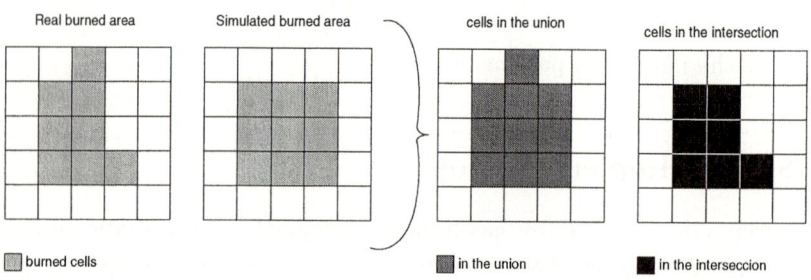

Fig. 2. Calculating the fitness for a 5 x 5 cell terrain

suitable to achieve this aim, because a main processor can calculate each combination of parameters and send them to a set of workers. These workers carry out the simulation and return the map to the master. This resulting map indicates which cells are burned and which are not.

Our system has a well defined structure. The Master process has a data reception stage (parameter files, terrain files, simulation time, etc.). After this there is an initialization stage for data structures. In the main loop, the Master process distributes scenarios to the workers, waits for results, receives results and distributes more data to idle workers (if there are more scenarios to simulate). Finally, it gives a graphical output.

The Worker structure is complementary. Each one has a data reception stage (to initialize terrain size, slope). Following this, it enters a loop to receive scenarios from the Master process to activate the simulation function for calculating fire spread.

4 Experimental Results

To test the system we used two experiments in the field. Both burns took place in Serra da Lousã (Gestosa, Portugal (40°15'N, 8°10'O)), at an altitude of between 800 and 950 m above sea level. The burns were part of the SPREAD project [15]. In the Gestosa field experiments [10], terrain was divided into dedicated plots in order to carry out different sorts of tests and measurements. In particular, we worked with plots 513 and 519, which had the following characteristics:

Experiment 1 (Plot 513): the plot was represented by means of a grid of 58 columns x 50 rows (each cell was 2.989 x 2.989 feet).

Experiment 2 (Plot 519): the plot was represented by means of a grid of 89 columns x 91 rows (each cell was 2.989 x 2.989 feet).

In order to gather as much information as possible about the fire-spread behavior, a camera recorded the complete evolution of the fire. The video obtained was analyzed and several images were extracted every 2 minutes in the first experiment and every 2.5 minutes in the second. From the images the corresponding fire contours were obtained and converted to cell format in order for S^2F^2M to interpret them.

4.1 Experiment 1

The first case is very complicated, because in a field experiment it is not possible to control environmental conditions. Nevertheless, we fixed certain known values (*slope* and *moisture* in 1, 10 and 100 hours) and let the others vary.

To make comparisons we fixed the initial time to 0 and a limit value of 12 minutes.

In table 1, we can see that the fitness for this experiment has values between 0.7 and 0.91. This indicates that our statistic output is very close to reality. It is important to note that in the Fitness table only those cells with 100% ignition probability (i.e. cells burned in 100% of the scenarios) are considered.

Table 1. Fitness of experiment 1 in each interval

Initial time	Final time	Fitness
0:00	2:00	0,749420
2:00	4:00	0,690152
4:00	6:00	0,864360
6:00	8:00	0,953166
8:00	10:00	0,826158
10:00	12:00	0,915669

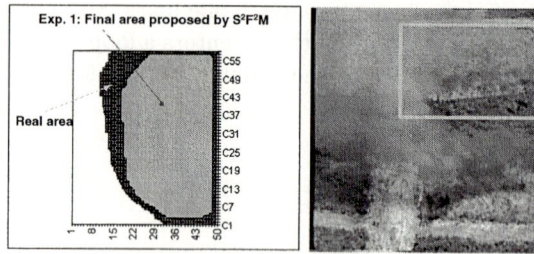

Fig. 3. S^2F^2M output for minute 12 of Experiment 1

In the figure 3 we can see that the S^2F^2M result is always inside reality, that is, the result does not exceed the real propagation. We only include the last step of the simulation (in this case at minute 12), as the previous steps are included in the real perimeter.

4.2 Experiment 2

The second experiment has a rank file equal to plot 513. This is because the plots are located very near each other, and therefore the terrain features can be taken to be equivalent. Table 2 shows the resulting fitness.

Finally, using the same criterion, we show the state proposed by S^2F^2M at minute 12.5 in figure 4. It is possible to identify clearly that the S^2F^2M area is inside the real perimeter.

4.3 Speed-Up Improvement

In a real case the system works under real time constraints and, therefore, it is necessary to analyze the speed-up obtained by using different numbers of

Table 2. Fitness of experiment 2 in each interval

Initial time	Final time	Fitness
2:30	5:00	0,451988
5:00	7:30	0,486521
7:30	10:00	0,425703
10:00	12:30	0,774615

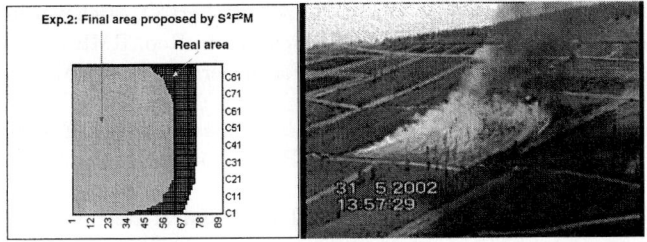

Fig. 4. S^2F^2M output for minute 12.5 of Experiment 2

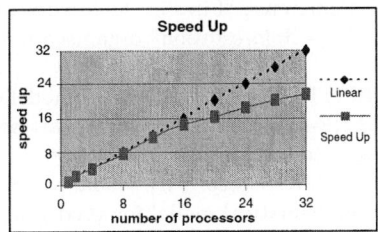

Fig. 5. Speed-up curve for experiment 1

processors. The number of processors used in the successive experiments were 1, 2, 4, 8, 12, 16, 20, 24, 28 and 32. Figure 5 shows the speed-up for a particular example compared with a linear speed-up (the ideal case).

It can be observed that the speed-up is close to being linear until 16 processors are used. From this point, an increase in the number of processors continues being profitable, but it can be observed that the speed-up is not so linear.

5 Conclusions

In this paper we have described a tool with the objective of offering an alternative to the normal use of a forest fire simulator. With this methodology we can obtain a prediction of the ignition probability of a terrain without knowing exact data about climatic factors, and without waiting for the fire to start.

From the experimental studies we can conclude that the area that S^2F^2M indicates with 100% probability of being reached by fire in a time interval is always included in the real burned area. Furthermore, since each output proposed by S^2F^2M needs a lot of calculations, we have used the parallel scheme of a master-worker programming paradigm in order to speed-up the whole process.

References

1. Andrews P. L. "BEHAVE: Fire Behavior prediction and modeling systems - Burn subsystem, part 1". General Technical Report INT-194. Odgen, UT, US Department of Agriculture, Forest Service, Intermountain Research Station; 1986.

2. Andrews, Patricia L.; Bevins, Collin D.; Seli, Robert C. "BehavePlus fire modeling system, version 2.0: User's Guide". Gen. Tech. Rep. RMRS-GTR-106WWW. Ogden, UT: Department of Agriculture, Forest Service, Rocky Mountain Research Station. 2003.
3. Baker Abdalhaq, G. Bianchini, Ana Cortés, Tomàs Margalef, Emilio Luque: "Improving Wildland Fire Prediction on MPI Clusters". LNCS 2840, pp. 520-528, 2003.
4. Collins D. Bevins, "FireLib User Manual & Technical Reference", 1996. http://www.fire.org.
5. E-FIS A ten telecom project. http://www.e-fis.org/
6. Eftichidi G., Varela V. 1999. SAFES: Safe Fire Expert System. Presentation in the International Scientific Conference "Fires in the Mediterranean forests: Prevention -Suppression - Soil Erosion - Reforestation" organised by UNESCO in Athens, 3-6 February 1999.
7. Finney, Mark A.. "FARSITE: Fire Area Simulator-model development and evaluation". Res. Pap. RMRS-RP-4, Ogden, UT: U.S. Department of Agriculture, Forest Service, Rocky Mountain Research Station. 47 p. 1998
8. ADAI Products: FIRESTATION http://www.adai.pt/products/firestation/
9. Rothermel, R. C., "A mathematical model for predicting fire spread in wildland fuels", USDA FS, Ogden TU, Res. Pap. INT-115, 1972.
10. ADAI - CEIF (Center of Forest Fire Studies) http://www.adai.pt/ceif/Gestosa/
11. MPI: The Message Passing Interface Standard http://www-unix.mcs.anl.gov/mpi/
12. Prometheus. http://kentauros.rtd.algo.com.gr/promet/schedule.htm
13. Reinhardt, E.D.; Keane, R.E.; Brown, J.K. "First Order Fire Effects Model: FOFEM 4.0, User's Guide". General Technical Report INT- GTR- 344. 1997.
14. Sixto Ríos Insua, Concepción Bielza Lozoya, Alfonso Mateos Caballero, "Fundamento de los Sistemas de Ayuda a la decisión", RaMa 2002 ISBN 84-7897-494-6
15. Project Spread, Forest Fire Spread Prevention and Mitigation http://www.adai.pt/spread/
16. Douglas C. Montgomery, George C. Runger, "Probabilidad y Estadística aplicada a la Ingeniería", Limusa Wiley 2002 ISBN: 968-18-5914-6

Concurrent Execution of Multiple NAS Parallel Programs on a Cluster

Adam K.L. Wong and Andrzej M. Goscinski

School of Information Technology, Deakin University, Geelong, Vic 3216, Australia
{aklwong, ang}@deakin.edu.au

Abstract. Currently, coordinated scheduling of multiple parallel applications across computers has been considered as the critical factor to achieve high execution performance. We claim in this report that the performance and costs of the execution of parallel applications could be improved if not only dedicated clusters but also non-dedicated clusters were used and several parallel applications were executed concurreontly. To support this claim we carried out experimental study into the performance of multiple NAS parallel programs executing concurrently on a non-dedicated cluster.

1 Introduction

Parallel processing has moved one step closer toward the computing mainstream by exploiting specialized dedicated clusters and MPI. These dedicated clusters are built using off-shelf elements (processors, memories and networks), which make them cost-effective computer systems.

The cost to performance ratio could be improved even further if non-dedicated clusters are used. These clusters are owned by many institutions, universities, business, and industry. They are made of not necessarily the fastest computers and networks, but still form a huge computational power that can be used to solve many problems that require parallel computing for high performance. Cluster's PCs in their working environments are on average idle for much more than 50% of time [2, 4, 12]. Therefore, a cluster has the potential of concurrently supporting a mixture of parallel and sequential applications of different users, which could lead to the improvement of the execution performance [14, 15, 7]. We claim that these computer systems can also be used cost-effectively to concurrently execute several parallel applications submitted by many competing users.

Parallel applications can share the computational resources of a cluster in two dimensions: space and time. Space-sharing is usually done by static allocation of processes to the available computers. Therefore, processes of different parallel applications can be mapped into different sets of computers of the cluster and the execution of those processes would not interfere with each other. On the other hand, time-sharing can be achieved if processes of parallel applications are mapped into the same computer and the local scheduler schedules these processes to share the CPU among them.

The time-sharing approach can provide good response time and good throughput for applications in a multi-user and multi-programs environment. However, it has been considered that uncoordinated scheduling of processes from different applications would seriously hurt the performance of a cluster [1, 3, 5, 17, 13]: the execution overhead can be up to over 15 times of the program execution time. To our knowledge, the only existing work which has shown an opposite result is [11]. Therefore, we believe that more experimental studies of this problem will provide not only a better understanding of the problem, lead industry, business and research institutions toward parallel processing on their existing clusters, but also form a background for the development of global scheduling facilities for computer clusters.

The aim of this paper is to confirm our claim by showing the results of our study into the performance of multiple parallel applications executing concurrently on a cluster. We also want to demonstrate that the results achieved by other researchers showing that uncoordinated scheduling of parallel applications would seriously hurt the performance of a cluster are unsubstantiated. For this purpose experiments were carried out using the well known and widely used NAS Parallel Benchmarks [9].

The paper is organized as follows. Section 2 presents the related work. Section 3 describes attributes of parallel applications required to carry out the proposed experiments on a cluster, and introduces the selected NAS benchmarks. Section 4 details the experiments carried out. Section 5 reports on the achieved results and presents the analysis of these results. Finally, the conclusion is presented in Section 6.

2 Related Work

Parallel computer systems such as the Massively Parallel Processors (MPPs) have a low communication overhead and therefore the effect of uncoordinated scheduling of processes from parallel applications executing concurrently on the systems is significant. Coordinated scheduling such as gang-scheduling [5] is normally used to alleviate the problem.

Time-sharing is intrinsically supported in a computer cluster via local scheduling. That means the local scheduler is responsible for time sharing of the CPU among all the processes which have been allocated to that computer. Processes from a parallel application can be placed into some or all of the computers in the cluster depending on the required parallelism. However, processes belonging to the same parallel application would not be guaranteed to execute at the same time across the computers in the cluster. Previous studies carried out using stimulation [1, 3, 17] have found that if the parallel application is communication intensive, uncoordinated scheduling of processes would lead to a great loss of performance because a process stalls when it communicates with a non-scheduled process.

[13, 11] present the results of co-scheduling of multiple parallel applications on a cluster using local scheduling, which are quite different. [13] shows that co-scheduling of parallel applications on a cluster worsens their execution performance. However, that result is difficult to assess as the experiment is not clearly described and applications used in the experiment are not defined. [11] shows that local scheduling can out-perform gang-scheduling of parallel applications executing on a Beowulf cluster with a slow network (100Mbps Ethernet).

3 The NAS Parallel Benchmarks (NPB)

We have selected to use the NAS Parallel Benchmarks (NPB) for this study because they have been widely used to objectively measure and compare the performance of parallel computer systems. In particular, the NAS Parallel Benchmarks 2.4 [16] can be used as parallel applications and run on non-dedicated clusters. Each NPB 2.x provides source codes written with MPI. Furthermore, these benchmarks can be coarsely divided into two major categories, computation-bound and communication-bound, which is important when carrying out experiments on commonly used clusters, with slow, 100Mbps, networks.

The NPB suite is a set of eight programs, which were derived from computational fluid dynamics (CFD) codes [9]. Each of these programs focuses on some important aspects of highly parallel supercomputing for aerophysics applications. There are two groups of these applications: five "kernels" and three "simulated computational fluid dynamics (CFD) applications". The kernels are relatively compact problems, easy to implement. They mimic the computational core of different numerical methods used by CFD applications – each of them emphasizes a particular type of numerical computation. They provide insight as to the general levels of performance that can be expected on these specific types of numerical computations. The simulated CFD applications are more difficult to implement – they indicate the types of actual data movement and computation required in state-of-the-art CFD application codes.

3.1 Attributes of the Benchmark Programs Affecting the Scheduling Behaviour

The behavior and scheduling study requirements led us to the specification of benchmark attributes that must be present because they influence the execution performance of a parallel application. These attributes form a basis of the selection of benchmarks for our experiments. They are as follows.

- *Computation attributes*: In general, the problem size of a parallel program is directly proportional to its execution time. Each of the programs of the NPB suite comes with several problem sizes (classes): A, B, C, W(orkstation) and S(ample). Excluding classes W and S, class A is the smallest whereas class C is the largest.
- *Communication attributes*: Different parallel program has different communication features. The communication features of the NAS programs can be considered in two aspects: communication volume and communication pattern. In respect to the communication volume, the programs can broadly be classified into three categories: low, medium and high. In respect to the communication pattern, two forms exist: point-to-point and collective.
- *Memory attributes:* The size of main memory of a program during execution affects the scheduling behaviour, in particular could lead to memory swapping. The memory scheduling behavior of the NAS programs depends on the program size, characterized by the program class.
- *Topology attributes:* The software topology of processes of a parallel program defines the size (number of processes) and the structure (the connections of processes) of the program. The NAS programs use various software topologies. FT, MG, CG, LU and IS run well with a power-of-2 number of processes; SP and BT run well with a square number of processes; and EP runs with any number of process.

Table 1. NAS programs with communication specific attributes

Program Name	Communication Volume	Communication Pattern
EP	Negligible	Negligible
LU	Low	Point-to-Point >>> Collective
BT	Medium	Point-to-Point >> Collective
MG	High	Point-to-Point > Collective

3.2 Selected NAS Benchmarks

To evaluate the impact of concurrent execution of multiple parallel applications on their performance, we carried out an analysis of the NAS programs to identify those that posses the attributes addressed in Section 3.1. We have found that four programs, EP, LU, BT and MG, of the NBS suite, represent a broad range of communication patterns of parallel applications that can commonly be found in real world. The communication features of these programs are summarized based on [6, 9, 10] and shown in Table 1.

We have chosen a problem size of class B for the programs listed above to make sure that (i) the execution time is long enough for the scheduling behaviour to be observed (only the number of iterations of execution in some programs have been altered), and (ii) memory swapping would not happen.

We could not satisfy fully the topology requirement. The reason is as follows. The MG and LU belong to the group of those programs that perform best (require) a power-of-2 number of processes, and EP can also be executed effectively with the same number of processes, BT performs best when it runs with a square number of processes, i.e., 1, 4, 9, 16. We assumed that the loss of performance for executing BT with eight (processes) rather than nine (processes) will not distort the experiment outcomes, and carried out the experiments using a power-of-2 number of processes for each selected NAS program. Thus, the four programs, two kernel (EP and MG) and two simulated (LU and BT) applications, which posses the attributes to carry out our study, are selected.

4 Experiments

All of the scheduling experiments were carried out on a cluster which consists of 16 Pentium III computers, each with 383 Mbytes of main memory. The computers are connected together by a 100Mbps Fast Ethernet network. Each computer runs the Red Hat Linux operating system with the parallel programming support of MPI.

We decided to use a two level scheduling system, where the upper level is responsible for allocation of processes of each parallel application to cluster computers and the lower level schedules local parallel processes (of one or more than one application) running on each local computer of the cluster. MPI provides initial allocation of parallel processes to cluster computers. On an individual computer these processes are scheduled by the Red Hat Linux operating system scheduler.

An MPI application usually can be executed by first constructing a network topology, which specifies in the topology configuration file (TCF) the number and the identity of the computers used for the execution. Such a network topology can then be booted up in the cluster, which basically starts a MPI daemon process on each of the computers specified in the TCF. The MPI daemons are responsible for handling communications among processes of a parallel application. We used in our experiment the LAM/MPI implementation [8] as it is one of the most popular implementations of the MPI specification, and its version is LAM/MPI-6.5.9.

First, we measured the execution times of the four selected NAS parallel programs (EP, LU, MG and BT) individually. Taking into account the topology attributes specified in [9]: a power-of-2 of processes for LU and MG, a square number of processes for BT and any number of processes for EP, we selected a power-of-2 of processes as the standard topology for all the four programs executing in this experiment. Since there are 16 computers in our cluster, each of the programs were run on 2, 4, 8 and 16 computers; all of the programs were compiled for using 16 processes. Then, we measured i) the execution time for each of the selected NAS benchmarks when multiple copies of itself were executed concurrently on a cluster of 16 computers, and ii) the execution time for each of the selected NAS benchmarks when one or multiple copies of EP were executed concurrently with itself on the same cluster. The influence of the local scheduling was then observed and measured.

5 Results and Analysis

Speedup of NAS Programs. Our experimental study was carried out using the programs of Class B of NPB.

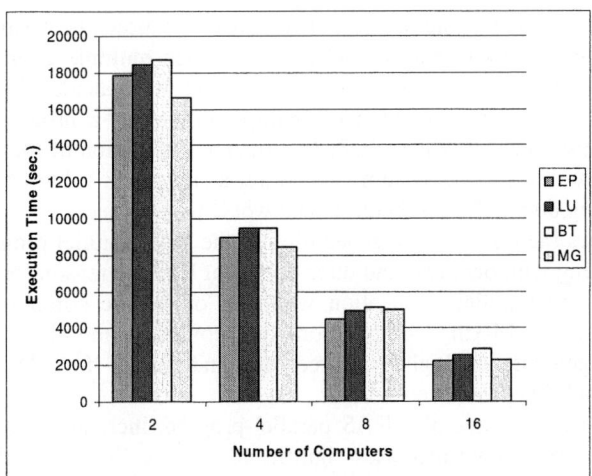

Fig. 1. Execution Time of NAS programs against different number of computers

Because the number of iterations of execution in the selected NAS parallel programs were increased, the execution time of each of parallel programs: EP, LU, BT and MG, against different number of computers: 2, 4, 8 and 16 of our cluster was measured. The result is presented in Figure 1.

The results we achieved are consistent with the benchmarking results listed in [9]. Besides, our results indicate that using a cluster of 16 Pentium III computers to execute each of the NAS parallel programs requires around 40 minutes of time. We believe that this time duration for an execution of a parallel program not only is realistic but also long enough for observing the scheduling behaviour when multiple copies of the programs are executed concurrently on a cluster.

Concurrent Execution of the NAS Benchmark Programs. The purpose of the experiment was to determine the effect of co-scheduling of multiple NAS programs on a cluster with particular attention paid to the execution performance of the programs. Table 2 shows the results of co-scheduling of two and three identical NAS parallel programs on our 16-computers cluster.

Table 2. Multi-program scheduling of parallel programs from the NBS

Program	Sequential Execution Time (SET) (sec.)			Concurrent Execution Time (CET) (sec.)			Slowdown = CET/SET		
	Number of copies			Number of copies			Number of copies		
	1	2	3	1	2	3	1	2	3
EP	2240	4480	6720	2240	4497	6711	1.00	1.00	1.00
LU	2522	5044	7566	2522	4942	7478	1.00	0.98	0.99
BT	2859	5718	8577	2859	5046	7561	1.00	0.88	0.88
MG	2279	4558	6837	2279	4550	6769	1.00	1.00	0.99

Expecting to observe some worse performance, we used the term slowdown to refer to the slowing down of a parallel program when multiple copies of the same program are running concurrently on the same system. It is calculated by dividing the time (CET) of concurrent execution of multiple copies of a program on a multi-programming system by the time (SET) of sequential execution of multiple copies of the program run on the same system. Therefore, a slowdown value of greater than 1 implies a slowing down of the execution, and worse performance.

Table 3 shows the results of co-scheduling of the NAS parallel programs: LU, BT and MG executing with one, two and three copies of EP. The reason for that is EP is a computation-bound parallel application while all others are communication-bound parallel applications. Mixing computation- and communication-bound applications together can generate a higher degree of randomness for the processes to communicate and synchronize each other.

As the number of copies of a NAS parallel program increases, the total memory required for the executions increases significantly, especially in the data-spaces. In both cases, the memory requirement for programs to execute is carefully controlled to make sure that no memory swapping would occur. Our results show that no slowdown is greater than 1 in all of the above scheduling cases which is different from the results presented in the majority of the current publications.

The results indicate that both computation-bound and communication-bound parallel applications are not sensitive to co-scheduling on a cluster with a slow network. From table 2, it can also be seen that the slowdown obtained in LU, BT and MG is slightly less than 1 which can be explained by the low CPU utilization of the programs. When multiple copies of those programs are scheduled together, a blocked process can be scheduled to keep the CPU as busy as possible.

Table 3. Mixed multi-program scheduling of parallel programs from the NBS

Program	Sequential Execution Time [SET] (sec.)			Concurrent Execution Time [CET] (sec.)			Slowdown = CET/SET		
	Number of copies of EP			Number of copies of EP			Number of copies of EP		
	1	2	3	1	2	3	1	2	3
LU	4762	7002	9242	4719	6951	9211	0.99	0.99	1.00
BT	5099	7339	9579	4752	6964	9188	0.93	0.95	0.96
MG	4519	6759	8999	4475	6705	8962	0.99	0.99	1.00

6 Conclusions

In this paper we have presented the results of the experimental study into the performance of multiple parallel applications executing concurrently on a cluster. The aim was to confirm our claim that the performance of the execution of parallel applications could be improved when several parallel applications are executed concurrently, and to show that the claim made by other researchers that uncoordinated scheduling of parallel applications would seriously hurt the performance of a cluster is unsubstantiated.

Contrary to the results presented in most of the current literature, we have found that even if a parallel application is communication intensive, there is no performance loss of the parallel application due to uncoordinated communications and synchronizations of processes. Our study of the concurrent execution of multiple parallel applications on a cluster does not confirm the simulation results reported in [1, 3, 17], which recommend synchronized scheduling of multiple parallel applications such as gang-scheduling. The results of our experimental study using real parallel applications are different from the result shown in [13]; the only result presented in the current literature which is inline with us is in [11].

We have demonstrated that concurrent execution of multiple parallel applications on a cluster did not make the execution performance of a parallel application worse. The execution performance was improved. It seems that this cost-effective scheduling scheme is particularly useful for computer cluster, especially with a slow network.

References

1. R.H. Arpaci, A.C. Dusseau, A.M. Vahdat, L.T. Liu, T.E. Anderson and D.A. Patterson. The Interaction of Parallel and Sequential Workloads on a Network of Workstations. In *Proceedings of 1995 ACM Joint International Conference on Measurement and Modeling of Computing Systems*, pages 267-278, May 1995.
2. A. Acharya, G. Edjlali and J. Saltz. The Utility of Exploiting Idle Workstations for Parallel Computation. In *Proceedings of 1997 ACM Sigmetrics International Conference on Measurement and Modeling of Computer Systems*, pages 225-236, May 1997.

3. C. Anglano. A Comparative Evaluation of Implicit Coscheduling Strategies for Networks of Workstations. In *Proceedings of 9th International Symposium on High Performance Distributed Computing (HPDC9)*, pages 221-228, August 2000.
4. W. Becker. Dynamic Balancing Complex Workload in Workstation Networks -- Challenge, Concepts and Experience. In *Proceedings High Performance Computing and Networking (HPCN) Europe Lecture Notes on Computer Science (LNCS)*, pages 407-412, 1995.
5. D.G. Feitelson and L. Rudolph. Gang Scheduling Performance Benefits for Fine-Grained Synchronization. *Journal of Parallel and Distributed Computing*, 16(4):306-318, 1992.
6. A. Faraj and X. Yuan. Communication Characteristics in the NAS Parallel Benchmarks. In *Proceedings of the 14th IASTED International Conference on Parallel and Distributed Computing and Systems (PDCS 2002)*, Nov. 2002.
7. A. M. Goscinski and A. K. L. Wong. Performance Evaluation of the Concurrent Execution of NAS Parallel Benchmarks with BYTE Sequential Benchmarks on a Cluster. Paper submitted to *the 11th International Conference on Parallel and Distributed Systems, ICPADS05*.
8. The LAM/MPI Homepage. URL: http://www.lam-mpi.org, lasted access: June 2004.
9. NAS Parallel Benchmarks. URL: http://www.nas.nasa.gov/Software/NPB/, last accessed: Nov. 2004.
10. J. Subhlok, S. Venkataramaiah and A. Singh. Characterizing NAS Benchmark Performance on Shared Heterogeneous Networks. In *11th International Heterogeneous Computing Workshop*, April 2002.
11. P. Strazdins and J. Uhlmann. Local Scheduling out-performs Gang Scheduling on a Beowulf Cluster. Technical Report TR-CS-04-01, The Australian National University, 2004.
12. F. Tandiary, S.C. Kothari, A. Dixit and E. W. Anderson. Batrun: Utilizing Idle Workstations for Large-scale Computing. *IEEE Parallel and Distributed Technology*, 4(2):41-48, 1996.
13. F.C. Wong, A.C. Arpaci-Dusseau and D.E. Culler. Building MPI for Multi-Programming Systems using Implicit Information. In *Proceedings of the 6th European PVM/MPI User's Group Meeting*, pages 215-222, 1999.
14. A. K. L. Wong and A. M. Goscinski. Scheduling of a Parallel Computation-Bound Application and Sequential Applications Executing Concurrently on a Cluster – A Case Study. In *Proceedings of the 2rd International Symposium on Parallel and Distributed Processing and Applications (ISPA04)*, Dec. 2004.
15. A.K.L. Wong and A.M. Goscinski. The Performance of a Parallel Communication-Bound Application and Sequential Applications Executing Concurrently on a Cluster – A Case Study. In *Proceeding of the 12th International Conference on Advanced Computing and Communication (ADCOM-2004)*, Dec. 2004.
16. Rob F. Van der Wijngaart. The NAS Parallel Benchmarks 2.4 NAS Technical Report NAS-95-020, NASA Ames Research Center, Moffett Field, CA, 1995.
17. B.B. Zhou, X. Qu and R.P. Brent. Effective Scheduling in a Mixed Parallel and Sequential Computing Environment. In *Proceedings of the 6th Euromicro Workshop of Parallel and Distributed Processing*, pages 32-37, Jan. 1998.

Model-Based Statistical Testing of a Cluster Utility

W. Thomas Swain[1] and Stephen L. Scott[2,*]

[1] Software Quality Research Laboratory,
University of Tennessee Department of Computer Science Knoxville, Tennessee 37996
[2] Network and Cluster Computing Group,
Computer Science and Mathematics Division,
Oak Ridge National Laboratory, Oak Ridge, Tennessee
swain@cs.utk.edu, scottsl@ornl.gov

Abstract. As High Performance Computing becomes more collaborative, software certification practices are needed to quantify the credibility of shared applications. To demonstrate quantitative certification testing, Model-Based Statistical Testing (MBST) was applied to *cexec*, a cluster control utility developed in the Network and Cluster Computing Group of Oak Ridge National Laboratory. MBST involves generation of test cases from a usage model. The test results are then analyzed statistically to measure software reliability. The population of *cexec* uses was modeled in terms of input selection choices. The J Usage Model Builder Library (JUMBL) provided the capability to generate test cases directly as Python scripts. Additional Python functions and shell scripts were written to complete a test automation framework. The resulting certification capability employs two large test suites. One consists of "weighted" test cases to provide an intensive fault detection capability, while the other consists of random test cases to provide a statistically meaningful assessment of reliability.

1 Introduction

The work described here had two primary objectives: (1) to certify the *cexec* command in the Cluster Command and Control (C3) tool suite [1,2], and (2) to demonstrate Model-Based Statistical Testing (MBST) as a certification methodology for computational software. Briefly C3 is a set of command line utilities to facilitate management of Linux clusters. The *cexec* command invokes a specified application on any combination of nodes in a Linux cluster.

MBST treats software testing as a statistical experiment. That is, each test is viewed as a sample from the population of all possible uses. This approach to testing involves six tasks: (1) usage model definition, (2) model analysis, (3) test automation, (4) test case generation, (5) test execution, and (6) results analysis

MBST is applied to systems by mapping input stimulus sequences to states of use and associated responses. However computational programs often have only two

---

* Research supported by the Mathematics, Information and Computational Sciences Office, Office of Advanced Scientific Computing Research, Office of Science, U. S. Department of Energy, under contract No. DE-AC05-00OR22725 with UT-Battelle, LLC.

states of use: pre-execution and post-execution. Such programs are further characterized by multiple input parameters, often with very large domains.

For *cexec*, usage was modeled in terms of the input selection process. The model analysis, performed using the J Usage Model Builder Library (JUMBL) [3], provided information needed to plan details of the testing strategy. Once a testing strategy was defined, a test automation framework was implemented using Python and *bash* shell scripts. From the model, JUMBL was used to generate test cases. Test cases for all paths through the model were generated in decreasing order of probability; and a set of random test cases were generated to be statistically representative of the usage profile embodied in the model. All test cases were executed using the test automation framework. The pass/fail outcome of all test cases was determined by automated inspection, supplemented by manual analysis of reported failures. The results were then analyzed statistically to provide a quantitative basis for certification.

2 Model Definition

Model definition requires four elements: (1) definition of the test boundary, (2) definition of a *use*, (3) identification of all input stimuli, and (4) identification of correct system responses.

The test boundary is defined by the list of interfaces where stimuli can be applied and responses can be observed. The interface list for *cexec* includes the following:

- The interface by which command line arguments are supplied on invocation.
- The interface by which environment variables are supplied.
- The file system interface by which files are read, written, or deleted.
- The interfaces to `stdout` and `stderr`.
- Other interfaces provided by UNIX system calls.

The definition of a use is simply execution of *cexec* by sending its command line to the operating system of the cluster head node. For each use, the command runs to completion with no intervening stimuli. Consequently sampling the *use* population is a matter of sampling the set of all possible input combinations. Modeling the input selection process as a discrete Markov chain allowed JUMBL to be used for test case generation and test analysis. With this approach, test cases are defined by input combinations obtained using the model as a statistical sampling mechanism.

The set of all possible combinations of the input parameters is quite large, even for a utility such as *cexec*. In addition to some discrete command line options, *cexec* input may include the following inputs as arbitrary character strings: Machine Definition expressions (to specify a subset of the cluster), UNIX command strings, configuration file names, and configuration file contents.

To cope with the large domains of these parameters, the following abstractions were used:

- Machine Definitions = [good, bad, none]
- UNIX commands = [good, bad]
- configuration file names = [good, bad, missing (when expected)]
- configuration files = [good, bad]

For test execution, sampling within these subdomains was built into the test automation framework.

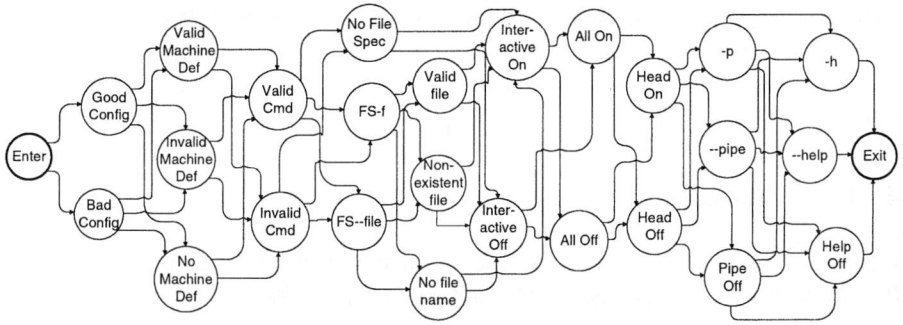

Fig. 1. *cexec* Usage Model

The resulting input selection model is illustrated graphically in Figure 1. The nodes of the graph represent input parameter values (or abstracted ranges), and the arcs represent individual value choices. Probabilities (not shown in the figure) are assigned to the arcs to indicate the relative likelihood of each value for a particular parameter. The probabilities used were based on the following assumptions:

- Except as noted below, all valid choices for a parameter are equally probable.
- Valid choices for a parameter are about 100 times more likely than invalid choices.
- A Machine Definition will be specified for 90% of all commands.
- About half of all commands will specify the configuration file on the command line instead of using the default configuration file.

3 Model Analysis

Analysis of the model as a Markov chain is used for model validation and test planning. Table 1 shows key results from the model analysis.

Table 1. *cexec* Usage Model Analysis

Node Count	27 nodes
Arc Count	60 arcs
Statistically Typical Sequences	$2^{6.65}$ cases
Maximum Mean First Passage	200 cases
Mean First Passage Variance	39,800

The model analysis results can be useful for both model validation and test planning. In this case the model structure is simple enough that the structure and probabilities can be validated by inspection.

For test planning, the analysis provides insights that are not available from inspection of the model. First the number of Statistically Typical Sequences given in Table 1 indicates that $2^{6.65}$ (~100) random test cases are needed to produce a test population that is typical of the usage profile embodied by the model. Additionally the maximum Mean First Passage indicates the average number of random test cases needed to encounter every node of the model is 200 with a variance of 39,800. The corresponding standard deviation σ is on the order of 200. Using the semi-quantitative argument that >3σ test cases would give high confidence of complete model coverage, 1000 random test cases were planned.

4 Test Automation

Figure 2 shows an overview of the test automation data flow. The processes in the figure are grouped into the following major automation tasks:

- Test case generation (*cexec* command line construction),
- Generation of configuration files and Machine Definitions,
- Results checking and comparison.

Most of the automation framework is reusable for other C3 commands or future versions of *cexec*. Only the usage model definition file and the Python functions used to check test results would need to be changed.

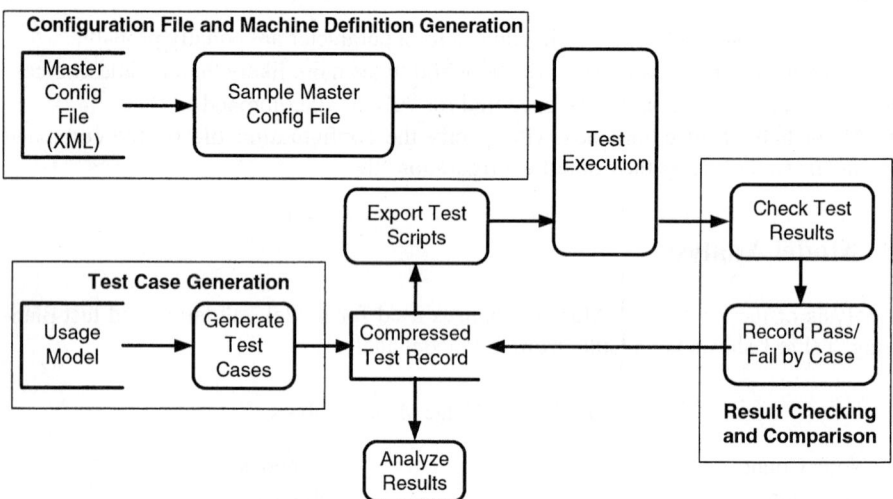

Fig. 2. Test Automation Overview

Automated Command Line Construction

To support test automation, the TML modeling language [4] allows each element (model, state, or arc) to be annotated with arbitrary text, referred to as a "label" in TML syntax. Test cases are initially generated and stored in a compressed Saved Test Record (STR) format. For test execution the JUMBL *managetest export* command [5] is used to convert STR to a human-readable form. The exported test case is a sequential listing of the labels associated with the model elements encountered.

Labels in the *cexec* TML file consist of Python statements. Initialization code is associated with the [Start] state. Python code in arc labels assigns values to *cexec* input parameters. Code associated with state labels records inputs for use in results checking. The label on the [Exit] state consists of code required to assemble the *cexec* command line, spawn the resulting command, and record and evaluate the output. Thus each exported test case is a short Python program that automates execution of the required command string and comparison of actual and expected output.

Configuration File and Machine Definition Generation

As mentioned above, four of the *cexec* input parameters are character strings and were each abstracted into two or three discrete values in the usage model. During test case generation, each of these parameters is assigned one of its abstract values. At test case run time, the assigned abstract value must be replaced with a specific value from the associated subdomain. For example, "good" file names are created as <test case name>.conf, and "bad" file names are always bogus.conf (a non-existent file).

Since configuration files and machine definitions are processed entirely within *cexec*, a much more varied sampling of their parameter spaces is desirable. Figure 3 shows an overview of this sampling process.

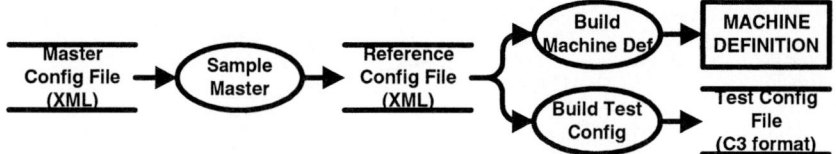

Fig. 3. Data Flow for Generating Configuration Files and Machine Definitions

To expedite data format conversion, an XML format was defined to represent the content of a C3 configuration file. The Master Config File in Figure 3 is the XML representation of the complete test system. At test case run time, a subset of clusters and nodes is selected from the Master Config File and used to create the Reference Config File. The Reference Config File is then transformed to the Test Config File in the standard C3 configuration file format.

If the test case calls for a "bad" configuration file, a single fault is inserted in the file by the Build Test Config process. The type of fault and the insertion location are selected randomly from among the syntactic elements that compose a C3 configuration file. If the test case requires a Machine Definition expression in the command line, the clusters and nodes to be included are selected and composed into the required string by the Build Machine Def process in Figure 3. When required, machine definition fault injection is handled in a manner similar to that for configuration files.

Results Checking

The standard output of *cexec* for each test case is redirected to a log file. The contents of the log file are then compared to the expected output. The graphs in Figures 4 and

5 depict the mapping of input combinations to responses. Each shaded leaf node in the graph represents a set of input combinations for which the expected output can be determined from a single rule.

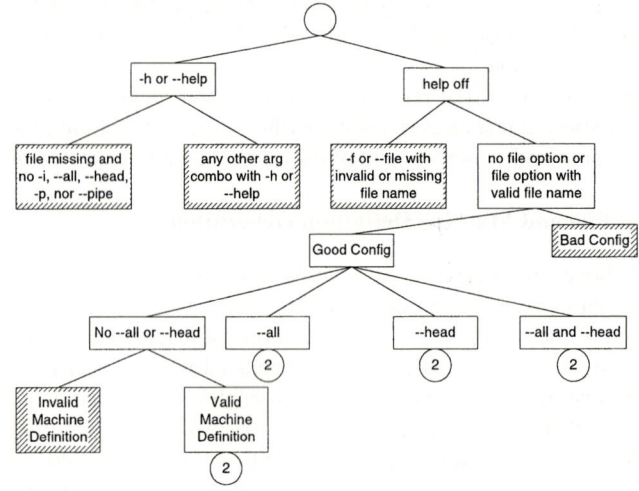

Fig. 4. Top Level *cexec* Response Tree

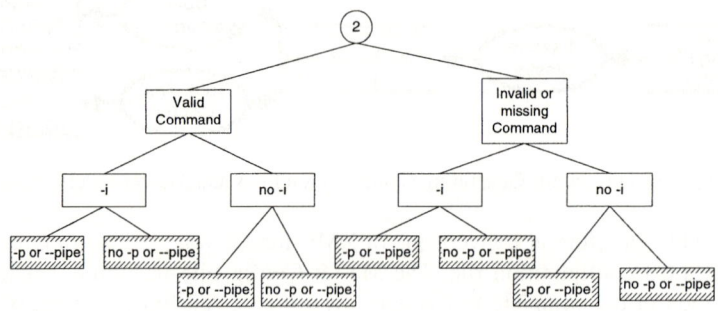

Fig. 5. Continuation of *cexec* Response Tree

Based on these rules for the expected output, a Python function was written to make a run-time pass/fail determination for each test case. For each test case, this function writes the test case name and a summary of the *cexec* input parameters to a file. If a test case failure is detected, additional diagnostic information is written to the file. If *cexec* terminates abnormally, the UNIX exit code is reported. If *cexec* terminates normally, the output line causing the failure determination is reported along with the expected result. Finally the test case is marked as a failure in the STR file.

A few cases involving invalid Machine Definition input were reported as failed by the run-time checking and required manual inspection to screen out the false failures. However these cases were easily recognized in the summary report.

5 Test Case Generation and Execution

The JUMBL provides five test case generation options: state and arc coverage, random, probability-weighted, cost-weighted, and manually crafted. The following factors were considered in selecting the test generation options to be used:

- Since the model involves relatively coarse abstractions for configuration file and Machine Definition inputs, minimal state and arc coverage is not adequate.
- With automation, up to several thousand test cases can be run within reasonable time and cost.
- Since quantitative certification is a primary objective, a subset of the test cases should represent a statistically typical sample (based on the model).
- Based on the model analysis above, the statistically typical sample requires 500 to 1000 random test cases.
- Since the model contains no loops or cycles, it is possible to test all possible paths through the model.

Based on these factors, the probability-weighted generation option was selected as a way to produce test cases for all possible paths through the model. This process generated 6048 test cases. To provide a statistically typical sample, 1000 random test cases were also generated.

The tests were executed on ORNL's XTORC cluster. As described above, each test case was exported as a Python program. One *bash* shell script was used to sequence the weighted test cases, and a second was used for random test cases. Execution of all 7048 test cases takes about four hours.

6 Results Analysis

Table 2 shows a summary of the test results. Sample Reliability is a simple ratio of the number of tests passed to the total for each type (*i.e.*, assumes Bernoulli sampling). Miller reliability is a Bayesian reliability estimate described in [6].

Table 2. *cexec* Certification Test Results

Case Type	Cases	Pass	Fail	Sample Reliability / Variance	Miller Reliability / Variance
Weighted	6048	5837	211	0.965/0.0337	0.965/0.00564
Random	1000	999	1	0.999/0.000999	0.998/0.000332
Combined	7048	6836	212	0.970/0.0292	0.970/0.00488

For weighted test cases, each path through the model is included exactly once in the test. This fact causes low probability test cases to be over-represented in the weighted sample relative to typical usage. The random test case sample, and therefore its reliabilities, reflects more realistic assumptions regarding actual usage.

The total of 212 failures observed among all test cases resulted from six distinct faults. All of the faults involved failure of *cexec* to handle input errors gracefully. The test results also included two significant observations regarding undocumented behavior of *cexec*: (1) Non-existent nodes in the Machine Definition are ignored, and (2) duplicate nodes in the machine definition are processed as many times as they appear.

7 Conclusions

The purpose of the work described above was twofold: (1) demonstration of MBST for computational software and (2) certification of the C3 *cexec* utility.

Using *cexec* as the system under test, this effort demonstrated that the methods and tools developed for MBST could be applied effectively to computational software. The key adaptation required is to model the input selection process instead of the discrete behavior. Using the test case generation and annotation features of the JUMBL tool set, it was possible to generate automatically a large number of test cases whose execution could be readily automated.

The following are key elements of the test automation framework:

- Direct generation of test cases in the form of Python programs,
- Python classes to generate run-time values for input parameters represented by abstractions in the model,
- Python utilities to capture *cexec* output, compare it to expected results, and record pass/fail information.

With this framework, *cexec* can be thoroughly tested in about four hours, and the automation framework is largely reusable.

In terms of *cexec* certification, the test cases generated using the "weighted" algorithm exercised every input combination at the defined level of abstraction. While the weighted test cases proved quite thorough in detecting even obscure software faults, the random test cases provide the primary basis for certification. The random test cases represent a statistically typical usage sample based on the model. The results of these tests indicate that users can expect the *cexec* program to function properly 99.8% of the time.

References

1. "Project C3 Cluster Command and Control," Network and Cluster Computing Group, Computer Science and Mathematics Division, ORNL, http://www.csm.ornl.gov/torc/C3.
2. M.Brim, R.Flanery, A.Geist, B.Luethke, and S.L.Scott, "Cluster Command and Control (C3) Tool Suite," pp. 381-399, *Parallel and Distributed Computing Practices Special Issue: Quality of Parallel and Distributed Programs and Systems*, Ed: P. Kacsuk and G. Kotsis, Vol. 4, No. 4, December 2001, issn 1097-2803, Nova Science Publishers Inc.
3. S. Prowell, "JUMBL: A Tool for Model-Based Statistical Testing", *Proceedings of the 36th Annual Hawaii International Conference on System Sciences (HICSS'03)*, January 2003.
4. S. Prowell, "TML: A description language for Markov chain usage models", *Information and Software Technology*, Vol. 42, No. 12, September 2000, 835--844.
5. JUMBL 4 User's Guide, Software Quality Research Laboratory, Department of Computer Science, University of Tennessee, November 22, 2002.
6. K. Sayre, J. Poore, "A Reliability Estimator for Model Based Testing", *Proceedings of the Thirteenth Symposium on Software Reliability Engineering*, November, 2002.

Accelerating Protein Structure Recovery Using Graphics Processing Units

Bryson R. Payne[1], G. Scott Owen[2], and Irene Weber[2]

[1] Georgia College & State University, Department of ISCM, Milledgeville, GA 31061
bryson.payne@gcsu.edu
[2] Georgia State University, Department of Computer Science, Atlanta, GA 30303
owen@siggraph.org, iweber@gsu.edu

Abstract. Graphics processing units (GPUs) have evolved to become powerful, programmable vector processing units. Furthermore, the maximum processing power of current generation GPUs is technically superior to that of current generation CPUs (central processing units), and that power is doubling approximately every nine months, about twice the rate of Moore's law. This research represents the first successful application of GPU vector processing to an existing scientific computing software package, specifically an application for computing the tertiary (3D) geometric structures of protein molecules from x-ray crystallography data. A framework for applying GPU parallel processing to other computational tasks is developed and discussed, and an example of the benefits of taking advantage of the visualization potential of newer GPUs in scientific computing is presented.

1 Introduction

The graphics processing units of the past four years have increased the capabilities of previous generations by a factor of two every six to nine months. Programmable graphics processing units (GPUs) are commonly included as hardware components in new computer workstations, including those workstations used for scientific computing. The current generation of GPUs is roughly equivalent to or greater than the processing power of current CPUs (central processing units), but that power is rarely used to its full capabilities. Numerous advances have been made recently in applying the GPU to non-graphical parallel matrix processing tasks, such as the Fast Fourier Transform (FFT) [6]. This research seeks to apply the vector processing power of a GPU to automatically computing the 3D geometry of proteins from x-ray crystallography data in an existing software package.

Our primary goal is to apply distributed GPU-CPU computation to automated tertiary structure fitting in protein crystallography, a process that normally takes several hours or even days to execute on a sequential CPU. We proposed to use existing software, ARP/wARP [11], that is already fully-featured and widely used in the industry rather than producing a competing package for several reasons. First, change is difficult to engender in any field, especially where the learning curve for effective utilization of new software is steep. Second, the software packages in existence are already well-suited to the needs of researchers in protein crystallography. Producing a complete product solely for the purpose of demonstrating GPU acceleration would have required hundreds of thousands of lines of programming to match the capabilities of packages already widely accepted and used.

Our secondary goal is to produce a reusable, portable framework for applying GPU computation to scientific computing problems in other fields. As GPUs gain acceptance as parallel vector processing units, a generalized framework for taking advantage of available GPU vector processing power in other scientific computing applications is needed. By documenting and examining the steps taken to add GPU parallel computation to an existing scientific computing package, it is hoped that future researchers will be able to reapply that framework to other existing and new software packages in other fields.

2 Literature Review

The first programmable consumer GPUs became available less than four years ago. The first generation of programmable GPUs was not well-suited to general-purpose computation for several reasons. First, they allowed access to only selected portions of the graphics pipeline and had no easily accessible off-screen rendering capabilities. Second, they had to be programmed in GPU-specific assembly code, with no standardization across manufacturers. Third, their limited accuracy of 8 bits-per-pixel combined with their slower clock speeds and memory accesses, as well as smaller memory sizes, compared to CPUs made them unattractive to general-purpose computing researchers who needed fast 32-bit floating point operations as a minimum point of entry.

By late 2002, however, a C-like programming language for the GPU, named Cg [5], had been developed for cross-platform GPU programming. Cg contained constructs for looping and conditional branching, required for most general-purpose computing, but GPU hardware took another two years to catch up to the capabilities provided for in the Cg language. Only in the NVIDIA GeForce FX6800 series GPU, released in late 2004, was it first possible to take advantage of true conditional branching on the GPU, as well as handle loops or programs that consisted of more than 1024 total instructions per pixel [8]. These advances enabled the research in this paper, and the previous work [9] upon which it is based.

2.1 The Bioinformatics Problem

X-ray crystallography is the most commonly used method for determining the 3D structure of proteins in bioinformatics. The 3D structure of a protein is valuable because it determines many of the protein's properties [4], making structure information useful in drug discovery research as well as many other fields in the biosciences. ARP/wARP [11] is the most popular, and most accurate, software package for automatically determining the structure of proteins from x-ray crystallography data [1]. ARP/wARP is used by researchers in our own Bioinformatics Lab so this program was the focus of our efforts.

ARP/wARP makes use of an iterative structure refinement process by means of a program called Refmac [7]. In run-time analysis, Refmac consumed as much as 83% of the run time of a typical 3D structure computation under ARP/wARP, and the source code for Refmac is freely available. Refmac was, therefore, selected as the test bed for our GPU acceleration research, as well as the subject of our investigation into visualization advantages of current GPUs in scientific computing.

3 Implementation

We first desired a proof of concept to test whether the stated superior performance of GPUs was possible to achieve in situations well-suited to the GPU. Our first implementation steps were to demonstrate that the GPU was at least as fast as the CPU at performing 2D convolutions like averaging filters, edge detection, and the Gaussian smoothing filters before implementing more complex algorithms for scientific computing. Straightforward algorithms for convolutions on the GPU and on the CPU were developed for comparison across a wide variety of matrix and image sizes (32x32 to 4096x4096).

3.1 Building GPU Code for ARP/wARP

In a trial run on a 247-residue protein molecule, ARP/wARP was able to fit 238 of the 247 peptides to a model with 98% connectivity in 18 hours on a 2.8 GHz CPU, using 1200 iterations of refinement. As previously mentioned, it was determined through run-time analysis that the most time-consuming process in ARP/wARP for our purposes is the Refmac refinement step, which uses an open-source molecular refinement program [7]. Therefore, we focused our attention on applying distributed GPU-CPU processing to the Refmac algorithms.

Run-time profiling with the GNU compiler tool `gprof` yielded two subprograms in Refmac that consumed over 30% of the total processing time of the program, `indens` and `prot_shrink`. Due to the fact that `indens` could take up to 25% of the total runtime, while `prot_shrink` accounted for 10% or less of the runtime across a sample set of three proteins of varying sizes, attention was given to `indens` first. We set out to translate `indens` from its native FORTRAN for the CPU to Cg on the GPU.

3.2 Toward a Framework

Integrating GPU-CPU distributed computation with existing bioinformatics software introduces two significant hurdles: commingling C and Cg code for the GPU with the native FORTRAN of Refmac, and re-mapping 1D, 2D, and 3D matrix computation from the CPU to 2D texture calculations on the GPU. Our goal with respect to these challenges is to produce a reusable framework for adding GPU acceleration of matrix computation to existing computational tools across any field of scientific computing. We also hope to demonstrate the advantages of adding 3D visualization, which the GPU handles optimally, to general purpose and scientific computing software packages.

4 Experimental Results

The GPU proved to be much faster at straightforward convolutions like Gaussian smoothing, averaging filters, and edge detection. In the case of a 3x3 convolution filter, the GPU (an NVIDIA GeForce FX6800) was from 10 to 90 times faster at high resolutions (2048x2048) than the CPU (a 2.8 GHz Intel Pentium 4), even after communication between main memory and the GPU was accounted for (see Figure 1).

Most image processing packages do not perform straightforward convolution algorithms, however. On the CPU, it is usually several times faster to use the Fast Fourier Transform (FFT) and perform a matrix multiplication, which is much faster on the CPU than a convolution, then use the inverse Fourier Transform to reacquire the resulting image. The GPU is still a full order of magnitude faster than the CPU even after Fourier optimization, which raises the possibility that Fourier processing for convolutions could become obsolete.

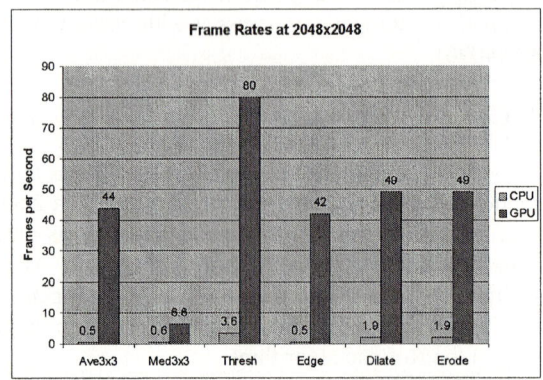

Fig. 1. Comparison of processing speed of various operations on CPU vs. GPU

The GPU is especially well-suited to performing 2D convolutions and morphological masking and filtering operations. Furthermore, programming the GPU version of these algorithms is a straightforward process, allowing the developer to access pixel neighborhoods using a relative indexing paradigm rather than a complicated modular arithmetic scheme for referencing 2D array elements in main memory. Figure 2 shows the GPU code in Cg for performing a 3x3 averaging filter. Figure 3 shows the same algorithm in a straightforward CPU implementation in C.

Notice that the Cg code shows relative texture lookups using vectors like (-1, 1) to denote left one, down one from the current pixel. The C version, on the other hand, is more difficult for several reasons. First, the C version must treat each color component (red, green, and blue, or *RGB*) as separate computations, while the GPU computes all three components simultaneously as three-component vectors of type `float3`. Second, the C method of computing array positions is one-dimensional rather than the 2D relative indexing of Cg. Therefore, non-intuitive modular arithmetic is necessary to resolve the location in main memory of a pixel that is one unit left and one down from the current pixel. More information on digital image processing on GPUs can be found in a related paper [10].

```
float3 a[9];
a[0] = texRECT(image, texCoord + float2(-1, 1));
a[1] = texRECT(image, texCoord + float2(0 , 1));
a[2] = texRECT(image, texCoord + float2(1 , 1));
a[3] = texRECT(image, texCoord + float2(-1, 0));
a[4] = texRECT(image, texCoord + float2(0 , 0));
a[5] = texRECT(image, texCoord + float2(1 , 0));
a[6] = texRECT(image, texCoord + float2(-1,-1));
a[7] = texRECT(image, texCoord + float2(0 ,-1));
a[8] = texRECT(image, texCoord + float2(1 ,-1));
color = (a[0]+a[1]+a[2]+a[3]+a[4]+a[5]+a[6]+a[7]+a[8])/9.0;
```

Fig. 2. A Cg 3x3 averaging filter function

```
int a1,a2,a3,a4,a5,a6,a7,a8,a9;
int wd=TexInfo->bmiHeader.biWidth;
int ht=TexInfo->bmiHeader.biHeight;
int m = wd*ht*3;
for (r=0; r<m;r++)
{
  a1=(r-(wd+1)*3+m)%m;
  a2=(r-(wd)*3+m)%m;
  a3=(r-(wd-1)*3+m)%m;
  a4=(r-3+m)%m;
  a5=r;
  a6=(r+3)%m;
  a7=(r+(wd-1)*3)%m;
  a8=(r+(wd)*3)%m;
  a9=(r+(wd+1)*3)%m;
  TexBits2[r] = (TexBits[a1]+TexBits[a2]+TexBits[a3]+TexBits[a4]+TexBits[a5]+
    TexBits[a6]+ TexBits[a7]+TexBits[a8]+TexBits[a9])/9;
}
```

Fig. 3. A simple subroutine in C for computing the 3x3 averaging filter

4.1 GPU Acceleration of Refmac

The GPU representation of the indens algorithm that operates on 1D arrays suffered from the same type of problem as the CPU operating on 2D arrays, only in reverse. Whereas the CPU version was able to use straightforward addressing to loop through the arrays, the GPU had to use more costly modular arithmetic to determine texture coordinates in 2D to correspond to the 1D positions of the array elements being processed. Because of the inefficiency of this additional computation, a speedup on the order of 90 times, as seen in straightforward convolutions, was not possible. However, due to the fact that the original FORTRAN code was not optimized for CPU caching, either, a significant advantage in speed was still afforded to the GPU version of the code.

In run-time analysis of the two algorithms across three proteins of varying sizes, the GPU version was 1.8 to 2.6 times faster than the GPU at the indens subprogram than the CPU. Figure 4 below shows the comparison, with the GPU version broken down into two components: CPU time (pre-processing and memory transfers from the CPU to the GPU and back again) and GPU time (the actual processing time of the algorithm on the GPU).

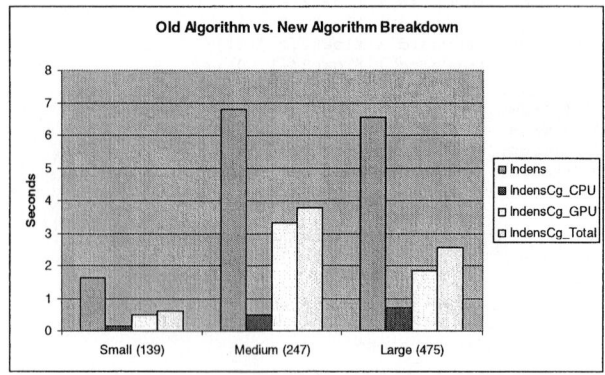

Fig. 4. Comparison of CPU and GPU versions of indens algorithm. The GPU version (indensCg) is broken down into CPU time (texture transfer, readback, etc.) and GPU processing time

While this demonstrated the significant potential for acceleration of a single algorithm, the 1.8 to 2.6 times speedup in the indens subprogram translated to only one to two hours of time savings on a 16 to 48-hour protein structure recovery. Even if the prot_shrink algorithm could yield an optimal 90:1 speedup, due to the fact that it consumes only 8-10% of each Refmac iteration, another hour or two would be the maximum time savings possible in this application. Clearly the acceleration potential of the GPU holds great promise in certain settings, but achieving significant gains in general-purpose applications consisting of hundreds of algorithms and subprograms may be achieved only by rewriting major portions of the code.

4.2 Adding Visualization to Refmac

The area of GPU potential still remaining to be explored was visualization. We had already observed, from poring through log files from ARP/wARP, that the program could resolve a great number of residue positions in a protein rather quickly, then reach a plateau after which few of the remaining residues could be placed no matter how many iterations were provided. However, ARP/wARP has no integrated visualization tool that would enable the researcher to visually inspect a protein for suitability and terminate the structure-building process early.

In the case of the smaller protein sample tested, a protein consisting of 139 amino acids, or residues, 134 of the residues had been correctly placed after only 300 refinement cycles, a plateau that went unsurpassed in the remaining 900 iterations of refinement. In other words, 12.6 hours of the 16.8 hour GPU-accelerated run time for the smaller protein were unnecessary; a 96% complete solution was achieved in only 4.2 hours.

By adding a system call once per Refmac iteration to a simple visualization program, RasMol [2], it was possible to view updated results of the previous iteration every one to four minutes, allowing the researcher to visually inspect the progress of the structure-building process from a different workstation across the room without having to sift through log or other intermediate files. Figure 5 shows some of the simple visualization styles provided by RasMol.

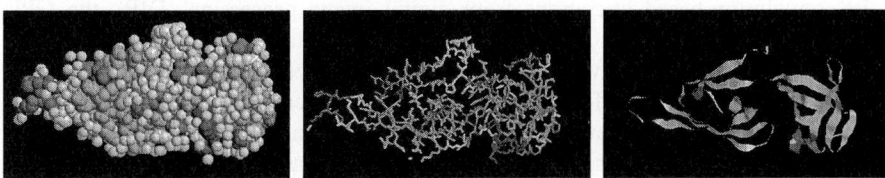

Fig. 5. Three visualization views of a protein with 134 residues in RasMol (space fill, stick, and ribbon views)

By providing this visual information automatically with no additional work on the part of the researcher, it was possible to halt each structure at 90% completion for a total time savings of 45 out of 80 hours of runtime, or 56%. It took only 34 minutes, 8 hours, and 26 hours, respectively, to resolve proteins with 139, 247, and 475 residues to 90% completion, versus a total of 80 hours and 50 minutes to allow them to run to 96% completion. For a researcher skilled at manually placing the remaining residues, the time saved by having automatic visualization built in to the model-building process would have been over 10 times greater than the time saved by accelerating the structure-building process using the GPU.

4.3 Developing a Framework

There were five main steps in the implementation used here that could be replicated for use in any existing or new software development in order to make use of GPU acceleration. First, run-time analysis, with a tool like `gprof`, is used to determine which subroutines consume the majority of processing time. Second, the source code of each high-use subroutine is examined for suitability for GPU implementation (small number of large vectors or matrices in the computation, etc.). Third, the selected algorithms are mapped to a 2D texture-rendering problem on a per-pixel basis, which is the MIMD-processing model of the GPU. Fourth, the source code is modified to include calls to the new GPU versions of the algorithms selected for acceleration. Finally, run-time testing is used to determine the speedup, if any, and adjustments are made, if needed, by revisiting steps two through four as needed.

While these steps are not trivial, they represent a model for future application of GPU vector processing to existing and new scientific and general-purpose computing packages. Possibilities for simplifying the framework in certain domains are given in the following section.

5 Conclusions and Future Work

This research presents the first successful application of GPU parallel vector processing to an existing scientific computing software package. In addition to implementing GPU-based acceleration of the given software, a visualization component was added to the process, resulting in a total time savings on the order of 60% over the CPU-only version with no visualization included.

In convolution-based operations from digital image processing (DIP), the GPU showed a speedup of up to 90:1 over the same implementation on the CPU, and a full order of magnitude of improvement over the FFT-optimized version of the algorithms on the CPU. Because 2D DIP is well-suited to GPU acceleration, a fruitful area for future research would be developing an image processing API at the same level of abstraction as the Brook language [3] for stream processing. An API or multi-step compiler system like Brook designed specifically for DIP operations could be a significant step toward general-purpose utilization of the GPU as a parallel vector processor.

While there exists significant potential for accelerating scientific computing by using GPUs for parallel vector computation, three main obstacles still exist. First, translating algorithms from the original implementations to 2D, GPU versions is not a trivial process. Second, not all vector processing algorithms are good candidates for GPU acceleration, including those with excessive branching and conditional logic. Finally, differences in graphics hardware and drivers still pose a problem in implementing the same algorithm across different platforms. A significant move toward standardization in programming langages for GPUs will be necessary if GPUs are to be used for scientific computing on a consistent basis.

References

1. Badger, J.: An evaluation of automated model-building procedures for protein crystallography, *Acta Crystallographica*. International Union of Crystallography (2003) 823-827.
2. Bernstein, H.J., Sayle, R.: RasMol Molecular Graphics Visualization Tool. Available online at http://openrasmol.com. (2000).
3. Buck, I., Foley, T., Horn, D., Sugerman, J., Fatahalian, K., Houston, M., Hanrahan, P.: Brook for GPUs: Stream Computing on Graphics Hardware, *Proceedings of ACM SIGGRAPH 2004*. ACM Press, (2004) 777-786.
4. Lunney, E.A.: Computing in Drug Discovery: The Design Phase, *Computing in Science & Engineering 3(5)*. IEEE, (2001) 105-108.
5. Mark, W.R., Glanville, R.S., Akeley, K., Kilgard, M.J.: Cg: a system for programming graphics hardware in a C-like language, *ACM Transactions on Graphics 22(3)*, pages 896-907. ACM Press, July 2003.
6. Moreland, K., Angel, E.: The FFT on a GPU, *Proceedings of the ACM SIGGRAPH/EUROGRAPHICS conference on Graphics hardware*, Eurographics Association (2003) 112-119.
7. Murshudov, G.N., Vagin, A.A., Dodson, E.J.: Refinement of Macromolecular Structures by the Maximum-Likelihood Method. *Acta Crystallographica*. International Union of Crystallography, (1997) 240-255.

8. NVIDIA Corporation: GeForce 6800 Product Overview. Available online at http://www.nvidia.com/object/IO_12464.html. (2004).
9. Payne, B.R.:Accelerating Scientific Computation in Bioinformatics by Using Graphics Processing Units as Parallel Vector Processors. (Doctoral dissertation, Georgia State University, 2004). *Dissertation Abstracts International* (UMI. No. pending)
10. Payne, B.R., Owen, G.S., Belkasim, S.O.: Digital Image Processing on GPUs. Submitted to the Fourth International Workshop on Computer Graphics and Geometric Modeling, CGGM'2005. Emory University, Atlanta, USA, May 22-25, 2005.
11. Perrakis A., Morris R., Lamzin V.S.: Automated protein model building combined with iterative structure refinement. *Nature Struct. Biol.*, (1999) 458-463.

A Parallel Software Development for Watershed Simulations

Jing-Ru C. Cheng[1], Robert M. Hunter[1], Hwai-Ping Cheng[2], and David R. Richards[3]

[1] Major Shared Resource Center,
Information Technology Laboratory
{ruth.c.cheng, robert.m.hunter}@erdc.usace.army.mil
[2] Coastal and Hydraulics Laboratory
hwai-ping.cheng@erdc.usace.army.mil
[3] Information Technology Laboratory,
U.S. Army Engineer Research and Development Center,
Vicksburg, MS 39180-6199, USA
david.r.richards@erdc.usace.army.mil

Abstract. A watershed software application is designed to model a coupled system of multiple physics on multiple domains. Tremendous computational resources are required to integrate the system equations on large spatial domains with multiple temporal scales among them. Supported by the Department of Defense Common High Performance Computing Software Initiative, the parallel WASH123D software development aims to efficiently simulate one aspect (i.e., soil and land) of the battlespace environment. Currently, the coupled two-dimensional overland and three-dimensional subsurface flows have been completed. Different numerical approaches are implemented to solve different components of the coupled system. The parallelization of such a complex system is developed on an IT-based approach—modular, hierarchical model construction, portable, scalable, and embedded parallel computational tools development and integration. Experimental results are presented to demonstrate the successful implementation of the parallel algorithms. Detailed profiling is also provided to show the imposed light-weight communication overhead.

1 Introduction

Watershed models simulate major hydrological processes on multiple spatial domains over varied temporal scales with interactions among them spanning from uncoupled to strongly coupled. Different numerical approaches for such a coupled nonlinear hydrologic process have been proposed to be efficient and affordable. Penn State Integrated Hydrologic Model (PIHM) [1] integrates hydrological models using the so-called "semi-discrete" method, which reduces the governing partial differential equations (PDE) to ordinary differential equations (ODE), while Yeh et al. [2] presented a first-principle, physics-based watershed model. According to Yeh's review [3], HSPF (Hydrologic Simulation Program—FORTRAN) and WASH123D are the only models that include complete media systems, i.e., stream/rivers, overland regimes, and subsurface media, and encompass the complete suite of fluid flows and thermal, salinity, sediment, and chemical transport processes. The difference between them is that HSPF employs the parametric approach, while WASH123D is based on a first-principle, physics-based approach. The drawback of the parametric approach is explained in Sect. 2.3.

The U.S. Army Corps of Engineers plays a critical role in the Nation's watershed management. In addition, supported by the Department of Defense (DoD) Common High Performance Computing Software Initiative (CHSSI), the parallel WASH123D software development aims to efficiently simulate one aspect of the battlespace environment, which includes space, weather, ocean, and soil, towards development of a complete battlespace environment. The current accomplishment, including the development of the numerical approaches, software designs, and parallel algorithms, is presented in this paper.

2 Numerical Algorithms in WASH123D

The governing equations of two-dimensional (2-D) overland flow and three-dimensional (3-D) subsurface flow, and the numerical approaches solving the system equations are described in the following subsections. The numerical methods that this paper presents are those used for demonstration in Sect. 4.

2.1 2-D Overland Flow

The 2-D overland flow is computed by solving the depth-averaged diffusive wave equation with the semi-Lagrangian finite element method (FEM). The governing equation can be written as

$$\frac{\partial h}{\partial t} + \nabla \cdot \mathbf{q} = S + R - E - I \quad \text{or} \quad \frac{\partial h}{\partial t} + \nabla \cdot [\mathbf{V}h] = S + R - E - I, \quad (1)$$

where h = overland water depth[L], t = time[t], $\mathbf{q}$ = overland flux[L^3/t/L], S = man-induced source[L^3/t/L^2], R = rainfall rate[L/t], E = evapotranspiration rate[L/t], I = infiltration rate [L/t], and $\mathbf{V}$ = overland flow velocity[L/t]. With the semi-Lagrangian FEM, (1) can be written in Lagrangian form, then integration along its characteristic line yields

$$\left(1 + \frac{\Delta\tau}{2} K_i^{(n+1)}\right) h_i^{(n+1)} = \left(1 - \frac{\Delta\tau}{2} K_i^*\right) h_i^* + \frac{\Delta\tau}{2} \left(S_i^{(n+1)} + S_i^*\right) +$$
$$\frac{\Delta\tau}{2} \left(R_i^{(n+1)} + R_i^*\right) - \frac{\Delta\tau}{2} \left(E_i^{(n+1)} + E_i^*\right) - \frac{\Delta\tau}{2} \left(I_i^{(n+1)} + I_i^*\right), \quad (2)$$

where $K = \nabla \cdot \mathbf{V}$, $\Delta\tau$ = the tracking time[t], which equals Δt (the time interval) when the backward tracking is carried out all the way to the root of the characteristic line, but is less than Δt when the backward tracking hits the boundary before Δt is completely consumed; $K_i^{(n+1)}$, $h_i^{(n+1)}$, $S_i^{(n+1)}$, $R_i^{(n+1)}$, $E_i^{(n+1)}$, and $I_i^{(n+1)}$ are the values of K, h, S, R, E, and I, respectively, at $\mathbf{x}_i$ at new time $t = (n+1)\Delta t$; and K_i^*, h_i^*, S_i^*, R_i^*, E_i^*, and I_i^* are the values at $\mathbf{x}_i^*$, i.e., where the backward tracking ends. Equation (2) is used to compute the water depth, h, at all nodes except for the upstream boundary nodes, where water depth is determined by applying adequate boundary conditions. Since the flow velocity can be computed and represented as a function of water depth, (2) is used in a nonlinear iteration loop until a convergent solution is obtained.

2.2 3-D Subsurface Flow

The governing equation of subsurface flow through saturated-unsaturated porous media can be derived based on the conservation law of water mass and can be written as follows.

$$\frac{d\theta}{dh}\frac{\partial h}{\partial t} = \nabla \cdot [\mathbf{K} \cdot (\nabla h + \nabla z)] + q ,\qquad(3)$$

where θ = moisture content$[L^3/L^3]$, h = pressure head [L], $\mathbf{K}$ = the hydraulic conductivity tensor$[L/t]$, z = the potential head[L], and q = man-induced source $[L^3/L^3/t]$. Equation (3), the well-known Richards equation, is solved with the Galerkin FEM that can be found elsewhere [2].

2.3 2- and 3-D Coupling

The fluxes between surface and subsurface media are computed by imposing continuity of fluxes and state variables (e.g., overland water depth and subsurface pressure head). If the state variables exhibit discontinuity, then a linkage term is used to simulate the fluxes. Considering the interaction between the 2-D overland and 3-D subsurface flows, the pressures in the overland flow (if present) and in the subsurface media must be continuous across the interface. Thus, the interaction must be simulated by imposing continuity of pressures and fluxes as

$$h^o = h^s \quad \text{and} \quad Q^o = Q^s \implies I = \mathbf{n} \cdot \mathbf{K} \cdot (\nabla h^s + \nabla z) ,\qquad(4)$$

where h^o is the water depth[L] in the overland if it is present, h^s is the pressure head[L] in the subsurface, Q^o is the flux $[L^3/L^2/t]$ from the overland to the interface, Q^s is the flux from the interface to the subsurface media $[L^3/L^2/t]$, and $\mathbf{n}$ is an outward unit vector of the ground surface. The use of a linkage term such as $Q^o = Q^s = k(h^o - h^s)$, while convenient, is not appropriate because it introduces a nonphysical parameter, k. The calibration of k to match simulation with field data renders the coupled model *ad hoc* even though the overland and subsurface models are each individually physics-based.

Algorithm 2.1 depicts the 2-D/3-D coupling algorithm used in WASH123D. Ideally, overland flow and subsurface flow should be strongly coupled within each time-step. However, this would introduce unaffordable computational characteristics because small time intervals may be required for solving nonlinear 2-D overland flow when a high-resolution mesh is employed. To make computation affordable, in WASH123D each 3-D flow-time interval may contain more than one 2-D flow-time interval. The fluxes through the surface-subsurface interface are updated using (4) for 2-D/3-D in each 3-D coupling/nonlinear iteration.

Algorithm 2.1 The 2-D/3-D Coupling Algorithm in WASH123D

Foreach 3D flow time step ($\triangle t_{3DF}$) do
    Foreach 3D coupling/nonlinear iteration do
        Foreach 2D flow time step ($\triangle t_{2DF}$) do
            Incorporate infiltration/seepage for 2D/3D coupling
            Foreach 2D coupling/nonlinear iteration do
                Solve linearized 2D flow equation
            Endfor
        Endfor
        Incorporate infiltration/seepage for 2D/3D coupling
        Solve linearized 3D flow equation
    Endfor
Endfor

3 Parallel Algorithms and Implementation

In WASH123D, different numerical approaches are implemented to solve different components of the coupled system. The parallelization of such a complex model starts with the data structure design and then tackles the programming paradigm. The original serial computational kernel is included in the parallel software without any changes to shorten the development time, because there is no parallelization involved. Therefore, the data structure design becomes very important if the goals of object orientation, parallelization, software integration, and language interoperability are to be reached.

3.1 Data Structure Design

To account for problem domains that may include 1-D river/stream network, 2-D overland regime, and 3-D subsurface media, three WashMesh objects are constructed in the object WashDomain, which embraces the computational domain as sketched in Fig. 1. These three objects describe the three subdomains, on which a set of governing equations is derived to mathematically describe the behavior of the component within the entire domain. Note that the WashDomain also includes a coupling object named WashCouple. Moreover, the WashGlobal object describes the common phenomena, and the WashProcinfo object sets up the parallel environment context. Each subdomain (i.e., WashMesh) is partitioned, based on users' partitioning criteria, to processors by DBuilder [4]. Hence, each WashMesh object may include vtxDomain and elementDomain, which are created and managed by DBuilder, to maintain coherent data structures among processors via ghost vertices/elements on a given mesh. The WashCouple object may include the coupler for (1-D, 2-D), (1-D, 3-D), and/or (2-D, 3-D) interactions. The coupler encapsulates all the implementation of a Message Passing Interface (MPI) scheme for communication/synchronization between different

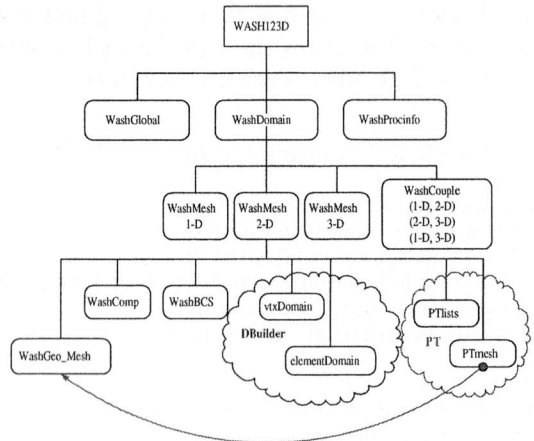

Fig. 1. Data structure design of the parallel WASH123D

`WashMesh` objects. This approach can partition each subdomain (i.e., `WashMesh`) independently. Therefore, this software tool can be reused to integrate two or more applications with different physics on multidomains.

3.2 Parallel Particle Tracking Software Integration

To solve the 2-D overland flow problem, with the Picard method solving the nonlinearity of the 2-D overland flow (2), the linearized equation can be solved by using particle tracking to compute the total time-derivative term in (1) and by manipulating the integration along the tracking path for the source/sink terms. Naturally, this application is perfectly suited for parallel implementation because the dependent variable, either water depth or water stage, can be obtained by solving the linearized equation independently. For such a purpose, the parallel particle tracking (PT) software [5] is facilitated with a new pathline computation kernel to accurately track particles under unsteady flow fields [6]. The design goal of the PT software development is to interface with different parallel or mesh programming environments. This goal is achieved through a software architecture specifying a lightweight functional interface [5].

3.3 DBuilder Software Toolkit Development

DBuilder [4] provides support of domain partitioning, parallel data management, coupling coordination, and parallel solver interface. This toolkit embeds all the MPI function calls required by the application codes and provides a set of user-interface functions to retrieve/modify parallel data. The basic concept of the development is based on a hierarchical modular design. To meet the requirements of different numerical methods, DBuilder can build a vertex domain with a distributed number of vertices, an element domain with a distributed number of elements, and a boundary element domain comprised of boundary elements in the element domain. DBuilder provides a default rule and a callback approach to users for the coordination between vertices and elements required by finite element applications.

The synchronization mechanism in DBuilder includes two steps. First, the local data that are shared on other processors are packed into a contiguous memory section, i.e., data are packed sequentially based on the receiving processor's rank. Within this set the data are then ordered based on global index values. Once the data have been packed, a single call to MPI_Alltoallv is made to update the ghost data on all processes. The computational cost of the pack routine for a given process is $O(N)$, where N is the number of local vertices. The low computational cost is achieved by construction and storage of a map of local data indices to the packed array indices. The communication cost of MPI_Alltoallv has a worst cost of $O(N^2)$, where N is the number of processors.

Multiphysics applications on multidomains have become a large focus. This requires executing multidomain integration of two or more applications. The spatial relationship between computational domains can be adjacent, partially/fully overlapped, or distinct. DBuilder allows for the building of a coupler object to avoid the dependency between meshes when partitioning. This functionality may extend the reuse of this software toolkit to coupling different areas of applications, e.g., ocean and atmosphere coupled systems. Details can be found in [4].

4 Experimental Results

Fort Benning military base is located in the Upatoi River watershed, west-central Georgia south of the city of Columbus, GA, and east of Phenix City, AL. Fort Benning is a major training area for the U.S. Army Infantry. The 2-D overland domain, which covers about 450 square miles, is discretized with 103,619 vertices and 206,167 elements. The underlying 3-D domain contains 3,092,505 elements and 1,657,904 vertices.

Figure 2 plots the wall clock time vs. number of processors for the coupled 2-D overland flow simulation and 3-D subsurface flow simulation on the Fort Benning site on three different architectures. At the Engineer Research and Development Center Major Shared Resource Center (ERDC MSRC), the Compaq AlphaServer SC45 machine is configured with 128 nodes. Each node has four 1-GHz processors connected by a 64-port, single-rail Quadrics high-speed interconnect switch. At the Naval Oceanographic Office MSRC, the IBM P4 machine has 168 nodes. Each node has eight 1.3-GHz CPUs and 8 GBs of memory. Nodes communicate through the IBM's Colony II switch. At the Army Research Laboratory (ARL) MSRC, the linux cluster, Evolocity II, has 256 processors. Each has a 3.06-GHz CPU using Myrinet interconnect for communication. Figure 3 shows the communication overhead taken up in the total wall clock time on these three different machine architectures.

From these figures, one can observe that ARL's linux cluster outperforms the others except when the simulation runs on 32 processors. For the 64-processor simulation, the parallel efficiency is around 74 percent on the Evolocity II linux

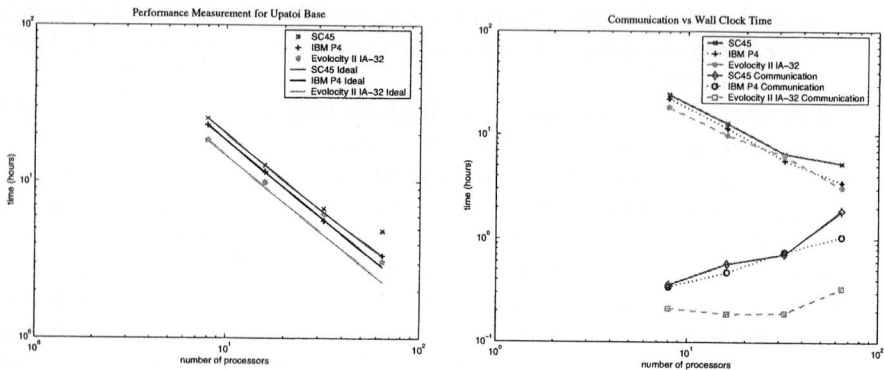

Fig. 2. Performance measurement on three machines

Fig. 3. Communication overhead compared with the total wall clock time

Table 1. Scalability on three machines

	Compaq SC45			IBM P4			Evolocity II Linux cluster		
nproc	Time (sec)	Speedup	PE	Time (sec)	Speedup	PE	Time (sec)	Speedup	PE
8	90978.28	—	—	82379.64	—	—	65983.72	—	—
16	45677.70	15.934	0.995	41497.16	15.882	0.993	35597.69	14.829	0.927
32	24316.91	29.931	0.939	20326.74	32.422	1.013	22360.68	23.608	0.738
64	17590.44	41.376	0.691	12224.27	53.912	0.842	11143.44	47.370	0.740

cluster, listed in Table 1, around 84 percent on the IBM P4 (see Table 1), and less than 70 percent on the SC45 (see Table 1). The main cause for such a difference is that the IBM P4 has better scalability with communication (see Fig. 3) and the Evolocity II has better CPU speed.

Figure 4 plots the high-water memory marks to show the memory scalability. While the parallel implementation imposes memory overhead, the memory requirements per processor are greatly reduced. The reduction of memory can then reduce the cache miss or page swapping, which can improve the performance significantly.

5 Summary and Future Plans

The software tool DBuilder has successfully embedded MPI routines so that application developers do not need to know the MPI library and parallel algorithms. The result shows that the implementation in DBuilder has successfully partitioned the domain, balanced the workload, and scaled the memory usage among processors. The following tasks for the software development have been completed: implementation of dynamic memory allocation, DBuilder functionality enrichment, parallel PT software integration, and the parallel performance

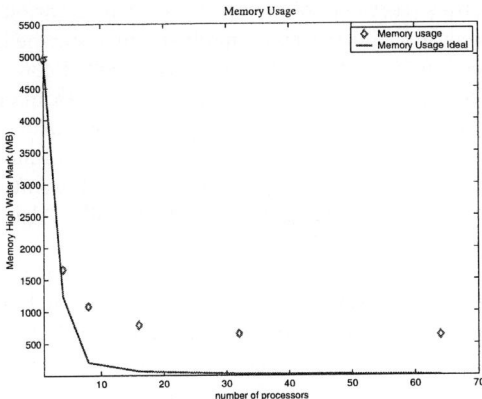

Fig. 4. Memory usage when run on different numbers of processors

evaluation. From the experimental demonstration, verification, fixed problem scalability, and memory scalability have been investigated. Further detailed profiling such as communication and memory overhead of each component will be performed. The WASH123D team has developed the 1-D module and is working on the 1-, 2-, and 3-D coupling module as well. Large-scale field problems are currently prepared for a large number of processors.

Acknowledgments

This work was supported in part by an allocation of computer time from the Department of Defense High Performance Computing Modernization Program at the U.S. Army Engineer Research and Development Center Major Shared Resource Center, Information Technology Laboratory, Vicksburg, Mississippi. The authors would like to thank their colleague Jerry Lin at the ERDC Coastal and Hydraulics Laboratory for preparing the data set for performance evaluation.

References

[1] Duffy, C.J.: Semi-discrete dynamical model for mountain-front recharge and water balance estimation, Rio Grande of southern Colorado and New Mexico. In: Groundwater Recharge in a Desert Environment: The Southwestern United States. Water Science and Applications Series, American Geophysical Union, D.C. (2004) 255–271

[2] Yeh, G.T., Cheng, H.P., Huang, G., Zhang, F., Lin, H.C., Edris, E., Richards, D.: A numerical model of flow, heat transfer, and salinity, sediment, and water quality transport in WAterSHed systems of 1-D stream-river network, 2-D overland regime, and 3-D subsurface media (WASH123D: Version 2.0). Technical Report Draft, Engineer Research and Development Center, U.S. Army Corps of Engineers, MS (2003)

[3] Yeh, G.T.: A rigorous treatment of interactions between various media in first principle, physics-based flow and transport modeling in watersheds. In: EOS Transac. Volume 83., American Geophysical Union, D.C. (2002) F436
[4] Hunter, R.M., Cheng, J.R.C.: DBuilder: A parallel data management toolkit for scientific applications. In: DoD High Performance Computing Modernization Program, 2004 Users Group Conference, Williamsburg, VA (June 7-11, 2004)
[5] Cheng, J.R.C., Jones, M.T., Plassmann, P.E.: A portable software architecture for mesh independent particle tracking algorithms. Parallel Algorithms and Applications (June-September 2004) 145–161
[6] Cheng, J.R.C., Plassmann, P.E.: Development of parallel particle tracking algorithms in large-scale unsteady flows. In: SIAM Conf. on Parallel Processing for Scientific Computing, San Francisco, CA (Feb. 25-27, 2004)

Design and Implementation of Services for a Synthetic Seismogram Calculation Tool on the Grid

Choonhan Youn, Tim Kaiser, Cindy Santini, and Dogan Seber

San Diego Supercomputer Center, University of California at San Diego,
9500 Gilman Drive,
La Jolla, CA 92093-0505
{cyoun, tkaiser, csantini, seber}@sdsc.edu

Abstract. We have built user environments that simplify and provide interactive access to data, models, and compute resources as well as integrate various distributed computational services to study earthquake waveforms utilizing 3D models and compute resources within the Geosciences Network (GEONgrid) and national computational grids, such as TeraGrid. These data and computing services are implemented using a Web services approach and are incorporated in a service-based portal architecture. We then illustrate how these data, models, and services can be used to build distributed, interactive scientific applications.

1 Introduction

Application-specific web-based scientific portals provide user-centric views of computational grid technologies based on global, large-scale distributed computing for scientific applications [1], [2]. Building on a foundation of distributed service components, we can typically construct a sophisticated, domain-specific portal system. These services may be built on top of Grid technologies such as Globus, Legion, and Condor, and web technologies and standards that are developed for Internet computing and are used to provide browser-based access to High Performance Computing (HPC) systems. Numerous such portals have been developed, with varying degrees of specializations. Some examples include NASA's Information Power Grid [3], NPACI's Hotpage [4] and application-specific Problem Solving Environments, PNNL's Extensible Computational Chemistry Environment (Ecce) system [5], UNICORE [6], Gateway System [7], and NMI's Open Grid Computing Environment (OGCE) portal [8].

As part of GEON's (Geoscience Network) computational and open grid computing environment (GEON project [9] funded by NSF), we have started developing a domain-specific application portal (SYNSEIS – SYNthetic SEISmogram generation tool) to help seismologists as well as any other researchers to calculate realistic 3D regional seismic waveforms using a well-tested, finite difference code, e3d. E3d was developed by the Lawrence Livermore National Laboratory [10]. This system is also designed to be used in day-to-day activities of researchers, especially EarthScope scientists who will be accessing data from hundreds of stations everyday and need to process the data in a timely fashion.

As the emergence of rich clients and rich Internet applications become important, we develop a user-friendly interface to SYNSEIS using Macromedia Flash MX [11]. Within our SYNSEIS user interface, we provide several components: an interactive mapping tool, event/station/waveform extraction tools that allow users to seamlessly access IRIS Data Management Center (DMC)'s remote archives [12], simulation of synthetic seismograms on the TeraGrid machines, and monitoring the job status.

This paper describes our design and initial efforts for building SYNSEIS application tool as one of the main GEON's science applications on the GEONgrid portal. The GEONgrid portal provides the unifying hosting environment for managing geoscience data and applications via the GEONgrid [13]. It is built out of portlet containers that access distributed data and computational resources via Web services, Grid services and other Internet computing technologies. Using GEON grid environments and national-scale TeraGrid supercomputer centers [14] for high performance computing, the SYNSEIS application is developed as a portlet object that can provide the hosting environments interacting with the codes and the data retrieval systems. It is also built using a service-based architecture for reusability and interoperability of each constituent service components which are exposed as Web services. This allows multiple use-case scenarios. Each service component can be reused by other researchers and portal developers.

2 SYNSEIS Architecture

The SYNSEIS architecture is based on a Web Services model which has received a great deal of attention from both the commercial and the Grid computing communities, the latter through the Globus group's proposed Open Grid Services Architecture (OGSA) [15], illustrated in Figure 1. Web Services [16] essentially do not present new concepts for distributed computing but rather implement important simplifying, standards-compliant ways for entities to find and invoke the appropriate remote service. These have important implications to the field of computational or science portals.

The user interacts with the SYNSEIS application tool through a web browser, which accesses a central user interface server that contains the collection of Web service clients in Web service connectors of Macromedia Flash MX [17]. The central goal of the SYNSEIS user interface (Figure 2) is to allow users to analyze seismic data hosted at the IRIS DMC which stored the recorded seismogram data, network and station data, and the earthquake event data [12] and calculate realistic 3D seismograms using the computational resources via the grid system. Using the data archival service, the user selects the bounding box area on the map and an event-station pair within the region, and then plots the waveform data which is accessed via IRIS DMC for the given event-station pairs. In order to run the application, e3d, the user selects the computational resource and creates the input data as an XML object using the input form provided by an interactive user interface for submitting the job. The input format for e3d application may be expressed in XML document which provides a useful data exchange language [18]. The XML object may be used to encode and provide structure to code input files and output data. We use XML descriptions as an independent format that may be used to generate input files for the codes in the proper legacy format.

Design and Implementation of Services for a Synthetic Seismogram Calculation Tool

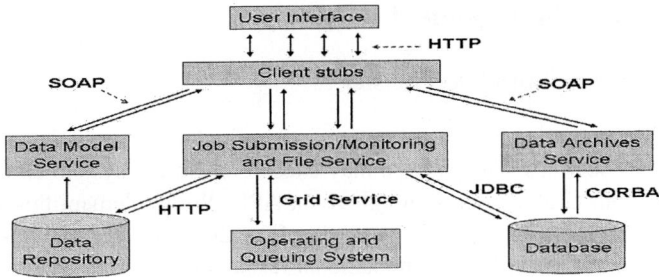

Fig. 1. SYNSEIS architecture with Web service invocations of remote services. Arrows indicate remote invocations with indicated protocols

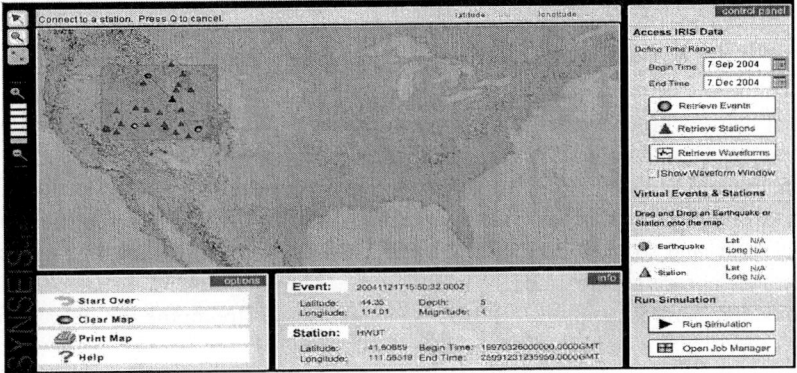

Fig. 2. SYNSEIS's interactive graphic user interface

A SYNSEIS user interface server maintains several client stubs that provide local interfaces to various remote services. These remote services are described in the WSDL format [19] and are invoked via SOAP [20] over HTTP. These services are invoked on various service-providing hosts, which may in turn interact with local or remote databases, data repositories, or remote queuing and system environments via grid services as shown in Figure 1.

The merit of the service-based architecture which we pursue in SYNSEIS tool allows the portal user to browse the geologic data model's service on a particular host, specify model parameters, and create the input data file and transfer it to the job submission service on a particular host. This service host maintains remote application executables, and the application may be run with the appropriate input data on a queuing system which may include a cluster or parallel computer. Following with the completion of the job execution, the user may transfer the simulation outputs back to another database or file system, or download the files to his/her desktop. Based on these service components that are described in Section 3, portal developers can also compose them in their portal environments easily.

3 Job and File Management

3.1 Job Submission with File Transfer

A job submission service is a basic component required to integrate the computational grid with Web Services. This service may execute operating system calls directly or may interact with Grid services through client APIs. We implement this service with file transfer on top of Grid technologies using Java CoG Kit [21] which is the Globus interface to run jobs on remote computational resources in a secure and authenticated manner. SOAP requests having the user input (XML string) and a compute host name as input parameters can be sent out to the targeting SOAP server which manages and interacts with computational grids.

Before submitting the job to the targeted host, this user XML document is stored into the data repository and is reused by the user later (for example, to modify some input parameters) for resubmitting the job, or getting the output file names.

The job RSL (Resource Specification Language) [22] script which contains the application metadata is needed to run the Globus job through the gatekeeper. It is created and integrated with Application Information Web Service developed by the Community Grids Lab, Indiana University, Bloomington (More detailed description is available from [23]) as the information repository describing e3d application for the seismic simulation and other system commands for dealing with the job files such as ls, rm, qstat, and so on. And the GASS (Global Access to Secondary Storage) server is created to receive the standard output/error. The GASS URL is also set as the stdout/stderr parameter in this job RSL script. This will stream the job output/error to GASS server that is redirected to the Job output listener.

Before submitting the job RSL script to the gatekeeper (GRAM server), the user input XML file that the application needs can be transferred to one of the hosts of computational grids (currently, we are using national computational grids, TeraGrid) through the GridFTP. After submitting the job RSL script, the output results describing the running job are caught from the GASS server. The information is parsed to get the job information to create the job table that specifies User ID, Job ID, Job Host Name, Job Error Handling message, user XML input's file name, Job directory name, Job submission time, and Job status. The job table is also stored into the job database. Finally, this job information is returned back to the client that invokes this job submission service as WSDL complex type.

3.2 Job Monitoring with File Transfer

A job monitoring service is also very basic service component in most of computational portals. Users are easily able to check the current job status from the queuing system about jobs on the high performance computing resources through the Web service client APIs. The job monitoring service is implemented using Java CoG Kit, like submitting the job. It is used by clients through a SOAP access to monitor the execution of a job running in a remote queuing system. Basically, a job submission service returns a unique job identifier (Job ID from the job information) that can be used for enquiry about the job status. If the job is submitted to a batch scheduler it is in the pending state while sitting in the queue waiting to be executed. The job may become suspended due to pre-emption mechanisms. In case of normal completion the job status is "Done", otherwise the job is "Failed". In our case, once the user's job is done, generated outputs are transferred to the data repository for allowing the user to access using the file transfer service.

The input parameters to this service consist of the user's job ID, the job output directory name, and the host name of the computational resource, such as, "SDSC", "NCSA". The service implementation is designed as a persistent data factory so that retrieval for a particular job can be performed easily without invoking the grid service depending on the job status. First of all, the user-saved job information is retrieved from the job database. If the job status column indicates "Done", the data are wrapped into URLs and returned to the client immediately. If the job status is not "Done", then a job RSL script, which specifies the execution of the job status, will be created to run the Globus job, integrating with Application Information Web Service [23]. For redirecting the standard output/error for the job running from the remote machine, the GASS server is started. The GASS URL is added into the generated job RSL script as the stdout/stderr parameter. Then the generated job script is submitted to the targeting gatekeeper (GRAM server).

The output results about the job status of the scheduler are transmitted to the GASS server that is running locally, and are then parsed for getting the job status. If the job is not completed yet, the job status of the particular submitted job is updated and saved into the job database. Otherwise, the user job's output files are transferred from the remote host, and the job status is updated and stored in the job database. For doing that, there is one more additional step. The job RSL script is generated for checking the list of the file names in the remote job directory, and submitted to the destination host (GRAM server) again. The output results which contain the list of the output file names are parsed to get the used XML file name and the generated output file names. This XML file is retrieved from the job data repository, and then unmarshalled into e3d's input XML schema [18] to extract the pieces of output file names. Based on that, the prefixes of the output file names are created for obtaining the targeted output files. This generated output list is used for transferring the output files to the job data repository via the GridFTP service. On the completion of file transferring, the URLs of the output files are returned to the clients for downloading.

4 Data Management

4.1 Data Model Service

SYNSEIS application allows users to select a region of interest on an interactive United States map and extract related geological data model which contains Moho and sediment thickness data via a Web service invocation which is remotely accessed through SOAP messages over HTTP. We have defined a WSDL interface for setting the boundary with geographic coordinates (longitude, latitude). This service may load the whole US Moho or sediment data and then calculate the location using the bounding box data provided by the input parameters. The selected region's data for the data model is returned to the users to construct a 3D geologic model. We implement this service in Java. Typically we use it to take a subset of the whole data, keeping the same data format. The different data models provided by the different institutions are also available to this service. In the SYNSEIS user interface, each data model can be viewed as a map after receiving the data model from the Web service.

4.2 Data Archival Service

IRIS (Incorporated Research Institutions for Seismology) is a university research consortium dedicated to exploring the Earth's interior through the collection and distribution of seismological data. IRIS DMC receives earthquake and seismic data from a variety of Data Collection Centers and is responsible for the long term archive and distribution of all IRIS generated data [12]. For accessing those data centers, the Data Handling Interface (DHI) servers at the FISSURES Project provide a framework for seismology software and data transmission based on CORBA [24] and Java Platforms that are developed in an open and cooperative fashion [25]. There are three servers at the DHI servers: the Network Server for providing information about networks, stations, sites, channels, and responses, the Event Server for accessing the event metadata including event origins, magnitudes, predicted arrival times and channels, and the Seismogram Server for providing for retrieval of seismograms consisting of the near real-time data and the archived data.

Based on the DHI clients that are implemented using Java CORBA IDL (Interface Definition Language), implementation, and utility tools provided by the FISSURES project, we implement the data archival Web service for easily integrating and reusing any applications. This service may be built on top of the DHI clients, which access the CORBA servers. We refer to the actual service interface for manipulating the data acquisition as the data retrieval manager. This is defined in WSDL and exposes the following methods for retrieving the archived data from the IRIS DMC:

- A user can retrieve one or more station data which contain the network code, the stations begin time and end time, the station channel data (BH*) from the network server within a given bounding box area (min, max point of the longitude and latitude).
- A user can search for one or more event data that have the origin time, magnitude, depth, longitude, and latitude from the event server within the same bounding box area in a region time period.
- A user can retrieve one or more seismogram waveform data from the seismogram server using a station and event pair and a time range. Depending on the event date, the seismogram data can be retrieved from the near real-time server, or from the archived server.

The above method calls are used for internally manipulating IRIS data. Externally, the seismogram data instance is represented as a set of URL wrapper classes. We store these instances from the seismogram server persistently on the file system.

5 Conclusion and Future Directions

We presented our initial architecture of an application-specific tool called SYNSEIS which was built using Web services, relevant job services and data management implementations. The architecture is composed of three major service components. First, we provide a job submission Web Service to run the relevant applications in TeraGrid or GEONgrid using Grid technologies, especially Globus, including a file transfer. After submitting the job that contains users' input parameters, the job information is saved into the job database. This enables us to extract out the job results and check out the job status. Next, we provide a means for obtaining a job status. This job monitoring Web Service is done through the execution of the command of queuing system for checking the job status using the grid services, providing the file transfer. Once the job is finished, the job outputs are transferred to the data repository which is running on a SOAP server. Finally, we must wrap the crustal data models in useful services that can be plugged into our SYNSEIS application tool. These modules must implement a set of specified interfaces for manipulating the bounding box on the US map. We provide data archival service interfaces for retrieving real seismogram waveform data from IRIS DMC as well.

There are several possible future revisions in our architecture. The first is to provide a secure role-based authorization control to fully integrate into the GEONgrid portal. In order to allow users to submit their jobs and check the job status, SYNSEIS SWF (the file format used by Macromedia Flash) code needs to use the user proxy credential generated by the user after logging on to the portal. The second possible improvement we plan to make is in the schema of the Application Information Web Service for integrating with the job script generator. The Application Descriptor schema contains the HostBinding element that indicates the Host Descriptor, which describes the hosting environment, especially the location, type, parameters of the queuing system. In our application tool, more system commands are required for doing the job and data handling. So, for a more general way, we will extend this schema to put the system environments. The third one is to add more user capabilities in the user interface for complex earthquake simulations. Currently, we provide simply the event-station pair and point source case. We will take into consideration providing event-multiple stations situations, a line source implementation, and multiple seismogram plots. The fourth possible modification is in the nature of the implementation for the effective Web service design. As shown in Figure 1, we have designed this application tool to compose of pure Web service components with Grids technologies. The job management Web services have been implemented on top of the system level Grid services in the back-end. And the data archive service has also been built on top of CORBA services in the back-end. However, we must also explore alternative stateful Web services mechanism that will implement applications that manage the state. This mechanism would remove some of the duplicate control capabilities and communications from the SOAP server. The definition of conventions for managing state may be handled through standard ways such as WSRF (Web Service Resource Framework) [26] or OGSI (Open Grid Services Infrastructure) [27] so that applications discover, bind, and communicate with stateful resources in standard and interoperable ways.

References

1. Foster, I., Kesselman, C. (ed.): The Grid: Blueprint for a New Computing Infrastructure. Morgan Kauffmann, 1999
2. Berman, F., Fox, G., and Hey, T.: Grid Computing: Making the Global Infrastructure a Reality. Wiley, 2003
3. W.E. Johnston, D. Gannon, B. Nitzberg. Grids as Production Computing Environments: The Engineering Aspects of NASA's Information Power Grid. Proceedings 8th IEEE International Symposium on High Performance Distributed Computing, 1999
4. M.P. Thomas, J.R. Boisseau. Building Grid computing portals: the NPACI Grid portal toolkit. In Grid Computing: Making the Global Infrastructure a Reality, F. Berman, G. Fox, and T. Hey. John Wiley & Sons, Chichester, England, March 2003
5. K. Schuchardt, B. Didier, G. Black. Ecce--a problem-solving environment's evolution toward Grid services and a Web architecture. Concurrency and Computation: Practice and Experience, Vol. 14, No. 13-15, pp 1221-1239 (2002)
6. D.W. Erwin. UNICORE—a Grid computing environment. Concurrency and Computation: Practice and Experience, Vol. 14, No. 13-15, pp 1395-1410 (2002)
7. Gateway Project. See http://www.gatewayportal.org
8. NMI Portal – Open Grid Computing Environment (OGCE). See http://www.collab-ogce.org/nmi/index.jsp
9. GEON (Cyberinfrastructure for the Geoscience) project. See http://www.geongrid.org
10. Larsen, S.: e3d: 2D/3D Elastic Finite-Difference Wave Propagation Code. Available from http://www.seismo.unr.edu/ftp/pub/louie/class/455/e3d/e3d.txt
11. Allaire, J.: Macromedia Flash MX—A next-generation rich client. March 2002. Available from http://www.macromedia.com/devnet/mx/flash/whitepapers/richclient.pdf
12. IRIS. See http://www.iris.washington.edu
13. GEONgrid portal. Available from https://geon01.sdsc.edu:8282/gridsphere/gridsphere
14. Teragrid project. Available from http://www.teragrid.org
15. I. Foster, et al. The Physiology of the Grid: an Open Grid Services Architecture for Distributed Systems Integration. See http://www.globus.org/research/papers/ogsa.pdf
16. Graham, S. et al.: Building Web Services with Java. SAMS, Indianapolis, 2002
17. Allaire, J.: Macromedia MX: Components and Web Services. April 2002. Available from http://www.macromedia.com/devnet/mx/coldfusion/whitepapers/components_ws.pdf
18. Input XML Schema for the e3d application. Available from http://geon01.sdsc.edu:8484/GEONappws/ApplSchema/event_sim_input_new.xsd
19. Christensen, E., Curbera, F., Meredith, G., Weerawarana, S.: Web Service Description Language(WSDL) version1.1. W3C Note 15 March 2001. See http://www.w3c.org/TR/wsdl
20. Gudgin, M., Hadley, M., Mendelsohn, N., Moreau, J., Nielsen, H. F.: SOAP Version 1.2 Part 1: Messaging Framework. W3C Recommendation 24 June 2003. Available from http://www.w3.org/TR/soap12-part1
21. Java CoG Kit. Available from http://www-unix.globus.org/cog/java/
22. RSL v1.0. See http://www.globus.org/gram/rsl_spec1.html
23. Youn, C., Pierce, M., and Fox, G., "Building Problem Solving Environments with Application Web Service Toolkits" ICCS 2003 Workshop on Complex Problem Solving Environments for Grid Computing, LNCS 2660, pp. 403-412, 2003
24. Object Management Group. Available from http://www.omg.org
25. Fissures project. Available from http://www.seis.sc.edu/software/Fissures/
26. The WS-Resource Framework. Available from http://www.globus.org/wsrf
27. Open Grid Services Infrastructure Working Group (OGSI-WG). Available from http://forge.gridforum.org/projects/ogsi-wg

Toward GT3 and OGSI.NET Interoperability: GRAM Support on OGSI.NET[*]

James V.S. Watson, Sang-Min Park, and Marty Humphrey

Department of Computer Science, University of Virginia, Charlottesville, VA 22904, USA

Abstract. OGSI.NET is the implementation of the Open Grid Services Infrastructure (OGSI) that leverages the Microsoft .NET Framework. OGSI.NET and the Globus Toolkit combine to create a comprehensive platform for computational science by supporting the emerging Grid protocols on Windows and Linux/UNIX, respectively. A significant challenge in building OGSI.NET is interoperability with the Globus Toolkit, both in terms of the rendering of individual services (OGSI-compliance and more recently WSRF-compliance) and also conformance to higher-level protocols developed in the Globus project and in the Global Grid Forum. This paper presents the design and experiences of implementing the Globus GRAM protocols on OGSI.NET. A major challenge was to easily and securely create processes as specific target users in the Windows environment. Differences between GT3 GRAM and OGSI.NET GRAM are described and an overview of WSRF.NET GRAM is presented.

1 Introduction

It is generally believed that the foundation of next-generation Grids will be the Open Grid Services Architecture (OGSA)[1], which is an overall vision for Grid computing that combines the strengths of projects such as Globus [2] and Legion [3] with Web Services. OGSA endorses the *service-oriented architecture* as the foundation of Grid Computing. The introduction of the largely abstract OGSA was accompanied by the Open Grid Services Infrastructure (OGSI)[4], which attempted to specify the low-level interfaces and—to a certain extent—the behaviors of the individual services in an OGSA-compliant Grid. OGSI defined a particular rendering of the service by using both standard and non-standard uses of the XML, SOAP, and the Web Services Description Language (WSDL). The Grid community actively contributed to the definition of OGSI through the Global Grid Forum (GGF) standardization process. OGSI.NET [5] is the implementation of the Open Grid Services Infrastructure (OGSI) that leverages the Microsoft .NET Framework. OGSI.NET and the Globus Toolkit combine to create a comprehensive platform for computational science by supporting the emerging Grid protocols on Windows and Linux/UNIX, respectively.

An important use of a computational grid is allowing users to submit jobs to the grid without needing explicit knowledge of which machines are being used or logging

---

[*] This work is supported in part by the US National Science Foundation under grants ACI-0203960 (Next Generation Software program), SCI-0438263 (NSF Middleware Initiative), the US Department of Energy through an Early Career Grant, and Microsoft Research.

onto each machine manually. For example, a user may desire to run a parameter-space problem in which hundreds of similar jobs can be executed in parallel. The user would prefer to authenticate himself only once to the computational grid, though many machines may be used to execute these parallel jobs. The user will also prefer to create these many job specifications with a minimum of effort and not have to configure the job specifications based on which machines are being utilized for each job.

A significant challenge in building OGSI.NET is interoperability with the Globus Toolkit, both in terms of the rendering of individual services (OGSI-compliance and more recently WSRF-compliance [6]) and also conformance to higher-level protocols developed in the Globus project and in the Global Grid Forum. Specifically with regard to remote execution described above, the challenge is to conform to the Grid Resource Allocation Manager (GRAM) [7] of the Globus Toolkit. This paper reports the implementation experiences of the development of GRAM-supporting services on OGSI.NET. A major challenge was to easily and securely create processes as specific target users in the Windows environment. This work is the *first* attempt to expand the Grid to include remote job execution on Windows machines via GRAM, as GT3 is only able to run on Windows machines now in a limited client-side mode via the Java Cog [8]. A .NET client was also written that runs on Windows and can submit jobs to machines running GT3 or OGSI.NET. These results provide the basis for remote execution currently being created in our implementation of WSRF on .NET [9] [10].

The next section reviews GRAM. Section 3 describes Windows security issues that came up as a result of trying to implement the same functionality of GRAM on Windows machines. Section 4 explains OGSI.NET GRAM, interoperability between Windows and Linux machines, and an example usage of this set of services. Section 5 gives a brief overview of WSRF.NET GRAM. Section 6 concludes this paper.

2 Review of GRAM

GRAM running on GT3 allows a user to easily submit a job and have it execute on the machine running the GRAM set of services. The Globus Resource Specification Language (RSL) [7] is an XML schema-defined language that is used by the user to specify the job submission. Many aspects of a job submission can optionally be included, such as the starting directory, user environment, and non-local input files. Substitution values are also allowed so that values can be better controlled and updated. For example, if the only changes in a parameter space problem were a single argument and related output filename, the user could specify the value like so: <rsl:substitutionDef name= "ArgValue"> <rsl:stringElement value="50"/> </rsl:substitutionDef>. The user would then be able to specify this value in multiple locations in the job submission like so: <rsl:substitutionRef name="ArgValue"/>.

Job submissions are authenticated by the use of a gridmap file located on each server, which contains a mapping between an X509 certificate DN and a local user name. The job submission is signed with the user's X509 certificate. The DN from the signature is then compared to the entries in the gridmap file. If a match is found, the job submission is accepted and the job will be run as the local user specified in the gridmap file. By running the job as a local user, the damage a malicious user can cause is limited to whatever access rights the local user already holds.

3 Windows Security

In this section, we describe the relevant aspects of Windows security that in some cases were leveraged and in some cases had to be overcome in order to implement GRAM in OGSI.NET. Implementing the GRAM set of services on Windows proved to be rather challenging because of the security mechanisms that are available in Linux and are not easily accessible in Windows. The major hurdle in Windows security is creating a new process (in this case, the actual job to run on the machine) as a different user than the user asking for the process to be created.

Security in Windows is based, in part, on passing user tokens to authentication methods. The well-documented methods available to create a new user token require a user name and password combination in clear text. If these well-documented methods are used, there are a number of possibilities for implementation. In one case, the user securely enters in the password for each job submission. Obviously, this is undesirable because it does not allow a user to sign-on once and for the grid services to authenticate the user on each machine. Another option is to include the user name and password pair in the gridmap file. This option has two major problems. If the file containing the passwords is read by a malicious user, that user can access the system as any user that is listed in the file. By mapping the DN to a local user name, the ability to read the gridmap file does not make it easier for a malicious user to gain access to the system by impersonating another user because it is assumed that the DN is hard to forge. Another problem is that the gridmap file will need to be changed every time the user changes his password on the machine, which would likely result in an updating nightmare for the system administrator.

UNIX and Linux systems allow a setuid bit to be set on files. When this bit is set on an executable, the executable may change the effective user id so that it can access resources that are inaccessible to the user that ran the executable. The call to change the effective user id only requires that the setuid bit be set on the executable.

There is a method on Windows called NtCreateToken() that is located in ntdll.dll. In order to use the NtCreateToken() method, the user running the executable that calls NtCreateToken() must be given the Create Token privilege. The call to NtCreateToken() is not as simple as the call to setuid() in *nix systems, which only requires the new user id. NtCreateToken() requires 12 input parameters that must be properly set for a new user token to be created. Since this method is undocumented in Windows help files, our code was influenced by other code that required the use of NtCreateToken(): CVSNT, Cygwin, and GUI-Based RunAsEx. Currently, the user token information also must be located on the machine where this code is running. This means that a user name cannot be intended to refer to an account name that is on a domain other than the local machine.

An important difference between the setuid bit and the NtCreateToken() method is the extent of the privileges. On *nix systems, the ability to impersonate other users is confined to the executable that has the setuid bit set, but any user with the proper permissions can run that executable. On Windows systems, the privilege to call

NtCreateToken() is given to users, instead of executables. This means that the user who runs any executable that eventually calls NtCreateToken() must have the Create Token privilege. This greatly increases the amount of code that is now susceptible to malicious users who would then gain the ability to run an executable as any user.

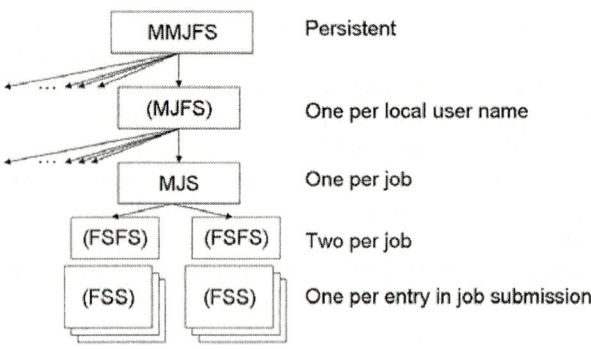

Fig. 1. OGSI.NET GRAM Architecture

4 OGSI.NET GRAM

The service architecture of OGSI.NET GRAM is based on GT3 GRAM and appears in Figure 1. The top service is the Master Managed Job Factory Service (MMJFS). This persistent service waits for new job submissions. When a job submission is received and the request is authenticated with the gridmap file, a Managed Job Factory Service (MJFS) may be created. In order to efficiently use local resources, at most one MJFS for each local user name that appears in the gridmap file will be running at any time. When the first job to run as a given local user name is submitted, a new MJFS is created and the job submission is forwarded to the new MJFS. Otherwise, the MMJFS will forward the job submission to the MJFS that already exists for the given local user name. When a MJFS receives a job submission, it creates a new Managed Job Service (MJS) that will actually execute and monitor the job as it runs locally on the system. The MJS creates two File Stream Factory Services (FSFS), one corresponding to standard output and one for standard error. Each FSFS will then create as many File Stream Services (FSS) as are specified in the job submission. The reason that a user can specify multiple locations is to make it easier for job results to be distributed. For example, a single user submits a job, but the user may wish all members of his group to receive the results. The RSL allows the user to not have to worry about sending the results out after the job has completed. The MJFS, FSFS, and FSS names appear in parentheses because the services are used by the system to satisfy the job submission, but it is not necessary for the user submitting the job to know that they exist. When a user submits a job, a reference to the newly created MJS is returned.

4.1 Differences Between OGSI.NET and GT3 GRAM

The GRAM set of services in GT3 also includes a Resource Information Provider Service (RIPS) that actually accesses the underlying operating system to get system-specific information, such as the scheduling and file systems. The MMJFS and MJS instances will subscribe to the RIPS to acquire information about the local system and job state changes, respectively. While we understand that this type of information is arguably necessary for OGSI.NET MMJFS and MJS instances, we do not believe that it is necessary to conform to the specific interfaces of the Globus RIPS at this time.

When the requested executable is actually run, the executable is called by invoking the Windows method CreateProcessAsUser() and passing in the user token created by NtCreateToken(). Except for the specified executable, everything in the OGSI.NET GRAM set of services run as the user with the Create Token privilege. The GRAM set of services in GT3, however, create each MJFS as the specified local user. This reduces the amount of security vulnerabilities that may allow a malicious user to gain access to the system as any user.

4.2 Interoperability

In many ways, the single criteria for success for OGSI.NET remote job execution is to faciliates clients on both Windows and Linux machines to send job submissions to both Windows and Linux machines with a minimum of user interaction.

By solely using GRAM in GT3, it is possible to take a Linux client and submit a job to a Linux machine running GT3. The Globus Alliance offers a Java CoG Kit [8] which offers some client functionality, but does not offer a client that can use the version of GRAM in GT3. Java CoG Kit version 1.1 only supports up to version 2.4.0 of the Globus Toolkit. Analysis of the client code offered in the GT3 source download revealed that it is possible to invoke a Java-only client that can use GRAM in GT3, thus allowing a Windows client to also submit jobs to a Linux machine running GT3.

The job submission sent from the client must be signed with the user's certificate in order for the server to properly authenticate the job submission request. In GT3, the versions of the WS-Security specification and other Web Services specifications are provided as part of the GT3 implementation, while OGSI.NET uses Microsoft's Web Services Enhancements 2.0 (WSE 2.0) [11]. In early development, we had difficulty getting our implementation of WS-Security to interoperable with the version of WS-Security in GT3. However, this has since been resolved as both implementations have recently conformed to the Basic Security Profile of the Web Services Interoperability group (WS-I [12]).

4.3 Example Usage

The prototypical use case is shown in Figure 2. At the time of this writing, due to incompatible security protocols inherited from tooling on both the Windows and Linux platforms, currently, there is no client that can run on Linux that can submit a job to OGSI.NET GRAM (although we expect this to change shortly). GT3 already provides a Linux client that can submit a job to GT3, so the arrow in Figure 2 indicating that functionality is not applicable to the work described in this paper. We have our client run from a Windows Server 2003 machine to show that the Windows client and server do not have to run the exact same operating system (i.e., Windows XP).

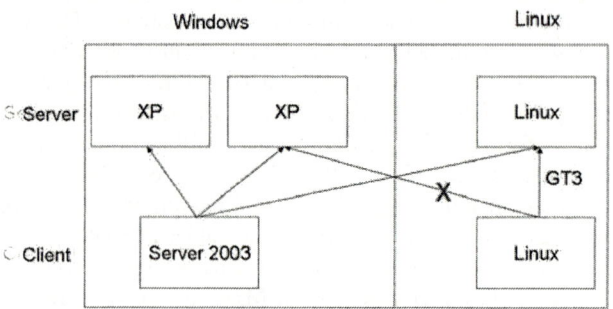

Fig. 2. Example Machine Configurations

The Windows client that we developed invokes either an OGSI.NET client or a Java GT3 client (via the Java Cog), depending on the machine that is being contacted. Other than the simple switch to know which client code to invoke, the arguments are exactly the same.

While GridFTP has been implemented as part of the OGSI.NET project, it is not currently fully-integrated into the OGSI.NET hosting environment such that a server can move files on behalf of a user automatically. Therefore, the client also examines the job submission file and downloads the first standard output and standard error files to the client's current running directory. This allows a client to easily view the results of running the job without having to manually check the output files on the server. This GridFTP functionality requires "grid-proxy-init" and "globus-url-copy" from the Java CoG Kit and GridFTP instances running on the servers.

In our testing, to ensure that the job was being run as the proper local user we set up the same file on both OGSI.NET servers that explicitly denied access to user Foo1. The same X509 certificate DN referred to different local user names on the two OGSI.NET servers (users Foo1 and Foo2). When the job submission included the DOS command "type" and specified the above file, one server displayed the contents of the file, while the other server denied the user access to that file and printed the error to standard error.

Running this client exposed the fact that Windows has a concept similar to real and effective user ids like *nix systems. For example, executing a .NET program as the job submission that displays the value System.Environment.UserName will show the name of the user that has the Create Token privilege. Running the Cygwin "whoami" command instead will show the local user name that is in the gridmap file.

5 WSRF.NET GRAM

In this section, we introduce a new GRAM implementation currently being developed on top of WSRF.NET. The Web Service Resource Framework (WSRF), announced in

January 2004, is an effort to create stateful web services by defining "WS-Resources" [6]. It provides an abstraction of stateful resources which could be identified by existing web service technologies (e.g., WS-Addressing). Globus Tookit version 4 (GT4) is now being developed based on the WSRF concepts. GRAM in GT4 has different features from GT3 GRAM. The Job (process) is now modeled as a WS-Resource, and a client can interact with a job through operations defined in WSRF portTypes and GRAM Services. In other words, a client can submit a job through an operation in the GRAM Factory Service, and query the job's state using one of the resource property operations defined in WSRF.

WSRF.NET is an implementation of full set of WSRF and WS-Notification specifications on Microsoft .NET [9]. WSRF.NET GRAM is an implementation of GRAM services using WSRF.NET libraries and tooling. GT4 GRAM defines two WSRF-compliant web services, ManagedJobFactoryService (MJFS) and ManagedJobService (MJS). It also defines an xml-based job description language. WSRF.NET GRAM exposes these two web services and accepts the job description. The ManagedJobFactoryService keeps information of hardware and software resources (e.g., localResourceManager) as WS-Resources and reflects their state as resource property document. WS-Resources in the ManagedJobService are information of process.

When a job-creation operation is called by a client, MJFS creates a new resource as a Windows process and the process is executed under the local user name which is different from the user name running ASP.NET infrastructure. The process handle is used to expose a process. WSRF.NET has a mechanism to save and retrieve the process handle to/from a database so that the process information can be sustained. The client may query MJS about the resource property of process (e.g., process state) using operations defined in the WSRF Resource Property portTypes. When the job's state changes, MJS notifies the client of new state using the WS-Notification libraries implemented in WSRF.NET. After receiving the 'done' or 'failed' notification, the client terminates.

The Globus Toolkit has been using the proxy certificate for WS-Security, and it has been the biggest challenge for interoperability. In WSRF.NET we support WS-Security processing with GT proxy certificate, thus GT4 GRAM and WSRF.NET GRAM are interoperable. In other words, users can submit the job to WSRF.NET GRAM services using GT4 client tools. In our future release, we are planning to provide a secure and reliable authorization mechanism, a delegation service, and file movement service all of which will be incorporated in WSRF.NET GRAM.

6 Conclusion

Remote execution is a fundamental operation for Grids. Prior to the work described in this paper, there was no Grid-standards-compliant mechanism for remote execution on the Windows platform. In this paper, we described the design and implementation of GRAM for OGSI.NET, focusing on the security model and issues that had to be overcome. We also described our prototype implement in WSRF.NET. We are currently building on this support to create the University of Virginia Campus Grid (UVaCG), using WSRF.NET and GT4 as the foundation of the Grid. By combining this new support for remote execution with our continuing support for remote data access via GridFTP and higher-level abstractions and GUIs, we have taken a significant step toward the realization of the Grid as a ubiquitous platform for computational science.

References

[1] Foster, C. Kesselman, J. Nick, and S. Tuecke. The Physiology of the Grid: An Open Grid Services Architecture for Distributed Systems Integration. Draft of 6/22/02. http://www.gridforum.org/ogsi-wg/drafts/ogsa_draft2.9_2002-06-22.pdf
[2] Globus Project. http://www.globus.org
[3] A.S. Grimshaw, A.J. Ferrari, F.C. Knabe and M.A. Humphrey, "Wide-Area Computing: Resource Sharing on a Large Scale," IEEE Computer, 32(5): 29-37, May 1999.
[4] S. Tuecke et. al. Open Grid Services Infrastructure (OGSI) Version 1.0. Global Grid Forum. GFD-R-P.15. Version as of June 27, 2003.
[5] G. Wasson, N. Beekwilder, M. Morgan, and M. Humphrey. OGSI.NET: OGSI-compliance on the .NET Framework. In *4th IEEE/ACM International Symposium on Cluster Computing and the Grid (ccGrid 2004)*. Chicago, Illinois. April 19-22, 2004
[6] K. Czajkowski., Ferguson, D., Foster, I., Frey, J., Graham, S., Sedukhin, I., Snelling, D., Tuecke, S., Vambenepe, W. 2004. The WS-Resource Framework. http://www-106.ibm.com/developerworks/library/ws-resource/ws-wsrf.pdf
[7] K. Czajkowski, I. Foster, N. Karonis, C. Kesselman, S. Martin, W. Smith, S. Tuecke. A Resource Management Architecture for Metacomputing Systems. *Proc. IPPS/SPDP '98 Workshop on Job Scheduling Strategies for Parallel Processing*, pg. 62-82, 1998.
[8] G. von Laszewski, I. Foster, J. Gawor, P. Lane. A Java Commodity Grid Toolkit. *Concurrency: Practice and Experience*, 13, 2001.
[9] M. Humphrey, G. Wasson, M. Morgan, and N. Beekwilder. An Early Evaluation of WSRF and WS-Notification via WSRF.NET. *2004 Grid Computing Workshop (associated with Supercomputing 2004)*. Nov 8 2004, Pittsburgh, PA.
[10] Web Services Resource Framework on the .NET Framework. http://www.ws-rf.net
[11] Microsoft. Web Services Enhancements (WSE). http://msdn.microsoft.com/webservices/building/wse/default.aspx
[12] Web Services Interoperability Organization (WS-I). http://www.ws-i.org

GEDAS: A Data Management System for Data Grid Environments

Jaechun No[1] and Hyoungwoo Park[2]

[1] Dept. of Computer Software,
College of Electronics and Information Engineering,
Sejong University, Seoul, Korea
[2] Supercomputing Center,
Korea Institute of Science and Technology Information,
Daejeon city, Korea

Abstract. In data grid environments, many large-scale scientific experiments and simulations generate very large amounts of data in the distributed storages, spanning thousands of files and data sets. In such environments, the replication technique for the fast data sharing between the community of researchers, and the high-performance I/O for the storage and efficient data accesses on heterogeneous resources present an extremely challenging task. Several data replication techniques have been developed to support high-performance data accesses to the remotely produced scientific data. However, most of those techniques were implemented with the assumption that the data being replicated is read-only so that it would not be modified once it has been generated. Furthermore, those techniques mainly focus on measuring up the network performance, but ignoring I/O overhead incurred during the data generation and replication. We have developed a software system, called Grid Environment-based Data Management System (GEDAS), that provides a high-level, user-friendly interface, while maintaining the consistent data replicas among the grid communities. We describe the design and implementation of GEDAS and present performance results on Linux cluster.

1 Introduction

In data grid environments, many large-scale scientific experiments and simulations generate very large amounts of data [1, 2, 3](on the order of several hundred gigabytes to terabytes) in the geographically distributed storages. Furthermore, these data are shared between the researchers and colleagues for data analysis, data visualization, and so forth. Several data replication techniques, including Globus toolkit [4, 5, 6], have been developed to support high-performance data accesses to the remotely produced scientific data.

However, most of those data replication techniques were implemented with the assumption that the data being replicated is read-only so that once it has been generated would it not be modified in any grid site. Furthermore, those techniques mainly focus on measuring up the network performance, but ignoring the I/O overhead incurred during the data generation and replication on the heterogeneous storages.

We have developed a software system, called Grid Environment-based Data Management System (GEDAS), that provides a high-level, user-friendly interface to share the remotely produced data among the grid communities. As a related work, we have implemented Scientific Data Manager (SDM) that combines the good features of both file I/O and databases [7, 8]. We, in GEDAS, extend the SDM capabilities to maintain the consistent data replicas among the grid communities and to support high-performance I/O using MPI-IO to store the real and replicated data. GEDAS interacts with database to store application-related and system-related metadata to support the integrated data replicas and high-performance I/O on the distributed resources, while taking advantage of various I/O optimizations available in MPI-IO, such as collective I/O and noncontiguous requests, in a manner that is transparent to the user.

The rest of this paper is organized as follows. In Section 2, we discuss our goals in developing GEDAS. In Section 3, we present the design and implementation of GEDAS. Performance results on the Linux cluster located at Sejong University are presented in Section 4. We conclude in Section 5.

2 Design Motivation

Our main objectives in developing GEDAS were to maintain consistent data replication among the geographically distributed sites, to provide high-performance parallel I/O, to provide a high-level application programming interface (API) and to support a convenient data-retrieval capability.

- **Consistent Data Replication.** In order to maintain the consistently integrated data replicas among the distributed resources, GEDAS uses the version checking method that is much similar to the locking mechanism of distributed file systems [9, 10]. Whenever an application running on a site modifies the replicated data, it informs the data modification to the sites that share the same data, including the owner that originally generated the data, using the version number, and thus would have them access the recently modified data.
- **High-Performance I/O.** To achieve high-performance I/O, GEDAS uses MPI-IO to access real data. MPI-IO, the I/O interface defined as part of the MPI-2 standard [11], is rapidly emerging as the standard, portable API for I/O in parallel applications. High-performance implementations of MPI-IO, both vendor and public-domain implementation, are available for most platforms. MPI-IO is specifically designed to enable the optimizations that are critical for high-performance parallel I/O. Examples of these optimizations include collective I/O, the ability to access noncontiguous data sets, and the ability to pass hints to the implementation about access patterns, file-striping parameters, and so forth.

- **High-Level API.** Our goal was to provide a high-level unified API for any kind of application (regular or irregular) while encapsulating the details of either MPI-IO or a database. The user can specify the data with a high-level description, together with annotations, and use a similar API for data retrieval. GEDAS internally translates the user's request into appropriate MPI-IO calls, including creating MPI derived data types for noncontiguous data. GEDAS also interacts with the database when necessary, by using embedded SQL functions.

3 Implementation Details

We describe the design and implementation of GEDAS.

3.1 GEDAS Metadata Structure

Figure 1 describes an overview of GEDAS. GEDAS stores application-related metadata and system-related metadata to *application_registry_table*, *data_registry_table*, *file_registry_table*, *performance_registry_table*, and *system_registry_table*. These data base tables are made for each application to provide a high-level remote data access abstraction, while supporting a transparent, user-friendly user interface.

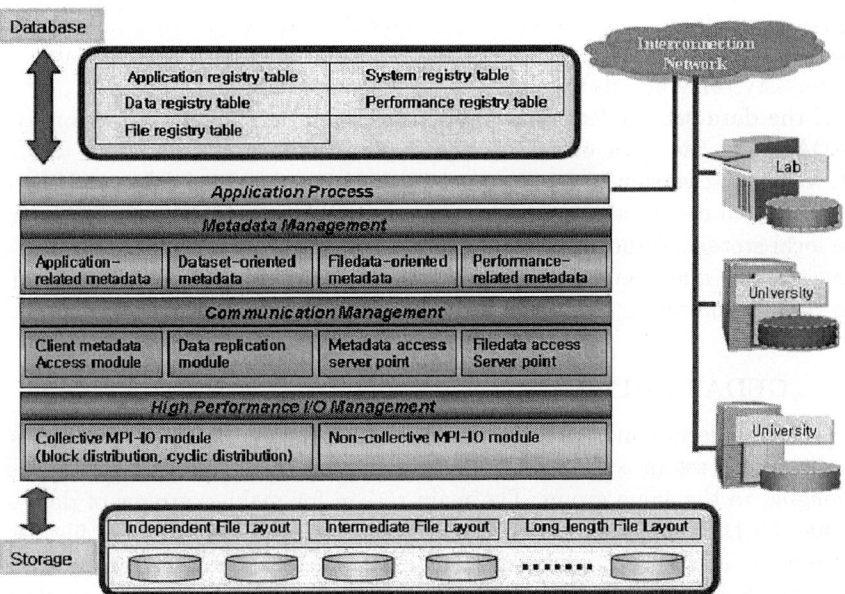

Fig. 1. GEDAS Architecture

The data_registry_tabe includes the properties of each data set, such as data type, storage order, data access pattern, and global size. Also, it contains data replication-related metadata, such as the owner ip, replication location ip, and version number to be used in checking the replica consistency.

The file_registry_table stores a globally determined file offset denoting the starting offset of the file of each data set. GEDAS uses this information to make appropriate MPI-IO calls to access the real data. The file_registry_table also includes the physical file name to be mapped to the data set.

The system_registry_table contains the system-related metadata, including filesystem type and striping unit size where the real data is stored. This information is used to choose the optimal file layout for the real data storage.

The performance_registry_table includes the performance values evaluated using buffered I/O and directed I/O on each filesystem type registered in the system_registry_table. In the buffered I/O, all the data transfers between the user buffer and the storage go through the kernel buffer. On the other hand, in the directed I/O, the data transfers bypass the kernel. These values are measured using a multiple of 2Mbytes of data size. When the data sets are accessed on a certain registered filesystem, GEDAS compares the performance values evaluated using buffered I/O and direct I/O to those that have been obtained using the closest data granularity in the performance_registry_table. If the direct I/O is selected as a better I/O option, then the direct I/O flag defined in the MPI-IO is turned on to bypass the kernel buffer during the data transfer.

GEDAS provides the capability to maintain the consistent data replication among the remote resources using the data version checking. When the data requested by an application could not be found on the local storage, GEDAS connects the data owner to replicate the data to the local storage. The associated metadata stored in the application_registry_table, data_registry_table, and file_registry_table are also replicated to the local database.

If the data set needed to the application has been replicated before, then GEDAS asks the data owner for the version number associated to the data set. When the version number stored in the local database table and the one received from the data owner are the same, GEDAS reads the data replica from the local storage. Otherwise, GEDAS invalidates the data replica stored in the local storage, and then receives the data from the data owner, while updating the version number in the local database to the highest one.

3.2 GEDAS API

In GEDAS, users can specify groups of data sets by assigning properties to the first data set in a group and by propagating them to the other data sets belonging to the same group. The main reason for making groups of data sets is that GEDAS can then use different ways of organizing data in files, with different performance implications. For example, each data set can be written in a separate file, or the data sets of a group can be written to a single file. The properties assigned to each data set are stored in the database by invoking GEDAS_set_attributes. In case of read operation, data from a specific run can be retrieved by specifying attributes of the data, such as the date of the run. Also, the properties of the data sets need not be specified because GEDAS retrieves this information from the database.

The main GEDAS functions for writing and reading data are GEDAS_write and GEDAS_read. Before calling these functions, the user must provide the information necessary for GEDAS to perform I/O, such as the starting points and sizes of the subarray in each dimension in the case of block distribution, or the size of process grids and distribution arguments in each dimension in the case of cyclic distribution.

4 Performance Evaluation

In order to measure the performance, we used two Linux clusters located at Sejong university. Each cluster consists of four nodes having Pentium3 866MHz CPU, 256 MB of RAM, and 100Mbps of Fast Ethernet each. The operating system installed on those machines was RedHat 9.0 with Linux kernel 2.4.20-8. Each cluster uses its own PostgreSQL to store the metadata.

The performance results were obtained using the template implemented based on the three-dimensional astrophysics application, developed at the University of Chicago. The application stores block-distributed data in each dimension. In the template, we tested GEDAS library for the cyclic-distributed data as well.

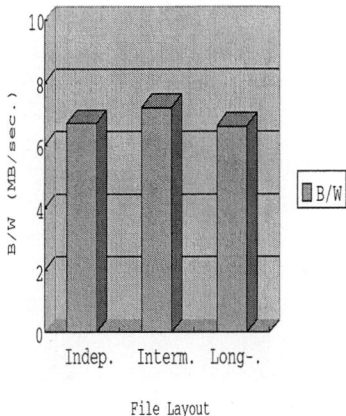

Fig. 2. Bandwidth to access the data from the remote owner in the block distribution. The accessed data is stored on the remote owner in three file layouts: independent file layout (Indep.), intermediate file layout (Interm.), and long-length file layout (Long-.)

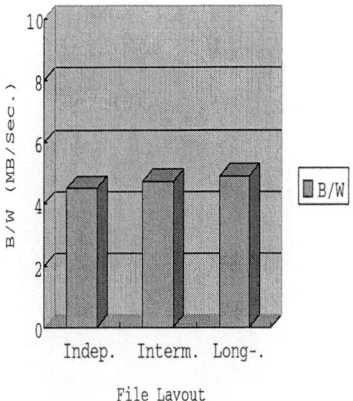

Fig. 3. Bandwidth to access the data from the remote owner in the cyclic distribution. The accessed data is stored on the remote owner in three file layouts: independent file layout (Indep.), intermediate file layout (Interm.), and long-length file layout (Long-.)

On one of the clusters, the template has generated six floating-point arrays for data analysis, another six floating-point arrays for restart, and seven character arrays for visualization. Once those data are generated, we accessed the data from the other cluster using the GEDAS library. The template grouped the data sets into three data groups, according to the usage.

Figures 2 and 3 describe the performance results when the data sets are accessed on one cluster, while the requested data sets are stored on the other cluster in three different ways of file layout. The data access patterns tested are block distribution in Figure 2 and cyclic distribution in Figure 3. The performance values include the database overhead to access the associated metadata.

As can be seen, the bandwidth measured in the block-distributed data access pattern is higher than that measured in the cyclic data access pattern due to the smaller communication overhead incurred in the MPI-IO's collective operation. In both distributions, the independent file layout shows the least bandwidth because it generates more files than the other two file layout methods, resulting in the higher file-open and close costs.

Figures 4 and 5 demonstrate the performance results obtained to access the replica of the desired data sets in the block distribution and cyclic distribution, respectively. The replicated data is stored in the independent layout, intermediate layout, and long-length layout. When compared to Figures 2 and 3, accessing the data replica shows almost as much as twice bandwidth accessing from the remote owner because of the less communication overhead.

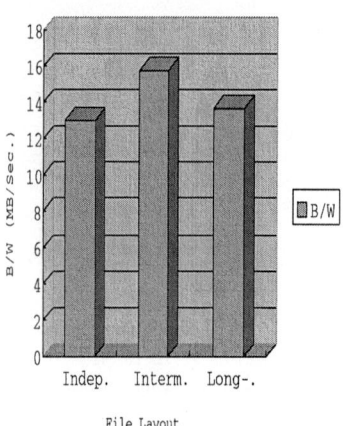
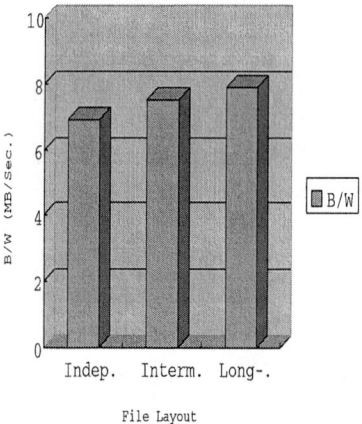

Fig. 4. Bandwidth to access the data replica in the block distribution, on top of three file layouts: independent file layout (Indep.), intermediate file layout (Interm.), and long-length file layout (Long-.)

Fig. 5. Bandwidth to access the data replica in the cyclic distribution, on top of three file layouts: independent file layout (Indep.), intermediate file layout (Interm.), and long-length file layout (Long-.)

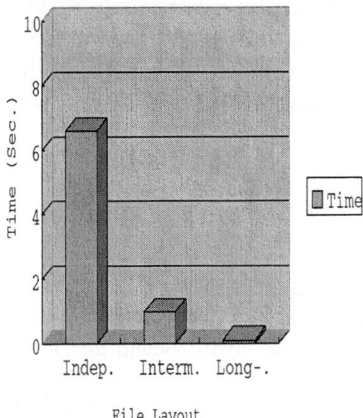
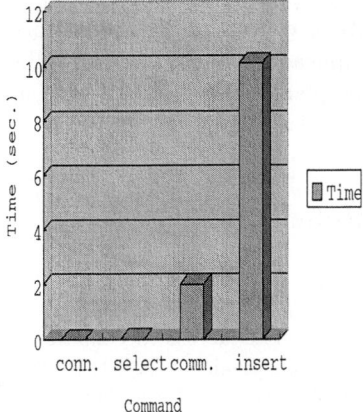

Fig. 6. Open and close overheads of three file layouts in time

Fig. 7. Database overhead in time for the data replication

Figure 6 shows the time overhead to open and close files, on top of Linux kernel 2.4.20-8. In the template, the independent file layout produces fifty-seven files: eighteen files for data analysis, twenty-one files for data visualization, and another eighteen files for restart. The intermediate file layout generates nineteen files: six files for data analysis, another six files for restart, and seven files for data visualization. The long-length file layout generates only three files, one for each data group.

Figure 6 shows that the less files are generated to store the data sets, the less file-open and close costs are incurred on Linux kernel 2.4.20-8.

Figure 7 shows the times to access the associated metadata from the remote database necessary for replicating the desired data set to the local storage. It demonstrates that the cost for inserting the metadata received to the local database is the highest overhead in the metadata communication.

5 Conclusion

We have presented the design and implementation of a toolkit, called GEDAS, for high-performance scientific data management in data grid environments. GEDAS provides a simple, high-level interface and performs all necessary I/O optimizations to access and replicate the remote data transparently to the user. We also experimented with different ways of organizing data in files, called independent file layout, intermediate file layout, and long-length file layout. In general, when file-open cost on a particular file system is high, long-length file layout performs well because it minimizes the number of files created. If the file-open cost is small, the performance of the three file layouts depends on how the number and size of files affect performance on the particular file system. An appropriate file layout policy can thereby be chosen for a particular file system. Also, we compared the performance produced in accessing the data sets from replicas to that in accessing the data sets from the remote owner. In the future, we plan to use GEDAS with more applications and evaluate both the usability and performance.

References

1. B. Allcock, I. Foster, V. Nefedova, A. Chervenak, E. Deelman, C. Kesselman, J. Leigh, A. Sim, A. Shoshani, B. Drach, and D. Williams. High-Performance Remote Access to Climate Simulation Data: A Challenge Problem for Data Grid Technologies. SC2001, November 2001
2. R. Moore, A. Rajasekar. Data and Metadata Collections for Scientific Applications. High Performance Computing and Networking (HPCN 2001), Amsterdam, NL, June 2001
3. A. Chervenak, E. Deelman, C. Kesselman, L. Pearlman, and G. Singh. A Metadata Catalog Service for Data Intensive Applications. GriPhyN technical report, 2002
4. I. Foster, C. Kesselman, J. Nick, and S. Tuecke. The Physiology of the Grid: An Open Grid Services Architecture for Distributed Systems Integration WG, Global Grid Forum, June 22, 2002
5. A. Chervenak, I. Foster, C. Kesselman, C. Salisbury, S. Tuecke. The Data Grid: Towards an Architecture for the Distributed Management and Analysis of Large Scientific Datasets. Journal of Network and Computer Applications, 23:187-200, 2001
6. I. Foster, C. Kesselman. Globus: A Metacomputing Infrastructure Toolkit. Intl J. Supercomputer Applications, 11(2):115-128, 1997.
7. J. No, R. Thakur, and A. Choudhary. High-Performance Scientific Data Management System. Journal of Parallel and Distributed Computing, (64)4:434-447, April 2003
8. X. Shen, A. Choudhary. A Multi-Storage Resource Architecture and I/O Performance Prediction for Scientific Computing. In 9th IEEE Symposium on High Performance Distributed Computing, 2000
9. C. A. Thekkath, T. Mann, and E. K. Lee. Frangipani: A Scalable Distributed File System. In Proceedings of the Symposium on Operating Systems Principles, 1997, pages 224–237
10. K. W. Preslan, A. P. Barry, J. E. Brassow, G. M. Erickson, E. Nygaard, C. J. Sabol, S. R. Soltis, D. C. Teigland, and M. T. O'Keefe. A 64-bit Shared Disk File System for Linux. In Proceedings of Sixteenth IEEE Mass Storage Systems Symposium Seventh NASA Goddard Conference on Mass Storage Systems & Technologies, March 15-18, 1999
11. R. Thakur and W. Gropp. Improving the Performance of Collective Operations in MPICH. In Proceedings of the 10th European PVM/MPI Users' Group Conference (Euro PVM/MPI 2003), September 2003

SPURport: Grid Portal for Earthquake Engineering Simulations

Tomasz Haupt, Anand Kalyanasundaram, Nisreen Ammari, Krishnendu Chandra, Kamakhya Das, and Shravan Durvasula

Mississippi State University, Center for Advanced Vehicular Systems,
Box 5405, Mississippi State, MS 39762, USA
{haupt, anand, ammari, krish, das, sharvan}@cavs.msstate.edu
http://www.cavs.msstate.edu

Abstract. This paper presents a successful implementation of the SPURport, a prototype Grid Portal for the earthquake engineering community. The portal eliminates the need for installing and maintaining simulation software by the end user. However, the portal infrastructure is application-neutral and can be used as a blueprint for developing portals for other communities. It provides seamless access to remote resources and supports the incorporation of legacy applications. It uses XML-based metadata extensively, which enables the introduction of high-level middle-tier services that aggregate and coordinate lower-level services provided by the Globus toolkit. For example, the high level Job Submission Service orchestrates resolution of logical entities, file transfers, and data streaming prior to actual job submission, hiding these activities from the end user.

1 Introduction

This paper describes a successful implementation of the SPURport - a prototype Grid Portal for the earthquake engineering community. The portal builds on and extends the current functionality of the NEESgrid [1], the cyberinfrastructure for the Network of Earthquake Engineering Simulations (NEES) [2]. The NEESgrid, in turn, is an application of OGSI/Globus Toolkit 3.0 [3]. The grid portal consists of three major components: a Web Portal providing the remote access to the SPURport functionality, the Grid Portal middleware that implements the portal specific services, and the back-end comprising the computational resources as well as data and metadata repositories.

The Web Portal is implemented using the CHEF[4] portal toolkit that adds collaborative features to the JetSpeed[5] portlet framework hosted by the Apache-Tomcat server. Since the portal front end is implemented as a portlet, it can be implemented using other portlet containers. Indeed, we have an alternative implementation using JSR-169 compliant GridSphere[6] toolkit. The use of CHEF is dictated by the integration with the NEESgrid. The portlet technology is widely used to implement Web Portals and it is recommended by the Open Grid Computing Environments consortium [7]. The portal toolkits, such as CHEF or GridSphere, provide a set of build-in portlets such as secure login (including access to myProxy server), user management and templates for the portal pages layout. They also provide a support for accessing remote services through Globus infrastructure (Java Cog [8]). However, access to the low level Globus functionality is not sufficient for implementing complex Grid Portals such as SPURport.

The actual functionality of SPURport is implemented by the SPURport middleware that creates a bridge between the required portal functionality and low level Globus services. Particular important is support for the incorporation of the legacy applications and orchestrating Globus services to perform complex tasks. Although driven by the specific requirements of earthquake engineering, the middleware is application-neutral, and can be reused for other application domains.

The rest of this paper is organized as follows: Section 2 briefly summarizes the goals of the SPUR project. Section 3 describes the requested functionality of the portal, and Section 4 explains the portal implementation. Due to space limitations, only selected portal features and the middle-tier implementation to support these features are discussed. Summary and conclusions are given in Section 5.

2 Large-Scale Seismic Performance of Urban Regions

The SPUR (*Seismic Performance of Urban Regions*) project is using the NEESgrid technology to link high-end simulations of earthquake ground motion, advanced models for building performance, large-scale databases, and visualizations to develop new knowledge about the spatial distribution of structural damage in a region and the effectiveness of building codes on controlling damage in a region. In this collaborative effort (between Mississippi State University, Carnegie Mellon University and Universities of California, Berkeley and Irvine) new visualization methods have been developed for understanding the dynamic processes in a large region, and new methods for communicating this information to scientists and engineers, as well as for conveying the effects of earthquakes to non-technical decision-makers. This is achieved through the use of the SPURport: a Grid-services-based middleware that provide access to computational resources, the databases for the regional models and simulations, as well as selecting scenarios and regional inventories for analysis.

3 Functionality of SPURport

The fundamental requirement for the SPURport is performing real time simulations of structural responses to ground motion caused by an earthquake. The implementation must provide an integrated simulation environment accessible anywhere through a Web Browser. Specifically, the portal must eliminate the need for explicit logging-in to remote systems (to retrieve, pre-process, transfer and post-process the data or submit jobs for execution); installation of any software; and the need for computer-related knowledge of the end user. For example, the user might be not familiar with the operating system of the remote computer, the user might not know how to query metadata servers to find the data or interpret the XML-encoded response of the server, the user might not know how to transfer data from one remote location (data repository) to another (compute servers), the user might not know or understand the data format which is needed to extract and format input data for the simulations, the user might be not familiar with OpenSees framework and consequently might not know how to prepare and run the simulations, the user might not know how to verify the consistency of the results., i.e., making sure that the right data are fed to the correctly prepared OpenSees script, etc.

The SPURport is required to be an integrated simulation environment; to provide an easy access to all services needed to specify, configure and perform simulation, as well as access the simulation results without the need of leaving the Web Browser. Furthermore, the portal should provide access to other NEESgrid services such as data and metadata repositories thus enabling collaborations between the users, data sharing and data dissemination to interested parities. The portal must be fault tolerant in the sense that work done by the end user cannot be lost as a result of interruption of connectivity (voluntary or not). At any time, the user should be able to reconnect to the servers and resume the broken session. Finally, the user should be able to reconstruct all circumstances that led to any particular simulation performed through the portal.

The context for a simulation is determined by three elements: (1) the earthquake model, (2) the structure model and (3) the inventory of structures. The earthquake model (or fault type) defines the ground motion. The simulations of earthquakes have been performed by our collaborators at Carnegie Mellon University [9]. The SPURport provides access to a repository of the results of these simulations. The structure models, as well as the OpenSees-based structure simulation framework, have been provided by our partners at University of California Berkeley [10]. An inventory of structures is defined as a collection of structures of a given types (for example, buildings of different sizes constructed using different materials according to specific building codes) distributed according to a selected pattern over the region of the simulated earthquake. The result of simulations of the seismic performance of the inventory is a distribution of potential damages in the region caused by an earthquake. At this stage of this project, the statistical analysis of the inventory performance has been done off-line because it is computationally very expensive and thus cannot be done interactively. Instead, the SPURport provides access to the repository of the results. We anticipate a parallel version of OpenSees running on high performance platforms to be developed by UC Berkeley outside this project. Once it is available, the simulations of the inventories will be incorporated into the SPURport.

Currently, a typical use of the SPURport is as follows. The user, guided by the metadata viewer and interactive visualizations, selects an earthquake model. Similarly, the user selects a structure type and sets its parameters, creating an instance of the structure to be simulated interactively. Next, the user chooses an inventory type and gets access to pre-computed results of the inventory simulation for the selected earthquake model. Finally, the user places the previously defined structure instance at a desired location on the distribution plot and runs the real-time simulation. The value of this scenario is that the user may try different what-if cases by placing different kinds of structures at locations known from the inventory simulations to be high- or low-risk locations.

4 SPURport Implementation

The Front End of SPURport is a CHEF teamlet, and thus it is seamlessly integrated with the rest of the NEESgrid user interface. This integration makes it possible to directly use NEESgrid/Chef user authentication mechanisms. The SPURport user interface is implemented as a single applet.

Figure 1 shows the layout of the SPURport GUI. The four text fields in the upper part of the SPURport applet display all currents selections that the user has made: Earthquake model, Structure Model, Inventory and Simulation (here, apparently at the beginning of a user session, only the Earthquake Model is selected). The rest of the applet frame is taken by the Swing Tabbed Pane: each tab corresponds to a particular function. The tabs are described in the following sections.

Access to Ground Motion Data Repository. Access to shared data repositories is a common use pattern for portals. SPURport provide support for this functionality through a combination of three web services: file repository service, metadata service and replica locator service. These three services are coordinated by a high-level data service provider, referred to as the data service façade. Each data set is uniquely identified by a logical Uniform Resource Identificator (URI), or a logical name, and is associated with its descriptor – a metadata record. The Replica Locator service resolves the URI into the physical location of the file – its Uniform Resource Locator (URL). Note that in general there is a one-to-many relationship between the data set URI and URL allowing for the creation file replicas to optimize access to it. This design follows a pattern employed by many other Grid-related efforts, and the implementation of the Replica Locator service is a part of the Globus distribution.

The typical use scenario is that user queries the metadata service to identify the data set, and the SPURport front end provides a graphical interface for making these queries and examining their results, as shown in Fig. 1). Once the file is selected, its

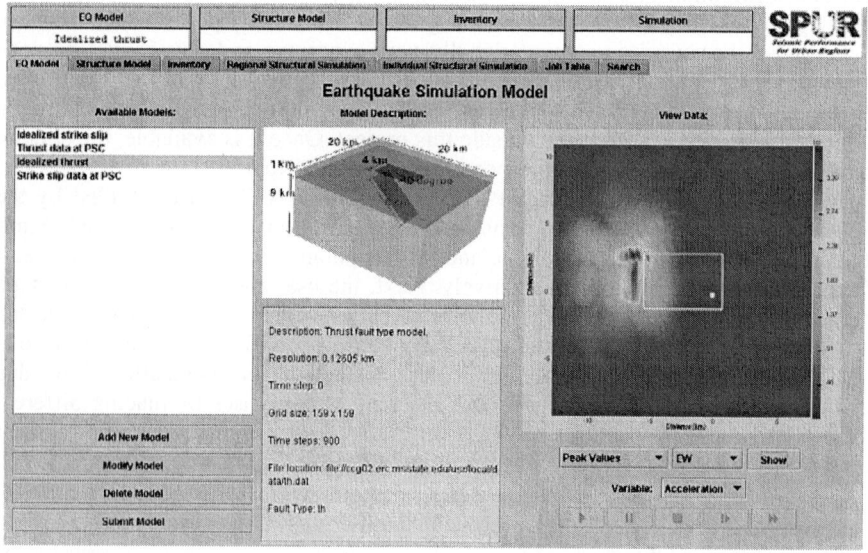

Fig. 1. SPURport GUI: Earthquake Model selection. The left panel shows a list of available datasets, the middle panel shows the metadata for the selected data set, and the right panel displays the contents of the retrieved file

URI is resolved by the replica locator and the resulting URL of the data set is used to actually retrieve the data. To simplify the front-end implementation and to reduce communication overhead, the interactions between the services are orchestrated by the façade. Thus, once the URI of the file is known, it is just a single request made by the applet to initiate the file transfer to the selected location, or streaming it to the applet for visualization. The applet call is converted by a JSP to an invocation of a corresponding façade's method that in turn makes all necessary calls to the services, as needed. The façade accesses the services through adapters, making it unaware and thus independent from the actual implementation of the back-end services. The current SPURport uses NEESgrid's OGSI/GT3-compliant NMDS [11] as metadata service and flat file system accessible through GridFTP as a file repository.

Incorporation of a Legacy Application. The SPURport simulation capabilities are based on the OpenSees framework, a Grid-unaware, legacy application. It has been brought to the portal using a similar approach as used in our previous portals, Gateway and Distributed Marine Environment Forecast System [12]. The key concept here is the employment of a metadata that describes the application. This metadata is referred to as "application descriptor". The application descriptor is an XML document comprised of three parts: application signature (name, credits, support, and the like), description of the application parameters including i/o specification, and instructions on how to run it (e.g., location of the executables, values of environmental variables, etc) that is automatically translated by the SPURport Job Submission Service into the Globus Resource Specification Language (RSL) at the submission time. This is the SPURport mechanism of hiding the complexity of the running applications in the Grid environment. Incorporating OpenSees, and other legacy applications, into the portal is accomplished by creating the XML application descriptor by filling out a web form. This is a task for the application expert who understands the application requirements (hardware, environment, input data, application parameters, etc) and thus not the end user. The support for the application metadata is provided by the SPURport Metadata Service.

Job Configuration – Structure Models. OpenSees applications, as many other applications, are controlled by input files. In the case of OpenSees, the controlling input file is written in the OpenSees script (an extension to the Tcl scripting language). In particular, the geometry and properties of the structures to be modeled are defined using the OpenSees scripts. The SPURport provides a library of such scripts, maintained by the data repository (in addition to files representing ground motion). Consequently, the user can browse the metadata to select and upload the script that represents the structure of interest. Furthermore, OpenSees scripting support parameters, and therefore the properties of a particular structure type can be adjusted. These parameters are captured in the corresponding script's metadata. The SPURport user selects thus a structure type (OpenSees script), and the portal seamlessly downloads the script's metadata and generates in-the-fly a graphical user interface revealing all adjustable parameters for the selected type of structure. The user sets the values of these parameters (or leaves the default values) and saves them as an instance of the structure. These values are used by the Job Submission Service.

Job Submission. The Globus toolkit provides all necessary services for submitting jobs on remote systems and the Java COG provides Java interfaces to these services making the invocation of the Globus services easy in the portal environment. However, the complete process of the job submission is a multistep process requiring coordination of several services. Therefore SPURport introduces its own, high-level Job Submission service that acts as a façade for the low-level Globus services and interacts with other SPURport services.

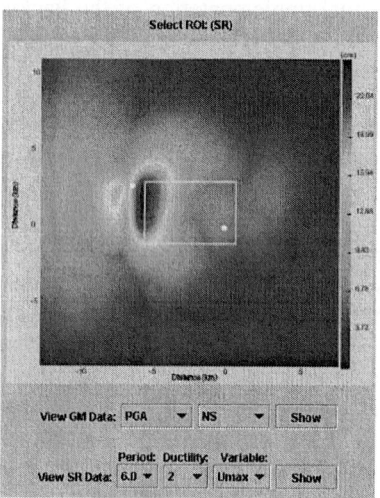

Fig. 2. To select the structure location the user displays the ground motion data or looks at a statistical distribution of previously run simulations. Here, the panel shows the distribution of U_{max} (maximum displacement of the top most floors) for the selected period of structure oscillations and ductility. The white rectangle in the central part of the plot shows the location of the fault, and the white dot inside the rectangle shows the epicenter of the simulated earthquake. The user selects the location by clicking on the panel (c.f. the second white dot on top of a dark region corresponding to high U_{max} values, left of the fault, which resulted by placing the cursor in that spot and clicking the left mouse button)

To submit a job, a user environment for running the job is created. This involves creating a working directory on the target system. Then input files are staged to the working directory. The OpenSees simulations require two input files. One is the driving OpenSees script that must be retrieved from the data repository and modified as specified by the user (changes of the structure parameters). The other is the acceleration time-history of the ground motion at the location where the simulated structure is located (see Fig. 2). To get this file, the data corresponding to the selected earthquake model in the data repository are fed to a filtering routine that extract data points needed for the simulation. That pruned data set is transferred to the target machine. Next, the job submission request is created as the Globus RSL document by combining the information stored in the application descriptor, parameters set by the user as well as information about the user environment. At this moment the job is ready for submission, and the history metadata is created, so that a record on how the simulation was run is preserved for a future reference. Then the job is submitted through Globus GRAM and the Job Service registers itself as a target for GRAM notifications. All changes in the job status reported by GRAM are forwarded to the SPURport monitoring services. Once the job is completed, its output files (including stdout and stderr) are staged out to the location specified by the user.

Monitoring Job Status and Access to the Results . All users jobs submitted through the SPURport are accessible through the Job Table interface (Fig. 3), until explicitly deleted by the user. The table displays where and when the job was submitted and its current status (PENDING, RUNNING, DONE). Once the job is completed the user can query its properties (e.g., input files, values of parameters, and history) and access – view or download - its results, including standard output, standard error and output data files.

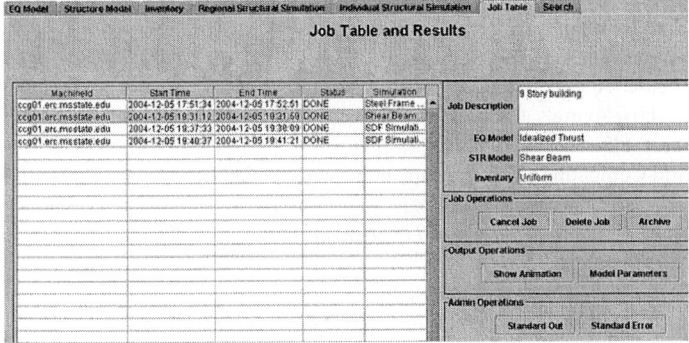

Fig. 3. SPURport Job Table interface allows monitoring the status of submitted jobs and provide access to their results. The jobs can be searched using metadata queries

5 Conclusions

This paper presents a successful implementation of a prototype Grid Portal for the earthquake engineering community. However, only the front-end has been customized for the particular domain. The portal infrastructure is application neutral and can be used as a blueprint for developing portals for other communities. It provides seamless access to remote resources (CPU and data) through OGSI/Globus 3.0 and provides support for easy incorporation of legacy applications. It extensively uses XML-based metadata for describing both data and applications. The metadata are used for automatic generation of the GUI, supporting the user with selection of applications and data. Furthermore, describing the application by metadata opens the opportunity to introduce the high-level middle-tier services that aggregate and coordinate lower-level services provided by the Globus toolkit.. Consequently, all the complexity of the heterogeneous, distributed Grid environment is hidden from the user. The applications and data are installed, maintained and upgraded by application experts. Therefore, the end user – the domain expert – can again concentrate on his or her research rather than acquiring computer wizardry.

Acknowledgements

This work presented in this paper is a part of the NSF-sponsored project Large-Scale Seismic Performance for Urban Region (grant EEC0121989). We would like to thank our collaborators, in particular Gregory Fenves and Jaesung Park of UC Berkeley, and Jacobo Bielak of Carnegie Mellon University for invaluable discussions and help with designing the functionality of the SPURport, as well as providing us with the ground motion data and structural models. Last but not least, we would like to thank the SPUR project Principal Investigator, Roger King of Mississippi State University for coordinating this successful collaborative effort.

References

1. NEESgrid, http://www.gridnees.org
2. NEES, http://www.nees.org
3. Globus Project, http://www.globus.org
4. Chef, http://chefproject.org
5. JetSpeed, http://portals.apache.org/
6. GridSphere, http://www.gridsphere.org
7. Open Grid Computing Environments, http://www.collab-ogce.org/nmi/index.jsp
8. Java CoG, http://www.globus.org/cog
9. J. Bielak et al, *"Visualizations of Large-Scale Earthquake Simulations"*, in the Proceedings of Supercomputing '03
10. OpenSees, http://opensees.berkeley.edu
11. NEESgrid Metadata Services, http://neesgrid.org/documents/TR_2004_38.pdf
12. T. Haupt, M. Pierce, *"Distributed object-based grid computing environments"*, book chapter in *"Grid Computing: Making the Global Infrastructure a Reality"*, Fran Berman, Geoffrey Fox and Tony Hey eds, John Wiley and Sons, 2003.

Extending Existing Campus Trust Relationships to the Grid Through the Integration of Pubcookie and MyProxy[*]

Jonathan Martin[1], Jim Basney[2], and Marty Humphrey[1]

[1] Department of Computer Science, University of Virginia, Charlottesville, VA 22904, USA
[2] National Center for Supercomputing Applications (NCSA),
University of Illinois at Urbana-Champaign,
Champaign, IL, 61820, USA

Abstract. In many ways, the ultimate success of the Grid will be highly dependent on the support for and integration with existing legacy infrastructure such as authentication infrastructure. This paper describes the design and implementation of the integration of Pubcookie, a popular Web-based authentication infrastructure used on campuses, and MyProxy, the on-line credential repository for the Grid. Specifically, we enable a valid Pubcookie credential to be dynamically exchanged for a Grid credential, without requiring the user to re-authenticate. Through this integration, this project makes an important contribution to an overall goal of single sign-on and more specifically the ability to "authenticate locally and act globally".

1 Introduction

Through a largely global effort, the last few years have seen a significant expansion and hardening of Grid technologies, particularly the maturation of the Globus Toolkit [1] and related technologies. Although the underlying software of the Grid can always be improved -- particularly the ability of the software to tolerate faults -- scientists are increasingly relying on Grid software to enable and manage their scientific explorations. The next version of the Globus Toolkit, based on WSRF [2], along with the WSRF-compliant hosting environment for the .NET Framework (WSRF.NET [3][4]), promises to expand these capabilities even further by creating uniform mechanism across Linux/UNIX and Windows, respectively.

However, one of the continuing challenges of the Grid software is to accommodate *legacy* mechanisms and policies. To make an existing scientific application "grid-aware", one must address issues related to security, I/O, scalability, licensing, etc., that often require non-trivial modifications. To truly be successful and a ubiquitous global computing infrastructure, the Grid must require fewer modifications to existing behaviors, particularly the manner in which scientists are accustomed to interacting

---

[*] J. Martin and M. Humphrey are supported in part by the US National Science Foundation under grants ACI-0203960 (Next Generation Software program), SCI-0438263 (NSF Middleware Initiative), and a subcontract from the San Diego Supercomputing Center (SDSC).

with local resources. When the local scientist *believes* she is performing experiments and manipulating data on a local set of resources when she is *actually* using non-local Grid resources, the Grid will be an unequivocal success.

This is particularly true with regard to security -- existing security policies, mechanisms, and trust relationships must be leveraged instead of being displaced by the Grid. For example, assume that a campus researcher has previously set up a secure Web portal for access either to local data or local compute cycles. To protect her resource, she has used Pubcookie [5], a popular open-source package for intra-institutional single-sign-on end-user Web authentication. Recently, she has been overwhelmed by local and non-local requests for this content and is investigating using the Grid (such as the TeraGrid [6] or the Open Science Grid [7]) as a back-end compute engine and data store as needed. However, she wants the back-end use of the Grid resource to be as seamless as possible and, unfortunately, she cannot afford to pay for the Grid resource usage on behalf of the requestors. Instead, these Grid computations and data request/storage must be performed by (and charged to) the requestor himself. In this situation, it is unacceptable to require the scientist to replace her existing local authentication with the Grid Security Infrastructure (GSI [8]), which is required of most Grid resources.

This paper describes our project to re-use existing campus trust relationships and authentication infrastructures when dynamically expanding to Grid computing resources on demand as in the scenario above. We bridge the campus environment of Pubcookie with the Grid environment of GSI through the MyProxy on-line credential repository [9][10]. In essence, the possession of a valid Pubcookie token is used as the basis for retrieving a Grid credential from the MyProxy server. In doing so, the user does not have to re-authenticate to the Grid -- the Pubcookie credential is *exchanged* for a valid credential to be used on the Grid. More broadly, this project makes an important contribution to an overall goal of single sign-on and more specifically the ability to "authenticate locally and act globally".

This paper is organized as follows. In Section 2, we present the existing projects that are related to this work. In Section 3, we describe Pubcookie and MyProxy in more detail and describe the integration design and implementation. Section 4 concludes and describes the future direction of this project.

2 Related Work

In this section, we describe the related work -- the Grid technologies in Section 2.1 and then the non-Grid technologies in Section 2.2. MyProxy and Pubcookie are presented separately in Section 3, along with the design for the integration of the two technologies.

2.1 Grid Technologies

There are a number of excellent projects that contribute pieces toward overall Grid security today. The Grid Security Infrastructure (GSI [8]) focuses on an authentication infrastructure for the Grid that is based on a Public-Key Infrastructure (PKI). GSI provides a standard programming interface for authentication, message integrity, and message confidentiality (GSSAPI), a mutual-authentication mechanism based on SSL/TLS, and a delegation protocol by which a user can temporarily empower a software service to act on her behalf. GSI also supports restricted delegation via proxy certificates; however, to our knowledge, to date, this capability is not used in most situations. The Community Authorization Service (CAS [11]) extends the base support of GSI so that a person can obtain and exercise authorization rights based on the group to which they belong. Additionally, two projects focus on the generation and creation of the *gridmap* file, which is used to both authorize users and specific the local account to which the global (grid) account is to be mapped: VOMS [12] manages a list of "Virtual Organizations" that define this mapping, while Walden [13] retrieves this list from a local authorization source such as an LDAP server.

2.2 Campus/Enterprise Technologies

Internet2 is leading the development of many important middleware projects aimed at the campus environment, focusing on directories and PKI. Shibboleth [14] is an open-source, privacy-preserving federating software project to support inter-institutional sharing of web resources subject to access controls. Essentially, when a user at one institution (the "Origin") tries to use a resource at another (the "Target"), Shibboleth sends attributes about the user to the Target institution without having to log into the target institution. The Target institution grants access based on the attributes. The user controls what attributes are given to the Target institution (for example, it is not strictly necessary that the attributes include the name of the user, if the Target institution bases its decision on, say, "member of UVa"). Stanford's Signet [15] is software to define and manage an organization's privilege system, with special emphasis on how to take a role-based organization and develop appropriate groups, policies, and attributes to operate an authority service. Signet is *not* an authorization service, but rather integrates with authorization services. An example use of Signet is a Web-based interface for assigning workers on campus the ability to make constrained purchasing decisions for a limited period of time. Grouper [16] offers support for basic group management, with subgroups and compound groups. To date, none of these projects have been made Grid-aware in the manner in which we have integrated Pubcookie with MyProxy as described in this paper (an effort to make Shibboleth Grid-aware has just commenced as of this writing).

3 Integration Design and Implementation

In this section, we first give the details of Pubcookie, and then MyProxy, and then we give the design for how we integrate Pubcookie and MyProxy to effectively connect the campus to the Grid without requiring a second sign-on.

3.1 Pubcookie

As shown in Figure 1, Pubcookie [5] provides single sign-on authentication to web sites using existing site-wide authentication services. In this way, a user who has already authenticated for access to one web site can access other protected sites without having to enter another password. For example, a user who has authenticated to a web mail system could browse a separate web database system without needing to log in again.

Pubcookie provides the glue between the authentication service and the web site and is not responsible for the actual authentication or authorization of users. Pubcookie does not verify who the user is – it hands this task off to a separate authentication service (like Kerberos [17], LDAP, or NIS). Likewise, Pubcookie does not decide if a user should be allowed to access a given page or resource. Control of access to the resource is left to the originating application.

This is accomplished, as the name implies, with cookies. When a user initially attempts to access a protected site (Arrow #1 of Figure 1), the Pubcookie module at that site automatically redirects her browser to a Pubcookie login page with a "granting request" cookie containing the request (the URL) and a random number (#2). The user enters her username and password on the login page (#3), which Pubcookie verifies using the configured authentication service (#4 and #5). Pubcookie then returns two cookies to the user's browser (#6) and redirects the user back to the original site (#7).

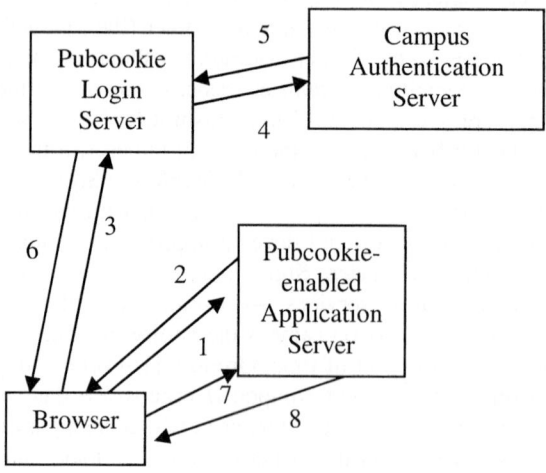

Fig. 1. Pubcookie Usage (without MyProxy integration)

The first of these cookies returned in step #6 of Figure 1 is a "granting cookie". This cookie authenticates the user to the target web site ("Pubcookie-enabled Application Server"). The cookie contains the information in the "granting request" cookie plus the authenticated username. When the Pubcookie login server redirects the browser back to the original web site (#7), the Pubcookie module at the web site verifies the "granting cookie" and determines if the user is authorized to access the web page. If the user is authorized, the web server returns the requested content (i.e., the web page) with a session cookie to authenticate subsequent requests at that site (#8). The second cookie returned by the login server is the Pubcookie "session cookie". As its name implies, this is the cookie that the user maintains throughout their Pubcookie session. When the user visits a new Pubcookie-enabled web site, it will again redirect her to the Pubcookie login server. However, now that the user has a Pubcookie "session cookie", she does not have to login again (Steps #4 and #5). The login server verifies the session cookie and automatically redirects the user back to the new web site with a new "granting cookie". Thus, the user has a single sign-on to all Pubcookie-enabled web sites.

There are two means by which the security of these cookies is guaranteed. First, they are encrypted. When Pubcookie is being set up, the Pubcookie login server (which issues the cookies) and the application server (i.e., the web server the user is trying to access) exchange a cryptographic key. This key is in turn used to encrypt and decrypt the data used by Pubcookie. The second method by which security is achieved is digital signatures. The Pubcookie login server digitally signs all cookies it issues, so that application servers can use the login server's public key to verify the authenticity of the cookies it issues.

3.2 MyProxy

MyProxy [9][10] is a service for securely storing GSI credentials. MyProxy uses the proxy delegation protocol to allow clients to retrieve short-lived proxy credentials without exporting long-lived keys from the MyProxy server. A dedicated MyProxy server provides more secure storage for user keys than a general-purpose workstation or file-server, and MyProxy can be integrated with cryptographic hardware to further protect user keys [18]. To retrieve a proxy credential from the MyProxy repository, the MyProxy client must first authenticate. Current versions of MyProxy support authentication via password, certificate, Kerberos, or PAM (Pluggable Authentication Modules). This allows users to retrieve proxy credentials from the MyProxy server whenever and wherever they are needed. Additionally, MyProxy can be used to renew credentials for trusted long-lived services.

MyProxy is widely used with grid portals, which provide a web interface to grid services. To allow users to interact with secure grid services through the portal interface, the portal must have access to the user's grid credentials, so it can perform secure operations on the user's behalf. To meet this requirement, users allow the grid portal to retrieve their credentials from the MyProxy server, by sending their MyProxy username and password to the portal, which it then uses to authenticate to MyProxy and retrieve the short-lived credentials. Given the security issues associated with web server platforms, this limits the vulnerability of user credentials on the portal by giving the portal access to credentials with a limited lifetime, with access logged at the MyProxy server, with the user able to change her MyProxy password at any time to deny the portal any further access to the credentials.

3.3 Integration

Because Pubcookie successfully implements single sign-on for the campus/enterprise environment, and MyProxy/GSI successfully implements single sign-on for the Grid environment, it is highly attractive to investigate the design, cost of implementation, and the implied security of using one credential as the basis for acquiring the *other* type of credential (i.e, Pubcookie-for-MyProxy or MyProxy-for-Pubcookie). Because our specific goal was to not disrupt the local campus environment in attempting to provide the ability to dynamically expand the local resources to a Grid such as the TeraGrid, we chose to use a valid Pubcookie as the basis for retrieving a GSI credential from MyProxy. More specifically, the user authenticates to the grid portal using the Pubcookie mechanism and the grid portal retrieve GSI credentials for the user from MyProxy by authenticating with the Pubcookie "granting cookie" without requiring the user to enter an additional password (for MyProxy). As shown in Figure 2, the system appears to the end-user as before (without MyProxy integration) but now has the option to expand onto the Grid, perhaps even without the local user realizing this is being done (Steps #8 through #11).

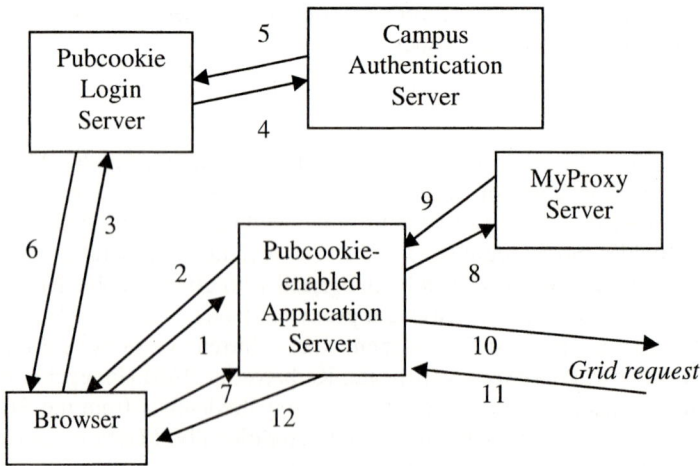

Fig. 2. Pubcookie Usage (with MyProxy integration)

To achieve this, we modified the Pubcookie-enabled Application Server software to send the signed authentication data from the decrypted cookie to MyProxy (#8), over an SSL authenticated and encrypted channel. The SSL channel is mutually authenticated, so the application server knows that it is communicating with the trusted MyProxy server, and the MyProxy server knows the identity of the application server, to be compared against the server name in the cookie, to ensure the cookie has not been stolen. For added security, the MyProxy server can be configured to only allow connections from specific trusted application servers (such as the specific Pubcookie-enabled Application Service shown in Figure 12). The MyProxy server then verifies the cookie's signature, verifies that the username in the cookie matches the MyProxy username, verifies that the server name in the cookie matches the portal's authenticated SSL name, and if verification succeeds, delegates the proxy credentials (#9). MyProxy uses the OpenSSL library to verify the signature; no Pubcookie code was added to the MyProxy software.

Our development was based on the widely-available source code for MyProxy from the MyProxy web page [10]. These modifications are being made as permanent additions to the MyProxy source code. The tests and verification were performed at the University of Virginia's Pubcookie installation, which is a modified version of Pubcookie 3.1.1. The modifications include the ability to use Radius authentication as the connection to the local authentication source and also SMB authentication.

It is important for the MyProxy username to match the Pubcookie username, so the portal uses the correct GSI credentials for each Pubcookie authenticated user. Multiple mechanisms are available to ensure this. First, the site's security administrators may load the MyProxy repository with user credentials under the correct usernames. This configuration relies on the MyProxy server, Pubcookie authentication service, and site Certification Authority (which creates the GSI credentials) being under the same administrative control. A second option is for the MyProxy administrator to maintain an access control list mapping Pubcookie usernames to MyProxy usernames. The mapping could support multiple Pubcookie servers by formatting Pubcookie usernames as *user@server*. The third option is for MyProxy users to set the Pubcookie authorization policy for their credentials when they upload them to the MyProxy repository, using MyProxy's existing per-credential authorization functionality. The MyProxy administrator must also configure the server to trust the Pubcookie login server signing key(s).

4 Conclusion

In many ways, the ultimate success of the Grid will be highly dependent on the support for and integration with existing legacy infrastructure. In this paper, we described our support for re-using existing campus/enterprise authentication infrastructure in combination with the Grid. That is, we have successfully and securely implemented modifications to MyProxy so that a valid Pubcookie-issued cookie could be used as the basis for acquiring a GSI credential.

Having successfully completed the modifications to MyProxy, we are currently rewriting our Grid portal that we have developed at the University of Virginia as part of the National Partnership for Advanced Computational Infrastructure (NPACI). The purpose of this portal is to provide an easy-to-use web interface for Computational Biophysics (focusing on applications such as CHARMM, Amber, and NAMD) [19]. This portal is currently protected via usernames and passwords. By making this portal protected by Pubcookie (while still supporting username/password when a user does not have the ability to be authenticated via Pubcookie), and making the portal connect to a Pubcookie-speaking Myproxy server, we significantly increase the ease-of-use for the computational scientist.

References

[1] Globus project, www.globus.org
[2] K. Czajkowski., Ferguson, D., Foster, I., Frey, J., Graham, S., Sedukhin, I., Snelling, D., Tuecke, S., Vambenepe, W. 2004. The WS-Resource Framework. http://www-106.ibm.com/developerworks/library/ws-resource/ws-wsrf.pdf
[3] WSRF.NET: The Web Services Resource Framework on the .NET Framework. http://www.ws-rf.net

[4] Humphrey, M., G. Wasson, M. Morgan, and N. Beekwilder (2004). An Early Evaluation of WSRF and WS-Notification via WSRF.NET. *2004 Grid Computing Workshop (associated with Supercomputing 2004).* Nov 8 2004, Pittsburgh, PA.
[5] Pubcookie project. www.Pubcookie.org
[6] TeraGrid project, www.teragrid.org
[7] Open Science Grid project, www.opensciencegrid.org
[8] Foster, C. Kesselman, G. Tsudik, S. Tuecke. A Security Architecture for Computational Grids. *Proc. 5th ACM Conference on Computer and Communications Security Conference*, pg. 83-92, 1998.
[9] J. Novotny, S. Tuecke, and V. Welch. An Online Credential Repository for the Grid: MyProxy. Proceedings of the Tenth International Symposium on High Performance Distributed Computing (HPDC-10), IEEE Press, August 2001.
[10] MyProxy on-line Credential Repository Project.
[11] L. Pearlman, V. Welch, I. Foster, C. Kesselman, S. Tuecke. A Community Authorization Service for Group Collaboration. *Proceedings of the IEEE 3rd International Workshop on Policies for Distributed Systems and Networks*, 2002.
[12] Virtual Organization Membership Service (VOMS). http://hep-project-grid-scg.web.cern.ch/hep-project-grid-scg/voms.html
[13] B. Kirschner, T. Hacker, W. Adamson, B. Athey. Walden: A Scalable Solution for Grid Account Management, *Proceedings of the 5th IEEE/ACM International Workshop on Grid Computing.* November 8, 2004. Pittsburgh, PA.
[14] Shibboleth. http://shibboleth.internet2.edu
[15] Internet2 MACE – Signet. http://middleware.internet2.edu/signet/
[16] T. Barton and B. Christiansen, eds. Grouper Phase 1 Specifications. Draft of 3 May 2004. Available at: http://home.uchicago.edu/~tbarton/draft-barton-christensen-grouper-phase1-specs-04.html
[17] B.C. Neuman, and T. Ts'o. Kerberos: An authentication service for computer networks. *IEEE Communications Magazine*, 32(9):33-38, September 1994.
[18] M. Lorch, J. Basney, and D. Kafura, "A Hardware-secured Credential Repository for Grid PKIs," 4th IEEE/ACM International Symposium on Cluster Computing and the Grid (CCGrid2004), Chicago, Illinois, April 19-22, 2004.
[19] NPACI Computational Biophysics Portal. https://wumpus.cs.virginia.edu/ NPACI-Computational BiophysicsPortal/

Generating Parallel Algorithms for Cluster and Grid Computing*

Ulisses Kendi Hayashida[1], Kunio Okuda[1], Jairo Panetta[2],
and Siand Wun Song[1]

[1] Universidade de São Paulo, Brazil
{ulisses, kunio, song}@ime.usp.br
http://www.ime.usp.br/~song/
[2] Instituto Nacional de Pesquisas Espaciais,
Centro de Previsão de Tempo e Estudos Climáticos, Brazil
panetta@cptec.inpe.br

Abstract. We revisit and use the dependence transformation method to generate parallel algorithms suitable for cluster and grid computing. We illustrate this method in two applications: to obtain a systolic matrix product algorithm, and to compute the alignment score of two strings. The product of two $n \times n$ matrices is viewed as multiplying two $p \times p$ matrices whose elements are $n/p \times n/p$ submatrices. For m such multiplications, using p^2 processors, the proposed parallel solution gives a linear speedup of $\frac{mp^3}{(m+2)p-2}$ or roughly p^2. The alignment problem of two strings of lengths m and n is solved in $O(p)$ communication rounds and $O(mn/p)$ local computing time. We show promising experimental results obtained on a 16-node Beowulf cluster and on an 18-node grid called InteGrade, consisting of desktop computers.

1 Introduction

The abundance of low cost computing resources in clusters and the increasing network bandwidth make it attractive to run parallel applications with the so-called grid-based computing. We attempt to use efficiently the computational resources that are idle to obtain a system of high computing power with the existing machines. Due to the high communication cost in cluster and grid computing, we are interested in designing parallel applications with low demand on communication. To this end, we propose a method to design parallel algorithms with the nice property of each processing having to communicate with only a few others. It is based on a modification of the dependence transformation method that was originally proposed to design systolic arrays for VLSI implementation on silicon chips. Given a sequential algorithm specified as nested loops, or more formally as a system of uniform recurrence equations, the specified computation

---

* Partially supported by CAPES, CNPq Grant Nos. 55.2028/02-9 and 30.5218/03-4.

can be transformed into a time-processor space domain adequate to be implemented on a VLSI chip [2, 3, 7, 6, 8, 10].

Some nice properties of systolic arrays include the regularity on the layout of the processing elements and local communication where each processing element communicates only with a few neighbor processors. These features make it suitable for implementation on a cluster where we wish to avoid costly global communication primitives. We illustrate the proposed method by deriving parallel algorithms for matrix multiplications and for computing the alignment score for string comparison. We then present implementation results on a 16-node Beowulf cluster and also on an 18-node grid, called InteGrade [4], consisting of desktop computers.

2 Dependence Transformation

Given a sequential algorithm expressed as nested loops or, more formally, as a system of uniform recurrence equations [5], the dependence transformation method [3, 10] transforms the computations involved into a time-space representation. We illustrate this method through an example.

Example 1. Given two $n \times n$ matrices $A = (a_{ij})$ and $B = (b_{ij})$, the matrix product can be expressed by a system of uniform recurrence equations, defined on the domain of all the points (i, j, k) for $0 \leq i, j, k < n$.

$0 \leq i < n, 0 \leq j < n, 0 \leq k < n,$
$C(i, j, k) = C(i, j, k-1) + A(i, j-1, k)B(i-1, j, k)$
$A(i, j, k) = A(i, j-1, k)$
$B(i, j, k) = B(i-1, j, k)$

$A(i, -1, k)$ and $B(-1, j, k)$ are the coefficients a_{ik} and b_{kj}, respectively. The values of $C(i, j, -1)$ are assumed to be zero. The output values $C(i, j, n-1)$, at the right side of the equations, give the desired product. To compute $C(i, j, k)$ we use the value $C(i, j, k-1)$ that needs to be computed earlier. We say there is a dependence vector for variable C, denoted by $\theta_c = \begin{pmatrix} 0 & 0 & 1 \end{pmatrix}^T$. Likewise we have the dependence vectors $\theta_a = \begin{pmatrix} 0 & 1 & 0 \end{pmatrix}^T$ and $\theta_b = \begin{pmatrix} 1 & 0 & 0 \end{pmatrix}^T$.

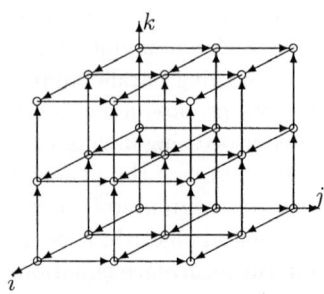

Fig. 1. Dependence graph assuming $n = 3$

Consider a *dependence graph* which has as vertices the points of the domain and directed edges from vertices y to z if z depends on y. See Figure 1 for an example. Each point of the dependence graph represents some elementary computation that needs to be calculated. Through a *time function* τ and a *processor allocation function* α, we can transform the dependences so that the points that do not depend on each other can be computed in parallel.

The time function maps each elementary computation of the dependence graph into a positive integer that represents the time unit it is executed. Consider a domain point $z = \begin{pmatrix} z_0 & z_1 & z_2 \end{pmatrix}^T$. The following is a time function: $\tau(z) = \lambda_0 z_0 + \lambda_1 z_1 + \lambda_2 z_2 + \delta$, where λ_i and δ are integers, and satisfying the condition: if z depends on y, then $\tau(z) > \tau(y)$.

Let $\lambda = \begin{pmatrix} \lambda_0 & \lambda_1 & \lambda_2 \end{pmatrix}^T$. Then $\tau(z) = \lambda^T z + \delta$. Suppose z depends on y by the dependence vector θ, that is, $z - y = \theta$. Then we have $\tau(z) > \tau(y)$ or, equivalently, $\tau(z) - \tau(y) > 0$. This can be written as $\lambda^T(z - y) > 0$, or $\lambda^T \theta > 0$, or $\lambda^T \theta \geq 1$. Thus for each dependence vector θ_i we have $\lambda^T \theta_i > 0$.

Consider Example 1, we have $\theta_c = \begin{pmatrix} 0 & 0 & 1 \end{pmatrix}^T, \theta_a = \begin{pmatrix} 0 & 1 & 0 \end{pmatrix}^T, \theta_b = \begin{pmatrix} 1 & 0 & 0 \end{pmatrix}^T$. It can be shown [10] that $\tau(z) = z_0 + z_1 + z_2$ is a time function. That is $\tau(z) = \lambda^T z$ where $\lambda = \begin{pmatrix} 1 & 1 & 1 \end{pmatrix}^T$.

The processor allocation function α maps each point z (of a domain of n dimensions) onto a point (of a space $n-1$ dimensions) $\alpha(z)$ where the elementary computation associated to z is performed. Different points of the domain that are computed at the same time instant should be mapped to different processing elements, that is, $\forall z, y$ in the domain, $\alpha(z) = \alpha(y) \Rightarrow \tau(z) \neq \tau(y)$.

Quinton and Robert [10] show how to obtain a processor allocation function α, given the time function λ, by projecting the domain points according to a direction given by a vector u that is not orthogonal to λ. Consider a domain of dimension 3. Let $u = (u_0\ u_1\ u_2)^T$ be a non-null vector such that $\lambda^T u \neq 0$. It can be shown [10] that $\alpha(z) = (\alpha_0(z), \alpha_1(z))$ is a processor allocation function where $\alpha_0(z) = u_2 z_0 - u_0 z_2$ and $\alpha_1(z) = u_2 z_1 - u_1 z_2$. In our example, we wish to obtain a processor allocation function α that maps each one of the dependence vectors onto either the null vector or $(0\ 1)^T$ or $(0\ 1)^T$. The following is a possible processor allocation function.

$$\alpha(z) = \begin{pmatrix} 1 & 0 & 0 \\ 0 & 1 & 0 \end{pmatrix} \begin{pmatrix} z_0 \\ z_1 \\ z_2 \end{pmatrix} = \begin{pmatrix} z_0 \\ z_1 \end{pmatrix}.$$

Figure 2 shows how this systolic array computes the matrix product of two matrices A and B. Each processing element receives an element of matrix A and an element of matrix B, computes the product of the two elements and adds it to the element of matrix C it stores. This parallel solution requires a total of $3n-2$ computing steps plus the same amount of communication steps. We obtain a speedup in terms of computing steps of $n^3/(3n-2)$ or $n^2/3$ for large n. Notice that we waste some time to fill in and flush out the pipeline. If we need to obtain m matrix products, then the parallel time is improved with the amortization of the costs of pipeline fill-in and flush-out. For m products of two $n \times n$ matrices,

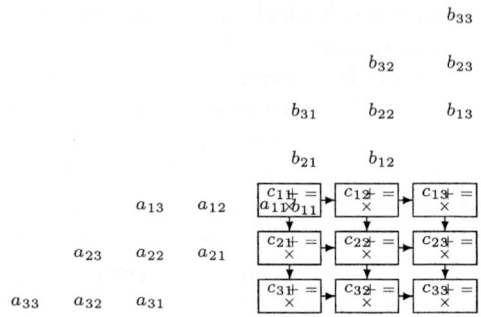

Fig. 2. Computing the product $A \times B$

we need a total of $(m+1)n - 1$ computing steps and communication steps. We now get a speedup in terms of computing steps of $(mn^3/((m+1)n-1)$ or n^2 for large n and m. Since we use n^2 processing elements, we get linear speedup.

Systolic arrays operate in a synchronous and lock-step fashion. Frequent synchronization of all the processor on a cluster or grid is too costly. However, synchronization of the *input, compute, output* steps of the parallel algorithm on the cluster can be achieved by using non-blocking sends and blocking receives.

The fine granularity of systolic algorithms is not suitable for cluster or grid computing due to the disparity between communication and computations speeds. To achieve a coarser grain, we can view each element of a matrix as a block or a submatrix. In other words, instead of each processor receives one single element of each matrix A and matrix B, it receives a submatrix of matrix A and matrix B. It can be shown that we still get linear speedup.

In the global atmospheric circulation model used for weather forecasting computation of Fourier and Legendre Transforms are needed [9]. A global model simulates the behavior of atmospheric fields along the discrete time. Partial differential equations are solved in two spaces: the grid space (here we use the term grid in the Mathematical sense) and the spectral space. The Fourier Transform moves a field between grid and Fourier representations while the Legendre Transform moves a field between Fourier and spectral representations. Fourier Transforms are computed by using the well-known FFT (fast Fourier Transform). Legendre Transforms constitute the most time-consuming part and can be computed as multiple matrix products, handled by the proposed algorithm.

3 A Parallel Algorithm for Alignment of Sequences

In [1] we have proposed a parallel algorithm for efficient sequence alignment. In this section we show that, by using the proposed dependence transformation method, it is straightforward to generate this algorithm. Consider two given strings A and C, where $A = a_1 a_2 \ldots a_m$ and $C = c_1 c_2 \ldots c_n$. To align the two strings, we insert spaces in the two sequences in such way that they become equal

A	a c t t c a - t
C	a t t c - a c g
Score	1 0 1 0 0 1 0 0	3

A	a c t t c a - t
C	a - t t c a c g
Score	1 0 1 1 1 1 0 0	5

Fig. 3. Examples of alignment

in length. See Figure 3 where each column consists of a symbol of A (or a space) and a symbol of C (or a space). An *alignment* between A and C is a matching of the symbols of A and of C in such way that if we draw lines between the matched symbols, these lines cannot cross each other. Figure 3 shows two simple alignment examples where we assign a score of 1 when the aligned symbols in a column match and 0 otherwise. The alignment on the right has a higher score (5) than that on the left (3).

A more general score assignment for a given string alignment considers insertion/deletion and match/mismatch of symbols, each with the respective scores. For example, a match has a positive score while the other operations negative scores. The similarity score S of the alignment between strings A and C can be computed by the following recurrence equation. For $0 < r \leq m, 0 < s \leq n$,

$$S(r, s) = \max \begin{cases} S[r, s-1] - t \\ S[r-1, s-1] + t & \text{(match) or} \\ S[r-1, s-1] - t & \text{(mismatch)} \\ S[r-1, s] - t \end{cases}$$

An $l_1 \times l_2$ *grid DAG* (Figure 4) is a directed acyclic graph whose vertices are the $l_1 l_2$ points of an $l_1 \times l_2$ grid, with edges from grid point $G(i,j)$ to the grid points $G(i, j+1)$, $G(i+1, j)$ and $G(i+1, j+1)$. In our terminology, the grid DAG is a dependence graph. Associate an $(m+1) \times (n+1)$ grid dag G with the alignment problem as follows: the $(m+1)(n+1)$ vertices of G are in one-to-one correspondence with the $(m+1)(n+1)$ entries of the S-matrix. It is easy to see that we need to compute a minimum source-sink path in the grid DAG. In Figure 4 the problem is to find the minimum path from (0,0) to (8,10).

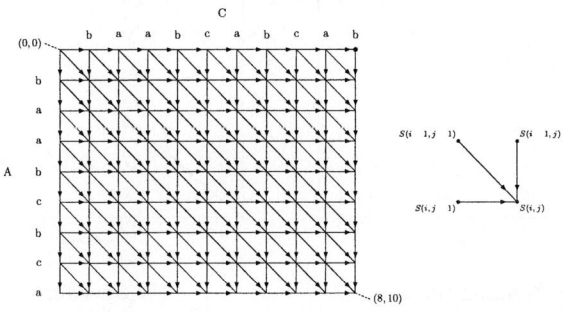

Fig. 4. Grid DAG G for $A =$ baabcbca and $C =$ baabcabcab

Let us now apply the dependence transformation method. From the recurrence equation, we obtain the dependence vectors: $d_1 = \begin{pmatrix} 0 & 1 \end{pmatrix}^T$, $d_2 = \begin{pmatrix} 1 & 1 \end{pmatrix}^T$, $d_3 = \begin{pmatrix} 1 & 0 \end{pmatrix}^T$. A possible time function is $\tau(z) = z_0 + z_1 = \lambda^T z$ where $\lambda = \begin{pmatrix} 1 & 1 \end{pmatrix}^T$. For the processor allocation function α, we can use $u = \begin{pmatrix} 1 & 0 \end{pmatrix}^T$.

This will result in the following scheduling of the tasks to be executed by p processors, each with $O(mn/p)$ local memory. The string A is stored in all processors, and the string C is divided into p pieces, of size $\frac{n}{p}$, and each processor P_i, $1 \leq i \leq p$, receives the i-th piece of C ($c_{(i-1)\frac{n}{p}+1} \ldots c_{i\frac{n}{p}}$). P_i^k denotes the work of Processor P_i at round k. Thus initially P_1 starts computing at round 0. Then P_1 and P_2 can work at round 1, P_1, P_2 and P_3 at round 2, and so on. In other words, after computing the k-th part of the sub-matrix S_i (denoted S_i^k), processor P_i sends to processor P_{i+1} the elements of the right boundary (rightmost column) of S_i^k. These elements are denoted by R_i^k. The systolic algorithm requires $O(p)$ communication rounds and $O(mn/p)$ local computing time.

4 Experimental Results

We ran our experiment on a 16-node Beowulf cluster, using the LAM-MPI interface. Each node has a 1.2GHz AMD Thunderbird Athlon processor, 256 KB L2 cache, 768 MB of RAM memory and 30.73 GB hard disk. The nodes are connected by a Switch 3COM 3300 FastEthernet 100Mb. We also used an 18-node grid using *InteGrade* (see [4]) middleware that provides efficient use of desktop microcomputers that are idle and available in a computer laboratory for graduate students, consisting of Pentium II 400 Mhz and 3 Athlon 1700+ desktop microcomputers interconnected in a local area network by 100Mb Ethernet. We have implemented and used some of the communication primitives of the Ox-

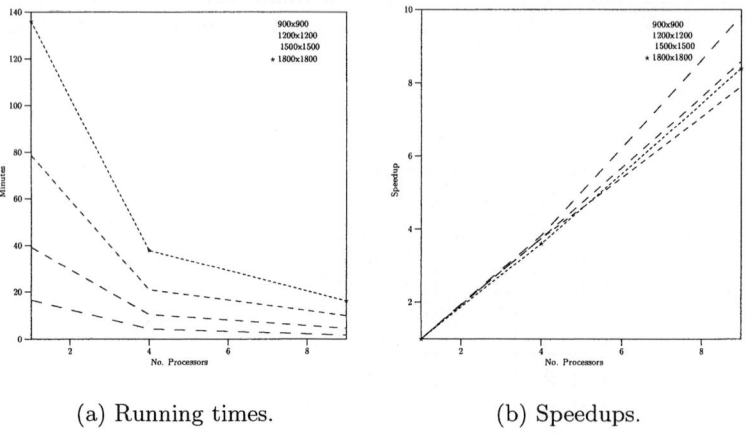

(a) Running times.   (b) Speedups.

Fig. 5. Matrix product results on a Beowulf cluster

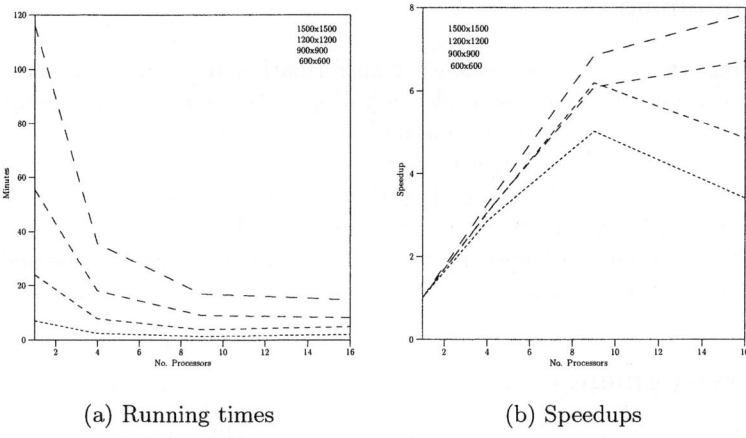

(a) Running times   (b) Speedups

Fig. 6. Matrix product results on the InteGrade grid

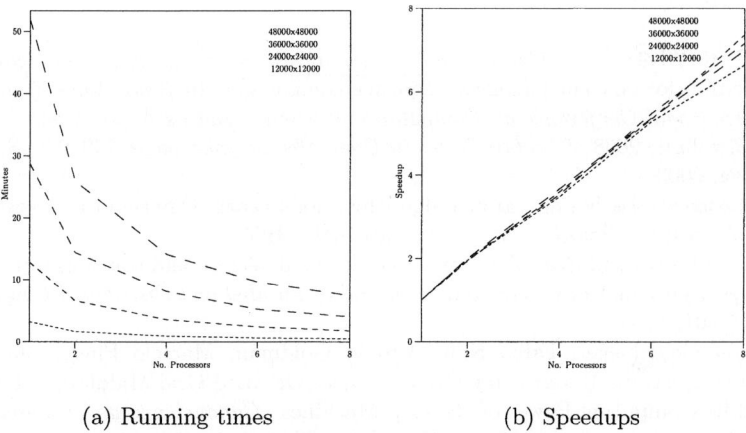

(a) Running times   (b) Speedups

Fig. 7. String alignment results on the InteGrade grid

ford BSPLib such as bsp_put() and bsp_get(). The number of processors used is always a perfect square (1, 4, 9, etc),

Figure 5 shows the results obtained on the Beowulf cluster, for 50 matrix products of sizes 900×900 to 1800×1800. The super-linear behavior on some of the speedup curves can be explained by the more efficient use of cache in the parallel program. The speedups start to decrease for large matrix sizes, when the communication costs start to dominate in relation to the overall computation time. Figure 6 shows the matrix product results on the InteGrade grid of desktop computers. Figure 7 show the results for the string alignment algorithm, also run on the InteGrade grid.

5 Conclusion

We propose to use the dependence transformation method, to generate parallel algorithms for cluster or grid computing, after some suitable adaptation. It should be observed that the method is general. In fact we have used this method to derive parallel algorithms for other problems, such as convolution, and transitive closure. All these algorithms share the desirable property of local communication with a few other processors, with no global communication. TCP/IP connection start-up is heavy. Thus an important issue to be addressed is the reuse of connections

Acknowledgments

We wish to thank the anonymous referees for their helpful comments.

References

1. C. E. R. Alves, E. N. Cáceres, F. Dehne, and S. W. Song. A parallel wavefront algorithm for efficient biological sequence comparison. In *Proceedings of the 2003 International Conference on Computational Science and its Applications - ICCSA 2003*, volume 2668 of *Lecture Notes in Computer Science*, pages 249–258. Springer Verlag, 2003.
2. M. Cosnard. Designing parallel algorithms for linearly connected processors and systolic arrays. *Parallel Computing*, 1:273–317, 1990.
3. J. A. B. Fortes and D. I. Moldovan. Parallelism detection and transformation techniques useful for VLSI algorithms. *Journal of Parallel and Distributed Computing*, 2:277–301, 1985.
4. Andrei Goldchleger, Fabio Kon, Alfredo Goldman, Marcelo Finger, and Germano Capistrano Bezerra. InteGrade: Object-Oriented Grid Middleware Leveraging Idle Computing Power of Desktop Machines. *Concurrency and Computation: Practice and Experience*, 16:449–459, March 2004.
5. R. M. Karp, R. E. Miller, and S. Winograd. The organization of computations for uniform recurrence equations. *Journal of the ACM*, 14:563–590, 1967.
6. G. M. Megson. *An Introduction to Systolic Algorithm Design*. Oxford University Press, 1992.
7. D. I. Moldovan. *Parallel Processing: from Applications to Systems*. Morgan Kaufmann Publishers, 1993.
8. K. Okuda. Cycle shrinking by dependence reduction. In *Proceedings 2nd International Euro-Par Conference*, volume 1123 of *Lecture Notes in Computer Science*, pages 398–401. Springer Verlag, 1996.
9. S. A. Orszag. Transform methods for calculation of vector coupled sums: Application to the spectral form of the vorticity equation. *Journal Atmospheric Science*, 27:890–895, 1970.
10. P. Quinton and Y. Robert. *Algorithmes et architectures systoliques*. Masson, 1989.

Relationship Networks as a Survivable and Adaptive Mechanism for Grid Resource Location

Lei Gao[1] and Yongsheng Ding[1,2,*]

[1] College of Information Sciences and Technology
[2] Engineering Research Center of Digitized Textile & Fashion Technology,
Ministry of Education, Donghua University,
Shanghai 200051, P. R. China
ysding@dhu.edu.cn

Abstract. Grid technologies have emerged to enable large-scale flexible resource sharing among virtual organizations. Resource location of next generation grid systems needs to be fully decentralized, more importantly, be survivable and adaptive to uncertain factors in grid environments. This paper applies the concepts and principles of relationship networks in human society to design a dynamic location approach. The simulation results have proved that our approach can: form relationship clusters to improve the location performance effectively; well adapt to different resource distributions and user request patterns; and survive from the changes of dynamic environments such as partial failure of agents.

1 Introduction

Grid systems have evolved over nearly a decade to enable large-scale flexible resource sharing among dynamic virtual organizations (VOs) [1]. The next generation grid systems [2] will exist in an unstable environment (where a large number of nodes join and leave the grids frequently). When this happens, new services that can survive and adapt to a user's environment have to be developed.

Resource location means that given a description of desired resources, a location mechanism can return a set of contact addresses of resources that match the description in a wide-area grid environment. As a critical activity in grid systems, it inevitably faces the challenges brought by next generation grid systems. It needs to be fully decentralized, more importantly, be survivable and adaptive to uncertain factors (such as different resource distribution and variable user request patterns) in grid environments. Unfortunately, existing grid location technologies such as Globus's MDS have not addressed the survivability and adaptability of dynamic location. So, it is valuable for us to look at the location approach in other fields. In the real life, it is a natural way for us to depend on our relationship networks for finding relevant information: our acquaintances, our acquaintances' acquaintances, and so on. These relationship entities and relationships among them can provide a fully decentralized, naturally adaptive and effective way to gather and disseminate the resource information.

Inspired by this, we study an approach that places the intelligence on the grid entities, enabling the users to locate desirable resources based on relationship networks. The work in this paper is the further results of our previous research [3], which viewed a grid as interacting agents and applied some key mechanisms of natural ecosystems to build a grid middleware named Ecological Network-based Grid Middleware (ENGM). The rest of this paper is organized as follows. Based on ENGM platform, Section 2 presents a basic framework for resource location. Relationship networks-based location mechanism is demonstrated in Section 3, which is followed by a simulation and corresponding analysis in Section 4. Section 5 briefly describes related work in P2P field and concludes our efforts.

2 A Wide-Area Resource Location Framework

2.1 Design of Ecological Network-Based Grid Middleware

From the bottom up, the architecture of ENGM system is presented as three-layers: Heterogeneous and Distributed Resources layer, ENGM layer, and Grid Applications for VOs layer. (1) *Heterogeneous and Distributed Resources* consist of different types of resources available from multiple service providers distributed in grids. Via a java virtual machine, an ENGM platform can run on a heterogeneous distributed system that established in a network node. (2) *ENGM* provides the services support a common set of applications in distributed network environments. It is made up by ENGM functional modules, ENGM core services, grid agent survivable environment, and emergent grid common service. (a) ENGM functional modules deal with the management of networks and systems. (b) ENGM core services layer provides a set of general-purpose runtime services that are frequently used by agents such as niche sensing service. (c) Grid agent survivable environment is runtime environment for deploying and executing agents that are characterized by high demand for computing, storage and network bandwidth requirements. (d) Emergent grid common services are kernel of middleware and responsible for resource allocation, information service, task assignment, and so on. These common services are emerged from the autonomous agents. (3) *Grid Applications for VOs* use developing kits and organize certain agents and common services automatically for special purpose applications.

2.2 Framework for Resource Location

Every grid resource supplier maybe has one or more servers that store and provide access to their resources. We regard these servers as nodes. An ENGM platform-based wide-area framework is set up on these nodes. Furthermore, some nodes can form a *niche* that refers to a logically defined area where agents can learn their surrounding environment. For instance, an agent may sense which agents are in the niche, what services they perform, and which resources it can access. Here, a niche can be regarded as a VO.

Agents in the framework fall into two categories: grid user agents (GUAs) and grid service agents (GSAs). A GUA represents a kind of user tasks. GSAs are used to comprise the main components of ENGM, such as grid information service agent (GISA). Agents are represented on their unit functions. Every unit function is defined as a metadata structure that enables collaboration among agents. The metadata of an agent consists of *agentID* (a global unique identifier of an agent), *agentAddress* (location of this agent), *serviceType* (service type of this agent), *serviceDescription* (description of service information), and *relationshipDescription* (information about its acquaintances, agents know each other are called acquaintances). Besides basic information of its acquaintances, *relationshipDescription* of an agent still consists *TrustCredit* (indication of reliability to acquaintance) and *collaborationRecord* (collaboration history records).

A GUA makes search instructions from a user into request messages and send them to GISA in local niche. The GISA responds with the matched resource descriptions if it has them locally, otherwise it forwards the requests to another GISA outside the niche until the request hit returns or request time exhausts. Note that there is only one GISA in a niche. Each GISA only knows about a subset of all GISAs, then it can select one to forward, and the selected one delivers the request to one of its acquaintances, and so on. Broadcast-type search strategies will lead to high bandwidth cost, scalability problem and congestion constraints. Instead of broadcasting a request to a large fraction of the network, a request is only passed onto one node at each step in our framework.

3 A Relationship Network-Based Location Approach

A three-phased location approach based on relationship networks is given out as follows: GISA relationship construction, request processing strategy and trust-based reconstruction of relationship. GISA relationship construction is responsible for collecting and updating information about the currently participating GISAs and for the forming relationship networks. Request processing strategy performs the search itself. Trust-based reconstruction of relationship makes the necessary preparations for a more efficient search.

1. GISA relationship construction. A GISA joins the grid by contacting a member GISA. Contact addresses of member GISA can be learned through sensing mechanism integrated in ENGM [3]. Once a nearby GISA is found, a relationship can be established. GISAs can be aware of their acquaintances and update the relationships of changed GISA.

2. Request processing strategy. A GUA receives a search instruction sent by a grid user in local niche, and then it creates a request message. A request message contains the message ID, information on request originator and the forwarder, a set of parameters to specify target resource attributes and a set of weights to describe the importance degree of each request resource attribute. When a GISA receive a request message, it first examines the message ID to check whether it has seen the message in the past. If has, it discards the message. Otherwise, it evaluates the request by its resource matching model and request forwarding

model. Then, the GISA decides to respond the request with its local resource, forward the request to its acquaintances, or reject the request. A GISA can both respond the request and forward the search request. Next, we give out resource matching model and request forwarding model as follows. The parameters used to describe the target resources are defined as the form of vector space. Grid resources provided in a peer are modeled as an attribute vector. The resources in the request message are also modeled as an attribute vector.

(a) Resource matching model: Given a request vector $R = \langle r_1, r_2, \ldots, r_n \rangle$, a weight vector $W = \langle w_1, w_2, \ldots, w_n \rangle$ (where $\sum_{t=1}^{n} w_t = 1$) that indicates the importance degree of each request attribute about R, and a grid resource vector $G = \langle g_1, g_2, \ldots, g_n \rangle$, the matching strength between and is defined as: $MS_{vec}(R, W, G) = \sum_{t=1}^{n} MS_{sin}(r_t, g_t) \cdot w_t$. If g_t satisfies r_t, $MS_{sin}(r_t, g_t) = 1$. Otherwise, $MS_{sin}(r_t, g_t) = 0$. For example, a single request attribute value r_t represents *more than 128 MB of available memory* and a single resource attribute value g_t stands for *256 MB of available memory*. So, we have $MS_{sin}(r_t, g_t) = 1$. In addition, the request originator will specify a threshold for resource matching denoted as ω_i ($0 \leq \omega_i \leq 1$). If $MS_{vec}(R, W, G) \geq \omega_i$, there is a matching between R and G.

(b) Request forwarding model: Assume that A_i is an any GISA and A_{i_j} is its one of acquaintances. The forwarding strength that A_i imposes on A_{i_j} is relative to *trustCredit* and *collaborationRecord*. The optimal matching strength on requests in *collaborationRecord* is defined as $MS_{opt}(R, W, R_{i_j}^k) = \max_{k=1}^{m} MS_{vec}(R, W, R_{i_j}^k) \cdot \delta_{i_j}^k$, where R and W have the same meaning in resource matching model, $R_{i_j}^k = \langle r_{i_j}^{k_1}, r_{i_j}^{k_2}, \ldots, r_{i_j}^{k_n} \rangle$ is a request vector A_{i_j} previously answered and/or forwarded and $\delta_{i_j}^k$ is a parameter in $[-\frac{1}{2}, 1]$ that indicates the user evaluation of location with $R_{i_j}^k$. Given a weight η to *trustCredit* and *collaborationRecord*, the forwarding strength that A_i imposes on A_{i_j} that is defined as $FS_{vec}(R, W, A_i, A_{i_j}) = \eta \cdot trust_{i,i_{i_j}} + (1 - \eta) \cdot MS_{opt}(R, W, R_{i_j}^k)$, where $trust_{i,i_j}$ is a number between $[0, 1]$ that represents *trustCredit* value that A_i has on A_{i_j}. If optimal forwarding strength $FS_{opt}(R, W, A_i, A_{i_k}) = \max_{k=1}^{n} FS_{vec}(R, W, A_i, A_{i_k})$ is obtained, a decision can be easily made by A_i regarding which acquaintance is more likely to lead towards desired resources. A_i can specify a threshold for request forwarding $\theta_i (0 \leq \theta_i \leq 1$ and usually $\theta_i \leq \omega_i)$. If $FS_{opt}(R, W, A_i, A_{i_k}) \geq \theta_i$, A_i will forward the request message. What's more, A_i will discard the message if it neither responses the request with its local resources nor forwards the request.

3. Trust-based reconstruction of relationship. This dynamic mechanism on reconstruction of relationship, can contribute to resource location. It updates and strengthens relationship networks by establishing and changing the *trustCredit* values among the GISAs. Reliability is expressed through a *trustCredit* value with which each GISA labels its acquaintances. On receiving a request hit, the request originator returns a defray message including a collaboration record and a credit that stands for user evaluation of request hit based on service defraying model. A credit could be a reward or a penalty, which indicates the degree of

its preference of the received request hit. This message is propagated through the same path where the location request has been originally forwarded. When an intermediate GISA on the path receives message, it adjusts the *trustCredit* value of the relationship that has been used to forward the original location request. *trustCredit* value is increased for a reward (i.e., high degree of the request originator's preference), and is decreased for a penalty (i.e., low degree of the request originator's preference). Service defraying model is given out as follows. Given a credit γ ($-\frac{1}{2} \leq \gamma \leq 1$) which stands for a reward or penalty contained in a defray message, A_i updates the *trustCredit* value of A_{i_j} using following method: if $\gamma \geq 0, trust^+_{i,i_j} = trust^-_{i,i_j}(1-\gamma^2) + \gamma^2$; if $\gamma < 0$, $trust^+_{i,i_j} = trust^-_{i,i_j}(2 - \frac{1}{1+\gamma})$, where $trust^-_{i,i_j}$ is a number in $[0,1]$ that represents *trustCredit* value which A_i imposes on A_{i_j} before updating. Correspondingly, $trust^+_{i,i_j}$ is the *trustCredit* value after updating. We have chosen the above method with the purpose of remaking ratings rise slowly and fall quickly.

To evaluate the approach, a simulator is developed upon ENGM platform. The simulation setup, results and analysis are described in next section.

4 Simulation Study

4.1 Simulation Setup

1. Initial relationship network topology. Define that a cycle starts from a request message sent by a GISA and ends till all the relevant messages disappear in the system, and 100 cycles is a generation in the simulation. We adopt the generation method of network topology proposed in [4] to establish relationship networks, whose nodes are GISAs and whose edges connect pairs of GISAs those know each other. The generated topology follows power law. It is pointed out emphatically that the statistic results from domain level and router level also follow above power law rule, and GISAs just play their roles on these two levels.

2. Resource distribution. Two involved thresholds, θ_1 and ω_i are predefined as fixed values. We make a set of common resources (contains 20000 different resource vectors) and a set of new-type resources (contains 2000 different resource vectors that are completely different from common resources). We experiment on two strategies. (a) Balanced distribution strategy: initially each GISA provides 3-5 resource vectors, which are randomly picked out from the common resource set. (b) Unbalanced distribution strategy: a few number of GISAs provide most of the resource vectors from common resource set, while the large number of GISAs share the small part of the resources.

3. User requests and user evaluation. Requests are initiated at a fixed percentage of randomly selected GISAs and contain resource vectors. The resource attributes of each request have the same weight. Two user request patterns are studied: (a) unbiased user request scenario (requesting all the vectors randomly) and (b) biased user request scenario (using a great probability to request a small part of and special vectors available in the simulation network). We randomly assign a fixed threshold to each GISA for indicating the satisfaction level of a

request originator. On receiving a request hit, the request originator will examine the match degree of responded resources. If the matched result is no greater than the threshold, the originator returns the rate message with a credit 0.1 to resource responder; otherwise, the credit will be -0.1. The GISAs in the credit-propagated chain will be given the same credit.

4.2 Simulation Results and Analysis

The simulation evaluates the survivability and adaptability from 3 aspects: resource distributions, user request patterns, and the reliability of GISAs. Also, we have a discussion of the performance distinctions affected by different η (0, 0.25, 0.5, 0.75, 1). Considering the randomness, we repeat the experiments multiple times. The results given out are average values of measurements.

A. Effects of Resource Distribution. Fig. 1 shows the performance results with different η in balanced and unbalanced resource environments. At the beginning of simulation, relationships are random, and location performs poorly. Many hops need to be visited to hit the target GISA. As more simulation cycles elapse, GISA gradually obtain many relationships similar to themselves, leading to improved performance in location process. We find that such improvement results from the clustering of vector-matched GISAs. The clusters have not formed in the process of random forwarding simulation, and the location will go aimlessly till it meets the matched GISA. So the random forwarding has the lowest efficiency, though it has lowest cost (no need to store any information in GISAs). The comparison between the results has proved that our approach can form clusters and improve the location performance adaptively.The condition with $\eta = 1$(locate totally according to *trustCredit* value) takes the minimum overhead costs, but it is with the lowest location efficiency in this scheme. While $\eta = 0$ (locate totally according to *collaborationRecord*), it is with the highest location efficiency, but not greater consumption. In unbalanced environments, the efficiency of the *trustCredit*-based location (with minor value of η) is much better than that in balanced environments. When a few number of GISAs hold most of the resources and so have rather more relationships, it is easier for them to form clusters. Compared to in a balanced environment, in a highly unbalanced environment, a GISA that had already responded some requests is more likely to have answers to other requests as well.

B. Influence of User Request Scenario. Comparing Fig. 1 with Fig. 2, we can see that, as a whole, the performance of our approach is better within biased request scenario than unbiased request scenario. Especially, the performance with $\eta = 1$ has evidently been improved, which means that location fully according to *trustCredit* value can adaptive to the relationship networks very well and can better support the special user request scenario. It shows very good efficiency in the unbalanced environments, response latency in this environment is almost half of that in the balanced environments. Similarly, in the unbalanced environments, the performance has been improved comparing to the balanced environments as η equals 0.75, 0.5 and 0.25. Relative to the unbalanced envi-

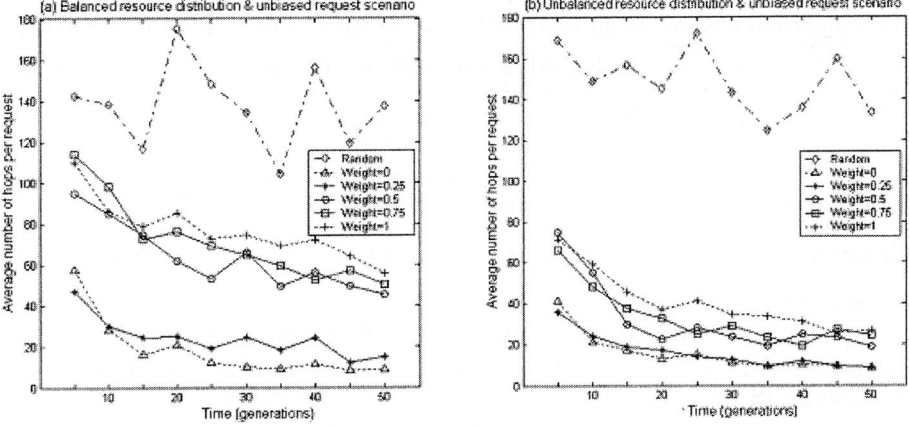

Fig. 1. Average number of hops per request as a function of simulation time, within unbiased user request scenario and 5000 GISAs, for two resource distribution(Left: balanced. Right: unbalanced)

ronments, when $\eta = 0$, the performance is better in the balanced environments, which demonstrates that the fair distribution of information can help to find a useful GISA as there is no relevant previous information of the current request.

C. Reliability of GISAs. We adopt the same simulation setup as the right of Fig. 2 (unbalanced and biased) to conduct a comparison when the GISAs are

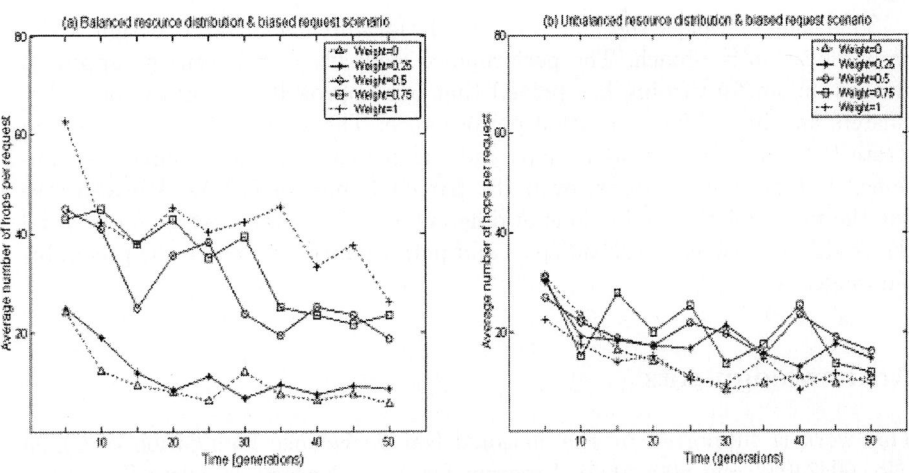

Fig. 2. Average number of hops per request as a function of simulation time, within biased user request scenario and 5000 GISAs in two environments(Left: balanced. Right: unbalanced)

not always available. At each generation, we chose 1% of GISAs randomly in the simulation and set them *unavailable* for location. The concept of difference $D = \frac{N_{hop}^{unreliable} - N_{hop}^{reliable}}{N_{hop}^{reliable}}$ is defined as a measure of the adaptability to unreliable GISAs, where $N_{hop}^{reliable}$ and $N_{hop}^{unreliable}$ are respectively the average number of hops per request for a certain number of GISAs in a static (all reliable) and dynamic scenario. The performance is compared between these two scenarios to determine how the approach is impacted by the unreliable GISAs. The average differences (equals to $\frac{1}{N} \sum_{i=1}^{N} D$, where $N = 5$ stands for 5 experiment points on the number of GISAs) with each η ($\eta = 0, 0.25, 0.5, 0.75, 1$) for all the numbers of GISAs are respectively 0.0373, 0.039, 0.0403, 0.0302, and 0.0429. Each average difference is very small, which suggests that the approach could be survivable to unreliable GISAs.

5 Related Work and Conclusions

Peer-to-peer technology are a very popular technology to deal with highly variable distributed environment. A distributed search approach in Gnutella is broadcasting a request to all peers in an unreasonable manner. CAN, Chord, and Pastry build distributed hashing structures that provide good scalability and search performance. However, the routing table in them for each node is fixed and thus the network is not reconfigurable. The location mechanism of Freenet is based on usage patterns, which make the popular files closer to users, while discard eventually the least popular files. Our design also proceed from this user-centered intention. However, it does not view that unpopular data is unimportant data.

This paper applies the concepts of relationship networks to design a dynamic location approach. The performance comparison between the approach and a random forwarding has proved that our approach can form relationship clusters and improve the location performance. The simulation results demonstrate the approach can well adapt and survive to different resource distributions, user request patterns, even the partial failure of GISAs. We also give out the performance distinctions of different η to help users better understand the tradeoffs between overhead costs and performance, and chose the preferable parameter.

Acknowledgments

This work is supported by the National Nature Science Foundation of China (No. 60474037 and 60004006), Program for New Century Excellent Talents in University, and Specialized Research Fund for the Doctoral Program of Higher Education from Educational Committee of China (No. 20030255009).

References

1. Foster, I., Kesselman, C., and Tuecke, S.: The anatomy of the grid: enabling scalable virtual organizations. Int. J. Supercomputers Applications.**15** (2001) 205–220
2. Cannataro, M. and Talia, T.: Semantics and knowledge grids: building the next-generation grid. IEEE Intelligent Systems. **19** (2004) 56–63
3. Gao, L., Ding, Y.-S., and Ren, L.-H.: A novel ecological network-based computation platform as grid middleware system. Int. J. Intelligent Systems. **19** (2004) 859–884
4. Barabsi, A.-L. and Albert, R.: Emergence of scaling in random networks. Science. **286** (1999) 509–512

Deployment-Based Security for Grid Applications

Isabelle Attali, Denis Caromel, and Arnaud Contes

INRIA Sophia Antipolis, CNRS - I3S - Univ. Nice Sophia Antipolis,
BP 93, 06902 Sophia Antipolis Cedex - France
`First.Last@inria.fr`

Abstract. Increasing complexity of distributed applications and commodity of resources through grids are making harder the task of deploying those applications. There is a clear need for a versatile deployment of distributed applications. In the same time, a security architecture must be able to cope with large variations in application deployment: from intra-domain to multiple domains, going over private, to virtually-private, to public networks. As a consequence, the security should not be tied up in the application code, but rather easily configurable in a flexible and abstract manner. To cope with those issues, we propose a high-level and declarative security framework for object-oriented Grid applications. This article presents the transparent deployment-based security we have developed. In a rather abstract manner, it allows one to set security policies on various security entities (domain, runtime, nodes, objects) in a way that is compatible with security needs according to a given deployment.

1 Introduction

This paper aims at introducing our security framework designed for distributed applications and optimized for grid infrastructure. We focus here on the transparent and deployment-based security part. It allows the secure deployment and the secure execution of security unaware distributed applications. Writing security policies for an application deployed within at least two different administrative domains is a challenge and leads to ad-hoc security solutions. Grid computation makes this challenge harder to overcome. The dynamic nature of grid resources enforces the use of a security framework that can be easily adapted to dynamically acquired resources. From one execution to the other, acquired resources could change due to concurrent access to the grid by external applications or node failures. Regarding this aspect, security features should be easily configurable and adaptable. There is also a strong need of security from users, but writing security-enabled program is often difficult and painful. Rather than letting programmers write security-related code and handle security concepts, we advocate that a middleware should provide an easy way to use security features. Our approach goes further and introduces a transparent security model for distributed applications. Our security framework focuses on authentication of users and processes; it supports user-to-process, process-to-user, process-to-process authentication. We provide authentication solutions that allow users, user's processes, and the resources used by those processes to verify each other's identity and to communicate securely. We assume that processes are running on a trusted environment, our framework is not intended to protect objects from malicious runtimes but applications from each other and network attacks.

The security model discussed in this paper is implemented on *ProActive* [1, 2]. Section 2 introduces *ProActive* and its descriptor-based deployment model. Section 3 introduces the security model associated to the deployment model. Section 4 presents the declarative language we created to store and express security rules. Then, section 5 gives some implementation details and section 6 presents some benchmarks. Section 7 compares with related work. Finally, section 8 concludes and presents future extensions.

2 The ProActive Middleware

ProActive is a LGPL Java library for concurrent, distributed and mobile computing. With a reduced set of simple primitives, *ProActive* provides a comprehensive API allowing to simplify the programming of applications that are distributed on Local Area Network, on clusters or on grids.

2.1 Base Model

A distributed application built using *ProActive* is composed of a number of medium-grained entities called *active objects*. Given a standard Java object, *ProActive* allows to transparently add behaviours to this java object. These transparent behaviours are, for example, location transparency, activity transparency or mobility. Each active object has one distinguished element, the *root*, which is the only entry point to the active object. It also contains a connected graph of standard java objects called *passive objects*. References are only possible onto root object and not onto passive objects. There is no sharing of passive objects. Each active object has its own thread of control.

2.2 Descriptor-Based Mapping and Deployment

Another extra service provided by *ProActive* is the capability to *remotely create remotely accessible objects*. For that reason, there is a need to identify *Runtimes*, and to provide them some services. A Runtime is a remotely accessible java object which offers a set of services needed by *ProActive* to access remote Java Virtual Machine. Offered services are the creation of local nodes, the creation of another VM (local or remote) and the creation of a local active object within an existing local node. At any time, a runtime hosts one or several nodes. A *Node* is an object defined in *ProActive* which gathers several active objects in a logical entity. Remote objects are identified by an URL, for example rmi://lo.inria.fr/Node1 identifies a node. However, the first step towards seamless deployment is to abstract away from hardware and software details as node or runtime URLs. To abstract away the underlying execution plate-form, and to allow a *source-independent deployment*, *ProActive* provides the following elements:

- *Virtual Nodes*: an abstraction of the distributed resources,
- *XML Descriptors*: a way to define the logical entities the application needs to run, the computing resources available and the mapping of those logical entities onto those hardware resources (i.e. real machines, using actual creation, registry, and lookup protocols).

Virtual Nodes (VNs) are identified as a name (a simple string) in the program source. They are defined and configured in the XML descriptor file. After activation, a VN is mapped to one or to a set of actual *ProActive* Nodes. Of course, distributed entities (active objects) are created on Nodes, not on Virtual Nodes. From application point of view, there is no notion of local or remote objects or remote runtimes, nor how the underlying architecture has been set up. As a consequence, applications cannot know if some security risks exist when an interaction is performed. The rest of the article presents our security model and how it ensures a secure seamless deployment and a secure application execution.

3 Generic Security Infrastructure Model

ProActive is used to deploy distributed applications as if they were deployed onto a big virtual computer. First, the underlying deployment mechanism starts or acquires remote or local execution places (runtimes or nodes). Then, once the deployment is over, the real program can be started. All these steps involve many different security features such as remote creation of node, of activities, method calls, etc. The first item, prior to seeing how the security infrastructure works, we have to solve is how to identify participating entities.

3.1 Security Entity Model

A Security Entity is an object which contains a wrapped object onto which it enforces a security policy (see figure 1). The wrapped object is seen as a black box. The security entity does not interact with its wrapped object internal code to handle security features. This allows to secure all kind of java objects even if the object has been loaded using Java Native Interface (JNI). The security entity is able to intercept incoming and outgoing calls thanks to the use of the proxy pattern. For each intercepted call, it performs needed security checks onto that call. Each Security Entity is uniquely identified by an EntityID which is, actually, a PKI certificate. Access control, communication privacy and integrity and specific *ProActive* features like migration or group communication are defined in Access Control Lists (ACLs) using our security language (see section 4). It is worth mentioning that, as no assumption is done onto the wrapped object, it can

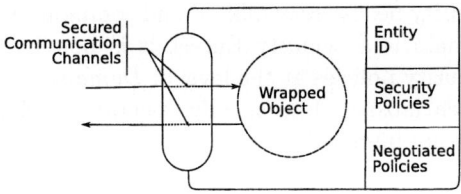

Fig. 1. A Security Entity

also be a security entity. This is an important point that allows an hierarchical organization of security entities.

3.2 Hierarchical Security Entities

A *ProActive* application is composed of several objects: runtimes, nodes and active objects. These objects are remotely accessible and must be secured. The *ProActive* infrastructure implies an ordered structure (see figure 2): runtimes contain nodes which contain active objects. This hierarchy is used to structure and to define our security hierarchy. Each part of this structure belongs to a specific security level. When a Security Entity is involved in an interaction with

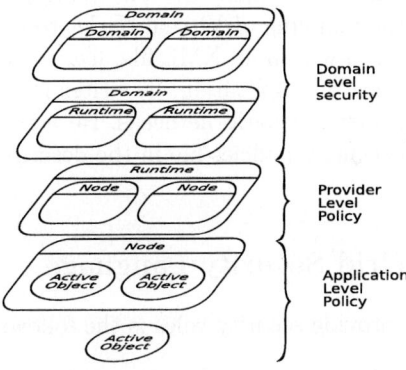

Fig. 2. Hierarchical Security Entities

an other, prior to performing the interaction, the Security Entity checks if this interaction is authorized by checking its owns security policy and by retrieving security rules from its wrapping security entities.

3.3 Domain Level Security

Domains are a standard way to structure (virtual) organizations involved in a Grid infrastructure. A Security Domain represents a distinct scope, within which common security needs are exhibited and common security rules observed. As a Security Domain is a Security Entity, it is possible to define fine grain and declarative security policies at the level of *Domains*. It allows organizations (companies, research labs, ...) to specify general security policies onto their resources. Domains are intended to be set by grid administrators.

3.4 Provider Level Security

A resource *Provider* is an individual, a research institute or an organization offering some resources under a certain security policy to a restricted set of users. A resource provider can set up a runtime with its own policy where clients will be able to perform computation. *ProActive* runtimes have been extended in order to become Security Entities. Alternatively, *ProActive* runtimes are directly deployed and set up by Grid users using deployment descriptors.

3.5 Application Level Security

Virtual Nodes are application abstractions, and nodes are only a run-time entity resulting from the deployment. A decisive feature allows one to define application-level security on those application-level abstractions. Application Security policies can contain references on virtual nodes used within the application source code. One can express which security attributes a communication between two Virtual Nodes requires. At execution, that security will be imposed on the Nodes resulting from the mapping of Virtual Nodes to runtimes.

Security policies are stored within an XML file, like the deployment descriptor. When a user want to start an application, he writes the deployment descriptor according to the deployment scenario he needs. He writes the security policies file and references this security policies file in the descriptor file.

4 Declarative Grid Security Language

The general syntax to provide security rules is the following[1]:

`Entities -> Entities: Interaction # [SecurityAttributes]`

An *entity* is a security entity onto which one want to define a policy rule. As an interaction always involves two entities, the first *Entities* part describes the entity which starts the interaction, the second one, the entity which receives the interaction. In order to express fine security rules, the *entities* part can be a set of entities. So, it is possible to specify that : (1) if the active object which starts the interaction belongs to Virtual Node A and if this Virtual Node is located within Domain D; (2) and the targeted active object belongs to Virtual Node B, then the communication must ciphered.

---

[1] For simplicity reasons, XML syntax has been removed

Interaction is a list of actions (communication, migration, creation of runtimes, nodes, active objects). Finally, security attributes specify how, if authorized, those interactions have to be achieved in terms of authentication, integrity and confidentiality. Each attribute can be required, optional or denied. This helps the security mechanism to compute a result security policy according to the security attributes computation algorithm (see section 5).

Our language also accepts joker rules. They are important to cover all cases, to specify default behaviors, and to provide a conservative security strategy for un-forecasted deployments.

In order to provide a flavor of the system, we consider the following example.

`Domain[inria.fr] -> Domain[ll.cnrs.fr] : Q,P # [+A,+I,?C]`

The rule specifies that between the domain *inria.fr* (identified by a specific certificate) and the parallel machine *ll.cnrs.fr*, all communications (reQuests, and rePlies) are authorized, they are done with *authentication* and *integrity*, *confidentiality* being accepted but not required.

5 Adaptative Security Policies

As we are in a distributed world, without a global administrator to handle all security policies, a given interaction could involve many security policies. The following protocol is used by security entities to compute a final security rule for a given interaction using all matching security policies. Indeed, the final policies being used are both dynamically computed, and possibly change during computation time.

1. The object acting as a client performs a method call onto the object acting as server.
   (a) The security entity intercepts the method call. The security mechanism contacts the callee security entity and requests callee entity location informations (callee security entity + encapsulating entities).
   (b) The client security entity collects all location informations. Once it has all location informations about caller and callee, it can retrieve all security rules matching to the interaction between callee and caller on the caller side.
   (c) The caller's security entity computes all matching rules to obtain a result rule and sends it to the callee security entity.
2. Callee security entity receives the requested policy rule by the caller security entity. The callee security entity :
   (a) retrieves all rules that match the requested interaction, computes them to find the result rule.
   (b) compares the rule requested by caller security entity with the locally computed rule. If they do not match, the interaction stops and the callee's security entity returns an exception to caller's security entity. Otherwise, an object matching that session is created and will be used to perform

all security related actions needed for this specific interaction. If needed, the caller generates a session key.
   (c) gets back to the callee security informations (Session ID, cyphered session key, computed rule).
3. The caller security entity checks that the returned policy rule matches its local policy, then creates a local session object. The method call is given to the Session object. The Session object performs security actions requested by the exchanged policy rule and gives the secured method call to the underlying transport layer.

6 Benchmarks

To perform benchmarks, we choose the *Jacobi iterations*, an algorithm to solve a linear matrix equation. It performs local computations and communications to exchange data. Benchmarks are executed on 5 Pentium IV@3.2Ghz, 1Go(DDR), 512 Kb L2 cache, Linux (2.4.22) computers interconnected by a 100Mb/s network. Java VM is Sun JVM 1.4.2. These computers are simultaneously used by their users. When activated, security policy requires that all interactions (communications and also deployment) are ciphered and authenticated. The left figure represents the average duration, in milliseconds, of a Jacobi iteration depending on the data. The right figure presents the duration of matrix initialization (data transfer) at the computation initialization step depending on the data. Benchmarks show that as soon as the size of computed data begins to grow the ratio of the overhead induced by the security strongly decreases, from 1.8 to 1.15.

7 Related Work

The .NET [3] framework provides security features allowing protection of a host against malicious code. Security system is based on user- and code-identity us-

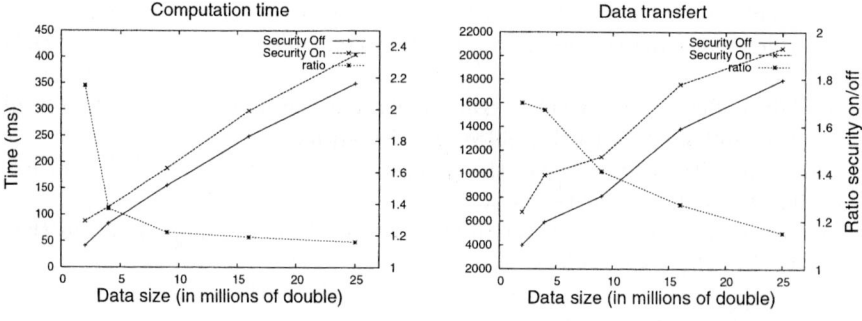

Fig. 3. Security Overhead

ing public/private keys, hash functions, digital signatures and security policies. There are four policy levels Enterprise, Machine, User and Application Domain. *ProActive* security levels are not restricted. Note that *ProActive* does not use code identity. Legion [4] encapsulates functionalities as objects and provide mechanisms for their location, migration, etc. Legion objects interact via remote method invocations and the main security objectives are authenticating the communicating parties, protecting traffic, enforcing access control, delegating rights and enforcing site-specific security concerns. Unlike *ProActive*, Legion does not support a hierarchical management of security policies. The Globus system relies on the Globus Security Infrastructure (GSI) [5] that supports, integrates and unifies popular security models. It supports an inter-domain and intra-domain security interoperability. The notion of virtual organization is defined as a set of individuals and/or institutions sharing resources and services under a set of rules and policies governing the extent and conditions for that sharing. *ProActive* is able to interact with GSI, it also goes further by proposing a dynamic and adaptive security which takes account computation mobility.

8 Conclusion

We have proposed a decentralized, declarative security mechanism for distributed systems that features interoperability with local policies, dynamically negotiated security policies, and multi-users systems. The use of a logical representation of the application (Virtual Nodes) allows to have a security policy adaptable to the deployment, a crucial feature for Grid applications. The security mechanism allows to express all security-related configurations within domain policy files, and application deployment descriptor, outside the source code of the application.

As a very short term perspective, the implementation of group communication security is in the process of being finalized; a single session key being used for all group members. This work is an attempt to contribute to the construction of flexible solutions that are very much needed for Grid deployment.

References

1. Caromel, D., Klauser, W., Vayssière, J.: Towards Seamless Computing and Metacomputing in Java. Concurrency Practice and Experience **10** (1998) 1043–1061
2. Attali, I., Caromel, D., Contes, A.: Hierarchical and declarative security for grid applications. In: High Performance Computing - HiPC 2003, 10th International Conference, Hyderabad, India, December 17-20, 2003, Proceedings. Volume 2913., Springer Verlag, Lecture Notes in Computer Science, LNCS (2003) 363–372
3. Wesley, A., ed.: .NET Framework Security. Addison Wesley Professional (2002)
4. Grimshaw, A., et al., W.W.: The Legion Vision of a World-wide Virtual Computer. Communications of the ACM **40** (1997)
5. Foster, I.T., Kesselman, C., Tsudik, G., Tuecke, S.: A Security Architecture for Computational Grids. In: ACM Conference on Computer and Communications Security. (1998) 83–92

Grid Resource Selection by Application Benchmarking for Computational Haemodynamics Applications

Alfredo Tirado-Ramos[1], George Tsouloupas[2], Marios Dikaiakos[2], and Peter Sloot[1]

[1] Faculty of Sciences, Section Computational Science,
University of Amsterdam,
Kruislaan 403, 1098 SJ Amsterdam, The Netherlands
{alfredo, sloot}@science.uva.nl

[2] Department of Computer Science,
University of Cyprus,
75 Kallipoleos St., P.O. Box 20537, 1678, Nicosia, Cyprus
{georget, mdd}@cs.ucy.ac.cy

Abstract. Grid benchmarking for improved computational resource selection can shed a light for improving the performance of computationally intensive applications. In this paper we report on a number of experiments with a biomedical parallel application to investigate the levels of performance offered by hardware resources distributed across a pan-European computational Grid network. We provide a number of performance measurements based on the iteration time per processor and communication delay between processors, for a blood flow simulation benchmark based on the lattice Boltzmann method. We have found that the performance results obtained from real application benchmarking are much more useful for running our biomedical application on a highly distributed grid infrastructure than the regular resource information provided by standard Grid information services to resource brokers.

1 Introduction

Resource selection in Grid environments is a crucial problem. Regardless of who performs the resource selection, be it users or automated systems (i.e. schedulers or resource brokers), the decision makers are faced with the difficult task of matching/mapping jobs to resources. Previous work on the specification of resources and services in complex heterogeneous computing systems and meta-computing environments in general [1] and, particularly, in grid environments, [2], has led to a better understanding of the issues. Nevertheless, the evolution of Grid architectures has underlined the need for addressing application-specific characterization of the resources available. Grid benchmarking, or the characterization of Grid computational resources for improving resource selection, can be used to help improving the performance of computationally intensive parallel applications by enhacing the resource selection process. Our application for preoperative support, a blood-flow simulation solver, is an implementation of the lattice Boltzmann method, a mesoscopic approach for simulating fluid flow based on the kinetic Boltzmann equation. The problem statement is straight-forward: how can we find the best resources to run the application at hand? Additionally, since the application is often run in multiple instaces using parameterised runs, it would be desirable to have acces to information that would help better schedule these jobs. In this article we discuss our results after performing a number of experiments to investigate the levels of performance offered by hardware resources distributed across the CrossGrid European computational Grid. We show how we can rank resources based on a benchmark derived from the blood-flow simulation kernel. In the remainder of this article, Section 2 describes the parallel solver we use as a benchmark in more detail. Section 3 lays out some specifications of the experimentation testbed used, the CrosGrid tesbed. In Section 4 we provide a short description of the GridBench framework which we used to perform our benchmarking experiments. In Section 5 we discuss some of the issues related to benchmarking on the Grid, particularly in light of resource characterization and ranking based on kernel performance, and offer our results. Finally, in Section 6 we briefly discuss our conclusions and relevant future work.

2 Non-invasive Vascular Reconstruction

For our benchmarking experiments we use a parallel solver from the Virtual Radiology Explorer (VRE) Grid-based Problem Solving Environment (PSE), a type of integrative collaborative environment [3] that includes simulation, interaction, and visualization components for pre-treatment planning in vascular interventional and surgical procedures. This PSE was developed by the University of Amsterdam and deployed within the European Crossgrid project [4]. The deployment of the interactive VRE system within Crossgrid resulted in a pan-European experimental PSE using resources from across Europe, exploiting available achievements from other European Grid projects such as European DataGrid[1] and the Large Hadron Collider Computing Grid[2]. We lay out a base architecture for PSEs using the Grid as a medium, with a validated case study in vascular reconstruction. For additional background, motivation, and the latest Grid-based results, see [6]. The VRE contains an efficient parallel computational hemodynamics solver [7] that computes pressure, velocities, and shear stresses during a full systolic cycle. The simulator is based on the Lattice-Boltzmann method (LBM), a mesoscopic approach for simulating fluid flow based on the kinetic Boltzmann equation [8]. The data used as input for the VRE can be obtained from several imaging techniques used to detect vascular disorders. For instance, 3D data acquired by Computed Tomography or Magnetic Resonance

---

[1] http://www.eu-datagrid.org
[2] http://lcg.web.cern.ch/LCG/Documents/default.htm

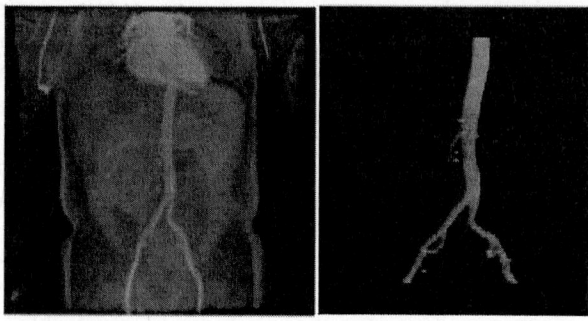

Fig. 1. Segmented medical data from the abdominal aorta, accesible via Grid Storage Elements functioning as medical repositories

Imaging, or particularly Magnetic Resonance Angiography for imaging blood vessels that contain flowing blood. To convert the medical scans into meshes our solver can work with, the raw medical data is first segmented so that only the arterial structures of interest remain in the data set (Figure 1).

Measurements are important for diagnoses. Clinical decision-making relies on evaluation of the vessels in terms of the degree of narrowing for stenosis and dilatation (increase over normal arterial diameter) for aneurysm. The selection of a bypass (its shape, length, and diameter) depends on sizes and geometry of an artery.

3 The CrossGrid Experimental Testbed

The CrossGrid distributed testbed[3] shares resources across 16 European sites. The sites range from relatively small computing facilities in universities, to large computing centers, offering an ideal mixture to test the possibilities of an experimental Grid framework. National research networks and the high-performance European network, Geant[4], assure interconnectivity between all sites. The network includes a local step, typically inside a University or Research Center, via Fast or Gigabit Ethernet, a jump via a national network provider at speeds that will range from 34 Mbit/s to 622 Mbit/s or even Gigabit, to the national node, and a link to the Geant network at 155 Mbit/s to 2.5 Gbit/s. Our experiments were conducted on the CrossGrid testbed, following the basic LCG-2 architecture. In this architecture, a Grid VO is made up of a set of geographically distributed sites (computer clusters) containing computational or storage resources. Each site contains a Computing Element, which manages a set of Worker Nodes. A site may also contain a Storage Element, which is an interface to mass storage.

---

[3] http://www.eu-crossgrid.org
[4] http://www.dante.net/geant/

4 The GridBench Tool

GridBench [9] is a tool for benchmarking Grids. It consists of a framework containing a set of tools that aim to facilitate the characterization of Grid nodes or collections of Grid resources. The framework has two main objectives: to generate metrics that characterize the performance capacity of resources belonging to a Virtual Organization (VO), and to provide a tool for researchers that wish to investigate various aspects of Grid performance using well-understood kernels that are representative of more complex applications deployed on the Grid. In order to perform benchmarking measurements in an organized and flexible way, GridBench provides a means for running benchmarks on Grid environments as well as collecting, archiving, and publishing the results. The framework allows for convenient integration of new and existing benchmarks into the suite, as well as the customization of existing benchmarks through parameters. We have used the tool to perform our biomedical application benchmarking experiments (Figure 2).

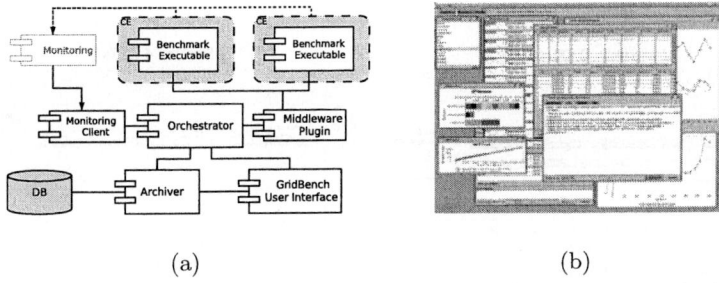

(a)   (b)

Fig. 2. GridBench: 2(a) shows the main components and services of the GridBench software architecture. 2(b) shows the GridBench GUI, used for defining and executing benchmarks, and browsing and analyzing results

5 Results and Discussion

Our application benchmark, the BStream kernel, is part of an interactive Grid application that involves processing of 3D data, which makes it computationally expensive. Shown here is the computationally intensive part of the application, which uses the Message Passing Interface (MPI) for parallelization. This code was instrumented to measure elapsed time for each iteration as well as the time spent on MPI communication, and integrated into GridBench[5]. As a dataset we used different sample files that represent our normal workload, from simple tube-like artery structures to aorta segments containing bifurcartions.

---

[5] The charts presented in this section were automatically generated using the GridBench Graphic User Interface.

5.1 Resource Comparison

Figure 3(a) shows the measured iteration times of the BStream kernel on 13 sites available in our testbed. In each case, the same workload was applied by using identical input data and parameters. Figure 3(a) shows the results obtained by using 2 CPUs in each measurement. For the 2 CPU measurements, using 2 CPUs on the same Worker Node was preferred over using two CPUs on two different Worker-Nodes. This is important, since it was found that this would seriously impact performance of this kernel (Figure 5). Resources *cluster.ui.sav.sk* and *loki01.ific.uv.es* employ single-CPU nodes while the majority of site employ dual-CPU Worker Nodes. In Figure 3(a) (as well as the rest of the charts in Figure 3) we observe that iteration times remain fairly constant throughout the duration of the computation. For our experiments, we have set the application kernel to run for 800 iterations, so it can be seen that right before the end of each run (at around 760 to 780 iterations) a jump in performance of about 30% larger time per iteration values is experienced in all nodes. This is mainly due to the design of the current version of the kernel, where the first processor that

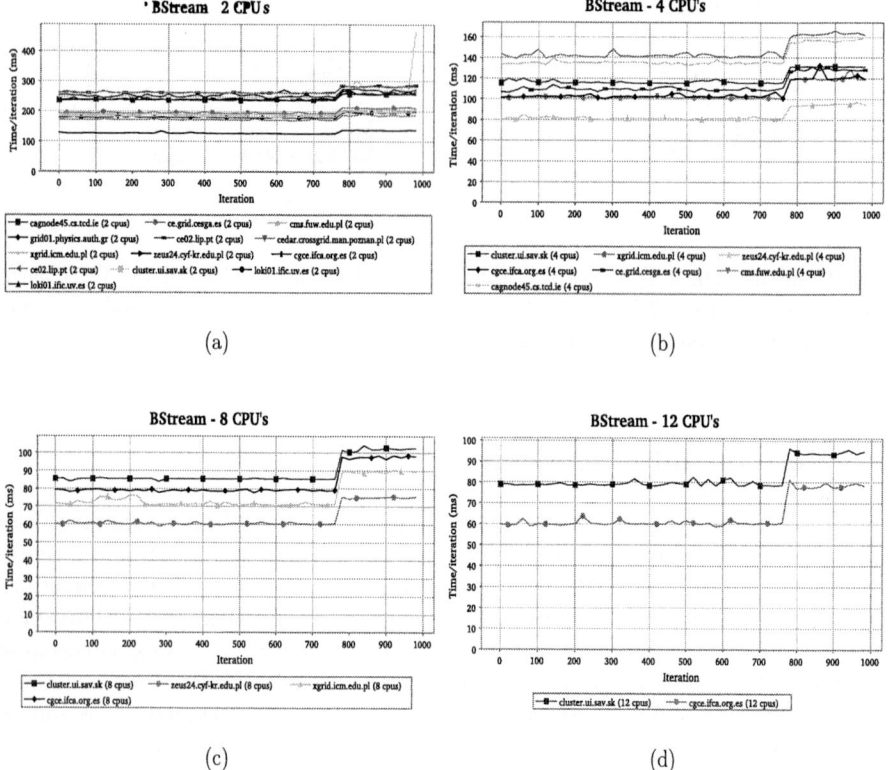

Fig. 3. The performance of the kernel at a set of sites using 2, 4, 8 and 12 CPU's

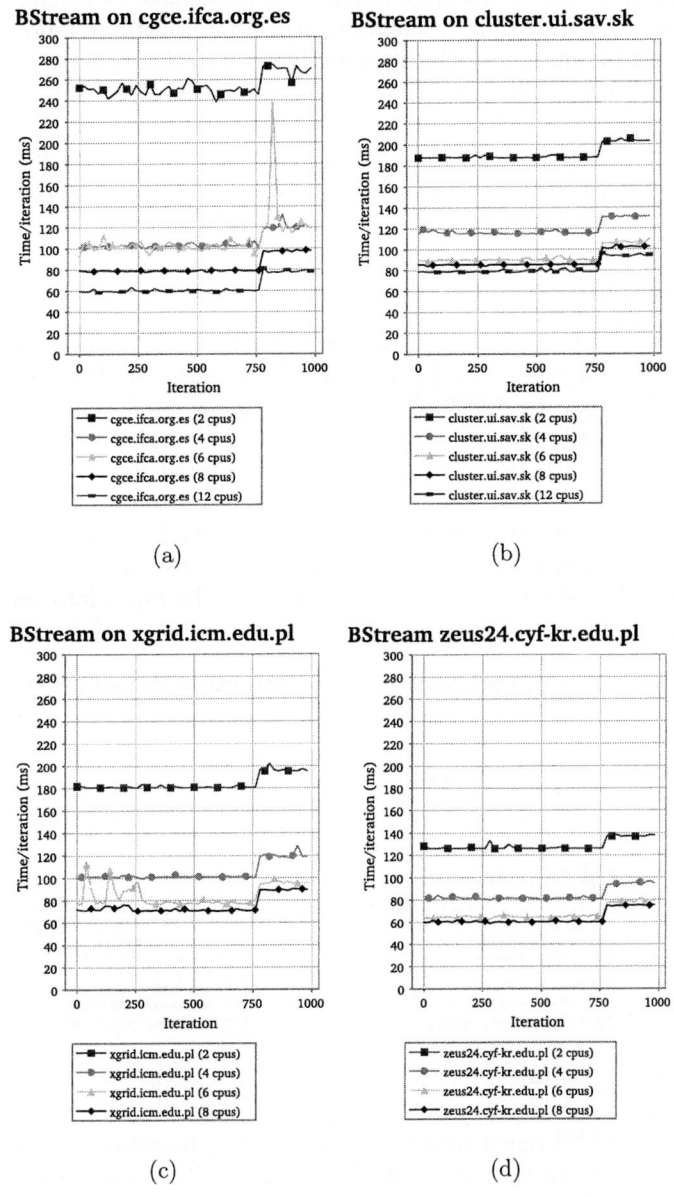

Fig. 4. Scalability as it is measured at four sites. Lower iteration times are better

started running gathers data from all other processors before producing the final output[6]. Nevertheless, iteration times remain relatively invariant regardless of

---

[6] The new version of the BStream kernel, which is work in progress at the time of writing this paper, has a different design that addresses this issue.

the number of iterations. For this reason it is reasonable to assume that short run-time experiments (using a small number of iterations) are representative of our real-life experiments, in which we use larger iteration counts.

Figures 3(b), 3(c) and 3(d) show the performance of the kernel at a set of sites using 4, 8 and 12 CPUs respectively. Generally we observe a "downward" trend indicating that the code is somewhat scalable, i.e. using a larger number of cpus at a given site will yield a faster run-time, but more on scalability will be given in the next sub-section.

5.2 Scalability

Figure 4 shows the scalability of the kernel as it is measured at four sites: *cgce.ifca.org.es* (up to 12 CPUs), *cluster.ui.sav.sk* (up to 12 CPUs), *xgrid.icm.edu.pl* (up to 8 CPUs) and *zeus24.cyf-kr.edu.pl* (up to 8 CPUs). It is quite interesting to observe that the different sites display a different scalability. For example, in Figure 4(a) the runtime is reduced to less than 30% when going from 2 CPUs to 8 CPUs, while in Figure 4(b) the improvement is only just under 50%. Similarly, while in 4(a) there is approximately a 25% improvement in runtime when going from 8 CPUs to 12 CPUs, in 4(b) there is only marginal improvement. Scalability *at each resource* needs to be taken into consideration for efficient resource selection.

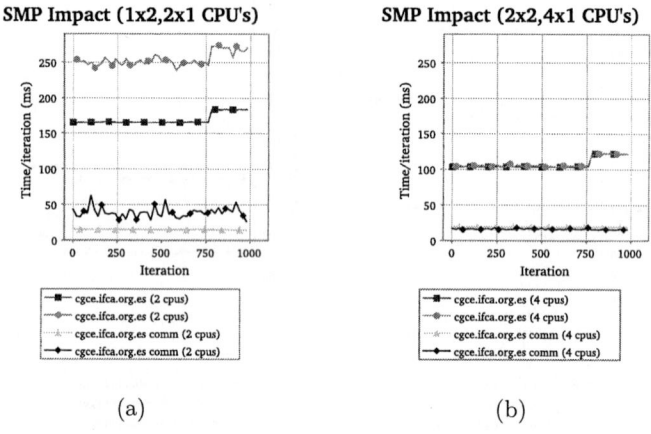

Fig. 5. Impact of MPI communication on runtime. 5(a) Iteration and communication times using 2 CPUs on the same (dual) Worker Node (1x2), and 1 CPU on each of 2 Worker Nodes. 5(b) Iteration and communication times using 2 CPUs on each of 2 (dual) Worker Nodes (2x2), and 1 CPU on each of 4 Worker Nodes (4x1)

5.3 Communication Measurements

The BStream kernel uses MPI for inter-proces communication, which we compiled using the MPICH4 device. The code is highly coupled and it is expected that the performance of the interconnect, i.e., the LAN connecting the cluster nodes will

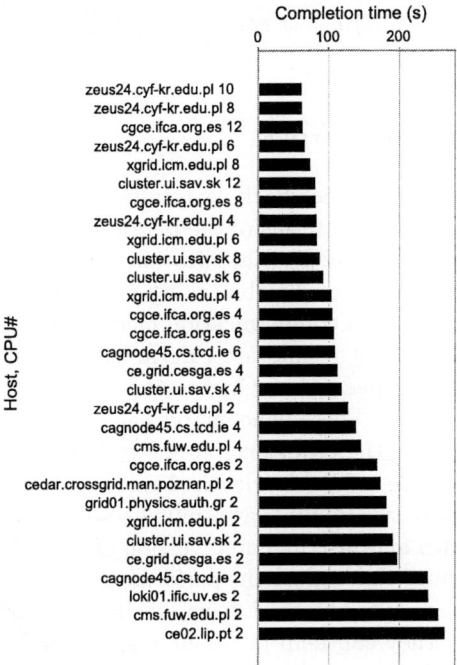

Fig. 6. Completion times of the BStream kernel using different numbers of CPUs on several resources

have a considerable impact on the performance of the kernel. To investigate this, the BStream code was instrumented to measure the time spent in communication[7]. To isolate the effect of the network we ran the code using just two CPUs on a dual-CPU Worker Node (1x2)[8], using two CPUs on two different (identical) Worker Nodes (2x1). This is shown in Figure 5(a). Figure 5(b) shows a similar experiment using 4 CPUs. In 5(a) we observe that there is considerable difference in communication performance which also impacts the time per iteration. On the other hand, in 5(b) we observe no significant difference when running in either mode, since the network is used is both cases (both in 2x2 and in 4x1).

5.4 Decision-Making

Figure 6 conveys a lot of usefull information since it provides a ranking of runtimes on all of the resources available. This ranking could be used directly in

---

[7] The impact of the instrumentation was measured and was found to be insignificant.
[8] The MPI library used was not optimized for SMP, and communication still went through the TCP/IP stack.

resource selection *especially* in cases where the relative CPU, Memory and network speeds at each resource (site) are not known. For example, it appears that it is better to run the code at *zeus24.cyf-kr.edu.pl* using 4 CPU's than at *cluster.ui.sav.sk* using 8 CPU's.

6 Conclusions and Future Work

We have presented a concise set of benchmarking results for the characterization of computational Grid resources in terms of the performance of CPU, main memory and interconnects, for a biomedical application for the simulation of blood flow. We have presented a set of benchmarking experiments to experiment with specific computation versus communication metrics. This small set of biomedical application benchmarking results are run on the highly distributed Grid resources offered by the European CrossGrid testbed with small overhead and with minimal effort by the user. These benchmarks can be invoked in a periodic and on demand manner using the GridBench framework, with the resulting measurements archived and made available via Grid services based on a web services framework. Furthermore, we found that ranking resources based on the performance of a stripped-down and instrumented version of an application can give us realistic resource rankings that reflect the performance of the application itself. We have shown how the results obtained using GridBench can be used for ranking of resources and how they can help in resource selection. In the future we plan to further investigate the relation between full-blown application performance and micro-benchmark performance in the context of highly distributed Grid environments.

Acknowledgments

We would like to acknowledge the meaningful contributions of R. Belleman, A. M. Artoli, and L. Abrahamyan to this work.

References

1. Matthias Brune, Jorn Gehring, Axel Keller, Burkhard Monien, Alexander Reinefeld, Specifying Resources and Services in Metacomputing Environments, Parallel Computing, 1998, vol. 24, pp. 1751–1776, Elsevier Science.
2. Greg Chun, Holly Dail, Henri Casanova, and Allan Snavely. Benchmark probes for grid assessment. Technical report, UCSD, 2003.
3. Houstis E.N., Rice J.R., Weerwarna S., Papachio P., Wang K. Yang and Gaitatzes M., Enabling Technologies for Computational Science Frameworks, Middleware and Environments, Chapter 14, pages 171-185. Kluwer Academic Publishers, 2000.
4. A. Tirado-Ramos; P.M.A. Sloot; A.G. Hoekstra and M. Bubak: An Integrative Approach to High-Performance Biomedical Problem Solving Environments on the Grid, Parallel Computing, (special issue on High-Performance Parallel Bio-computing) vol. 30, nr 9-10 pp. 1037-1055. Chun-Hsi Huang and Sanguthevar Rajasekaran editors, 2004.

5. W. Hoschek, J. Jaen-Martinez, A. Samar, H. Stockinger, K. Stockinger, Data Management in an International Data Grid Project, IEEE/ACM International Workshop on Grid Computing Grid'2000, 17-20 December 2000 Bangalore, India "Distinguished Paper" Award.
6. http://www.science.uva.nl/research/scs/HotResults/
7. A.M. Artoli, A.G. Hoekstra, P.M.A. Sloot, Simulation of a systolic cycle in a realistic artery with the Lattice Boltzmann BGK method., Int. J. Mod. Phys. B, Vol. 17, Nos. 1-2 (2003) 95-98.
8. S. Succi, The Lattice Boltzmann Equation for fluid dynamics and beyond, Oxford Science Publications, Clarendon Press, 2001.
9. George Tsouloupas, Marios D. Dikaiakos, GridBench: A Tool for Benchmarking Grids, Proceedings of the 4th International Workshop on Grid Computing (GRID2003), Phoenix, AZ, November 2003, pp. 60-67, IEEE.

AGARM: An Adaptive Grid Application and Resource Monitor Framework*

Wenju Zhang, Shudong Chen, Liang Zhang, Shui Yu, and Fanyuan Ma

Shanghai Jiaotong University, Shanghai, P.R.China, 200030
{zwj03, chenshudong, zhang_liang, merlin, fyma}@sjtu.edu.cn

Abstract. Most people assume that grid computing is the mainstream high performance and throughput computing platform in the future. To monitor these grid environment and provide grid information service become extremely important for using grid computing resource efficiently. In this paper we propose an adaptive grid application and resource monitor framework named AGARM. The AGARM framework can be used not only to monitor resources but also to monitor applications in the grid environment. Moreover, we show how the monitor activity is optimized by the adaptive pulling techniques, which is based on the moving average estimation algorithm.

1 Introduction

A grid [3] system is an environment in which users, such as scientists, can access resources in a transparent and secure manner. Many grid middleware have been developed to facilitate people unitize those distributed resources including grid security, grid resource monitor and management, meta-scheduler, user interface, etc. It is extremely important to monitor these grid computing resources and provide grid information service for using grid computing resource efficiently. There are some grid middleware have the capability to do that to some extent [2, 4, 8, 9, 10, 11]. But, we need a monitoring system that is scalable, flexible, adaptable, modular, open standard, and OGSA-compatible.

However, due to the complexity and diversity of the applications and resources in the grid computing environment, existing grid monitoring architecture can not monitor all of the resource belonging to the grid. When the size of computing facility grows, existing monitoring strategy will significantly increase system overhead. The dynamic characteristics of grid resources allows the grid computing resource participate and withdraw from the resource pool constantly. Only a few existing monitoring system address this characteristics partially.

This paper proposes an adaptive grid application and resource monitor framework named AGARM to solve grid application and resource monitor problems.

---

* This research work is supported in part by the the National High Technology Research and Development Program of China (863 Program), under Grant No. 2002AA104270.

It leverages grid services and adaptive pulling techniques which based on estimation algorithm to monitor the grid applications and resources. The primarily evaluation result confirms that the AGARM framework is scalable, effective and flexible.

The rest of the paper is organized as follows. Section 2 describes overview of the AGARM. Section 3 presents the architecture of the AGARM. Section 4 introduces the prototype implementation and evaluation result; Section 5 discusses the relevant grid application and resource monitor systems. We conclude in Section 6.

2   AGARM Overview

2.1   Service and Layered Model

The number one element of SOA is the flexibility. Essentially, an SOA is a set of loosely coupled services. Each service is relatively inexpensive to build and replace if necessary. Loose coupling further allows the architecture to adapt to changes unlike the traditional tightly coupled architectures that are more brittle. In an SOA you can replace one service with another without having to worry about the underlying technology. The interface is what matters, and it is defined in the universal standards of OGSA/OGSI, Web services and XML. That is flexibility through interoperability.

The grid system usually spans different continents and administrative domains. So we keep the layer architecture in mind in the design and implementation of the AGARM.

2.2   Pull and Push Event Model

Grid application and resource status information can be divided into two categories: static and dynamic. Static status information keeps almost the same in the lifetime of the grid computing environment, such as the number of CPU of a grid computing node. Dynamic status information fluctuates with the time, such as the CPU utilization of a grid computing node. We monitor this two categories status information by means of two different event models: *push* event model for static status information and *pull* event model for dynamic status information. The evaluation result of AGARM shows that this model has the minimum overhead. The essential problem of traces of dynamic status information is how to deal with the stale information. We introduce the EWMA (Exponential Weighted Moving Average) algorithm [10, 12] to estimate the time of the next significant update. The EWMA algorithm is a very simple but effective time series estimation instrument.

3   Architecture

The AGARM framework is illustrated in Figure 1. In virtual organization A, there are many resources (compute, storage, instruments, etc.) to be monitored.

We collect and present those resources status information through application and resource monitor service components. Those grid status information of virtual organization are stored in XML database and can be utilized by other management applications such as scheduling and performance tuning services. The virtual organization is the base unit we monitor. In order to monitor different virtual organization, we just only point to them through virtual organization switcher.

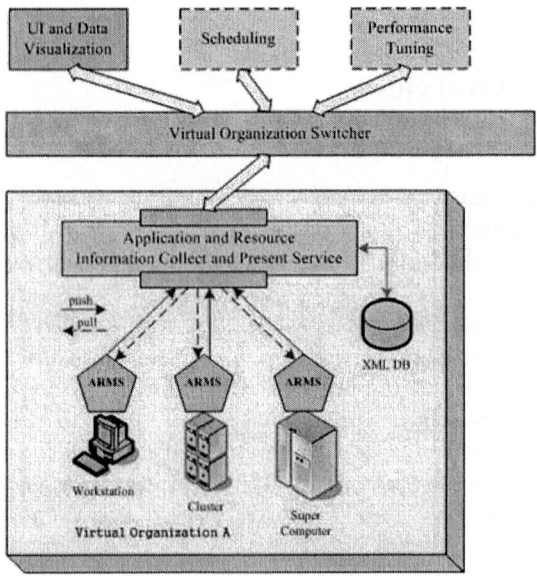

Fig. 1. AGARM architecture

3.1 Components

The AGARM framework is composed of five key components: UI and Data Visualization, Virtual Organization Switcher, Application and Resource Information Collect and Present Service (ARICPS), Application and Resource Monitor Service (ARMS) and XML Database.

UI and Data Visualization: This is a web portal or a standalone application function as the user interface and data visualization. Usually, this component is integrated with resource scheduling components.

Virtual Organization Switcher: Virtual Organization (VO) is "abstract entity grouping Users, Institutions and Resources (if any) in the same administrative domain"[3]. VO switcher acts as a bridge between ARICPS in different VO and UI component.

Application and Resource Information Collect and Present Service (ARICPS): This is the collect layer in the AGARM framework. The ARICPS provides the functionality within which Service Data Elements can be collected, aggregated, and queried; data feeds can be monitored; and Service Data Elements can be created dynamically on demand.

Application and Resource Monitor Service (ARMS): This is a grid service component function as the sensor in some resource monitoring systems. The ARMS provides data to the ARICPS either in *active* or *passive* mode.

XML Database: All the data we collect are stored in XML databases. It is very convenient to exchange data in XML format. Our prototype implementation is based on open source Berkeley DB XML.

3.2 Adaptive Pull and Estimation Mechanism

It is very essential to keep the status information updated with the minimize overhead for grid monitoring systems. The Moving Average estimation algorithm is a very effective but simple time series estimation instrument [10, 12]. The service maintains a window over the timestamps of significant measurements from the monitor service. The Moving Average estimation algorithm does an average of the time interval between two consecutive updates. At any time instance t, the estimation of the time to next pull Z_t is given by

$$Z_t = \frac{1}{N}\sum_{i=1}^{N} x_i, \qquad (1)$$

where x_i is the ith update interval value. The effectiveness of this scheme depends on the window size that is used. Larger window sizes are more suitable for usage patterns that have a gradual or long-term fluctuation patterns.

The Exponential Weighted Moving Average (EWMA) operates on the principle of assigning exponentially lower weights to older values of the time series. There are different types of EWMA estimation algorithm. The Single EWMA algorithm is effective in estimating traces which do not exhibit well defined trends or patterns. At any time instance t, the estimation of the time to next pull Z_t can be reduced by formula (1) as follow

$$Z_t = \lambda \overline{x_t} + (1-\lambda) Z_{t-1}, \quad 0 < \lambda \leq 1 \qquad (2)$$

In this definition $\overline{x_t}$ is the sample mean from time period t, λ is the weight assigned to the current observation. Smaller the value of λ is, the more the weight is given to the older values in the series. Larger values of λ result in more weight being given to more recent observations.

It is essential to choose the right value for λ. That requires some amount of knowledge of the usage patterns on the series. This implies that the value of λ should be chosen dynamically using the observed time series characteristics.

4 Prototype Implementation and Evaluation

We have implemented a prototype of the AGARM framework on top of Globus Toolkit version 3. Our evaluation preliminary proved the effectiveness of we

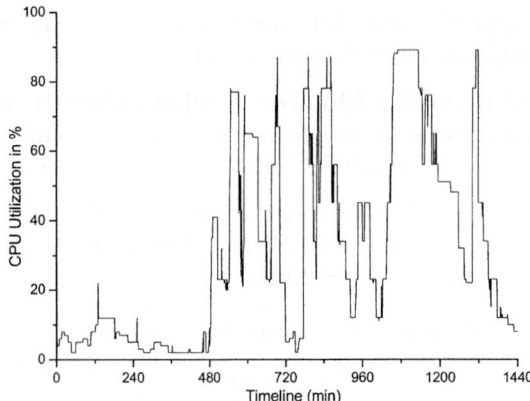

Fig. 2. CPU utilization of our sample application server

proposed AGARM framework and the estimation algorithm. The experiment data was collected over a three week period with a periodicity of 60 seconds on the sample application server with in the four nodes Linux cluster. As shown in Figure 2, the CPU utilization exhibits a flat behavior with only occasional changes in a small period and a large amount of fluctuation with sharp changes in a large period.

4.1 Application and Resource Monitor Techniques

In traditional systems, after a user initiates the jobs, she or he has the ability to monitor the jobs and the corresponding resource consumption as they execute. Queue systems have been used to manage job scheduling on a cluster of nodes. Most queues operate in the "batch" mode. In other words, when a user submits a job, the submission program immediately returns the user to the prompt and provides her with a ticket or job ID or token, which can be used to monitor the job at any later time.

However, there are a number of disadvantages with respect to application and resource monitoring in traditional and queue systems. For example, the monitoring requires the user have an account on each machine, requires knowing on which nodes the jobs are executing, only monitor limited status of application and resource in queue systems, etc. Those tools are well designed but do not meet the needs of grid systems.

Typically, in a grid system, when a user submits a job, the job runs with the permissions of an ordinary user. Grid systems typically span multiple organizations and administrative domains. Often grid systems run on machines that are controlled by queuing systems. Currently, there are no standard methods for monitoring the progress of jobs executed by grid systems. Different grid systems use different techniques for monitoring applications and resources.

In the AGARM framework, our Application and Resource Monitor Service (ARMS) component solves this problem practically. The ARMS component leverages the grid services and traditional operating system provided tools to monitor the applications and resources of grid systems.

4.2 The Performance of the Estimation Algorithm

Single EWMA estimation algorithm is characterized by its parameter λ. A lower value of λ results in larger weight given to the oldest historical information, a higher value of λ implies greater weight for the more recent observations. As shown in Figure 3, the out of sync period varies along with λ. What is the optimization λ for CPU utilization usage pattern of our sample application server? We conclude that λ_{opt} ($\lambda = 0.8$) is the optimization λ from Figure 3. Does this conclusion hold for all usage patterns? This paper does not cover this problem. We will study the optimization λ of different usage pattern. This is our future work.

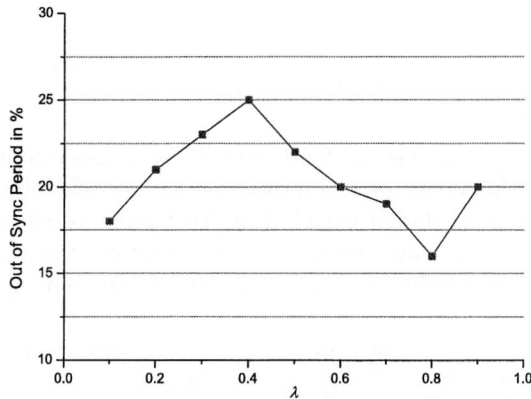

Fig. 3. Out of sync period at different λ

The performance of monitoring is also influenced by the window size. The window size determines the significance attached to older observations in the current estimation. Figure 4 illustrates the effect of varying window sizes on the number of pulls and out of sync period. Those data were collected at optimization λ_{opt} ($\lambda = 0.8$) for CPU utilization usage pattern of our sample application server. Intuitively, one might expect that nodes which exhibit a larger non-periodic rate of change will require a smaller window size for better response times, while nodes which exhibit a smooth fluctuation or infrequent changes will require a larger window size in order to capture this behavior.

5 Related Work

GMA (Grid Monitoring Architecture) [6] is the grid monitor architecture proposed by Global Grid Forum (GGF). It consists of three components: consumers,

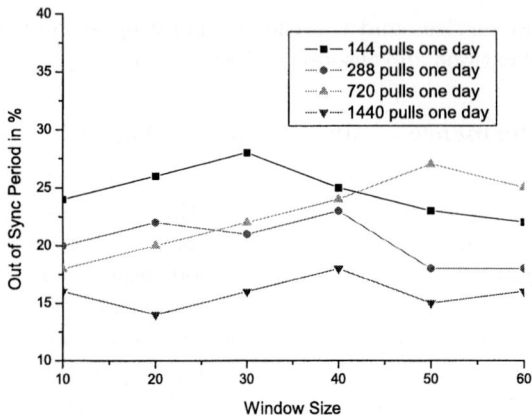

Fig. 4. Effect of window size at different number of pulls

producers and a directory service. R-GMA (Relational Grid Monitoring Architecture) [1] is based on the GMA from GGF. Its strong comes from relational model.

Ganglia (http://ganglia.sourceforge.net) [2] is an open-source project that grew out of the University of California, Berkeley Millennium Project. It is based on a hierarchical design targeted at federations of clusters. It relies on a multicast-based listen/announce protocol to monitor state within clusters and uses a tree of point-to-point connections amongst representative cluster nodes to federate clusters and aggregate their state.

Globus MDS [4] aims to provide a standard mechanism for publishing and discovering resource status and configuration information. It have been designed to cater for the OGSA/OGSI service architecture. It is the most resemble architecture to our AGARM framework hitherto.

NWS (Network Weather Services) [7] provides accurate forecasts of dynamically changing performance characteristics from a distributed set of grid computing resources. It can produce short-term performance forecast based on historical performance measurement.

The above systems solve the grid monitor problem to some extent. However, none of these systems provides any mechanism for monitoring the grid applications execution. None of these systems is OGSA/OGSI compatible with the exception of Globus MDS3. Some of these systems, such as Globus MDS and NWS, impose heavy overhead on the being monitored resources.

6 Conclusions

This paper has discussed how the services provided by the AGARM framework help constructing monitor services for grid computing environments. A novel monitor service for grid applications and resources has been presented. We have identified the main components of their design, and evaluated the effect of each components on the overall performance of the models. Furthermore, the single EWMA estimation algorithm and the adaptive pulling mechanism have been introduced. Therefore, the AGARM framework has its practical significance for the grid applications and resources monitor.

References

1. Andy Cooke, et al. "The Relational Grid Monitoring Architecture: Mediating Information about the Grid". Kluwer Academic Publishers. April 2004.
2. Ganglia: http://ganglia.sourceforge.net
3. Foster, I., and C. Kesselman and S. Tuecke, "The Anatomy of the Grid", International Journal of High performance Computing Applications, 15, 3 2001.
4. Globus MDS3: http://www.globus.org/mds
5. X. Zhang, J. Freschl, and J. Schopf. "A Performance Study of Monitoring and Information Services for Distributed Systems". Proceedings of HPDC, August 2003.
6. Tierney, B., R. Aydt, D. Gunter, W. Smith, V. Taylor, R. Wolski, and M. Swany, "A Grid Monitoring Architecture". The Global Grid Forum GWD-GP-16-2, January 2002.
7. NWS: http://nws.cs.ucsb.edu
8. Anand Natrajan, Michael P. Walker, "Monitoring Remote Jobs in a Grid System", http://www.cs.virginia.edu/ mpw7t/papers/NatrajanWalker01.pdf
9. Plale, B., P. Dinda, and G. Laszewski, "Key Concepts and Services of a Grid Information Service", ISCA 15th International Parallel and Distributed Computing Systems (PDCS), 2002.
10. R. Sundaresan, M. Lauria, T. Kurc, S. Parthasarathy, and Joel Saltz, "Adaptive Polling of Grid Resource Monitors using a Slacker Coherence Model", Proceedings of HPDC-12, Seattle, June 2003.
11. Ken'ichiro Shirose, Satoshi Matsuoka, Hidemoto Nakada, and Hirotaka Ogawa, "Autonomous Configuration of Grid Monitoring Systems", In Proceedings of the 2004 International Symposium on Applications and the Internet (SAINT 2004 Workshops) pp. 651-657, 2004.
12. Hunter, J. S., "The Exponential Weighted Moving Average", Journal of Quality Technology, 18, 203C210, 1986.

Reducing Transaction Abort Rate of Epidemic Algorithm in Replicated Databases

Huaizhong Lin[1,*], Zengwei Zheng[1,2], and Chun Chen[1]

[1] College of Computer Science, Zhejiang University, 310027 Hangzhou, China
{linhz, zhengzw, chenc}@zju.edu.cn
[2] City College, Zhejiang University, 310015 Hangzhou, China
zhengzw@zucc.edu.cn

Abstract. Easy to deploy, robust, and highly resilient to failures, epidemic algorithms are a potentially effective mechanism for propagating information in large peer-to-peer systems deployed on Internet or ad hoc networks. In the paper, we explore the epidemic algorithms used for transaction processing in replicated databases that reside in weakly connected environments. We concentrate on the transaction commit voting process of the epidemic algorithms and suggest a new voting method, which takes an optimistic approach in conflict reconciliation. The optimistic voting protocol decreases abort rate and improves average response time of transactions.

1 Introduction

In recent years, the wireless communication and wide area network technologies, especially Internet, evolve rapidly. Weakly connected environments, which are characterized by low bandwidth, excessive latency, instability of connection, and constant disconnection, are used more and more frequently. Data replication is the common approach to improve system performance and availability. But due to the massive communication overhead in weakly connected environments, eager replication may bring about unacceptable number of failed or blocked transactions, and result in dramatic drop of system performance [1,2].

Epidemic algorithms [3], which mimic the spread of a contagious disease, have recently gained popularity as a potentially effective solution for disseminating information in large-scale systems, particularly P2P systems deployed on Internet or ad hoc networks. In addition to their inherent scalability, they are easy to deploy, robust, and resilient to failure. It is possible to adjust the parameters of an epidemic algorithm to achieve high reliability despite process crashes and disconnections, packet losses, and a dynamic network topology.

Epidemic algorithms can be used for managing replicated data [4-9]. In an epidemic approach, sites perform update operations locally and communicate peer-to-peer in a lazy manner to propagate updates. Transactional consistency is achieved by

---

* Supported by the Natural Science Fundation of Zhejiang Province, China (Grant no. M603230) and the Research Fund for Doctoral Program of Higher Education from Ministry of Education, China (Grant no. 20020335020).

decentralized conflict detect and reconciliation. Sites communicate in a way that maintains the causal order of updates and the communication can pass through one or more intermediate sites. Therefore, the epidemic model provides an environment that is tolerant of communication failures and doesn't require continuous connection between sites. Epidemic model is suitable for transaction processing of replication systems in weakly connected environments.

Several protocols have been proposed for implementing epidemic model in replicated databases, like ROWA (Read-One Write-All) protocol [4], quorum protocol [5], voting protocol [7,8], etc. In this paper, we describe the optimistic voting protocol, which introduces *condition* and *order* vote in the election process in transaction commitment. *Condition* vote postpones the final decisions on conflicting transactions and therefore improves the chances for transactions to get *yes* vote. *Order* vote prescribes the commit order of transactions that have read-write and write-write conflicts and eliminates transaction aborts due to these kinds of data conflicts. Optimistic voting protocol reduces abort rate and improves average response time of transactions when compared to other protocols.

The rest of the paper is organized as follows. In section 2, we develop the necessary background and introduce the epidemic model used in replicated databases. In section 3, we describe the optimistic voting protocol. In section 4, we perform the performance evaluation. We conclude the paper in section 5.

2 Epidemic Model

We consider a distributed system consisting of n sites labeled $S_1, S_2, ..., S_n$ and data items replicated fully or partially at all sites. Epidemic model assumes a fail-stop model of site failures and an unreliable communication medium. Sites communicate each other through messages passing in a pair-wise manner. Messages can arrive in any order, take an unbounded amount of time to arrive, or may be lost entirely, however, messages will not arrive corrupted. For this reason, timeout is not used in the protocols to detect conflicts and deadlocks.

Epidemic model is based on the causal delivery of log records where each record corresponds to one transaction instead of one operation. An event model [5] is used to describe the system execution, $(E, \rightarrow)$, where E is a set of transaction events and $\rightarrow$ is the happened-before relation which is a partial order on all events in E. The partial order $\rightarrow$ satisfies the following two conditions:

(1) Events occurring at the same site are totally ordered;
(2) If e is a sending event and f is the corresponding receiving event, then $e \rightarrow f$.

Vector clocks are used to ensure the property that if two events are causally ordered, their effects should be applied in that order at all sites. Each site S_k keeps a two-dimension time-table, which corresponds to S_k's most recent knowledge of the vector clocks at all sites. Upon communication, S_k sends a message including its own time-table and all records that receiving site hasn't received. Then the receiving site processes the events according to causal order and incorporates the time-table in an atomic step to reflect the new information from S_k.

The site S_k determines the records that receiving site S_j hasn't received according the following predicate [5] :
HasRecvd(T_k, t, S_j) ≡ T_k[j, Site(t)]≥Time(t)
Where t is an event, Site(t) is the site at which t occurred, and Time(t) is the local time at Site(t) when t occurred.

Upon completion of operations, a read-only transaction can be committed locally whereas an update transaction pre-commits and becomes a candidate. The read set, write set, and the update values of the candidate are recorded in log. Then sites exchange their respective log records to detect global conflicts and propagate values written by the transaction. A candidate is voted on and is eventually either committed (if it wins a plurality of the total system votes) or aborted.

When a transaction pre-commits, it is attached with a global distinct timestamp denoted by (local_ts, site_index), which is composed of a local timestamp and a distinct site index. Formally, we define a total order < on timestamps as follows. Suppose two timestamps ts(T_1)=(local_ts$_1$, site_index$_1$) and ts(T_2)=(local_ts$_2$, site_index$_2$), then ts(T_1)<ts(T_2) if and only if:

(1) local_ts$_1$<local_ts$_2$, or
(2) local_ts$_1$=local_ts$_2$ and site_index$_1$<site_index$_2$.

The information of local timestamp is piggybacked in the usual epidemic messages and a site adjusts its local timestamp as follows [10]: when site A receives a message from site B, it advances its local timestamp to max{ local_ts$_A$, the local_ts$_B$ carried by message}. If there are no communications between sites, their local timestamps will drift apart. But this doesn't matter since, in the absence of such communications, there is no need for synchronization in the first place and the drift will not affect the correctness of the protocol.

3 Optimistic Voting Protocol

3.1 Condition and Order Vote

Suppose two conflicting transactions T_i and T_j are issued by two sites concurrently. To maintain serializability, previous epidemic protocols consider that there is only one transaction can be committed and each site can only cast *yes* vote to one transaction in election, for example T_i. In optimistic voting protocol, to increase the chances to get *yes* vote for transaction T_j, sites can cast *condition* vote on it (whereas it is cast *no* vote in quorum or voting protocols). The *condition* vote on T_j can be transformed to *yes* vote if T_i is aborted. The use of *condition* vote postpones the final vote decision on transactions.

Definition 1. When voting on transaction T, suppose C={T_1,...,T_p} is the set in which each transaction conflicts with T, the *condition* vote *cond*(C) means that it can be transformed to *yes* vote in case each transaction in C is aborted, otherwise to *no* vote. The transform rules of *condition* vote are as follows:

(1) If $\exists T_i \in C$, T_i has been aborted, then *cond*(C) → *cond*(C-T_i);
(2) If $\forall T_i \in C$, T_i has been aborted, then *cond*(C) → *yes*;
(3) If $\exists T_i \in C$, T_i has been committed, then *cond*(C) → *no*.

For two transactions T_i and T_j that only have read-write and write-write conflicts, if the correct order can be preserved at all sites, e.g. T_i is committed before T_j, then the two conflicting transactions T_i and T_j can all be committed maintaining consistency. *Order* vote prescribes the commit order of these kinds of conflicting transactions. Additionally, it is easily observed that *condition* and *order* vote can coexist on one transaction T.

Definition 2. When voting on transaction T, suppose $C=\{T_1,\ldots,T_p\}$ is the set in which each transaction has only read-write and write-write conflicts with T, the *order* vote $order(C)$ means that it can be transformed to *yes* vote when all transactions in C have been committed or aborted at one site.

The transform rule of *order* vote is as follows:

If at one site, $\forall T_i \in C$, T_i has been committed or aborted, then $order(C) \rightarrow yes$.

Each site S_k maintains a list of candidates by the receiving order. Let $list_k$ denote the candidate set in which the vote on each transaction by the site is not *no* vote. When S_k receives a new candidate T, it votes on T according to the following rules. For convenience of description, Let

cond_set={ $T_i | T_i \in list_k$, wr_conflict(T_i,T) is true },

order_set={ $T_i | T_i \in list_k$, rw_conflict(T_i,T) or ww_conflict(T_i,T) is true, and wr_conflict(T_i,T) is false }.

(1) If $\exists x \in$ ReadSet(T), ReadVN(T,x)<CurrVN(S_k,x), it means that the value read by T has been overwritten, then vote *no*;
(2) If cond_set=∅ and order_set=∅, it means that there are no transactions in $list_k$ that have conflict with T, then vote *yes*;
(3) If $\exists T_i \in$ cond_set∪order_set, ts(T_i)>ts(T), then vote *no*;
(4) If $\forall T_i \in$ cond_set∪order_set, ts(T_i)<ts(T), then vote *cond*(cond_set) + *order*(order_set). The '+' denotes that the vote is transformed to *yes* vote if and only if both the *condition* and *order* vote are transformed to *yes*, otherwise to *no* vote.

The correctness proof of optimistic voting can be found in [9].

The votes collected in optimistic voting protocol can be viewed as optimistic quorum. The optimistic quorum differs from ordinary quorum in replicated databases in that the quorum is conditional and can only be transformed to really quorum based on the results of other transactions. This optimistic quorum increases the chance for a transaction to win a majority of sites, thus reducing the transaction abort rate.

3.2 An Example

We explain the optimistic voting protocol with an example. Suppose three transactions T_1, T_2 and T_3 (ts(T_1)<ts(T_2)<ts(T_3)).

Fig. 1 shows the voting process of voting protocol. Because three transactions have data conflict with each other, only one transaction can be committed. Fig. 2 shows optimistic voting process, which avoid the abort of T_3 by use of *order* vote. From the figures, we observe that T_2 in optimistic voting can be committed earlier than in voting protocol by use of *order* vote. For clarity, we omit some unimportant information exchanges in Fig. 1 and Fig. 2.

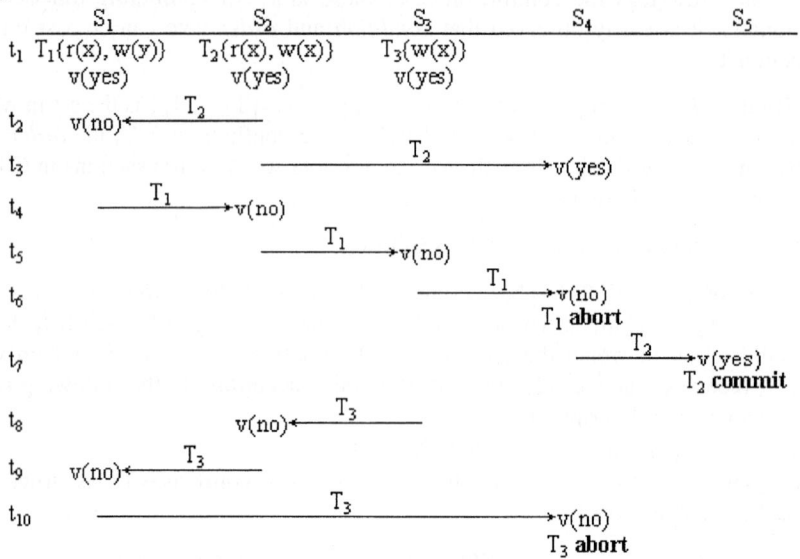

Fig. 1. Voting protocol

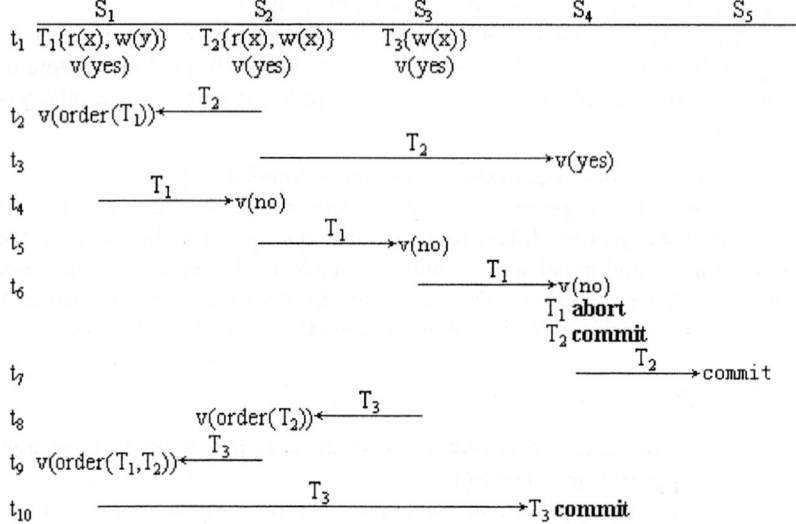

Fig. 2. Optimistic voting protocol

4 Performance Evaluation

We perform experiments to show performance improvement attained by optimistic voting (OV) protocol. Additionally, we investigate two representative epidemic replication schemes from the literature, ROWA protocol [4] and voting protocol [7] (quorum protocol [5] is similar to voting protocol). The evaluations are done at 10 desktops connected via a 10Mbps Ethernet network.

The simulation assumes that data items are fully replicated at all sites and tickets are uniformly distributed among sites. Each site generates transactions randomly according to a global transaction generation rate. Data items are accessed uniformly by transactions. Each site periodically initiates a synchronization session with a given synchronization interval by sending a pull request to another randomly selected site.

Since we focus on the transaction abort rate and commit delay of different protocols, we don't model any read-only transactions. Each transaction read 5-10 data items and write 5 data items that are in the read set, so there are no blind writes. The main parameters and settings used in the experiments are summarized in Table 1.

Table 1. Experimental parameters

Parameters	Descriptions	Values
N	Site number	10
Sync. interval	Average synchronization interval	1~5s
Trans. rate	Average generation rate of update transactions	0.2~20/s
Data items	Total data item number	500

Fig.3 illustrates the transaction abort rate of three protocols for various values of transaction generation rate. From the figures, it is obvious that optimistic voting protocol outperforms the other two protocols.

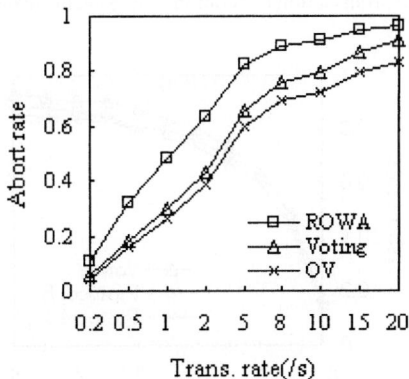

Fig. 3. Abort rate vs. transaction generation rate (Synchronization interval=1.0s)

In optimistic voting protocol presented above, the *condition* and *order* vote of a transaction is dependent on other transactions. This dependency relation in optimistic voting is one-way, i.e. a transaction can only depend on transactions that have smaller global timestamp than it (noted as protocol A). The one-way dependency ensures that there are no cycles among transactions and therefore no global deadlocks. This one-way dependency can be converted to depending on transactions that have larger global timestamp (noted as protocol B). We explore the impacts on performance of different dependency direction by experiments. Fig.4 and Fig.5 illustrate the transaction abort rate of protocol A, protocol B, and voting protocol for various values of transaction generation rate and synchronization interval. From the figures, it is obvious that protocol A is better than protocol B with average 5.4% performance gain. It is the natural direction for a transaction to depend on other transactions with smaller global timestamp. The transaction with smaller timestamp has been stay in the system for a longer time span and will be committed or rollbacked much earlier, which makes the transform of *condition* and *order* vote more quickly, thus reduce the delay of a transaction in the system. Additionally, we can notice that both protocol A and B outperforms the voting protocol.

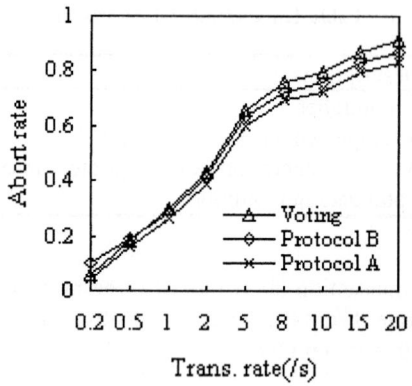

Fig. 4. Abort rate vs. transaction generation rate (Synchronization interval=1.0s)

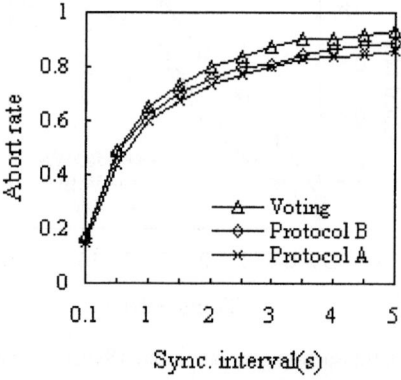

Fig. 5. Abort rate vs. synchronization interval (Transaction generation rate=5.0/s)

5 Conclusion

Epidemic replication schemes are used extensively in transaction processing in weakly connected environments. Some continuously connected systems also use epidemic model to improve system efficiency. The optimistic voting protocol presented in this paper improves system performance in epidemic model and is of high practical values.

References

1. J. Gray, P. Helland, P. O'Neil, and D. Shasha. The dangers of replication and a solution. In Proceedings of ACM SIGMOD International Conference on Management of Data. Montreal, Canada, 1996. 173-182.
2. T. Anderson, Y. Breitbart, H. F. Korth, and A. Wool. Replication, consistency, and practicality: are these mutually exclusive. In Proceedings of ACM SIGMOD International Conference on the Management of Data. Seattle, Washington, 1998. 484-495.
3. Patrick T. Eugster, Rachid Guerraoui, Anne-Marie Kermarrec, and Laurent Massoulié. Epidemic Information Dissemination in Distributed Systems. IEEE Computer, 2004, (5): 60-67.
4. D. Agrawal, A. El Abbadi, and R. Steinke. Epidemic algorithms in replicated databases. In Proceedings of 16th ACM SIGACT-SIGMOD-SIGART Symposium on Principles of Database Systems. Tucson, Arizona, 1997. 161-172.
5. J. Holliday, R. Steinke, D. Agrawal, and A. El Abbadi. Epidemic quorums for managing replicated data. In Proceedings of 19th IEEE International Performance, Computing, and Communications Conference. Phoenix, Arizona, 2000. 93-100.
6. K. Petersen, M. J. Spreitzer, D. B. Terry, M. M. Theimer, and A. J. Demers. Flexible update propagation for weakly consistent replication. In Proceedings of 16th ACM Symposium on Operating System Principles. St. Malo, France, 1997. 288-301.
7. U. Çetintemel, P. J. Keleher, and M. J. Franklin. Support for speculative update propagation and mobility in Deno. In Proceedings of 21st International Conference on Distributed Computing Systems. Phoenix, Arizona: IEEE Computer Society Press, 2001. 509-516.
8. P. J. Keleher. Decentralized replicated object protocols. In Proceedings of 18th Annual ACM Symposium on Principles of Distributed Computing. Atlanta, Georgia, 1999. 143-151.
9. Huaizhong Lin and Chun Chen. Optimistic voting for managing replicated data. Journal of Computer Science and Technology, 2002, 17(6): 874-881.
10. M. M. Deris, A. Mamat, and M. P. Hamzah. Replicated data management for transactions sharing in distributed database. In Proceedings of 4th International Conference/Exhibition on High Performance Computing in Asia-Pacific Region. Beijing, China: IEEE Computer Society Press, 2000. 836-841.

Snap-Stabilizing k-Wave Synchronizer

Doina Bein[1], Ajoy K. Datta[2], Mehmet H. Karaata[3], and Safaa Zaman[4]

[1] School of Computer Science, University of Nevada Las Vegas, NV
siona@cs.unlv.edu
[2] School of Computer Science, University of Nevada Las Vegas, NV
datta@cs.unlv.edu
[3] Department of Computer Engineering, Kuwait University
karaata@cairo.eng.kuniv.edu.kw
[4] Department of Computer Engineering, Kuwait University

Abstract. We propose a *snap-stabilizing* k-wave algorithm for routed trees, called kWA, which implements k consecutive waves, $k > 1$. A *snap-stabilizing* algorithm, starting from an arbitrary system configuration, always behaves according to its specification. The maximum number of states per process is $2k+1$, better than the previous result of at most $3k$ states per process. A *snap-stabilizing* algorithm guarantees that, starting from an arbitrary system configuration, always behaves according to its specification.

Keywords: Distributed computing, fault tolerance, multi wave algorithm, routed tree, snap stabilization, synchronizer.

1 Introduction

Solutions to a large number of fundamental problems in distributed computing require the execution of a constant number of consecutive distinct waves with corresponding feedbacks [1] (e.g. broadcasting of information, distributed infimum computation, leader election, mutual exclusion, spanning tree construction problems, clock synchronization, etc.). Consecutive distinct broadcasts with the corresponding feedbacks are implemented by the *multi-wave* algorithms, also referred to as the *asynchronous clock synchronizers*.

In the literature, PIF (propagation of information with feedback) algorithms are also referred to as *echo algorithms* [2]. Such wave algorithms are essential components of some distributed algorithms: leader election [3], snapshot [4], global reset [4,5,6,7], synchronization [6,8,9].

A *self-stabilizing* system automatically recovers to normal behavior in case of transient faults, without a centralized control. Regardless of the current configuration, the system reaches a legal state in finite number of steps and the system state remains legal thereafter [10], being able to withstand *transient failures* (a fault that perturbs the state of the system but not the program). Dolev provides a detailed survey on self-stabilization in [11].

If a certain stabilizing algorithm always behaves according to its specification, then is said to be *snap-stabilizing*, therefore it has optimal stabilization time. The notion of snap-stabilization was introduced in 1999 in [12]. The first snap-stabilizing PIF algorithm for arbitrary networks is proposed in [13]. A snap-stabilizing implementation of the PIF with cleaning called the PFC scheme (*propagation of information with feedback and cleaning*) is proposed by Bui et. al in [12], in which are three states in every wave. Therefore, a straight forward solution for a k-wave snap-stabilizing algorithm will have $3k$ states per process.

One of the main advantages of the PIF scheme to be snap-stabilizing is that the arbitrary initial state has limited or no effect on the pace of the broadcast propagation, so it implements the propagation with the optimal delay. Using the snap-stabilizing PIF algorithm of Bui et. al [12], distinct consecutive waves can be implemented by means of a counter per each process incremented upon receipt of each broadcast message. However, notice that the resulting algorithm is no longer stabilizing due to the need to initialize the counters. Therefore, the adaptation of the snap-stabilizing wave algorithm of Bui et. al to multiple consecutive distinct waves with feedback is not straight forward. The primary challenge stems from the introduction of new states representing the new phases while maintaining the snap-stabilization property.

Contributions. In this paper, we propose a *snap-stabilizing* k-wave algorithm (kWA), $k > 1$, for routed trees. The algorithm runs forever and the delay to start a kWA cycle is $O(h)$ *rounds*, which is optimal, where h is the height of the tree (the delay represents the time taken by the algorithm to start a kWA cycle beginning in an arbitrary initial state). A *round* refers to a minimum execution sequence in which each enabled action is performed at least once. The maximum number of states per process is $2k+1$, which is better than the previous implementations of $3k$ states per process and the same delay. A round refers to a minimum execution sequence in which each enabled action is taken at least once. The algorithm is optimal with respect to its time complexity and can be generalized to general networks using any of the self-stabilizing spanning tree construction algorithms proposed in [5, 6, 14].

Outline of the paper. Section 2 describes the distributed model used and snap-stabilization. Section 3 describes the specification of propagation of information with feedback (PIF). The k-wave snap-stabilizing algorithm is presented in the Section 4. Because of lack of space, the complete correctness proof of kWA algorithm is omitted, only few important remarks are given. The paper ends with concluding remarks and discussions on future work in Section 5.

2 Stabilization in Distributed Systems

A tree can be represented by an undirected graph $T = (V, E)$ with vertex (node) set V and edge set E, where $|V| = n$, and $|E| = n - 1$. We assume that each vertex of T is a process with a unique ID. We denote the set of leaf processes in

T by L and the set of internal nodes by I. For each process i, $N.i$, $D.i$, and $p.i$ denote the set of its neighbors, the set of its children and its parent, respectively.

Each process maintains a set of local variables whose values can only be updated by the process itself after inspecting them and the local variables of its neighbors.

The distributed program of some process can be expressed as a finite set of *guarded commands* or *guarded actions* [15]:

* [$G[1] \to A[1]$ ▯ $G[2] \to A[2]$ ▯ ... ▯ $G[k] \to A[k]$],

In the *guarded action* $G[] \to A[]$, the *guard* $G[]$ is a Boolean function of the variables of the process and its neighbors, and *action* $A[]$ can read the variables of the process and its neighbors, and update only the process variables. If $G[]$ is *true*, then the guarded action is called *enabled*. A process that has at least one guard enabled is called *enabled* too. *[S] corresponds to the repeated execution of S while there exists enabled guards. ▯ is called the *nondeterminism* symbol, and means that one of the guarded actions separated by those symbols is selected nondeterministically in each iteration and executed.

We assume the presence of a *distributed daemon* or *scheduler*: if more than one process is enabled, then a subset of enabled processes is selected, and exactly one enabled guard of each selected process is executed atomically and concurrently. In addition, we assume that the scheduler is *weakly fair*, i.e., any action remaining enabled is eventually executed. A self-stabilizing algorithm implementing this can be adapted from [16, 17]. Therefore, an implementation of this is omitted.

The *state (configuration)* of a process is the set of variables used by the process. The *system state* is the cartesian product of the states of the processes in the system. The set of all the possible system states is denoted by $\mathcal{C}$. An execution is a maximal sequence of states $e = c_1, c_2, ...$ such that $\forall c_i \in \mathcal{C} : c_{i-1} \mapsto c_i$ by executions some guarded action, for each $i > 0$. Given P a predicate defined over $\mathcal{C}$, the set of all states that satisfy P, $\mathcal{L}_P$, is called the set of all *legitimate states with respect to P*, or simply, the set of all *legitimate states*. Let P be a predicate defined over $\mathcal{C}$. The notation $c \vdash P$ means that an element $c \in \mathcal{C}$ satisfies the predicate P defined on the set $\mathcal{C}$. Given a task $\mathcal{T}$, the specification of $\mathcal{T}$ is a predicate called $\mathcal{SP}_\mathcal{T}$, and the distributed protocol to accomplish the task $\mathcal{T}$ is denoted by $\mathcal{P}_\mathcal{T}$. The set of all possible computations of $\mathcal{P}_\mathcal{T}$ in the system is denoted by $\mathcal{E}$.

Definition 1 (Snap-stabilization). *Let $\mathcal{T}$ be a task, and $\mathcal{SP}_\mathcal{T}$ the specification of $\mathcal{T}$. The program $\mathcal{P}_\mathcal{T}$ is snap-stabilizing for $\mathcal{SP}_\mathcal{T}$ on $\mathcal{E}$ if and only if $\forall e \in \mathcal{E} :: e \vdash \mathcal{SP}_\mathcal{T}$.*

3 Propagation of Information with Feedback (*PIF*)

Wave algorithms are often implemented using the *propagation of information with feedback (PIF) algorithms* [2, 18]. A snap-stabilizing implementation of the *PIF* with cleaning called the *PFC scheme* (*propagation of information with feedback and cleaning*) is proposed by Bui et. al in [12]. The value C, called the *cleaning state*, denotes the initial state of a process before it participates in a *PIF*-cycle, or the ending state after it participates in a *PIF* cycle.

Each process i maintains a variable S_i called the S-value of i. The root can use only two state values: B (broadcast) and C (cleaning). When executing its B-action and C-action, the root changes its state from C to B and B to C, respectively. The leaf processes use only F (feedback) and C states: the F-action changes the state from C to F, and the C-action changes the state from F to C. An internal process can be in C, B, and F states: after executing a B-action, F-action, respectively C-action, the S-value changes from C to B, from B to F and F to C, respectively.

According to the PFC cycle specification, a normal broadcast phase is followed by a feedback phase which is initiated by the leaf processes. After initiating the feedback phase, in the next round, the leaf processes can initiate the cleaning phase by changing their states from F to C. So, the feedback and the cleaning phases run concurrently. Eventually, the feedback phase reaches the descendants of the root. The root now executes its C-action, i.e., changes its state from B to C, and then waits until all its descendants are in cleaning phase. Then it changes its state from C to B. This marks the end of the current PIF cycle and the start of the next PIF cycle. Since this solution requires three phases (*broadcast, feedback,* and *cleaning*) in the PIF cycle, the method is called the *propagation of information with feedback and cleaning (PFC)* and the corresponding cycle the *PFC cycle*.

We say that a process p is B-*done*, F-*done*, or C-*done* to indicate that p has executed its corresponding actions (B, F, or C, respectively).

4 k-Wave Snap-Stabilizing Algorithm (kWA)

We present the k-wave algorithm, $k > 1$, to implement k consecutive distinct waves scheme on tree structured networks.

A kWA *cycle* consists of k consecutive PIF cycles where each PIF has its own broadcast and feedback phase with a specific task. Starting from an arbitrary configuration where no message has yet been broadcasted, the root (r) initiates the *first broadcast phase* by sending a *first broadcast message* to its children. The internal processes forward the broadcast message to their children upon receipt of the first broadcast message. Once the first broadcast phase reaches the leaf processes, they notify their parents of the termination of the broadcast phase by initiating the *first feedback phase* by sending *first feedback messages*. After receiving first feedback messages from all its children, a process sends a first feedback message to its parent. When all the neighbors of the root are in the first feedback phase, the second wave will be initiated by the root through the *second broadcast* phase, which acts similarly. Following the k feedback phase, a *clear phase* is initiated.

These k waves can overlap in a manner similar to those of the PFC cycle. Unlike the PFC cycle, in a kWA cycle the feedback phase of each wave and the cleaning phase do not execute concurrently. Instead, upon termination of the l feedback phase, the $l+1$ wave is initiated, $1 \le l < k$.

The state value C denotes the initial state of any process before it participates in a kWA cycle. The state values B^l, respectively F^l denotes the broadcast, respectively feedback state of a process, $1 \le l < k$. The root uses $k+1$ state values: C, B^l, $1 \le l < k$. An internal process can have $2k+1$ different state values C, B^l and F^l, $1 \le l < k$. A leaf process uses only $k+1$ state values, C, F^l, $1 \le l < k$.

The root executes its rB^1-action, rB^l-action, $1 < l < k$, and rC-action by changing its state from C to B^1, B^{l-1} to B^l, and B^k to C, respectively. An internal process executes its iB^1-action, iF^1-action, iB^l-action, iF^l-action, $1 < l \le k$, and iC-action, by changing its S-value from C to $B1$, from B^1 to F^1, from F^{l-1} to B^l, from B^l to F^l, and from F^k to C respectively. A leaf process executes its lF^1-action, lF^l-action, $1 < l \le k$, and lC-action respectively by changing its state from C to F^1, from F^{l-1} to F^l, and from F^k to C respectively.

Based on the above description, we define the kWA cycle in terms of the following kWA-actions: B^1-action, F^1-action, B^2-action, F^2-action, ..., B^k-action, F^k-action, C-action. The B^l-action, F^l-action, with $1 \le l \le k$, refer to the actions executed during l-th broadcast, l-th feedback phase respectively. The C-action is executed in order to terminate the current kWA cycle.

Definition 2 (*kWA cycle*). *A finite computation $e \in \mathcal{E}$ is called a* kWA *cycle, denoted by* $e \vdash kWA$-*cycle, if and only if the following conditions hold:*

L1 At least one process r, called the initiator, sends a message or completes the kWA cycle ($rB1$).

L2 If an internal process i ($i \in I$) receives a first/second/.../ k-th broadcast message from its parent $p.i$, then i eventually sends this message to all its descendants ($iB1$, iBl).

L3 If a leaf process i ($i \in L$) receives a first/second/.../ k-th broadcast message from its parent $p.i$, then i eventually sends a first/second/.../k-th feedback message to $p.i$ ($lF1$, lFl).

L4 If an internal process i ($i \in I$) receives the first/second/... k-th feedback message from all of its descendants, then i eventually sends this message to its parent $p.i$ ($iF1$, iFl).

L5 If the initiator is in l-th broadcast phase and receives l-th feedback message from all of its descendents, then it sends the $(l+1)$-th broadcast message to all of its descendents (rBl).

L6 If the root r receives the k-th feedback message from all of its descendants, then r eventually terminates the current kWA cycle (rC).

S The root r cannot terminate the kWA cycle (rC) more than once.

Conditions [L1] to [L6] are called the liveness properties, and condition [S] is called the safety property. By Condition [L1], we know that there exists at least one initiator in a kWA cycle.

The algorithm implementing the above mechanism is given in Figure 1. The value of l is $1 < l \le k$.

Variables
  $S_i(i=r) \in \{B^1, B^2, ..., B^k, C\}$ for the root
  $S_i(i \in I) \in \{B^1, F^1, B^2, F^2, ..., B^k, F^k, C\}$ for internal processes
  $S_i(i \in L) \in \{F^1, F^2, ...F^k, C\}$ for the leaf processes

Actions
  {Program for root process i }

rB1 $\quad * \Big[S_i = C \land \forall_{j \in N.i} S_j = C \quad\longrightarrow\quad S_i := B^1;$
rBl $\quad \Box\ S_i = B^{l-1} \land \forall_{j \in D.i} S_j = F^l \quad\longrightarrow\quad S_i := B^l;$
rC $\quad \Box\ (S_i = B^k \land (\forall_{j \in N.i} S_j = F^k)) \lor$
$\quad\quad (S_i = B^1 \land \exists_{j \in D.i} S_j \notin \{B^1, F^1, C\}) \lor$
$\quad\quad (1 < l < k \land S_i = B^l \land \exists_{j \in D.i} S_j \notin \{B^{l-1}, F^{l-1}, B^l, F^l\}) \quad\longrightarrow\quad S_i := C; \Big]$

  {Program for internal process i }

iB1 $\quad * \Big[S_i = C \land S_{p.i} = B1 \land \forall_{j \in D.i} S_j = C \quad\longrightarrow\quad S_i := B^1;$
iF1 $\quad \Box\ S_i = B1 \land S_{p.i} = B1 \land \forall_{j \in D.i} S_j = F^1 \quad\longrightarrow\quad S_i := F^1;$
iBl $\quad \Box\ S_i = F^{l-1} \land S_{p.i} = B^l \land \forall_{j \in D.i} S_j = F^{l-1} \quad\longrightarrow\quad S_i := B^l;$
iFl $\quad \Box\ S_i = B^l \land S_{p.i} = B^l \land \forall_{j \in D.i} S_j = F^l \quad\longrightarrow\quad S_i := F^l;$
iC $\quad \Box\ (S_i = F^k \land S_{p.i} = F^k \land \forall_{j \in D.i} S_j \in \{F^k, C\}) \lor$
$\quad\quad (1 < l \leq k \land S_i = B^{l-1} \land (S_{p.i} \notin \{B^{l-1}, F^{l-1}, B^l\} \lor$
$\quad\quad \exists_{j \in D.i} S_j \notin \{B^{l-1}, F^{l-1}, F^l\}) \lor$
$\quad\quad (1 < l \leq k \land S_i = F^{l-1} \land (S_{p.i} \notin \{B^{l-1}, F^{l-1}, B^l\} \lor$
$\quad\quad \exists_{j \in D.i} S_j \neq F^{l-1}) \quad\longrightarrow\quad S_i := C; \Big]$

  {Program for leaf process i }

lF1 $\quad * \Big[S_i = C \land S_{p.i} = B^1 \quad\longrightarrow\quad S_i := F^1;$
lFl $\quad \Box\ S_i = F^{l-1} \land S_{p.i} = B^l \quad\longrightarrow\quad S_i := F^l;$
lC $\quad \Box\ (S_i = F^k \land S_{p.i} = F^k) \lor$
$\quad\quad (1 < l \leq k \land S_i = F^{l-1} \land S_{p.i} \notin \{B^{l-1}, F^{l-1}, B^l\}) \lor$
$\quad\quad (S_i = F^k \land S_{p.i} \notin \{B^k, F^k, C\}) \quad\longrightarrow\quad S_i := C; \Big]$

Fig. 1. The Snap-Stabilizing k-Wave Algorithm (Algorithm kWA)

Let $Init$ be the set of processes that initiate kWA cycles. We will call a kWA cycle as a *normal kWA cycle* if $Init = \{r\}$ and the first action of r is a rB^1-action.

Definition 3 (kWA scheme). *We define computation e as a kWA scheme, denoted by $e \vdash \mathcal{SP}_{kWA}$, as an infinite sequence of kWA cycles, $e = e_0, e_1, ...$ such that $\forall i \geq 1$, e_i is a complete kWA cycle.*

The following proposition follows from the definition of the kWA cycle.

Proposition 1. *In a normal kWA cycle, every process executes a B^1-action, a F^1-action, ..., a B^k-action, and a F^k-action at most once.*

From Definition 2, it is obvious that once p executes one of the kWA-actions, p will not execute that action anymore in the same kWA cycle.

In any configuration, either the root is enabled to execute and the other processes have no enabled actions, or the root has no enabled actions and the other processes are enabled to execute. The following proposition follows from Definition 2 and Proposition 1.

Proposition 2. *In any configuration of the kWA cycle, either one of the following conditions holds:*
  (i) The root r is enabled to start the lth wave, $1 \leq l \leq k$, by executing its B^l-action.
  (ii) The root r is surrounded by a maximal set of processes all of which are enabled to execute a kWA-action corresponding to either l^{th} wave, $1 \leq l \leq k$, such that each process between r and a process in this set (including r) is B^l-done.

Definition 4 (Snap-Stabilizing kWA). *A kWA algorithm is snap-stabilizing if and only if the root executes a kWA-action infinitely often, and the kWA cycle satisfies $\mathcal{SP}_{kWA}$.*

Theorem 1. *Algorithm* kWA *is snap-stabilizing.*

The optimal delay of starting a kWA cycle of any implementation of the kWA cycle is shown to be $O(h)$ rounds in [6]. We can show that the delay to start the first kWA cycle using Algorithm kWA is bounded by $O(h)$ rounds. Therefore, the algorithm kWA is optimal with respect to its delay to start a kPW cycle.

5 Conclusions

In this paper, we propose a *snap-stabilizing k-wave algorithm for routed trees*, called kWA, with $k > 1$. The algorithm ensures that a cycle may start after $O(h)$ *rounds*, which is optimal, where h is the height of the tree. The maximum number of states per process is $2k + 1$, which is better than the previous implementations of $3k$ states per process and the same delay. We intend to apply the concept of kWA scheme in devising optimal snap-stabilizing algorithms for some fundamental problems in distributed computing.

References

1. G. Tel. Total algorithms. *In: Vogt F.H. (ed) Concurrency 88, Springer-Verlag LNCS 335*, pages 277–291, 1988.
2. E.J. Chang. Echo algorithms: depth parallel operations on general graphs. *IEEE Transactions on Software*, SE-8:391–401, 1982.

3. S. Dolev, A. Israeli, and S. Moran. Uniform dynamic self-stabilizing leader election. *IEEE Transactions on Parallel and Distributed Systems*, 8(4):424–440, 1997.
4. G. Varghese. Self-stabilization by counter flushing. *SIAM Journal on Computing*, 30(2):486–510, 2000.
5. A. Arora and M.G. Gouda. Distributed reset. *IEEE Transactions on Computers*, 43:1026–1038, 1994.
6. B. Awerbuch, S. Kutten, Y. Mansour, B. Patt-Shamir, and G. Varghese. Time optimal self-stabilizing synchronization (extended abstract). In *Proceedings of the Twenty-Fifth Annual ACM Symposium on the Theory of Computing*, pages 652–661, San Diego, California, 16–18 May 1993.
7. B. Awerbuch, B. Patt-Shamir, and G. Varghese. Self-stabilization by local checking and correction. In *Proceedings of the 32th Annual IEEE Symposium on Foundations of Computer Science*, pages 268–277, 1991.
8. L.O. Alima, J. Beauquier, A.K. Datta, and S. Tixeuil. Self-stabilization with global rooted synchronizers. In *ICDCS98 Proceedings of the 18th International Conference on Distributed Computing Systems*, pages 102–109, 1998.
9. B. Awerbuch and G. Varghese. Distributed program checking: a paradigm for building self-stabilizing distributed protocols. In *FOCS91 Proceedings of the 31st Annual IEEE Symposium on Foundations of Computer Science*, pages 258–267, 1991.
10. E.W. Dijkstra. Self-stabilizing systems in spite of distributed control. In *EWD 391, In Selected Writings on Computing: A Personal Perspective*, pages 41-46,1973.
11. S. Dolev. *Self-Stabilization*. MIT Press, Cambridge, MA, 2000. Ben-Gurion University of the Negev, Israel.
12. A. Bui, A.K. Datta, F. Petit, and V. Villain. State-optimal snap-stabilizing PIF in tree networks. In *Proceedings of the Third Workshop on Self-Stabilizing Systems (published in association with ICDCS99 The 19th IEEE International Conference on Distributed Computing Systems)*, pages 78–85. IEEE Computer Society, 1999.
13. A. Cournier, A.K. Datta, F. Petit, and V. Villain. Snap-stabilizing PIF algorithm in arbitrary networks. In *ICDCS02 The 22nd IEEE International Conference on Distributed Computing Systems*, pages 199–206, 2002.
14. S.T. Huang and N.S. Chen. A self-stabilizing algorithm for constructing breadth-first trees. *Information Processing Letters*, 41:109–117, 1992.
15. E.W. Dijkstra. Guarded commands, nondeterminacy, and formal derivation of programs. *Communications of the ACM*, 18(8):453–457, 1975.
16. M.G. Gouda and F. Haddix. The alternator. In *Proceedings of the Third Workshop on Self-Stabilizing Systems (published in association with ICDCS99 The 19th IEEE International Conference on Distributed Computing Systems)*, pages 48–53. IEEE Computer Society, 1999.
17. M. Mizuno and M. Nesterenko. A transformation of self-stabilizing serial model programs for asynchronous parallel computing environments. *Information Processing Letters*, 66(6):285–290, 30 June 1998.
18. A. Segall. Distributed network protocols. *IEEE Transactions on Information Theory*, IT-29(1):23–35, January 1983.

A Service Oriented Implementation of Distributed Status Monitoring and Fault Diagnosis Systems

Lei Wang[1], Peiyu Li[2], Zhaohui Wu[1], and Shangjian Chen[1]

[1] College of Computer Science, Zhejiang University,
Hangzhou, R.P.China, 310027
{alwaysbeing, wzh, ipipip}@cs.zju.edu.cn
[2] College of Mechanical and Energy Engineering, Zhejiang University,
Hangzhou, R.P.China, 310027
lipeiyu@zju.edu.cn

Abstract. The service-oriented architecture (SOA) can greatly improve the collaboration of devices in a distributed environment. With this architecture, human can be liberated from the burdensome configuration and management of distributed systems, especially when the nodes of the systems are dynamically changed. In this paper, we present a method to implement service-oriented distributed systems in industry. As a case study, we present a service-oriented Status Monitoring and Fault Diagnosis Systems (SMFDS) for large equipment. Our work is based on the UPnP (Universal Plug and Play) technology. Through abstracting the interfaces of devices and providing service middleware in an independent thread or process, the SOA can be easily implemented on an existing system with the least modifications. This characteristic is very useful when upgrading an existing system, for cost and risk consideration.

1 Introduction

The increase in amount of computing devices and the diversity of them will enhance the complexity to configure and utilize them, which is becoming an urgent issue. The traditional computer-centered model cannot solve this problem. Human beings need to be liberated from the burdensome work of devices management and maintenance.

An emerging trend in distributed application development is that of a SOA. It is defined as [1]: "SOA takes the existing software components residing on the network and allows them to be published, invoked and discovered by each other. SOA allows a software programmer to model programming problems in terms of services offered by components to anyone, anywhere over the network. In other words, any application residing anywhere on any computer system would be able to interact with any service anywhere over the network." This architecture regards an application as a federation of services, where a service represents a logical concept such as a printer or chat-room service. Services are provided and used by components and they are discovered dynamically and used according to a mutual agreed contract between providers and users. With this scheme, the interoperability of a distributed system can be greatly improved, and the devices in the system will be more autonomous. This new approach makes it possible to reduce the configuration and management efforts and is considered to be one of the solutions of human-centered computing.

In this paper, we focus on the implementation of SOA distributed systems in industrial environments. As a case study, we will present how to implement a SMFDS for large equipment. SMFDS is a kind of data stream centered distributed system, which is representative in actual industrial applications. Our work concentrates on implement the service-oriented model based on existing systems. The remainder of paper is structured as follows: Section 2 introduces the SOC base on UPnP. Section 3 describes the traditional implementation of SMFDS. Section 4 presents our method to develop the service-oriented SMFDS. Finally, we conclude our paper and describe a program of further work in section 5.

2 SOA on UPnP

SOA can be implemented in many ways. Four common implementation technologies are CORBA [2], JINI [3], Web Services [4] and UPnP [5]. Since these implementation technologies provide the service-oriented architecture for up-level applications and are independent of any particular operating system, they are often called service middleware. As UPnP needs small footprint software components on devices, it is quite applicable to resource-constrained embedded systems. In this paper, we choose UPnP to implement the status monitoring and fault diagnosis system.

UPnP technology is a distributed, open networking architecture that employs TCP/IP and other Internet technologies to enable seamless proximity networking, in addition to control and data transfer among networked devices in the home, office, and public spaces. It is independent of any particular operating system, programming language, or physical medium. It is designed to support zero-configuration, "invisible" networking, and automatic discovery for a breadth of device categories from a wide range of vendors.

The basic building blocks of an UPnP network are devices, services and control points. An UPnP device is a container of services and nested devices. Different categories of UPnP devices will be associated with different sets of services and embedded devices. A service is the smallest unit of control in an UPnP network. A service exposes actions and models its state with state variables. A service in an UPnP device consists of a state table, a control server and an event server. The state table models the state of the service through state variables and updates them when the state changes. The control server receives action requests, executes them, updates the state table and returns responses. The event server publishes events to interested subscribers anytime the state of the service changes. A control point in an UPnP network is a controller capable of discovering and controlling other devices. After discovery, a control point could retrieve service descriptions for interesting services and invoke actions to control the service. It can also subscribe to a service's event source. Anytime the state of the service changes, the event server will send an event to the control point.

3 Traditional Implementation of SMFDS

Large equipment, such as water pumps and dynamotors, are widely used in industrial environments. This equipment is so large and heavy that they have to be assembled and fixed in workshops. Usually, this equipment keeps on working day and night and makes a very important role in industrial manufacture. If they fail to work, the production will be blocked and the safety of production may even be threatened. To avoiding the economic loss caused by equipment failures, it is needed to continuously monitor the status of the pivotal equipment, detect their abnormality, locate the faults and take measures to prevent failures. This is the aim to develop SMFDS.

A SMFDS usually works as a distributed system. As shown in Fig.1, it usually consists of sensors, data collectors, monitoring computers and a data storage server. The sensors collect equipment status signals, such as librations signals, temperatures, and pressures, and send them to a data collector, which performs the A/D conversion and corresponding digital signal processing. The processed data is then packed and sent to the monitoring computers and the data storage server. Since the distance between sensors and the data collector is limited, the collector is generally fixed near the monitored equipment. If there is a great deal of equipment in a workshop, multiple data collectors are required.

The monitoring computers and data storage server are located in a monitoring center. They connect with the data collectors through a LAN or WAN. The data storage server stores and manages the data streams from data collectors. The monitoring computers process, analyze and display the real-time data streams from data collectors. They can also access the historical data in the data storage server. With the monitoring computers, operators can monitor the running status of all equipment. An expert system running on monitoring computers can detect the equipment abnormality and locate the faults based on its current and historical data.

When the status of equipment is abnormal, the faults usually need to be eliminated manually. It will be very useful for a maintainer to refer to the real-time analyzed results of the equipment status. For the cost and space considerations, most data collectors cannot provide the functions of monitoring computers. In this case, PDAs with the same functions of monitoring computers can be use in the workshop. The PDAs are connected to the network through wireless access points.

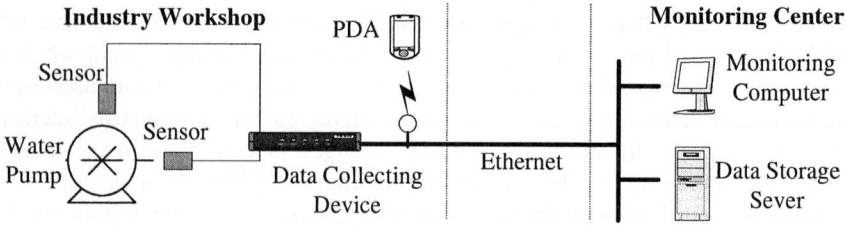

Fig. 1. Topology of traditional SMFDS

In this type of systems, all the devices connected to the network have fixed addresses or the operators can configure their addresses. Any device has to know the addresses of others to communicate with them.

4. Implementation of Service-Oriented SMFDS

The traditional implementation of SMFDS described in Section 3 has several drawbacks. Firstly, when a PDA is brought into the workshop, it has to be configured to make it connected with the data collectors and the data storage server. Secondly, when new equipment with a new data collector is assembled in the workshop, many configurations have to be done on the data collector, all the monitoring devices and the data storage server to make the whole system work correctly. Finally, to guarantee the integrality and availability of history data, more than one data storage servers are required. If one of the servers breaks down, other devices should find and access the backup server automatically. Current SMFDS has difficulty in meeting this requirement. In this section, we will describe how to implement the SOA in SMFDS, which overcomes the drawbacks of the traditional implementations.

4.1 Service-Oriented Architecture of SMFDS

According to the description of SMFDSs in Section 3, we can determine that there are four types of devices related to services in a SMFDS. They are the data collectors, the PDAs, the remote monitoring computer and the data storage server. Based on their actions and the principle of UPnP, the SOA of this system can be abstracted.

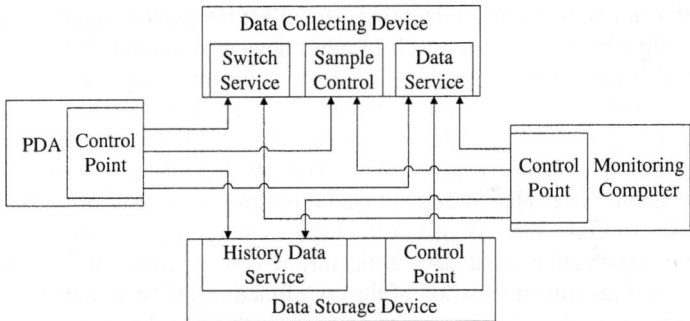

Fig. 2. The service-oriented architecture of SMFDS

As shown in Fig. 2, the data collectors provide three services. The switch service is to start and shut down the data collector, the sample control service is to control the sampling frequency of the data collector and the data service is provided for accessing the latest equipment status data. The data storage server provides a service named historical data access service. Both the PDAs and the remote monitoring computers work as control points. They use the services provided by the data collectors and the data storage server. Besides providing a service, the data storage server also works as a control point, since it uses data service provided by the data collectors to obtain the equipment status data. Obviously, the actions of the PDAs and the remote monitoring computers are quite similar, except that the PDAs are mobile devices.

The service interfaces in this system is described in Table 1. The status variables of a service model the state of the service. When the state is update, an event will be published. Control points can subscribe to the service's event source. In our implementation of SMFDS, for example, the PDAs subscribe the event notification from the switch status variable in the data collectors. When a data collector is started or shut down, the PDAs can capture this event and take corresponding actions.

Table 1. Services provided by devices in SMFDS

Device	Service	Interface	Type	Description
Data collector	Switch service	Start	Action	Start data collection
		ShutDown	Action	Stop data collection
		GetStatus	Action	Get the switch status
		SwitchStatus	State	Switch status: running or stopped
	Data service	GetData	Action	Get the latest equipment status data
		DataID	State	Get ID of the latest status data
	Sampling control service	GetFrequency	Action	Get the sampling frequency
		SetFrequency	Action	Set the sampling frequency
		Frequency	State	The sampling frequency
Data storage server	History data service	GetData	Action	Get a historical data record
		DataStatus	State	Denote the update of data

4.2 Implementation of Data Collectors

The essential function of data collectors is to collect the analog signals from sensors, process the signals and send the result to other devices requiring the status data. The traditional software implementations of data collectors may be quite different: component-based or not, in a single thread or in multiple threads. To implement service-oriented data collectors, the software can be completely rewritten based on the UPnP standards. This policy is acceptable when develop a bran-new system. But if a mature and stable system has been deployed after long-term development, this policy will be costly and take the risk of crashing the existing system. When exploring the software implementation of a data collector, it can be found that what limits it autonomy is not the implementation of the core functions, such as signal process, data management, but its interfaces to other devices, such as how to notify other devices that new data is ready. An alternative method is to insert a service-oriented middleware between the existing data collector software and the control points using these services. The middleware provides the service interfaces to the control points on other devices. When the control points invoke the actions exposed by the middleware, the middleware will perform the service actions by invoking the corresponding operations in the data collection software. Fig. 3 illustrates this scheme considering only the implementation of switch service.

In this figure, data collection software denotes the core functions implementation of a data collector. It provides two access interfaces: start and shut down the data collector. The switch service is exposed by the switch service middleware, which will accept the action invocations from the control points on other devices. The action starts and shutdowns in the middleware are implemented by invoking the corresponding interfaces in the data collection software. Since the middleware has to continually handle the requirements from control points and provide an event server, it is usually implemented in an independent process. With the consideration of efficiency, an asynchronous communication scheme is needed to invoke functions and transfer data between the middleware and the data collection software process. When the Start operation is completed in data collection software, it will invoke a StartReturn operation in the UPnP middleware by sending a SRSIG signal. The operation will change the state of the middleware from SwitchOff to SwitchOn. When the state of the middleware changed, a status change notification will be sent to the control points that have subscribed this event.

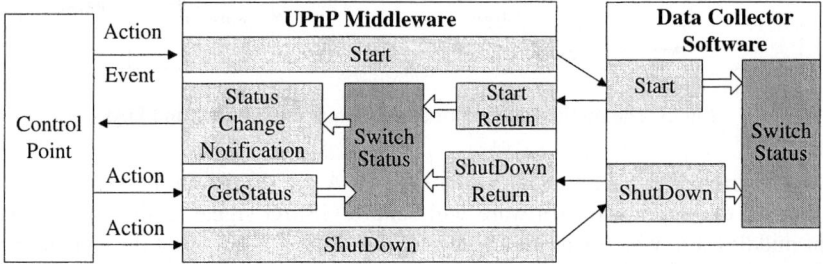

Fig. 3. The implementation of switch service in data collectors

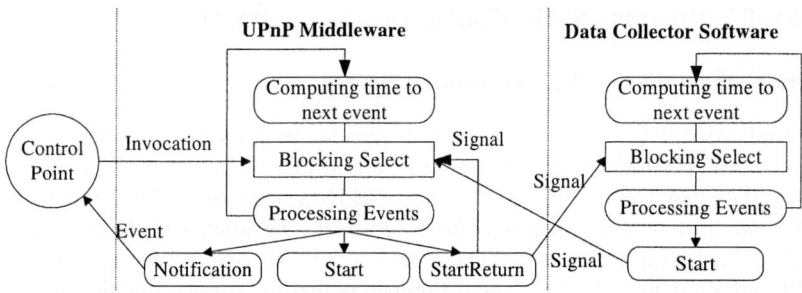

Fig. 4. The thread model of switch service implementation

The thread model corresponding to the implementation of the switch service is shown in Fig. 4. Both the middleware and data collection software work in a blocking select manner. When an event arrives, its corresponding action will be invoked. The actions must be completed quickly, so that other events can be responded immediately. If time-consuming operations have to be performed in an action, another thread or process can be created to execute it. The interfaces between the middleware and the data collection software are event notifications and data transfer, which can be implemented through IPC. In the same way, the data service and sample control service in data collectors can also be implemented.

4.3 Implementation of the Control Point on PDAs

The implementation of a control point is much like that of a service. It works as a middleware to provide application software with the channels for interacting with external service providers. In order to build a control point on an existing device with the least modification, the control point middleware is implemented in an independent thread or process. The application software can interoperate with it through IPC.

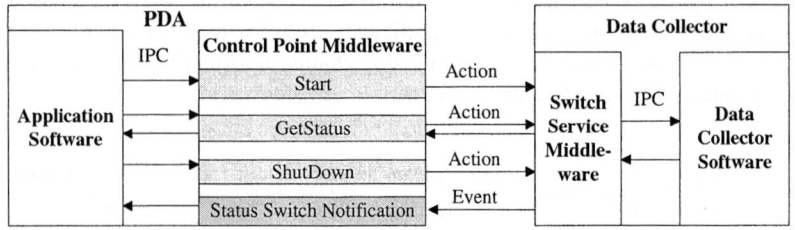

Fig. 5. The implementation of control point middleware on PDAs

As shown in Fig. 5, if an application on a PDA aim to obtain the switch status of equipment, the application software will send a GSSIG signal to the switch control point middleware, which will in turn invoke the GetStatus action of the switch service middleware on the corresponding data collector. The interactions between control points and service providers are based on UPnP specification. If the switch status of a data collector is changed, an event will be send to the control point middleware, which will in turn send a SSSIG signal to application software.

4.4 Implementation of the Data Storage Server

The implementation of control points in remote monitoring computers and data storage server is similar to the implementation on PDAs. On the data storage server, the core data management function is provided by a database. A history data service middleware provides an interface for interactions between the database and the control points using the history data access service. It works very like the service middleware on data collectors. The difference is that the service middleware on data storage server passively wait the invocations from the control points. If there are several data storage servers on the network, each server receives the data stream from data collectors through its control point middleware. Although each monitoring device usually fetches historical data from one server, it maintains a list of all the existing servers. Once its current server cannot provide data normally, a backup server will be chosen from the list for historical data service.

4.5 Prototypes and Application

A SMFDS prototype for large pumps has been developed. The data collector is implemented on an ARM9 single-board embedded computer with Linux running on it. The monitoring device is implemented on both HP 5500h PDAs and PCs. The operating systems are Linux and Windows respectively. The data storage server runs Windows 2000 Server and SQL server. Three data collectors, two data storage servers and one monitoring computer are connected with an Ethernet network. Two PDAs are connected to the network through a wireless access point. The services and the control point middleware are implemented through the Intel® NMPR v2.0 Device Enabling

Kit [6], which can generate the UPnP framework codes for several OS platform. When an operator holding a PDA enters the workshop, the PDA can discover all the equipment and display their status. It can also obtain the history data from the data storage server and compare their difference. If one of the data storage servers breaks down, all the monitoring devices can automatically fetch data from the backup server.

5 Conclusion

The SOA can provide a flexible cooperative scheme for distributed heterogeneous computing devices. Taking advantage of this architecture in industry can greatly reduce the work of the configuration and maintenance of computing devices. We present a method to implement the SOA on SMFDS for large equipment based on UPnP technology. In this implementation, it will incur zero configurations for equipment or computing devices to join or quit the system, as well as to dynamically change the historical data source. Through providing service and control point middleware in an independent thread or process, the service-oriented architecture can be easily implemented on existing systems with the least modifications. A prototype SMFDS for large pumps is also developed based on this method. In this prototype, devices and equipment can be automatically discovered and cooperate with other devices in a predefined rule, which will help maintainers concentrate on the equipment check-up and fault elimination. Dependability is very critical in industry. Current UPnP technology cannot provide sufficient support to implement dependable service-oriented industrial applications. Our future work will focus on the solutions of dependable SOA, especially for embedded distributed real-time systems.

Acknowledgements

This research was supported by 863 National High Technology Program under Grant No. 2002AA1Z2308, No. 2003AA1Z2130 and No. 2004AA1Z2060.

References

1. Introduction to UDDI and Service Oriented Architecture (SOA). http://archive.devx.com/xml/articles/sm100901/sidebar1.asp retrieved on 8 Oct. 2003.
2. The Object Management Group (OMG). The Common Object Request Broker: Architecture and Specification. OMG Document formal/99-10-07, Revision 2.3.1, 1999.
3. K. Arnold et al..The Jini Specification. Addison-Wesley Longman, Reading, Mass., 1999.
4. W3C Web Services Activity. Web Service Architecture. W3C Working Group Note, 2004.
5. Univeral Plug and Play Forum, http://www.upnp.org/
6. Intel® NMPR v2.0 Device Enabling Kit, Available: http://intel.com/technology/upnp/download.htm

Adaptive Fault Monitoring in Fault Tolerant CORBA*

Soo Myoung Lee[1], Hee Yong Youn[1], and We Duke Cho[2]

[1] School of Information and Communications Engineering
SungKyunKwan University, 440-746, Suwon, Korea
{lifetime, youn}@ece.skku.ac.kr
[2] CUCN, Suwon, Korea
chowd@ajou.ac.kr

Abstract. A number of different kinds of applications developed on CORBA framework need fault tolerance in asynchronous distributed system or network environment, and it is important to quickly detect the faults. There exist various fault monitoring and detection algorithms that employ a timeout-based mechanism. However, they are occasionally inaccurate in unstable or overloaded system. The goal of the proposed algorithm is to enhance the accuracy of fault monitoring. This is achieved by promptly adjusting the timeout interval using the past elapsed time values accumulated. Additionally, we use asynchronous invocation to call 'is_alive()' method of monitorable object with a sequence number. Experiment on CORBA-compliant Orbix ORB confirms the effectiveness of the proposed scheme compared to the existing one.

Keywords: Adaptation, elapsed time, fault monitoring and detection, fault-tolerant CORBA, timeout.

1 Introduction

A number of different kinds of applications developed by the members of the OMG and users of CORBA [1] need fault tolerance features. Fault-tolerant CORBA [2] aims to provide robust support for the applications requiring high reliability. Here it is very important to quickly detect a fault if it occurs. We call the mechanism detecting any fault occurrence as fault monitoring and detection. The mechanism is primarily based on timeout event occurring when a request message sent from a monitor to the monitorable object does not return in time.

In asynchronous distributed systems it is hard to differentiate between real crash and network overload. The design of fault detector must be careful because a timeout value much smaller than the average transmission time will cause false alarm, while a timeout value much greater than the average transmission time will delay the fault detection when a crash occurs.

There are various fault detection models such as Crash-Recovery Model [3] and the model of Chandra and Toueg [4]. They suggest a way that increases the timeout

---

* This research was supported by the Ubiquitous Autonomic Computing and Network Project, 21st Century Frontier R&D Program in Korea and the Brain Korea 21 Project in 2004. Corresponding author: Hee Yong Youn

value when a false fault or crash alarm occurred. However, this strategy allows a scenario where the timeout value is increased several times due to overload in the network. It can lead to an unnecessarily long timeout value even after the network returns to normal workload condition. The ADAPTATION-Algorithm proposed by Sotoma and Mauro Madeira [5] presents a fault detection mechanism that aims at accurate failure detection in spite of network overloads by means of periodic increases and decreases of the timeout value. However, it may deduce incorrect timeout values and monitoring intervals in unstable system or network environment because it regulates timeout values only after timeout or response event occurs and initializes several factors used for calculating the mean values.

We thus propose a solution that can solve the problem above. First of all, it employs an asynchronous invocation approach in which 'is_alive()' message is transmitted with a sequence number to the monitorable object and buffer is used that can reserve a number of measured time values for calculating several factors. Additionally, it uses some equations in the algorithm for achieving better result. Experiment on an actual client and server system reveals that the proposed algorithm allows more accurate prediction of monitoring interval than the ADAPTATION algorithm regardless of load condition of the system.

The rest of the paper is organized as follows. Section 2 presents an overview of Fault-tolerant CORBA and related work. Section 3 describes the proposed scheme. Section 4 evaluates the performance of the proposed scheme and compares it with the ADAPTATION algorithm. Finally, Section 5 concludes the paper.

2 Background

In fault-tolerant CORBA, fault management encompasses the following activities; **fault detection:** detecting the presence of a fault in the system and generating a fault report, **fault notification:** propagating fault reports to the entities registered for such notifications, **fault analysis/diagnosis:** analyzing a (potentially large) number of related fault reports and generating condensed or summary reports.

2.1 The Models for Fault Monitoring

Most implementations of fault detectors are based on timeout, and use either pull or push-based monitoring. Because push-based monitoring depends on the characteristics of the application, it is not defined in Fault-tolerant CORBA specification.

- **Pull model**: The fault monitor periodically calls 'is_alive()' method of the monitorable object asking whether it is alive. If the monitorable object does not reply within some time interval, then the fault monitor suspects it is faulty. The main advantage is that it allows a status check only when it is needed by the application.
- **Push model**: The monitorable object periodically calls 'i_am_alive()' method of the fault monitor informing that it is still alive. If the monitorable object has not called the 'i_am_alive()' method within some time interval, then the fault monitor suspects it is faulty. The main advantage is fast detection of the failure of the monitorable object.

2.2 Related Works

The design and verification of fault-tolerant distributed applications are generally viewed as complex endeavor. In recent years, several paradigms have been identified which simplify the task. The model using the weakest failure detector [6] determines what information about failures is necessary and sufficient to solve the Consensus problem in asynchronous distributed systems subject to crash failures. In the model any failure detector has to provide at least as much information as $\Diamond W$. Thus, $\Diamond W$ is indeed the weakest failure detector for solving the Consensus problem in asynchronous systems with a majority of correct processes.

The hybrid approach [7] presents a consensus algorithm that combines randomization and unreliable failure detection, two well-known techniques solving the Consensus problem. This hybrid algorithm combines the advantages of both the approaches; it guarantees deterministic termination if the failure detector is accurate and probabilistic termination otherwise.

The model using the heartbeat failure detector [8] considers partitionable networks with process crashes and lossy links, and focuses on the problems of reliable communication and consensus for such networks. The model solves the problem by using $\Diamond S$ and the *quiescent*, i.e., algorithms that eventually stop sending messages.

In addition to the fault monitoring mechanisms, a framework called DOORS [9] was proposed, which was developed prior to the FT-CORBA standard as an experimental Fault-tolerant CORBA middleware. It was implemented as a CORBA service to provide end-to-end application-level fault tolerance.

2.3 Detection Problem

To accurately design a fault detection mechanism, we should avoid some situations. Figure 1 shows two examples of inaccurate timeout.

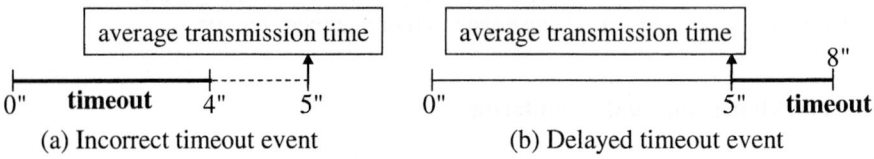

Fig. 1. Examples of inaccurate fault monitoring

As shown in Figure 1(a), if timeout occurs in 4 seconds when the average transmission time from a monitor to the monitorable object is 5 seconds, the system assumes a fault. On the other hand, as shown in Figure 1(b), if timeout occurs in 8 seconds, the system has 3 second latency in detecting a fault and this may cause a lot of fatal problems. Thus, Chandra and Toueg suggest an increase on the timeout value after a false crash alarm. However, this strategy may cause the timeout value to be increased several times due to overload in the network. It can lead to an unnecessarily long timeout value too long even after the network returns to normal workload condition.

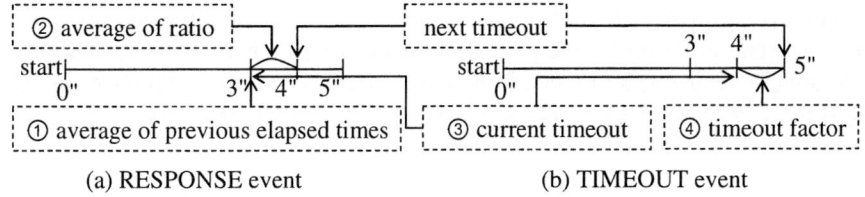

Fig. 2. The operational mechanism of the ADAPTATION algorithm

Figure 2 shows the solution of the ADAPTATION algorithm toward the problem mentioned above. It regulates the timeout value using the RESPONSE and TIMEOUT event. If the RESPONSE event occurs, the algorithm decreases the next timeout value by the product of ① and ② (the average of the ratios of the two successive elapsed time pairs). On the other hand, with TIMEOUT event, the algorithm sets the next timeout value by the product of ③ and ④ (the timeout factor set by the user). This algorithm also shows slow reaction time and possibly incorrect timeout value because it regulates timeout value only after a timeout event occurs or a few response events occur. We next present the proposed scheme solving these problems.

3 The Proposed Scheme

The proposed scheme employs some new features to solve the problems mentioned above. First of all, it uses asynchronous invocation to transmit 'is_alive()' message with a sequence number to the monitorable object. Figure 3 presents how to judge normal or abnormal state such as loss or delay of the request message using the sequence number. Here successful and failed transmission/reception of a message are marked as black and white dot, respectively.

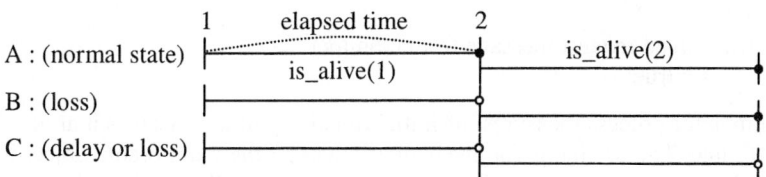

Fig. 3. Judging the state of a monitorable object

Here Case A is normal state because 'is_alive(2)' is transmitted after 'is_alive(1)' arrives. However, Case B presents loss of 'is_alive(1)' because 'is_alive(2)' arrives before 'is_alive(1)' arrives. Hence there is a need to regulate the monitoring interval and notify the state to the Notifier. At last, nobody can judge the state of the monitorable object for Case C because both 'is_alive(1)' and 'is_alive(2)' have not arrived. This is due to delay or loss.

Another feature of the proposed scheme is to promptly adjust the timeout value using the past elapsed times. For example, assume that there is a system with capriciously varying workload. To distinguish failed object from overly workload when an abnormal state is identified, the proposed monitor keeps a number of sequence numbers, round-trip times (RTT), etc. in its buffer. Here the size of a buffer is 140bytes and there exist 1000 of them. The asynchronous invocation and buffer allow the proposed scheme to outperform the existing schemes including the ADAPTATION algorithm.

The following variables are used in the proposed algorithm. **seqNum:** the sequence number attached to each packet; **monInter:** predicted monitoring interval time; **sTime:** the time the message is sent; **rTime:** the time the response is received; **eTime:** the elapsed time obtained by subtracting sTime from rTime; **message:** the character type pointer variable with "SET" or "CHANGE" value; **ratio:** the ratio of the current elapsed time to the previous elapsed time; **pCount:** the number of previous contiguously increasing or decreasing ratio values; **weight:** the weight of a ratio; **avg_ratio:** the means of ratio values. The followings are the monitor and monitorable object process.

```
process monitor object:
01   monInter[0] = the initial monitoring interval;
02   while(do monitoring){
03     check if all previous packets have returned;
04     seqNum = seqNum + 1;
05     sTime[seqNum] = timer;
06     if(call asynchronously an 'is_alive(seqNum)'){
07       rTime[seqNum] = timer;
08       eTime[seqNum] = rTime[seqNum] – sTime[seqNum];
09       MonitoringInterval(seqNum, "SET or CHANGE");
10     }else{
11       send a notification to Notifier;}
12   }}
```

```
process monitorable object:
is_alive():
01   if(received request message from monitor)
02     return true;
```

The monitor process measures the initial round trip time and saves it at monInter[0] in the 1st line. The 3rd line is for checking the state of the monitorable object if it is in state A, B, or C of Figure 3. In the 6th line, the monitor calls 'is_alive()' method asynchronously. If any response does not return, it sends a notification to its notifier. The MonitoringInterval() function at the 9th line regulates the monitoring interval. Here two messages are used, one for calculating the next packet's monitoring interval and another for recalculating it according to the status of returning message. The MonitoringInterval() function is as follows.

```
process MonitoringInterval(int seqNum, char* message):
01   if(message == "SET"){
02     set next monitoring interval;
03   }else if(message == "CHANGE"){
04     if(increase or decrease continually more twice){
```

```
05     if(pCount/2 == 0) weight[n-1] = (1);
06     else weight[n-1] = (2);
07     total = 1;
08     for(int i = n-2; i >= 0; i--) {
09         total = total – weight[i+1];
10         if(i != 0) weight[i] = total * weight[i-1];
11         else weight[i] = total;}}
12     monInter[seqNum] = eTime[seqNum-1]*(3)
13  }else if(after once increased, once decreased or vice versa){
14     monInter[seqNum] = Regulating equation(4);
15     ratio[seqNum] = eTime[seqNum] – eTime[seqNum-1];}
```

The MonitoringInterval process has two parameters. The sequence number identifies the packet called the 'is_alive()' method, and the message parameter of a character type pointer variable indicates the condition whether it establishes next monitoring interval or updates current monitoring interval. The message that has a "SET" value sets next monitoring interval, while "CHANGE" value asynchronously reflects the result calculated by the following equations to the current monitoring interval. This process is to distinguish the case that the elapsed time increases or decreases continuously from the case that the elapsed time increases and then decreases next or vice versa.

$$\frac{n+4}{4n}, \ (n = 2n) \tag{1}$$

$$\frac{n+1}{2n}, \ (n = 2n-1) \tag{2}$$

$$\text{avg\_ratio} = \frac{\sum_{i=0}^{pCount-1} eTime_i \times weight_i}{pCount}, \ \text{(The pCount is an integer larger than 1.)} \tag{3}$$

$$\text{Regulating} = eTime_0 + \frac{\sum_{i=0}^{1} eTime_i}{2} \tag{4}$$

Equation (1) and (2) are used when the last elapsed time increases and decreases continuously more than twice, respectively. Equation (3) is used to get the average ratio. Equation (4) is to regulate the monitoring interval in the case that the elapsed time increases and then decreases next or vice versa.

4 Performance Evaluation

In this section the proposed scheme is evaluated by experiment and compared with the ADAPTATION algorithm to display what factor allows better monitoring interval according to the system workload changed per second. The client and server system used for the experiment have Intel Pentium 4 processor and 1GB main memory. The OS is Windows XP with established CORBA-compliant Orbix Enterprise Version 6.1 [11, 12] and Visual C++ 6.0. The monitor and monitorable object program reside in different systems. We use two sets of delay values {0, 0.01, 0.1, 1} and {0, 0.01, 0.1, 1, 1.5, 2.5, 5, 10} that represent the response delay time for the request message caused by changing the system workload for comparing the performances of the algorithms in stable and unstable condition, respectively. We assume that initial monitoring interval is 1.2 microsecond. Also, in order to cause changing elapsed time, the workload is gradually increased in the first 4 second period and then decreased in the next 4 second period repeatedly.

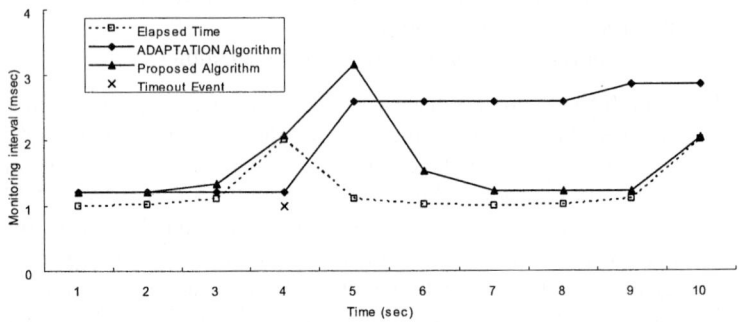

Fig. 4. The monitoring intervals as time moves in stable condition

Figure 4 displays the changing monitoring intervals of the two algorithms as the system workload changes in stable system. Notice that the line of elapsed time (the dotted line) does not vary a lot in stable condition, while the monitoring interval regulated by the proposed algorithm (the line with triangular dots) chases the elapsed time trajectory quite fast. However, the monitoring interval regulated by the ADAPTATION algorithm(the line with lozenge dots) is not as prompt as the proposed algorithm. Particularly, it has incorrect monitoring interval when time is 4 (in other words the interval is shorter than the elapsed time). In spite the monitoring interval should decrease if the elapsed time decreases, the monitoring interval regulated by the ADAPTATION algorithm shows little change in contrast with the proposed algorithm.

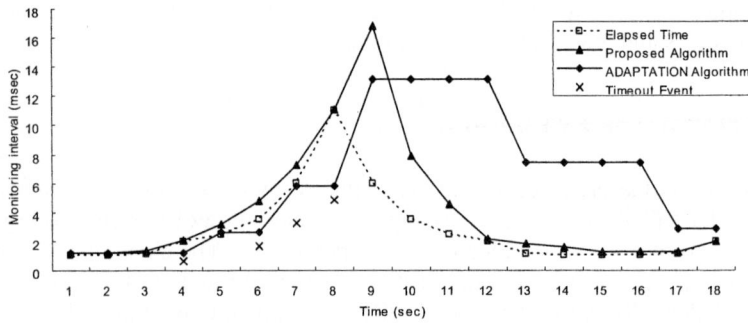

Fig. 5. The monitoring intervals as time moves in unstable condition

Figure 5 presents the changing monitoring intervals of the two algorithms as the system workload changes in unstable system or network environment. The experiment results of the two algorithms are similar to Figure 4. Notice that, in spite of the system workload increasing rapidly, the proposed algorithm chases the trajectory of the elapsed time fast. The ADAPTATION algorithm is worse in the unstable condition.

5 Conclusion

A number of different kinds of applications developed on CORBA framework in asynchronous distributed system or network environment need fault tolerance. Particularly, in real-time distributed processing systems, it is important to quickly detect the fault and properly adjust the monitoring interval in which the 'is_alive()' message is predicted to return. This paper has presented a new fault monitoring algorithm that uses elapsed time values to decide the monitoring interval. Also, it has an important feature of asynchronous invocation with sequence number.

Experiment was performed on a client and server system presenting varied monitoring interval according to the changed system workload in stable and unstable condition, respectively. The proposed algorithm allows monitoring interval close to the elapsed times while the ADAPTATION algorithm does not. Also, few timeout events occurred with the proposed algorithm in unstable system or network situation unlike the ADAPTATION algorithm. The monitoring interval must be larger than the elapsed time and the proposed algorithm allows it. A more fine tuning of the proposed approach will be carried out in the future.

References

1. Object Management Group, The Common Object Request Broker Architecture: Core Specification *Version 3.0.3-Editorial changes formal/04-03-12*, 2004
2. Object Management Group, Fault Tolerant CORBA, *OMG document: formal/04-03-12*, 2004
3. M. K. Aguilera, W. Chen and S. Toueg. Failure Detection and Consensus in the Crash-Recovery Model. *Technical Report 98-1676, Department of Computer Science, Cornell University*, 1998.
4. T. D. Chandra and S. Toueg. Unreliable failure detectors for reliable distributed systems. *Journal of the ACM*, 43(2):225-267, 1996
5. Sotoma, I. and Mauro Madeira, E.R. ADAPTATION-Algorithms to ADAPTive FaulT MonItOriNg and Their Implementation on CORBA. *Distributed Objects and Applications*, 2001.
6. T. D. Chandra, V. Hadzilacos and S. Toueg. The Weakest Failure Detector for Solving Consensus. *Journal of the ACM,* 43(4):685-722, 1996
7. M. K. Aguilera and S. Toueg, Randomization and Failure Detection: A Hybrid Approach to Solve Consensus. *SIAM Journal on Computing*, 28(3):890-903, 1999
8. M. K. Aguilera, W. Chen and S. Toueg. Using the Heartbeat Failure Detector for Quiescent Reliable Communication and Consensus in Partitionable Networks, *Theoretical Computer Science, Volume 220(1): 3-30*, 1999

9. B. Natarajan, A. Gokhale and D. C. Schmidt, DOORS: Towards High-performance Fault-Tolerant CORBA, *Proceedings of the 2nd International Symposium on Distributed Objects and Applications (DOA2000)*, 2000
10. D. Szentivanyi and S. Nadjm-Tehrani, Building and Evaluating a Fault-Tolerant CORBA Infrastructure, *in Proceedings of the Workshop on Dependable Middleware-Based Systems (WDMS'02) - part of the International Conference on Dependable Systems and Networks (DSN)*, 2002
11. IONA, Orbix ORB Enterprise Version 6.1 with Visual C++ Development Toolkits, *http://www.iona.com/downloads/*, 2004
12. IONA, "Orbix CORBA Programmer's Guide C++ Edition, Version 6.1", 2003

Simulated Annealing Based-GA Using Injective Contrast Functions for BSS

J.M. Górriz, C.G. Puntonet, J.D. Morales, and J.J. delaRosa

Facultad de Ciencias, Universidad de Granada,
Fuentenueva s/n, 18071 Granada, Spain
gorriz@ugr.es

Abstract. In this paper we present a novel GA-ICA method which converges to the optimum. The new method for blindly separating unobservable independent component signals from their linear mixtures (Blind Source Separation BSS), uses genetic algorithms (GA) to find the separation matrices which minimize a cumulant based contrast function. The paper also include a formal prove on the convergence of the proposed algorithm using guiding operators, a new concept in the genetic algorithms scenario. This approach is very useful in many fields such as biomedical applications i.e. EEG which usually use a high number of input signals. The Guiding GA (GGA) presented in this work converges to uniform populations containing just one individual, the optimum.

1 Introduction

The starting point in the Independent Component Analysis (ICA) research can be found in [1] where a principle of redundancy reduction as a coding strategy in neurons was suggested, i.e. each neural unit was supposed to encode statistically independent features over a set of inputs. But it was in the 90´s when Bell and Sejnowski applied this theoretical concept to the blindly separation of the mixed sources (BSS) using a well known stochastic gradient learning rule [2] and originating a productive period of research in this area [3]. In this way ICA algorithms have been applied successfully to several fields such as biomedicine, speech, sonar and radar, signal processing, etc. and more recently also to time series forecasting [4], i.e. using stock data.

In general, any abstract task to be accomplished can be viewed as a search through a space of potential solutions and whenever we work with large spaces, GAs are suitable artificial intelligence techniques for developing this optimization [4]. Such search requires balancing two goals: exploiting the best solutions and exploring the whole search space. In this work we prove how GA-ICA algorithms converge to the optimum. They work efficiently in the search of the separation matrix (i.e. EEG and scenarios with the BSS problem in higher dimension) proving the convergence to the optimum. We organize the essay as follows. In section 2 and 3 we give a brief overview of the basic ICA and GA theory. Then we introduce a set of genetic operators in sections 3 and 4 and prove convergence. Finally state some conclusions in section 6.

2 ICA and Statistical Independence Criterion

We define ICA using a statistical latent variables model (Jutten & Herault, 1991). Assuming the number of sources n is equal to the number of mixtures, the linear model can be expressed, using vector-matrix notation and defining a time series vector $\mathbf{x} = (x_1, \ldots, x_n)^T$, $\mathbf{s}$, $\tilde{\mathbf{s}}$ and the matrix $\mathbf{A} = \{a_{ij}\}$ and $\mathbf{B} = \{b_{ij}\}$ as:

$$\tilde{\mathbf{s}} = \mathbf{B}\mathbf{x} = \mathbf{B}\mathbf{A}\mathbf{s} = \mathbf{G}\mathbf{s} \tag{1}$$

where we define $\mathbf{G}$ as the overall transfer matrix. The estimated original sources will be, under some conditions included in Darmois-Skitovich theorem [5], a permuted and scaled version of the original ones. The statistical independence of a set of random variables can be described in terms of their joint and individual probability distribution. This is equivalent to [6]:

$$\Pi = \sum_{\{\lambda,\lambda^*\}} \beta_\lambda \beta_{\lambda^*}^* \cdot \Gamma_{\lambda,\lambda^*} \qquad |\lambda| + |\lambda^*| < \tilde{\lambda} \tag{2}$$

where the expression defines a summation of cross cumulants [6] and is used as a fitness function in the GA. The latter function satisfies the definition of a contrast function Ψ defined in [7] as can be seen in the following generalized proposition given in [8].

Proposition 1. *The criterion of statistical independence based on cumulants defines a contrast function Ψ given by:*

$$\psi(\mathbf{G}) = \Pi - \log|\det(\mathbf{G})| - h(\mathbf{s}) \tag{3}$$

where $h(\mathbf{s})$ is the entropy of the sources and G is the overall transfer matrix.

Prove: To prove this proposition see Appendix A in [8] and apply the multi-linear property of the cumulants.

3 Genetic Algorithms: A Theoretical Background

Let $\mathcal{C}$ the set of all possible creatures in a given world and a function $f : \mathcal{C} \to R^+$, namely fitness function. Let $\Xi : \mathcal{C} \to \mathcal{V}_\mathbf{C}$ a bijection from the creature space onto the free vector space over $\mathcal{A}^\ell$, where $\mathcal{A} = \{\overline{a}(i), \ 0 \leq i \leq a - 1\}$ is the alphabet which can be identified by $\mathcal{V}_1$ the free vector space over $\mathcal{A}$. Then we can establish $\mathcal{V}_\mathbf{C} = \otimes_{\lambda=1}^\ell \mathcal{V}_1$ and define the free vector space over populations $\mathcal{V}_\mathcal{P} = \otimes_{\sigma=1}^N \mathcal{V}_\mathbf{C}$ with dimension $L = \ell \cdot N$ and a^L elements. Finally let $S \subset \mathcal{V}_\mathcal{P}$ be the set of probability distributions over $\mathcal{P}_\mathbf{N}$, that is the state which identifies populations with their probability value.

Definition 1. *Let $S \subset \mathcal{V}_\mathcal{P}$, $n, k \in \mathcal{N}$ and $\{P_c, P_m\}$ a variation schedule. A Genetic Algorithm is a product of stochastic matrices (mutation, selection, crossover, etc..) act by matrix multiplication from the left:*

$$\mathbf{G^n} = \mathbf{F_n} \cdot \mathbf{C^k_{P_c^n}} \cdot \mathbf{M_{P_m^n}} \tag{4}$$

where $\mathbf{F_n}$ is the selection operator, $\mathbf{C_{P_c^n}^k} = C(K, P_c)$ is the simple crossover operator and $\mathbf{M_{P_m^n}}$ is the local mutation operator (see [6] and [10])

In order to improve the convergence speed of the algorithm we could include another mechanisms such as elitist strategy (a further discussion about reduction operators, can be found in [11]). Another possibility is:

4 Guided GAs

In order to include statistical information into the algorithm we define an hybrid statistical genetic operator as follows. The value of the probability to go from individual p_i to q_i depends on contrast functions (i.e. based on cumulants) as: $P(\xi_{n+1} = p_i | \xi_n = q_i) = \frac{1}{\aleph(T_n)} \exp\left(-\frac{\Psi(p_i)+\Psi(q_i)}{T_n}\right)$; $p_i, q_i \in \mathbf{C}$ where $\aleph(T_n)$ is the normalization constant depending on iteration n; temperature follows a variation decreasing schedule, that is $T_{n+1} < T_n$ converging to zero, and $\Psi(q_i)$ is the value of the selected contrast function over the individual (an encoded separation matrix). This sampling (Simulated Annealing -SA- law) is applied to the population and offspring emerging from the canonical genetic procedure.

Proposition 2. *The guiding operator can be described using its associated transition probability function (t.p.f.) by column stochastic matrices $\mathbf{M_G^n}$, $n \in \mathcal{N}$ acting on populations.*

1. *The components are determined as follows: Let p and $q \in \wp_N$, then we have*

$$\langle q, \mathbf{M_G^n} p \rangle = \frac{N!}{z_{0q}! z_{1q}! \ldots z_{a^L-1q}!} \prod_{i=0}^{a^L-1} \{P(i)\}^{z_{iq}}; \quad p, q \in \mathcal{P_N} \quad (5)$$

where z_{iq} is the number of occurrences of individual i on population q and $P(i)$ is the probability of producing individual i from population p given above. The value of the guiding probability $P(i) = P(i, \Psi)$ depends on the fitness function used:[1] $P(i) = \frac{z_{ip} \exp\left(-\frac{\Psi(p_i)+\Psi(q_i)}{T_n}\right)}{\sum_{i=0}^{a^L-1} z_{ip} \exp\left(-\frac{\Psi(p_i)+\Psi(q_i)}{T_n}\right)}$
2. *For every permutation $\pi \in \Pi_N$, we have $\pi \mathbf{M_G^n} = \mathbf{M_G^n} = \mathbf{M_G^n} \pi$.*
3. *$\mathbf{M_G^n}$ is an identity map on $\mathbf{U}$ in the optimum, that is $\langle p, \mathbf{M_G^n} p \rangle = 1$; and has strictly positive diagonals since $\langle p, \mathbf{M_G^n} p \rangle > 0$ $\forall p \in \mathcal{P_N}$.*
4. *All the coefficients of a GA consisting of the product of stochastic matrices: the simple crossover $\mathbf{C_{P_c}^k}$, the local multiple mutation $\mathbf{M_{P_m}^n}$ and the guiding operator $\mathbf{M_G^n}$ for all $n, k \in \mathcal{N}$ are uniformly bounded away from 0.*

---

[1] The condition that must be satisfied the transition probability matrix $P(i, f)$ is that it must converge to a positive constant as $n \to \infty$ (since we can always define a suitable normalization constant). The fitness function or selection method of individuals used in it must be injective.

Proof: (1) *follows from the transition probability between states.* (2) *is obvious and* (3) *follows from [7] and checking how matrices act on populations.* (4) *follows from the fact that* $\mathbf{M^n_{P_m}}$ *is fully positive acting on any stochastic matrix* $\mathbf{S}$.

It can be viewed as a suitable fitness selection and as a certain Reduction Operator, since it preserves the best individuals into the next generation using a non heuristic rule, unlike the majority of GAs used.

The convergence and strong and weak ergodicity of the proposed algorithm can be proved using several ways. A MC modelling a CGA has been proved to be strongly ergodic (hence weak ergodic, see [10]). So we have to focus our attention on the transition probability matrix that emerges when we apply the guiding operator. We can write the overall process as:

$$\langle q, \mathbf{G}^n p \rangle = \sum_{v \in \wp_N} \langle q, \mathbf{M^n_G} v \rangle \langle v, \mathbf{C}^n p \rangle \quad (6)$$

where $\mathbf{C}^n$ is the stochastic matrix associated to the CGA and $\mathbf{M^n_G}$ is given by equation 5.

Proposition 3. Weak Ergodicity
A MC with transition probability function associated to guiding operators that converges to uniform populations (populations with the same individual) satisfies weak ergodicity.

Prove: If we define a GGA on CGAs, the ergodicity properties depends on the new defined operator since they satisfy them as we said before. To prove this proposition we just have to check the convergence of the t.p.f. of the guiding operator on uniform populations. If the following condition is satisfied:

$$\langle u, \mathbf{G}^n p \rangle \to 1 \quad u \in \mathbf{U} \quad (7)$$

Then we can find a series of numbers which satisfies:

$$\sum_{n=1}^{\infty} \min_{n,p}(\langle u, \mathbf{G}^n p \rangle) = \infty \leq \sum_{n=1}^{\infty} \min_{q,p} \sum_{v \in \wp_N} \min\left(\langle v, \mathbf{M^n_G} p \rangle \langle v, \mathbf{C}^n q \rangle\right) \quad (8)$$

which is equivalent to weak ergodicity [9].

Proposition 4. Strong Ergodicity
Let $\mathbf{M^n_{P_m}}$ describe multiple local mutation, $\mathbf{C^k_{P_c^n}}$ describe a model for crossover and $\mathbf{F^n}$ describe the fitness selection. Let $(P^n_m, \hat{P}^n_c)_n \in \mathcal{N}$ be a variation schedule and $(\phi_n)_{n \in \mathcal{N}}$ a fitness scaling sequence associated to $\mathbf{M^n_G}$ describing the guiding operator according to this scaling. [2] Let $\mathbf{C}^n = \mathbf{F^n} \cdot \mathbf{M^n_{P_m}} \cdot \mathbf{C^k_{P_c^n}}$ represent the first n steps of a CGA. In this situation,

---

[2] A scaling sequence $\phi_n : (\mathcal{R}^+)^N \to (\mathcal{R}^+)^N$ is a sequence of functions connected with a injective fitness criterion f as $f_n(p) = \phi_n(f(p))$ $p \in \wp_N$ such that $\mathbf{M^\infty_G} = \lim_{n \to \infty} \mathbf{M^n_G}$ exist.

$$v_\infty = \lim_{n\to\infty} \mathbf{G}^n v_0 = \lim_{n\to\infty} (\mathbf{M}_\mathbf{G}^\infty \mathbf{C}^\infty)^n v_0 \qquad (9)$$

exists and is independent of the choice of v_0, the initial probability distribution. Furthermore, the coefficients $\langle v_\infty, p \rangle$ of the limit probability distribution are strictly positive for every population $p \in \wp_N$.

Prove: The demonstration of this proposition is rather obvious using the results of Theorem 16 in [10] and the point 4 in Proposition 2. In order to obtain the results of the latter theorem we only have to replace the canonical selection operator $\mathbf{F_n}$ with our guiding selection operator $\mathbf{M_G^n}$ which has the same essential properties.

Proposition 5. Convergence to the Optimum
Under the same conditions of propositions 3, 4 the GGA algorithm converges to the optimum.

Prove: To reach this result, one has to prove that the probability to go from any uniform population to the population containing only the optimum is equal to 1 when $n \to \infty$:

$$\lim_{n\to\infty} \langle p^*, \mathbf{G}^n u \rangle = 1 \qquad (10)$$

since the GGA is an strongly ergodic MC hence any population tends to uniform in time. If we check this expression we finally have the equation 10. In addition we have to use point 3 in Proposition 2 to make sure the optimum is the convergence point. Thus any guiding operator following a simulated annealing law converges to the optimum uniform population in time.

5 Simulations

At the first step, we compare the previous canonical method for apply GAs to ICA [12] with the GGA version for a reduced input space dimension ($n = 3$). The Computer used in these simulations was a PC 2 GHz, 256 MB RAN in the case of a low number of signals and the software used is an extension of ICATOOLBOX2.0 in MatLab code, protected by the Spanish law N° CA-235/04. We test these two algorithms for a set of independent signals plotted in figure 2(a) using 50 randomly chosen mixing matrices (50 runs); i.e. using the mixing matrix: $B = \{1.0000, -0.9500, 0.5700; -0.5800, 1.0000, 0.0900; 0.6300, -0.0100, 1.0000\}$, we get the signals shown in figure 2(b). We have chosen two super-gaussian signals and one bimodal signal for the first attempt ($n_inps = 3$). The order of the statistics used is the same in both methods (cumulants of 4th order)[3] and the size

---

[3] Based on section 2, we can define the fitness function approach for BSS as:

$$f(p_o) = \sum_{i,j,\ldots} ||Cum(\overbrace{y_i, y_j, \ldots}^{stimes})|| \qquad \forall i,j,\ldots \in [1,\ldots,n] \qquad (11)$$

where p_o is the parameter vector (individual) containing the separation matrix and $||\ldots||$ denotes the absolute value.

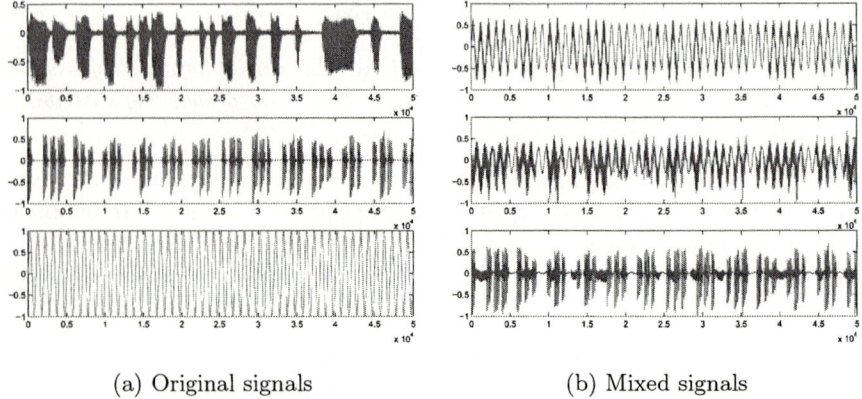

(a) Original signals (b) Mixed signals

Fig. 1. Set of independent series used in the comparison GA-GGA and a mixed case

Table 1. Mean and deviation of the parameters in the separation over 50 runs for the cost function of 4th order by the GA-method, GGA method and the FASTICA method. 1st row GA-ICA, 2nd row GGA-ICA and 3th row FATICA

param	a_{11}	a_{12}	a_{13}	a_{21}	a_{22}	a_{23}	a_{31}	a_{32}	a_{33}
mean	-0.2562	0.1473	-0.1657	-0.0647	-0.1393	0.2475	-0.4910	0.0998	-0.0350
dev.(%)	≤ 5	≤ 5	≤ 5	≤ 5	≤ 5	≤ 5	≤ 5	≤ 5	≤ 5
mean	-0.1481	0.1647	-0.2564	0.1401	-0.2464	-0.0649	-0.1003	0.0345	-0.4914
dev.(%)	≤ 6.5	≤ 6.5	≤ 6.5	≤ 6.5	≤ 6.5	≤ 6.5	≤ 6.5	≤ 6.5	≤ 6.5
mean	0.0756	-0.1099	0.2271	-0.0715	0.1648	0.0659	0.0512	-0.0226	0.4435
dev.(%)	≤ 10	≤ 10	≤ 10	≤ 10	≤ 10	≤ 10	≤ 10	≤ 10	≤ 10

of population was 100. In this way we can compare the search efficiency of both methods. Later we will focus our attention with a third statistical algorithm for ICA, the well-known FastICA [3]. This method uses the same level of information in its contrast function (4th order) thus the comparison is significant.

Results obtained from simulations are conclusive. We find out how the number of iterations (CPU time) needed to reach convergence is higher using the proposed method in [12]. This is due to blind search strategy used in the latter reference unlike the guided strategy proposed in this paper. We measure convergence by means of the well-known methods: Crosstalk (between original and recovered signals) and Normalized Round Mean Square Error (NRMSE). The set of recovering signals using GGA method can be found in figure 2(b). In the case of the three method comparison we observe how the efficiency of the FastICA in low dimension is better than the genetic approaches (see figure 2 somehow the standard deviation in time and error measure in higher than the genetic methods (see tables 1 and 2) since it suffers the local minima effect. As is shown in the latter tables the genetic procedures are slower but finally reach a better solution (the new proposed method is faster than the method in [12]).

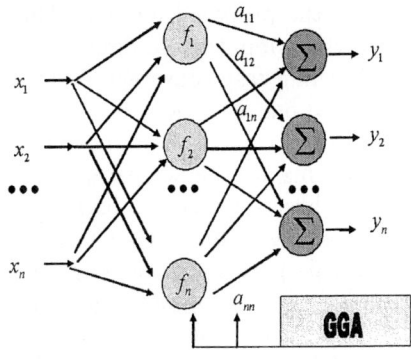
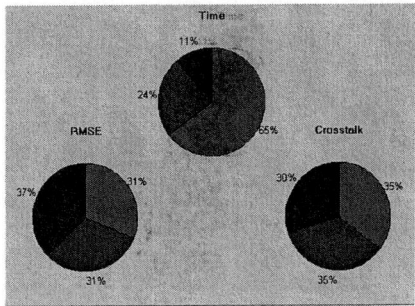

(a) Schematic Representation of the Separation System in ICA-GA

(b) Comparison for number of inputs equal to 3 GGA (red) ,GA (light blue) and FastICA (blue). Observe how GA methods obtain the same level of recovery but the time efficiency is quite different

Fig. 2. Schematic representation and comparison of the 3 method

Table 2. Mean and deviation of the parameters in the separation over 50 runs for the cost function of 4th order by the GA-method, GGA-method and the FASTICA method (cont)

Method	param.	Comp. Time(s)	NRMSE	Crosstalk(dB)
GA-ICA	mean	10.21	1.5635^{-4}	-34.709
	dev.(%)	≤ 2	≤ 1	≤ 1
GGA-ICA	mean	3.3	1.5408^{-4}	-37.7507
	dev.(%)	≤ 2	≤ 1	≤ 1
FastICA	mean	1.64	1.6355^{-4}	-29.663
	dev.(%)	≤ 5	≤ 2	≤ 4

Finally, we checked the performance of the proposed hybrid algorithm in a high dimensional scenario [6]. The results for the crosstalk were conclusive: FASTICA convergence rate decreases as dimension increases whereas GA approaches work efficiently. Of course we used the number of starting points equal to the number of individuals in the genetic generation.

6 Conclusions

A GGA-based BSS method has been developed to solve BSS problem from the linear mixtures of independent sources. The proposed method obtain a good performance overcoming the local minima problem over multidimensional domains.

Extensive simulation results prove the ability of the proposed method. This is particular useful in some medical applications where input space dimension increases and in real time applications where reaching fast convergence rates is the major objective. In this work we have focussed our attention to linear mixtures. The nonlinear problem can be interpreted as a piece-wise linear model and is expected that results improve even more since the higher parameters to encode the better results we obtain. GAs are the best strategies in high dimensional domains so it would be interesting how these algorithms (non CGAs) face the nonlinear ICA. The experimental work on this part is on the way. In the theoretical section we have prove the convergence of the proposed algorithm to the optimum unlike the ICA algorithms which usually suffer of local minima and non-convergent cases. Any injective contrast function can be used to build a guiding operator, as a elitist strategy i.e. the Simulated Annealing function defined in section 4. The convergence is shown under little restrictive conditions for the guiding operator: its effect must disappear in time like the simulated annealing.

References

1. Barlow, H.B, Possible principles underlying transformation of Sensory messages. Sensory Communication, MIT Press, New York, (1961).
2. Bell,A.J. et al. An Information-Maximization Approach to BSS and Blind Deconvolution. Neural Comp., 7, 1129-1159 (1995).
3. Hyvärinen, A. et al. A fast fixed point algorithm for ICA Neural Comp., 9: 1483-1492
4. Górriz, J.M. et al. New Model for Time Series Forecasting using rbfs and Exogenous Data. Neural Comp. and Appl., 13/2 (2004)
5. Cao, X.R. et al. General Approach to BSS. IEEE Trans. on Signal Proc., 44/3, 562-571 (1996)
6. Górriz J.M. et al. Hybridizing GAs with ICA in Higher dimension LNCS 3195,414-421, (2004)
7. Comon, P., ICA, a new concept? Signal Proc. 36 (1994) 287-314
8. Cruces, S. et al. Robust BSS algorithms using cumulants. Neurocomp. 49 (2002) 87-118
9. Isaacson, D.L. et al. *MCs: Theory and Appli.*, Wiley, 1985.
10. Schmitt, L.M. et al. *Linear Analysis of GAs*, Theoretical Computer Science, 200, pp 101-134, 1998.
11. Rudolph, G., *Convergence Analysis of CGAs*, IEEE Trans. on NN, 5/1,(1994) 96-101.
12. Tan, Y. et al. Nonlinear BSS Using HOS and a GA. IEEE Trans. on Evol. Comp., 5/6 (2001)

A DNA Coding Scheme for Searching Stable Solutions

Intaek Kim[1], HeSong Lian[1], and Hwan Il Kang[2]

[1] Department of Communication Eng., Myongji University,
449-728, Yongin, South Korea
kit@mju.ac.kr, hslian@hotmail.net
[2] Department of Information Eng., Myongji University,
449-728, Yongin, South Korea
hwan@mju.ac.kr

Abstract. This paper presents a novel method for searching stable solutions using a DNA coding scheme. Often there is more than one solution that satisfies the system requirements. These solutions can be viewed as extremes in the multimodal function. All extremes are not the same in that some of them are sensitive to noise or perturbation. This paper addresses the method that selects a solution that meets the system requirements in terms of output performance and is tolerant to the perspective noise or perturbation. A new method called a gradient DNA Coding is proposed to achieve such objectives. A numerical example is presented and comparing DNA coding with genetic algorithm is also given.

Keywords: DNA coding, genetic algorithms, gradient DNA coding.

1 Introduction

The genetic algorithm (GA) has been widely used in many optimization problems. The GA offers an efficient way to search a global solution in the multimodal function [1][2]. The multimodal function may have several solutions, but some of them are very sensitive to small perturbations of their parameter values. They may be not good solutions in certain situations.

In many optimization tasks, there is a need to find solutions whose performance will not change much due to small variation of the parameter values. In this paper, we define a stable solution as one whose variation results in a small amount of change in output performance that satisfies the system requirement. We propose a new coding method for searching stable solutions. It is based on the biological DNA and a mechanism of artificial DNA [4][5]. A gradient information is utilized to find solutions and it is named a gradient DNA coding method. In addition, the comparison between methods using the simple DNA coding and the gradient DNA coding is presented. In the next section, the DNA coding method is described. The proposed algorithm is given in section 3 and it is followed by simulation to show the effectiveness of the proposed method. The final section is followed for the conclusion and future work.

2 DNA Coding Method

The biological DNA consists of nucleotides which have four bases, Adenine(A), Guanine(G), Cytosine(C) and Thymine(T)[4][5][6]. A messenger RNA (mRNA) is first synthesized from the DNA. In the synthesis of RNA, each base is translated into the complementary base and the unused parts are cut out. This operation is a splicing. After this splicing the mRNA is completed. Three successive bases called codons are allocated sequentially in the mRNA. These codons are the codes for amino acids. 64 kinds of codons correspond to 20 kinds of amino acids as shown in Table 1. The details of translation into amino acid from codons are omitted here. This allocation of amino acid makes proteins, and proteins make up cells.

A Figure 1 shows an example of the DNA chromosome and its translation mechanism. A gene begins with the start codon ATG, and closes with end codons TAG, TAA or TGA [6]. The Figure 1 indicates that Gene1 consists of eight condons: CGG, CGT, ..., TCC and they are translated into amino acids: Arg, Arg, ..., Ser, respectively. Each amino acid has a number from 0 to 9 as shown in Table 2. The sum of the acids value represents gene's value and it is plugged in as a parameter value. For example, Gene2 consists of acids: Arg, Gly, Phe, Leu, Ala, Ser and Gly and its value becomes 26.25 by adding the value of each acids.

Table 1. RNA(DNA) Codon and amino acid

	T		C		A		G		
T	TTT	Phe	TCT	Ser	TAT	Tyr	TGT	Cys	T
	TTC		TCC		TAC		TGC		C
	TTA	Leu	TCA		TAA	Stop	TGA	Stop	A
	TTG		TCG		TAG		TGG	Trp	G
C	CTT	Leu	CCT	Pro	CAT	His	CGT	Arg	T
	CTC		CCC		CAC		CGC		C
	CTA		CCA		CAA	Gln	CGA		A
	CTG		CTG		CAG		CGG		G
A	ATT	Ile	ACT	Thr	AAT	Asn	AGT	Ser	T
	ATC		ACC		AAC		AGC		C
	ATA		ACA		AAA	Lys	AGA	Arg	A
	ATG	Met	ACG		AAG		AGG		G
G	GTT	Val	GCT	Ala	GAT	Asp	GGT	Gly	T
	GTC		GCC		GAC		GTC		C
	GTA		GCA		GAA	Glu	GGA		A
	GTG		GCG		GAG		GGG		G

Table 2. The value for each amino acid

Phe	0.25	Pro	0.50	His	0.75	Glu	1.00
Leu	1.25	Thr	1.50	Gln	1.75	Cys	2.00
Ile	3.25	Ala	2.50	Asn	2.75	Trp	3.00
Met	3.25	Tyr	3.50	Lys	3.75	Arg	4.00
Ser	4.25	Val	4.50	Asp	4.75	Gly	5.00

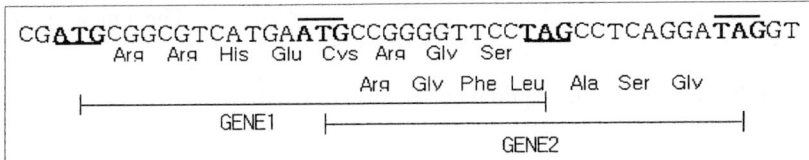

Fig. 1. The example of genes overlapping and chromosome translation mechanism

3 Algorithm

In this paper, the proposed algorithm can be summarized in 7 steps as follows:

Step 1: Initialization
Determine the population size, the chromosome length, and probabilities of crossover and mutation. The simple GA (genetic algorithm) is initialized with the binary number (0, 1), but the DNA coding is initialized with four bases: A(adenine), T(thymine), G(guanine) and C(cytosine).

Step 2: Addition of the noise
The noise is generated and added to parameters in the system. If a given population size is n, the parameter can be expressed as $X = (x_1, x_2, \cdots x_n)$ and the noise is given by $\Delta = (\delta_1, \delta_2, \cdots \delta_n)$ where δ_k has a random variable from the interval [0, 0.02]. Either $X + \Delta$ or $X - \Delta$ is considered as a noise-added parameter. This noise vector will be used in next step.

Step 3: Calculation of the average value of gradients
In this procedure, the gradients $S(x_i, x_i + \delta_i)$ and $S(x_i, x_i - \delta_i)$ will be obtained, so is the average value of absolute gradient, that is, $(|S(x_i, x_i + \delta_i)| + |S(x_i, x_i - \delta_i)|)/2$. This value will be used in calculating the fitness function.

Step 4: Calculation of the fitness
The above average value will be taken into account in this step. A Gene with a larger gradient value has a smaller fitness value so that it can be degenerated in the next

generation. Fitness values also include the object function's value as well as one obtained from the gradient information. The optimal fitness function requires two conditions: the high object function value and the low average absolute gradient value.

Step 5: Selection
The Roulette Wheel selection method is used. It is better than Tournament selection method to avoid the object function converging into local optimums.

Step 6: Crossover and mutation
The Elitist strategy is adopted as a crossover operator. It saves the best gene in every chromosome so that the best solution will never degenerate. The mutation is performed randomly by the given mutation probability.

Step 7: go to step 2 until the required conditions are satisfied.

4 Simulation

In order to demonstrate the capability of the proposed algorithm, we applied the algorithm to a 1-D function that is the multimodal function. In simulation, the parameters were kept constant with the mutation probability $p_m = 0.02$, the crossover probability $p_c = 0.6$, the population size N=60, the maximum number of generation 50. we performed 50 simulations for each experiment with randomly initializing the population.

1) Function y_1: Consider a function y_1 [Fig. 2(a)], which has five unequal peaks in the range $0 \le x \le 1$ and is a variant of the function used in [3]. It is defined as

$$Y_1(x) = \begin{cases} e^{-2\ln 2(\frac{x-0.1}{0.8})^2} |\sin(5\pi x)|^{0.5} & 0.4 < x \le 0.6 \\ e^{-2\ln 2(\frac{x-0.1}{0.8})^2} \sin^6(5\pi x) & \text{otherwise}. \end{cases}$$

As shown in Fig. 2(a) the global optimum is located at $x = 0.1$ with the function value 1.0. There are four sharp peaks. The third peak is broad compared to others and is located at $x = 0.486$ with function value 0.715.

Fig.2 (b), (c) shows a typical distribution of the individuals in the after 50 generations for the simple DNA coding method and the gradient DNA coding method. Fig.2 (d) shows a convergence process of the mean value of parameter x in the population with trials. The simple DNA coding method converged at $x = 0.1$, the center of the highest peak. The gradient DNA coding method converged to the stable peak ($x = 0.486$) zone. Indeed, we can observe from Fig.2 (d) that it approached the broad peak.

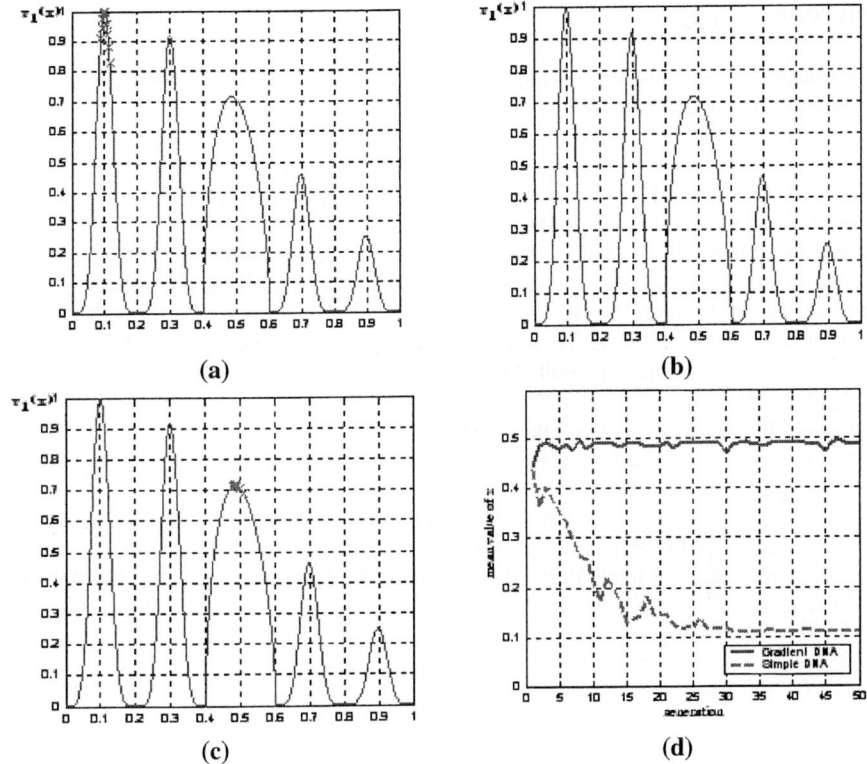

Fig. 2. (a) The original function y_1 (b) A typical distribution of the individuals in function y_1 after 50 generation for the simple DNA coding method (c) A typical distribution of the individuals in function y_1 after 50 generation for the gradient DNA coding method (d) The variation of mean (over the population) value of x with function evaluations

5 Conclusions

This paper proposed a gradient DNA coding method, which extends the application of GA's to domains that require detection of stable solutions. The gradient DNA coding method was found that this approach can be effective when we want to detect more than one stable solutions on different peaks.

The future work will focus on analyzing the behavior of gradient DNA coding method on more complicated problems where many peaks interact, evaluating the gradient DNA coding method on real-world problems.

Acknowledgment

This work was supported by grant No. (R01-1999-000-00226-0) from the Basic Research Program of the Korea Science & Engineering Foundation.

References

1. Zhijiang Guo, Hongtao Zheng, Jinping Jiang,: A powerful modified genetic algorithm for multimodal function optimization, , Proceedings of the American Control Conference, Vol. 4 , (2002) 3168 -3173
2. Park Chang-Su, Lee Hungu, Bang Hyo-Choong, Tahk Min-Jea: Modified Mendel operation for multimodal function optimization, Evolutionary Computation, Proceedings of the 2001 Congress on , Vol. 2 , (2001) 1388 -1392
3. Nasraoui, O.; Krishnapuram, R.: A novel approach to unsupervised robust clustering using genetic niching, Fuzzy Systems, 2000. FUZZ IEEE The Ninth IEEE International Conference on , Vol. 1 (2000) 170 -175
4. Wasiewicz, Piotr, Janczak, Tomasz, JMulaka, J.: the Inference via DNA Computing, Evolutionary Computation, 1999. CEC 99. Proceedings of the 1999 Congress on , Vol. 2,(1999) 988-993
5. Deaton, R., and et. Al: A DNA Based Implementation of an Evolutionary Search for Good Encodings for DNA Computation, Proc. IEEE *Int. Conf. Evolution computation*, Indianapolis, IN, USA, April, (1997) 267-271
6. Yoshikawa, Tomohiro, Furuhashi, Takeshi, Uchikawa Yoshiki: DNA Coding Method and a Mechanism of Development for Acquisition of Fuzzy Control Rules, Fuzzy Systems, 1996., Proceedings of the Fifth IEEE International Conference on , Vol. 3 , (1996) 2194 - 2200

Study on Asymmetric Two-Lane Traffic Model Based on Cellular Automata*

Xianchuang Su[1], Xiaogang Jin[2,**], Yong Min[2], and Bo Peng[2]

[1] College of Software Engineering, Zhejiang university, Hangzhou 310027, China
[2] AI Institute, College of Computer Science, Zhejiang university,
Hangzhou 310027, China
xiaogangj@cise.zju.edu.cn

Abstract. With the consideration of driver's velocity preference, we redefine the velocity updating rules of NaSch model and extend them to an asymmetric two-lane cellular automaton model with a recently proposed lane changing rule set. The analysis focuses on the reproduction of empirically observed results and the relations between empirical results and vehicle plugs. Vehicle plug is a structured vehicle set which slows down the traffic flow and it is also a bridge from microscopic to macroscopic level. Simulation results show that encouraging lane changes have little effect on decomposing vehicle plugs. In order to decrease plugs we should keep slow vehicles out of expressway or make a flexible overtaking ban of slow vehicle according to practical situations.

1 Introduction

In recent years, more and more cellular automata (CA) models of traffic flow have been proposed. CA model formulate simple rules which mimic the behavior of the drivers as simply as possible, yield reasonable results compared with empirical findings and can be easily modified to study various instances. The basic model has been introduced by Nagel and Schreckenberg(NaSch model) [NS]. And a more sophisticated CA model [KSS1] has been developed which is capable of reproducing all of the empirically observed traffic states in single-lane traffic, i.e. free flow, wide moving jams and especially synchronized traffic.

In an asymmetric two-lane traffic only a few empirical results exist which help to specify lane changing rule[GY][HL]. Lane change frequency should increase with vehicle density, shows a maximum in the vicinity of the flow maximum and then decreases with increasing density. And a special feature of a highway with a right-lane preference is the empirically observed lane usage inversion. There are more vehicles distributed on the left than on the right lane, and the flow is larger for the left than for the right lane while vehicle density is over the vicinity

---

* Supported by Natural Science Foundation of China (NSFC) grant 60103015 and The Project-sponsored by SRF for ROCS, SEM.
** Corresponding author.

of the flow maximum. In [KSS2][KSS3][KSS4][RN] the two-lane extensions of NaSch model reproduce the density dependence of lane change frequency and the lane usage inversion quite well.

While the NaSch model only ensures the avoidance of crashes and every driver is only confined with a maximum velocity. For the desire of the drivers for smooth driving, we introduce ideal velocity with the consideration of individual velocity preference into NaSch model. Each driver takes speed at the vicinity of individual ideal velocity and moves on with the anticipation. Also the lane changing rules in [KSS3] could be mended by anticipation method which is capable of improving the usage of two-lane highway space and making the lane change process more neatly.

In the next section a modified asymmetric two-lane model is proposed not only for reproducing the empirical results on macroscopic level, i.e. density dependence of lane change frequency and the lane usage inversion, but also for letting the drivers in model have more human natures by reconstructing the local rule on microscopic level. In section 3 we introduce $U(t)$ which represent the average unhappiness of all drivers and show that by encouraging lane changing, $U(t)$ could stay at a low degree and the flow of traffic goes up linearly with the increasing density when the traffic is in free flow state. The influence of slow vehicle such as truck is remarkable in inhomogeneous asymmetric two-lane highway. Without the overtaking ban on truck, the flow would go down steeply with the increasing rate of trucks in all vehicles at the early stage and retain at a low level afterwards. The negative impact of truck rate on flow is vanished with the increasing vehicle density. We catch different segments of the highway and find that vehicle plug plays a main role in the negative impact. Once a vehicle plug is formed by a pair of trucks which occupy both two-lanes at short distance, the vehicles behind the plug are difficult to get through it. The relations between vehicle density, rate of truck and the formation of plug are discussed in detail. The plug is easy to form and spread all over but the measures that can dissipate it are difficult to find. Here we make some tries that may be helpful. In the last section a short summary and a discussion are followed.

2 Models

In this inhomogeneous asymmetric two-lane traffic model, a system update is performed in two sub-steps. In the first step the vehicles change lanes according to the lane changing rules and do not move. In the second step, the cars move according to the calculated velocity. Both sub-steps are performed in parallel for all vehicles. Before we present the lane changing rules, we briefly define the single-lane motion rules. For the sake of completeness we briefly recall the definition of the NaSch model. The NaSch model is a discrete model for traffic flow. The road is divided into cells which can be either empty or occupied by a car with a velocity $v = 0, 1, \cdots, v_{\max}$. The vehicles move from left to right on single lane with periodic boundary conditions and the system update is performed in parallel for all vehicles according to the following rules (see table 1 for a summary of the parameters and variables of models):

Table 1. Summary of the variables and parameters used in the model definition

Variable Parameter		Variable Parameter	
x	Position	R_s	Rate of slow vehicle
v	Velocity	ρ	Vehicle density
v_a	Anticipation velocity	$v_{\max}$	Maximum velocity
d	Distance headway	d_s	Safe gap
d_e	Effective distance headway	v_i	Ideal velocity
d_{ofe}	Effective distance headway other lane	P_b	Deceleration probability
d_{ose}	Effective distance behind other lane	P_a	Acceleration probability

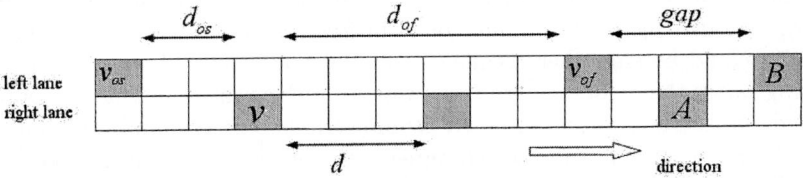

Fig. 1. Sketch of a road segment. Here take right lane as main lane and left lane as overtake lane. The hatched cells are occupied by vehicles

(1) Acceleration and deceleration:

$$v(n, t+1) = \begin{cases} \min\left(v(n,t)+1, v_{\max}\right), & v(n,t) < d \\ d, & v(n,t) \geq d \end{cases}$$

(2) Decelerated by noise: if $rand() < P_b$ then $v(n, t+1) = \max\left(v(n,t+1) - 1, 0\right)$
(3) Motion: $v(n, t+1) = x(n,t) + v(n, t+1)$
where $v(n,t)$ and $x(n,t)$ denotes the velocity and the position of number n-th vehicle at system time t, respectively. With the deceleration probability P_b, various motions of drivers could be described in a random way. For the desire of the drivers for smooth and comfortable driving, Schneider et al[SE] and Knospe et al[KSS3] proposed a more realistic model based on NaSch model by introducing the effective distance headway d_e:

$$\begin{cases} d_e(n,t) = d(n,t) + \max\left(v_a(n,t) - d_s, 0\right) \\ v_a(n,t) = \min\left(d(n+1, t), v(n+1, t)\right) \end{cases}$$

where vehicle $n+1$ denotes the leading vehicle of n. The velocity of leading vehicle is anticipated and $d_e(n,t)$ allows for smaller gaps, larger velocity and higher lane space usage.

In real world, each driver has one's velocity preference. Here, we introduce the ideal velocity $v_i(v_i < v_{\max})$ for each driver and the rules are designed to let the velocity varying around v_i. The status of the brake light B is assigned to 1 when a deceleration occurred, otherwise B is assigned to 0.

The update rules for motion are then as follows, which should be done step by step:

(1) Acceleration and deceleration: if $d_e(n,t) > v_i(n) > v(n,t)$ then $v(n,t+1) = v(n,t)+1$; if $(d_e(n,t) > v_i(n)$ and $v(n,t) \geq v_i(n)$ and $rand() < P_a)$ or $d_e(n,t) \leq v_i(n)$ then $v(n,t+1) = \min(v(n,t)+1, d_e(n,t))$; otherwise $v(n,t+1) = v(n,t)$;
(2) Decelerated by noise: if $rand() < P_b$ then $v(n,t+1) = \max(v(n,t)-1, 0)$;
(3) Upper limit: $v(n,t+1) = \min(v(n,t+1), v_{\max})$;
(4) Brake light: $B(n,t+1) = \begin{cases} 1, v(n,t+1) < v(n,t) \\ 0, v(n,t+1) \geq v(n,t) \end{cases}$;
(5) Motion: $x(n,t+1) = x(n,t) + v(n,t+1)$.

The acceleration and deceleration rules of step one is designed for keeping the drivers driving at the vicinity of v_i in free flow state, and the changing of velocity could be smoother. While in the state of synchronized traffic or wide moving jams, the velocities are constrained by the vehicle density. In order to extend the single-lane model to an asymmetric two-lane model, we should introduce lane changing rules which agree with the two mechanisms used in real life. The one is right-lane preference that the driver should use the right lane as often as possible. The other one is the right-lane overtaking ban. In [KSS3] a sophisticated lane changing rule set has been proposed, which agrees with the two mechanisms and is capable of reproducing the density dependence of lane change frequency and the lane usage inversion. In order to keep the model simply we restrict the lane interaction to vehicles which have to brake ($B = 1$) in the next time step due to an insufficient gap in front. The lane changing rules are as follows (see table 1 and figure 1 for the parameters and variables of rules):

(1) From right to left lane:
  (i) Incentive criterion: $B = 0$ and $v > d_e$
  (ii) Safety criterion: $d_{ofe} > v$ and $d_{ose} > v_{os}$
(2) From left to right lane:
  (i) Incentive criterion: $(B = 0$ and $\frac{d_{ofe}}{v} > 2.0$ and $(\frac{d_e}{v} > 4.0))$ or $(v > d_e)$;
  (ii) Safety criterion: $d_{ofe} > v$ and $d_{ose} > v_{os}$

where d_{ofe} denotes the effective gap to the leading vehicle on the destination lane, d_{ose} denotes the effective gap to the succeeding vehicle on the destination lane, and they are defined as follows:

$$\begin{cases} d_{ofe} = d_{of} + \max(\min(gap, v_{of}) - d_s, 0) \\ d_{ose} = d_{os} + \max(\min(d_{of}, v) - d_s, 0) \end{cases}$$

It is obviously that the difference between two incentive criterions agrees with the right-lane preference mechanism. Since the velocity of the vehicle is taken into account in the incentive criterion that from left to right, slow vehicles are allowed to change lane even at small distances.

3 Results and Discussions

The numerical simulations are started with randomly generated initial configurations, and evolve to steady terminal states over a long time. On a 2×20000 lattice as figure 1, trucks and cars are randomly distributed. We distribute the v_i of trucks according to a Gaussian profile with different variances and a mean $\overline{v}_i = 5$. While the $\overline{v}_i$ of cars is set to 8, and $v_{\max} = 12$. The sum of left and right lane usages is 2. In order to describe the emotions of drivers, we introduce $U(t)$ which reflects the unhappiness of all drivers who are not at their ideal velocities at system time t:

$$U(t) = \frac{1}{n} \sum_{j=1}^{n} \frac{|v(j,t) - v_i(j)|}{v_i(j)}$$

In figure 2 the variables of traffic flow varying with the increasing vehicle density are depicted (the units only have a relative meaning) and agree with the empirically observed results well. With the increasing density at early stage, traffic flow goes up linearly, mean velocity keeps on a high level and unhappiness stays at a low level while the lane change frequency goes up steeply. And we

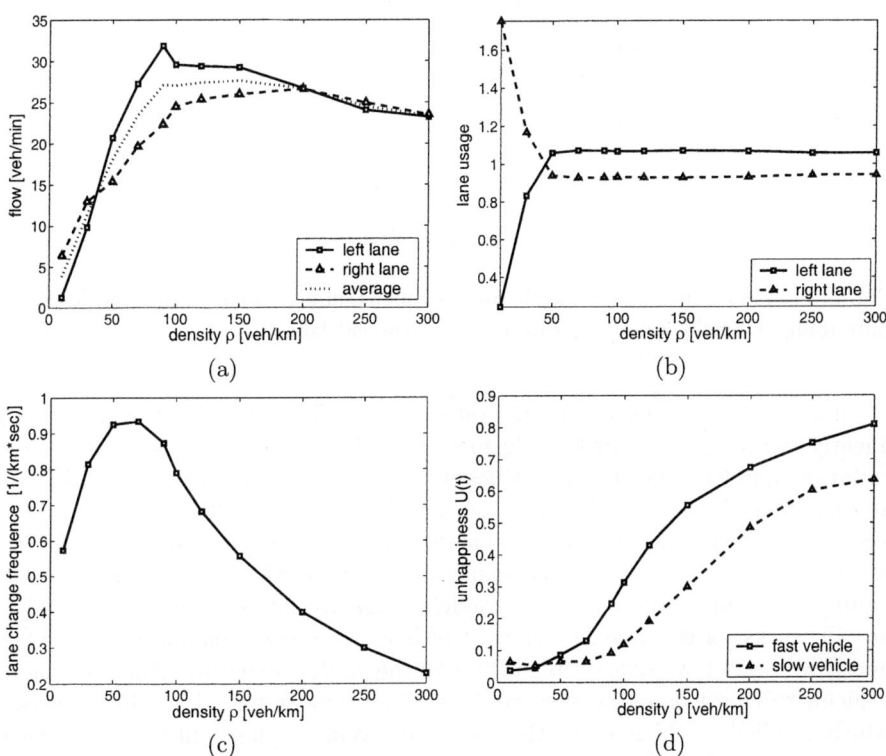

Fig. 2. Fundamental diagrams of asymmetric two-lane model with 10% slow vehicles. The slow vehicles are not allowed to change to left lane

can see that the active lane changes provide more chances for driver to drive at a high and satisfying velocity at free flow phase. Meanwhile, most of the vehicles are distributed on right lane for the right-lane preference. However, the left lane usage will exceed the right lane and then achieve its maximum in the vicinity of flow maximum, and the flow of left lane is larger than the right at the same time. A simple explanation for these is that the trucks or some slow cars dominate the right lane and the cars would like to take the left lane for more satisfying velocities. There are many factors in producing these phenomena, and a detail analysis is presented in [KSS3]. With the increasing density, the velocity predominance of cars against trucks is decreased, and the free flow phase turns into synchronized flow phase. In this stage the mean velocity of all drivers is determined only by density but not at the wills of drivers.

In real world, the entry of slow vehicle into highway is forbidden or the overtaking of slow vehicle is banned. Here the numerical simulations below will show us how the performances will be if we did not follow these instructions.

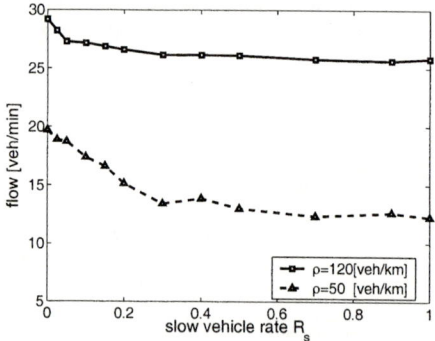

Fig. 3. Flow statistics of asymmetric tow-lane model under two different flow densities. Note that the trucks are allowed to change to the left lane

In figure 3, when $\rho = 50$ and the rate of trucks $R_s \in [0, 0.27)$, flow goes down linearly and steeply. Meanwhile, the mean velocity of all drivers also has the same variation, and the lane change frequency goes up steeply with the increasing R_s at this stage. However, when $\rho = 50$ and $R_s \in [0.27, 1]$, flow maintains at a low level, and so does the mean velocity. The lane change frequency decreases at the same time. All these phenomena are related to the slow vehicle plugs. In figure 1, we can see how the slow vehicle plugs are formed. Suppose A and B are all trucks, in the vehicle structure of figure 1, the succeeding vehicles of A and B will have few chances to go through the vehicle structure. And then more vehicles will be formed as slow vehicles at a segment with local high density which is called a vehicle plug. More and more vehicle plugs will be formed with the increasing $R_s \in [0, 0.27)$ until they have been distributed everywhere. We tried to decompose the vehicle plugs by encouraging lane changing, i.e. tuning the numerical parameters in incentive criterion of lane changing rule that from left to right lane. But the result was even worse, since the flow enhanced only a little while the lane changing frequency increased doubly which would cause more crashes and make drivers tired and depressed. Therefore, it is important to keep slow vehicles out of expressway under these circumstances.

In figure 3, when $\rho = 120$, flow declines linearly but more steeply than the flow with $\rho = 50$ and then it reaches a stationary value over a short range of $R_s(R_s \in [0, 0.05])$. Due to the vehicle plugs are easy to be formed than to be decomposed under higher density (the balance between plug forming and decomposing is interesting), under $\rho = 120$ a small addition of R_s can generate more plugs than under $\rho = 50$. The velocity predominance of cars against trucks is declined under the phases of synchronized traffic or wide moving jams. At these phases, flow and mean velocity are determined mainly by vehicle density. Therefore, the range of R_s that flow varies over to reach its stationary value under $\rho = 120$ is shorter than that under $\rho = 50$. We can unchain the slow vehicle entry ban under the phases of synchronized traffic or wide moving jams for that the ban has little effect on flow increasing. And the constraints on slow vehicles can be applied flexibly according to different situations. For example, it is unpractical to keep all slow vehicles out of highways in the region with road network underdeveloped. In this situation, we can make a flexible overtaking ban of slow vehicles to achieve a higher flow and make drivers happy.

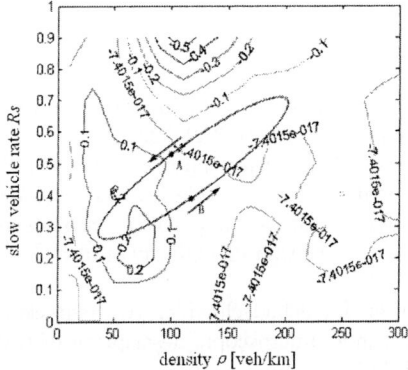

Fig. 4. Contour of $(Flow_{forbid} - Flow_{allow})/Flow_{forbid}$ as a function of flow density ρ and slow vehicle rate R_s. Here $Flow_{forbid}$ and $Flow_{allow}$ represent the traffic flow when slow vehicles are forbidden to change to the left lane or allowed respectively

Suppose that the ellipse in figure 4 represents density and R_s of a two-lane highway as a function of time in one day. The arrow represents the direction that time goes by. The A and B are the moments of day. In order to achieve higher flow, slow vehicles should observe the overtaking ban from moment A to moment B. However, from moment B to moment A the slow vehicle overtaking ban should be cancelled.

4 Conclusion

We redefined the velocity updating rules of NaSch model with the consideration of velocity preference of driver, and extended it with a sophisticated lane changing rule set to an asymmetric two-lane model. The model reproduced empirical results on macroscopic level, i.e. density dependence of lane change frequency and the lane usage inversion. And we found that active lane changes provide more chances for driver to drive at a high and satisfying velocity at free flow phase with the slow vehicle overtaking ban. However, the effect of lane change will vanished without the ban since the vehicle plugs are easier to be formed by structured slow vehicles than to be decomposed only by active lane changes. And even a low fraction of slow vehicle can make the mean velocity and flow go down steeply. In order to weaken the negative impact of vehicle plugs and achieve higher flow, we should keep slow vehicles out of expressway or make a flexible overtaking ban of slow vehicle according to practical situations.

With a simple local interaction set, CA model can reproduce most empirically observed results of complex traffic system and present microscopic causes of formulations of the empirical results. Since the traffic simulations based on CA are at the stage of developing and most analyzes are qualitative. There are many works could be done, for example, clarifying the quantitative relations between simulated and realistic results for quantitative analysis.

References

[NS] Nagel K., Schreckenberg M.: A cellular automaton model for freeway traffic. Journal de Physique I. **2** (1992) 2221-2229
[SE] Scheider J., Ebersbach A.: Anticipatory drivers in the Nagel-Schreckenberg model. International Journal of Modern Physics C. **13** (2002) 107-113
[KSS1] Knospe W., Santen L., Schadschneider A., Schreckenberg M.:Towards a realistic microscopic description of highway traffic. J. Phys. A. Math. Gen. **33** (2000) 477-485
[KSS2] Knospe W., Santen L., Schadschneider A., Schreckenberg M.. Single-vehicle data of highway traffic: microscopic description of traffic phases . Phys. Rev. E. **65** (2002) 056133
[KSS3] Knospe W., Santen L., Schadschneider A., Schreckenberg M.. A realistic two-lane traffic model for highway traffic. J. Phys. A **35** (2002) 3369-3388
[KSS4] Knospe W., Santen L., Schadschneider A., Schreckenberg M.. Disorder effects in cellular automata for two-lane traffic. Physica A. **265** (1999) 614-633
[RN] Rickert M., Nagel K., Schreckenberg M., Latour A.: Two lane traffic simulations using cellular automata. Physica A. **231** (1995): 534-550
[GY] Ganglen C.,Yangming K. An empirical investigation of macroscopic lane-changing characteristics on uncongested multilane freeways. Transportation Research Part A. **25** (1991) 375-389
[HL] Hall F. L., Lam T. N.: The characteristics of congested flow on a freeway across lanes, space, and time. Transportation Research Part A. **22** (1988) 45-56

Simulation of Parasitic Interconnect Capacitance for Present and Future ICs

Grzegorz Tosik, Zbigniew Lisik, Malgorzata Langer,
and Janusz Wozny

Institute of Electronics, Technical University of Łódź,
91-924 Lodz, Wolczanska 223
{Pgrzegorz.tosik, lisikzby, malanger,
jwozny}@p.lodz.pl

Abstract. The performance of modern integrated circuits is often determined by interconnect wiring requirements. Moreover, continuous scaling of VLSI circuits leads to an increase in the influence of interconnects on system performance. It is desired therefore, to calculate accurately its parasitic components, particularly wiring capacitance. In order to recognize which one from the most popular empirical approaches gives the evaluation of the total capacitance that suits to the real capacitance of the interconnect line, the numerical simulations based on the numerical solving of Maxwell equations have been employed.

1 Introduction

Due to continually shrinking feature sizes, higher clock frequencies, and the simultaneous growth in complexity, the role of interconnections in determining circuit performance is growing in importance. This trend has led to the increasing dominance of interconnect delay over logic propagation delay – even with new metal technologies such as copper or new low-k dielectrics [1]. Additionally, increasingly large chip dimensions result in the longer interconnect lines, which give considerable contributions to the total power dissipation. Since the metallic interconnections are the crucial design issue for current and future generation of IC's, it is increasingly important to compute accurately all its parasitic components, particularly wiring capacitances. Their evaluation is a non-trivial task and it is a subject of many investigations [2],[3],[4]. There are two major approaches to calculate the parasitic capacitance. The first type is to use a numerical simulation, often based on Finite Difference Time Domain or on Finite Element Method [5],[6]. The numerical methods have good accuracy, however are too time-consuming, when applied to the whole integrated circuit. The second approach uses analytic formulations, derived from the equations of electromagnetism. These methods have a sufficient accuracy and a simulation speed, but they can be considered to simulate a few physical configurations, only. Empirical 2- and 3D capacitance models have been reported extensively in the literature. In order to recognize which one from the most popular approaches [7],[8],[9],[10] gives the evaluation of the total capacitance that suits to the real capacitance of the interconnect line, the numerical simulations based on the numerical solving of Maxwell equations have been employed. The simulations were performed using the commercial software package OPERA [11], which uses finite element techniques to analyze the electromagnetic problems.

2 Electrical Models of On-chip Interconnections

Initially, interconnect has been modeled as a single lumped capacitance in the analysis of the performance of on-chip interconnects. Currently, at low frequency, the lumped RC models are used for high-resistance nets (Fig.1a) and capacitance models are used for less resistive interconnect [8],[12]. To represent the distributed nature of the wire, the interconnect is broken down into n smaller lumped section (Fig.1b). The simulation accuracy increases with increasing n. If the signal rise time is too short or the wire is very long, the inductance must also be included and an RLC network (Fig.1c), or the transmission line model (Fig.1d) must be used [13],[14]. The possible representations of interconnect line models include the Π, L and T networks.

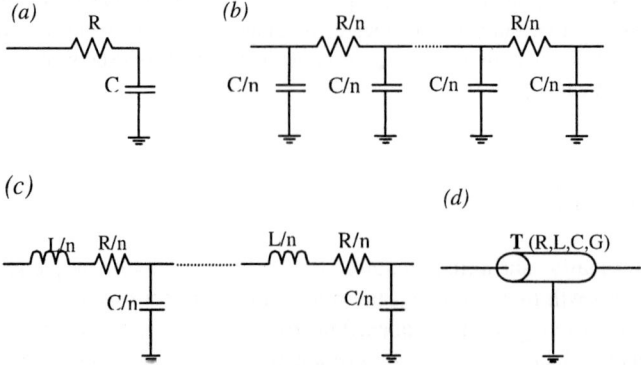

Fig. 1. Models of interconnections. (a) Lumped RC line (b) Distributed RC line. (c) Distributed RLC line (d) Lossy transmission line.(n: number of distributed cells)

3 Analytical Models of Parasitic Interconnect Capacitance

An accurate model for the crossover capacitance is essential for estimating the interconnect circuit performance. To get an accurate interconnect capacitance electric field solvers (2D or 3D) should be used. It is, however, so huge task that it would take ages to estimate the capacitance of the whole chip. Therefore, various assumptions and approximations are used to get quick estimates. The empirical formulas, derived from the equations of electromagnetism have the sufficient accuracy and the simulation speed. The simplest equation for interconnect capacitance is given by.

$$C = \varepsilon_o \varepsilon_{SiO2} \frac{W}{H} L_{Int} \qquad (1)$$

where H is the interlayer dielectric (ILD) thickness, ε_{SiO2} is dielectric constant, ε_o is the permittivity of free space, W and L_{Int} are the width and the length of the interconnect respectively.

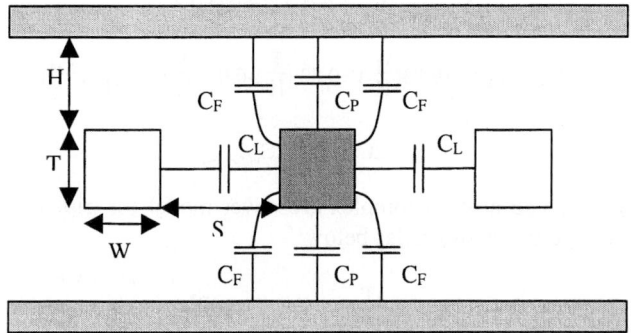

Fig. 2. Interconnect capacitance components

For interconnections in modern integrated circuits a large part of the capacitance comes from lateral coupling between adjacent wires placed on the same metal level and from fringing fields This means that the values of extracted capacitance using simple parallel plate capacitor approximations are extremely inaccurate. In reality, the total interconnect capacitance is a sum of a few different capacitances resulting from the particular design of interconnect system and cannot be treated as the simple plane described by (1). As it is shown in Fig.2, one can distinguish the parallel plate capacitance component, C_P, the fringing field component, C_F, and lateral coupling capacitance component, C_L. It should be noted, that if the neighbor line switches in the opposite direction, the effective lateral capacitance doubles but if the neighbor line switches in the same direction the effective lateral capacitance equals to zero.

There are several approaches to identify the total capacitance C_{Total}, which use different models to define the components capacitances The most popular of them are presented below. One of the first one was developed by Sarasvat [7]. He considered the structure shown in Fig.2 and described the component capacitances as follows:

$$C_P = \varepsilon_{ox}\varepsilon_o \frac{W}{H} \tag{2}$$

$$C_L = \varepsilon_{ox}\varepsilon_o \frac{T}{S} \tag{3}$$

$$C_{Total} = k(2C_P + 2C_L) \tag{4}$$

where T - line thickness, S - distance between two adjacent wires and k - the factor which takes into account the fringing fields, which value can be calculated using two-dimensional analysis of Dang and Shigyo [15].

Another model was presented by Sakurai [8] who derived a simple analytical formula for the total capacitance for symmetrical interlevel dielectric thickness. However, this formula takes into account a basic wire structure with one ground plane only. Because of the symmetrical nature of the interconnect structure considered in this work, like is shown in Fig.2 the capacitance between wire and ground $C_{vertical}$ and capacitance between adjustment wires $C_{horizontal}$ are multiplied by two.

$$C_{vertical} = \varepsilon(1.15(\frac{W}{H}) + 2.80\,(\frac{T}{H})^{0.222}) \tag{5}$$

$$C_{horizontal} = \varepsilon(0.03(\frac{W}{H}) + 0.83\,(\frac{T}{H}) - 0.07\,(\frac{T}{H})^{0.222}(\frac{S}{H})^{-1.34}) \tag{6}$$

$$C_{Total} = 2C_{vertical} + 2C_{horizontal} \tag{7}$$

Chern in [9] presented a more complex crossover model for triple-level metal layer, which was the base for the formulas below:

$$C_{vertical} = \varepsilon(\frac{W}{H} + 1.086\,(1 + 0.685 e^{\frac{-T}{1.343S}} - 0.9964\,e^{\frac{-S}{1.421H}})(\frac{S}{S+2H})^{0.0476}(\frac{T}{H})^{0.337}) \tag{8}$$

$$C_{horizontal} = \varepsilon(\frac{T}{S}(1 - 1.897 e^{\frac{-H}{0.31S}} - e^{\frac{-T}{2.474S}} + 1.302\,e^{\frac{-H}{0.082S}} - 0.1292 e^{\frac{-T}{1.421S}})$$
$$+ 1.722(1 - 0.6548 e^{\frac{-W}{0.3477H}}) e^{\frac{-S}{0651H}}) \tag{9}$$

$$C_{Total} = 2C_{vertical} + 2C_{horizontal} \tag{10}$$

The last formula that will be presented here is taken from Wong work [10]. According to his model, the interconnect capacitance is described as:

$$C_{vertical} = \varepsilon(\frac{W}{H} + 2.217(\frac{S}{S+0702H})^{2.193} + 1.17(\frac{S}{S+1.510H})^{0.7642}(\frac{T}{T+4.532H})^{0.1204}) \tag{11}$$

$$C_{horizontal} = \varepsilon(1.412\frac{T}{S}\exp(\frac{-4S}{S+8.014H}) + 2.3704(\frac{W}{W+0.3078S})^{0.25724}$$
$$(\frac{H}{H+8.961S})^{0.7571}\exp(\frac{-2S}{S+6H})) \tag{12}$$

$$C_{Total} = 2C_{vertical} + 2C_{horizontal} \tag{13}$$

4 Numerical Model of Parasitic Interconnect Capacitance

In order to recognize which from the above formulas gives the evaluation of the total capacitance that suits to the real capacitance of the interconnect line, the numerical simulations based on the numerical solving of Maxwell equations have been employed. The simulations were performed using the Vector Field commercial software package OPERA [11], which uses finite element method to analyze the electromagnetic problems in 2D domain. This method divides the studied structure into sub-domains. Then, it is possible, with this tool, to fit any polygonal shape by choosing element shapes and sizes.

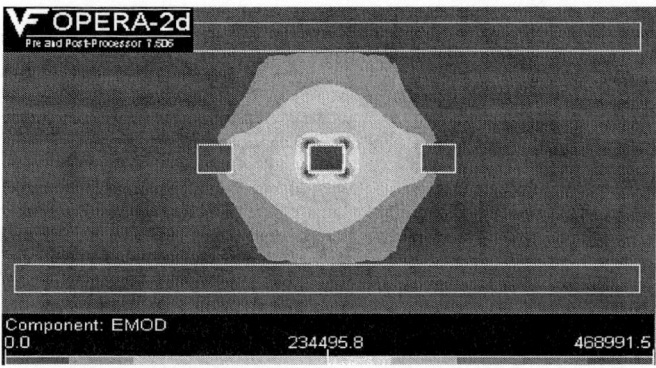

Fig. 3. Interconnect structure used in numerical capacitance calculation

The test interconnect structure is shown schematically in Fig.3. It contains three wires sandwiched between a two ground planes. Three copper lines are imbedded in a dielectric matrix (with the permittivity dependent on technology). The electric potentials and the currents have been fixed to arbitrary values. The cross-section dimension of investigated structure has been changed according to the technology. Typical views of the field distributions are represented in Figure 3. The shades give the intensity of the electric potential. This gives qualitative indications on the intensity of the coupling strengths. The software reports the total charge on the strip in Coulombs per meter. Since the total charge equals capacitance times voltage ($CV = Q$) then we can interpret the results from the software directly as capacitance per meter.

5 Comparison Between Analytical and Numerical Models

The total capacitance was calculated numerically and using four formulas mentioned above. Such a procedure has been repeated for several structures fabricated in 0.13, 0.10 and 0.05µm CMOS processes, which design parameters are collected in Table.1.

Table 1. Cross-section interconnect dimensions							
Technology	Layer	w	t	s	h	Eps	V
		[m]	[m]	[m]	[m]	-	[V]
130	1	1.46E-07	1.08E-07	1.54E-07	3.55E-07	2.00	1.20
	6	2.35E-06	2.00E-06	2.53E-06	6.67E-06	2.00	1.20
100	4	1.18E-07	2.04E-07	1.22E-07	3.55E-07	1.50	0.90
	7	2.22E-06	2.00E-06	4.98E-06	6.67E-06	1.50	0.90
50	7	1.20E-06	1.50E-06	1.20E-06	9.00E-07	1.50	0.60
	9	2.00E-06	2.50E-06	2.00E-06	1.40E-06	1.50	0.60

Eps – dielectric constant

The discrepancies between the total capacitance evaluation by means of analytical formulas and the OPERA simulations are shown in Table.2. The error used as the discrepancy measure is defined as follows:

$$Error = \frac{Analytical\ calculation - Numerical\ calculation}{Numerical\ calculation} 100\%. \quad (14)$$

Table 2. Error table of the analytical formulas compared with numerical simulator

Tech.	Layer	OPERA	Mod.1	error	Mod.2	error	Mod.3	error	Mod.4	error
130	1	7.36E-14	7.84E-14	7%	8.26E-14	12%	1.16E-13	58%	3.94E-14	-46%
	6	7.39E-14	7.88E-14	7%	8.46E-14	14%	1.17E-13	58%	4.05E-14	-45%
100	4	7.63E-14	8.07E-14	6%	9.19E-14	20%	1.23E-13	61%	5.32E-14	-30%
	7	4.37E-14	4.56E-14	4%	5.23E-14	20%	7.51E-14	72%	1.95E-14	-55%
50	7	8.61E-14	9.22E-14	7%	1.06E-13	23%	1.48E-13	72%	6.86E-14	-20%
	9	8.93E-14	9.39E-14	5%	1.09E-13	22%	1.52E-13	70%	7.11E-14	-20%

where Mod.1 is the Chern [9] formula, Mod.2 is the Wong [10] formula and Mod.3 and Mod.4 are the Sakurai [8] and Sarasvat [7] approaches respectively. Based on this comparison one can judge that the empirical Chern's formula that error value is less then 8% over a wide range of interconnect parameters can be treated as the more realistic one. It should be noted that the valid range of interconnect dimensions for this formula is [9]:

$$0.3 \leq \frac{W}{H} \geq 10; \quad 0.3 \leq \frac{H}{T} \geq 10 \quad (15)$$

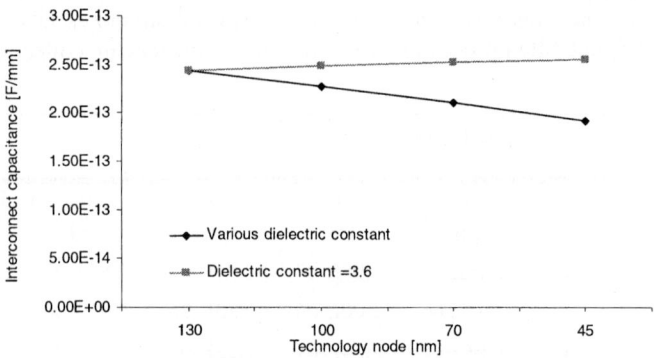

Fig. 4. Global interconnects capacitance versus technology node

Fig.4 shows the wiring capacitance per unit length C_o calculated by Chern's formula as a function of technology nodes. The calculations have been done for global interconnect with minimum cross-section dimensions (minimum wire pitch). It can be noticed that the interconnect capacitance calculated for the present value of dielectric constant 3.6 remains almost the same, while the capacitance calculated for the values predicted by ITRS [1] tends to lessen.

6 Conclusion

The crossover parasitic interconnect capacitance is essential for estimating the interconnect circuit performance. To get an accurate interconnect capacitance electric field solvers should be used. Since it is too time-consuming, the empirical approximations derived from the equations of electromagnetism are used. Unfortunately these methods can be considered to simulate a few physical configurations, only. In order to recognize which from the most popular approaches gives the evaluation of the total capacitance that suits to the real capacitance of the considered interconnect structure (Fig.2), the numerical simulations based on the numerical solving of Maxwell equations have been employed. The presented comparison showed that only for the Chern formula the error value is less then 7% over a wide range of ITRS parameters. Therefore one can conclude that the Chern formula is the most realistic one for the considered interconnect structure.

References

[1] The International Technology Roadmap for Semiconductors 2001, http ://public.itrs.net
[2] A.Ruehli P.Brennan "Capacitance Models for Integrated Circuit Metalization Wires "Journal of Solid-State Integrated Circuits Vol.10 No.6 pp.530-536 1975.
[3] M.Lee "A Multilevel Parasitic Interconnect Capacitance Modeling and Extraction for Reliable VLSI On-Chip Clock Delay Evaluation" Journal of Solid-State Integrated Circuits Vol.33 No.4 pp.657-661 1998.
[4] E.Barke "Line-to-Ground Capacitance Calculation for VLSI: A Comparison" IEEE Transaction of Computer Added Design Vol.7 No.2 pp.295-298 1988
[5] J.C.Chen B.McGaughy D.Sylvester C.Hu "An On-Chip Atto-Farad Interconnect Charge-Based Capacitance Measurement Technique" IEEE Tech. Dig. Int. Electron device Meeting 1996.
[6] O.P.Jensen "Calculating Wire Capacitance in Integrated Circuits" IEEE Circuits and Devices No.3 pp.36-40 1994.
[7] K.C.Saraswat F.Mohammadi Effect of Scaling of Interconnections on the Time Delay of VLSi Circuits Journal of Solid-State Circuits Vol.17 No.2 pp.275-280, 1982.
[8] T. Sakurai K. Tamaru Simple Formulas for Two- and Three –Dimensional Capacitances IEEE Tran. On Electron Devices Vol.30 No.2 pp.183-185, 1983.
[9] J.H. Chern, J. Huang, L. Arledge, P.C. Li, and P. Yang, Multilevel metal capacitance models for CAD design synthesis systems, IEEE Electron Device Letters, Vol.13, pp. 32-34, 1992.
[10] Shyh-Chyi Wong T.G.Y.Lee D.J.Ma CH.J.Chao An Empirical Three-Dimensional Crossover Capacitance Model for Multilevel Interconnect VLSI Circuits IEEE Trans. Semiconductor Manufacxturing Vol.13 No.2 pp.219-223, 2000.
[11] Vector Fields http://www.vectorfields.com/op2d

[12] J.Rubinstein P.Penfield M.A.Horowitz "Signal Delay in RC Tree Networks" IEEE Transactions on Computer-Aided Design Vol.2 No.3 pp.202-211 1983
[13] J.A.Davis J.D.Meindl "Compact Distributed RLC Interconnect Models, Part II- Coupled line transient expressions and peak crosstalk in multilevel networks" IEEE Transactions on Electron Devices Vol.47 No.11 pp.2078-2087 2000.
[14] Y.I.Ismail E.G.Friedman "Equivalent Elmore Delay for RLC Trees" IEEE Transaction on Computer-Aided Design of Integrated Circuits and systems Vol.19 No.1 Pp83-96 Jan 2000.
[15] R.L Dang N.Shigyo A two-dimensional simulations of LSI interconnect capacitance IEEE Electron Device Lett. EDL-2 pp.196-197, Aug 1981.

Self-optimization of Large Scale Wildfire Simulations

Jingmei Yang[1], Huoping Chen[1], Salim Hariri[1], and Manish Parashar[2]

[1] University of Arizona,
{jm_yang, hpchen, hariri}@ece.arizona.edu
[2] Rutgers, The State University of New Jersey,
parashar@caip.rutgers.edu

Abstract. The development of efficient parallel algorithms for large scale wildfire simulations is a challenging research problem because the factors that determine wildfire behavior are complex. These factors make static parallel algorithms inefficient, especially when large number of processors is used because we cannot predict accurately the propagation of the fire and its computational requirements at runtime. In this paper, we propose an Autonomic Runtime Manager (ARM) to dynamically exploit the physics properties of the fire simulation and use them as the basis of our self-optimization algorithm. At each step of the wildfire simulation, the ARM decomposes the computational domain into several natural regions (e.g., burning, unburned, burned) where each region has the same temporal and special characteristics. The number of burning, unburned and burned cells determines the current state of the fire simulation and can then be used to accurately predict the computational power required for each region. By regularly monitoring and analyzing the state of the simulation, and using that to drive the runtime optimization, we can achieve significant performance gains because we can efficiently balance the computational load on each processor. Our experimental results show that the performance of the fire simulation has been improved by 45% when compared with a static portioning algorithm.

1 Introduction

For over fifty years, attempts have been made to understand and predict the behavior of wildfires. However, the factors that determine wildfire behavior are complex and the computational loads associated with regions in the domain vary greatly both in time and space. Load balancing and efficient parallel execution of these simulations on large numbers of processors present significant challenges.

Optimizing the performance of parallel applications through load balancing is well studied and can be classified as either static or dynamic. The static approaches [3][4] assign work to processors before the computation starts and can be efficient if we know how the computations will progress a priori. If the workload cannot be estimated beforehand, dynamic load balancing strategies have to be used [5][6][7][8]. Some global schemes [9][10] predict future performance based on past information or based on some prediction tools such as Network Weather Service (NWS)[11]. Other optimization techniques are based on application-level scheduling [12][13]. AppLeS

[12] assumes the application performance model is static and provided by users and GHS system[13] assumes the applications computation load is a constant.

There are a few techniques that assume adaptive applications [14][15][16]. However, the wildfire simulation is a continuously changing application and requires adaptive and efficient runtime optimization techniques. In this paper, we present an Autonomic Runtime Manager (ARM) that continuously monitoring the computing requirements of the application, analyzing the current state of the application as well as the computing and networking resources and then making the appropriate planning and scheduling actions at runtime. The ARM control and management activities are overlapped with the application execution to minimize the overhead incurred.

The reminder of this paper is organized as follows: Section 2 gives a brief overview of the ARM system and a detailed analysis of the wildfire simulation. Results from the experimental evaluation of the ARM system are presented in Section 3. A conclusion and outline of future research directions are presented in Section 4.

2 Autonomic Runtime Manager (ARM) Architecture

The Autonomic Runtime Manager(ARM) is responsible for controlling and managing the execution for large-scale applications at runtime. The ARM main modules include (Fig. 1): 1) Online Monitoring and Analysis Module and 2) Autonomic Planning and Scheduling Module. The online monitoring and analysis module monitors the state of the application and underlying system and determines whether the online planning engine should be invoked. The planning and scheduling engine uses the resource capability models as well as performance models associated with the computations, and the knowledge repository to select the appropriate models and partitions for each region and then decompose the computational workloads into schedulable Computational Units (*CUs*). In this paper, we will use the wildfire simulation as a running example to explain the main operations of the ARM modules.

2.1 An Illustrative Example - Wildfire Simulation

In the wildfire simulation model, the entire area is represented as a 2-D cell-space composed of cells of dimensions length x breadth. For each cell, there are eight major wind directions as shown in Fig. 2. When a cell is ignited, its state will change from "unburned" to "burning". During its "burning" phase, the fire will propagate to its eight neighbors. The direction and the value of the maximum fire spread rate within the burning cell can be computed using Rothermel's fire spread model [2]. When the simulation time advances to the ignition times of neighbors, the neighbor cells will ignite. In a similar way, the fire would propagate to the neighbors of these cells. With different terrain, vegetation and weather conditions, the fire propagation could form very different spread patterns within the entire region.

Our wildfire simulation model is based on fireLib [1], which is a C function library for predicting the spread rate and intensity of free-burning wildfires. We parallelized the sequential fire simulation using MPI. This parallelized fire simulation divides the entire cell space among multiple processors such that each processor works on its own

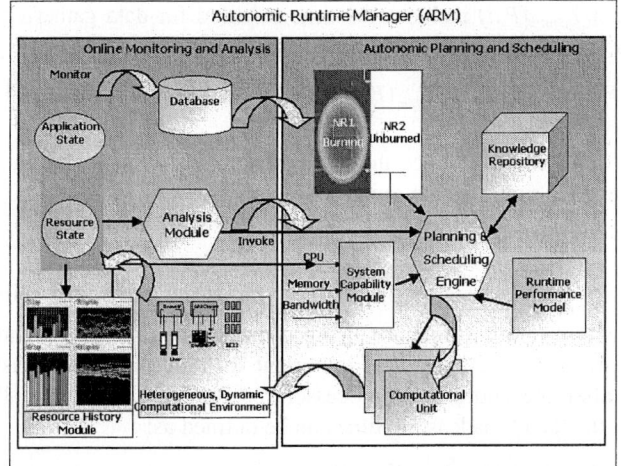
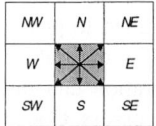

Fig. 1. Autonomic Runtime Manager (ARM) architecture

Fig. 2. Fire direction after ignition

portion and exchanges the necessary data with each other's after each simulation time step. At each time step, each processor computes and maintains the ignition maps of the 8 neighbors of the current ignited cell. Then the ignition map changes are exchanged between processors.

In our current implementation, a coordinator processor gathers the ignition map changes from each worker processor and then broadcasts them to all processors. Since there are only a few cells whose ignition times are changed at each time step, we believe the communication overhead with the coordinator is low. Thus the estimated execution time at time t for processor P_i can be defined as follows:

$$T_i(t) = T_{comp}(P_i,t) + T_{comm}(P_i,t) \tag{1}$$

where $T_{comp}(P_i,t)$ and $T_{comm}(P_i,t)$ are the computation and communication time at step t for processor P_i, respectively.

The application computational workload (ACW) of the simulation is defined as:

$$ACW(t) = N_B(t)T_B + N_U(t)T_U \tag{2}$$

where $N_B(t)$ and $N_U(t)$ are the number of burning and unburned cells at time t; T_B and T_U are the estimated computation times of each burning and unburned cell. $T_B > T_U$ because burning cells are more computation intensive than unburned cells, which contribute significantly to the imbalance conditions at runtime. Let α_i be the fraction of the workload assigned to processor P_i, it will be given a workload of $\alpha_i \times ACW(t)$. Therefore, the expected computation time for processor P_i can be defined as follows:

$$T_{comp}(P_i,t) = \alpha_i(N_B(t)T_B + N_U(t)T_U) = N_B(P_i,t)T_B + N_U(P_i,t)T_U \tag{3}$$

where $N_B(P_i, t)$ and $N_U(P_i, t)$ are the number of burning cells and unburned cells assigned to processor P_i at time step t.

The communication cost $T_{comm}(P_i,t)$ includes the time required for data gathering, synchronization and broadcasting, which can be defined as follows:

$$T_{comm}(P_i,t) = T_{gather}(P_i,t) + T_{sync}(P_i,t) + T_{bcast}(t) \quad (4)$$

Data gathering operation can be started once the computation is finished. The data gathering time of processor P_i at time step t is given by:

$$T_{gather}(P_i,t) = mT_{Byte}N_c(P_i,t) \quad (5)$$

where m is the message size in bytes sent by one cell, $N_c(P_i,t)$ is the number of cells assigned to processor P_i whose ignition time are changed during the time step t, and T_{Byte} is the data transmission time per byte. It is important to notice that broadcast operation can only start after the coordinator processor receives the data from all processors. Consequently, the data broadcasting time can be defined as:

$$T_{bcast}(t) = mT_{Byte}\sum_{i=0}^{P-1} N_c(P_i,t) \quad (6)$$

Then, the estimated execution time of the wildfire simulation on processor i can be computed as:

$$T_{total_i} = \sum_{t=1}^{N_t} T_i(t) \quad (7)$$

where N_t is the number of time steps performed by the wildfire simulation.

2.2 Online Monitoring and Analysis

The online monitoring module collects the information about the wildfire simulation state, such as the number and the location of burning cells and unburned cells, and the computation time for the last time step. At the same time, it monitors the states of the underlying resources, such as the CPU load, available memory, network load etc. The runtime state information is stored in a database. The online analysis module analyzes the load imbalance of the wildfire simulation and then determines whether or not the current allocation of workload needs to be changed.

Figure 3 shows the breakdown of the execution time and type of activities performed by four processors. Processor P_0 has the longest computation time because it is handling a large number of burning cells. Consequently, all the other three processors have to wait until processor P_0 finishes its computation and then the data broadcasting can be started. To balance the workload, the online analysis module should quickly detect large imbalance and invoke the repartitioning operation. To quantify the imbalance, we introduce a metric, Imbalance Ratio (*IR*) that can be computed as:

$$IR(t) = \frac{Max_{i=0}^{P-1}(T_{comp}(P_i,t)) - Min_{i=0}^{P-1}(T_{comp}(P_i,t))}{Min_{i=0}^{P-1}(T_{comp}(P_i,t))} \times 100\% \quad (8)$$

We use a predefined threshold $IR_{threshold}$ to measure how severe the imbalance is. If $IR(t) > IR_{threshold}$, the imbalance is considered severe and repartitioning is required. Then the automatic planning and scheduling module will be invoked to carry the appropriate actions to reparation the simulation workload.

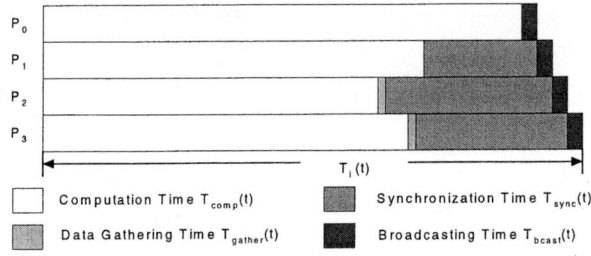

Fig. 3. The breakdown of the processor execution time at time step t

The selection of the threshold $IR_{threshold}$ can significantly impact the effectiveness of the self-optimization approach. If the threshold chosen is too low, too many load repartitioning will be triggered and the high overhead produced outweigh the expected performance gains. On the other hand, when the threshold is high, the imbalance conditions cannot be detected quickly. In the experimental results subsection, we show how we can experimentally choose this threshold value.

2.3 Autonomic Planning and Scheduling

The autonomic planning and scheduling module partitions the whole fire simulation domain into several natural regions (burning, unburned) based on its current state and then assigns them to processors by taking into consideration the states of the processors involved in the fire simulation execution. To reduce the rescheduling overhead, we use a dedicated processor to run the ARM self-optimizing algorithm and overlap that with the worker processors that compute their assigned workloads. Once the new partition assignments are finalized, a message is sent to all the worker processors to read the new assignments once they are done with the current computations. Consequently, the ARM self-optimization activities are completely overlapped with the application computation and the overhead is very minimum less than 4% as will be discussed later.

3 Experimental Results

The experiments were performed on two problem sizes for the fire simulation. One is a 256*256 cell space with 65536 cells. The other is a 512*512 cell domain with 262144 cells. To introduce a heterogeneous fire patterns, the fire is started in the southwest region of the domain and then propagates northeast along the wind direction. To make the evaluation accurate, we maintain total number of burning cells during the simulation is about 17% of the total cells for both problem sizes.

We begin with an examination of the effects of the imbalance ratio threshold on application performance. We ran the fire simulation with a problem size of 65536 on 16 processors and varied the $IR_{threshold}$ values to determine the best value that minimizes the execution time. The results of this experiment are shown in Fig. 4. We observed that the best execution time, 713 seconds, was achieved when the $IR_{threshold}$

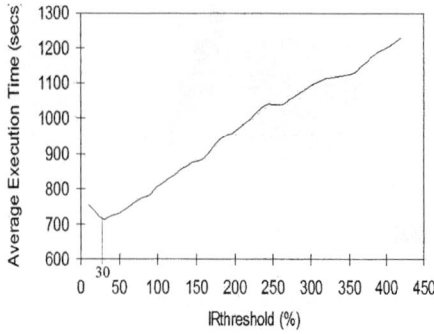 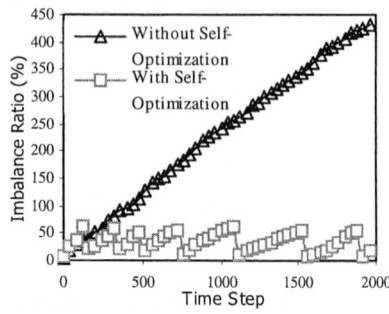

Fig. 4. The sensitivity of the fire simulation to the $IR_{threshold}$ value

Fig. 5. Imbalance ratios for 2000 time steps of the fire simulation, problem size = 65536, number of processors = 16, $IR_{threshold}$ = 50%

is equal to 30%. Figure 5 shows how the imbalance ratio increases as the simulation progresses using static partitioning algorithm and compares that with our self-optimization algorithm. For example, at time step 2000, the imbalance ratio in the static parallel algorithm is about 450% while it is around 25% in our approach. Using our approach, the imbalance ratio is kept bound within a small range.

Figure 6 shows the computation time for each processor at time steps 1, 300 and 600 with and without the ARM self-optimization. For example, at time step 1, the computation load is well balanced among most processors for both static partitioning and self-optimization. However, as shown in Fig. 6(a), at time step 300, processor P_0 and P_1 experience longer computation times while other processors keep the same computation time as before. This is caused by having many burning cells assigned to these two processors P_0 and P_1. At time step 600, more and more cells on processor P_0 and P_1 are burning and the maximum computation time of 0.24 seconds is observed for P_1. However, if we apply the ARM self-optimization algorithm, all processors finish their computations around the same time for all the simulation time steps (see Fig. 6 (b)). For example, the maximum execution time of 0.1 seconds is observed for processor P_2 at time step 600, which is 58% reduction in execution time when compared to the 0.24 seconds observed for the static portioning algorithm.

Tables 1 and 2 summarize the comparison of the execution time of the fire simulation with and without our self-optimization algorithm. Our experimental results show that the self-optimization approach improves the performance by up to 45% for a problem size of 262144 cells on 16 processors. We expect to get even better performance as the problem size increases because it will need more simulation time and will have more burning cells than smaller problem sizes.

In our implementation, one processor is dedicated to the autonomic planning and scheduling operations while all the worker processors are running the simulation loads assigned to them. Consequently, our self-optimization algorithm will not have high overhead impact on the fire simulation performance. The only overhead incurred is the time that ARM sensors collect the runtime information and the time that worker processors read new assigned simulation loads. To quantify the overhead on the whole system, we conducted experiments to measure the overhead. Based on our

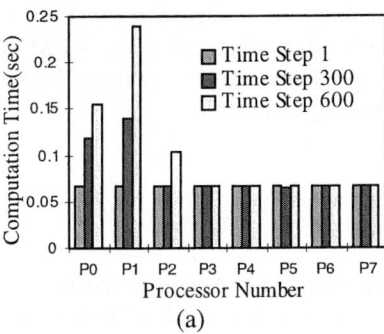 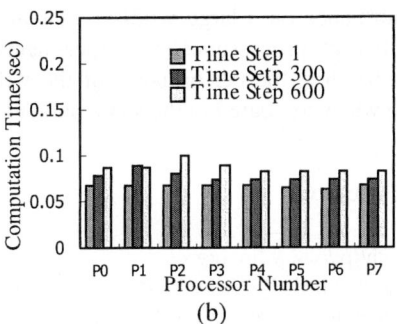

Fig. 6. Computation times of different time steps on 8 processors. Each group of adjacent bars shows the computation time of time step 1, 300 and 600, respectively. (a) Without self-optimization (b) With self-optimization

Table 1. Performance comparison for the fire simulation with and without self-optimization for different number of processors, problem size = 65536, and $IR_{threshold} = 30\%$

Number of Processors	Execution Time without static partitioning (sec)	Execution Time with Self-Optimization (sec)	Performance Improvement
8	2232.11	1265.94	43.29%
16	1238.87	713.17	42.43%

Table 2. Performance comparison for the fire simulation with and without self-optimization for different number of processors, problem Size = 262144, and $IR_{threshold} = 30\%$

Number of Processors	Execution Time without Self-Optimization (sec)	Execution Time with Self-Optimization (sec)	Performance Improvement
16	17276.02	9486.3	45.09%
32	9370.96	5558.55	40.68%

experiments, we observed that the overhead cost is less than 4% of the total execution time for both problem sizes of the fire simulation.

4 Conclusions and Future Work

In this paper, we described an Autonomic Runtime Manager that can self-optimize the parallel execution of large-scale applications at runtime by continuously monitoring and analyzing the state of the computations and the underlying resources, and efficiently exploit the physics of the problem being optimized. In our approach, the physics of the problem and its current state are the main criterion used to in our self-optimization algorithm. The activities of the ARM modules are overlapped with the algorithm being self-optimized to reduce the overhead. We show that the overhead of our self-optimization algorithm is less than 4%. We have also evaluated the ARM performance on a large wildfire simulation for different problem sizes and different number of processors. The experimental results show that using the ARM self-optimization, the performance of the wildfire simulation can be improved by up to 45% when compared to the static parallel partitioning algorithm.

References

1. <http://www.fire.org>
2. Rothermel, R. C.: A Mathematical Model for Predicting Fire Spread in Wildland Fuels. Research Paper INT-115. Ogden, UT: U.S. Department of Agriculture, Forest Service, Intermountain Forest and Range Experiment Station(1972)
3. Ichikawa, S., Yamashita, S.: Static Load Balancing of Parallel PDE Solver for Distributed Computing Environment. Proc. 13th Int'l Conf. Parallel and Distributed Computing Systems (2000) 399-405
4. Cierniak, M., Zaki, M. J., Li, W.: Compile-Time Scheduling Algorithms for Heterogeneous Network of Workstations. Computer J., vol. 40, no. 6(1997) 256-372
5. Willebeek-LeMair, M., Reeves, A.P.: Strategies for Dynamic Load Balancing on Highly Parallel Computers. IEEE Trans. Parallel and Distributed Systems, vol.4, no. 9 (1993) 979-993
6. Lin, F. C. H., Kelle, R. M. r: The Gradient Model Load Balancing Method, IEEE Trans. on Software Engineering, vol. 13, no. 1 (1987) 32-38
7. Cybenko, G.: Dynamic Load Balancing for Distributed Memory Multiprocessors. J. Parallel and Distributed Computing, vol. 7, no.2 (1989) 279-301
8. Horton, G.: A Multi-Level Diffusion Method for Dynamic Load Balancing. Parallel Computing, vol.19 (1993) 209-229
9. Nedeljkovic, N., Quinn, M. J.: Data-Parallel Programming on a Network of Heterogeneous Workstations. 1st IEEE HPDC (1992) 152-160
10. Arabe, J., Beguelin, A., Lowekamp, B., Seligman, E., Starkey, M., Stephan, P.: Dome: Parallel Programming in a Heterogeneous Multi-User Environment. Proc. 10th Int'l Parallel Processing Symp. (1996) 218-224
11. Wolski, R., Spring, N., Hayes, J.: The Network Weather Service: A Distributed Resource Performance Forecasting Service for Metacomputing. Journal of Future Generation Computing Systems (1998) 757-768
12. Berman, F., Wolski, R., Casanova, H., Cirne, Dail, W., H., Faerman, M., Figueira, S., Hayes, J., Obertelli, G., Schopf, J., Shao, G., Smallen, S., Spring, N., Su, A., Zagorodnov, D.: Adaptive Computing on the Grid Using AppLeS. IEEE Trans. on Parallel and Distributed Systems, vol. 14, no. 4(2003) 369--382
13. Sun, X.-H., Wu, M.: Grid Harvest Service: A System for Long-Term, Application-Level Task Scheduling. Proc. of 2003 IEEE International Parallel and Distributed Processing Symposium (IPDPS 2003)(2003)
14. Oliker, L., Biswas, R.: Plum: Parallel Load Balancing for Adaptive Unstructured Meshes", J. Parallel and Distributed Computing, vol. 52, no. 2(1998) 150-177
15. Walshaw, C., Cross, M., Everett, M.: Parallel Dynamic Graph Partitioning for Adaptive Unstructured Meshes. J. Parallel and Distributed Computing, vol. 47(1997)102-108
16. Zhang, Y., Yang, J., Chandra, S., Hariri, S., Parashar, M.: Autonomic Proactive Runtime Partitioning Strategies for SAMR Applications. Proceedings of the NSF Next Generation Systems Program Workshop, IEEE/ACM 18th International Parallel and Distributed Processing Symposium (2004)

Description of Turbulent Events Through the Analysis of POD Modes in Numerically Simulated Turbulent Channel Flow

Giancarlo Alfonsi[1] and Leonardo Primavera[2]

[1] Dipartimento di Difesa del Suolo, Università della Calabria,
Via P. Bucci 42b, 87036 Rende (Cosenza), Italy
alfonsi@dds.unical.it
[2] Dipartimento di Fisica, Università della Calabria,
Via P. Bucci 33b, 87036 Rende (Cosenza), Italy
lprimavera@fis.unical.it

Abstract. The flow of a viscous incompressible fluid in a plane channel is simulated numerically with the use of a parallel computational code for the numerical integration of the Navier-Stokes equations. The numerical method is based on a mixed spectral-finite difference algorithm and the approach of the Direct Numerical Simulation (DNS) is followed in the calculations. A turbulent-flow database is constructed, including 500 non-dimensional time steps of the turbulent statistically steady state flow field at Reynolds number $Re_\tau = 180$. The coherent structures of turbulence are extracted from the fluctuating portion of the velocity field with the use of the Proper Orthogonal Decomposition technique (POD). Turbulent events occurring in the flow are described in terms of temporal dynamics of the coherent turbulent structures, the flow modes educed with the POD technique.

1 Introduction

The properties of turbulence in wall bounded flows have been investigated with the use of a variety of techniques and methods. Accurate and extensive works in which a considerable amount of results have been reviewed are due to Robinson [1] and Panton [2]. Vortical structures of various kind have been observed in the inner region of wall-bounded flows, following different mechanisms of self-sustainment (Panton [2]). A fist type of mechanism involves the generation of a streamwise vortex from an already existing streamwise vortex, near either end of the original vortex. A second type of mechanism involves the generation of an horseshoe vortex from a parent horseshoe vortex and the related interaction of the vortex legs, that includes the lift-up of the vortex head. A third mechanism is the bridging between two streamwise fingers of vorticity. Other phenomena occur in the outer region, like the regeneration of hairpin vortices into packets with the vortical structures of the outer region interacting with those of the inner region.

In spite of the large amount of scientific work accomplished, still there are no definite conclusions on the character of the phenomena occurring in the near-wall region of a wall-bounded turbulent flow. This is due to the complexity of turbulence itself and to the quality of the scientific information that is possible to gather with the use of the current research techniques.

Contemporary times are characterized by the fact that high-performance computers are frequently used in turbulence research. In computational fluid dynamics a relevant issue of is that of the method that has to be followed to take into account the phenomenon of turbulence in the numerical simulations. The most rigorous approach is that of the Direct Numerical Simulation (DNS) according to which the objective of calculating all turbulent scales is pursued and the Navier-Stokes equations are numerically integrated without modifications. The crucial aspect of this method is the accuracy of the calculations that in theory should to be sufficiently high as to resolve the Kolmogorov microscales in space and time. DNS results for the case of the plane channel have been reported, among others, by Kim *et al.* [3] and Moser *et al.* [4].

Modern techniques for the numerical integration of the Navier-Stokes equations in conjunction with the increasing power of computers have the ability of greatly increasing the amount of data gathered during a research of computational nature. Moreover, the effort of studying turbulence in its full complexity has brought to the condition of managing large amounts of data. A typical turbulent-flow database includes all three components of the fluid velocity in all points of a three-dimensional domain, gathered for an adequate number of time steps of the turbulent statistically steady state. Such a database contains a considerable amount of information about the physics of the flow and, in the formation of the instantaneous value of each variable, all turbulent scales have contributed, being the effect of each scale nonlinearly combined with all others. Methods can be applied in order to extract from a turbulent-flow database only the *relevant* information, by separating the effects of appropriately-defined modes of the flow from the background flow, i.e. extract the coherent structures of turbulence.

In this work the issue of the coherent structures of turbulence in the wall region of a turbulent channel flow is addressed. The coherent turbulent motions are educed with the technique of the Proper Orthogonal Decomposition (POD) from a numerical database that has been built with the use of a parallel, three-dimensional, time-dependent computational code for the numerical integration of the Navier-Stokes equations. The DNS approach has been followed in the calculations.

2 Numerical Techniques

The numerical simulations have been performed with a parallel computational code based on a mixed spectral-finite difference algorithm. The system of the unsteady Navier-Stokes equations for incompressible fluids in three dimensions and non-dimensional conservative form is considered ($i \ \& \ j = 1,2,3$):

$$\frac{\partial u_i}{\partial x_i} = 0 \qquad (1a)$$

$$\frac{\partial u_i}{\partial t} + \frac{\partial}{\partial x_j}(u_i u_j) = -\frac{\partial p}{\partial x_i} + \frac{1}{Re_\tau}\frac{\partial^2 u_i}{\partial x_j \partial x_j} \qquad (1b)$$

where $u_i(u,v,w)$ are the velocity components in the cartesian coordinate system $x_i(x,y,z)$. Equations (1) are nondimensionalized by the channel half-width δ for lenghts, wall shear velocity $u_\tau = \sqrt{\tau_w/\rho}$ for velocities, ρu_τ^2 for pressure and δ/u_τ for time, being $Re_\tau = (u_\tau \delta/\nu)$ the friction Reynolds number. The fields are admitted to be periodic in the streamwise (x) and spanwise (z) directions, and equations (1) are Fourier transformed accordingly. The nonlinear terms in the momentum equation are evaluated pseudospectrally by anti-transforming the velocities back in physical space to perform the products (FFTs are used). A dealiasing procedure is applied to avoid errors in transforming the results back to Fourier space. In order to have a better spatial resolution near the walls, a grid-stretching law of hyperbolic-tangent type has been introduced for the grid points along y, the direction orthogonal to the walls. For the time advancement, a third-order Runge-Kutta algorithm has been implemented and the time marching procedure is accomplished with the fractional-step method (Kim et al. [3]). No-slip boundary conditions at the walls and cyclic conditions in the streamwise and spanwise directions have been applied to the velocity. More detailed descriptions of the numerical scheme, of its reliability and of the performance obtained on the parallel computers that have been used, can be found in Alfonsi et al. [5] and Passoni et al. [6],[7],[8].

3 Turbulent-Flow Database

By recalling the wall formalism, one has: $x_i^+ = x_i u_\tau/\nu = x_i/\delta_\tau$, $t^+ = tu_\tau^2/\nu = tu_\tau/\delta_\tau$, $\delta^+ = \delta/\delta_\tau$, $u^+ = \bar{u}/u_\tau$, $Re_\tau = u_\tau\delta/\nu = \delta/\delta_\tau = \delta^+$, where $\bar{u}$ is streamwise velocity averaged on a x-z plane and time, $\delta_\tau = \nu/u_\tau$ is the viscous length and δ/u_τ the viscous time unit. In Table 1 the characteristic parameters of the numerical simulations are reported.

Table 1. Characteristic parameters of the numerical simulations

Computing domain			Computational grid			Grid spacing		
L_x	L_y	L_z	N_x	N_y	N_z	Δx^+	Δy^+_{wall}	Δz^+
$2\pi\delta$	2δ	$\pi\delta$	96	129	64	11.8	0.87	8.8

It can be verified that in the present work there are 8 grid points in the y direction, within the viscous sublayer ($y^+ \leq 7$). The Kolmogorov spatial microscale, estimated using the criterion of the average dissipation rate per unit mass across the width of the channel, results $\eta^+ \approx 1.8$. After the insertion of appropriate initial conditions, the initial transient of the flow in the channel has been first simulated, the turbulent statistically steady state has been reached and then calculated for a time $t = 10\delta/u_\tau$ ($t^* = 1800$) to form a 500 time-step database with a temporal resolution of $\Delta t = 2 \times 10^{-2} \delta/u_\tau$ ($\Delta t^* = 3.6$) between each instant. In Table 2 predicted and computed values of a number of mean-flow variables are reported.

Table 2. Predicted vs. computed mean-flow variables

Predicted variables						
Re_τ	Re_b	Re_c	U_b/u_τ	U_c/u_τ	U_c/U_b	C_{fb}
180	2800	3244	15.56	18.02	1.16	8.44×10^{-3}

Computed variables						
Re_τ	Re_b	Re_c	U_b/u_τ	U_c/u_τ	U_c/U_b	C_{fb}
178.74	2786	3238	15.48	17.99	1.16	8.23×10^{-3}

In Table 2, U_b and U_c are the bulk mean velocity and the mean centerline velocity respectively, while Re_b and Re_c are the related Reynolds numbers. The *predicted* values of U_c/U_b and C_{fb} are obtained from the correlations suggested by Dean [9]:

$$\frac{U_c}{U_b} = 1.28(2Re_b)^{-0.0116} ; \quad C_{fb} = 0.073(2Re_b)^{-0.25}. \tag{4}$$

The *computed* skin friction coefficient is defined as:

$$C_{fb} = \frac{\tau_w}{\frac{1}{2}\rho U_b^2} \tag{5}$$

and is evaluated from the actual shear stress at the wall obtained in the computations. procedure is followed for the evaluation of the friction velocity in the computed Re_τ.

4 Proper Orthogonal Decomposition

The Proper Orthogonal Decomposition is a technique that can be applied for the extraction of the coherent structures from a turbulent flow field (Berkooz et al. [10], Sirovich [11]). By considering an ensemble of temporal realizations of a velocity field $u_i(x_j,t)$ on a finite domain D, one wants to find which is the most similar function to the elements of the ensemble, on average. This problem corresponds to find a deterministic vector function $\varphi_i(x_j)$ such that (i & $j=1,2,3$):

$$\max_{\psi} \frac{\left\langle \left| \left(u_i(x_j,t), \psi_i(x_j) \right) \right|^2 \right\rangle}{\left(\psi_i(x_j), \psi_i(x_j) \right)} = \frac{\left\langle \left| \left(u_i(x_j,t), \varphi_i(x_j) \right) \right|^2 \right\rangle}{\left(\varphi_i(x_j), \varphi_i(x_j) \right)}. \quad (6)$$

A necessary condition for problem (6) is that $\varphi_i(x_j)$ is an eigenfunction, solution of the eigenvalue problem and Fredholm integral equation of the first kind:

$$\int_D R_{ij}(x_l, x'_l)\varphi_j(x'_l)dx'_l = \int_D \left\langle u_i(x_k,t)u_j(x'_k,t) \right\rangle \varphi_j(x'_k)dx'_k = \lambda \varphi_i(x_k) \quad (7)$$

where $R_{ij} = \left\langle u_i(x_k,t)u_j(x'_k,t) \right\rangle$ is the two-point velocity correlation tensor. To each eigenfunction $\varphi_i^{(n)}(x_j)$ is associated a real positive eingenvalue $\lambda^{(n)}$ and every member of the ensemble can be reconstructed by means of a modal decomposition in the eigenfunctions themselves:

$$u_i(x_j,t) = \sum_n a_n(t) \varphi_i^{(n)}(x_j) \quad (8)$$

that can be seen as a decomposition of the originary random field into deterministic structures with random coefficients.

In the present work the POD technique is applied for the analysis of the fluctuating portion of the velocity field in a plane channel at $Re_\tau = 180$. Besides the time-averaged quantities outlined above, the decomposition properties are used to compute two time-dependent quantities (Webber et al. [12]), the kinetic energy of the fluctuations $E(t)$ and the representational entropy $S(t)$, respectively:

$$E(t) = \int_D u_i(x,y,z,t)u_i(x,y,z,t)dx_i = \sum_n a^n(t)a^{-n}(t) \quad (9)$$

$$S(t) = -\sum_n p^n(t)\ln(p^n(t)) \quad (10)$$

where:

$$p^n(t) = a^n(t)a^{-n}(t)/E(t), \quad (11)$$

to be used to follow the temporal dynamics of the coherent structures of the flow. A small value of $S(t)$ indicates that the corresponding energy is contained in few modes, while a large value of $S(t)$ indicates that the energy is distributed over many modes. A more detailed description of the POD algorithm that has been developed and previous results can be found in Alfonsi & Primavera [13] and Alfonsi et al. [14].

5 Results

1,820,448 eigenfunctions $(3 \times N_x \times N_y \times N_z)$ and correspondent eigenvalues are determined as a result of the decomposition. Figure 1 shows the fluctuating energy content $E(t)$ (9) of the first 5,188 eigenfunctions of the decomposition with time $(0 \leq t^+ \leq 1800)$, that alone incorporate 95% of the total energy content. The plot of $E(t)$ shows a sharp peak at $t^+ = 414$, besides some other smaller spikes. The peak of

$E(t)$ reflects a sharp rise and subsequent fall of the turbulent activity in the interval $129.6 \leq t^+ \leq 615.6$ and reveals the occurrence of a *turbulent event*. The correspondent plot of $S(t)$ (not shown here) shows a sharp fall and rise during the event in a time frame practically concident with that of $E(t)$. The minimum value of $S(t)$ occurs at $t^+ = 284.4$ indicating that during the growing-energy/falling entropy part of the event, the energy arrives to be distributed over a rather low number of flow modes. After $t^+ = 284.4$ the entropy starts growing while the energy is still rising until $t^+ = 414$, the instant corresponding to the peak of the energy within the event cycle, actually followed by another smaller peak at $t^+ = 489.6$. This indicates that for a short time interval $\Delta t^+ = 75.6$ the energy remains high and then rapidly decreases in the interval $489.6 \leq t^+ \leq 615.6$, redistributing over many modes (the entropy is still rising). The event cycle actually ends at $t^+ = 615.6$. The turbulent event in the whole lasts for $\Delta t^+ = 486$ ($129.6 \leq t^+ \leq 615.6$), 2.7 of $10\,\delta/u_\tau$ time units computed.

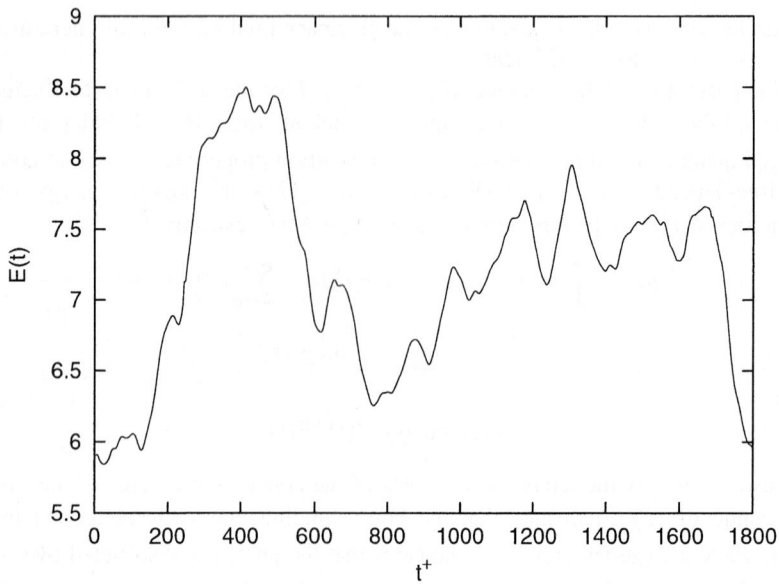

Fig. 1. Fluctuating energy content $E(t)$ (9) (first 5,188 eigenfunctions) with time

Figure 2 shows a visualization of the flow field generated by the first 5,188 eigenfunctions of the decomposition, in terms of isosurfaces of streamwise vorticity. The isovorticity surfaces are represented at $t^+ = 414$, when the energy reaches its maximum. A remarkably large number of flow structures is visible in the wall region of the flow.

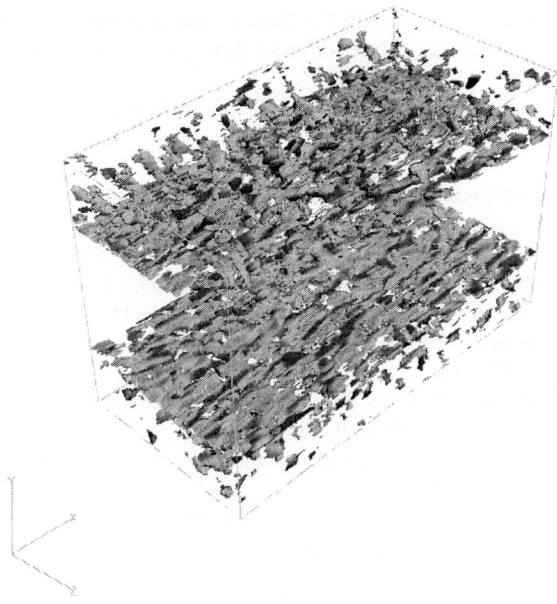

Fig. 2. Isosurfaces of streamwise vorticity at $t^+ = 414$ (first 5,188 eigenfunctions)

6 Concluding Remarks

The analysis of the flow field of a numerically simulated turbulent channel flow is performed in terms of flow modes determined with the POD technique. A peak of the kinetic energy of the velocity fluctuations reflects a sharp rise and subsequent fall of the turbulent activity in the interval $129.6 \leq t^+ \leq 615.6$ and reveals the occurrence of a turbulent event.

References

1. Robinson S.K.: Coherent motions in the turbulent boundary layer. *Ann. Rev. Fluid Mech.* **23** (1991) 601
2. Panton R.L.: Overview of the self-sustaining mechanisms of wall turbulence. *Prog. Aero. Sci.* **37** (2001) 341
3. Kim J., Moin P. & Moser R.: Turbulence statistics in fully developed channel flow at low Reynolds number. *J. Fluid Mech.* **177** (1987) 133
4. Moser R.D., Kim J. & Mansour N.N.: Direct numerical simulation of turbulent channel flow up to Re= 590. *Phys. Fluids* **11** (1999) 943
5. Alfonsi G., Passoni G., Pancaldo L. & Zampaglione D.: A spectral-finite difference solution of the Navier-Stokes equations in three dimensions. *Int. J. Num. Meth. Fluids* **28** (1998) 129
6. Passoni G., Alfonsi G., Tula G. & Cardu U.: A wavenumber parallel computational code for the numerical integration of the Navier-Stokes equations. *Parall. Comp.* **25** (1999) 593

7. Passoni G., Cremonesi P. & Alfonsi G.: Analysis and implementation of a parallelization strategy on a Navier-Stokes solver for shear flow simulations. *Parall. Comp.* **27** (2001) 1665
8. Passoni G., Alfonsi G. & Galbiati M.: Analysis of hybrid algorithms for the Navier-Stokes equations with respect to hydrodynamic stability theory. *Int. J. Num. Meth. Fluids* **38** (2002) 1069
9. Dean R.B.: Reynolds number dependence of skin friction and other bulk flow variables in two-dimensional rectangular duct flow. *J. Fluids Eng.* **100** (1978) 215
10. Berkooz G., Holmes P. & Lumley J.L.: The Proper Orthogonal Decomposition in the analysis of turbulent flows. *Ann. Rev. Fluid Mech.* **25** (1993) 539
11. Sirovich L.: Turbulence and the dynamics of coherent structures. Part I: coherent structures. Part II: symmetries and transformations. Part III: dynamics and scaling. *Quart. Appl. Math.* **45** (1987) 561
12. Webber G.A., Handler R.A. & Sirovich L.: The Karhunen-Loéve decomposition of minimal channel flow. *Phys. Fluids* **9** (1997) 1054
13. Alfonsi G. & Primavera L.: Coherent structure dynamics in turbulent channel flow. *J. Flow Visual. Imag. Proc.* **9** (2002) 89
14. Alfonsi G., Restano C. & Primavera L.: Coherent structures of the flow around a surface-mounted cubic obstacle in turbulent channel flow. *J. Wind Eng. Ind. Aerodyn.* **91** (2003) 495

Computational Modeling of Human Head Conductivity

Adnan Salman[1], Sergei Turovets[1], Allen Malony[1], Jeff Eriksen[2], and Don Tucker[2]

[1] NeuroInformatics Center, 5219 University of Oregon, Eugene, OR 97403, USA
malony@cs.uoregon.edu
[2] Electrical Geodesic, Inc., 1600 Millrace Dr, Eugene, OR 97403, USA
dtucker@egi.com

Abstract. The computational environment for estimation of unknown regional electrical conductivities of the human head, based on realistic geometry from segmented MRI up to 256^3 resolution, is described. A finite difference alternating direction implicit (ADI) algorithm, parallelized using OpenMP, is used to solve the forward problem describing the electrical field distribution throughout the head given known electrical sources. A simplex search in the multi-dimensional parameter space of tissue conductivities is conducted in parallel using a distributed system of heterogeneous computational resources. The theoretical and computational formulation of the problem is presented. Results from test studies are provided, comparing retrieved conductivities to known solutions from simulation. Performance statistics are also given showing both the scaling of the forward problem and the performance dynamics of the distributed search.

1 Introduction

Tomographic techniques determine unknown complex coefficients in PDEs governing the physics of the particular experimental modality. Such problems are typically non-linear and ill-poised. The first step in solving such an inverse problem is to find a numerical method to solve the direct (forward) problem. When the physical model is three-dimensional and geometrically complex, the forward solution can be difficult to construct and compute. The second stage involves a search across a multi-dimensional parameter space of unknown model properties. The search employs the forward problem with chosen parameter estimates and a function that determines the error of the forward calculation with an empirically measured result. As the error residuals of local inverse searches are minimized, the global search determines convergence to final property estimates based on the robustness of parameter space sampling.

Fundamental problems in neuroscience involving experimental modalities like electroencephalography (EEG) and magnetoencephalograpy (MEG) are naturally expressed as tomographic imaging problems. The difficult problems of source localization and impedance imaging require modeling and simulating the associated bioelectric fields. Forward calculations are necessary in the computational formulation of these problems. Until recently, most practical research in this field has opted for analytical or semi-analytical models of a human head in the forward calculations [1, 2]. This is in contrast to approaches that use realistic 3D head geometry for purposes of significantly improving the accuracy of the forward and inverse solutions. To do so, however, requires that the geometric information be available from MRI or CT scans. With such image data, the tissues of the head can be better segmented and more accurately represented in the computational model. Unfortunately, these realistic modeling techniques have intrinsic computational complexities that grow as the image resolution increases.

In source localization we are interested in finding the electrical source generators for the potentials that might be measured by EEG electrodes on the scalp surface. Here, the inverse search is looking for those sources (their position and amplitude) on the cortex surface whose forward solution most accurately describes the electrical potentials observed. The computational formulation of the source localization problem assumes the forward calculation is without error. However, this assumption in turn assumes the conductivity values of the modeled head tissues are known. In general, for any individual, they are not known. Thus, the impedance imaging problem is actually a predecessor problem to source localization. In impedance imaging, the inverse search finds those tissue impedance values whose forward solution best matches measured scalp potentials when experimental stimuli are applied. In either problem, source localization or impedance imaging, solving the inverse search usually involves the large number of runs of the forward problem. Therefore, computational methods for the forward problem, which are stable, fast and eligible for parallelization, as well as intelligent strategies and techniques for multi-parameter search, are of paramount importance.

To deal with complex geometries, PDE solvers use finite element (FE) or finite difference (FD) methods [3, 4]. Usually, for the geometry with the given complexity level, the FE methods are more economical in terms of the number of unknowns (the size of the stiffness matrix A, is smaller, as homogeneous segments do not need a dense mesh) and resulting computational cost. However, the FE mesh generation for a 3D, highly heterogeneous subject with irregular boundaries (e.g., the human brain) is a difficult task. At the same time, the FD method with a regular cubed grid is generally the easiest method to code and implement. It is often chosen over FE methods for simplicity and the fact that MRI/CT segmentation map is also based on a cubed lattice of nodes. Many anatomical details (e.g., olfactory perforations and internal auditory meatus) or structural defects in case of trauma (e.g., skull cracks and punctures) can be included as the computational load is based on the number of elements and not on the specifics of tissues differentiation. Thus, the model geometry accuracy can be the same as the resolution of MRI scans (e.g., $1 \times 1 \times 1mm$).

In the present study we adopt a model based on FD methods and construct a distributed and parallel simulation environment for conductivity optimization through inverse simplex search. FE simulation is used to solve for relatively simple phantom geometries that we then apply as "gold standards" for validation.

2 Mathematical Description of the Problem

The relevant frequency spectrum in EEG and MEG is typically below $1kHz$, and most studies deal with frequencies between 0.1 and $100Hz$. Therefore, the physics of EEG/MEG can be well described by the quasi-static approximation of Maxwell's equations, the Poisson equation. The electrical forward problem can be stated as follows: given the positions and magnitudes of current sources, as well as geometry and electrical conductivity of the head volume Ω calculate the distribution of the electrical potential on the surface of the head (scalp) Γ_Ω. Mathematically, it means solving the linear Poisson equation [1]:

$$\nabla \cdot \sigma(x,y,z) \nabla \phi(x,y,z) = S, \tag{1}$$

in Ω with no-flux Neumann boundary conditions on the scalp:

$$\sigma(\nabla \phi) \cdot n = 0, \tag{2}$$

on Γ_Ω. Here $\sigma = \sigma_{ij}(x,y,z)$ is an inhomogeneous tensor of the head tissues conductivity and S is the source current. Having computed potentials $\phi(x,y,z)$ and current densities $J = -\sigma(\nabla \phi)$, the magnetic field B can be found through the Biot-Savart law. We do not consider anisotropy or capacitance effects (the latter because the frequencies of interest are too small), but they can be included in a straightforward manner. (Eq.(1) becomes complex-valued, and complex admittivity should be used.)

We have built a finite difference forward problem solver for Eq. (1) and (2) based on the multi-component alternating directions implicit (ADI) algorithm [7, 8]. It is a generalization of the classic ADI algorithm as described by Hielscher et al [6], but with improved stability in 3D (the multi-component FD ADI scheme is unconditionally stable in 3D for any value of the time step [8]). The algorithm has been extended to accommodate anisotropic tissues parameters and sources. To describe the electrical conductivity in the heterogeneous biological media within arbitrary geometry, the method of the embedded boundaries has been used. Here an object of interest is embedded into a cubic computational domain with extremely low conductivity values in the external complimentary regions. This effectively guarantees there are no current flows out of the physical area (the Neuman boundary conditions, Eq.(2), is naturally satisfied). The idea of the iterative ADI method is to find the solution of Eq. (1) and (2) as a steady state of the appropriate evolution problem. At every iteration step the spatial operator is split into the sum of three 1D operators, which are evaluated alternatively at each sub-step. For example, the difference equations in x direction is given as [8]

$$\frac{\phi_i^{n+1} - \frac{1}{3}(\phi_i^n + \phi_j^n + \phi_k^n)}{\tau} + \delta_x(\phi_i^{n+1}) + \delta_y(\phi_i^n) + \delta_z(\phi_i^n) = S, \tag{3}$$

where τ is a time step and $\delta_{x,y,z}$ is a notation for the appropriate $1D$ spatial difference operator (for the problems with variable coefficients it is approximated on a "staggered" mesh). Such a scheme is accurate to $O(\tau^2) + O(\Delta x^2)$. In contrast with the classic ADI method, the multi-component ADI uses the regularization (averaging) for evaluation of the variable at the previous instant of time.

Parallelization of the ADI algorithm is straightforward, as it consists of nests of independent loops over "bars" of voxels for solving the effective 1D problem (Eq. (3)) at each iteration. These loops can be easily unrolled in a shared memory multiprocessor environment. It is worth noting, that the ADI algorithm can be also easily adapted for solving PDEs describing other tomographic modalities. In particular, we have used it in other related studies, for example, in simulation of photon migration (diffusion) in a human head in near-infrared spectroscopy of brain injuries and hematomas.

The inverse problem for the electrical imaging modality has the general tomographic structure. From the assumed distribution of the head tissue conductivities, σ_{ij}, and the given injection current configuration, S, it is possible to predict the set of potential measurement values, ϕ^p, given a forward model F (Eq. (1), (2)), as the nonlinear functional [5, 6]:

$$\phi^p = F(\sigma_{ij}(x,y,z)). \qquad (4)$$

Then an appropriate objective function is defined, which describes the difference between the measured, V, and predicted data, ϕ^p, and a search for the global minimum is undertaken using advanced nonlinear optimization algorithms. In this paper, we used the simple least square error norm:

$$E = \left(\sum_{i=1}^{N} (\phi_i^p - V_i)^2 \right)^{1/2}, \qquad (5)$$

where N is a total number of the measuring electrodes. To solve the nonlinear optimization problem in Eq.(5), we employed the downhill simplex method of Nelder and Mead as implemented by Press et al[3]. In the strictest sense, this means finding the conductivity at each node of the discrete mesh. In simplified models with the constrains imposed by the segmented MRI data, one needs to know only the average regional conductivities of a few tissues, for example, scalp, skull, cerebrospinal fluid (CSF) and brain, which significantly reduces the demensionality of the parameter space in the inverse search, as well as the number of iterations in converging to a local minimum. To avoid the local minima, we used a statistical approach. The inverse procedure was repeated for hundreds sets of conductivity guesses from appropriate fisiological intervals, and then the solutions closest to the global minimum solutions were selected using the simple critirea $E < E_{threshold}$.

3   Computational Design

The solution approach maps to a hierarchical computational design that can benefit both from parallel parametric search and parallel forward calculations. Fig. 1 gives a schematic view of the approach we applied in a distributed environment of parallel computing clusters. The master controller is responsible for launching new inverse problems with guesses of conductivity values. Upon completion, the inverse solvers return conductivity solutions and error results to the master. Each inverse solver runs on a compute server. Given N compute servers, N inverse solves can be simultaneously active, each generating forward problems that can run in parallel, depending on the number of processors available. The system design allows the number of compute servers and the number of processors per server to be decided prior to execution, thus trading off inverse search parallelism versus forward problem speedup.

At the University of Oregon, we have access to a computational systems environment consisting of four multiprocessor clusters. Clusters *Clust1*, *Clust2*, and *Clust3* are

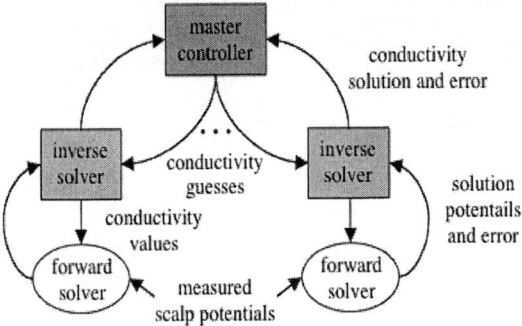

Fig. 1. Schematic view of the parallel computational system

8-processor IBM p655 machines and cluster *Clust4* is a 16-processor IBM p690 machine. All machines are shared-memory multiprocessors running the Linux operating system. The clusters are connected by a high-speed gigabit Ethernet network. In our experiments below, we treated each machine as a separate compute server running one inverse solver. The forward problem was parallelized using OpenMP and run on eight (*Clust1-3*) and sixteen (*Clust4*) processors. The master controller can run on any networked machine in the environment. In our study, the master controller ran on Clust2.

4 Computational Results

The forward solver was tested and validated against a 4-shell spherical phantom, and low ($64 \times 64 \times 44$) and high ($256 \times 256 \times 176$) resolution human MRI data. For comparison purposes, the MRI data where segmented into only four tissue types and their values were set to those in the spherical model (cl. Table 1). When we computed potentials at standard locations for the 129 electrodes configuration montage on the spherical phantom and compared the results with the analytical solution [2] available for a 4-shell spherical phantom we observed good agreement, save for some minor discrepancies (average error is no more than a few percents) caused by the mesh orientation effects (the cubic versa spherical symmetry).

Similarly, we found the good agreement for spherical phantoms between our results and the solution of the Poisson equation using the standard FEM packages such as FEMLAB. Also, we have performed a series of computations for electric potentials and currents inside a human head with surgical or traumatic openings in the skull. We

Table 1. Tissues parameters in 4-shell models[2]

Tissue type	$\sigma(\Omega^{-1}m^{-1})$	Radius(cm)	Reference
Brain	0.25	8	Geddes(1967)
Csf	1.79	8.2	Daumann(1997)
Skull	0.018	8.7	Law(1993)
Scalp	0.44	9.2	Burger(1943)

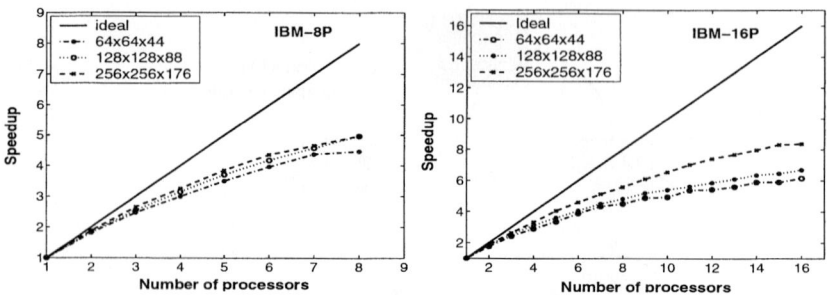

Fig. 2. Speed-up of the forward solver for different problem sizes at an 8-processor (left) and a 16-processor (right) IBM machines

found that generally low resolution (64 × 64 × 44 voxels) is not enough for accurate description of the current and potentials distribution through the head, as the coarse discretization creates artificial shunts for currents (mainly in the skull). With increased resolution (128 × 128 × 88 or 256 × 256 × 176 voxels) our model has been shown to be capable to capture the fine details of current/potential redistribution caused by the structural perturbation. However, the computational requirements of the forward calculation increase significantly.

The forward solver was parallelized using OpenMP. The performance speedups for 64 × 64 × 44, 128 × 128 × 88 and 256 × 256 × 176 sized problems on the IBM p655 (8 processors) and p690 (16 processors) machines are shown in Fig. 2. The performance is reasonable at present, but we believe there are still optimizations that can be made. The importance of understanding the speedup performance on the cluster compute servers is to allow flexible allocation of resources between inverse and forward processing.

In the inverse search the initial simplex was constructed randomly based upon the mean conductivity values (cl. Table 1) and their standard deviations as it is reported in the related biomedical literature. In the present test study we did not use the real experimental human data, instead, we simulated the experimental set of the reference potentials V in Eq. 5 using our forward solver with the mean conductivity values from Table 1, which had been assumed to be true, but not known a priory for a user running the inverse procedure. The search was stopped when one or two criteria were met. The first is when the decrease in the error function is fractionally smaller than some tolerance parameter. The second is when the number of steps of the simplex exceeds some maximum value. During the search, the conductivities were constrained to stay within their pre-defined plausible ranges. If the simplex algorithm attempted to step outside of the acceptable range, then the offending conductivity was reset to the nearest allowed value. Our procedure had the desired effect of guiding the search based on prior knowledge. Some number of solution sets included conductivities that were separated from the bulk of the distribution. These were rejected as outliers, based on the significant larger square error norm in Eq. (5) (i.e., the solution sets were filtered according to the criteria $E < E_{threshold}$). We have found empirically that setting $E_{threshold} = 1\mu V$ in most of our runs produced a fair percentage of solutions close to the global minimum.

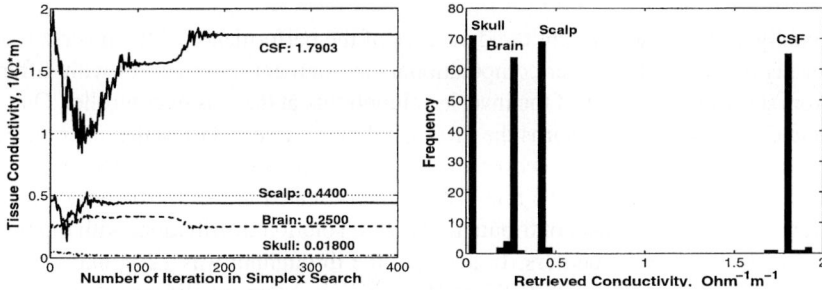

Fig. 3. Results of the inverse search. Dynamics of the individual search (left) and statistics of the retrieved conductivities for about 200 initial random guesses. The actual number of the solutions shown is 71, their error function is less than 1 microvolt

The distribution of the retrieved conductivities is shown in Fig. 3 (right). The fact that the retrieved conductivities for the intracranial tissues (CSF and brain) have wider distributions is consistent with the intuitive physical explanation that the skull, as having the lowest conductivity, shields the currents injected by the scalp electrodes from the deep penetration into the head. Thus, the deep intracranial tissues are interrogated less in comparison with the skull and scalp. The dynamics of an individual inverse search convergence for a random initial guesses is shown in Fig. 3 (left). One can see the conductivities for the extra cranial tissue and skull converging faster than the brain tissues, due to the better interrogation by the injected current.

After filtering data according to the error norm magnitude, we fitted the individual conductivities to the normal distribution. The mean retrieved conductivities $\sigma(\Omega^{-1} m^{-1})$ and their standard deviations $\Delta\sigma(\Omega^{-1} m^{-1})$ are: Brain (0.24 / .01), CSF (1.79 / .03), Skull (0.0180 / .0002), and Scalp (0.4400 / .0002) It is interesting to compare these values to the "true" conductivities from Table 1. We can see excellent estimates for the scalp and skull conductivities and a little bit less accurate estimates for the intracranial tissues. Although we have not yet done runs with the realistic noise included, the similar investigation in Ref. 2 for a spherical phantom suggests that noise will lead to some deterioration of the distributions and more uncertainty in the results. In general, it still will allow the retrieval of the unknown tissue parameters.

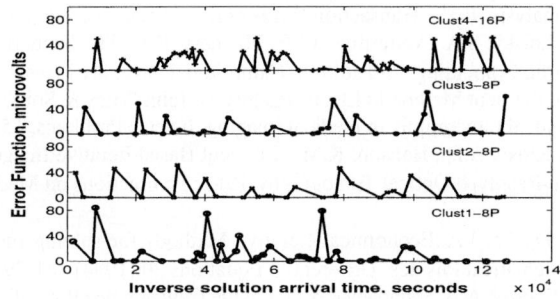

Fig. 4. Solution flow at the master controller. Inverse solution arrival to the controller are marked

Finally, in Fig. 4 we present the dynamics of the performance of the inverse search in our distributed multi-cluster computational environment. Four curves with different markers show the dynamics of the inverse solution flux at the master controller. One can see that Clust4 on average returns the inverse solution twice as fast as the other clusters, as would be expected. Note, however, the time to inverse solution also depends on both forward speed and convergence rate. The markers seated at the "zero" error function line represent solutions that contribute to the final solution distribution, with the rest of the solutions rejected as outliers. In average, the throughput was 12 minutes per one inverse solution for $128 \times 128 \times 88$ MRI resolution. More intelligent schemes of the search with intermediate learning from the guiding process with smaller resolution to control (narrow) the range of the initial guesses in simulation with the higher resolution are under investigation.

5   Conclusion

We have built an accurate and robust 3D Poisson solver based on a FDM ADI algorithm for modeling electrical and optical problems in heterogeneous biological tissues. We focus in particular on modeling the conductivity properties of the human head. The computational formulation utilizes realistic head geometry obtained from segmented MRI datasets. The results presented here validate our FDM approach for impedance imaging and provide a performance assessment of the parallel and distributed computation.

In the future, we will enhance the computational framework with additional cluster resources that the naturally scalable inverse search can use. Our intent is to evolve the present interprocess communication (IPC) socket-based code to one that uses grid middleware support, allowing the impedance imaging program to more easily access available resources and integrate with neuroimaging workflows.

The authors wish to thank Dr. V.M. Volkov, of Institute of Mathematics, Belarus Academy of Sciences, for providing many ideas and fruitful discussions on the multicomponent ADI algorithm.

References

1. Gulrajani, R.M.: Bioelectricity and Biomagnetism. John Wiley & Sons, New York (1998)
2. Ferree, T. C., Eriksen, K. J., Tucker, D. M.: Regional head tissue conductivity estimation for improved EEG analysis. IEEE Transactions on Biomedical Engineering 47(2000) 1584-1592
3. Press, W.H., Teukolsky, S.A., Vetterling, W.T., Flannery, B.P.: The Numerical Recipes in C: The art of Scientific Computing. 2nd edition. Cambridge University Press, New York (1992)
4. Jin, J.: The Finite Element Method in Electromagnetics. John Wiley & Sons, New York(1993)
5. Arridge, S.R.: Optical tomography in medical imaging. Inverse Problems, 15 (1999) R41-R93
6. Hielscher, A.H., Klose, A.D., Hanson, K.M.: Gradient Based Iterative Image Reconstruction Scheme for Time-Resolved Optical Tomography. IEEE Transactions on Medical Imaging. 18 (1999) 262-271
7. Abrashin, V.N., Dzuba, I.A.:Economical Iterative Methods for solving multi- dimensional problems in Mathematical Physics. Differential Equations 30 (1994) 281-291
8. Abrashin, V.N., Egorov, A.A., Zhadaeva, N.G. On the Convergence Rate of Additive Iterative Methods. Differential Equations. 37 (2001) 867-879

Modeling of Electromagnetic Waves in Media with Dirac Distribution of Electric Properties

André Chambarel and Hervé Bolvin

UMR A 1114 Climate, Soil, Environment,
33, rue Louis Pasteur, F-84000 AVIGNON - France

Abstract. We develop a new numerical approach of the electromagnetic wave propagation in a complex media model. We use a new Finite Element Method for space discretization and a finite difference method in time domain. This study is applied to TDR (Time Domain Reflectometry) which appears as a wave guide. In this context, the experimental approach is very uneasy and the numerical study brings about very interesting results, particularly for local values of the electromagnetic field. Moreover we test models of an heterogeneous medium. Dielectric properties of the medium are represented by an homogeneous component associated with Dirac distribution. We present wave front profiles with useable signals simulation.

Keywords: Finite Element Method, Backward Difference Method, Maxwell's equations.

1 Introduction

We present a model of electromagnetic wave propagation in a wave guide containing a complex medium. This work is a contribution to the interpretation of the TDR (Time Domain Reflectometry) probe's signals. The TDR probe [1] is a wave guide constituted by two parallel metallic rods. At initial time, a step of electric field is applied between the electrodes and intensity is measured at the entrance of the wave guide. Measurement and processing of the signal obtained by reflection should theoretically allow the determination of medium properties. An interesting application of this technology concerns soil science. In practice, a soil is composed of a complex mixture with inhomogeneous electric permittivities.

The theoretical approach is formulated using Maxwell's equations. They are solved in the time domain using the Finite Element Method. Moreover we consider an homogeneous medium in which random space distribution of electric properties is added to some finite element nodes with Dirac distribution. Each numerical simulation is performed using a given ratio r of Dirac dedicated nodes. The main objective is then to study the influence of ratio r on the electric signal.

2 General Presentation of the Model

2.1 Maxwell Equations

The 2D wave guide is represented by two parallel plate electrodes whose electric conductivity is infinite. The free space around the electrodes is constituted by a complex medium. With the usual notations, the dimensionless Maxwell equations are formulated as follows [2]:

$$\mu_r . \frac{\partial \vec{H}}{\partial t} = \overrightarrow{curl} \, \vec{E} \qquad \varepsilon_r . \frac{\partial \vec{E}}{\partial t} = \overrightarrow{curl} \, \vec{H} \qquad (1)$$

$$div\left(\mu_r . \vec{H}\right) = 0 \qquad div\left(\varepsilon_r . \vec{E}\right) = 0$$

2.2 Finite Element Formulation

The Finite Element method is applied to Maxwell's equations. In our formulation, we use the weak formulation, presented by Chambarel et al [3], with Galerkin's ponderation [4], in order to obtain integral forms for the electromagnetic equations. Then, with the Finite Element formulation of the integral forms for a first order form of Maxwell's equations, we can get an approximate solution of the weak formulation, with discretization of the electric and the magnetic fields. The weighted residual method can be written:

$$\int_{(\Omega)} \overrightarrow{\delta H} . \mu_r . \frac{\partial \vec{H}}{\partial t} . d\Omega = - \int_{(\Omega)} \overrightarrow{\delta H} . \overrightarrow{curl} \, \vec{E} . d\Omega \qquad (2)$$

$$\int_{(\Omega)} \overrightarrow{\delta E} . \varepsilon_r . \frac{\partial \vec{E}}{\partial t} . d\Omega = \int_{(\Omega)} \overrightarrow{\delta E} . \overrightarrow{curl} \, \vec{H} . d\Omega$$

In formula (2), ε_r incorporates the Dirac distribution.

The medium arises as a medium whose permittivity is homogeneous but in which a distribution of singularities is inserted. For the Dirac formulation of electric permittivity we define the following density θ :

$$\varepsilon_r = \frac{d\theta}{d\Omega} \qquad \varepsilon_r = \varepsilon_H + \sum_p \theta_p . \delta(x - x_p) \qquad (3)$$

where index H denotes the homogeneous component of the medium. So the weak formulation becomes :

$$\int_{(\Omega)} \overrightarrow{\delta H} . \mu_r . \frac{\partial \vec{H}}{\partial t} . d\Omega = - \int_{(\Omega)} \overrightarrow{\delta H} . \overrightarrow{curl} \, \vec{E} . d\Omega$$

$$\int_{(\Omega)} \overrightarrow{\delta E} . \varepsilon_H . \frac{\partial \vec{E}}{\partial t} . d\Omega + \sum_p \left(\overrightarrow{\delta E} . \theta_p . \frac{\partial \vec{E}}{\partial t} \right)_{x=x_p} = \int_{(\Omega)} \overrightarrow{\delta E} . \overrightarrow{curl} \, \vec{H} . d\Omega \qquad (4)$$

In that way, we can define the discretization. Let $n_i(x)$ be a base of work space and index n denote the elementary nodal values. For an isoparametrical element the polynomial interpolation can be written:

$$x_i^h = n_j . x_{ij}^n \quad , \quad H_i^h = n_j . H_{ij}^n \quad and \quad E_i^h = n_j . E_i^n$$

The discretization of the electric fields is as follows :

$$\{E_h\} = \begin{bmatrix} \langle n_i \rangle & \langle 0 \rangle & \langle 0 \rangle \\ \langle 0 \rangle & \langle n_i \rangle & \langle 0 \rangle \\ \langle 0 \rangle & \langle 0 \rangle & \langle n_i \rangle \end{bmatrix} \begin{Bmatrix} E_x^n \\ E_y^n \\ E_z^n \end{Bmatrix} = [N]\{E^n\} \quad (5)$$

So the discretization of the curl operator is easy and we have :

$$\overrightarrow{\text{curl }} \overrightarrow{E_h} = \begin{bmatrix} \langle 0 \rangle & -\left\langle \frac{\partial n_i}{\partial z} \right\rangle & \left\langle \frac{\partial n_i}{\partial y} \right\rangle \\ \left\langle \frac{\partial n_i}{\partial z} \right\rangle & \langle 0 \rangle & -\left\langle \frac{\partial n_i}{\partial x} \right\rangle \\ -\left\langle \frac{\partial n_i}{\partial y} \right\rangle & \left\langle \frac{\partial n_i}{\partial x} \right\rangle & \langle 0 \rangle \end{bmatrix} \begin{Bmatrix} E_x^n \\ E_y^n \\ E_z^n \end{Bmatrix} = [R]\{E^n\} \quad (6)$$

We can write similar relations for the magnetic field. The weak formulation can be written in matricial form:

$$\int_{(\Omega)} \overrightarrow{\delta H} . \mu_r . \frac{\partial \overrightarrow{H}}{\partial t} . d\Omega = \langle \delta H^n \rangle . \int_{(\Omega)} N^T . \mu_r . N . d\Omega . \left\{ \frac{\partial H^n}{\partial t} \right\} \quad (7)$$

$$\int_{(\Omega)} \overrightarrow{\delta H} . \overrightarrow{\text{rot }} \overrightarrow{E} . d\Omega = \langle \delta H^n \rangle . \int_{(\Omega)} N^T . R . d\Omega . \{E^n\} \quad (8)$$

$$\int_{(\Omega)} \overrightarrow{\delta E} . \varepsilon_r . \frac{\partial \overrightarrow{E}}{\partial t} . d\Omega = \langle \delta E^n \rangle . \left[\int_{(\Omega)} N^T . \varepsilon_H . N . d\Omega . . + \sum_{k \in (\Omega_e)} \{n(x_k)\} . \theta_k . \langle n(x_k) \rangle \right] \left\{ \frac{\partial E^n}{\partial t} \right\} \quad (9)$$

$$\int_{(\Omega)} \overrightarrow{\delta E} . \overrightarrow{\text{rot }} \overrightarrow{H} . d\Omega = \langle \delta E^n \rangle . \int_{(\Omega)} N^T . R . d\Omega . \{H^n\} \quad (10)$$

2.3 Matricial Formulation

$$\sum_{i=1}^{n} \langle \delta H^n, \delta E^n \rangle . \left(\begin{bmatrix} m_\mu & 0 \\ 0 & m_\varepsilon \end{bmatrix} . \begin{Bmatrix} \frac{\partial H^n}{\partial t} \\ \frac{\partial E^n}{\partial t} \end{Bmatrix} + \begin{bmatrix} 0 & k_i \\ -k_i & 0 \end{bmatrix} \begin{Bmatrix} H^n \\ E^n \end{Bmatrix} \right) = 0 \quad (11)$$

or $\sum_{N_e} \langle \delta ue \rangle . \left([me_H + me_{Dirac}] \left\{ \frac{due}{dt} \right\} + [ke]\{ue\} - \{fe\} \right) = 0$

After an assembling process, we obtain :

$$[M_H + M_{Dirac}] \frac{d}{dt} \{U\} = \{F\} - [K]\{U\} \quad \text{with} \quad \{U\} = \{H, E\} \quad (12)$$

3 Numerical Resolution

We obtain a large size time-differential system. By inversion of the mass matrix [M], the general formulation of this differential system is :

$$\frac{d}{dt}\{U\}=[M]^{-1}.\{\Psi(U,t)\} \quad \text{where} \quad \{\Psi(U,t)\}=\{F\}-[K].\{U\} \tag{13}$$

For the numerical quadrature of formula (11) we choose the nodes of the element as integration points [3]. Consequently, the mass matrix [M] is diagonal and this inversion is an easy procedure. It is a necessary condition for the efficiency of the methods described below. For numerical time-resolution, we modify the Backward Difference Method [5]. With the propose formulation one can choose the time order of discretization and the upward scheme. Under these conditions we can test a semi-implicit method. With the matrix-free technique, the mass matrix and the stiffness matrix are never built: only the elementary matrices in (11) are calculated. So we note a high performance level both for the CPU and the storage costs.

3.1 Semi-implicit Method

The corresponding k order algorithm is as follows :

$t_n = 0$
while $(t_n \leq t_{max})$

$$\left\{ \begin{array}{l} \{\Delta U_n^i\} = \Delta t_n . \sum_{j=0}^{k-1} \lambda_j . [M_{n-j}^i]^{-1} . \{\Psi_{n-j}(U_{n-j} + \alpha_j . \Delta U_{n-j}^{i-1}, t_n + \alpha_j . \Delta t_n)\} \\ i = 1, 2, \ldots \quad \text{until} \quad \left\| \Delta U_n^i - \Delta U_n^{i-1} \right\| \leq tolerence \end{array} \right\} \tag{14}$$

$\{U_{n+1}\} = \{U_n\} + \{\Delta U_n\}$
$t_{n+1} = t_n + \Delta t_n$
end while

where α is the upward time-parameter.

This method requires inner iterations for each time-step with index I, for ΔU determination, and the convergence criteria can be written for example at the first order [5]:

$$\alpha . \Delta t . \left\| \frac{\partial}{\partial U}\{[M]^{-1}.\Psi(U,t)\} \right\| < 1 \quad \text{let} \quad \Delta t < \Delta t_0 \tag{15}$$

But we always find a time-step value for the convergence of the process. We note the good stability of the scheme in the implicit case. If $\alpha < 0.5$ a CFL condition is also required [6]. In this way we choose a time step as follows [7]:

$$\Delta t \leq \text{Min}(\Delta t_0, \Delta t_{CFL})$$

The advantage of this method is the matrix free technique associated with an iterative method. The initial solution of each time step is close to the next solution. So the number of inner iterations is very low, two or three in practice. Initially the electromagnetic field is zero in full domain. For time t_k, the time step Δt_k can be changed at each step with a CFL condition.

3.2 The Software

For this method to work with efficacy, we must adapt each formulation in accordance with the boundary conditions. For easy adaptation of the code to the different formulations described above we must have a dedicated software. We use efficient C++ Objects-Oriented Programming for the Finite Element code called FAFEMO (Fast Adaptive Finite Element Modular Object). All details are given in reference [3].

Figure 1 shows the general structure of the compact code. It is organized in three classes corresponding to the functional blocks of the FEM's different stages. With these classes we built three objects that are connected by a single heritage. So the transmission of the parameters between the objects is defined by a list technique. We adapt the "class elementary matrices" for Dirac distribution.

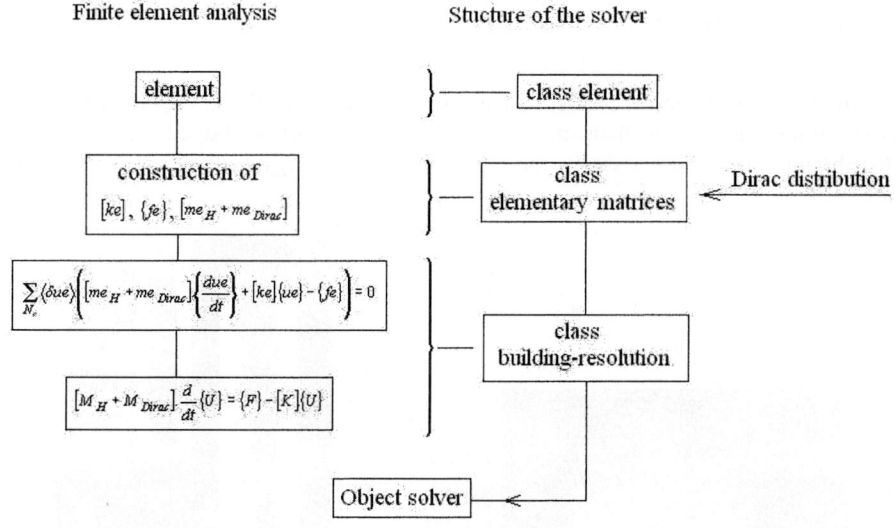

Fig. 1. Structure of the software

This constitutes a general Finite Element solver for the Maxwell equations.

4 Numerical Results

We study the transverse magnetic mode -T.M.- in the case of a 2D model. Figure 2 shows the meshing of the domain. We use here a high density of elements because the wave front propagates in the whole domain, but does not return significantly in the wave guide. In this context we develop stationary waves and we notice at each period that a part of the electromagnetic energy is scattered in the cavity. So the amplitude of the signal decreases (Fig. 3). After the electromagnetic field computation, we can deduce the value of the signal by Ampere's theorem. For TDR technology only intensity at the entrance of the wave guide is measurable. So we simulate the useable signal.

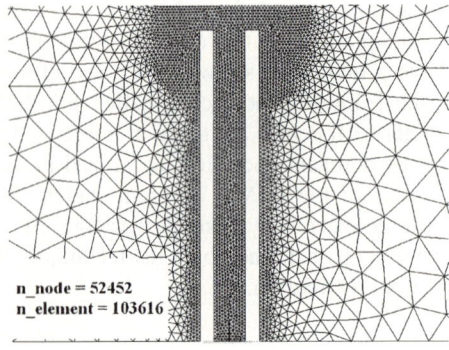

Fig. 2. The meshing

Fig. 3. The useable signal

Another test studies the effect of a single Dirac distribution. Each distribution concentrates the electromagnetic energy. Figure 4 gives the details around a distribution and Figure 5 shows the electromagnetic energy in the wave guide with several distributions.

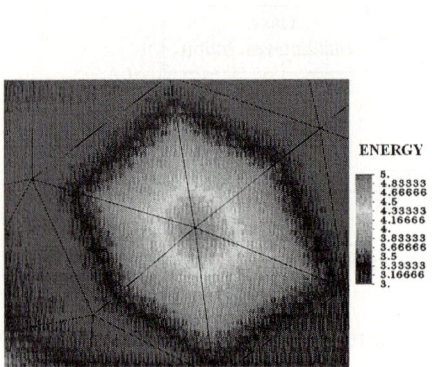
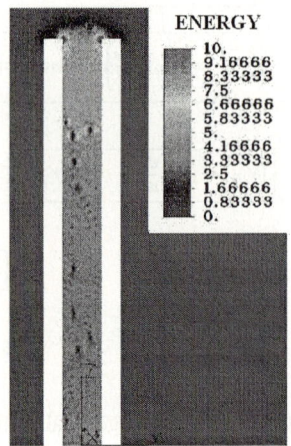

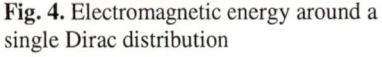

Fig. 4. Electromagnetic energy around a single Dirac distribution

Fig. 5. Electromagnetic energy in the wave guide

The curve in Figure 6a shows the differential signal obtained by a single Dirac distribution visible in the wave guide (Fig.6b). The progressive wave reflects on the Dirac distribution and a signal returns towards the entrance. In practice our numerical model is constituted by many distributions with random localizations (Fig. 7a) (Fig. 7b). The superposition of the multiple signals associated to each distribution is associated to a mean electric property of the complex medium. So it is possible to choose a random profile of Dirac distributions according to a complex medium model as a soil.

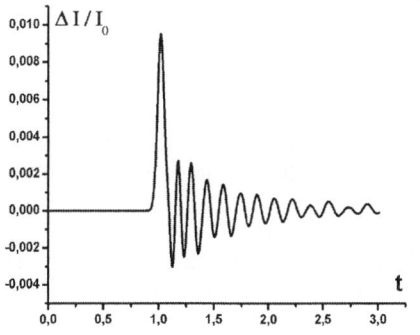

Fig. 6a. Signal at the entrance of the wave guide obtained with a single Dirac distribution

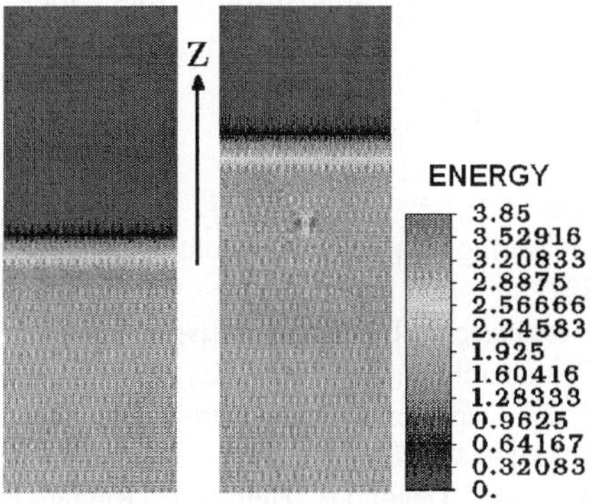

Fig. 6b. Electromagnetic energy in the guide around a single Dirac distribution

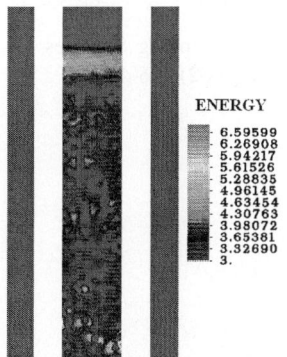

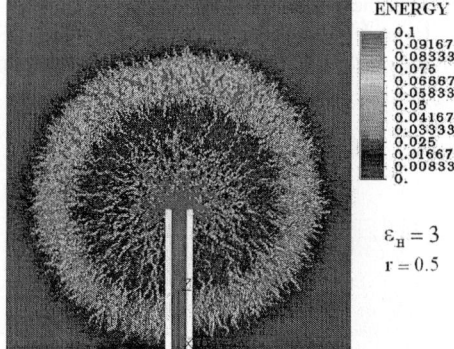

Fig. 7a. Electromagnetic energy in the wave guide in the case of some Dirac distributions with random localizations

Fig. 7b. Electromagnetic energy out of the wave guide

5 Conclusion and Further Works

We have presented a numerical study of the propagation of electromagnetic waves in a complex medium. First we have developed a set of numerical techniques applied to Maxwell's equations. A special algorithm has been developed for wave front study. Secondly we have tested a new model of complex medium by introduction of Dirac distributions in the physical properties. Also we hope to modelize the electric properties of a soil. This numerical simulation is performed by Object-Oriented Programming. In this context we obtain efficient and low cost calculus. This set of techniques constitutes a very interesting tool for soil science.

Acknowledgement. The authors would like to thank Ralph Beisson for his assistance with the English composition of this paper.

References

1. Spaans E.J.A. and Baker J.: Examining the Use of Time Domain Reflectometry for Measuring Liquid Water-Content in Frozen Soil. Water Resour. Res., Vol. 31 (1995), 2917.
2. Fleckinger E.: Electromagnétisme, Editions Masson, Paris (1991).
3. Chambarel A. and Ferry E.: Finite Element Formulation for Maxwell's Equations with Space Dependent Electric Properties. Revue Européenne des Eléments Finis, Vol. 9 (8) (2000) 941-967.
4. Assous F., Degond P., Heintze E., Raviard P.A.. and Segre J.: On a Finite Element Method for Solving the Three-Dimensional Maxwell Equations. J. Comp. Phys., Vol.109 (1993) 222-237.
5. Anderson D.A., Tannehill J.C. and Pletcher R.H.: Computational Fluid Mechanics and Heat Transfer. Hemisphere Publishing Corporation Editor (1984).
6. Dhatt G. and Touzot G.: Une Présentation de la Méthode des Eléments Finis, Editions Maloine S.A ,Paris (1981).
7. Rylander T. and Bondeson A.: Stability of Explicit-Implicit Time-stepping Schemes for Maxwell's Equations. Journal of Computational Physics, Vol. 179 (2002) 426-438.

Simulation of Transient Mechanical Wave Propagation in Heterogeneous Soils

Arnaud Mesgouez, Gaëlle Lefeuve-Mesgouez, and André Chambarel

UMR A Climate, Soil and Environment,
Université d'Avignon et des Pays de Vaucluse,
INRA, Domaine St-Paul, F-84914 Avignon Cedex 9, France
arnaud.mesgouez@univ-avignon.fr

Abstract. A Finite Element Method using a C++ code is developed to study the mechanical wave propagation in saturated porous soils. The modelization uses the complete Biot theory including the couplings between the solid and fluid phases. A matrix-free algorithm and a selection data technique are implemented in the code. Researches are focused on homogeneous and heterogeneous semi-infinite media in the case of transient regimes. The time domain results present the displacements over and within the half-space. In particular, we will see that the fluid wave front is strongly dependent on the proportion of heterogeneities in the ground.

1 Introduction

The study of the mechanical wave propagation in porous media is a subject of great interest in diverse scientific fields ranging from environmental engineering or vibration isolation to geomechanics. A saturated porous medium is a medium that presents on the microscopic spatial scale a solid deformable skeleton and a porous space filled with a viscous fluid. A macroscopic formulation can be deduced from the microscopic approach by homogenization: in this case, the medium is considered as a two-phase continuum. Biot's articles (see [1] and [2]) and Bourbié et al.'s review [3] are works of reference for the macroscopic mechanical wave propagation theory. In such a medium, three body waves exist: the P1 and P2 compressional waves and the S shear wave. The first compressional wave is a wave said to be quick whereas the second compressional wave is said to be slow and strongly attenuated. Moreover, for a semi-infinite medium, a surface wave also exists denoted as the Rayleigh R wave. Theoretical works are restricted to simple geometries. Consequently, they have to be completed by numerical approaches such as Finite Element or Boundary Element Methods, allowing the study of more complex problems to better modelize the ground. The difficult study of transient regimes has been treated numerically for specific cases by several authors: Zienkiewicz and Shiomi [4], Simon et al. [5] and Gajo et al. [6] for instance. In this paper, the authors propose a Finite Element Method of the whole Biot's equations in the case of homogeneous and heterogeneous soils. An efficient and accurate C++ code is developed to deal with this problematic.

2 Mechanical Model

2.1 Governing Equations

The theoretical approach is formulated using the Biot model: writing u_i and U_i respectively the solid and fluid displacement components, with usual notations, Biot's equations can be written as follows

$$\sigma_{ij,j} = (1-\phi)\rho_s \ddot{u}_i + \phi \rho_f \ddot{U}_i \qquad (1)$$

$$p_{,i} = -\frac{\phi}{K}(\dot{U}_i - \dot{u}_i) + \rho_f(a-1)\ddot{u}_i - a\rho_f \ddot{U}_i \qquad (2)$$

$$\sigma_{ij} = \lambda_{0v}\varepsilon_{kk}\delta_{ij} + 2\mu_v \varepsilon_{ij} - \beta p \delta_{ij} \qquad (3)$$

$$-\phi(U_{k,k} - u_{k,k}) = \beta u_{k,k} + \frac{1}{M}p \qquad (4)$$

The soil's characteristics are: λ_{0v} and μ_v (drained Lamé constants for the purely viscoelastic equivalent porous media), ρ_s and ρ_f (solid grains and fluid densities), ϕ (porosity), K (hydraulic permeability), a (tortuosity), M and β (Biot coefficients).

(1) and (2) are the motion equations, (3) and (4) are the constitutive relationships including a physical viscoelastic hysteretic Rayleigh damping of the soil.

2.2 Finite Element Formulation and Numerical Resolution

In order to obtain a first order time differential system, the solid and fluid particles' velocities are introduced: $v_i = \dot{u}_i$ and $V_i = \dot{U}_i$. Then, some algebraic manipulations are done to yield in the end a diagonal mass matrix. The weighted residual method with Galerkin ponderation gives integral forms (5) and (6) defined in the vector space $V(f) = [0,T] \times \{\frac{\partial f}{\partial t} \in L^2(\Omega), f \in H_1(\Omega)\}$ where T and (Ω) are respectively the study time and the study space.

$$\int_\Omega a\sigma_{ij,j}\delta v_i d\Omega + \int_\Omega \phi p_{,i}\delta v_i d\Omega = \int_\Omega [a(1-\phi)\rho_s + \phi(a-1)\rho_f]\dot{v}_i \delta v_i d\Omega$$
$$+ \int_\Omega \frac{-\phi^2}{K}V_i \delta v_i d\Omega + \int_\Omega \frac{\phi^2}{K}v_i \delta v_i d\Omega \qquad (5)$$

$$\int_\Omega \rho_f(a-1)\sigma_{ij,j}\delta V_i d\Omega - \int_\Omega (1-\phi)\rho_s p_{,i}\delta V_i d\Omega = \int_\Omega \frac{(1-\phi)\rho_s \phi}{K}V_i \delta V_i d\Omega$$
$$+ \int_\Omega \frac{-(1-\phi)\rho_s \phi}{K}v_i \delta V_i d\Omega + \int_\Omega [\rho_f^2(a-1)\phi + \rho_s \rho_f a(1-\phi)]\dot{V}_i \delta V_i d\Omega \qquad (6)$$

The weak formulation is then transformed using the Green theorem that introduces the boundary conditions. Then, for respectively two-dimensional and three-dimensional problems, triangular and tetrahedral linear isoparametric elements are used to mesh the (Ω) space. Afterwards, we can get an approximative solution of the second formulation with analytical discretization of the displacement and velocity components.

For an elementary space (subscript $()_e$), the discrete system can be synthesized in the following form

$$\sum_e \langle \delta W_e \rangle \left\{ \begin{bmatrix} [m_e] & 0 & 0 & 0 \\ 0 & [m_e] & 0 & 0 \\ 0 & 0 & [m_{se}] & 0 \\ 0 & 0 & 0 & [m_{fe}] \end{bmatrix} \{\dot{W}_e\} \right. \\ \left. + \begin{bmatrix} 0 & 0 & -[m_e] & 0 \\ 0 & 0 & 0 & -[m_e] \\ [k_{se}] & [k_{sfe}] & [c_{se}] & -[c_{se}] \\ [k_{fse}] & [k_{fe}] & [c_{fe}] & -[c_{fe}] \end{bmatrix} \{W_e\} \right\} = \sum_e \langle \delta W_e \rangle \{F_e\} \quad (7)$$

where $\{W_e\} = \langle \{u_i\}^n \ \{U_i\}^n \ \{v_i\}^n \ \{V_i\}^n \rangle^t$ is the elementary vector of the variables and $\{F_e\}$ the elementary vector of the load.

The assembling procedure leads to the following global differential system (superscript $()^{(G)}$) with a diagonal global mass matrix

$$[M]\frac{d}{dt}\{W^{(G)}\} + [K]\{W^{(G)}\} = \{F^{(G)}\} \quad (8)$$

The time integration algorithm is a backward difference method modified with an upward time parameter α, and is presented as follows for a k-order, Ψ being the global residuum

$$\begin{cases} \text{While}(t_n \leq t_{max}) \\ \quad \{\Delta W_n^i\} = \Delta t_n \sum_{j=0}^{k-1} \lambda_j \left[M_{n-j}^i\right]^{-1} \times \\ \qquad \{\Psi_{n-j}(\{W_{n-j}\} + \alpha_j \{\Delta W_{n-j}^{i-1}\}, t_n + \alpha_j \Delta t_n)\} \\ \quad i = 1, 2... \text{until } \| \{\Delta W_n^i\} - \{\Delta W_n^{i-1}\} \| \leq \text{tolerance} \\ \{W_{n+1}\} = \{W_n\} + \{\Delta W_n\} \\ t_{n+1} = t_n + \Delta t_n \\ \text{end while} \end{cases} \quad (9)$$

It requires inner iterations for each time step until the tolerance criterium is reached.

2.3 Structure of the Code

The Finite Element C++ code is organized in three classes: *element*, *elementary matrices* and *building-resolution* classes. Thus, three objects connected by a single heritage are constructed and they form a solver. In practise, we obtain very low sized solvers (less than 1000 C++ lines).

In this code, the mass and stiffness matrices are never built because a matrix-free technique is used. As the global mass matrix is diagonal, its inversion is an easy process. The size of the global vector of unknowns is optimized by the use of an expert multigrid system called AMS (see [7] and [8]). On the whole, a high performance level is obtained both in terms of CPU and storage costs.

3 Results

3.1 Half-Space Model

A vertical impulse (y axis) load acts over the free surface (x axis) of a two-dimensional half-space. The soil characteristics are relative to a porous stiff ground: the dimensionless values are given in Table 1. In all the following section, the numerical values are dimensionless, using three independent mechanical quantities. Dirichlet's conditions corresponding with the zero displacements for both phases are imposed at depth at the border of the medium. The (Ω) space is meshed with 50 626 triangular elements and 25 617 eight-degrees-of-freedom nodes.

Table 1. Dimensionless soil parameters and associated celerity values

λ_0	μ	K_s	K_f	M	β	ρ_s	ρ_f	a	K	ϕ	η	c_{P1}	c_{P2}	c_S	c_R
0.56	0.83	4	0.22	0.53	0.72	1	0.39	1.5	0.26	0.4	0.01	1.9	0.65	1.1	1.05

3.2 Homogeneous Case

Figure 1(a) presents the horizontal solid displacements on the surface of the half-space ($-3 < x < 3, y = 0$) versus time variable ($0 < t < 2$). The P1 (respectively R) contribution is focused on a line the slope of which corresponds to the P1 (resp. R) celerity. The S wave is lost to the eye due to the dominating Rayleigh wave: its contribution could be seen when in vertical displacements. The P2 wave carries very little energy and is mixed with the whole surface displacement which here prevents us from seeing it. The P1 and R waves give opposite contributions with a preponderant part to Rayleigh's. Moreover, in the case of a vertical sollicitation as studied here, the horizontal displacement is an odd function of x variable, whereas the vertical displacement is even. Also note, from the same figure, that the displacement decreases geometrically in addition to the physical damping. In Figure 1(b), which shows a section of Figure 1(a) for $x = 1.5$, a visualization of the vertical displacement is added. The theoretical arrival times are underlined and the contribution of the S wave is clearly perceptible on the vertical displacement.

Figure 2(a) shows the variation of the fluid vertical displacements in depth under the load ($x = 0$) versus time variable. The Rayleigh wave decreases exponentially with depth. Consequently, the displacements underline the two compressional waves P1 and P2. On Figure 2(b), the two theoretical lines relative to each wave speed are plotted: a good agreement is obtained with the contour levels.

Figures 3 and 4 are plotted for $t = 1.5$. The P1 wave front corresponds to a half-circle the radius of which equals 2.8: this is clearly perceptible on Figure 3. From the same figure, note that the Rayleigh wave gives a predominant contribution on the surface: this corresponds to the red areas beginning at $x = 1.6$. Moreover, with the chosen parameters, the fluid and solid phases are strongly

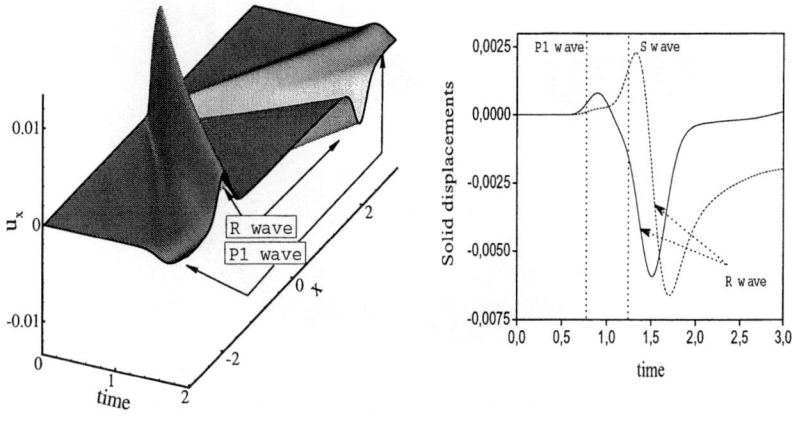

(a) Surface horizontal displacements

(b) Observational point $x = 1.5$, horizontal (solid line) and vertical (dotted line) displacements

Fig. 1. Surface solid displacements versus time

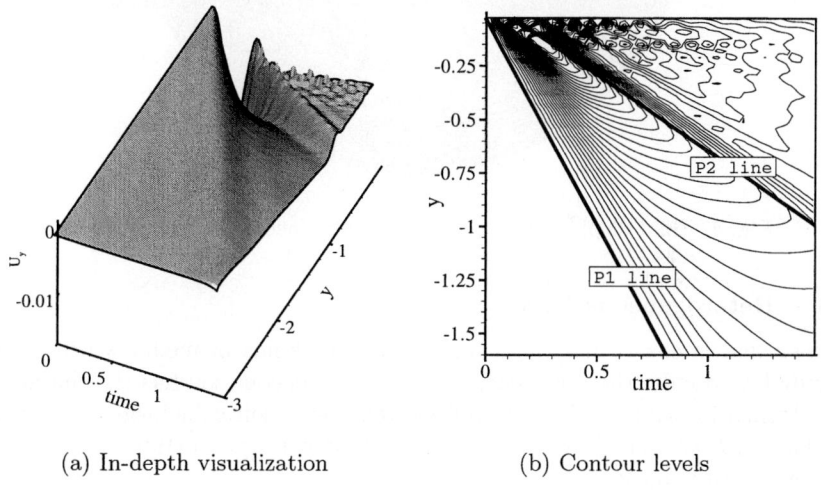

(a) In-depth visualization

(b) Contour levels

Fig. 2. In-depth vertical fluid displacements versus time

uncoupled, thus the S wave exists only in the solid phase ($x = 1.7$). Due to the competition between the R and S waves which have neighboring speeds, the S wave front is more perceptible in depth.

For the fluid phase, Figure 4, red zones are caused by the P1 wave which presents a more widely spread time wave front. This analysis has been confirmed by a study of the displacements of time-related individual points positioned under the load.

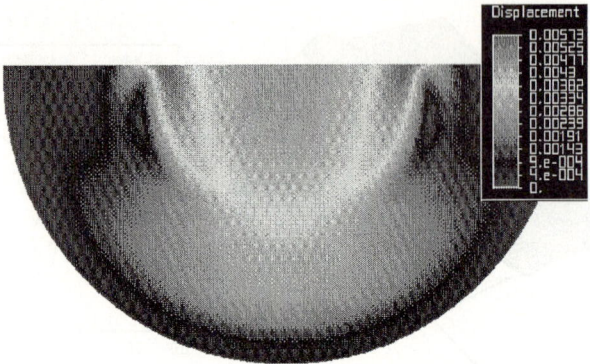

Fig. 3. Solid displacement field for a homogeneous medium at $t = 1.5$

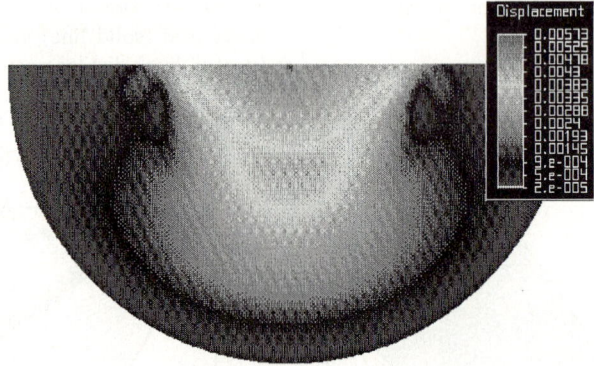

Fig. 4. Fluid displacement field for a homogeneous medium at $t = 1.5$

3.3 Heterogeneous Case

The ground is constituted by a mixture of two kinds of media which are distributed randomly: the main part of this heterogeneous soil has the characteristics defined in Table 1. The second one represents softer inclusions (λ_0=0.011, μ=0.011, M= 0.271, β=0.995, ϕ=0.6, η=1) and its proportion ranges from 1 percent to 40 percent.

In these cases, the solid wave fronts are on the whole less modified than the fluid ones, (see Figures 5 and 6). They remain well-ordered up to at least a 20 percent distribution. For a higher proportion, the solid wave fronts are dismembered and the contour levels are not concentric any more. Moreover, the inclusions focus the mechanical deformations over a smaller area in which the displacement values are higher. The presence of soft elements slows down the progression of the solid and fluid deformations in comparison with the purely homogeneous case.

As for the fluid phase, even for a low proportion of inclusions (1 or 5 percent), the wave fronts are disordered. Some areas present higher deformations due to

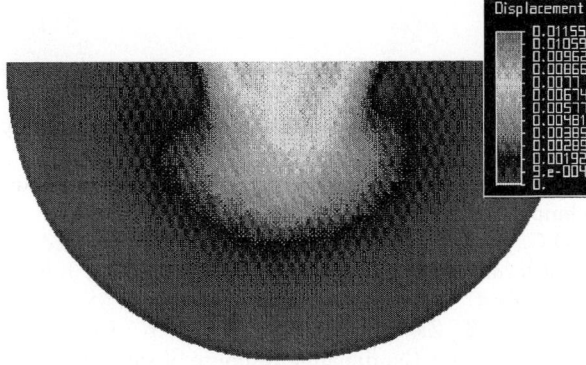

Fig. 5. Solid displacement field for a heterogeneous medium with 40% of inclusions at $t = 1.5$

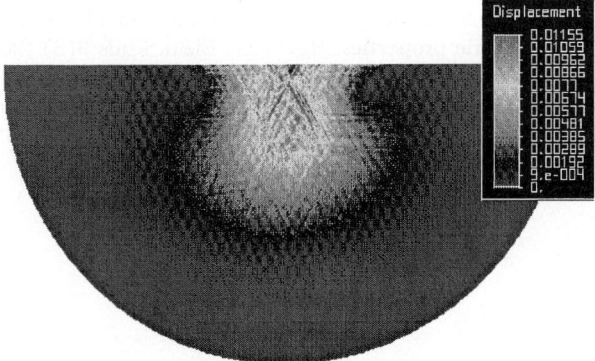

Fig. 6. Fluid displacement field for a heterogeneous medium with 40% of inclusions at $t = 1.5$

the higher proportion of fluid in the softer inclusions. Moreover, fluid amplitudes, which are lower than solid ones in the homogeneous case, become equal if not higher with the inclusions. In conclusion, a deeper change occurs for the fluid phase.

4 Conclusion and Further Works

The study of the mechanical wave propagation has been carried out using a finite element method and a modified backward difference method for the time integration algorithm. A matrix-free algorithm and a selection data technique are implemented in the C++ objet-oriented program. Different examples of homogeneous and heterogeneous semi-infinite media have been presented. Further works will include the parallelisation of the code and the study of three-dimensionnal geometries.

References

1. Biot, M.A.: Generalized theory of acoustic propagation in porous dissipative media. J. Acoust. Soc. Am. **34(9)** (1962) 1254–1264
2. Biot, M.A.: Mechanics of deformation and acoustic propagation in porous media. J. Appl. Phys. **33(4)** (1962) 1482–1498
3. Bourbié, T., Coussy, O., Zinszner, R.: Acoustics of Porous Media. (1987) Paris: Editions Technip
4. Zienkiewicz, O.C., Shiomi, T.: Dynamic behaviour of saturated porous media: the generalized Biot formulation and its numerical solution. Int. J. Numer. Anal. Methods Geomech. **8** (1984) 71–96
5. Simon, B.R., Wu, J.S.S., Zienkiewicz, O.C., Paul, D.K.: Evaluation of u-w and u-π finite element methods for the dynamic response of saturated porous media using one-dimensional models. Int. J. Numer. Anal. Methods Geomech. **10** (1986) 461–482
6. Gajo, A., Saetta, A., Vitaliani, R.: Evaluation of three and two field finite element methods for the dynamic response of saturated soil. Int. J. Numer. Anal. Methods Geomech. **37** (1994) 1231–1247
7. Chambarel, A., Ferry, E.: Finite Element formulation for Maxwell's equations with space dependent electric properties. Rev. Eur. Elém. Finis **9(8)** (2000) 941–967
8. Chambarel, A., Bolvin, H.: Simulation of a compressible flow by the finite element method using a general parallel computing approach. Lect. Notes Comput. Sc., Springer-Verlag, **Vol. 2329** (2002) 920–929

Practical Modelling for Generating Self-similar VBR Video Traffic

Jong-Suk R. Lee[1] and Hae-Duck J. Jeong[2]

[1] Grid Technology Research Department, Supercomputing Centre,
Korea Institute of Science and Technology Information,
Daejeon, South Korea
jsruthlee@kisti.re.kr
[2] Department of Information Science, Korean Bible University,
Seoul, South Korea
joshua@bible.ac.kr

Abstract. Teletraffic in the Internet is rapidly growing and diversifying, and there is a strong need for QoS support in high-speed communication networks. There are a number of research issues concerning the transmission of JPEG/ MPEG video over modern high speed computer networks. These problems have been studied intensively over the last ten years in order to provide a consistent and desirable QoS for JPEG/MPEG video traffic, construct accurate models for JPEG/MPEG video traffic and utilise efficient resource allocation techniques. In the paper we show that synthetically generated streams of VBR video, compressed under such standards as JPEG, MPEG-1 and MPEG-2, can be statistically equivalent to real video traces. We also investigate how compression algorithms on correlation structure of compressed teletraffic influence real video traffic.

1 Introduction

Teletraffic in the Internet is rapidly growing and diversifying, and there is a strong need for QoS (Quality of Service) support in high-speed communication networks [1], [6]. The introduction of many new multimedia services requires a high bandwidth to transport data such as real-time digital video. Modern computer networks can no longer cope with uncompressed multimedia traffic, resulting in the development of several image and video compression standards such as JPEG and MPEG. In this paper we focus on VBR (Variable Bit Rate) JPEG/MPEG video, i.e., on video streams compressed according to JPEG and MPEG standards and transmitted as VBR components of an ATM network.

There are a number of research issues concerning the transmission of JPEG/ MPEG video over modern high speed computer networks, such as the dimensioning of multiplexer buffers and monitoring of video cell streams. These problems have been studied intensively over the last ten years in order to provide a consistent and desirable QoS for JPEG/MPEG video traffic, construct accurate models for JPEG/MPEG video traffic and utilise efficient resource allocation techniques. We look at the influence of compression algorithms on correlation structure of compressed teletraffic, see Section 4, where results of compression of *Star Wars* video under JPEG and MPEG-1 are discussed.

Applicability of gamma/Pareto model as marginal distributions of compressed video streams is discussed in Section 4, where we look at accuracy of this approximation in relation to data coming from different videos (*Star Wars* and *Titanic*) compressed under three different algorithms (JPEG, MPEG-1 and MPEG-2). We also show that synthetically generated streams of VBR video, compressed under such standards as JPEG, MPEG-1 and MPEG-2, are statistically similar to real video traces [7].

A number of researchers tried to fit a specific mathematical model to traces of real VBR video traffic. For example, several models (based on gamma [3], lognormal [10], and combined gamma/Pareto [2], [9]) have been suggested for VBR video traffic. Heyman et al. [3] used a 30-minute compressed video-teleconferencing sequence for simulation studies using the gamma model. Krunz et al. [10] used a 23-minute movie, *The Wizard of Oz*, to study statistical characteristics of VBR MPEG-coded video streams using the lognormal model. The gamma model for video traffic became inaccurate in the tail of distribution, and the lognormal model was too heavy-tailed at first and then fell off too rapidly. Garrett and Willinger [2] used a two-hour VBR video, *Star Wars*, and proposed a hybrid gamma/Pareto model based on the F-ARIMA process [4]. They found that the tail behaviour of the marginal distribution can be accurately described using the heavy-tailed Pareto distributions.

Huang et al. [5] presented a unified approach to modelling VBR video traffic using both SRD (Short-Range Dependent) and LRD (Long-Range Dependent) empirical ACFs (Auto-correlation Functions). They applied this approach to 2 hours' trace of *Last Action Hero* video. Their approach is potentially accurate, but establishing an automatic search for the best background ACF remains an open problem. Lombardo et al. [12] proposed the generation of pseudo-MPEG video traffic with a specific correlation structure based on FFT [14] and an ICDF transformation, assuming an arbitrary marginal distribution of the output process. The proposed algorithm has been used to generate a sequence with the same statistical characteristics as those of the movie "The Simpsons", however, the robustness of this algorithm remains an issue open to further investigation.

2   JPEG/MPEG Video Compression

Several algorithms have been developed to compress video data, in order to reduce the memory required for their storage, the time or bandwidth necessary for their transmission, and the effective data access or transfer rate. We focus on MPEG-1 and MPEG-2 of the MPEG standard family. MPEG-2 uses encoders from the MPEG-1 scheme, and in the case of multi-layer encoding, the statistical properties of its base layer are almost identical to MPEG-1. A video sequence is simply a series of pictures taken at closely spaced time intervals starting with a sequence header. The sequence header is followed by one or more group(s) of

Table 1. Parameters for generating the *Titanic* video sequence

Parameters	Values
Coding algorithm	DCT
Duration	3 hours
Video frames	285,890 (I-, P- and B-frames only)
Frame dimensions	720 x 576 pixels
Pixel resolution	24 bits/pixel (colour)
Frame rate	29.97/second
Average bandwidth	692,150 bytes/second
Average compression rate	53.87
A group of pictures	15 frames (IBBPBBPBBPBBPBB)

pictures (GOP) and ends with a sequence end code. Additional sequence headers may appear between any GOP within the video sequence. This is achieved by using three types of frames: Intra-coded picture (I-frame), Predictive-coded picture (P-frame) and Bi-directionally predictive-coded picture (B-frame) [13].

Only I- and P-frames can be used as a reference for past and/or future prediction. An ordered collection of I-, P- and B-frames is called a group of pictures. The proportion of I-, P- and B-frames is application-dependent and is left to the user. For example, for many scenes, spacing the reference frames at about one-twelfth of a second interval seems appropriate, i.e., IBBPBBPBBPBB.... The MPEG GOP pattern was used to encode the MPEG-1 version of *Star Wars* by Garrett and Willinger [2].

We encoded three hours of *Titanic* video to obtain a realistic full-length trace of video traffic. This will be used as a control reference self-similar trace in our investigations. Parameters of the sequence are summarised in Table 1. We chose MPEG-2 to obtain encoded frame sequences of our trace. In this paper we will use the following three self-similar sequences: (i) two hours of *Star Wars* video encoded by JPEG [2], (ii) two hours of *Star Wars* video encoded by MPEG-1 [2], and (iii) three hours of *Titanic* video encoded by MPEG-2 [7]. The last trace was obtained by taking a sample that was approximately 60% longer than Sequence (i) and (ii).

3 Modelling for Self-similar VBR Video Traffic

Following the recommendation of Garrett and Willinger [2], we chose to use the combined gamma/Pareto model for VBR video traffic. They along with Krunz and Makowski [9], showed that the gamma distribution can be used to capture the main part of the empirical distribution, but is inappropriate for the tail. Addition of a heavy-tailed Pareto distribution corrects this, as shown in Figure 1.

Let F_Γ and F_P be the CDF (Cumulative Distribution Function) for the gamma and Pareto distributions, respectively. Note that F_Γ has no closed form

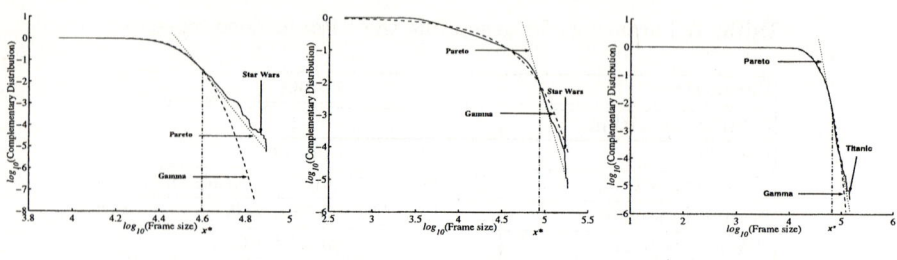

Fig. 1. Complementary cumulative distributions of real video traffic and gamma/Pareto model

of the CDF when α_Γ, the shape parameter of the gamma distribution assumes non-integer values.

If α_Γ is a positive integer, then the CDF for the gamma distribution is given by

$$F_\Gamma(x) = \begin{cases} 0, & \text{for } x \leq 0, \\ 1 - e^{-x/\beta_\Gamma} \sum_{j=0}^{\alpha_\Gamma - 1} \frac{(x/\beta_\Gamma)^j}{j!}, & \text{for } x > 0, \end{cases} \quad (1)$$

where α_Γ is the shape parameter, $\alpha_\Gamma > 0$, and β_Γ is the scale parameter, $\beta_\Gamma > 0$.

The CDF $F_P(x)$ of the Pareto distribution is given as:

$$F_P(x) = \begin{cases} 0, & \text{for } x < 1, \\ 1 - \left(\frac{b_P}{x}\right)^{\alpha_P}, & \text{for } 1 \leq x \leq \infty, \end{cases} \quad (2)$$

where α_P is the shape parameter, $\alpha_P > 0$, and b_P is the minimum allowed value of x, $0 < b_P \leq x$.

Thus, the combined gamma/Pareto distribution is determined by

$$F_{\Gamma/P}(x) = \begin{cases} 0, & \text{for } x \leq 0, \\ F_\Gamma(x), & \text{for } 0 < x \leq x^*, \\ F_P(x), & \text{for } x > x^*. \end{cases} \quad (3)$$

The complementary CDFs of $F_\Gamma(x)$ and $F_P(x)$ can be used to determine x^* in Equation (3). The parameters of the gamma distribution are obtained by matching the first and second moments of the empirical sequence to those of a gamma random variate. x^* can be obtained graphically by inspecting the tail behaviour of the empirical distribution, and determining where it starts to deviate from the tail of the gamma curve. The values of b_P and α_P for the estimated Pareto distribution can be obtained by finding $x = x^*$ for which the least-square fit of the Pareto tail gives $F_\Gamma(x) = F_P(x)$. Figure 1 shows log-log plots of gamma and Pareto complementary CDF for real VBR video traffic. While the gamma curve fits the main part of the empirical video traffic well, the Pareto curve closely fits its tail part. Applying this method, we have determined values of x^* for all three samples; see Table 2.

Table 2. Estimated parameter values obtained from *Star Wars* and *Titanic* video traffic utilising the combined gamma/Pareto model

Parameters	Estimated values		
	Star Wars JPEG	*Star Wars* MPEG-1	*Titanic* MPEG-2
Length (frames)	171,000	174,136	285,890
Duration	2 hours	2 hours	3 hours
Compression algorithm	Intra-frame	MPEG-1	MPEG-2
Sample mean	27,791	15,598	26,353
Standard dev.	6,254	18,165	11,600
Maximum	78,459	185,267	146,608
Minimum	8,622	476	12
Gamma α_Γ	25.8	0.737	5.16
Gamma β_Γ	1,100	21,154	5,106
Pareto α_P	12.42	9.19	10.06
Pareto b_P	30,000	51,500	37,800
x^*	39,810	86,003	57,280

Given a self-similar sequence of the FGN-DW (Fractional Gaussian Noise-Daubechies Wavelets) process **X** [8], we can transform the marginal distribution by mapping each point as

$$Z_i = F_{\Gamma/P}^{-1}(F_N(X_i)), \quad i = 1, 2, \ldots, \qquad (4)$$

where $F_N(\cdot)$ is the CDF of the normal distribution and $F_{\Gamma/P}^{-1}(\cdot)$ is the inverse CDF of the combined gamma/Pareto model given by

$$F_{\Gamma/P}^{-1}(y) = \begin{cases} F_\Gamma^{-1}(y), & \text{for } y \leq 1 - (b_P/x^*)^{\alpha_P}, \\ F_P^{-1}(y) = b_P/(1-y)^{1/\alpha_P}, & \text{for } y > 1 - (b_P/x^*)^{\alpha_P}. \end{cases} \qquad (5)$$

Note that for computing $F_\Gamma^{-1}(y)$, we used the Newton-Raphson technique [11], [15]. The procedure for the MPEG video consists of I-, P- and B-frames of sequences generated from FGN-DW [8], which are then combined in I-, B- and P-frame order (e.g., IBBPBBPBBPBBPBBI ...) before transforming the corresponding time series into time series with the gamma/Pareto marginal distributions defined in Equation (4). The procedure for the JPEG synthetic sequence generated from FGN-DW is simple. We used the sequences obtained from the previous procedure for simulation studies of VBR video traffic, which we describe in the next section.

4 Numerical Results

4.1 Analysis of Hurst Parameter Estimates for VBR Video Traffic

The Hurst parameter estimates obtained from the most efficient estimators (i.e., the wavelet-based H estimator and Whittle's MLE) [7], have been used to anal-

Table 3. Estimates of the Hurst parameter obtained from the wavelet-based H estimator and Whittle's MLE for *Star Wars* JPEG, *Star Wars* MPEG-1 and *Titanic* MPEG-2 video traffic. We give 95% confidence intervals for the means of two estimators in parentheses

Estimators	*Star Wars* JPEG	*Star Wars* MPEG-1	*Titanic* MPEG-2
Wavelet-based	.8841(.879, .889)	.8634(.859, .868)	.9034(.895, .911)
Whittle's MLE	.8997(.887, .912)	.8680(.855, .880)	.8999(.886, .914)

yse *Star Wars* JPEG, *Star Wars* MPEG-1 and *Titanic* MPEG-2 video sequences. Table 3 shows the estimates of the Hurst parameter for the three video sequences. Comparing *Star Wars* after JPEG and MPEG-1, we can formulate hypothesis that JPEG produces stronger dependent output video sequences. Our results show 2% difference in H parameter when using the wavelet-based H estimator, and 11% difference when using Whittle's MLE, see Table 3. On the other hand, two different videos (*Star Wars* and *Titanic*) show that regardless of compression algorithm resulted processes have the same marginal distribution well approximated by our gamma/Pareto model. This can be regarded as generalisation of a finding by Garrett and Willinger [2] who showed that gamma/Pareto model is a good approximation of marginal distributions for *Star Wars* compressed under JPEG.

The wavelet-based H estimator of three hours of real *Titanic* video traffic calculates $\hat{H} = 0.9034$, shown in Table 3. Estimate of the Hurst parameter $\hat{H}$ obtained from Whittle's MLE is 0.8999. The Hurst parameter estimates for the *Star Wars* JPEG and *Star Wars* MPEG-1 video sequences are also given in Table 3.

4.2 Simulation Results of the VBR Video Traffic Model

Figure 2 shows quantile-quantile plots for the distribution of frame sizes in bytes of real VBR video traffic, (i.e., *Star Wars* JPEG, *Star Wars* MPEG-1, and *Titanic* MPEG-2), and the combined gamma/Pareto model based on FGN-DW. We observed that although the compression algorithms used for encoding the various videos were different, the combined model fits the real traffic statistic well. Note especially that the distribution of the gamma/Pareto model matches the real *Star Wars* JPEG video traffic well.

Figure 3 shows that the ACF of the combined gamma/Pareto model also fit the empirical video traffic statistic well. The ACF curve of the gamma/Pareto model at large lags (i.e., lags > 1,800) fit the real *Star Wars* JPEG video traffic well, but the model slightly underestimated at small lags. The ACF curves in Figures 3 (b) and (c) also oscillated more than the one in Figure 3 (a), due to the MPEG format. Furthermore, the autocorrelation structure in Figure 3 (b) oscillated more than that in Figure 3 (c) because they use different frame formats (i.e., while every 12th frame in Figure 3 (b) is an I-frame, every 15th frame in Figure 3 (c) is an I-frame). In addition, we found strong evidence of LRD, as all ACF curves obtained from the real video traffic and the gamma/Pareto model

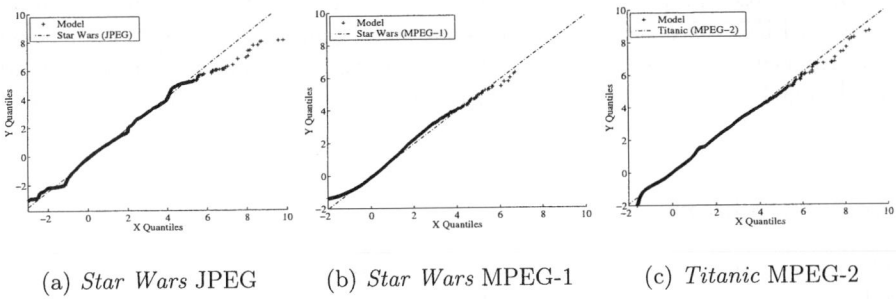

Fig. 2. Distributions of real video traffic and traffic from the gamma/Pareto models

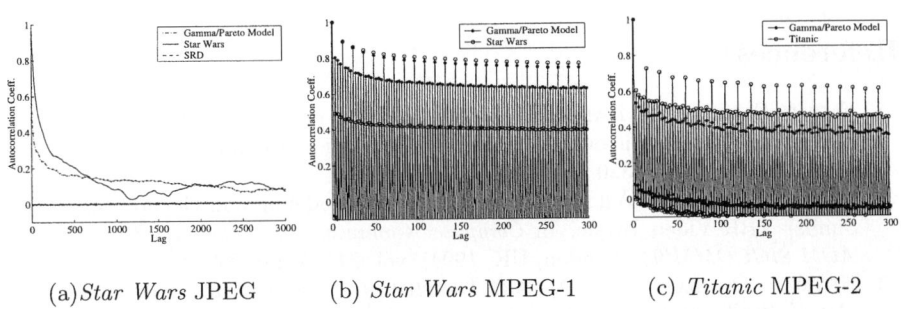

Fig. 3. ACF of real video traffic and traffic from the gamma/Pareto models

decayed slowly, while the SRD (i.e., Poisson model) in Figure 3 (a) decayed quickly.

5 Conclusions

We showed how pseudo-random self-similar sequences can be applied to produce a model of teletraffic associated with the transmission of VBR JPEG/MPEG video. A combined gamma/Pareto model based on the application of the FGN-DW generator was used to synthesise VBR JPEG/MPEG video traffic.

In the paper we showed that synthetically generated streams of VBR video, compressed under such standards as JPEG, MPEG-1 and MPEG-2, can be statistically equivalent to real video traces shown in Section 4 (see also Figures 2 – 3). We investigated how compression algorithms on correlation structure of compressed teletraffic influence real video traffic. Generalisation of findings of Garrett and Willinger, showing that video compression algorithms (MPEG-1 and MPEG-2) lead to self-similar processes was studied. We considered outcomes of MPEG-1 and MPEG-2 in addition to previously studied outcomes of JPEG [2], to show that the results of (Garrett and Willinger [2]) do not depend on the compression algorithms.

Better GOP and cell-layered modelling is needed for development of integrated MPEG video traffic models. While some general assessment of queueing performance can be obtained from single-streams, more universal results could be obtained from the queueing performance analysis of multiplexed streams of video traffic. These issues await further investigations.

Acknowledgements

The authors thank Dr. Don McNickle, Dr. Krzysztof Pawlikowski, Dr. Manfred Jobmann, Dr. Matthew Roughan and three anonymous referees for their valuable comments. The authors also wish to thank the financial support of Korean Bible University, and Korea Institute of Science and Technology Information, Korea.

References

1. FAHMY, S., JAIN, R., RABIE, S., GOYAL, R., AND VANDALORE, B. Quality of Service for Internet Traffic over ATM Service Categories. *Computer Communications 22*, 14 (1999), 1307–1320.
2. GARRETT, M., AND WILLINGER, W. Analysis, Modeling and Generation of Self-Similar VBR Video Traffic. In *Computer Communication Review, Proceedings of ACM SIGCOMM'94* (London, UK, 1994), vol. 24 (4), pp. 269–280.
3. HEYMAN, D., TABATABAI, A., AND LAKSHMAN, T. Statistical Analysis and Simulation Study of Video Teleconference Traffic in ATM. *IEEE Transactions on Circuits and Systems for Video Technology 2*, 1 (1992), 49–59.
4. HOSKING, J. M. Fractional Differencing. *Biometrika 68*, 1 (1981), 165–176.
5. HUANG, C., DEVETSIKIOTIS, M., LAMBADARIS, I., AND KAYE, A. Modeling and Simulation of Self-Similar Variable Bit Rate Compressed Video: A Unified Approach. *Computer Communication Review, Proceedings of ACM SIGCOMM'95 25*, 4 (1995), 114–125.
6. JAIN, R. Current Issues in Telecom Networks: QoS, Traffic Engineering and DWDM. Keynote speaker, http://www.cis.ohio-state.edu/~jain/talks/icon99.htm, 2000.
7. JEONG, H.-D. J. *Modelling of Self-Similar Teletraffic for Simulation*. PhD thesis, Department of Computer Science, University of Canterbury, 2002.
8. JEONG, H.-D. J., MCNICKLE, D., AND PAWLIKOWSKI, K. Fast Self-Similar Teletraffic Generation Based on FGN and Wavelets. In *Proceedings of the IEEE International Conference on Networks, ICON'99* (Brisbane, Australia, 1999), pp. 75–82.
9. KRUNZ, M., AND MAKOWSKI, A. A Source Model for VBR Video Traffic Based on $M/G/\infty$ Input Processes. In *Proceedings of IEEE INFOCOM'98* (San Francisco, CA, USA, 1998), pp. 1441–1448.
10. KRUNZ, M., SASS, R., AND HUGHES, H. Statistical Characteristics and Multiplexing of MPEG Streams. In *Proceedings of IEEE INFOCOM'95* (Boston, Massachusetts, 1995), pp. 455–462.
11. LAW, A., AND KELTON, W. *Simulation Modeling and Analysis*. 2nd ed., McGraw-Hill, Inc., Singapore, 1991.
12. LOMBARDO, A., MORABITO, G., PALAZZO, S., AND SCHEMBRA, G. MPEG Traffic Generation Matching Intra- and Inter-GoP Correlation. *Simulation 74*, 2 (2000), 97–109.

13. MITCHELL, J., PENNEBAKER, W., FOGG, C., AND LEGALL, D. *MPEG Video Compression Standard*. Chapman and Hall, New York, 1997.
14. PAXSON, V. Fast, Approximate Synthesis of Fractional Gaussian Noise for Generating Self-Similar Network Traffic. *Computer Communication Review, ACM SIGCOMM 27*, 5 (1997), 5–18.
15. PRESS, W., TEUKOLSKY, S., VETTERLING, W., AND FLANNERY, B. *Numerical Recipes in C*. Cambridge University Press, Cambridge, 1999.

A Pattern Search Method for Image Registration

Hong Zhou[1] and Benjamin Ray Seyfarth[2]

[1] Saint Joseph College,
West Hartford, CT 06117, USA
hzhou@sjc.edu
[2] The University of Southern Mississippi,
Hattiesburg, MS 39406, USA

Abstract. This paper presents a pattern search method for transformation function search in automatic image registration. In this search method, affine transformation parameters are represented as a 5-parameter vector and the search proceeds toward the direction resulting in higher similarity values between the reference and sensed images. Experiments show that this method can successfully find an optimal affine transformation function for edge images generated by computing the local standard deviation of the images. In addition, a series of overlapped aerial images of the Mississippi Delta area are successfully registered into a large reference image automatically by employing this search method.

1 Introduction

Image registration is a process of aligning two different images of the same object such that corresponding points in the two images represent the same physical location. The image to be registered is usually called the sensed image, while the image to which the sensed image is registered is commonly called the reference image. In general, image registration involves three steps including feature detection, feature matching and transformation function construction, and image transformation. Feature detection is a step to identify salient and distinctive objects. In the second step, feature matching and transformation function construction, the correspondence between the features detected in the sensed image and those detected in the reference image is established in terms of similarity measurement. Based on the correspondence, transformation functions (also called mapping functions) which are used to align the two images are constructed. In the third step, by means of the transformation function, the sensed image is transformed to register with the reference image.

The key work in image registration is to find the optimal transformation function f such that most of the points of the sensed image can be accurately mapped to the reference image by means of f [5]. However, reaching the optimal transformation function has never been a trivial task because of the large search space and possible image distortions. The search space is characterized by the transformation model underlying each registration process. Affine transformation is the most frequently used global transformation model [3, 4, 13, 15, 16], since it is reliable, efficient and can correct some global distortions. Global transformation methods are typically either a search for the allowable transformation which maximizes the used metric, or a search for the optimal parameters of the transformation function [3]. Though most of the current research on search strategies are related to concepts of multiresolution, wavelet, relaxation matching, and Hausdorff distance [2, 3, 4, 6, 8, 10, 11, 16], direct search related techniques are still in active use because of their easy implementation and freedom from derivative computation [9, 12].

Direct search methods started from the work of Hooke and Jeeves [7]. The idea can be explained as the following. Starting from a base point p, repeatedly search for a new point p' from various directions until either a p' is found such that p' is a better point than p and therefore p is replaced with p', or no better p' can be found. If no better p' can be found, the search is finished. One key step in direct search methods is how to select a new point. Based on the tactic of the new point selection, direct search methods are usually categorized into

1. pattern search methods,
2. simplex search methods,
3. and methods with adaptive sets of search directions.

In this work, a pattern search method that works with the affine transformation model is presented for image registration. It is demonstrated that with the classic similarity metric, normalized cross-correlation (NCC), this method could successfully find optimal affine transformation function parameters for edge-images generated by computing the local standard deviation of the images.

2 The Search Method

To precisely address the proposed search method, it is necessary to explain the edge-image used in the method first. The edge-image is generated by computing the local standard deviation of the image.

2.1 Local Standard Deviation for Edge Detection

Normalized cross correlation is one of the most used similarity metrics in image registration. However, edge-based correlation, which is computed on the edges extracted from the images rather than on the original images, is advantageous over the general correlation method in that it is less sensitive to the intensity differences between the reference and sensed images [1, 14, 16]. In the proposed search method, the NCC value is computed between two edge-images.

While there are a large number of edge detection methods, the local standard deviation (LSD) method in which edges are detected by computing the standard deviation inside a local small window shows some important properties and advantages:

1. With proper window sizes, LSD can be used for successful line edge detection. Sharp lines can be detected as line edges by LSD with small window sizes. Wider lines require larger window sizes.
2. LSD widens the detected edges. The width of the detected edge is proportional to the size of the local window used. This property plays an important role in the proposed search method.

2.2 The Search Method

An affine transformation is precisely a function of 6 parameters

$$x' = a_0 + a_1 x + a_2 y,$$
$$y' = b_0 + b_1 x + b_2 y. \tag{1}$$

An affine transformation without shear can also be written as:

$$\begin{vmatrix} x' \\ y' \end{vmatrix} = \begin{vmatrix} s_x & 0 \\ 0 & s_y \end{vmatrix} \begin{vmatrix} \cos\theta & -\sin\theta \\ \sin\theta & \cos\theta \end{vmatrix} \begin{vmatrix} x \\ y \end{vmatrix} + \begin{vmatrix} t_x \\ t_y \end{vmatrix}. \tag{2}$$

Clearly, an affine transformation without shear can be fully determined by 5 parameters: s_x and s_y which are the x and y scale factors, t_x and t_y which are the x and y translations, and θ which is the rotation angle. These 5 parameters constitute a vector fully describing an affine transformation composed of rotation, translation and scaling. Each vector element determines the amount of transformation of a specific type. Thus, the search for an optimal affine transformation becomes the task of the search for an optimal vector function, and the key is the search direction. The search method can be detailed as the following:

1. Start with the affine transformation defined by the vector $A = <A_0, A_1, A_2, A_3, A_4>$. The values of A must be somewhat close to optimal and might be determined by tie points selected from each image or by other methods.
2. Start with a small *step* size.
3. For each element of A, determine whether increasing or decreasing that element by the *step* size improves the transformation. The reference image is transformed to match the sensed image and the NCC between the warped reference image and the sensed image is computed. Improvement is defined to be an increase in the NCC value. Record the amount of NCC improvement for this element and its direction.
4. Based on the amount of improvement and the direction for each vector element of A, construct a 5-dimensional vector $a = <a_0, a_1, a_2, a_3, a_4>$ that determines the search direction. The value of each element in a is proportional to the amount of NCC value improvement incurred by its corresponding element in A. The sum of the absolute values of the elements in a equals 1. However, if there is no element in A whose variation results in any improvement, *step* is divided by two and the process repeats from step 3 until either at least one element incurs some improvement or the *step* is smaller than a predefined limit. In case that *step* is smaller than this predefined limit, the search is finished.
5. The search starts toward the direction determined by the vector a. The vector A is recomputed, i.e. $A = A + step \times a$. If the newly computed A generates a better NCC value, it is kept and the old vector is discarded.
6. Repeat step 5 until it fails to generate a higher NCC value. Then *step* is divided by two. If *step* is smaller than a predefined limit, the search toward this vector direction is over and the search restarts from step 2. Otherwise, the search repeats from step 5 with the decreased *step*.

In brief, for the current optimal direction, the search continues until there is no gain in this direction. At this point, a new direction is computed and expressed in the vector *a*. The search stops when there is no direction resulting in a larger NCC value. The computation of the vector direction can be done in other ways. The way presented here is a very simple method, in which each vector element is weighted equally.

The proposed search method is basically the pattern search method described by Hooke and Jeeves with some variations [7]. A critical factor in pattern search methods is the *step* value. Large *step* sizes may result in quickly finding the optimal solution, but may seriously degenerate the search result [9]. It turns out that for different image registration cases, the optimal *step* size should be different. As a general rule, the *step* size is small. However, a small *step* size is more likely to run into a local maximum when the starting point is relatively far from the optimal solution.

To successfully minimize the local maximum problem, the LSD edge detection technique is employed to generate line edges of different widths in the method. In the beginning when the search is relatively far from the optimal parameters, the LSD technique is applied with larger window sizes to generate wider edges. Images with wide line edges have less content details, thus the search is focused on major features of the image, i.e. the search is on a coarse level. As the solution approaches the optimum, the LSD processing is performed with smaller window sizes to generate sharp edges, i.e. more image details are taken into consideration. Using a large initial window size for LSD processing has another valuer, which can be explained in Fig. 1. In Fig. 1, (*a*) and (*b*)

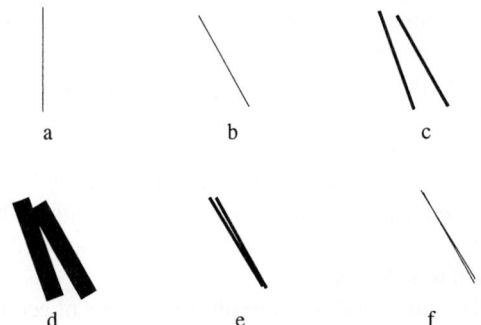

Fig. 1. The significance of the width of line edges in image registration

are the reference and sensed images each with only one line. If the width of the detected line edges are small at the initial search step as shown by (*c*), the NCC metric between the warped reference image and the sensed image generates no significant result (0, in this case) to guide any proper search direction. However, if the width of the two line edges are large enough as shown in (*d*), then there is a meaningful NCC value that could be used to guide the next step search. As the search is closer to the optimal parameters as shown by (*e*) and (*f*), the width of the two line edges are cut down for finer alignment. It is important to observe the limit on the window sizes for LSD processing. Large window sizes can significantly diminish prominent features. They can make close edges indistinguishable. Thus, a proper starting LSD window size should be large enough to account for the distance between the initial warped reference image and the sensed image, but still leave separate edges distinguishable.

2.3 Refinement

Shear is an important component of affine transformation and can be used to correct some global distortions. An affine transformation without shear is not complete. Thus, shear is applied after the search is finished to refine the search transformation function parameters. As there exist both x and y shears, we take the shear as a vector of 2 elements and repeat the above process again specifically for the shear components.

The iterative refinement procedure is based on the fact that different search starting points usually lead to slightly different optimal solutions. In the iterative refinement procedure, after the first iteration of the search and shear refinement, the searched transformation parameters are slightly modified and another iteration of the search and shear refinement is carried out. However, in this iteration, the LSD is always processed with a small window size.

3 Experimental Results

In order to test the method and demonstrate its feasibility, the proposed search method was implemented and aerial images of the Mississippi Delta area were used as test images. The reference and sensed images were taken seven years apart and have significant differences. First, we artifically generated a number of rotated, translated, and scaled images from the original aerial images to perform some experiments to determine the impact of the initial LSD window size on the registration quality. The results showed that when the *step* sizes are small, proper LSD window sizes could significantly improve the registration quality. Therefore, in all the experiments presented below, the initial and final window sizes for LSD processing are 21×21 and 5×5, the initial *step* values applied to rotation, x-translation, y-translation, x-scaling, and y-scaling are 0.2, 1.0, 1.0, 0.01, and 0.01 respectively. The similarity metric NCC was computed between the warped reference image and the sensed image with both images being LSD processed and smoothed. In addition, shear refinement is applied after the search was finished to improve the transformation.

Figures 2 and 3 show the registration results of two sets of aerial images of the Mississippi Delta area. Both figures are rotated 90° counterclockwise. In Fig. 2 from left to right, the first image is the reference image, the middle image is a combination of the reference image and the unregistered sensed image in such a way that the unregistered image lies in the middle surrounded by the reference image. The right image is also a combination of the reference and the registered sensed images arranged as the middle image. The purpose of such two combined images is to demonstrate the registration quality visually. Clearly, the proposed search method can render good registration quality for aerial images of the Mississippi Delta area.

We performed another set of experiments in which a series of the individual aerial sensed images were registered to a large map automatically based on the proposed search method. The Mississippi Delta area is primarily an agricultural area with many fields and ponds. To successfully monitor the variation of the ground cover in the Delta area, aerial images are taken periodically and these aerial images are registered to construct an integrated overview image of the area. These aerial images were taken in order. They have similar sizes, and are overlapped sequentially. We designed an algorithm

Fig. 2. Registration of a sample aerial image

Fig. 3. Automatic registration of 12 aerial images of the Mississippi Delta area

in which the first sensed image was manually aligned with the large reference image roughly well, then the proposed search method was applied to register it. After the first image was registered, based on the overlapping properties between the sensed images, the search starting point of the second sensed image could be obtained automatically, and the proposed registration procedure was applied to register the second image. This process was repeated for 12 partially overlapped aerial images. In addition, 3 levels of iterative refinement were applied to the registration process of the 12 aerial images, and the result is shown in Fig. 3. This result shows that the proposed search method can find an optimal transformation function for the registration of each individual aerial image in the automatic registration procedure.

4 Conclusion

The proposed search method is a pattern search method in which the affine transformation parameters are represented as a 5-parameter vector. Applied to edge images generated by LSD, the proposed search method is demonstrated to be able to find optimal parameters for affine transformation. We plan to extend the search method to higher-order polynomial transformation models in the future.

References

1. Anuta, P.: Spatial registration of multispectral and multitemporal digital imagery using fast Fourier transform. IEEE Transactions on Geoscience Electronics. **8** (1970) 353–368
2. Baker, S., Matthews, I.: Equivalence and efficiency of image alignment algorithms. In Proceedings of the 2001 IEEE Conference on Computer Vision and Pattern Recognition. (2001) 1090–1097
3. Brown, L.: A survey of image registration techniques. ACM Computing Surveys. **24** (1992) 326–376
4. Chalermwat, P.: High performance automatic image registration for remote sensing. PhD thesis, George Mason University. (1999)
5. Chmielewski, L., Kozinska, D.: Image registration. In Proceedings of the 3rd Polish Conference on Computer Pattern Recognition Systems, Poland. (2001) 163–168
6. Fonseca, L., Costa, M.: Automatic registration of satellite images. http://www.inf.uni-konstanz.de/cgip/lehre/ss03-proj/papers/papers/FoCo97.pdf
7. Hooke, R., Jeeves, T.: Direct search solution of numerical and statistical problems. Journal of the Association for Computing Machinery. **8** (1961) 212–229
8. Huttenlocher, D., Klanderman, G., Rucklidge, W.: Comparing images using the Hausdorff distance. IEEE Transactions on Pattern Analysis and Machine Intelligence. **15** (1993) 850–863
9. Isaacs, A.: Direct-search methods and dace. Seminar Report 02401002, Indian Institute of Technology Bombay. (2003)
10. Kruger, S., Calway, A.: Image registration using multiresolution frequence domain correlation. In British Machine Vision Conference. (1998) 316–325
11. Lai, S., Fang, M.: Robust and efficient image alignment with spatially varying illumination models. http://www.cs.nthu.edu.tw/~lai/CVPR99_Lai_Fang.pdf
12. Lewis, R.: Direct search methods: then and now. ICASE Report 2000-06, Institute for Computer Applications in Science and Engineering, National Aeronautics and Space Administration Langley Research Center. (2000)
13. Image Fusion Systems Research: Transformation functions for image registration. http://www.imgfsr.com/ifsr_tf.pdf. (2003)
14. Wie, P., Stein, M.: A landsat digital image rectification system. IEEE Transaction on Geoscience Electronics. **15** (1977) 130–136
15. Wang, X., Feng, D., Jin, J.: Elastic medical image registration based on image intensity. In Pan-Sydney Area Workshop on Visual Information Processing (VIP2001), Sydney, Australia. (2001) 139–142
16. Zitová, B., Flusser, J.: Image registration methods: a survey. Image and Vision Computing. **21** (2003) 977–1000

Water Droplet Morphing Combining Rigid Transformation*

Lanfen Lin, Shenghui Liao, RuoFeng Tong, and JinXiang Dong

State Key Laboratory of CAD and CG,
Department of Computer Science and Engineering,
Zhejiang University, China
liaoshenhui@zju.edu.cn

Abstract. This paper presents a plausible method for large water droplet morphing, taken into account the rigid transformation. The droplet on a plane was represented by the contact area, which was described via "a variational implicit function" in 2D, together with a profile curve. Then we made use of distance fields generated from the boundary of contact area to drive the morphing. The deforming direction and speed were well controlled to give a very smooth and stable deformation. The morphing procedure was combined with rigid transformation synchronously, which yield more natural effect. Finally, the ray-tracing method was employed for rendering the realistic scene.

1 Introduction

Several different methods addressing water modeling and animation have been developed since the 1980's. Most of them concerned the motion of water in forms of waves and other connected fluids, for example, waves approaching and braking on a beach [1]. Realistic and practical animation of liquids [2] has also been made. Only a few methods proposed during the 1990's addressed the problems of the water droplets.

Kaneda et al [3] developed methods for realistic animation of water droplets on a glass plate, on curved surfaces, and meandering down a transparent surface. The main purpose is to generate a realistic animation, taken into account gravity of water droplets, inter-facial tensions, and so on. Fournier et al [4] presented a model that focuses on the simulation of large liquid droplets as they travel down a surface. The aim is to simulate the visual contour and shape of water droplets when affected by the underlying surface and other force fields. Malin [5] gave a method for animation of water droplets flowing on structured surfaces, the droplets in this method were affected by underlying bump mapped surface.

---

* Project supported by National Grand Fundamental Reasearch 973 (No.2002CB312106) of China.
* Project supported by Natural Science Foundation (No.M603129) of Zhejiang Province, China.

All these methods are physical based. It is quite difficult to simulate the flow of water droplets for the purpose of high-precision engineering, due to the diversiform shape representation of the droplet and the complicated flow process. While, our method is not to make a simulation physically correct, but to make an effective and physically plausible droplets morphing on a plane.

2 The Main Idea of Our Approach

2.1 Droplet Representation

First, the droplet on a plane is represented by contact area between the droplet and the plane, together with a profile curve.

To describe the contact area between the droplet and the plane, such as Fig.1 (b), we employ "a variational implicit function" in 2D, proposed in Turk and O'Brien's paper [6], which can model arbitrary 2D shape of general topology.

Yu et al [7] discussed the droplets' shape on a simple plane in the gravitational field. We take advantage of their result and use a profile curve to describe the height of droplet within the contact area, as Fig.1 (a) illustrates (The x coordinate of profile curve means the distance from a point to the boundary of contact area, and y coordinate gives the height of droplet at this point).

When the contact area as well as the profile curve is ready, the ray-tracing method is employed for rendering the scene, to produce a realistic effect of the water droplet. Such as Fig.2 shows.

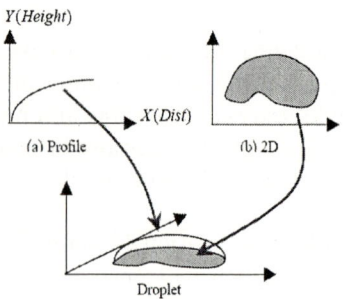

Fig. 1. Droplet representation

Fig. 2. Rendering scene

2.2 Initial Method

Our morphing between two (or more) droplets consists of blending between two profile curves and morphing between two contact areas. Blending between two profile curves is simple, thus we will concentrate on morphing between two contact areas, that is, a kind of 2D volume morphing.

There are several existing approaches for volume morphing. Pasko simply interpolated the values of corresponding nodes [8]. This may cause unnecessary distortion and change in topology. Other techniques are based on a decomposition of the discrete values. Hughes considered Fournier decomposition [9], and T. He decomposed the functions with a wavelet transform [10]. Lerios proposed a feature-based morphing [11]. Cohen presented a method that the interpolation of distance field was guided by a warp function controlled by anchor points [12]. But there is more or less artificial interference in these methods.

Our initial idea is using the intensity fields of implicit functions to produce potential fields, with which we can change the source intensity field to the target. This approach is somewhat like Whitaker's Level-Set Model [13].

Assume $F_1(X) = \delta$ is the source intensity field and $F_2(X) = \delta$ is the target, where $X = (x, y)$. The morphing can be generated by the movement of X, as illustrated in Fig.3.

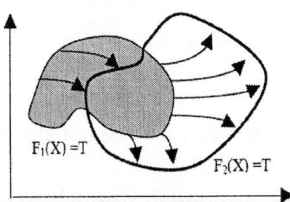 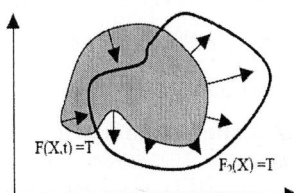

Fig. 3. Potential Field   **Fig. 4.** Movement direction

The problems are:
1. How does X move (The direction and speed)?
2. How does the movement affect the intensity field function F?

Obviously, the movement of the X can be described by regarding X as a function of time t. Thus the whole morphing procedure can be represented by $F(X(t),t)= \delta$, $t \in (0,1)$, and $F(X(0),0)=F_1(X)$, $F(X(1),1)=F_2(X)$. Because F equals to δ over time, the time derivation should be zero:

$$\frac{dF}{dt} = \frac{\partial F(X(t),t)}{\partial t} + \nabla F(X(t),t) \bullet \frac{\partial X(t)}{\partial t} = 0 \; where \nabla F(X,t) = \left(\frac{\partial F}{\partial x}, \frac{\partial F}{\partial y}\right) \quad (1)$$

Thus,

$$\frac{\partial F(X(t),t)}{\partial t} = -\nabla F(X(t),t) \bullet \frac{\partial X(t)}{\partial t} \quad (2)$$

Eq.2 describes how the movement of X (denoted by $\frac{\partial X(t)}{\partial t}$) affects the intensity function F. Now the only problem is how does X move.

First of all, when t changes from t_0 to t_n, $\frac{\partial X(t)}{\partial t}$ must changes F(X(t),t) from $F_1(X)$ to $F_2(X)$, otherwise it will not be a morphing.

Rewrite Eq.2 in a discrete way:

$$\frac{F(X,t_{k+1}) - F(X,t_k)}{\Delta t} = -\nabla F(X,t_k) \bullet \left.\frac{\partial X}{\partial t}\right|_{t_k} \quad (3)$$

To make F(X(t),t) change from $F_1(X)$, just set $F(X, t_0) = F_1(X)$. The problem is how to ensure the deformation end at $F_2(X)$. Note that the intensity field at next time is controlled by Eq.3, whose parameter can be used to control the deformation is $\frac{\partial X(t)}{\partial t}$. One choice of $\frac{\partial X(t)}{\partial t}$ that may satisfy this demand is

$$\frac{\partial X(t)}{\partial t} = -\left(F_2(X) - F(X,t)\right) \frac{\nabla F(X,t)}{\|\nabla F(X,t)\|} \qquad (4)$$

The geometric explanation of Eq.4 is the movement speed of X is $|F_2(X) - F(X,t)|$. And the direction is along the current normal($-\frac{\nabla F(X,t)}{\|\nabla F(X,t)\|}$), if X is inside the target object; else, the direction is the reverse of the normal, as illustrated in Fig.4. Substitute Eq.4 into Eq.3, get the iteration equation:

$$F(X, t_{k+1}) = F(X, t_k) + \|\nabla F(X, t_k)\| \left(F_2(X) - F(X, t_k)\right)\Delta t \qquad (5)$$

where $\nabla F \approx \left(\frac{F(x_{i+1}, y_j, t_k) - F(x_{i-1}, y_j, t_k)}{2h}, \frac{F(x_i, y_{j+1}, t_k) - F(x_i, y_{j-1}, t_k)}{2h}\right)$.

The ending condition of the iteration of Eq.5 is $\max_{X \in \Omega} |F_2(X) - F(X,t)| < \varepsilon$, where Ω is the boundary of F(X,t), and ε is a minimal positive constant.

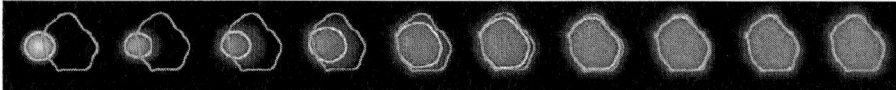

T=0.0 T=0.05 T=0.1 T=0.2 T=0.3 T=0.4 T=0.5 T=0.6 T=0.8 T=1.0

Fig. 5. Morphing using Eq.5

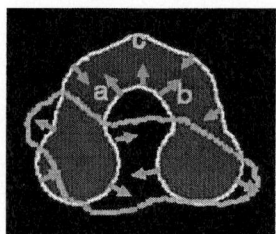 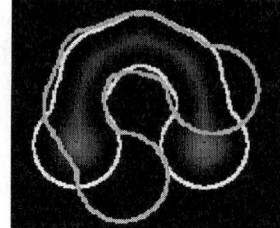

Fig. 6. Direction and distance problem   **Fig. 7.** Rigid deformation problem

By iterating Eq.5, F(X(t),t) changes gradually from $F_1(X)$ into $F_2(X)$. Unfortunately, there are several unpleasant phenomena.

First problem: as Fig.5 shows, the morphing shape changes quickly at fist, and then becomes very slow. Eq.4 gives cause for it, as the pixels move at $|F_2(X) - F(X, t_k)|$. When $F(X, t_k)$ is approaching $F_2(X)$, the speed will get slower and slower. Problem 2: every pixel outside the target shrinks, as in Fig.6. Because according to Eq.4, every pixel's movement direction is reverse of the normal when out of the target. Actually, we expect the pixels between **a** and **b** expend instead of shrink. Problem 3: at the place far enough from the boundary of target (such as **c** in Fig.6), the intensity value may be all zero, this prevents us from controlling the movement. Problem 4: when the shape of source and target are the same, such as Fig.7, we expect the morphing changes by rigid deformation composed of pure rotation and translation. If they are different, we want least distortion of the in-between objects. Eq.5 can not gratify this demand.

To get a natural and pleasant morphing effect, these problems should be addressed. Section 3 gives a solution to the first 3 problems; and problem 4 is handled in section 4.

3 Distance Field Morphing

First, we generate the distance field from the boundary of contact area. Then, use the distance field instead of intensity field to control the deformation.

3.1 Distance Field Transformation

The distance inside the area of object is positive, outside the area is negative, and zero on the boundary. Assume dist(X) is the Euclidean distance from point X to the boundary of contact area, define distance field as:

$$D(X) = \begin{cases} +dist(X) & \text{if X is inside object;} \\ 0 & \text{if X is on the boundary;} \\ -dist(X) & \text{if X is outside object;} \end{cases} \quad (6)$$

Denote the source distance field as S(X), the target field as T(X), and the current deforming distance field as D(X(t),t). Using the same general spirit as the previous study, we rewrite Eq.4 and the iteration Eq.5 as:

$$\frac{\partial X(t)}{\partial t} = -T(X) \frac{\nabla D(X(t),t)}{\|\nabla D(X(t),t)\|} \quad (7)$$

$$D(X, t_{k+1}) = D(X, t_k) + T(X) \bullet \|\nabla D(X, t_k)\| \quad (8)$$

Now, problem 3 is handled naturally, as T(X) still works at the place far enough from the target object.

3.2 Control Morphing Speed and Direction

In the first problem, we want a stable morphing, that is, the maximum speed of the pixels at every time step should be identical. Assume N is the number of total morphing step. We give the stable iteration Equation:

$$D(X, t_{k+1}) = D(X, t_k) + \frac{T(X)}{N-k} \bullet \|\nabla D(X, t_k)\| \quad (9)$$

Morphing in Fig.8 is generated by applying Eq.9. As expected, the morphing speed is very stable. What's more, the number of total morphing step N, that is, the total morphing frames number, is now under control.

Now let's consider the morphing direction in problem 2. Assume $\mathbf{N}_T(X)$ is the unit normal vector of target distance field, and $\mathbf{N}_D(X)$ is that of current distance field. Note that in the area where **a** and **b** belong to (in Fig.6), the angle between $\mathbf{N}_T(X)$ and $\mathbf{N}_D(X)$ is an obtuse angle; while in the area where **c** belongs to, it is an acute angle. So, the dot production $\mathbf{N}_T(X) \bullet \mathbf{N}_D(X)$ can be employed to decide the morphing direction of the outside part, where T(X)<0.

Thus we get the last iteration Equation:

$$D(X, t_{k+1}) = \begin{cases} D(X, t_k) + \frac{T(X)}{N-k} \bullet \|\nabla D(X, t_k)\| \bullet (\mathbf{N}_T(X) \bullet \mathbf{N}_D(X)), T(X) < 0 \\ D(X, t_k) + \frac{T(X)}{N-k} \bullet \|\nabla D(X, t_k)\|, T(X) > 0 \end{cases}$$
(10)

Applying Eq.10, Fig.9 gives a smooth and natural morphing. And the pixels between **a** and **b** all expend to the target rather than shrink.

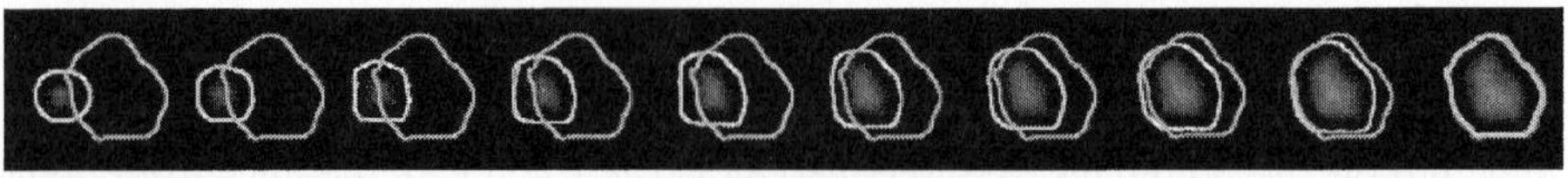

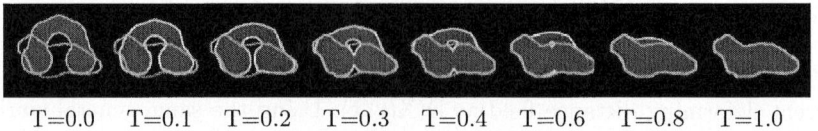

T=0.0 T=0.05 T=0.1 T=0.2 T=0.3 T=0.4 T=0.5 T=0.6 T=0.8 T=1.0

Fig. 8. Morphing using Eq.9

T=0.0 T=0.1 T=0.2 T=0.3 T=0.4 T=0.6 T=0.8 T=1.0

Fig. 9. Morphing using Eq.10

4 Combining Rigid Transformation

As in problem 4, in some cases, to get better morphing effect, we must transform (warp) source object O_s, to another object $W(O_s)$, which approximates the target object O_t, as well as possible. The transformation W is composed of a pure translation C and a rotation R.

The translation is easily obtained by comparing the center of the source and that of the target. The problem is how to define the rotation to make $W(O_s)$ approximate O_t.

First, in the space O_s, emit a line in each direction of $\theta_i = i\Delta\theta$ (i=0, ... N-1) from the center, calculate the distance to the farthest boundary. Store the distances in array $D_s[i]$. Define $D_j[i] = \begin{cases} D_S[i+j], i+j < N \\ D_S[i+j-N], i+j \geq N \end{cases}$ (i,j=0, ... N-1). Do the same thing to O_t to generate $D_t[i]$. Then calculate the correlation between $D_j[i]$ and $D_t[i]$, and select the j which maximizes the correlation, that is, the j which minimizes $\sum_{i=0}^{N-1}(D_t[i] - D_j[i])^2$, to get the rotation angle $\theta = j\Delta\theta$.

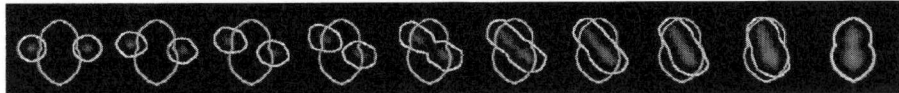

Fig. 10. Combine rigid transformation

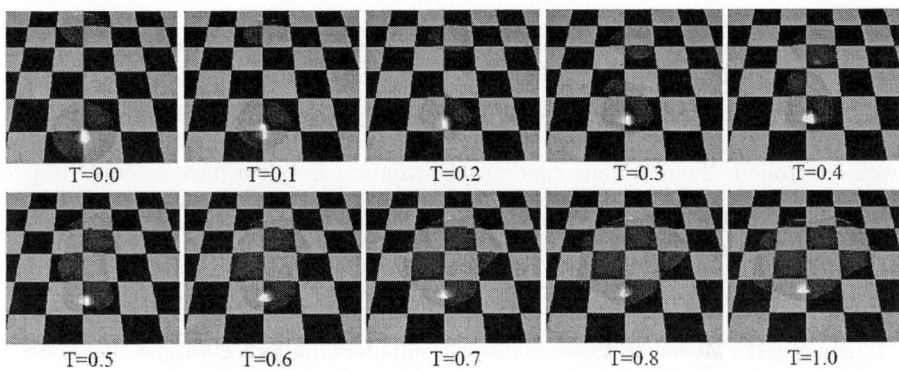

Fig. 11. Complete morphing Procedure

So, the rigid transformation used to rotate and translate the source object to match the coarse features of the target object is $W(X) = R_\theta(X) + C, (\theta = j\Delta\theta)$.

One thing must be pointed out is: when combine the morphing with the rigid transformation, the exact number of frames is pre-requisite. This is because the morphing speed must keep pace with the rigid transformation. If it is slower, it has not arrived the target when the rotation and translation have ended, and, contrariwise. Recall Eq.9 (also Eq.10), the number of total morphing step N is under control. So, they can be combined synchronously precisely.

Now, we can organize complete distance field morphing procedure as follows:

Step 1: Generate the transformation W, exerting on the source object O_s, which approximates the target object O_t, denoted as $W(O_s)$. The transformation W is composed of a pure translation C and a rotation R_θ.

Step 2: Generate a new distance field W(S) according to the transformation W, by assigning the value of point X in field S to the point W(X) in field W(S), where $W(X) = R_\theta(X) + C$.

Step 3: Generate a series of intermediate fields M(W(S),T,t) by morphing W(S) to T, using Eq.10.

Step 4: Transform every field M(W(S),T,t) to field $W_t^{-1}M(W(S),T,t))$, by assigning the value at point X in field M(W(S),T,t) to the point $W_t^{-1}(X)$ in field $W_t^{-1}M(W(S),T,t))$, where $W_t^{-1}(X) = R_{-(1-t)\theta}(X) - (1-t)C$, $t \in [0, 1]$.

The series of $W_t^{-1}M(W(S),T,t))$ is the in-between fields from S to T.

Fig.10 shows a morphing combining rigid transformation. As expected, the morphing procedure combines rigid transformation synchronously smoothly.

Then, we can calculate the contact area between the droplet and the plane, as well as take advantage of the profile curve directly from these intermediate distance fields, which are used to represent the water droplet. Subsequently, the scene is rendered by ray-tracing method. Fig.11 is our final example, which shows a complete water droplets morphing procedure. We can see that the plausible morphing procedure gives a very smooth and natural effect.

5 Conclusion

We have presented a metamorphosis method between two (or more) large water droplets with arbitrary shape and topology, without any artificial interaction. The morphing procedure is not physical based, but driven by distance field. Various instances are considered, thus the morphing speed and direction are well controlled. Synchronous rigid transformation is taken into account to give a more natural effect.

References

1. Peachey D.: Modeling Waves and Surf. Computer Graphics, 20(4), pp. 65-74, 1986.
2. Foster N, Fedkiw R.: Practical Animation of Liquids. Proceedings of the 28th annual conference on Computer graphics and interactive techniques, 2001.
3. Kaneda K, Shinya I, Yamashita H.: Animation of Water Droplets Moving Down a Surface. The Journal of Visualization and Computer Animation 10, 1999.
4. Fournier P, Habibi A, Poulin P.: Simulating the flow of liquid droplets. Proceedings of Graphics Interface 98 pp.133-42, 1998.
5. Malin Jonsson, Anders Hast.: Animation of Water Droplet Flow on Structured Surfaces. Conference: Sigrad 2002, NorrkÖping, pp. 17-22, 2002
6. Turk G., O'Brien J.: Shape transformation using variational implicit functions. Proceedings of ACM SIGGRAPH 99, pages 335-342, 1999.
7. Yu Y-J, Jung H-Y, Cho H-G.: A new water droplet model using metaball in the gravitational field. Computer & Graphics 23, pp. 213-222, 1999.
8. Alexander Pasko, Vladimir Savchenko.: Constructing functionally defined surfaces. Implicit Surfaces'95, Grenoble, France, pp. 97-106, April, 1995.
9. J. F. Hughes.: Scheduled Fourier volume morphing. SIGGRAPH Computer Graphics, 26(2), pp. 43-45, 1992.
10. T. He, S. Wang and A. Kaufman.: Wavelet-based volume morphing. In Proceedings of IEEE Visualization '94, Washington, D.C., pp. 85-92, 1994.
11. A. Lerios, C. D. Garfinkle and M. Levoy.: Feature-based volume metamorphosis. Proc. SIGGRAPH 95, Los Angeles, California, pp. 449-456 , August 6-11, 1995.
12. D. Cohen-Or, D. Levin, and A. Solomivici.: Three-dimensional distance field metamorphosis. ACM Transactions on Graphics, 17(2), pp. 116-141, 1998.
13. D. Breen, R. Whitaker.: A level set approach for the metamorphosis of solid models. IEEE Transactions on Visualization and Computer Graphics, 7(2):173–192, 2001.

A Cost-Effective Private-Key Cryptosystem for Color Image Encryption

Rastislav Lukac and Konstantinos N. Plataniotis

The Edward S. Rogers Sr. Dept. of Electrical and Computer Engineering,
University of Toronto, 10 King's College Road, Toronto, M5S 3G4, Canada
{lukacr, kostas}@dsp.utoronto.ca
http://www.dsp.utoronto.ca/~lukacr

Abstract. This paper presents a cost-effective private-key cryptosystem for color images. The scheme allows for secret sharing of the color image by generating two color shares with dimensions identical to those of the original. Encryption is performed via simple binary operations realized at the image bit-levels altering both the spectral correlation among the RGB color components and the spatial correlation between the neighboring color vectors. The decryption procedure uses both noise-like color shares as the input and recovers the original image with perfect reconstruction.

1 Introduction

Image secret sharing techniques [1],[2] represent a popular encryption tool used to secure transmission of personal digital photographs [3],[4] and digital documents [5] via public communication networks. The so-called $\{k, n\}$-threshold scheme [1],[4]-[9] encrypts the input image by splitting the original content into n noise-like shares. The secret information is recovered only if k or more shares are available for decryption [6]-[9].

Among the various $\{k, n\}$-threshold schemes, a simple $\{2, 2\}$ solution, seen as a private-key cryptosystem [8],[9], takes a great popularity due to its simplicity and adequate information security. The encryption process splits the secret image into two noise-like shares which are delivered to the end-user independently. To recover the actual information the end-user should be in possession of both shares.

Although numerous secret sharing solutions have been proposed for encryption of binary and gray-scale images [1],[6]-[11], secret sharing of color images has become increasingly important in recent times. The growing interest in the development the color image encryption techniques [3]-[5],[12]-[17] can be attributed primarily to the proliferation of color imaging systems and imaging-enabled consumer electronic devices such as digital cameras and mobile phones [18]. End-users and system developers have to protect personal digital photographs and scanned digital documents in emerging applications such as digital photo archiving and image transmission through wireless (mobile) networks. The surge of emerging applications and the proliferation of color imaging systems and imaging-enabled consumer electronic devices suggests that the demand for color image encryption solutions will continue.

2 Color Imaging Basics

Let us consider a $K_1 \times K_2$ Red-Green-Blue (RGB) image $\mathbf{x} : Z^2 \rightarrow Z^3$ representing a two-dimensional matrix of three-component vectorial inputs. In a given $K_1 \times K_2$ RGB color image $\mathbf{x}$, pixel $\mathbf{x}_{(p,q)} = [x_{(p,q)1}, x_{(p,q)2}, x_{(p,q)3}]$ denotes the color vector occupying the spatial location (p, q), with $p = 1, 2, ..., K_1$ and $q = 1, 2, ..., K_2$ denoting the image row and column, respectively. The component $x_{(p,q)i}$, for $i = 1, 2, 3$, of the RGB vector $\mathbf{x}_{(p,q)}$ denotes the vector's spectral component, namely: $x_{(p,q)1}$ denotes the R component, $x_{(p,q)2}$ denotes the G component, and $x_{(p,q)3}$ indicates the B component.

Following the tristimulus theory of color representation, each color pixel $\mathbf{x}_{(p,q)}$ is a three-dimensional vector, uniquely defined by its length (magnitude) $M_\mathbf{x} : Z^2 \rightarrow R^+$ and orientation (direction) $D_\mathbf{x} : Z^2 \rightarrow S^2$ in the vector space, [18]. The magnitude

$$M_{\mathbf{x}_{(p,q)}} = \|\mathbf{x}_{(p,q)}\| = \sqrt{x_{(p,q)1}^2 + x_{(p,q)2}^2 + x_{(p,q)3}^2} \qquad (1)$$

of the color vector relates to the luminance, whereas the vectors' directionality

$$D_{\mathbf{x}_{(p,q)}} = \frac{\mathbf{x}_{(p,q)}}{\|\mathbf{x}_{(p,q)}\|} = \frac{\mathbf{x}_{(p,q)}}{M_{\mathbf{x}_{(p,q)}}} \qquad (2)$$

with S^2 denoting a unit ball in R^3 and $\|D_{\mathbf{x}_{(p,q)}}\| = 1$, relates to the chromaticity characteristics of the pixel.

Since both measures are essential for human perception [19], any color image encryption solution should alter both the magnitude and the orientation characteristics of the original color vectors. Using secret sharing principles for color image encryption [3]-[5],[13], the cryptographic solution generates color shares which contain noise-like, seemingly unrelated information.

3 Overview of Color Image Secret Sharing Solutions

Popular visual secret sharing (VSS) schemes, such as those proposed in [1],[6]-[9], allow for visual recovery of the encrypted images [1]. By utilizing the transparent/frosted representation of the shares (usually printed on transparencies) and color mixing principles, the color VSS schemes such as those listed in [12],[14],[15] use the properties of the human visual system (HVS) to force the recognition of a secret message from the required set of shares without additional computations or any knowledge of cryptography [1]. On the other hand, VSS schemes can also be implemented via simple logical operations and used in a computer-centric environment. Although these features make the VSS decryption system

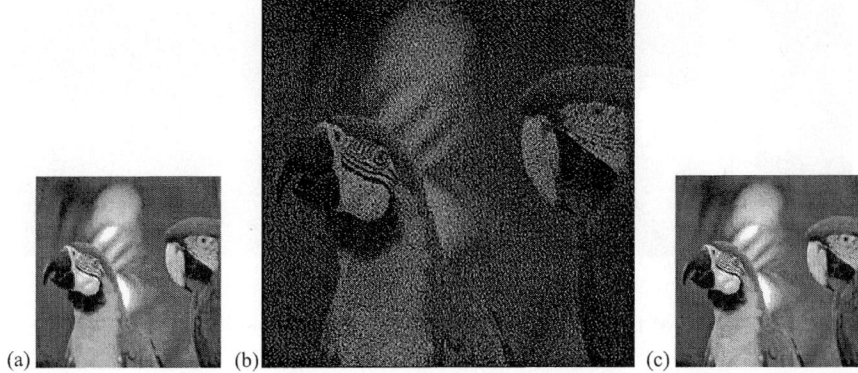

Fig. 1. Obtained results: (a) color test image Parrots, (b) the decrypted output obtained using the VSS scheme, (c) the decrypted output obtained using the bit-level processing based secret sharing scheme

cost-effective, such an approach is not well suited for natural color images. This is due to the fact that the procedure increases the spatial dimension of both shares and output image and introduces a number of visual impairments to the output image (*Fig. 1b*).

The secret sharing schemes of [3]-[5] operate on the bit planes of the digital input and generate the shares with a bit representation identical to the one of the input image. Although the input image and the produced shares have different spatial dimensions due to encryption of each pixel into the blocks of share pixels, the decryption procedure produces an output image (*Fig. 1c*) which is identical to the input (*Fig. 1a*). It was explained in [13] that such a cryptographic solution satisfies the so-called perfect reconstruction property. Since both bit-level processing based $\{k, n\}$-solutions [3]-[5], and $\{2, 2\}$ private key cryptosystem of [13] increase the dimensions of the shares compared to those of the original (secret) image, the produced shares with the enlarged spatial resolution may become difficult to transmit via wireless mobile networks.

4 Private-Key Cryptosystem for Color Image Encryption

To avoid this drawback, our new private-key cryptosystem for color image encryption reduces the block-based share operations to the pixel-based operations [16]. Thus, the spatial resolution of the shares remain unchanged during processing. Moreover, since the scheme performs cryptographic processing on bit planes, it: i) preserves the bit-representation of the shares, and ii) produces an output image identical to the input image. This suggests that the scheme recovers the input image with perfect reconstruction.

Assuming the conventional RGB color image, each color component $x_{(p,q)k}$ of the color vector $\mathbf{x}_{(p,q)}$ is coded with $B = 8$ bits allowing $x_{(p,q)k}$ to take an

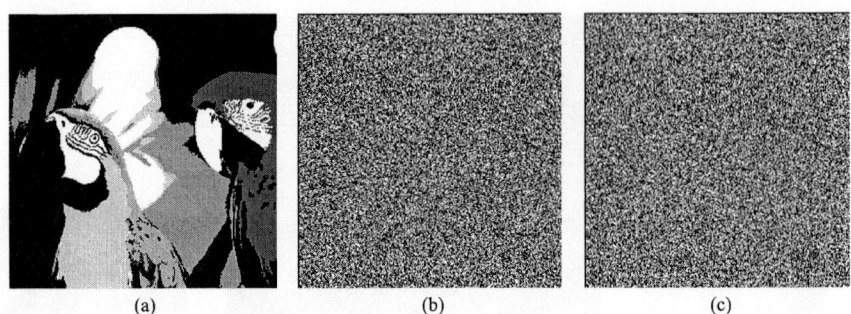

Fig. 2. Encryption of the MSB ($b = 1$) plane: (a) the original input image $\mathbf{x}^b$, (b) the share $\mathbf{s}'^b$, (c) the share $\mathbf{s}''^b$

integer value between 0 and $2^B - 1$. Using a bit-level representation [13], the color vector $\mathbf{x}_{(p,q)}$ can be equivalently expressed in a binary form as follows:

$$\mathbf{x}_{(p,q)} = \sum_{b=1}^{B} \mathbf{x}^b_{(p,q)} 2^{B-b} \tag{3}$$

where $\mathbf{x}^b_{(p,q)} = [x^b_{(p,q)1}, x^b_{(p,q)2}, x^b_{(p,q)3}] \in \{0,1\}^3$ denotes the binary vector at the b-bit level, with $b = 1$ denoting the most significant bit (MSB).

Since the proposed private-key cryptosystem is a simple $\{2,2\}$-secret sharing scheme, binary vectors $\mathbf{x}^b_{(p,q)}$, for $b = 1, 2, ..., B$, are encrypted into two binary share vectors $\mathbf{s}'^b_{(p,q)} = [s'^b_{(p,q)1}, s'^b_{(p,q)2}, s'^b_{(p,q)3}]$ and $\mathbf{s}''^b_{(p,q)} = [s''^b_{(p,q)1}, s''^b_{(p,q)2}, s''^b_{(p,q)3}]$, whose components $s'^b_{(p,q)k}$ and $s''^b_{(p,q)k}$ are generated as follows [16]:

$$[s'^b_{(p,q)k}\ s''^b_{(p,q)k}] \in \begin{cases} \{[0\ 1], [1\ 0]\} & \text{if } x^b_{(p,q)k} = 1 \\ \{[0\ 0], [1\ 1]\} & \text{if } x^b_{(p,q)k} = 0 \end{cases} \tag{4}$$

where the binary set $[s'^b_{(p,q)k}\ s''^b_{(p,q)k}]$ is obtained from the basis elements 0 and 1.

To ensure the random nature of the encryption (4), a random process should be guided by a random number generator. In this paper, we use the conventional rand function that is built in to common C++ programming tools to determine $[s'^b_{(p,q)k}\ s''^b_{(p,q)k}]$ from the sets $\{[0\ 1], [1\ 0]\}$ and $\{[0\ 0], [1\ 1]\}$ via (4). By repeating the process at each binary level $b = 1, 2, ..., B$ and each vector component $k = 1, 2, 3$, the procedure generates two color share vectors $\mathbf{s}'_{(p,q)} = [s'_{(p,q)1}, s'_{(p,q)2}, s'_{(p,q)3}]$ and $\mathbf{s}''_{(p,q)} = [s''_{(p,q)1}, s''_{(p,q)2}, s''_{(p,q)3}]$ defined as follows:

$$\mathbf{s}'_{(p,q)} = \sum_{b=1}^{B} \mathbf{s}'^b_{(p,q)} 2^{B-b},\ \mathbf{s}''_{(p,q)} = \sum_{b=1}^{B} \mathbf{s}''^b_{(p,q)} 2^{B-b} \tag{5}$$

where $\mathbf{s}'^b_{(p,q)}$ and $\mathbf{s}''^b_{(p,q)}$ denote the binary share vectors obtained in (4). As shown in *Fig. 2* the formation of the binary share vector arrays increases the degree

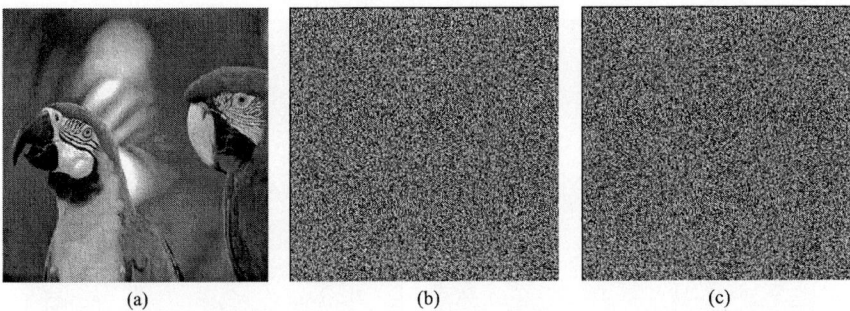

Fig. 3. Proposed color image encryption: (a) the input (original) color image **x**, (b) the color share **s'**, (c) the color share **s''**

of protection from two possible options in (4) to eight options observed for the binary share vectors $\mathbf{s}'^b_{(p,q)}$ and $\mathbf{s}''^b_{(p,q)}$ at each bit level b.

By performing the processing operations (3)-(5) in each spatial location (p,q), for $p = 1, 2, ..., K_1$ and $q = 1, 2, ..., K_2$, the procedure produces two $K_1 \times K_2$ color shares $\mathbf{s}' : Z^2 \rightarrow Z^3$ and $\mathbf{s}'' : Z^2 \rightarrow Z^3$, with the share vectors $\mathbf{s}'_{(p,q)}$ and $\mathbf{s}''_{(p,q)}$ denoting the color pixels located at (p,q) in $\mathbf{s}'$ and $\mathbf{s}''$, respectively. This increases protection of both the individual color channels (2^B levels) and the color shares (2^{3B} possible color vectors) depicted in *Figs.3b,c* since the random binary vectors $\mathbf{s}'^b_{(p,q)}$ and $\mathbf{s}''^b_{(p,q)}$ are weighted by 2^{B-b} needed in the determination of the word-level color image model.

It is shown in [16] that due to the randomness in the encryption process defined in (4) the color share vectors $\mathbf{s}'_{(p,q)}$ and $\mathbf{s}''_{(p,q)}$ differ both in magnitude ($M_{\mathbf{s}'_{(p,q)}} \neq M_{\mathbf{s}''_{(p,q)}}$) and in direction ($D_{\mathbf{s}'_{(p,q)}} \neq D_{\mathbf{s}''_{(p,q)}}$). In addition, they differ in both magnitude and direction from those ($M_{\mathbf{x}_{(p,q)}}, D_{\mathbf{x}_{(p,q)}}$) observed for the original color inputs $\mathbf{x}_{(p,q)}$. As it can be seen in *Fig.3* the encryption process changes the spectral correlation characteristics of the input vectors and alters the spatial correlation characteristics of the input image. The utilization of the complete RGB color gamut ($256^3 = 16,777,216$ colors) in the proposed encryption procedure i) ensures adequate protection against attacks in the color share domain, and ii) prohibits unauthorized access to the original bit information.

To faithfully decrypt the secret color image from the color shares, the decryption function must satisfy the perfect reconstruction property. This can be obtained when the encryption and decryption operations are reciprocal [13]. Therefore, the original color/structural information is recovered by processing the share vector arrays at the binary level, and the decryption function follows the encryption mechanism in (4) resulting in the following definition [16]:

$$x^b_{(p,q)k} = \begin{cases} 0 \text{ if } s'^b_{(p,q)k} = s''^b_{(p,q)k} \\ 1 \text{ if } s'^b_{(p,q)k} \neq s''^b_{(p,q)k} \end{cases} \quad (6)$$

Based on the reciprocal concept between (4) and (6), the original binary component $x^b_{(p,q)k}$ is recover as $x^b_{(p,q)k} = 1$ if the binary share components

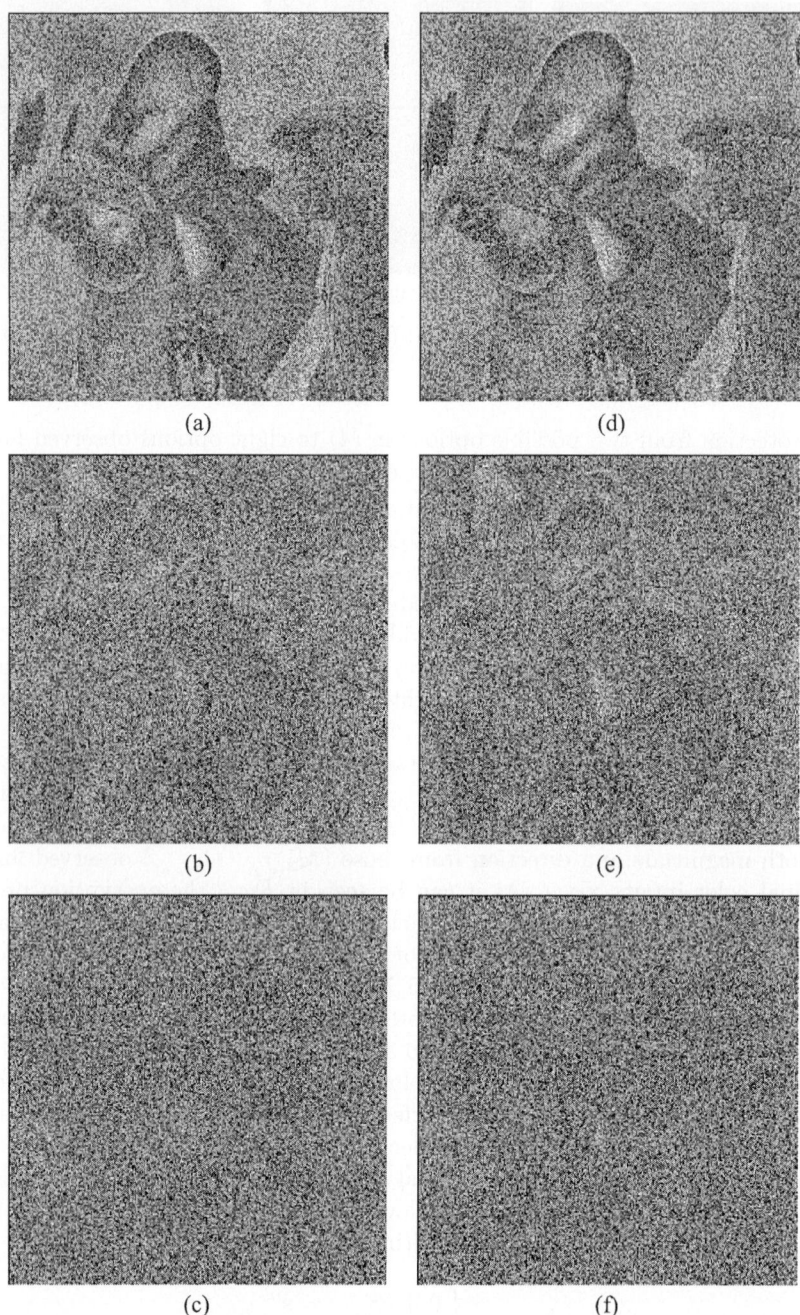

Fig. 4. Color shares s' (a-c) and s'' (d-f) obtained using the color image Parrots when cryptographic processing is performed for the reduced set of binary levels: (a,d) $b = 1$, (b,e) $b = 1, 2$, (c,f) $b = 1, 2, 3$

$s'^b_{(p,q)k}$ and $s''^b_{(p,q)k}$ are not equal or $x^b_{(p,q)k} = 0$ if $s'^b_{(p,q)k}$ and $s''^b_{(p,q)k}$ are identical. Stacking together the recovered bits $x^b_{(p,q)k}$ using the spectral relationship $\mathbf{x}^b_{(p,q)} = [x^b_{(p,q)1}, x^b_{(p,q)2}, x^b_{(p,q)3}]$ the binary vector $\mathbf{x}^b_{(p,q)}$ is formed. Application of (3) results in the recovered original color vector $\mathbf{x}_{(p,q)}$, which suggests that the proposed method satisfies the perfect reconstruction property.

Figs.3b,c show the color shares obtained when the cryptographic operations are applied to all bit planes ($b = 1, 2, ..., B$). The complexity of the solution can be reduced by performing the encryption/decryption operations for a reduced set of bit planes. Visual inspection of the color shares depicted in *Fig.4* reveals that the encryption of the MSB (*Figs.4a,d*) or the two most significant bits (*Figs.4b,e*) only, fine details are sufficiently encrypted, however, large flat regions can be visually revealed. The results depicted in *Figs.4c,f* indicate that a sufficient level of protection is achieved by cryptographically processing the first three most significant bits ($b = 1, 2, 3$). The remaining bits of the original image vectors can be simply copied into the shares unchanged. If this option is selected, image decryption has to be performed only for $b = 1, 2, 3$.

5 Conclusion

A private-key cryptosystem for color image encryption was presented. The solution can be seen a $\{2,2\}$-secret sharing scheme which satisfies the perfect reconstruction property. By applying simple logical cryptographic operations at the bit-level, the encryption procedure: i) changes both the magnitude and the orientation of the color vectors generating color noise-like shares, and ii) it produces random-like color vectors which differ significantly in both magnitude and orientation from the original color inputs. Thus, the generated shares which can be transmitted over unsecured public channels with reasonable overhead. Since the method performs the pixel-based share operations instead of the usual block-based operations, the produced color hares have the same spatial resolution as the original image. The input color image is perfectly reconstructed from the share vector arrays using elementary bit-level logical functions. The perfect reconstruction allowed by the procedure makes it ideal for cost-effective dissemination of digital imaging material over untrusted communication channels.

References

1. Naor, M., Shamir, A.: Visual Cryptography. Proc. EUROCRYPT'94, LNCS **950** (1994) 1–12
2. Yang, C.N., Laih, C.S.: New colored visual secret sharing schemes. Designs Codes and Cryptography **20** (2000) 325–336
3. Lukac, R., Plataniotis, K.N., Smolka, B., Venetsanopoulos, A.N.: A new approach to color image secret sharing. Proc. XII European Signal Processing Conference (EUSIPCO'04) in Vienna, Austria, (2004) 1493–1496

4. Lukac, R., Plataniotis, K.N.: A color image secret sharing scheme satisfying the perfect reconstruction property. Proc. 2004 IEEE International Workshop on Multimedia Signal Processing (MMSP'04) in Sienna, Italy, (2004) 351–354
5. Lukac, R., Plataniotis, K.N.: Document image secret sharing using bit-level processing. Proc. 2004 IEEE International Conference on Image Processing (ICIP'04) in Singapore, (2004) 2893–2896
6. Lin, C.C., Tsai, W.H., Visual cryptography for gray-level images by dithering techniques. Pattern Recognition Letters **24** (2003) 349–358
7. Chang, C.C, Chuang, J.C.: An image intellectual property protection scheme for gray-level images using visual secret sharing strategy. Pattern Recognition Letters, **23** (2002) 931–941
8. Lukac, R., Plataniotis, K.N.: Bit-level based secret sharing for image encryption. Pattern Recognition **38** (2005) 767–772
9. Ateniese, G., Blundo, C, de Santis, A., Stinson, D.G.: Visual cryptography for general access structures. Information and Computation **129** (1996) 86–106
10. Yang, C.N.: New visual secret sharing schemes using probabilistic method. Pattern Recognition Letters, **25** (2004) 481-494
11. Yang, C.N., Chen, T.S: Aspect ratio invariant visual secret sharing schemes with minimum pixel expansion. Pattern Recognition Letters, **26** (2005) 193-206.
12. Hou, J.C.: Visual cryptography for color images. Pattern Recognition **36** (2003) 1619–1629
13. Lukac, R., Plataniotis, K.N.: Colour image secret sharing. IEE Electronics Letters **40** (2004) 529-530
14. Koga, H., Iwamoto, M., Yakamoto, H.: An analytic construction of the visual secret sharing scheme for color images. IEICE Transactions on Fundamentals **E84-A** (2001) 262–272
15. Ishihara, T., Koga, H.: A visual secret sharing scheme for color images based on meanvalue-color mixing. IEICE Transactions on Fundamentals **E86-A** (2003) 194–197
16. Lukac, R., Plataniotis, K.N.: A new encryption scheme for color images. Computing and Informatics, submitted (2004)
17. Sudharsanan, S.: Visual cryptography for JPEG color images. Proceedings of the SPIE, Internet Multimedia and Management Systems V, **5601** (2004) 171-178
18. Lukac, R., Smolka, B., Martin, K., Plataniotis, K.N., Venetsanopoulos, A.N.: Vector filtering for color imaging. IEEE Signal Processing Magazine, Special Issue on Color Image Processing **22** (2005) 74–86
19. Sharma, G., Trussell, H.J.: Digital color imaging. IEEE Transactions on Image Processing **6** (1997) 901–932

On a Generalized Demosaicking Procedure: A Taxonomy of Single-Sensor Imaging Solutions

Rastislav Lukac and Konstantinos N. Plataniotis

The Edward S. Rogers Sr. Dept. of Electrical and Computer Engineering,
University of Toronto, 10 King's College Road, Toronto, M5S 3G4, Canada
{lukacr, kostas}@dsp.utoronto.ca
http://www.dsp.utoronto.ca/~lukacr

Abstract. This paper presents a generalized demosaicking procedure suitable for single-sensor imaging devices. By employing an edge-sensing mechanism and a spectral model, the proposed demosaicking framework preserves both the spatial and spectral characteristics of the captured image. Experimental results reported in this paper indicate that the solutions designed within the proposed framework produce visually pleasing full color, demosaicked images.

1 Introduction

Color filter array (CFA) interpolation or demosaicking is an integral step in single-sensor imaging solutions such as digital cameras, image-enabled wireless phones, and visual sensors for surveillance and automotive applications, [1]-[6]. The CFA is used to separate incoming light into a mosaic of the color components (*Fig.1a*). The sensor, usually a charge-coupled device (CCD) or complementary metal oxide semiconductor (CMOS) sensor, is essentially a monochromatic device [1],[7], and thus, the raw data that acquires in conjunction with the CFA constitute a $K_1 \times K_2$ gray-scale image z with scalar pixels $z_{(p,q)}$, with $p = 1, 2, ..., K_1$ and $q = 1, 2, ..., K_2$ denoting the image row and column, respectively. The two missing color components are estimated from the adjacent pixels using the demosaicking process to produce the full-color demosaicked image [8]-[11].

Although a number of CFA have been proposed, the three-color Red-Green-Blue (RGB) Bayer CFA pattern (*Fig.1a*) [12] is the most commonly used due to the simplicity of the subsequent demosaicking procedure. Assuming the GRGR phase in the first row, a Bayer CFA image z, depicted in *Fig.2a*, can be transformed to a $K_1 \times K_2$ three-channel image $\mathbf{x}$ (*Fig.2b*) as follows [1],[13]:

$$\mathbf{x}_{(p,q)} = \begin{cases} [z_{(p,q)}, 0, 0] & \text{for } p \text{ odd and } q \text{ even,} \\ [0, 0, z_{(p,q)}] & \text{for } p \text{ even and } q \text{ odd,} \\ [0, z_{(p,q)}, 0] & \text{otherwise.} \end{cases} \quad (1)$$

where $\mathbf{x}_{(p,q)} = [x_{(p,q)1}, x_{(p,q)2}, x_{(p,q)3}]$ denotes the color vector. The values $x_{(p,q)k}$ indicate the R ($k = 1$), G ($k = 2$), or B ($k = 3$) CFA components. Since the sensor image z is a mosaic-like gray-scale image, the missing components in $\mathbf{x}_{(p,q)}$ are set equal to zero to indicate their portion to the coloration of $\mathbf{x}$.

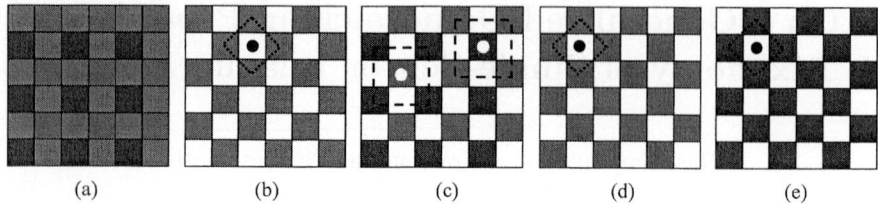

Fig. 1. (a) Bayer CFA pattern with the GRGR phase in the first row, (b-e) spatial arrangements of the four-neighboring color components observed during the proposed demosaicking procedure: (b,d,e) $\zeta = \{(p-1,q),(p,q-1),(p,q+1),(p+1,q)\}$, (c) $\zeta = \{(p-1,q-1),(p-1,q+1),(p+1,q-1),(p+1,q+1)\}$

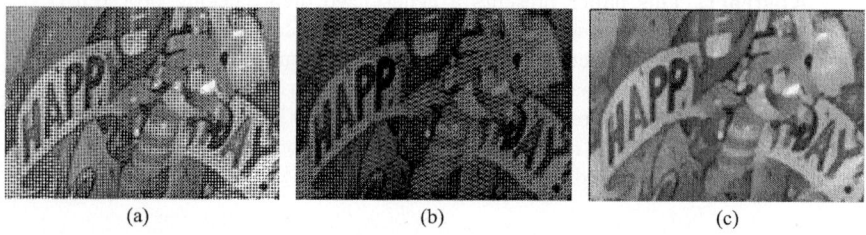

Fig. 2. Single-sensor imaging: (a) a gray-scale Bayer CFA sensor image, (b) a Bayer CFA image arranged as a color image, (c) a full-color, demosaicked image

2 A Generalized Demosaicking Procedure

Due to the dominance of the G component in the Bayer CFA pattern, most demosaicking procedures, for example those listed in [1]-[4],[7]-[9], start the process by interpolating the G color plane. In order to quantify the contribution of the adjacent samples, the missing component $x_{(p,q)k}$ is calculated as follows:

$$x_{(p,q)k} = \sum_{(i,j)\in\zeta} \{w'_{(i,j)} x_{(i,j)k}\} \qquad (2)$$

where $x_{(i,j)k}$ denotes the k-th components of the color vector $\mathbf{x}_{(i,j)} = [x_{(i,j)1}, x_{(i,j)2}, x_{(i,j)3}]$, with $(i,j) \in \zeta$ denoting the spatial location arrangements on the image lattice (*Figs. 1b-e*).

The normalized weighting coefficients $w'_{(i,j)}$ used in (2) are defined as

$$w'_{(i,j)} = w_{(i,j)} / \sum_{(i,j)\in\zeta} w_{(i,j)} \qquad (3)$$

where $w_{(i,j)} \geq 0$ is the so-called edge-sensing weight. The weights $w_{(i,j)}$ are used to regulate the contribution of the available color components inside the spatial arrangements shown in *Figs. 1b-e*. To ensure that the demosaicking procedure is an unbiased solution, the condition $\sum_{(i,j)\in\zeta} w'_{(i,j)} = 1$ must be satisfied, [1].

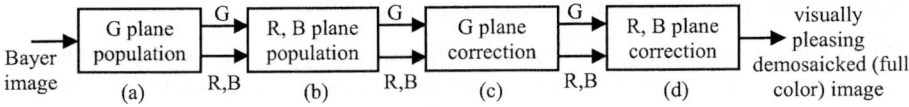

Fig. 3. Block scheme diagram of a generalized demosaicking procedure: (a,b) mandatory steps, (c,d) recommended (optional) steps

By populating the G color plane in *Fig. 3a* via (2) with $k = 2$ and $\zeta = \{(p-1,q), (p,q-1), (p,q+1), (p+1,q)\}$ *(Fig. 1b)*, the missing R (or B) components of **x** can be obtained through the use of the spectral correlation that exists between the G and R (or B) components of a natural image. Adopting the notation and concept introduced in [13], the R ($k = 1$) or B ($k = 3$) components $x_{(p,q)k}$ are calculated in *Fig. 3b* as follows:

$$x_{(p,q)k} = x_{(p,q)2} \bar{\oplus} \sum_{(i,j) \in \zeta} \{w'_{(i,j)}(x_{(i,j)k} \oplus x_{(i,j)2})\} \quad (4)$$

where $\oplus$ and $\bar{\oplus}$ denote the spectral quantity formation and normalization operations, respectively. The procedure first produces the R and B components $x_{(p,q)k}$ located in the center of the shape-masks $\zeta = \{(p-1,q-1), (p-1,q+1), (p+1,q-1), (p+1,q+1)\}$ *(Fig. 1c)*, and then those located in the center of the shape-masks $\zeta = \{(p-1,q), (p,q-1), (p,q+1), (p+1,q)\}$ observed for the updated planes *(Figs. 1d,e)*.

Since the G color plane was populated without the utilization of the essential spectral characteristics, the demosaicked G components obtained using (2) should be re-evaluated in *Fig. 3c* as follows [13]:

$$x_{(p,q)2} = x_{(p,q)k} \bar{\oplus} \sum_{(i,j) \in \zeta} \{w'_{(i,j)}(x_{(i,j)2} \oplus x_{(i,j)k})\} \quad (5)$$

where $\zeta = \{(p-1,q), (p,q-1), (p,q+1), (p+1,q)\}$, as shown in *Fig. 1b*.

Finally, the proposed demosaicking procedure completes by correcting the demosaicked R and B components *(Fig. 3d)*. This demosaicking step is realized using (4) with $k = 1$ for R and $k = 3$ for B components. As before, the spatial arrangements of the adjacent samples are described using $\zeta = \{(p-1,q-1), (p-1,q+1), (p+1,q-1), (p+1,q+1)\}$ *(Fig. 1c)* and $\zeta = \{(p-1,q), (p,q-1), (p,q+1), (p+1,q)\}$ *(Figs. 1d,e)*.

3 Taxonomy of Demosaicking Solutions

Within the proposed generalized demosaicking framework, numerous demosaicking solutions may be constructed by changing the form of the spectral model, as well as the way the edge-sensing weights are calculated. The choice of these two construction elements essentially determines the characteristics and the performance of the single-sensor imaging solution, [1],[13],[14].

3.1 Non-adaptive Versus Adaptive Solutions

Based on the nature of the determination of $w_{(i,j)}$ in (3), the demosaicking solutions can be differentiated as i) non-adaptive, and ii) adaptive demosaicking schemes.

Non-adaptive demosaicking schemes such as those listed in [15]-[18] use a simple linear averaging operator (fixed weights $w_{(i,j)} = 1$) without considering any form of adaptive weighting, [1],[13]. Since non-adaptive schemes do not utilize structural information of the captured image to direct the demosaicking process, they produce the full-color images with blurred edges and fine details.

To restore the demosaicked image in a sharp form, adaptive demosaicking solutions use the edge-sensing weights $w_{(i,j)}$ to emphasize inputs which are not positioned across an edge and to direct the demosaicking process along the natural edges in the captured image, [1],[19],[20]. In most available designs, such as those listed in [13],[14],[21]-[25], the edge-sensing coefficients $w_{(i,j)}$ use some form of inverse gradients. In order to design a cost-effective and robust solution, the following form of $w_{(i,j)}$ defined using inverse gradients [13],[26] is used throughout the paper:

$$w_{(i,j)} = \left\{1 + \sum\nolimits_{(g,h)\in\varsigma} \left|x_{(i,j)k} - x_{(g,h)k}\right|\right\}^{-1} \qquad (6)$$

3.2 Component-Wise Versus Spectral Model-Based Solutions

Based on the use of the essential spectral characteristics of a captured image in the demosaicking process, the demosaicking schemes can be divided into the following two classes: i) component-wise, and ii) spectral model based solutions.

The component-wise processing solutions do not use the spectral correlation that exists between the color channels in a natural image. Such a demosaicking procedure uses (2) to fully populate R ($k = 1$), G ($k = 2$), and B ($k = 3$) color planes. It has been widely observed [1],[20]-[28] that the omission of the spectral information in the component-wise demosaicking process in [3],[16],[17] leads to a restored output which contains color artifacts and color moire noise.

The use of the spectral model preserves the spectral correlation that exists between the color components. Since natural RGB images exhibit strong spectral correlation characteristics [1],[6],[18], both researchers and practitioners in the camera image processing community rely on spectral models to eliminate spectral artifacts and color shifts. A commonality of the currently used spectral models of [1],[15],[27],[28] is that they incorporate RG or BG spectral characteristics into the demosaicking process. The spectral model based demosaicking procedure, such as those used in [2],[18],[22],[23],[25], first populates the G color plane (*Fig.3a*) via (2), and then use the spectral characteristics in the demosaicking steps (*Figs.3b-d*) defined via (4), (5). It has been shown in [13] that the use of $\oplus$ and $\bar{\oplus}$ in (4)-(5) generalize the previous spectral models. Assuming for the simplicity the color-difference based modelling concept, the spectral modelling operators $\oplus$ and $\bar{\oplus}$ denote the addition and subtraction operations, respectively, and these modelling operations are used throughout the paper.

Fig. 4. Test images: (a) Snake, (b) Girls, (c) Butterfly

Table 1. Obtained objective results

Image	Snake			Girls			Butterfly		
Method	MAE	MSE	NCD	MAE	MSE	NCD	MAE	MSE	NCD
NCS	12.446	906.1	0.2648	2.456	35.1	0.0503	3.184	70.4	0.0449
ACS	10.729	859.4	0.2473	2.279	30.6	0.0479	2.868	59.1	0.0420
NSMS	9.103	525.5	0.1832	1.867	16.4	0.0420	1.768	12.8	0.0309
ASMS	7.806	460.0	0.1590	1.742	13.8	0.0399	1.614	10.5	0.0281

4 Experimental Results

To examine the performance of the basic demosaicking solutions designed within the proposed generalized framework, a number of test images have been used. Examples such as the 512 × 512 images Snake, Girls, and Butterfly are depicted in *Fig.4*). These test images, which vary in color appearance and complexity of the structural content (edges), have been captured using three-sensor devices and normalized to 8-bit per channel RGB representation.

Following common practices in the research community [1],[2],[6],[18], mosaic versions of the original color images are created by discarding color information in a GRGR phased Bayer CFA filter (*Fig.1a*) resulting in the CFA image z. The demosaicked images are obtained from applying the demosaicking solution designed within the proposed framework (*Fig.3*) to process the CFA image. Comparative evaluations are performed by comparing, both objectively and subjectively, the original full color images to demosaicked images. To facilitate the objective comparisons [1], the mean absolute error (MAE), the mean square error (MSE) and the normalized color difference (NCD) criterion are used. While the MAE and MSE criteria are defined in the RGB color space which is conventionally used for storing or visualization purposes, the perceptual similarity between the original and the processed image is quantified using the NCD criterion expressed in the CIE LUV color space [29].

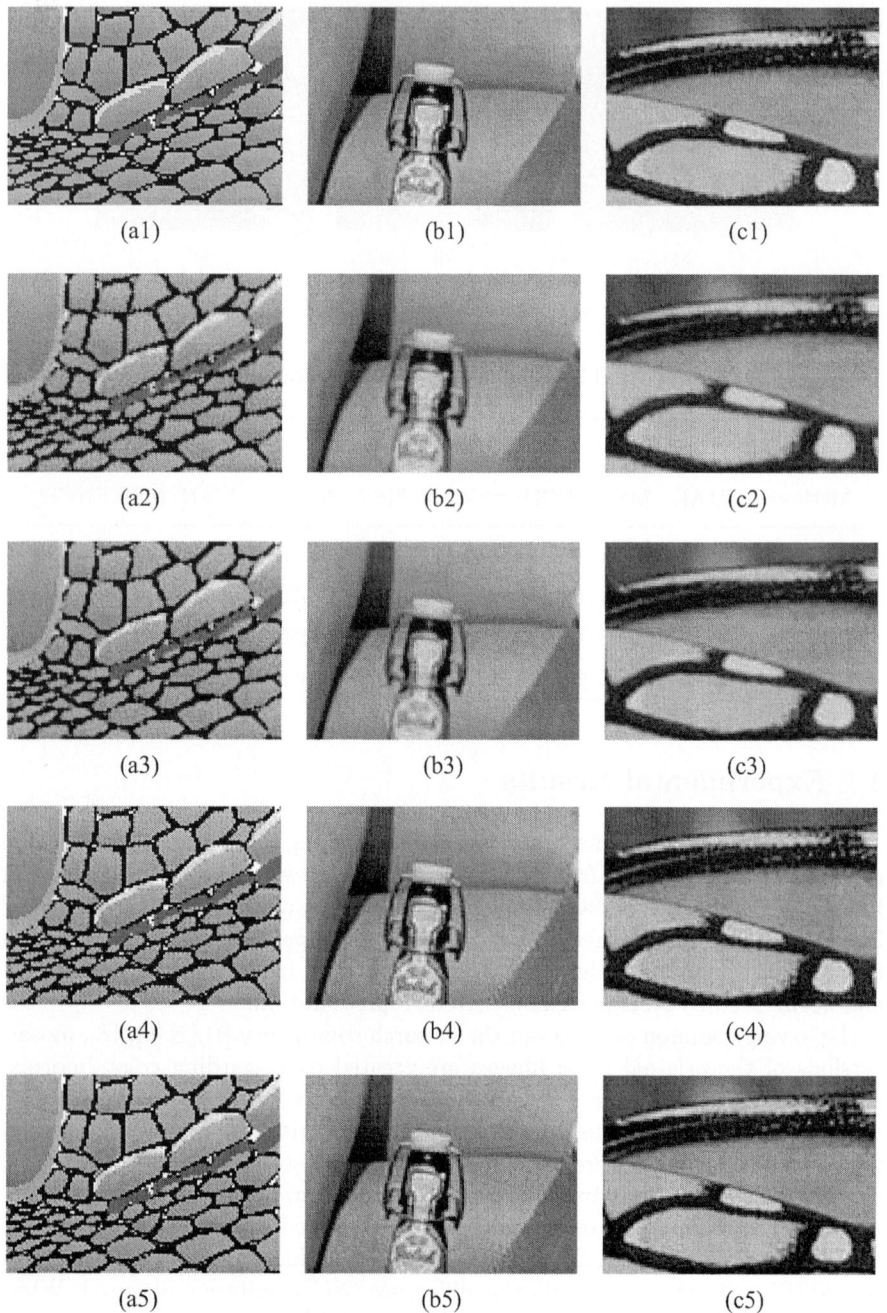

Fig. 5. Enlarged parts of the images: (a) Snake, (b) Girls, (c) Butterfly; (1) original image, (2) NCS, (3) ACS, (4) NSMS, (4) ASMS

To demonstrate the importance of the edge-sensing mechanism and the spectral model, the four solutions designed within the proposed framework (*Fig.3*) defined in (2),(4), and (5) are considered. Namely, the selected demosaicking schemes include the non-adaptive component-wise scheme (NCS), the adaptive component-wise scheme (ACS), the non-adaptive, spectral model-based scheme (NSMS), and the adaptive, spectral model based scheme (ASMS).

Table 1 summarizes the objective results obtained by comparing the different solutions designed within the proposed demosaicking framework. It can be easily seen that the NCS scheme is the worst performing method among the tested schemes. This should be attributed to its non-adaptive and component-wise nature. The use of the adaptive its adaptive ACS variant improves the result in terms of all objective criteria. However, the significant improvement of the performance of the demosaicking process is observed when the processing solution employs both the spectral model and the edge-sensing mechanism.

Figs.5 depicts enlarged parts of the test images cropped in edge areas which are usually problematic for Bayer CFA demosaicking schemes. The results show that NCS and ACS solutions blur edges and produce a number of color shifts in the demosaicked image, while the ASMS solution produces the highest visual quality among the tested schemes.

5 Conclusion

A generalized demosaicking framework for single-sensor imaging was presented. The framework allows for the utilization of both the spatial and spectral characteristics during the demosaicking process. Experimentation performed here suggests that both the spectral model and the edge-sensing mechanism should be used in the demosaicking pipeline.

References

1. Lukac, R., Plataniotis, K.N.: Normalized color-ratio modelling for CFA interpolation. IEEE Transactions on Consumer Electronics **50** (2004) 737–745
2. Lukac, R., Plataniotis, K.N., Hatzinakos, D., Aleksic, M.: A novel cost effective demosaicing approach. IEEE Transactions on Consumer Electronics **50** (2004) 256–261
3. Ramanath, R., Snyder, W.E., Bilbro, G.L., Sander, W.A.: Demosaicking methods for Bayer color arrays. Journal of Electronic Imaging **11** (2002) 306–315
4. Wu, X., Zhang, N.: Primary-consistent soft-decision color demosaicking for digital cameras. IEEE Transactions on Image Processing **13** (2004) 1263–1274
5. Lukac, R., Martin, K., Plataniotis, K.N.: Digital camera zooming based on unified CFA image processing steps. IEEE Transactions on Consumer Electronics **50** (2004) 15–24
6. Gunturk, B., Altunbasak, Y., Mersereau, R.: Color plane interpolation using alternating projections. IEEE Transactions on Image Processing **11** (2002) 997–1013
7. Adams, J., Parulski, K., Spaulding, K.: Color processing in digital cameras. IEEE Micro **18** (1998) 20–30

8. Freeman, W.T.: Median filter for reconstructing missing color samples. U.S. Patent 5 373 322, (1988)
9. Cai, C., Yu, T.H., Mitra, S.K.: Saturation-based adaptive inverse gradient interpolation for Bayer pattern images. IEE Proceedings - Vision, Image, Signal Processing **148** (2001) 202–208
10. Lukac, R., Plataniotis, K.N.: Digital camera zooming on the colour filter array. IEE Electronics Letters **39** (2003) 1806–1807
11. Hur, B.S., Kang, M.G.: High definition color interpolation scheme for progressive scan CCD image sensor. IEEE Trans. Consumer Electronics **47** (2001) 179–186
12. Bayer, B.E.: Color imaging array. U.S. Patent 3 971 065 (1976)
13. Lukac, R., Plataniotis, K.N.: Data-adaptive filters for demosaicking: a framework. IEEE Transactions on Consumer Electronics, submitted (2004)
14. Lukac, R., Plataniotis, K.N., Hatzinakos, D.: Color image zooming on the Bayer pattern. IEEE Transactions on Circuit and Systems for Video Technology **15** (2005)
15. Cok, D.R.: Signal processing method and apparatus for producing interpolated chrominance values in a sampled color image signal. U.S. Patent 4 642 678, (1987)
16. Sakamoto, T., Nakanishi, C., Hase, T., Software pixel interpolation for digital still cameras suitable for a 32-bit MCU. IEEE Transactions on Consumer Electronics **44** (1998) 1342–1352
17. Longere, P., Zhang, P., Delahunt, P.B., Brainard, D.H.: Perceptual assessment of demosaicing algorithm performance. Proceedings of the IEEE **90** (2002) 123–132
18. Pei, S.C., Tam, I.K.: Effective color interpolation in CCD color filter arrays using signal correlation. IEEE Trans. Circuits and Systems for Video Technology **13** (2003) 503–513
19. Ramanath, R., Snyder, W.E.: Adaptive demosaicking. Journal of Electronic Imaging **12** (2003) 633–642
20. Kakarala, R., Baharav, Z.: Adaptive demosaicing with the principal vector method. IEEE Transactions on Consumer Electronics **48** (2002) 932–937
21. Hamilton, J.F., Adams, J.E.: Adaptive color plane interpolation in single sensor color electronic camera. U.S. Patent 5 629 734, (1997)
22. Kimmel, R.: Demosaicing: image reconstruction from color CCD samples. IEEE Transactions on Image Processing **8** (1999) 1221–1228
23. Lu, W., Tang, Y.P.: Color filter array demosaicking: new method and performance measures. IEEE Transactions on Image Processing **12** (2003) 1194–1210
24. Kehtarnavaz, N., Oh, H.J., Yoo, Y.: Color filter array interpolation using color correlations and directional derivatives. Journal of Electronic Imaging **12** (2003) 621–632
25. Chang, L., Tang, Y.P., Effective use of spatial and spectral correlations for color filter array demosaicking. IEEE Trans. Consumer Electronics **50** (2004) 355–365
26. Lukac, R., Plataniotis, K.N.: A Robust, Cost-Effective Postprocessor for Enhancing Demosaicked Camera Images. Real-Time Imaging, Special Issue on Spectral Imaging II, **11** (2005)
27. Adams, J., Design of practical color filter array interpolation algorithms for digital cameras. Proceedings of the SPIE **3028** (1997) 117–125
28. Lukac, R., Martin, K., Plataniotis, K.N.: Demosaicked image postprocessing using local color ratios. IEEE Transactions on Circuit and Systems for Video Technology **14** (2004) 914–920
29. Plataniotis, K.N., Venetsanopoulos, A.N.: Color Image Processing and Applications. Springer Verlag, Berlin, (2000)

Tile Classification Using the CIELAB Color Model

Christos-Nikolaos Anagnostopoulos[1], Athanassios Koutsonas[2],
Ioannis Anagnostopoulos[3], Vassily Loumos[2], and Eleftherios Kayafas[2]

[1] University of the Aegean, Cultural Technology & Communication Dpt.,
Mytilene, Greece, GR 81100
canag@ct.aegean.gr
http://www.aegean.gr/culturaltec/canagnostopoulos
[2] National Technical University of Athens, Electrical & Computer Engineering School,
Athens, Greece, GR 15773
{loumos,kayafas}@cs.ntua.gr
[3] University of the Aegean, Information & Communication System Engineering Dpt.,
Karlovassi, Samos, GR 83200
janag@aegean.gr

Abstract. An image processing algorithm was developed for tile shade classification on the basis of quantitative measurements in CIELAB color space. A total of 50 tile images of 10 types were recorded, and evaluated with the proposed algorithm in comparison with the conventional classification method. The objectivity of the method is based on the fact that it is not subject to inter- and intra-observer variability arising from human's profile of competency in interpreting subjective and non-quantifiable descriptions.

1 Introduction

The objective of this paper is the presentation of a method that can be used for the surface inspection and shade classification of ceramic tiles. The shade of a tile is the combination of a number of visual characteristics of the tile surface, including its color and the distribution pattern of color or decoration over the tile surface. Due to the nature of the process, with the use of natural materials, intentionally variable decoration effects and high temperature firing, it is usually not possible to guarantee the production is of a single shade. What is needed is a rational method of establishing a shade classification system for a tile product and then applying it in a consistent way in order to avoid or minimize these problems and so increase profit. Sorting and classification of tiles by shade is a challenging and complicated subject. It is also a critical one for tile manufacturers.

Until now this work is performed almost exclusively by specialized workers, causing several problems to the chain of production. Despite the fact that automated sorting and packing lines have been in existence for a number of years, the complexity of shade classification of tiles has meant that, until recently, automated classification systems have not been possible [1]. However, based on the science of color measurement, attempts are being made to introduce an automatic tile shade classification system [2],[3].

2 Shade Measurements – CIELAB Color Space

Until now the discussions have been equally applicable to manual shade sorting and automatic shade classification. However, a reliable shade measurement metric should be considered and defined.

Appearance measurement instruments and systems take objective measurements which correspond with these descriptions. The tile surface should be illuminated with one or more light sources of controlled intensity. Then, the light reflected from the tile surface will be acquired, converted to digital signals and sent for further analysis by an appropriate software module.

To be useful for tile shade classification an individual shade measurement should:

- measure an appearance variation that is clearly visible to the human eye, in a way that correlates with human perception.
- give reliable and repeatable, measurement values.
- measure shade variations which occur or could occur during the production of the tile.
- be linear with human perception – a 1 unit difference at one part of the measurement scale should appear to a person to be just as large as a 1 unit difference at another part of the measurement scale.

Moreover, the set of shade measurements selected for tile shade classification of a particular type should:

- be able to detect and measure all of the shade variations that occur or could occur.
- be orthogonal to each other – a change in the appearance characteristic measured by one measurement should not affect any of the other measurements.
- be isotropic – a 1 unit difference of one of the measurements should be perceived as being just as large as 1 unit difference of any of the other measurements.

The work in science of color measurement has been directed at finding ways of measuring and specifying color and color variations. The progress in this field resulted in several well documented and understood color measurement system, defined in international standards under the name of CIELAB.

3 Algorithmic Description

This part of the paper constitutes the presentation of the complete software algorithm that will be used for the estimation of shade variation in tile surface.

3.1 Transformation from RGB to CIELAB

For the development of such application, it is essential to use a color appearance chromatic model such as the CIE $L^*a^*b^*$. Consequently, the first, and relatively independent part of application, is the implementation of formulas that make possible the transformation in the CIELAB color space. For fluorescent light the D65 model values were used [5].

3.2 Cluster Segmentation of the Reference Image Using K-Means

Clustering is a way to separate groups of objects. K-means clustering treats each object as having a location in space. It finds partitions such that objects within each cluster are as close to each other as possible, and as far from objects in other clusters as possible. K-means clustering requires a number of clusters to be partitioned and a distance metric to quantify how close two objects are to each other. Since from step 1, the color information exists in the 'a*b*' space, the objects are pixels with 'a*' and 'b*' values. For the proposed algorithm K-means was used to cluster the objects into n clusters according the Euclidean distance metric, where n is the number of basic colors in every type of tile.

Using the above method, the user specifies the number of the basic colors without taking into consideration the effect of luminosity. This step was designed in such a way to handle variations of illumination, watersheds and spots that are presented in tile surface. An example is shown in Figure 1, where a picture in Lab -was segmented in 2 sub-images according the Euclidean distance. The basic colors were orange and beige. An algorithm for partitioning (or clustering) N data points into K disjoint subsets S_j containing N_j data points so as to minimize the sum-of-squares criterion as shown in Equation 1:

$$J = \sum_{j=1}^{K} \sum_{n \in S_j} |x_n - \mu_j|^2 \qquad (1)$$

where x_n is a vector representing the nth data point and μ_j is the geometric center of the data points in S_j. In general, the algorithm does not achieve a global minimum of J over the assignments. In fact, since the algorithm uses discrete assignment rather than a set of continuous parameters, the "minimum" it reaches cannot even be properly called a local minimum. Despite these limitations, the algorithm is used fairly frequently as a result of its ease of implementation.

The algorithm consists of a simple re-estimation procedure as follows. First, the data points are assigned at random to the K sets. Then the center is computed for each set. These two steps are alternated until a stopping criterion is met, i.e., when there is no further change in the assignment of the data points.

The clustering step returns the centers of each segmented color in a*, b* values. Supposing that we have n basic colors in an image the result of K-means clustering will be the following matrix:

$$K = \begin{bmatrix} color\ 1 & a_1 & b_1 \\ color\ 2 & a_2 & b_2 \\ color\ 3 & a_3 & b_3 \\ \ldots & \ldots & \ldots \\ color\ n & a_n & b_n \end{bmatrix}$$

Where a_n, b_n are the centers of cluster n in the a* b* space.

3.3 Calculation of Cluster Luminosity

In the second procedure for K-means clustering, placement of the K centers can be done by the following procedure, which is similar to the one proposed in [6].

- Place K points into the space represented by the objects that are being clustered. These points represent the reference centers found previously in the step of cluster segmentation. Starting with those values the K-means algorithm needs minimum effort to find the new centers according the Euclidean distance.
- Assign each object to the group that has the closest center.
- When all objects have been assigned, recalculate the positions of the K centers.
- Repeat the steps of cluster segmentation and calculation of luminosity until the centers no longer move. This produces a separation of the objects into groups from which the metric to be minimized can be calculated.

As the centers of the clusters become stable for every tile, the calculation of the mean luminosity (L*) of the pixels belonging to each cluster follows. Based on the luminosity values, the tiles will be then classified in comparison to the reference L* values.

4 Experimental Results

Tile samples obtained from a Greek industrial company (FILKERAM/JOHNSON) form the training set of the algorithm. Following shade classification by an expert engineer, each tile was labeled with an identifier code of its corresponding type and the appropriate scale shade from 1 to 15.

The images of the above tiles were then digitized and stored for later retrieval and processing. The hardware for image acquisition, and digitization was a digital camera SONY DSC-V1 with a resolution of 2592x1944 pixels. The white balance of the camera was set to fluorescent light, as image acquisition was performed in indoor conditions and stable illumination conditions in the laboratory. The distance between the camera and the tile was set equal to 30 cm with automatic focusing of the camera.

Image analysis was performed by using specialized software (Matlab 6.0 by Math-Works Inc). Following acquisition of the tile image in TIFF format, a file name was given according its type and the shade classification scale (e.g tile5_12.tif corresponds to tile type 5 and shade scale 12). Measurement data were sent by a DDE (Dynamic Data Exchange) method to be saved in a spreadsheet (Excel® by Microsoft) for analysis.

4.1 Experimental Example

A small sub-set of 3 tiles of type 1 was considered for demonstration purposes of the developed algorithm. The data set consists of tile1_1, tile1_4, tile1_8, all belonging to tile type 1 and shade scale 1, 4 and 8 respectively, where scale 1 indicates darker shade and scale 15 the lighter.

The image tile1_1 is set as the reference tile image. The algorithmic sequence as well as the results for every step are the following:

a. Transformation from RGB to CIEXYZ.
b. Transformation from CIEXYZ to CIELab.
c. Cluster segmentation of the reference image using K-means

As shown in Figure 1a, tile type1 has a pattern of 2 basic colors, beige and light orange. Therefore, the K-mean software module was activated for K=2 clusters. This step returns, the centers of the two clusters in terms of a* and b* values.
Using a simple AND masking technique, the two clusters are now clearly visible in Figures 1b, 1c. Following the masking technique, the mean luminosity value for every cluster is calculated with Equation 2:

$$\overline{L_c} = \frac{\sum_{i=1}^{2592_c} \sum_{j=1}^{1944} L_c(i,j)}{N_c} \quad (2)$$

where N_c equals the number of the pixel that belong to cluster c.

Specifically for the tile type 1, the values a*, b* for every cluster and their average luminosity L* are:

$$K = \begin{bmatrix} cl1 & a_1 & b_1 \\ cl2 & a_2 & b_2 \end{bmatrix} = \begin{bmatrix} beige & 2.2704 & 21.1612 \\ orange & 6.3916 & 25.3697 \end{bmatrix}, \quad L = \begin{bmatrix} cl1 & 66.1546 \\ cl2 & 61.1420 \end{bmatrix} = \begin{bmatrix} beige & 66.1546 \\ orange & 61.1420 \end{bmatrix}$$

d. Cluster classification

The above centers (2.2704, 21.1612) and (6.3916, 25.3697) are now considered to be the reference centers for further cluster processing of similar tile type, while the pair (66.1546, 61.1420) corresponds to the reference luminosity values for shade classification.

As the centers of the two clusters are recalculated for every tile of type 1, the calculation of the mean luminosity for each cluster follows. On the basis of the luminosity values, the tile will be then classified in comparison to the reference values.
As a result, the K matrix of tile1_4 and tile1_8 are:

$$K_{1\_4} = \begin{bmatrix} cl1 & a_1 & b_1 \\ cl2 & a_2 & b_2 \end{bmatrix} = \begin{bmatrix} beige & 5.9694 & 19.5014 \\ orange & 11.5815 & 24.3607 \end{bmatrix}, \quad K_{1\_8} = \begin{bmatrix} cl1 & a_1 & b_1 \\ cl2 & a_2 & b_2 \end{bmatrix} = \begin{bmatrix} beige & 6.1521 & 18.8210 \\ orange & 11.0485 & 24.5022 \end{bmatrix}$$

Using the same AND masking technique, the new two clusters for every tile are now depicted in Figures 2b, 2c, 3b and 3c. Using Equation 2 the average luminosity value for each new cluster is given in matrix L as follows:

$$L_{1\_4} = \begin{bmatrix} c1 & 65.4438 \\ c2 & 59.2922 \end{bmatrix} = \begin{bmatrix} beige & 65.4438 \\ orange & 59.2922 \end{bmatrix}, \quad L_{1\_8} = \begin{bmatrix} c1 & 62.9300 \\ c2 & 58.4622 \end{bmatrix} = \begin{bmatrix} beige & 62.9300 \\ orange & 58.4622 \end{bmatrix}$$

Comparing both L_{1-4} and L_{1-8} with the reference matrix L, it is well shown that there is a slight difference in tile illumination values, which is in complete agreement with the experts decision. Therefore, it is assumed that the luminosity measurements in CIELAB space can give a reliable metric for tile classification according their shade, providing that the illumination conditions remain the same for the whole process.

As shown in Figures 1a, 2a and 3a the shade of tile type 1 is the combination of two equally distributed colors over the tile surface. Therefore, no one can safely decide which color plays the most important role for shade classification and it is quite rational to consider the average value of both clusters as an adequate shade metric. However, in other tile types the decoration over the surface consists mainly of one color.

Depending on the tile pattern decoration the user may choose which cluster (color) is the most important for luminosity measurements. For instance, for of tile type 1, the manufacturer may consider cluster 1 (beige's) measurement as the most important for shade classification. Consequently, the average luminosity of the basic color should be taken into consideration.

4.2 Overall Results

A total of 50 images of 50 individual tiles in various shades belonging to 10 types were recorded. This sample was already classified by conventional classification methods by an expert and labeled accordingly (e.g. tile2_11 for tile type 2 and shade scale 11). Following image processing with the proposed algorithmic procedure those images were classified according the criterion of average luminosity described in step d. Deviations in classification from the expert's decision occurred when the classification metric of average luminosity of cluster 1 and cluster 2 was considered. However, when the average luminosity of both clusters (whole image) is considered to be the classification metric, there is a complete agreement between the algorithmic results and the expert's decision.

5 Discussion

An automatic tile inspection system implementing digital cameras and image processors could be used for tile classification. It would be able to calculate shade measurements and exploit these measurements in a trustworthy and repeatable way. It is proved herein, that it is possible to utilize image processing methods in CIELAB color space to build up a reliable shade classification scheme compatible to human color perception. This would possible, however, if the algorithm developer has a clear understanding of the requirements and theory of tile shade classification, and experience of what shade measurements are required for different tile types. Potential applications of such systems include quality control and shade-variation monitoring in the industry of ceramic materials. The benefits introduced are cost reduction and quality improvement as also shown in [7].

References

1. Vincent LEBRUN,"Quality control of ceramic tiles by machine vision," Flaw Master 3000, Surface Inspection Ltd. 2001.
2. C. Boukouvalas, F. De Natale, G. De Toni, J. Kittler, R. Marik, M. Mirmehdi, M. Petrou, P. Le Roy, R. Salgari and G. Vernazza. *An Integrated System for Quality Inspection of Tiles.* Int. Conference on Quality Control by Artificial Vision, QCAV 97 , pp 49-54, France, 1997.
3. Murat Deviren, M.Koray Balci, U. Murat Leloglu, Mete Severcan, "A Feature Extraction Method for Marble Classification", *Third International Conference on Computer Vision, Pattern Recognition & Image Processing (CVPRIP 2000)*, Atlantic City, pp 25-28. (HSI)

4. A. D. H. Thomas, M. G. Rodd, J. D. Holt, and C. J. Neill. Real-time industrial inspection: A review. *Real-Time Imaging*, 1:139–158, 1995.
5. Mark D. Fairchild, Color Appearance Models, Addison-Wesley, Reading, MA (**1998**).
6. Park., S.H., Yun, I.D., Lee, S.U., "Color Image Segmentation Based on 3-D Clustering: Morphological Approach", *Pattern Recognition*, Vol. 31. No. 8, pp. 1061-1076, 1998.
7. Martin Coulthard, "Tile shade classification strategies for maximum profit", Tile & Brick international journal, Vol.: 16, N°.: 3, June, 2000.

Appendix: Example Screenshots

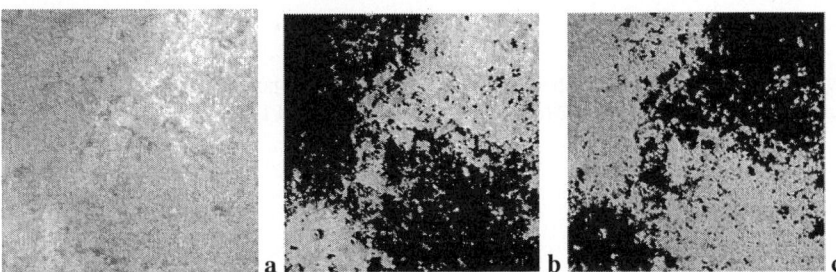

Fig. 1. a) Tile1_1, b) Cluster 1 (beige): a*=2.2704, b*=21.1612, L*=66.1546, c) Cluster 2 (orange): a*=6.3916, b*=25.3697, L*=61.1420

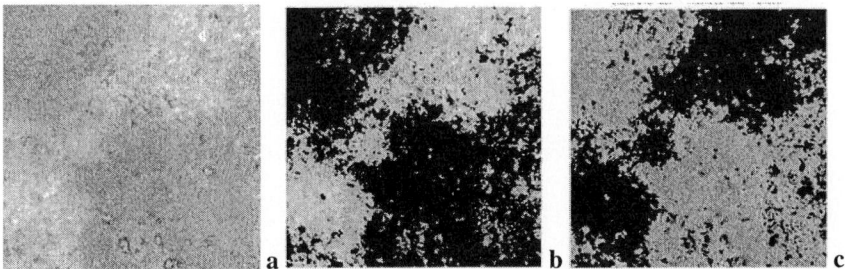

Fig. 2. a) Tile1_4, b) Cluster 1 (beige): a*=5.9694, b*=19.5014, L*=65.4438, c) Cluster 2 (orange): a*=11.5815, b*=24.3607, L*=59.2922

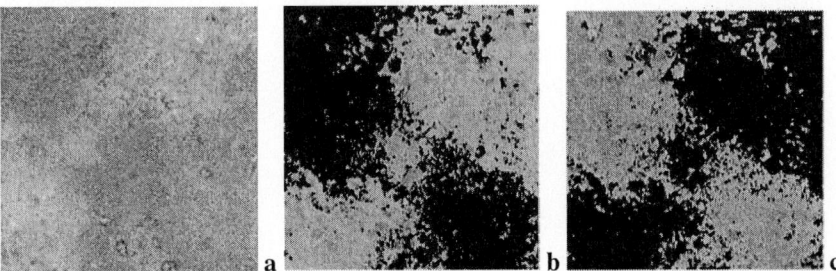

Fig. 3. a) 1_8, b) Cluster 1 (beige): a*=6.1521, b*=18.8210, L*=62.9300, c) Cluster 2 (orange): a*=11.0485, b*=24.5022, L*=58.4622

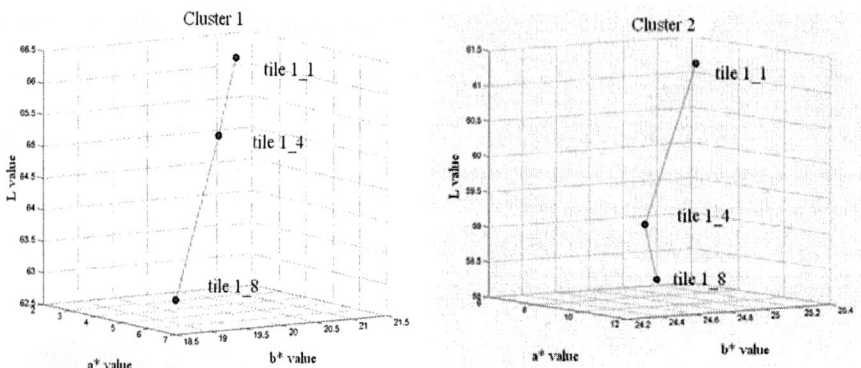

Fig. 4. Visual representation of how values L*, a* and b* are positioned in the CIELAB color space for clusters 1 (beige) and 2 (orange) in tiles 1_1, 1_4 and 1_8

A Movie Is Worth More Than a Million Data Points

Hans-Peter Bischof and Jonathan Coles

Rochester Institute of Technology,
102 Lomb Memorial Dr., Rochester, NY 14623
{hpb, jpc1870}@cs.rit.edu

Abstract. This paper describes a distributed visualization system called Spiegel, which displays and analyzes the results of N-body simulations. The result of a N-body simulation is an enormous amount of data consisting of information about how the bodies interact with each other over time. The analysis of this data is difficult because it is not always clear ahead of time what is important. The visualization system allows a user to explore the simulation by moving through time and space in a 3-dimensional environment. Because of its flexible architecture, the visualization system can be easily extended to add new features.

1 Introduction

The N-body problem is the problem of calculating gravitational force vectors between N particles over a given time period. Since each particle affects all other particles the complexity of this problem is $O(2^N)$. This becomes challenging to compute for astrophysicists who want to build models of entire galaxies where the number of particles is often in the millions.

To overcome this exponential growth problem, researchers in Tokyo, Japan developed specialized hardware called GRAPEs (GRAvity PipElines) [8]. A GRAPE board is solely designed to calculate force vectors on a given set of particles using a highly parallel architecture. Recently, newer models have been developed which are smaller and can easily be installed in computer clusters [2].

The Department of Physics at the Rochester Institute of Technology (RIT) has a 1 teraflop, 8 dual-Xenon node GRAPE cluster that can run simulations up to one million particles. The result of a typical run of a simulation is a file on the order of 20 Gigabytes. These files contain the information of how the particles interacted with each other over time. The difficulty is to interpret this vast amount of information.

One of the best ways to analyze this information is through a visualization system. Such a system, named Spiegel, has been developed as a joint project between the Department of Physics and the Department of Computer Science at RIT. The system projects the simulation data into a 3-dimensional world and allows the user to explore the world by moving through space and time. The basic functionality for the Spiegel system is to:

- Provide multiple viewpoints by having virtual cameras positioned in the simulation universe. The particles are represented as points colored based on physical properties.
- Allow a camera position and angle to be easily adjusted by using standard GUI widgets, the mouse, joystick, or other specialized input hardware such as Flock of Birds [3].
- Allow the current time index to be adjusted. The software should be able to create a movie of the simulation. Sample movies can be found on the project website [4].
- Provide a scripting language to control the system and allow new features to be added on easily.
- Allow collaborators across the Internet to connect together to view and manipulate the same simulation.

2 System Architecture

In designing Spiegel, one primary goal was flexibility. The system should be able to work on one machine or across many machines and allow new functionality, such as interface extensions or visualization techniques, without having to modify any of the code base.

The general system design consists of three major components: the Switchboard, the Feeder, and the ViewController. Each of these components talks to the others using a simple scripting language called Sprache. This language controls the creation, movement, and positioning of the cameras, and includes commands for loading particle information and changing the virtual time frame that is being viewed. Using this language, complex sequences of commands can be written without having to modify the system. These scripts can be saved and loaded on demand to be played repeatedly. An simple example is given in Figure 1.

```
camera create Camera0    # create a camera view labelled Camera0
camera 0 moveto 2 3 4    # move it to x=2 y=3 z=4
camera 0 rotate 4 3 2    # rotate it along the x,y,z axes by 4,3,2 degrees
updateview               # commit the changes to the view
```

Fig. 1. Simple example of the Sprache scripting language

The Switchboard is the module responsible for maintaining the current state. All commands that affect the position of the cameras, or anything else in the system are first sent to the Switchboard. When commands are sent to the Switchboard, the Switchboard distributes them to all of the different displays. There is no limit to the number of displays that can be connected. Before commands can be sent to the Switchboard, however, they must first be sent to the Feeder.

The Feeder acts as a central point to which other pieces of the system send commands. The commands are then ordered and sent to the Switchboard one at a time. The Feeder also allows an additional layer of interpretation. If a different, more complex, language were to be developed, the Feeder could provide a translation mechanism to convert the higher level language into Sprache commands. The rest of the system would be unaffected, but complex sequences could more easily be programmed, either directly by the user, or through an interface of some sort.

The ViewController is the visual part of the system that is connected to the Switchboard and it receives instructions from the Switchboard on how to update the display. Within the ViewController is the MapView and the CameraViews. The MapView offers a global view of the positions of the cameras relative to each other and the coordinate system of the virtual space. The CameraViews mechanism manages the different CameraViews. Each CameraView shows the particles as they appear from the point of view of one camera. The two pieces of the ViewController also act as an interactive interface. The user can manipulate the cameras by directly clicking and dragging the cameras in the map view, or by clicking and dragging in one of the CameraView displays.

3 Component Interaction

The graphical user interface provided by the ViewController controls the main functionality of the system. Interaction with the UI generates a sequence of Sprache commands that are sent to the Feeder and then to the Switchboard. Once the Switchboard has sent the actual movement commands to the ViewController, the ViewController updates its display with the current state of the system. It is important to understand this sequence of events. The ViewController will not act on a user command unless instructed to do so by the Switchboard. This is to prevent the state in the Switchboard from became unstable.

The three components communicate using one of two mechanisms. Either they can talk over an TCP/IP socket connection, or, if they are running on the same machine, they can be connected directly in memory. The advantage of communicating in memory is one of speed. However, allowing for network communication yields two powerful features. First, any language that supports a network library can be used to design any portion of the system. If one ViewController would be better written in C++ or Python then it would be easy to plug in to the system. The code would simply have to understand how to talk to the Feeder and Switchboard using Sprache commands over a TCP/IP connection.

If there is a new input device that is better suited to manipulate three dimensional images, such as a glove, then the driver would only have to support talking to the Feeder. This allows for an incredible degree of freedom for anyone wishing to add to the system. Second, it means that the ViewControllers, for example, can be spread across the Internet, allowing researchers to view, discuss, and manipulate the data simultaneously. In order for researchers to do this, however, they must have access to the same data.

There is currently only a tentative design which allows the ViewController to retrieve particle data by talking to a module called the ParticleView. The

ParticleView would have direct access to the data and would allow the ViewController to query for particle information at a given point in time. The data is stored in a format which gives initial values and then incremental changes over time. The time intervals for each change may differ between particles, but using some simple algorithms it is possible to extrapolate the position, velocity, etc. of every particle at a given time. The ParticleView performs the extrapolation at the request of a ViewController. A picture of the design is in Figure 2.

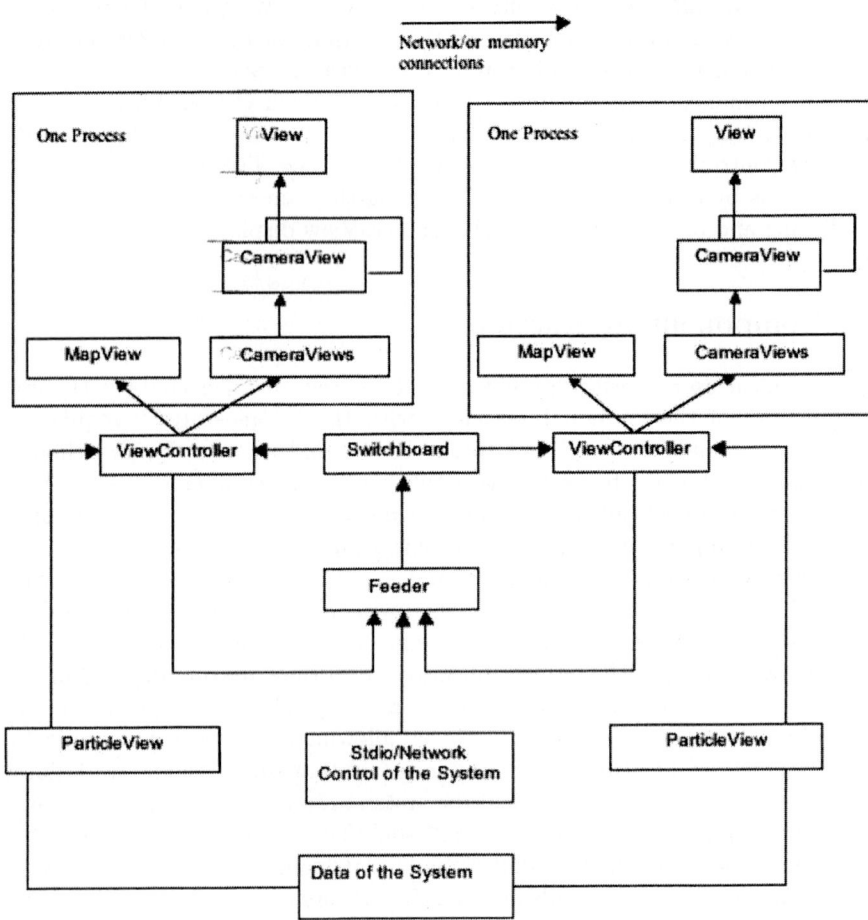

Fig. 2. System design for Spiegel

4 Source Data

The simulation is executed on a 1 teraflop GRAPE cluster system that creates a file containing the state of particles over time. This file is typically very large and in a plain text format. Before it can be used by the visualization system it must be converted into a denser binary format that is more efficient and can be loaded quickly.

The plain text file is divided into three parts. The first part of the file contains three lines holding header information:

- Number of particles used for this specific simulation n
- Time units for this simulation t_{max}
- Number of black holes involved in this simulation nBH

Each of the remaining lines represents the state of a particle, i, at a given moment in time, t. A line has the following format, for $1 \leq i \leq n$:

$$i\ t_i\ x_i\ y_i\ z_i\ v_{x_i}\ v_{y_i}\ v_{z_i}\ a_{x_i}\ a_{y_i}\ a_{z_i}\ \left[\frac{da_x}{dt}\right]_i \left[\frac{da_y}{dt}\right]_i \left[\frac{da_z}{dt}\right]_i \left[\frac{d^2a_x}{dt^2}\right]_i \left[\frac{d^2a_y}{dt^2}\right]_i \left[\frac{d^2a_z}{dt^2}\right]_i.$$

Initially, all particles are listed with the state at time $t = 0$. The first nBH lines are the black holes and the remaining $n-nBH$ lines are the other particles. After the initial conditions, particle information is present for some $0 < t \leq t_{max}$ only when that particle information has changed. Some particles are noted more often then others, because the energy level for these particles is extremely high. If the exact position is not known at time t, the position of the particle is interpolated. The following property is guaranteed: t_k on line k is t_l on line l if $k < l$. Pseudo-code for displaying the particles is given in Figure 3.

```
while (t < tmax) do
    collectionOfStars = findStarsInTime(t)
    collectionOfStars.display()
    t = t + deltaT
```

Fig. 3. Pseudo-code for extracting, interpolating, and displaying particles

The original interpolation predicted where a particle will be in the future based on state of the particle in the past. Given the number of time units in the future to predicate, Δt, and the first and second derivatives of acceleration from the input file, the equation for the future x coordinate is:

$$x_{future} = x + v_x \Delta t + \left(\frac{\Delta t^2}{2} \cdot \frac{da_x}{dt}\right) + \left(\frac{\Delta t^3}{6} \cdot \frac{da_x}{dt}\right) + \left(\frac{\Delta t^4}{24} \cdot \frac{d^2 a_x}{dt^2}\right).$$

Similar equations calculate y_{future} and z_{future}.

This worked well for stable galaxies, but for more interesting galaxies that are more active, the calculations became too imprecise. The solution for this problem is to find a particle's state s_1 and s_2 where $t_{s_1} \leq t < t_{s_2}$ and use linear interpolation based on time. This also decreased the amount of information stored for each particle because only the position information is important.

The original design did not taken into account the complexity involved in processing the particle information. Disk I/O speed was the major problem. More than 24 frames per second (FPS) are required in order to get a smooth visualization. The initial version could only read enough data to achieve 1 FPS. Preprocessing, removing non-essential data, and compressing the file increased the speed by 500%. The system can now display at 5 FPS.

5 The System in Use

The system was successfully used to visualize the collision of two galaxies, which resulted in the creation of a new, single galaxy. The astrophysicists created an initial situation where they assumed one galaxy would be stable. The visualization system proved that this assumption was not correct. Figure 4 shows the system at three different points in time. The black holes are the three dark masses at the center of the three galaxies. The stars are single points. The picture on the left shows the initial situation. Over time, the upper galaxy becomes unstable and eventually captured by the center galaxy.

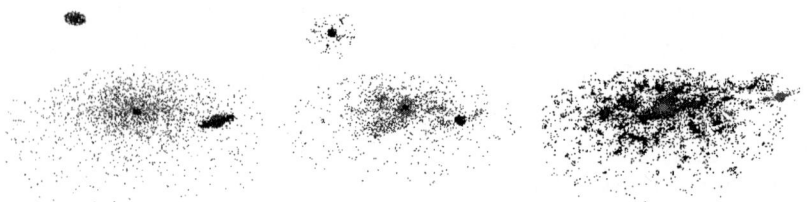

Fig. 4. Three snapshots of three galaxies merging

6 Creating a Movie

It is relatively simple to create a movie. First, a system must be defined that controls the transition through space and time. This systems output are simple Spiegel commands, which are sent to the Feeder.

The rendering is done off-screen and each individual image is stored in a file. The individual images are later assembled into a movie using the Java Media Framewor

7 Related Work

Spiegel was designed with minimum hardware in mind–no special hardware or software is required to run the system. Hut [7], followed a different approach by using the planetarium of the American Museum for Natural History (AMNH)

in New York City as a display device. The user gets a very realistic visual experience. This system is based on partiview [1], which uses OpenGL [9] to render images. This allows the program to interact with the museums projection system, an Onyx2 with 28 CPUs, 14GB of memory, 2TB disk space, and 7 graphics pipes. The user interaction with partiview requires some practice; it does not allow interpolation of the position of stars over time. Partiview is very well suited for display existing galaxies and not very well suited to visualize the results of simulations.

Most visualization systems, like [6] provide a generic framework for information specialization which is mainly focused on dynamic hierarchy computation and user controlled refinement of those hierarchies for preprocessing unstructured information space. Spiegel has a very structured information space and would not benefit, and therefore does not need, the immense overhead required by the framework described in [6].

8 Future Work

Spiegel is work in progress. The decision to use the scripting language Sprache turned out to be a great idea. However, Sprache is a small, simple language which does not include any kind of control structure. The system would benefit if the language would be extensible and act like a macro processor interpreter [5]. This would allow new, often used, commands to be to added without changing the language or the systems that depend on it.

The Feeder is designed to listen on a network port for connections from programs that supply Sprache commands. With this, the system can be easily extended to accomplish tasks which have not been programmed into the system. Smarter, better suited tools can be programmed and attached to Spiegel in order to perform more complicated tasks, like a specialized fly-through path that follows a particle, or connection and disconnection of cameras in a sophisticated time driven fashion. Some of these ideas are planned for future development.

The visualization part would benefit from the study of perception. It would be beneficial to choose colors, fog effects, and other perceptual characteristics in order to give a better visual presentation of the physical reality. No human has seen the collision of galaxies in a short period of time. Therefore, what can be visualized is not based on experience; it must be based on perception.

Since the system is extensible it could visualize gas clusters if a plug-in were available. These parts have to be developed in order to prove that the framework is flexible enough.

The current system can render an animation sequence at 5 FPS. More than 24 FPS is required in order to see a flicker-free movie. The major performance bottleneck of the current system is file I/O. Around 80% of the CPU cycles are used to read the data from disk, around 5% are used for data interpolation and graphic rendering. At the moment, the file is a compressed. Clearly, the goal is to decrease the amount of data that describes the system. Each particle follows a 3-dimensional spline curve. Therefore, it should be possible to describe paths of the particles as a function of time and not with discrete points. More research needs to be done to answer this question.

Acknowledgements

We would like to thank the following Computer Science Students at RIT for their contribution: Andrew Bak, Meng Jiang, Raun Krisch, Andrew Rader, Prachi Shinde, Christopher Stelma, and Peter Weisberg. We would also like to express our thanks to David Merritt, Andras Szell, and Peter Berczik from the Department of Physics at RIT. They provided the actual data and the interpretation of the visual representation.

References

1. Brian Abbott. partiview. http://www.haydenplanetarium.org/hp/vo/du/partiview.html.
2. Ernst Nils Dorband, Marc Hemsendorf, and David Merritt. Systolic and hypersystolic algorithms for the gravitational N-body problem, with an application to Brownian motion. J. Comput. Phys., 185(2):484C511, 2003.
3. Flock of birds. http://www.ascension-tech.com/products/flockofbirds.php.
4. The grapecluster project: A dedicated parallel platform for astrophysical dynamics. http://www.cs.rit.edu/grapecluster.
5. Brian W. Kernighan and Dennis M. Ritchie. The m4 macroprocessor. 1977.
6. Matthias Kreuseler, Norma Lpez, and Heidrun Schumann. A scalable framework for information visualization. Electronic Edition (IEEE Computer Society DL), 27.
7. Peter Teuben, Piet Hut, Stuart Levy, Jun Makino, Steve McMillan, Simon Portegies Zwart, Mike Shara, and Carter Emmart. Immersive 4-d interactive visualization of large-scale simulation. volume 238, page 499. Astronomical Society of the Pacific, 2001.
8. The grape project. http://grape.astron.s.u-tokyo.ac.jp/grape/.
9. Mason Woo, Jackie Neider, and Tom Davis. OpenGL Programming Guide: The Official Guide to Learning Opengl, Version 1.1. Addison-Wesley.

A Layout Algorithm for Signal Transduction Pathways as Two-Dimensional Drawings with Spline Curves[1]

Donghoon Lee, Byoung-Hyon Ju, and Kyungsook Han*

School of Computer Science and Engineering, Inha University,
Inchon 402-751, Korea
khan@inha.ac.kr

Abstract. As the volume of the biological pathway data is rapidly expanding, visualization of pathways is becoming an important challenge for analyzing the data. Most of the pathways available in databases are static images that cannot be refined or changed to reflect updated data, but manual layout of pathways is difficult and ineffective. There has been a recent interest in the use of the three-dimensional (3D) visualization for signal transduction pathways due to the ubiquity of advanced graphics hardware, the ease of constructing 3D visualizations, and the common perception of 3D visualization as cutting-edge technology. However, our experience with visualizing signal transduction pathways concluded that 3D might not be the best solution for signal transduction pathways. This paper presents an algorithm for dynamically visualizing signal transduction pathways as 2D layered digraphs.

1 Introduction

Recently a number of biological pathway databases have been developed, and visualization of biological networks is crucial to the effective analysis of the data. There are several types of biological networks, such as signal transduction pathways, protein interaction networks, metabolic pathways, and gene regulatory networks. Different types of network represent different biological relationships, and are visualized in different formats in order to convey their biological meaning clearly. The primary focus of this paper is the representation of signal transduction pathways.

A signal transduction pathway is a set of chemical reactions in a cell that occurs when a molecule, such as a hormone, attaches to a receptor on the cell membrane. The pathway is a process by which molecules inside the cell can be altered by molecules on the outside [1]. A large amount of data on signal transduction pathways is available in databases, including diagrams of signal transduction pathways [2, 3, 4]. However, most of these are static images that cannot be changed to reflect updated data. It is increasingly important to visualize signal transduction pathways from databases.

---

[1] This study was supported by the Ministry of Health & Welfare of Korea under grant 03-PJ1-PG3-20700-0040.
* Correspondence Author.

Signal transduction pathways are typically visualized as *directed graph* (*digraph* in short) in which a node represents a molecule and an edge between two nodes represents a biological relation between them. Signal transduction pathways convey their meaning best when they are visualized as layered digraphs with uniform edge flows. Therefore, the problem of visualizing signal transduction pathways can be formulated as a graph layout problem. There has been a recent interest in the use of the three-dimensional (3D) visualization for signal transduction pathways due to the ubiquity of advanced graphics hardware, the ease of constructing 3D visualizations, and the common perception of 3D visualization as cutting-edge technology. However, our experience with visualizing signal transduction pathways concluded that 3D might not be the best solution for signal transduction pathways. Because most 3D visualization techniques have a 2D visualization counterpart, a question arises with respect to the appropriateness of 3D visualization as opposed to 2D visualization for signal transduction pathways. This paper presents an algorithm for automatically visualizing signal transduction pathways as 2D layered digraphs.

2 Layout Algorithm

To discuss the layout algorithm, a few terms should be defined. Suppose that $G=(V, E)$ is an acyclic digraph. A layering of G is a partition of V into subsets $L_1, L_2, \ldots, L_h$, such that if $(u, v) \in E$, where $u \in L_i$ and $v \in L_j$, then $i > j$. The *height* of a layered digraph is the number h of layers, and the *width* of the digraph is the number of nodes in the largest layer. The *span* of an edge (u, v) with $u \in L_i$ and $v \in L_j$ is $i - j$.

We visualize signal transduction pathways as layered digraphs. The visualization algorithm is composed of 3 steps at the top level: (1) layer assignment and cycle handling, (2) crossing reduction, and (3) placement of edges with span > 1.

2.1 Layer Assignment and Cycle Handling

This step assigns a y-coordinate to every node by assigning it to a layer. Nodes in the same layer have the same y-coordinate values. It first places all the nodes with no parent in layer L_1, and then each remaining node n in layer L_p+1, where L_p is the layer of n's parent node. When the layer of a node is already determined, the larger value of layers is assigned to the node. Node L in the middle graph of Fig. 1, for example, is assigned to layer 4 from the path (A, E, I, L), but 3 from the path (B, G, L). The larger value of 4 becomes the layer number of node L.

The main drawback of this layering is that it may produce a too wide digraph and that an edge may have a span greater than one. The number of edges whose span > 1 should be minimized because they cause the subsequent steps (steps 2-3) of the algorithm take long [9]. We place the source node of an edge whose span > 1 to higher layers so that the span of the edge becomes one.

Fig. 1 shows an example of the initial layer assignment for the input data of signal transduction below. The initial layer assignment is adjusted to minimize edge spans, as shown in Fig. 2. This step also handles cycles in signal transduction pathways. Starting with a node with no parent node, it assigns a layer to every node along the edges connected to the node. When it encounters a node with a layer assigned to it, it has found a cycle. It starts with a new node no parent node and repeats the same thing. When every node has been assigned a layer, it goes on to step 2.

A	E
B	F
B	G
C	I
D	I
E	I
F	H
G	J
G	L
I	K
I	L
K	D

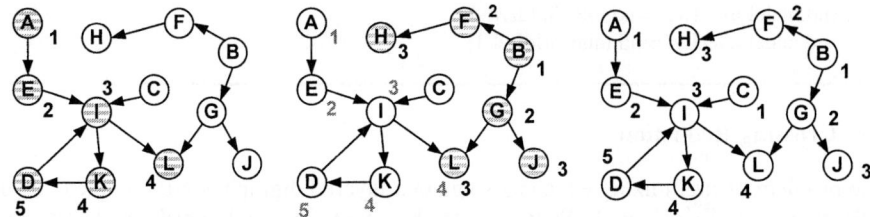

Fig. 1. An example of assigning nodes to layers. The layer numbers in grey indicate the previously assigned numbers. There is a cycle (D, I, K) in the graph

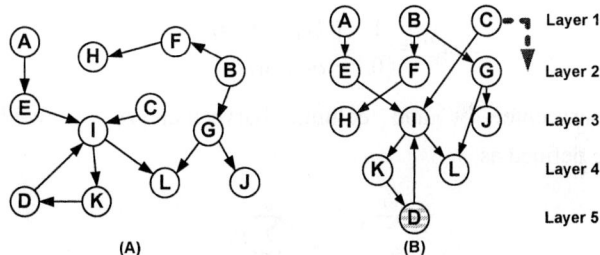

Fig. 2. (A) Initial digraph. (B) Layered digraph. Node D has an upward edge due to a cycle (I, K, D, I)

Algorithms 1-3 describe step 1 in detail

Algorithm 1 AssignLevel ()
---
foreach(nd ∈ G)
  nd.nodeLayerLvl=-1;
foreach(nd ∈ G)
  if(not (nd has owner))
    AssignNodeLevel(nd, 0);
AlignBottom();

Algorithm 2 AssignNodeLevel (node, nodeLvl)
___
if(node.nodeLayerLvl < nodeLvl) {
  node.nodeLayerLvl = nodeLvl;
  foreach(eg ∈ Node)
    if (eg is node's child) {
      if(node's child is not count) {
        AssignNodeLevel(eg, nodeLvl+1);
      }
      else return;
}

Algorithm 3 AlignBottom ()
___
foreach(nd ∈ G) {
  if(nd.nodeLayerLvl != nd.minChildLvl-1)
    nd.nodeLayerLvl = nd.minChildLvl-1;
}

2.2 Crossing Reduction

The problem of minimizing edge crossings in a layered digraph is NP-complete, even if there are only two layers [10]. We use the barycenter method to order nodes at each layer [11, 12]. In the barycenter method, the x-coordinate of each node is chosen as the barycenter (average) of the x-coordinates of its neighbors. Since two adjacent layers are considered in this method, edges whose span >1 are ignored in this step.

Suppose that the element $m_{kl}^{(i)}$ of incidence matrix $M(i)$ is given by

$$m_{kl}^{(i)} = \begin{cases} 1, & if\ (v_k, v_l) \in E \\ 0, & otherwise \end{cases} \quad (1)$$

The row barycenter γ_k and column barycenter ρ_l of incidence matrix $M = (m_{kl})$ are defined as

$$\gamma_k = \sum_{l=1}^{q} l \cdot m_{kl} / \sum_{l=1}^{q} m_{kl} \quad (2)$$

$$\rho_l = \sum_{k=1}^{p} k \cdot m_{kl} / \sum_{k=1}^{p} m_{kl} \quad (3)$$

The number C of crossings of the edge between v_k and v_l is given by

$$C(v_k, v_l) = \sum_{\alpha=k+1}^{p} \sum_{\beta=1}^{l-1} m_{\alpha\beta} \cdot m_{kl} \quad (4)$$

When rearranging the order of rows (columns), the row (column) barycenters are computed and arranged in increasing order with the order of columns (rows) fixed. By repeatedly alternating row and column barycenter ordering, the total number of crossings is reduced. The algorithm for reducing the total number of crossings is given in Algorithms 5-7, and Fig. 3A shows an example of computing the initial row and column barycenters, and the total number of crossings for a graph in Fig. 2B. Fig 3B shows the final row and column barycenters, and the total number of crossings after applying Algorithms 5-7, and the final layout obtained is displayed in Fig. 4A.

Algorithm 4 baryCenter(tarLayer)

foreach(nd in curLayer)
   nd.calculateBCValue(tarLayer);
if(isNeedSort) {
   curLayer.sortbyBCValue;
   return true;
} else return false;

Algorithm 5 calculateBCValue(tarLayer)

BCValue=0; node_layer_cntSub=0;
   foreach(nd in tarLayer) {
     if(this is nd's neighbor) {
       BCValue += nd.LayerIdx;
       node_layer_cntSub++;
     }
   }
BCValue/=node_layer_cntSub;

Algorithm 6 calcBaryCenter()

do {
   bWork=true;
   foreach(layer in G) bWork &= layer.nextLayer.baryCenter(layer);
   foreach(layer in G) bWork &= layer.baryCenter(layer.preLayer);
} while (bWork)

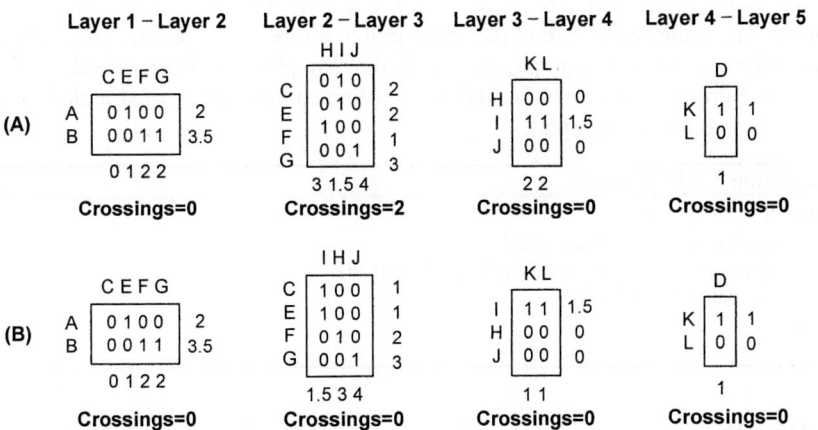

Fig. 3. (A) The initial row and column barycenters, and the total number of crossings for a graph in Fig. 3B. (B) The final row and column barycenters, and the total number of crossings of a graph in Fig. 4A. When rearranging the order of rows and columns based on the row barycenter γ_k and column barycenter ρ_l, nodes with $\gamma_k=0$ or $\rho_l=0$ need not be rearranged

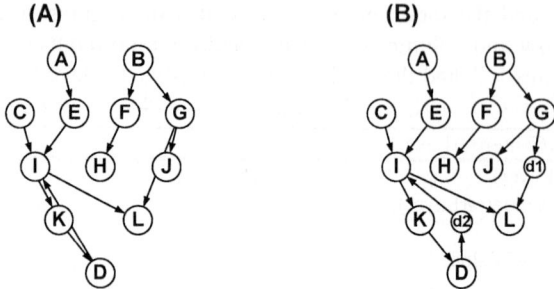

Fig. 4. (A) Digraph with an optimized placement for all nodes and edges except the edges with span > 1. (B) Graph with dummy nodes introduced for edges with span > 1. Both nodes d1 and d2 are inner dummy nodes

2.3 Placement of Edges with Span > 1

This step places long edges whose edge span > 1, which have been ignored in the previous steps. Every edge with span=2 is placed using an inner dummy node. Dummy nodes that can be included inside the current graph are called inner dummy nodes, and are considered when computing barycenters (Fig. 4B). On the contrary, dummy nodes that cannot be included inside the current graph are called outer dummy nodes and excluded when computing barycenters. For edges with span > 2, it first computes possible edge crossings caused by inner dummy nodes and outer dummy nodes and selects whichever with fewer edge crossings. When one or more inner dummy nodes are created, the crossing reduction step is performed. Otherwise, there is no further crossing reduction.

Inner dummy nodes are created at each layer and the inner dummy nodes are included when computing barycenters. Therefore, their positions are not fixed until the last step. On the other hand, a pair of outer dummy nodes are created for an edge, one below the source node and the other above the sink node. When the source node is located in the left of the center line of the entire graph, all the outer dummy nodes are placed in the left of the current graph; otherwise, they are placed in the right side. Edges connecting dummy nodes are displayed using spline curves instead of straight line segment (see Fig. 6 for an example).

Algorithm 7 AddDummy()

```
for each node n ∈ V {
    for each downward edge e of n {
        if (n.nodeLayerLvl+1 < child(n).nodeLayerLvl)
            CreateDummyNodes();
    }
}
```
---

3 Results

The algorithms were implemented in a web-based program called PathwayViewer. PathwayViewer runs on Windows 2000/XP/Me/98/NT 4.0 systems. An example of the user interface of the program is shown in Fig. 5. Fig. 6 shows the actual signal transduction pathway of mitogen-activated protein kinase. Note the two spline curves for the edges with span > 1, which have been placed using dummy nodes.

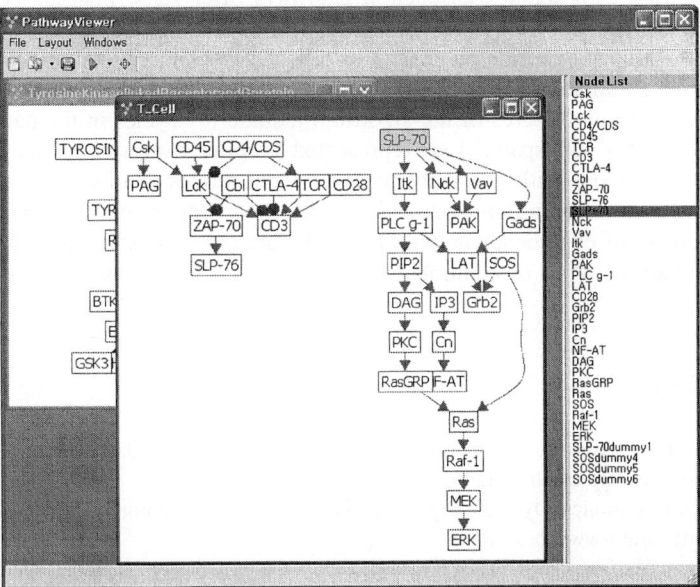

Fig. 5. Example of the user interface of the program. Red arrows indicate the source nodes activate the sink nodes. Blue lines ended with filled circles indicate that source nodes inhibit the sink nodes. The node selected by a user (the yellow node in the signal transduction pathway) is also highlighted in the node list window

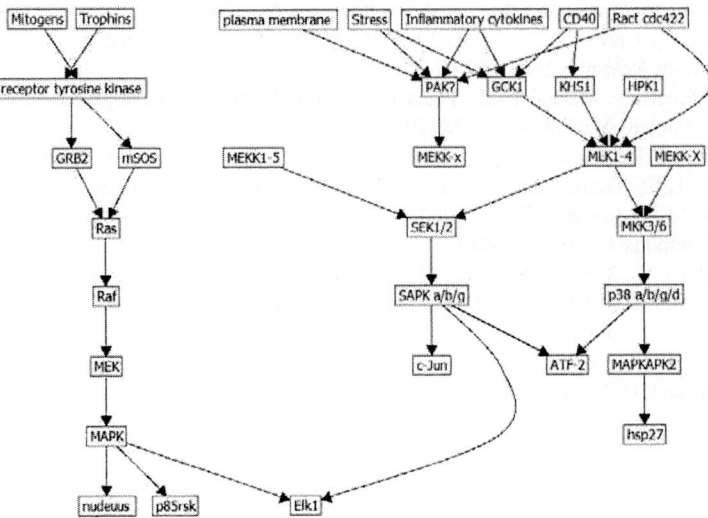

Fig. 6. The signal transduction pathway of mitogen-activated protein kinase, visualized by our algorithm

4 Conclusion

Most databases with signal transduction pathways offer static images. These static images are esthetic since they are hand-made and are fast in loading. However, static images are difficult to refine or change to reflect new data. We have developed an algorithm for automatically representing signal transduction pathways from databases or text files. Unique features of the algorithm include (1) cycles in the pathways are handled; (2) edges with span > 1 are represented as spline curves; (3) it does not place all sink nodes (nodes with no parent) in layer 1 but moves them to a lower layer so that edge spans can be minimized; and (4) edge bends occur only at dummy nodes and the number of edge bends is minimized. We are currently extending the program to overlay various types of additional information onto the signal transduction pathways.

References

1. Kanehisa, M., Goto, S., Kawashima, S., Nakaya, A.: The KEGG databases at GenomeNet. Nucleic Acids Research 30 (2002) 42-46
2. Hippron Physiomics, Dynamic Signaling Maps. http://www.hippron.com/products.htm
3. BioCarta. http://www.biocarta.com
4. Kanehisa, M., Goto, S.: KEGG: Kyoto encyclopedia of genes and genomes. Nucleic Acids Research 28 (2002) 27-30
5. Wackett, L., Ellis, L., Speedie, S., Hershberger, C., Knackmuss, H.J., Spormann, A., Walsh, C., Forney, L., Punch, W., Kazic, T., Kaneshia, M., Berndt, D.: Predicting microbial biodegradation pathways. ASM News 65 (1999) 87-93
6. Overbeek, R., Larsen, N., Pusch, G., D'Souza, M., Selkov, E., Kyrpides, N., Fonstein, M., Maltsev, N.: WIT: integrated system for high-throughput genome sequence analysis and metabolic reconstruction. Nucleic Acids Research 28 (2000) 123-125
7. Selkov, E., Grechkin, Y., Mikhailova, N.: MPW: the metabolic pathways database. Nucleic Acids Research 26 (1998) 43-45
8. Karp, P., Riley, M., Saier, M., Paulsen, I., Paley, S. M.: The EcoCyc and MetaCyc databases. Nucleic Acids Research 28 (2000) 56-59
9. Gansner, E.R., Koutsofios, E., North, S.C., Vo, K.-P.: A technique for drawing directed graphs. IEEE Transactions on Software Engineering 19 (1993) 214-230
10. Garey, M.R., Johnson, D.S.: Crossing Number is NP-Complete. SIAM J. Algebraic Discrete Methods 4 (1983) 312-316
11. Sugiyama, K., Tagawa, S., Toda, M.: Method for visual understanding of hierarchical system structures, IEEE Transaction on Systems, Man, and Cybernetics SMC-11 (1981) 109-125
12. Sugiyama, K.: Graph Drawing and Applications for Software and Knowledge Engineering. Singapore (2002)

Interactive Fluid Animation and Its Applications

Jeongjin Lee[1], Helen Hong[2], and Yeong Gil Shin[1]

[1] School of Electrical Engineering and Computer Science,
Seoul National University,
{jjlee, yshin}@cglab.snu.ac.kr
[2] School of Electrical Engineering and Computer Science BK21,
Information Technology, Seoul National University,
San 56-1 Shinlim 9-dong Kwanak-gu, Seoul 151-742, Korea
hlhong@cse.snu.ac.kr

Abstract. In this paper, we propose a novel technique of fluid animation for interactive applications. We have incorporated an enhanced particle dynamics simulation method with pre-integrated volume rendering. The particle dynamics simulation of fluid flow can be conducted in real-time using the Lennard-Jones model. The computational efficiency is enhanced since a small number of particles can represent a significant volume. To get a high-quality rendering image with small data, we use the pre-integrated volume rendering technique. Experimental results show that the proposed method can be successfully applied to various fluid animation applications at interactive speed with acceptable visual quality.

1 Introduction

The demand for interactive fluid animation has increased recently for 3D computer games and virtual reality applications. However, it is very difficult to animate natural fluid phenomena at interactive speed, because their motions are so complex and irregular that intensive simulation and rendering time is needed.

In the previous work, only off-line fluid animation methods have been reported [1-3]. In general, fluid animation is carried out by physical simulation immediately followed by visual rendering. For the physical simulation of fluids, the most frequently used practices are the particle dynamics simulation of isolated fluid particles and the continuum analysis of flow via the Navier-Stokes equation. Miller et al. [4] proposed a spring model among particles to represent viscous fluid flow. Terzopoulos et al. [5] introduced molecular dynamics to consider interactions between particles. In these approaches, when the number of particles increases significantly, the number of related links between particles exponentially increases. Therefore, it takes too much time for realistic fluid simulation due to the large number of particles for describing complex fluid motions. Stam [1] proposed a precise and stable method to solve the Navier-Stokes equations for any time step. Foster [3] applied a 3D incompressible Navier-Stokes equation. Above methods using the Navier-Stokes equations yield a realistic fluid motion when properly conditioned, but still need huge calculations of complex equations. The second limitation is the time complexity of visual rendering. Global illumination has been widely used for natural fluid animation. Jensen et al. [6] proposed a photon mapping method currently used in many applications. Global illumination is generally successful in rendering premium-quality images, but too slow to be used in interactive applications.

In this paper, we propose a novel technique for interactive fluid animation and its applications. For rapid analysis of the motion of fluids, we use a modified form of particle dynamic equations. The fluid interaction is approximated by the attractive and repulsive forces between adjacent particles using the Lennard-Jones model to emulate fluid viscosity. To get a high quality rendering image with a smaller volume data, we use a pre-integrated volume rendering method [7]. Experimental results show that our method is successfully applied to various interactive fluid animation applications.

The organization of the paper is as follows. In Section 2, we discuss the particle dynamics simulation of our method, and describe how the simulation data are rendered. In Section 3, experimental results show various kinds of interactive fluid animation applications. This paper is concluded with brief discussions of the results in Section 4.

2 Interactive Fluid Animation Methodology

2.1 Dynamics Simulation of Fluid Particles

Two approaches, particle dynamics and continuum dynamics, have been widely used for fluid simulation. The continuum dynamics approach is not suitable for interactive applications due to its high time complexity of calculating the Navier-Stokes equation [1-3]. In our approach, a simple particle dynamics approach is chosen since it is much faster than a continuum dynamics approach based on the Navier-Stokes equation.

In particle dynamics, a spherical particle is assumed to be the basic element that makes an object such as for solid, liquid and gas, and used for calculating interactions between particles. For N spherically symmetric particles, the total inter-particle potential energy $E(\mathbf{r}^N)$ is the sum of isolated pair interactions according to pair-wise addition.

$$E(\mathbf{r}^N) = \sum_{i=1}^{N} \sum_{j=1}^{N} u(r_{ij}), \ i \neq j, \tag{1}$$

where $\mathbf{r}^N$ is the set of vectors that locate centers of mass, i.e. $\mathbf{r}^N = \{\mathbf{r}_1, \mathbf{r}_2, \mathbf{r}_3, ..., \mathbf{r}_N\}$ and r_{ij} is the scalar distance between particles i and j.

The elementary potential energy $u(r_{ij})$ is taken from the Lennard-Jones (LJ) potential model [8]. For two particles i and j separated by a distance r_{ij}, the potential energy $u(r_{ij})$ between the both can be defined as

$$u(r_{ij}) = 4\varepsilon \left(\left(\frac{\sigma}{r_{ij}}\right)^{12} - \left(\frac{\sigma}{r_{ij}}\right)^{6} \right). \tag{2}$$

The force field f_{ij} created by two particles i and j can be given as

$$f_{ij} = -\frac{du(r_{ij})}{dr_{ij}} = \left(\frac{48\varepsilon}{\sigma^2}\right)\left[\left(\frac{\sigma}{r_{ij}}\right)^{14} - \frac{1}{2}\left(\frac{\sigma}{r_{ij}}\right)^{8}\right]r_{ij}. \quad (3)$$

Since the inter-particle potential forces are conservative within a given potential field, the overall potential force $F_{i,p}$ acting on particle i is related to the potential by

$$F_{i,p} = -\frac{\partial E(\mathbf{r}^N)}{\partial \mathbf{r}_i} = m_i\ddot{\mathbf{r}}_i, \quad (4)$$

where m_i is the mass of particle i. Fig. 1 illustrates the Lennard-Jones potential and the force extended over a modest range of pair separations. The critical distance at which the positive sign of the inter-particle force becomes negative can be considered as the particle radius.

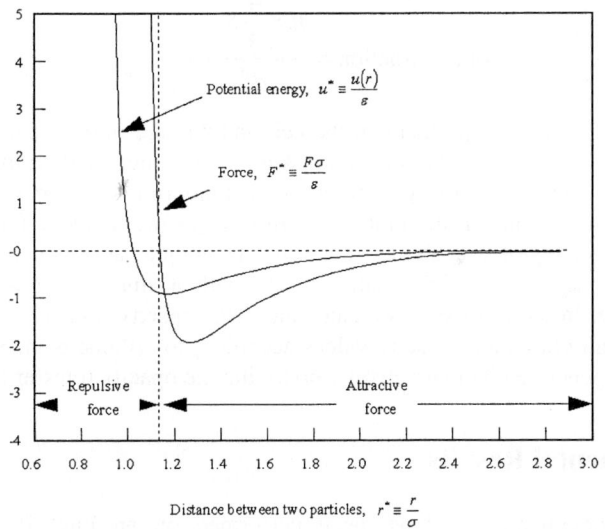

Fig. 1. The Lennard-Jones potential and the force

The friction force $F_{i,f}$ on particle i can be given as

$$F_{i,f} = -\zeta \dot{\mathbf{r}}_i, \quad (5)$$

where ζ is the friction coefficient. The governing equation of the force balance on particle i can hence be written as

$$-\frac{\partial E(\mathbf{r}^N)}{\partial \mathbf{r}_i} + \zeta \ddot{\mathbf{r}}_i = 0 \ . \tag{6}$$

Eq. (6) can be used to calculate the interaction between particles by assuming them as slightly deformable soft spheres. The soft sphere collision method is useful for treating interactions between particles with the separation distance within a moderate multiple of the critical distance.

2.2 Realistic Rendering of Simulation Data

A photon mapping method [6] accomplishes the global illumination effect using ray tracing, which is excessively slow for interactive applications. For interactive rendering without the loss of image quality, we use a pre-integrated volume rendering on graphics hardware. Rendering of simulation data is accomplished in the following three steps. In the first step, we transform the simulation data into volume data. We divide the 3D space, in which particles of simulation data are stored, into regular cells having unit length d. The density of each cell is determined by the volume fraction value as Eq. (7). These density values are used for volume rendering.

$$\text{Volume fraction} = \frac{n \times \frac{4}{3}\pi r^3}{d^3} \ , \tag{7}$$

where n is the number of particles in the cell and r is the radius of particles. In the second step, we visualize volume data using a pre-integrated volume rendering technique. The color and opacity between two neighboring slices of volume data are pre-integrated and stored in the graphics hardware texture memory for acceleration. Using the pre-integrated volume rendering technique accelerated by graphics hardware we can get the high quality image with a smaller volume data at the interactive rate. In the third step, we can control the opacity transfer function, which assigns different colors and opacity values according to volume data. Various visual effects can be generated by interactively modifying the opacity transfer function.

3 Experimental Results

All of the implementations have been performed on an Intel Pentium IV PC containing 2.4 GHz CPU with GeForce FX 5800 graphics hardware. Fig. 2 and 3 show the animation of water flowing from a bottle to a cup. The opacity transfer function of this animation is given in Table 1. Between two control points Vi and Vj, color and opacity are determined using linear interpolation. As shown in Fig. 2, when the number of particles (n) is 2000 and the radius of particles (r) is 0.002 [m], particle characteristics of the water are emphasized. This animation occurs in 7 ~ 8 fps at 730 x 520 resolution. Fig. 3 shows that when the number of particles (n) is 10000 and the radius of particles (r) is 0.003 [m], continuum characteristics of the water are emphasized. This animation occurs in 3 ~ 4 fps at 730 x 520 resolution.

Table 1. Opacity transfer function for water animation

d1 / V1(R, G, B, A)	0.21 / (0, 0, 255, 0.0)
d2 / V2(R, G, B, A)	0.23 / (36, 65, 91, 1.0)
d3 / V3(R, G, B, A)	0.28 / (154, 208, 228, 1.0)
d4 / V4(R, G, B, A)	0.30 / (0, 0, 160, 0.0)

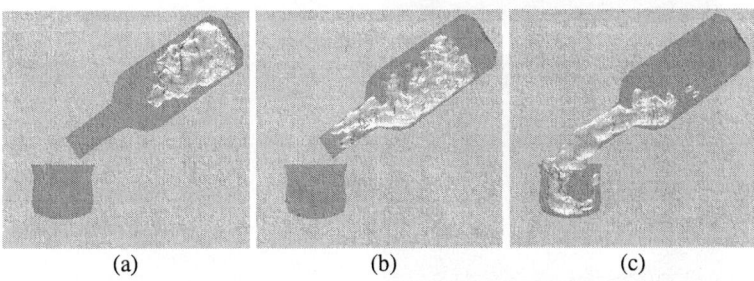

(a)   (b)   (c)

Fig. 2. The animation of pouring water from a bottle into a cup (n = 2000, r = 0.002 [m]) (a) t = 0.12 [s] (b) t = 0.33 [s] (c) t = 1.02 [s]

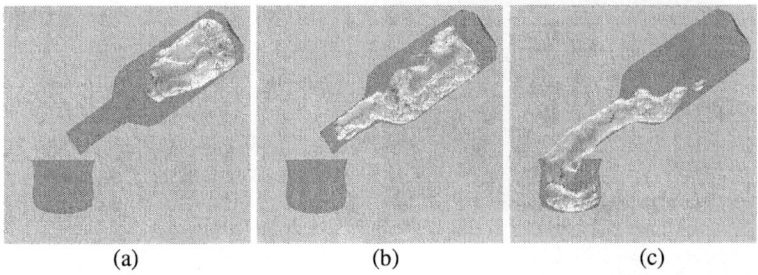

(a)   (b)   (c)

Fig. 3. The animation of pouring water from a bottle into a cup (n = 10000, r = 0.003 [m]) (a) t = 0.12 [s] (b) t = 0.33 [s] (c) t = 1.02 [s]

Fig. 4 shows the comparison of our method with RealFlow, which is the widely used commercial software for fluid animation. The average number of particles used for the animation of RealFlow is 1870000 while that of our method is 8000. For the generation of 10 seconds' animation in Fig. 4(e), (f), RealFlow computes during 4 hours 30 minutes whereas our method finishes within 150 seconds. The experimental results show that our method gives similar visual quality comparing with RealFlow using much smaller number of particles. In addition, total processing time of our method is dramatically faster than that of RealFlow.

Fig. 5 shows water and artificial fluid flow using our fluid animation method. Fig. 6(a) shows a conceptual overview of fluidic shadow dance application. The shadow of the viewer is projected onto a screen by a light. Two columns to the left and right of the viewer are equipped with a vertical array of switches. When the viewer's hand

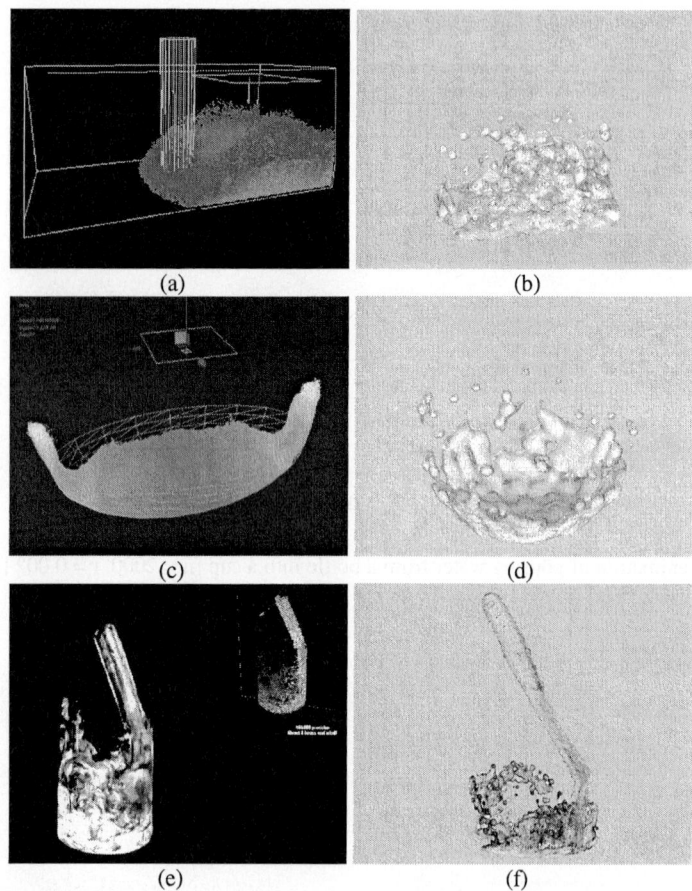

Fig. 4. Comparison of our method with RealFlow (a) RealFlow (n = 160000) (b) our method (n = 8000) (c) RealFlow (n = 5000000) (d) our method (n = 6000) (e) RealFlow (n = 450000) (f) our method (n = 10000)

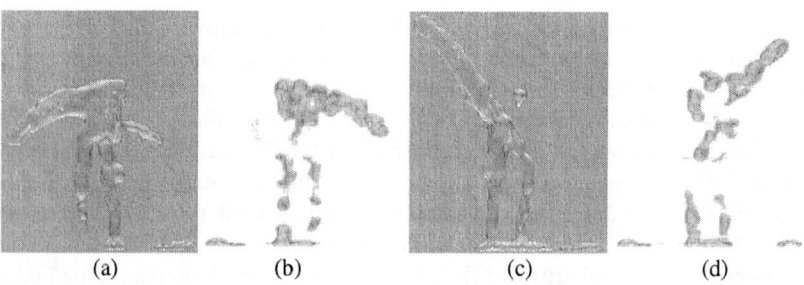

Fig. 5. Fluid animation for fluidic shadow dance application (a) water flow from left-lower side (b) artificial fluid flow from right-lower side (c) water flow from left-upper side (d) artificial fluid flow from right-upper side

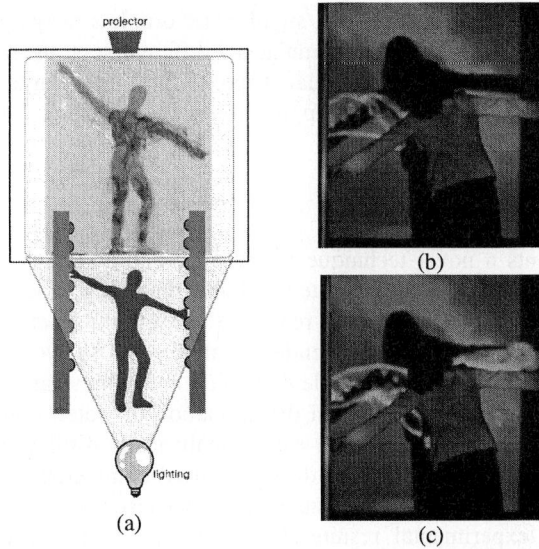

Fig. 6. The fluidic shadow dance application (a) conceptual overview (b) flow from left column (c) flow from both columns

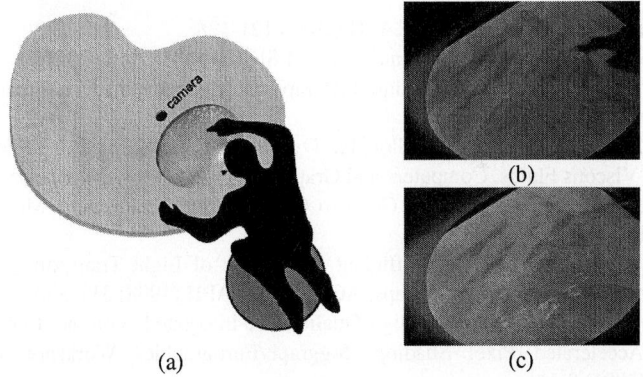

Fig. 7. The fluid portrait application (a) conceptual overview (b) by fingertip (c) by stick

comes in contact with a specific location, the corresponding switch is inputted into our program, and the image of a fluid flowing out from that location is animated. Fluids coming out of the two columns are ejected towards the center of the viewer's body, subsequently fall towards the bottom simulating gravity, collide into one another, and then splash upwards resulting in a free and irregular motion. The demonstration of this interactive application is shown in Fig. 6(b), (c).

Fig. 7(a) shows a conceptual overview of fluid portrait application. A camera projects the viewer's face onto the surface of a touch screen. Interaction is created between the viewer's hand movements and rendered fluid in real-time. If the viewer continuously stimulates a certain spot, small water droplets merge together to form a large water drop. When the viewer stimulates a different spot, the merged water drop disperses and moves to the other stimulated spot to form into a large water drop again. The demonstration of this interactive application is shown in Fig. 7(b), (c).

4 Conclusion

This paper presents a novel technique of fluid animation, which integrates particle dynamics simulation and pre-integrated volume rendering. The particle dynamics simulation can be conducted in real-time using the Lennard-Jones model. Furthermore, pre-integrated volume rendering allowed us to avoid the unnaturalness of images usually obtained with particle dynamics, and achieve an image quality good enough for interactive applications. In the animation of water, both the particle and continuum characteristics of the water can be realistically displayed by manipulating simulation parameters. Comparing with widely used commercial software, our fluid animation method is performed at dramatically faster speed with comparable visual quality. Various experimental results show that our method can be successfully applied to interactive fluid animation applications.

References

1. Stam, J., Stable Fluids, ACM SIGGRAPH (1999) 121-128.
2. Foster, N., Practical Animation of Liquids, ACM SIGGRAPH (2001) 23-30.
3. Foster, N., Realistic Animation of Liquids, Graphical Models and Image Processing (1996) 204-212.
4. Miller, Gavin S. P., Pearce, A., Globular Dynamics: A Connected Particle System for Animating Viscous Fluids, Computers and Graphics Vol. 13, No. 3 (1989) 305-309.
5. Terzopoulos, Platt, Fleischer, From Goop to Glop: Melting Deformable Models, Graphics Interface (1989).
6. Jensen, H.W., Christensen, P.H., Efficient Simulation of Light Transport in Scenes with Participating Media using Photon Maps, ACM SIGGRAPH (1998) 311-320.
7. Engel, K., Kraus, M., Ertl, T., High-Quality Pre-Integrated Volume Rendering Using Hardware-Accelerated Pixel Shading, Siggraph/Eurographics Workshop on Graphics Hardware (2001) 9-16.
8. McQuarrie, D.A., Statistical Mechanics, Harper and Row (1976).

ATDV: An Image Transforming System

Paula Farago[1], Ligia Barros[1], Gerson Cunha[2], Luiz Landau[2], and Rosa Maria Costa[3]

[1] Universidade Federal do Rio de Janeiro – UFRJ
Instituto de Matemática - Dept de Ciência da Computação
Centro de Tecnologia – Bl. C
Rio De Janeiro - Rj – Brasil
farago@lamce.ufrj.br, ligia@nce.ufrj.br
[2] Universidade Federal do Rio de Janeiro - UFRJ
COPPE - Programa de Eng. Civil
Caixa Postal 68552
CEP 21949-900 -Rio de Janeiro - RJ – Brasil
{gerson, landau}@lamce.ufrj.br
[3] Universidade do Estado do Rio de Janeiro – UERJ
IME – Dept. de Informática e Ciência da Computação
Rua São Francisco Xavier 524- 6º B
CEP 20550-013 - Rio de Janeiro - RJ – Brasil
rcosta@ime.uerj.br

Abstract. Some macular pathology produces degeneration, which causes a perception loss in specific areas of the visual field. In this case, people visualize "blind spots" that hide part of the images captured by the impaired eye.

The goal of this paper is to present an image distortion system that explores the Virtual Reality technology potential. To implement it we developed a system that is able to obtain the data generated by the visual field exam, and decode it. A camera captured the real images and the Convex Hull algorithm and Isoperimetric Mapping were used to identify the spots limits generating a divergent distortion in the images situated inside of the blind spot.

The modified images can be visualized in a Virtual Reality head mounted display with eye tracking. This system offers opportunities to people with this visual problem to get a general perception of the covered area, reducing the visual loss.

1 Introduction

Nowadays, the sense of vision is one of the most required and necessary for humans because we have a wide range of visual illustrated information that is presented on the Internet, on television, in cinema, etc.

Some people that had toxoplasmosis, glaucoma or senile macular degeneration can get impairments causing an area of diminished vision within the visual field. In general, they see a "blind spot", denominated scotoma, which continually hides some parts of the central and/or peripheral viewing area. The scotoma position is fixed, independent of the vision focus or angle. In the majority of cases, the person can have a perception of the general scene, because our eyes interchange central and peripheral vision all the time, but have great difficulty in visualizing objects in movement.

The goal of this paper is, therefore, to present the ATDV (Apoio Tecnológico aos Deficientes Visuais - Technological Support for Visual Impairments) system that aims at temporarily supporting the vision of a person with this pathology. We assume that there is only one scotoma in each eye and we will treat the worst. This system uses Virtual Reality and Augmented Reality techniques and will generate a deformation in the captured images. The objects that are hidden will be seen on the outside limits of the spot. It permits them to have the general perception of the visualized image.

The visual field (VF) exam measures and defines the spot extension of each user. A micro camera captures the mobile images (streams) and transforms them according to the VF exam results. An Isoperimetric Mapping and a Convex Hull algorithm are applied to the VF data. A specific programming language transforms the images that are projected in real time, by a server, on a Head Mounted Display (HMD) connected to a PC or handheld computer.

In the following sections we will present the basic characteristics of Virtual Reality technology, and some medical applications. Next, the ATDV system is explained: the main concepts of the visual field exam, the module that transforms the images, and the Isoperimetric Mapping technique. An example illustrates the algorithm application. In the Conclusions we summarize and draw final remarks.

2 Virtual Reality Applications

Virtual Reality (VR) includes advanced technologies of interface, immersing the user in environments that can be actively interacted with and explored [1]. Moreover, the user can accomplish navigation and interaction in a three-dimension synthetic environment generated by computer using multi-sensory channels. In this case, diverse stimuli can be transmitted by specific devices and perceived by one or more user senses.

There are some devices to reproduce generated images from the head-and-eye movements. The first is the Head Mounted Display (HMD), which provokes an immersion in the user.

Some kinds of VR simulations, such as Augmented Reality (AR), have specific classifications. Augmented Reality attempts to superimpose graphics over a real environment in real-time, and also changes those graphics to accommodate a user's head-and-eye movements. Therefore, AR supplements reality, rather than completely replacing it. Virtual Reality and Augmented Reality have a wide scope of application domains like medicine [2], entertainment [3], design [4], robotics and telerobotics [5].

These technologies are being broadly applied in the medical area. Much research has tested developed and tested virtual environments for treating a wide variety of mental and physical disorders [6], [7], [8], [9].

In the specific literature, there are many works describing image transformations. However, they do not explain in details the main technologies being used. In 1986 [10], NASA engineers proposed some ideas to reduce vision loss caused by different kinds of visual deficits. They developed a simple approach exploring a bi-prism arrangement in which text is viewed through two prisms placed base to base. Dagnellie [11] describes some vision rehabilitation projects, using intraocular prosthesis and mentions an application for people with scotoma. Starner [12] remapped images that are presented in a wearable computer with a head-mounted display. He uses a simple 2D 'hyper fisheye' coordinate transformation to magnify texts.

The ATDV presents some differences from these systems. The aim of the software introduced in this article is to provide an integrated platform for the visualization of transformed images in an integrated way. The patient can put the diskette with the results of his VF exam (made previously at doctor's office) in the library's PC, the home's PC or a handheld portable; a program decodes these data, generating the inputs to the Isoparametric module. In such a manner, the patient can put the HMD with camera and eye tracking attached and read or see with less difficulty in real time. The system uses the mesh transformation points and translates them dynamically, producing altered images in real-time.

Next, we present the ATDV system that is composed by modules which generates a personalized image transformation.

3 ATDV - Technological Supporting for Visual Impairments

The ATDV system integrates different input-output devices, a micro camera, a video server, an HMD with eye tracking and a Handheld [18], or a Personal Computer.

The ATDV system is considered the Central Module, and is responsible for processing data from the VF exam and for generating the transformed images, as presented in Figure 1.

The Central Module is divided in three stages: "Decoder", "VRML generator" and "Isoperimetric Mapping". All these stages compose a single program implemented in C++ language, and will be described next.

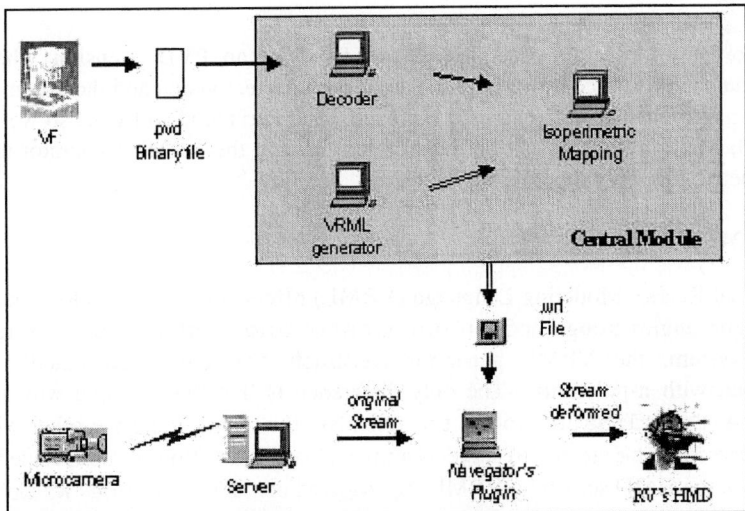

Fig. 1. ATDV system structure

3.1 Decoder

Some diseases such as Toxoplasmosis, Syphilis and Disease of Harada [13] can damage the external and internal part of the Ocular Globe and the optic nerve. This pathological process decreases the visual acuity and may cause lesions denominated scotomas. In general, it is seen as a "black spot" over the visualized images and generates a severe visual field loss.

The scotoma can be evaluated, among other manners, through a visual field exam. It determines the monocular field of view of a person, identifying vision and damaged areas.

The VF exam, from Octopus [17], is made separately on the right and on the left eye. The results are represented by graphics, which include only the 30°/120° central angle of the total vision field. Various kinds of computerized exam graphics exist, and the Gray Tone Scale, as seen in Figures 2A and 2B, is an example of it. The areas of the same visual perception are shown with the same shade in the gray scale. The area of low visual acuity is the darkest, and the blind refers to the area totally blind, the scotoma.

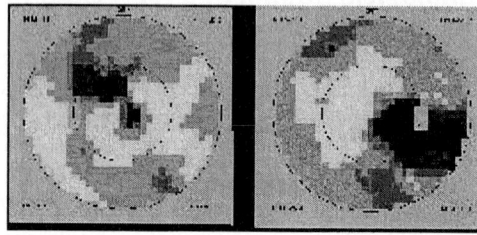

Fig. 2A and 2B. Graphics of Gray tone scale corresponding a 30°/120° central angle (respectively, left eye and right eye)

The computerized exam generates a binary file (with an .PVD extension) where the relative data is registered, informing the areas with perfect vision and those containing the scotomas. The VF data can only be read by the PeriTrend software, so we had to decode it. Next, these data will be transformed through the VRML Generator and the Isoperimetric Mapping modules.

3.2 VRML Generator

The Virtual Reality Modeling Language (VRML) offers some facilities to manipulate the polygon angles trough nodes (Coordinate and Texture Coordinate). In this system, the VRML generator constructs two equal reticulated screens overlapped with n-polygons. The only difference is that the first one will use the Texture Coordinate node to reticulate the texture and the other will use the Coordinate node to serve as a mirror-base flat. Both can be resized to be adapted on the HMD screen by VRML (Navegation Info {type"Walk"}).

The captured images are associated to the Coordinate mesh, and the reticulate angles are warped according to the scotoma dimensions by the Isoperimetric Mapping techniques. In this way, the information that is inside the scotoma area can be mapped away from the blind retinal area (Figures 6A and 7A).

3.3 Isoperimetric Mapping and Convex Hull

The Isoperimetric Mapping in mesh generation was first described by Zienkiewicz and Philips[14]. This technique causes a distortion in images, and uses polynomial interpolation functions to promote a mapping between Cartesian and curvilinear coordinates.

Consider a particular parabolic quadrilateral shape, presented in Figure 3, with eight nodes, in which associated coordinates X, Y (and Z) are well defined. In this case, $X = \sum N_i x_i$, $Y = \sum N_i y_i$ and $Z = N_i z_i$, where $I = 1...8$. Each N_i is defined in terms of a curvilinear coordinates system ξ and η, that have values between 1 and -1 at the opposed sides.

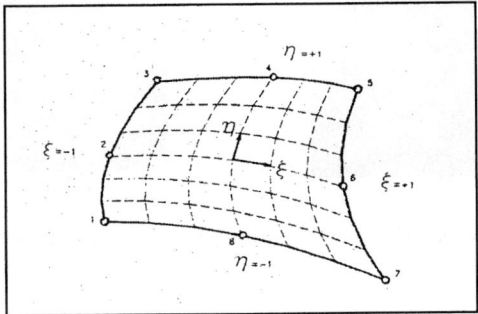

Fig. 3. Parabolic Quadrilateral

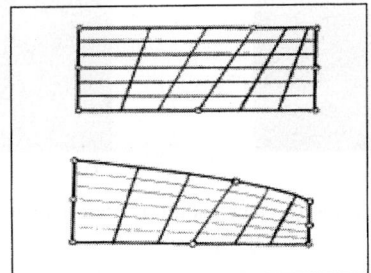

Fig. 4. Mesh transformation

It is used to map the scotoma shape, as shown in Figure 4.

Since the scotoma has a very irregular boundary we have to apply a Convex Hull Algorithm to adjust the transformed mesh to its limits. We used the Graham's scan algorithm [15] and the vectorial product.

To do this, we use the points from the binary .PDV file. Figure 5 presents the adopted technique.

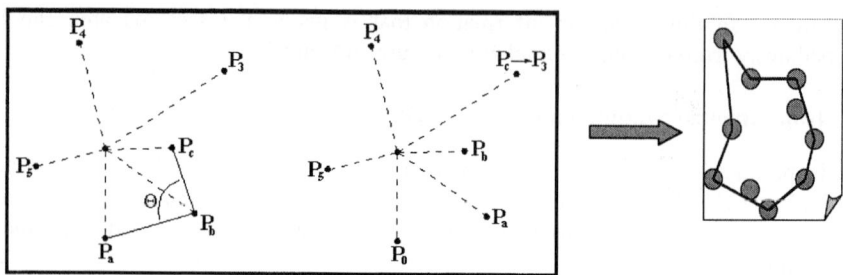

Fig. 5. The algorithm eliminates and translates some points according to scotoma limits. When Pb = P0, the process stops, and only the points around the scotoma limits remain

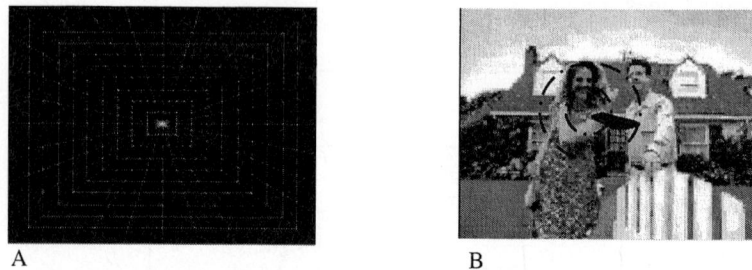

Fig. 6. Original coordinate mesh (A) and original texture (B)

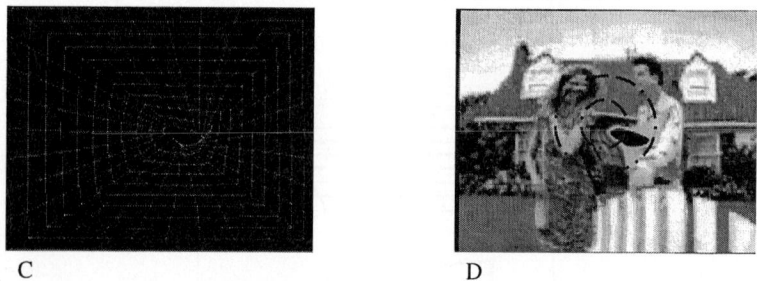

Fig. 7. Coordinate mesh (C) and texture (D) deformed by the ATDV system

The information generated by the Central Module are transformed in a .WFL extension file, that keep the mesh that will be used to transform the images that are captured by a micro camera coupled on a head-mounted display (HMD). The Video Server processes these streams, in real time, according to this .WFL file. It distributes these images to the HMD, in unicast model, using a VRML plug in.

An example of this system can be seen in Figures 6 and 7, which present a distortion based on the 30°/120° center right eye scotoma, showed in Figure 2. As a result, the objects in the warped area will look "stretched" to the outside border of the scotoma

This system has preliminarily been tested with only two people. The initial results show that it is useful to read movie captions and watch some games on television. As mentioned in the NASA projects [10] it seems that patients must have time to become accustomed to the devices.

4 Conclusions

This article presents an image distortion system that offers new opportunities to people with diminished vision caused by scotomas. The users of this system are persons that have a visual impairment that causes a vision loss in the central field of view, specifically, with only one scotoma per eye.

The system integrates the Isoperimetric and Convex Hull techniques, obtaining a deformed stream that can be visualized in real time in a Virtual Reality device.

We consider that this approach has important technical implications for the development of this kind of Augmented Reality applications.

To spread the use of VR in this domain, we are verifying the possibility of simultaneous images projection to both eyes. This will demand a larger processing capacity in order not to delay the modified images generation. Future experiments should be designed in an attempt to overcome this obstacle.

Lastly, we would like to emphasize that the advances in wearable computers and monocular screens will open new possibilities to the use of this kind of system in this specific domain.

References

1. Burdea, G., Coiffet, P.(ed): Virtual Reality Technology, 2nd edition, Wiley-Interscience, (2003).
2. Barfield, W. , Caudell, T.: Basic Concepts in Wearable Computers and Augmented Reality, In: Fundamentals of Wearable Computers and Augmented Reality, W. Barfield, T. Caudell (eds.), Mahwah, New Jersey, (2001), 3-26.
3. Stapleton, C., Hughes, C., Moshell, M., Micikevicius, P., Altman, M.: Applying Mixed Reality to Entertainment, In:Computer, 35(12), (2002), 122-124.
4. Hoff, W., Vincent, T.: Analysis of Head Pose Accuracy in Augmented Reality, In: IEEE Trans. Visualization and Computer Graphics, 6(4), (2000).
5. Wheeler, A., Brujic-Okretic, V., Hitchin, L., Parker, G.: Augmented Reality Enhancements for the Teleoperation of Remote Mobile Robots, In: Second International Workshop VR-MECH'2001, Brussels, (2001).
6. Costa R.M.; Carvalho, L. A.: The Acceptance of Virtual Reality Devices for Cognitive Rehabilitation: a report of positive results with schizophrenia, In: Computer Methods and Programs in Biomedicine, 73, , (2004), 173-182.
7. Matheis, R.; Schulteis, M., Rizzo, A.:Learning and Memory in a virtual office environment, In:Proceedings of Second International Workshop on Virtual Rehabilitation, (2003), 48-54.
8. Broeren, J., Lundberg, M., Molén, T., Samuelsson, C., Bellner, A., Rydmak, M.:Virtual Reality and Hapitcs as an assessment tool for patients with visuospatial neglect: a preliminary study, In:Proceedings of Second International Workshop on Virtual Rehabilitation, (2003), 27-32.

9. Niniss, H.; Nadif, A.: Simulation of the behavior of a Powered Wheelchair using Virtual Reality, In:3rd International Conference on Disability, Virtual Reality and Associated Technologies, (2000), 9-14.
10. NASA: Low Vision Tests and devices, In: www.ski.org/rerc/5yrReport/General/V.html. Visited on 2004.
11. Dagnellie, G.: Virtual Technologies aid tin restoring sight to the blind, In: Communication Through virtual technology: Identity, Community and Technology in the Internet age, G. Riva and F. Davide (eds), IOS Press, (2003), 247-272.
12. Staner, T., Mann, S., Rhodes, B., Levine, J.: Augmented Reality Through Wearable Computing, In Presence, 6(4), (1997), 386-389.
13. Miller, S.: Enfermidades dos olhos e Afecções da retina (Eye Illnesses and Retina Diseases). Ed. Artes Médicas, (1991), 247–269 (in Portuguese).
14. Haber, C.: A General two-dimensional, graphical finite element processor utilizing discrete transfinite mappings. In: International Journal Numerical Methods in Engineering, 17, (1981),1015 – 1044.
15. 15.Riot. In: http://riot.ieor.berkeley.edu/riot/Applications/ConvexHull/CHDetails.html. Visited on 2003.
16. IIS: Inside Information Systems, In: http://www.iis.com.br. Visited on 2001.
17. Octopus. In: http://www.octopus.ch/. Visited on 2005.
18. Poma, In: http://www.xybernaut.com/Solutions/product/downloads/XYBpoma.pdf. Visited on 2005.

An Adaptive Collision Detection and Resolution for Deformable Objects Using Spherical Implicit Surface

Sunhwa Jung[1], Min Hong[2], and Min-Hyung Choi[1]

[1] Department of Computer Science and Engineering,
University of Colorado at Denver and Health Sciences Center,
Campus Box 109, PO Box 173364, Denver, CO 80217, USA
sjung@ouray.cudenver.edu, Min-Hyung.Choi@cudenver.edu
[2] Bioinformatics, University of Colorado at Denver and Health Sciences Center,
4200 E. 9th Avenue Campus Box C-245, Denver, CO 80262, USA
Min.Hong@UCHSC.edu

Abstract. A fast collision detection and resolution scheme is one of the key components for interactive simulation of deformable objects. It is particularly challenging to reduce the computational cost in collision detection and to achieve the robust treatment at the same time. Since the shape and topology of a deformable object changes continuously unlike the rigid body, an efficient and effective collision detection and resolution is a major challenge. We present a fast and robust collision detection and resolution scheme for deformable objects using a new enhanced spherical implicit surface hierarchy. The penetration depth and separating distance criteria can be adjusted depending on the application specific error tolerance. Our comparative experiments show that the proposed method performs substantially faster than existing algorithms for deformable object simulation with massive element-level collisions at each iteration step. Our adaptive hierarchical approach enables us to achieve a real-time simulation rate, well suited for interactive applications.

1 Introduction

Physically based simulation for deformable objects is indispensable for many modern character animation and medical simulation. Deformable object is usually discretized into a set of basic meshed elements to model the geometry and behavior of the object. It often consists of thousands of nodes to avoid undesirable sharp folds and to better represent their detailed dynamic behaviors. The collision detection and contact resolution often cost more than 90% of the total simulation time per iteration and it is considered as a major bottleneck for any deformable object simulation. Although many researchers achieved robust collision detection schemes recently [4,5,7], they are often not applicable to the interactive applications due to the complexity of their algorithms. Recently stochastic collision detection methods [17] have been proposed to achieve the interactive rate of simulation. These approaches guarantee the desirable computation time for the collision detection, but they miss significant amounts of collisions, thus the robust behavior can not be expected. In addition, more critical issue is that the collision resolution scheme can not be separated from the collision detection methods. If the collision is not resolved properly in current step, the collision detection query in the next step will start with unacceptable initial conditions and consequently the collision resolution will fail. Therefore the collision detection method must guarantee the comprehensive collision check and it has to provide enough information about the colliding and penetrating objects so that the collision resolution scheme can handle the collisions properly. So the collision detection and resolution is inseparable and the accurate collision response can only be achieved with a well coordinated detection and resolution scheme. In addition there must be a way to adjust the error tolerance and a strategy to resolve a degenerate penetration situation that guarantees the penetration free status after the collision resolution, since unavoidable numerical drift and other geometric errors could result in element level intersections.

Our proposed method is based on a spherical implicit surface that provides fast and robust collision detection and the resolution approximation between deformable objects. It extends the hierarchical spherical surface to the sub-divided triangle level adaptively to detect and resolve collisions within a tight error bound. It is designed to handle massive collisions at a given iteration step so its applicability is wide across variety of deformable structures as long as their surface is meshed with triangles. Due to the controllability of collision tolerance, the view dependent adaptive application of collision resolution can be applied to further reduce the computational cost.

2 Related Works

To reduce collision detection query space, several bounding volume hierarchies and spatial subdivision schemes have been proposed [10,18,19,20]. Most of them require pre-computation to generate efficient tree structures, and they are designed to perform a series of quick rejection tests to reduce the collision query space. But they can not provide enough collision information to resolve the collisions, i.e. proximity information, separating and penetrating depth, and involved features, for an accurate collision resolution. Therefore after finding the primitives that violate the bounding volume conditions, a geometry based exact collision detection method should be applied. For an exact collision detection between surface geometric primitives, the continuous collision detection using a coplanar condition is a de-facto standard. Originally the continuous collision detection for a moving triangle and a point was proposed by Moore [1]. It required solving a fifth polynomial, but Provot [8] reformulated the problem with a cubic equation and they extended it to include edge-edge cases. Bridson et al. [7] further addressed the numerical error handling method to achieve robust collision detection. But these methods are still expensive and have no ability to reduce the computation time due to the nature of the triangle-triangle geometric computation. Moreover the numerical error is inevitable in degenerated cases such as severely stacked and pinched triangles, often found in complex cloth patches [4].

Hubbard [14] proposed a method to approximate polyhedra using a hierarchical bounding sphere for collision detection between rigid polyhedra. This method achieved the accuracy through searching down the collision spheres in the hierarchical structure and it provided the ability to control the computational cost by adjusting the level of detail of the bounding spheres. But building a hierarchy that fits the model with minimal overestimation is computationally expensive. Bradshaw et al. [16] proposed methods that generate better fit sphere trees for rigid polyhedra. However deformable objects are difficult to generate proper sphere trees because fixed size spheres can not cover the changing geometry of the deformed structure effectively. Recently James et al. [21] proposed a method to build an efficient bounding tree for reduced deformable objects but it can not be used for general deformable objects with large deformation since it only works for simple and limited deformation based on modal analysis.

Collision resolution is also a crucial component to achieve accurate post-collision behavior. Baraff et al [3] applied the penalty force to resolve cloth and cloth collision. But it could be difficult to find the correct magnitude of forces for the collision resolution, and when the contact spring force is too strong or weak it will generate oscillations between collided regions. The other popular method is to resolve geometrically after the penetration happens. Volino [5] and Provot [8] used this method and most deformable objects with volume can utilize this method. Bridson et al [7] combined these two methods to take advantages of them. These approaches require accurate collision information from the previous iteration steps. Baraff et al [4] proposed a method that does not require historical information but it has a limitation when the colliding objects do not have closed surfaces.

3 Collision Detection

Conventional continuous collision detection [5,6,7,8] in triangle/node and edge/edge cases requires solving a cubic equation of the coplanar condition and solving one 2 by 2 linear system. Although the individual computation is simple, massive collision often creates a big numerical system and it becomes a bottle neck of the simulation. In addition, the collision result has numerical error because of many numerical operations (320 multiplications and 156 subtractions or additions) to generate the cubic equations. When a triangle is relatively small, the relative error can be a serious problem. Our method can substitute this method with quick distance calculation in an adaptive fashion.

In our method we used Axis Aligned Bounding Box (AABB) hierarchy for reducing the collision detection query space. AABB works efficiently for deformable objects because of the quick updating time. After searching the collision through AABB, the collision detection algorithm will find the two triangles which are close enough to require an exact collision check. In this chapter, our method to decide the collision state of the two triangles is described.

3.1 Approximation Using Spherical Implicit Surface

Triangle and node collision test. First we calculate the proper radius of the sphere (d) to generate the spherical implicit surface of the triangle for collision detection. d is the 2/3 of the distance between the triangle's center of mass (C4) to one of its nodes. The distance fields (spheres S1-S3) are created from each triangle's node using radius d and S4 is created from the triangle's center of mass. In figure 1 (level 0) four spheres

(S1-S4) represent the spherical implicit surface of the triangle. The spherical implicit surface of the triangle is the sum of the distance field from the three nodes of the triangle and the triangle's center of the mass. Node P is the one of three nodes in the other triangle in the exact collision detection. After performing the distance calculations from the center of each sphere to the node P, we can conclude whether node P is inside of the spherical implicit surface or not. Then we can check the distance between the closest location of the triangle and node P. If node P is in the one of spheres S1, S2, S3, and S4, the approximated distance between the triangle and the node is the distance from node P to one of the closest sphere centers. The collision normal for collision resolution is the vector from the selected center of the sphere to the node P. But if node P is in the two spheres, the approximated distance between triangle and node P is the distance between the mid point of the center of the two involved spheres and node P. If node P is in the three spheres, the approximation can be done from the three involved spheres' center. This collision result can be used to decide the collision resolution direction. The key idea of our method is using the result of the quick distance check to determine the collision resolution information (direction and distance). If more accuracy for the exact collision location and distance is required and the error tolerance (floating distance between triangles) is set below the size of the radius of the spherical implicit surface, we can perform the same algorithm recursively using the subdivided triangles.

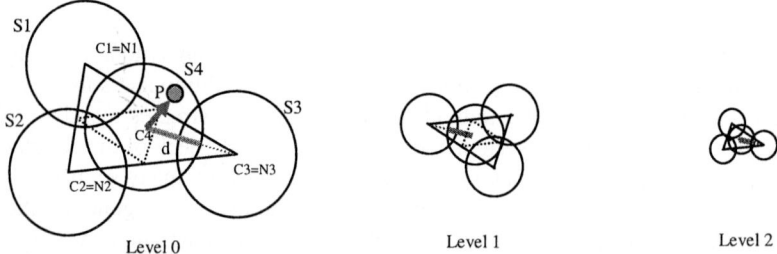

Fig. 1. Level 0 collision detection in triangle C1, C2, and C3 and with the node P. N1, N2, and N3 are the node positions of the triangle. Level 1 and Level 2 illustrates the detailed levels

Edge and edge collision test. After checking triangle and point collision, the possibly missed collision might be the edge and edge case which the points of the edge are outside of the implicit surface. To check these collisions, the distance between two mid points of the edges is checked. If they are overlapped, the distance between two mid points of the edge is tested whether it is within the threshold. Figure 2 illustrates that the collision detection can be performed in different levels. To select the collision check points (C3 and C4) at the next level, we use a binary search method. For example, in figure 2 (level 0) we select the minimum distance between four nodes (AC, AD, BC, and BD). In this case distance between BC is the minimum distance, thus the next level sets are edge C1 and B and edge C2 and C. We can pre-store the distance data in the previous triangle and node collision test and can reduce the computation time.

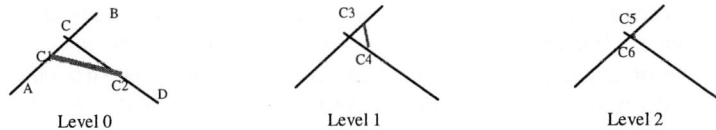

Fig. 2. Level 0 collision detection between two edges. C1 and C2 are the mid points of the edges. If the distance between C1 and C2 is less than the threshold, the collision is determined

3.2 Adaptive Re-sampling

Typically we found that the level 0 collision result would be sufficiently accurate enough for the robust collision handling. If we use the relatively finer mesh, we do not need any further collision detection query. But when the triangle is too coarse, we need to adaptively search down to the next level. The collision detection algorithm on level 0 can be applied recursively until we get a satisfying result. No extra storage is required for the substructure to achieve this. Since we have already four distance calculations at the upper level, we need three more distance calculations for the next lower level collision detection. The beauty of our method is that we can have enough collision resolution information whenever we stop collision detection processing. The decision of further searching can be done using human collision perception, which O'Sullivan [15] researched with psychophysical experiment. We changed the threshold according to the distance between the camera and the object to reduce the computational burden. If the user wants to achieve more accuracy, it can be done by simply expanding the tree structure but the cost is not substantial since the involved triangles will be subdivided and each operation only includes a quick distance comparison.

4 Collision Resolution

Figure 3 illustrates the collision results in level 0 when we consider that the distance is less than the threshold. We can use the distance vector as the collision normal to apply the collision resolution method. From the figure 1, the node P was detected as collided with sphere 4. Figure 3 shows the normal of the collisions in triangle-point case and edge and edge case. Using these collision reports, geometry method and penalty force method are used for the collision resolution. Our method utilizes both methods to handle the collisions depending on the collision situation. Penalty force

Fig. 3. The red arrow in (A) is the collision normal in the point and triangle collision detection result from figure 1. The red arrow in (B) is the collision normal in the edge and edge collision detection result from figure 2 (A)

method reduces the number of collisions and geometry method guarantees the collision-free states. For the penalty force method, we sum up the mass of the object (edge, node or triangle) and calculate the repulsion force for two objects and the forces are distributed to each node according to their masses. For the geometry method, we modified the velocity to the collision response velocity and moved the position to the legal location.

5 Experiments

In the comparative experiments we performed the collision detection at the level 0 with threshold 1, and the collision resolution scheme was applied differently to each object. Since cloth model is thin and generates numerous self-collisions, we need to treat collisions carefully utilizing penalty force method and geometry method. We applied the penalty force after calculating the internal force to reduce the number of collision and the geometry method applied after finishing the simulation and before displaying the scene to enforce the collision free. For the two cylinders simulation, the geometry method is enough to handle the collision because there is no chance to miss the collision detection completely and the number of collisions is small.

Cloth simulation. To compare the performance between existing continuous collision detection method using the coplanar conditions and our proposed method, we applied both methods to a patch of cloth simulation to check self-collision. The conventional method requires solving cubic equations for each coplanar condition. Figure 4 is the snapshot of the cloth piling up simulation. The cloth model contains 3000 (30 by 100) particles. This illustrates the robust collision resolution with a predefined cloth thickness. We used a 2.4 GHz Pentium 4 processor computer and it took about 0.5 second per iteration. The collision detection threshold which we provide became the thickness of the cloth model. Figure 5 shows the comparative performance in the piling cloth simulation. At the beginning of the simulation, less exact collision detections is required due to the low curvature of the surface and less self collision occasions. After the cloth model begins to fold, the computation time grows as the number of colliding elements increases and more cases for exact element level collision resolution are occurred. The conventional method takes approximately in the range of 0.3-1.8 seconds to resolve all collision conditions. However, our method takes in the range of 0.015-0.5 second. Throughout the simulation, our method shows a significant speedup, about 20 to 200 times faster. We can achieve more accuracy by applying the small threshold with adaptive re-sampling distance checking.

Cylinder and cylinder simulation. We applied our method to a collision simulation between two deformable cylinders in figure 4. These cylinders are coarsely meshed and they are modeled with soft material property using linear finite element method [21]. Although these cylinders are under large deformation, the result demonstrates that our proposed method handles robustly the collision detection and resolution in real-time. The computational cost of our collision detection and resolution is negligible for the total simulation time due to the small number of nodes and heavy computation time in FEM analysis. You can see the result animation from [23].

Fig. 4. Left two figures are the snapshots of the cloth simulation falling down and folded and crumpled. Due to the size of meshed triangles, our approximation in level 0 gives reasonably accurate collision detection result for the collision resolution. Right figure shows the deformation between two cylindrical volumetric models meshed with tetrahedra in contact

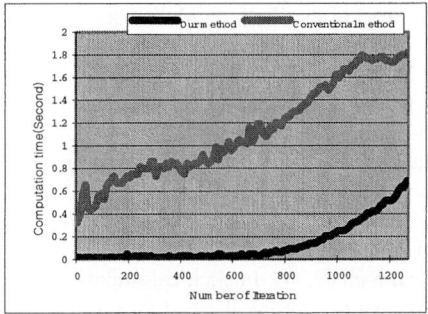

Fig. 5. The performance comparison between the conventional method and our method

6 Conclusion

We propose a fast and robust collision detection and contact resolution for deformable objects. We have utilized the human perception to determine the necessary accuracy of collision detection. Our experimental results demonstrate that the massive collisions between deformable objects can be effectively handled. Our method substantially reduces the complexity of the collision management and alleviates the numerical error in calculation. Due to the simplicity of our method, it is possible to achieve even better performance when it employs a GPU-based hardware acceleration. For coarsely meshed objects, the proposed method may have to go down to the numerous levels of details to achieve the necessary accuracy of a collision, and the depth of the tree is inversely proportional to the size of a triangle mesh. However, since deformable models should be meshed reasonably to generate plausible motion, the levels of detail are limited within a few steps.

References

1. Moore, M. and Wilhelms, J.: Collision detection and response for computer animation. In SIGGRAPH 1998, ACM Press / ACM SIGGRAPH, Computer Graphics Proc., 289-298.

2. Baraff, D. and Witkin, A.: Physically based modeling. SIGGRAPH 03' course notes, 2003
3. Baraff, D. and Witkin, A.: Large steps in cloth simulation. In Proc. of SIGGRAPH 1998, ACM Press / ACM SIGGRAPH, Computer Graphics Proc., 1-12
4. Baraff, D., Kass, M. and Witkin, A.: Untangling cloth. In Proc. of SIGGRAPH 2003, ACM Press / ACM SIGGRAPH, Computer Graphics Proc., 862-870.
5. Volino, P. and Thalmann, N. Magnenant: Efficient self-collision detection on smoothly discretized surface animations using geometrical shape regularity. In Proc. of Eurographics, vol. 13 of Computer Graphics Forum, Eurographics Association, C–155–166, 1994.
6. Volino, P., Courchesne, M. and Thalmann, N. Magnenant: Accurate collision response on polygonal meshes. In Proc. of Computer Graphics, 179–188, 2000.
7. Bridson, R., Fedkiw, R. and Anderson, J.: Robust treatment of collisions, contact and friction for cloth animation, In Proc. of SIGGRAPH 2002, ACM SIGGRAPH, 594-603.
8. Provot, X.: Collision and self-collision handling in cloth model dedicated to design garment. In Graphics Interface, 177–89, 1997.
9. Lafleur, B., Thalmann, N. Magnenat and Thalmann, D.: Cloth animation with self-collision detection. In Proc. of the Conf. on Modeling in Comp. Graphics, Springer, 179–187, 1991.
10. Zhang, D. and Yuen, M.: Collision detection for clothed human animation. In Proceedings of the 8th Pacific Graphics Conference on Computer Graphics and Application (PACIFIC GRAPHICS-00), 328–337, 2000.
11. Lin, M. and Gottschalk, S.: Collision detection between geometric models: A survey. In Proc. of IMA Conf. on Mathematics of Surfaces, 1998.
12. Choi, K. and Ko, H.: Stable but responsive cloth. In Proc. of SIGGRAPH 2002, ACM SIGGRAPH, Comput. Graphics Proc., 604–611.
13. Choi, M., Hong, M. and Samuel, W.: Implicit constraint enforcement for stable and effective control of cloth behavior. A technical report form Univ. of Colorado at Denver, Computer Science and Engineering Department TR-03-01, 2004.
14. Hubbard, P. M.: Approximating polyhedra with spheres for time-critical collision detection. ACM Trans. Graph., 15(3):179-210, 1996.
15. O'Sullivan, C.: A model of collision perception for real-time animation. In Computer Animation and Simulation'99, pages 67–76.
16. Bradshaw, G. and O'Sullivan C.: Sphere-tree construction using dynamic medial axis approximation. In ACM SIGGRAPH Symposium on Computer Animation, pages 33-40. ACM Press, July 2002
17. Kimmerle, S., Nesme, M. and Faure, F.: Hierarchy Accelerated Stochastic Collision Detection. Vision, Modeling, and Visualization 2004
18. Gottschalk, S., Lin, M. C. and Manocha, D.: OBBTree: A hierarchical structure for rapid interference detection. In Proc. of SIGGRAPH 1996. ACM SIGGRAPH, 1996.
19. Klosowski, J. T., Held, M., Mitchell, J. S. B., Sowizral, H., and Zikan, K.: Efficient collision detection using bounding volume hierarchies of k-DOPs. IEEE Transactions on Visualization and Computer Graphics, v.4 n.1, p.21-36, 1998
20. Palmer, I. J., and Grimsdale, R. L.: Collision detection for animation using sphere-trees." Computer Graphics Forum, 14(2):105-116, 1995.
21. James, D. L. and Pai, D. K.: BD-Tree: Output-Sensitive Collision Detection for Reduced Deformable Models. In Proc. of SIGGRAPH 2004, ACM SIGGRAPH, 2004
22. Hong, M., Choi, M. and Lee, C.: Constraint-Based Contact Analysis between Deformable Objects. In Proceedings of International Conference on Computational Science, 2004.
23. http://graphics.cudenver.edu/~lab/iccs05.htm

Automatic Categorization of Traditional Chinese Painting Images with Statistical Gabor Feature and Color Feature

Xiaohui Guan, Gang Pan, and Zhaohui Wu

Dept. of Computer Science, Zhejiang University, Hangzhou, China
{hxgy, gpan, wzh}@zju.edu.cn

Abstract. This paper presents an automatic statistical approach to categorize traditional Chinese painting (TCP) images according to subject matter into three major classes: figure paintings, landscapes, and flower-and-bird paintings. A simple statistical Gabor feature is presented to describe the local spatial configuration of the image, which is then integrated with color histogram that represents the global visual characteristic to build the feature subspace. A relative-distance based voting rule is proposed for final classification decision. The effectiveness of the proposed scheme is demonstrated by the comparable experimental results.

1 Introduction

Traditional Chinese painting (TCP) is highly regarded throughout the world for its theory, expression, and techniques. As an important part of Chinese cultural heritage, the traditional Chinese painting is distinguished from Western art in that it is executed on xuan paper (or silk) with the Chinese brush, Chinese ink and mineral and vegetable pigments. Currently more and more museums and artists like to present the paintings on the Web. This paper addresses the automatic categorization of the traditional Chinese painting images. The potential applications include Web searching and browsing, digital art libraries etc.

Traditional Chinese painting can be divided according to subject matter into three major categories [15]: landscape paintings, flower-and-bird paintings and figure paintings. Landscape paintings mainly depict the natural scenery of mountains and rivers. Flower-and-bird paintings concentrate on the plants and small animals. Figure paintings mainly portray humans and their activities, generally along with the plant, small animal and scenery. Although the categorization is usually an easy task for humans, it is an extremely difficult problem for machines. The major difficulties lie in its wide variation in subject's shape, color, and painting style. The subjects' appearance in TCP is usually highly abstracted from the real word.

There are many researchers made efforts for image classification and indexing. As one of the simplest representations of digital images, color histogram [1] has been widely used for various image categorization problems. Color histogram is insensitive to image resolution and rigid transform. Pass [2] presents color coherence vector (CCV) as image index. CCV is the statistical pixels' convergence measurement in space. Stricker and Orengo think that the color information centralizes in the low rank color moments, so they use the statistical moment to characterize the color [3]. The above methods attempt to capture the global color features and discard the spatial configuration. The reference [4] utilizes spatial information to improve the histogram method.

Besides histogram, a number of block-based methods have been proposed to employ local and spatial properties by dividing an image into rectangular sub-images. [5] exploits the wavelet coefficients in high frequency bands for classification. Ma and Manjunath evaluate various wavelet forms to reach a conclusion that Gabor wavelet transforms get the best results in texture retrieval methods [6]. Yu [7] uses 1D HMM to classify indoor and outdoor scene. Carson et al [8] utilizes EM algorithm to segment the image into several "blobworld" which are coherent in color and texture. Huang [9] relies on the banded color correlogram as image features and then hierarchical classification tree is construct to classify new image. Jiang [10] presents a scheme to classify the TCP images from non-TCP images.

This paper proposes a statistical method to automatically classify the traditional Chinese paintings according to subject matter into three major categories: landscape paintings, flower-and-bird paintings and figure paintings. We use a statistical Gabor feature as the representation of local texture of TCP, and color histogram as the global representation of color information.

2 Statistical Feature Representation

In our method, we present a statistical feature integrating local texture and color information. Firstly, the TCP image is filtered using a bank of Gabor functions with multiple scales v and orientations u. Secondly, with the filtered image, we construct a statistical Gabor feature such as average, variance. And then, we combine the Gabor feature with the color histogram of the TCP.

2.1 Statistical Gabor Feature

Gabor wavelets are biologically motivated because they model the response properties of human cortical cells. They have been widely adopted to extract features in image retrieval, face recognition, texture classification [11, 12] and got wonderful results. Basically, Gabor filters are a group of wavelets, with each wavelet capturing the variation at a specific frequency and direction. It has been shown to correspond to human visual system.

Gabor filter family is similar to the simple visual perceptive cells, which are characterized as localized orientation selective and frequency selective. The kernel is the product of a Gaussian envelope and a wave function, as shown in Equ (1), where the term $\exp(-\frac{\delta^2}{2})$ is subtracted to make the filter insensitive to the overall level of illumination.

$$\Phi_j(\vec{x}) = \frac{\|\vec{k_j}\|}{\delta^2} \exp(-\frac{\|\vec{k_j}\| \cdot \|\vec{x}\|}{\vec{x}})[\exp(i\vec{k_j}\vec{x}) - \exp(-\frac{\delta^2}{2})] \qquad (1)$$

Where

$$\vec{x} = I(x,y) \quad \vec{k_j} = \begin{bmatrix} k_\nu \cos \varphi_\mu \\ k_\nu \sin \varphi_\mu \end{bmatrix}$$

$$k_\nu = \frac{\Pi}{2} * \frac{1}{2^\nu} \quad \varphi_\mu = \mu \frac{\Pi}{6} \quad \nu = 0 \cdots 3, \mu = 0 \cdots 5$$

On basis of [13], the parameters setting of Gabor filter are 4 scales, 6 orientations, and filter mask size is 13×13. So we get 24 functions determined by the parameters ν, μ. This bank of filters is applied to the TCP image, shown in Equ (2). This means Gabor wavelet transformation, that is, every input image is convoluted with the 24 Gabor filters.

$$G_{\nu\mu}(\vec{x}) = \int I(\vec{y})\Phi_{\nu\mu}(\vec{x} - \vec{y})d^2\vec{y} \qquad (2)$$

The resulted Gabor coefficient has two parts: real part $Re(G_{\nu\mu})$ and imaginary part $Im(G_{\nu\mu})$. From the filtered image, we could compute the average $\varrho_{\nu\mu}$ and variance $\sigma_{\nu\mu}$ of the magnitude $\sqrt{Re(G_{\nu\mu})^2 + Im(G_{\nu\mu})^2}$, then obtain the statistical Gabor feature, a 48-dimensional vector:

$$\vec{g} = \{\varrho_{00}, \sigma_{00}, \varrho_{01}, \sigma_{01}...\varrho_{\nu\mu}\sigma_{\nu\mu}|\nu = 3, \mu = 5\} \qquad (3)$$

2.2 Statistical Color Feature

Color is the most intuitive characteristic of human vision. Every kind of subjects has its different intrinsic color. We employ the color histogram [1] as our statistical feature for color information, which has been used in the image retrieval and indexing. It is invariant to translation and rotation around the viewing axis and varies slowly with changes of view angle, scale and occlusion.

Assume colors in an image are mapped into a discrete color space containing n colors, a color histogram of image I is n-dimensional vector, where each element represents the number of pixels with color index j. Each element of a histogram is normalized so that the histogram represents the image without regarding to the image size. The element of the normalized color histogram $h(I)$ is defined as:

$$h_j(I) = H_j(I)/\sum_{i=1}^{n} H_i(I) \qquad (4)$$

Where $H_i(I), H_j(I)$ is the number of pixels with color i,j in image I. In our experiment, we select $8 \times 8 \times 8$ RGB color space. So we get a color feature vector:

$$\vec{h} = \{h_j(I)|j = 1 \cdots 512\} \qquad (5)$$

We combine the color histogram $\vec{h}$ with the Gabor feature $\vec{g}$ into a longer vector $\{\vec{g}|\vec{h}\}$, which together characterizes the texture and color of TCP image.

3 Categorization

3.1 Voting with Relative-Distance in Feature Space

To classify a new input image I in a feature space, we exploit voting with sample images. In many articles, Euclidean distance is used. However, for the traditional Chinese painting images, works in different style may have different dispersibility. The Euclidean distance in the feature space is often not enough, since the scatter information of the same class should be taken into account.

Here we take the intra-class variance into consideration. Suppose that $X = \{x_1, x_2 \cdots x_N\}$, $Y_{mk} = \{y_1, y_2 \cdots y_N\}$, where X is the feature vector of the input image I, Y_{mk} is kth sample feature vector of class m. K_m is the total number of training images of class m. We use the Equ (6) to determine the distance between the input image and the class m.

$$\bar{d}_m = \sum_{i=1}^{K_m} d_{mk}/K_m \qquad v_m = \sqrt{\sum_{k=1}^{K_m}(d_{mk} - \bar{d}_m)^2/K_m} \qquad (6)$$

And finally the relative-distance between the input image I and the kth sample is:

$$Rd_{mk} = d_{mk}/v_m \qquad d_{mk} = d(X, Y_{mk}) = \sqrt{\sum_{n=1}^{N}(x_i - y_i)^2} \qquad (7)$$

We make the relative distance to decide whether to cast a vote of a certain class to the input image. In most cases the relative distance performs better than the common Euclidean measure.

The voting policy is according to the following rule:

$$V_m = \sum_{k=1}^{K_m} V_{mk} \qquad V_{mk} = \begin{cases} 1 & \text{if } Rd_{mk} < Threshold \\ 0 & \text{otherwise} \end{cases} \qquad (8)$$

V_m is the total number of votes received by the class m. Assuming C_m is the total samples of the class m. Thus, maximum V_m/C_m determines the pattern of test image.

3.2 Sample Selection

In order to refine the training samples and also to improve the classification accuracy, sample selection procedure is implemented. The selection of training samples may affect completely the categorization results, because images to classify need be compared with all the training samples in the reference database. $Fig.1.(a)$ shows two classes of traditional Chinese painting image, where blue star denotes the flower-and-bird paintings and red point denotes figure paintings. Each painting is represented by a point in the feature space.

We use the k-nearest neighbor editing algorithm [14] to select the registered samples from a set of reference samples. The algorithm is as follows:

1. Divide randomly reference sample set into several subsets, suppose: $\Theta = \{S_0, S_1 \cdots S_{n-1}\}$, $n \geq 3$,
2. With $S_{(i+1)mod(n)}$ as reference samples, Use K-nearest classification method to classify S_i samples, which $i = 0, 1 \cdots n - 1$.
3. Discard classified falsely samples, the residual samples constitute new sample set Θ^N.
4. If there are no samples to be discarded, stop. Otherwise, go to: 1.

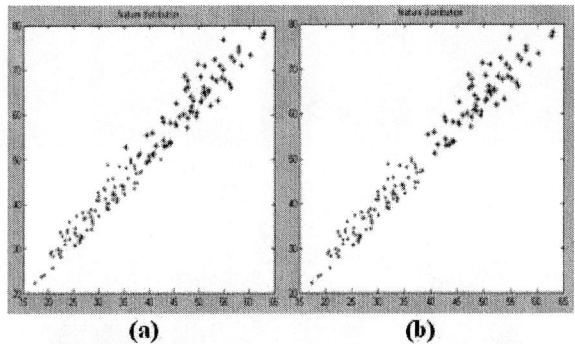

Fig. 1. The illustration of sample selection: (a) before and (b) after sample selection. Where x-axis: $\bar{\varrho}$ of $\nu = 0$, $\mu = 1$ and y-axis: $\bar{\varrho}$ of v=3, u=5

$Fig.1.(b)$ shows the result of $Fig.1.(a)$ by the editing algorithm. From this figure we can see that the intersectional points are discarded. This decreases the affect of ambiguous samples for voting procedure. In the next section, we will show that after sample selection the categorization accuracy has been improved to a certain extent.

4 Experimental Results

In our experiment, we use a database with 392 traditional Chinese painting images, which includes 132 flower-and-bird paintings, 127 figure paintings and 133 landscapes. These images are painted by different painters in different dynasty. Some works by modern painters are also included. The image size varied from 200 × 200 to 500 × 500. Some samples are shown in Fig.2.

We use the leave-one-out rule of cross validation to test our approach. Every time we select one as the test image and the others are as reference samples. This process repeats throughout the whole experimental data. During relative-distance based voting, we need to determine the threshold for voting. The threshold will decide whether the sample casts a vote of class m to the test image. We compute average dispersivity of the selected training samples as the threshold, i.e.

$$Threshold = d/v \qquad (9)$$

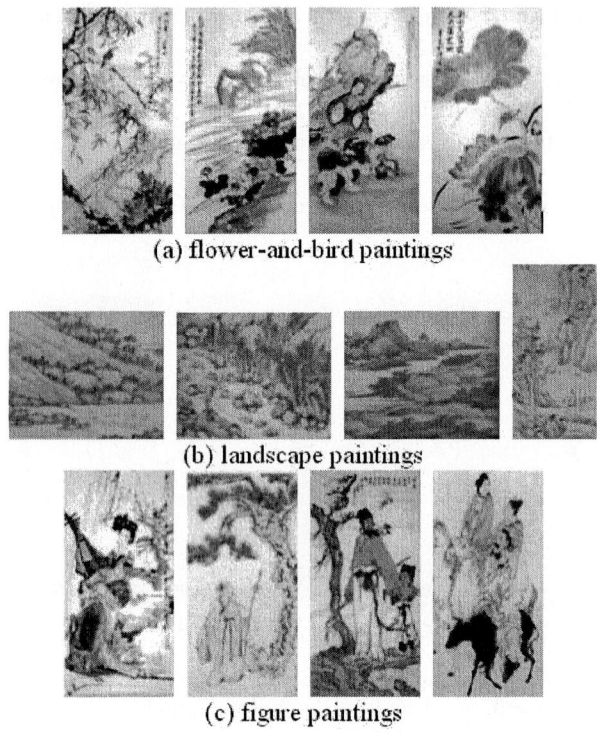

Fig. 2. Some samples of the experimental data

Where d and v are for the average mutual distance and variance of samples feature vectors.

The comparison with different features is carried out. The categorization results are shown in $Table.1$. presented in the confusion matrix form. In the class confusion matrix $A = (a_{ij})$, a_{ij} means the samples number of class i that predicted into class j. The diagonal elements are the correct classification numbers. The non-diagonal elements are the numbers of false categorization. $Table.1$. demonstrates that the scheme combining the statistical Gabor feature and color feature (GC) has improved the categorization accuracy comparing with the single feature (G or C). The sample selection ($GC+$) farther enhances the classification result.

5 Conclusion

In this paper, we present an automatic classification scheme. The statistical Gabor feature and color histogram are combined to classify the traditional Chinese painting images. And we use relative-distance based voting rule to finally categorize the images. The Gabor feature, description of local texture information of image, is compensated with the absence of spatial information, the drawback

Table 1. The confusion matrix of landscape, figure and flower-and-bird paintings using various features.   C: color histogram;   G: Gabor feature;   GC: combining Gabor feature and color histogram without sample selection;   GC+: combining Gabor and color histogram after sample selection

	Feature Used	Landscape	Figure	Flower-and-bird
Landscape	C	110	23	0
	G	111	21	1
	GC	119	14	0
	GC+	124	9	0
Figure	C	1	92	34
	G	4	92	31
	GC	1	100	26
	GC+	0	106	21
Flower-and-bird	C	1	36	95
	G	3	35	94
	GC	3	24	105
	GC+	1	19	112

of global color histogram. The preliminary experimental results have shown the superiority of our method.

References

1. M. Swain, D.Ballard. Color indexing. International Journal of Computer Vision, 7(1): 11-32, 1991.
2. G.Pass, R.Zabih, J.Miller. Comparing images using color coherence vectors. ACM Multimedia. Boston, MA, 1996.
3. M.Striker, M.Orengo. Similarity of color images, SPIE proceedings, 1995.
4. Y.L.Ho, K.L.Ho, H.H.Yeong. Spatial color descriptor for image retrieval and video segmentation. IEEE Transaction on multimedia. 5(3), 2003.
5. J.Z.Wang, J.Li, G.Wiederhold. SIMPLIcity: Semantics-sensitive integrated matching for picture libraries. IEEE PAMI, 23(9):947-963, 2001.
6. W.Y.Ma, B.S.Manjunath. Texture features and learning similarity. IEEE CVPR, 1996.
7. H.Yu, W.Wolf. Scenic classification methods for image and video databases. SPIE Proceeding, vol.2606, 363-371, October 1995.
8. S.Belongie, C.Carson, H.Greenspan, J.Malik. Color and texture-based image segmentation using EM and its application to content-based image retrieval. IEEE ICCV. pp.675-682, Jan. 1998.
9. J.Huang, S.Ravikumar, R.Xabih. An automatic hierarchical image classification scheme. Proc. ACM multimedia, Sep. 1998.
10. S.Jiang, W.Gao, W.Wang. Classifying traditional Chinese painting images. IEEE PCM, pp.1816-1820, Dec. 2003.
11. C.Liu, H.Wechsler. Gabor feature based classification using the enhanced fisher linear discriminant model for face recognition. IEEE Trans. IP, 11(4):467-476, 2002.
12. G.M.Haley, B.S.Manjunath. Rotation-invariant texture classification using modified Gabor filters. Proceedings of the International Conference on Image Processing. Vol.1, 1995.

13. L.Chen, G.Lu, D.Zhang. Effects of different gabor filter parameters on image retrieval by texture. Proc. 10th Int'l Conf. Multimedia Modeling, 2004.
14. R.O.Duda, P.E.Hart, D.G.Stork, Pattern classification, Wiley, October 2000.
15. W. Liu ed. Traditional Chinese paintings thesaurus, pp.392-399, Huawen publisher, 1990 (in Chinese).

Nonlinear Finite Element Analysis of Structures Strengthened with Carbon Fibre Reinforced Polymer: A Comparison Study

X.S. Yang, J.M. Lees, and C.T. Morley

Department of Engineering, University of Cambridge
Trumpington Street, Cambridge CB2 1PZ, UK
xy227@eng.cam.ac.uk

Abstract. Modelling crack propagation in fracture mechanics is a very challenging task. Different methods are usually robust under different conditions and there is no universally efficient numerical method for dynamic fracture simulations. Most available methods are computationally extensive and usually require frequent remeshing. This comparison study focuses on three major methods: the discrete element method, the adaptive fixed crack method and the element-free Galerkin method. By implementing these methods to study a 2D concrete beam with reinforcement of carbon-fibre reinforced polymer straps, we have shown that for simulations of a limited number of cracks, fixed crack method gives the best results. For multiple crossover cracks, the discrete element method is more suitable, while for moderate number of elements, the element-free Galerkin method are superior. However, for large number of elements, fixed crack method is most efficient. Comparisons will be given in details. In addition, new algorithms are still highly needed for the efficient simulations of dynamic crack propagations.

1 Introduction

The technique of strengthening reinforced concrete structure using carbon fibre-reinforced polymer (CFRP) is promising and can have important applications in many areas. However, an understanding of the behaviour and influence of the FRP composite system is crucial for the proper design of structural reinforcement strategies in infrastructure such as bridges and buildings [6, 7]. The computer simulation is an cost-effective way of improving such an understanding.

In simulating the fracture and crack growth, there are three major paradigms: discrete element methods, smeared crack methods and element-free Galerkin methods. The discrete element method has the advantage of tracing each individual crack with an irregular geometry. In order to do this, substantial remeshing is required to accommodate the arbitrary geometry and boundaries along the newly created cracks and thus extensive computation is required which is usually time-consuming even with the modern fast computers. The adaptive fixed crack concept or smeared crack model is very efficient in computation but cannot directly deal with the crack propagation, especially in the case of multiple cracks.

This method uses the equivalent material properties to simulate the effect of the cracks in the context of continuum-based finite element analysis [4, 5]. On the other hand, the new promising element-free Galerkin method has the ability of dealing with an arbitrary geometry but it is still computationally extensive compared with the conventional finite element methods, especially in the interface regions and crack zones [1, 10, 11]).

In this paper, we will compare different numerical approaches to the nonlinear finite element simulations of concrete structures with CFRP reinforcement and analyse the appropriate conditions in detail. The main focus is to find an appropriate way to simulate the nonlinear behaviour of concrete beams strengthened with CRFP straps. In addition, we will also discuss the implication of simulation paradigms in engineering applications.

2 Models and Their Formulations

In order to compare different numerical methods, we first outline their formulations briefly, and then we implement them to simulate the same 2D concrete beam with CFRP reinforcement.

2.1 Discrete Element Method

The discrete element method for rigid granular materials was developed by Cundall and Strack [3] and this method was later incorporated into a combined finite-discrete element method [9]. There are some variations of this method, and the main procedure takes the whole domain as a multibody system and each domain is considered as a finite element continuum where the usual finite element analysis applies. The total strain increment ε_{ij} consists of the elastic strain ε_{ij}^e and the inelastic strain ε_{ij}^p, or

$$\varepsilon_{ij} = \varepsilon_{ij}^e + \varepsilon_{ij}^p. \tag{1}$$

The general stress-strain relation and strain energy density U are

$$\sigma_{ij} = C_{ijkl}\varepsilon_{kl}, \quad U = \frac{1}{2}C_{ijkl}\varepsilon_{ij}\varepsilon_{kl}, \tag{2}$$

where the elastic constants satisfy certain symmetry conditions such as $C_{ijkl} = C_{klij} = C_{ijlk}$ and $i,j = 1,2,3$. The interactions between different domains will be computed using Hertz kinetics.

2.2 Smeared Crack Model

In the smeared crack model, the total strain increment is decomposed into a concrete strain increment $d\varepsilon^{co}$ and a crack/fracture strain increment $d\varepsilon^f$, in a similar fashion as the decomposition of the total strain into an elastic strain increment and a plastic strain increment in elasto-plastic formulation. One can further decompose the concrete strain increment into an elastic part $d\varepsilon^e$ and a plastic part $d\varepsilon^p$. Thus, the decomposition becomes

$$d\varepsilon_{ij} = d\varepsilon_{ij}^e + d\varepsilon_{ij}^p + d\varepsilon_{ij}^f. \tag{3}$$

The transformation is necessary from the local coordinate system $\mathbf{d\varepsilon}$ which is aligned with the crack to the global coordinate system $(\mathbf{d\varepsilon})$. We use the same conventional notation as de Borst [4] and Kesse [8] where the subscript n means normal to the local crack plane and t means tangent to the crack plane. After some tedious but straightforward calculations, we have

$$d\sigma = [\mathbf{I} - \mathbf{N}(\mathbf{D}^{cr} + \mathbf{N}^T\mathbf{D}\mathbf{N})^{-1}\mathbf{N}^T]\mathbf{D}d\varepsilon. \tag{4}$$

The matrix $\mathbf{D}^{cr}$ can be considered as a normal crack stiffness $\mathbf{D}_n^{cr}$ and a shear crack stiffness $\mathbf{D}_t^{cr}$, i.e.,

$$\mathbf{D}^{cr} = \mathbf{D}_n^{cr} + \mathbf{D}_t^{cr}, \tag{5}$$

where

$$\mathbf{D}_n^{cr} = \begin{pmatrix} k_c & 0 \\ 0 & 0 \end{pmatrix}, \quad \mathbf{D}_t^{cr} = \begin{pmatrix} k_{nn} & k_{nt} \\ k_{tn} & k_{tt} \end{pmatrix}, \tag{6}$$

where k_c, k_{nn} etc are the constants determined from experiments. It is worth pointing out that the choice of the values for these parameters is relatively arbitrary and detailed studies are highly needed.

2.3 Element-Free Galerkin Method

The element-free Galerkin method was developed by Belytschko and his colleagues [1, 2], and a coupled finite element-element-free Galerkin (EFG) method was applied by Sukumar et al to simulate a fracture problem [10]. The main advantage of an EFG method is that it requires only nodal data for constructing approximate functions and no element structure is necessary. In this sense, it is not a method based on interpolants but a method using moving least square approximations. The moving least square approximations are constructed in the following manner. Let the m−term approximation be

$$\hat{u}(\mathbf{x}) \equiv \mathbf{p}^T(\mathbf{x})\mathbf{a}(\mathbf{x}) = \sum_{j=1}^{m} p_j(\mathbf{x})a_j(\mathbf{x}), \tag{7}$$

where $p_j(\mathbf{x})$ are basis functions. The associated coefficients $a_j(\mathbf{x})$ in the approximation are determined by minimizing the quadratic form $\Phi(\mathbf{x})$

$$\Phi(\mathbf{x}) = \sum_{i}^{n} w_i(\mathbf{x})[\mathbf{p}^T(\mathbf{x})\mathbf{a}(\mathbf{x}) - u_i]^2, \quad w_I(\mathbf{x}) = w(\mathbf{x} - \mathbf{x}_i) \geq 0, \tag{8}$$

where w_I is a weighting function and u_i is a nodal parameter. This is equivalent to define a EFG shape function

$$\phi_i(\mathbf{x}) = \sum_{j}^{m} p_j(\mathbf{x})[\mathbf{Q}]_{ji}, \tag{9}$$

where $\mathbf{Q}$ is a matrix function of $\mathbf{w}_i(\mathbf{x})$ and $\mathbf{p}(\mathbf{x}_i)$. It is worth pointing out that $u_i \neq \hat{u}(x_i)$ as $\hat{u}(\mathbf{x})$ is only an approximant.

3 Simulations and Results

By implementing these three major numerical methods, we can now start to simulate crack propagation for various structures with CFRP reinforcement. For simplicity of comparison, we use the same fundamental structure of a 2D concrete beam for all simulations.

3.1 Discrete Element Method: Fracture of Concrete Beam

In order to demonstrate the simulation capability of the discrete element method, we first show the fracturing of a concrete beam due to an impact of an elastic block moving downward with an initial velocity 5m/s. The dimensions for the beam are 2200mm (length), 200mm (width) and 100mm (thickness). Figure 1 shows the different stages of the fracture patterns. The visualization has been carried out using Abaqus. The values used are $E = 22$ GPa, $\nu = 0.2$, $\sigma_Y = 3.36$ MPa, and $\rho = 2400$kg/m^3. We can see that discrete element method is very powerful in fracture simulations and different fragments are computed in a realistic manner. Fracture patterns are traced as the crack propagates in each time step with local remeshing.

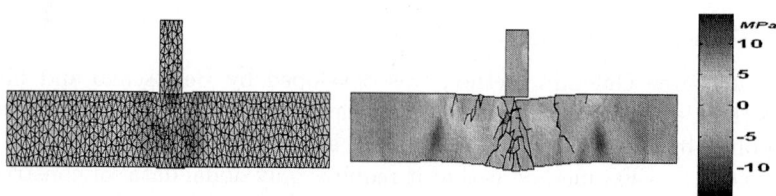

Fig. 1. The impact of a drop load on a 2D concrete beam and the fracture pattern at $t = 0.1$ s and $t = 0.5$ s

3.2 Adaptive Fixed Crack Model

We now proceed to investigate the four-point bending of the same 2D concrete beam in the plane strain condition. The loading is applied as a linear function of time with a peak value of $f = 100$kN, and the maximum time step $t = 10,000$ (pseudo time). We can see from Figure 2 that the vertical cracks start to form and grow gradually as the loading force increases, and shear cracks start to

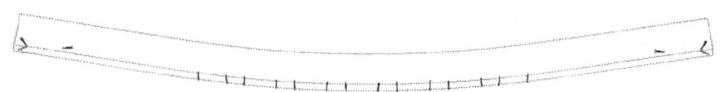

Fig. 2. The crack pattern for a 2D concrete beam with four-point bending conditions. Crack pattern are mainly vertical in the flexural section that is consistent with experimental results

appear once the loading force reaches 40kN. However, the crack pattern is just an indication and does not correspond to the real fracture although the overall patterns of the fracture are similar to that derived from experiments.

3.3 Element-Free Galerkin Method

A typical distribution of nodal points for the analysis of a 2D beam is shown in Figure 3. As no elements are requried in the element-free Galerkin method, only adaptive points are shown. The location of these points can be distributed in an adaptive manner where the density of the points varies with the potential density of cracks. The stress distribution σ_{xx} at time $t = 10,000$ s is also shown. Red color corresponds to high shear stress and light green corresponds to compression. However, although the crack geometry are still limited by the density of meshless points, this method is much efficient compared to the discrete element method.

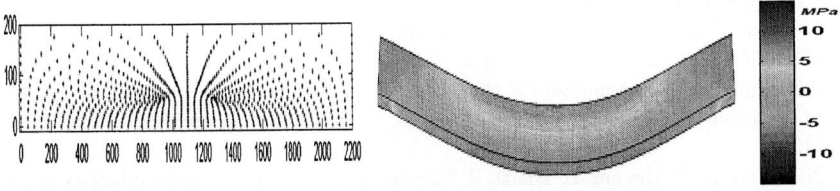

Fig. 3. Random points or locations for the EFG analysis, stress σ_{xx} and the crack patterns at $t = 10,000$ s and $f = 50$ kN

4 Conclusion

By implementing three major numerical methods for simulating crack formation in a concrete beam, we have carried out a comparison study. A comprehensive comparison of three major methods for modelling the nonlinear behaviours of structures shows that for a limited number of cracks (less than 40) and elements (< 2000), the smeared crack model gives the best results. For multiple cracks (up to 100 cracks), the discrete element give the best results, while for moderate large number of elements with arbitrary geometry, the element- free Galerkin method seems superior. However, the number of cracks may be limited in this case. Simulations also suggest that for the choice of numerical methods shall take the type of problem into consideration even though material properties are the same. In order to get a sharp crack indication or trace multiple cracks, fine or dense meshes are recommended for all methods. In addition, the development of new efficient algorithms for fracture analysis is still highly needed.

Acknowledgement

The authors thank the financial support by EPSRC.

References

1. T Belytschko, Y Y Lu, L Gu, Element-free Galerkin method, *Int. J. Num. Meth. Eng.*, **37** (1994) 229-256.
2. T Belytschko, L Gu and Y Y Lu, Fracture and crack growth by element free Galerkin methods, *Modelling Simul. Mater. Sci. Eng.*, **2** (1994) 519-534.
3. P A Cundall and O D L Strack, A discrete element model for granular assemblies, *Geotechnique*, **29**, 47-65 (1979).
4. R de Borst, Smeared cracking, plastity, creep and thermal loading - a unified approach, *Comp. Meth. Appl. Mech. Eng.*, **62** (1987) 89-110.
5. R de Borst and A H van den Boogaard, Finite element modeling of deformation and cracking in elary-age concrete, *J. Eng. Mech. Div.*, ASCE **120** (1994) 2519-2534.
6. J M Lees, A U Winistoerfer and U Meier, External prestressed CFRP straps for the shear enhancement of concrete, *ASCE J Composites for Construction*, **6** (2002).
7. N A Hoult and J M Lees, Shear retrofitting of reinfoced concrete beams using CFRP straps, *Proc. Advanced Composite Materials in Bridges and Structures*, 20-23 July, Calgary, (2004).
8. G Kesse, *Concrete beams with external prestressed carbon FRP shear reinforcement*, PhD Thesis, Cambridge University, (2003).
9. A Munjiza, D R J Owen, and N Bicanic, A combined finite-discrete element method in transient dynamics of fracturing solids, *Engineering Computation*, **12**, 145-174 (1995).
10. N Sukumar, B Moran, T. Black, T Belytschko, An element-free Galerkin method for three-dimensional fracture mechanics, *Computaional Mech.*, **20** (1997) 170-175.
11. Zienkiewicz O. C. and Taylor R. L., *The Finite Element Method*, Vol. I/II, McGraw-Hill, 4th Edition, 1991.

Machine Efficient Adaptive Image Matching Based on the Nonparametric Transformations

Bogusław Cyganek

AGH – University of Science and Technology
Al. Mickiewicza 30, 30-059 Kraków, Poland

Abstract. The paper presents machine efficient area-based image matching method that is based on a concept of matching-regions that are adaptively adjusted to image contents. They are in a form of square windows which grow to convey enough information for reliable matching. This process is controlled by local image contents. The images, however, are transformed into nonparametric representation. Such a liaison of information-theoretic models with nonparametric statistics allows for compensation for noise and illumination differences in the compared images, as well as for better discrimination of compared regions. This leads to more reliable matching in effect. Machine efficient implementation is also discussed in the paper. Finally the experimental results and conclusions are presented.

1 Introduction

The area-based image matching algorithms are ubiquities in image processing systems. They all operate in the similar fashion: the image regions of the same shape and size are compared by means of some measure. Then based on a result of this comparison the regions are pronounced to be matched or not. There are also many improvements to this basic scheme, since in practice the method is not always reliable. One of the most cumbersome problems is choice of the matching regions beforehand. Their shape and size determine quality of matching and unfortunately depend greatly on image contents, as well as on noise, texture, and image distortions. One of such improvements consists of adaptively adjusted matching regions to accommodate diverse image contents. There are different strategies that control this process. One of the very original solutions with a statistical model was proposed by Kanade and Okutomi [8]. The appropriate window is selected by evaluating the variations in intensity and disparity. Their method is based on the observation that at discontinuities the intensity and disparity variations are larger, unlike at the positions of surfaces. However, the implementation of this method is quite complicated.

The adaptive window technique developed by Lotti [9] for matching aerial images bases on window adjusting that is limited by edges and statistical contents of the matching regions. However, this technique is quite time consuming and rather limited to special class of images.

There are also many other techniques [3], such as multiple windowing [6] or a variable window concept for integral images proposed by Veksler [10].

In this paper we present an extension to the adaptive window growing technique (AWG) originally proposed in [3]. This method tries to adjust size of the matching window in order to conveyed *enough* information for more reliable matching. At the same time size of a window is kept as small as possible. However, in respect to [3], the technique has been simplified for easier implementation and speed up.

It can be shown that the AWG technique is equivalent to computation of signal entropy in a window, however in a more efficient way. This is achieved thanks to the prior image transformation into nonparametric representation which is more resistive to noise and local lighting incompatibilities. Such domain allows for fast information assessment since this nonparametric signal representations conveys *mutual relation* among neighboring pixels.

2 The Adaptive Window Concept for Image Matching

Zabih and Woodfill proposed the nonparametric *Rank* and *Census* transforms for computation of correspondences. The *Census*, used in the presented method, returns an ordered stream of bits, defined as follows: a bit at a given position is '1' if and only if an intensity value at a pixel is greater or equal to the central pixel; otherwise a bit is set to '0'. Local neighborhood of pixels is usually 3×3 and 5×5 square window [11].

The consecutive image matching, in the nonparametric domain, is usually done by means of Hamming measure used to compare different image areas [11][1], although other measures are possible as well [5]. The AWG technique presented in this paper allows for better control of the size of matching windows (whereas size of the *Census* transform is fixed to 3×3 neighborhoods and should not be confused hereafter).

For a given central pixel at *(i,j)* and surrounding $n \times n$ square neighbourhood, the corresponding *Census* measure *IC(i,j)* can be defined as a series of bits [3][5]:

$$IC(i,j) = b_{n^2-1} \ldots b_k \ldots b_3 b_2 b_1 b_0, \text{ where } k \in [0, \ldots, n^2-1] \; / \; \left\{ \left\lfloor \frac{n^2}{2} \right\rfloor \right\}. \quad (1)$$

The b_k parameter can be expressed as follows:

$$b_k = 1 \text{ if } I\left(i - \left\lfloor \frac{n}{2} \right\rfloor + \left\lfloor \frac{k}{n} \right\rfloor, j - \left\lfloor \frac{n}{2} \right\rfloor + k \bmod n\right) \geq I(i,j) \text{, 0 otherwise,} \quad (2)$$

where $I(i,j)$ denotes the intensity value for an image at a point at *(i,j)* in image coordinates, $\lfloor k/n \rfloor$ integer division of *k* by *n*, k **mod** n modulo *n* division.

Let us examine some 3×3 pixel neighbourhoods and their *Census* representations (Fig. 1). Not knowing values of pixels surrounding this neighbourhood, the complete *Census* value can be estimated only for the central pixel from the neighbourhood of interest. Therefore all the other pixels have their *Census* values assessed only in respect to their closest neighbouring pixels belonging to this 3×3 neighbourhood. All other values are left undefined and marked as 'x' (don't care). Such bits can be assigned '0' or '1' depending on intensity values of other pixels, lying in other 3×3 neighbourhoods, where the given pixel in turn lies in their centre. The function *q('b')* in Fig. 1 returns number of bits with value 'b' (i.e. number of '0s' or '1s').

The question arises on distribution of the *Census* values, computed in local 3×3 neighborhoods for the whole image, in respect to the distribution of intensities in the input image. It is well known that the latter greatly influences any subsequent matching, no matter what matching criteria are taken into consideration. The answer to this question is fundamental to the AWG technique described in this paper.

Based on definition (1) we observe that neighborhoods with constant intensity produce bit streams with all bits set to '1' (Fig. 1a). That is data has zero entropy and the conveyed information is too unreliable for matching. However, in Fig. 1b and Fig. 1c we notice that with an increase of spatial distribution – counted as an entropy of intensity values – $q('0')$ increases as well. For exemplary values in Fig. 1 the entropy of data in selected 3×3 regions increases conveying more information that in consequence allows for more reliable matching.

For any $n \times n$ neighborhood it is evident that the following relation holds:

$$q('0') + q('1') + q('x') = Bn^2, \qquad (3)$$

where B stands for number of bits assigned for *Census* representation of a pixel. Therefore maximization of $q('0')$ can be achieved by keeping $q('1')$ at minimum and when undefined bits 'x' become '0'. This feature is used in implementation.

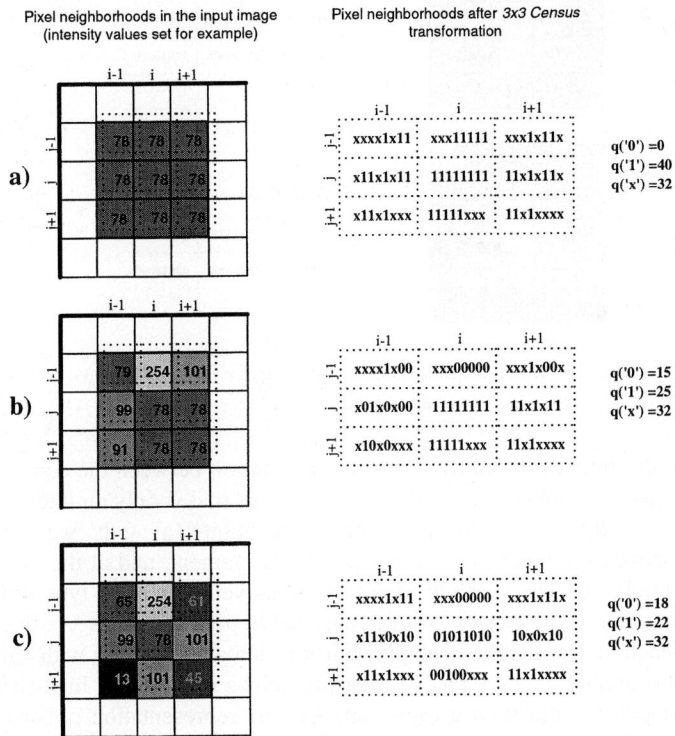

Fig. 1. Some 3×3 neighborhoods and their *Census* representations. $q('b')$ returns number of bits with value '*b*'. Intensity values set only as an example – relevant are only their mutual relations

Disambiguation of all *Census* values in any 3×3 neighborhood requires knowledge of intensity values of all pixels surrounding this neighborhood. Such a situation with some exemplary intensity values is presented in Fig. 2. We notice that an increase of the $q('0')$ entails a further increase of the entropy of data in the pixel neighbourhoods of interest. For 3×3 neighborhoods around a point at indexes *(i,j)* in Fig. 2, the entropies are respectively (from the top): 0.29, 0.41, and 0.55.

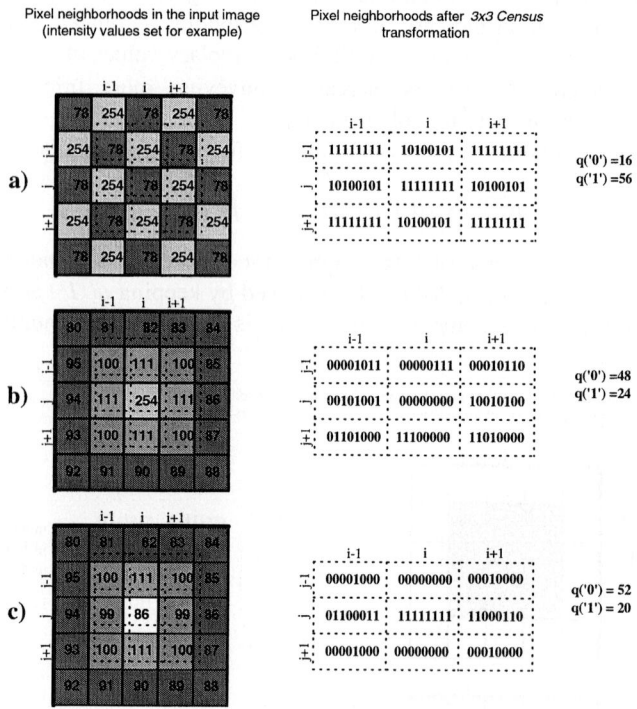

Fig. 2. The specific 3×3 pixel neighborhoods and their *Census* transformations. Relevant are relations of intensity values (set for example here)

From the definition (1) and from the mutual relation between the central point 'X' and all its closest neighbors 'a' – 'h' in Fig. 3 we see that if only an intensity value of the point 'X' is different from all values of the points 'a' – 'h' we have met the condition for maximum entropy for this point arrangement, and at the same time the maximum number of '0's is achieved. There exists yet an another type of the mutual pixel relation – the relation among three consecutive pixels, such as 'a', 'b', and 'd' in Fig. 3. For each pair of neighboring pixels from such a triple, a bit with value '0' can be assigned if and only if any pair of intensity values is different. In such a case, for each pair of pixels from such a triple, one *Census* representation obtains '1' at the corresponding bit position, while the same bit position – but in the complementary pixel from this pair – holds '0'. This is clear from (2) since there is an exclusive relation. Thus, if only the abutted pixels have different values, at the pertaining bit position the only possibility is: bit '1' in the first *Census* pixel of a pair and '0' for the second one.

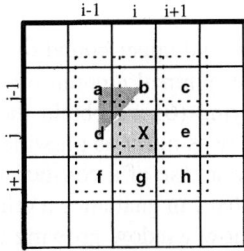

Fig. 3. Relations among pixel values in the *Census* transformation

The rules for the maximum number of '0s' – and in consequence entropy increase – in a *Census* representation of any 3×3 neighborhood (see Fig. 3) are as follows:

1. A value of the central pixel **X** is irrelevant provided that it is different than any other pixel from its closest neighborhood (i.e. from the all pixels **a, b, c, d, e, f, g,** and **h** in Fig. 3):

$$\forall p \in N - \{X\}: I(p) \neq I(X), \qquad (4)$$

where *I(p)* is an intensity value of a pixel at index *p* (see Fig. 1b) from the 3×3 neighborhood *N*.

2. Pixel values of any corner-triple (such as e.g. **a, b, d** in Fig. 3) must be different, i.e.:

$$I(a) \neq I(b) \neq I(c). \qquad (5)$$

3. All other pixels bordering with pixels from the neighborhood *N* must have their intensity values less than their direct pixels-neighbors from *N*. For example, for the pixel **b** in Fig. 3 these would be pixels at indexes: (j-2,i-1), (j-2,i), and (j-2,i+1).

The upper bound for any 3×3 *Census* neighbourhood $q_{max}('0') = 52$ can be easily found from the conditions 1-3 and after observing Fig. 3. This with (3) corresponds with lower bound $q_{min}('1') = 20$. The lower bound for data entropy (LBE) in any 3×3 *Census* neighbourhood that preserves the conditions 1-3 is obtained by taking only four different values and can be easily found from (4) to be: LBE=-4/9*log4/9-2*2/9*log2/9-1/9*log1/9≈0.55. The upper bound for the entropy in any 3×3 neighbourhood is MBE=log(9)≈0.95 (i.e. nine different values). Thus, any 3×3 neighbourhood of pixels that preserve conditions 1-3 guarantee almost 58% of maximum possible entropy in this neighbourhood. Concluding we can state that the set of conditions 1-3 guarantying the maximum number of '0's (or equivalently the minimal number of '1's) in any 3×3 *Census* neighborhood leads to at least LBE. Therefore the sought size *n* of a matching square window can be found as to maximize the *averaged* (per pixel) value of $q_{av}('0')$ in a window of interest with a constraint $q_{tot}('0') > \vartheta_0 = 0$:

$$n: \max_{n_{min} \le n \le n_{max}} q_{av}('0') \wedge q_{tot}('0') > \vartheta_0 = 0 \qquad (6)$$

where n_{min} and n_{max} set the lower and upper bound of size of the matching window, θ_0 is a certain threshold for a total minimal amount of '0s' in a window of interest; θ_0 usually is set to 0. Thus, applying (6) we obtain the *minimal size* of the matching window that is *adapted* to convey as much as possible of information to allow a more reliable matching. Otherwise, in a case of a constant intensity we obtain information that there is no such a window that can guarantee a reliable matching.

The other possibility is to allow window growing up to such a *minimal* size n for which the achieved *average* $q_{tot}('0')$ reaches a predefined threshold value ϑ_1:

$$\min(n) : q_{av}('0') > \vartheta_1 \qquad (7)$$

Similarly, one can search for such a *minimal* size n for which the *total* $q_{tot}('0')$ reaches a predefined threshold value ϑ_2 as follows:

$$\min(n) : q_{tot}('0') > \vartheta_2 \qquad (8)$$

Computations of (7) and (8) are simpler and in consequence faster than (6). However they require an arbitrary setting of the threshold values.

3 Experimental Results

Experimental results and discussion of the image matching based on AWG operating in accordance with (6) can be found in [3]. In this section we present results that help compare different versions of the AWG given by (6)-(8). The computational platform consists of the IBM PC with Pentium 4 with 3.4 GHz, 2GB, implementation in C++,.

Fig. 4 presents results of the AWG technique applied to match the "PCB" stereo pair of size 253×193 (a). The disparity map obtained in accordance with (6) is

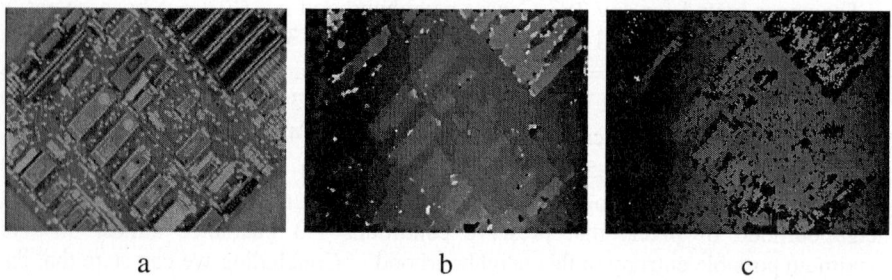

a   b   c

Fig. 4. The AWG technique applied to the "PCB" stereo image (253×193): left image of a pair (a), disparity map from formula (6) (b), disparity map from formula (7) and θ_1=2 (c)

depicted in (b) while disparity from (7) with $\theta_1=2$ in (c). Execution times are 1.75 and 1.68 s, respectively. Parameter θ_1 was chosen based on prior experiments. Allowable window growing spanned: $n_{min}=3$ to $n_{max}=30$.

Fig. 5 contains results of the AWG applied to the "Mask" 450×375 pair (a). Disparity map from (6) is shown in (b) and with formula (7) with $\theta_1=2$ in (d), formula (8) with $\theta_2=15$ in (e), and $\theta_2=150$ in (f). Execution times are as follows: 2.05 for (b), 1.99 for (d), 0.83 for (e), and 4.22 for (f). The latter caused by the fact of much higher matching windows due to high θ_2. However, execution of (e) compared to (d) is match faster, because total amount of '0s' is easier to compute than their average.

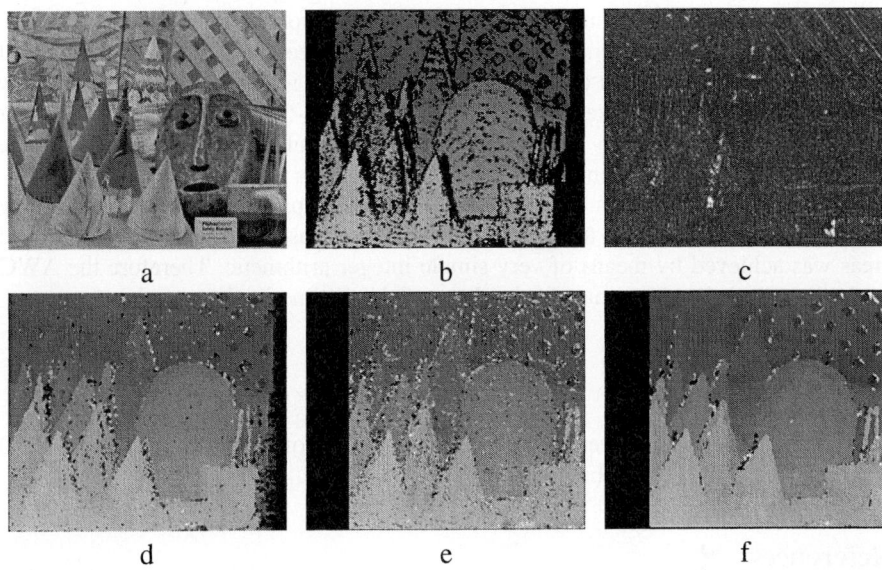

Fig. 5. The AWG technique applied to the "Mask" pair (450×375): left image of a pair (a), disparity map (6) (b), visualization of adaptive windows (lither places denote bigger windows) (c), disparity map with formula (7) and $\theta_1=2$ (d), disparity map with formula (8) and $\theta_2=15$ (e), disparity map with formula (8) and $\theta_2=150$ (f)

For each pixel location the size of a matching window is determined with the AWG technique defined by (6). Disparity values are found for each pixel by matching square windows of sizes appropriately set in the previous step. As a matching measure the Hamming distance is computed from the *Census* representation of each pair of candidate windows. Matching is done with a fast block matching algorithm based on the winner-update strategy with hashing table [2]. The visible limited occlusions are due to a simple stereo algorithm without the cross-checking [12]. In Fig. 5(d-f) we notice also many occluding areas due to large displacement between images of this stereo pair. However, the details were preserved due to adaptively chosen matching windows.

The worst case time complexity of the AWG algorithm (6)-(8) can be estimated as follows:

$$O(NMn_{max}^2) \tag{9}$$

for image of size $N{\times}M$ pixels, where n_{max} is the maximum allowable matching window. However, the arithmetic operations are limited only to *integer comparisons and summations*. Therefore the technique fits well hardware implementations.

4 Conclusions

The paper presents the image matching technique with adaptively changed matching areas. The whole matching is performed after changing the original images into nonparametric domain by means of the *Census* transformation. The basics of the AWG technique were presented which can be summarized as follows: an increase of the zero-value-bits $q('0')$ in any *Census* neighborhoods indicates increase of entropy in this neighborhood of pixels. Different mutations (6)-(8) of the method allow for adjustment of the matching region at different computational cost. In effect, the subsequent matching is much more reliably than in a case of fixed windows [3].

The experimental results with the AWG technique applied to the stereo matching showed robustness of this technique. The main purpose of avoiding low-textured areas was achieved by means of very simple integer arithmetic. Therefore the AWG method seems to be appropriate for hardware and real-time implementations.

Acknowledgement

This paper has been sponsored by the Polish Committee of Scientific Research (KBN) grant number: 3T11C 045 26.

References

1. Banks J., Bennamoun M., Corke P.: Non-Parametric Techniques for Fast and Robust Stereo Matching. CSIRO Manufacturing Science and Technology, Australia (1997)
2. Chen, Y-S., Hung, Y-P., Fuh, C-S.: Fast Block Matching Algorithm Based on the Winner-Update Strategy. IEEE Trans. On Image Processing, Vol. 10, No. 8, (2001) 1212-1222
3. Cyganek, B.: Adaptive Window Growing Technique for Efficient Image Matching. Technical Report, AGH-University of Science and Technology, Krakow, Poland (2004)
4. Cyganek, B., Borgosz, J.: Maximum Disparity Threshold Estimation for Stereo Imaging Systems via Variogram Analysis. ICCS 03, Russia/Australia (2003) 591-600
5. Cyganek, B.: Comparison of Nonparametric Transformations and Bit Vector Matching for Stereo Correlation. Springer LNCS 3322 (2004) pp. 534-547
6. Fusiello, A. et.al.: Efficient stereo with multiple windowing. CVPR 858–863 (1997)
7. Haykin, S.: Neural Networks. A Comprehensive Foundation. Prentice-Hall (1999)
8. Kanade, T., Okutomi, M.: A stereo matching algorithm with an adaptive window: Theory and experiment. PAMI, 16(9) (1994) 920–932
9. Lotti, J-L., Giraudon, G.: Adaptive Window Algorithm for Aerial Image Stereo. INRIA Technical Report No 2121 (1993)

10. Veksler, O.: Fast Variable Window for Stereo Correspondence using Integral Images. Computer Vision and Pattern Recognition (2003)
11. Zabih, R., Woodfill, J.: Non-Parametric Local Transforms for Computing Visual Correspondence. Proc. Third European Conf. Computer Vision (1994) 150-158
12. Zhengping, J.: On the Mutli-Scale Iconic Representation for Low-Level Computer Vision. Ph.D. Thesis. The Turing Institute and University of Strathclyde (1988) 114-118

Non-gradient, Sequential Algorithm for Simulation of Nascent Polypeptide Folding

Lech Znamirowski

Institute of Informatics, Silesian University of Technology,
ul. Akademicka 16, 44-100 Gliwice, Poland
lznamiro@top.iinf.polsl.gliwice.pl

Abstract. In the paper, the method for determining of the conformation of nascent protein folding based on two-phase, sequential simulation approach is presented. In both phases the potential energy of molecule under construction is minimized, however in the first phase the minimization is performed for the structure {*new amino acid*}/{*existing amino acid chain*} and in the second phase, the {*new existing amino acid chain*} conformation is "tuned" to reach the minimal potential energy of growing chain. The formed, nascent conformation of protein determines initial condition for a future conformation modifications and plays a crucial role in fixing the biological and chemical features of created protein. The simulation of conformation process is illustrated through the numerical example of nascent protein folding for a selected protein.

1 Introduction

The chain of amino acids synthesized in a ribosome forms a polypeptide and in case of a long chain, a protein. For the conformation simulation tasks, the synthesis programs generating the conformation of the main chain of amino-acid residues in a polypeptide, use the amino acids library and need the information on the angles between α-carbons bonded with their side chains and contiguous peptide groups [1, 2, 3]. The backbone conformation of the chain of amino-acid residues can be determined by the set of torsion angles [5] φ (rotation around the nitrogen-α-carbon bond), and ψ (rotation around the α-carbon-carbon bond in the chain of the polypeptide).

1.1 Torsion Angles

The relationship between the peptide groups, α-carbons and torsion angles can be expressed in the following form

$$\rightarrow PB \rightarrow \phi_1 \rightarrow C_{\alpha 1} \rightarrow \psi_1 \rightarrow PB \rightarrow \phi_2 \rightarrow C_{\alpha 2} \rightarrow \psi_2 \rightarrow PB \rightarrow \qquad (1)$$

where PB denotes the peptide bond and $C_{\alpha i}$ is the i-th α-carbon atom. The torsion angles play a crucial role in the conformation of proteins (Fig. 1) because the three-dimensional structure of protein determines its biological functions including its activity in a signal transduction cascades [6, 7, 8].

On the other hand not all combinations of torsion angles φ_i and ψ_i in (1) for each amino acid are possible, as many leads to collision between atoms in adjacent residues exposed by Ramachandran restrictions (map) [7, 9]. By assuming that atoms behave as hard spheres, allowed ranges of torsion angles can be predicted and visualized in contour of Ramachandran plots based on X-ray crystallography [10]. Nearly every polypeptide or protein in a cell is chemically altered after its synthesis in a ribosome. Many proteins are later modified in a reversible manner, fulfilling its functions after the activation forced by the messenger molecules. Such modifications expressed in the conformational changes [11, 12, 13], may alter the biological activity, life span, or cellular location of proteins, depending on the nature of alteration.

1.2 Conformational Energy

The force powered changes of spatial shape is a tendency of the chain of polypeptide to get minimal potential energy. The potential energy of nanostructure defined also as conformational energy, according to the principles of *molecular mechanics* [17, 21, 23] accepts the general form

$$E(\mathbf{x}) = \sum_j U_j(\mathbf{x}) \ , \qquad (2)$$

where the expression $U_j(\mathbf{x})$ describes the proper type of potential energy (bonds stretching energy around equilibrium length of bonds, angles bending energy, multipole interactions energy, Lennard-Jones or generally Mie potentials etc.) of the intermolecular interactions there be a function of degrees of freedom and the physicochemical constants determined for the specified types of atoms, bonds, and in molecule configurations. The entries x_1, x_2, x_3, ... of a vector $\mathbf{x}$ represent the degrees of freedom relating with distances and proper characteristic angles.

The spatial structure of molecule is described in one of the accepted standard formats: PDB [28] or mmCIF [27]. Dependent on type of atoms, type of bonds and the configuration of atoms in a molecule, the physicochemical constants are collected in tables (so called *Force Field*) e.g. AMBER [26], ENCAD [22], CHARMM [25] and others (depending on type of chemical compounds), applied in computations [16].

Minimizing the expression (2) which describes a conformational energy of structure for vector $\mathbf{x}$ with constraints (Ramachandran plots), it is possible to find a stationary point defining the stable structure in a given conditions. However, the complex form of expression (2), usually leads to the problem related with multiple local minima or saddle points. In the saddle points of the surface of energy for which the gradient $E(\mathbf{x})$ reaches zero the crucial role play the eigenvalues of hessian matrix built for function $E(\mathbf{x})$, which allow to define the farther run of the settlement of nanostructure conformation. And so simple algorithms of optimization e.g. gradient-based, Newton's or Fletcher-Powell have to be modified to find the global minimum. The implementation of algorithms of dynamic programming [3, 19, 20] or Monte Carlo methods [4, 15, 24] is better solution.

Using new, two-phase sequential simulation approach, we will find the free conformation (initial conformation or nascent protein folding) for the exemplary protein through the minimization of the energy [16] as a function of the torsion angles [3, 16, 17], however, we do not look for the global minimum of conformational energy, rather we perform the sequence of optimization corrections [14] for the growing chain. The simulation process retains the same sequence of events as a real process of nascent protein elongation.

2 Two-Phase Sequential Simulation Algorithm

The basic assumption for the construction of simulation algorithm is an observation, that the new amino-acid appearing from the ribosome rotates the existing chain of amino acids in a such manner, that the torsion angles on the new peptide bond, accordingly with minimization of potential energy of the structure {*new amino acid*} /{*existing amino acid chain*} are determined, and next, the conformation of the whole structure is somewhat modified to minimize the potential energy of {*new existing amino acid chain*}. The first minimization plays a crucial role because of dimensions of peptide chain in comparison with the new molecule attached.

2.1 Initialization

The space defining available regions of φ and ψ angles (Fig. 1) is quantized and translated to the linear vector of available coordinates. The two first amino acids adopt the conformation minimizing expression

$$\min_{\varphi_0, \psi_0, \varphi_1, \psi_1} E_2(\varphi_0, \psi_0, \varphi_1, \psi_1), \qquad (3)$$

where E_2 is a potential energy of the system of two amino acids, φ_0 and ψ_0 are the torsion angles of the first amino acid, as well as φ_1 and ψ_1 are the torsion angles of the second amino acid (Fig. 1a). In the Initialization and First Phase of simulation the search for energy decreasing is performed in the discretized, allowed ranges of torsion angles for glycine, proline and all 18 remaining amino acids specifically, predicted as modified Ramachandran plots based on X-ray crystallography [10]. In the Second Phase of simulation during the chain "shaking" process, only the generated torsion angles which remain in the allowed ranges are used. The energy computations apply the AMBER force field and the Tinker subroutines [16].

2.2 First Phase of Simulation

When the third residue appears (#2 in Fig. 1b), the following expression is minimized

$$\min_{\varphi_2, \psi_2} E_3(\varphi_2, \psi_2, \varphi_{0f}, \psi_{0f}, \varphi_{1f}, \psi_{1f}), \qquad (4)$$

where E_3 is a potential energy of the system of three amino acids, $\varphi_{0f}, \psi_{0f}, \varphi_{1f}, \psi_{1f}$ are the torsion angles determined in the initialization.

The first phase for fourth amino acid is preceded by the second phase which tunes the existing, whole chain to get the minimal potential energy (point 2.3).

Accordingly, for the fourth amino acid we minimize the expression

$$\min_{\varphi_3,\psi_3} E_4(\varphi_3,\psi_3,\varphi_{2s},\psi_{2s},\varphi_{0fs},\psi_{0fs},\varphi_{1fs},\psi_{1fs}),\quad (5)$$

where E_4 is a potential energy of the system of four amino acids, and φ_{2s}, ψ_{2s}, φ_{0fs}, ψ_{0fs}, φ_{1fs}, ψ_{1fs} are the torsion angles determined in the second phase for the three amino acids (Fig. 1c).

2.3 Second Phase of Simulation

For the consecutive residues, bonded to the growing chain, the global tuning after the first phase takes place. Along the backbone of the chain (Fig. 1b, c), the torsion angles are changed in the four directions: $+k\Delta\varphi_i$, $-k\Delta\varphi_i$, $+k\Delta\psi_i$, and $-k\Delta\psi_i$, including the current φ_i and ψ_i ($i = 0, 1, \ldots n$) angles. Values of $\Delta\varphi_i$ and $\Delta\psi_i$ determine the amplitudes of reversible changes of torsion angles along the chain, and k is a positive and decreasing parameter during the consecutive steps of the minimization of potential energy of the chain.

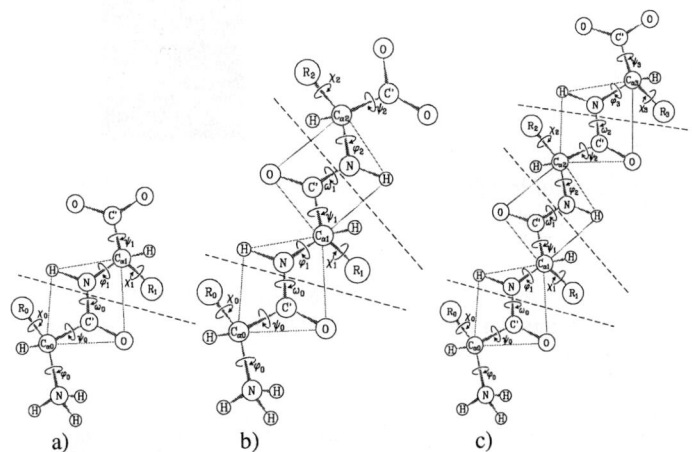

Fig. 1. Process of polypeptide elongation: a) initialization of simulation, b) the three residues (first phase of simulation), c) the four residues (first phase of simulation)

The parameter k can be constant in one iteration step, as well as may be a function of selected parameters. In the proposed approach, we omit the problem of hydrogen and disulfide bonds creation.

3 Numerical Example

An illustration of presented two-stage simulation of nascent protein folding are the intermediate and final results in simulation of elongation process and conformational changes of the small protein containing thirty eight amino acids:

asn – ser – tyr – pro – gly – cys – pro – ser – ser – tyr – asp – gly – tyr – cys – leu –
asn – gly – gly – val – cys – met – his – ile – glu – ser – leu – asp – ser – tyr – thr –
cys – asn – cys – val – ile – gly – tyr – ser.

The initialization phase of simulation based on the minimizati on of expression (3), gives the dipeptide asn-ser with four torsion angles. The conformation of dipeptide and the set of torsion angles is presented in Fig. 2.

When the third amino acid appears i.e. tyrosine, the first phase of simulation for the elongated tripeptide gives the results presented in Fig. 3a, b. The expression (4) is minimized over the φ_2 and ψ_2 angles with remaining angles unchanged.

In the second phase, the tripeptide chain is "shaked" along the backbone with $\Delta \varphi_i = \Delta \psi_i = 32°$ and k decreasing twice per step until resolution 1°, better resolution usually needs parallel processing.

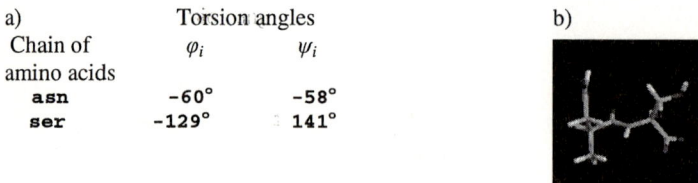

a)
Chain of amino acids	Torsion angles φ_i	ψ_i
asn	-60°	-58°
ser	-129°	141°

b)

Fig. 2. The spatial shape of folded dipeptide asn-ser: a) the chain of torsion angles, b) RasMol [18] visualization of a molecule

The iterations stop when there is no decreasing value of the potential energy of tripeptide. As a result, the conformation of nascent tripeptide asn-ser-tyr gets the form presented in Fig. 3c, d.

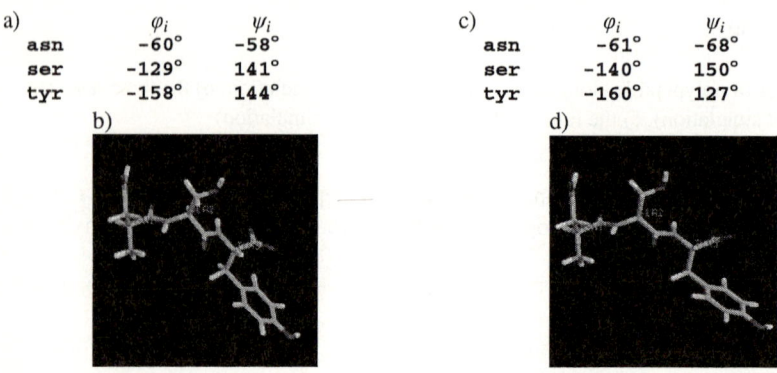

a)
	φ_i	ψ_i
asn	-60°	-58°
ser	-129°	141°
tyr	-158°	144°

c)
	φ_i	ψ_i
asn	-61°	-68°
ser	-140°	150°
tyr	-160°	127°

b)   d)

Fig. 3. The initial shape of folded tripeptide asn-ser-tyr: a) the chain of torsion angles, b) visualization of a molecule, and the final conformation of nascent tripeptide: c) torsion angles, d) visualization of a molecule

Continuing this process, the eight amino acids chain obtains the conformation after the first and second phase of simulation presented in Fig. 4. Finally, the whole 38 amino acids chain adopts the conformation after first and second phase of simulation presented in Fig. 5. The results of computations have been compared with the results obtained in different discretization grid of the permissible torsion angles, and the differences were acceptable.

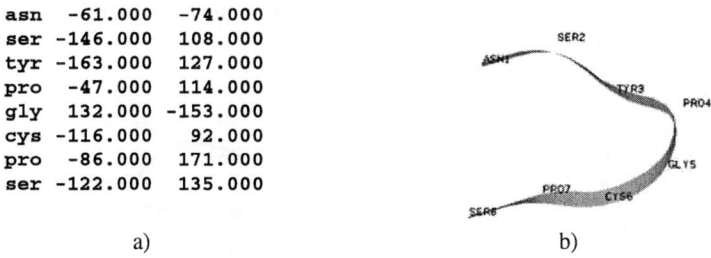

```
asn  -61.000   -74.000
ser -146.000   108.000
tyr -163.000   127.000
pro  -47.000   114.000
gly  132.000  -153.000
cys -116.000    92.000
pro  -86.000   171.000
ser -122.000   135.000
```

a)   b)

Fig. 4. Final conformation of eight amino acids nascent protein after the second phase of simulation: a) torsion angles, b) backbone of molecule (ribbons visual representation [18])

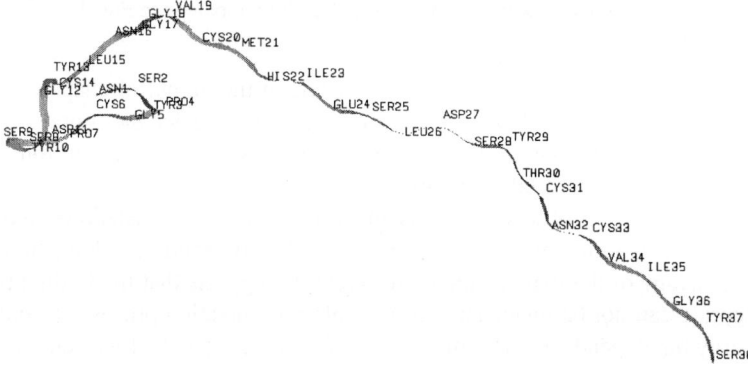

Fig. 5. Final conformation of 38 amino acids nascent protein after the second phase of simulation

The interesting observation can be do refer to the conformations of the same polypeptide sub-chain during the elongation phase and post-translational modifications namely cleavage of the main chain. In Fig. 6 the conformation of the initial eight amino acids cleaved from the whole chain has been presented. Comparing this conformation with the conformations of the same sub-chain of polypeptide in elongation phase (Fig. 4) we see the difference in the spatial shapes of these nanostructures. It suggests that the folding process of polypeptide depends on the sequence of external events preceding the final conformation.

4 Conclusions

The paper presents two-phase algorithm for modeling and simulation of composite phenomena in protein processing: model building and simulation of initial phase of establishing the conformation of nascent protein folding.

The introductory experiments reveal, that the crucial role plays the phase when the new, consecutive amino acid bonds to the existing chain of polypeptide. This phenomenon is represented in simulation by the *first phase*, underlying fact that the new amino acid is partly fixed in the ribosome.

```
asn   -58.000   -75.000
ser   -89.000    82.000
tyr  -165.000   128.000
pro   -47.000   114.000
gly   126.000  -157.000
cys  -115.000    92.000
pro   -86.000   171.000
ser   -94.000   144.000
         a)
```

b)

Fig. 6. Conformation of the initial eight amino acids cleaved from the whole 38 amino acids nascent protein: a) torsion angles, b) backbone of molecule (compare with Fig. 4)

We can expect, that the similar situation appears at the output of chaperones [29]. In the *second phase* of simulation, the whole structure is under the influence of annealing process reaching the nascent conformation. This phase of protein folding can be treated as a correction of initial conformation.

The very interesting observation was made refer to post-translational modifications. The conformations of the same sub-chain of polypeptide in elongation phase and after cleavage of the main chain are different. It suggests that the folding process of polypeptide can not be understand as a simple minimization process of conformational energy but depends on the sequence of events preceding the final state.

Developed simulation software implementing mainly paradigms of molecular mechanics will be enriched in the scope of detection of stabilizing bonds as well as in the analysis tools for tracking the environmental modifications of protein conformation.

References

1. Berman, H. M., et al.: The Protein Data Bank. Nucleic Acid Res. 1 (2001) 235-242
2. "Klotho: Biochemical Compounds Declarative Database – Alphabetical Compound List". (2001), http://www.ibc.wustl.edu/moirai/klotho/compound_list.html
3. Znamirowski, L., Zukowska, E. D.: Simulation of Environment-Forced Conformations in the Polypeptide Chains. Proc. of the ICSEE'03, 2003 Western MultiConf., The Soc. for Mod. and Sim. Int., Orlando, Florida, January 19-23 (2003) 87-91
4. Zieliński, R.: Monte Carlo Methods. WNT Publisher, Warsaw (1970) (in Polish)

5. Liebecq, C. (ed.): The White Book. Biochemical Nomenclature and Related Documents. IUPAC-IUBMB Joint Commission 2nd edn. Portland Press, London (1992)
6. "Cell Signaling Technology Database: Pathways". Cell Sign. Techn. Inc., Beverly, MA. (2003) info@cellsignal.com
7. Stryer L.: Biochemistry. W. H. Freeman and Company, New York (1994)
8. Heater, S. H., Zukowska, E. D., Znamirowski, L.: Information Processing in the Signal Cascades of Nanonetworks. ICCE-10, Tenth Ann. Intern. Conf. on Composites/Nano Engr., Intern. Comm. for Composite Engr., New Orleans, Louisiana, July 20-26 (2003) 233-234
9. Ramachandran, G. N., Sasisekharan, V.: Conformation of polypeptides and proteins. Advanced Protein Chem., 23 (1968) 506-517
10. Hovmoller, S, Zhou, T., Ohlson, T.: Conformations of amino acids in protein. Acta Cryst., 5 (2002) 768-776
11. Baker, D.: A surprising simplicity to protein folding. Nature, 406 (2000) 39-42
12. Okamoto, Y.: Protein folding simulations and structure predictions. Computer Physics Comm., 142 (2001) 55-63
13. Wei, C., Zhang, Y., Yang, K.: Chaperone-Mediated Refolding of Recombinant Prochymosin. Journal of Protein Chemistry, 6 (2000) 449-456
14. Znamirowski, A. W., Znamirowski, L.: Two-Phase Simulation of Nascent Protein Folding. Proc. of the Fourth IASTED Intern. Conf. on MODELLING, SIMULATION, AND OPTIMIZATION MSO 2004, Kauai, Hawaii, August 17-19, 2004, ACTA Press, Anaheim-Calgary-Zurich (2004) 293-298
15. Warecki, S., Znamirowski, L.: Random Simulation of the Nanostructures Conformations. Intern. Conference on Computing, Communication and Control Technology, Proceedings Volume I, The Intern. Institute of Informatics and Systemics, Austin, Texas, August 14-17 (2004) 388-393
16. Ponder, J.: TINKER – Software Tools for Molecular Design. Dept. of Biochemistry & Molecular Biophysics Washington University, School of Medicine, St. Louis (2001) http://dasher.wustl.edu/tinker
17. Hinchliffe, A.: Modelling Molecular Structures. 2nd edn. John Wiley & Sons, Ltd, Chichester (2000)
18. Sayle, R.: RasMol, Molecular Graphics Visualization Tool. Biomolecular Structures Group, Glaxo Wellcome Research & Development, Stevenage, Hartfordshire (1998) H. J. Bernstein v.2.7.1.1, rasmol@bernstein-plus-sons.com
19. Bellman, R.: Dynamic Programming. Princeton University Press, Princeton, N. J. (1957)
20. Węgrzyn, S., Winiarczyk, R., Znamirowski, L.: Self-replication Processes in Nanosystems of Informatics. Int. Journ. of Applied Math. and Comp. Science, Vol. 13, 4 (2003) 585-591
21. Atkins, P. W.: Physical Chemistry, 6th edn. W. H. Freeman & Co., New York (1998)
22. Levitt, M., Hirshberg, M., Sharon, R., Daggett, V.: Potential Energy Function and Parameters for Simulation of the Molecular Dynamics of Protein and Nucleic Acids in Solution. Comp. Physics Comm., 91 (1995) 215-231
23. McMurry, J.: Organic Chemistry, 4th edn. Brooks/Cole Publ. Co., Pacific Grove (1996)
24. Metropolis, N., Ulam, S.: The Monte Carlo Method. Journal of the American Stat. Assoc., Vol. 44, 247 (1949) 335-341
25. MacKerrell, A. D. et al.: All-Atom Empirical Potential for Molecular Modeling and Dynamics Studies of Proteins. Journ. Phys. Chem. B, 102 (1998) 3586-3616
26. Cornell, W. D. et al.: A Second Generation Force Field for the Simulation of Proteins, Nucleic Acids, and Organic Molecules. Journ. Am. Chem. Soc., Vol. 117 (1995) 5179-5197

27. "The Nucleic Acid DataBase Project". Rutgers, The State University of New Jersey, Camden, Newark, New Brunswick/Piscataway (2001), http://ndb-mirror-2.rutgers.edu/mmcif/index.html
28. Callaway, J., Cummings, M. et al.: Protein Data Bank Contents: Atomic Coordinate Entry Format Description. Federal Govern. Agency (1996), http://www.rcsb.org/pdb/docs/ format/ pdbguide2.2/
29. Lodish, H., Berk, A., Zipursky, S. L., et al.: Molecular Cell Biology, 4th edn. W. H. Freeman and Comp., New York (2001)

Time Delay Dynamic Fuzzy Networks for Time Series Prediction

Yusuf Oysal

Anadolu University, Computer Engineering Department, Eskisehir, Turkey
yoysal@anadolu.edu.tr

Abstract. This paper proposes a Time Delay Dynamic Fuzzy Network (TDDFN) that can be used for tracking and prediction of chaotic time series. TDDFN considered here has unconstrained connectivity and dynamical elements in its fuzzy processing units with time delay state feedbacks. The minimization of a quadratic performance index is considered for trajectory tracking applications. Gradient with respect to model parameters are calculated based on adjoint sensitivity analysis. The computational complexity is significantly less than direct method, but it requires a backward integration capability. For updating model parameters, Broyden-Fletcher-Golfarb-Shanno (BFGS) algorithm that is one of the approximate second order algorithms is used. The TDDFN network is able to predict the Mackey-Glass chaotic time series and gives good results for the nonlinear system identification.

1 Introduction

Some of the nonlinear dynamical systems produce chaotic time series outputs that are highly depend on initial conditions. If the initial condition does not specified within a suitable precision range, it is very difficult to predict the long time future behavior of these time series. But the short time behavior can be exactly encapsulated. There are various types of evolutionary systems and neural networks with time delays to solve time series prediction in short time. For example in [1], an adaptive-network-based fuzzy inference system (ANFIS) was used to identify nonlinear components on-line in a control system to predict a chaotic time series. In another study, a genetic fuzzy predictor ensemble (GFPE) was proposed for the accurate prediction of the future in the chaotic or nonstationary time series [2]. Moreover, an evolutionary system, i.e., EPNet was used to produce very compact artificial neural networks (ANNs) to predict the Mackey-Glass time series prediction with generalization ability in comparison with some other algorithms [3] and in [4] a hybrid approach to fuzzy supervised learning was applied through software called GEFREX for approximation problems, classification problems, and time series prediction.

Another approach for time series prediction is to provide dynamical neural network structures. The typical workaround is the usage of a large parallel input vector consisting of a number of states or past samples of process data. This "tapped delay line" approach has proven successful for chaotic time series prediction ([5],[6]), but it has the drawback of the curse of dimensionality: the number of parameters in the units increases exponentially and parameters can get larger values.

This work focuses on modeling and prediction of nonlinear systems with time delay dynamic fuzzy networks (TDDFNs) to overcome these drawbacks. TDDFNs are continuous-time recurrent neural networks that contain dynamical elements such as integrators in their fuzzy processing units and time delayed feedbacks. Successful control and modeling applications of DFNs without time delay elements can be found in [7] and [8]. In this study an approximate second order gradient algorithm based on adjoint theory [9]-[12] which is faster than the direct method is used for training the TDDFNs to obtain the appropriate parameters. Given a desired trajectory, a nonlinear optimization problem is solved to determine appropriate values for network parameters.

2 Time Delay Dynamic Fuzzy Network Architecture

The dynamic fuzzy network considered here represents the biological neuron that is constrained to be feedforward with dynamic elements in its fuzzy processing units, and with time delay state feedbacks. The processing unit is called "feuron" (stands for fuzzy neuron) [7],[8]. It represents a biological neuron that fires when its inputs are significantly excited through a lag dynamics (i.e. Hopfield dynamics).

The feuron's activation model which resembles the receptive field units found in the visual cortex, in parts of the cerebral cortex and in outer parts of the brain is a standard fuzzy system with Gaussian membership functions, singleton fuzzifier, product inference engine and a center average defuzzifier [13].

The activation function of the ith feuron can be expressed as:

$$\phi_i(x_i) = \frac{\sum_{k=1}^{R_i} b_{ik}\mu_k(x_i)}{\sum_{k=1}^{R_i} \mu_k(x_i)} = \frac{\sum_{k=1}^{R_i} b_{ik} \exp\left(-\frac{1}{2}\left(\frac{x_i-c_{ik}}{\sigma_{ik}}\right)^2\right)}{\sum_{k=1}^{R_i} \exp\left(-\frac{1}{2}\left(\frac{x_i-c_{ik}}{\sigma_{ik}}\right)^2\right)} \quad (1)$$

where c_{ik} is the center and σ_{ik} is the spread of the kth receptive field unit of the its ith feuron.

The membership functions of the feuron are assumed to be normal and orthogonal with the boundary membership functions (the lower and upper membership functions of the universe of discourse) represented by hard constraints, i.e., it is assumed that membership value is equal one at out of range.

An example of the computational model for TDDFNs with two-feuron and two-inputs/two-outputs that is used in this study is shown in Fig.1.

The general computational model that we have used for TDDFN is summarized in the following equations:

$$z_i = \sum_{j=1}^{n} q_{ij} y_j, \quad i=1,2......M \quad (2)$$

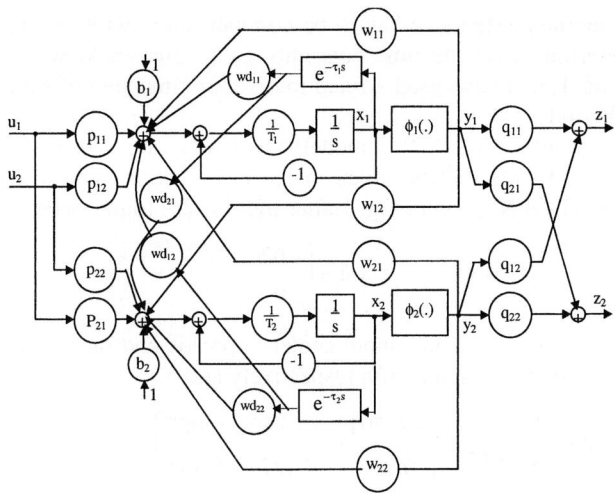

Fig. 1. The state diagram of a TDDFN with two-feurons two-inputs/two-outputs

$$y_i = \phi_i(x_i, \pi_i) \qquad (3)$$

$$\dot{x}_i = f_i(x_i, p) = \frac{1}{T_i}[-x_i + \sum_{j=1}^{n} wd_{ij} x_j(t-\tau_j) + \sum_{j=1}^{n} w_{ij} y_j + \sum_{j=1}^{L} p_{ij} u_j + r_i]; \; x_i(0) = x_{i0}, i=1..n \qquad (4)$$

where w, p, q, r are the interconnection parameters of a TDDFN with **n** units and **L** input signals, T is the time constants and π is the parameter sets (centers **c**, spreads σ, output centers **b**) corresponding to fuzzy SISO activation functions of the feurons.

In general, the units have dynamics associated with them, as indicated, and they receive inputs from themselves and delayed-themselves and all other units. The output of a unit y_i is a standard fuzzy system $\phi(x_i, \pi_i)$ of a state variable x_i associated with the unit. The output of the network is a linear weighted sum of the unit outputs. Weights p_{ij} are associated with the connections from input signals **j** to units **i**, w_{ij} with interunit connections from **j** to **i**, wd_{ij} with delay-interunit connections from **j** to **i** and q_{ij} is the output connection weights from j^{th} feuron to i^{th} output. T_i is the dynamic constant, r_i is the bias (or polarization) term and τ_i is the time-delay of ith feuron.

3 Illustrative Examples of Some Dynamical Behaviors of TDDFN

This model (TDDFN) can be used to approximate many of the behaviors of nonlinear dynamical systems with time delay. In this section, examples are given in which TDDFN converges to a point attractor, a perodic attractor (limit cycle). For this aim, given a set of parameters, initial conditions, and input trajectories, the set of equations (2), (3) and (4) can be numerically integrated from t=0 to some desired final time t_f. This will produce trajectories overtime for the state variables x_i (i=1...n). We have used Adams-Bashforth predictor method, extended with trapezoidal corrector in some cases. The integration step size has to be commensurate with the temporal scale of dynamics, determined by the time constants T_i. In our work, we have specified a lower bound on T_i and have used a fixed integration time step of some fraction (e.g., 1/10) of this bound.

As a first example, a TDDFN is modeled as a point attractor system by a training algorithm whose details will be given in the next section. The interconnection parameters of the TDDFN given in Fig. 1 after training are found to be:

$$w = \begin{bmatrix} 0 & 2 \\ -1 & 0 \end{bmatrix}, wd = \begin{bmatrix} -0.5 & 0 \\ 0 & -.5 \end{bmatrix}, r = \begin{bmatrix} 0.1 \\ 0.1 \end{bmatrix}$$

And the data for time constants, input centers, spreads and output centers with three membership functions in each feuron respectively are:

$$T = \begin{bmatrix} 1 & 0 \\ 0 & 1 \end{bmatrix}, c = \begin{bmatrix} -0.791 & -1.356 & 0.087 \\ 0.996 & -0.974 & 0.785 \end{bmatrix},$$

$$\sigma = \begin{bmatrix} 1.412 & 0.770 & 1.283 \\ 0.837 & 1.116 & 0.875 \end{bmatrix}, b = \begin{bmatrix} 0.542 & -0.764 & 1.211 \\ 0.898 & 0.176 & 1.176 \end{bmatrix}$$

In this case initial conditions are chosen as $x(0) = [\ 1\ 1]^T$. Fig. 2a shows the example of zero-input state space trajectories for two-feuron network with time delay (19 seconds) that converges to a point attractor.

As a second example, a TDDFN is modeled as a periodic attractor system. In this case the interconnection parameters of the TDDFN are calculated as:

$$w = \begin{bmatrix} 0 & 1 \\ 2 & 0 \end{bmatrix}, wd = \begin{bmatrix} -0.85 & 0 \\ 0 & -0.85 \end{bmatrix}, r = \begin{bmatrix} 0.1 \\ 0.1 \end{bmatrix}$$

The data for time constants, input centers, spreads and output centers with three membership functions in each feuron respectively are:

$$T = \begin{bmatrix} 2 & 0 \\ 0 & 1 \end{bmatrix}, c = \begin{bmatrix} -4.174 & 1.418 & -2.987 \\ -0.470 & -0.177 & 0.374 \end{bmatrix},$$

$$\sigma = \begin{bmatrix} 1.988 & 2.296 & 2.167 \\ 1.015 & 1.041 & 0.877 \end{bmatrix}, b = \begin{bmatrix} 0.001 & 1.000 & -0.007 \\ -0.551 & 1.152 & 1.206 \end{bmatrix}$$

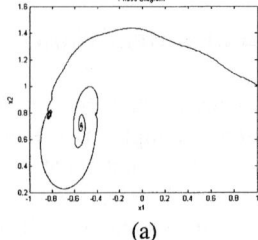
(a)

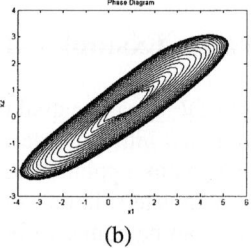
(b)

Fig. 2. State space trajectory of TDDFN a) Point attractor, b) periodic attractor

In this application trajectory tracking performance is excellent with initial conditions $x(0) = [\ 1\ 1]^T$. Fig. 2b shows the example of zero-input state space trajectories for two-feuron network with time delay (19 seconds).

4 Training of TDDFN Based on Adjoint Sensitivity Analysis

In this section, we consider a particular application of the TDDFN: trajectory tracking for process modeling. That is, we desire to configure the network so that its output and input trajectories have been specified, but all the parameters w, w_d, r, τ and π, T are adjustable. This is done by minimizing the cost functional. We associate a performance index of the network for this task as given by

$$J = \frac{1}{2}\int_0^{t_f} [z(t) - z^d(t)]^T [z(t) - z^d(t)] dt \tag{5}$$

where $z^d(t)$ and $z(t)$ are actual and modeled process responses respectively.

Our focus in this paper has been on gradient-based algorithms for solving this computational problem. We require the computation of gradients or sensitivities of the performance index with respect to the various parameters of the TDDFN:

$$\frac{\partial J}{\partial w}, \frac{\partial J}{\partial w_d}, \frac{\partial J}{\partial \tau}, \frac{\partial J}{\partial T}, \frac{\partial J}{\partial r}, \frac{\partial J}{\partial c}, \frac{\partial J}{\partial \sigma}, \frac{\partial J}{\partial b}] \tag{6}$$

In this study above gradients are calculated by "adjoint" method which is based on the use of calculus of variations [9]-[12], [14]. In this method, a set of dynamical systems is defined with adjoint state variables λ_i:

$$-\dot{\lambda}_i = -\frac{\lambda_i}{T_i} + \frac{1}{T_i}\sum_j wd_{ij} x'_j(t-\tau_j)\lambda_j \frac{1}{T_i}\sum_j w_{ij} y'_j \lambda_j + e_i(t)\sum_j q_{ij} y'_j;\ \lambda_i(t_f) = 0 \tag{7}$$

where $y'_j = \dfrac{\partial \phi_j(x_j)}{\partial x_j}$ and can be computed by partial differentiation of (3):

$$\frac{\partial \phi_j(x_j)}{\partial x_j} = \frac{\sum_{k=1}^{R_j}(\phi_j - b_{jk})\exp\left(-\frac{1}{2}\left(\frac{x_j - c_{jk}}{\sigma_{jk}}\right)^2\right)\left(\frac{x_j - c_{jk}}{\sigma_{jk}^2}\right)}{\sum_{i=1}^{R_j}\exp\left(-\frac{1}{2}\left(\frac{x_j - c_{jk}}{\sigma_{jk}}\right)^2\right)} \tag{8}$$

The size of the adjoint vector is thus **n** and is independent of the number of DFN parameters. The computation of sensitivities using the adjoint method requires the solution of **n** differential equations. This is a significant savings for real-time applications. Then, the cost gradients with respect to TDDFN parameters are given by the following quadratures;

$$g = \frac{\partial J}{\partial p} = \int_0^{t_f}\left(\frac{\partial f}{\partial p}\right)^T \lambda\, dt \tag{9}$$

The cost gradients as in [7], [8], [12] can be easily computed. We assume that at each iterations, gradients of the performance index with respect to all TDDFN parameters are computed. Once g is computed several of gradient-based algorithms can be used to update parameter values of the TDDFN. Here for updating model parameters, Broyden-Fletcher-Golfarb-Shanno (BFGS) algorithm [15] that is one of the approximate second order algorithms is used.

5 Mackey-Glass Time Series Prediction with TDDFNs

This section deals with a complex problem of approximating a nonlinear dynamical time series using TDDFN. Here, a benchmark chaotic time series first investigated by Mackey and Glass [16] which is a widely investigated problem in the fuzzy-neural literature [1], [2] is considered. The time series is generated by the following differential equation:

$$\frac{dx}{dt} = \frac{0.2x(t-\tau)}{1+x^{10}(t-\tau)} - 0.1x(t) \tag{10}$$

As τ in this equation varies, the system can exhibit either fixed point, limit cycle or chaotic behavior. For $\tau = 17$ the systems response is chaotic and we attempt the problem of approximating time series function of (10) for this value. The fifth order Runge-Kutta method was used to obtain simulation data with the following initial conditions $x(0)=1.2$ and $x(t-\tau)=0$ for $0 \le t < \tau$.

A TDDFN with one feuron was used in the simulations. Time delay of the feuron was taken to be the fixed value as the same as the Mackey-Glass, $\tau=17$. The other weights are adjusted as presented previously. The first 100 data points were used to train the TDDFN. The prediction performance of the TDDFN was tested after 200th data points. Fig. 3 shows the result of the test with 200 training points. As seen in Fig. 4, the neural network prediction capability is excellent. Table 1 compares the performance of TDDFN with various classical models, neural networks, and fuzzy neural networks. The comparison is based on normalized root mean square error (NRMSE), which is defined as the RMSE divided by the standard deviation of the target series.

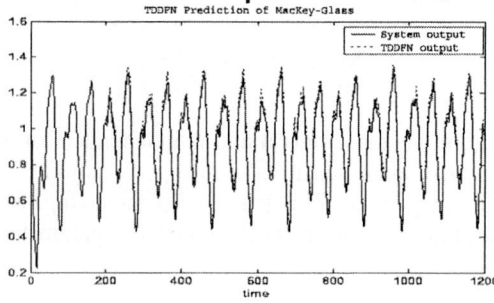

Fig. 3. TDDFN prediction result (solid-line: Mackey-Glass, dotted-line: TDDFN output)

Note from the Table 1 that both ANFIS [1] and GEFREX [4] outperform all other model in terms of NRMSE. However, ANFIS has the drawback that it has less interpretability in terms of learned information, and the implementation of the GEFREX is difficult. Excluding ANFIS, GEFREX and EPNet [3], TDDFN performs the best with an NRMSE of 0.024. In comparison to other models, the proposed TDDFN model is much less complex with easy implementation, and less number of parameters to be calculated. These mean that TDDFN can be easily used for real-time applications.

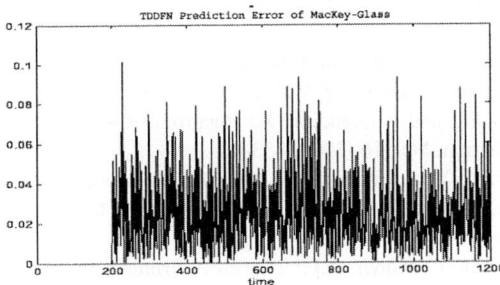

Fig. 4. TDDFN prediction error

Table 1. Comparison of TDDFN with other models for Mackey-Glass time series prediction problem (*: Results adapted from [2])

Method	NRMSE
GEFREX [4]	0.0061
ANFIS [1]	0.0074
EPNet [3]	0.02
TDDFN	0.025
GFPE*	0.026
6th order polynomial*	0.04
Cascade Correlation-NN Model*	0.06
Auto Regressive Model*	0.19
Linear Predictive Model*	0.55

6 Conclusion

The work reported here has concentrated on laying the theoretical and analytic foundations for training of TDDFN. The TDDFN presented in this paper was successfully applied to time series prediction. They are to be used in real-time applications with process modeling and advanced control.

The significance of this work is that efficient computational algorithms have been developed for parameter identification and training of fully nonlinear dynamical systems with time delay. The gradients can be computed by adjoint sensitivity analysis methods. Another time-delay approach for neural networks should be investigated to improve the prediction capability.

References

1. Jang, J.S.R.: ANFIS: Adaptive-Network-Based Fuzzy Inference System. IEEE Trans. Syst. Man, Cybern., Vol. 23 (1993) 51-63
2. Kim, D., Kim, C.: Forecasting Time Series with Genetic Fuzzy Predictor Ensemble. IEEE Trans. Fuzzy Systems, Vol. 5 (1997) 523-535
3. Yao, X., Lin, Y.: A New Evolutionary System for Evolving Artificial Neural Networks. IEEE Trans. Neural Networks, Vol. 8 (1997) 694-713
4. Russo, M.: Genetic Fuzzy Learning. IEEE Trans. Evolutionary Computation, Vol. 4 (2000) 259-273
5. Alex, A.: Dynamic Recurrent Neural Networks Towards Prediction and Modeling of Dynamical Systems. Neurocomputing, Vol. 28 (1999) 207-232
6. Ryad, Z., Daniel, R., and Noureddine, Z.: Recurrent Radial Basis Function Network for Time series Prediction, Vol. 16 (2003) 453-463
7. Oysal, Y., Becerikli, Y. and Konar, A.F.: Generalized Modeling Principles of a Nonlinear System with a Dynamic Fuzzy Network. Computers & Chemical Engineering, Vol. 27 (2003) 1657-1664
8. Becerikli, Y., Oysal, Y., Konar, A.F.: Trajectory Priming with Dynamic Fuzzy Networks in Nonlinear Optimal Control. IEEE Trans on Neural Networks, Vol.15 (2004) 383-394
9. Pearlmutter, B.: Learning State Space Trajectories in Recurrent Neural Networks. Neural Computation, Vol. 1 (1989) 263-269
10. Barhen, J., Toomarian, N., and Gulati, S.: Adjoint Operator Algorithms for Faster Learning in Dynamical Neural Networks. In Advances in Neural Information Processing Systems 2, D.S. Touretzky (Ed.) San Mateo, Ca.: Morgan Kaufmann, (1990)
11. Leistritz, L., Galicki, M., Witte, H., and Kochs, E.: Training Trajectories by Continuous Recurrent Multilayer Networks. IEEE Trans. on Neural Networks, Vol. 13(2) (2002)
12. Becerikli, Y., Konar, A.F. and Samad, T.: Intelligent Optimal Control with Dynamic Neural Networks. Neural Networks, Vol.16(2) (2003) 251-259
13. Passino, K. M., & Yurkovich, S. : Fuzzy control, Menlopark, Cal.: Addison-Wesley (1998)
14. Bryson, A.E. and Ho, Y.C.: Applied Optimal Control. Hemisphere Publishing Corporation (1975)
15. Edgar, T. F., & Himmelblau, D. M.: Optimization of Chemical Processes. McGraw-Hill (1988)
16. Mackey, M., and Glass, L.: Oscillation and Chaos in Physiological Control Systems. Science, Vol. 197 (1977) 287-289

A Hybrid Heuristic Algorithm for the Rectangular Packing Problem

Defu Zhang[1], Ansheng Deng[1], and Yan Kang[2]

[1] Department of Computer Science, Xiamen University, 361005, China
dfzhang@xmu.edu.cn
[2] School of Software, Yunnan University, Kunming, 650091, China

Abstract. A hybrid heuristic algorithm for the two-dimensional rectangular packing problem is presented. This algorithm is mainly based on divide-and-conquer and greedy strategies. The computational results on a class of benchmark problems have shown that the performance of the heuristic algorithm can outperform that of quasi-human heuristics.

1 Introduction

Packing problems have found many industrial applications, with different applications incorporating different constraints and objects. For example, in wood or glass industries, rectangular components have to be cut from large sheets of material. In warehousing contexts, goods have to be placed on shelves. In newspapers paging, articles and advertisements have to be arranged in pages. In the shipping industry, a batch of object of various sizes have to be shipped as many as possible in a larger container, a bunch of optical fibers have to be accommodated in a pipe with as small as possible. In VLSI floor planning, VLSI has to be laid. These applications can formalize as bin packing problems [1]. For more extensive and detailed descriptions of packing problems, the reader is referred to [1,2,3].

In this paper, two-dimensional rectangular packing problem is considered. This problem belongs to a subset of classical cutting and packing problems and has been shown to be NP hard [4,5]. Optimal algorithms for orthogonal two-dimension cutting are proposed in [6,7]. However, they might not be practical for large problems. Hybrid algorithms combining genetic with deterministic methods for the orthogonal packing problem are proposed [8, 9, 10]. An empirical investigation of meta-heuristic and heuristic algorithms of the orthogonal packing problem of rectangles is given by [11]. However, generally speaking, those non-deterministic algorithms are more time consuming and are less practical for problems having a large number of rectangles. Recently, an effective quasi-human heuristic, Less Flexibility First, for solving the rectangular packing problem is presented [12]. Namely, the rectangle with less flexibility should be packed earlier. This heuristic is fast and effective. Recently, several researchers have started to apply evolutionary algorithms to solve rectangular packing problem ([11, 13, 14, 15]. Based on the previous studies [16, 17], a rather fast and effective hybrid heuristic algorithm for the rectangular packing problem is presented.

Computational results have shown that the performance of the hybrid heuristic algorithm can outperform that of quasi-human heuristics.

2   Mathematical Formulation of the Problem

Given a rectangular empty box and a set of rectangles with arbitrary sizes, we want to know if all rectangles can be packed into the empty box without overlapping. This problem can also be stated as follows.

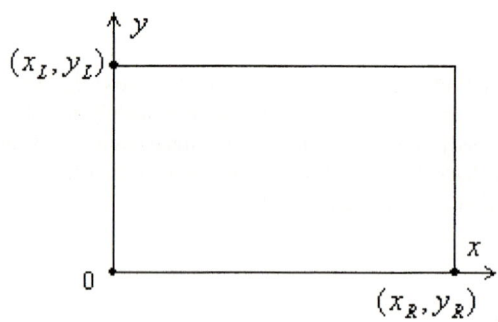

Fig. 1.

Given an empty box with length L and width W, and n rectangles with length l_i and width w_i, $1 \le i \le n$, take the origin of two dimensional Cartesian coordinate system at the left-down corner of the empty box, (x_L, y_L) denotes the left-up corner coordinates of the rectangular empty box and (x_R, y_R) denotes the right-down corner coordinates of this box (See Fig. 1.). Does there exist a solution composed of n sets of quadruples

$$P = \{\langle (x_{li}, y_{li}), (x_{ri}, y_{ri}) \rangle \mid 1 \le i \le n, x_{li} < x_{ri}, y_{li} > y_{ri}\},$$

where, (x_{li}, y_{li}) denotes the left-up corner coordinates of rectangle i, and (x_{ri}, y_{ri}) denotes the right-down corner coordinates of rectangle i. For all $1 \le i \le n$, and the coordinates of rectangle i satisfies the following conditions:

1. $x_{ri} - x_{li} = l_i \wedge y_{li} - y_{ri} = w_i$ or $x_{ri} - x_{li} = w_i \wedge y_{li} - y_{ri} = l_i$.
2. for all $1 \le j \le n, j \ne i$, and rectangle i and j cannot overlap, namely, $x_{ri} \le x_{lj}$ or $x_{li} \ge x_{rj}$ or $y_{ri} \ge y_{lj}$ or $y_{li} \le y_{rj}$.
3. $x_L \le x_{li} \le x_R, x_L \le x_{ri} \le x_R$ and $y_R \le y_{li} \le y_L, y_R \le y_{ri} \le y_L$.

If there is no such a solution, then obtain a partial solution which minimizes the total area of the unpacked space.

It is noted that the packing process has to ensure the edges of each rectangle are parallel to the $x-$ and $y-$ axes respectively.

3 Hybrid Heuristic Algorithm

Many useful algorithms have recursive structure: to solve a given problem, they call themselves recursively one or more times to deal with closely related subproblems, so these algorithms are simple and effective. These algorithms typically follow a divide-and-conquer approach: they break the problem into several subproblems that are similar to the original problem but smaller in size, solve the subproblems recursively, and then combine these solutions to create a solution to the original problem.

The divide-and-conquer paradigm involves three steps at each level of the recursion [18]:

(1) Divide the problem into a number of subproblems.
(2) Conquer the subproblems by solving them recursively. If the subproblem sizes are small enough, however, just solve the subproblems in a straightforward manner.
(3) Combine the solutions to the subproblems into the solution for the original problem.

Intuitively, we can construct a divide-and-conquer algorithm for the rectangular packing problem as follows:

(1) Pack a rectangle into the space to be packed. Divide the unpacked space into two subspaces (see Fig. 2).
(2) Pack the subspace by packing them recursively. If the subspace sizes are small enough to only pack a rectangle, however, just pack this rectangle into the subspace in a straightforward manner.
(3) Combine the solutions to the subproblems into the solution for the rectangular packing problem.

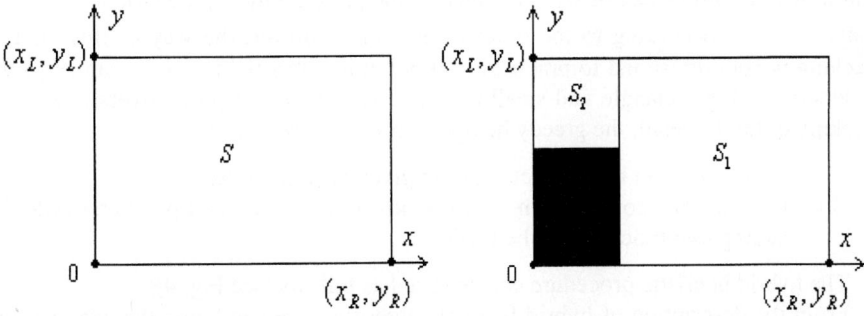

Fig. 2. Pack a rectangle into S and divide the unpacked space into S_1, S_2

It is noted that how to select a rectangle to be packed into S is very important to improve the performance of divide-and-conquer algorithm. In addition, two subspaces

S_1 and S_2, which to be first packed recursively also affect the performance of algorithm. In this paper, we present a greedy heuristic for selecting a rectangle to be packed, namely a rectangle with the maximum area or perimeter is given priority to pack. In detail, unpacked rectangles should be sorted by non-increasing order of area or perimeter size. The rectangle with maximum one should be selected to pack if it can be packed into the unpacked subspace.

Thus the divide-and-conquer procedure can be stated as follows (Fig.3):

Pack(S)
    if no rectangle can be packed into S
    then return;
    else
        Select a rectangle and pack it into S ;
        Divide the unpacked space into S_1, S_2 ;
        if the perimeter of $S_1 >$ the perimeter of S_2
            Pack(S_1);
            Pack(S_2);
        else
            Pack(S_2);
            Pack(S_1);

Fig. 3. The divide-and-conquer procedure

For this recursive procedure, the average running time of the recursive procedure is $T(n) = \theta(n \lg n)$.

Due to the packing orders affect the packed ratio of the divide-and-conquer algorithm, several orders can be tried to enhance the packed ratio. If we strictly pack rectangles into S according to the order of area or perimeter, the way of this kind of packing is not correspond to practical packing in industry fields, so we can swap the orders of the big rectangle and small rectangle in order to keep the diversification of packing order. In detail, the greedy heuristic strategies are as follows:

1) Swap the order of two rectangles in given packing order;
2) Consider the computation in two kinds of area order and perimeter order by calling two times hybrid heuristic ().

The hybrid heuristic procedure can be stated as follows (see Fig. 4):
From the description of hybrid heuristic algorithm, we can know the average running time of this algorithm is $T(n) = \theta(n^3 \lg n)$. However, the worst running time of Heuristic1 [12] is $T(n) = \theta(n^5 \lg n)$, the worst running time of Heuristic2 [12]

is $T(n) = \theta(n^4 \lg n)$. Therefore, the computational speed of the hybrid heuristic algorithm is faster than that of Heuristic 1 and Heuristic 2.

```
Hybrid heuristic ()
   Repeat do
         for i=1 to n
            for j=i to n
               Swap the order of rectangle i and j in current order;
               Pack(S);
               Save the best order of rectangles and the best packed area so far;
               Current order =best order;
   Until (the best packed area has no improvement)
```

Fig. 4. The hybrid heuristic algorithm

4 Computational Results

Performance of the hybrid heuristic (HH) has been tested with seven different sized test instances ranging from 16 to 197 items [11]. These test instances have optimal solutions. The computational results are reported in Table 1. In order to understand HH more easily and compare HH with Heuristic 1 and Heuristic 2, we give their unpack ratio and running time comparison in Fig. 5, Fig. 6. Here, Heuristic 1 and Heuristic 2 are not implemented in this paper, they are run on a SUN Sparc20/71, a machine with a 71 MHz SuperSparc [12], so the % of unpacked area and running time are directly taken from [12]. Our experiments are run on a Dell GX260 with a 2.4GHz CPU. It is noted that the computational results of C7 are not reported in [12], and so is Heuristic 1 for C43. So in Fig. 5 and Fig. 6, % of unpacked area and the running time for Heuristic 1 are shown to be empty. In addition, we give four packed results on test instances C1 and C73 for HH in Fig. 7 and Fig. 8 respectively.

On this test set, as shown in Table 1, HH runs much faster for all test instances than Heuristic1 and Heuristic 2. For C1-C6, the average running time of HH is less than 1.5s, for C7, the average running time of HH is less than 13.47s. It has shown that the actual running time of HH accords with its computational complexity. The unpacked area for HH ranges from 0% to 3.5% with the average unpacked ratio of 0.88%. The average unpacked ratio of Heuristic 1 and Heuristic 2 is 0.92 and 3.65 respectively. The average unpacked ratio of HH is lower than that of Heuristic 1 and Heuristic 2. From Fig. 5 and Table 1, we can observe that the packed density increases with the increasing of the number of rectangles. The running time of Heuristic 1 increases more badly. With the increasing of the number of rectangles, it is imaginable that the computational speed of Heuristic 1 for practical applications is unacceptable.

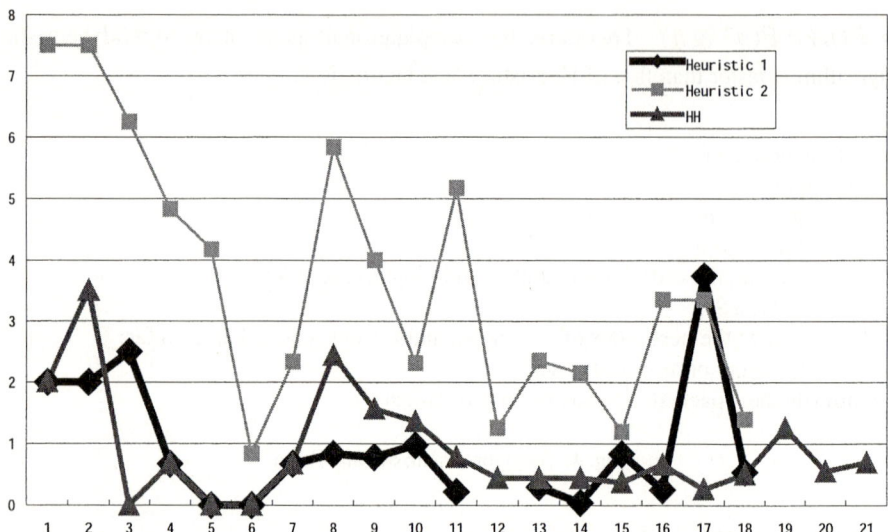

Fig. 5. % of unpacked area for Heuristic 1, Heuristic 2 and HH

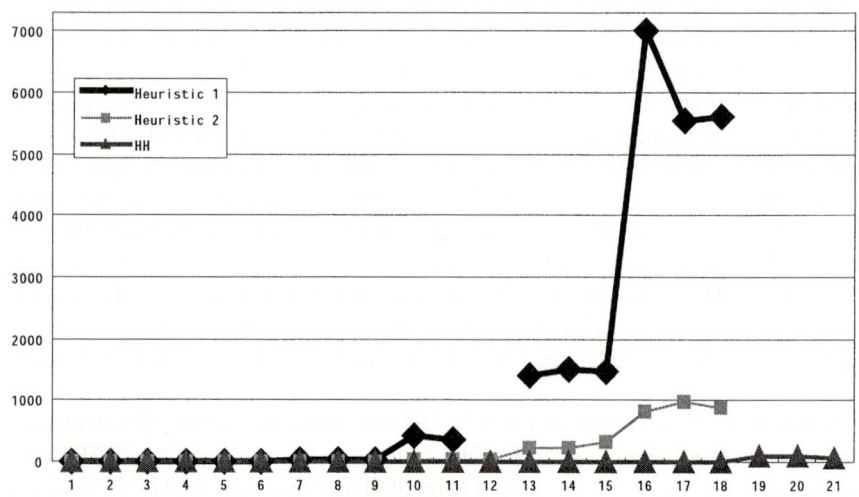

Fig. 6. CPU time comparison of Heuristic 1, Heuristic 2 and HH

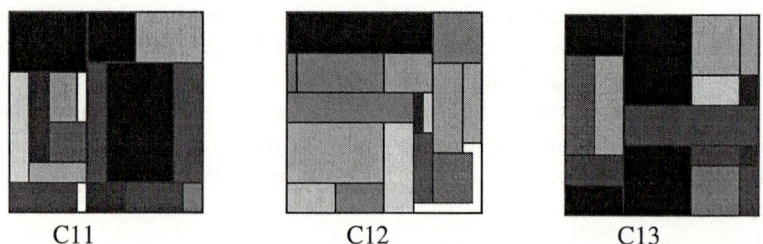

Fig. 7. Packed results of C1 for HH

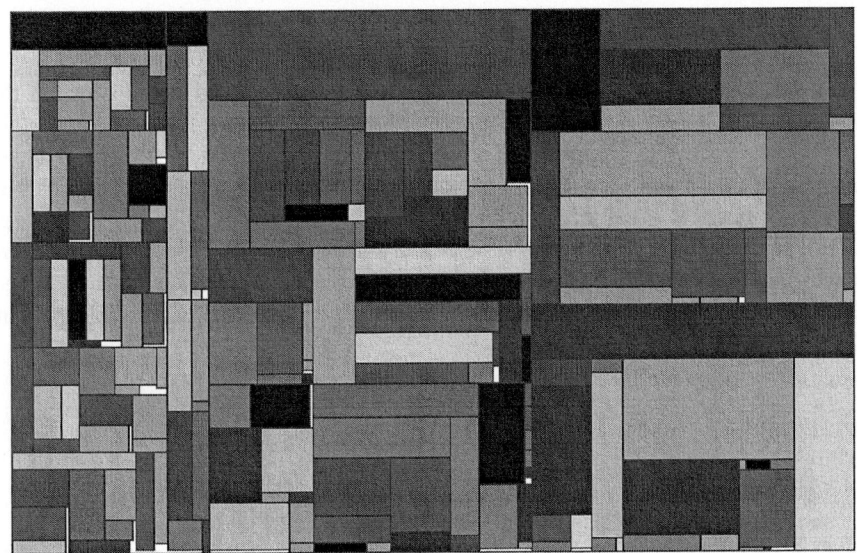

Fig. 8. Packed result of C73 for HH

Table 1. Experimental results on Heuristic 1, Heuristic 2 and HH

Test instances		n	L×W	CPU time (seconds)			% of unpacked area		
				Heuristic1	Heuristic2	HH	Heuristic1	Heuristic2	HH
C1	C11	16	20×20	1.48	0.10	0.0	2.00	7.50	2
	C12	17	20×20	2.42	0.18	0.0	2.00	7.50	3.5
	C13	16	20×20	2.63	0.17	0.0	2.50	6.25	0
C2	C21	25	15×40	13.35	0.80	0.05	0.67	4.83	0.67
	C22	25	15×40	10.88	0.98	0.05	0.00	4.17	0.0
	C23	25	15×40	7.92	7.92	0.0	0.00	0.83	0.0
C3	C31	28	30×60	23.72	2.07	0.05	0.67	2.33	0.67
	C32	29	30×60	34.02	1.98	0.05	0.83	5.83	2.44
	C33	28	30×60	30.97	2.70	0.05	0.78	4.00	1.56
C4	C41	49	60×60	438.18	32.90	0.44	0.97	2.31	1.36
	C42	49	60×60	354.47	33.68	0.44	0.22	5.17	0.78
	C43	49	60×60		29.37	0.33		1.25	0.44
C5	C51	73	90×60	1417.52	234.52	1.54	0.30	2.35	0.44
	C52	73	90×60	1507.52	237.58	1.81	0.04	2.15	0.44
	C53	73	90×60	1466.15	315.60	2.25	0.83	1.19	0.37
C6	C61	97	120×80	7005.73	813.45	5.16	0.25	3.35	0.66
	C62	97	120×80	5537.88	975.30	5.33	3.74	3.35	0.26
	C63	97	120×80	5604.70	892.72	5.6	0.54	1.39	0.5
C7	C71	196	240×160			94.62			1.25
	C72	197	240×160			87.25			0.55
	C73	196	240×160			78.02			0.69
Average time and % of unpacked area				1379.97	199.00	13.47	0.92	3.65	0.88

5 Conclusions

A novel hybrid heuristic algorithm for the rectangular packing is presented in this paper. This algorithm is very simple and intuitional, and can fast solve the rectangular packing problem. The computational results have shown that HH outperform heuristic 1 and Heuristic 2 in two ways of % of unpacked area and the CPU time. So HH may be of great practical value to the rational layout of the rectangular objects in the engineering fields, such as the wood-, glass- and paper industry, and the ship building industry, textile and leather industry. Further improve the performance of hybrid heuristic algorithm and extend this algorithm for three-dimensional rectangular packing problems are the future work.

References

1. Andrea Lodi, Silvano Martello, Michele Monaci. Two-dimensional packing problems: A survey. European Journal of Operational Research 141 (2002) 241–252
2. K.A. Dowaland, W.B. Dowsland. Packing problems. European Journal of Operational Research 56 (1992) 2–14
3. David Pisinger. Heuristics for the container loading problem. European Journal of Operational Research 141 (2002) 382–392
4. D.S. Hochbaum Wolfgang Maass. Approximation schemes for covering and packing problems in image processing and VLSI. Journal of the Association for Computing Machinery 32 (1) (1985) 130–136
5. J. Leung, T. Tam, C.S. Wong, Gilbert Young, Francis Chin. Packing squares into square. Journal of Parallel and Distributed Computing 10 (1990) 271–275
6. J.E. Beasley. An exact two-dimensional non-guillotine cutting tree search procedure. Operations Research 33 (1985) 49–64
7. E. Hadjiconstantinou, N. Christofides. An optimal algorithm for general orthogonal 2-D cutting problems. Technical report MS-91/2, Imperial College, London, UK
8. S. Jakobs. On genetic algorithms for the packing of polygons. European Journal of Operational Research 88 (1996) 165–181
9. D. Liu, H. Teng. An improved BL-algorithm for genetic algorithm of the orthogonal packing of rectangles. European Journal of Operational Research 112 (1999) 413–419
10. C.H. Dagli, P. Poshyanonda. New approaches to nesting rectangular patterns. Journal of Intelligent Manufacturing 8 (1997) 177–190
11. E. Hopper, B.C.H. Turton. An empirical investigation of meta-heuristic and heuristic algorithms for a 2D packing problem. European Journal of Operational Research 128 (2001) 34–57
12. Yu-Liang Wu, Wenqi Huang, Siu-chung Lau, C.K. Wong, Gilbert H. Young. An effective quasi-human based heuristic for solving the rectangle packing problem. European Journal of Operational Research 141 (2002) 341–358
13. Andrea Lodi, Silvano Martello,and Daniele Vigo. Heuristic and Metaheuristic Approaches for a Class of Two-Dimensional Bin Packing Problems, INFORMS Journal on Computing 11 (1999) 345–357
14. E. Hopper. Two-Dimensional Packing Utilising Evolutionary Algorithms and other Meta-Heuristic Methods, PhD Thesis, Cardiff University, UK. 2000

15. J. Puchinger and G. R. Raidl. An evolutionary algorithm for column generation in integer programming: an effective approach for 2D bin packing. In X. Yao et. al, editor, Parallel Problem Solving from Nature - PPSN VIII, volume 3242, pages 642–651. Springer, 2004
16. De-fu Zhang, An-Sheng Deng. An effective hybrid algorithm for the problem of packing circles into a larger containing circle. Computers & Operations Research 32(8) (2005) 1941–1951
17. Defu Zhang, Wenqi Huang. A Simulated Annealing Algorithm for the Circles Packing Problem. Lecture Notes in Computer Science (ICCS 2004) 3036 (2004) 206–214
18. Thomas H. Cormen, Charles E. Leiserson, Ronald L. Rivest and Clifford Stein. Introduction to Algorithms. Second Edition, The MIT Press (2001)

Genetically Dynamic Optimization Based Fuzzy Polynomial Neural Networks

Ho-Sung Park[1], Sung-Kwun Oh[2], Witold Pedrycz[3], and Yongkab Kim[1]

[1] Department of Electrical Electronic and Information Engineering, Wonkwang University,
344-2, Shinyong-Dong, Iksan, Chon-Buk, 570-749, South Korea
[2] Department of Electrical Engineering, The University of Suwon, San 2-2 Wau-ri,
Bongdam-eup, Hwaseong-si, Gyeonggi-do, 445-743, South Korea
ohsk@suwon.ac.kr
[3] Department of Electrical and Computer Engineering, University of Alberta,
Edmonton, AB T6G 2G6, Canada
and Systems Research Institute, Polish Academy of Sciences, Warsaw, Poland

Abstract. In this paper, we introduce a new architecture of genetically dynamic optimization based Fuzzy Polynomial Neural Networks (gdFPNN) and discuss its comprehensive design methodology involving mechanisms of genetic optimization, especially genetic algorithms (GAs). The proposed gdFPNN gives rise to a structurally and parametrically optimized network through an optimal parameters design available within FPN. Through the consecutive process of such structural and parametric optimization, an optimized and flexible gdFPNN is generated in a dynamic fashion. The performance of the proposed gdFPNN is quantified through experimentation that exploits standard data already used in fuzzy modeling. These results reveal superiority of the proposed networks over the existing fuzzy and neural models.

1 Introduction

The challenging quest for constructing models of the systems that come with significant approximation and generalization abilities as well as are easy to comprehend has been within the community for decades [1], [2], [3], [4]. The most successful approaches to hybridize fuzz systems with learning and adaptation have been made in the realm of CI [5]. As one of the representative design approaches which are advanced tools, a family of fuzzy polynomial neuron (FPN)-based SOPNN(called "FPNN" as a new category of neuro-fuzzy networks)[6] were introduced to build predictive models for such highly nonlinear systems. The FPNN algorithm exhibits some tendency to produce overly complex networks as well as a repetitive computation load by the trial and error method and/or the repetitive parameter adjustment by designer like in case of the original GMDH algorithm.

In this study, in addressing the above problems with the conventional SOPNN (especially, FPN-based SOPNN called "FPNN" [6], [7]) as well as the GMDH algorithm, we introduce a new genetic design approach; as a consequence we will be referring to these networks as genetically dynamic optimization based FPNN (to be called "gdFPNN"). The determination of the optimal values of the parameters available within an individual FPN (viz. the number of input variables, the order of the polynomial, input variables, the number of membership function, and the apexes of membership function) leads to a structurally and parametrically optimized network.

2 The Architecture and Development of Fuzzy Polynomial Neural Networks (FPNN)

2.1 FPNN Based on Fuzzy Polynomial Neurons (FPNs)

The FPN consists of two basic functional modules. The first one, labeled by **F**, is a collection of fuzzy sets that form an interface between the input numeric variables and the processing part realized by the neuron. The second module (denoted here by **P**) is about the function – based nonlinear (polynomial) processing. This nonlinear processing involves some input variables.

2.2 The Review of Conventional FPNN Architecture

Proceeding with the conventional FPNN architecture as presented in [6], [7], essential design decisions have to be made with regard to the number of input variables and the order of the polynomial occurring in the conclusion part of the rule. The overall selection process of the conventional FPNN architecture is shown in Fig. 1

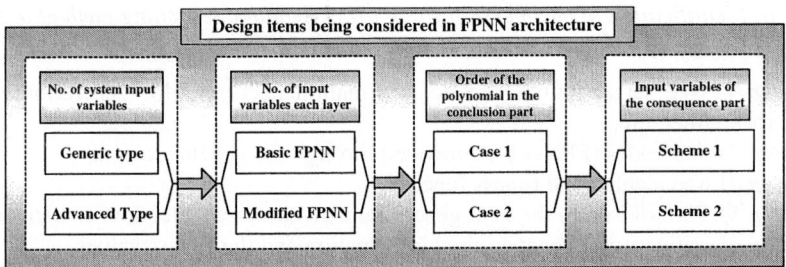

Fig. 1. Taxonomy of the conventional FPNN architecture

3 The Algorithms and Design Procedure of Genetically Dynamic Optimization Based FPNN

3.1 Genetic Optimization of FPNN

GAs is optimization techniques based on the principles of natural evolution. In essence, they are search algorithms that use operations found in natural genetics to guide a comprehensive search over the parameter space. GAs has been theoretically and empirically demonstrated to provide robust search capabilities in complex spaces thus offering a valid solution strategy to problems requiring efficient and effective searching [8]. In this study, for the optimization of the FPNN model, GA uses the serial method of binary type, roulette-wheel used in the selection process, one-point crossover in the crossover operation, and a binary inversion (complementation) operation in the mutation operator. To retain the best individual and carry it over to the nest generation, we use elitist strategy [9].

3.2 Design Procedure of Genetically Dynamic Optimization Based FPNN

[Step 1] *Determine system's input variables*
[Step 2] *Form training and testing data*

The input-output data set $(x_i, y_i)=(x_{1i}, x_{2i}, ..., x_{ni}, y_i)$, $i=1, 2, ..., N$ is divided into two parts, that is, a training and testing dataset.

[Step 3] *Decide initial information for constructing the gdFPNN structure*
[Step 4] *Decide FPN structure using genetic design*

We divide the chromosome to be used for genetic optimization into four sub-chromosomes. The 1st sub-chromosome contains the number of input variables, the 2nd sub-chromosome involves the order of the polynomial of the node, the 3rd sub-chromosome contains input variables, and the 4th sub-chromosome (remaining bits) involves the number of MF coming to the corresponding node (FPN).

[Step 5] *Design of structurally optimized gdFPNN*

In this step, we design the structurally optimized gdFPNN by means of FPNs that obtained in [Step 4].

[Step 6] *Identification of membership value using dynamic searching method of GAs*
[Step 7] *Design of parametrically optimized gdFPNN*

Sub-step 1) We set up initial genetic information necessary for generation of the gdFPNN architecture.
Sub-step 2) The nodes (FPNs) are generated through the genetic design.
Sub-step 4) we calculate the fitness function.
Sub-step 5) To move on to the next generation, we carry out selection, crossover, and mutation operation using genetic initial information and the fitness values.
Sub-step 6) We choose optimal gdFPNN characterized by the best fitness value in the current generation. For the elitist strategy, selected best fitness value used.
Sub-step 7) We generate new populations of the next generation using operators of GAs obtained from ***Sub-step 2***. We use the elitist strategy. This sub-step carries out by repeating ***sub-step 2-6***.
Sub-step 8) Until the last generation, this sub-step carries out by repeating ***sub-step 2-7***.

4 Experimental Studies

The performance of the gdFPNN is illustrated with the aid of well-known and widely used dataset of a gas furnace process utilized by Box and Jenkins [10].

We try to model the gas furnace using 296 pairs of input-output data. The total data set consisting 296 input-output pairs was split into two parts. The first one (consisting of 148 pairs) was used for training. The remaining part of the series serves as a testing set. In order to carry out the simulation, we use six-input [$u(t-3)$, $u(t-2)$, $u(t-1)$, $y(t-3)$, $y(t-2)$, $y(t-1)$] and one-output ($y(t)$). To come up with a quantitative evaluation of the network, we use the standard MSE performance index.

Table 1 summarizes the performance index of gdFPNN when using dynamic searching method.

Table 1. Performance index of gdFPNN for the gas furnace process data (3rd layer)

Max	(a) Selected input variables				(b) Entire system input variables			
	Triangular MF		Gaussian-like MF		Triangular MF		Gaussian-like MF	
	PI	EPI	PI	EPI	PI	EPI	PI	EPI
2	0.019	0.102	0.013	0.101	0.011	0.105	0.012	0.102
3	0.015	0.115	0.011	0.116	0.011	0.110	0.007	0.106

Fig. 2 illustrates the detailed optimal topologies of gdFPNN for 3 layer when using Max=2 and Gaussian-like MF: the results of the network have been reported as PI=0.012 and EPI=0.102.

As shown in Fig 2, the proposed network enables the architecture to be a structurally and parametrically more optimized and simplified network than the conventional FPNN.

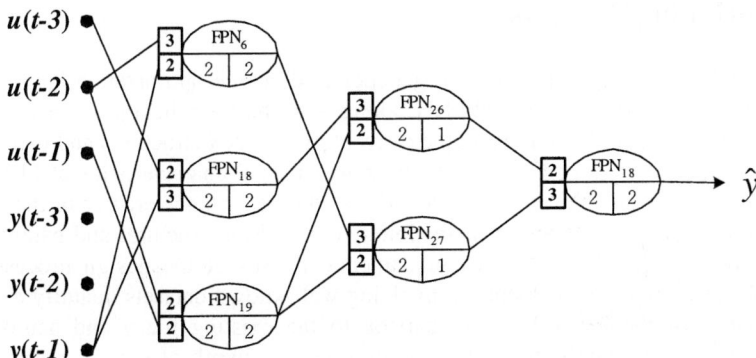

Fig. 2. Genetically dynamic optimization based FPNN (gdFPNN) architecture

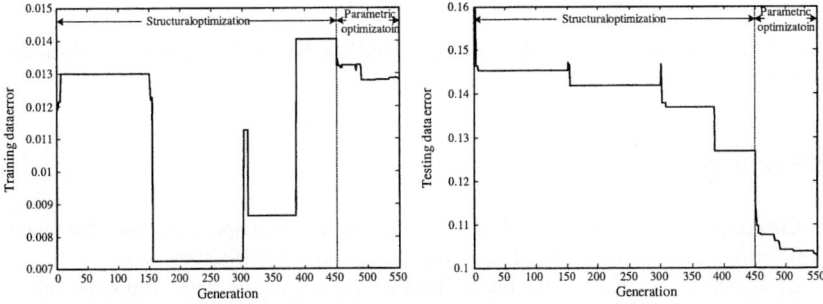

Fig. 3. The optimization process by the genetic algorithms

Fig. 3 illustrates the optimization process by visualizing the values of the performance index obtained in successive generations of GA.

Table 2 gives a comparative summary of the network with other models.

Table 2. Comparative analysis of the performance of the network; considered are models reported in the literature

Model				PI	PI_s	EPI_s
Box and Jenkin's model [10]				0.710		
Tong's model [11]				0.469		
Sugeno and Yasukawa's model [12]				0.190		
Xu and Zailu's model [13]				0.328		
Oh and Pedrycz's model [14]				0.123	0.020	0.271
Lin and Cunningham's model [15]					0.071	0.261
FPNN [16]	CASE I				0.016	0.116
FPNN [16]	CASE II				0.012	0.125
gFPNN [17]	Triangular MF	3rd layer	Max=4		0.018	0.122
gFPNN [17]	Gaussian-like MF		Max=4		0.020	0.104
Proposed gdFPNN	Triangular MF		Max=2		0.011	0.105
Proposed gdFPNN	Gaussian-like MF		Max=3		0.007	0.106

5 Concluding Remarks

In this study, the design procedure of genetically dynamic optimization based Fuzzy Polynomial Neural Networks (gdFPNN) along with their architectural considerations has been investigated. In contrast to the conventional FPNN structures and their learning, the proposed model comes with a diversity of local characteristics of FPNs that are extremely useful when coping with various nonlinear characteristics of the system under consideration. The design methodology comes as a structural and parametrical optimization being viewed as two fundamental phases of the design process. The comprehensive experimental studies involving well-known datasets quantify a superb performance of the network in comparison to the existing fuzzy and neuro-fuzzy models. Most importantly, through the proposed framework of genetic optimization we can efficiently search for the optimal network architecture (structurally and parametrically optimized network) and this becomes crucial in improving the performance of the resulting model.

Acknowledgement. This work was supported by Korea Research Foundation Grant (KRF-2004-002-D00257)

References

1. Cherkassky, V., Gehring, D., Mulier, F.: Comparison of adaptive methods for function estimation from samples. IEEE Trans. Neural Networks. **7** (1996) 969-984
2. Dickerson, J. A., Kosko, B.: Fuzzy function approximation with ellipsoidal rules. IEEE Trans. Syst., Man, Cybernetics. Part B. **26** (1996) 542-560

3. Sommer, V., Tobias, P., Kohl, D., Sundgren, H., Lundstrom, L.: Neural networks and abductive networks for chemical sensor signals: A case comparison. Sensors and Actuators B. **28** (1995) 217-222
4. Kleinsteuber, S., Sepehri, N.: A polynomial network modeling approach to a class of large-scale hydraulic systems. Computers Elect. Eng. **22** (1996) 151-168
5. Cordon, O., et al.: Ten years of genetic fuzzy systems: current framework and new trends. Fuzzy Sets and Systems. 2003(in press)
6. Oh, S.K., Pedrycz, W.: Self-organizing Polynomial Neural Networks Based on PNs or FPNs : Analysis and Design. Fuzzy Sets and Systems. **142**(2) (2003) 163-198
7. Oh, S.K., Pedrycz, W.: Fuzzy Polynomial Neuron-Based Self-Organizing Neural Networks. Int. J. of General Systems. **32** (2003) 237-250
8. Michalewicz, Z.: Genetic Algorithms + Data Structures = Evolution Programs. 3rd edn. Springer-Verlag, Berlin Heidelberg New York. (1996)
9. Jong, D., K. A.: Are Genetic Algorithms Function Optimizers?. Parallel Problem Solving from Nature 2, Manner, R. and Manderick, B. eds., North-Holland, Amsterdam. (1992)
10. Box, D.E., Jenkins, G.M.: Time Series Analysis, Forcasting and Control, California, Holden Day. (1976)
11. Tong, R.M.: The evaluation of fuzzy models derived from experimental data. Fuzzy Sets and Systems. **13** (1980) 1-12
12. Sugeno, M., Yasukawa, T.: A Fuzzy-Logic-Based Approach to Qualitative Modeling. IEEE Trans. Fuzzy Systems. **1** (1993) 7-31
13. Xu, C.W., Xi, T.G., Zhang, Z.J.: A clustering algorithm for fuzzy model identification. Fuzzy Sets and Systems. **98** (1998) 319-329
14. Oh, S.K., Pedrycz, W.: Identification of Fuzzy Systems by means of an Auto-Tuning Algorithm and Its Application to Nonlinear Systems. Fuzzy Sets and Systems. **115** (2000) 205-230
15. Lin, Y., Cunningham III, G. A.: A new approach to fuzzy-neural modeling. IEEE Trans. Fuzzy Systems. **3** (1995) 190-197
16. Park, H.S., Oh, S.K., Yoon, Y.W.: A New Modeling Approach to Fuzzy-Neural Networks Architecture. Journal of Control, Automation and Systems Engineering. **7** (2001) 664-674(in Koreans)
17. Oh, S.K., Pedrycz, W., Park, H.S.: Genetically Optimized Fuzzy Polynomial Neural Networks. IEEE Trans. Fuzzy Systems. (2004) (submitted)
18. Park, B.J., Lee, D.Y., Oh, S.K.: Rule-based Fuzzy Polynomial Neural Networks in Modeling Software Process Data. International journal of Control, Automations, and Systems. **1**(3) (2003) 321-331

Genetically Optimized Hybrid Fuzzy Neural Networks Based on Simplified Fuzzy Inference Rules and Polynomial Neurons

Sung-Kwun Oh[1], Byoung-Jun Park[2], Witold Pedrycz[3], and Tae-Chon Ahn[2]

[1] Department of Electrical Engineering, The University of Suwon, San 2-2 Wau-ri, Bongdam-eup, Hwaseong-si, Gyeonggi-do, 445-743, South Korea
ohsk@suwon.ac.kr
[2] Department of Electrical Electronic and Information Engineering, Wonkwang University, 344-2, Shinyong-Dong, Iksan, Chon-Buk, 570-749, South Korea
[3] Department of Electrical and Computer Engineering, University of Alberta, Edmonton, AB T6G 2G6, Canada
and Systems Research Institute, Polish Academy of Sciences, Warsaw, Poland

Abstract. We introduce an advanced architecture of genetically optimized Hybrid Fuzzy Neural Networks (gHFNN) and develop a comprehensive design methodology supporting their construction. The gHFNN architecture results from a synergistic usage of the hybrid system generated by combining Fuzzy Neural Networks (FNN) with Polynomial Neural Networks (PNN). As to the consequence part of the gHFNN, the development of the PNN dwells on two general optimization mechanisms: the structural optimization is realized via GAs whereas in case of the parametric optimization we proceed with a standard least square method-based learning.

1 Introductory Remarks

The models should be able to take advantage of the existing domain knowledge and augment it by available numeric data to form a coherent data-knowledge modeling entity. The omnipresent modeling tendency is the one that exploits techniques of Computational Intelligence (CI) by embracing fuzzy modeling [1], [2], [3], [4], [5], [6], neurocomputing [7], and genetic optimization [8].

In this study, we develop a hybrid modeling architecture, called genetically optimized Hybrid Fuzzy Neural Networks (gHFNN). In a nutshell, gHFNN is composed of two main substructures driven to genetic optimization, namely a fuzzy set-based fuzzy neural network (FNN) and a polynomial neural network (PNN). The role of the FNN is to interact with input data, granulate the corresponding input spaces. The role of the PNN is to carry out nonlinear transformation at the level of the fuzzy sets formed at the level of FNN. The PNN that exhibits a flexible and versatile structure [9] is constructed on a basis of Group Method of Data Handling (GMDH [10]) method and genetic algorithms (GAs). The design procedure applied in the construction of each layer of the PNN deals with its structural optimization involving the selection of optimal nodes (polynomial neurons; PNs) with specific local characteristics (such as the number of input variables, the order of the polynomial, and a collection of the specific subset of input variables) and addresses specific aspects of parametric optimization.

2 Conventional Hybrid Fuzzy Neural Networks (HFNN)

The architectures of conventional HFNN [11], [12] result as a synergy between two other general constructs such as FNN and PNN. Based on the different PNN topologies, the HFNN distinguish between two kinds of architectures, namely basic and modified architectures. Moreover, for the each architecture we identify two cases. In the connection point, if input variables to PNN used on the consequence part of HFNN are less than three (or four), the generic type of HFNN does not generate a highly versatile structure. Accordingly we identify also two types as the generic and advanced. The topologies of the HFNN depend on those of the PNN used for the consequence part of HFNN. The design of the PNN proceeds further and involves a generation of some additional layers. Each layer consists of nodes (PNs) for which the number of input variables could the same as in the previous layers or may differ across the network. The structure of the PNN is selected on the basis of the number of input variables and the order of the polynomial occurring in each layer.

3 Genetically Optimized HFNN (gHFNN)

3.1 Fuzzy Neural Networks Based on Genetic Optimization

We consider two kinds of FNNs (viz. FS_FNN and FR_FNN) based on simplified fuzzy inference. The fuzzy partitions formed for each case lead us to the topologies visualized in Fig. 1.

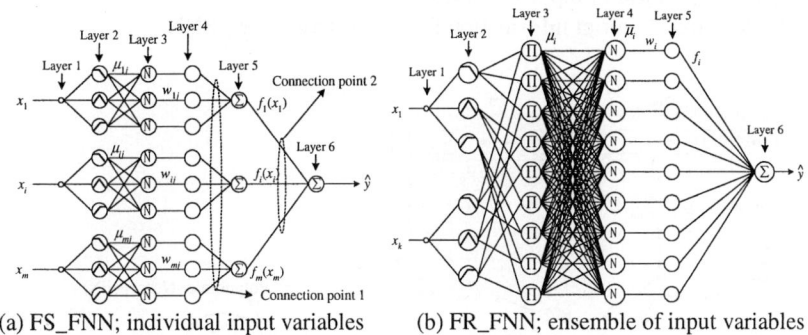

(a) FS_FNN; individual input variables   (b) FR_FNN; ensemble of input variables

Fig. 1. Topologies of FNN

The learning of FNN is realized by adjusting connections of the neurons and as such it follows a BP algorithm [14]. GAs are optimization techniques based on the principles of natural evolution. In essence, they are search algorithms that use operations found in natural genetics to guide a comprehensive search over the parameter space [8]. In order to enhance the learning of the FNN and augment its performance of a FNN, we use GAs to adjust learning rate, momentum coefficient and the parameters of the membership functions of the antecedents of the rules.

3.2 Genetically Optimized PNN (gPNN)

When we construct PNs of each layer in the conventional PNN [9], such parameters as the number of input variables (nodes), the order of polynomial, and input variables available within a PN are fixed (selected) in advance by the designer. This could have frequently contributed to the difficulties in the design of the optimal network. To overcome this apparent drawback, we introduce a new genetic design approach; especially as a consequence we will be referring to these networks as genetically optimized PNN (to be called "gPNN").

4 The Algorithms and Design Procedure of gHFNN

The premise of gHFNN: FS_FNN (Refer to Fig. 1)
[Layer 1] Input layer.
[Layer 2] Computing activation degrees of linguistic labels.
[Layer 3] Normalization of a degree activation (firing) of the rule.
[Layer 4] Multiplying a normalized activation degree of the rule by connection.
[Layer 5] Fuzzy inference for the fuzzy rules.
[Layer 6; Output layer of FNN] Computing output of a FNN.

The design procedure for each layer in FR_FNN is carried out in a same manner as the one presented for FS_FNN.

The consequence of gHFNN: gPNN (Refer to Fig. 2)
[Step 1] Configuration of input variables.
[Step 2] Decision of initial information for constructing the gPNN.

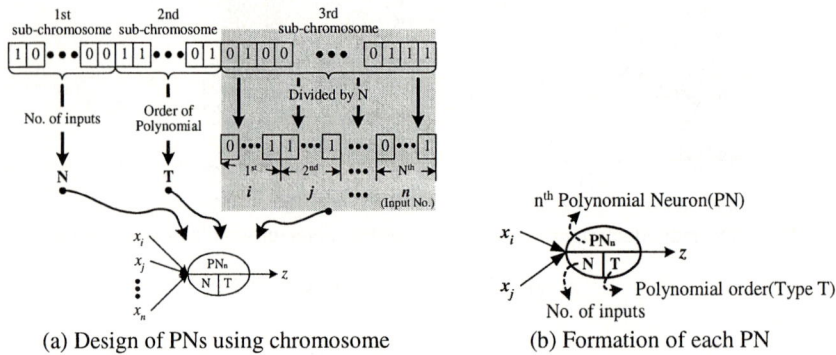

(a) Design of PNs using chromosome   (b) Formation of each PN

Fig. 2. The PN design using genetic optimization

[Step 3] Initialization of population.
[Step 4] Decision of PNs structure using genetic design. as shown in Fig. 2.
[Step 5] Evaluation of PNs.
[Step 6] Elitist strategy and selection of PNs with the best predictive capability.
[Step 7] Reproduction.
[Step 8] Repeating Step 4-7.
[Step 9] Construction of their corresponding layer.
[Step 10] Check the termination criterion (performance index).
[Step 11] Determining new input variables for the next layer.
   The gPNN algorithm is carried out by repeating Steps 4-11.

5 Experimental Studies

The performance of the gHFNN is illustrated with the aid of a time series of gas furnace [14].

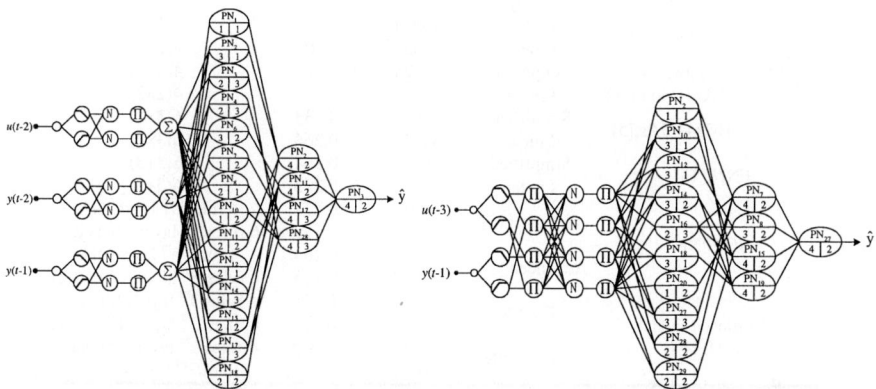

(a) In case of using FS_FNN with Type II   (b) In case of using FR_FNN with Type I

Fig. 3. Optimal topology of genetically optimized HFNN for the gas furnace

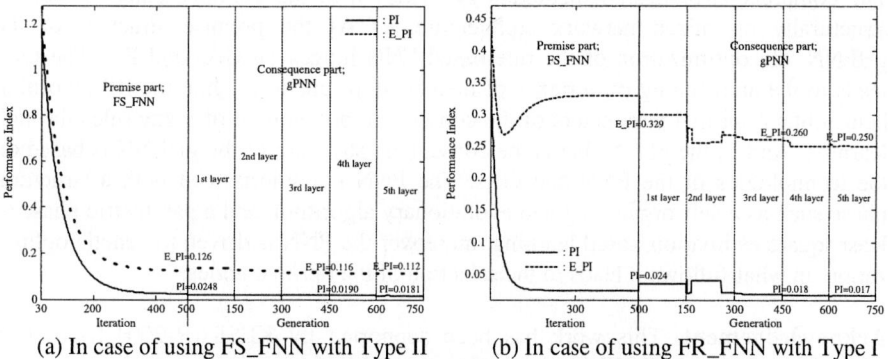

(a) In case of using FS_FNN with Type II   (b) In case of using FR_FNN with Type I

Fig. 4. Optimization procedure of gHFNN by BP learning and GAs

We use two types of system input variables of FNN structure, Type I and Type II to design an optimal model from gas furnace data. Type I utilize two system input variables such as $u(t-3)$ and $y(t-1)$ and Type II utilizes 3 system input variables such as $u(t-2)$, $y(t-2)$, and $y(t-1)$. The output variable is $y(t)$.

The optimal topology of gHFNN is shown in Fig. 3. Fig. 4 illustrates the optimization process by visualizing the performance index in successive cycles. Table 1 contrasts the performance of the genetically developed network with other fuzzy and fuzzy-neural networks studied in the literatures.

Table 1. Comparison of performance with other modeling methods

Model			PI	EPI	No. of rules
	Box and Jenkin's model [14]		0.710		
	Pedrycz's model [1]		0.320		
	Xu and Zailu's model [2]		0.328		
	Sugeno and Yasukawa's model [3]		0190		
	Kim, et al.'s model [15]		0.034	0.244	2
	Lin and Cunningham's mode [16]		0.071	0.261	4
Fuzzy	Complex [4]	Simplified	0.024	0.328	4(2×2)
		Linear	0.023	0.306	4(2×2)
	Hybrid [6] (GAs+Complex)	Simplified	0.024	0.329	4(2×2)
		Linear	0.017	0.289	4(2×2)
	HCM+GAs [5]	Simplified	0.022	0.333	6(3×2)
		Linear	0.020	0.264	6(3×2)
	FNN [13]	Simplified	0.043	0.264	6(3+3)
		Linear	0.037	0.273	6(3+3)
SOFPNN		Generic [11]	0.023	0.277	4 rules/5th layer (NA)
			0.020	0.119	6 rules/5th layer (22 nodes)
		Advanced [12]	0.019	0.264	4 rules/5th layer (NA)
			0.017	0.113	6 rules/5th layer (26 nodes)
Proposed model (gHFNN)		FS_FNN	0.020	0.265	4 rules/3rd layer (12nodes)
			0.019	0.116	6 rules/3rd layer (19 nodes)
		FR_FNN	0.018	0.260	4 rules/3rd layer (15nodes)
			0.018	0.114	7 rules/3rd layer (9 nodes)

6 Concluding Remarks

The comprehensive design methodology comes with the parametrically as well as structurally optimized network architecture. 1) As the premise structure of the gHFNN, the optimization of the rule-based FNN hinges on GAs and BP: The GAs leads to the auto-tuning of vertexes of membership function, while the BP algorithm helps obtain optimal parameters of the consequent polynomial of fuzzy rules through learning. And 2) the gPNN that is the consequent structure of the gHFNN is based on the technologies of the PNN and GAs: The PNN is comprised of both a structural phase such as a self-organizing and evolutionary algorithm, and a parametric phase of least square estimation-based learning, moreover the PNN is driven to genetic optimization, in what follows it leads to the selection of the optimal nodes.

Acknowledgement. This work has been supported by KESRI(R-2004-B-133-01), which is funded by MOCIE(Ministry of commerce, industry and energy).

References

1. Pedrycz, W.: An identification algorithm in fuzzy relational system. Fuzzy Sets and Systems. **13** (1984) 153-167
2. Xu, C.W., Zailu, Y.: Fuzzy model identification self-learning for dynamic system. IEEE Trans. on Syst. Man, Cybern. SMC-**17**(4) (1987) 683-689
3. Sugeno, M., Yasukawa, T.: A Fuzzy-Logic-Based Approach to Qualitative Modeling. IEEE Trans. Fuzzy Systems. **1**(1) (1993) 7-31
4. Oh, S.K., Pedrycz, W.: Fuzzy Identification by Means of Auto-Tuning Algorithm and Its Application to Nonlinear Systems. Fuzzy Sets and Systems. **115**(2) (2000) 205-230
5. Park, B.J., Pedrycz, W., Oh, S.K.: Identification of Fuzzy Models with the Aid of Evolutionary Data Granulation. IEE Proceedings-Control theory and application. **148**(5) (2001) 406-418
6. Oh, S.K., Pedrycz, W., Park, B.J.: Hybrid Identification of Fuzzy Rule-Based Models. International Journal of Intelligent Systems. **17**(1) (2002) 77-103
7. Narendra, K.S., Parthasarathy, K.: Gradient Methods for the Optimization of Dynamical Systems Containing Neural Networks. IEEE Transactions on Neural Networks. **2** (1991) 252-262
8. Michalewicz, Z.: Genetic Algorithms + Data Structures = Evolution Programs. Springer-Verlag, Berlin Heidelberg. (1996)
9. Oh, S.K., Pedrycz, W., Park, B.J.: Polynomial Neural Networks Architecture: Analysis and Design. Computers and Electrical Engineering. **29**(6) (2003) 653-725
10. Ivahnenko, A. G.: The group method of data handling: a rival of method of stochastic approximation. Soviet Automatic Control. **13**(3) (1968) 43-55
11. Park, B.J., Oh, S.K., Jang, S.W.: The Design of Adaptive Fuzzy Polynomial Neural Networks Architectures Based on Fuzzy Neural Networks and Self-Organizing Networks. Journal of Control, Automation and Systems Engineering. **8**(2) (2002) 126-135 (In Korean)
12. Park, B.J., Oh, S.K.: The Analysis and Design of Advanced Neurofuzzy Polynomial Networks. Journal of the Institute of Electronics Engineers of Korea. **39**-CI(3) (2002) 18-31 (In Korean)
13. Oh, S.K., Pedrycz, W., Park, H.S.: Hybrid Identification in Fuzzy-Neural Networks. Fuzzy Sets and Systems. **138**(2) (2003) 399-426
14. Box, D. E. P., Jenkins, G. M.: Time Series Analysis, Forecasting, and Control, 2nd edition Holden-Day, SanFransisco. (1976)
15. Kim, E., Lee, H., Park, M., Park, M.: A Simply Identified Sugeno-type Fuzzy Model via Double Clustering. Information Sciences. **110** (1998) 25-39
16. Lin, Y., Cunningham III, G. A.: A new Approach to Fuzzy-neural Modeling. IEEE Transaction on Fuzzy Systems. **3**(2) 190-197
17. Park, H.S., Park, B.J., Kim, H.K., Oh, S,K,: Self-Organizing Polynomial Neural Networks Based on Genetically Optimized Multi-Layer Perceptron Architecture. International journal of Control, Automations, and Systems. **2**(4) (2004) 423-434

Modelling and Constraint Hardness Characterisation of the Unique-Path OSPF Weight Setting Problem

Changyong Zhang and Robert Rodosek

IC-Parc, Imperial College London, London SW7 2AZ, United Kingdom
{cz, r.rodosek}@icparc.imperial.ac.uk

Abstract. Link weight is the primary parameter of OSPF, the most commonly used IP routing protocol. The problem of setting link weights optimally for unique-path OSPF routing is addressed. A complete formulation with a polynomial number of constraints is introduced and is mathematically proved to model the problem correctly. An exact algorithm is thereby proposed to solve the problem based on the analysis of the hardness of problem constraints.

1 Introduction

Open Shortest Path First (OSPF) [13] is the most widely deployed protocol for IP networks. As with most other conventional IP routing protocols [6], OSPF is a shortest path routing protocol, where traffic flows between origin and destination nodes are routed along the shortest paths, based on a shortest path first (SPF) algorithm [5]. Given a network topology, the SPF algorithm uses link weights to compute shortest paths. The link weights are hence the principal parameters of OSPF.

A simple way of setting link weights is the hop-count method, assigning the weight of each link to one. The length of a path is thereby equal to the number of hops. Another default way recommended by Cisco is the inv-cap method, setting the weight of a link inversely proportional to its capacity, without taking traffic into consideration. More generally, the weight of a link may depend on its transmission capacity and its projected traffic load. Accordingly, a task is to find an optimal weight set for OSPF routing, given a network topology, a projected traffic matrix [8], and an objective function. This is known as the OSPF weight setting problem.

The problem has two instances, depending on whether multiple shortest paths or only a unique one from an origin to a destination is allowed. For the first instance, a number of heuristic methods have been developed, based on genetic algorithm [7] and local search method [9]. For the second instance, Lagrangian relaxation method [12], local search method [15], and sequential method [2] have been proposed to solve the problem. With these heuristic methods, the problem is not formulated completely or explicitly and so generally is not solved optimally.

From a management point of view, unique-path routing requires much simpler routing mechanisms to deploy and allows for easier monitoring of traffic flows [3]. Therefore, this paper focuses on the unique-path instance. The problem is referred as the unique-path OSPF weight setting (1-WS) problem. It is a reduction of the NP-complete integer multicommodity flow problem [16].

With the aim of developing a scalable approach to solve the 1-WS problem optimally, a complete formulation with a polynomial number of constraints is introduced and is mathematically proved to model the problem correctly in Section 2. The hardness of problem constraints is studied in Section 3. Based on the analysis of constraint hardness, an exact algorithm is proposed in Section 4. Conclusions and further work are presented in Section 5.

2 A Complete Formulation

2.1 Problem Definition

The unique-path OSPF weight setting problem is defined as follows. Given

- A network topology, which is a directed graph structure $G=(V, E)$ where V is a finite set of nodes and E is a set of directed links. For each $(i, j) \in E$, i is the starting node, j is the end node, and $c_{ij} \geq 0$ is the capacity of the link.
- A traffic matrix, which is a set of demands D. For each demand $k \in D$, $s_k \in V$ is the origin node, $t_k \in V$ is the destination node, and $d_k \geq 0$ is the demand bandwidth. Accordingly, S is the set of all origin nodes.
- Lower and upper bounds of link weights, which are positive real numbers w_{min} and w_{max}, respectively.
- A pre-specified objective function, e.g., to maximise the residual capacities.
- Find an optimal weight w_{ij} for each link $(i, j) \in E$, subject to
- Flow conservation constraints. For each demand, at each node, the sum of all incoming flows (including demand bandwidth at origin) is equal to the sum of all outgoing flows (including demand bandwidth at destination).
- Link capacity constraints. For each link, the traffic load over the link does not exceed the capacity of the link.
- Path uniqueness constraints. Each demand has only one routing path.
- Path length constraints. For each demand, the length of each path assigned to route the demand is less than that of any other possible and unassigned path to route the demand.
- Link weight constraints. For each link $(i, j) \in E$, the weight w_{ij} is within the weight bounds, i.e., $w_{min} \leq w_{ij} \leq w_{max}$.

2.2 Mathematical Modelling

According to the requirements of the 1-WS problem, the routing path of a demand is the shortest one among all possible paths. For each link, the routing path of a demand either traverses it or not. Based on this observation and the relationship between the length of a shortest path and the weights of links that it traverses, the problem can be formulated by defining one routing decision variable for each link and each demand, which results in the following model.

Routing decision variables:

$$x_{ij}^k \in \{0,1\}, \forall k \in D, \forall (i,j) \in E \qquad (1)$$

is equal to 1 if and only if the path assigned to route demand k traverses link (i,j).

Link weight variables:

$$w_{ij} \in [w_{\min}, w_{\max}], \forall (i,j) \in E \qquad (2)$$

represents routing cost of link (i,j).

Path length variables:

$$l_i^s \begin{cases} = 0, i = s \\ \in [0, +\infty), i \neq s \end{cases}, \forall s \in S, \forall i \in V \qquad (3)$$

represents the length of the shortest path from origin node s to node i.

Flow conservation constraints:

$$\sum_{h:(h,i) \in E} x_{hi}^k - \sum_{j:(i,j) \in E} x_{ij}^k = b_i^k, \forall k \in D, \forall i \in V \qquad (4)$$

where $b_i^k = -1$ if $i = s_k$, $b_i^k = 1$ if $i = t_k$, and $b_i^k = 0$ otherwise.

Link capacity constraints:

$$\sum_{k \in D} d_k x_{ij}^k \leq c_{ij}, \forall (i,j) \in E \qquad (5)$$

Path length constraints:

$$\left. \begin{array}{l} x_{ij}^k = 0 \wedge \sum_{h:(h,j) \in E} x_{hj}^k = 0 \Rightarrow l_j^{s_k} \leq l_i^{s_k} + w_{ij} \\ x_{ij}^k = 0 \wedge \sum_{h:(h,j) \in E} x_{hj}^k = 1 \Rightarrow l_j^{s_k} < l_i^{s_k} + w_{ij} \\ x_{ij}^k = 1 \Rightarrow l_j^{s_k} = l_i^{s_k} + w_{ij} \end{array} \right\}, \forall k \in D, \forall (i,j) \in E \qquad (6')$$

The above logic constraints can be linearised by introducing appropriate constants ε and M with $0 < \varepsilon \ll M$.

$$\begin{cases} l_j^{s_k} \leq l_i^{s_k} + w_{ij} - \varepsilon(\sum_{h:(h,j) \in E} x_{hj}^k - x_{ij}^k) \\ l_j^{s_k} \geq l_i^{s_k} + w_{ij} - M(1 - x_{ij}^k) \end{cases}, \forall k \in D, \forall (i,j) \in E \qquad (6)$$

Objective function: to maximise the residual capacities, alternatively, to minimise the throughput:

$$\min \sum_{(i,j) \in E} \sum_{k \in D} d_k x_{ij}^k \qquad (7)$$

Accordingly, the complete model is presented as follows:

1-WS 0: Optimise (7) Subject to (4), (5), (6), (1), (2), (3)

2.3 Proof of Correctness

A relaxation of the 1-WS problem is the integer multicommodity flow problem [1], a recognised correct model of which is presented as follows:

1-WS I: Optimise (7) Subject to (4), (5), (1)

Apparently, the difference between 1-WS 0 and 1-WS I are path length constraints (6) and the resulting additional link weight variables (2) as well as path length variables (3). In order to ensure that 1-WS 0 formulates the 1-WS problem correctly, constraints (6) are proved to represent correctly the additional path length as well as path uniqueness constraints in the following. As the initial logic constraints are identical to the linearised constraints (6), the following proof is based on the initial constraints (6').

Proposition 1. *The path length constraints in 1-WS 0 restrict that each routing path is a shortest path.*

Proof. Assume for demand k, $P_j = (j_1, j_2) \rightarrow (j_2, j_3) \rightarrow ... \rightarrow (j_{n-1}, j_n)$, $j_1 = s_k, j_n = t_k$ is the assigned routing path and $P_i = (i_1, i_2) \rightarrow (i_2, i_3) \rightarrow ... \rightarrow (i_{m-1}, i_m)$, $i_1 = s_k, i_m = t_k$ is one of any other possible and non-assigned paths. Then, according to the definition of routing decision variables, $x^k_{j_l j_{l+1}} = 1, l = 1,...,n-1$ and $\exists (i_q, i_{q+1}) \in P_i$, $x^k_{i_q i_{q+1}} = 0$, $q \in \{1,2,...,m-1\}$.

As a result, according to (6'), on one hand, since $x^k_{ij} = 1 \Rightarrow l^{s_k}_j = l^{s_k}_i + w_{ij}$,

$$l^{s_k}_{t_k} = l^{s_k}_{j_n} = l^{s_k}_{j_{n-1}} + w_{j_{n-1} j_n} = l^{s_k}_{j_{n-2}} + w_{j_{n-2} j_{n-1}} + w_{j_{n-1} j_n} = ... = l^{s_k}_{j_1} + w_{j_1 j_2} + ... + w_{j_{n-1} j_n} = l_{P_j}$$

On the other hand, since $x^k_{ij} = 0 \Rightarrow l^{s_k}_j \leq l^{s_k}_i + w_{ij}$,

$$l^{s_k}_{t_k} = l^{s_k}_{i_m} \leq l^{s_k}_{i_{m-1}} + w_{i_{m-1} i_m} \leq l^{s_k}_{i_{m-2}} + w_{i_{m-2} i_{m-1}} + w_{i_{m-1} i_m} \leq ... \leq l^{s_k}_{i_1} + w_{i_1 i_2} + ... + w_{i_{m-1} i_m} = l_{P_i}$$

Therefore, $l_{P_j} \leq l_{P_i}$. It is proved that path P_j is a shortest path. □

Lemma 1. *The path uniqueness constraints are satisfied by 1-WS I.*

Proposition 2. *The path length constraints in 1-WS 0 restrict that the resulting routing path of each demand is a unique shortest path.*

Proof. As 1-WS 0 is a reduction of 1-WS I, the solution to routing decision variables x^k_{ij} of 1-WS 0 is a solution to 1-WS I.

According to Lemma 1, there is only one path to route each demand. Suppose for demand k, $P_j = (j_1, j_2) \rightarrow (j_2, j_3) \rightarrow ... \rightarrow (j_{n-1}, j_n), j_1 = s_k, j_n = t_k$ is the assigned routing path, and $P_i = (i_1, i_2) \rightarrow (i_2, i_3) \rightarrow ... \rightarrow (i_{m-1}, i_m), i_1 = s_k, i_m = t_k$ is one of any other possible and non-assigned paths to route demand k. Then, according to the definition of routing decision variables, $x^k_{j_l j_{l+1}} = 1, l = 1,...,n-1$.

As a result, according to (6'), since $x^k_{ij} = 1 \Rightarrow l^{s_k}_j = l^{s_k}_i + w_{ij}$,

$$l^{s_k}_{t_k} = l^{s_k}_{j_n} = l^{s_k}_{j_{n-1}} + w_{j_{n-1} j_n} = l^{s_k}_{j_{n-2}} + w_{j_{n-2} j_{n-1}} + w_{j_{n-1} j_n} = ... = l^{s_k}_{j_1} + w_{j_1 j_2} + ... + w_{j_{n-1} j_n} = l_{P_j}$$

As both P_i and P_j are paths between s_k and t_k, they finally merge at one node. Assume it is node r and $r = j_p = i_q, p \in \{2,3,...,n-1,n\}, q \in \{2,3,...,m-1,m\}$. Then, according to the definition of routing decision variables, $x^k_{j_{p-1}r} = 1$ and $x^k_{i_{q-1}r} = 0$, and hence $\sum_{h:(h,r)\in E} x^k_{hr} = 1$. As a result, according to (6'),

$$l^{s_k}_{t_k} = l^{s_k}_{i_m} = l^{s_k}_{i_{m-1}} + w_{i_{m-1}i_m} = ... = l^{s_k}_r + w_{ri_{q+1}} + ... + w_{i_{m-1}i_m}$$
$$< l^{s_k}_{i_{q-1}} + w_{i_{q-1}r} + w_{ri_{q+1}} + ... + w_{i_{m-1}i_m} \leq ... \leq l^{s_k}_{i_1} + w_{i_1i_2} + ... + w_{i_{m-1}i_m} = l_{P_i}$$

Therefore, $l_{P_j} < l_{P_i}$. It is proved that path P_j is the unique shortest path to route demand k. □

3 Constraint Hardness Characterisation

There are three types of constraints in 1-WS 0, flow conservation constraints, link capacity constraints, and path length constraints. Among them, flow conservation constraints are the basic and core constraints of the problem. In order to compare the hardness of the other two types of constraints, two relaxed problems are studied.

First, path length constraints are relaxed from the 1-WS problem, which results in the integer multicommodity flow problem 1-WS I, as introduced in Section 2.3.

Second, link capacity constraints are relaxed from the 1-WS problem. This results in the un-capacitated unique-path OSPF weight setting problem:

1-WS II: Optimise (7) Subject to (4), (6), (1), (2), (3)

Forty-eight data sets with combinations of different parameter scenarios were generated for empirical study. In Table 1, Nds, Lnks, and Dmnds denote the numbers of nodes, links, and demands, respectively. All the three problems 1-WS 0, 1-WS I, and 1-WS II were implemented in ECLiPSe [11] and solved using CPLEX 6.5 [10] on all data sets generated. The timeout was set to be *3600 seconds* for each data instance. The following analyses are thereby based on the performance of using CPLEX.

Table 1. Details of data sets tested

ID	Nds	Lnks	Dmnds	ID	Nds	Lnks	Dmnds	ID	Nds	Lnks	Dmnds
1	10	22	3	17	30	78	60	33	50	130	49
2	10	26	5	18	30	78	375	34	50	128	50
3	10	24	9	19	30	136	3	35	50	130	100
4	10	24	10	20	30	144	15	36	50	136	788
5	10	26	20	21	30	142	29	37	50	238	3
6	10	24	50	22	30	144	30	38	50	238	25
7	10	46	3	23	30	142	60	39	50	238	49
8	10	46	5	24	30	142	450	40	50	242	50
9	10	46	9	25	30	236	3	41	50	240	100
10	10	46	10	26	30	234	15	42	50	238	1000
11	10	48	20	27	30	236	29	43	50	644	3
12	10	44	50	28	30	234	30	44	50	648	25
13	30	80	3	29	30	236	60	45	50	642	49
14	30	78	15	30	30	234	450	46	50	642	50

15	30	82	29	31	50	128	3	47	50	646	100
16	30	76	30	32	50	132	25	48	50	642	1000

Consider the OSPF weight setting problem, it is shown that routing performances resulting from the proposed complete formulation are much better than those from using the default methods. The resulting average maximum utilisation is *28.79%* of that from using the hop-count method and *40.68%* of that from using the inv-cap method, which demonstrates the significant gains achieved by formulating the problem completely and solving it optimally.

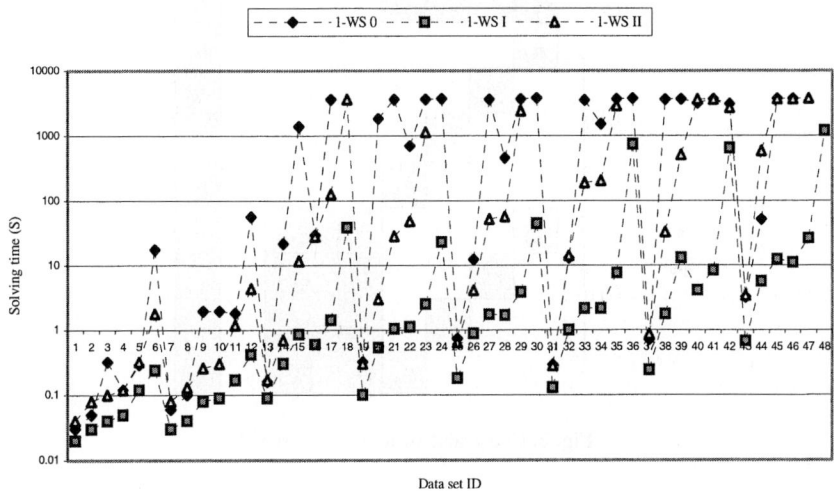

Fig. 1. Solving times of 1-WS 0, 1-WS I, and 1-WS II

Figure 1 compares the solving times of 1-WS 0 with those of 1-WS I and 1-WS II. It can be noted that 1-WS I is solved optimally within *1000 seconds* for all instances, except the last one, which is detected infeasible in *1191 seconds*. Meanwhile, it takes more time to solve 1-WS 0 than to solve 1-WS I on all instances. For most large-scale instances, it even cannot be solved when timeout. It is thus shown that path length constraints are very hard constraints for the 1-WS problem. It can be further seen that, although it takes less time to solve 1-WS II than the initial problem on most data instances, the difference is not so significant. The relaxed problem still cannot be solved when timeout on a few data instances. It is therefore indicated that the link capacity constraints are not the hardest constraints.

In addition, it can be observed that between the two relaxed problems, 1-WS I is much easier to solve than 1-WS II. Therefore, path length constraints, which are relaxed in 1-WS I, are the hardest constraints for the 1-WS problem.

In order to investigate further the reason behind the above observations, the constraint structure of the 1-WS problem is shown in Figure 2. The first row represents link capacity constraints (5), the next four rows correspond to flow conservation constraints (4), and the last four rows represent path length constraints (6). As it can be seen, among the three types of constraints, flow conservation constraints and link capacity constraints contain only routing decision variables, while path length constraints couple routing decision variables with link weight variables and path length variables, which makes the problem more complicated than the integer multicommodity flow problem. This observation can also be used to explain why path length constraints are the hardest constraints for the 1-WS problem, instead of link capacity constraints, which are the hardest for the integer multicommodity flow problem [14].

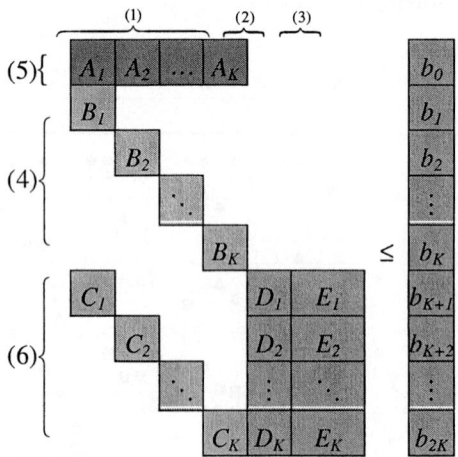

Fig. 2. Constraint structure of 1-WS 0

4 Proposed Algorithm

Based on the above study of constraint hardness, a proposed algorithm to solve the 1-WS problem is Benders decomposition method [4], which decomposes the problem into an integer multicommodity flow master problem and a linear programming (LP) subproblem. The master problem deals with flow conservation constraints and link capacity constraints, and so contains routing decision variables only. Accordingly, the LP subproblem deals with the hardest constraints, path length constraints. Compared with the initial mixed integer programming (MIP) problem, the resulting master problem has a much smaller model size. Therefore, instead of solving a larger and more complicated MIP problem in one step, the proposed algorithm solves the problem by dealing with a smaller and simpler master problem and an LP subproblem iteratively. For the integer master problem, Lagrangian relaxation method has been demonstrated to be an appealing algorithm [14].

It was shown be preliminary results that, for small data instances, MIP solver solves the problem slightly faster than Benders decomposition method. However, when data instances get larger, the latter takes the advantage.

5 Conclusion

In order to develop a complete solution approach to the unique-path OSPF weight setting problem, the problem has been explicitly formulated as a complete model, the correctness of which is mathematically proved. The model has three types of constraints, flow conservation constraints, link capacity constraints, and path length constraints. Among them, path length constraints have been identified to be the hardest constraints for the problem. Based on the study of constraint structure of the formulation, Benders decomposition method, embedded with Lagrangian relaxation method for the integer master problem, has been proposed to solve the problem.

Our future work includes developing the proposed algorithm completely and investigating possible improvements to both model formulation and solution algorithm to accelerate the convergence rate of the solution approach.

References

1. Ahuja, R. K., Magnanti, T. L., and Orlin, J. B.: Network Flows: Theory, Algorithms, and Applications. Prentice-Hall (1993)
2. Ameur, W. B., Bourquia, N., Gourdin, E., and Tolla, P.: Optimal Routing for Efficient Internet Networks. In Proc. of ECUMN (2002) 10-17
3. Ameur, W. B., and Gourdin, E.: Internet Routing and Related Topology Issues. SIAM J. on Discrete Mathematics, 17(1) (2003) 18-49
4. Benders, J.: Partitioning Procedures for Solving Mixed-Variables Programming Problems. Numerische Mathematik, 4 (1962) 238-252
5. Bertsekas, D., and Gallager, R.: Data Networks. Second Edition, Prentice Hall (1992)
6. Black, U.: IP Routing Protocols: RIP, OSPF, BGP, PNNI and Cisco Routing Protocols. Prentice Hall, (2000)
7. Ericsson, M., Resende, M. G. C., and Pardalos, P. M.: A Genetic Algorithm for the Weight Setting Problem in OSPF Routing. J. of Combinatorial Optimization, 6 (2002) 299-333
8. Feldmann, A., Greenberg, A., Lund, C., Reingold, N., Rexford, J., and True, F.: Deriving Traffic Demands for Operational IP Networks: Methodology and Experience. IEEE/ACM Transactions on Networking, 9(3) (2001) 265-279
9. Fortz, B., and Thorup, M.: Internet Traffic Engineering by Optimizing OSPF Weights. In Proc. of INFOCOM (2000) 519-528
10. ILOG Inc.: ILOG CPLEX 6.5 User's Manual. (1999)
11. Imperial College London: ECLiPSe 5.7 User's Manual. (2003)
12. Lin, F. Y. S., and Wang, J. L.: Minimax Open Shortest Path First Routing Algorithms in Networks Supporting the SMDS Services. In Proc. of ICC, 2 (1993) 666-670
13. Moy, J.: OSPF Anatomy of an Internet Routing Protocol. Addison-Wesley (1998)
14. Ouaja, W., and Richards, B.: A Hybrid Multicommodity Routing Algorithm for Traffic Engineering. Networks, 43(3) (2004) 125-140
15. Ramakrishnan, K. G., and Rodrigues, M. A.: Optimal Routing in Shortest-Path Data Network. Bell Labs Technical Journal, January-June (2001) 117-138
16. Wang, Y., and Wang, Z.: Explicit Routing Algorithms for Internet Traffic Engineering. In Proc. of ICCCN (1999) 582-588

Application of Four-Dimension Assignment Algorithm of Data Association in Distributed Passive-Sensor System

Li Zhou[1,2], You He[1], and Xiao-jing Wang[3]

[1] Research Institute of Information Fusion, Naval Aeronautical,
Engineering Institute, Yantai, 264001, P. R. China
zxyzlzwh@vip.sina.com
[2] Math & Information College, Yantai Teachers' University,
Yantai, 264025, P. R. China
[3] Department of Basic Science,
Beijing Institute of Civil Engineering and Architecture,
Beijing, 100044, P.R. China

Abstract. The disadvantage of multi-dimension assignment algorithm of data association in distributed passive-sensor system is the lower processing speed. The modified optimal assignment algorithm is presented in this paper. The new algorithm avoids a large quantity of calculation of cost function of 4-tuple measurements and removes the disturbance of some false location points. Simulation results show that by using the algorithm discussed in this paper, not only the calculation burden is reduced greatly but also the accuracy of data association is improved correspondingly.

1 Introduction

Because passive-sensor doesn't radiate any electromagnetism signal, and has the advantage of both better concealment and less interference of the enemy, passive-sensor location has become a hot research point which is studied by many scholars. When the target's position is estimated by bearing-only measurements from passive sensors, the better method of data association is to transform the problem to the optimal assignment problem of operational research [1], [2]. The disadvantage of the optimal assignment algorithm of multi-passive-sensor multi-target data association is the long processing time caused by the heavier calculation burden of cost function [3]. In the case of the four bearing-only sensors and multi-target are in the same plane, this paper proposes a modified algorithm based on direction-finding cross location. The essence of the modified algorithm is a two-stage association algorithm. In the first stage, we eliminate some false intersection points by using a cross location technology, and in the second stage, only those points which have passed through the gating of the correlation test in the first stage can be permitted to join the assignment process. This reduces the computation burden from the calculation of cost function for assignment problem. In the meantime, with the removing of the large number of false location points, the effect of data association is improved.

2 Data Association

2.1 The Traditional Optimal Algorithm

Supposed four bearing-only sensors are used to locate targets as illustrated in Fig.1. The positions of four sensors are $p_s = (x_s, y_s)^T$, $s = 1,2,3,4$; suppose there are N targets in Surveillance view, and the position of target t is $p_t = (x_t, y_t)^T$; Suppose the number of measurements of sensor s is n_s, $s = 1,2,3,4$; the bearing-only sensor only measures the azimuth of target t, and it is denoted as Z_{si_s}, $i_s = 1,2...n_s$; if the measurement is from a real target, it is the true observable θ_{st} plus the Gaussian noise $N(0, \sigma_s^2)$, and if it is from spurious measurement, suppose it follows the uniform distribution in the field of view of sensor s.

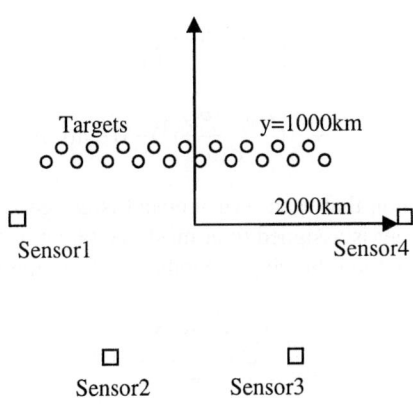

Fig. 1. Position of sensors

The maximum likelihood function of 4-tuple $Z_{i_1 i_2 i_3 i_4} = \{Z_{1i_1}, Z_{2i_2}, Z_{3i_3}, Z_{4i_4}\}$ coming from the same target t is [1], [2]

$$\Lambda(Z_{i_1 i_2 i_3 i_4} | \omega_t) = \prod_{s=1}^{4} [P_{ds} \cdot p(Z_{si_s} | \omega_t)]^{u(i_s)} [1 - P_{ds}]^{[1-u(i_s)]} \quad (1)$$

Where P_{ds} is the detect probability of sensor s, $u(i_s)$ is a binary indicator function, if sensor s missed the detection from target t, i.e., $i_s = 0$, then $u(i_s) = 0$. Otherwise, $u(i_s) = 1$. $p(Z_{si_s} | \omega_t)$ is the probability density function of Z_{si_s} being from target t, The likelihood that the measurements are all spurious or unrelated to target t, i.e., $\omega_t = \Phi$ is

$$\Lambda(Z_{i_1i_2i_3i_4}|t=\Phi) = \prod_{s=1}^{4}[\frac{1}{\Psi_s}]^{u(i_s)} \qquad (2)$$

where Ψ_s is the field of view of sensor s, the cost of associating the 4-tuple to target t is given by

$$c_{i_1i_2i_3i_4} = -\ln\frac{\Lambda(Z_{i_1i_2i_3i_4}|t)}{\Lambda(Z_{i_1i_2i_3i_4}|t=\Phi)} \qquad (3)$$

As ω_t in (1) is unknown, it can usually be replaced by its maximum likelihood or least-square estimation as

$$\hat{\omega}_t = \arg\max_{\omega_t} \Lambda(Z_{i_1i_2i_3i_4}|t) \qquad (4)$$

Hence, the cost of associating $Z_{i_1i_2i_3i_4}$ with target t can be induced as [1], [2]

$$c_{i_1i_2i_3i_4} = \sum_{s=1}^{4}[u(i_s)(\ln(\frac{\sqrt{2\pi}\cdot\sigma_s}{P_{ds}\cdot\Psi_s}) \\ + \frac{1}{2}(\frac{Z_{si_s}-\hat{\theta}_{st}}{\sigma_s})^2) - (1-u(i_s))\cdot\ln(1-P_{ds})] \qquad (5)$$

With the assumption that each measurement is assigned to a target or declared false, and each measurement is assigned to at most one target, the problem of data association can be transformed to the following generalized 4-D assignment problem [2]

$$\min_{P_{i_1i_2i_3i_4}} \sum_{i_1=0}^{n_1}\sum_{i_2=0}^{n_2}\sum_{i_3=0}^{n_3}\sum_{i_4}^{n_4} c_{i_1i_2i_3i_4}\cdot P_{i_1i_2i_3i_4} \qquad (6)$$

subject to

$$\begin{cases} \sum_{i_2=0}^{n_2}\sum_{i_3=0}^{n_3}\sum_{i_4=0}^{n_4} P_{i_1i_2i_3i_4} = 1; & \forall i_1 = 1,2\cdots n_1 \\ \sum_{i_1=0}^{n_1}\sum_{i_3=0}^{n_3}\sum_{i_4=0}^{n_4} P_{i_1i_2i_3i_4} = 1; & \forall i_2 = 1,2\cdots n_2 \\ \sum_{i_1=0}^{n_1}\sum_{i_2}^{n_2}\sum_{i_4}^{n_4} P_{i_1i_2i_3i_4} = 1; & \forall i_3 = 1,2\ldots n_3 \\ \sum_{i_1=0}^{n_1}\sum_{i_2=0}^{n_2}\sum_{i_3=0}^{n_3} P_{i_1i_2i_3i_4} = 1; & \forall i_4 = 1,2\ldots n_4 \end{cases} \qquad (7)$$

The solution of 4-D assignment problem can be shown to be NP-hard [2], [5]. The optimal technique requires unacceptable time and is of little practical value. Instead, fast and near the optimal solutions are most desirable. Among various heuristic algorithms of multi-dimension assignment problem, Lagrangian relaxation algorithm has a dominant role owing to its satisfying result in application. It relaxes the 4-D assignment problem to a series of 2-D assignment problem to solve, which can be resolved by various algorithms in polynomial time [4]. The advantage of this algorithm compared with the other modern optimal algorithm is that not only can we obtain a suboptimal solution which is near to the optimal solution, but also can obtain a measure of the quality of this solution. When the density of targets and false alarms in surveillance is higher, the run time of Lagrangian relaxation algorithm is still too long to satisfy the need of engineering because of the large quantity of calculation burden from association cost, so we present a modified algorithm in next section.

2.2 The Modified Algorithm

The description of the cross location for four bearing-only sensors are shown in Fig.2. θ_i ($i=1,2,3,4$) denote the bearing measurements, The positions of four sensors are denoted by (x_i, y_i), $i=1,2,3,4$, and the Cartesian position of A, B, C is denoted by (x_A, y_A), (x_B, y_B), (x_C, y_C). Then from the formula

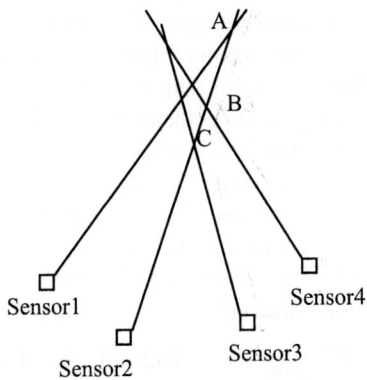

Fig. 2. Location of four sensors

$$tg\theta_1 = \frac{y_A - y_1}{x_A - x_1}, \quad tg\theta_2 = \frac{y_A - y_2}{x_A - x_2} \tag{8}$$

We can get [3],

$$x_A = \frac{y_2 - y_1 + x_1 tg\theta_1 - x_2 tg\theta_2}{tg\theta_1 - tg\theta_2} \tag{9}$$

$$y_A = \frac{y_2 tg\theta_1 - y_1 tg\theta_2 + (x_1 - x_2) tg\theta_1 tg\theta_2}{tg\theta_1 - tg\theta_2} \tag{10}$$

Let

$$d_{AB} = \sqrt{(x_B - x_A)^2 + (y_B - y_A)^2} \tag{11}$$

Again from

$$tg\theta_2 = \frac{y_B - y_2}{x_B - x_2}, \quad tg\theta_3 = \frac{y_B - y_3}{x_B - x_3},$$

and

$$tg\theta_3 = \frac{y_C - y_3}{x_C - x_3}, \quad tg\theta_4 = \frac{y_C - y_3}{x_C - x_3}$$

we can get the Cartesian position of B, C and the distance function d_{BC} and d_{AC} in the same way.

Let

$$d = d_{AB} + d_{BC} + d_{AC} \tag{12}$$

In actuality, the Cartesian position of A, B and C usually don't coincide because of the sensor's measurement error. In consideration of the special position of sensor 2, the probability that the four measurements come from a same target can be estimated by the value of d. In general, the smaller the value is, the more possible that they come from a same target has. We can use the value of d to be the statistics to determine if we should calculate the association cost of 4-tuple of measurements. This can avoid the large operation quantity from statistic test aiming to each 4-tuple of measurements one by one. The disadvantages using distance function to construct test statistics of χ^2 distribution to carry on the data correlation test, one is the heavy burden of calculation, the other is inferior data association result caused by the reason that the statistic may be not obey to the χ^2 distribution strictly sometimes. According to the large number of numeral simulations with the various measurement errors, the gating of statistics d can be determined by $\tau = 500 \times 180 \times \sigma / \pi$, where σ denotes the bearing measurement error of each sensor, if it is different from various sensors, σ can be replaced by the average of four error values.

As the association cost of 4-D assignment problem is defined as formula (3), so the corresponding 4-tuple of measurements $Z_{i_1 i_2 i_3 i_4}$ may be considered a candidate association if and only if $c_{i_1 i_2 i_3 i_4} < 0$, and all the 4-tuple of measurements $Z_{i_1 i_2 i_3 i_4}$ with $c_{i_1 i_2 i_3 i_4} > 0$ can be eliminated from the list of candidate associations by a certain measure. The sparsity of candidate association is defined as the ratio of the number of potential measurement-target association in the 4-D assignment problem to the number of a fully connection. In the former algorithm, sparsity s_0 is defined as the ratio of the number of association with negative cost value of the 4-tuple of measurements to the total association number. In the modified method, sparsity s_τ is defined as the ratio of the number of association having passed the gating of the rough correlation test to the number of the fully association. In order to remove some false location points whose association cost are smaller relatively, s_τ is adopted as lower than s_0 or nearer to it. This can decrease the system error of the model of assignment problem. Therefore, the accuracy of data association of multi-sensor multi-target can be improved correspondingly.

The experimental function τ which is given above is just be taken as a approximate superior gating, we can also adjust the size of the gating by observing the number of cost function in the pages of two-dimension at different circumstance timely. In fact, sparsity of candidate target in 4-D assignment problem is usually lower. According to the current accuracy of direction-finding cross location technique, the ratio of the number of cost function need to be calculated to the total number of cost function is about 10-15% in the two-dimension pages of association cost.

From above analysis, the modified algorithm can be regarded as a two-stage correlation process constituted by both rough correlation process based on verdict function and the accurate correlation process based on the optimal assignment.

3 Performance of Simulation

3.1 Simulation Model

Suppose the sensors' position are shown in Fig.1, the position of four sensors are (-2000, 0), (-1000, -1000 $\sqrt{3}$), (1000, -1000 $\sqrt{3}$) and (2000, 0). The bearing measurement error of various sensor is the same value, we use them, σ_θ of $0.4^0, 0.8^0$. The detection probability of each sensor is assumed to be 1, and the false alarm rate is assumed 1/rad.

3.2 Analysis of Simulation Results

As shown in table 1, in the modified algorithm, either the sparsity s_τ is adopted as $s_\tau \approx s_0$ or adopted as $s_\tau < s_0$, both the run times reduced largely compared with the former algorithm. This is because that a large quantity of calculation from the false location points is cut. The difference between the two case is that the procedure time of $s_\tau < s_0$ is little lower than the result of the case of $s_\tau \approx s_0$. In addition, an obvious result is that the association accuracy correlated to $s_\tau < s_0$ is advantage of the result of $s_\tau \approx s_0$. This result mainly because that with the appropriate limitation of the gating of verdict function in the case of $s_\tau < s_0$, some false location points with lower cost value are removed from the candidate association, and this causes the accuracy of data association and location to be improved by 7~8 percents.

As the measurement error is supposed 0.02^0, the relations between the proceeding time of 4-D assignment algorithm and the number of targets are given in Fig.3. It is obvious that the run time growths rapidly with the increase of the number of targets when the former algorithm is adopted, but it increases slowly when the modified algorithm ($s_\tau < s_0$) is used. This result shows that the more the number of targets is, the more obvious the advantage of the modified algorithm demonstrates. In other words, the new algorithm is more applicable to solve the data association problem in high density environment of targets and false alarms.

Table 1. Average results (25 runs) of 4-D assignment algorithmin (number of targets =20)

		Number joining accurate process	association accuracy	Run time (s)	RMS error (km)
0.4^0	Former	1405	85.7%	75.37	1.61
	Later ($s_\tau \approx s_0$)	1386	86.2%	33.72	1.56
	Later ($s_\tau < s_0$)	715	93.1%	31.96	1.07
0.8^0	former	4776	76.7%	84.58	2.07
	Later ($s_\tau \approx s_0$)	4765	77.4%	37.63	2.03
	Later ($s_\tau < s_0$)	2974	85.6%	34.41	1.58

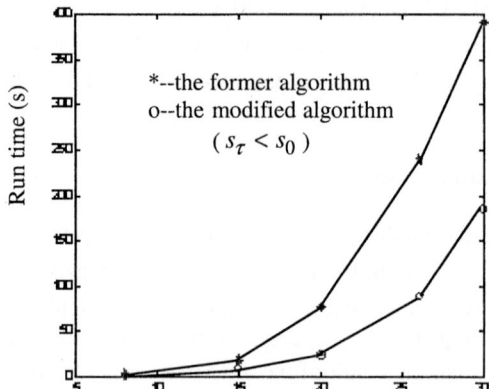

Fig. 3. Run time of 4-Dassignment algorithm versus number of targets

4 Conclusions

This text mainly studies data association algorithm in multi-sensor multi-target location system. A modified algorithm based on Lagarangian relaxation algorithm of 4-D assignment problem is presented. The gating of the verdict function in the first stage correlation process is discussed and given. The new algorithm reduces the run time of the global optimal algorithm of 4-D assignment problem, and improves the effect of data association and location. This idea can be developed to the situation that the sensors and targets are not in the same plane, and it has widely application in other fields of the social activities.

References

1. Pattipati, S.Deb, K.R, and Bar-shalom, Y. et al.: A new Relaxation Algorithm and Passive Sensor Data Association. IEEE Transactions on Automatic Control, 37(1992), 198-213
2. Yeddanapudi, S.Deb, M. Pattipati, K. Bar-shalom, Y.: A Generalized S-D Assignment Algorithm for Multisensor-Multitarget State Estimation. IEEE Transactions on Aerospace and Electronic Systems, 33(1997) 523-537
3. Jianjuan, X.: Study on Multi-target Algorithm in Two Direction-finding Location System. Acta Electronic Sinica, Beijing, China 30, 12 (2002) 1763-1767
4. Bertsekas, D P.: The Auction Algorithm: A Distributed Relaxation method for the Assignment Problem. Annals of Operat Res, 14 (1988) 105-123
5. You, H.: Multisensor Information Fusion with Application.Publishing House of Electronics Industry. Beijing, China (2000)
6. Wenxun, X.: The Modern optimal Algorithm. Publishing House of Tsinghua University, Beijing, China (1999) 20-28
7. Deb S, Pattipati K.R, and Bar-shalom,Y.: A Multisensor-Multitarget Data Association Algorithm for Heterogeneous sensors. IEEE Transactions on Aerospace and Electronic Systems, 29, 2 (1993) 560-568

Using Rewriting Techniques in the Simulation of Dynamical Systems: Application to the Modeling of Sperm Crawling

Antoine Spicher and Olivier Michel

LaMI, umr 8042 du CNRS, Université d'Évry – GENOPOLE,
Tour Evry-2, 523 Place des Terrasses de l'Agora,
91000 Évry, France
{aspicher, michel}@lami.univ-evry.fr

Abstract. Rewriting system (RS) are a formalism widely used in computer science. However, such a formalism can also be used to specify executable models of dynamical systems (DS) by allowing the specification of the evolution laws of the systems in a local manner.

The main drawback of RS is that they are well understood and well known only for terms (a tree-like structure) and that their expressivity is not enough for the representation of complex organizations that can be found in DS.

We propose a framework based on topological notion to extend the notion of RS on more sophisticated structures; the corresponding concepts are validated through the development of an experimental programming language, MGS, dedicated to the simulation of DS. We show how the MGS rewriting system can be used to specify complex dynamical systems and illustrate it with the simulation of the motility of the nematode's sperm cell.

1 Introduction

In this paper, we advocate the use of rewriting techniques for the simulation of complex dynamical systems. The systems we are interested in, are often systems with a dynamical structure [1]. They are difficult to model because their state space is not fixed *a priori* and is jointly computed with the current state during the simulation. In this case the evolution function is often given through local rules that drive the interaction between some system components.

These rules and their application are reminiscent of rewriting rules and their strategy. As a programming language, rewriting systems have the advantage of being close of the mathematical formalism (transparencial referency and declarativeness).

The aim of the MGS project is to develop new rewriting techniques on data structures beyond tree-like organization, and to apply these techniques to the modeling and simulation to various dynamical systems with a dynamical structure in biology. The key idea used here to extend rewriting systems to more general data structures is a topological point of view: a data structure is a set of elements with neighborhood relationship that specifies which elements of the data structure can be accessed from a given one.

This paper is organized as follows. Section 2 recalls the basic notions of rewriting system and sketches its application to the simulation of dynamical systems. Then we present the MGS programming language. An example illustrates the introduced notion: the MGS simulation approach on a dynamical system with a dynamical structure. The system to be modeled is the motility of a cell, inspired by a previous work [2].

2 Rewriting and Simulations

2.1 A Computational Device

A *rewriting system* [3] (RS) is a device used to replace some part of an entity by another. In computer science, the entities subject to this process are usually expressions represented by formal trees. A RS is defined by a set of rules, and each rule $\alpha \to \beta$ specifies how a subpart that matches with the pattern α is substituted by a new part computed from the expression β. We call the pattern α the *left hand side* of the rule (*l.h.s*), and β the *right hand side* (*r.h.s*).

We write $e \to^* e'$ to denote that an expression e is transformed by a series of rewriting in expression e'. It is called a *derivation* of e. The transformation of e into e' can be seen as the result of some computations defined by the rewriting rules and the derivation corresponds to the intermediate results of the computation.

2.2 Rewriting and Simulation

We will see how rewriting can be used for the simulation of *dynamical systems* (DS), *i.e.*, systems described by a state that changes with the time. Using RS for the simulation of DS means:

– the state of the DS is represented by an expression,
– its evolution is specified by a set of rewriting rules defining local transformation.

Then, given an initial state e, a derivation of e following a RS corresponds to a possible trajectory of the DS.

The role of a rule is to specify an interaction between different parts (atomic or not) of the system, or the answer of the system to an exterior message. So, at a cellular scale, $c + s \to c'$ means a cell c that receives a signal s, will change its state to c' ; $c \to c' + c''$ specifies a cell division and $c \to$. represents apoptosis. In these examples, operator $+$ denotes the composition of entities into subsystems. The formalism of RS has consequences of the properties of DS taken in considerations, especially on the management of time and space.

Discretized Time. An important point in the modeling of a DS is the handling of time. Clearly the model of time naturally supported by the framework of rewriting is a discrete, event based, model of time: the application of a rule corresponds to some event in the system and this event corresponds to an atomic instantaneous change in the system state.

Locality of Space. The previous operator + that joins entities and messages expresses the spatial and/or the functional organization of the modeled system and is used to denotes interacting parts of the system and the composition of entities into a subsystem. So, on the first hand, the l.h.s and the r.h.s of a rule specify a local part of the system where an interaction occurs. As a consequence, rules represent local evolution laws of the DS. On the other hand, the organization structures specified in the l.h.s and the r.h.s can differ to generate a modification of the structure. This allows the modeling of a special and difficult to represent kind of DS, the *dynamical systems with a dynamical structure* or $(DS)^2$ (see [4]).

3 MGS: A Framework for Modeling and Simulating Dynamical Systems Using RS

MGS is a project that aims at integrating the formalism of RS in a programming language dedicated to the modeling and the simulation of $(DS)^2$. In this section, we will present this language. MGS embeds a complete, impure, dynamically typed, strict, functional language.

3.1 Topological Collections

One of the distinctive features of the MGS language is its handling of entities structured by *abstract topologies* using *transformations* [5]. The notion of data structures is unified in the notion of *topological collection*, a set of entities organized by an abstract topology. Topological means here that each collection type defines a neighborhood relation inducing a notion of *subcollection*.

Topological Collection and the Representation of a DS State. Topological collections are well-fitted to represent the complex states of DS at a given time. The elements of the topological collection are the atomic elements of DS and each element has a value.

3.2 Transformations

Topological collections represent a possible framework for an extension of RS. Indeed, the neighborhood relationship provides a local view of the structural organization of elements. *Transformations* extends the notion of RS to structures other than trees and they are used to specify evolution functions of modeled DS. A *transformation* of a topological collection S consists in the *parallel application* of a set of *local rewriting rules*. A local rewriting rule r specifies the replacement of a subcollection by another one. The application of a rewriting rule $\sigma \Rightarrow f(\sigma, ...)$ to a collection S (1) selects a subcollection S_i of S whose elements match the *pattern* σ, (2) computes a new collection S'_i as a function f of S_i and its neighbors, and (3) specifies the insertion of s'_i in place of s_i into s.

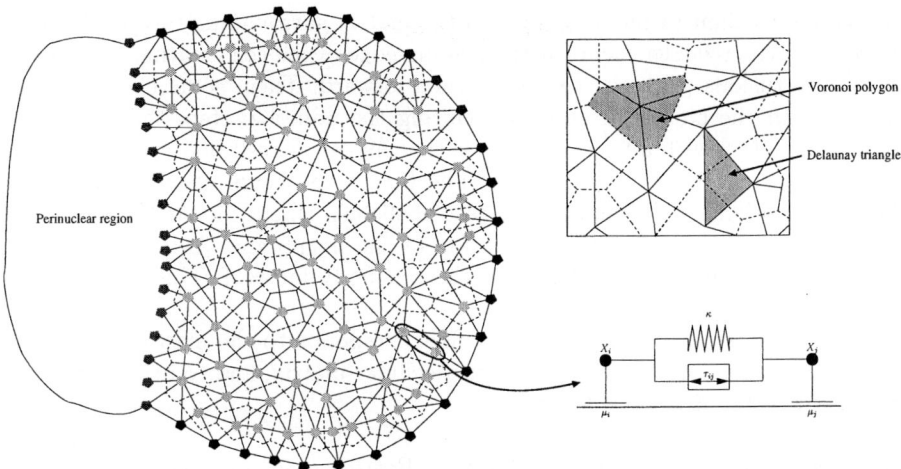

Fig. 1. The nematode sperm cell. At the left, a schematic diagram showing the cell organization: on the left, the nuclear region is found, on right is the lamellipodium. Its discretization is done by the nodes. The plain edges correspond to the Delaunay neighborhood. The dashed edges are the boundaries of the Voronoi polygons. At top right, a figure of a zoomed part of the mesh is given. At bottom right, a Delaunay edge links two nodes with a spring of modulus κ in parallel with a tensile element of stress τ. A friction of coefficient μ appears when a node is in contact with the exterior tissue (these diagrams are inspired by figures from [2])

Path Pattern. A pattern σ in the l.h.s of a rule specifies a subcollection where an interaction occurs. This subcollection can have an arbitrary shape, making it very difficult to specify. Thus, it is more convenient (and not so restrictive) to enumerate sequentially its elements. Such enumeration will be called a *path*.

Replacement. The right hand side of a rule specifies a collection that replaces the subcollection matched by the pattern in the left hand side.

4 Application to the Simulation of Nematode Sperm Crawling

In this part, we are interesting in implementing a complex biological model proposed by Bottino *et al* [2]. This model simulates the motility of the sperm cell of the nematode *Ascaris suum*. We first describe the model and its discretization in 2D. Then we see how this model can be translated in the MGS formalism.

4.1 Description of the Model

The sperm of *Ascaris suum* crawls using a lamellipodial protusion, adhesion and retraction cycle. The chemical mechanisms of motility are located in the front of the cell called *lamellipodium*. In this model, the system corresponds to the lamellipodium membrane stuck to the matrix surrounding the cell. First, a fibrous polymerization occurs at the leading edge of the cell creating protusions. These protusions push the cell membrane forward. Then, some elastic energy is stored in the created fibrous gel. During the adhesion step, the protusions stick the matrix with a traction process that makes the cell body traveling. The fibrous gel undergoes a contraction. The final step occurs near the boundary between the lamellipodium and the rest of the cell where the depolymerization of the fibrous gel causes the deadhesion of the membrane. As a consequence, the stored energy is released to pull the cell body forward. This mechanics is moderated by a pH gradient.

The considered continuous equations corresponds to the elastic and tensile stress in the membrane fixed to the extracellular matrix, and to pH distribution to deal with the pH dependence.

Mechanical Forces. The equation given by Bottino et al. governing the mechanical forces is:

$$\mu(u)\frac{\partial u}{\partial t} = \nabla.\sigma(u)$$

where u is a position vector. The l.h.s corresponds to the drag force due to the contact between the membrane and the matrix. The r.h.s computes the mechanical forces from the stress given by $\sigma =$ Elastic Stress $-$ Tensile Stress. All the coefficients depend on distribution of the pH.

pH Distribution. The pH distribution follows a diffusion equation with a leak. But, this distribution is done in a shorter time scale. Therefore, considering a quasi-static approximation, we obtain:

$$D\nabla^2[H^+] = P([H^+] - [H^+]_{\text{ext}})$$

where $[H^+]$ is the proton concentration at a given position, $[H^+]_{\text{ext}}$ is the external proton concentration, D and P are properties of the cell. The l.h.s represents the diffusion and the r.h.s is the leakage.

4.2 The Finite Element Model

This 2D surface is divided into finite elements in order to approximate the previous continuous differential equations. Each element corresponds to a node of a mesh (see figure 1). A node represents a Voronoi tessellated cell are represented by dashed edges on figure 1. The neighborhood of each Voronoi polygon is given by a Delaunay triangulation and is figured as plain edges.

There are three kinds of element: (1) the lamellipodial boundary ($Bnodes$) where the protusion occurs (in black), (2) the interface ($NRnodes$) between the lamellipodium and the cell body (in dark grey) where the retraction is done, and (3) the interior ($Inodes$) of the lamellipodium (in light grey).

Polymerization and Depolymerization. The polymerization and the depolymerization of the gel respectively correspond to the creation and the deletion of

Inodes. Two thresholds give the upper and the lower lengths of a Delaunay edge. Let X_i and X_j be two nodes and l_{ij} the length of the Delaunay edge linking X_i and X_j. If $l_{ij} > l_{\max}$, a node is created is the middle of the edge with a pH being the average between the pH at X_i and X_j. In practice, nodes are created near the boundary. On the opposite, if $l_{ij} < l_{\min}$ and X_i is a $NRnode$, X_j is deleted.

Discretization of the Continuous Equations. The previous equations are translated; for a node X_i:

$$\frac{\partial u_i}{\partial t} = \frac{1}{\mu_i} \sum_j (\kappa |X_i - X_j| - \tau_{ij}) C_2^{ij} \frac{X_j - X_i}{|X_j - X_i|} \quad \text{(mechanical forces)}$$

$$\sum_j C_1^{ij}([\text{H}^+]_i - [\text{H}^+]_j) = \frac{\text{P}}{\text{D}}([\text{H}^+]_i - [\text{H}^+]_{\text{ext}}) \quad \text{(pH distribution)}$$

In these two equations, the ∇ operators of the continuous one are replaced by a finite iteration over the neighbors X_j of the node X_i. The computation is local and well-suited to a rewriting framework. The coefficients C_k^{ij} depend on geometrical properties of the Voronoi/Delaunay triangulation (such as the area of Voronoi polygons). In the first equation, the term $\kappa |X_i - X_j|$ corresponds to the elastic force between X_i and X_j (on bottom right of figure 1). In fact, the Delaunay edges are considered as an elastic element (emulated by a spring of modulus κ) in parallel with a tensile element (with the stress τ_{ij}). The coefficient μ_i represents the drag effect.

4.3 MGS Implementation

The translation of the model in terms of transformations and topological collections is straightforward.

Data Structures. First, we have to represent a node of the Delaunay graph. We use an MGS *record* (a data structure equivalent to a C struct) composed by 8 fields:

```
record Node = { px:float, py:float, vx:float, vy:float
                H:float, pH:float, Bflag:bool, NRflag:bool };
```

Fields px and py represent the position of the node, vx and vy the speed vector, and H and pH the proton concentration and the pH. We also define three predicates Inode, Bnode and NRnode, to determine the type of the node. They use the booleans Bflag and NRflag of a node.

A Delaunay graph is a predefined type of a topological collection available in MGS. This type of collection is parameterized by a function that returns the position of an element in space to automatically compute the neighborhood. So we define this function for our example:

```
delaunay(2) D2 = fun elt ->
    if Node(elt) then (elt.px, elt.py)
                 else error("bad element type") fi ;;
```

Evolution Laws. Now that we have a representation for the data, we have to specify the evolution laws from the discrete equations. We start with checking the structure to create or delete nodes. The following MGS rules compute both polymerization and depolymerization:

```
polymerization:   Xi, Xj / (length(Xi,Xj) > lmax) =>
                  Xi, {pH=(Xi.pH+Xj.pH)/2,...}, Xj
depolymerization: Xi:Inode, Xj / (length(Xi,Xj) < lmin) => Xj
```

where function length returns the length between two nodes, and Xi:Inode specifies that the node Xi must be a Inode. As soon as these rules are applied, the Delaunay neighborhood is automatically updated.

After that, the pH distribution has to be updated to take account of the new or the deleted nodes. The equation provides the value of the proton concentration of a node as a function of the proton concentration of its neighbors:

```
trans update_pH = {
  Xi:Bnode => ...; Xi:NRnode => ...;
  Xi => let num = neighborsfold(
                    (fun Xj acc -> C1(Xi,Xj) * Xj.H + acc),
                    0, Xi) + (P/D) * H_ext
        and den = neighborsfold(
                    (fun Xj acc -> C1(Xi,Xj) + acc),
                    0, Xi) + (P/D)
        in Xi + {H = num/den, pH = -log10(num/den) }
}
```

The transformation update_pH is composed by 3 rules. The two first deal with the boundary conditions of the equation, and the last one applies the equation. The function neighborfold is used to evaluate the sum of proton concentration of the neighbors of Xi balanced by the coefficient C_1^{ij}. neighborfold corresponds to a basic fold on the sequence of the neighbors of Xi. Finally, Xi is replaced by Xi+{H = num/den, pH = -log10(num/den)} that denotes the new value of Xi where the fields H and pH are updated. To deal with the quasi-static approximation, this transformation is iterated until a fixpoint is reached. This iteration corresponds to the resolution of inverting a matrix as Bottino *et al.* do.

To end one step of the simulation, the force equation has to be computed and the velocities and positions of the nodes updated. The implementation of this transformation is quite similar to update_pH.

5 Discussion and Conclusion

The simulation developed here mimics in MGS the initial model developed by Bottino *et al.* and implemented in Matlab [6]. One of the main motivations for the development of this example, was to compare the conciseness and the expressivity of the MGS programming style compared to a more traditional programming language. Our opinion (which is subjective) is that the developed code is more concise and more readable, for instance because the management of the Voronoi tessellation and the Delaunay triangulation is completely transparent to the programmer. From the point of view of the performance, our approach is comparable (with respect to the few indications available into the articles of Bottino *et al.*) despite that the current MGS interpreter is a prototype version.

Acknowledgments

The authors would like to thank J.-L. Giavitto and J. Cohen at LaMI, D. Boussié, F. Jacquemard at INRIA/LSV-Cachan and the members of the "Simulation and Epigenesis" group at Genopole for technical support, stimulating discussions and biological motivations. This research is supported in part by the CNRS, GDR ALP, IMPG, University of Évry and Genopole/Évry.

References

1. Giavitto, J.L.: Invited talk: Topological collections, transformations and their application to the modeling and the simulation of dynamical systems. In: Rewriting Technics and Applications (RTA'03). Volume LNCS 2706 of LNCS., Valencia, Springer (2003) 208 – 233
2. Bottino, D., Mogilner, A., Roberts, T., Stewart, M., Oster, G.: How nematode sperm crawl. Journal of Cell Science **115** (2002) 367–384
3. Dershowitz, N., Jouannaud, J.P.: Rewrite systems. In: Handbook of Theoretical Computer Science. Volume B. Elsevier Science (1990) 244–320
4. Giavitto, J.L., Godin, C., Michel, O., Prusinkiewicz, P.: "Computational Models for Integrative and Developmental Biology". In: Modelling and Simulation of biological processes in the context of genomics. Hermes (2002).
5. Giavitto, J.L., Michel, O.: The topological structures of membrane computing. Fundamenta Informaticae **49** (2002) 107–129
6. Bottino, D.: *Ascaris suum* sperm model documentation (2000)

Specifying Complex Systems with Bayesian Programming. An Alife Application*

Fidel Aznar, Mar Pujol, and Ramón Rizo

Department of Computer Science and Artificial Intelligence,
University of Alicante
{fidel, mar, rizo}@dccia.ua.es

Abstract. One of the most important application areas of Artificial Life is the simulation of complex processes. This paper shows how to use Bayesian Programming to model and simulate an artificial life problem: that of a worm trying to live in a world full of poison. Any model of a real phenomenon is incomplete because there will always exist unknown, hidden variables that influence the phenomenon. To solve this problem we apply a new formalism, Bayesian programming. The proposed worm model has been used to train a population of worms using genetic algorithms. We will see the advantages of our method compared with a classical approach. Finally, we discuss the emergent behaviour patterns we observed in some of the worms and conclude by explaining the advantages of the applied method. It is this characteristic (the emergent behaviour) which makes Artificial Life particularly appropriate for the study and simulation of complex systems for which detailed analysis, using traditional methods, is practically non-viable.

Keywords: Bayesian Programming, Complex Systems Modeling, Artificial Life Formalization Model.

1 Introduction

Initially, Artificial Life was defined as a broad field of work in which attempts are made to simulate or recreate one or more natural processes using artificial methods. Nevertheless, applications in this field have quickly exceeded purely biological applications. The immediate applications of Artificial Life are in the simulation of complex processes, chemical synthesis, multivariate phenomena, etc. Very complex global behaviour patterns can be observed, initiated by simple local behaviour. It is this characteristic (sometimes called emergent behaviour) which makes Artificial Life particularly appropriate for the study and simulation of complex systems for which detailed analysis, using traditional methods, is practically non-viable. Nevertheless, it is necessary to bear in mind that any model of a real phenomenon will always be incomplete due to the permanent

---

* This work has been financed by the Generalitat Valenciana project GV04B685.

existence of unknown, hidden variables that will influence the phenomenon. The effect of these variables is malicious since they will cause the model and the phenomenon to have different behavioural patterns. In this way both artificial systems and natural systems have to solve a common problem: how each individual within the system uses an incomplete model of the environment to perceive, infer, decide and act in an efficient way.

Reasoning with incomplete information continues to be a challenge for artificial systems. Probabilistic inference and learning try to solve this problem using a formal base. A new formalism, the Bayesian programming (BP) [1], based on the principle of the Bayesian theory of probability, has been successfully used in autonomous robot programming. Bayesian programming is proposed as a solution when dealing with problems relating to uncertainty or incompleteness. Certain parallelisms exist between this kind of programming and the structure of living organisms as shown in a theoretical way in [2].

We will see a simple example of how to apply BP formalism to a specific artificial life problem. We will define a virtual world, divided into cells, some of which contain poison. In this world lives a worm with only one purpose, to grow indefinitely. In order to grow the worm must move through a certain number of non poisonous cells in its world. If the worm moves into a poisonous cell then it will die. The worm has a limited vision of the world, provided by its sensorial organs, found in its head. These sensors allow the worm to see no further than the adjacent cell. We believe that this is one of the first approaches that uses Bayesian programming for the formalization of an artificial life problem as we haven't found any evidence of it's application in this field.

2 Bayesian Programming

Using incomplete information for reasoning continues to be a challenge for artificial systems. Probabilistic inference and learning try to solve this problem using a formal base. Bayesian programming has been used successfully in autonomous robot programming [2], [1], [3], [4], [5]. Using this formalism we employ incompleteness explicitly in the model and then, the model's uncertainty chosen by the programmer, are defined explicitly too.

A Bayesian program is defined as a mean of specifying a family of probability distributions. There are two constituent components of a Bayesian program. The first is a declarative component where the user defines a description. The purpose of a description is to specify a method to compute a joint distribution. The second component is of a procedural nature and consists of using a previously defined description with a question (normally computing a probability distribution of the form $P(Searched|Known)$). Answering this question consists in deciding a value for the variable $Searched$ according to $P(Searched|Known)$ using the Bayesian inference rule:

$$P(\text{Searched}|\text{Known} \otimes \delta \otimes \pi) = \frac{1}{\Sigma} \times \sum_{\text{Unknown}} P(\text{Searched} \otimes \text{Unknown} \otimes \text{Known}|\delta \otimes \pi) \qquad (1)$$

It is well known that a general Bayesian inference is a very difficult problem, which may be practically intractable. However, for specific problems, it is assumed that the programmer would implement an inference engine in an efficient manner. More details about BP can be found in [1],[2].

3 Specifying the Problem Using Bayesian Programming

We commented above on the existence of a world, composed of $n \times m$ cells, where each cell C_{ij} could be in any one of four different states: empty, containing poison, part of the wall which surrounds this artificial world or it could be hidden from view beneath the worm's tail. In this way a cell $C_{ij} = \{\emptyset, V, M, L\}$. The wall configuration is uniform for each generated world, however, in contrast, the distribution of poison is random and varies from world to world. Initially, we asume the amount of poisonous cells to be between 5%-10% of the total. Within each world lives only a single worm which only objective is to move and to grow. A worm grows and increases its length by one unit every time it moves through d cells inside its world. If the worm moves to a cell that is not empty then it will die. The only information about the world available to the worm is provided by its sensors, located in its head (see figure 1a). A sensor is only able to see the state of cells adjacent to and in front of the worm's head, no further.

We assume that each worm has a certain knowledge represented as states. In this way each worm can stay in one state E_t given a reading and a previous state. Furthermore, a worm could obtain a reading of the world L_t represented as a binary triplet which specifies if the cell in the position of its components is occupied '1' or not '0'. Finally, a worm could execute three actions. Go straight ahead, turn left or turn right $A_t = \{u, l, r\}$ the actions will be guided only by a reading and the actual state of the worm. Once the action A_t has been executed the worm can change to a new state E_{t+1}.

The first part of a Bayesian program is to define the pertinent variables of the problem. To develop a movement in the world, the worm only needs to know the reading L_t of it's sensor and the actual state E_t, in addition to the set of actions A it could develop in the world. As we commented previously, an action A_t must be followed by an instant change in state $t + 1$. In this way we define the following variables for each instant t:

$$L_t = \{000, 001, 010, ..., 111\}, \lfloor L_t \rfloor = 8 \\ E_t = \{0, 1, 2, ..., k\}, \lfloor E_t \rfloor = k + 1 \\ A_t = \{u, l, r\}, \lfloor A \rfloor = 3 \quad (2)$$

The second part of a BP is to define a decomposition of the joint probability distribution $P(L_t \otimes E_{t-1} \otimes E_t \otimes A | \pi_W)$ as a product of simpler terms. This distribution is conditioned by the previous knowledge π_w we are defining.

$$P(L_t \otimes E_{t-1} \otimes E_t \otimes A_t | \pi_W) = P(L_t|\pi_W) \times P(E_{t-1}|L_t \otimes \pi_W) \times \\ \times P(E_t|E_{t-1} \otimes L_t \otimes \pi_W) \times P(A_t|E_t \otimes E_{t-1} \otimes L_t \otimes \pi_W) = \\ = P(L_t|\pi_W) \times P(E_{t-1}|L_t \otimes \pi_W) \times \\ \times P(E_t|E_{t-1} \otimes L_t \otimes \pi_W) \times P(A_t|E_t \otimes L_t \otimes \pi_W) \quad (3)$$

The second equality is deduced from the fact that an action only depends on the actual state and the reading taken.

Next, in order to be able to solve the joint distribution, we need to assign parametrical forms to each term appearing in the decomposition:

$$\begin{aligned} P(L_t|\pi_W) &\equiv Uniform \\ P(E_{t-1}|L_t \otimes \pi_W) &\equiv Uniform \\ P(E_t|E_{t-1} \otimes L_t \otimes \pi_W) &\equiv G\left(\mu(E_{t-1}, L_t), \sigma(E_{t-1}, L_t)\right) \\ P(A_t|E_t \otimes L_t \otimes \pi_W) &\equiv G\left(\mu(E_t, L_t), \sigma(E_t, L_t)\right) \end{aligned} \quad (4)$$

We assume that the probability of a reading is uniform because we have no prior information about the distribution of the world. In the same way we consider that all possible worm states can be reached with the same probability. Give a state E_{t-1} and a lecture L_t we believe that only one state E_t would be preferred. In this way the distribution $P(E_t|E_{t-1} \otimes L_t \otimes \pi_W)$ is unimodal. However, depending on the situation, the decision to be made may be more or less certain. This behaviour is resumed by assigning a Gaussian parametrical form to $P(E_t|E_{t-1} \otimes L_t \otimes \pi_W)$. In the same way, given a state and a reading we suppose that an action with more or less intensity would be prepared.

We show a set of free parameters which define the way the worm moves. These free parameters (that must be identified), derived from the parametrical form (means and standard deviations of all the Gaussians $\lfloor E_{t-1} \rfloor \times \lfloor L_t \rfloor$ and $\lfloor E_t \rfloor \times \lfloor L_t \rfloor$), would be the ones to be learned.

Finally we specify the steps that the worm needs to move (using the joint distribution): to obtain a reading L_t from the worm's sensors, to answer the question $Draw(P(A_t|E_t \otimes L_t \otimes \pi_W))$, then the worm will execute the movement command A and will answer the question $Draw(P(E_{t+1}|E_t \otimes L_T \otimes \pi_W))$, finally the worm will change to the state E_{t+1}.

4 Genetic Algorithms

Genetic algorithms (GA) are a global search technique which mimic aspects of biological evolution. We initially assume that the worm's parameters are generated randomly. The worm only has previous knowledge provided by its knowledge decomposition. The learning process would be produced generation after generation, where the longest living worms in the world would be those most enabled and adapted to reproduce and to maintain their *intelligence*. Next we will describe the main parts of our genetic algorithm.

Chromosome Codification. A chromosome is represented using two tables. The first one is formed by $2 \cdot k \cdot 8$ components specifying the Gaussians $\lfloor E_{t-1} \rfloor \times \lfloor L_t \rfloor$ which represent $P(E_t|E_{t-1} \otimes L_t \otimes \pi_W)$. The second table is formed by the same component numbers specifying the Gaussians $\lfloor E_t \rfloor \times \lfloor L_t \rfloor$ which represent $P(A_t | E_t \otimes L_t \otimes \pi_W)$. In this way, each chromosome contains $32 \cdot k$ gens. In the described experiments, the initial chromosome population is obtained by randomly initializing the Gaussian parameters.

Fitness Function. We want to reward the worms that live the longest time in the world. In this way we describe the fitness function as the number of iterations that a worm lives in a randomized generated world. In order to avoid the situation where a simple world produces an overvalued worm, we generate w random worlds to evaluate each worm's fitness. All worlds are the same size and have the same wall disposition, only the quantity and position of poisonous cells varies, being selected randomly and comprising between 5% and 10% of the total cells.

Selection, Crossover an Mutation Operators.
- Selection operator. We used a stochastic remainder sampling selector (SRS) with a two-staged selection procedure. In addition we use elitism (the best individual from each generation is carried over to the next generation).
- Crossover operator. We use an asexual two-point crossover operator. In this way the mother genes will be selected until the crossover point where the father genes will be copied. This process will be done for the two tables (see figure 1b) that describe the chromosome.
- Mutation operator. We define an incremental mutation operator for states, in this way given a gene x we define a mutation as: $x \in [0, k]$, $mut(x) = x + 1 \, MOD \, k$. Suppose we have four states, and that $q_2 = 3$, if we mutate this element we will obtain $q_2 = 3 + 1 \, MOD \, 4 = 0$. A random mutation scheme is used to choose the directions for the worm to take. A new direction is generated randomly and then substitutes the original gene.

4.1 Used Parameters

Using the operators presented in the previous section we obtain an evolutive learning process for a worm. Developing empirical tests we arrive at the conclusion that a number of states (k) greater than five complicates the learning process of the worm and does not improve the movements made by the worm. For this reason, in the rest of the experiments, we use a fixed number of states equal to five.In addition, for the remainder of tests we use $d = 5$ (for each five cells the worm moves through it will increase it's size by one unit) and $w = 6$ (six random worlds will be generated in order to evaluate each worm).

4.2 Worms Evolution

In order to obtain the best individual we use 100 executions for each state with a population of 250 individuals and 500 generations using the operators specified in the previous section. In figure 1c an example is shown of the executions showing the fitness of the worst and best performing individual as well as the average results obtained, illustrating the algorithm convergence. For each algorithm execution the evaluation took about 2 minutes using a Pentium IV running at 2Ghz.

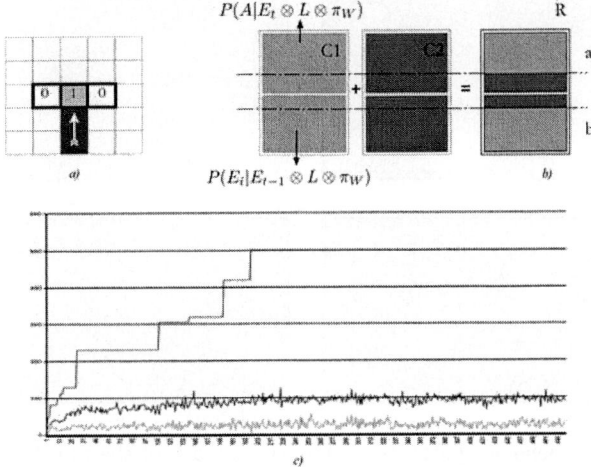

Fig. 1. a) Worm's vision relating to its head and the direction it has in the world. b) Asexual two point crossover operator for the worm's chromosome. c) Evolution of the worst (bottom line), the average (in the middle) and the best performing individual (top). The y axis represents the worm's fitness and the x axis the actual generation. Until the first 50 iterations an improvement is produced in the medium and the worst individual. Then the graph tends to oscillate although a slight increase is produced (because the best case increases and maintains it's level through elitism)

5 Survival Behaviours and Experimentation

In this section we will analyze some characteristics and emergent behaviours that were observed in the worms. Readers of this paper are invited to test our simulator at the following address http://www.dccia.ua.es/~fidel/worm.zip. Using Bayesian Programming the worm's previous knowledge is defined and mechanisms are given to provide new knowledge to the worm. This data is represented using two sets of discrete Gaussians which were learned using genetic algorithms. However, we should remember that to get the information of the learned distributions we use the *Draw* function which randomly extracts a value for the distribution. In this way we obtain a non-deterministic behaviour, which is more adaptable to variations in complex worlds. After training the worm population we simulate, in a graphical way, the best individual found. It is curious to see different behaviour patterns, which provide more survival opportunities. Some of these patterns even seem to imitate natural behaviour developed in some animals.

One of the most common patterns is to follow the edge of the world while no poison is found near it (see figure 2a). This is a good way to move if the proportion of poison is low near the edges and configurations don't exist that trap the worm between the perimeters and the poison. Another curious behaviour is the development of a zigzag movement emulating the way some snakes move (see figure 2b) so reducing the area that the worm occupies in the world. In addition it is quite common for the worm to move up and down like a ping-pong ball

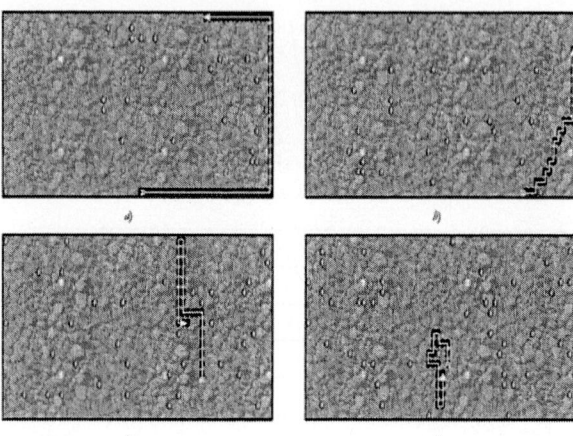

Fig. 2. Different behavior patterns. *a*) Follow the edge of the world. *b*) Zigzag movement. *c*) *Ping-pong* behaviour. *d*) In the movement the worm seems to follow its tail. The arrow points to the next worm displacement

(see figure 2c). Finally, we underline the movement of some worms which seem to move as if trying to reach their tails, so forming a spiral (see figure 2d). The behaviour described above (and some others) are repeated and combined with the obtained worms. These behaviour are not programmed implicitly, they have been obtained using the proposed *brain* model and selected using an evolving process in a population.

5.1 Comparing Our Model with a Clasical One

We can see some advantages of our method if we compare it to a more classical model approach, the finite states machine (FSM) [6]. This approach has various drawbacks. First, it assumes a perfect world model, which is false. It is necessary to know that any model of a real phenomenon is incomplete because there will always exist non-considered, hidden variables, that will influence the phenomenon. The effect of these variables is malicious since they will cause the model and the phenomenon to show different behavioural patterns. Second, a FSM develop a deterministic behaviour therefore in certain world configurations it will fail. On the other hand, a Bayesian model has not a deterministic behaviour and two different executions in the same world may have different results, which provide greater adaptability to changes in the environment configuration.

6 Conclusions

In this paper we have seen an application of Bayesian programming in an artificial life system. The formalism of the artificial life models is a continuous field of investigation because of the complexity of the systems we work with

[7],[8]. In addition, we have the added difficulty of working with uncertainty and include it into the model we want to use. The Bayesian programming brings up a formalism where implicitly, using probabilities, we work with the uncertainly.

In a world with randomly distributed poison lives a worm, which main purpose is to grow up. We propose a formalization of the virtual worm using a decomposition in terms of a joint probability distribution of their knowledge. In this way, applying the Bayesian Programming we obtain a versatile behaviour adaptable to changes and what is more a mathematical description of the probabilistic environment model. We have seen some advantages of our method comparing it to a more classical model approach, the finite states machine (FSM [6]) (see section 5.1).The learning process, given a worm population, has been developed with evolving techniques, using genetic algorithms. The principal reason for using GA was because they are a global search technique which mimic aspects of biological evolution even though other search techniques could be picked to select the worms. Each used chromosome is the codification of the two distributions obtained with the previous Bayesian formalism. Satisfactory results were obtained that prove the validity of the proposed model. Relatively complex and elaborate behavioural patterns were observed in the movements of the most highly adapted worms. These behaviour patterns were not implicitly programmed but were obtained in an emergent way using the proposed model.

Bayesian programming is, therefore, a promising way to formalize both artificial and natural system models. In this example, we have seen how this paradigm can be adapted to a simple, artificial life problem. Future studies will try to model different artificial life systems using this new formalism.

References

1. Lebeltel, O., Bessire, P., Diard, J., Mazer, E.: Bayesian robots programming. Autonomous Robots **16** (2004) 49–79
2. Bessire, P., Group, I.R.: Survei:probabilistic methodology and tecniques for artefact conception and development. INRIA (2003)
3. Koike, C., Pradalier, C., Bessiere, P., Mazer, E.: Proscriptive bayesian programming application for collision avoidance. Proc. of the IEEE-RSJ Int. Conf. on Intelligent Robots and Systems (IROS); Las Vegas, USA (2003)
4. C. Cou, Th. Fraichard, P.B., Mazer, E.: Using bayesian programming for multi-sensor data fusion in automotive applications. IEEE Intelligent Vehicle Symposium (2002)
5. Bellot, D., Siegwart, R., Bessire, P., Cou, C., Tapus, A., Diard, J.: Bayesian reasoning for real world robotics: Basics, scaling and examples. Book Chapter in LNCS/LNAI, http://128.32.135.2/users/bellot/files/David_Bellot_LNCS_LNAI.pdf (2004)
6. Dysband, E.: Game Programming Gems. A finite-state machine class (237-248). Charles River Media (2000)

7. Agre, P., Horswill, I.: Lifeworld analysis. Journal of Artificial Intelligence Research **6** (1997) 111–145
8. Rasmussen, S., L. Barrett, C.: Elements of a theory of simulation. ECAL1995 (Advances in Artificial Life, Third European Conference on Artificial Life, Granada, Spain). (1995) 515–529

Optimization Embedded in Simulation on Models Type System Dynamics – Some Case Study

Elżbieta Kasperska and Damian Słota

Institute of Mathematics, Silesian University of Technology, Kaszubska 23,
44-100 Gliwice, Poland
{e.kasperska, d.slota}@polsl.pl

Abstract. The problem of optimization embedded in simulation on models type System Dynamics is rather new for field modelers. In model DYNBALANCE(2-2-c) authors link the algorithm of solving the pseudosolution of overdetermined system of equations to classical structure of type System Dynamics – to minimize the specific Euclidean norm which measures some aspect of the dynamical behaviour of system.

1 Introduction

The classical concept of system Dynamics [1,2,3,4,10,11] assumes, that, during the time horizon of the simulation, the structure (given a-priori) will remain constant. Last couple of years, some ideas of structural evolution has occurred in System Dynamics modelling. First prof. Coyle [1] took the problem of, so called "simulation during optimization". The question was: how to "optimize" the structure in order to achieve the desired behaviour? Kasperska and Słota have gone the opposite way. The idea was: how to embedding the optimization in simulation on System Dynamics models? In work [7], the attention has been focused on two methods of embedding the optimization in simulation. One of them took advantage from linear programming. However the second considered the idea of Legras (for pseudosolution of overdetermined system). Such "hybrid" ideas has enriched the possibilities of classical System Dynamics, specially in context of measuring the interesting aspect of dynamical behaviour of system. Now authors present extended version of model DYNBALANCE(2-2) (named: DYNBALANCE(2-2-c)) and some results of experiments undertaken by applying language Professional Dynamo 4.0.

2 Mathematical Model of the System

On Figure 1 the reader can study the main structure of the model, in Lukaszewicz symbols [10]. Below the meanings of matrixes: A, x and b can be recognize in presented matrix equation $Ax = b$:

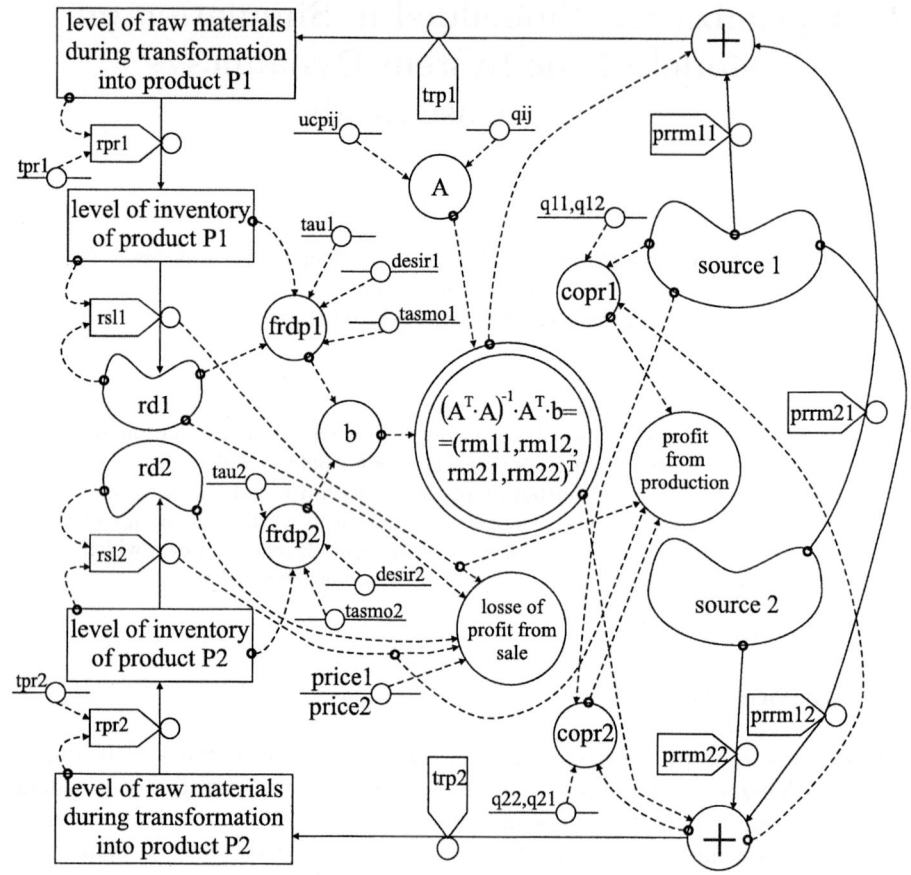

Fig. 1. Structure of model DYNBALANCE(2-2-c)

$$\begin{pmatrix} q11 & 0 & q12 & 0 \\ 0 & q21 & 0 & q22 \\ 1 & 1 & 0 & 0 \\ 0 & 0 & 1 & 1 \\ ucp11 \cdot q11 & 0 & ucp21 \cdot q12 & 0 \\ 0 & ucp12 \cdot q21 & 0 & ucp22 \cdot q22 \\ 1 & 0 & 0 & 0 \\ 0 & 1 & 0 & 0 \\ 0 & 0 & 1 & 0 \\ 0 & 0 & 0 & 1 \end{pmatrix} \begin{pmatrix} rm11 \\ rm12 \\ rm21 \\ rm22 \end{pmatrix} = \begin{pmatrix} frdp1(t) \\ frdp2(t) \\ sourc1(t) \\ sourc2(t) \\ b5 \\ b6 \\ b7 \\ b8 \\ b9 \\ b10 \end{pmatrix} \quad . \quad (1)$$

The idea of solving the matrix equation (1) uses the method of Legras [9] by finding the so called "pseudosolution", which minimize the norm of overdetermined equation (1). This is the Euclidean norm, so it is the square root of sum of squares of discrepancies $(Ax - b)_i$, for $i = 1, 2, \ldots, 10$. So, the found solution is that give the best "fitting" of balance (modelling by equation (1)).

Technically speaking, the solving of matrix equation (1) is the problem of programming specific works on matrixes, in order to attain x:

$$x = \begin{pmatrix} rm11 \\ rm12 \\ rm21 \\ rm22 \end{pmatrix} = \left(A^T \cdot A\right)^{-1} \cdot A^T \cdot b \ .$$

The formula was examined by authors and for given number values we have gotten results, that will be described in next section.

3 The Assumptions for Experiments

The table 1 presents in synthetical form the elements of scenario of simulations experiments.

Table 1. The main assumptions for experiments on model DYNBALANCE(2-2-c) (where $\mathbf{1}(t)$ is the unit step function, equal to 0 for $t < 0$ and 1 for $t \geq 0$)

No.	rd1	rd2
1	Sinusoidal: $rd1(t) = 100 + 30\sin(\pi t/52)$	Sinusoidal: $rd2(t) = 100 + 30\sin(\pi t/52)$
2	Step function (increase): $rd1(t) = 100 + 300\,\mathbf{1}(t-10)$	Sinusoidal: $rd2(t) = 100 + 30\sin(\pi t/52)$
3	Step function (decrease): $rd1(t) = 100 - 100\,\mathbf{1}(t-10)$	Sinusoidal: $rd2(t) = 100 + 30\sin(\pi t/52)$
4	Linear decrease: $rd1(t) = \begin{cases} 1000 & t < 10 \\ 1100 - 10t & t \geq 10 \end{cases}$	Sinusoidal: $rd2(t) = 100 + 30\sin(\pi t/52)$
5	Step function (increase): $rd1(t) = 100 + 300\,\mathbf{1}(t-10)$	Step function (decrease): $rd2(t) = 100 - 100\,\mathbf{1}(t-10)$

Moreover the values of chosen parameters are as follows:

- prices: $price1 = 300$, $price2 = 350$;
- sources of raw materials: $sourc1 = 20$, $sourc2 = 20$;
- initial levels of inventory of product $P1$ and $P2$: $lin1 = 350$, $lin2 = 350$;
- initial levels of raw materials during transformation into product $P1$ and $P2$: $lmt1 = 200$, $lmt2 = 400$;
- parameters of technology: $q11 = 1$, $q12 = 2$, $q21 = 1$, $q22 = 2$.

The possibilities of experimentations on model DYNBALANCE(2-2-c) are practically unlimited. For example: the prices can become random as sources of raw materials too. Below, in the next section, the chosen results of experiments 1–5 will be presented.

4 The Results of Experiments

On Figures 2–6 the main characteristic of some variables from the model DYNBALANCE(2-2-c) are presented. The Reader, who will be interested in details, has possibilities to compare such important (in sense of "quality" of fitting the balance: $Ax = b$, and in sense of dynamics of whole system) variables, like: *norm*, *lin*1, *lin*2, *prof*1, *prof*2 and the others.

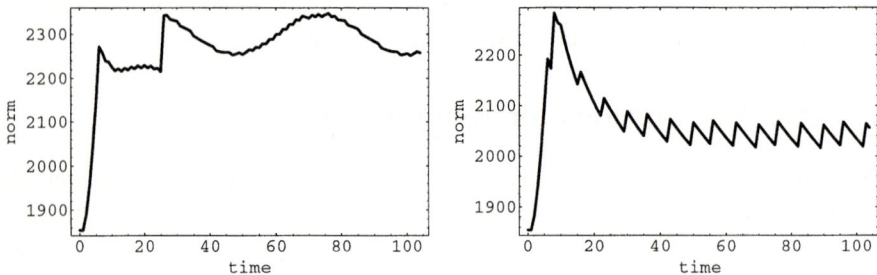

Fig. 2. The characteristic of variable *norm* (the "fitting of balance $Ax = b$ in whole horizon of simulation; left figure – experiment number 1, right figure – experiment number 5)

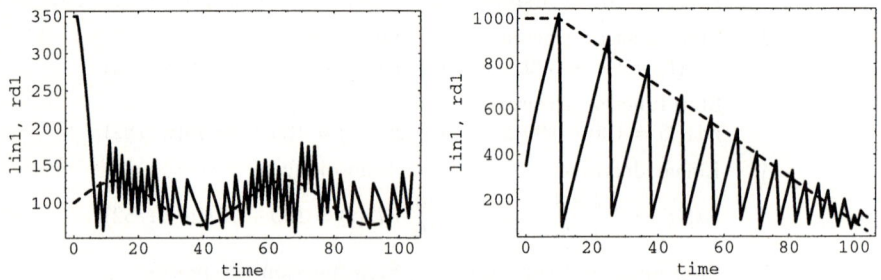

Fig. 3. The characteristic of level of inventory of product $P1$ (*lin*1) and rate of demand $rd1$ (left figure – experiment number 1, right figure – experiment number 4; solid line – *lin*1, dash line – $rd1$)

Under the different condition about the demands for product $P1$ and $P2$ system reacts in such a way that minimize the Euclidean norm. So, the rates of raw materials: $rm11$, $rm12$ and $rm22$, are optimal to fit the balance $Ax = b$ (see Figure 1) and the "quality" of fitting, variable *norm* has optimal value in each step of simulation (in whole horizon two years). So, the embedding optimization in simulation, has dynamical character and is "compatible" with idea of dynamical modelling in Forrester's way. Precisely speaking, it's some extension of System Dynamics method possibilities.

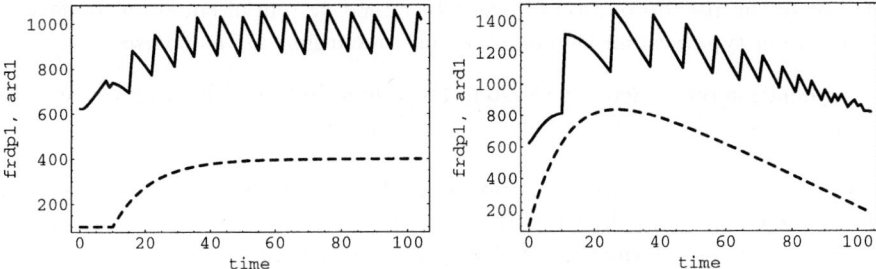

Fig. 4. The characteristic of plan of production of product $P1$ ($frdp1$) and averaging rate of demand $ard1$ (left figure – experiment number 2, right figure – experiment number 4; solid line – $frdp1$, dash line – $ard1$)

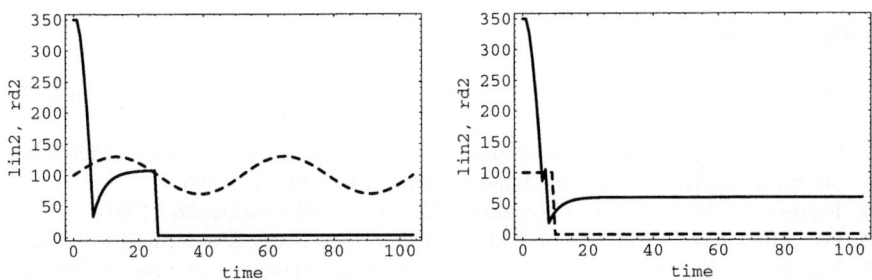

Fig. 5. The characteristic of level of inventory of product $P2$ ($lin2$) and rate of demand $rd2$ (left figure – experiment number 3, right figure – experiment number 5; solid line – $lin2$, dash line – $rd2$)

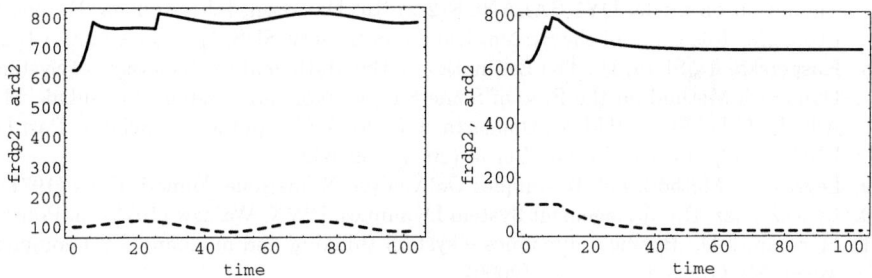

Fig. 6. The characteristic of plan of production of product $P2$ ($frdp2$) and averaging rate of demand $ard2$ (left figure – experiment number 4, right figure – experiment number 5; solid line – $frdp2$, dash line – $ard2$)

5 Final Remarks and Conclusions

The purpose of the paper was to present some results of experiments undertaken on model DYNBALANCE(2-2-c). Such experiments consider the problem of optimization embedded in simulation. The authors have linked the algorithm of solving the pseudosolution of overdetermined system of equation to classical structure of type System Dynamics. Final conclusions are as follows:

- simulation on models type System Dynamics, linked with optimization technique, is a good tool for analysing the dynamical, complex, nonlinear, multilevel system;
- the locally optimal decisions can be confront with dynamic of a system like a whole – its give the supportive tool for decision-makers and for the projectors of the information- decision systems as well;
- if seems that such "hybrid" ideas, like presented in paper, has a future – that enrich the possibilities of famous Forrester's method and makes them adequate to different situations in real systems.

References

1. Coyle, R.G.: System Dynamics Modelling. A Practical Approach. Chapman & Hall, London (1996)
2. Coyle, R.G.: The practice of System Dynamics: milestones, lessons and ideas from 30 years experience. System Dynamics Rev. **14** (1998) 343–365
3. Forrester, J.W.: Industrial Dynamics. MIT Press, Massachusetts (1961)
4. Forrester, J.W.: Principles of Systems. Cambridge Press, Massachusetts (1972)
5. Kasperska, E., Mateja-Losa, E., Słota, D.: Some extension of System Dynamics method – practical aspects. In: Deville, M., Owens, R. (eds.): Proc. 16th IMACS World Congress. IMACS, Lausanne (2000) 718-11 1–6
6. Kasperska, E., Słota, D.: Mathematical Method in the Management in Conceiving of System Dynamics. Silesian Technical University, Gliwice (2000) (in Polish)
7. Kasperska, E., Słota, D.: Two different methods of embedding the optimization in simulation on model DYNBALANCE(2-2). In: Davidsen, P.I., Mollona, E. (eds.): Proc. 21st Int. Conf. of the System Dynamics Society. SDS, New York (2003) 1–23
8. Kasperska, E., Słota, D.: The Estimation of the Mathematical Exactness of System Dynamics Method on the Base of Some Simple Economic System. In: Bubak, M., Albada, G.D., Sloot, P.M.A., Dongarra, J.J. (eds.): Computational Science, Part II. LNCS 3037, Springer-Verlag, Berlin (2004) 639–642
9. Legras, J.: Methodes et Techniques De'Analyse Numerique. Dunod, Paris (1971)
10. Łukaszewicz, R.: Management System Dynamics. PWN, Warsaw (1975) (in Polish)
11. Sterman, J.D.: Business dynamics – system thinking and modeling for a complex world. Mc Graw-Hill, Boston (2000)

A High-Level Petri Net Based Decision Support System for Real-Time Scheduling and Control of Flexible Manufacturing Systems: An Object-Oriented Approach

Gonca Tuncel and Gunhan Mirac Bayhan

Department of Industrial Engineering,
University of Dokuz Eylul,
35100 Bornova-Izmir, TURKEY
{gonca.tuncel, mirac.bayhan}@deu.edu.tr

Abstract. Petri nets (PNs) are powerful tools for modeling and analysis of discrete event dynamic systems. The graphical nature and mathematical foundation make Petri net based methods appealing for a wide variety of areas including real-time scheduling and control of flexible manufacturing systems (FMSs). This study attempts to propose an object-oriented approach for modeling and analysis of shop floor scheduling and control problem in FMSs using high-level PNs. In this approach, firstly, object modeling diagram of the system is constructed and a heuristic rule-base is developed to solve the resource contention problem, then the dynamic behavior of the system is formulated by high-level PNs. The methodology is illustrated by an FMS.

1 Introduction

Due to high complexity of flexible manufacturing systems, some mathematical programming related methods such as integer programming, linear programming, and dynamic programming often lack to describe the practical constraints of complex scheduling problems. On the other hand, classical scheduling techniques such as branch-and bound and neighborhood search techniques cause substantial increase in state enumeration, and thus computation time grows exponentially as the problem size increases. Recently, either existing techniques are improved or new scheduling approaches are developed to deal with the practical constraints of real-world scheduling problems. These methods include simulation-based approaches, expert systems, PNs based methods, AI-based search techniques, and hybrid methods [1]. Although the first applications of PNs were in the field of communication protocols and computer systems, they have been extended and applied to the broader range of problems including the scheduling of production systems by means of their modeling capabilities and formulation advantages. A significant advantage of PNs is its representation capability. PNs can be explicitly and concisely model concurrent, asynchronous activities, multi-layer resource sharing, routing flexibility, limited buffers, and precedence constraints [2], [3]. However, modeling and analysis of complex or large size systems require too much effort, since considerable number of states and transition requirements cause state explosion problem. Furthermore, PNs are system dependent and lack some features like modularity, maintability, and reusability, which are the properties of Object-Oriented approach [4]. Therefore, recently, there has been a growing interest in merging PNs and Object-oriented approach to combine graphical representation and mathematical foundation of PNs with the modularity, reusability, and maintability features of Object-orientation. Here, we present a PN based Decision Support System (DSS) for real-time shop floor scheduling and control of FMSs. High-level PNs and Object Oriented Design (OOD) approach were used for system modeling, and a heuristic rule (knowledge) based approach was employed for scheduling /dispatching in control logic. The presented DSS helps managers to take control decisions effectively and efficiently by considering the current status of the shop floor. The proposed methodology is illustrated on an example FMS. Many researchers have employed PNs for scheduling and control of manufacturing systems [5], [6], [7], [8]. However, Petri-net based dynamic scheduling of FMSs dealing with production of several product types with flexible routes, setup times, material handling system, and operator constraints has not been given much attention. In Section 2, modeling methodology is given. The paper is concluded by Section 3.

2 Modeling Methodology for Real-Time Shop Floor Scheduling and Control Problem of FMSs

FMSs are characterized by concurrency, resource sharing, routing flexibility, limited buffer sizes, and variable lot sizes. For interaction activities, and the coordination of individual units, a descriptive and dynamic modeling tool is required to model in detail the concurrency, and synchronization with respect to time [1]. The units in this system can be considered as objects interacting with each other, and they can be added, removed or modified. Therefore, OOD concepts can be employed to design of shop floor scheduling and control systems in FMSs, and PNs can be used to model dynamic behavior of the system. The properties and behavior of the objects were modeled by the data/attribute and methods /operations. OOD methodology used for development of the DSS is similar to those given in [9], [10].

2.1 Illustration of the Methodology with an FMS with Alternative Operations and Setup Times

An FMS with alternative operations and setup times is employed to explain the modeling methodology of the proposed decision support system for real-time shop floor scheduling and control problem. The physical layout of the system considered is shown in Figure 1. The system consists of a loading and unloading

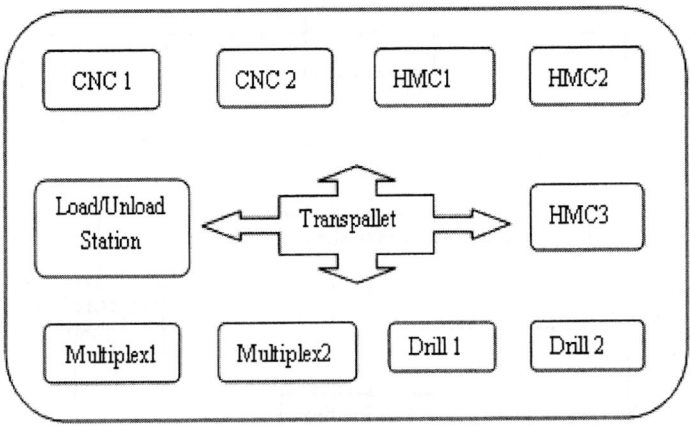

Fig. 1. Layout of the FMS

station, 9 workstations (CNC1, CNC2, Multiplex1, Multiplex 2, HMC1, HMC2, HMC3, Drill 1, and Drill 2), a material-handling system, and several operators. Each workstation has one machine with input and output buffers with limited capacity. Each machine can only process a part at a time, and once a part is processed the machine setup is needed to process a different part type. Each workstation is allowed to contain two pallets at most at a time. The FMS is designed to produce a variety of products simultaneously. Each product has alternative routes for some operations (i.e. two or more machines are enable to perform same operations). The operational policy is under push paradigm. Each job has a best operation sequence that determines the order in which resources must be assigned to the job. But the efficient utilization of resources on real time basis requires a real time resource allocation policy to assign resources to jobs as they advance through the system. Since each machine is capable of processing a variety of part types and each part has to visit a number of machines, there is often a conflict when more than one part is contending for the same machine or material handling system. The problem is the allocation of resources to a set of tasks, that is, determination of the best route of each task in the system according to the current shop-floor condition (due dates, release dates, order quantities, tardiness penalties, inventory levels, and setup status).

Object Modeling of the System. The object modeling technique (OMT) that is the most widely used OOD methodology is employed to describe and analyze the object classes. It divides the system considered into object classes, and it is used to model the static relations among FMS objects by the class structure. Each individual FMS object is derived as an instance of CPN class module. The behavior of the objects is described by operations associated with object classes. Figure 2 shows the OMT diagram corresponding to the DSS. This diagram captures the relevant properties of the FMS objects and their functions

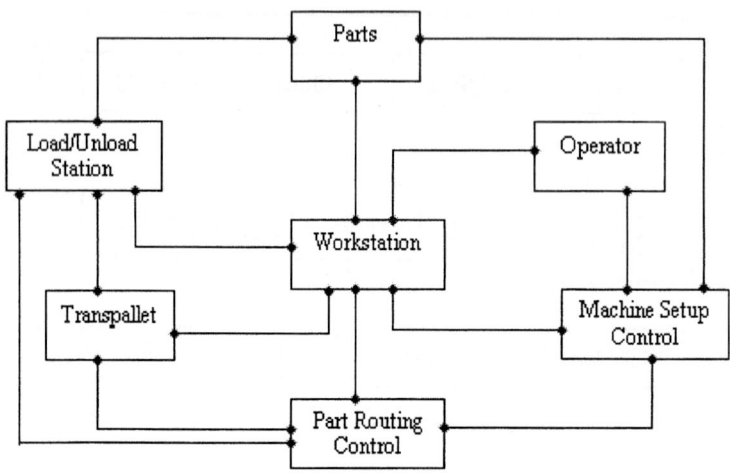

Fig. 2. OMT Diagram of the FMS

in the DSS. The links between objects with black dots at each end represent the associations. Individual classes of FMS objects are defined as follows:

Load and Unload Station Class: raw parts with the attributes such as products type, due date, and process plan, are set up on pallets and stored to the load storage buffer, and a route request is sent to the Scheduler Class for the pallet stored. When information on next destination for a pallet is received from the Scheduler Class, the pallet is moved to the active buffer, loaded to the transpallet, and sent to the destination workstation. If all operations of the parts on a pallet are completed, the pallet returns to the load/unload (L/UL) station and it is transferred to unload storage buffer. Then the finished products are unloaded from the pallet. Thus, pallets become available and are sent to the free pallet buffer to be used again.

Workstation Class: parts are processed at machining stations. Once a transpallet arrives at a workstation, the pallet is moved to the input buffer of the workstation and the transpallet becomes free. When machining station is idle, the pallet goes from the input buffer to the machining buffer, and one operator request is sent to the Operator Object. Machine setup control is also performed in WS class. If the machine operates the same part type with the previous part type, it doesn't need a setup operation. When all the parts on the pallet are processed, the pallet is sent to the output buffer of the workstation, and a routing request is sent to the Part Routing Control object. When a routing request is replied for a pallet, it is transferred to the active buffer, and then sent to the destination station by transpallet. Part transport class: parts are transported by transpallet between stations. The Part Transport Object accepts a transport request from the Scheduler Class, and forwards a transpallet to the station, which required it. Thus ready pallet is loaded to the transpallet, and transferred to the destination station.

Operator class: machine setup operation, loading and unloading operations are performed by operators. Operators working at workstations are represented by Operator Class.

Scheduler class: part flow between stations are controlled and managed by Scheduler Class. Part Routing Control Object is an instance of Scheduler Class. In a manufacturing system, concurrent flow of parts competing for sharing limited resources causes resource contention problem. Resource sharing in any manufacturing system often leads to conflicts. The system controller must be capable of resolving these conflicts effectively. From the standpoint of the PNs, resource sharing increases the complexity in scheduling. The proposed DSS use a heuristic rule based approach to solve resource contention problems and to determine the best route(s) of the parts, which have routing flexibility. The following heuristics are used to solve the resource contention problem:

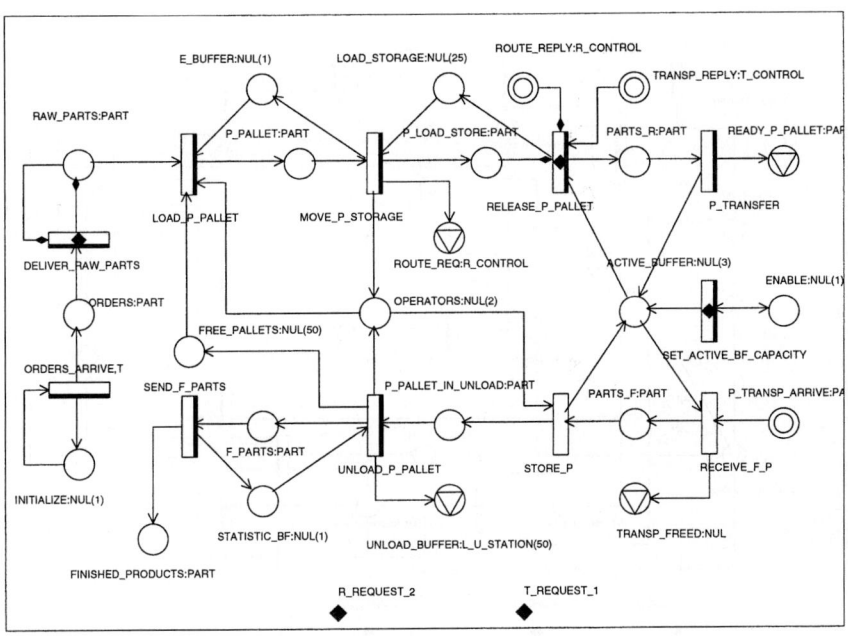

Fig. 3. CPN Model of Load / Unload Station Class

- Finished products have priority for transportation.
- Semi-finished or raw parts are ranked according to their next operation number (highest is first), or due dates (lowest is first) for the parts with the identical operation number.
- Alternative machines are ranked according to their operation time (shortest is first), and the pallet is routed to the available machine among the alternative machines which is ranked first and doesn't require setup operation.

- If there is no available machine, which doesn't need setup operation, the route request is accepted regarding the order's critical ratio.
- After all the route requests are checked under the current system status, the route requests, which are not critical, are reevaluated to be routed.

CPN Models of FMS Objects. Dynamic behavior/control logic of the FMS is formulated by constructing high-level PN model of the system based on the static relations of the OMT model. For this purpose, each CPN class model is first constructed to capture concurrency and synchronization of the system. Internal places and transitions are used to model the operations and dynamic behavior inside the Object Class, and input and output places are used to model the interface of the objects. Then all the related PN models of classes are connected through the input and output places to obtain the complete model and the control logic of the system. In Figures through 3-7, the CPN class models are displayed.

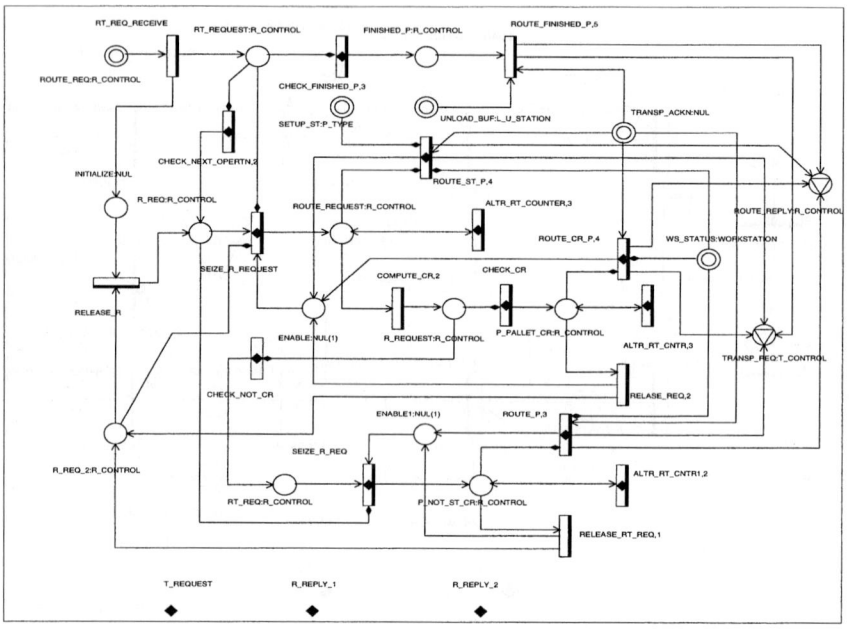

Fig. 4. CPN Model of Scheduler Class

Since the standard PNs are insufficient to model the complex and large size systems, they have been extended to High-level PNs, which allow arbitrary complex data types for tokens. Thus, transitions and places can be constructed by using a special programming language. In this study, Artifex which is a modeling and simulation environment supporting the design of discrete event systems is employed to model CPN classes by using a high level language C/C++.

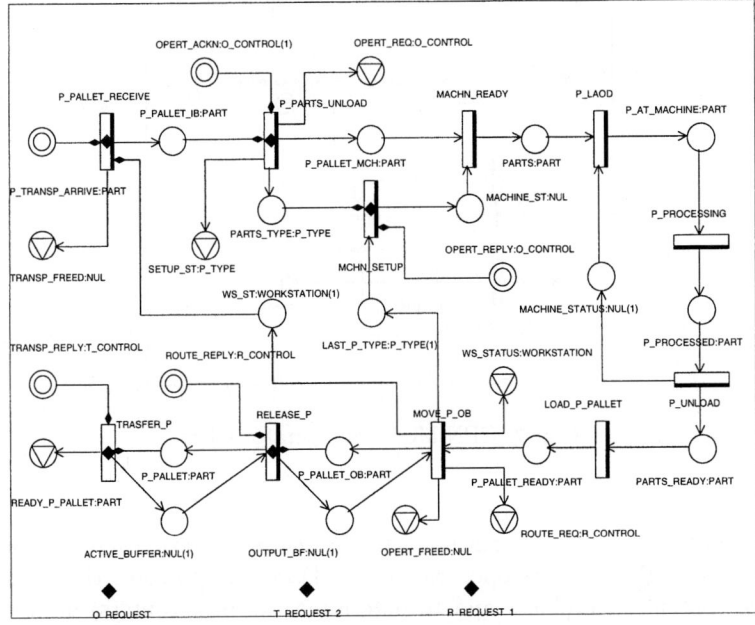

Fig. 5. CPN Model of Workstation Class

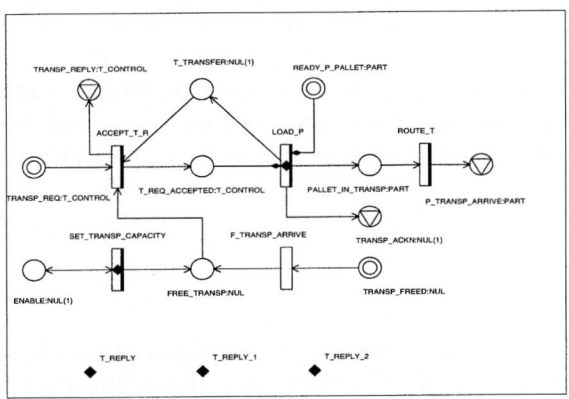

Fig. 6. CPN Model of Part Transport Class

3 Conclusion

In recent years, there has been a growing interest in applying PN theory for scheduling of production systems by means of their modeling capabilities and formulating advantages. In this study, a PN based DSS for shop floor schedul-

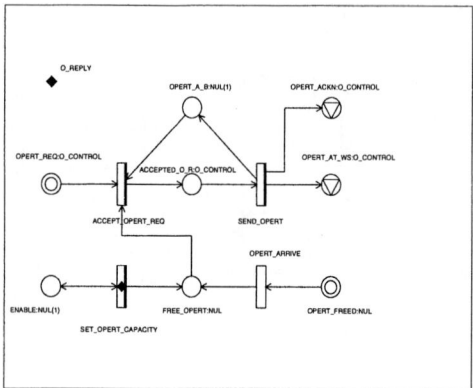

Fig. 7. CPN Model of Operator Class

ing and control of FMSs is presented. In the modeling process, object-oriented approach is used. The developed rule-based DSS aims to solve the problem of resource contention problem and to determine the best route(s) of the parts, which have routing flexibility. Decisions are taken on real-time basis by checking product types, due dates, alternative process plans, next possible destination resource, setup status, and resource utilizations rates. The DSS takes a global view of the system state before making decision about resource assignment, and proposes a dynamic approach to solve the conflict problems. It has adaptive and autonomous ability for obtaining intelligent control, and can be used for different production systems by only changing some system parameters (i.e. number of operators and stations, types of products, and process plan information) in Object Classes. Performance analysis can be performed under different system configurations.

References

1. Lin, J.T., Lee, C.C.: Petri net-based integrated control and scheduling scheme for flexible manufacturing cells. Computer Integ. Manuf.Systems, 10 (1997)109-122.
2. Murata, T.: Petri Nets: Properties, Analysis and Applications. Proceedings of the IEEE, vol.77/4 (1989) 541-580.
3. Zurawski, R., Zhou, M.C.: Petri Nets and Industrial Applications: A Tutorial. IEEE Transactions on Industrial Electronics, 4 (1994) 567-582.
4. Wang, L.: Object-oriented Petri nets for modelling and analysis of automated manufacturing systems. Computer Integ. Manuf. Systems, 26 (1996) 111-125.
5. Hatano, I., Yamagata, K., Tamura, H.: Modeling and on-line scheduling of flexible manufacturing systems using stochastic Petri nets. IEEE Trans. Software Eng., 17 (1991) 126-133.
6. Lee, D.Y., DiCesare, F.: Scheduling flexible manufacturing systems using Petri nets and heuristic search. IEEE Trans. Robot. Automat., 10 (1994) 123-133.

7. Chetty, O.V.K., Gnanasekaran, O.C.: Modelling, simulation and scheduling of flexible assembly systems with coloured petri nets. Int. J. Adv. Manufacturing Technology, 11 (1996) 430-438.
8. Jain, P.K.: Solving resource contention problem in FMS using Petri nets and a rule-based approach. Int. J. of Production Research, 39 (2001) 785-808.
9. Chen, J., Chen, F.F.: Performance modelling and evaluation of dynamic tool allocation in flexible manufacturing systems using coloured Petri nets: An object-oriented approach. Int. J. of Adv. Manufacturing Technology, 21 (2003) 98-109.
10. Venkatesh, K., Zhou, M.C.: Object-oriented design of FMS control software based on object modeling techniques diagrams and Petri nets. J. of Manufacturing Systems, 17 (1998) 118-136.

Mesoscopic Simulation for Self-organization in Surface Processes

David J. Horntrop

New Jersey Institute of Technology,
Department of Mathematical Sciences,
Newark, NJ 07102 USA

Abstract. The self-organization of particles in a system through a diffusive mechanism is known as Ostwald ripening. This phenomenon is an example of a multiscale problem in that the microscopic level interaction of the particles can greatly impact the macroscale or observable morphology of the system. The mesoscopic model of this physical situation is a stochastic partial differential equation which can be derived from the appropriate particle system. This model is studied through the use of recently developed and benchmarked spectral schemes for the simulation of solutions to stochastic partial differential equations. The results included here demonstrate the effect of adjusting the interparticle interaction on the morphological evolution of the system at the macroscopic level.

1 Introduction

The coarsening phenomenon by which mass is transported via diffusion in order to reduce the overall interfacial area is known as Ostwald ripening. In general, larger droplets grow at the expense of smaller droplets–the number of droplets decreases while the average size of the remaining droplets increases because of the conservation of mass in the system. This phenomena has received a great deal of study in the literature from theoretical, experimental, and computational points of view. ([1], [2], and [3] are just a few examples.) However, Ostwald ripening is also a very difficult problem to study due to the inherently multiscale nature of the problem. In this note, a mesoscopic model which can be related to many of the commonly studied models of phase separation is used in a computational study of domain coarsening. To begin, the model under consideration here will be described.

2 Description of Mesoscopic Model

Mesoscopic (or local mean field) models are designed to bridge the gap between microscopic (molecular) and macroscopic (observable) scales by incorporating microscopic level behavior in the macroscopic level. Mesoscopic models can be derived through a coarse graining of the underlying microscopic system in a rigorous fashion without the introduction of artificial cutoffs. Details of the derivation of mesoscopic models are given in [4], [5], and [6]. Here, this modeling approach will be briefly described and the mesoscopic model appropriate for studying Ostwald ripening will be given.

In the mesoscopic framework, each micromechanism is incorporated through an evolution equation of the average coverage which is derived from the specific micromechanism under consideration. These evolution equations contain stochastic forcing terms as a result of a direct derivation rather than being included in an ad hoc manner as a means of describing microscale behavior. Thus, the mesoscopic model equations are stochastic partial integrodifferential equations which incorporate the molecular interactions explicitly through convolutions.

The mesoscopic model that would be an appropriate model of Ostwald ripening arises from the spin exchange (surface diffusion) mechanism and can be given as follows:

$$u_t - D\nabla \cdot [\nabla u - \beta u(1-u)\nabla J_m * u] + \gamma \operatorname{div}\left[\sqrt{2Du(1-u)}\, dW(x,t)\right] = 0 \quad (1)$$

where u is the concentration of the particles on the surface, D is the diffusion constant, β is proportional to the inverse of the temperature of the underlying Ising model, J_m is the migration potential, γ is proportional to the interaction length of the particles, and dW represents a process that is delta correlated (white noise) in both space and time $\langle dW(x,t)\rangle = 0$ and $\langle dW(x,t)dW(x',t')\rangle = \delta(x-x')\delta(t-t')$ where the angular brackets are used to denote mean values. It is important to observe that the noise in (1) is multiplicative rather the additive noise that is commonly added to deterministic models in an ad hoc manner.

3   Description of Simulation Method

The numerical scheme that is used to study the mesoscopic model in (1) is based upon a generalization of spectral schemes for deterministic partial differential equations to the stochastic setting. Spectral schemes are a particularly attractive way to study (1) due to the simplicity of calculating convolutions spectrally. Just as in the deterministic setting, spatial variables and derivatives are treated spectrally while the time evolution is calculated using finite differences. A detailed description of this scheme is given in [7] where some exactly solvable problems are used as computational benchmarks to validate this method. The emphasis here will be on those aspects of the scheme which are unique to the stochastic system, such as the spectral treatment of the multiplicative noise term in (1) and the use of a suitable time discretization technique. For the sake of simplicity in the description of the spectral method in this section, it will be assumed that the stochastic partial differential equation has only one spatial dimension. The method straightforwardly generalizes to higher spatial

3.1 Treatment of Noise Term

One of the most important steps for developing a spectral scheme for a stochastic partial differential equation such as the mesoscopic model in (1) is to determine a spectral representation for the spatial component of the noise term. It is well-known in the stochastic processes literature that a stationary, isotropic, Gaussian random field $v(x)$ can be represented by the following stochastic integral in Fourier space:

$$v(x) = \int e^{2\pi i x \cdot \xi} S^{\frac{1}{2}}(\xi) \, dW(\xi) \tag{2}$$

where W is Brownian motion and S is the spectral density function of the random field. Most consistent numerical schemes are based upon discretizations of this stochastic integral ([8],[9]). The simplest such discretization uses equispaced nodes and is known as the Fourier method and is essentially a Fourier series. This approximation is

$$v(x) \approx \sum_{j=1}^{M} a_j \cos(2\pi \xi_j x) + b_j \sin(2\pi \xi_j x) \tag{3}$$

where a_j and b_j are independent Gaussians with mean zero and variance $S(\xi_j)\Delta\xi$. Due to the periodicity of the approximation in (3), the Fourier method would not be suitable in applications in which the desired random field has long range correlations [8]. However, since the random field that is needed for the numerical study of (1) has delta correlation (lacks long range correlations), the Fourier method should give a good approximation. The Fourier method is also computationally attractive since an FFT can be used directly to evaluate (3) at all physical space lattice sites.

The next issue in the treatment of the noise term in (1) is determining how to use (3) to obtain a spectral representation of the multiplicative noise term. Because each realization of the noise at a given time is approximated by a Fourier series representation (3), the noise term in (1) can be calculated by completing all multiplications in physical space and all differentiations in Fourier space, passing back and forth between physical space and Fourier space as necessary using an FFT. Thus, the existence of the Fourier series representation of the noise allows the treatment of the stochastic term in the same basic fashion as for deterministic terms in partial differential equations.

3.2 Time Discretization

At this point, the solution of the stochastic partial differential equation in (1) effectively has been reduced to the solution of a stochastic ordinary differential equation which can be written in the form

$$u_t = a(u)\, dt + b(u)\, dW(t) \tag{4}$$

where a is the drift coefficient and b is the diffusion coefficient. Note that it is not appropriate to directly apply schemes that were derived for deterministic ordinary differential equations to (4) since the additional stochastic corrections from the Ito calculus must be included. For instance, it has been shown in [10] that the Euler method applied to a stochastic ordinary differential equation has a strong (pathwise) order of convergence of $\frac{1}{2}$ rather than order 1 as in the deterministic setting. Thus, it is essential to use schemes that are derived directly from suitable truncations of Taylor-Ito series. However, such schemes typically include derivatives of the drift and the diffusion coefficients. Given the highly nonlinear nature of these terms in (1), it is especially useful to consider derivative-free Runge-Kutta type schemes [10]. One choice of a first order strong scheme is the following:

$$u_{n+1} = u_n + a(u_n)\Delta t + b(u_n)\Delta W + \frac{1}{2\sqrt{\Delta t}}\left[b(\tilde{u}_n) - b(u_n)\right]\left[(\Delta W)^2 - \Delta t\right] \quad (5)$$

with supporting value $\tilde{u}_n$ given by

$$\tilde{u}_n = u_n + b(u_n)\sqrt{\Delta t} \quad (6)$$

and ΔW is a Gaussian with mean 0 and variance Δt. In practice, the time stepping is usually done in Fourier space so that terms, such as the standard diffusion, which are linear in Fourier space may be treated exactly through the use of an integrating factor. The integrating factor also allows the use of larger times steps than are numerically stable if the linear terms were not calculated exactly. Of course, since the solution of (1) is real, the ΔW_n's in the time stepping scheme (when taking time steps in Fourier space) must be selected to respect the symmetries in Fourier coefficients that are present in spectral representations of real fields.

4 Simulation Results

In this section, some simulation results for (1) in two space dimensions using the numerical scheme described in the preceding section are given. The migration potential is chosen to be

$$J_m(r) = \frac{1}{\sqrt{\pi r_0^2}} \exp\left(\frac{-|r|^2}{r_0^2}\right) \quad (7)$$

with r_0 a parameter that describes the interaction length. The other parameters in (1) are chosen to be $D = 0.1, \beta = 6$, and $\gamma = r_0^2$. Simulation parameters include 64 wave numbers in each direction and a step size of $\Delta t = 0.00001$, which insures numerical stability of the time evolver. The computational domain is a unit square with periodic boundary conditions. The system is initialized with two circular regions of high concentration surrounded by regions of low concentration. A rapidly decaying exponential is used to "connect" these two regions to insure the continuity and differentiability of the initial concentration field. The centers of the circles are a distance 0.5 apart. The upper circle initially has a radius of 0.1, while the lower circle has an initial radius of 0.09. This sort of initialization is commonly used in studies of Ostwald ripening as it is the simplest physical case in which this phenomenon can be observed ([11], [12]). In this system, the larger (upper) circle is expected to grow while the smaller (lower) circle shrinks since lower curvatures are preferred; as the smaller circle disappears, the larger region should be distorted away from its circular shape in the locations nearest the small circle but eventually become circular again. The deformation in shape would be expected to occur at both ends of the upper circle in this case due to the symmetries in the initial state and the periodic boundary conditions used.

Figure 1 contains contour plots from simulations of (1) using the spectral scheme described in Section 3 for the case $r_0 = 0.05$ for times 5, 10, 13.5, and 15. The light areas represent regions of high concentration while the dark areas represent regions of low concentration. At the earlier times presented in Figure 1,

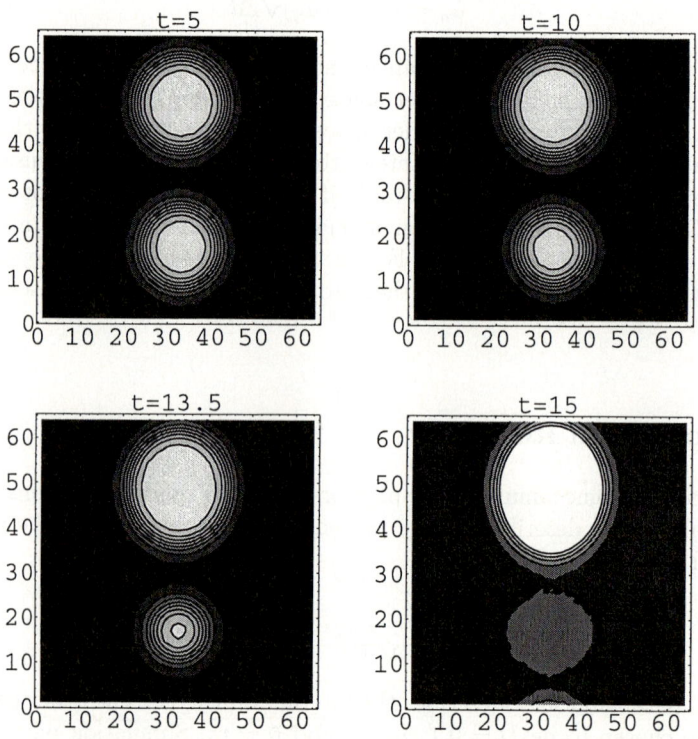

Fig. 1. Contour plots showing the time evolution of the concentration field obtained from mesoscopic simulations. The lighter shades represent regions of higher concentration while the darker shades represent regions of lower concentration. The expected Ostwald ripening is observed

the shrinking of the lower circle and the growth of the upper circle are already evident. As time continues, the rate of shrinking of the smaller circle increases until at time 13.5, the region of highest concentration has almost completely disappeared from the lower circle; the deformation of the upper region into an elliptical shape is also becoming quite evident. By time 15, the lower region has essentially diffused away with the majority of particles appearing in the upper region which has become quite elliptical in shape. In later time results not shown here, this upper region eventually evolves to a circular shape again. The total concentration in the system was observed to be conserved for all times considered, including this final state. All of these results are in qualitative agreement with those expected; in addition, these results are also quite similar in nature those obtained in prior studies of others using different mathematical models and numerical techniques, for example, boundary integral techniques are used in [12].

The ease of changing the interparticle interaction potential in the mesoscopic model in (1) straightforwardly allows for many computational studies. The effect

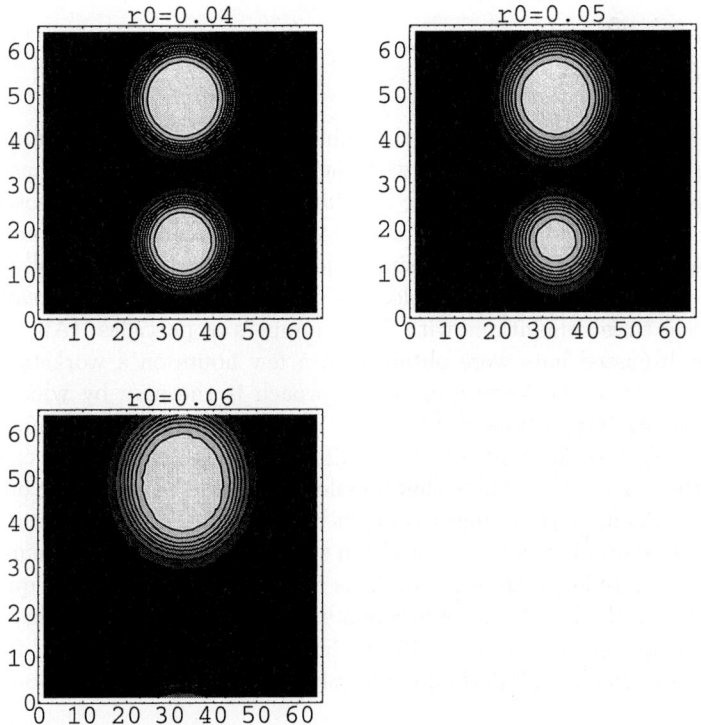

Fig. 2. Mesoscopic simulation results at time $t = 10$ which demonstrate the effect of adjusting the particle interaction length on the time evolution of the system. Longer interaction lengths lead to faster evolution while shorter interaction lengths lead to slower evolution

of adjusting the interaction length r_0 in (7) can be observed in Figure 2 where simulation results are given at time $t = 10$ for $r_0 = 0.04, 0.05$, and 0.06. All other simulation parameters are unchanged from those used for Figure 1; the initial concentration field is also unchanged. Recall that increasing the value of r_0 increases the interaction length of the underlying particles of the system. For the smallest interaction length shown here ($r_0 = 0.04$), the system has evolved very little from the initial state; on the other hand, for the largest interaction radius shown here ($r_0 = 0.06$), not only has the smaller (lower) circle completely diffused away but also the upper region has already evolved back to a nearly circular shape. These results are in good agreement with physical intuition since one would expect particles in systems with longer range interactions to more readily "find" each other. In fact, the smaller circle has essentially diffused away by time 8 when $r_0 = 0.06$ but remains until time 26 for the much smaller interaction radius of $r_0 = 0.04$. Also, just as for the simulation results in Figure 1, the total concentration was conserved throughout the entire simulation. Thus, the time scale of the evolution is quite sensitive to the interaction length of the particles but the overall qualitative behavior of the evolution remains the same.

5 Conclusions

This brief note can only indicate the value and power of using the mesoscopic modeling approach in conjunction with the spectral scheme for stochastic partial differential equations ("mesoscopic simulation"). For instance, mesoscopic simulation can be applied to physical situations with underlying micromechanisms other than diffusion. Mesoscopic simulation is more computationally tractable than other approaches such as molecular dynamics, especially for results which require long time simulations with large numbers of particles. (All results presented or discussed here were obtained in a few hours on a workstation.) One particularly attractive feature of this approach is the ease by which different sorts of particle interactions can be used. While Figure 2 certainly indicates one interesting class of adjustments to the interaction potential that can be considered, there are many others that should be considered including interactions with repulsive interaction ranges as might be encountered with charged particles. Anisotropic potentials would be useful in studies of coarsening systems in which there is a preferred growth direction. It is also important that more quantitative comparisons with theory and experiments be made. As a step in this direction, results of this type are given in [13], including simulations starting from a disordered state which exhibit the Lifshitz-Slyozov growth law as self-organization proceeds.

Acknowledgments

This research is partly supported by NSF-DMS-0219211, NSF-DMS-0406633, and an NJIT SBR grant.

References

1. Voorhees, P.: The theory of Ostwald ripening. J. Statis. Phys. **38** (1985) 231–252.
2. Yao, J., Elder, K., Guo, H., Grant, M.: Late stage droplet growth. Physica A **204** (1994) 770–788.
3. Niethammer, B., Pego, R.: The LSW model for domain coarsening: asymptotic behavior for conserved total mass. J. Statis. Phys. **104** (2001) 1113–1144.
4. Hildebrand, B., Mikhailov, A.: Mesoscopic modeling in the kinetic theory of adsorbates. J. Phys. Chem. **100** (1996) 19089–19101.
5. Giacomin, G., Lebowitz, J.: Exact macroscopic description of phase segregation in model alloys with long range interactions. Phys. Rev. Letters **76** (1996) 1094–1097.
6. Vlachos, D., Katsoulakis, M.: Derivation and validation of mesoscopic theories for diffusion-reaction of interacting molecules. Phys. Rev. Letters **85** (2000) 3898–3901.
7. Horntrop, D.: Spectral schemes for stochastic partial differential equations. (submitted).
8. Elliott, F., Horntrop, D., Majda, A.: Monte Carlo methods for turbulent tracers with long range and fractal random velocity fields. Chaos **7** (1997) 39–48.
9. Elliott, F., Horntrop, D., Majda, A.: A Fourier–wavelet Monte Carlo method for fractal random fields. J. Comp. Phys. **132** (1997) 384–408.
10. Kloeden, P., Platen, E.: *Numerical Solution of Stochastic Differential Equations.* Springer, Berlin (1992).
11. Imaeda, T., Kawasaki, K.: Theory of morphological evolution in Ostwald ripening. Physica A **186** (1992) 359–387.
12. Voorhees, P., McFadden, G., Boisvert, R., Meiron, D.: Numerical simulation of morphological development during Ostwald ripening. Acta Metall. **36** (1988) 207–222.
13. Horntrop, D.: Mesoscopic simulation for domain coarsening. (in preparation).

Computer Simulation of the Anisotropy of Fluorescence in Ring Molecular Systems

Pavel Heřman[1] and Ivan Barvík[2]

[1] Department of Physics, University of Hradec Králové, Rokitanského 62,
CZ-500 03 Hradec Králové, Czech Republic,
pavel.herman@uhk.cz

[2] Institute of Physics of Charles University, Faculty of Mathematics and Physics,
CZ-12116 Prague, Czech Republic

Abstract. The time dependence of the anisotropy of fluorescence after an impulsive excitation in the molecular ring (resembling the B850 ring of the purple bacterium *Rhodopseudomonas acidophila*) is calculated. Fast fluctuations of the environment are simulated by dynamic disorder and slow fluctuations by static disorder. Without dynamic disorder modest degrees of static disorder are sufficient to cause the experimentally found initial drop of the anisotropy on a sub-100 fs time scale. In the present investigation we are comparing results for the time-dependent optical anisotropy of the molecular ring for four models of the static disorder: Gaussian disorder in the local energies (Model A), Gaussian disorder in the transfer integrals (Model B), Gaussian disorder in radial positions of molecules (Model C) and Gaussian disorder in angular positions of molecules (Model D). Both types of disorder - static and dynamic - are taken into account simultaneously.

1 Introduction

We are dealing with the ring-shaped units resembling those from antenna complex LH2 of the purple bacterium *Rhodopseudomonas acidophila* in which a highly efficient light collection and excitation transfer towards the reaction center takes place. Due to a strong coupling limit (large interaction J between bacteriochlorophylls) our theoretical approach considers an extended Frenkel exciton states model.

Despite intensive study, the precise role of the protein moiety in governing the dynamics of the excited states is still under debate [1]. At room temperature the solvent and protein environment fluctuate with characteristic time scales ranging from femtoseconds to nanoseconds. The dynamical aspects of the system are reflected in the line shapes of electronic transitions. To fully characterize the line shape of a transition and thereby the dynamics of the system, one needs to know not only the fluctuation amplitude (coupling strength) but also the time scale of each process involved. The observed linewidth reflect the combined influence of static disorder and exciton coupling to intermolecular, intramolecular, and solvent nuclear motions. The simplest approach is to decompose the line profile into homogeneous and inhomogeneous contributions of the dynamic and static disorder. Yet, a satisfactory understanding of the nature of the static disorder in light-harvesting systems has not been reached [1]. In the site excitation basis, there can be present static disorder in both diagonal and off-diagonal elements. Silbey pointed out several questions: It is not clear whether only the consideration of the former is enough or the latter should be included as well. If both are considered, then there remains a question about whether they are independent or correlated.

Time-dependent experiments of the femtosecond dynamics of the energy transfer and relaxation [2, 3] led for the B850 ring in LH2 complexes to conclusion that the elementary dynamics occurs on a time scale of about 100 fs [4, 5, 6]. For example, depolarization of fluorescence was studied already quite some time ago for a model of electronically coupled molecules [7, 8]. Rahman et al. [7] were the first who recognize the importance of the off-diagonal density matrix elements (coherences) [9] which can lead to an initial anisotropy larger than the incoherent theoretical limit of 0.4. Already some time ago substantial relaxation on the time scale of 10-100 fs and an anomalously large initial anisotropy of 0.7 was observed by Nagarjan et al. [4]. The high initial anisotropy was ascribed to a coherent excitation of a degenerate pair of states with allowed optical transitions and then relaxation to states at lower energies which have forbidden transitions. Nagarjan et al. [5] concluded, that the main features of the spectral relaxation and the decay of anisotropy are reproduced well by a model considering decay processes of electronic coherences within the manifold of the excitonic states and thermal equilibration among the excitonic states. In that contribution the exciton dynamics was not calculated explicitly.

In several steps [10, 11, 12, 13] we have recently extended the former investigations by Kumble and Hochstrasser [14] and Nagarjan et al. [5]. For a Gaussian distribution of local energies in the ring units we added the effect of dynamical disorder by using a quantum master equation in the Markovian and non-Markovian limits. We also investigated influence of static disorder in transfer integrals [15, 16].

In our present investigation we are comparing results for the time-dependent optical anisotropy of the molecular ring for four models of the static disorder: Gaussian disorder in the local energies, Gaussian disorder in the transfer integrals, Gaussian disorder in radial positions of molecules and Gaussian disorder in angular positions of molecules.

2 Model

In the following we assume that only one excitation is present on the ring after an impulsive excitation [14]. The Hamiltonian of an exciton in the ideal ring coupled to a bath of harmonic oscillators reads

$$H^0 = \sum_{m,n(m\neq n)} J_{mn} a_m^\dagger a_n + \sum_q \hbar\omega_q b_q^\dagger b_q + \frac{1}{\sqrt{N}} \sum_m \sum_q G_q^m \hbar\omega_q a_m^\dagger a_m (b_q^\dagger + b_{-q})$$

$$= H_{\text{ex}}^0 + H_{\text{ph}} + H_{\text{ex-ph}}. \tag{1}$$

H_{ex}^0 represents the single exciton, i.e. the system. The operator $a_m^\dagger$ (a_m) creates (annihilates) an exciton at site m. J_{mn} (for $m \neq n$) is the so-called transfer integral between sites m and n. H_{ph} describes the bath of phonons in the harmonic approximation. The phonon creation and annihilation operators are denoted by $b_q^\dagger$ and b_q, respectively. The last term in Eq. (1), $H_{\text{ex-ph}}$, represents the exciton–bath interaction which is assumed to be site–diagonal and linear in the bath coordinates. The term G_q^m denotes the exciton–phonon coupling constant.

Inside one ring the pure exciton Hamiltonian H_{ex}^0 (Eq. (1)) can be diagonalized using the wave vector representation with corresponding delocalized "Bloch" states and energies. Considering homogeneous case with only nearest neighbor transfer matrix elements $J_{mn} = J_{12}(\delta_{m,n+1}+\delta_{m,n-1})$ and using Fourier transformed excitonic operators (Bloch representation)

$$a_k = \sum_n a_n e^{ikn}, \quad k = \frac{2\pi}{N} l, \quad l = 0, \pm 1, \ldots \pm N/2, \tag{2}$$

the simplest exciton Hamiltonian in k representation reads

$$H_{\text{ex}}^0 = \sum_k E_k a_k^\dagger a_k, \quad \text{with} \quad E_k = -2 J_{12} \cos k. \tag{3}$$

Influence of static disorder is modelled by a Gaussian distribution

A) for the uncorrelated local energy fluctuations ϵ_n (with a standard deviation Δ)

$$H_s^A = \sum_m \epsilon_n a_m^\dagger a_n$$

B) for the uncorrelated transfer integral fluctuations δJ_{nm} with a standard deviation Δ_J

$$H_s^B = \sum_{m,n(m\neq n)} \delta J_{mn} a_m^\dagger a_n.$$

We are using nearest neighbor approximation.

C) for the uncorrelated fluctuations of radial positions of molecules (with standard deviation Δ_r and $\langle r_n \rangle = r_0$)

$$r_n = r_0(1 + \delta r_n)$$

leading to H_s^C.

D) for the uncorrelated fluctuations of positions of molecules on the ring without the changing of orientations of transition dipole moments (with standard deviation Δ_φ)

$$\varphi_n = \varphi_n^0 + \delta\varphi_n$$

leading to H_s^D.

Hamiltonian of the static disorder adds to the Hamiltonian of the ideal ring

$$H = H^0 + H_s^X. \tag{4}$$

All of the Q_y transition dipole moments of the chromophores (bacteriochlorophylls (BChls) B850) in a ring without static and dynamic disorder lie approximately in the plane of the ring and the entire dipole strength of the B850 band comes from a degenerate pair of orthogonally polarized transitions at an energy slightly higher than the transition energy of the lowest exciton state.

The dipole strength μ_a of eigenstate $|a\rangle$ of the ring with static disorder and the dipole strength μ_α of eigenstate $|\alpha\rangle$ of the ring without static disorder read

$$\mu_a = \sum_{n=1}^{N} c_n^a \mu_n, \qquad \mu_\alpha = \sum_{n=1}^{N} c_n^\alpha \mu_n, \tag{5}$$

where c_n^α and c_n^a are the expansion coefficients of the eigenstates of the unperturbed ring and the disordered one in site representation, respectively. In the case of impulsive excitation the dipole strength is simply redistributed among the exciton levels due to disorder [14]. Thus the impulsive excitation with a pulse of sufficiently wide spectral bandwidth will always prepare the same initial state, irrespective of the actual eigenstates of the real ring. After impulsive excitation with polarization $\mathbf{e}_x$ the excitonic density matrix ρ [11] is given by [5]

$$\rho_{\alpha\beta}(t=0;\mathbf{e}_x) = \frac{1}{A}(\mathbf{e}_x \cdot \boldsymbol{\mu}_\alpha)(\boldsymbol{\mu}_\beta \cdot \mathbf{e}_x),$$

$$A = \sum_\alpha (\mathbf{e}_x \cdot \boldsymbol{\mu}_\alpha)(\boldsymbol{\mu}_\alpha \cdot \mathbf{e}_x). \tag{6}$$

The usual time-dependent anisotropy of fluorescence

$$r(t) = \frac{\langle S_{xx}(t)\rangle - \langle S_{xy}(t)\rangle}{\langle S_{xx}(t)\rangle + 2\langle S_{xy}(t)\rangle}, \tag{7}$$

$$S_{xy}(t) = \int P_{xy}(\omega,t)d\omega$$

is determined from

$$P_{xy}(\omega,t) = A\sum_a \sum_{a'} \rho_{aa'}(t)(\boldsymbol{\mu}_{a'} \cdot \mathbf{e}_y)(\mathbf{e}_y \cdot \boldsymbol{\mu}_a)[\delta(\omega - \omega_{a'0}) + \delta(\omega - \omega_{a0})]. \tag{8}$$

The brackets $\langle\rangle$ denote the ensemble average and the orientational average over the sample.

The crucial quantity entering the time dependence of the anisotropy in Eq. (7) is the exciton density matrix ρ. The dynamical equations for the exciton density matrix obtained by Čápek[17] read

$$\frac{d}{dt}\rho_{mn}(t) = \sum_{pq} i(\Omega_{mn,pq} + \delta\Omega_{mn,pq}(t))\rho_{pq}(t). \tag{9}$$

In long time approximation coefficient $\delta\Omega(t \to \infty)$ becomes time independent.

All details of calculations leading to the time-convolutionless dynamical equations for the exciton density matrix are given elsewhere [13] and we shall not repeat them here. The full time dependence of $\delta\Omega(t)$ is given through time dependent parameters [17]

$$\mathcal{A}_{mn}^p(t) = \int_0^t \frac{i\hbar}{N} \sum_k \omega_k^2 (G_{-k}^m - G_{-k}^n) \sum_r G_k^r \sum_{\alpha,\beta} \langle\beta|r\rangle\langle r|\alpha\rangle\langle\alpha|m\rangle\langle p|\beta\rangle \times$$
$$e^{-\frac{i}{\hbar}(E_\alpha - E_\beta)\tau} \left\{[1 + n_B(\hbar\omega_k)] e^{i\omega_k \tau} + n_B(\hbar\omega_k) e^{-i\omega_k \tau}\right\} d\tau. \tag{10}$$

Obtaining of the full time dependence of $\delta\Omega(t)$ is not a simple task. We have succeeded to calculate microscopically full time dependence of $\delta\Omega(t)$ only for the simplest molecular model namely dimer [18]. In case of molecular ring we should resort to some simplification [13].

In what follows we use Markovian version of Eq. (10) with a simple model for correlation functions C_{mn} of the bath assuming that each site (i.e. each chromophore) has its own bath completely uncoupled from the baths of the other sites. Furthermore it is assumed that these baths have identical properties [2, 19]. Then only one correlation function $C(\omega)$ of the bath is needed

$$C_{mn}(\omega) = \delta_{mn}C(\omega) = \delta_{mn}2\pi[1 + n_B(\omega)][J(\omega) - J(-\omega)]. \tag{11}$$

Here $J(\omega)$ is the spectral density of the bath [19] and $n_B(\omega)$ the Bose-Einstein distribution of phonons. The model of the spectral density $J(\omega)$ often used in literature is

$$J(\omega) = \Theta(\omega) j_0 \frac{\omega^2}{2\omega_c^3} e^{-\omega/\omega_c}. \tag{12}$$

Spectral density has its maximum at $2\omega_c$. We shall use (in agreement with [2]) $j_0 = 0.4$ and $\omega_c = 0.2$.

3 Results and Conclusions

The anisotropy of fluorescence (Eq. (7)) has been calculated using dynamical equations for the exciton density matrix ρ to express the time dependence of the optical properties of the ring units in the femtosecond time range. Details are the same as in Ref. [13, 15, 16].

For the numerical time propagation of the density matrix ρ (Eq. 9) the short iterative Arnoldi method [20] as well as the standard Runge-Kutta scheme have

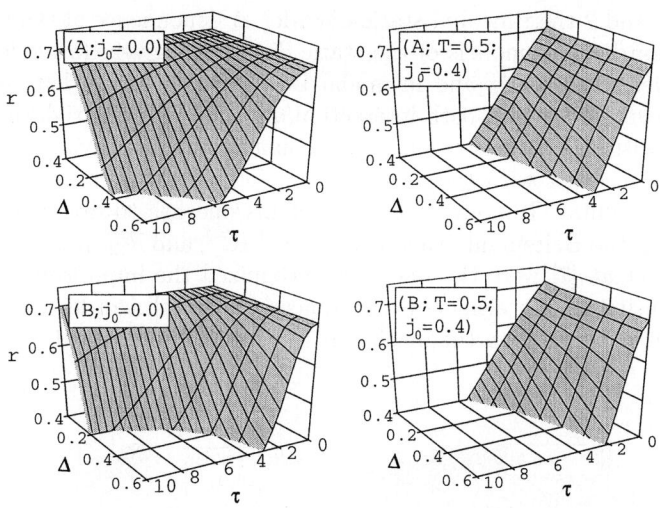

Fig. 1. The time and Δ dependence of the anisotropy depolarization for two models (A) and (B) of the static disorder is given. In the left column the results without the exciton-bath interaction are shown, in the right column the interaction with the bath is taken into account in the Markovian treatment of the dynamic disorder with the $j_0 = 0.4$ and for temperature $T = 0.5$ (in dimensionless units)

been used. An advantage of the short iterative Arnoldi method with respect to the standard Runge-Kutta scheme is the low computational effort for moderate accuracy [21]. Furthermore, the expansion coefficients are adapted at each time to a fixed time step with a prespecified tolerance in contrast to the Runge-Kutta scheme in which the time step is adapted. An uniform time grid is important for averaging of various realizations at the same time points without interpolation. The realization averaging and the orientational averaging can easily be parallelized by means of *Message passing interface* (MPI). Some computations were performed on a PC cluster. So instead of running about 10 000 realizations on one node, 312 realizations can be calculated on each of the 32 nodes (or 52 realizations on each of 192 nodes).

In Ref. [14], which does not take the bath into account, the anisotropy of fluorescence of the LH2 ring decreases from 0.7 to $0.3 - 0.35$ and subsequently reaches a final value of 0.4. One needs a strength of static disorder of $\Delta \approx 0.4-0.8$ to reach a decay time below 100 fs.

Results of our simulations are presented graphically in Fig. 1. and Fig. 2. We use dimensionless energies normalized to the transfer integral J_{12} and the renormalized time τ. To convert τ into seconds one has to divide τ by $2\pi c J_{12}$ with c being the speed of light in cm s^{-1} and J_{12} in cm^{-1}. Estimation of the transfer integral J_{12} varies between 250 cm^{-1} and 400 cm^{-1}. For these extreme values of J_{12} our time unit ($\tau = 1$) corresponds to 21.2 fs or 13.3 fs.

In Fig. 1 and 2 the time and static disorder Δ dependence of the anisotropy depolarization for four models of the static disorder is given. In model B) $\Delta = \Delta_J$, in model C) $\Delta = \Delta_r$ and in model D) $\Delta = \Delta_\varphi$. In the left column the results without the exciton-bath interaction are shown, in the right column the interaction with the bath is taken into account in the Markovian treatment of the dynamic disorder with the $j_0 = 0.4$ and for the temperature $T = 0.5$ (in dimensionless units). To convert T into kelvins one has to divide T by k/J_{12} with k beeing the Boltzmann constant in cm^{-1} K^{-1} and J_{12} in cm^{-1}.

Rahman et al. [7] were the first who recognized the importance of the off-diagonal density matrix elements (coherences) [9] which can lead to an initial anisotropy $r(0)$ larger than the incoherent theoretical limit of 0.4.

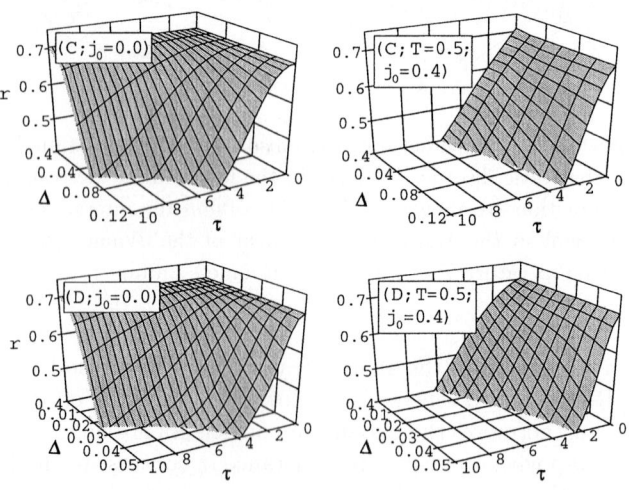

Fig. 2. The same as in Fig. 1 but for the models (C) and (D)

Without dynamic disorder modest degrees of static disorder are sufficient to cause the experimentally found initial drop of the anisotropy on a sub-100 fs time scale. Difference between the Gaussian static disorder in the local energies (Model A) and the Gaussian static disorder in the transfer integrals (Model B) calculations expressed by the time interval in which the anisotropy depolarization reaches $r = 0.4$ (the incoherent theoretical limit) is almost as much as 100 % for the same value of the static disorder $\Delta = \Delta_J$. It means that the same drop of the anisotropy may be caused even by the diagonal static disorder (model A) with Δ or by the static disorder in the transfer integrals with $\Delta_J = 0.5\Delta$. This difference between the Model A and the Model B calculations is still present also in the case, when the exciton interaction with the bath is taken into account. In model C) the strength of the static disorder $\Delta_r = 0.12$ and in model D) the strength of the static disorder $\Delta_\varphi = 0.04$ have practically the same effect as $\Delta = 0.6$ in model A).

Acknowledgement

This work has been funded by the project GAČR 202-03-0817.

References

1. Jang S., Dempster S. F., Silbey R. J.: J. Phys. Chem. **B 105** (2001) 6655
2. Sundström V., Pullerits T., van Grondelle R.: J. Phys. Chem. **B 103** (1999) 2327
3. Novoderezhkin V., van Grondelle R.: J. Phys. Chem. **B 106** (2002) 6025
4. Nagarjan V., Alden R. G., Williams J. C., Parson W. W.: Proc. Natl. Acad. Sci. USA **93** (1996) 13774
5. Nagarjan V., Johnson E. T., Williams J. C., Parson W. W.: J. Phys. Chem. **B 103** (1999) 2297
6. Nagarjan V., Parson W. W.: J. Phys. Chem. **B 104** (2000) 4010
7. Rahman T. S., Knox R. S., Kenkre V. M.: Chem. Phys. **44** (1979) 197
8. Wynne K., Hochstrasser R. M.: Chem. Phys. **171** (1993) 179.
9. Kühn O., Sundström V., Pullerits T.: Chem. Phys. **275** (2002) 15
10. Heřman P., Kleinekathöfer U., Barvík I., Schreiber M.: J. Lumin. **94&95** (2001) 447
11. Heřman P., Kleinekathöfer U., Barvík I., Schreiber M.: Chem. Phys. **275** (2002) 1
12. Barvík I., Kondov I., Heřman P., Schreiber M., Kleinekathöfer U.: Nonlin. Opt. **29** (2002) 167
13. Heřman P., Barvík I.: Czech. J. Phys. **53** (2003) 579
14. Kumble R., Hochstrasser R.:, J. Chem. Phys. **109** (1998) 855
15. Reiter M., Heřman P., Barvík I.: J. Lumin. **110**(2004) 258
16. Heřman P., Barvík I., Reiter M.:, J. Lumin., in press.
17. Čápek V.: Z. Phys. **B 99** (1996) 261
18. Barvík I., Macek J.: J. Chin. Chem. Soc. **47** (2000) 647
19. May V., Kühn O.: Charge and Energy Transfer in Molecular Systems, Wiley-WCH, Berlin, 2000
20. Pollard W. T., Friesner R. A.: J. Chem. Phys. **100** (1994) 5054
21. Kondov I., Kleinekathöfer U., Schreiber M.: J. Chem. Phys. **114** (2001) 1497

The Deflation Accelerated Schwarz Method for CFD

J. Verkaik[1], C. Vuik[2,*], B.D. Paarhuis[1], and A. Twerda[1]

[1] TNO Science and Industry, Stieltjesweg 1, P.O. Box 155, 2600 AD Delft,
The Netherlands
[2] Delft University of Technology, Faculty of Electrical Engineering, Mathematics and
Computer Science, Mekelweg 4, 2628 CK Delft, The Netherlands
Phone: +31 15 27 85530, Fax: +31 15 27 87209
c.vuik@ewi.tudelft.nl.

Abstract. Accurate simulation of glass melting furnaces requires the solution of very large linear algebraic systems of equations. To solve these equations efficiently a Schwarz domain decomposition (multi-block) method can be used. However, it can be observed that the convergence of the Schwarz method deteriorates when a large number of subdomains is used. This is due to small eigenvalues arising from the domain decomposition which slow down the convergence. Recently, a deflation approach was proposed to solve this problem using constant approximate eigenvectors. This paper generalizes this view to piecewise linear vectors and results for two CFD problems are presented. It can be observed that the number of iterations and wall clock time decrease considerably. The reason for this is that the norm of the initial residual is much smaller and the rate of convergence is higher.

Keywords: efficiency; computational fluid dynamics; domain decomposition; deflation; Krylov subspace acceleration.

1 Introduction

Simulation by Computational Fluid Dynamics (CFD) is important for the design, optimization and trouble shooting of glass melting furnaces. It gives engineers in the glass industry great insight into the occurring transport phenomena. At TNO Science and Industry, a CFD simulation package called GTM-X is being developed for simulating gas- and oil-fired glass-furnaces. This is a complete model for simulating glass furnaces, describing the glass melt and combustion space simultaneously, and predicting the effects on melting performance and glass quality.

A domain decomposition (DD) approach is applied within GTM-X, for which the spatial domain is decomposed into subdomains (blocks). A DD (or multi-block) approach has several important advantages for simulating glass-melting

---

* Corresponding author.

furnaces. Since a glass-furnace geometry is often complex, this enables us to easily describe the geometry using blocks (subdomains). Furthermore, parallel computing can be done and variables for each block can be simultaneously computed on different processors. This is an advantage, because simulation of glass-melting furnaces often results in very large computation times.

A disadvantage of multi-block solvers is that the convergence behavior deteriorates significantly when a large number of blocks is used. This is especially the case for solving elliptic equations. In [11], a reason for this problem is given which relates the loss of convergence to the presence of small eigenvalues arising from the domain decomposition. These eigenvalues can be 'eliminated' by applying a deflation operator using constant vectors for approximating the corresponding eigenvectors. The authors of [11] present convergence rates for solving Poisson's equation which are independent on the number of blocks.

In this paper we extend and apply this idea to solve general linear systems of equations. The deflation method is implemented in GTM-X and both GCR and CG Krylov subspace acceleration is used to solve the resulting linear system of equations. Both constant (CD) and combined constant and linear deflation (CLD) vectors are considered. The research in this paper focusses on solving the singular pressure-correction system in the SIMPLE method, which is used in GTM-X to solve the Navier-Stokes equations. However, the solver can be used for general linear systems and is therefore applicable to a wide range of problems.

2 Description of the Mathematical Model

Besides the incompressible Navier-Stokes equations and the energy equation, GTM-X has dedicated models for turbulence, combustion, glass melting and chemical vaporization. The user can apply different models in different domains.

The equations arising from the physical models are discretised with the finite volume method. Several schemes can be used for discretization: upwind, central, TVD higher-order schemes, and blending of schemes can be done using deferred correction. The grid is boundary fitting and colocated, meaning that all variables are located in the volume cell centers. The discretised non-linear Navier-Stokes equations are solved by the SIMPLE method ([6, 1, 12]), using pressure-weighted (Rhie & Chow) interpolation to exclude checkerboard pressure modes. The SIMPLE method is an iterative method in which the system in each iteration (outer iteration) is splitted up into linear systems for the pseudo-velocities and pressure-correction. So-called SIMPLE Stabilization Iterations (SSI) are applied, which can be seen as additional outer iterations for solving the pressure system without solving the velocities. Linear systems of equations are solved with a domain multi-block approach (inner iteration). In GTM-X, an additive Schwarz DD method with minimal overlap is applied, in combination with inaccurate subdomain solutions. Stone's SIP-solver [7], which is based on an incomplete LU decomposition, is used for obtaining the subdomain solutions. Local grid refinement at block level can be done, as well as parallel computing using MPI libraries. Furthermore, solutions on coarse grids can be used as starting solutions for finer grids by using interpolation. The reader is referred to [10] for more details on the code.

3 Deflation and Domain Decomposition

Consider a decomposition of the entire computational domain consisting of n grid points into L nonoverlapping subdomains. Discretization of the partial differential equations to be solved and grouping together the unknowns per subdomain, results in the block system

$$\begin{bmatrix} A_{11} & \cdots & A_{1L} \\ \vdots & \ddots & \vdots \\ A_{L1} & \cdots & A_{LL} \end{bmatrix} \begin{bmatrix} y_1 \\ \vdots \\ y_L \end{bmatrix} = \begin{bmatrix} b_1 \\ \vdots \\ b_L \end{bmatrix},$$

which we will denote simply as $Ay = b$. The diagonal blocks of A are the coefficients for the interior of the subdomains, the off-diagonal blocks represent coupling across subdomain boundaries. This system is solved with the deflation method combined with a Krylov subspace method.

Let the matrix Z be $n \times m$, where $m \leq n$. Furthermore, let Z be of rank m, i.e. Z has linear independent columns. The columns of Z, the so-called deflation vectors, span the deflation subspace, i.e., the space that approximates the eigenspace belonging to the smallest eigenvalues and which is to be projected out of the residual. To do this we define the projectors

$$P = I - AZE^{-1}Z^T , \quad Q = I - ZE^{-1}Z^T A ,$$

where the $m \times m$ matrix E is defined as $E = Z^T AZ$, and I is the identity matrix of appropriate size. To solve $Ay = b$ with the deflation technique, we write $y = (I - Q)y + Qy$, and since

$$(I - Q)y = Z(Z^T AZ)^{-1}Z^T Ay = ZE^{-1}Z^T b , \tag{1}$$

can be computed immediately, we only need to compute Qy. In light of the identity $AQ = PA$ we can solve the system

$$PA\tilde{y} = Pb \tag{2}$$

for $\tilde{y}$, premultiply this by Q and add it to (1).

Since $PAZ = 0$, this system is obviously singular, and therefore no unique solution exists. However, it can be shown that Qy is unique [9]. In [2] and [8–p. 147], it is noted that a positive semidefinite system can be solved as long as it is consistent, i.e., as long as its right-hand side does not have components in the null space (column space of Z) of PA. This assumption holds for (2), because

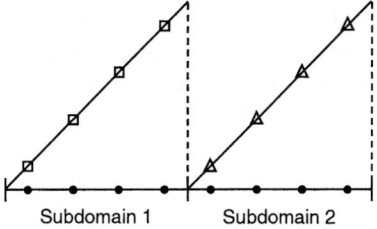

Fig. 1. Linear deflation vectors for the 1-D example of two subdomains with 4 grid cells per subdomain

the same projection is applied to both sides of $Ay = b$. In [8–p. 147], it is shown that a Krylov subspace method could be used to solve (2), because the null-space never enters the iteration, and therefore the corresponding zero-eigenvalues do not influence the convergence. The reader is referred to [4] for more details on the spectrum of PA.

When system (2) is solved, this involves the computation of E^{-1}, for example statements like $f = E^{-1}g$ have to be evaluated. In [4–Section 3], it is shown that deflation works correctly as long as E^{-1} is computed with a high accuracy. Since we never determine E^{-1} explicitly, this means that $Ef = g$ is solved by a direct method, for example by an LU factorization.

Deflation can also be combined with preconditioning. Suppose K is a suitable preconditioner of A, then solving $Ay = b$ can be replaced by: solve $\tilde{y}$ from the left-preconditioned system $K^{-1}PA\tilde{y} = K^{-1}Pb$, and form $Q\tilde{y}$, or solve $\tilde{y}$ from the right-preconditioned system $PAK^{-1}\tilde{y} = Pb$, and form $QK^{-1}\tilde{y}$. Both systems can be solved by a one's favorite Krylov subspace solver, for example by the GCR method. In our implementation we take K the block Gauss-Jacobi preconditioner (block diagonal of A), corresponding to the additive Schwarz iteration. This preconditioner lends itself well for parallel computing. We will refer to the deflation method combined with Krylov subspace acceleration in a domain decomposition context as the deflated Krylov-Schwarz method.

4 Approximating Eigenvectors by Constant and Linear Functions

Many authors [3, 5, 11] consider constant deflation vectors for approximating the eigenvectors corresponding to the small eigenvalues which slow down the convergence. For each subdomain, exactly one deflation vector is defined having elements that are constant in the grid points on the corresponding subdomain, and zero elements in the grid points on the other subdomains. We will refer to this as CD deflation. Generalization to the 2-D and 3-D case is straightforward: for a subdomain, we simply take the elements of the deflation vector to be constant in the grid nodes.

This idea is generalized to approximating the eigenvectors using piecewise linear vectors in the grid directions [1]. By this we augment the space of constant vectors with linear vectors. Figure 1 shows the linear deflation vectors. A linear deflation vector is defined as a polynomial of degree one on each subdomain and zero on the other subdomains. If the deflation vectors consist of both the constant and the linear vectors, we will denote it by Constant Linear Deflation (CLD). For the case of Figure 1, we have two deflation vectors per subdomain: one constant and one linear. Generalization to 2-D and 3-D is not so straightforward anymore, compared to CD deflation. However, it appears that for the 2-D case three vectors are connected to each subdomain: one constant, and one linear vector in each of the two grid directions. For the 3-D case, it appears that we need four deflation vectors: one constant and one linear vector in each of the three grid directions.

When the coefficient matrix A is singular then the deflation vectors has to be chosen such that $Z^T A Z$ is non-singular, since break-down can occur with a direct solution method. Two options seem satisfactory to overcome this problem. The first one is to remove a constant deflation vector for one subdomain; the second is to adjust one entry in a constant deflation vector for one subdomain.

5 Numerical Experiments

The deflated GCR method is compared to the SIP method for solving the singular pressure-correction system arising in the SIMPLE method applied to the stationary incompressible Navier-Stokes equations. Two test cases are considered: a buoyancy-driven cavity flow for a high Rayleigh number, and a glass tank. In both test cases the stationary energy equation is solved besides the Navier-Stokes equations. Furthermore, the flow is assumed to be laminar and buoyant. One constant deflation vector is removed in order to overcome the singularity. Furthermore, one SIP iteration is used for obtaining the subdomain solution and the GCR algorithm is truncated for one search direction. In this paper we will restrict ourselves to the results for the inner iterations.

The first test case is a 2-D buoyancy-driven cavity flow in a unit square consisting of a 60×60 uniform grid and a decomposition of 4×4 subdomains, see Figure 2. The Rayleigh number is chosen to be large, i.e. $Ra = 1.0 \cdot 10^6$, meaning that the flow is highly buoyant but still laminar. Figure 3 shows the residuals for solving the pressure-correction system. Clearly, the CLD deflation performs best, followed by CD deflation and the GCR without deflation. The SIP method performs very disappointing for this case. Note that a large jump in the initial residual can be observed for CLD deflation.

---

[1] In this paper we let deflation vectors in the coordinate directions out of consideration. However, numerical experiments show that choosing linear vectors in grid- or coordinate directions can result in different convergence behavior.

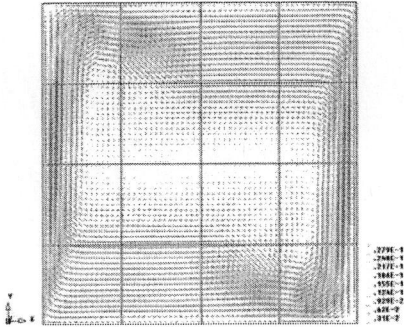

Fig. 2. Velocity field for the buoyancy-driven cavity flow

In the second test case, a glass tank is considered having dimensions 7 [m] × 3 [m] ×1 [m], see Figure 4. Glass fractions are injected (left of figure) from above and melted glass with desired properties leaves the tank at the outlet (right of figure). For this case the grid is uniform consisting of 10,500 grid cells, and the domain is decomposed into 18 subdomains. Figure 5 shows the results for the inner iteration residuals. Clearly, CLD deflation performs best concerning convergence behavior.

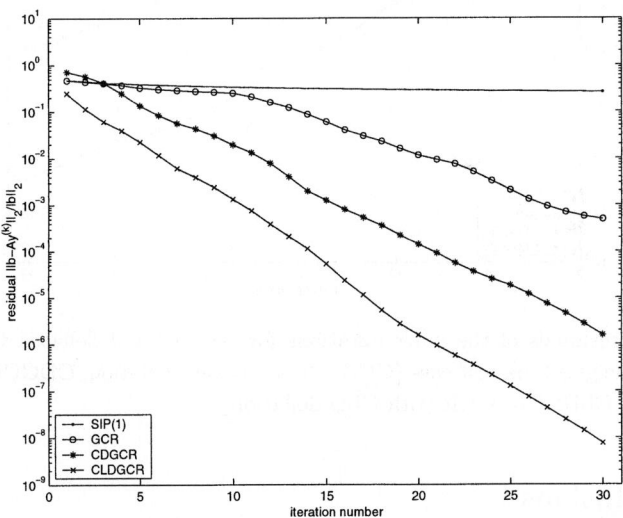

Fig. 3. The inner iteration residuals for the SIP and deflated GCR method considering the buoyancy-driven test case (GCR: GCR without deflation; CDGCR: GCR with CD deflation; CLDGCR: GCR with CLD deflation)

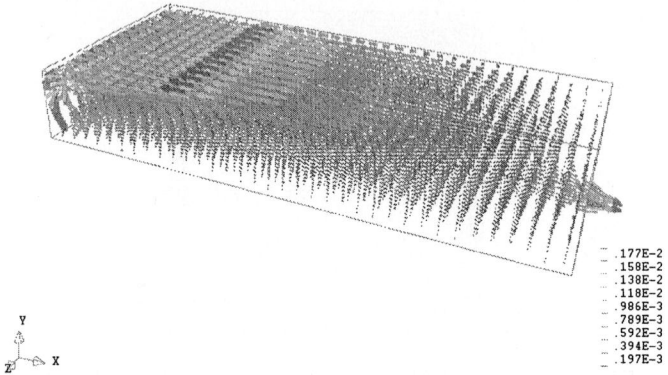

Fig. 4. Velocity field for the glass tank test case

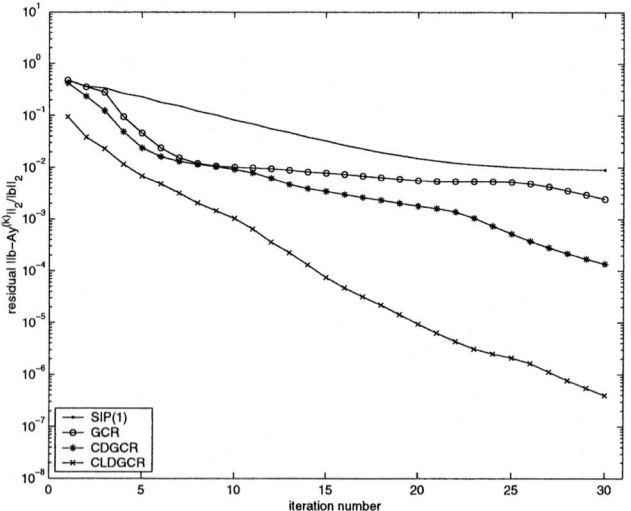

Fig. 5. The residuals of the inner iterations for the SIP and deflated GCR method considering the glas tank test case (GCR: GCR without deflation; CDGCR: GCR with CD deflation; CLDGCR: GCR with CLD deflation)

6 Conclusions

From the experiments presented in this paper, we conclude that the deflation accelerated Schwarz method is a very efficient technique to solve linear systems of equations arising from domain decomposition. It seems that the combination of constant and linear deflation vectors is most efficient. A large jump in the norm of the initial residual can be observed as well as a higher convergence rate. Moreover, it should be noted that the deflation method can be implemented in existing software with relatively low effort.

References

1. J.H. Ferziger and M. Perić. *Computational Methods for Fluid Dynamics*. Springer, Heidelberg, second edition, 1999.
2. E.F. Kaasschieter. Preconditioned conjugate gradients for solving singular systems. *J. Comp. Applied Math.*, 41:265–275, 1988.
3. L. Mansfield. On the conjugate gradient solution of the Schur complement system obtained from domain decomposition. *SIAM J. Numer. Anal.*, 7(6):1612–1620, 1990.
4. R. Nabben and C. Vuik. A comparison of deflation and coarse grid correction applied to porous media flow. Report 03-10, Delft University of Technology, Department of Applied Mathematical Analysis, Delft, 2003.
5. R.A. Nicolaides. Deflation of conjugate gradients with applications to boundary value problems. *SIAM J. Numer. Anal.*, 24(2):355–365, 1987.
6. S.V. Patankar. *Numerical Heat Transfer and Fluid Flow*. McGraw-Hill, New York, 1980.
7. H.L. Stone. Iterative solution of implicit approximations of multi-dimensional partial differential equations. *SIAM J. Numer. Anal.*, 5:530–558, 1968.
8. H.A. van der Vorst. *Iterative Krylov Methods for Large Linear systems*. Cambridge University Press, Cambridge, 2003.
9. F. Vermolen, C. Vuik, and A. Segal. Deflation in preconditioned conjugate gradient methods for finite element problems. In M. Křižek, P. Neittaanmäki, R. Glowinski, and S. Korotov, editors, *Conjugate Gradient and Finite Element Methods*, pages 103–129. Springer, Berlin, 2004.
10. R.L. Verweij. *Parallel Computing for furnace simulations using domain decomposition*. PhD thesis, Delft University of Technology, Delft.
11. C. Vuik and J. Frank. Coarse grid acceleration of a parallel block preconditioner. *Future Generation Computer Systems*, 17:933–940, 2001.
12. P. Wesseling. *Principles of Computational Fluid Dynamics*. Springer, Heidelberg, 2001.

The Numerical Approach to Analysis of Microchannel Cooling Systems

Ewa Raj, Zbigniew Lisik, Malgorzata Langer,
Grzegorz Tosik, and Janusz Wozny

Institute of Electronics, Technical University of Lodz,
223 Wolczanska Str, 90-924, Lodz, Poland
{ewaraj, lisikzby, malanger,
grzegorz.tosik, jwozny}@p.lodz.pl

Abstract. The paper deals with microchannel cooling where water is the cooling liquid. ANSYS software and CFDRC-ACE one were used to analyse the flows and the origin of large amount of heat that can be overtaken from the chip when microchannels are applied. The concept of microscale heat transfer coefficient is discussed. The phenomena taking place in microchannel flows are simulated and some conclusions are introduced to explain the results met in many references but still unexplained. In contrast to existing, standard methods, the new approach describes the local phenomena and is used in the further investigation of the cooling microstructure. An effect of its geometry on the total overtaken heat flux is analysed with respect to optimal conditions as well as to technological restrictions.

1 Introduction

The fast development of nowadays electronics induces the increase of heat dissipation inside semiconductor structures. In case of power module, the total power can exceed 1kW [1, 2] and the heat flux that needs to be overtaken from the device reaches several MW/m^2. The greatest problem nowadays is not only the huge amount of heat dissipation but mainly its density at the surface of the structure. Therefore, one has revealed the challenging task: to design as effective heat exchanger as only possible with regard to microelectronic dimensions restrictions.

A forced cooling system with the coolant characterised by the large enough heat capacity and thermal conductivity could meet these demands only. For example, it could be a liquid cooling system with water as a cooling medium, and such systems already exist. Their effectiveness can be improved when one introduces the coolant stream as close to the heat source as possible. This idea has been employed in a new solution of a liquid cooling system that is based on microchannels formed in the substrate of semiconductor device [3-5], or in cooling microstructure placed at the chip directly [6, 7]. Unfortunately, such a huge heat transfer capacity has been observed at very large inlet pressure, e.g. above 200kPa in [3], that is unacceptable because of the reliability and the life of the electronic equipment.

Although the differences in the behaviour of liquid flow in the micro- and macrostructures have been already reported [3-12], no coherent explanation of their origin exists. In [11], they have noticed that the flow character in microchannels changes for much smaller values of Reynolds numbers. Others observed in [12] the slip of water flowing in microchannels while there is no possibility to observe the phenomenon in macrochannels under these particular conditions. It indicates that the transition from the macroscale, corresponding to the typical liquid cooling systems, to the microscale, when the considered thermal phenomena take place, is not the scaling problem only but it creates quite new problems in heat and mass transport. The lack of their satisfactory description as well as the contradictory opinions presented by different authors encouraged us to analyse the problems from the microchannel cooling efficiency aspect. Some of our results are presented in the paper.

2 Heat and Mass Transfer in Microchannels

The rise of turbulent flow should be described from the origin; when the fluid makes contact with the surface, viscous effects become significant. As a result, boundary layers develop with increasing distance from the inlet, until they merge at the centreline. Then the flow is fully developed and the distance at which this condition is achieved is called hydrodynamic entrance length. Within the fully developed region one can distinguish a turbulent core and a laminar sublayer. In the first region the heat is transferred by the mass transport whereas in the second one, the heat transport is dominated by diffusion. In consequence, this area is responsible for heat exchange at the solid-fluid border mainly.

2.1 One Channel Approach

At first, our interest has been devoted to analyse the heat overtaking process by the water flow through a single channel. We have based it on simulations of water cooling conducted with the aid of ANSYS and CFDRC-ACE software. The heat transfer process has been investigated with use of the heat transfer coefficient that represents the amount of heat overtaken by the coolant that flows along the cooled walls with respect to assumed temperature difference between the wall temperature and the reference ambient one. When one considers liquid heat exchange systems, the heat transfer coefficient is commonly considered as the ratio of the heat flux overtaken from the wall at some particular point and the difference of the point temperature and some reference one that usually is the liquid inlet temperature, the same for the whole heat exchange system [13]. Such an attitude to the analysis of heat exchange problems can be called the macroscale one with the macroscale heat transfer coefficient that refers the local thermal phenomena to one arbitrary, defined temperature. This feature of the macroscale coefficient allows to treat the system as the whole and to estimate the influence of the system design on the entire heat exchange process. Its value does not, however, depend on the real heat exchange ability at the particular point only and does not characterise the local heat exchange phenomenon. It is evident when we consider the changes of the macroscale coefficient along the channel shown in Fig.1. Its value decreases although no changes in the mechanics of the heat exchange take place.

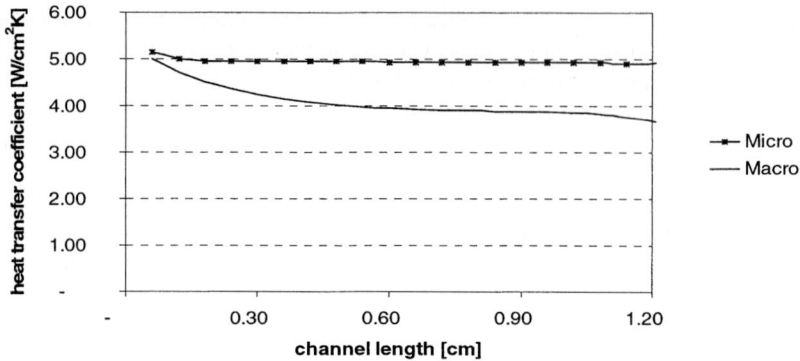

Fig. 1. Micro and macro heat transfer coefficient versus the channel length

The macroscale approach is not sufficient in case of microscale problems, like the microchanel cooling. Therefore, another definition of the heat transfer coefficient, called microscale one, has been proposed. It introduces a new reference temperature connected directly with the local heat exchange process instead of the inlet temperature. It seems that the temperature at the turbulent core – the laminar sublayer border meets this demand the best. The efficiency of the heat overtaking process depends on the heat diffusion that depends directly on the laminar sublayer thickness. The thinner laminar sublayer the smaller difference between the wall temperature and the temperature just on the border, and in consequence the better cooling abilities are. The right of above argument has been confirmed by the results of single channel simulations performed for homogeneous heat dissipation on one wall (Fig.1). While the macroscale coefficient changes along the channel the microscale one remains constant what is in agreement with the mechanics of the heat exchange phenomenon. Its a little larger value at the inlet is obvious when we take into account the laminar sublayer that starts to create itself at the inlet. At the beginning it is very narrow and achieves the final thickness on some distance. At the inlet the laminar sublayer is narrower what leads to more intensive heat exchange.

2.2 Multi-channel Approach

In Fig. 2, one can find the outline of the copper structure that has been investigated as an example of multichannel cooler [8, 9]. It contains several microchannels, with the dimensions w_{CH}xh; separated by the walls (columns) of the thickness w_{COL}. Since the heat exchange processes are homogeneous along the channel if the microscale heat transfer coefficient is used, the 3D analysis can be simplified to the investigation of 2D model that is a crosscut in the direction perpendicular to the water flow.

Once again the incoherent reports [3,5,7] forced us to look for the optimal geometry of the presented above structure. The series of numerical simulations lead us to the conclusion that in the microstructure, one can find two competitive phenomena: the enhancement and the chocking effect. The total heat removal increases when the number and the height of channels increase since the larger area of the water-heat sink contact is observed. On the other hand, the overtaken heat flux decreases when the column width decreases due to the chocking effect that results in reduction of the temperature at the bottom wall of the channel.

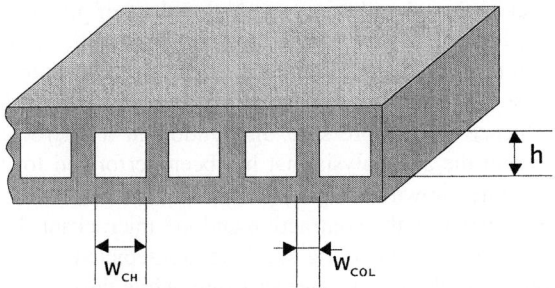

Fig. 2. The 3D multichannel structure

With the aid of numerical analysis, an influence of the channel width and height, the column width as well as the heat transfer coefficient on total heat overtaken from the structure have been examined. The considerations above are illustrated graphically in Fig. 3. The chart presents two curves for the constant column and channel width (2mm) and for two chosen heights, one twice as high as the other. The simulations have been performed for the heat transfer coefficient equal to $10W/cm^2K$. This value has been settled with the aid of one channel simulations. One can easily notice the peaks in the curves that are created by the two phenomena. The most crucial conclusion is that the optimal geometry of the structure from Fig. 2 exists. There are certain ranges of parameters deviations from the most preferable values where changes have no influence on the efficiency of the whole system.

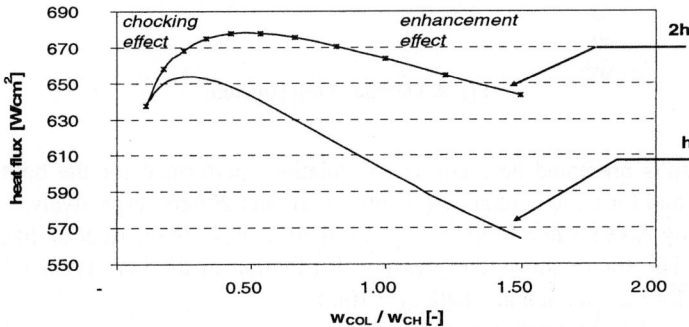

Fig. 3. Two competitive phenomena in the characteristic of heat flux versus the ratio of the column and the channel

3 Inlet Elements

All the above considerations deal with the fully developed turbulent flow inside the microchannel cooler. Its efficiency depends, however, on the flow velocity inside the channels and the input pressure that are strongly combined each with the other. Since the pressure across the liquid cooling unit is limited in microelectronic applications the pressure losses in inlet elements are of large importance for the final effectiveness of the microchannel coolers. In addition, too high values of pressure and subpressure can be very destructive, shorten the life and lower the reliability of the whole system. Therefore, one must take into considerations the hydrodynamic entrance phenomena.

In this chapter some results of the numerical investigations of the inlet element shape and its influence on the pressure distribution in a microchannel cooler are reported. They present the 2D analysis that has been performed for the cross-section of the cooling structure shown in Fig.3. It corresponds to the water flow path that consists of the inlet chamber, the contraction and the microchannel section. In Fig. 3 two considered solutions of the cooling structure are depicted - the dotted line in the contraction section shows the sharp edges structure, while the continuous one presents the smooth structure. During the simulations, the contraction length, d_k, has varied in the range 5 ÷ 20mm. The other dimensions have been settled on the basis of the design and technological restrictions as well as the results of earlier numerical analysis. The simulations have been conducted with the aid of ANSYS software for similar boundary conditions as the previous ones. The inlet pressure has been assumed as equal to 40kPa and the outlet pressure has been kept at 0Pa. It has been established on the basis of earlier considerations of mass flow resulting from the pump efficiency and additional limits for allowable pressure in microstructures.

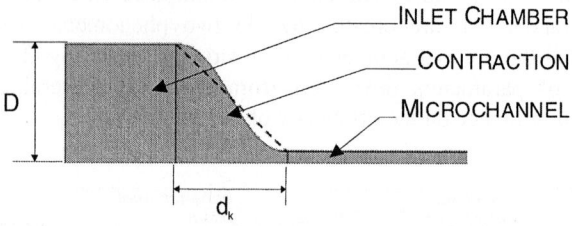

Fig. 4. Outline of test structure

The results presented here concern simulations performed for the both types of structures and for three contraction lengths 5, 10 and 20mm, respectively. They were delivered by ANSYS in a form of maps presenting pressure distribution like the ones in Fig. 5. The figure shows the pressure distribution in the both types of analyzed structures for the contraction length d_k = 10mm.

The basic data characterising the hydrodynamics processes in the considered inlet elements, like the highest static subpressure p_{min} and the highest static pressure p_{max}, have been extracted from the pressure maps and are collected in Table 1 together with additional characteristic parameters discussed below.

Comparing the values of maximal and minimal pressure, one can notice a surprising large magnitude of the subpressure in the sharp edge design contrary to the smooth edge one, where the subpressure area almost does not exist.

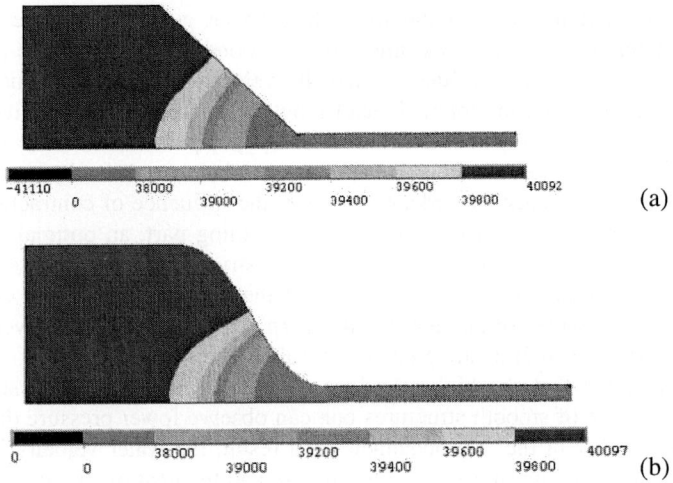

Fig. 5. Pressure distribution in the structures with $d_k = 10$mm and with (a) sharp edges and (b) smooth edges

Table 1. Characteristic parameters for various inlet structures

contraction length d_k	p_{min} [kPa]	p_{max} [kPa]	p_{3mm} [kPa]	d_{opt} [mm]
	sharp edge structures			
5mm	- 47.4	40.09	4.05	6.50
10mm	- 41.1	40.09	5.39	3.00
20mm	- 20.2	40.09	5.86	1.50
	smooth edge structures			
5mm	- 1.8	40.11	5.92	1.40
10mm	0	40.10	5.96	1.20
20mm	0	40.09	5.82	1.00

It indicates that the additional efforts to get the smooth edge contraction are worth to be undertaken. It can result in the higher reliability and the considerably longer microstructure life. One can strengthen this result increasing the contraction length d_k. There are some natural limits arisen from the permissible length of the cooling structure but they apply to sharp designs rather. In case of smooth inlets the subpressure becomes negligible small at $d_k = 10$ mm. The pressure gradients generated in contraction segment penetrate the entrance of the microchannel disrupting the pressure distribution in that region and changing the heat transfer process inside the whole microstructure. Therefore, it is impossible to use an input microchannel pressure to evaluate the microchannel cooling efficiency. One needs the input pressure independent on the local disturbance generated by the contraction segment for this goal. One has assumed that this condition is met at the place where the pressure starts to change linearly and the velocity profile is symmetrical. This distance measured from the microchannel input is denoted as d_{opt}, and its values for considered structures are gathered in Table 1. They depend on the contraction shape and dimensions. The smaller the value the lower turbulence and subpressure occur in the microchannel entrance. In order to evaluate the influence of contraction segment design on the phenomena in the microchannel cooling part, an optimal distance for microchannel pressure determination for all the structures should be settled. It has been chosen as equal to 3mm on the base of the d_{opt} values. Such a value allows comparing the pressures of all smooth and sharp edge structures. The average values of the static pressure at 3mm are gathered in Table 1.

One can notice that the highest values of p_{3mm} occur for smooth structures. It means that in case of smooth structures one can observe lower pressure drop on inlet element and higher in the microchannel. As a result, the water velocity is higher in the microchannel, the turbulence is more intense and the heat overtaking efficiency is better. On the other hand, the sharp edges introduce strong turbulent eddies at the entrance to the channel that are suppressed at quite a long distance. This kind of turbulences result in the high subpressure and can be very destructive for microchannel structure what is an additional disadvantage of this design contrary to the smooth one. The av. pressure at the 3mm from the inlet vs. contraction length for the smooth structure is graphically presented in Fig. 3. The maximum pressure is obtained for $d_k = 8mm$. For the values of contraction length in the range from 5 to 10mm, the differences in pressure are lower than 1%. Hence one can choose an arbitrary d_k value from the given set. An area of low eddy turbulences (subpressure) is created even for the smooth but too abrupt contraction. The elongation of d_k causes the shrinkage and finally diminishing of the subpressure region. It results in a peak at the curve presented in Fig. 3.

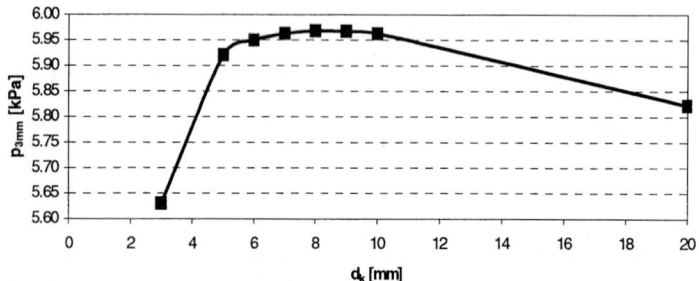

Fig. 6. The average pressure at the 3mm for smooth edge structures

4 Conclusions

The paper aims at the problem of microchannel liquid cooling system analysis. As an necessary element of the analysis, a new method for heat transfer coefficient calculation that is dedicated for microscale analysis has been introduced. Contrary to existing, standard methods, the new approach describes the local phenomena and is used in the further investigation of the cooling microstructure. An effect of its geometry on the total overtaken heat flux is analysed with respect to optimal conditions as well as to technological restrictions. Furthermore, an influence of the shape of the inlet element on pressure distribution in the structure is presented. It is proven that the contraction length as well as the edge design is very essential for life and reliability of the whole system and in consequence for thermal efficiency of the microstructure. The main aims of the inlet element optimisation process are to minimise maximum pressure and subpressure values, to reduce the subpressure area and to decrease the pressure drop losses.

References

1. Capriz, C.: Trends in Cooling Technologies for Power Electronics. Power Electronics. no 1/2 (1999) 22-24
2. Haque, S. et all: Thermal Management of High-Power Electronics Modules Packaged with Interconnected Parallel Plates. SEMI-THERM, San Diego, USA, (1998) p. 111
3. Tuckerman, D.B., Pease, R.F.W.: High Performace Heat Sinking for VLSI. IEEE Electron Devices Lett. EDL-2 (1981) 126-129
4. Goodson, K.: Electroosmotic Microchannel Cooling System for Microprocessor. Electronics Cooling. Vol.8. (2002)
5. Harms, T.M., Kazimierczak, M.J., Gerner, F.M.: Developing Convective Heat Transfer in Deep Rectangular Microchannels. Int. J. Heat Fluid Flow. Vol.20. (1999) 149-157
6. Sittig, R., Steiner, T.:Vision of Power Modules. Proc. Inter. Conf. CIPS 2000, Bermen, Germany (2000) 134-139
7. Gillot, C.H., Bricard, A., Schaeffer, C.H.: Single and Two-phase heat exchangers for Power Electronics. Int. J. Therm. Sci. Vol.39. (2000) 826-832
8. Raj, E., Langer, M., Lisik, Z.: Numerical Studies for Jet Liquid Cooling in Electronics. Proc. Int. Conf. Thermic'2000. Zakopane, Poland (2000) 68-72
9. Langer, M., Lisik, Z., Raj, E.: Optimising of Microchannel Cooling. Proc. Int. Conf. ICSES'2001. Łódź, Poland (2001) 383-387
10. Pfahler, J. et all: Liquid Transport in Micron and Submicron Channels. J. Sensors Actuators. Vol.21. (1990) 431-434
11. Peng, X.F., Peterson, G.P.: Convective Heat Transfer and Flow Friction for Water in Microchannel Structures. Int. J. Heat Mass Transfer . Vol.39. (1996) 2599-2608
12. Tretheway, D.C., Meinhart, C.D.: Apparent Fluid Slip at Hydrophobic Microchannel Walls. Physics of Fluids. Vol.14. (2002)
13. Incropera, F.P., DeWitt, D.P.: Fundamentals of Heat and Mass Transfer. 3rd ed. Wiley-Interscience, New York (1990)

Simulation of Nonlinear Thermomechanical Waves with an Empirical Low Dimensional Model

Linxiang Wang[1] and Roderick V.N. Melnik[2]

[1] MCI, Faculty of Science and Engineering,
University of Southern Denmark,
Sonderborg, DK-6400, Denmark
[2] Centre for Coupled Dynamics & Complex Systems,
Wilfrid Laurier University,
75 University Avenue West,
Waterloo, ON, Canada, N2L 3C5

Abstract. In this paper we analyse the performance of a low dimensional model for the nonlinear thermo-mechanical waves. The model has been obtained by using proper orthogonal decomposition methods combined with a Galerkin projection. First, we analyse the original PDE model in order to obtain the system states at many time instances. Then, by using an empirical orthogonal basis extracted from our numerical results, we construct an empirical low dimensional model. Finally, we compare the results obtained with the original PDE model and those obtained with our low-dimensional model. These comparisons are carried out for mechanically induced phase transformations in a shape-memory alloy rod.

Keywords: Nonlinear waves, thermo-mechanical dynamics, proper orthogonal decompositions, Galerkin projections.

1 Introduction

The field of smart material systems is rapidly developing. Due to their unique properties, the smart materials have attracted an increasing attention from mathematicians, physicists, control theorists, and engineers. Smart materials such as piezoelectrics, shape memory alloys, and magnetostrictive materials can sense and respond to external stimuli. They can also be used as actuators. Ultimately, one would like to achieve a certain degree of control over phenomena associated with the complex behaviour of these materials ([17, 6] and references therein). As a result, there is an increasing number of efforts to construct simple and robust mathematical models describing the dynamics of smart materials and such that would be relatively easy to amend to control.

Control strategies for many dynamic systems described by nonlinear ODEs are well developed, in particular in those cases where the dimension of the controlled system is not large [2, 13]. At the same time, existing mathematical models for the dynamical behaviour of smart materials are typically based on a set of coupled nonlinear PDEs. For example, the mathematical models for the shape memory alloys are formulated as a system of PDEs that couples thermal and elastic fields, while the models for the piezoelectric materials are formulated as a system of PDEs that couples electric and elastic fields [11, 19]. All such models are infinite dimensional and one of the ways to deal with such models is to discretize them in space and apply the method of lines to the result. However, due to the coupling between multi-physics fields and system nonlinearities, the number of nodes for the spatial discretization needs to be sufficiently large. This leads to computational difficulties due to the fact that the resultant large system of ODEs is usually stiff. Even in the case of linear PDEs, control issues of the resulting models are highly non-trivial [18, 14, 4, 9]. Hence, it seems natural to try to approximate PDE systems such as those arising in the description of smart material systems by a lower dimensional ODE systems. One way to do that is to use the Proper Orthogonal Decomposition (POD). This is a very efficient tool for this purpose as soon as a collection of system states is available. This idea has been applied to many active control problems involving fluid flows (see [1, 14, 5, 15] and references therein). In its essence, the POD methodology is analogous to the principal component analysis, techniques based on the singular value decomposition, or the Karhunen-Loeve decomposition [8, 5, 15].

In what follows, we propose a low dimensional model for the nonlinear thermomechanical waves, describing the dynamics of shape memory alloys. The model is constructed on the basis of the numerical results obtained from the original system of coupled nonlinear PDEs. The dynamical behavior of the considered system is simulated by the empirical low dimensional model, and the performance of the low dimensional model is compared to the original PDE model. It is shown that the empirical low dimensional captures all the characteristic features of the material.

2 The Original PDE Model

Many smart materials encountered in applications have been extensively investigated by both experimentalists and theoreticians. Today, mathematical models for the 1D shape memory alloys rods (or wires) are well established on the basis of the modified Ginzburg-Landau theory. The well known Falk model has been constructed on the basis of conservation laws for linear momentum and energy, and thermo-dynamical consistency. To model the coupled thermo-mechanical wave interactions and the first order phase transitions in the shape memory alloys, we use the following 1D mathematical model [3, 11, 19]:

$$\begin{aligned}
\rho \frac{\partial^2 u}{\partial t^2} &= \frac{\partial}{\partial x}\left(k_1\left(\theta - \theta_1\right)\frac{\partial u}{\partial x} - k_2(\frac{\partial u}{\partial x})^3 + k_3(\frac{\partial u}{\partial x})^5\right) + F, \\
c_v \frac{\partial \theta}{\partial t} &= k \frac{\partial^2 \theta}{\partial x^2} + k_1 \theta \frac{\partial u}{\partial x}\frac{\partial v}{\partial x} + G,
\end{aligned} \quad (1)$$

where u is the displacement, θ is the temperature, ρ is the density, k_1, k_2, k_3, c_v and k are re-normalized material-specific constants, θ_1 is the reference temperature for 1D martensitic transformations, and F and G are distributed mechanical and thermal loadings.

It is well known that even in this one-dimensional case, the analysis of the system is far from trivial due to the strong nonlinear coupling between thermal and elastic fields. Thermal and mechanical hysteresis effects and complicated phase transformations are the phenomena that need to be dealt with. Following our previous works [10, 11, 19], we re-write our original system in a way convenient for computational implementation:

$$c_v \frac{\partial \theta}{\partial t} = k \frac{\partial^2 \theta}{\partial x^2} + k_1 \theta \epsilon \frac{\partial v}{\partial x} + G, \quad \frac{\partial \epsilon}{\partial t} = \frac{\partial v}{\partial x}$$
$$\rho \frac{\partial v}{\partial t} = \frac{\partial}{\partial x} \left(k_1 (\theta - \theta_1) \epsilon - k_2 \epsilon^3 + k_3 \epsilon^5 \right) + F, \quad (2)$$

where $\epsilon = \partial u / \partial x$, $v = \partial u / \partial t$.

3 The Construction of a Low Dimensional Model

3.1 Orthogonal Basis

The first step in constructing a low dimensional model is to construct an effective basis for the approximation of the system states. This step is equivalent to the one described in [12] where Eq.(2) is rewritten in a general dynamical system form

$$\frac{\partial \mathcal{U}(x, t)}{\partial t} = \mathcal{F}(\mathcal{U}(x, t)), \quad (3)$$

and $\mathcal{U}(x, t)$ is the sought-for vector function with components ϵ, v and θ. This function depends on the spatial position continuously in a given domain Ω and $\mathcal{F}$ is a nonlinear function of $\mathcal{U}$, and the first, second order derivatives of $\mathcal{U}$. At this stage, loadings are not included in the above model.

The POD is concerned with the possibility to find a set of orthonormal basis functions $\phi_j(x), j = 1, \ldots, P$ which are optimal in the sense that the P dimensional approximation

$$\mathcal{U}_P(x, t) = \sum_{i=1}^{P} a_i(t) \phi_i(x) \quad (4)$$

gives the best approximation to the function $\mathcal{U}(x, t)$ among all those P dimensional approximations, in the least square sense [8, 5, 16]. As usual, here a_i are the general Fourier coefficients for ϕ, that are functions of time. In other words, the idea of POD applied here is based on a choice of the basis functions ϕ to maximize the mean projection of the function $\mathcal{U}(x, t)$ on ϕ

$$\max_{\phi \in L_2(\Omega)} \frac{E\left(|\langle \mathcal{U}, \phi \rangle|^2\right)}{\|\phi\|^2} \quad (5)$$

where $E(\cdot)$ denotes the mean value functional, and $\langle \cdot \rangle$ is the inner product [5, 16]. Finally, the maximization problem leads to the following eigenvalue problem

$$\int_\Omega E(\mathcal{U}(x)\mathcal{U}(x'))\phi(x')dx' = \lambda\phi(x), \tag{6}$$

whose kernel $\mathcal{K} = E(\mathcal{U}(x)\mathcal{U}(x'))$ can be interpreted as the auto-covariance function of the two points x and x' (see [5, 16] and references therein).

In practice, the states of the above system could be obtained by either experimental measurements or numerical simulation with suitable initial and boundary conditions, so $\mathcal{U}(x,t)$ will be a discrete function in both time and space. If we assume that the system states are available at N different time instances, we will call the system state at the ith time instance U^i as the ith snapshot. In the discrete form, each snapshot can be written as a column vector with M entries, where M is the number of nodes for spatial discretization.

In order to construct the orthonormal basis, all the snapshots need to be collected in one matrix $U = \{U^i, i = 1, \ldots, N\}$, and we can put each snapshot as one column in the collection matrix so that the collection U will be a $M \times N$ matrix. Then, the orthonormal basis for the given collection U can be calculated by the singular value decomposition as follows

$$U = LSR^T \tag{7}$$

where L is $M \times M$ orthonormal matrix, R is a $N \times N$ orthonormal matrix, S is a $M \times N$ matrix with all elements zero except along the diagonal and those non-zero elements are arranged in a decreasing order along the diagonal (the singular values of U with associated eigenvectors in L and R).

If we let $SR^T = Q^T$ in the singular value decomposition, then $U = LQ^T$. Let then ϕ_k be the kth column of L and a_k be the kth row of Q, so that the matrix U's singular value decomposition can be rewritten as

$$U = \sum_{k=1}^{m} a_k \phi_k \tag{8}$$

where $m = \min(N, M)$ is the rank of the collection matrix U. This approximation is what we are looking for in Eq.(4). Following [7, 5], we note that the lower dimensional approximation of the matrix U can be easily obtained by just keeping the first few largest singular values and their associated eigenvectors in L and Q, and the number of eigenvectors should be determined by compromising between the dimension number of the resultant system and the approximation accuracy. Furthermore, the basis vectors obtained from the singular value decomposition are orthonormal, which gives us the following relations

$$\langle \phi_i, \phi_j \rangle = \begin{cases} 1 & \text{if } i = j \\ 0 & \text{if } i \neq j \end{cases} \tag{9}$$

and the general Fourier coefficients in this case could be calculated as

$$a_k = \langle U^k, \phi_k \rangle. \tag{10}$$

3.2 Galerkin Projection

The idea of constructing a lower dimensional dynamic system from a given higher dimensional system is to replace the dynamics of the given system by a lower dimensional subspace of the origin state space [7, 5, 12]. Following the standard procedure, we substitute the approximation Eq.(4) into the given dynamic system Eq.(3), with the basis vectors extracted from the collection matrix U, and write it in the residual form

$$r(x,t) = \frac{\partial \mathcal{U}(x,t)}{\partial t} - \mathcal{F}(\mathcal{U}(x,t)) = \sum_{k=1}^{M} \frac{\partial a_k}{\partial t}\phi_k - \mathcal{F}(\sum_{k=1}^{M} a_k \phi_k). \qquad (11)$$

Then, we use a Galerkin projection so that we approximate the system by a lower dimensional subspace approximation, and set the residual, induced by this approximation, to be orthogonal to all the basis functions

$$(r, \phi_k) = \int_{\Omega} r(x)\phi(x)dx = 0. \qquad (12)$$

In the discrete case, we set the inner product between the residual vector and all the basis vectors to zero, so that for the given dynamic system, it is easy to get that

$$\langle \sum_{k=1}^{M} \frac{\partial a_k}{\partial t}\phi_k, \phi_j \rangle = \langle \mathcal{F}(\sum_{k=1}^{M} a_k \phi_k), \phi_j \rangle. \qquad (13)$$

Since the ϕ_j produces a set of orthonormal basis vectors, the system can be re-cast into the following set of ODEs

$$\frac{\partial a_j}{\partial t} = \langle \mathcal{F}(\sum_{k=1}^{M} a_k \phi_k), \phi_j \rangle, \quad k, j = 1, \ldots, M. \qquad (14)$$

Since the basis vectors are extracted from the snapshots, all the boundary conditions are "embedded" into the basis vectors which help avoid a stiff system of equations. The resulting system of ODEs is integrated given the initial conditions projected into the orthonormal basis, and the loadings are applied.

4 Numerical Results

In what follows we demonstrate the technique described above on the empirical low dimensional model by simulating the dynamical behavior of a SMA rod. The dynamical behavior of the SMA rod is strongly nonlinear with coupling effects between the elastic and thermal fields, first order martensitic transformations, and hysteresis [3, 11, 10, 19]. The simulation is performed for a $Au_{23}Cu_{30}Zn_{47}$ rod of length $L = 1$cm based on Eq.(1). All the physical parameters for this specific material are taken here the same as in [11]. We use the following procedure.

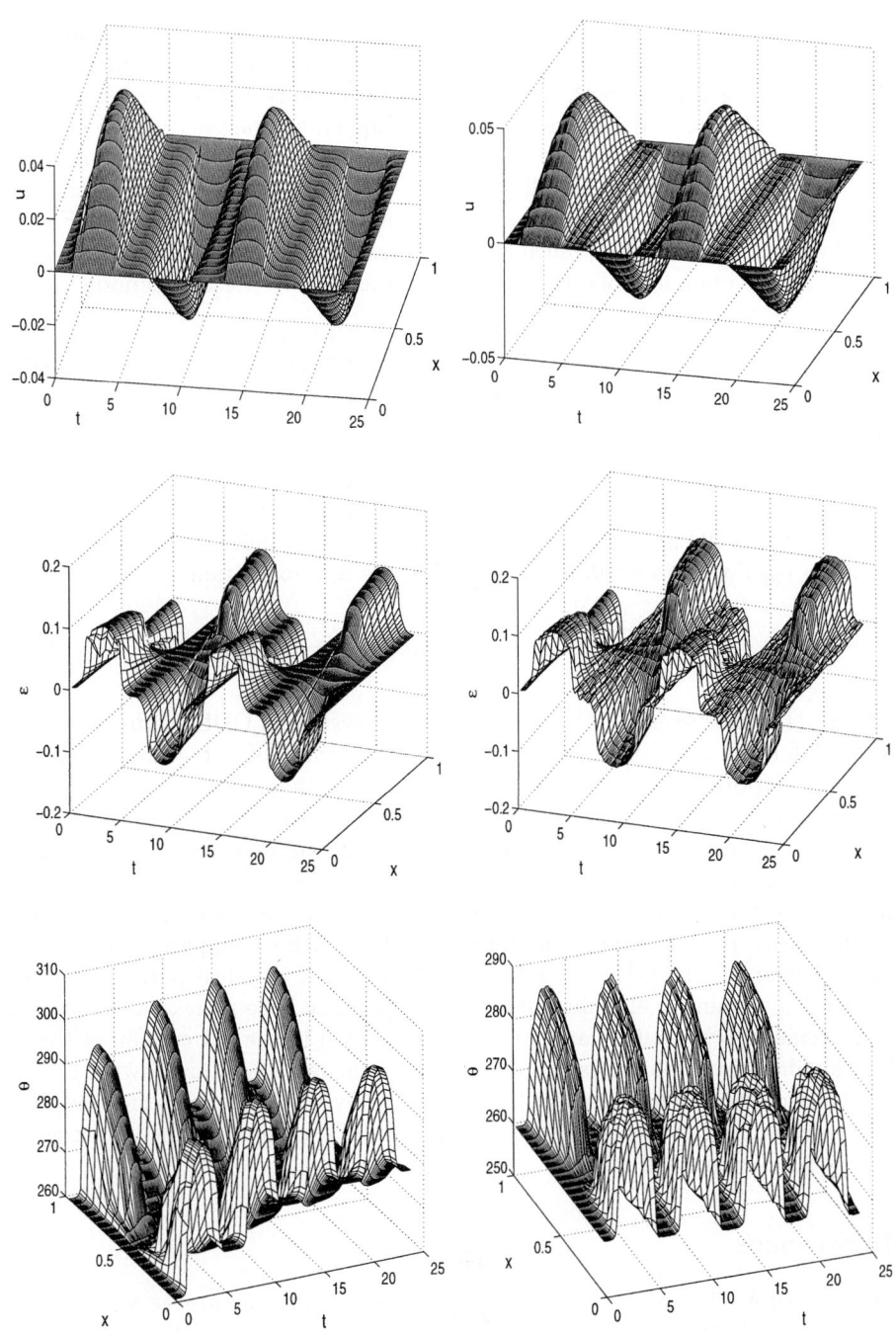

Fig. 1. Comparison of the nonlinear thermo-mechanical behavior of a SMA rod: the full PDE model (left) and the empirical low dimensional model (right)

First, we perform the numerical simulation using Eq.(1) with a representative mechanical load. As a result, the collection matrix will be constructed and the empirical orthonormal basis vectors can be extracted. Then, we simulate the dynamics using Eq.(14) with different mechanical loads. By comparing the numerical results from Eq.(14) with those from Eq.(1), we analyse the performance of the low dimensional model in reproducing the main features of the dynamics.

The initial conditions for all simulations are set the same. In particular, $\theta(x,0) = 260^\circ K$ and $\epsilon(x,0) = v(x,t) = 0$. Boundary conditions for Eq.(1) are taken as mechanically pinned-end and thermally insulated. The distributed mechanical loading for collecting snapshots is as follows (for one period)

$$F = 7000 \begin{cases} t/3, & 0 \leq t \leq 3, \\ (6-t)/3, & 3 \leq t \leq 9, \\ (t-12)/3, & 9 \leq t \leq 12. \end{cases} \quad (15)$$

We follow here ideas reported in [11, 10, 19]. There are 18 nodes used for ϵ and θ discretization and 19 nodes used for v. Two periods of loadings are performed ($t \in [0, 24]$) and totally 201 evenly distributed snapshots are sampled.

After the POD is applied, there are 9 basis vectors that are kept for ϵ, 9 for v, and 8 basis vectors for θ. By substituting the approximation

$$u = \sum_{i=1}^{9} \epsilon_i \phi_i^\epsilon, \quad v = \sum_{i=1}^{9} v_i \phi_i^v, \quad \theta = \sum_{i=1}^{8} \theta_i \phi_i^\theta \quad (16)$$

into Eq.(1), the system will be converted into a system of ODEs with dimension of 26 (9 for ϵ, 9 for v, and 8 for θ). Then, a standard ODE integrator (ode23 in Matlab) is applied to simulate the state evolution, without changing any physical parameters, except the mechanical loadings that is now $F = 6000 \sin(\pi t/6)^3$.

The numerical results obtained with the low dimensional model are presented in the right column of Fig.1. The displacements (displacement is calculated by integrating ϵ along x after simulation), temperature and strain distributions in the entire rod are plotted as functions of time. For comparison purposes, the behavior of the SMA rod with exactly the same physical parameters and mechanical loading is also simulated using Eq.(1), and the numerical results for displacement, temperature and strain are presented in the left column of Fig.1, in a similar way. The validation of the numerical results obtained with Eq.(1) has been previously discussed in [19, 11]. By analysing Fig. 1, we conclude that all the characteristic features of the material are captured by the constructed here lower dimensional model.

References

1. Atwell,J.A., Borggard,J.T., King,B.B.: Reduced Order Controller for Burgers' Equation With a Nonlinear Observer. Int.J.Appl.Math.Comput.Sci. **6** 11 (2001) 1311-1330.
2. Beeler,S.C., Tran.H.T., Banks.H.T.: Feedback control methodologies for nonlinear systems, Journal of Optimization Theory and Application, **1** 107 (2000) 1-33

3. Falk, F., Konopka, p.: Three-dimensional Landau theory describing the martensitic phase transformation of shape memory alloys. J.Phys.:Condens.Matter **2** (1990) 61-77.
4. Fattorini,H.O.: Infinite Dimensional Optimization and Control Theory. Cambridge University Press, (1999) Cambridge.
5. Holmes,P., Lumley,J.L., Berkooz,G.: Turbulence, Coherent Structures, Dynamical systems and Symmetry. Cambridge University Press, (1996) Cambridge.
6. Hu,M, Du,H.J., Ling,S.F., Zhou,Z.Y., Li,Y.: Motion Control of Electrostrictive Actuator. Mechatronics, **2** 14 (2004) 153-161
7. Kerschen,G., Colinval,J.C.: Physical Interpretation of the Proper Orthogonal Modes Using the Singular Value Decomposition. Journal of Sound and Vibration. **5** 249 (2002) 849-865.
8. Liang,Y.C., Lee,H.P., Lim,S.P., Lin,W.Z., Lee,K.H., Wu,C.G.: Proper Orthogonal Decomposition And Its Applications-Part I:Theory. Journal of Sound and Vibration, **3** 252 (2002)527-544.
9. Lurie,K.A., Applied Optimal Control Theory of Distributed Systems. Plenum Press, (1993) New York.
10. Melnik, R., Roberts, A., Thomas, K.: Coupled thermomechanical dynamics of phase transitions in shape memory alloys and related hysteresis phenomena. Mechanics Research Communications **28** 6 (2001) 637-651.
11. Melnik, R., Roberts, A., Thomas, K.: Phase transitions in shape memory alloys with hyperbolic heat conduction and differential algebraic models. Computational Mechanics **29** (1) (2002) 16-26.
12. Melnik, R.V.N. and Roberts, A.J., Modelling nonlinear dynamics of shape-memory alloys with approximate models of coupled thermoeleasticity. ZAMM: Zeitschrift fur Angewandte Mathematik Mechanik **83** (2) (2003) 93–104.
13. Meyer,M., Matthies,H.G.:Efficient Model Reduction in Nonlinear Dynamics Using the Karhunen-Loeve Expansion and Dual-Weighted Residual Methods. Computational Mechanics **31** (2003) 179-191
14. OR,A.C., Kelly,R.E.: Feedback Control of Weakly Nonlinear Rayleigh Bernard Marangoni Convection. J.Fluid.Mech., **440** (2001) p27-47.
15. Rowley,C.W, Colonius,T., Murray,R.M.: Model Reduction for Compressible Flows Using POD and Galerkin Projection. Physic D **189** (2004) 115-129.
16. Rowley,C.W., Marsden,J.E.: Reconstruction Equations and Karhunen-Loeve Expansions for Systems with Symmetry. Physica.D **142** (2000) 1-19.
17. Sood,D.K, Lawes,R.A., Varadan,V.V (Eds): Smart Structures and Devices, Proceedings of SPIE, December 2000, Melbourne, Australia.
18. Teman,R.: Infinite Dimensional Dynamical Systems in Mechanics and Physics. Springer-Verlag (1998) New York
19. Wang, L., Melnik, R.: Nonlinear coupled thermomechanical waves modelling shear type phase transformation in shape memory alloys. in Mathematical and Numerical Aspects of Wave Propagation, Eds.G.C.Cohen, et al,Springer,723-728 (2003).

A Computational Risk Assessment Model for Breakwaters

Can Elmar Balas

Gazi University, Faculty of Engineering and Architecture,
Civil Engineering Department, 06570 Ankara, Turkey
cbalas@gazi.edu.tr

Abstract. In the reliability-risk assessment, the second order reliability index method and the Conditional Expectation Monte Carlo (CEMC) simulation were interrelated as a new Level III computational approach in order to analyse the safety level of the vertical wall breakwaters. The failure probabilities of sliding and overturning failure modes of the Minikin method for breaking wave forces were forecasted by approximating the failure surface with a second-degree polynomial having an equal curvature at the design point. In this new computational approach, for each randomly generated load and tide combination, the joint failure probability reflected both the occurrence probability of loading condition and the structural failure risk at the limit state. This new approach can be applied for the risk assessment of vertical wall breakwaters in short CPU durations of portable computers.

1 Introduction

In the structural design of vertical wall breakwaters, two methods have been widely applied in European countries. The first method is the First Order Mean Value Approach (FMA) [1], and the second one is the Hasofer-Lind second order reliability (HL) index. The partial coefficient system utilizes the former and the latter has been employed to compare risk levels of rubble mound and vertical wall structures [2]. Goda and Tagaki [3] suggested a reliability design criteria in which the Monte Carlo simulation of expected sliding distance was carried out for caisson breakwaters.

The reliability-risk assessment of Ereğli harbor main breakwater involves the second order reliability index (β_{II}) method interrelated with CEMC simulation as a Level III method. In this technique, uncertainties that affected most of the variables in the design were incorporated throughout the lifetime of structures by the use of the simulation of design conditions, i.e. the water level change due to tidal action and the random wave action. This proposed Level III computational methodology was compared with the individual application of β_{II} (Level II) method.

2 Computational Risk Model

The safety of vertical wall breakwater was evaluated by modelling random resistance and load variables with common probability distributions at their limit-state. The primary variable vector **z** in the normalized space indicates these random variables.

The functional form of the basic variables consistent with the limit state is the failure function denoted by: $g(\mathbf{z})=(z_1,z_2,...,z_n)$. The safety of the structure can be assured by designating an admissible value of the probability of achieving the limit state defined by: $g(\mathbf{z})=0$. In the reliability-based study, the second-order reliability index method was utilized, in which the failure surface was approximated by a rotational parabolic surface. The parabolic limit state surface in standard normal space, $g(\mathbf{z})$ [4] was taken in the model as follows:

$$g(\mathbf{z}) \approx a_0 + \sum_{i=1}^{n} b_i z_i + \sum_{i=1}^{n} c_i z_i^2 \tag{1}$$

where a_0, b_i, and c_i are the regression coefficients of the second-order polynomials; z_i are the standardized normal random variables and n is the number of random variables. Regression coefficients were obtained by using the response surface approach in standard normal space [5]. The positive sum of the principle curvatures of limit state surface at the design point ($\mathbf{z}^*$) was expressed as:

$$K_s = \frac{2}{|\nabla g|} \sum_{i=1}^{n} c_i \left[1 - \frac{1}{|\nabla g|^2} \left(b_i + 2c_i z_i^* \right)^2 \right] \tag{2}$$

$$|\nabla g| = \sqrt{\sum_{i=1}^{n} \left(b_i + 2c_i z_i^* \right)^2} \tag{3}$$

$$\beta_{II} = -\phi^{-1} \left[\phi(-\beta_I) \left(1 + \frac{\varphi(\beta_I)}{R\phi(-\beta_I)} \right)^{-\frac{n-1}{2} \left(1 + \frac{2K_s}{10(1+2\beta_I)} \right)} \right] \tag{4}$$

where, β_{II} is the second-order reliability index, R is the average principal curvature radius expressed as $R=(n-1)/K_s$, β_I is the first order reliability index $\beta_I = \mathbf{a}^T \mathbf{z}^*$; Φ is the standard normal distribution function, φ is the standard normal probability density function, $\mathbf{a}$ is the directional vector at the design point. The structural performance of the breakwater under the affect of wave loading was investigated by utilizing the Conditional Expectation Monte Carlo (CEMC) simulation. The exceedance probability (P_f) of failure damage level was obtained by utilizing the control random variable vector of $\mathbf{z}_i = (z_{i1}, z_{i2}, ..., z_{ik})$ as follows:

$$P_f = E_{z_j : j=1,2,...,n \text{ and } j \neq i} [P_f(\mathbf{z}_i)] \tag{5}$$

where, $E[.]$ is the conditional expectation (mean) and $P_f(\mathbf{z}_i)$ is the failure probability evaluated for $z_{i1}, z_{i2}, ..., z_{ik}$, by satisfying the conditional term in eqn (6) for the last control variable as follows:

$$P_f(z_{ik}) = \Pr[z_{ik} < g_{ik}(z_j : j=1,2,...,n \,\&\, j \neq i)] \quad (6)$$

where, k is the number of control variables in the simulation. A computer program was developed for the simulations that repetitively reproduced breakwater performance at the limit state condition until the specified standard mean error of convergence (ε) was satisfied. The limit state equations for breaking wave forces acting on the vertical wall breakwaters were derived in this study from the Minikin's method [6] as illustrated in Figure (1). For sliding failure mode, the limit state equation utilized in the model was [7]:

$$g = \mu_f \left[Bh_s\gamma_c - Bd_s\gamma_o - \frac{1}{4}\gamma_o BH_b \right] - \frac{101}{3}\gamma_o \frac{d_s(d_s+d)}{dL_d} H_b^2 - \frac{1}{2}\gamma_o d_s H_b$$

$$- \frac{1}{8}\gamma_o H_b^2 = 0 \quad (7)$$

The limit state equation for the overturning failure mode was obtained as:

$$g = \left(\frac{h_s B^2}{2}\right)\gamma_c - \left[\left(\frac{1}{4}d_s^2\gamma_o + \frac{1}{6}\gamma_o B^2\right)H_b + \left(\frac{101}{3}\frac{\gamma_o}{L_d d}d_s^3 + \frac{101}{3}\frac{\gamma_o}{L_d}d_s^2\right)\right.$$

$$\left. + \frac{1}{8}\gamma_o d_s \right)H_b^2 + \left(\frac{1}{48}\gamma_o\right)H_b^3 + \left(\frac{1}{2}d_s B^2 \gamma_o\right) \right] = 0 \quad (8)$$

In eqns (7) and (8), d_s is the depth from still water level, h_s is the height of vertical wall breakwater, B is the width of wall, μ_f is the coefficient of friction, γ_o is the weight per unit volume of seawater, γ_c is the weight per unit volume of concrete, H_b is the breaking wave height, h_c is the breaker crest taken as $H_b/2$, P_m is the maximum pressure acting at the SWL, d is the depth at a distance one wavelength seaward of the structure, L_d is the wavelength at the water of depth d.

In the application of the suggested Level III method, the offshore wave height was randomly generated and a linear wave transformation was carried out to obtain the design load of the structure. Then, the reliability of the structure was investigated (on average 30,000 times) by the β_{II} method at the limit state. As a result, the joint failure risk reflected the occurrence probabilities of wave loading and the limit state for each random load combination generated in the simulation. Then, the β_{II} method was applied individually to the case study as a Level II approach and the results obtained from these methods were compared with each other.

3 Model Application

A commercial harbor will be constructed in Marmara Ereğlisi on the inland Sea of Marmara of Turkey (Figures 2 and 3). The basic parameters in the design are listed in Table (1) with the mean (μ) and standard deviation (σ) of normally distributed random variables.

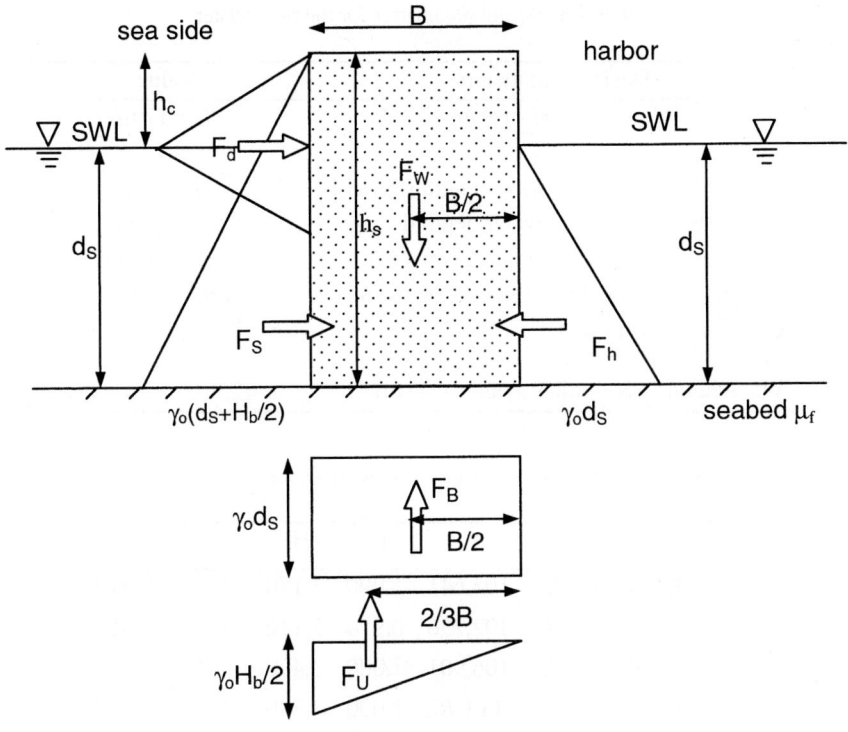

Fig. 1. Breaking wave forces

Fig. 2. Locations of recently planned harbours in Turkey

Table 1. Design parameters for the breakwater

Design Parameters	Value	
Height of the structure	$h_s = 17$ m	
Depth at the toe of the wall	$d_s = 6.5$ m	
Depth at distance one wave length seaward	$d = 8$ m	
Wave length at water depth d	$L_d = 76$ m	
Weight per unit volume of sea water	$\gamma_o = 1.02$ t/m^3	
Weight per unit volume of concrete (t/ m^3)	$\mu=2.4$ t/m^3	$\sigma=0.1$ t/m^3
Coefficient of friction (normally distributed)	$\mu= 0.64$	$\sigma= 0.1$
Wave height (Weibull-Rayleigh) H (m)	$\alpha =2.97$	$\beta =1.55$

Table 2. Annual maximum significant wave characteristics of the site

M	Hs	Ts	L_o	Kr	H_o'	H_b/H_o'	H_b
1	3,170	8,072	101,641	1,000	3,170	1,11	3,519
2	3,200	8,096	102,256	0,974	3,116	1,11	3,459
3	3,430	8,278	106,901	1,000	3,430	1,1	3,773
4	3,780	8,539	113,76	1,000	3,780	1,09	4,120
5	3,820	8,568	114,528	1,000	3,820	1,09	4,164
6	3,900	8,625	116,058	0,973	3,793	1,09	4,134
7	4,540	9,055	127,911	0,972	4,411	1,08	4,764
8	4,880	9,267	133,962	0,984	4,802	1,08	5,186
9	4,900	9,279	134,313	0,984	4,822	1,08	5,207
10	5,070	9,381	137,277	0,984	4,989	1,08	5,388
11	5,520	9,640	144,955	0,984	5,432	1,07	5,812
12	6,020	9,911	153,227	0,984	5,924	1,06	6,279
13	6,650	10,231	163,305	0,984	6,544	1,05	6,871

The wave height was modeled by a joint Weibull-Rayleigh probability distribution with the scale (α) and shape (β) parameters listed in Table (1) by using the wave characteristics listed in Table (2) [8]. In Table (2), H_b is the breaker height at construction depth, H_o' is the unrefracted deep-water wave height, M is the plotting position, Kr is the linear refraction coefficient, Hs and Ts are the annual maximum significant deep-water wave height and period, respectively. The new Level III reliability approach, in which the second order reliability index (β_{II}) and the Conditional Expectation Monte Carlo (CEMC) simulation were interrelated, was suggested to handle the uncertainties inherent in wave data and design methodology

[9]. Wave characteristics of the site were randomly generated by simulation. Afterwards, the failure mode probability was predicted by the parabolic limit state surface having the identical curvature at the design point with the higher degree failure surface. The mean (μ) and the standard deviation (σ) of the wave height distribution were μ_H= 4.77m and σ_H = 0.95m, respectively. Probabilities of failure of the Ereğli vertical wall breakwater in 50 years of lifetime obtained by the suggested Level III approach for both sliding and overturning failure criteria are given in Figure (4) in which sliding criterion governs the design. The sensitivity study carried out by using a rank correlation method reveals that the overturning failure function is sensitive to the wave height with a correlation coefficient of Rc=-0.84 (load variable) obtained by Level II and Level III methods for sliding criterion under breaking wave forces and to the weight per unit volume of concrete with Rc=0.46 (resistance variable). The sliding failure function is sensitive to the wave height with Rc=-0.73 (loading variable), to the weight per unit volume of concrete Rc=0.22 (resistance variable), and to the coefficient of friction with Rc=0.57.

For sliding failure criterion, the probability of failure determined from Level II reliability method is lesser than the probability of failure obtained from Level III reliability method with a root mean square error of 0.25 and a bias of –0.17 (Figure 5). For overturning failure criterion, the probability of failure obtained from Level III reliability method is greater than the probability of failure obtained from Level II method with a root mean square error of 0.32 and a bias of –0.17.

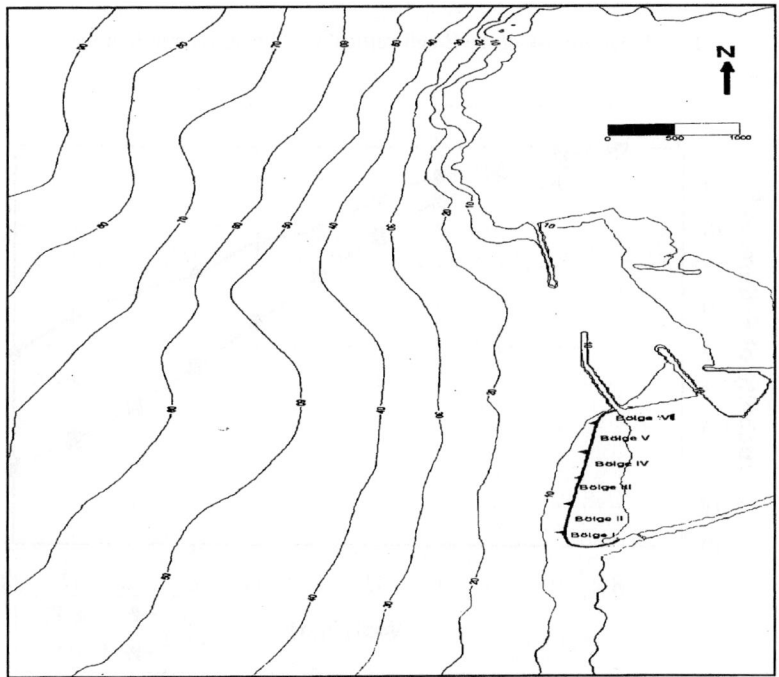

Fig. 3. Ereğli (Marmara) harbour

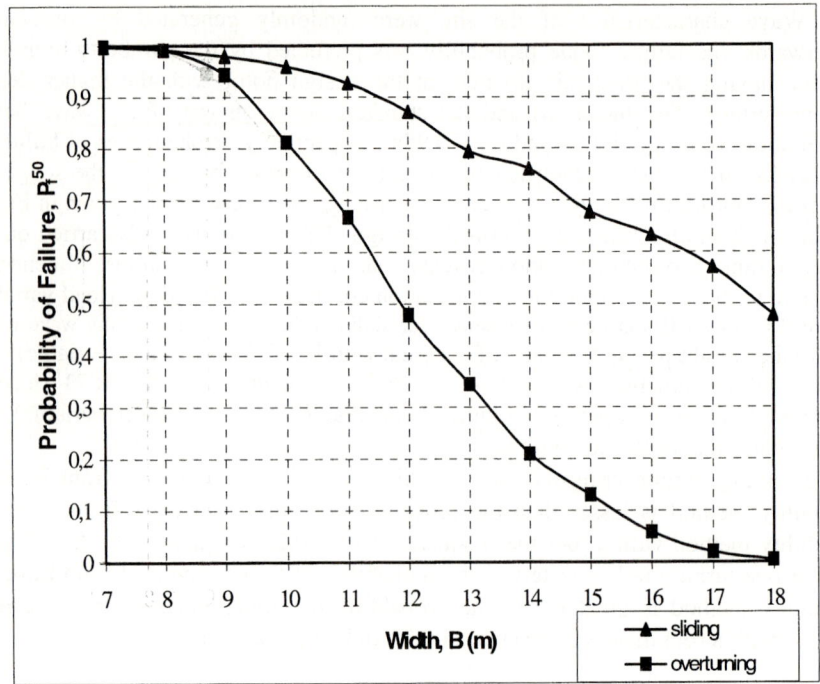

Fig. 4. Failure risk of the Ereğli vertical wall breakwater in lifetime

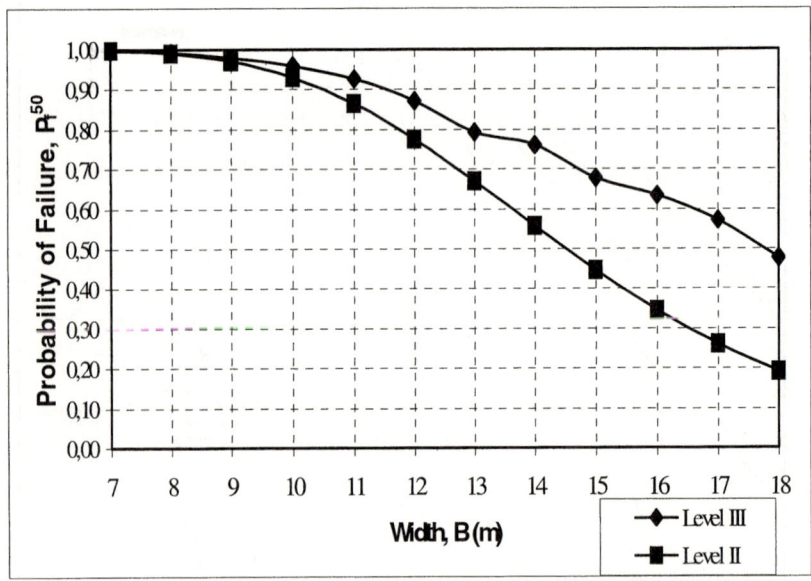

Fig. 5. Comparison of failure probability of vertical wall breakwater in a lifetime of 50 years obtained by Level II and Level III methods for sliding criterion under breaking wave forces

4 Conclusions

A new computational simulation methodology was presented for the design of breakwaters. For the case study at Ereğli, results obtained from Level II method deviated from results obtained by simulation. Therefore, structural reliability evaluated by using Level II method was considered as approximate, when compared to the Level III method presented in this paper. The type of distribution (Normal or extreme value) in design parameters of the failure function also effected the reliability evaluations irrespective of the design level. As a result, the reliability of vertical wall structures was highly variable and depended upon the unpredictable nature of coastal storms, the reliability method and distributions utilized in the design.

This new computational method has advantages in practical design applications, since the random behaviour of structural performance in lifetime can be estimated at the planning stage. The new computational approach applied in this paper within few minutes of CPU time in a portable computers, was recommended for the risk assessment of vertical wall breakwaters.

References

1. Burcharth, H.F. and Sorensen J.D., Design of vertical wall caisson breakwater using partial safety factors. Proc. of the 26th International Conference on Coastal Engineering, ASCE, Copenhagen, Denmark (1998) 2138-2151.
2. Burcharth, H.F. and Sorensen J.D., The PIANC safety factor system for breakwaters. Proc. of the Coastal Structures'99, ed. I.J. Losada, Balkemare, Spain, (1999) 1125-1144.
3. Goda Y. and Tagaki H., A Reliability design method of caisson breakwaters with optimal wave heights. Coastal Engineering Journal, 42 (4), 357-387 (2000).
4. Balas C.E. & Ergin A., A sensitivity study for the second order reliability based design model of rubble mound breakwaters, Coastal Engineering Journal, 42, (1) 57-86 (2000).
5. Hong, H.P., Simple approximations for improving second-order reliability estimates, Journal of Engineering Mechanics, 125 (5), 592-595 (1999).
6. CERC, Shore Protection Manual, Coastal Engineering Research Center, U.S. Army, Corps of Engineers, USA (1984).
7. İçmeli, F., Risk assessment of vertical wall breakwaters using sliding and overturning design criteria for breaking waves, MSc. Thesis, Middle East technical University, Ankara, Turkey (2002).
8. Özhan, E., Ergin, A., Ergun, U., Yalçıner, A.C., Abdalla, S., & Yanmaz, M., Hydraulic Research on the Harbor of Marmara Ereğlisi, Middle East Technical University Technical Report, Ankara, Turkey, 57 (1998).
9. Balas, C.E. & Ergin, A.E. Reliability-Based Risk Assessment in Coastal Projects: A Case Study in Turkey, Journal of Waterway, Port, Coastal and Ocean Engineering, ASCE, (2002).

Wavelets and Wavelet Packets Applied to Termite Detection

Juan-José González de-la-Rosa[1], Carlos García Puntonet[2],
Isidro Lloret Galiana, and Juan Manuel Górriz

[1] University of Cádiz, EPS-Electronics Instrumentation Group, Electronics Area, Av.
Ramón Puyol S/N. 11202, Algeciras-Cádiz, Spain
`juanjose.delarosa@uca.es`
[2] University of Granada, Department of Architecture and Computers Technology,
ESII, C/Periodista Daniel Saucedo. 18071, Granada, Spain
`carlos@atc.ugr.es`

Abstract. In this paper we present an study which shows the possibility of using wavelets and wavelet packets to detect transients produced by termites. Identification has been developed by means of analyzing the impulse response of three sensors undergoing natural excitations. Denoising exhibits good performance up to SNR=-30 dB, in the presence of white Gaussian noise. The test can be extended to similar vibratory or acoustic signals resulting from impulse responses.

1 Introduction

In acoustic emission (AE) signal processing a customary problem is to extract some physical parameters of interest in situations which involve join variations of time and frequency. This situation can be found in almost every nondestructive AE tests for characterization of defects in materials, or detection of spurious transients which reveal machinery faults [1]. The problem of termite detection lies in this set of applications involving nonstationary signals [2].

When wood fibers are broken by termites they produce acoustic signals which can be monitored using *ad hoc* resonant AE piezoelectric sensors which include microphones and accelerometers, targeting subterranean infestations by means of spectral and temporal analysis. The drawbacks are the relative high cost and their practical limitations due to subjectiveness [2].

Second order methods (spectra) failure in low SNR conditions even with *ad hoc* piezoelectric sensors. Bispectrum have proven to be a useful tool for characterization of termites in relative noisy environments using low-cost sensors [3],[4]. The computational cost could be pointed out as the main drawback of the technique. This is the reason whereby diagonal bispectrum have to be used.

Numerous wavelet-theory-based techniques have evolved independently in different signal processing applications, like wavelets series expansions, multiresolution analysis, subband coding, etc. The wavelet transform is a well-suited technique to detect and analyze events occurring to different scales [5]. The idea of decomposing a signal into frequency bands conveys the possibility of extracting subband information which could characterize the physical phenomenon under study [6].

In this paper we show an application of wavelets' de-noising possibilities for the characterization and detection of termite emissions in low SNR conditions. Signals have been buried in Gaussian white noise. Working with three different sensors we find that the estimated signals' spectra match the spectra of the acoustic emission whereby termites are identified.

The paper is structured as follows: Section 2 summarizes the problem of acoustic detection of termites; Section 3 remembers the theoretical background of wavelets and wavelet packets. Experiments and conclusions are drawn in Section 4.

2 Acoustic Detection of Termites

2.1 Characteristics of the AE Signals

Acoustic Emission(AE) is defined as the class of phenomena whereby transient elastic waves are generated by the rapid (and spontaneous) release of energy from a localized source or sources within a material, or the transient elastic wave(s) so generated (ASTM, F2174-02, E750-04, F914-03 [1]).

Figure 1 shows one impulse in a burst produced by termites and its power spectrum. Significant drumming responses are produced over the range 200 Hz-10 kHz. The carrier (main component) frequency of the drumming signal is around 2600 Hz. The spectrum is not flat as a function of frequency as one would expect for a pulse-like event. This is due to the frequency response of the sensor (its selective characteristics) and also to the frequency-dependent attenuation coefficient of the wood and the air.

2.2 Devices, Ranges of Measurement and HOS Techniques

Acoustic measurement devices have been used primarily for detection of termites (feeding and excavating) in wood, but there is also the need of detecting termites in trees and soil surrounding building perimeters. Soil and wood have a much longer coefficient of sound attenuation than air and the coefficient increases with frequency. This attenuation reduces the detection range of acoustic emission to 2-5 cm in soil and 2-3 m in wood, as long as the sensor is in the same piece of material [7]. The range of acoustic detection is much greater at frequencies <10 kHz, and low frequency accelerometers have been used to detect insect larvae over 1-2 m in grain and 10-30 cm in soil [8].

It has been shown that ICA succeeded in separating termite emissions with small energy levels in comparison to the background noise. This is explained away by

---

[1] American Society for Testing and Materials. F2174-02: Standard Practice for Verifying Acoustic Emission Sensor Response. E750-04: Standard Practice for Characterizing Acoustic Emission Instrumentation. F914-03: Standard Test Method for Acoustic Emission for Insulated and Non-Insulated Aerial Personnel Devices Without Supplemental Load Handling Attachments

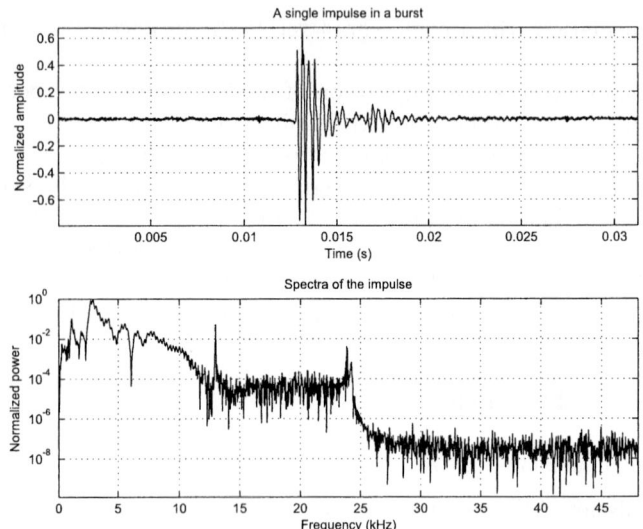

Fig. 1. Normalized power spectrum of a single pulse

statistical independence basis of ICA, regardless of the energy associated to each frequency component in the spectra [4]. The same authors have proven that the diagonal bispectrum can be used as a tool for characterization purposes [3]. With the aim of reducing computational complexity wavelets transforms have been used in this paper to de-noise corrupted impulse trains.

3 Wavelet Packets (WP)

3.1 Wavelet Bases

The WP method is a generalization of wavelet decomposition that offers more possibilities of reconstructing the signal from the decomposition tree. If L is the number of levels in the tree, WP methods yields more than $2^{2^{L-1}}$ ways to encode the signal. The wavelet decomposition tree is a part of the complete binary tree.

When performing a split we have to look at each node of the decomposition tree and quantify the information to be gained as a result of a split. An entropy based criterion is used herein to select the optimal decomposition of a given signal. We use an adaptative filtering algorithm, based on the work by Coifman and Wickerhauser [9].

Any finite energy signal $s(t)$ can be decomposed over a wavelet orthogonal basis [5] [2] of $\mathbf{L}^2(\Re)$ according to:

$$s(t) = \sum_{j=-\infty}^{+\infty} \sum_{k=-\infty}^{+\infty} \langle s, \psi_{j,k} \rangle \psi_{j,k} \qquad (1)$$

---

[2] $\left\{ \psi_{j,k}(t) = \frac{1}{\sqrt{2^j}} \psi\left(\frac{t - 2^j k}{2^j}\right) \right\}_{(j,k) \in \mathbb{Z}^2}$

Each partial sum can be interpreted as the details variations at the scale $a = 2^j$:

$$d_j(t) = \sum_{k=-\infty}^{+\infty} \langle s, \psi_{j,k} \rangle \psi_{j,k} \qquad s(t) = \sum_{j=-\infty}^{+\infty} d_j(t) \qquad (2)$$

The approximation of the signal $s(t)$ can be progressively improved by obtaining more layers or levels, with the aim of recovering the signal selectively. For example, if $s(t)$ varies smoothly we can obtain an acceptable approximation by means of removing fine scale details, which contain information regarding higher frequencies or rapid variations of the signal. This is done by truncating the sum in 1 at the scale $a = 2^J$:

$$s_J(t) = \sum_{j=J}^{+\infty} d_j(t) \qquad (3)$$

3.2 Multiresolution and Tree Decomposition

We consider the resolution as the time step 2^{-j}, for a scale j, as the inverse of the scale 2^j. The approximation of a function s at a resolution 2^{-j} is defined as an orthogonal projection on a space $\mathbf{V}_j \subset \mathbf{L}^2(\Re)$. $\mathbf{V}_j$ is called the scaling space and contains all possible approximations at the resolution 2^{-j}.

Let us consider a scaling function ϕ. Dilating and translating this function we obtain an orthonormal basis of $\mathbf{V}_j$:

$$\left\{ \phi_{j,k}(t) = \frac{1}{\sqrt{2^j}} \phi\left(\frac{t - 2^j k}{2^j}\right) \right\}_{(j,k) \in \mathbb{Z}^2} \qquad (4)$$

The approximation of a signal s at a resolution 2^{-j} is the orthogonal projection over the scaling subspace $\mathbf{V}_j$, and is obtained with an expansion in the scaling orthogonal basis $\{\phi_{j,k}\}_{k \in \mathbb{Z}}$:

$$P_{\mathbf{V}_j} s = \sum_{k=-\infty}^{+\infty} \langle s, \phi_{j,k} \rangle \phi_{j,k} \qquad (5)$$

The inner products

$$a_j[k] = \langle s, \phi_{j,k} \rangle \phi_{j,k} \qquad (6)$$

represent a discrete approximation of the signal at level j (scale 2^j). This approximation is low-pass filtering of s sampled at intervals 2^{-j}.

A fast wavelet transform decomposes successively each approximation $P_{\mathbf{V}_{j-1}} s$ into a coarser approximation $P_{\mathbf{V}_j} s$ (local averages) plus the wavelet coefficients carried by $P_{\mathbf{W}_j} s$ (local details). The smooth signal plus the details combine into a multiresolution of the signal. Averages come from the scaling functions and details come from the wavelets.

$\{\phi_{j,k}\}_{k \in \mathbb{Z}}$ and $\{\psi_{j,k}\}_{k \in \mathbb{Z}}$ are orthonormal bases of $\mathbf{V}_j$ and $\mathbf{W}_j$, respectively, and the projections in these spaces are characterized by:

$$a_j[k] = \langle s, \phi_{j,k} \rangle \qquad d_j[k] = \langle s, \psi_{j,k} \rangle \qquad (7)$$

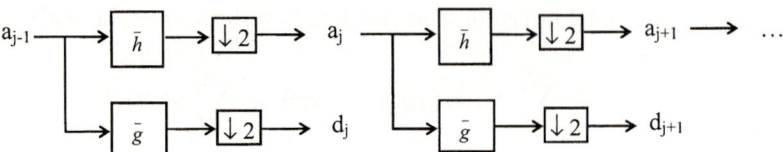

Fig. 2. Cascade of filters and subsampling

A space $\mathbf{V}_{j-1}$ is decomposed in a lower resolution space $\mathbf{V}_j$ plus a detail space $\mathbf{W}_j$, dividing the orthogonal basis of $\mathbf{V}_{j-1}$ into two new orthogonal bases:

$$\{\phi_j(t - 2^j k)\}_{k \in \mathbb{Z}} \quad \text{and} \quad \{\psi_j(t - 2^j k)\}_{k \in \mathbb{Z}} \tag{8}$$

$\mathbf{W}_j$ is the orthogonal complement of $\mathbf{V}_j$ in $\mathbf{V}_{j-1}$, and $\mathbf{V}_j \subset \mathbf{V}_{j-1}$, thus:

$$\mathbf{V}_{j-1} = \mathbf{V}_j \oplus \mathbf{W}_j. \tag{9}$$

The orthogonal projection of a signal s on $\mathbf{V}_{j-1}$ is decomposed as the sum of orthogonal projections on $\mathbf{V}_j$ and $\mathbf{W}_j$.

$$P_{\mathbf{V}_{j-1}} = P_{\mathbf{V}_j} + P_{\mathbf{W}_j}. \tag{10}$$

The recursive splitting of these vector spaces is represented in the binary tree. This fast wavelet transform is computed with a cascade of filters $\overline{h}$ and $\overline{g}$, followed by a factor 2 subsampling, according with the scheme of figure 2.

Functions that verify additivity-type property are suitable for efficient searching of the tree structures and node splitting. The criteria based on the entropy match these conditions, providing a degree of randomness in an information-theory frame. In this work we used the entropy criteria based on the p-norm:

$$E(s) = \sum_{i}^{N} \|s_i\|^p; \tag{11}$$

with p≤1, and where $s = [s_1, s_2, \ldots, s_N]$ in the signal of length N. The results are accompanied by entropy calculations based on Shannon's criterion:

$$E(s) = -\sum_{i}^{N} s_i^2 \log(s_i^2); \tag{12}$$

with the convention $0 \times log(0) = 0$.

4 Experiments and Conclusions

Two accelerometers (KB12V, seismic accelerometer; KD42V, industrial accelerometer, MMF) and a standard microphone have been used to collect data (alarm signals from termites) in different places (basements and subterranean wood structures and roots) using the sound card of a portable computer and a

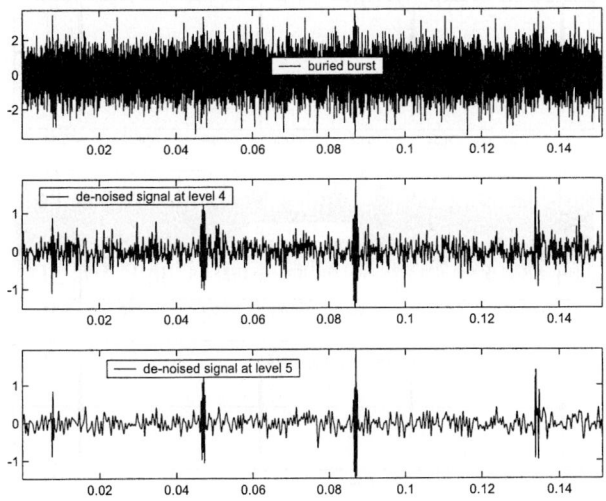

Fig. 3. Limit situation of the de-noising procedure using wavelets (SNR=-30 dB). From top to bottom: a buried 4-impulse burst, estimated signal at level 4, estimated signal at level 5

sampling frequency of 96000 (Hz), which fixes the time resolution. These sensors have different sensibilities and impulse responses. This is the reason whereby we normalize spectra.

The de-noising procedure was developed using a *sym*8, belonging to the family *Symlets* (order 8), which are compactly supported wavelets with least

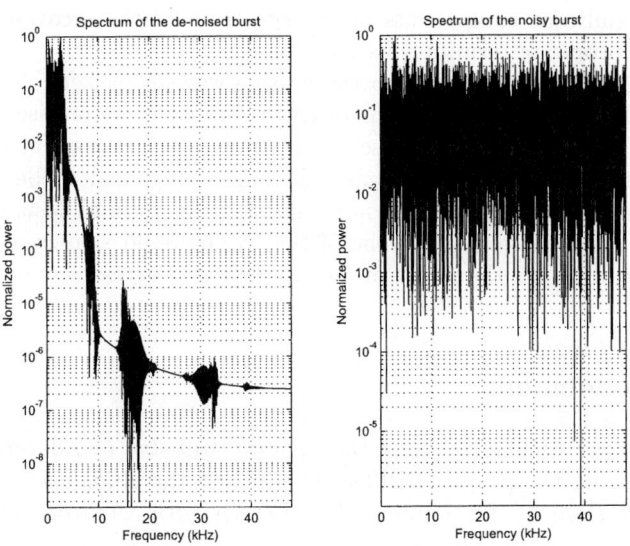

Fig. 4. Spectra of the estimated signal and the buried burst

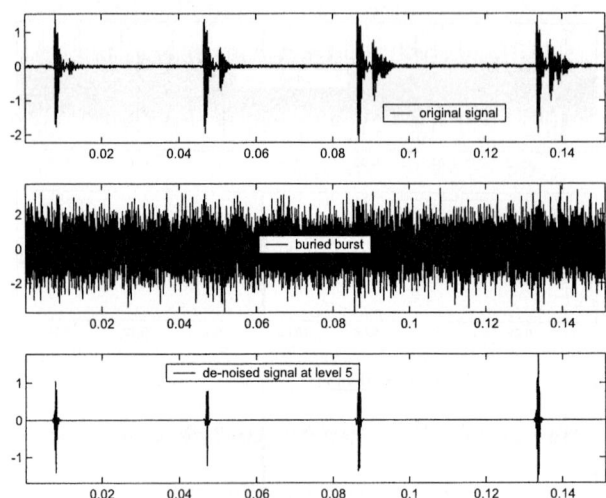

Fig. 5. Limit situation of the de-noising procedure using WP (SNR=-30 dB). From top to bottom: original signal, a buried 4-impulse burst, estimated signal at level 5

asymmetry and highest number of vanishing moments for a given support width. We also choose a soft heuristic thresholding.

We used 15 registers (from *reticulitermes grassei*), each of them comprises a 4-impulse burst buried in white gaussian noise. De-noising performs successfully up to an SNR=-30 dB. Figure 3 shows a de-noising result in one of the registers. Figure 4 shows a comparison between the spectrum of the estimated signal at level 4 and the spectrum of the signal to be de-noised, taking a register as an example. Significant components in the spectrum of the recovered signal are found to be proper of termite emissions.

The same 15 registers were processed using wavelet packets. Approximation coefficients have been thresholded in order to obtain a more precise estimation of the starting points for each impulse. Stein's Unbiased Estimate of Risk (SURE) has been assumed as a principle for selecting a threshold to be used for denoising. A more thorough discussion of choosing the optimal decomposition can be found in [5]. Figure 5 shows one of the 15 de-noised signals using wavelets packets. It can be see the result of reconstructing progressively each a_j by the filter banks.

Acknowledgement

The authors would like to thank the *Spanish Ministry of Education and Science* for funding the project DPI2003-00878, and the *Andalusian Autonomous Government Division* for funding the research with *Contraplagas Ambiental S.L.*

References

1. Lou, X., Loparo, K.A.: Bearing fault diagnosis based on wavelet transform and fuzzy inference. Mechanical Systems and Signal Processing **18** (2004) 1077–1095
2. de la Rosa, J.J.G., Puntonet, C.G., Górriz, J.M., Lloret, I.: An application of ICA to identify vibratory low-level signals generated by termites. Lecture Notes in Computer Science (LNCS) **3195** (2004) 1126–1133 Proceedings of the Fifth International Conference, ICA 2004, Granada, Spain.
3. de la Rosa, J.G., Lloret, I., Puntonet, C.G., Górriz, J.M.: Higher-order statistics to detect and characterise termite emissions. Electronics Letters **40** (2004) 1316–1317 Ultrasosics.
4. de la Rosa, J.J.G., Puntonet, C.G., Lloret, I.: An application of the independent component analysis to monitor acoustic emission signals generated by termite activity in wood. Measurement **37** (2005) 63–76 Available online 12 October 2004.
5. Mallat, S.: A wavelet tour of signal processing. 2 edn. Academic Press (1999)
6. Angrisani, L., Daponte, P., D'Apuzzo, M.: A method for the automatic detection and measurement of transients. part I: the measurement method. Measurement **25** (1999) 19–30
7. Mankin, R.W., Osbrink, W.L., Oi, F.M., Anderson, J.B.: Acoustic detection of termite infestations in urban trees. Journal of Economic Entomology **95** (2002) 981–988
8. Robbins, W.P., Mueller, R.K., Schaal, T., Ebeling, T.: Characteristics of acoustic emission signals generated by termite activity in wood. In: Proceedings of the IEEE Ultrasonic Symposium. (1991) 1047–1051
9. Coifman, R.R., Wickerhauser, M.: Entropy-based algorithms for best basis selection. IEEE Trans. on Inf. Theory **38** (1992) 713–718

Algorithms for the Estimation of the Concentrations of Chlorophyll A and Carotenoids in Rice Leaves from Airborne Hyperspectral Data

Yanning Guan, Shan Guo, Jiangui Liu, and Xia Zhang

State Key Laboratory of Remote Sensing Science,
Jointly Sponsored by the Institute of Remote Sensing Applications,
Chinese Academy of Sciences, and Beijing Normal University,
P. Box 9718, Beijing 100101, China
guan@irsa.ac.cn

Abstract. Algorithms based on reflectance band ratios and first derivative have been developed for the estimation of chlorophyll a and carotenoid content of rice leaves by using airborne hyperspectral data acquainted by Pushbroom Hyperspectral Imager (PHI). There was a strong R680/R825 and chlorophyll a relationship with a linear relationship between the ratio of reflectance at 680nm and 825nm. The first derivative at 686 nm and 601 nm correlated best with carotenoid. The relationship between the ratio of R680/R825 and chlorophyll a relationship, the first derivative at 686 nm and carotenoid concentration were used to develop predictive regression equations for the estimation of canopy chlorophyll a and carotenoid concentration respectively. The relationship was applied to the imagery, where a chlorophyll a concentration map was generated in XueBu, which is one of the sites for rice.

1 Introduction

Hyperspectral remote sensing exploits the fact that all material reflect, absorb, and emit electromagnetic energy, at specific wavelengths, in distinctive patterns related to their molecular composition. Hyperspectral imaging sensors in the reflective regions of the spectrum acquire digital images in many contiguous and very narrow spectral bands that typically span the visible, near infrared, and mid-infrared portions of the spectrum. This enables the construction of an essentially continuous radiance spectrum for every pixel in the scene. Thus, Hyperspectral imaging data exploitation makes possible the remote identification of ground materials-of –interest based on their spectral signatures. (D.G.Manolakis & G.Shaw, 2002)

Hyperspectral algorithms for the estimation of the concentrations of chlorophyll A and carotenoids can be develop using statistical approaches. Spectral band in the visible and near-infrared regions of the spectrum have been used to develop a number of indices for estimating chlorophyll. The commonly used normalized difference vegetation index (NDVI) was developed by contrasting the chlorophyll absorption in the red wavelengths. The NDVI has been found to be insensitive to medium and high chlorophyll concentrations.(Buschman & Nagel,1993) Miller found the concentration of chlorophyll within a vegetation canopy is related positively to the point of maximum slope at wavelengths between 690 and 740 nm in reflectance.(Miller et al.,1990) This point, Known as the Red edge of plant reflectance.(Curran et al.,1991) The Recent studies have developed several new vegetation indices based on other visible wavelength bands instead of the red wavelengths near 670-680nm.The spectral derivative technique has been used to improve characterizing the spectra and estimating accuracy of vegetation biochemical concentrations (Demerriades-Shah & Steven.M.D. et al., 1990, Pu & Gong 1997). Maximum sensitivity of reflectance to chlorophyll content was found in the green wavelengths region at 550 nm and at 708 nm in the far-red wavelengths. The reflectance in the main chlorophyll absorption regions in 400-500nm and 660-690nm proved to be insensitive to variation in chlorophyll content (Bisun.D.,1998). Strong correlation were observed between red edge position and canopy chlorophyll concentration using airborne image (Rosemary.A.J. et al.,1999).

2 Data Acquisition

2.1 Airborne Radiance Data

The hyperspectral image data in this paper are acquired from two airborne flights in 1999 in Changzhou area, Jiangsu province of East China, by using the Push-broom Hyperspectral Imager (PHI), which is developed by the Shanghai Institute of Technical Physics (SITP), Chinese Academy of Sciences (CAS). Table 1 shows the specification of PHI. The flights were carried out in September 9 and October 18 when the rice was in the broom and ripe stages, respectively. 80 bands were selected for recording in a range of 400-850nm for the 244-band PHI with the spectral resolution <5nm. The pixel size or spatial resolution is 2.25m with a flight altitude about 1500m. A 3D stabilized platform was applied for PHI and GPS data was recorded with the image data.

Table 1. Specification of the PHI

FOV	0.36rad(21°)
IFOV	1.0mrad (vertical to flight line)
S/N	300
No. of bands	244
Spectral Range	VIS-NIR(400nm--850nm)
Spectral Resolution	<5nm
Number of Pixels/scan	367pixels/line
Weight	9Kg
Frame Frequency	60Fr/sec
Data Rate	7.2Mb/Sec.

The field spectra are measured by the 252-channel SE590 spectrometer with a spectral resolution about 2-3nm in the range of 0.4-1.1μm.

The PHI image was pre-processed for radiometric, spectral and geometric calibration then was transformed to a reflectance image by field calibration using empirical line method.

2.2 Field Site

Night field sites were selected to analyzed chlorophyll concentration. There are four sites for rice within the nine. Unfortunately the PHI imagery was acquired on 18 October 1999 was deflected one field site for rice. Three sites were used to analyze rice samples. Nine study plot were located three sites which to be ensure were larger than one 2.25m*2.25m PHI pixel. Field reflectance and samples of different rice species for biochemical analyses were collected at the nine study areas.

2.3 Canopy Biochemical Concentration Data

Chlorophyll a, chlorophyll b, and carotenoid of rice leaves have been measured in Sep.1 1999 and Oct.18, 1999 in the study area of Changzhou.

The concentration of chlorophyll a, chlorophyll b, and total chlorophyll was calculated by the following equations. The unit is microgram per milliliter.

$$C_a = 12.7 A_{663} - 2.69 A_{645} \tag{1}$$

$$C_b = 22.9 A_{645} - 4.68 A_{663} \tag{2}$$

$$C_{a+b} = 20.2 A_{645} + 8.02 A_{663} \tag{3}$$

$$C_K = 4.7 A_{440} - 0.27 C_{A+B} \tag{4}$$

Where C_a is the concentration of chlorophyll a; C_b is the concentration of chlorophyll b, C_{a+b} is the concentration of the total chlorophyll and C_K is the concentration of carotenoid (mg/ml) A_{663} and A_{645} are the absorption value of leaves where wavelengths are 663nm and 645nm respectively. These values were measured by 721 spectrophotometer.

From the C_a, C_b, and C_{a+b}, the concentration of chlorophyll a, chlorophyll b, and carotenoid can be got from the following equations.

$$X_a = C_a \times V/W \tag{5}$$

$$X_b = C_b \times V/W \tag{6}$$

$$X = C_{a+b} \times V/W \tag{7}$$

$$Y = (4.7 A_{440} - 0.27 C_{a+b}) \times V/W \tag{8}$$

Where X_a, X_b, X, and Y are the concentration of the chlorophyll a, chlorophyll b, chlorophyll a and b, and carotenoid respectively. (mg/g). And A_{440} is the absorption value of the leaf where wavelength equals 440 nm. W (g) is the weight of the samples and V (ml) is the volume of the sample liquor.

3 Develops Algorithms for Estimating the Concentrations of Vegetation Biochemical

The image, which transformed into reflectance was used to produce the first derivative image. In order to develop better algorithms for estimating the concentrations of vegetation biochemical, the wavelength bands with maximum and minimum sensitivities to the concentrations of vegetate were identified from correlogram plots

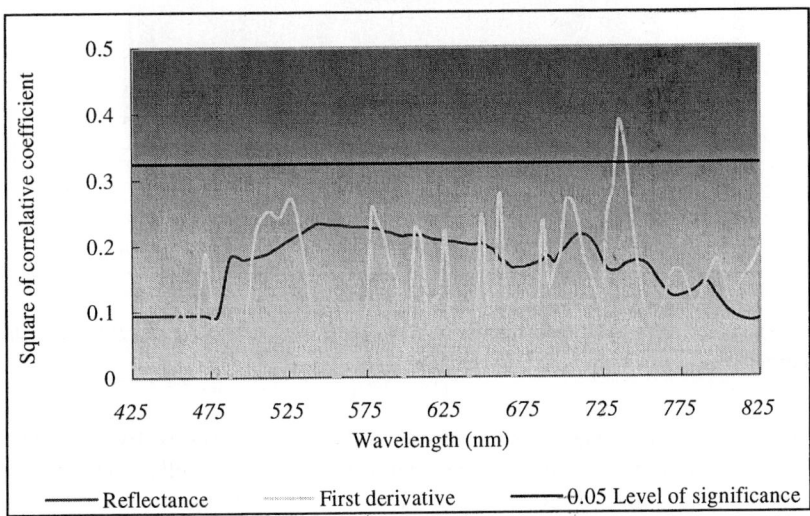

Fig. 1. The correlation coefficients between reflectance, first derivative of reflectance and the concentrations of chlorophyll at all wavelengths

Table 2. Correlation of vegetation indices with pigment content

	Chlorophyll a	Chlorophyll b	Chlorophyll a+b	Carotenoids
R680/R550	-0.633286(S)	-0.027519	-0.567520	0.303747
R680/R700	-0.597131(S)	0.012407	-0.527937	0.356874
R680/R825	-0.901990(S)	-0.376614	-0.871511(S)	-0.148244
R760/R550	-0.101527	0.146617	-0.062699	-0.380614
R760/R700	-0.240858	0.214545	-0.173705	-0.401802
R825/R550	0.791722 (S)	0.406786	0.779243 (S)	0.058519
NDVI	-0.051386	0.175014	-0.012855	-0.286543
GNDVI	-0.099975	0.135721	-0.063361	-0.352727
Red edge Position	-0.121271	0.444960	-0.024358	-0.231216
Red edge slope	0.497118	0.266054	0.491274	0.625607 (S)

(S) = significant at 0.05 level

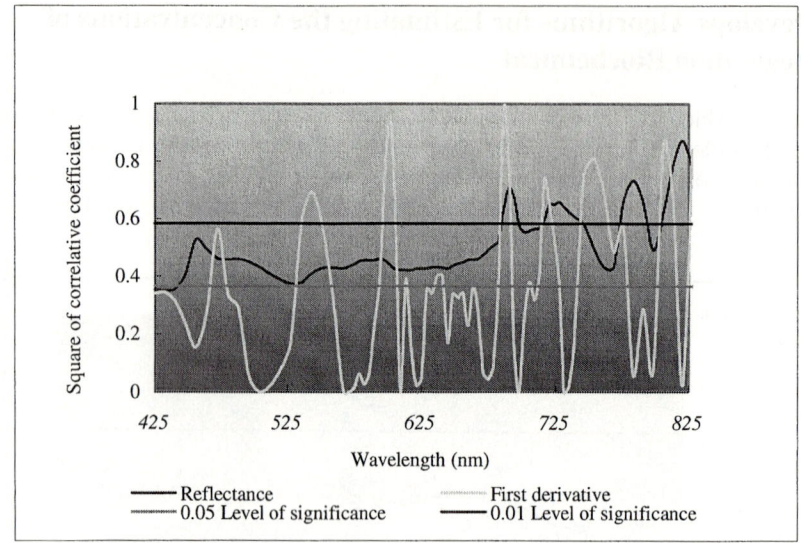

Fig. 2. The correlation coefficients between reflectance, first derivative of reflectance and the concentrations of carotenoid at all wavelengths

showing the correlation coefficients between reflectance, first derivative of reflectance and the concentrations of vegetation biochemical at all wavelengths. The correlogram for chlorophyll is shown in Figure 1. The similar correlogram for cartooned is demonstrated in Figure 2. The reflectance is insensitive to chlorophyll. The first derivative at 735 nm correlated best with chlorophyll. The reflectance and the first derivative were well correlated with the carotenoid contents at many wavelength bands. The first derivative at 601 nm and 686 nm correlated best with carotenoid.

Chlorophyll a, b, a+b, and carotenoid content were related to the vegetation indices R680/R550, R680/R700, R680/R825, R760/R550, R760/R700, R825/R550, NDVI, GNDVI, the Red edge Position and the Red edge slope. The correlation coefficients between these indices and pigment are given in Table 2.

The NDVI is normalized difference vegetation index, defined as (R760-R670)/(R760+R670).

A "green" NDVI defined as GNDVI=(R760-R550)/(R760+R550), by using the green wavelength band near 550nm.

Red edge position is the point of maximum slope at wavelengths between 690 nm and 740 nm in reflectance. This point characterizes the effective boundary between the strong absorption of red radiation by chlorophyll and the increased multiple scattering of radiation in near-infrared wavelengths (Curran et al., 1991).

Red edge slope is the slope of red edge and is the maximum slope at wavelength between 690 nm and 740 nm.

R680/R550, R680/R700, R680/R825 and R825/R550 are sensitive to chlorophyll a, and R680/R825 is correlated best with chlorophyll a.

R680/R825 and R825/R550 are well correlated with chlorophyll a+b. The red edge slope is sensitive to carotenoid.

4 Algorithms for Estimates of Canopy Chlorophyll and Carotenoid from Image Spectra

There was a strong R680/R825 – chlorophyll a relationship with a linear relationship between the ratio of reflectance at 680nm and 825nm.. There was a statistically significant correlation between R680/R825 and chlorophyll a at the 99% confidence level.(r = -0.90)

The relationship between R680/R825 and chlorophyll a concentration was used to develop a predictive regression equation for the estimation of canopy chlorophyll a concentration. The algorithms equation had the form of Equation (9).

$$\text{Chl a} = -2.657 \text{R680/R825} + 1.8029 \tag{9}$$

Where Chl a is the concentration of chlorophyll a (mg/g). R680 and R825 are the reflectance at 680nm and 825nm.

This relationship was applied to the imagery, where a chlorophyll a concentration map (Figure 3) was generated in Xuebu, which is one of the sites for rice.

The first derivative at 686 nm correlated best with. There was a very strong first derivative-carotenoid relationship with a linear relationship between the first derivative at 686nm. There was a statistically significant correlation between the first derivative at 686 nm and carotenoid at the 99.9% confidence level. (r = 0.99)

The relationship between the first derivative at 686 nm and carotenoid concentration was used to develop a predictive regression equation for the estimation of canopy carotenoid concentration. The algorithm equation was the Equation (10).

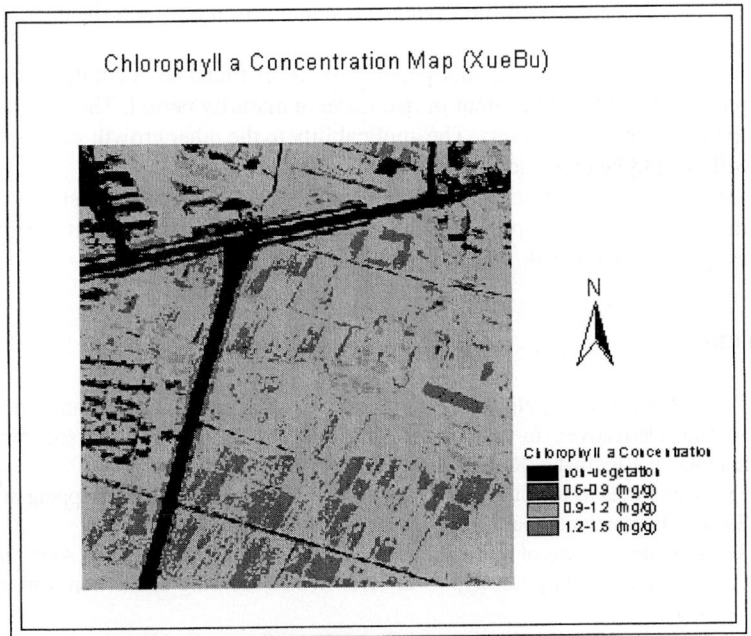

Fig. 3. Chlorophyll concentration map (Xuebu, Oct.18)

$$CK = 0.9527D_{686} + 0.1955 \tag{10}$$

Where CK is the concentration of carotenoid (mg/g). D_{686} is the first derivative at 686nm.

5 Discussion

Analysis of visible/infrared reflectance and biochemical concentration data for rice leaves has revealed some new information on the quantification of chlorophyll and carotenoid by airborne hyperspectral data. PHI data, by means of its continuous, high resolution spectral imagine, have been found useful for the estimation of chlorophyll a and carotenoid content of rice leaves. It was found that the indices R680/R550, R680/R700, R680/R825 and R825/R550 correlate well with chlorophyll a. R680/R825 and R825/R550 showed the more sensitivity to chlorophyll a+b. It was also found red edge slope correlate well with carotenoid. The new index R680/R825 is the best indicator of chlorophyll a. The reflectance ratio is the simple transformation, which removes irrelevant information from reflectance spectra, and highlights subtle variations in reflectance cause by chemical absorptions (B.Datt, 1999). The R680/R825 is calculated from wavelengths that is not affected by leaf structure (M.L.Adams, W.D.Philpot and W.A.Norvell, 1999), and also be unaffected by leaf water content (Bowman, W.D.1989).

The reflectance and the first derivative of reflectance at many wavelengths are more sensitive to the carotenoid content of rice leaves. The phase of the airborne hyperspectral data acquisition is October 18, which is complete maturity period of rice. The rice color is yellow and the carotenoid content is more than the others growth stages of rice.

The algorithms devolved in this paper have been found to accurately predict the chlorophyll and carotenoid content in rice leave in maturity period. They are needed to be compared with more data sets. The applicability to the other growth stages and other crops will need to be evaluated by further experiments.

The employed ground-truth was found to be limited due to some constrains. Despite these limitations however, optimistic chlorophyll and carotenoid content of rice for the PHI was clearly established.

References

Adams M.L., Philpot W.D., Norvell W.A.: Yellowness Indexes: an Application of Spectral Second Derivatives to Estimate Chlorophyll of Leaves in Stressed Vegetation. International Journal of Remote Sensing, 20(1999) 3663-3675

Bagheri S., Stein M., Dios R.: Utility of Hyperspectral Data for Bathymetric Mapping in a Turbid Estuary. International Journal of Remote Sensing, 19(1998) 1179-1188

Bisun, D.: Remote Sensing of Chlorophyll A, Chlorophyll B, Chlorophyll A+B, and Total Carotenoid Contenting Eucalyptus Leaves. Remote Sensing of Environment, 66(1998) 111-121

Bowman, W.D.: The Relationship between Leaf Water Status, Gas Exchanges, and Spectral Reflectance in Cotton Leaves. Remote Sensing of Environment, 30(1989) 249-255

Buscchman, C., Nagel, E.: In Vivo Spectroscopy and Internal Optics of Leaves as a Basis for Remote Sensing of Vegetation. International Journal of Remote Sensing, 14(1993) 711-722

Curran, P.J., Dungan, J.L., Macler, B.A., Plummer, S.E.: The Effect of a Red Leaf Pigment on the Relationship between Red Edge and Cholorophyll Concentration. Remote Sensing of Environment, 35(1991) 69-76

Datt B.: Visible/Near Infrared Reflectance and Chlorophyll Content in Eucalyptus Leaves. International Journal of Remote Sensing, 20(1999) 2741-2759

Demetriades-Shah, T.H., Steven, M.D., Clark, J.A.: High Resolution Derivative Spectra in Remote Sensing. Remote Sensing of Environment, 33(1990) 55-64

Manolakis D.G., Shaw G.: Detection Algorithms for Hyperspectral Imaging Applications. IEEE Signal Processing, 19(2002) 29-43

Miller, J.R., Hare, E.W., Wu, J.: Quantitative Characterization of the Vegetation Red Edge Reflectance I. An Inverted-Gaussian Reflectance Model. International Journal of Remote Sensing, 11(1990) 1755-1773

Pu, R., Gong, P.: Relationships between Forest Biochemical Concentrations and CASI Data along the Oregon Transect. Journal of Remote Sensing, 1(1997) 115-123 (in Chinese)

Rosemary, A.J., Mark, E.J.C., Paul, J.C.: Estimating Canopy Chlorophyll Concentration from Field and Airborne Spectra. Remote Sensing of Environment, 68(1999) 217-224

Multiresolution Reconstruction of Pipe-Shaped Objects from Contours

Kyungha Min[1] and In-Kwon Lee[2],*

[1] Center for Computer Graphics and Virtual Reality,
Ewha Womans Univ., Seoul, Korea
minkh@cs.rutgers.edu
[2] Dept. of Computer Science,
Yonsei Univ., Seoul, Korea
iklee@yonsei.ac.kr

Abstract. We reconstruct pipe-shaped objects from a set of contours, each of which is extracted from an image representing a slice sampled from 3D volume data. The contours are formed by connecting the intersection points between rays cast from a central pixel of an image slice and the boundary of the shape. The edges on the contours are classified into several types, which are exploited in triangulating the contours, thus eliminating most of the floating-point computation from the tiling. Initially, contours of lowest resolution are extracted to reconstruct a lowest-resolution object, which is refined by adding points to the contours.

1 Introduction

Reconstruction of a pipe-shaped object from contours is an important problem in many fields, such as medical imaging, computer-aided design, and reverse engineering. The reconstruction problem is composed of three subproblems [11]: the correspondence problem, the branch problem, and the tiling problem. In this paper, we concentrate on the tiling problem.

We propose a multiresolution approach that yields a polygonal surface from a set of contours sampled from 3D volume data. The proposed algorithm is outlined in Fig. 1. Contours of the lowest resolution are extracted on image slices sampled from the 3D volume data, and an initial pipe-shaped object is reconstructed by tiling the contours. This object is brought to a higher resolution through refinement of the contours.

Our first key idea is a search algorithm for corresponding pairs of points or edges on the successive contours that does not require floating-point computations, except for a few degenerate cases. Note that the conventional tiling algorithms search the corresponding pairs by means of geometric tests that would

---

* This work was supported (in part) by the Ministry of Information & Communications, Korea, under ther Information Technology Research Center (ITRC) Support Program at Ewha Womans Univ. and Yonsei Univ.

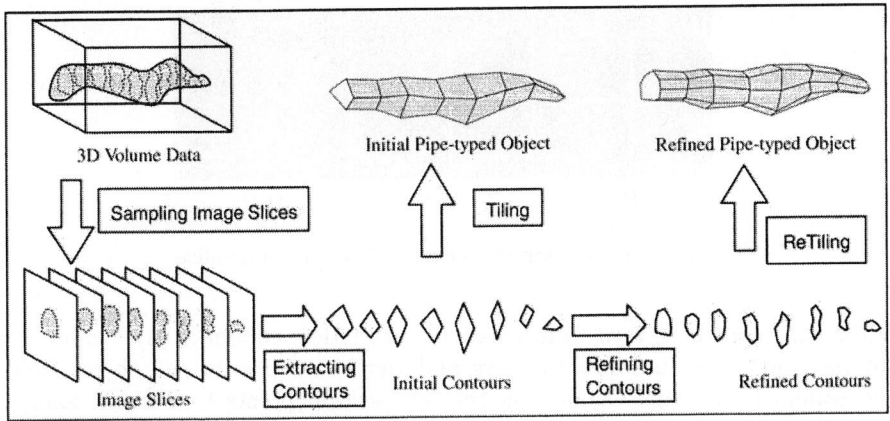

Fig. 1. Overview of the proposed algorithm

involve floating-point operations. The second key idea of this paper is the reconstruction of objects by a stepwise refinement. An object is refined by refinement of its constituent contours, which is achieved by casting additional rays, which increases the number of points on the contours in a stepwise fashion.

In Section 2, we review related works on the reconstruction of pipe-shaped objects from contours. We provide a scheme for extracting contours in Section 3, and a tiling algorithm in Section 4. In Section 5, we provide details of our implementation and results. Finally, in Section 6, we draw the conclusions and suggest future work.

2 Related Works

Keppel [9] originally proposed a scheme for reconstructing a surface from 2D contours. He built a toroidal graph between points on the contours and proposed a search using the graph that would generate polygons from the points. Many researchers improved Keppel's work by a divide and conquer approach [7], a greedy search algorithm [3], Delaunay triangulation [2], decomposing non-convex contours into convex sub-contours [5], tiling using ellipses [11], approximating the contours using discrete field functions [8], a constraint-based approach [1], a generalized Voronoi diagram [13], solving PDE [4], and a morphological dilation operators [10]. Some researchers have developed multiresolutional approaches for tiling surface from contours using wavelets [12], medial axis and simplification [15], and a combined scheme of wavelet and dynamic programming [6].

3 Extraction of Contours

Extracting Contours. Our reconstruction process starts with the extraction of a contour from an image slice. Fig. 2 illustrates the three steps of the ex-

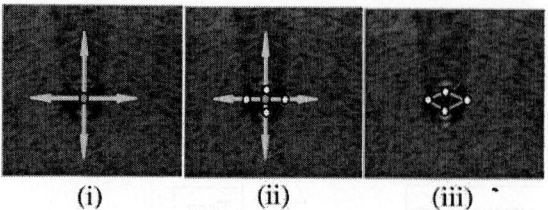

Fig. 2. Extraction of a contour from an image slice

traction: (i) Casting rays from a central pixel (left column), (ii) Computing intersection points between the rays and the boundaries of the object (middle column), and (iii) Connecting the intersection points to build a contour (right column). Depending on the relationship between the ray and the boundary of the object, we can classify an intersection point into two types: either IN-OUT (ray passing out of the object) or OUT-IN (ray coming in the object). In order to connect the intersection points, we apply the well-known chain code algorithm [14] to traverse the boundary pixels in counterclockwise order. The algorithm is slightly modified to traverse the border edges of the boundary pixels.

Classifying Edges. We classify edges of the contours into the following categories: Forward (F), Backward (B), Down (D), and Up (U). A D edge is further classified into Down and In (D_I) and Down and Out (D_O), while a U edge is into Up and In (U_I), and Up and Out (U_O). The classification is based on the following three factors:

i. **The relation between the ray and the endpoints.** An F edge is an edge with a start point is determined by the x-th ray and end point by the $(x+1)$-th ray, A B edge has its start point determined by the $(x+1)$-th ray and its endpoint by the x-th ray. Note that the start and end points of the U and D edges lie on the same ray.
ii. **The direction of the edge and of the ray.** We distinguish a U edge from a D edge based on the direction of the edges. The direction of U edge is identical to the ray, while the direction of D edge is reverse to the ray.

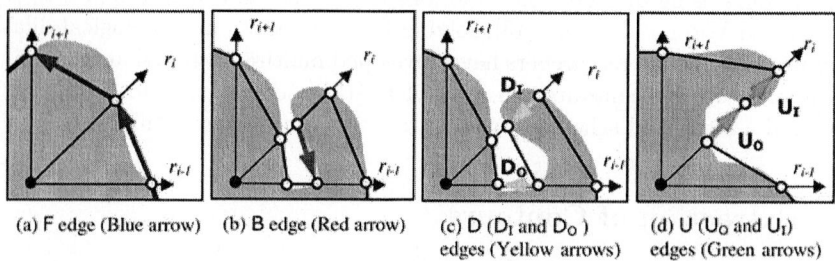

(a) F edge (Blue arrow) (b) B edge (Red arrow) (c) D (D_I and D_O) edges (Yellow arrows) (d) U (U_O and U_I) edges (Green arrows)

Fig. 3. Types of edges

iii. **The types of the endpoints of an edge.** D_I and U_I edges connect IN-OUT and OUT-IN endpoints, while D_O and U_O edges connect OUT-IN and IN-OUT endpoints.

Fig. 3 illustrates the types of the edges on a contour.

4 Tiling Algorithm

4.1 Building a Contour Tree

Description of a Contour Tree. A contour tree represents a contour based on the types of the edges in that contour. It is composed of three types of nodes.

Root node. Initially, the root node of a contour tree has n_0 child nodes, where n_0 denotes the number of initial rays.
Level-1 node. A level-1 node is a child of a root node. The x-th level-1 node contains an F edge determined by the x-th ray and the $(x + 1)$-th ray.
Edge node. An edge node is a node that contains one of the edge types.

Note that the leaf nodes and level-1 nodes contain the information about the edges of the contour.

Building Rules. A contour is unambiguous if n rays create exactly n edges of type F. If n rays determine more F edges, then the contour is said to be ambiguous. The rules for building a contour tree for an unambiguous contour are listed as follows:

[Building rule for level-1 nodes]

[1] An x-th level-1 node stores the F edge determined by the x-th ray and the $(x + 1)$-th ray.
[2] A D edge node is attached to the right child of a level-1 node and a U edge node to the left child.

[Building rule for leaf nodes]

[3] For all non-F edges, we build edge nodes that contain the geometry of the edge. These edge nodes are the leaf nodes of a contour tree.

[Building rule for edge nodes]

[4] If a D_I (D_O) edge and a D_O (D_I) edge are incident, we build a new D (D_R) internal node and assign the nodes as the child nodes of the new node.
[5] If a U_O (U_I) edge and a U_I (U_O) edge are incident, we build a new U (U_R) internal node and assign the nodes as the child nodes of the new node.
[6] If two D (U) internal nodes are incident, we build a new D (U) internal node and assign both of the D (U)-typed nodes as the child node of the new node.
[7] If a D internal node and a U internal node are incident, we build a new D (or U) internal node and assign both of the nodes as child nodes of the new internal node.

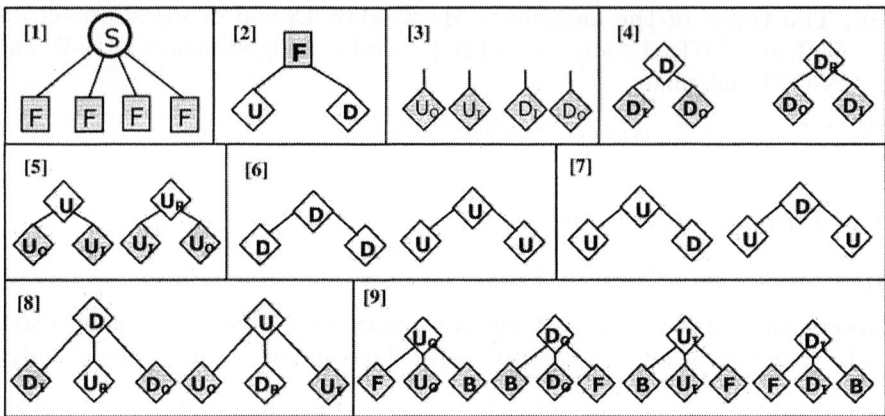

Fig. 4. Examples of the building rules

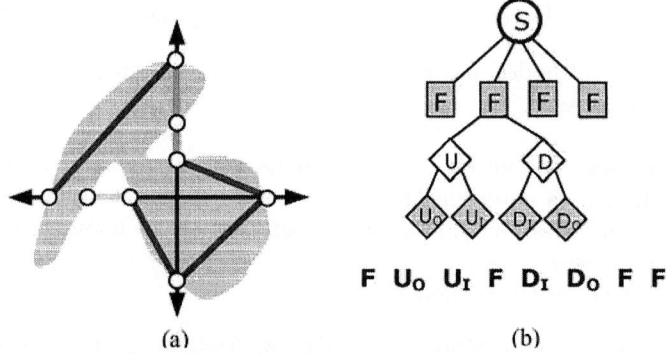

Fig. 5. An example of an unambiguous contour tree: (a) illustrates an example contour and its edges, and (b) illustrates the corresponding contour tree. The string in (b) denotes the edges represented its types in the counterclockwise order

[8] If D_I (U_O) and U_R (D_R) and D_O (U_I) internal nodes are incident, then we build a new D (U) internal node and assign three internal nodes are the child nodes of the new node.

[9] For D_I (D_O, U_I, or U_O) edge, which is encapsulated by F edge and B edge, we build a new internal node of the identical type and assign the three internal nodes that contain the three edges as child nodes.

Fig. 4 illustrates the examples of the building rules [1] - [9]. Fig. 5 illustrates an example of a contour tree for an unambiguous contour.

An ambiguous contour can correspond to more than two contour trees, each of which is defined from a contour. In this case, we choose a contour tree based on the tree of the neighboring contour. In case that a contour is ambiguous, more than two contour trees can be derived from a contour. Among the contour trees, we select one tree based on the contour tree of a neighboring contour. The building rule [1] is rewritten for the level-1 node. Other building rules for an unambiguous case can be applied for the ambiguous case.

[1'] There are t edges of type F determined by x-th ray and $(x+1)$-th ray, where $t > 1$. We choose one of them and store it in the x-th level-1 node. The F edge to will be stored in the level-1 node should be the one that is geometrically closest to the F edge stored in the x-th level-1 node of the corresponding contour tree.

4.2 Tiling by Traversing Contour Trees

The tiling step generates triangles between contours by searching corresponding edges and vertices on the contours. We determine corresponding pairs on the contours by traversing their contour trees simultaneously. The rules for traversal are described as follows:

1. At the root nodes
   From the root nodes, the level-1 nodes at the same position on each tree are visited simultaneously.
2. At level-1 nodes
   Child nodes of the same type are traversed simultaneously.
3. At the internal edge nodes, we have three different cases (See Fig. 6):
   Case 1. Both of the nodes have an identical set of child nodes. All the child nodes of the same type on each tree are traversed simultaneously.
   Case 2. Both of the nodes have at least one child node. The same-typed child nodes on each of the tree are traversed simultaneously. If more than two child nodes are of the same type with a child node in the corresponding tree, we choose one of them arbitrarily.
   Case 3. The nodes have no children of the same type. The child node whose type is identical to the parent node is traversed, while the edge node without a child of the same type is not traversed.
4. At the leaf edge nodes, the edges stored in the nodes become corresponding edges to each other, and the basis of tile pairs.

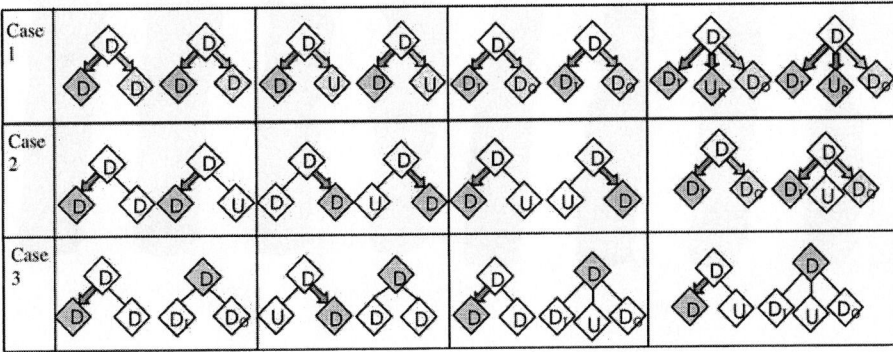

Fig. 6. Traversing rules for D-typed edge node. Note that the child nodes of the identical colors are traversed simultaneously

4.3 Refinement

We refine a contour by casting new rays between the existing rays. The new intersection points sampled on the boundary of the object are inserted into the existing contour. The contour tree follows the refinement of its contour as follows:

Refinement of a level-1 node. Suppose a new ray is cast between the x-th ray and the $(x+1)$-th ray. The x-th level-1 node is divided into two level-1 nodes, and new F edges determined by the new ray are stored in the new level-1 nodes.

Refinement of an edge node. All the existing edges interrupted by new intersection points are removed. Consequently, the nodes that store these edges are removed from the contour tree recursively. To add the new edges to the contour, we apply the building rules described in Section 3 to build a contour tree, and attach the resulting fragment as a child of a level-1 node.

5 Implementation and Results

We implemented our algorithm on an 800 MHz Pentium III CPU PC with 256 MB of RAMs. For testing, we used a phantom object and an artery at the neck. Both of the objects are 3-D volume with 256 X 256 X 512 resolutions. To capture the volume from the source object, we exploited 3-D ultrasonic scanner developed for medical purpose. Fig. 7 (a) illustrates four levels of detail for the phantom object, and Fig. 7 (b) illustrates the artery at four levels of details. For comparison, we also implemented the well-known toroidal search algorithm [2, 3, 7]. The resulting computation time, number of triangles are itemized in Table 1.

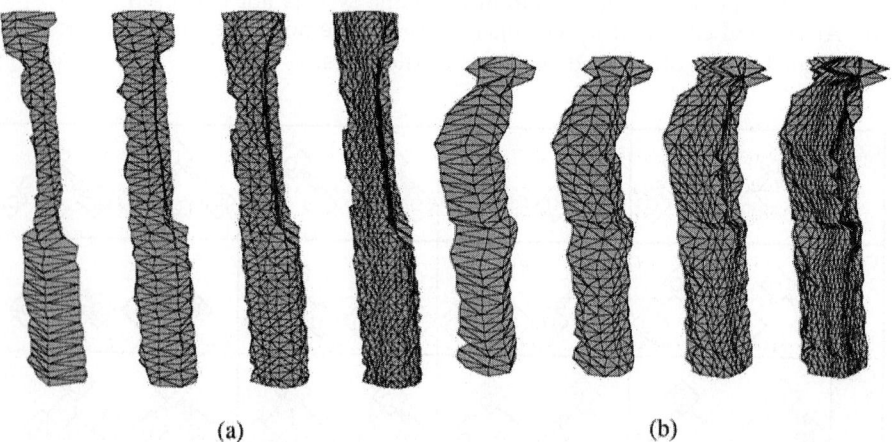

Fig. 7. Reconstruction of pipe-shaped objects in four resolutions (from left to right)

Table 1. Comparison of the proposed scheme to the conventional scheme

Resolution	Applied Algorithm	Time (sec)	No. of Triangles	Size of data (Kbytes)
1	Toroidal search	0.340	444	6.384
	Proposed	0.241	412	2.684
2	Toroidal search	0.761	1,061	15.126
	Proposed	0.550	867	6.054
3	Toroidal search	1.442	2,424	34.032
	Proposed	0.992	1,754	12.652
4	Toroidal search	2.904	4,791	136.178
	Proposed	2.062	3,392	50.354

6 Conclusion and Future Work

We have presented a new multiresolution scheme to reconstruct a pipe-shaped object from contours, which are extracted from a set of image slices by connecting the intersection points between the rays cast from a central pixel and the boundary of the shape. The edges on the contour are classified into several types, and this classification is exploited in searching for correspondence between adjacent contours. Efficiency is improved by eliminating most of the floating-point computations from the tiling process. A pipe-shaped object at the lowest resolution is refined in a stepwise way by refinement of the contours, which is achieved by casting new rays on the image slices.

We aim to develop a reconstruction scheme that addresses the branching problem based on the approach reported in this paper. We also intend to apply our scheme in areas such as medical imaging.

References

1. Bajaj, C. L., Coyle, E. J., and Lin, K. N., "Arbitrary Topology shape reconstruction from planar cross sections,", GMIP, 58(6), pp. 524-543, 1996.
2. Boissonnat, J. D., "Shape reconstruction from planar cross sections," CVGIP, 44, pp. 1-29, 1988.
3. Christiansen, H. N. and Sederberg, T. W., "Conversion of complex contour line definitions into polygonal element mosaics," SIGGRAPH 78, pp. 187-192, 1978.
4. Cong, G. and Parvin, B., "An algebraic solution to surface recovery from cross-sectional contours," GMIP, 61(4), pp. 222-243, 1999.
5. Ekoule, A. B., Peyrin, F. C., and Odet, C., L., "A triangulation algorithm from arbitrary shaped multiple planar contours," ACM ToG, 10(2), pp. 182-199, 1991.
6. Fix, J. D. and Ladner, R. E., "Multiresolution based refinement to accelerate surface reconstruction from polygons," Computational Geometry, 13(1), pp. 49-64, 1999.
7. Fuchs, H., Kedem, Z. M., and Uselton, S. P., "Optimal surface reconstruction from planar contours," Communications of ACM, 20(10), pp. 693-702, 1977.
8. Jones, M. W. and Chen, M., "A new approach to the construction of surfaces from contour data," Computer Graphics Forum, 13(3), pp. 75-84, 1994.

9. Keppel, E., "Approximating complex surfaces by triangulation of contour lines," IBM Journal of Res. Dev. 19, pp. 2-11, 1975.
10. Marsan, A. L. and Dutta, D., "Computational techniques for automatically tiling and skinning branched objects," Computers & Graphics, 23(1), pp. 111-126, 1999.
11. Meyers, D., Skinner, S., and Sloan, K., "Surfaces from contours," ACM ToG, 11(3), pp. 228-258, 1992.
12. Meyers, D., "Multiresolution tiling," Computer Graphics Forum, 13(5), pp. 325-340, 1994.
13. Oliva, J. M., Perrin, M., and Coquillart, S., "3D reconstruction of complex polyhedral shapes from contours using a simplified generalized Voronoi diagram," Computer Graphics Forum, 15(2), pp. 397-408, 1996.
14. Pitas, I., Digital Image Processing Algorithms, Prentice Hall, 1993.
15. Schilling, A. and Klein, R., "Fast generation of multiresolution surfaces from contours," Eurographics Workshop on Visualization, pp. 35-46, 1998.

Multi-resolution LOD Volume Rendering in Medicine

Kai Xie[a,*], Jie Yang[a], and Yue Min Zhu[b]

[a] Inst. of Image Processing & Pattern Recognition,
Shanghai Jiaotong Univ., 200030 Shanghai,China
[b] CREATIS - CNRS research unit 5515 & INSERM unit 630,
69621 Villeurbanne, France
xie_kai2001@sjtu.edu.cn

Abstract. This paper presents a level of detail (LOD) selection algorithm for multi-resolution volume rendering using 3D texture mapping. It uses an adaptive scheme that renders the volume in a region-of-interest at a high resolution and the volume away from this region at progressively lower resolutions. The algorithm is based on several important criteria, rendering is done adaptively by selecting high-resolution cells close to a center of attention and low-resolution cells away from this area. In addition, our hierarchical level-of-detail representation guarantees consistent interpolation between different resolution levels. Experiments have been applied to a number of large medical data and produced high quality images at interactive frame rates using standard PC hardware.

1 Introduction

Volume rendering of large data sets is a very common task in many areas in medicine. To address the challenge, researchers have proposed various algorithms. Among the existing techniques, hierarchical rendering algorithms can effectively control the tradeoff between quality and speed, and thus show a great potential. In essence, hierarchical methods first create a multi-resolution representation for the volume. Data of different resolutions in different regions are chosen for rendering. The effectiveness of hierarchical algorithms relies on their ability to adaptively simplify rendering in regions where data are unimportant or uninteresting, so that both the memory and computational cost can be reduced without significantly affecting the rendering quality.

Hierarchical volume rendering algorithms [1][2][3] can accelerate the speed of rendering. The selection of appropriate volume resolutions, or levels of detail (LOD), is often done for the different applications. Although various LOD selection algorithms for polygon rendering systems [5][6] are available, LOD selection algorithms for volume rendering, however, are still scarce. Frequently used approaches such as selecting the volume resolutions based on user specified error tolerances [4], based on the volume block's distance to the viewpoint and the angle to the view vector [1][2] are difficult to guarantee the rendering quality and the continuity at level boundaries while reducing the rendering time.

This paper presents a LOD selection algorithm for rendering hierarchical volumes using 3D texture hardware. The main focus of our algorithm is adaptive LOD selection based on several important criteria: Maximum opacity, Distance to the view point, Projection area and Gaze distance. In addition, we reduce the run-time performance of volume rendering and insure interpolation consistency between levels.

The rest of the paper is organized as follows. In section 2, we present our multi-resolution rendering algorithm in detail. Experimental results are discussed in section 3. Conclusions are discussed in section 4.

2 Multi-resolution LOD Volume Rendering

We develop a multi-resolution hierarchy that allows consistent interpolation between levels and give a solution to the rendering artifacts that still persist when rendering two differing but adjacent levels. In addition, to make sure that regions of interest are rendered at a higher quality, our algorithm allocates the different hierarchy to different subvolumes based on several important criteria, which can be simple heuristics such as the distance to the view point, the volume opacity, or other application-specific criteria. In the following, we describe our algorithm in detail.

2.1 The Multi-resolution Texture Hierarchy

Prior to the rendering, the volume data set is subdivided into multiple subvolumes of smaller size, which get assigned the chunk of texture that is necessary to render the subvolume at the original resolution. Each subvolume builds its own local hierarchy by constructing copies of the original texture at ever coarser resolution. These different levels-of-detail are stored in additional texture maps as shown in Figure 1. On every level the size of texture elements increases by a factor of two. Note that at subvolume boundaries on the same level, texture elements have to be included in multiple subvolumes in order to guarantee continuous texture interpolation (filled areas).

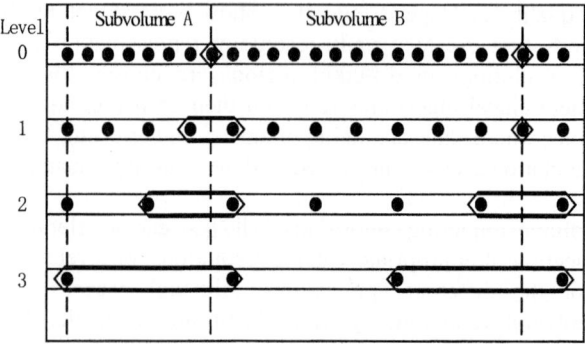

Fig. 1. One-dimensional of a data set with four levels of detail

The original data set is represented by textures of level 0. The width of texture elements on level k is twice the width of elements on level $k-1$, whereas the texture size is reduced by a factor of two. Texture elements on different resolution levels are aligned in such a way, that their centers on a certain level correspond to the center of an even element on the next finer level. Since the geometry or shape of bricks will be retained throughout the hierarchy, the domain of the underlying texture function, necessary to compute appropriate texture coordinates, has to be adapted accordingly.

On the first level, the domain of texture coordinates ranges from the center of the first element to the center of the last one. Due to the alignment of voxels on different levels, borders of the texture function domain on coarser levels often fall between two adjacent voxels. In this case additional voxels are needed in order to guarantee correct interpolation. On the other hand, since the width of texture elements on each level is known and because the shape of subvolumes is not going to be modified, offsets for correct calculation of texture coordinates can always be determined.

We should note that it is always necessary to expand textures in order to cope with the power-of-two restriction imposed by the OpenGL implementation. However, a hierarchical technique will be outlined below that allows us to effectively avoid unnecessary texture elements.

2.2 Continuous Level Transitions

If local texture hierarchies are constructed as described, interpolation artifacts at the boundaries between adjacent bricks at the same level do not appear. This is because boundary voxels are shared by adjacent subvolumes. However, this does not apply to adjacent bricks rendered on different levels.

Therefore, it is necessary to slightly modify the treatment of level transitions to meet the continuity requirements: in the level-of-detail representation the continuity at level transitions can be established by letting the finer cells to the left and to the right of the subvolume boundary interpolate the scalar field from the coarser level (Figure 2). For cells with even index in each dimension this is equivalent to a copy operation as they correspond exactly to a cell on the next coarser level. This procedure only adapts the subvolume textures on the finer level. Thus, with the combination of the proposed multi-resolution representation together with appropriately re-sampled values at level transitions we guarantee the continuity of the 3D scalar field.

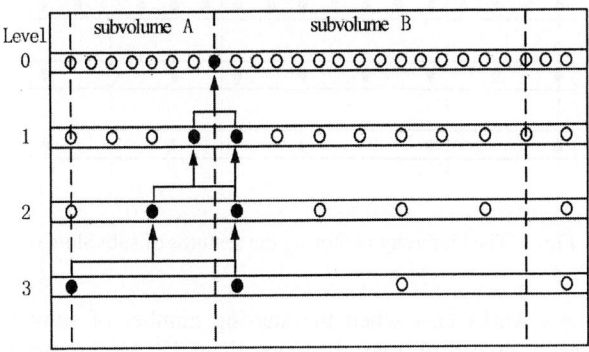

Fig. 2. Adaption of adjacent subvolumes with different level-of-detail to ensure consistent texture interpolation

Note in particular, that we restrict transitions to differ by at most one level in order to maintain the continuity between levels. However, this does not impose a real restriction as higher transitions can be achieved by consecutive transitions of one level.

The number of voxels that has to be adapted depends on the position of the brick boundary relative to the voxel coordinates. Either the boundary is located at the center of one voxel, as is the case on level 0, or between two voxels, as is the case on other levels. In the first case only one voxel needs to be adapted while both voxels have to be modified in the latter case. Due to the overlap between bricks, adaption has to be performed whenever a subvolume is adjacent to a coarser one.

2.3 Level-Wise Texture Merging

The enlargement of textures in order to guarantee continuous level transitions leads to significant overhead in the texture memory that is used. Refer to the example shown in Figure 2. If we want to render the information represented by the first 31 voxels on level 0, we need two textures of size 16 for the finest level and two textures of size 16 on level 1 whereas only nine texture elements are necessary to store the information. Effectively there is no saving of texture memory when switching from level 0 to level 1. Considering higher levels leads to similar results, as only half of the voxels of every texture – plus one or two for overlap – contain non-redundant information.

This overhead can be minimized by merging the texture data of adjacent subvolumes into a single texture as shown in Figure 3. This figure demonstrates the one-dimensional analogue. For each level-of-detail the same texture size is used. In level 0 each brick has a texture of its own. As in any direction only half the voxels of a particular level are needed on the next coarser level, eight adjacent subvolumes can be represented by the same texture map, with appropriate texture coordinates. On higher levels, the number of subvolumes sharing the same texture map is multiplied by a factor of 8, which finally results in an octree-like hierarchy of texture maps.

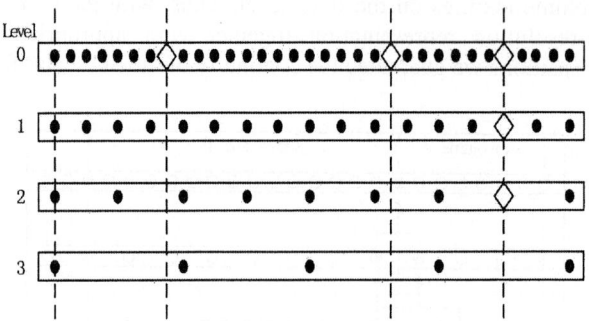

Fig. 3. The hierarchy of storing the textures of subvolumes

Merging textures works best when the starting number of subvolumes in every direction matches a power of two. If more than 2^{l-1} bricks are created by the subdivision of the initial data set, where l is the number of levels used for rendering, additional subvolumes will be generated that build their own set of textures as described in subsection 3.1. If fewer subvolumes are generated, then the depth of the texture hierarchy has to be reduced by storing textures on the coarsest level individually. We utilize a texture manager object for administration of the texture hierarchy. This object creates the desired texture maps and assigns appropriate sub-textures to the subvolumes requesting texture memory.

2.4 Opacity Correction

When rendering subvolumes at different levels of hierarchy, the opacity properties of the subvolumes are different. The classical rendering algorithms depend on using the same sampling along rays for each pixel. But in the context of a multi-resolution format, the volume is sampled in different ways, and at varying resolutions. To preserve optical properties between tiles of different resolutions, we must modify the transfer functions for those subvolumes generated by subsampling the original texture.

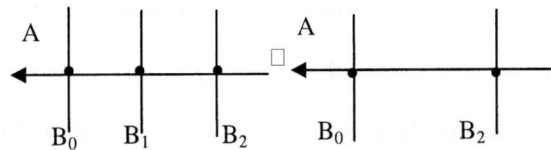

Fig. 4. Sampling a texture at two different resolutions

Figure 4 shows an example where we have sampled a texture at two different resolutions – one is half the resolution of the other. Each sample B_i has an associated color c_i and an opacity value α_i. By considering only the first three samples, the resulting colors A and A' are given by

$$A = \alpha_0 c_0 + (1-\alpha_0)\alpha_1 c_1 + (1-\alpha_0)(1-\alpha_1)C_2 \tag{1}$$

$$A' = \alpha'_0 c_0 + (1-\alpha'_0)C'_2 \tag{2}$$

Where C_2 is the incoming color from samples $B_2, B_3 \cdots$, C'_2 is the color calculated as a result of the samples $B_4, B_6 \cdots$

However, if we compute the total-accumulated opacities D and D', we obtain

$$D = \alpha_0 + (1-\alpha_0)\alpha_1 + (1-\alpha_0)(1-\alpha_1)D_2 \tag{3}$$

$$D' = \alpha'_0 + (1-\alpha'_0)D'_2 \tag{4}$$

Assuming that the accumulated opacities are equal at the even samples, it follows that $D_2 = D_2'$ and $D = D'$, i.e.,

$$\alpha_0 + (1-\alpha_0)\alpha_1 + (1-\alpha_0)(1-\alpha_1)D_2 = \alpha_0' + (1-\alpha_0')D_2' \qquad (5)$$

Solving this equation for α_0', one obtains

$$\alpha_0' = 1 - (1-\alpha_0)(1-\alpha_1) \qquad (6)$$

By assuming that $\alpha_1 = \alpha_0 + \varepsilon$ (where ε is a very small number), we obtain the equation

$$\alpha_0' = 1 - (1-\alpha_0)^2 + O(\varepsilon) \qquad (7)$$

Therefore, we modify the transfer function of the parent (coarser) texture by

$$\alpha' = 1 - (1-\alpha_0)^2 \qquad (8)$$

for all opacity values in the subsampled texture to minimize the artifacts between subvolumes. This formula is used when applying the transfer function to a level of the texture hierarchy.

2.5 Adaptive Level-of-Detail

The initial size of each subvolume has to be specified in advance before the multi-volume representation is constructed. Multi-criteria are applied to LOD selection.

- Maximum opacity: The maximum opacity of a subvolume is determined by the highest opacity of all the voxels in the subvolume. The rationale behind is that a more opaque region in the volume should be rendered in a higher accuracy.
- Distance to the view point: Here the distance is calculated from the center of the subvolume to the view point. For those subvolumes that are closer to the view point, as they have high importance value, they will be rendered in a high quality. For other subvolumes, as they are farther away and may be occluded, will be rendered in a low quality.
- Projection area: Projection area is calculated based on the bounding box of the subvolume. For those subvolumes with high projection area, a higher quality rendering should be assigned to them.
- Gaze distance: This parameter is especially useful for gaze-directed rendering. Gaze distance is the distance between the center of the gaze area and the center of the projected area of a subvolume. For those regions that are closer to the gaze area, we ensure a higher image quality. For those regions that are farther away from the gaze area, the image becomes blur because lower resolution data are used.

Furthermore, as long as no level transitions greater than one are specified, our approach allows us to use arbitrary texture resolution for every subvolume.

2.6 Temporal Coherence Consideration

The above LOD selection procedure will be executed at every frame, which could cause the subvolume's LOD to change frequently. However, frequent changes of the volume LODs can cause flickering when the user changes views. To solve this problem, we take special care to maintain the temporal coherence in consecutive rendering frames. We realize temporal coherence by recording the importance value of each subvolume from the last rendering. When a new frame is being rendered, if the user changes the viewing direction or gaze area, the importance value of each subvolume will be recalculated. For those subvolumes whose importance values do not differ too much (we use a threshold of percentage to decide this), we do not change its LOD in the new frame.

3 Results and Discussion

We have implemented our LOD volume rendering algorithm on a Windows 2000 PC with a 1.7 GHz AMD processor and NVIDIA GeForceFx graphics accelerator with 128 MB of Video RAM. Three data sets were used for the experiments.

Table 1. Rendering speed for the various data sets

Model	Data Set Size	Texture memory	Rendering Speed(fps)		
			our algorithm	ray casting	shear-warp
MRIBrain	$256^2 \times 109$	100%	17.8	6.4	21.5
		54%	21.4	-	-
		33%	24.2	-	-
Foot	102x247x200	100%	18.6	5.7	20.1
		53%	22.5	-	-
		27%	27.4		
Spine	$256^2 \times 299$	100%	14.3	5.2	18.4
		57%	16.8	-	-
		29%	19.1	-	-

Abbreviations: fps = frames per second.

In Figure 5, results of the proposed level-of-detail rendering technique applied to MRI-Brain of size $256^2 \times 109$ are shown. In the leftmost image, all subvolumes were rendered with full resolution. In the middle image only 54% of the original texture memory was used, while on the right, only 33% was used (Table 1). The number of texture lookups was reduced, from 59% to 35%. This leads to improved rendering performance. Whereas the frame rate is 17.8 frames/s with the full resolution data set, the level-of-detail representation used for the rightmost image allows us to create a frame rate is 24.2 frames/s (Table 1). The rendering speed of our algorithm is quick compared to other volume rendering techniques, such as ray casting and shear-warp.

In Figure 6, we use gaze-directed rendering to show the results when the gaze area is placed at different parts of the viewport.

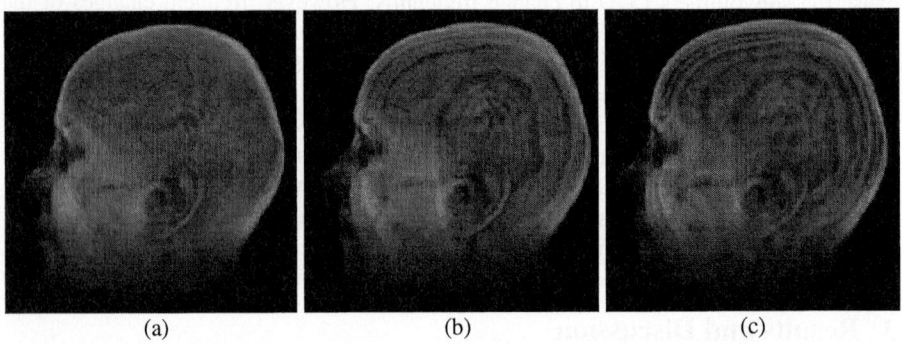

Fig. 5. On the left, the data set is displayed with full resolution. In the middle, two different levels of detail are used with lower resolution in the back. This reduces texture memory consumption to 54%. On the right, the adaptive representation using four levels of detail from front-to-back requires only 33% of the original texture memory

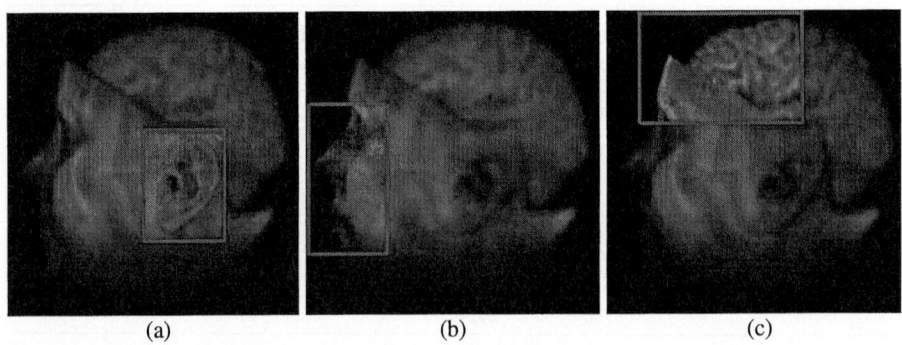

Fig. 6. Gaze-directed rendering. (a)(b)(c) show the results when the gaze area is placed at different parts of the viewport

4 Conclusions

In this work we have emphasized a multi-resolution approach for the rendering of large-scale volume data via 3D textures. The major contribution here is that we entirely avoid artifacts that occur due to incorrect texture interpolation and opacity correction at subvolume boundaries. In this respect, we have developed two beneficial extensions that guarantee continuous transitions between different levels of detail, and yield correct pixel opacities.

References

1. E. LaMar, B. Hamann, and K. Joy. Multiresolution techniques for interactive texture-based volume visualization. In Proceedings of Visualization '99, pages 355–361. IEEE Computer Society Press, Los Alamitos, CA,1999.
2. M. Weiler, R. Westermann, C. Hansen, K. Zimmerman,and T. Ertl. Level-of-detail volume rendering via 3d textures. In Proceedings of 2000 Symposium on Volume Visualization, pages 7–13. ACM SIGGRAPH, 2000.
3. D. Ellsworth, L. Chiang, and H.-W. Shen. Accelerating time-varying hardware volume rendering using tsp trees and color-based error metrics. In Proceedings of 2000 Symposium on Volume Visualization. ACM SIGGRAPH,2000.
4. D. Laur and P. Hanrahan. Hierarchical splating: A progressive refinement algorithm for volume rendering. In Proceedings of SIGGRAPH 91, pages 285–287. ACM SIGGRAPH, 1991.
5. T. Funkhouser and C. Sequin. Adaptive display algorithms for interactive frame rate during visualization of virtual environment. In Proceedings of SIGGRAPH 93, pages 247–254. ACM SIGGRAPH, 1993.
6. E. Gobbetti and E. Bouvier. Time-critical multiresolution scene rendering. In Proceedings of 1999 Symposium on Volume Visualization, pages 123–130. ACM SIGGRAPH, 1999.

Automatic Hepatic Tumor Segmentation Using Statistical Optimal Threshold

Seung-Jin Park[1], Kyung-Sik Seo[2], and Jong-An Park[3]

[1] Dept. of Biomedical Engineering,
Chonnam National University Medical School, Gwangju, Korea
sjinpark@jnu.ac.kr
[2] Dept. of Electrical & Computer Engineering,
New Mexico State University, Las Cruces, NM, USA
nmsu2@hanmail.net
[3] Dept. of Information & Communications Engineering,
Chosun University, Gwangju, Korea
japark@chosun.ac.kr

Abstract. This paper proposes an automatic hepatic tumor segmentation method of a computed tomography (CT) image using statistical optimal threshold. The liver structure is first segmented using histogram transformation, multi-modal threshold, maximum a posteriori decision, and binary morphological filtering. Hepatic vessels are removed from the liver because hepatic vessels are not related to tumor segmentation. Statistical optimal threshold is calculated by a transformed mixture probability density and minimum total probability error. Then a hepatic tumor is segmented using the optimal threshold value. In order to test the proposed method, 262 slices from 10 patients were selected. Experimental results show that the proposed method is very useful for diagnosis of the normal and abnormal liver.

1 Introduction

Liver cancer, which is the fifth most common cancer, is more serious in areas of western and central Africa and eastern and southeastern Asia [1]. The average incidence of liver cancer in these areas is 20 per 100,000, and liver cancer is the third highest death cause from cancer [1]. In Korea, the incidence of liver cancer is quite high at 19% for males and 7% for females [2].. In order to improve the curability of liver cancer, early detection is critical. Liver cancer, like other cancers, manifests itself with abnormal cells, conglomerated growth, and tumor formation. If the hepatic tumor is detected early, treatment and curing of a patient may be easy, and human life can be prolonged.

Liver segmentation using CT images has been vigorously performed because CT is a very conventional and non-invasive technique. Bae et al. [3] used priori information about liver morphology and image processing techniques. Gao et al. [4] developed automatic liver segmentation using a global histogram, morphologic operations, and the parametrically deformable contour model. Park et al. [5] built a probabilistic atlas of the brain and extended abdominal segmentation including the liver, kidneys, and spinal cord. Tsai [6] proposed an alternative segmentation method using an artificial neural network to classify each pixel into three categories. Also, Husain et al. [7] used neural networks for feature-based recognition of liver region. Pan et al. [8] presented a level set technique for the automatic liver segmentation by proposing a novel speed function. Seo et al. [9] proposed fully automatic liver segmentation based on the spine. However, most previous research has been concentrated on only liver segmentation and volume construction. In this paper, a simple automatic hepatic tumor segmentation method using statistical optimal threshold (SOT) is proposed. An automatic hepatic tumor segmentation method is presented in the following section. Experiments and analysis of results are described in the next section. Finally, the conclusion will be drawn in the last section.

2 Hepatic Tumor Segmentation

In this section, an automatic hepatic tumor segmentation method is presented. A liver structure is first segmented and then vessels in the liver are removed. Statistical optimal threshold (SOT) is found by transformed mixture probability density (MPD) and minimum total probability error (MTPE). A region of interest (ROI) of a hepatic tumor is segmented and estimated.

2.1 Liver Segmentation

The first important work to segment a hepatic tumor is to segment a liver boundary. The ROI of the liver is extracted using histogram transformation [10], multi-modal threshold [11], and maximum a posteriori decision [12]. In order to eliminate other abdominal organs such as the heart and right kidney, binary morphological (BM) filtering is performed by dilation, erosion, closing, and filling [13, 14, 15]. Fig. 1(a) shows an abnormal CT image with a tumor. Fig. 1(b) shows the ROI of the liver. Also, Fig. 1(c) shows the segmented liver image using BM filtering.

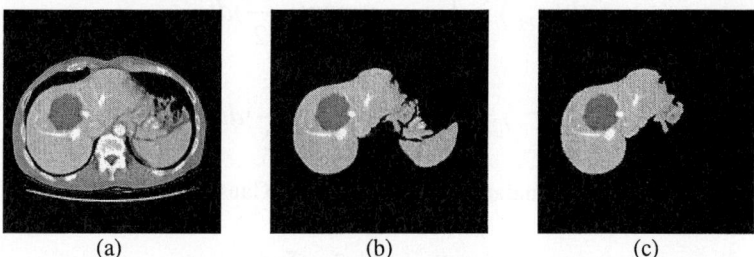

Fig. 1. Liver segmentation: (a) CT image, (b) ROI of the liver, (c) liver image segmented by BM filtering

2.2 Vessel Elimination

The liver image with a tumor obtained by BM filtering consists of the pure liver, tumor, and vessels. As vessels of the liver have no influence on tumor segmentation, vessels are eliminated from the liver. Histogram transformation for better histogram threshold is first performed to reduce histogram noises. Then the left and right valleys, called object ranges, are calculated using a piecewise linear interpolation method [11]. The vessel range is located in the rightmost side of the histogram because pixel values are higher than other objects. Therefore, the vessel range is decided easily. Fig. 2(a) shows the liver image after vessel elimination.

2.3 Statistical Optimal Threshold

After eliminating vessels, the histogram has only two peaks, and the liver image consists of the pure liver and tumor region. Therefore, the gray-level value thresholding two regions is easily calculated by using multi-modal threshold method. However, we do not know this threshold value is optimal. In order to find the SOT value, $T_{optimal}$, the histogram of the liver is transformed. The histogram with two peaks as the mixture probability density (MPD) is expressed as

$$p(x) = \frac{P_{PL}}{\sqrt{2\pi}\sigma_{PL}}\exp\left(-\frac{(x-\mu_{PL})^2}{2\sigma_{PL}^2}\right) + \frac{P_T}{\sqrt{2\pi}\sigma_T}\exp\left(-\frac{(x-\mu_T)^2}{2\sigma_T^2}\right) \quad (1)$$

where P_{PL}, μ_{PL}, and σ_{PL} are pixel occurrence, mean, variance of the pure liver and P_T, μ_T, and σ_T are pixel occurrence, mean, and variance of the tumor. Using four parameters such as $\mu_{PL}, \sigma_{PL}, \mu_T$, and σ_T, the transformed MPD [16] is created

- Generate numbers, $\Phi(z_{PL})$ and $\Phi(z_T)$, from a uniform distribution, $U \sim [0, 1]$.
- Find the inverse, z_{PL} and z_T, from the standard normal distribution function defined as

$$\Phi(z_{PL}) = \int_{-\infty}^{z_{PL}} \frac{1}{\sqrt{2\pi}}\exp\left(-\frac{t^2}{2}\right)dt \quad (2)$$

$$\Phi(z_T) = \int_{-\infty}^{z_T} \frac{1}{\sqrt{2\pi}}\exp\left(-\frac{t^2}{2}\right)dt. \quad (3)$$

- Converse the standard normal data to the Gaussian observations defined as

$$\hat{x}_{PL} = \mu_{PL} + z_{PL}\sigma_{PL} \quad (4)$$

$$\hat{x}_T = \mu_T + z_T\sigma_T. \quad (5)$$

- The transformed MPD is created defined as

$$p(\hat{x}) = \frac{P_{PL}}{\sqrt{2\pi}\sigma_{PL}}\exp\left(-\frac{(\hat{x}-\mu_{PL})^2}{2\sigma_{PL}^2}\right) + \frac{P_T}{\sqrt{2\pi}\sigma_T}\exp\left(-\frac{(\hat{x}-\mu_T)^2}{2\sigma_T^2}\right) \quad (6)$$

The $T_{optimal}$ is found by calculating the minimum total probability error. The total probability error (TPE) $E(T)$ from $p(\hat{x})$ is calculated as [12]

$$E(T) = P_{PL} \int_{-\infty}^{T} p(\hat{x}_{PL})d\hat{x} + P_T \int_{T}^{\infty} p(\hat{x}_T)d\hat{x} \tag{7}$$

where T is the threshold value. Then the $T_{optimal}$ is selected by the threshold value calculating the minimum TPE. Fig. 2(b) shows the segmented tumor using the $T_{optimal}$.

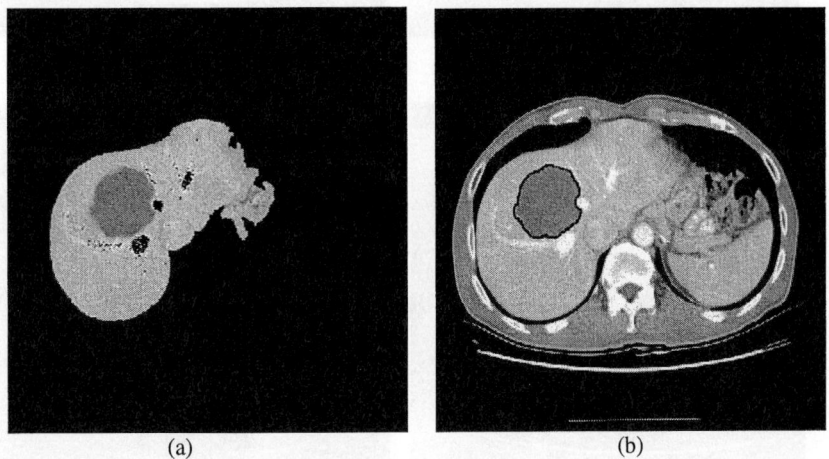

Fig. 2. Tumor segmentation: (a) liver image after hepatic vessel elimination, (b) segmented tumor using the optimal threshold value

3 Experiments and Analysis

CT images to be used in this research were provided by Chonnam National University Hospital in Kwangju, Korea. The CT scans were obtained by using a LightSpeed Qx/i, which was produced by GE Medical Systems. Scanning was performed with intravenous contrast enhancement. Also, the scanning parameters used a tube current of 230 mAs and 120 kVp, a 30 cm field of view, 5 mm collimation and a table speed of 15 mm/sec (pitch factor, 1:3).

Ten patients were selected for testing the new proposed method to segregate a hepatic tumor. Five people had normal livers and the other five people had abnormal livers. 262 total slices from ten patients were used. One radiologist took part in this research in order to evaluate liver status. Fig. 3 shows examples of segmented tumors. Table 1 shows the data of evaluated slices followed by slice numbers, true negative (TN), false positive (FP), false negative (FN), and true positive (TP) [17].

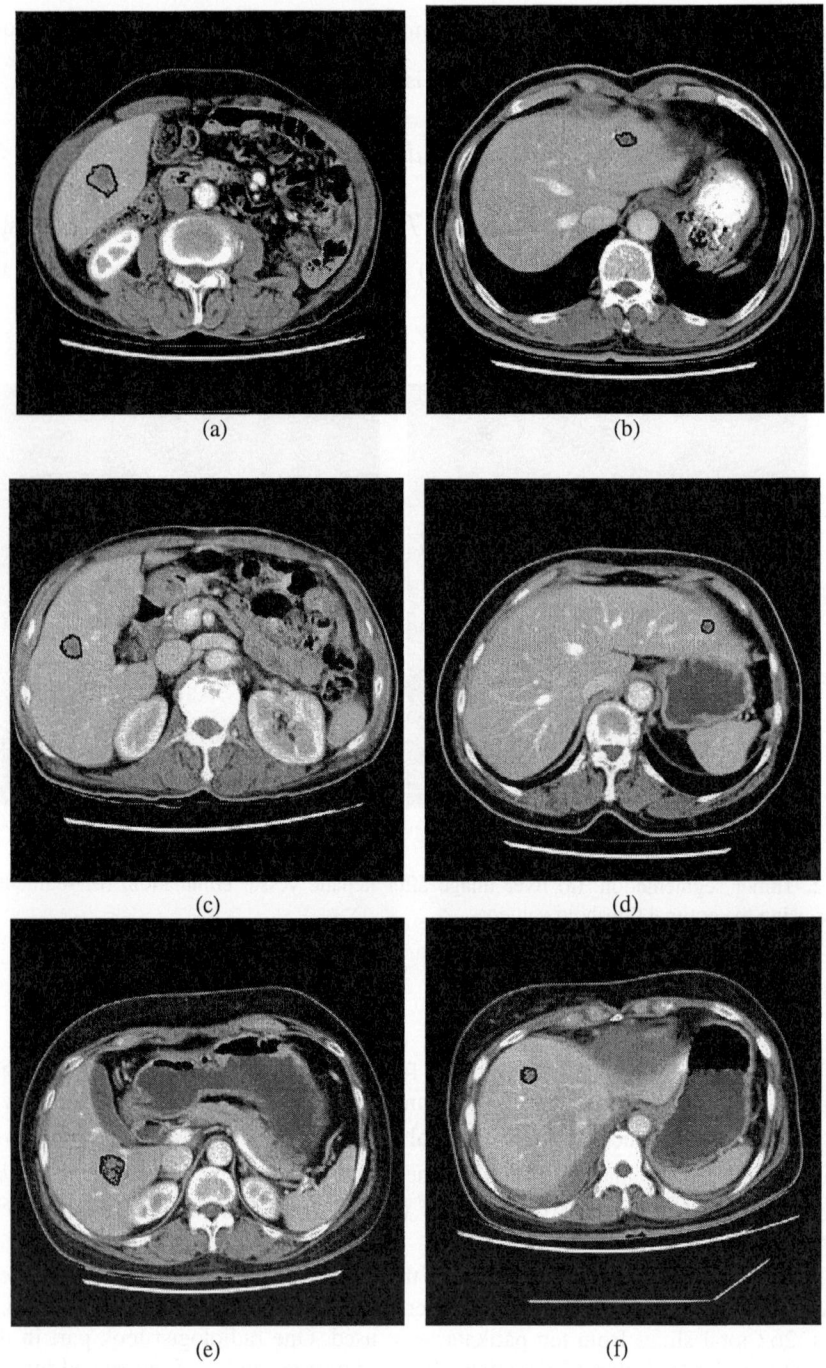

Fig. 3. Examples of tumor segmentation

Table 1. Data of evaluated slices

PATIENTS	SLICES TAKEN	FREQUENCY OF TN	FREQUENCY OF FP	FREQUENCY OF FN	FREQUENCY OF TP
PAT. 01	23	21	2	0	0
PAT. 02	31	30	1	0	0
PAT. 03	24	21	3	0	0
PAT. 04	26	25	1	0	0
PAT. 05	28	28	0	0	0
PAT. 06	34	30	2	1	1
PAT. 07	23	21	1	0	1
PAT. 08	23	19	3	0	1
PAT. 09	26	16	0	3	7
PAT. 10	24	17	4	2	1
TOTAL NUM.	262	228	17	6	11

As the evaluation measure, sensitivity, specificity, and accuracy were calculated. As sensitivity represents the fraction of patients with disease who test positive, sensitivity is defined as

$$Sensitivity = \frac{TP}{TP + FN}. \tag{8}$$

As specificity represents the fraction of patients without disease who test negative, specificity is defined as

$$Specificity = \frac{TN}{TN + FP}. \tag{9}$$

Also, accuracy is defined as

$$Accuracy = \frac{TP + TN}{TP + TN + FP + FN}. \tag{10}$$

In this research, we had 0.6471 of sensitivity, 0.9306 of specificity, and 0.9122 of accuracy. These results show the proposed method is very useful for diagnosis of the normal liver. Values of FP and FN are high for tumors located in the left portal branch and tumors with a diameter less than 2 cm.

4 Conclusions

In this paper, an automatic hepatic tumor segmentation method using statistical optimal threshold was proposed. The liver structure was first segmented in order to remove other abdominal organs. Hepatic vessels were removed from the liver because hepatic vessels were not related to tumor segmentation. Then statistical optimal threshold was calculated by a transformed mixture probability density and minimum total probability error. Finally, a hepatic tumor was segmented using the optimal threshold value. In order to evaluate the proposed method, 262 slices from 10 patients were selected. From the evaluation results, we had 0.6471 of sensitivity, 0.9306 of specificity, and 0.9122 of accuracy. These results show that the proposed method is very useful for diagnosis of normal and abnormal livers. In the future, algorithms for reducing false positives of the left portal branch will be developed.

References

1. Parkin, D. M.: Global cancer statistics in the year 2000. Lancet Oncology, Vol. 2. (2001) 533-54
2. Lee H.: Liver cancer. The Korean Society of Gastroenterology, Seoul Korea (2001)
3. Bae, K. T., Giger, M. L., Chen, C. T., Kahn, Jr. C. E.: Automatic segmentation of liver structure in CT images. Med. Phys.,Vol. 20. (1993) 71-78
4. Gao, L., Heath, D. G., Kuszyk, B. S., Fishman, E. K.: Automatic liver segmentation technique for three-dimensional visualization of CT data. Radiology, Vol. 201. (1996) 359-364
5. Park, H., Bland, P. H., Meyer, C. R.: Construction of an abdominal probabilistic atlas and its application in segmentation. IEEE Trans. Med. Imag., Vol. 22. No. 4. (2003) 483-492
6. Tsai, D.: Automatic segmentation of liver structure in CT images using a neural network. IEICE Trans. Fundamentals, Vol. E77-A. No. 11. (1994) 1892-1895
7. Husain, S. A., Shigeru, E.: Use of neural networks for feature based recognition of liver region on CT images. Neural Networks for Sig Proc.-Proceedings of the IEEE Work., Vol.2. (2000) 831-840
8. Pan, S., Dawant, B. M.: Automatic 3D segmentation of the liver from abdominal CT images: a level-set approach. Proceedings of SPIE, vol. 4322. (2001) 128-138
9. Seo, K., Ludeman, L. C., Park S., Park, J.: Efficient liver segmentation based on the spine. LNCS, Vol. 3261. (2004) 400-409
10. Orfanidis, S. J.: Introduction to signal processing. Prentice Hall, Upper Saddle River NJ (1996)
11. Schilling, R. J., Harris, S. L.: Applied numerical methods for engineers. Brooks/Cole Publishing Com., Pacific Grove CA (2000)
12. Ludeman, L. C.: Random processes: filtering, estimation, and detection. Wiley & Sons Inc., Hoboken NJ (2003)
13. Gonzalez, R. C., Woods, R. E.: Digital image processing. Prentice Hall, Upper Saddle River NJ (2002)
14. Shapiro, L. G., Stockman, G. C.: Computer vision. Prentice-Hall, Upper Saddle River NJ (2001)
15. Parker, J.R.: Algorithms for image processing and computer vision. Wiley Computer Publishing, New York (1997)
16. Hines, W. W., Montgomery, D. C., . Goldsman, D. M, Borror, C.M.: Probability and statistics in engineering. Wiley, Hoboken NJ (2003)
17. Rangayyan R.M.: Biomedical signal analysis. Wiley, New York NY (2002)

Spatio-Temporal Patterns in the Depth EEG During the Epileptic Seizure*

Jung Ae Kim[1], Sunyoung Cho[2], Sang Kun Lee[3], Hyunwoo Nam[3], and Seung Kee Han[1,2]

[1] Department of Physics, Chungbuk National University,
Cheongju, Korea,
[2] Basic Science Research Institute, Chungbuk National University,
Cheongju, Korea,
[3] Departments of Neurology, College of Medicine, Seoul National University,
Seoul, Korea,
sycho@chungbuk.ac.kr

Abstract. We analyzed the spatio-temporal patterns of the depth EEG recorded from a patient with lateral temporal-lobe epilepsy. Statistical analysis based on the Jensen-Shannon entropy (JS-E) as well as linear power spectral analyses were performed. The spatio-temporal patterns from JS-E and β rhythm successfully detected the onset timing of the seizure and revealed the temporal topology of the epileptic focus. The robustness of these patterns was proved by inter-trial consistency. As the patterns are well matched with the clinical diagnosis, it could be used for the identification of onset time and focus region of epileptic seizure.

1 Introduction

The epilepsy is characterized by a disturbance on the electrochemical activity of the brain, producing an excessive and hypersynchronous activity of the neuron. Epileptic seizures reflect this sudden and recurrent malfunction of the brain areas. Partial (focal) seizures originate from a restricted region and remain to this region (epileptic focus), while generalized seizures involve almost the entire brain. Partial seizures generally originate at a cortical area; most of them begin focally at the temporal lobes.

Exact localization of the epileptic focus and its delineation from functionally relevant areas are required for successful surgical treatment and understanding the basic mechanism of the seizures. The primary tool for this purpose is the EEG due to high temporal resolution and close relationship to physiological and pathological function of the brain. Various spectral and nonlinear analyses of epileptic EEG have been used to detect the epileptic focus and to predict the seizure onset [1~4]. Investigators have extracted specific quantitative features from the EEG, which would be characteristics for the discharge patterns during seizure [5].

---

* This work was supported by Korean Research Foundation, KRF 2002-075-H0007 to S.Y. Cho, and a grant (M103KV010011-03K2201-01130) from Brain Research Center of the 21st Century Frontier Research Program funded by the Ministry of Science and Technology of Republic of Korea to S.K. Han.

We analyzed the spatio-temporal patterns of the epileptic EEG recorded intracranial for the temporal topology and source localization of seizure initiation and propagation. Statistical analysis based on the Jensen-Shannon entropy (JS-E) and linear power spectral analyses were performed [6].

2 Method

2.1 Intracranial Recording

Our data was from a patients presenting lateral temporal epilepsy clinically diagnosed at Epilepsy Center in Medical collage of Seoul National University. Intracranial recording was required to confirm the exact site of the structures generating seizure onsets before surgical operation. The EEG recording were performed on 30 sites as depicted in Fig 1 and 2 (Telefactor Beehive telemonitoring system, sampling rate 200Hz, bandwidth filtering 0.1~70Hz). The data set analyzed in this study had duration of 3-5 min including 1 min before the seizure. The seizure onset was determined by electrographic criteria as localized, sustained rhythmic discharges (burst of rhythmic spikes, low-voltage fast rhythms, etc.) and associated with subsequent clinical seizure activity [7].

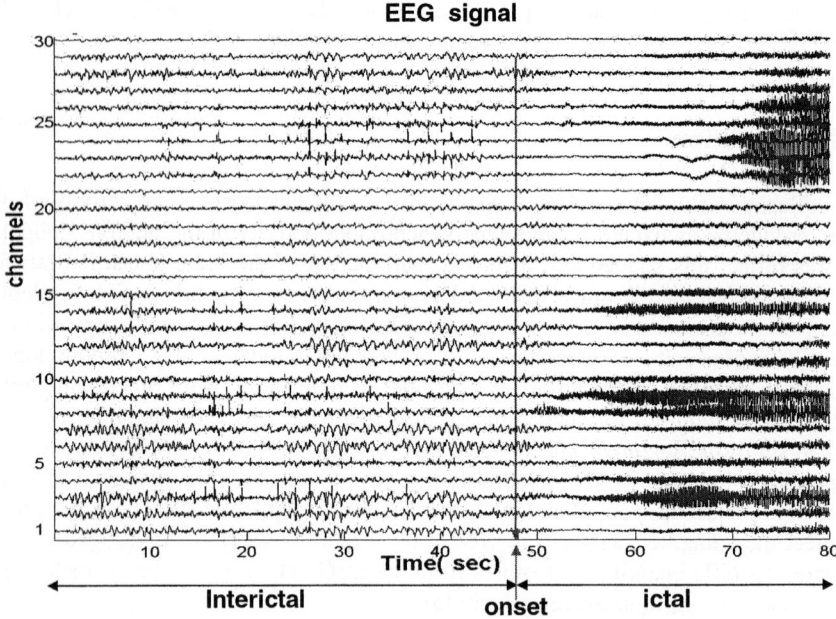

Fig. 1. Example of depth EEG raw data recorded from 30 sites on the temporal lobe. The red arrow (onset) points the EEG onset of seizure activity diagnosed clinically

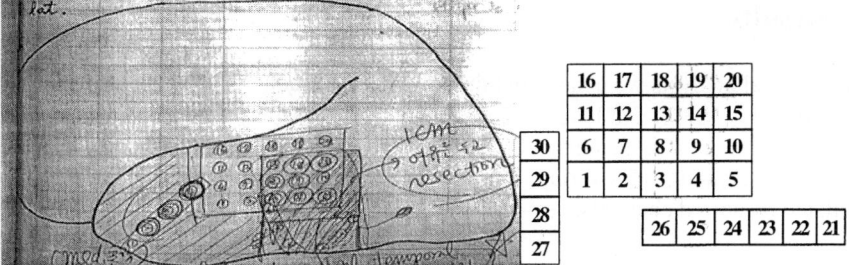

Fig. 2. Schematic view of the recording sites on the temporal lobe. 21-26 sites were covered on the ventral area of inferial temporal region. The numbers of channels are shown in the left boxes

2.2 Jensen-Shannon Entropy

We applied the Jensen-Shannon entropy to analyze spatio-temporal patterns of the EEG time series data. Jensen-Shannon entropy (JS-E) is a statistical measurement that quantifies the difference between two (or more) probability distributions [8]. This value would be maximized when two distributions reveal the most prominent difference in the statistical characteristics.

In case of epileptic EEG, we can find the abrupt changing points of the time series data using JS-E. When total series S is divided to two subset, S_1 and S_2 of lengths l_1 and l_2, JS-E is calculated from the entropies of total time range (S) and two subsets (S_1 and S_2) by equations below. With moving the dividing point by 1.5 sec, the distribution of JS-E was obtained. The EEG data in whole trial could be separated into sub periods using JS-E peak points (see Fig 6).

$$z(t) = x(t+1) - x(t)$$

$$JS_2(S_1, S_2) = H[S] - (\frac{l_1}{L}H[S_1] + \frac{l_2}{L}H[S_2]) \geq 0$$

$$L = l_1 + l_2, \quad S = S_1 \oplus S_2$$

$$H[\bullet] = -\Sigma\, p \log_2 p$$

where H denotes Shannon entropy of the probability distribution {p}.

3 Results

For spectral analysis we considered the following frequency bands: theta (θ: 3 ~ 7Hz), alpha (α: 8 ~ 13Hz) and beta (β: 20 ~ 30Hz). Fig. 3 displays the mean time course of powers (averaged in 30 channels) of total and three frequency bands with window size 10.24sec and 50% overlap. Fig. 4 displays temporal patterns of each band accumulated by sorted channels.

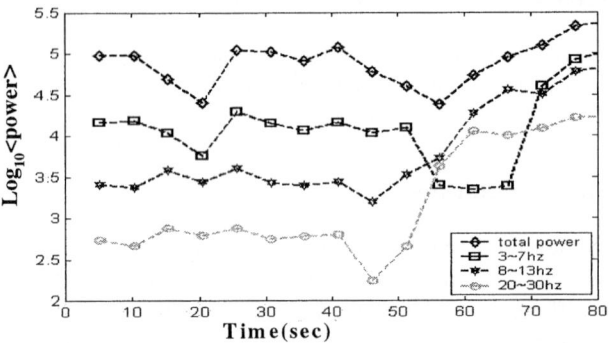

Fig. 3. The mean time course of powers of total and three frequency bands

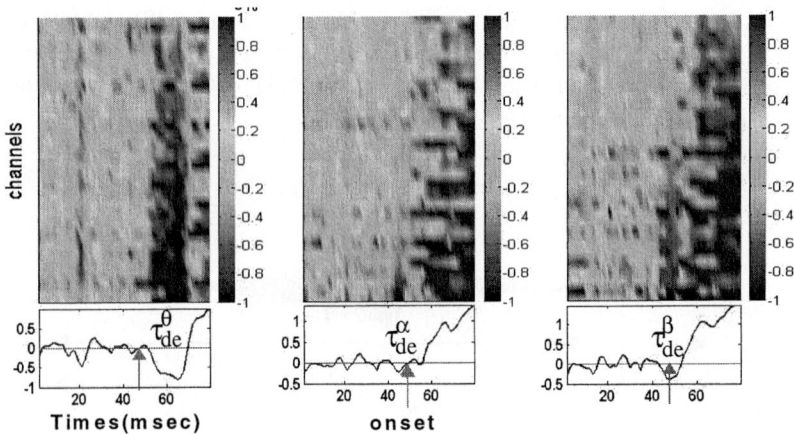

Fig. 4. Power spectral density accumulated by channels sorted by the amount of decrease after seizure onset. The averages on the channels were shown in the boxes below. The red arrow (onset) points the EEG onset of seizure diagnosed clinically

Near the onset time of seizure, β rhythm started to decrease for a short time (about 5-10 sec) and followed by an increase as the seizure evolved. Fig 5 shows the change of power spectral topography of β rhythm over time. These activities are well matched to the location of seizure focus defined clinically and reflect the progression of the seizure. Compare to β rhythm, θ it started to decrease later and sustained longer.

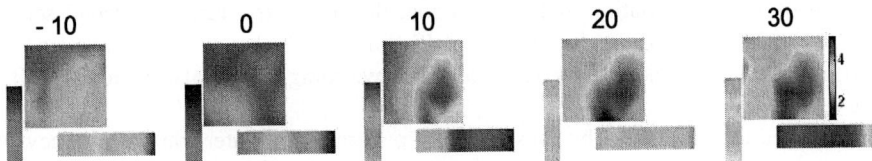

Fig. 5. Temporal evolution of power spectral topography of β rhythm. The onset time is scaled zero and the window size is 10.24sec and 50% overlap

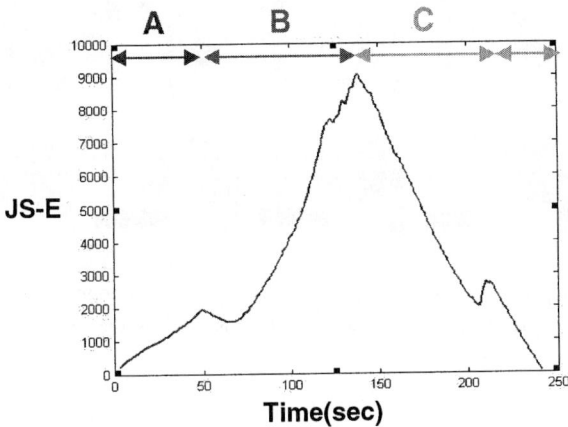

Fig. 6. JS-E mean distribution over time. Sub periods (A: inter-ictal, B: ictal, C: post-ictal period) could be identified using the peaks

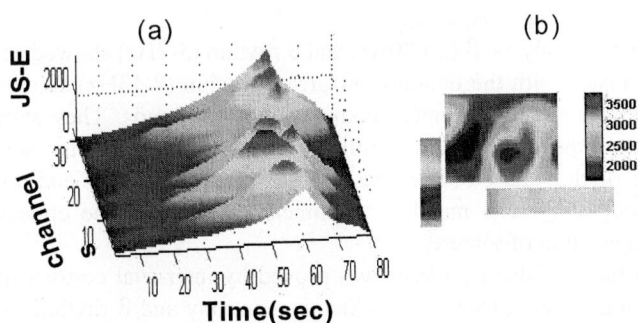

Fig. 7. JS-E distribution in each channel during early 80 sec (a) and the peak entropy topography (b)

In Fig 6, a whole trial EEG data were divided into sub periods using JS-E peak points. Those sub periods fitted the evolving stage of seizure (inter-ictal / ictal / post-ictal) very well. We analyzed further the data for the early period for the source localization of seizure initiation and propagation, that is 80 sec-duration around seizure onset. Fig 7(a) showed the distribution of JS-E across 30 channels over time, and Fig 7(b) depicted the peak entropy value on each recording site. It also matched well to the seizure focus.

We tried to prove the robustness of these patterns using inter-trial consistency. As shown in Fig 9, the JS-E peak entropy topography and β power topography at the 200% power- increasing point were consistent across the trials.

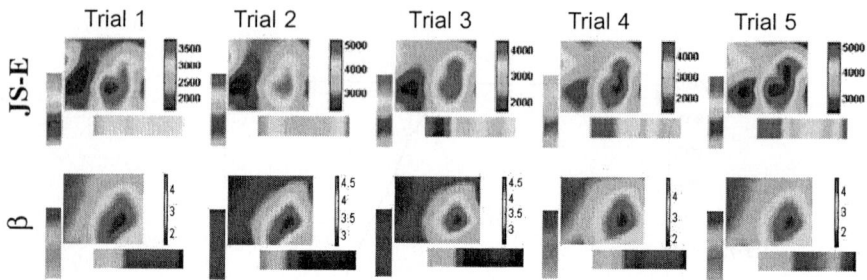

Fig. 8. JS-E peak entropy topography and β power topography at the 200% power- increasing point in each trial, that were consistent across the trials as well as the exact localization of epileptic focus

4 Discussion

The spatio-temporal analysis using the Jensen-Shannon entropy successfully detected the onset timing of the seizure and revealed the topographic information of ongoing seizure. Besides the localization of epileptic focus it could be extended to the temporal profiles of seizure propagation.

In the spectral analysis, β (20-30Hz) and θ rhythm (3-7Hz) showed spatio-temporal patterns correlated with the ongoing seizure. α rhythm (8-13Hz) showed similar patterns even much more moderately compare to other bands. They showed decrease near the onset time of seizure and followed by an increase as the seizure evolved. Especially β rhythm started to decrease before the seizure onset and revealed spatio-temporal topography well matched to clinical diagnose of the epileptic focus and temporal propagation of seizure.

The robustness of these patterns was proved by inter-trial consistency, that is the spatio-temporal patterns from Jensen-Shannon entropy and β rhythm were consistent across the trials. As the patterns are well matched with the clinical diagnosis, it could be used for the identification of onset time and focus region of epileptic seizure.

References

1. Lerner, D. E. : Monitoring changing dynamics with correlation integrals:Case study of an epileptic seizure.Phys. D 97(1996) 563-576
2. Martinerie, J., Adam,C., Quyen, M. L. V. , Baulac,M., Clemenceau, ., Renault,B. and Varela, F.,J.: Epileptic crisis can be anticipated by non-linear analysis, Nature Medicine vol.4 (1998) 1173-1176
3. Pereda, E., Rial, R., Gamundi, A., Gonzalez, J.: Assessment of changing interdependencies between human electroencephalograms using non-linear methods, Phys. D 148 (2001) 147.
4. Quyen, M. L. V., Martinerie, J, Baulac, M. Varela, F.: Anticipating epileptic seizures in real time by a non-linear analysis of similarity between EEG recordings, NeuroReport 10, (1999) 2149-2155
5. Mormann, F., Elger,C.E. and Lehenrtz, K.: Automatic detection of a preseizure state based on a decrease in synchronization in intracranial electroencephalogram recordings from epilepsy patients, Phys. Rev. E 67, (2003) 021912-1-10
6. Galvan, P. B., Grosse, I., Carpena, P., Oliver, J. L., Roldan, R.R. Stanley, H.E.: Finding borders between coding and noncoding DNA regions by an entropy segmentation method, Phys. Rev. Lett. 85(6), (2000) 1342-1345
7. Engel, C. E.; Seizure and epilepsy: contemporary neurology series. Philadelphia: FA Davis, (1989)
8. Grosse, I., Galvan, P. B., Carpena, P., Roldan, R.,R., Oliver, J., and Stanley, H., E.,: Analysis of symbolic sequences using the Jensen-Shannon divergence, Phys. Rev. E 65,(2002) 041905-1-16

Prediction of Ribosomal Frameshift Signals of User-Defined Models

Yanga Byun, Sanghoon Moon, and Kyungsook Han[*]

School of Computer Science and Engineering,
Inha University, Inchon 402-751, Korea
{quaah, jiap72}@hanmail.net,
khan@inha.ac.kr

Abstract. Programmed ribosomal frameshifts are used frequently by RNA viruses to synthesize a single fusion protein from two or more overlapping open reading frames. Depending on the number of nucleotides shifted and the direction of shifting, frameshifts are classified into –n and +n frameshifts, n being the number of nucleotides shifted. Computational identification of frameshift sites is in general very difficult since the sequences of frameshifting cassettes are diverse and highly dependent on the organism. Most current computational methods focus on predicting -1 frameshift sites only, and cannot handle other types of frameshift sites. We previously developed a program called FSFinder for predicting -1 and +1 frameshifts. As an extension of FSFinder, we now present FSFinder2, which is capable of predicting frameshift sites of general type, including user-defined models. We believe FSFinder2 is the first program capable of predicting frameshift signals of general type, and that it is a powerful and flexible tool for predicting genes that utilize alternative decoding, and for analyzing frameshift sites.

1 Introduction

Programmed ribosomal frameshifting is involved in the expression of certain genes in a wide range of organisms such as viruses, bacteria and eukaryotes, including humans [1-4]. In this process the ribosome shifts to an alternative reading frame by one or a few nucleotides at a specific site in a messenger RNA [5]. The most common of these events requires the ribosome to shift to a codon that overlaps a codon in the existing frame [6]. Frameshifts are classified into different types depending on the number of nucleotides shifted and the shifting direction. The most common type is a -1 frameshift, in which the ribosome slips a single nucleotide in the upstream direction. -1 frameshifting requires a slippery site and a stimulatory RNA structure. A spacer of 5-9 nucleotides separating the slippery site from the stimulatory RNA structure also affects the probability of frameshifting. The slippery site generally consists of a heptameric sequence of the form X XXY YYZ in the incoming 0-frame where X, Y and Z can be the same nucleotide [7], but other slippery sequences differ from this

---

[*] Correspondence Author.

motif. The ribosome changes reading frame on these sequences. The stimulatory RNA structure forms a secondary structure such as a pseudoknot or stem-loop. Ribosomal pausing is generally believed to account for programmed frameshifting, but Kontos et al. [8] consider that pausing is not sufficient to cause frameshifting.

+1 frameshifts are much less common than -1 frameshifts, but have been observed in diverse organisms [6]. The *prfB* gene encoding release factor 2 in *E.coli* is a well-known example [9, 10]. In RF2 frameshifting, a Shine-Dalgarno (SD) sequence is observed upstream of the slippery sequence CUU URA C where R can be adenine or guanine. Among other +1 frameshift sites, the frameshift signal of the ornithine decarboxylase antizyme (*oaz*) gene encoding antizyme 1 consists of a slippery sequence and a downstream RNA secondary structure such as a pseudoknot [11].

In previous work we developed a program called FSFinder (Frameshift Signal Finder) for predicting -1 and +1 frameshift sites [12, 13]. Trials of FSFinder on ~190 genomic and partial DNA sequences showed that it predicted frameshift sites efficiently and with greater sensitivity and specificity than other programs [14, 15, 16]. Although -1 and +1 frameshifts are the most frequently found frameshifts, other types occur. This paper presents an extension of FSFinder that can handle frameshifts of any type, including user-defined types. We believe this is the first program capable of predicting frameshift signals of general type.

2   A Computational Model of Frameshifts

2.1   Basic Models of Frameshifts

Three types of frameshifts are considered as basic frameshifts and their models are predefined: the -1 and +1 frameshifts for the peptide chain release factor B gene encoding release factor 2 (RF2), and the +1 frameshift for the ornithine decarboxylase antizyme (ODC antizyme). The models for these frameshifts consist of four components: Shine-Dalgarno sequence, frameshift site, spacer and downstream secondary structure (Fig. 1). FSFinder2 extends the previous models used in FSFinder to incorporate a user-defined model. For the upstream Shine-Dalgarno sequence, FSFinder2 considers AGGA, GGGG, GGAG, AGGG and GGGA as well as classical Shine-Dalgarno sequences such as GGAGG and GGGGG. For the slippery site of the +1 frameshift, the sequence CUU URA C, where R is a purine (that is, either adenine or guanine), is considered. For the downstream structure, H-type pseudoknots as well as stem-loops are considered.

2.2   User-Defined Frameshift Models

As shown in Fig. 1, the three basic models can be defined by a combination of a few components. We classify the components into four types.

- The *pattern* component represents a pattern of nucleotide characters like the slippery site of the -1 frameshift model (first box in Fig. 1A). The pattern characters are defined first, followed by the nucleotide characters that represent the pattern characters.

- The *signal* component represents a nucleotide string such as Shine-Dalgarno sequences, stop codons, or CUU URA C, UUU strings in a +1 frameshift model (first and third boxes in Fig. 1B, and first and second boxes in Fig. 1C).
- The *secondary structure* components are simple stem-loops or pseudoknots, in which only canonical pairs are considered (third box in Fig. 1A, and third box in Fig. 1C).
- The *counter* component represents the number of nucleotide characters in the specified region. This component is useful for finding regions with specific nucleotide content such as those with high GC content.

Unlike the signal component, the pattern component can specify the number of occurrences of a specific nucleotide. For example, the pattern of -1 frameshift signal is X XXY YYZ in which X can be N (A,G,C,T), Y can be W (A or C), and Z can be H (A, C, T). But, the first three nucleotides must be the same to each other, and the next three nucleotides must be the same. If this pattern is defined using a signal component N NNW WWH instead, there is no way of avoiding unwanted matches such as C AGA TTA.

Flexible spacers are inserted between the components. With a combination of the components of the four types users can define not only the basic models but also their own models. An arbitrary number of the components in any order can appear in a user-defined model.

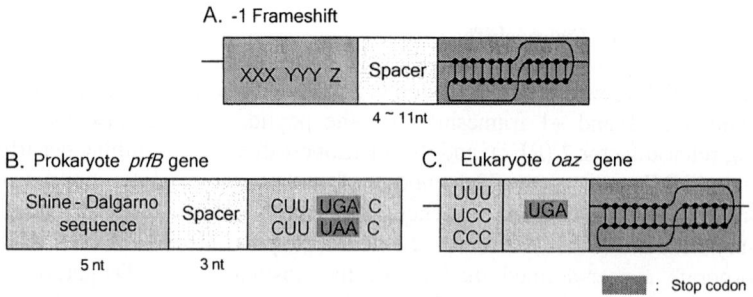

Fig. 1. The three basic frameshift models. (A) -1 frameshift. (B) +1 frameshift model of the *prfB* gene in *E.coli*. (C) +1 frameshift model of the *oaz* gene in eukaryotes

2.3 Algorithm for Predicting User-Defined Models

FSFinder2 has an algorithm to handle various user-defined models. Since an arbitrary number of components can be included, the most important component should be specified as a pivot by the user. Based on the user's choice, FSFinder2 first finds matches with the pivot component. It then finds matches to other components in each direction from the pivot component, starting with the closer one to the pivot component. For example, for a user-defined model in which components 1, 2, 3, 4, and 5 appear in this order, and component 3 is the pivot, it finds components either in the order of 3, 2, 1, 4, 5 or 3, 4, 5, 2, 1. Algorithm 1 shows a way of locating a user-defined model in genomic sequences.

Algorithm 1 Find a user-defined model

```
Length(A) is the length of array A.
Firstof(match) is the first index of a match.
Lastof(match) is the last index of a match.

Set F be an array of components in the user-defined model.
Set M be a 2-dim array that will save all matches of a component.
Set 1-dim of M as Length(F), and the size of M is flexible.
pi ← index of pivot model
Set M[pi] an array of matches with F[pi], sorted in increasing
order of the first indices of matches.

for i ← pi-1 to 0 do
    count ← 0
    for mi ← 0 to length(M[i+1]) do
        if mi ≠ 0 and Firstof(M[i, mi])= Firstof(M[i, mi-1]) then
            go to next step.
        end if
Set FM be an array of matches with F[i] in upstream of M[i+1,
mi]. Sort FM in increasing order of the first indices of matches.
        for fmi ← 0 to Length(FM)-1 do
            M[i, count] ← FM[fmi]
            Count ← count + 1
        end for
    end for
end for

for i ← pi+1 to Length(F)-1 do
    count ← 0
    for mi ← 0 to length(M[i-1]) do
        if mi ≠ 0 and Lastof(M[i, mi])= Lastof(M[i, mi-1]) then
            go to next step.
        end if
Set FM be an array of matches with F[i] in downstream of M[i-1,
mi]. Sort FM in increasing order of the last indices of matches.
        for fmi ← 0 to Length(FM)-1 do
            M[i, count] ← FM[fmi]
            count ← count + 1
        end for
    end for
end for
```

3 Results and Discussion

Fig. 2 shows default models for the three basic frameshifts. These models can be redefined in the edit panels of the four components. Each component has its own edit panel that helps users easily define and modify their frameshift models (Fig. 3). The basic models as well as user-defined models can be defined with any combination of the four components. Fig. 4 shows an example of the graphical user interface of FSFinder2 for finding a user-defined frameshift signal. User-defined models can be saved in an XML file and loaded later as desired.

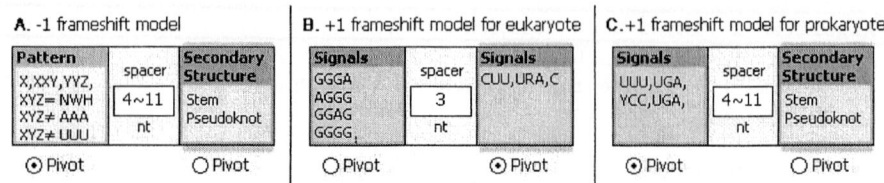

Fig. 2. Three basic models for frameshift mutation defined in FSFinder

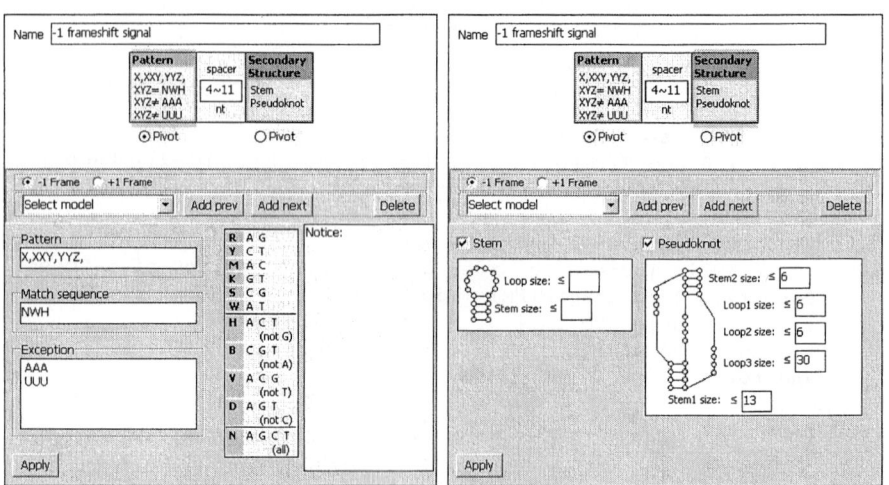

Fig. 3. Edit panels for user-defined frameshift models. The left panel is for the pattern component, and the right, for secondary structure components

FSFinder2 focuses on finding frameshift sites in the overlapping regions of open reading frames (ORF) since most known frameshift cassettes are found in the overlapping regions of two ORFs [13]. Consider two ORFs of frames 0 and -1 in Fig. 4. The starting positions of the two ORFs are extended from their original start codons to upstream stop codons (positions a and c in Fig. 4A). The extended regions a-b and c-d of the two ORFs partially overlap at their termini if position a of frame -1 is to the left of position d of frame 0 and there exists a start codon in frame 0. The highlighted region e of Fig. 4A is the overlapping region of the two ORFs. Our definition of an overlapping region identifies a wider region than an actual overlapping region since it is extended to the upstream stop codon. The reason for the extended overlapping region of ORFs is to avoid missing possible frameshift sites.

We tested FSFinder2 on *Shewanella algae*, for which no frameshift signal is known. A *Shewanella algae* is a mesophilic marine bacterium and plays an important role in the turnover of inorganic material. It may also cause disease in humans [17]. We selected 11 contigs out of 113 contigs in the *Shewanella algae*, which are longer than 100 Kb. The frameshift signals of *Shewanella algae* were divided into 6 types with the help of expert biologists, and searched from the 11 contigs. As shown in Table 1, FSFinder2 found 28 frameshift sites that exactly match to our defined frameshift model (model not shown here) and 915 frameshift sites that partially match to the defined model in the overlapping regions of open reading frames. There were many other frameshift sites (955 exact matches and 24,572 partial matches to the defined models) that were found in non-overlapping regions of open reading frames, but there are likely to be false positives. Considering the shape and length of the overlapping regions of open reading frames, at least 12 out of the 28 exact matches in the overlapping regions are strong candidates of frameshift sites and the remaining 16 sites are also good candidates.

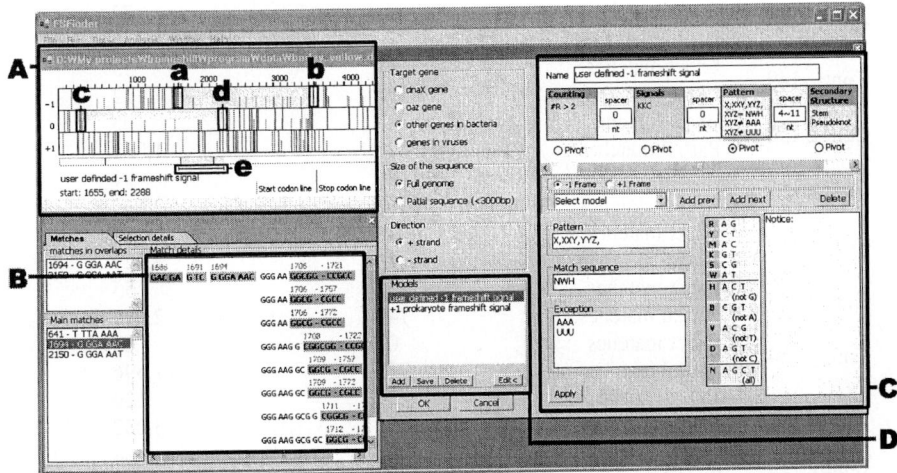

Fig. 4. Graphical user interface of FSFinder2. (A) The display window shows the positions of start codons, stop codons and matches with the user-defined model. Reading frames a–b and c–d partially overlap at their termini. FSFinder2 focuses on finding frameshift sites in the overlap region e. (B) The match details show all match results. (C) The edit panel shows a user-defined model with four components. (D) List of user-defined models

4 Conclusion

Understanding programmed ribosomal frameshifts is important because they are related to biological phenomena such as fidelity of mRNA-tRNA binding, and some genetic controls and enzyme activities. They are also involved in the expression of certain genes in a wide range of organisms. However, identifying programmed frameshifts is very difficult due to the diverse nature of the frameshift events. Existing computational approaches focus on a certain type of frameshift only and cannot handle frameshifts of variable type.

We have developed a program called FSFinder2 for predicting frameshift sites of any type, including user-defined types. A user can define his or her frameshift model with a combination of four components. The user-defined model can be saved in an XML file and loaded later. FSFinder2 is currently being used to find unknown frameshift sites in the *Shewanella* genome. *Shewanella* is a metal reducing bacterium that can be used for cleaning-up polluted environments. We believe FSFinder2 is the first program capable of predicting frameshift signals of general type.

Table 1. Predicted frameshift signals (FS) in the *Shewanella algae* contigs

Contig number	Type of match	FS in overlapping region	FS in non-overlapping region
Contig 147	Exact matches	0	57
	Partial matches	63	1,834
Contig 148	Exact matches	3	72
	Partial matches	78	1,455
Contig 149	Exact matches	1	53
	Partial matches	57	1,122
Contig 150	Exact matches	2	105
	Partial matches	82	1,981
Contig 151	Exact matches	8	91
	Partial matches	100	2,643
Contig 152	Exact matches	1	81
	Partial matches	53	2,027
Contig 153	Exact matches	0	55
	Partial matches	74	2,027
Contig 154	Exact matches	0	69
	Partial matches	77	1,596
Contig 155	Exact matches	3	120
	Partial matches	103	3,052
Contig 156	Exact matches	2	84
	Partial matches	77	2,021
Contig 157	Exact matches	8	168
	Partial matches	151	4,314
Total number of exact matches		28	955
Total number of partial matches		915	24,572

Acknowledgements

This work was supported by the Korea Science and Engineering Foundation (KOSEF) under grant R01-2003-000-10461-0.

References

1. Namy, O., Rousset, J., Napthine, S., Brierley, I.: Reprogrammed genetic decoding in cellular gene expression. Mol. Cell 13 (2004) 157-169
2. Stahl, G., McCarty, G.P., Farabaugh, P.J.: Ribosome structure: revisiting the connection between translational accuracy and unconventional decoding. Trends Biochem. Sci. 27 (2002) 178-183
3. Dinman, J.D., Icho, T., Wickner, R.B.: A -1 ribosomal frameshift in a double-stranded RNA virus of yeast forms a gag-pol fusion protein. Proc. Natl Acad. Sci. USA 88 (1991) 174-178

4. Licznar, P., Mejlhede, N., Prere, M., Wills, N., Gesteland, R.F., Atkins, J.F.: Programmed translational -1 frameshifting on hexanucleotide motifs and the wobble properties of tRNAs. EMBO J. 22 (2003) 4770-4778
5. Baranov, P.V., Gesteland, R.F., Atkins, J.F.: Recoding: translational bifurcations in gene expression. Gene 286 (2002) 187-201
6. Farabaugh, P.J.: Programmed translational frameshifting. Ann. Rev. Genetics 30 (1996) 507-528
7. Jacks, T., Varmus, H.E.: Expression of the Rous sarcoma virus pol gene by ribosomal frameshifting. Science 230 (1985) 1237-1242
8. Kontos, H., Napthine, S., Brierley, L.: Ribosomal pausing at a frameshifter RNA pseudoknot is sensitive to reading phase but shows little correlation with frameshift efficiency. Mol. Cell. Biol. 21 (2001) 8657-8670
9. Weiss, R.B., Dunn, D.M., Atkins, J.F., Gesteland, R.F.: Slippery runs, shifty stops, backward steps, forward hots: -2, -1, +1, +2, +5, and +6 ribosomal frameshifting. Cold Spring Harb. Symp. Quant. Biol. 52 (1987) 687-693
10. Baranov, P.V., Gesteland, R.F., Atkins, J.F.: Release factor 2 frameshifting sites in different bacteria. EMBO Rep. 3 (2002) 373-377
11. Ivanov, I.P., Gesteland, R.F., Atkins, J.F.: Antizyme expression: a subversion of triplet decoding, which is remarkably conserved by evolution, is a sensor for an autoregulatory circuit. Nucleic Acids Research 28 (2000) 3185-3196
12. Sanghoon, M., Yanga, B., Kyungsook, H.: Computational identification of -1 frameshift signals. LNCS. 3036 (2004) 334-341
13. Sanghoon, M., Yanga, B., Hong-jin, K., Sunjoo, J., Kyungsook, H.: Predicting genes expressed via -1 and +1 frameshifts. Nucleic Acids Research 32 (2004) 4884-4892
14. Hammell, A.B., Taylor, R.C., Peltz, S.W., Dinman, J.D.: Identification of putative programmed -1 ribosomal frameshift signals in large DNA databases. Genome Research 9 (1999) 417-427
15. Bekaert, M., Bidou, L., Denise, A., Duchateau-Nguyen, G., Forest, J., Froidevaux, C., Hatin, Rousset, J., Termier, M.: Towards a computational model for -1 eukaryotic frameshifting sites. Bioinformatics 19 (2003) 327-335
16. Shah, A.A., Giddings, M.C., Parvaz, J.B., Gesteland, R.F., Atkins, J.F., Ivanov, I.P.: Computational identification of putative programmed translational frameshift sites. Bioinformatics 18 (2002) 1046-1053
17. Gram, L., Bundvad, A., Melchiorsen, J., Johansen, C., Vogel, B.F.: Occurrence of Shewanella algae in Danish coastal water and effects of water temperature and culture conditions on its survival. Applied and Environmental Microbiology 65 (1999) 3896-3900

Effectiveness of Vaccination Strategies for Infectious Diseases According to Human Contact Networks

Fumihiko Takeuchi and Kenji Yamamoto

Department of Infection Control Science, Juntendo University, 113-8421, Tokyo, Japan
Research Institute, International Medical Center of Japan, 162-8655, Tokyo, Japan
fumihiko@takeuchi.name

Abstract. A 'contact network' modeling infection transmission comprises of nodes (or individuals) that are linked when they are in contact that possibly transmits an infection. We here studied infection transmission on contact networks of various degree distributions—scale-free, exponential and constant—under SIRV model assuming susceptible, infected, removed and vaccinated statuses of nodes. Aiming for infectious disease containment within the very early stage of spreading, we computed the minimum transmissibility at which an infectious disease epidemic begins to emerge, and its change according to mass preventive and ring post-outbreak vaccination. In the most degree-heterogeneous scale-free network, the 'super-spreading' by the hubs, or high-degree nodes, allowed epidemics even for low transmissibility. In compensation, vaccination was much more efficient for the scale-free network. We also found that basic reproductive number R_0 defines a measurement of epidemic emergence universally applicable to networks of various degree distributions. These results are significant for public health design.

1 Introduction

A 'contact network' modeling infection transmission comprises of nodes (or individuals) that are linked when they are in contact that possibly transmits an infection. Here, the magnitude of the spreading of infection is determined not solely by the infectiousness of the pathogen but also by the structure of the contact network. In particular, a major factor is the distribution of each node's 'degree,' which is the number of nodes linked to it. If all nodes have the same degree and the links between nodes are random, there exists a threshold value in 'transmissibility,' the probability that an infected node transmits the infection to a susceptible node in contact, less than which an outbreak immediately extinguishes as an endemic [1]. On the other hand, if there are a significant number of high-degree nodes, or hubs, such nodes can become super-spreaders, and allow an outbreak even under weak transmissibility. In fact, in scale-free networks, which only have a power decrease in the number of high-degree nodes, there remains a marginal number of stationary infected nodes under SIS model (nodes transit between susceptible and infected), even for pathogens of infinitely small transmissibility [2]. This phenomenon is typically observed as computer virus infections in the Internet.

Vulnerability to infection attributable to the heterogeneous degree distribution is applicable to contact networks of infectious diseases as well, thus can become a concern for public health. Although our knowledge on the contact network of infectious diseases is limited, social networks can give some insight. Sexual contacts that follow the scale-free degree distribution [3] must be those with largest degree-heterogeneity, whereas friendship relations that follow Gaussian distribution [4] must include much less number of hubs. Studies on such infections and their containment are important both for existing diseases such as AIDS or SARS or for those introduced deliberately by bioterrorism such as smallpox.

The primary measure for containing infection is vaccination, either preventive or post-outbreak. The epidemic mentioned above under SIS model in scale-free networks cannot be stopped by preventive mass vaccination of randomly selected nodes even of a large proportion, but can be halted by prioritized vaccination of hub nodes [5, 6]. However, for vaccination in case of infectious diseases in human, the latter hub vaccination is difficult to implement, because the contact network is not apparent and potential hubs are not evident. On the other hand, among post-outbreak vaccination strategies, the one important in practice is the ring vaccination, in which the susceptible individuals in contact with an infected individual are vaccinated. Yet, there has been no study evaluating the effectiveness of ring vaccination or its combination with preventive mass vaccination, the two practical containment strategies. Moreover, for the study of infectious diseases in human a modeling more realistic than the SIS is necessary.

Thus, we here studied infection transmission on contact networks of various degree distributions under SIRV model assuming not only susceptible and infected nodes, but also those removed (by death or acquiring immunity) and vaccinated, and evaluated the effectiveness of mass preventive and ring post-outbreak vaccinations.

2 Results

Aiming for infectious disease containment within the very early stage of spreading, we studied the minimum transmissibility at which an infectious disease epidemic begins to emerge, and its change according to contact networks or vaccination strategies. We generated random contact networks comprising of $n=100,000$ nodes having average degree $\langle k \rangle=10$ with high-degree nodes decreasing by power (scale-free network), exponentially (exponential network), or with all nodes having degree $\langle k \rangle$ thus with no high-degree nodes (constant network) (Figure 1). Under no vaccination, the scale-free network that has the largest heterogeneity of degree allowed epidemic emergence even for transmissibility $T=0.032$, which was followed by $T=0.087$ for the less heterogeneous exponential network, and then by $T=0.111$ for the homogeneous constant network (Figure 2, Table 1). This exhibits the vulnerability of degree-heterogeneous contact networks to infection.

Although the transmissibilities causing epidemic emergence were different among the networks, the corresponding values of basic reproductive number R_0 were consistently around one (Table 1). This number R_0 is defined as the expected number of secondary infections among nodes in contact with a primary infected node and is evaluated as $R_0 = (\langle k^2 \rangle / \langle k \rangle - 1)T$ [7]. Whereas the transmissibility T basically defines the pathogen's biological strength for transmission, R_0 indicates the strength of spreading in a specific contact network, and reflects the degree distribution of the network. In particular, the value R_0 can become large even under a small T, when the degree distrubution is heterogeneous and $\langle k^2 \rangle / \langle k \rangle$ is large. Furthermore, we observed the epidemic emergence to occur around $R_0=1$ independently of the number of nodes (see the case $n=10,000$ in Table 1) or the average degree (see the case $\langle k \rangle=100$). These results indicate that basic reproductive number serves as a faithful indicator for the emergence of epidemic independently of the degree distribution.

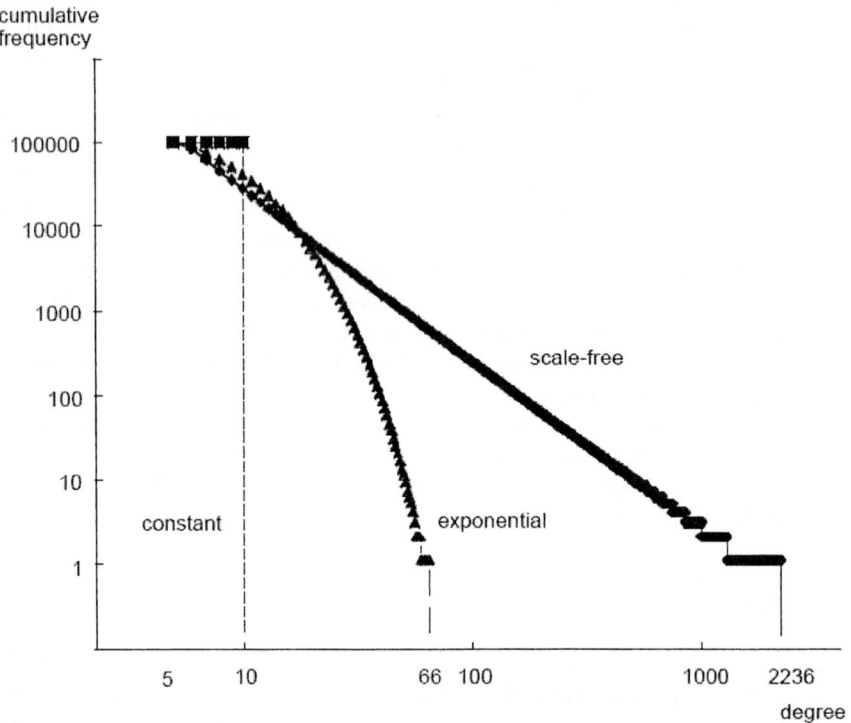

Fig. 1. Cumulative frequency of nodes according to degree for scale-free, exponential and constant networks of $n=100,000$ nodes with average degree $\langle k \rangle = 10$

Compensating the vulnerability of scale-free networks to weak transmissibility, vaccination worked much effectively for such networks compared to the exponential and constant networks. We simulated for various ratio u of randomly selected nodes preventively vaccinated (mass preventive vaccination) and ratio v of susceptible status nodes in contact with an infected node to be vaccinated (ring vaccination). In exponential or constant networks, the transmissibility necessary to cause emergence of epidmic was enlarged by the factor of $1/[(1-u)\cdots(1-v)]$ compared to the case without vaccination (Table 1). This can be explained as vaccinations decreasing the number of (susceptible) nodes around an infected node approximately by the factor of $(1-u)\cdots(1-v)$. In scale-free networks, on the other hand, the transmissibility necessary for epidemic was observed to increase by a much larger factor of 16^{u+v} (Table 1). For example, a pathogen of transmissibility $T=0.018$ could emerge an epidemic under no vaccination, whereas the value arose to $T=0.123$ under $u+v=0.5$.

Effectiveness of Vaccination Strategies for Infectious Diseases 959

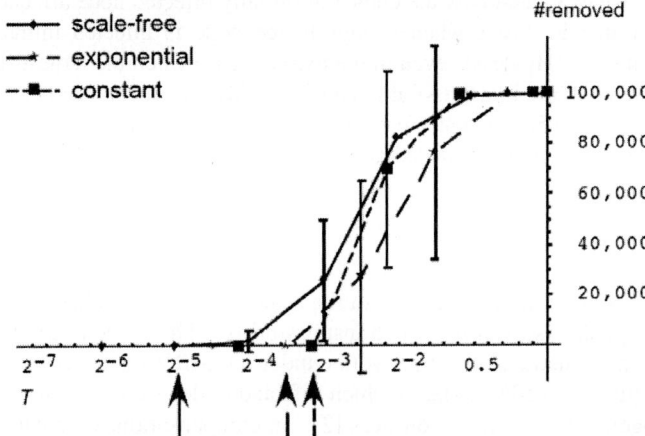

Fig. 2. The number of removed nodes after infection simulation for scale-free, exponential and constant networks of $n=100{,}000$ nodes with average degree $\langle k \rangle=10$ under various transmissibility, and without vaccination. Horizontal axis indicates the transmissibility T, and vertical axis indicates the final number of removed nodes. Data points for the scale-free network are diamonds connected by lines, those for the exponential network are stars connected by sparsely dashed lines, and those for the constant network are squares connected by densely dashed lines. Standard deviation of five experiments is indicated by a vertical error bar centered at the mean point. The transmissibility (interpolated in log scale) for which 100 nodes are removed was defined as the epidemic emerging transmissbility, and the values for the three networks are indicated by arrowed vertical lines of the same type as those connecting data points

Table 1. Transmissibility T and basic reproductive number R_0 causing emergence of epidemic for various contact networks

n	$\langle k \rangle$	initial infection	vaccination		scale-free		exponential		constant	
			mass preventive	ring	T	R_0	T	R_0	T	R_0
100,000	10	random	0%	0%	0.032	1.04	0.087	1.00	0.111	1.00
100,000	10	random	0%	25%	0.064	2.08	0.088	1.01	0.111	1.00
100,000	10	random	0%	50%	0.123	4.03	0.174	2.01	0.222	2.00
100,000	10	random	0%	75%	0.251	8.21	0.350	4.03	0.446	4.01
100,000	10	random	25%	0%	0.062	2.03	0.088	1.01	0.111	1.00
100,000	10	random	25%	25%	0.123	4.04	0.095	1.09	0.115	1.04
100,000	10	random	25%	50%	0.128	4.17	0.176	2.03	0.224	2.01
100,000	10	random	25%	75%	0.492	16.09	0.353	4.07	0.447	4.02
100,000	10	random	50%	0%	0.123	4.02	0.174	2.01	0.223	2.00
100,000	10	random	50%	25%	0.246	8.03	0.175	2.02	0.223	2.01
100,000	10	random	50%	50%	0.254	8.32	0.349	4.02	0.445	4.00
100,000	10	random	50%	75%	0.499	16.33	0.700	8.06	0.891	8.02
100,000	10	random	75%	0%	0.248	8.12	0.349	4.02	0.446	4.01
100,000	10	random	75%	25%	0.263	8.60	0.353	4.06	0.447	4.02
100,000	10	random	75%	50%	0.354	11.58	0.699	8.05	0.896	8.06
100,000	10	random	75%	75%	0.761	24.90	-	-	-	-
10,000	10	random	0%	0%	0.075	2.03	0.088	1.02	0.112	1.01
100,000	100	random	0%	0%	0.003	1.04	0.008	1.00	0.012	1.18
100,000	10	hub	0%	0%	0.018	0.58	0.057	0.66	0.111	1.00

Dashes indicate absence of epidemic (even under $T=1$).
For each contact network, R_0 is calculated as $R_0 = (\langle k^2 \rangle / \langle k \rangle - 1)T$ (see text).

Finally, we evaluated how the choice of initially infected node affects epidemics. An infection transmits faster when a high-degree node is infected initially (say by hub targeted attack). However, even in the extreme case when the largest-degree node was initially infected, the transmissibility causing epidemic emergence became smaller only by the factor of 0.56 for scale-free network and 0.66 for the exponential, compared to the case when an initially infected node was selected randomly (Table 1).

3 Discussion

The 'super-spreading' by the hubs in the scale-free networks had significant effect of allowing epidemics even for low transmissibility. The demonstrated vulnerability of degree-heterogeneous contact networks under the SIRV model was in parallel to their vulnerability under SIS model in which infected nodes were stationarily reserved even for pathogens of weak infectiousness [2]. In compensation, vaccination turned out to be much more efficient for the scale-free network thus still enabling infectious disease containment in the very early stage of epidemics. In addition, the consistent measurement of epidemic emergence by basic reproductive number R_0 among the three types of networks indicated the appropriateness of the evaluation of infectiousness by R_0, which takes the degree heterogeneity into account. Our results could explain super-spreading caused simply by the structure of contact networks, but other factors such as pathogens changing transmissbility by recombination within a host, also can give accounts for super-spreaders.

4 Methods

4.1 Generating Contact Networks

We generated random contact networks of either $n=10,000$ or $100,000$ nodes with average degree either $\langle k \rangle=10$ or 100 for three types of degree distributions in order to adjust the amount of hub nodes. In the scale-free degree distribution, the number of high-degree nodes decreased only by the power: the proportion of degree k nodes among all nodes was set to be $p(k) = (m^2 k^{-3})/2$ for $k \geq m/2$, and $p(k) = 0$ otherwise. In the exponential degree distribution, the high-degree nodes decreased exponentially: $p(k) = (2e \exp(-2k/m))/m$ again for $k \geq m/2$, and $p(k) = 0$ otherwise. As for the other extremity, we tested the constant degree case, where all nodes had degree m, and there were no hubs at all. For each case, a contact network was generated by random connection of edges: firstly nodes with various degrees according to the distribution were listed, then 'untied links' emanating from each node by the number of its degree were generated, and finally the 'untied links' were connected randomly. The scale-free network had the largest number of high-degree nodes, the exponential network was in the medium, and the constant network had no such hubs (Figure 1). The mean squared degree $\langle k^2 \rangle = \Sigma_k k^2 p(k)$ of the generated scale-free network was 279.5 for ($n=10,000$, $m=10$), 336.9 for ($n=100,000$, $m=10$) and 33,701.1 for ($n=100,000$, $m=100$), that of the exponential network was 125.0 for ($n=10,000$, $m=10$), 125.1 for ($n=100,000$, $m=10$) and 12,499.1 for ($n=100,000$, $m=100$), and that of the constant network was 100 for ($m=10$) and 10,000 for ($m=100$).

4.2 SIRV Model

The nodes in our simulation had four possible statuses: susceptible, infected, removed or vaccinated. The simulation proceeds stepwise. In each step, a portion of 'susceptible' nodes that are vaccinated change to 'vaccinated.' Then, each 'infected' node transmits the infection to a 'susceptible' node in contact, by converting its status to 'infected' (in the next step) with probability T, the transmissibility. Meanwhile, all 'infected' nodes are changed to 'removed' in the next step. (The infected period is one step, which corresponds to various periods in reality depending on infectious diseases.) Thus, nodes in status 'removed' or 'vaccinated' do not change their status further. The simulation becomes stable and terminates when there are no more nodes in status 'infected.'

4.3 Simulation of Infection and Vaccination

For various contact networks and values of transmissibility, we performed simulations parameterized by the implemented rate of mass preventive and ring post-outbreak vaccinations. In the first step, 0, 0.25, 0.5 or 0.75 of the population was randomly assigned the 'vaccinated' status (mass vaccination), one node among the remaining was either randomly selected (random initial infection) or the highest-degree node was selected (hub initial vaccination) to become an 'infected' node, and all of the remaining nodes were set as 'susceptible.' In each of the following step, all of the 'susceptible' nodes in contact with an 'infected' node were listed, and either 0, 0.25, 0.5 or 0.75 of them were assigned the 'vaccinated' status (ring vaccination). Each set of simulation was repeated five times. The parameters used for our simulation was $R_0 = 0.5, 1, 2, ..., 64$, and the corresponding values for T, caclulated from the above mentioned relation between R_0 and T. (The values of R_0 differ widely according to diseases: influenza has 1.7, SARS has 1.2–3.6, smallpox has 4–10, and measles have 17 [1].)

Acknowledgments

This study was partially supported by the 'Special Coordination Funds for Promoting Science and Technology' from the Ministry of Education, Culture, Sports, Science and Technology.

References

1. Anderson, R.M. and R.M. May, *Infectious diseases of humans : dynamics and control.* Oxford science publications. 1991, Oxford ; New York: Oxford University Press. viii, 757 p.
2. Pastor-Satorras, R. and A. Vespignani, *Epidemic spreading in scale-free networks.* Physical Review Letters, 2001. **86**(14): p. 3200-3203.
3. Liljeros, F., et al., *The web of human sexual contacts.* Nature, 2001. **411**(6840): p. 907-8.
4. Amaral, L.A., et al., *Classes of small-world networks.* Proc Natl Acad Sci U S A, 2000. **97**(21): p. 11149-52.

5. Pastor-Satorras, R. and A. Vespignani, *Immunization of complex networks.* Physical Review E, 2002. **65**: p. 036104.
6. Cohen, R., S. Havlin, and D. ben-Avraham, *Efficient immunization strategies for computer networks and populations.* Physical Review Letters, 2003. **91**(24): p. 247901.
7. Meyers, L.A., et al., *Network theory and SARS: predicting outbreak diversity.* J Theor Biol, 2005. **232**(1): p. 71-81.

A Shape Constraints Based Method to Recognize Ship Objects from High Spatial Resolution Remote Sensed Imagery

Min Wang, Jiancheng Luo, Chenghu Zhou, and Dongping Ming

State Key Laboratory of Resources & Environmental Information System,
Institute of Geographical Sciences and Natural Resources Research,
Chinese Academy of Sciences, A11,Datun Road,
Anwai, Beijing, 100101, P.R. China
wangm@lreis.ac.cn

Abstract. Automatically extracting moored ships from high spatial resolution remote sensed imagery is more difficult than offshore ships because their spectrum values and textures are very close to the harbor. For this reason, different routes are designed and applied to extract the two kinds of ships. Our whole method can be divided into three main steps: 1) extract water polygons with histogram segmentation, 2) extract holes in the water polygons with morphological operations as the possible offshore ships, and extract possible moored ship with the identification of the salients to sea along the water boundary, and then 3) screen real ships out of these possible ships with more shape constraints. A case study is carried out on Spot 5 imagery to validate our method.

1 Introduction

Automatically recognizing and extracting man-made objects from remote sensed imagery has always been an important research task in the fields such as computer vision, pattern recognition and application of remote sensing, etc. Recently, information extraction and object recognition from high spatial resolution remote sensed imagery become one hot spot with the launch of many high resolution spaceborne sensors, such as Ikonos, QuickBird, Spot 5, etc[1][2].

The resolution of high spatial resolution remote sensing can reach less than 10m, which gives very fine details of the earth surface. It can undertake some tasks in a much cheaper spaceborne remote sensing way which can only be done by aerial remote sensing before, thus has a very wide application scope in photogrammetry, city planning, transportation, military, agriculture, forestry, etc. But, with the improving of resolution, data volume increases remarkably, and spectral distribution become more complicated (e.g., more prevailing substantial spectral confusion), all these bring greater difficulties to information extraction work [3][4].

This paper focuses on automatically extracting one kind of special objects, which are large ships from high resolution remote sensed imagery. To the author's knowledge, most work on ship recognition are on radar imagery (e.g., SAR imagery, see literature [5][6][7][8] for examples). In them, for example, Zhou et al. [5] separate sea and land with gray level histogram segmentation, and detect ships in sea with moment invariant threshold from SAR imagery. Eldhuset et al. [6] detect ship targets with moving window filtering on SAR imagery. Besides, Zhang et al. [9] first get edge enhanced image with fuzzy theory, binarize it with segmenting threshold, separate targets and background with filtering operations, and then recognize running ships with combing pixel values and shape features on Spot optical imagery.

We can find that the common ground of most ship target recognition work, which is also adopted in this paper, is firstly to distinguish land and water (if without land in image, this step can be omitted), and then distinguish ships with surrounding water with their pixel values and some shape features. However, the main contributions of our work are:

- We use high spatial resolution optical imagery, a high resolution and good complementarity to radar imagery;
- Our method can recognize moored ships conglutinated to shoreline, which are much difficult to be separated automatically from harbors, for their pixel values and textures are much close between each other.

In this paper, we design different algorithm routes for the recognition and extraction of offshore and moored ships, which combine the shape knowledge of harbor and different running statuses of ships.

The rest of this paper is organized as follows. Section 2 describes our shape constraints based ship recognition method. Section 3 is a case study on Spot 5 imagery. Section 4 sums up our paper.

2 Shape Constraints Based Method

In common circumstances, water in high spatial resolution panchromatic imagery has relatively lower pixel values than other objects. For this reason, we can separate offshore ships with surrounding water directly with pixel values. We first extract water with image segmentation, detect those holes in these water areas with mathematics morphology operators (we call those holes "possible ship areas"), and then screen the real ships out from these possible ship areas with more shape constraints.

But for moored ships, different routes should be carried out because simple morphological operations only give very poor results. Our core idea is: because coastline of most harbors is very straight in general, the elliptical or semi-elliptical salients to sea along their coastline are often caused by moored ships. We first detect those parts, and then screen the real ships out from them with more shape constraints.

2.1 Water Extraction

The first step of this method is to extract water. For this, we can select histogram segmentation or Markov random field texture classification. We choose histogram segmentation, which automatically get the first valley after the first summit in the gray level histogram as the segmentation threshold T.

The principle of histogram segmentation is:

Regarding the histogram $h(z)$ as a curve, we can then get the histogram valleys with calculating the minima, which satisfy that the first order difference=0 and the second order difference >0. Then, T can be gotten by [10]:

1. Calculate $h(z)$ of the image $f(x,y)$, and segment it to N levels;
2. Remove burrs with median filtering;
3. Remove possible errors from local dithering with re-smoothing;
4. Calculate the first order difference $h'(z)=h(z+1)-h(z)$, and detect from left to right the first position which transits from negative to positive as T.

We then segment the image with T with the following rules: water $\leq T$ and non-water $> T$.

2.2 Screening Out Ships with Shape Constraints

We vectorize the water areas and get their boundaries. The work of moored ships extraction would be carried out along their boundaries.

We first set an a priori length range of possible ships (L). It is used to remove those salients whose lengths fall out of L and are impossible to be ships, which can make the following searching and judgment quicker and more precise.

After that, starting from the first node of a water polygon as the begin vertex, we search an end vertex which should satisfies:

- The length of the straight line connecting the two vertices falls in the length range L;
- All the other vertices between the two vertices are on the same side of the connecting line, and the area closed by this boundary segment and connecting line protrudes to water.

This area can be regarded as a 'possible ship area'. If this area does exist, then the next searching begins from the end vertex. Otherwise, we move the searching position to the next vertex, and repeat the above searching process until all of the possible ship areas are extracted.

After extracting all these possible offshore and moored ship areas, we screen out the real ships with more shape constraints. We know that real large ships are relatively symmetric between their head and tail, like long and narrow ellipses. But, with many shape influences, which include image precision selection (to accelerate the segmentation of water, we often re-sample the image), outstanding backstays, and conglutination to harbor, the extracted ship areas are often not very regular in shape, and then can not be recognized only by means of 'like a ellipse'.

From several experiments, we define the following shape indices to validate these areas:

1. Simple length index
   They are: length range (L) and width range of ship (W).
2. Length to width ratio range (LW)

For moored ships, LW is defined as the ratio of the length of the connecting line of the start to end vertices (l) and the max distance of the other vertices to the line (d). The reason to define LW is because the long axes of the enveloping ellipses of moored ships are usually parallel to shoreline (see figure 1(a)), and figure 1(b) often represents natural salients. With *LW*, these salients can easily be kicked out from the following searching.

For offshore ships, *LW* is simply defined as the length ratio of long, short axes.

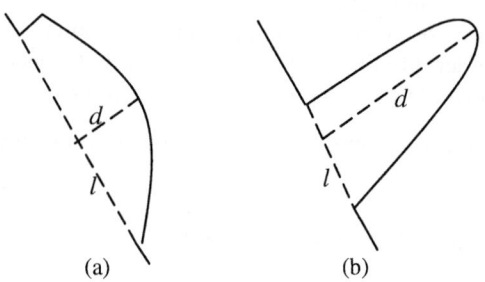

Fig. 1. Two illustrations of seashore salient

3. Shape regularity index (*SR*)

If the possible ship area has much irregular shape (for example, zigzag shape), it should be kicked out. We should find one relatively robust index which can identify irregularity and has the position, size and rotation invariant properties. After many experiments, we choose the third values of the 7 Hu moment invariants [11] as *SR*.

The principle of Hu moments can briefly described as:

Given object boundary curve *C* and its vertex coordination *f(x, y)*, the *(p + q)*th order boundary moments are defined as:

$$m_{pq} = \oint_C x^p y^q ds, p,q = 0,1,\dots . \quad (1)$$

The central boundary moments are defined as:

$$\mu_{pq} = \oint_C (x-\bar{x})^p (y-\bar{y})^q ds, p,q = 0,1,\dots , \quad (2)$$

where

$$\begin{cases} \bar{x} = m_{10}/m_{00} \\ \bar{y} = m_{01}/m_{00} \end{cases}. \quad (3)$$

In discrete cases, for example, digital imagery, we have

$$\begin{cases} m_{pq} = \sum_{x,y \in C} x^p y^q \\ \mu_{pq} = \sum_{(x,y) \in C} (x-\bar{x})^p (y-\bar{y})^q \end{cases}. \quad (4)$$

The normalized boundary moments are:

$$\eta_{pq} = \mu_{pq} / \mu_{00}^{p+q+1} \quad p,q = 0,1,\ldots \quad (5)$$

The third moment invariant can then be obtained by:

$$\phi_3 = (\eta_{30} - 3\eta_{12})^2 + (3\eta_{21} + \eta_{03})^2 \quad (6)$$

From experiments we found that ϕ_3 is often rather small for irregular shapes (e.g., zigzag shape), and it satisfies our need of independent of position, size and orientation, so we use ϕ_3 as *SR*, which can remove all these irregular shapes difficulty to be deleted with other indices.

4. Symmetric regularity index (*SMR*)

$$SMR = 1 - \left|(A_{0-l/2} - A_{l/2-l})\right| / A \quad (7)$$

We know that real large ships are commonly near symmetric between head and tail, like narrow ellipses. Considering that possible ship areas are often not very regular in shape, we use area to define symmetry. In formula 7, A is the total area of the possible ship area, $A_{0-l/2}$ and $A_{l/2-l}$ are the partial areas before and after the midpoint of the long axis. *SMR* can be used to index the symmetry of the possible ship's head and tail.

3 A Case Study

Several experiments have been carried out to validate our method. Due to the page limits, here we only give one typical example on Spot 5 image.

The Inputs are:

1. Data source: Spot 5 panchromatic image, with spatial resolution 2.5m, one harbor in south of China
2. Experiment inputs:

- *L*: [100-350]meters
- *W*: [20-60] meters
- *LW*: [3-10]
- *SR*: <=1.0
- *SMR*:>=0.6

These indices are obtained with visual interpretation and a priori knowledge and as the default inputs.

In this experiment, we get the water threshold with histogram segmentation. From figure 2 we can know that the distribution of pixel values is much concentrated, which makes it difficulty in separating sea and land. What's more, the shape of land, running statuses of ships are much complicated. For example, one ship is not moored directly on the shore, but on a bracket. It can be extracted as offshore ship in our method however.

Figure 3(b) show the water segmentation result with the automatically identified threshold 105. We can see that because the pixel values of some parts of the harbor is very close to water, many small land areas are misclassified into water. To these areas, we can use one water area size threshold (for example, 1000 pixels in size) to remove them from the following searching process.

We first extract the offshore ships (in this experiment, the ship on the bracket). With the minimum value of W 15, we can determine the temperate size of morphological operator closing, which first dilate then erode the water area. After closing, we then get the possible offshore ships with the result minus the original water area image. From figure 3(c) we can see that many possible offshore ships are conglutinated to shoreline, but most real possible moored ships have not been extracted, which indicates that we can not extract ships simply with morphological operations.

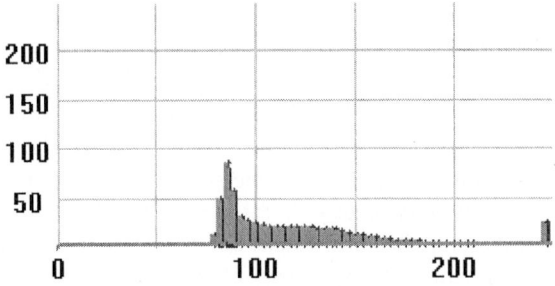

Fig. 2. Gray level histogram of this experiment area

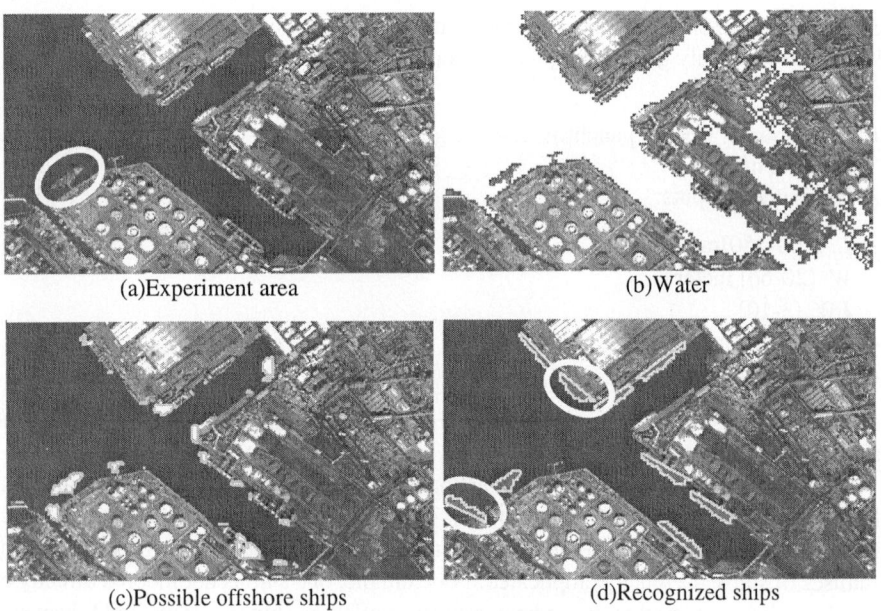

(a)Experiment area (b)Water

(c)Possible offshore ships (d)Recognized ships

Fig. 3. A Case Study on Spot 5 imagery

Fig. 4. *SR* values

We then vectorize the water areas, and extract all these possible moored ships with salient identification. Finally, we extract the real large ships with our shape constraints in all these offshore and moored possible ships.

In this experiment, there exist many irregular possible ship areas, which can not be kicked out with the other shape constraints. From figure 4, it's obvious that the irregular shapes give relatively smaller *SR* values to the other regular shapes, which indicates *SR* can be used to denote the regularity of ship areas. With the condition of $SR \leq 1.0$, these irregular shapes can be removed successfully.

The final result is showed in figure 3(d). We can find that all the 7 large ships are extracted, but with 2 misrecognized areas (see the two ellipses). It's reasonable and difficulty to be avoided with our shape constraints based method. For further studies, we can reselect these areas with combining more features of pixel values, textures, which are believed to can further improve the recognizing precision.

From this experiment, we can see that under the complicated distribution of pixel values, harbor shapes and running statuses of the ships, our method get good result however, which indicates that our method is robust and applicable.

4 Conclusions

In this paper, a shape constraints based method to recognize ships from high spatial resolution remote sensed imagery is proposed. Because large ships are totally similar in shapes, the inputs of *LW*, *LR* and *LMR* need not be changed frequently, but *L*, *W* should be revised a little according to data sources sometimes. Totally speaking, our method is rather simple in use.

In further study, with combing a priori knowledge of 'moored many ships' and shape features of harbors (for example, with relatively straight shoreline and narrow exit), we can try to recognize harbors, the complex man-made objects. We will give detailed discussion in other paper.

Acknowledgment. This work is supported by Chinese National Natural Science Foundation under grant No.40401039, Chinese National Programs for High Technology Research and Development under grand No. 2002AA135230, and Chinese Postdoctoral Foundation.

References

[1] Proc. of the processing and application of high spatial resolution remote sensed satellite imagery. Lanzhou, (2003)
[2] Proc. Of IGRASS, Toulouse, France, (2003)
[3] Ma Tin: Technological Model for High Resolution Satellite Images and Information Processing. Information of Remote Sensing, 3(2001)6-10
[4] Cheng Chenqi, Ma Tin: Automatically Extraction of Linear Features from High Resolution Remotely Sensed Imagery. Journal of Remote Sensing,7(2003)26-30
[5] Zou Huanxin, Kuang Gangyao, Jiang Yongmei, Zheng jian: Algorithm of ship targets detection in SAR Ocean Image based on the moment invariant. Computer engineering, 29(2003)114-116
[6] Eldhuset K: An Automatic ship and ship wake detection system for spaceborne SAR images in Coastal Regions. IEEE transactions on GeoScience and Remote Sensing, 34(1996)1010-1018
[7] Q. Jiang, S. Wang, and D. Ziou: Ship Detection in RADARSAT SAR Imagery. In Proc. of CMS, San Diego, (1998)
[8] Qingshan Jiang, Shengrui Wang, Djemel Ziou, and Ali El Zaart: Automatic Detection for Ship Targets in RADARSAT SAR Images from Coastal Regions, In Proc. Of Vision Interface, (1999)131-137
[9] Zhang Yutian, Zeng Anjun: Moving maritime targets detection technology on high spatial resolution remote sensed imagery. Research on telecom technology, 9(2003)1-8
[10] Zhang Yujing: Image segmentation. Beijing: Science press, (2001)
[11] Ming-Kuei Hu: Visual pattern recognition by moment invariants. IRE Transactions on Information Theory, IT-8(1962)179-187

Statistical Inference Method of User Preference on Broadcasting Content

Sanggil Kang[1], Jeongyeon Lim[2], and Munchurl Kim[2]

[1] Department of Computer Science, College of Information Engineering,
The University of Suwon, Hwaseong, Gyeonggi-do, Korea
sgkang@suwon.ac.kr
[2] Laboratory for Multimedia Computing, Communications and Broadcasting,
Information and Communications University, Daejeon, Korea
{jylim,mkim}@icu.ac.kr

Abstract. This paper proposes a novel approach for estimating the statistical multimedia user preference by providing weights to multimedia contents with respective to their consumed time. The optimal weights can be obtained by training the statistical system in the sense that the mutual information between old preference and current preference is maximized. The weighting scheme can be done by partitioning a user's consumption history data into smaller sets in a time axis. With developing a mathematical derivation of our learning method, experiments were implemented for predicting the TV genre preference using 2,000 TV viewers' watching history and showed that the performance of our method is better than that of the typical method.

1 Introduction

With the flood of multimedia content over the digital TV channels, the internet, and etc., users sometimes have a difficulty in finding their preferred content, spend heavy surfing time to find them, and are even very likely to miss them while searching. By predicting or recommending the user's preferred content, based on her/his usage history in content consumptions, the problems can be solved to some extent.

Various preference recommendation techniques can be classified into three possible categories such as the rule-based, collaborative filtering, and inference method. The rule-based recommendation is usually implemented by a predetermined rule, for instance, if -then rule. Kim et al. [1] proposed a marketing rule extraction technique for personalized recommendation on internet storefronts using tree induction method [2]. As one of representative rule-based techniques, Aggrawall et al. [3, 4] proposed a method to identify frequent item sets from the estimated frequency distribution using association-rule mining algorithm [5]. Collaborative filtering (CF) technique recommends a target user the preferred content of the group whose content consumption mind is similar to that of the user. Because of the nature of the technique, CF has been attractive for predicting various preference problems such as net-news [6, 7], e-commerce [8, 9], digital libraries [10, 11], digital TV [12, 13]. In general, rule-based and CF techniques need expensive effort, time, and cost to collect a large number of users' consumption behavior due to the nature of their methodologies. However, inference is the technique that a user's content consumption behavior is predicted based on the history of personal content consumption behaviors. Ciaramita et al. [14] presented a Bayesian network [15], the graphical representation of probabilistic relationship among variables which are encoded as nodes in the network, for verb selectional preference by combining the statistical and knowledge-based approaches. The architecture of the Bayesian network was determined by the lexical hierarch of Wordnet [16]. Lee [17] designed an interface agent to predict a user's resource usage in the UNIX domain by the probabilistic estimation of behavioral patterns from the user behavior history.

From the literatures mentioned above, there is one thing not to be overlooked, which is that all data are equally weighted in computing the statistical preference. In this case, recently collected data may not be appreciated because the size of the user's usage history data usually dominates over that of the new data. In general, the recent usage history will give more impact on predicting the future preference than old one. In order to take into the consideration, we provide weights to data with respect to their collected or consumed time.

The objective of our work is to find the optimal weights bringing better performance than the typical methods. In this paper, a new adaptive learning method is proposed for obtaining the optimal weights. Our method executes by partitioning the usage history data into smaller sets in a time axis, on which the weights are provided. Thus, the weighted data can differently reflect their significance on predicting the preference. We utilize a supervised learning technique commonly used in neural network for estimating the weights using the mutual information , which is an optimality index to be maximized during the learning process. Also, the weights are updated whenever a predetermined amount of data is collected.

The remainder of this paper is organized as follows. Section 2 describes a window weighing scheme. Section 3 describes our learning method. In Section 4, we show the experimental results performed on a realistic set of data. We then conclude our paper in Section 5.

2 Window Weights

For predicting users' preference, the time that content is consumed can be a critical factor. In order to consider this, we partition the usage history data stored in a chronicle into smaller sets in a time axis and give weights to the frequencies of content in the partitioned datasets. The smaller dataset, named as *window* in our paper, resides within a predetermined non-overlapped time interval L. Usually, the content in the recent windows will give a potent influence on predicting the future preference. In case the data size in those windows is not big, it can not be enough to represent the statistical future preference, which makes us reluctant to use only the latest window. To compensate this problem, we group the windows into two regions: old preference region (OPR) and current preference region (CPR). CPR includes the latest window and is used as a reference for predicting future preference. The OPR has enough windows in order for the estimated conditional probabilities in the region to be able to represent the future preference (FP), compared to CPR. The frequencies of content can be modified by assigning weights to the windows in OPR. A set of the window weights can be denoted as $\underline{w} = [w_1 \ w_2 \ ... \ w_M]$, here M is the total number of windows in the OPR. The statistical preference of the ith content x_i in OPR can be expressed as

$$\hat{\theta}_{x_i} = p(X = x_i \mid E) = \sum_{m=1}^{M} w_m n_{i,m} / \sum_{m=1}^{M} w_m N_m \qquad (1)$$

where X is a set of consumed content, and N_m and $n_{i,m}$ is the sample number of whole content and content x_i within the mth window in OPR, respectively. Also, $\hat{\theta}_{x_i}$ is the estimated statistical preference of x_i for FP with given evidence E. From Equation (1), we can see that the conditional probability is a function of the window weights, which means the accuracy of the preference prediction depends upon the values of the weights. As addressed, the latest window can give a big impact on predicting the future preference (FP) so the weights are adjusted in the sense that a set of the content in the OPR, denoted as $\tilde{X}_O$, is getting correlated with a set of weighted content in the CPR, denoted as X_C. Here, $\tilde{X}_O$ is the weighted version of X_O which is a set of content in the OPR.

In some typical methods, the statistical preference is computed using the entire history dataset, which causes the new coming content not be appreciated in estimating the preference because the size of the user's usage history data usually dominates over that of the new content.. In order to complement this weak point, we rule out the most outdated window when new data is filed up during the next time interval L from the last window in the usage history data. It can be operated by shifting (or sliding) a window at a time and continuing the same processing to find the optimal weights and so on. This scheme can allow an on-line prediction. From the following section, s with the parenthesis indicate the sth window shifting process.

3 Determination of Optimal Window Weights

Our learning method is to determine an optimal set of the window weights in the sense that the mutual information, denoted as $I(\tilde{X}_O(s); X_C(s))$, between $X_C(s)$ and $\tilde{X}_O(s)$ is maximized at the sth shift. At each window shifting, the weight updates are done based on a gradient ascent algorithm. The weight update continues until the mutual information (MI) reaches the maximum value. The mathematical derivation of our learning method starts with the definition of the MI such as

$$\begin{aligned} I(\tilde{X}_O(s); X_C(s)) &= \log(p(X_C(s) \mid \tilde{X}_O(s)) / p(X_C(s))) \\ &= \log(p(X_C(s) \mid \tilde{X}_O(s))) - \log(p(X_C(s))). \end{aligned} \qquad (2)$$

From Equation (2), the larger the value of the MI, the more $X_C(s)$ is correlated with $\tilde{X}_O(s)$. We assume that $X_C(s)$ includes at least a content to exclude the extreme

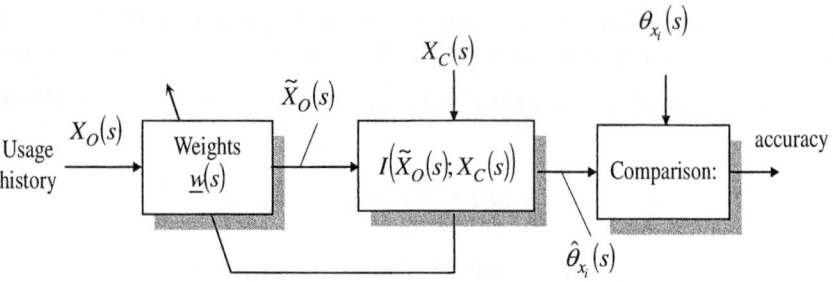

Fig. 1. The schematic of learning method

case in which the equation can not be appropriately applied. For instance, if $X_C(s) = \Theta$ then the $\log(p(\Theta | \tilde{X}_O(s)))$ can not be calculated, where Θ means the empty set. As searching the optimal weights in the weight space, the partial derivative of the mutual information with respective to the weight $w_m(s)$ is calculated as following:

$$\partial I(\tilde{X}_O(s); X_C(s))/\partial w_m(s) \\ = \partial \log(p(X_C(s)|\tilde{X}_O(s)))/\partial w_m(s) - \partial \log(p(X_C(s)))/\partial w_m(s) \quad (3)$$

where, the MI is composed of the logarithmic functions so it is differentiable at a concerning point in the weight space. Thus, it is feasible to use the gradients. Since the prior probability $p(X_C(s))$ on the right hand side is not a function of $\underline{w}(s)$, Equation (3) can be simplified as

$$\partial I(\tilde{X}_O(s); X_C(s))/\partial w_m(s) = \partial \log(p(X_C(s)|\tilde{X}_O(s)))/\partial w_m(s) \quad (4)$$

In order to compute the partial derivative, we need to formulize $p(X_C(s)|\tilde{X}_O(s))$ in terms of the weights. Let's consider $X_C(s) = [x_1(s) \ x_2(s) \cdots x_i(s) \cdots x_J(s)]$, here J is the number of attributes in $X_C(s)$ and assume that all elements in $X_C(s)$ are contained in $\tilde{X}_O(s)$. In order to obtain $p(X_C(s)|\tilde{X}_O(s))$, we first need to determine the conditional probability $p(x_i(s)|\tilde{X}_O(s))$ which means the probability of occurring attribute x_i given $\tilde{X}_O(s)$. The conditional probability $p(x_i(s)|\tilde{X}_O(s))$ is given by

$$p(x_i(s)|\tilde{X}_O(s)) = \sum_m w_m(s) n_{i,m}(s) / \sum_m w_m(s) N_m(s) \quad (5)$$

The conditional probability $p(X_C(s)|\tilde{X}_O(s))$ can be obtained by multiplications of the conditional probabilities of all content in $X_C(s)$ under the assumption that the consumption of the contents is independent among them.

$$p(X_C(s) \mid \tilde{X}_O(s)) = \prod_i p(x_i(s) \mid \tilde{X}_O(s))^{n_{x_i}(s)} \qquad (6)$$

$$= \prod_i \left(\sum_m w_m(s) n_{i,m}(s) / \sum_m w_m(s) N_m(s) \right)^{n_{x_i}(s)}$$

where $n_{x_i(s)}$ is the number of x_i in $X_C(s)$. Thus, $\log(p(X_C(s) \mid \tilde{X}_O(s)))$ in (2) is

$$\log(p(X_C(s) \mid \tilde{X}_O(s))) = \log\left(\prod_i \left(\sum_m w_m(s) n_{i,m}(s) / \sum_m w_m(s) N_m(s) \right)^{n_{x_i}(s)} \right) \qquad (7)$$

$$= \sum_i n_{x_i}(s) \left(\log\left(\sum_m w_m(s) n_{i,m}(s) \right) - \log\left(\sum_m w_m(s) N_m(s) \right) \right)$$

Therefore,

$$\partial I(\tilde{X}_O(s); X_C(s)) / \partial w_m(s) \qquad (8)$$

$$= \sum_i n_{x_i}(s) \left(n_{i,m}(s) / \sum_m w_m(s) n_{i,m}(s) - N_m(s) / \sum_m w_m(s) N_m(s) \right)$$

The weights are updated every epoch defined as one sweep for all data included in the current windows. The amount of update $\Delta w_m(s)$ can be obtained using the delta rule such as

$$\Delta w_m(e,s) = \eta \cdot \partial I(\tilde{X}_O(s); X_C(s)) / \partial w_m(s) \qquad (9)$$

where notation e is the number of epochs and η is the learning rate which determines the degree of searching step in the weight space during the learning process. We can express the weight update in every epoch e as

$$w_m(e,s) \leftarrow w_m(e-1,s) + \Delta w_m(e,s). \qquad (10)$$

However, the point we should not overlook is that the weights may trace to the negative weight space during the learning process according to the initial location of the weights. The negative weights cause a problematic situation in calculating the conditional probabilities which should always be positive. Thus, the weight searching can be caught in a vicious circle as the epoch runs. To avoid this situation, we put restrictions on weight update in order to force weights in only positive space during training. If a weight move to the negative space, we do not update the weight, as seen in Equation (11).

$$\begin{cases} w_m(e,s) \leftarrow w_m(e-1,s) + \Delta w_m(e,s), & \text{if } w_m(e,s) > 0 \\ w_m(e,s) \leftarrow w_m(e-1,s), & \text{otherwise} \end{cases} \qquad (11)$$

Now, we have a question about stopping criteria like "When do we have to stop learning?" In our method, the learning process continues until the predetermined number of epochs is reached, or until the maximum MI is reached. In the case that MI reaches the saturation much faster than the predetermined epoch, we pay the overtraining time and effort for the large epoch. To avoid this situation, we stop the training when the MI does not increase any more for some predetermined epoch.

4 Experimental Results and Analysis

In this section, we show the numerical accuracy of our learning method by comparing with that of the typical method, that is, no-training method in which no weights are assigned. We applied our method to the Digital TV genre recommendation problem.

For the training and test data set, we used a large set of TV watching history data collected from December 1, 2002 to May 31, 2003, which is provided by AC Nielsen Korea, one of the authorized market research company in Korea. For 2,000 TV viewers, the TV watching history was collected by a set-top box installed in their houses, which can record login and logout time, broadcasting time and day, watched program genre, etc. From the data, we can extract only two evidences such as TV watching time and day for computing the statistical preference of each genre. The genre includes eight attributes such as Education, Drama & Movie, News, Sports, Children, Entertainment, Information, Others. The considered watching time is only from 6 p.m. to 12 p.m. in our experiment because the user barely watched TV during other time periods. The watching time period was slotted by every two hours for each day. Thus, the set of evidence can be expressed as E = {(6 p.m. ~ 8 p.m., Monday), (6 p.m. ~ 8 p.m., Monday), . . . , (10 p.m. ~ 12 p.m., Sunday)} with 21 elements. For each case in E, we first extracted and arranged the genres of the watched TV programs from each viewer's watching history and then partitioned them by every one week. If there is any missing week, the window of the week was replaced with the next week in order to render training to be possible.

The accuracy is evaluated in terms of the error between the estimated statistical preference and the true preference for the future preference with $\eta = 0.1$ and the initial weight vector whose elements are all ones.

$$Error(s) = \sum_{i=1}^{J} \left| \hat{\theta}_{x_i}(s) - \theta_{x_i}(s) \right| \qquad (13)$$

where $\hat{\theta}_{x_i}(s)$ and $\theta_{x_i}(s)$ is the estimated statistical preference and true preference of x_i for the future preference at shift s, respectively. As sliding one window at a time until approaching to the last window, we repeated the processes of training and calculating the errors by varying M, the number of windows in the OPR, for 2,000 TV viewers. By gender and age group of the viewers, we tabulated the mean error over the viewers, for the typical and our method as shown in the table.

From the table, it is shown that the performances of our method were better than those of the typical method for 10s, 20s, 30s, and 40s. For 10s, around 50% at maximum improvement was made, for 20s and 30s, around 37%, for 40s, around 12%. However, for 50s, it can be stated that there was no improvement. It can be induced that the trend of the preference of 40s and 50s viewers usually is steadier than

10s, 20s, and 30s. It is hard to provide a unique number of windows in OPR for obtaining optimal performance for all age and gender identity, for instance, the number of window 3 or 4 for 10s, 4 or 5 for 20s, etc.

Table 1. The mean errors of the typical method and our method by varying the value of M

Age (Gender)	Method	The number of windows in the OPR, M					
		3	4	5	6	7	8
10s (male)	Typical	0.24	0.21	0.22	0.2	0.24	0.23
	Our	0.12	**0.1**	0.13	0.14	0.16	0.15
10s (female)	Typical	0.22	0.2	0.21	0.24	0.23	0.25
	Our	**0.13**	0.15	0.16	0.15	0.14	0.15
20s & (male)	Typical	0.22	0.21	0.19	0.21	0.2	0.22
	Our	0.14	0.15	**0.12**	0.14	0.16	0.17
20s & female	Typical	0.2	0.19	0.22	0.19	0.21	0.22
	Our	0.14	**0.11**	0.13	0.15	0.13	0.12
30s & (male)	Typical	0.22	0.23	0.2	0.22	0.22	0.23
	Our	0.13	0.14	**0.12**	0.13	0.14	0.13
30s &(female)	Typical	0.21	0.19	0.2	0.19	0.22	0.22
	Our	0.14	0.15	0.11	**0.11**	0.15	0.13
40s & (male)	Typical	0.23	0.22	0.23	0.21	0.19	0.22
	Our	**0.16**	0.17	0.17	0.17	**0.16**	0.18
40s& (female)	Typical	0.19	0.18	0.19	0.18	0.2	0.19
	Our	0.17	**0.16**	0.17	**0.16**	0.17	0.18
50s & (male)	Typical	0.14	0.15	0.14	0.15	0.14	0.14
	Our	0.15	0.14	0.13	0.13	0.13	**0.12**
50s& (female)	Typical	0.13	0.14	0.15	0.14	0.14	0.15
	Our	0.15	0.13	0.15	0.13	**0.12**	**0.12**

5 Conclusion

In this paper, we presented a new system for estimating the statistical user preference with introducing the window weights and developing the optimal index, the mutual information, used as a teacher when training the system. By forcing to have old preference be correlated with current preference, the prediction can be good for the near future preference which is correlated with the current preference. From the experimental results, the training speed, which is less than 100 epochs when the initial weighs are all ones, can be acceptable in the practical situation. Also, it was shown that our method was outperformed to the typical method for the 10s, 20s, 30s, and 40s age.

However, we determined the optimal values of the parameters from the exhaustive empirical experience using 2,000 TV viewers' watching information. The 2,000 viewers might not be enough for the exhaustive experiment. It is needed to collect more viewers' information. Also, we need to do further study for developing an automatic algorithm to estimate the optimal values of parameters for each TV viewer when training our system.

Acknowledgements

This research work was carried out at Information and Communications University under the Project titled by "Development of Intelligent Agent and Metadata Management Technology in SmarTV" in 2004 funded by Ministry of Information and Communication in Korean government.

References

1. J.W. Kim, B.H. Lee, M.J. Shaw, H.L. Chang, M. Nelson, "Application of decision-tree induction techniques to personalized advertisements on internet storefronts," International Journal of Electronic Commerce, vol. 5, no. 3, pp. 45-62, 2001
2. J.R. Quinlan, Induction of decision trees, "*Machine Learning*," vol. 1, pp. 81-106, 1986
3. R. Aggrawall, T. Imielinski, A. Swami, "Mining association rules between sets of items in large databases," *Proc. ACM SIGMOD Int'l Conference on Management of Data*, pp. 207-216, 1994
4. R. Aggrawall, R. Srikant, "Fast algorithms for mining association rules," *Proc. 20th Int'l Conference on Very Large Databases*, 478-499, 1994
5. M.Z. Ashrafi, D. Tanizr, K. Smith, "ODAM: An optimized distributed association rule mining algorithm," *IEEE Distributed Systems Online*, vol. 3, no. 3, pp. 1-18, 2004
6. P. Resnick, N. Lacovou, M. Suchak, P. Bergstrom, J. Riedl, "GroupLens: an open architecture for collaborative filtering of netnews," *Internet Research Report*, MIT Center for Coordination Science, 1994, http://www-sloan.mit.edu/ccs/1994wp.html
7. D.A. Maltz, "Distributing information for collaborative filtering on Usenet net news," *SM Thesis*, Massachusetts Institute of Technology, Cambridge, MA, 1994
8. J.B. Schafer, J. Konstan, J. Riedl, "Recommender systems in e-commerce," *ACM Conference on Electronic Commerce*, pp. 158-166, 1999
9. G. Linden, B. Smith, J. York, "Amazon.com recommendations: item-to-item collaborative filtering," *IEEE Internet Computing*, pp. 76-80, 2003
10. K.D. Bollacker, S. Lawrence, C.L. Giles, "A system for automatic personalized tracking of scientific literature on the web," *Proc. ACM Conference on Digital Libraries*, pp. 105-113, 1999
11. R. Torres, S.M. McNee, M. Abel, J.A. Konstan, J. Riedl, "Enhancing digital libraries with TechLens+", *ACM/IEEE-CS Joint Conference on Digital Libraries*, pp. 228-236, 2004
12. P. Cotter, B. Smyth, "Personalization techniques for the digital TV world," *Proc. European Conference on Artificial Intelligence*, pp. 701-705, 2000
13. W.P. Lee, T.H. Yang, "Personalizing information appliances: a multi-agent framework for TV programme recommendations," *Expert Systems with Applications*, vol. 25, no. 3, pp. 331-341, 2003
14. M. Ciaramita, M. Johnson, "Explaining away ambiguity: Learning verb selectional preference with Bayesian networks," *Proc. Intl. Conference on Computational Linguistics*, pp. 187-193, 2000
15. F. V. Jensen, *Bayesian Networks and Decision Graphs*, Springer, 2001.
16. G. Miller, R. Beckwith, C. Fellbaum, D. Gross, K.J. Miller, "Wordnet: An on-line lexical database," *International Journal of Lexicography*, vol. 3, no. 4, pp. 235-312, 1990
17. J.J. Lee, "Case-based plan recognition in computing domains," *Proc. The Fifth International Conference on User Modeling*, pp. 234-236, 1996

Density-Based Spatial Outliers Detecting

Tianqiang Huang[1], Xiaolin Qin[1], Chongcheng Chen[2], and Qinmin Wang[2]

[1] Department of Computer Science and Engineering,
Nanjing University of Aeronautics and Astronautics, Nanjing, 210016, China
tianqianghuang@163.com
[2] Spatial Information Research Center in Fujian Province,
Fuzhou, 350002, China
http://www.sirc.gov.cn/

Abstract. Existing work in outlier detection emphasizes the deviation of non-spatial attribution not only in statistical database but also in spatial database. However, both spatial and non-spatial attributes must be synthetically considered in many applications. The definition synthetically considered both was presented in this paper. New Density-based spatial outliers detecting with stochastically searching approach (*SODSS*) was proposed. This method makes the best of information of neighborhood queries that have been detected to reduce many neighborhood queries, which makes it perform excellently, and it keeps some advantages of density-based methods. Theoretical comparison indicates our approach is better than famous algorithms based on neighborhood query. Experimental results show that our approach can effectively identify outliers and it is faster than the algorithms based on neighborhood query by several times.

1 Introduction

A well-quoted definition of outliers is the Hawkin-Outlier [1]. This definition states that an outlier is an observation that deviates so much from other observations as to arouse suspicion that it was generated by a different mechanism. However, the notion of what is an outlier varies among users, problem domains and even datasets[2]: (i) different users may have different ideas of what constitutes an outlier, (ii) the same user may want to view a dataset from different "viewpoints" and, (iii) different datasets do not conform to specific, hard "rules" (if any).

We focus on outlier in spatial database, in which objects have spatial and non-spatial attributions. Such datasets are prevalent in several applications. Existing work of Multidimensional outlier detection methods can be grouped into two sub-categories, namely homogeneous multidimensional and bipartite multi- dimensional methods [3]. The homogeneous multidimensional methods model data sets as a collection of points in a multidimensional isometric space and provide tests based on concepts such as distance, density, and convex hull depth. These methods do not distinguish between spatial dimensions and attribute dimensions (non-spatial dimensions), and use all dimensions for defining neighborhood as well as for comparison. Another multidimensional outlier detection method is bipartite multidimensional test which is designed to detect spatial outliers. They differentiate between spatial and non-spatial attributes. However, they defined outlier as "spatial outlier is spatially referenced objects whose non-spatial attribute values are significantly different from those of other spatially referenced objects in their spatial neighbor- hoods [3,4]", which emphasizes non-spatial deviation and ignores spatial deviation.

In some application, domain specialist needs detect the spatial objects, which have some non- spatial attributes, deviation from other in spatial dimension. For example, in image processing, detecting a certain type vegetable is anomaly in spatial distribution. The vegetable type is non-spatial attribute, and the vegetable location means spatial attributes. As another example, government wants to know middle incoming residents distribution in geo-space. To detect outliers in these instances, spatial and non-spatial attributes may be synthetically taken into account. For example, there are two type objects in Fig. 1. The solid points and rings respectively represent two objects with different non-spatial attribute, such as the solid objects represent one vegetable and the rings are the other. All objects in Fig. 1 are one cluster when we didn't consider non-spatial attribute, but they would have different result when we took spatial and non-spatial attribute into account. Apparently, when we focus solid objects, the solid objects in *C1* and *C2* are clusters, and object *a* and *b* are outliers.

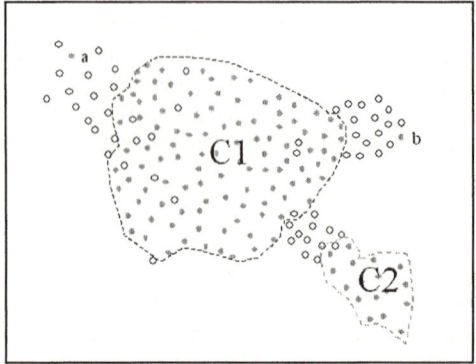

Fig. 1. An illumination example

We took into account of spatial and non-spatial attributes synthetically to define the outliers. If the objects that have some non-spatial attributes are keep away from their neighbor in spatial relation. We defined them outliers.

The main contributions of this paper are: (1) we propose a novel density-based algorithm to detect it, which is the quicker than existing algorithms based on neighborhood query. (2) We evaluate it on both theory and experiments, which demonstrate that algorithm can detect outlier successfully with better efficiency than other algorithms based on neighborhood query.

The remainder of the paper is organized as follows: In section 2, we discuss formal definition of outliers. Section 3 presents the *SODSS* algorithm. Section 4 evaluates performance of *SODSS*. Section 5 reports the experimental evaluation. Finally, Section 6 concludes the paper.

2 Density-Based Notion of Spatial Outliers

In this section we present the new definition of outlier, in which spatial and non-spatial attributes were synthetically taken into account.

Given a dataset D, a symmetric distance function *dist*, parameters *Eps* and *MinPts*, and variable *attrs* indicates the non-spatial attributes.

Definition 1. The **impact neighborhood** of a point p, denoted by $IN_{Eps}(p)$, is defined as $IN_{Eps}(p) = \{q \in D \mid \text{dist}(p, q) \leq Eps \text{ and } q.attrs \text{ satisfy } C\}$.

Definition 2. The **Neighbor** of p is any point in impact neighborhood of p except p.

Definition 3. If a point's impact neighborhood has at least *MinPts* points, the impact neighborhood is **dense**, and the point is **core point**.

Definition 4. If a point's impact neighborhood has less than *MinPts* points, the impact neighborhood is **not dense**. If a point is a neighbor of core point, but his neighborhood is not dense, the point is **border point**.

Definition 5. If a point is core point or border point, and it near a border point p, the point is **near-border point** of p.

Definition 6. A point p and a point q are **directly density-reachable** from each other if (1) $p \in IN_{Eps}(q)$, $|IN_{Eps}(q)| \geq MinPts$ or (2) $q \in IN_{Eps}(p)$, $|IN_{Eps}(p)| \geq MinPts$.

Definition 7. A point p and a point q are **density-reachable** from each other, denoted by $DR(p, q)$, if there is a chain of points $p_1,\ldots,p_n$, $p_1=q$, $p_n=p$ such that p_{i+1} is directly density-reachable from p_i for $1 \leq i \leq n-1$.

Definition 8. A **cluster** C is a non-empty subset of D satisfying the following condition: $p, q \in D$: if $p \in C$ and $DR(p, q)$ holds, then $q \in C$.

Definition 9. **Outlier** p is not core object or border object, i.e., p satisfying the following conditions: $P \in D$, $|IN(p)| < MinPts$, and $\forall q \in D$, if $|IN(q)| > MinPts$, then $p \notin IN(q)$.

3 *SODSS* Algorithm

In *DBSCAN* [5] or *GDBSCAN* [6], to guarantee finding density-based clusters or outliers, determining the directly density-reachable relation for each point by examining the neighborhoods is necessary. However, performing all the region queries to find these neighborhoods is very expensive. Instead, we want to avoid finding the neighborhood of a point wherever possible. In our method, the algorithm discards these dense neighborhoods in first, because these objects in it are impossibly outliers. The algorithm stochastically researched in database but not scan database one by one to find the neighborhood of every point like *DBSCAN*, so the algorithm outperform famous algorithms based on neighborhood query, such as *DBSCAN* [5], *GDBSCAN* [6], *LOF* [7].

In the following, we present the density-based Spatial Outlier Detecting with Stochastically Searching (*SODSS*) algorithm. *SODSS* is consisted of three segments. The first (lines 3~17) is *Dividing Segment*, which divide all object into three parts, cluster set, candidate set or outlier; The second (lines 19~23) is *Near-border Detecting*

Segment, which detect and record the near-border objects of candidate, i.e., the neighbors of these border objects that may be labeled candidate, which would be used to detect these border objects in the third segment; The third (lines 24~31) is *Fining Segment*, using the near-border objects to find these border objects and remove them.

SODSS starts with an arbitrary point *p* and Examine its impact neighborhood *NeighborhoodSet* with *D.Neighbors(p, Eps)* in line 5. If the size of *NeighborhoodSet* is at least *MinPts*, then *p* is a core point and its neighbors are belong to some clustering, to put them into clustering set list; otherwise, if the size is 0, *p* is outlier, so put them into outlier set; or else *p* and his neighbor may be outliers, so put them into candidate set. Lines 19~23 detect neighbors of these that were labeled candidates in *Dividing Segment* and include them into candidate set. These objects would be used to detect border objects that are not outliers from candidate set. Lines 24~31 check every object in candidate set to remove the border objects.

SODSS algorithm

```
Algorithm SODSS(D, Eps, MinPts)
1. CandidateSet = Empty;
2. ClusteringSet = Empty;
3. While ( !D.isClassified( ) )
4.    {Select one unclassified point p from D;
5.     NeighborhoodSet = D.Neighbors(p, Eps);
6.     if ( | NeighborhoodSet | > MinPts )
7.         ClusteringSet = ClusteringSet ∪ NeighborhoodSet
8.     else
9.            if( | NeighborhoodSet | > 0 )
10.              {NeighborhoodSet.deleateCluserLabledPoit;
11.               CandidateSet = CandidateSet ∪
                                NeighborhoodSet ∪ p
12.              }
13.         else
14.              OutlierSet = OutlierSet ∪ p
15.         endif;
16.    endif;
17.   } // While !D.isClassified
18. Borders = Empty;
19. While ( !CandidateSet.isLabel )
20.     { Select one point q from CandidateSet;
21.       q.isLabel;
22.       Borders = Borders ∪ CluseringSet.Neighbors(q,
                             Eps);
23.     } // While !CandidateSet.isLabel
24. While ( !Borders.isLabel )
25.     { Select one point b from CandidateSet;
26.       b.isLabel;
27.       Bord_NB = D.Neighbors( b );
28.       if ( | Bord_NB | > MinPts )
29.           CandidateSet.delete (Bord_NB);
30.       OutlierSet = OutlierSet ∪ CandidateSet;
31.     } // While !Borders.isLabel
```

To understand this algorithm, we give example as Fig. 2. There are two type objects in Fig. 2. The solid point represented one-type objects and the ring represented the other type objects. Supposing we focus on solid objects. Apparently, there are two clusters and two outliers in solid objects in the figure. Clusters are located in center and right down, and outliers are object a and object d. when algorithm run lines 3~17 to divide spatial objects to three parts, cluster set, outlier or candidate set. Algorithm may select object a, and calculate neighborhood A. Supposing object b and c have not been labeled in any dense neighborhood. They are the neighbors in neighborhood A, and neighborhood A is sparse, so they are labeled to candidate. When object b and c is included in candidate set, the near-border objects near b and c, which include in the red polygon P in Fig. 2., are also included in candidate set through the *Near-border Detecting Segment* in line 19~23. Some of near-border objects in red polygon P are dense, so object b and c would be removed from candidate set. So *SODSS* can identify real outlier.

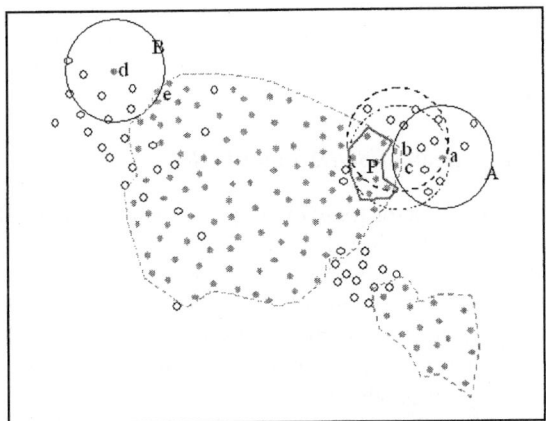

Fig. 2. Object a and d are outliers. Object b and c would be labeled candidates. The objects in red polygon P are border objects that are put into candidate set in *Near-border Detecting Segment*

4 Theoretical Performance Comparison of *SODSS* and the Other Density-Based Algorithm

There are many density-based algorithm that were proposed to detect outliers, but calculation efficiency is not obviously improved. In the worst case, the time cost of the algorithms are $O(n^2)$. *SODSS* outperform existing algorithms in calculation efficiency.

The neighborhood query $D.Neighbors(p, Eps)$ in line 5 is the most time-consuming part of the algorithm. A neighborhood query can be answered in $O(logn)$ time using spatial access methods, such as R*-trees [8] or SR-trees [9]. When any clusters are found, their neighbors would not be examine by *SODSS* again, so *SODSS* will perform fewer neighborhood queries and save much time. Clustering objects must much more than outlier objects, so *SODSS* can reduce much neighborhood query and then have good efficiency. Supposing *SODSS* performs k neighborhood queries, its time complexity is $O(klogn)$, which k is much smaller than n. In the second and third segment algorithm must query neighborhood again, but these operation are in candidate set and the number of candidate is very few. The k is related to *Eps*, so the time complexity is related to *Eps*. With increasing of *Eps* time cost decreases in certain range, however, the candidates would increase greatly when *Eps* exceeds the threshold and the time cost would increase obviously.

4.1 Performance Comparison of *SODSS* and *GDBSCAN*

GDBSCAN [6] extended the famous algorithm *DBSCAN* to apply to spatial database. *GDBSCAN* identify spatial outlier through detecting cluster, i.e., the noises are outliers. This algorithm scans database and examine all objects neighborhoods.

Eps-Neighborhood of *GDBSCAN* corresponds to impact neighborhood of *SODSS*, which is expensive operation. One crucial difference between *GDBSCAN* and *SODSS* is that once *SODSS* has labeled the neighbors as part of a cluster, it does not examine the neighborhood for each of these neighbors. This difference can lead to significant time saving, especially for dense clusters, where the majority of the points are neighbors of many other points.

4.2 Performance Comparison of *SODSS* and *LOF*

LOF [7] calculates the outlier factor for every object to detect outliers. It is the average of the ratio of the local reachability density of p and those of p's *MinPts*-nearest neighbors. The local reachability density is based on *MinPts*-nearest neighbors. *LOF* must calculate k-distance neighborhoods of all objects, which time costs are equal to impact neighborhoods query. Calculating k-distance neighborhoods is the main expensive operation. *SODSS* detect outlier by removing cluster objects with stochastically researching. All neighbors in dense neighborhood would not calculate their neighborhood again, so the region query of *SODSS* must be less than *LOF*'s. Accordingly, *SODSS* have better efficiency than *LOF*.

5 Experimental Evaluation

The criteria evaluating outlier detection approaches can be divided into two parts: efficiency and effectiveness. Good efficiency means the technique should be applicable not only to small databases of just a few thousand objects, but also to larger databases with more than hundred thousand of objects. As for effectiveness, a good approach should have ability to divide exactly outliers from clusters. We have done many experiments to examine the efficiency and effectiveness, but here limiting to extension we only presented two. In first, we use synthetic data to explain effectiveness of our approach. Secondly, we use large database to verify the efficiency. Experiments showed that our ideas can be used to successfully identify significant local outliers and performance outperforms the other density-based approaches. All experiments were run on a 2.2 GHz PC with 256M memory.

5.1 Effectiveness

To compare *SODSS* with *GDBSCAN* [6] and *LOF* [7] in terms of effectiveness, we use the synthetic sample databases which are depicted in Fig. 3. In these datasets, the non-spatial property for the points is depicted by different symbol, rings and solid points. Experiment focus on solid objects, and set *q.attrs = solid*. Fig. 4 shows the outliers and clusters identified by *SODSS*. The radius set to 2.1 in *SODSS*, MinPts set to 3. *SODSS* and *GDBSCAN* can identify outliers correctly, because they consider non-spatial attribute. As shown in Fig. 5, *LOF* does not find the outliers because it ignores non-spatial attributes and considers all objects are cluster.

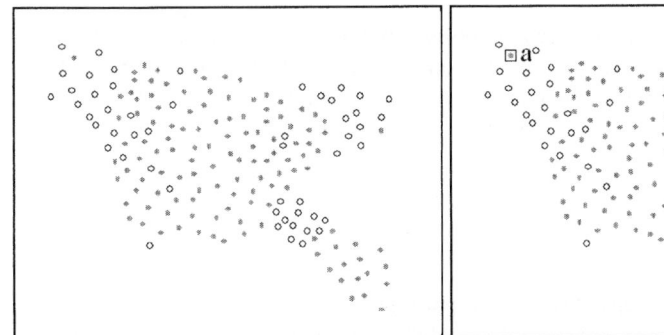

Fig. 3. Synthetic sample databases

Fig. 4. Outlier *a* and *b* identified by *SODSS* or *SDBDCAN*

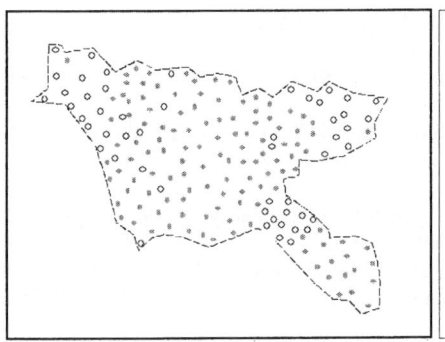

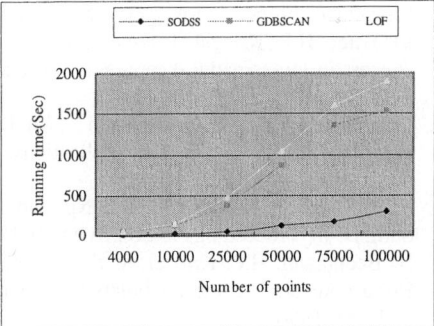

Fig. 5. *LOF* can't identify outliers

Fig. 6. Time efficiency comparisons between *GDBSCAN*, *LOF* and *SODSS*

5.2 Efficiency

For comparison computational efficiency of *SODSS* and *GDBSCAN* and *LOF*, we used synthetic datasets that are consisted of points from 4000 to 100,000. The *Eps* is 5, and *MinPts* is 10, when *SODSS* query the neighborhood. They are the same when

GDBSCAN run. We set *MinPts* = 30 and *LOF* > 1.5. Fig. 6. shows the running time for *SODSS* increases with the size of the datasets in an almost linear fashion, and the performance is obviously better than the other two.

6 Conclusion

In this paper, we formulated the problem of one-type spatial outlier detection and presented effective and efficient *SODSS* algorithms for spatial outlier mining. This algorithm does not calculate neighborhood of very objects but stochastically research. It discards much region query of cluster, and gained good efficiency.

Acknowledgements. This research supported by the National Nature Science Foundation of China (No. 49971063), the National Nature Science Foundation of Jiangsu Province (BK2001045), the High-Tech Research of Jiangsu Province (BG2004005) and the National High-Tech Research and Development Plan of China (No. 2001AA634010-05).

References

1. D. Hawkins. Identification of Outliers. Chapman and Hall, London, 1980
2. H. Dai, R. Srikant, and C. Zhang. OBE: Outlier by Example. In: Proceedings of PAKDD 2004, Sydney, Australia, May 26-28, 2004, LNAI 3056, pages: 222-234, 2004
3. S. Shekhar, C.T. Lu, and P. Zhang. A unified approach to detecting spatial outliers. GeoInformatica, 7(2): 139-166, 2003
4. C.T. Lu, D. Chen, and Y. Kou. Algorithms for spatial outlier detection. In Proceedings of the 3rd IEEE International Conference on Data Mining (ICDM 2003), December 19-22, 2003, Melbourne, Florida, USA, pages: 597-600. IEEE Computer Society, 2003
5. M. Ester, H.P. Kriegel, J. Sander, and X. Xu. A density-based algorithm for discovering clusters in large spatial databases. In: Proceedings of KDD'96, Portland OR, USA, pages: 226-231, 1996
6. J. Sander, M. Ester, H. Kriegel, and X. Xu. Density-based Clustering in Spatial Databases: the algorithm GDBSCAN and its applications. Data Mining and Knowledge Discovery, val. 2, no. 2, pages: 169-194, 1998
7. M.M. Breunig, H.P.Kriegel, R.T.Ng, and J. Sander. LOF: Identifying density-based local outliers. In: Proceedings of SIGMOD'00, Dallas, Texas, pages: 427-438, 2000
8. N. Beckmann, H.P. Kriegel, R. Schneider, and B. Seeger. The R*-Tree: An Efficient and Robust Access Method for Points and Rectangles. SIGMOD Record, vol. 19, no. 2, pages: 322-331, 1990
9. N. Katayama and S. Satoh. The SR-tree: An Index Structure for High-Dimensional Nearest Neighbor Queries. SIGMOD Record, vol. 26, no. 2, pages: 369-380, 1997

The Design and Implementation of Extensible Information Services

Guiyi Wei[1], Guangming Wang[1], Yao Zheng[2], and Wei Wang[2]

[1] Zhejiang Gongshang University, Hangzhou, 310035, P. R. China
weiguiyi@tom.com, hucl@mail.hz.zj.cn
[2] Center for Engineering and Scientific Computation, Zhejiang University,
310027, P. R. China
{yao.zheng, weiw}@zju.edu.cn

Abstract. High-performance distributed computing often requires careful selection and configuration of computers, networks, application protocols, and algorithms. In grid environment, the participators of domains provide site information by default that may be not adequate for resources manager to achieve optimal performance. In this paper, an Extensible Information Service Architecture (EISA) is described to resolve this problem. EISA supports a virtual organization participator's autonomous control of local resources. It consists of three layers, the registration and deployment layer, normalization and organization layer, and information collection layer. Dynamic information services in EISA could be deployed and extended easily. EISA improves the extensibility of grid information services and the efficiency of resources utilization.

1 Introduction

The term "Grid" denotes a distributed computing infrastructure for advanced science and engineering [1] and is distinguished from conventional distributed computing by its focus on large-scale resource sharing, innovative applications, and, in some cases high-performance orientation [2]. The real and specific problem underlies the Grid concept is coordinated resources sharing in dynamic, multi-institutional virtual organizations. As Grid becoming a viable high performance computing alternative to the traditional supercomputing environment, various aspects of effective Grid resources, such as proper scheduling and efficient resource utilization scheme across the Grid can lead to improve system and a lower turn-around time for jobs.

High-performance distributed computing often requires careful selection and configuration of computers, networks, application protocols, and algorithms. These requirements do not arise in traditional distributed computing. The situation is also quite different in traditional high-performance computing where systems are usually homogeneous and hence can be configured manually [4]. In grid computing, neither defaults nor manual configuration is acceptable. Using default configurations often can not result in acceptable performance, and manual configuration requires steering of remote systems that is usually impossible. We need an information-rich approach to configuration in which decisions are made, whether at compile-time, link-time, or run-time [5], based upon information about the structure and state of the site on which a program is to run.

In grid environment, the participators of some special domains provide site information that is discovered by default. In some cases, default information may be not adequate for the scheduler to achieve optimal result. The most important part of a good scheduler is an excellent algorithm. The input arguments and objectives of the algorithm are subject to different applications. The computation intensive applications, such as Monte-Carlo simulations, need high speed computing power. The I/O intensive applications, such as distributed data sharing system, need large capacity storage. The communication intensive applications, such as computational fluid dynamics simulations, need large bandwidth and low latency. To achieve the optimal objectives of these applications, adequate information about distributed nodes in the grid environment must be acquired. Nowadays the information is discovered and retrieved by grid information services, such as MDS, LDAP, etc., which is provided by grid middleware. Typically, the original information is generated by many different elementary programs, which are always encapsulated as web services or grid services. According to the difference of operation systems, networks, storage management and other local resource management policies, some programs may not always be resided in all of the domain sites. It is possible that the schedule result is not optimal because of the information scarcity. In this paper, an Extensible Information Service Architecture (EISA) is described to resolve above problems, which support dynamic deployment of scalable information service.

2 Related Work

The Multidisciplinary ApplicationS-oriented SImulation and Visualization Environment (MASSIVE) project adopted EISA in its information service infrastructure. The grid created for the MASSIVE utilizes resources located at the CESC and relevant departments of ZJU. The resources at other institutions of high performance computing are to be added to it recently. The grid currently uses Globus Toolkit 2.x as the middleware to enable the resources to be accessed after authentication and authorization.

Using OGSA architecture, the MASSIVE project enables grid-based CFD and CSM by some specified services. Services for geometry pre-processing and mesh generation, migration and execution of application programs on remote platforms, collaborative visualization, and data analysis, form the basis of a globally distributed virtual organization. A typical usage would be the generation of a mesh using a mesh generation service on an IRIX platform, the solution of CFD and CSM problems with the meshes previously created on a PC cluster, and the collaborative visualization of the numerical results with equipments such as a stereo tiled display wall and a BARCO stereo projection system at the CESC.

To minimize the risks of remote execution of simulations, the MASSIVE Grid makes it possible for the user to keep the codes and data at a local machine, and to transfer them to a remote machine only when it is to be executed, and it is then erased from the remote site after execution. Services will be provided to support the secure migration, remote execution, deletion of the application program, and the secure return of the application results to the user [6].

Visualization of large scaled scientific data sets using the Grid is an important aspect of the MASSIVE project. To make effective use of all of the devices, a powerful information service faculty is needed.

3 EISA

The grid participator provides its local static resources information and dynamical status of devices' activities with core resource information service supported by the adopted grid middleware. MDS [4] is the information service provided by Globus Toolkit in grid environment.

The variety of resources leads to the variety of Information Providers (IP). EISA operates over the wide area and supports a VO participator's autonomous control of local resources. It consists of three layers, a registration and deployment layer, normalization and organization layer, and information collection layer (as depicted in figure 1).

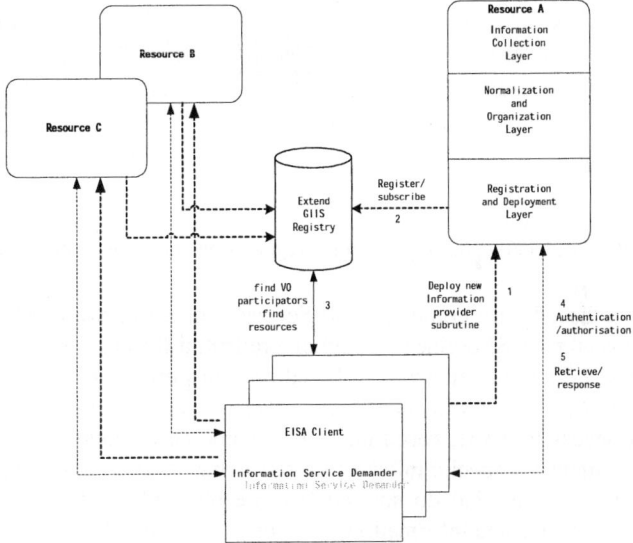

Fig. 1. The EISA Architecture

The registration and deployment layer, which provides Virtual Organization interaction, is interconnected with a Grid Information Index Service (GIIS) server. The normalization and organization layer consists of two components, information normalization module and IPs management module. At the information normalization module, information is collected from native or temporarily deployed IPs, and normalized into annotated resource information that is tailored to a client's needs. At the IP management module each IP is migrated, configured and activated within its control. The information collection layer is the IPs' container, which is extensible and capable of adopting external information discovery programs and encapsulating them into grid services.

3.1 The Registration and Deployment Layer

The registration and deployment layer provides mechanisms to support interaction among GIIS servers, local host and EISA clients. It can also enforce the site's security policies. Logically the registration and deployment layer is depicted in Figure 2, where a range of clients interacts with it to request resource information and deploy new IPs. The registration and deployment layer register the site's contact details, and the categories of resource they manage to the GIIS servers of the VO. Clients locate the site by searching GIIS registry. Thereafter, clients directly interact with registered sites, query it, and subscribe for reservation and notification when conditions changed.

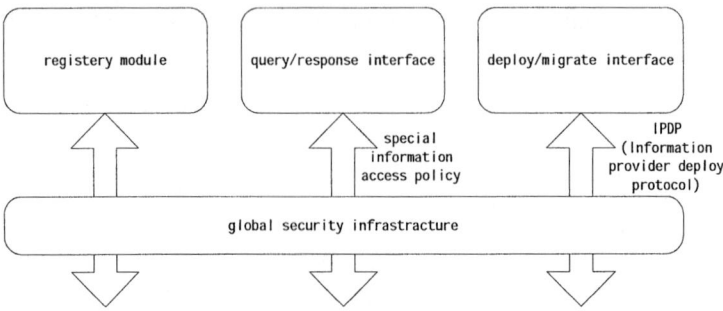

Fig. 2. Logical representation of the registration and deployment layer

The EISA is based on a common Grid Security Infrastructure (GSI) that provide security functionality. GSI defines a common credential format based on X.509 identity certificates and a common protocol based on transport layer security (TLS, SSL) [7]. The registration and deployment layer security can be configured with GSI to restrict client access to a site, based on VO. In some cases, there may be additional security requirement for special information retrieving. A site policy may decree that the requests from clients that do not satisfied are blocked, or the requests are only permitted to retrieve cached information for a subset of resources.

3.2 The Normalization and Organization Layer

The information normalization module and IPs management module are two major components of the layer.

3.2.1 Normalization
Normalization is the process of transforming raw data from diverse different IPs into information that is well defined and described in a consistent and standard format (as

depicted in figure 3). Normalization is used to provide a homogeneous view of resources [3]. The raw data from IPs may be in many different formats. They could be: (1) name-value pairs of device's attributes and its value; (2) encoded in binary, ASCII, Unicode, or other formats; (3) represented using comma-delimited fields, or a metadata language like XML; (4) Context requiring knowledge to understand it.

XML naming schemas provide a way to enforce normalization requirements. A naming schema is an abstract description of the name, unit, meaning, organization and relation of attributes that compose a given entity within a domain of reference [3]. For example, the GLUE [8] naming schema provides a common conceptual model used for grid resource monitoring and discovery. GLUE provides common terms of reference and semantics that define how computing, storage and network resources, and their attributes, should be labeled, described and related.

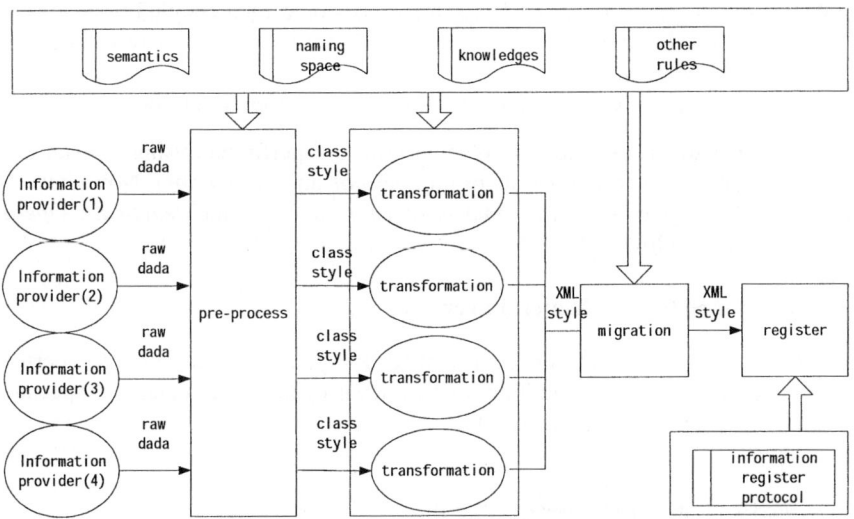

Fig. 3. The normalization process

The normalization process includes four steps described as follow.

1. Discover raw data of the resources.
2. Encapsulate raw data into class style.
3. Transform information in class style to XML format and migrate them into a single XML document.
4. Register the migrated information to reservations and cache it.

3.2.2 Information Provider Management

The IP management module controls the deployment of each IP. At this module, we define a deployment in this way:

$$D < P, IN, OUT, S > \qquad (1)$$

Here:

- **D** is the name of the deployment that is a character string and identifies itself exclusively.
- **P** is the name of information provider. P includes the executable program, its requirements of running environment (such as OS type, OS version, and necessary library), and its version. P is given in XML style. It can be denoted as:

$$P = (exe\_name, env, ver) \tag{2}$$

- **IN** is the input parameters set of P. It includes parameters name, parameters data type, and query rules. It can be denoted as:

$$IN = (<para\_name, data\_type>, <query\_rules>) \tag{3}$$

- **OUT** is the output raw data of P. Each data is stored by a couple of value as <attribute, value>. For example, <FreeMemory, 54.6> means that there is 54.6 megabytes free RAM memory.
- **S** is the collection of semantics, knowledge, and transformation rules.

At IP management module, management jobs are performed using a serial of instructions. Each instruction consists of an *operator* and an *operand*. Nowadays, *Operator* includes *import, upgrade, remove, register, activate,* and *deactivate*. Operand is the name of a deployment.

3.3 The Information Collection Layer

The information collection layer consists of IPs for querying local resources, performing reservation, concurrent handling, and retrieval of historical data, and special information access policies (as depicted in figure 4).

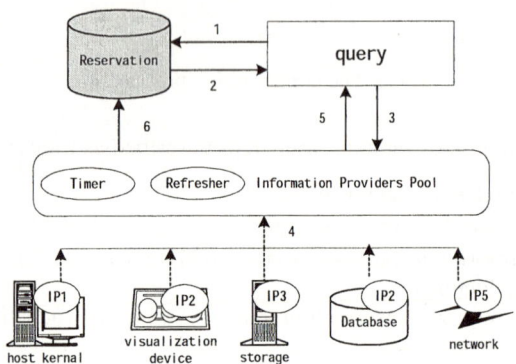

Fig. 4. The information collection layer

A key element of the information collection layer is the IP pool that is used for interacting with a variety of IPs and maintaining a reservation refresher with the help of time rules. Clients can query an arbitrary resource in a standard way and retrieve information described in a standard format. Clients do not require prior knowledge of resources; IPs can be queried to return details of naming schemas that are appropriate to the selected resource.

Figure 4 shows the steps that occur when a client queries a resource:

- Step 1: When query come, information collection layer retrieve the reservation.
- Step 2: Return information hold in reservation. If it fulfills requirement, results will be returned to the upper layer. Otherwise it continues retrieval with step 3.
- Step 3-5: Run proper IPs to perform discovering. Then IPs return results.
- Steps 6: Perform reservations.

4 Implementation

Besides the elementary information, EISA should provide additional information of running program and their result. For example, network bandwidth, network data flow, remote mapped user, remote mapped working directory and visualization devices information are additional that should not be provided by default information service of VO sites. The programs have been developed for discovering information above, encapsulated into OGSA grid services, and can be deployed in the EISA.

The implementation of these additional information services should fulfill the following requirements.

- The IP's output format should be compatible to LDIF of LDAP.
- EISA uses security component of Globus Toolkit. It is also compatible with OGSA standard, so that they can be integrated seamlessly into Globus-based grid application system.
- User-defined IPs should not conflict with those existing services, or not it will work improperly.
- New IPs should be flexible enough. Both the information and the way they integrated into EISA system can be customized.

Each attribute item in a resource object is stored by a couple of values as <attribute, value>. All information is encapsulated as <node_id, res_info>, where the "node_id" is a host name or IP address of the host, and the "res_info" is an instance of class "MdsResult". The C++ class definition of "MdsResult" is shown below:

```
typedef map<const string , string> InfoMap;
class MdsResult {
  map <string , InfoMap *> entry
  public;
    string getValue(string node, string attr);   }
class VGridMds {
  MdsResult result;
  LDAPServer * server;
  public:
  search();
  search(string filter, int scope)
  search(string basedn, string filter, int scope)
  search(string basedn, string filter, string[]
attrToReturn, int scope)
```

```
connect();
disconnect();
update();
getResult();   }
```

5 Conclusions

Resource information service is a vital part of any Grid software. The variety of resources leads to the variety of Information Providers. This paper describes an Extensible Information Service Architecture that operates over the wide area and supports a VO participator's autonomous control of local resources. EISA consists of three layers, a registration and deployment layer, normalization and organization layer, and information collection layer. It implements dynamic deployments of information services for additional information search with high-level extensibility and security.

Acknowledgements

The authors wish to thank the National Natural Science Foundation of China for the National Science Fund for Distinguished Young Scholars under grant Number 60225009. We would like to thank the Center for Engineering and Scientific Computation, Zhejiang University, for its computational resources, with which the research project has been carried out.

References

1. I. Foster, and C. Kesselman (eds.): *the Grid: Blueprint for a New Computing Infrastructure*, Morgan Kaufmann, (1999)
2. Ian Foster, Carl Kesselman, Steven Tuecke: the Anatomy of the Grid: Enabling Scalable Virtual Organizations, *Int. J. Supercomputing Applications*, 15(3), (2001)
3. Mark Baker, Garry Smith: Ubiquitous Grid Resource Monitoring, *Proceedings of the UK e-Science All Hands Meeting 2004*, 61-68, (2004)
4. Steven Fitzgerald, Ian Foster, Carl Kesselman, et al.: A Directory Service for Configuring High-Performance Distributed Computations, *Proceedings of the 6th International Symposium on High Performance Distributed Computing (HPDC '97)*, p.365, (1997)
5. D. Reed, C. Elford, T. Madhyastha, E. Smirni, et al.: The Next Frontier: Interactive and Closed Loop Performance Steering, *Proceedings of the 1996 ICPP Workshop on Challenges for Parallel Processing*, 20-31, (1996)
6. David W. Walker, Jonathan P. Giddy, Nigel P. Weatherill, Jason W. Jones, Alan Gould, David Rowse, and Michael Turner: GECEM: Grid-Enabled Computational Electromagnetics, *Proceedings of UK e-Science All Hands Meeting*, 2003
7. V. Welch, F. Siebenlist, I. Foster, et al.: Security for Grid Services, *Twelfth International Symposium on High Performance Distributed Computing (HPDC-12)*, IEEE Press, 2003
8. GLUE-Schema, http://www.hicb.org/glue/glueschema/schema.htm, December 2004

Approximate B-Spline Surface Based on RBF Neural Networks

Xumin Liu[1,2], Houkuan Huang[1], and Weixiang Xu[3]

[1] School of Computer and Information Technology, Beijing Jiaotong University, 100044
[2] School of Information Engineering, Capital Normal University, Beijing 100037
[3] School of Traffic and Transportation, Beijing Jiaotong University, 100044
liuxmxxxy@263.net

Abstract. Surface reconstruction is a key technology in geometric reverse engineering. In order to get the geometric model of object we need a large number of metrical points to construct the surface. According to the strong points of RBF network such as robust, rehabilitating ability and approximating ability to any nonlinear function in arbitrary precision, we presented a new method to reconstruct B-spline surface by using RBF. Simulation experiments were made based on the theoretical analysis. The result indicated that this model could not only efficiently approximate incomplete surface with noise, automatically delete and repair the input wrong points through self-learning, but has a rapid learning speed, which improves the reconstructing efficiency and precision of dilapidation incomplete surface. The surface obtained by this model has a good smooth character.

1 Introduction

Reverse Engineering (RE), which means the designing of products is based on practicality instead of conception, has recently become a researching hotspot in the field of CAD/CAM. It's widely applied in the industrial formative field, which relates to freeform surface, such as aviation, automobile and shipping, etc. Based on the attainable object model, RE can be used for the construction of the design model. Using the adjustment and modification of character parameter of the reconstructed model, the result is to approach or modify the object model, so that it can meet many kinds of requirements. After the data attained from scanning and measure is processed, surface reconstruction become a key technology of RE. It denotes the surface of the measured object in the form of some piecewise smooth, and unites continued global continuous surfaces at definite range of error through some scatter data points acquired by measure.

As geometrical scanning machines are more widely used, the number and complexity of scanned objects increase. It is inevitable to attain 3D model with noise when scanned by the scanning machine. Therefore, it is important to remove the noise data out of all the data points and meanwhile keep the original shape of the object. With the development of Artificial Neural Networks (ANN) theory, neural networks surface reconstruction method is brought forward by a great deal of home and broad literatures. They put forward scatter data points surface modeling method from a new point of view. ANN is a large-scale, distributing and parallel non-linear information processing system. It has powerful learning capability. It can learn and gain knowledge from samples. If we use the powerful non-linear approach ability of ANN to start measured scatter data points surface modeling, the result model will not only have higher approach precision, but also be somewhat antinoise capability.

There is primary research in the surface reconstruction in intelligent ways, such as the reconstruction for partial disrepaired surface by improved BP neural networks. Presently, the ANN aimed to modeling often adopts three layered feedforward networks as its network topology structure, and adopts algorithm of Back-Propagation [1] as its learning algorithm. Because the algorithm uses negative gradient descent algorithm to guide the search optimize process of weight and threshold of network, it has some disadvantages such as making search optimize process get into local minimum and slow convergence speed, etc. Although Simulated-Annealing method can overcome the BP algorithm's shortcoming of local optimize [2], it is at the price of sacrificing convergence speed, so its application has some limits. The Genetic Algorithms neural network has the specialty of search traverse [3]. It is based on search optimize chromosome by evolution of population, so it has some discomfort when modeling.

The Radial Basis Function (RBF) adopted in this paper is better than the above networks no matter it is approaching ability, classifying ability, convergence speed, or search traversal. So it is applied widely.

This paper put forward the reconstruction method of surface, which come from the standard geometrical description model of industrial product. The method combines with the advantages of Neural Network surface reconstruction, such as precision, high efficiency, and anti-noise. It aims at requirement of the crust surface reconstruction in industrial product. The paper adopts RBF network to pre-fit scatter measure data points from B-spline surface. We use RBF network to obtain the mapping relationship of the surface and store this relationship in the weight and threshold. This mold of holographic memory has strong capacity of fault-tolerant and association, so it won't bias the output because of the break of a few neuron or slight lack of input data, hence it has a strong robust. Better precision of reconstruction can be obtained by using the nonlinear approaching ability of the RBF network to construct unorganized points.

2 RBF Neural Network

RBF neural network (RBFNN) is a kind of efficient Multi-Layer forward network after Multi-Layer Perceptions (MLP) in recent years. It is a partially approaching network, which originates from Radius Basis Function method of multivariable interpolation in numerical analysis. The basic structure includes input layer, hidden layer and output layer, all of which has different functions. The input layer is composed of some source point (perceptive unit), which connect the network with the outside environment. The function of the hidden layer is to transits between input space and hidden space. In most cases, hidden space has a higher dimension. The output layer, which is linear, has the function of providing echoing for the stimulated mold (information) of the function and the input layer. It has been proved that a RBF neural network can approach any non-linear function at any precision, and it has the ability of optimizing functional approach ability[4], in case of the concealed nodes are adequate, and the studying is sufficient.

2.1 RBF Neural Network Structure and Learning Way

RBF neural network is a kind of network of three layers feedforward. The number of each layer's node respectively is p, n, m. It's structure is showed in Fig.1.

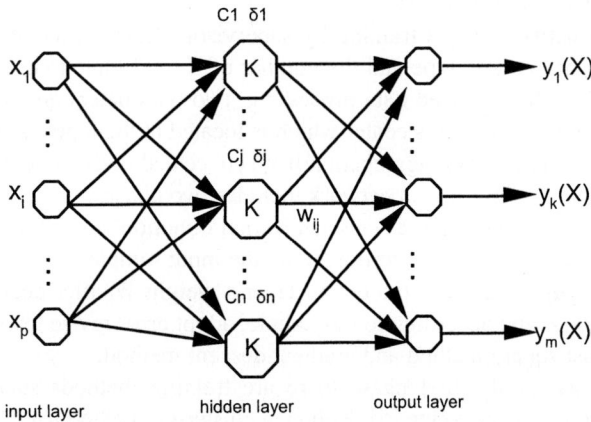

Fig. 1. Construction of RBF neural network

The weight from input layer to concealed layer is fixed in 1. Only the weight $w_{kj}(j=1,2,\ldots,n; k=1,2,\ldots,m)$ coming from concealed layer to output layer can be adjusted.

The concealed layer output function (namely Kernel Function) is often defined as Gaussian Kernel Function. It produces a localized response to input prompting. This character makes Gaussian concealed layer have a clustering impact on input sample. The nodes concealed represent the class numbers of clustering. The center of concealed layer is the agglomerate center point of this kind. The Kernel Function's form is usual as follows:

$$\phi_j(X) = \exp[-\frac{\|X - C_j\|^2}{2\delta_j^2}] \quad (j = 1,2,\cdots,n) \tag{1}$$

In the expression (1), $\Phi_j(X)$ is the output of jth node in concealed layer. $X=(x_1,x_2,\ldots,x_p)^T$ is network input vector, where $\|*\|$ represents Euclidean normal data. Let $X-C_j=(X-C_j)^T(X-C_j)$; C_j be the center vector of jth Gaussian unit in concealed layer; δ_j be radius. From formula (1), we can see that: The function transform is from input layer to output layer. The output of concealed layer nodes ranges from 0 to 1, and the input sample is closer to the center of the nodes, the output value is larger.

RBF neural network's output is the linear combination of concealed nodes output. The output form is represented as follows:

$$y_k = \sum_{j=1}^{n} w_{kj}\phi_j(X) \quad (k=1,2,\cdots,m) \qquad (2)$$

The formula (2) is written in the form of matrix.

$$Y = W\phi \qquad (3)$$

In the formula (3), $W=(w_1,w_2,\ldots,w_q,\ldots,w_m)^T$ is the weight matrix ($w_q=(w_{q1},w_{q2},\ldots,w_{qn})$) between concealed layer to output layer. $\Phi=(\Phi_1,\Phi_2,\ldots,\Phi_n)^T$ is the output matrix of the concealed.

RBF neural network do its training by supervisory learning method. For a neural network which has certain type and the number of Kernel Function, in order to make it rightly perform the expected data processing, two parameters are needed to be ensured: one is Kernel Function's center which is located at the input sample data space, the other is the weight vector aggregation from concealed layer to output layer. So the learning process of a RBF neural network includes two phases:

The first stage is to decide the Gaussian Kernel Function's center C_j and radius δ_j of each node of concealed layer, according to all the input samples.

The second stage is to seek the output layer's weight W_i after deciding concealed layer's parameters, and according to the sample, adopt error revise learning algorithm, for instance least square method and gradient descent method.

At present, as for the first phase there are training methods such as clustering method based on k-means value [5], Kohonen clustering method [6], Orthogonal least squares learning algorithm [7], gradient descent method [8], and for each class sample covariance's Gram-Schmidt Orthogonal method [9] etc. The training method in the second phase is mainly least gradient descent method [8], which is used to seek the target function. Recently, Kaminski has put forward the standard of Kernel Function Gram-Schmidt Orthogonal training method [10].

RBF neural network can make certain the Gaussian Kernel Function's parameters C_j and δ_j by means of Clustering algorithms. The most simple and effective way is K-means [11] method. The detailed steps are as follows:

1) Initialize all the clustering center C_j. The number of the center is usually equal to the number of concealed layer's nodes n.

2) Classify all the input samples X according to the closest center C_j, from criterion (4), classify the sample as class i; G is the sum of sample.

$$\|X_g - C_i\| = \min\|X_g - C_i\| \quad (g=1,2,\cdots,G) \qquad (4)$$

3) Calculate the mean of each kind of samples.

$$E(C_j) = \frac{1}{N_j}\sum_{i=1}^{N_j} X_i \qquad (5)$$

N_j is the number of sample of each class. We substitute $E(C_j)$ for C_j.

4) Repeat the process (2), (3), until $E(C_j)-C$ inclines to zero, and confirm C_j as the center of hidden layer activation function.

5) Each kind of radius can be defined as the average distance from the training sample, which belongs to this class, to the clustering center. For Gaussian Radius-Basis-Function, usually δ_j can simply have the value of 1.

2.2 Use RBF Network to Do Pre-fit Scatter Measure Data Points

Every point of the free-form surface of the three-dimensional space can be described by the mapping relation $z=f(x,y)$. So the essential of using RBF neural network to pre-fit scatter measure data points is a optimize process of neural network parameter for precisely mapping the antetype surface S. Firstly, input the data of measure points into RBF neural network. Then the neural network begins to learn from the known information. When the error between the neural network output and the real value is less than the fixed maximal error, the training of neural network is finished. The pre-fitting to archetypal surface is achieved by RBF neural network.

3 The Transform to B-Spline Surface

These are the most widely used class of approximating splines. B-splines have two advantages: (1) the degree of a B-splines polynomial can be set independently of the number of control points (with certain limitations), and (2) B-splines allow local control over the shape of a spline curve or surface.

We can obtain a vector point function over a B-spline surface using the Cartesian product of B-spline basis functions. The Bicubic B-spline surface's mathematic description is as follows:

$$S(u,v) = \sum_{s=1}^{m+2}\sum_{t=1}^{n+2} d_{st} B_{s,3}(u,v) \cdot B_{t,3}(u,v) \tag{6}$$

Formula (6) satisfy $S(u_i,v_j)=P_{ij}(i=1,2,\ldots,m; j=1,2,\ldots,n)$, where d_{st} $(s=1,2,\ldots, m+2; t=1,2,\ldots,n+2)$ is the control vertexes of B-spline surface. $B_{s,3}(u, v)$ and $B_{t,3}(u, v)$ are Bicubic B-spline basis function.

Append with standardized condition, the basic function of cubic B-spline curve can be written as

$$\begin{aligned}B_{3,3}(u) &= 1/6 \cdot u^3\\B_{2,3}(u) &= 1/6 \cdot (1+3u+3u^2-3u^3)\\B_{1,3}(u) &= 1/6 \cdot (4-6u^2+3u^3)\\B_{0,3}(u) &= 1/6 \cdot (1-3u+3u^2-u^3)\end{aligned} \tag{7}$$

Where u is a parameter variable between two controlling vertexe ($0 \leq u \leq 1$).

Suppose the known data points mesh is P_{ij} (namely the standard mesh defined by data point set P_{ij}, $i=1,2,\ldots,m; j=1,2,\ldots,n$), m and n are the u orientation and v orientation of the number of data points. Now reverse seeks the Bicubic B-spine surface.

B-spline interpolation includes two steps. Firstly, calculate every control polygon's vertex according to the data points at the direction of u (or v). Then calculate the polygon mesh at the direction of v (or u) according to the obtained polygon vertex.

Finally, the B-spline surface defined by the mesh can interpolate the original data points P_{ij}. The essential of resolving is seeking the control vertex mesh d_{st} of the target B-spline surface.

The main steps of the transformation from RBF neural network mapping model $z=f(x,y)$ to B-spline surface representation are as follows[11].

1) Picking up boundary points concentrated from antetype surface S, and then fit into four boundary B-spline curve C_1,C_2,C_3,C_4 according to the measured object's specific geometrical character. Using bilinear interpolation method to gain the boundary interpolation B-spline surface as the basis surface S_b.

2) Doing uniform parameter segmentation to the basis surface S_b. We get the mesh data point set of the parameter region.
$D=\{(u_i,v_j):i=1,2,\ldots,m; j=1,2,\ldots,n\}$
Seek the mesh data point set D's mapping in the XYZ space.
$D'=\{(x(u,v),y(u,v),z(u,v))\}$
Map the D' to the XOY projection surface to project, get the basic interpolation mesh.
$D''=\{(x(u,v),y(u,v))\}$. The result is show as Fig. 2.

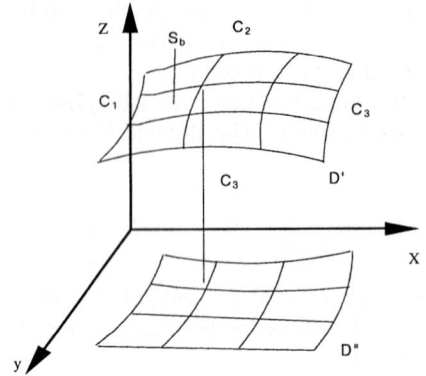

Fig. 2. The formation of basic interpolation mesh

3) Input each node of basic interpolation mesh D'' into the trained perfectly RBF neural network at measure data points pre-fitting stage. Then get the standard mesh topology point set of surface S.
$P=\{(x,y,z):((x,y)\in D'', z=f(x,y))\}$
Among above formula, $z=f(x,y)$ if output information of RBF neural network.

4) Following standard mesh topology point set P to interpolation B-spline surface, we can achieve antetype surface's transformation from RBF neural network mapping model $z=f(x,y)$ to B-spline surface representation.

5) If the reverse resolution precision does not meet the requirement, we regard interpolation B-spline surface as new basis surface S_b, and return to 2). Then we start the next turn calculation for interpolation.

4 Simulation Experiment

Experiment of fitting B-splines surface by RBF network.

The control vertex coordinates of input as follow:

d_{11}= (10,150); d_{12}=(40,150); d_{13}=(50,60); d_{14}=(80,130);
d_{21}=(20,150); d_{22}=(60,130); d_{23}=(80,110); d_{24}=(120,170);
d_{31}=(40,250); d_{32}=(70,200); d_{33}=(120,150); d_{34}=(150,250);
d_{41}=(60,200); d_{42}=(90,150); d_{43}=(130,100); d_{44}=(150,200);

In experiment, B-Splines surface is drawn according to control vertexes firstly. Then noise is added, that range of noise is 0.5. The data with noise is training set. The step is 0.01.

Control point coordinates are network inputs. Data points produced by B-splines surface are expected output. We train the noise data by the above RBF and reconstruct the surface by using the original data set, noise contaminated data set and data set obtained by the trained RBF. The simulation result is depicted as fig. 3.

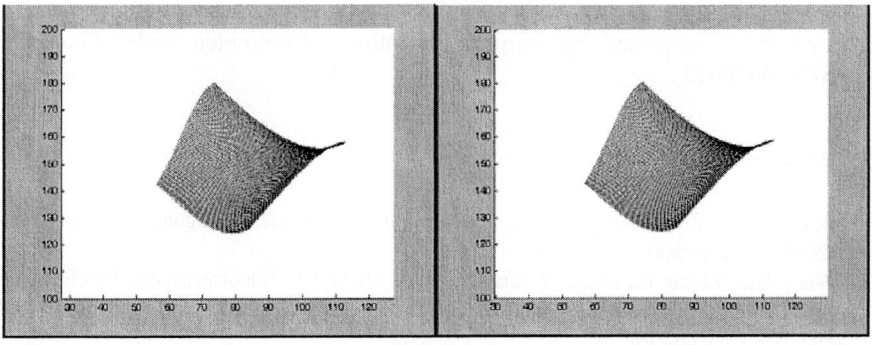

(a) Surface obtained original data set   (b) Noise is added in x and y coordinate

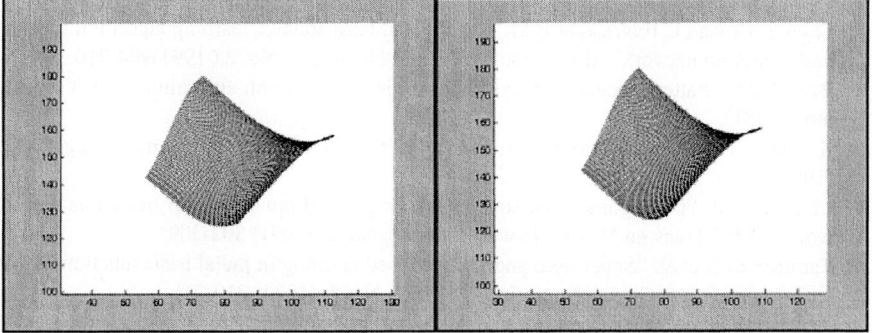

(c) Noise is added in x coordinate   (d) Surface obtained by the data set after fitting

Fig. 3. B-splines surface by RBF network fitting noise

The experiment result denotes that the fitting precision of RBF is quite good and the training speed of network is fast. The quality and quantity of training sample influence directly the fitting precision and training speed. The less the training sample's variance, the higher is the convergence precision and the convergence speed.

5 Conclusions

This paper presented a RBF model to reconstruct surface according to the characteristics of RBF such as good approximating ability, good generalizability, fast convergence speed and strong robust ability. Theoretic analysis and experiment result both denote that by this model, we can construct very smooth surface and raise the efficiency and precision of reconstructing dilapidation incomplete surface. This paper gives a new method of surface modeling with dense and messy data set in a new viewpoint.

Acknowledgements

This paper is supported by Beijing Educational Committee under Grant No. KM200410028013

References

1. Bian Zhaoqi, Zhang Xuegong, Pattern Recognition Principles, Tsinghua University Press, (2000). (in Chinese)
2. Wang Kai, Zhang Caiming, "Neural Network Method to Reconstruct the Freeform Surface", Journal of Computer-Aided Design & Computer Graphics, Vol. 10, No. 3, (1998) 193-199. (in Chinese)
3. Gao hongfeng et al. "Neural Network Trained with Genetic Algorithms and Its Application for Non-Linear System Identification", Journal of Luoyang Institute of Technology, Vol. 19, No. 1, (1998) 74-77. (in Chinese)
4. Chen S, Cowan C F N, Grant P M, "Orthogonal least squares learning algorithm for radial basis function network", IEEE Trans on Neural Net works, No. 2,(1991)904-910.
5. Bezdek J C, Pattern recognition with fuzzy objective function algorithm, New York: Plenum (1981).
6. Kohonen T, Self-organization and associative memory, Berlin, Germany Springer Velag (1989).
7. Chen S et al, "Orthogonal least squares learning algorithm for radial basis function network", IEEE Trans on Neural Networks, Vol. 2, No. 2, (1991) 302-309.
8. Tarassenko B et al, "Supervised and unsupervised learning in radial basis function classifiers", Vision Image Signal Processing, Vol. 141, No. 4, (1994) 210-216.
9. Musavi M F et al, "On the training of radial basis function neural networks ", Neural Networks, Vol. 5, No. 3, (1992) 595-603.
10. Kaminski W et al, "Kernel orthonormalization in radial basis function neural networks", IEEE Trans on Neural Networks, Vol. 8, No. 5, (1997) 1177-1183.
11. Zhou Jinyu, Xie Liyang, "B-spline Surface Reconstruction Based on RBFNN Pre fitting", Journal of Northeastern University, Vol. 24, No. 6, (2003)556-559. (in Chinese)

Efficient Parallelization of Spatial Approximation Trees

Mauricio Marín[1] and Nora Reyes[2]

[1] Computer Cs. Department, University of Magallanes, Chile
[2] Computer Cs. Department, University of San Luis, Argentina
Mauricio.Marin@umag.cl

Abstract. This paper describes the parallelization of the Spatial Approximation Tree. This data structure has been shown to be an efficient index structure for solving range queries in high-dimensional metric space databases. We propose a method for load balancing the work performed by the processors. The method is self-tuning and is able to dynamically follow changes in the work-load generated by user queries. Empirical results with different databases show efficient performance in practice. The algorithmic design is based on the use of the bulk-synchronous model of parallel computing.

1 Introduction

The Spatial Approximation Tree (SAT) is a recent data structure devised to support efficient search in high-dimensional metric spaces [5, 6]. It has been compared successfully against other data structures devised for the same purpose [2, 3] and update operations have been included in the original design [1, 7].

The typical query for this data structure is the *range query* which consists on retrieving all objects within a certain distance from a given query object. From this operation one can construct other ones such as the nearest neighbors. The distance between two database objects in a high-dimensional space can be very expensive to compute and in many cases it is certainly the relevant performance metric to optimize; even over the cost secondary memory operations [1]. For large and complex databases it then becomes crucial to reduce the number of distance calculations in order to achieve reasonable running times. The SAT is able to achieve that goal but still range query operations can be very time consuming. This makes a case for the use of parallelism.

In this paper we propose efficient parallel algorithms for range query operations upon the SAT data structure. The model of parallel computing is the so-called BSP model [10] which provides independence of the computer architecture and has been shown to be efficient in applications such as text databases and others [4, 8]. The proposed algorithms can be implemented using any modern communication library such as PVM, MPI or special purpose libraries such as BSPlib or BSPpub.

The main contribution of this paper is a low-cost method for load balancing the number of distance calculations performed by the processors. It works with any strategy of tree node distribution onto the processors. The method is self-tuning and is able to adapt itself to the evolution of the work-load generated by the stream of queries being submitted to the SAT (we assume a high-traffic client-server setting in which the SAT is used as an index structure to efficiently solve user's range queries). Experimental results with different databases show that the method is able to achieve efficient performance.

2 SAT and BSP

The SAT construction starts by selecting at random an element a from the database S. This element is set to be root of the tree. Then a suitable set $N(a)$ of neighbors of a is defined to be the children of a. The elements of $N(a)$ are the ones that are closer to a than any other neighbor. The construction of $N(a)$ begins with the initial node a and its "bag" holding all the rest of S. We first sort the bag by distance to a. Then, we start adding nodes to N(a) (which is initially empty). Each time we consider a new node b, we check whether it is closer to some element of $N(a)$ than to a itself. If that is not the case, we add b to N(a). We now must decide in which neighbor's bag we put the rest of the nodes. We put each node not in $a \cup N(a)$ in the bag of its closest element of $N(a)$. The process continues recursively with all elements in $N(a)$.

The resulting structure is a tree that can be searched for any $q \in S$ by spatial approximation for nearest neighbor queries. Some comparisons are saved at search time by storing at each node a its covering radius, i.e., the maximum distance $R(a)$ between a and any element in the subtree rooted by a.

Range queries q with radius r are processed as follows. We first determine the closest neighbor c of q among $\{a\} \cup N(a)$. We then enter into all neighbors $b \in N(a)$ such that $d(q,b) \leq d(q,c) + 2r$. This is because the result elements q^* sought can differ from q by at most r at any distance evaluation, so it could have been inserted inside any of those b nodes. In the process, we report all the nodes q^* we found close enough to q. Finally, the covering radius $R(a)$ is used to further prune the search, by not entering into subtrees such that $d(q,a) > R(a)+r$, since they cannot contain useful elements.

In the bulk-synchronous parallel (BSP) model of computing [10, 8], any parallel computer (e.g., PC cluster, shared or distributed memory multiprocessors) is seen as composed of a set of P processor-local-memory components which communicate with each other through messages. The computation is organized as a sequence of *supersteps*. During a superstep, the processors may perform sequential computations on local data and/or send messages to other processors. The messages are available for processing at their destinations by the next superstep, and each superstep is ended with the barrier synchronization of the processors.

We assume a server operating upon a set of P machines, each containing its own memory. Clients request services to a *broker* machine, which in turn distribute those requests evenly onto the P machines implementing the server. Requests are queries that must be solved with the data stored on the P machines. We assume that under a situation of heavy traffic the server start the processing of a batch of Q queries in every superstep.

Every processor has to deal with two kind of messages, those from newly arriving queries coming from the broker, in which case a range search is started in the processor, and those from queries located in other processors that decided to continue their range search in a subtree of this processor (in this case the query is sent packed into a message, thus an initial query can give place to a number of additional query messages).

In the following discussion the main metric used to measure performance is the load balance of the computations effected by the processors. This is defined by the ratio A/B where A is the average of a given measure across processors, and B is the maximum of the measure in any processor. The average A/B is taken over all the supersteps. This is called efficiency E_f and the value $E_f = 1$ indicates the optimal. Speed-up is defined by PE_f. We measure computation by considering the number of distance calculations among objects during query operations and communication is the number of query messages among processors.

In the experiments below we use a 69K-words English dictionary and queries are composed by words selected uniformly at random. The distance between two objects is the edit distance, that is, the minimum number of character insertions, deletions, and replacements to make the two strings equal. We assume a demanding case in which queries come in pairs that are a range query for the same object but with two different radius (large and small) and the broker distribute them circularly among the processors. We use the values 1 (small) and 2 (large) for the dictionary database. The SAT is initialized with the 90% of the database and the remaining 10% are left as query objects (randomly selected from the whole database).

3 Range Queries in Parallel

A first point to emphasize is that the SAT structure contains nodes of very diverse number of children. Every child node causes a distance comparison, so it is relevant to be able to balance the number of distance comparisons performed in every processor per superstep. Thus it is desirable to map the tree nodes onto the processors by considering the number of distance comparisons that can be potentially performed in every sub-tree rooted at the children of the SAT's root. That is, the sub-trees associated with nodes b in $N(a)$ where a is root and $N(a)$ is the set defined in the previous section.

We count the total number of nodes $C(b)$ in each sub-tree with $b \in N(a)$. We then sort the $C(b)$ values and collapse these values onto the P processors. To this end, we define an array U of size P to sum up the $C(b)$ values. For every $b \in N(a)$ such that b is in decreasing order of $C(b)$ values, the node b goes to processor i such that $U[i]$ has the minimum sum among the P elements of U.

We can improve efficiency by setting upper limits V to the number of distance comparisons that are performed per processor in every superstep. During a superstep, every time any processor detects that it has effected more than V distance comparisons, it suspends query processing and waits until the next superstep to continue with this task. "Suspending" in our BSP realization means that all queries going down recursively in the tree in every processor k are sent to processor k as a message exactly as if they found out that the search has to continue in another processor (no communication cost is involved for these extra messages). Also the processors stop extracting new queries from their messages input queues. Figure 1.a shows the effect of this strategy in load balance; average efficiency improves to $E_f = 0.95$ in this case and communication is less than 0.1%.

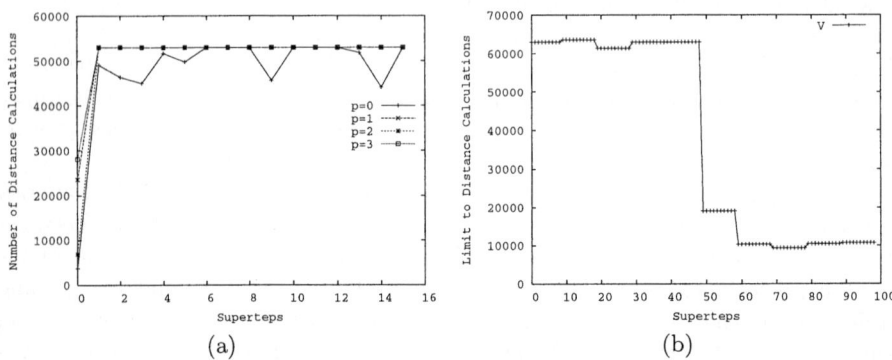

Fig. 1. (a) SAT distributed onto the processors. Number of distance calculations (comparisons) per processor versus supersteps. Demanding case for queries: Circular distribution and every object is queried with two radius. (b) Automatic and adaptive calculation of limits V. Halfway the work-load changes suddenly to queries with one small radius

The value V should be able to adapt itself to follow changes in the workload produced by the flow of queries arriving constantly to the processors. We propose a simple and low-cost solution to this problem. The key idea is to periodically, e.g., every s supersteps, collect statistics that are used to define the value of V for the next sequence of s supersteps. The key is to make these statistics independent of the current value of V and thereby of the s supersteps used to calculate them. Because of the limits V, supersteps can be truncated before processing all the available queries. Therefore real supersteps are not a reliable measure of the real average number of supersteps required to complete a query.

However, for each query we can know exactly when it is necessary to cross over to another processor. We can equip every query q with a counter of "virtual" supersteps $q.w$. We distinguish these virtual supersteps from the real supersteps being executed by the BSP computer. We also keep counters of virtual supersteps $S_v(k)$ in each processor k. Every time a new query q is initiated in a processor

k we set $q.w$ to be equal the batch number at which it belongs to. The broker (or the processors themselves) can assign to every query the batch number in $q.w$ before sending it to one of the processors. In a situation with $V = \infty$ the reception of every new batch of queries marks the beginning of a new superstep (in this case, virtual and real supersteps are the same). Thus every time a new query is received we set $S_v(k) = \max(S_v(k), q.w)$.

We further refine the $S_v(k)$ values by considering that every time a query q has to migrate to another processor we must set $q.w = q.w + 1$ because it takes one virtual superstep to get there. Thus every time one of such queries q arrives to a processor k we also set $S_v(k) = \max(S_v(k), q.w)$, and from this point onwards this query q takes the value $q.w = S_v(k)$. This ensures that queries traveling through several processors will account for the necessary (minimal) number of virtual supersteps to get their target processors and they will reflect this fact in the counters $S_v(k)$, a fact that will also be reflected in all other queries visiting the processor k.

In addition we keep counters $D(k)$ that maintain the total number of distance calculations that has been performed in every processor k. Thus after the processing of the s real supersteps has been completed, the total number of virtual supersteps u that has been completed is given by the maximum among the P values $S_v(k)$, i.e., $u = \max_{0 \leq k \leq P-1}\{S_v(k)\}$. Thus the limit V set in every processor k for the next s real supersteps is given by $V = d/u$ with $d = \operatorname{avg}_{0 \leq k \leq P-1}\{D_v(k)\}$. Figure 1.b shows the effectiveness of this method.

Figures 2.a and 2.b show further experiments. In the first figure it is seen the positive effects of limits V for the milder workloads which do not mix two radius for the same query object. In contrast, we show in the second figure results for a naive strategy of duplicating the SAT in every processor and distributing queries evenly for the same workloads. The results show that even in the case with no

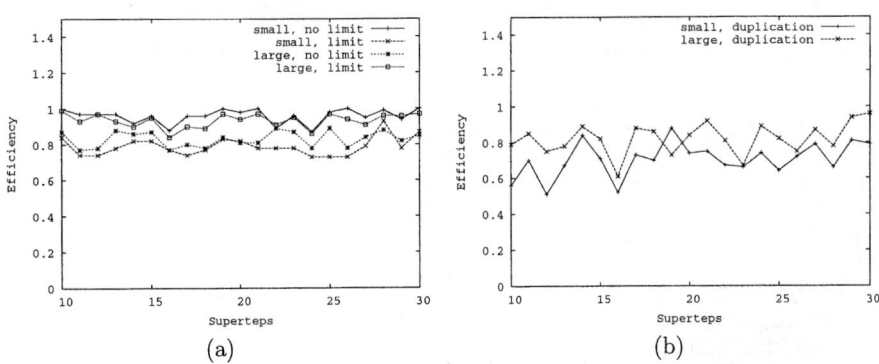

Fig. 2. (a) Efficiencies per supersteps. Proposed strategy of SAT distribution onto the processors and limits to the number of distance calculations per superstep. Two separate workloads; small and large radius. (b) Efficiencies per supersteps. Case in which a copy of the whole SAT is kept in each processor. Two separate workloads; small and large radius

limits V, our proposed strategy for range queries outperform the alternative approach. Speed-ups are shown in table 1. Column "serial" is a measure of the effect of suspending queries in every superstep, i.e., the increase in supersteps (no more than 20%).

3.1 Other Databases and Nodes Mapping

In figure 3.a we show results for the amount of communication demanded by an alternative mapping of SAT nodes to processors. In this case, nodes are

Table 1. Comparing different strategies

Case	work-load	ssteps	serial	speed-up
Duplicated	small	432	–	2.85
Duplicated	large	432	–	3.33
Distributed	small	433	–	3.18
Distributed	large	433	–	3.43
Dist., Limits	small	519	0.17	3.62
Dist., Limits	large	482	0.10	3.53

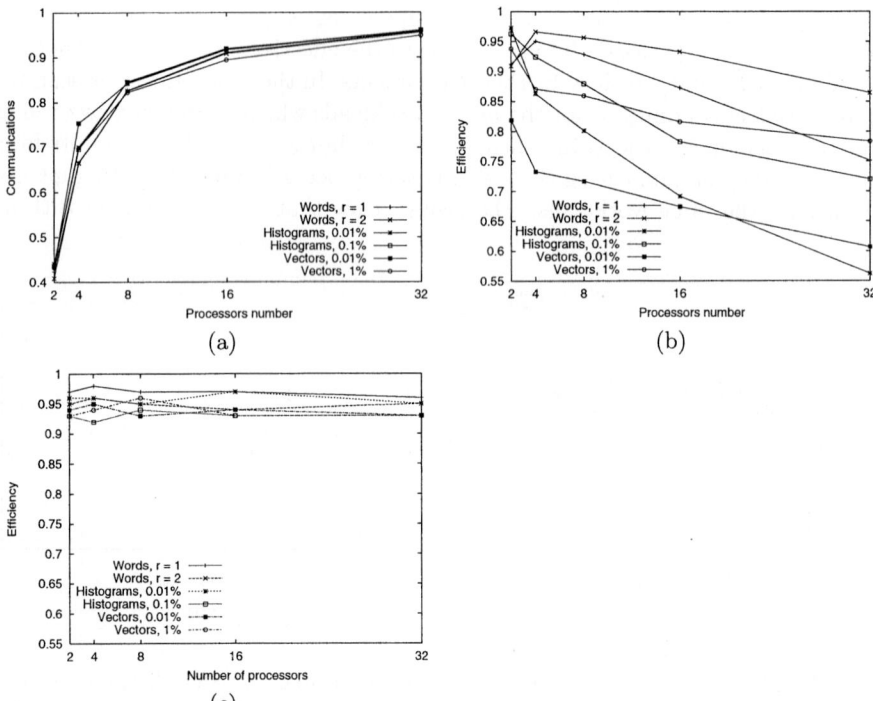

Fig. 3. (a) Effect in communication of multiplexed node distribution. (b) Efficiency achieved by the multiplexed node distribution. (c) Efficiency achieved by the multiplexed node distribution with upper limits to the number of distance calculations

placed in a circular manner independently of their father-child-brother relations. The results were obtained with the above dictionary words and two additional databases. The new data sets are a 20-dimensional cube with points generated uniformly at random and a set of images represented by histograms defined by values of frequency amplitude. Every curve is a value for a different range query radious. Each point indicate the ration A/B where A is the total number of messages sent between processors and B the total number of times that the function *range-search* was called to complete the processing of all queries. These results are in contrast with the mapping based on distance calculation suggested in this paper as the values for this case are all less than 0.1 for all databases. As expected, the amount of communication in the multiplexed approach increases significantly with the number of processors.

The figure 3.b shows results for the efficiency E_f with the same node distribution. Note that the vectors and histograms databases are more demanding in terms of load balance. Similar efficiency values are observed for the distribution method proposed above (selecting the least loaded processor to place a new sub-tree). The actual difference is in the reduced communication. Figure 3.c shows the improvement in efficiency E_f as a result of setting upper limits V.

4 Final Comments

We have proposed parallel algorithms for range queries on a distributed SAT structure. We have focused on balancing the number of distance calculations across supersteps because this is most relevant performance metric to optimize. Distance calculation between complex objects are known to be expensive in running time. On the other hand, we have observed that the amount of communication and synchronization is indeed extremely small with respect to the cost of distance calculations. We have observed that the number of message transmissions is well below 1% with respect to the number of distance calculations. This does not consider the cost of sending solution objects as this cost has to be paid by any strategy. Also, less than 500 supersteps for processing 13K queries is also a very modest amount of synchronization.

Note that we have mapped SAT nodes by considering the sub-trees belonging to the children of the root. It may happen that we have more processors than those children. This case cannot be treated by considering the mapping of sub-trees one or more levels downwards the tree. Those sub-trees actually generate too few distance comparisons. We believe such case can be treated by simply using less processors than available (after all that particular SAT does not admit more parallelism) or by resorting to duplication of some sub-trees in other processors. Here we select the ones which generates more distance comparisons in an attempt to further increase parallelism by ways of dividing the query flow for that sub-trees in two or more processors. This is the subject of our future investigation.

We also tried a multiplexed approach in which every node is circularly distributed onto the processors. Efficiency is good but the cost of communication is extremely high.

Acknowledgements

This work has been partially funded by research project Fondecyt 1040611 and RIBIDI VII.19.

References

1. D. Arroyuelo, F. Mu noz, G. Navarro, and N. Reyes. Memory-adaptative dynamic spatial approximation trees. In *Proceedings of the 10th International Symposium on String Processing and Information Retrieval (SPIRE 2003)*, LNCS 2857, pages 360–368. Springer, 2003.
2. C. Bohm, S. Brchtold, and D. Kein. Searching in high-dimensional spaces: Index structures for improving the performance of multimedia databases. *ACM Computing Surveys*, 33(3):322–373, 2001.
3. V. Gaede and O. Gnnther. Multidimensional access methods. *ACM Computing Surveys*, 30(2):170–321, 1998.
4. M. Marín and G. Navarro. Distributed query processing using suffix arrays. In *Proceedings of the 10th International Symposium on String Processing and Information Retrieval (SPIRE 2003)*, LNCS 2857, pages 311–325. Springer, 2003.
5. G. Navarro. Searching in metric spaces by spatial approximation. *The Very Large Databases Journal (VLDBJ)*, 11(1):28–46, 2002.
6. G. Navarro and N. Reyes. Fully dynamic spatial approximation trees. In *Proceedings of the 9th International Symposium on String Processing and Information Retrieval (SPIRE 2002)*, LNCS 2476, pages 254–270. Springer, 2002.
7. G. Navarro and N. Reyes. Improved deletions in dynamic spatial approximation trees. In *Proc. of the XXIII International Conference of the Chilean Computer Science Society (SCCC'03)*, pages 13–22. IEEE CS Press, 2003.
8. D.B. Skillicorn, J.M.D. Hill, and W.F. McColl. Questions and answers about BSP. Technical Report PRG-TR-15-96, Computing Laboratory, Oxford University, 1996. Also in *Journal of Scientific Programming*, V.6 N.3, 1997.
9. URL. BSP PUB Library at Paderborn University, http://www.uni-paderborn.de/bsp.
10. L.G. Valiant. A bridging model for parallel computation. *Comm. ACM*, 33:103–111, Aug. 1990.

The Visualization of Linear Algebra Algorithms in Apt Apprentice

Christopher Andrews, Rodney Cooper, and Ghislain Deslongchamps,
and Olivier Spet

University of New Brunswick, P.O. Box 4400 Fredericton NB E3B 5A3, Canada
c.andrews@unb.ca,
http://www.cs.unb.ca/research-groups/mmsdt/

Abstract. The development of tools that can increase the productivity of computational chemists is of paramount importance to the pharmaceutical industry. Reducing the cost of drug research benefits consumer and company alike. Apt Apprentice is a visual programming paradigm designed to reduce the overhead associated with creating software to implement algorithms in the data analysis phase of rational drug design. It draws on both standard programming language environments and programming by demonstration. The approach of Apt Apprentice and an example of its use in implementing a linear algebra routine are described.

1 Introduction

Chemical modelling *in silico* is essential to rational drug design. Quantum mechanics, molecular mechanics, and Newtonian dynamics are first used to analyze the chemical structure of compounds found in a training set of molecules all of which exhibit a specific biological activity to some degree. The structurally derived descriptions of these compounds are then statistically analyzed to find predictors of the specific biological activity exhibited by elements of the training set. The subject area of this type of research is called Quantitative Structure Activity Relationships or QSAR. If a predictive relationship between structure and activity can be discovered, other compounds with known structural properties can be analyzed to determine their likelihood of exhibiting specific biological activity. Testing of any promising compounds can then proceed to the laboratory. These compounds are called *drug leads*.

Testing an eventual drug can cost upwards of one billion dollars and take up to 15 years to complete. Since not all promising compounds will pass the tests required by regulatory agencies risk avoidance is paramount.

It is well understood in the pharmaceutical industry that chemical modelling must be done by chemists as the modelling algorithms are not foolproof and can easily be wrongly applied. Apt Apprentice is a visual programming tool designed to facilitate algorithm development in chemical modelling software. It is hoped that this will lead to faster implementation of new algorithms in existing software, aid in the training of chemists to perform this task and facilitate modifications to software in the light of chemical knowledge.

2 The Apt Apprentice Model

The next section describes some of the unique characteristics of Apt Apprentice along with an example of its use.

2.1 Fundamental Concept

At the University of New Brunswick we teach Biochemistry majors computer assisted drug design. Our chemical modelling tool of choice is the Molecular Operating Environment, MOE™, produced by the Chemical Computing Group in Montreal, Canada. MOE is based on the Scientific Vector Language which is a *Collection Oriented Language*. Collection Oriented Languages are programming languages that operate on aggregate structures rather than single elements. Such aggregates can contain many atomic elements and generally allow for parallel processing. APL is one early implementation of this type of language which allows operations on aggregates themselves such as summing, multiplying, and reversing their elements and implements functionality like *apply to each* allowing for operations on each element of a collection. Since MOE permits the insertion of modules designed by the user in SVL it is essential our students are provided with tools that can reduce the challenge learning numerical algorithms and SVL have for non-programmers. While text-based languages may be suitable for regular programmers, they do not necessarily provide an adequate solution for non-programmers. One of the largest hurdles for non-programmers is learning language syntax. Non-programmers do not want to concern themselves with semicolons, comma placement, or for that matter, any other issue related to syntax. In addition, a text-based language mentally places the programmer into "editor" mode, rather than "creator" mode. This means that programmers are often more concerned with how an algorithm is implemented, rather than simply focusing on what the computer needs to do. This situation is akin to an author who writes with a computer, rather than with pen and paper. Using the computer, the author switches into "editor" mode, hindering creativity with its logical structure and as a result, causes a decrease in productivity. That is, the writer spends time trying to do things like craft perfect sentences, instead of concentrating on getting the main ideas out. Of course, it is reasonable to assume that a professional programmer will want to consider all aspects of an implementation, but in the research community, it is often better to get something up-and-running as soon as possible. In lead development, it is the *result* of the computation that is generally of most interest.

The paradigm that *Apt Apprentice* uses is designed to address these issues. It employs a direct manipulation interface, drawing on techniques from both text-based programming and programming by demonstration. The semantics of the environment attempt to mirror those found in a standard text editor, but also leverage the graphical interface to remove many syntax-related problems and facilitate program construction. It uses ideas from programming by demonstration to reduce the more technical aspects of algorithm implementation, allowing the programmer to concentrate on more abstract details. The environment makes use of XML to produce a language independent description of the graphically inputted algorithm. XML serves as the input to an interpreter, which translates the XML into a standard text-based language in this case SVL (see figure. 1). XML is used because of its extensibility and its self-descriptive nature. Interpreters can be built that will not be broken by future additions to the XML schema. The programming environment can be improved while maintaining backwards compatibility.

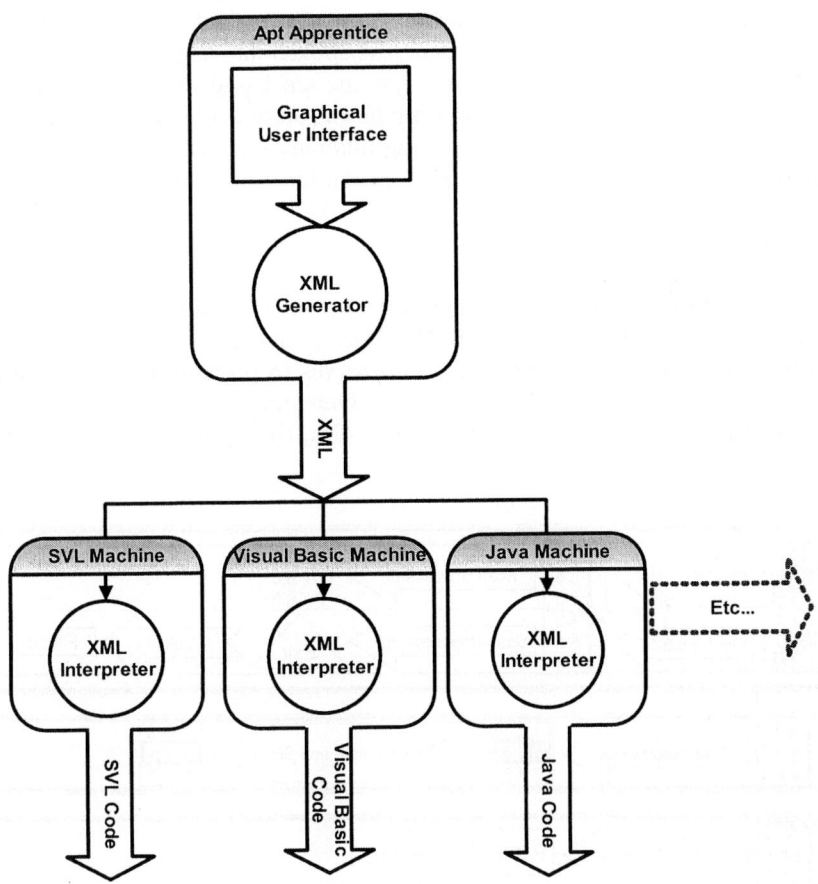

Fig. 1. XML view of the Apt Apprentice Architecture

Apt Apprentice, unlike many other end-user programming environments, is intended to serve as a procedure level programming tool. As such, it assumes a more sophisticated end-user than many of its counterparts, and therefore does not attempt to remove all aspects of programming from the end-user. By allowing the programmer to partake in the programming process, Apt Apprentice is able to offer the power of a standard text-based language, while at the same time using the graphical interface to keep this process at a higher level of abstraction. However, the design of such an environment requires the solution to its own set of problems.

2.2 Program Design

The first problem to be addressed in the design of Apt Apprentice is the issue of program flow, which is simply the order in which program statements are executed. In a text-based system, one expects a compiler (or person) to read a source code listing from top to bottom, moving from left to right. This is the natural way to read any document written using the spoken language of programmers[1]. Unfortunately, as direct manipulation systems are based in graphical environments, this natural restriction on program flow is often lost. Many attempts have been made to indicate program flow using different symbols connected together by arrows in some form of constructed diagram, but the syntax and semantics of these are more learned than they are natural. The largest obstacles are: where to begin and end the path through the diagram, and in some cases, which arrow to take at any given time. Apt Apprentice overcomes this limitation by introducing the concept of *rows*. The idea is straightforward; based on the contents of a given row, the program either progresses to the right within the row, or drops down to the row directly below (see figure 2). This serves to promote the same natural flow that is found in standard source code. One simply starts at the first row, traverses that row according to the proper rules, then proceeds on to the next row.

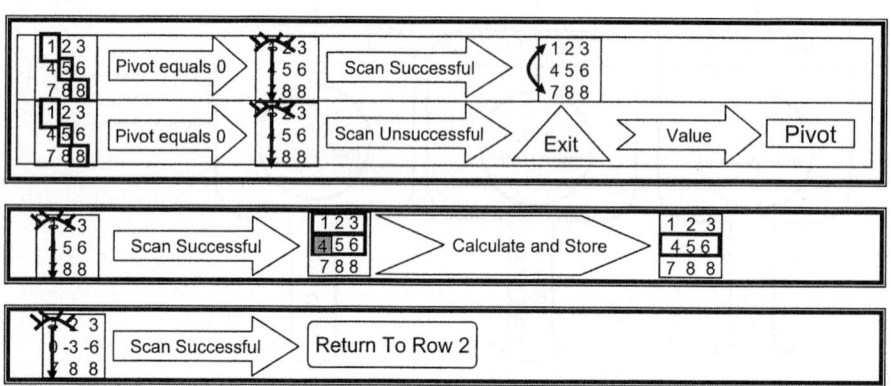

Fig. 2. Elements of Apt Apprentice depicted to show three complete rows of the determinant algorithm, including an explicit loop

Of course, once the concept of flow is introduced, it must be controlled. In most text-based systems this is accomplished through the use of *if-else* statements. This is where the difficulty level rises for the novice programmer in a

---

[1] Assuming a European based language is used.

text-based system, since the program statements are no longer executed in a strictly top-to-bottom order. The implementation of conditional statements has also proved to be a difficult task in direct manipulation interfaces, with many of them omitting them entirely [9]. Within Apt Apprentice, it is *conditionals* that regulate the rightward flow within a particular row. Conditionals are represented as right pointing arrows. In other words, as long as the conditionals evaluate to a value of true, flow continues on to the right. A conditional that evaluates to false causes execution to drop down to the next row. To supply the notion of an *else* statement there needs to be a plausible state between not moving right and dropping down to the next row. This is accomplished through the use of a *sub-row* notation. A sub-row is a further subdivision of a standard row. These divisions are color coded as a means to differentiate them. They are also subject to the same rules as a standard row. That is, the algorithm continues to the right if the conditional evaluates to true. This criterion will ensure that at any given time only a single sub-row can be completely traversed to the right. This notation is similar to a *case* statement in a traditional text-based language. The flow of control in the program is thus readily apparent to anyone viewing the algorithm, and does not change even when the nesting of conditionals occurs. This idea can be seen in figure 2.

In the development of Apt Apprentice the focus to this point has been primarily in the area of Linear Algebra. The use of matrices is often a requirement in implementing statistical procedures for QSAR. This has constrained the exploration of various notations to a limited subject matter, while still maintaining the relevance to computational chemistry.

To illustrate various features the example of calculating the determinant of a matrix using Gaussian elimination will be used. This algorithm can be found in any standard textbook on Linear Algebra. The user begins describing the algorithm by entering an example matrix, say: [[1,2,3],[4,5,6],[7,8,8]]. The first step of the algorithm involves examining the element in position (1,1) to determine that it is non-zero. If it is non-zero the element becomes the pivot element for this iteration. If it is zero an element below must satisfy the condition; or execution stops. The process is repeated for each pivot element found on the main diagonal of the matrix. Row 1 of figure 2 is Apt Apprentice's graphical representation of this process, with the exception of the stopping condition when the final pivot element is reached. To tell Apt Apprentice the sequence of pivots the programmer simply clicks on each element on the main diagonal. This guides Apt Apprentice in choosing the next pivot in each iteration. The idea of a guide is not new, and has been used in several implementations of programming by demonstration [9]. To our knowledge however, it has not been used before to depict a sequence of elements operated on by an algorithm. The *pivot* element is defined by the user and given a name. The system is apprised of the conditions regarding this element (i.e. that it not be equal to zero). After defining the guide, the system generates an arrow pointing to the right, which represents the situation when the pivot is equal to zero. The algorithm traverses to the right if and only if the pivot element is zero, otherwise either the next sub-row is chosen, or if there are no more sub-rows, the next main row. If the pivot is zero, a non-zero element must be found below in the same column. This is achieved in Apt Apprentice through the *scan* operation. Scanning may be performed in many different directions, but in this example we confine ourselves to downwards only. According to the algorithm, the scan result must also be non-zero. The system automatically adds a new right arrow to direct the algorithm if the scan is successful. The final step, once a non-zero element has been found, is to swap the original pivot row with the one uncovered by the scan. The *swap* operation is used to perform this task. The operation is automatically associated with the proceeding *scan*, and the user only need specify the other row involved in the swap. Dialog boxes are used to allow the user to select the appropriate item. The use of dialog boxes to aid in programming by demonstration systems has been well documented [9],[6],[2]. Finally, the system realizes that there is still one condition unspecified, that is, what should the system do if the *scan* is unsuccessful. The second sub-row of row 1 gives the proper course of action. For this condition to be evaluated, the pivot must be equal to zero; thus the system automatically replicates this portion of the upper sub-row. The user is then only required to show the system how to manage the case when the entire column is zeroes (i.e. there is no row available for a swap). In this case, the user indicates that the algorithm has finished through the use of the *exit* icon. The exit icon behaves like a standard *return* statement, allowing the user to indicate the value to produce at this point. The rightward *value* arrow is notational in nature and only serves to remind the user to progress to the right within the row. The user then provides the exit value, which in this example is simply the value contained in the pivot variable. Upon completion of the first row the pivot elements of the matrix have been defined and the algorithm has been told how to handle exceptional values. In essence, a nested conditional statement has been created.

To continue on to row 2 the pivot element must be non-zero. In other words, at least one conditional arrow in each sub-row of the proceeding row has evaluated to false. Row 2 describes the process of eliminating the remaining elements in the pivot column. According to the standard algorithm this is carried out through elementary row operations. In Apt Apprentice a *scan* operation is first used to find an element in the pivot column that is non-zero, and also not located at the position of pivot. Every variable has a position associated with it by default. Positions may be compared with operators such as @ and *not* @. If an element is found that matches this criteria, the algorithm continues to the right.

At this point Apt Apprentice must be shown how to carry out the required calculations. The notation used for this operation is a different style arrow which again points to the right. The user demonstrates the calculation to Apt Apprentice using appropriate drag and drop operations. The user shows how the element found by the *scan* is reduced to zero using an elementary row operation. A series of dialog boxes is used to connect the proper rows to the position of the pivot element and the scan result.

In row 3, a loop is introduced using a *return* operation. A *return to* works in exactly the way one would expect. The icon is placed on a row; then the row that the algorithm is to *return to* is selected using the mouse. This is the familiar **GOTO**. Forward **GOTO**s are not allowed. The responsibility falls to Apt Apprentice to generate code that avoids the usual pitfalls of using a **GOTO** statement.

To begin, a scan is performed to determine if there are any remaining non-zero elements in the column (excluding the pivot value). If there are non-zero elements, the algorithm returns to row 2. At the point when the scan is unsuccessful, the condition associated with the arrow evaluates to false and algorithm execution moves on to row 4. The remainder of the algorithm can be seen in figure 3. Row 4 defines the *finalpivot* position and a condition that *pivot* is not @ *finalpivot*. If this is the case, then the algorithm has not visited every pivot element, and so, should return to row 1. Once the *finalpivot* has been reached execution drops down to row 5. Here Apt Apprentice is once again directed to exit, however this time the exit value demonstrated is the product of the diagonal elements. Apt Apprentice maintains properties associated with specific operations such as the number of times the swap is called. Swapping rows of a determinant changes its sign so the pivot value is multiplied by (-1) raised to the power of the swapcount value. The result of this expression is the determinant of this particular matrix.

3 Conclusions

Apt Apprentice presents a hybrid method for end-user programming. It attempts to merge the feel of a standard text-based programming language with the ab-

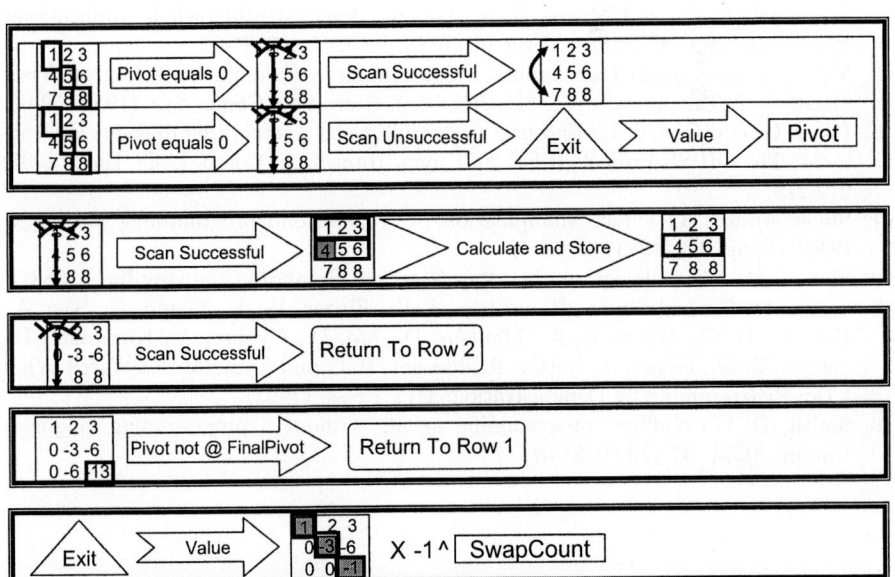

Fig. 3. Depiction of Apt Apprentice showing the complete determinant algorithm

straction offered by direct manipulation environments. The ultimate goal of Apt Apprentice is to be a system that enables computational chemists to extend applications, without requiring the activation of a computer science team. Until now, the focus for the programming capabilities of Apt Apprentice has been on standard computer science algorithms directed towards classroom use. Apt Apprentice assumes a more sophisticated end-user than many of its peers and so does not attempt to hide all aspects of the programming task. In the future, it will be necessary to develop a larger repertoire of graphical constructs as well as submit the system to extensive user testing.

Acknowledgments

Student PGSB support provided by the Natural Sciences and Engineering Research Council of Canada is gratefully acknowledged.

References

1. Amant, R. St., Lieberman, H., Potter, R., and Zettlemoyer, L.: Programming by example: visual generalization in programming by example. Comm. ACM. **43** (2000) 107-114.
2. Bocionek, S. and Sassin, M.: Dialog-Based Learning (DBL) for Adaptive Interface Agents and Programming-by-Demonstration Systems. CMU-CS-93-175, Carnegie Mellon University (1993) 1-51.
3. Chang, S., Korfhage, R. R., Levialdi, S., and Ichikawa, T.: Ten years of visual languages research. Proc. IEEE Sym. Vis. Lang. (1994) 196-205.
4. McDaniel, R. G. and Myers, B. A.: Gamut: demonstrating whole applications. Proc. 10th Ann. ACM Sym. User Int. Soft. Tech. (1997) 81-82.
5. Myers, B. A.: Visual Programming, Programming by Example, and Program Visualization: A Taxonomy. Proc. SIGCHI Conf. Hum. Fac. Comp. Sys. (1986) 59-66.
6. Patry, G. and Girard, P.: End-user programming in a structured dialogue environment: The GIPSE project. IEEE 2001 Sym. Hum. Cen. Comp. Lang. Env. (2001) 212-219.
7. Shneiderman, B.: Direct Manipulation: A step beyond programming languages. IEEE Comp. **16** (1983) 57-69.
8. Shu, N. C.: Visual Programming. Van Nostrand Reinhold Company Inc. (1988).
9. Smith, D. C., Lieberman, H., Witten, I. H., Finzer, W. F., Gould, L., Kay, A., Halbert, D. C., Myers, B. A., Maulsby, D., Mo, D., Cypher, A., Kurlander, D., Jackiw, R. N., Feiner, S., Potter, R., Piernot, P. P, and Yvon, M. P.: Watch What I Do: Programming by Demonstration. MIT Press. (1993).
10. Smith, D. C.: KidSim: programming agents without a programming language. Comm. ACM. **37** (1994) 54-67.

A Visual Interactive Framework for Formal Derivation

P. Agron, L. Bachmair, and F. Nielsen

[1] Department of Computer Science,
Stony Brook University, Stony Brook, New York
[2] Sony Computer Sciences Laboratory Inc,
Tokyo, Japan

Abstract. We describe a visual interactive framework that supports the computation of syntactic unifiers of expressions with variables. Unification is specified via built-in transformation rules. A user derives a solution to a unification problem by stepwise application of these rules. The software tool provides both a debugger-style presentation of a derivation and its history, and a graphical view of the expressions generated during the unification process. A backtracking facility allows the user to revert to an earlier step in a derivation and proceed with a different strategy.

1 Introduction

Kimberly is a software tool intended to be used in college-level courses on computational logic. It combines results from diverse fields: mathematical logic, computational geometry, graph drawing, computer graphics, and window systems. A key aspect of the project has been its emphasis on visualizing formal derivation processes, and in fact it provides a blueprint for the design of educational software tools for similar applications.

Applications of computational logic employ various methods for manipulating syntactic expressions (i.e., terms and formulas). For instance, *unification* is the problem of determining whether two expressions E and F can be made identical by substituting suitable expressions for variables. In other words, the problem requires one to syntactically *solve* equations $E \approx F$. For example, the two expressions $f(x, c)$ and $f(h(y), y)$, where f and h denote functions, c denotes a constant, and x and y denote variables, are *unifiable*: substitute c for y and $h(c)$ for x. But the equation $f(x, x) \approx f(h(y), y)$ is not solvable.[1]

Logic programming and automated theorem proving require algorithms that produce, for solvable equations $E \approx F$, a *unifier*, that is, a substitution σ such that $E\sigma = F\sigma$. Often one is interested in computing *most general unifiers*, from which any other unifier can be obtained by further instantiation of variables. Such unification algorithms can be described by collections of *rules* that are

---

[1] Note that we consider *syntactic* unification and do not take into account any semantic properties of the functions denoted by f and h.

designed to transform a given equation $E \approx F$ into *solved form*, that is, a set of equations, $x_1 \approx E_1, \ldots, x_n \approx E_n$, with variables x_i on the left-hand sides that are all different and do not occur in any right-hand-side expression E_j. The individual equations $x_i \approx E_i$ are called *variable bindings*; collectively, they define a unifier: substitute for x_i the corresponding expression E_i.

For example, the equation $f(x,c) \approx f(h(y),y)$ can be *decomposed* into two equations, $x \approx h(y)$ and $c \approx y$. Decomposition preserves the solutions of equational systems in that a substitution that *simultaneously* solves the two simplified equations also solves the original equation, and vice versa. We can *reorient* the second equation, to a variable binding $y \approx c$, and view it as a partial specification of a substitution, which may be applied to the first equation to yield $x \approx h(c)$. The two final equations, $x \approx h(c)$ and $y \approx c$, describe a unifier of the initial equation. The example indicates that in general one needs to consider the problem of simultaneously solving several equations, $E_1 \approx F_1, \ldots, E_n \approx F_n$.

A *derivation* is a sequence of sets of equations as obtained by repeated applications of rules. If transformation rules are chosen judiciously, the construction of derivations is guaranteed to terminate with a final set of equations that is either in solved form (and describes a most general unifier) or else evidently unsolvable. Unsolvability may manifest itself in two ways: (i) as a *clash*, i.e. an equation of the form $f(E_1, \ldots, E_n) \approx g(F_1, \ldots, F_k)$, where f and g are different function symbols, or (ii) as an equation $x \approx E$, where E contains x (but is different from it), e.g., $x \approx h(x)$. The transformation rules we have implemented in *Kimberly* are similar to those described in [8].

In Section 2 we describe the visual framework for derivations. The current version of *Kimberly* provides graphical representation of terms and formulas as *trees*, but has been designed to be augmented with additional visualization functionality for representation of *directed graphs*, as described in Section 3. We discuss some implementation details in Section 4 and conclude with plans for future work.

2 Visual Framework for Derivations

Kimberly features a graphical user interface with an integrated text editor for specifying sets of equations. The input consists of a set of (one or more) equations, to be solved simultaneously. Equations can be entered via a "source" panel, see Figure 1. They can also be saved to and loaded from a text file. For example, the upper section of the window in Figure 1 contains the textual representation of three equations to be solved, $f(a, h(z), g(w)) \approx f(y, h(g(a)), z)$, $g(w) \approx g(y)$, and $s(z, s(w, y)) \approx s(g(w), s(y, a))$. Note that in *Kimberly* we distinguish between variables and function symbols by enclosing the former within angle brackets.

Once the initial equations have been parsed, the user may begin the process of transforming them by selecting an equation and a rule to be applied to it. The transformation rules include decomposition, orientation, substitution, elimination of trivial equations, occur-check, and detection of clash (cf. [8]), each of

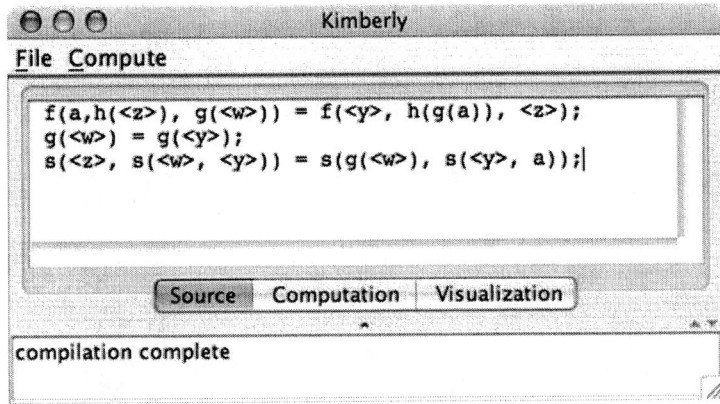

Fig. 1. The upper section of the window contains a tabbed view to manage panels named according to their functionality. The source panel, shown, provides text editing. The lower section is used to display messages generated by the application

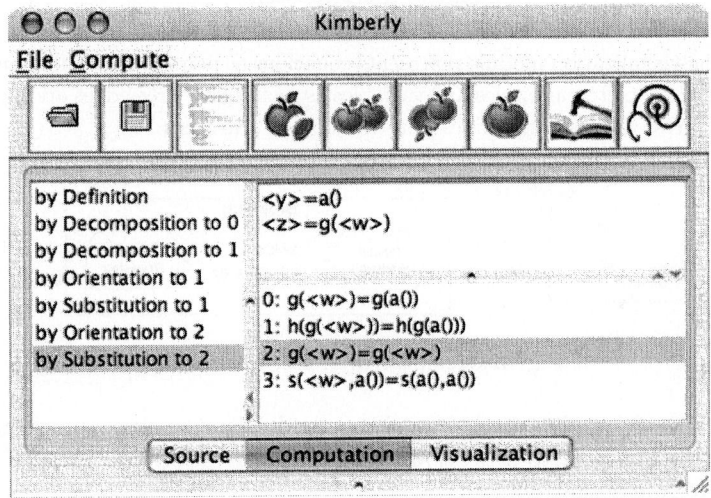

Fig. 2. The computation panel presents a debugger-style view of a derivation. The left section of the panel shows the history of rule applications. The right section lists the current equations, with variable bindings listed on top and equations yet to be solved at the bottom

which is represented by a button in the toolbar of the "computation" panel, see Figure 2. If an invoked rule is indeed applicable to the selected (highlighted) equation, the transformed set of equations will be displayed and the history of the derivation updated.

The derivation process is inherently nondeterministic, as at each step different rules may be applicable to the given equations. The software tool not only keeps

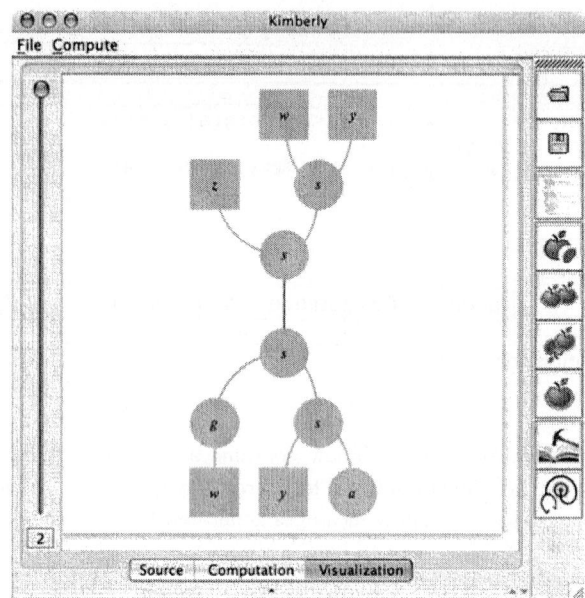

Fig. 3. The visualization module represents equations as pairs of vertically joined trees. Functions are depicted as circles and variables as rectangles. Tree edges are represented by elliptic arcs, while straight lines join the roots of two trees

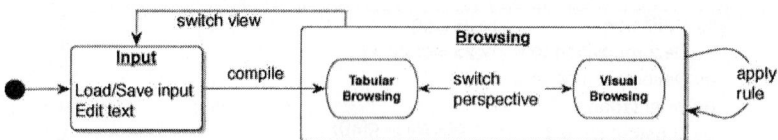

Fig. 4. Organization of user interaction

track of all rule applications, but also provides backtracking functionality to allow a user to revert to a previous point in a derivation and proceed with an alternative rule application.

Kimberly features a "visualization" module that supplies images of a trees representing terms and equations, and the user may switch between a textual and a graphical view of the current set of equations. The browser allows a user to view one equation at a time, see Figure 3.

The design of the user interface has been guided by the intended application of the tool within an educational setting. User friendliness, portability, and conformity were important considerations, and the design utilizes intuitive mnemonics and hints, regularity in the layout of interface elements, and redundant keyboard/mouse navigation. The transition diagram in Figure 4 presents a high-level view of user interaction.

Kimberly is a single document application, with all the user interface elements fixed inside the main window. Addition of a new browser to the application in-

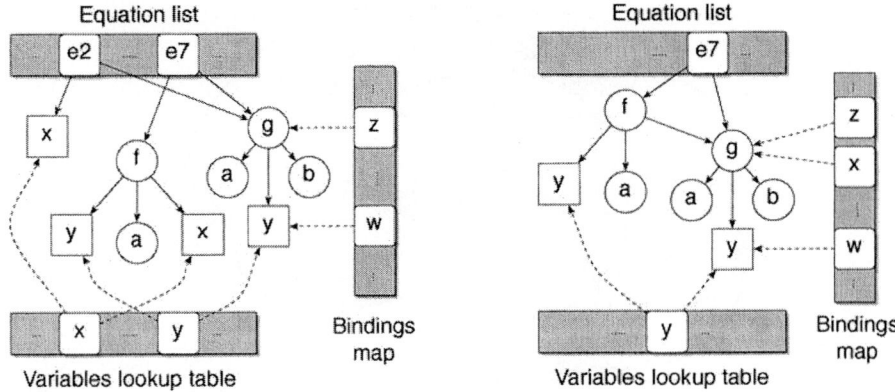

Fig. 5. Application of the substitution rule

volves only the introduction of a panel to host the new browser. The browsing architecture is designed to reflect interaction with the active browser, e.g., backtracking to an earlier state, to the remaining browsing modules.

For efficiency reasons we internally represent terms and equations by directed acyclic graphs, and utilize various lookup tables (dictionaries) that support an efficient implementation of the application of transformation rules. Specifically, we keep those equations yet to be solved in an "equation list," separate from variable bindings, which are represented by a "bindings map." An additional *variables lookup table* allows us to quickly locate all occurrences of a variable in a graph. The latter table is constructed at parse time, proportional in size to the input, and essential for the efficient implementation of applications of the substitution rule.

Figure 5 demonstrates the effect of an application of the substitution rule on the internal data structures. The rule is applied with an equation, $x \approx g(a, y, b)$, that is internally represented as *e2* on the equation list in the left diagram. The effect of the substitution is twofold: (i) the equation *e2* is removed from the equation list and added (in slightly different form) to the bindings map and (ii) all occurrences of x are replaced by $g(a, y, b)$, which internally causes several pointers to be redirected, as shown in the right diagram.

3 Visualization of Directed Graphs

Many combinatorial algorithms can be conveniently understood by observing the effect of their execution on the inherent data structures. Data structures can often be depicted as graphs, algorithmic transformations of which are naturally suited for interactive visualization. Results on graph drawing [1] can be combined with work on planar labeling for animating such dynamics. At the time this article was written no public domain software packages were available that were suitable for our purposes, therefore we decided to develop a novel method for browsing animated transformations on *labeled* graphs.

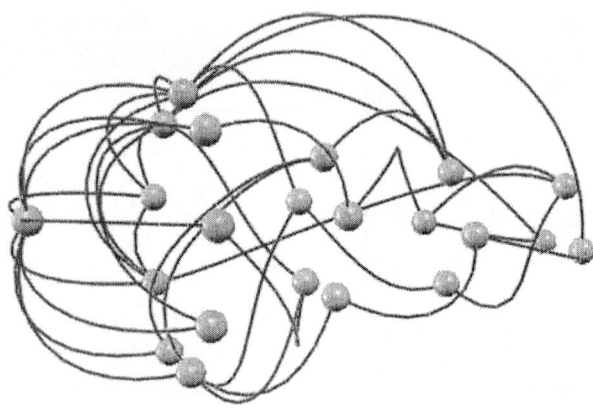

Fig. 6. A 3D layout of a randomly generated directed acyclic graph

We conducted a number of experiments in interactive graph drawing to equip *Kimberly* with an adequate visualization module for the representation of directed graphs. In particular, we studied issues of graph layout and edge routing (and plan to address labeling and animation aspects in future work).

Our graph drawing procedure is executed in two stages. The *layering step*, which maps nodes onto *vertices* (points in $\mathbb{R}^3$), is followed by the *routing step*, which maps edges onto *routes* (curves in $\mathbb{R}^3$).

In [2] it was shown that representing edges as curved lines can significantly improve readability of a graph. In contrast to the combinatorial edge-routing algorithm described in [2] (later utilized in [5] to address labeled graphs) our algorithm is iterative and consequently initialization sensitive. We treat vertices as sources of an outward irrotational force field, which enables us to treat routes as elastic strings in the solenoidal field (routes are curves that stay clear of the vertices). A route is computed by minimizing an energy term that preserves both length and curvature. Our approach is similar to the "active contours" methods employed for image segmentation [7, 9].

The layering step involves computing a clustering of a directed acyclic graph via topological sorting. We sorted with respect to both minimum and maximum distance from a source vertex and found that the latter approach typically produces significantly more clusters than the former. We map clusters onto a series of parallel circles centered on an axis, see Figure 6.

4 Implementation

The design of *Kimberly* is centered around providing multiple views of the derivation process. A flexible architecture allows for straightforward introduction of new visualization modules (*perspectives*). To achieve the flexibility we have followed the canonical MVC (Model-View-Controller) paradigm [4–pp 4-5]. The *model* object of the application keeps the history of rule applications and information on the currently selected step in the derivation and the selected equation. The *controller* object inflicts rule-firings on the model and notifies the perspectives of the changes. Perspectives are automatically kept in synch with the model; A perspective may, for example, change selection parameters of the model which would mechanically cause other perspectives to reflect the changes.

Images generated during the graph drawing stages are cached, as in-time image generation may take unreasonably long and adversely affect user experience. Memory requirements for storing the images may be very high, and hence images are stored compressed. An alternative approach would be to rely on a customized culling scheme under which only the elements in the visible region (intersecting the viewport) are drawn, using fast rectangle intersection algorithms and other computational geometry techniques.

For the 3D experiments we built an application with animation and simulation processes running in separate threads, enabling us to adjust simulation parameters such as step size while observing the effects in real time. Routes are implemented as polylines and computed by iteratively displacing the joints (mass points) under the sum of internal and external forces until the displacements become negligible. Motion of mass points is considered to be in a viscous medium and is therefore calculated under the assumption of $F = m\frac{dx}{dt}$. Segments of the polyline are treated as strings with *non-zero* rest length to encourage an even distribution of the mass points. At each step forces are scaled appropriately to ensure there are no exceedingly large displacements. Once the visualization scheme is fully understood we plan to base a *perspective* on it and make it an integral part of *Kimberly* .

5   Future Work

Further experimentation is needed to address the labeling and animation of graph transformations. We plan to use the same edge-routing algorithm to compute trajectories for animation of graph transformations. An advantage of drawing graphs in 3D is that edge intersections can always be avoided. But cognitive effects of shading and animation on graph readability need to be further investigated. Although the extra-dimensional illusion is convenient for setting up interaction, it is apparent that bare "realistic" 3D depictions are not sufficient.

Herman et al. [6] point out that while its desirable to visualize graphs in 3D because the space is analogous to the physical reality, it is disorienting to navigate the embeddings. Our research suggests that highly structured 3D layouts are more beneficial in this sense than straightforward extensions of the corresponding 2D methods because they are less disorienting. While our layout possesses axial symmetry, one can imagine a decomposition into spherical or toroidal clusters. The authors of [6, 3] corroborate our conviction that there is no good method for viewing large and sufficiently complex graphs all at once, and that interactive exploration of a large graph is indispensable. A key issue is how to take advantage of the graphics capabilities common in today's computers to increase the readability of graphs.

Once the directed-graph browser has been integrated in the existing tool, the next step would be to consider applications to other derivation-based methods. Possible educational applications include grammars, as used for the specification of formal languages. The current system focuses on the visualization of derivation-based methods; future versions are expected to feature more extensive feedback to a user during the derivation process.

References

1. G. Di Battista, P. Eades, R. Tamassia, and I. Tollis. *Graph Drawing*. Prentice Hall, 1998.
2. David P. Dobkin, Emden R. Gansner, Eleftherios Koutsofios, and Stephen C. North. Implementing a general-purpose edge router. In *Proceedings of the 5th International Symposium on Graph Drawing*, pages 262–271. Springer-Verlag, 1997.
3. Irene Finocchi. *Hierarchical Decompositions for Visualizing Large Graphs*. PhD thesis, Universita degli Studi di Roma "La Sapienza", 2002.
4. E. Gamma, R. Helm, R. Johnson, and J. Vlissides. *Design Patterns: Elements of Reusable Object-Oriented Software*. Addison-Wesley, 1995.
5. Emden R. Gansner and Stephen C. North. Improved force-directed layouts. In *Proceedings of the 6th International Symposium on Graph Drawing*, pages 364–373. Springer-Verlag, 1998.
6. I. Herman, G. Melançon, and M. S. Marshall. Graph visualization and navigation in information visualization: A survey. *IEEE Transactions on Visualization and Computer Graphics*, 6(1):24–43, 2000.
7. M. Kass, A. Witkin, and D. Terzopoulos. Snakes: Active contour models. *International Journal of Computer Vision*, 1(4):321–331, 1987.
8. R. Socher-Ambrosius and P. Johann. *Deduction Systems*. Springer, 1996.
9. C. Xu and J. Prince. Snakes, shapes, and gradient vector flow. *IEEE Trans. Image Processing*, 7(3):359–369, 1998.

ECVlab: A Web-Based Virtual Laboratory System for Electronic Circuit Simulation

Ouyang Yang, Dong Yabo, Zhu Miaoliang,
Huang Yuewei, Mao Song, and Mao Yunjie

Lab of Networking Center, Collage of Computer Science and Technology
Zhejiang University, Hangzhou 310027, P.R.China
{lily, dongyb, zhum, yuewei, mason, myjesky}@zju.edu.cn

Abstract. In this paper, we describe the design and implementation of a web-based electronic circuit simulation system named ECVlab. This system combines technologies as rich client technology, XML, and circuit simulation, and provides the user with vivid interface, convenient operation and powerful simulation capability. At the moment, ECVlab has been implemented and successfully applied in an undergraduate course in Zhejiang University and it was evaluated by undergraduate students. Students made statistically significant learning gains as a result of using ECVlab, and rated them positively in terms of improving operation capability, enhancing interests for digital circuit and offering opportunity to inspire innovation and exploration.

1 Introduction

Nowadays, there is an increasing interest in web-based education for its prominent advantages of system opening and resources reusability. At universities, electronic circuit experiment is a vital part of higher education since it is a valuable method of learning which gives a learner the feeling of involvement. It can practice students' operation ability and also provide a good opportunity for them to put into practice what they've learned in class. And the learning process can be best facilitated if tools and models are freely and widely available, not just in the dedicated laboratories. Web-based simulation environments, combining distance education, group training and real-time interaction, can serve as a good approach.

Simulation is a process, during which experiments are conducted on a designed real system model for the purpose of understanding the behavior of the system or evaluating various strategies for the operation of the system. The power of simulation is the ability to model the dynamics of a real system and to analyze the results [6]. Meanwhile, the availability and interactive nature of web-based simulation serves as a good medium for students to experience the complexity of collaborative work.

This paper is to present a web-based virtual laboratory system for electronic circuit simulation (ECVlab). In this system, students can conduct experiment through web browser using virtual apparatuses like multimeter and equipments like resistance and tolerance.

The rest of this paper is organized as follows. In section 2, we list some related Web-based simulations. The design and structure of the vlab system is specified in section 3. In section 4, we demonstrate an example of applying our system. Later, an analysis of survey in undergraduate students is presented in section 5. Finally, in section 6, we draw conclusions of this work and point out some directions of future work.

2 Related Works

Web-based simulation represents the combination of rapidly-developing internet & multimedia technology and simulation science. The ability of web technologies enhances the power of simulation in that web can service large simulation communities and allow developers to distribute models and simulations to end users. Since Fishwick [4] started to explore Web-based simulation as a new research area in the field of simulation, many Web-based simulation models and support systems have been developed.

Kuljis and Paul [2] introduced a variety of new technologies for discrete-event Web-based simulation technologies and reviewed related environments and languages too. John C. Waller and Natalie Foster [1] designed a virtual GC-MS (Gas Chromatography / Mass Spectrometry) by means of copying presentation of monitor to the screen. Students can operate this virtual instrument via web and the real instrument can be used to take more valuable experiment. Cheng K.W.E. et.al [3] demonstrated how a power electronics experiment is programmed in a remotely controlled laboratory setup. Huang and Miller [5] presented a prototype implementation of a Web-based federated simulation system using Jini and XML.

Some of the above simulation systems operate the experimental equipment through remote control and monitor by web-based tools. Due to the involvement of real equipment, limitation of cooperation and risk of damaging certain equipment still exist. The application of Java development tools and JavaBeans technology are widely used in several simulation systems. However, the interface developed by java applet is lack of intuitive and vivid expression. The major contribution of our work is to design a rich client simulation system which is totally independent of real experiment environment.

3 ECVlab System

3.1 Design Objectives

The ECVlab system is based on the project supported by a grant from the National Key Technologies R & D Program of China. As an assistant experiment education system which plays an important role in distance-learning, it should be provided with characteristics such as relatively low requirement of system resources, powerful collaboration capability between user and server, platform flexibility, high performance of transform and communication on internet and so on. Furthermore, the system should provide students with real expression.

3.2 System Architecture

The ECVlab system uses an Browser/Server system architecture, which is shown in figure 1.

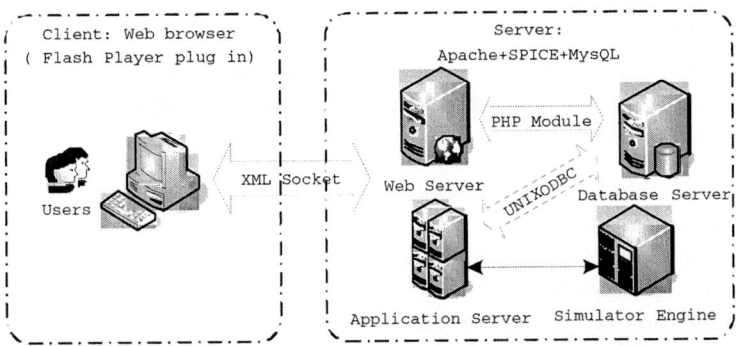

Fig. 1. ECVlab System Architecture

The server side runs on Linux system and can be divided into 3 parts: web server, application server and database server. The web server is responsible for user authentication, establishment of the virtual experiment framework, maintenance of user information; the application server manages the communication with multiple online flash client users, transfers parsed or encapsulated data to the spice simulator and return the results to client side. In ECVlab, information about user and experiments are stalled in database server, while the simulation logic is contained in application server.

On the server side, simulator engine plays an important roll in the whole system and we utilize SPICE (Simulation Program with Integrated Circuit Emphasis) to meet our simulation requirements. SPICE is a general-purpose circuit simulation program for nonlinear dc, nonlinear transient, and linear ac analysis which enables server to simulate analog circuit and print output in ASCII text.

The client side consists of the web browser and the embedded flash. Flash is a multimedia graphics program especially for use on the web. Compared with Animated images and Java applets, which are often used to create dynamic effects on Web pages, the advantages of Flash are:

– Flash loads much faster than animated images
– Flash allows interactivity, whereas animated images do not
– Flash can create a real scene , whereas Java applets can't
– Flash edition 2004 supports the object-oriented technology which enables us to develop powerful interactive web animation

With these features, we use Flash to construct the virtual experiment environment. More detailed discussion about the interface will be shown in section 4. When a user starts an experiment, the client side firstly logins to the server through web browser and after certification, the user logins to the ECVlab system. After entering certain experiment, virtual experiment environment downloads automatically from the web browser and runs. During the experiment process, the environment consistently communicates with the application server to exchange data and display the simulation results.

3.3 Communication Protocol

An expandable and flexible communication protocol between server and client plays an important role in a well-designed web-based Vlab system. In our system, the protocol is used to transport following information:

- 1) Information about experiment and user, such as user name and password, experiment id and so on,
- 2) Parameters of components like value of resistance and apparatuses like pace of oscillograph,
- 3) Simulation results returned from simulator engine.

We use XML to construct the communication protocol. XML is a markup language for documents containing structured information and is designed to describe data that allows the author to define his own tags document structure. With this powerful data-exchange language, we can construct our own protocol by defining tags related to parameters of apparatus and simulation results. The form of ECVlab communication protocol is shown in figure 2.

The communication model of the ECVlab is shown in figure 3. The ECVlab system can be conveniently layered into four abstract levels. When the circuit

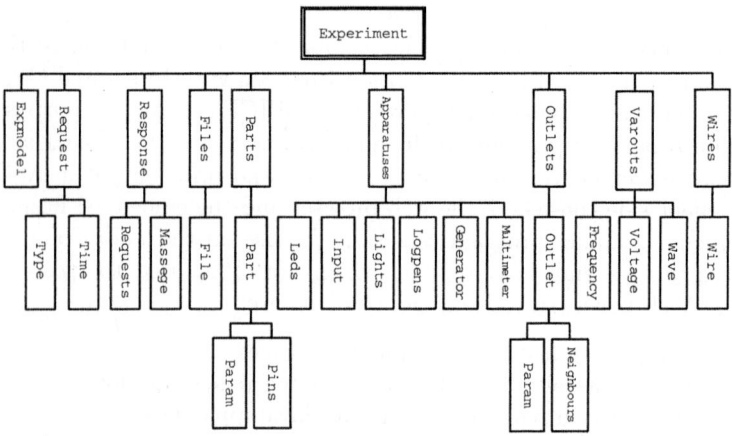

Fig. 2. ECVlab Communication Protocol

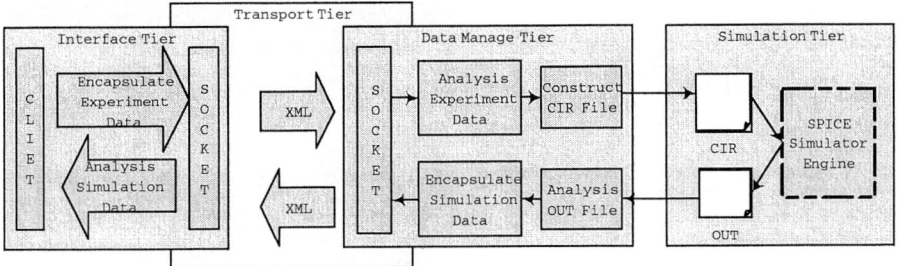

Fig. 3. Communication Mechanism of ECVlab

began to simulate, the circuit description file which contained circuit parameters and control information was encapsulated in XML file and transferred from interface layer to the data manage layer by socket. Later the submitted data was transferred into the formatted data needed by simulator engine and created a SPICE supported input file (.cir). After the simulator engine worked out the results and stored them in a .out file, the data management layer was responsible for analyzing the output and encapsulating required information into XML file similarly. Finally, the client converted the results into the form shown in apparatus such as the waveform shown in oscillograph and the voltage value in multimeter.

4 An Example of Using the ECVlab

We select one experiment of analog circuit in the undergraduate courses named Operation Amplifier (OPAMP). Through this experiment, students learn to use operation amplifier, resistance and capacitor to construct analog operation circuit and analyze the function of this circuit. In figure 4, the schematic diagram of OPAMP is shown in a), and b) demonstrates the equivalent circuit built by user in the virtual circuit environment.

Figure 5 shows a snapshot of the interface in ECVlab and demonstrates simulation results of the OPAMP. On the left hand, the system displays the tool bar which contains buttons for different operations. When a user starts to construct a circuit, he can add necessary components with appointed value to the circuit from the tool bar. For example, in OPAMP, user can add one operation amplifier and several resistances. After that, user needs to connect lines between these components. ECVlab enables the user to add or delete lines from this virtual circuit board freely and change the color of lines to differentiate between each others. In order to modify the circuit parameters and obtain the results of the experiment, user should use several apparatuses such as multimeter, oscillograph, and signal generator. In our system, the user can choose which of these apparatuses to be displayed on screen, and can zoom out or zoom in or move around at will. Furthermore, all these virtual apparatuses are highly similar to the real ones and the user can operate the buttons on them to modify parame-

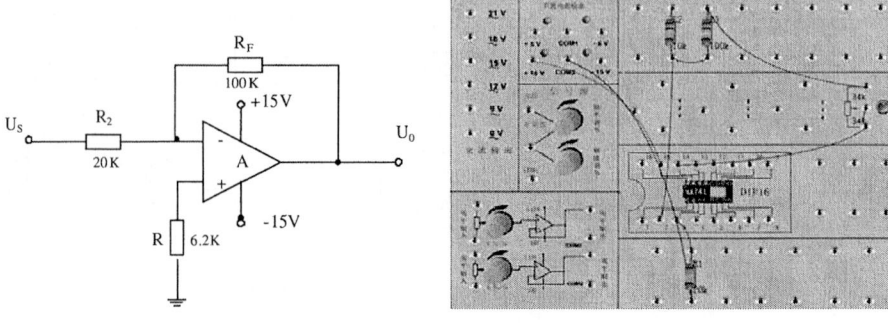

(a) Schematic Circuit Diagram (b) Equivalent Virtual Circuit

Fig. 4. OPAMP Electronic Circuit

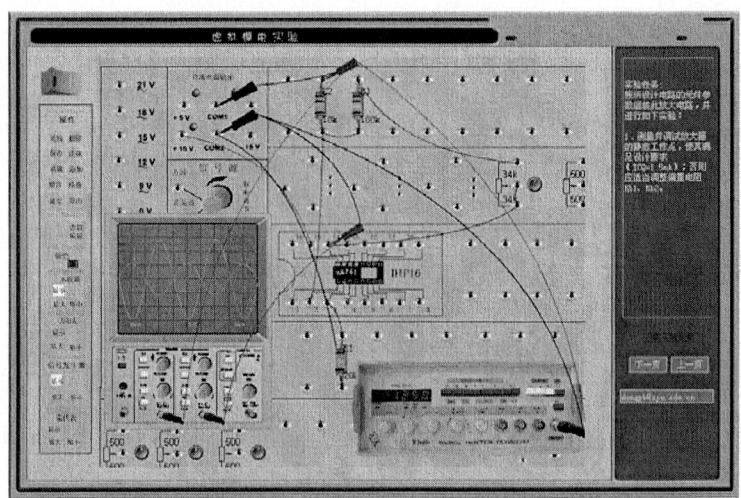

Fig. 5. Interface of ECVlab

ters of the circuit. On the right hand lists the guidance about this experiment, the user can check it anytime during the process.

Once the circuit has been constructed, user can click the power button on the left corner. The ongoing status and results sent back from server will be displayed in related virtual apparatuses. As shown in figure 5, we can clearly see the wave of the input signal and the amplified output signal on the screen of oscillograph as a result of the OPAMP. If the user adjusts the parameters of circuit or apparatus, for example, changes the signal frequency of the signal generator, the displayed result will be refreshed in real time.

The ECVlab also provides the user with convenient load and save mechanism to keep the status of the circuit which facilitates the experiment process a lot.

5 Implementation

Presently, the ECVlab has been implemented and successfully applied to the students as an undergraduate course in Zhejiang University. In order to evaluate the effects, we conduct a survey among the students who have used the system for a semester. According to the results, around 87% of the students think the ECVlab is helpful to their study, concerning both theory and experiments. Figure 6 demonstrates the analysis of the survey about how does the ECVlab facilitate in Electronic Circuit study.

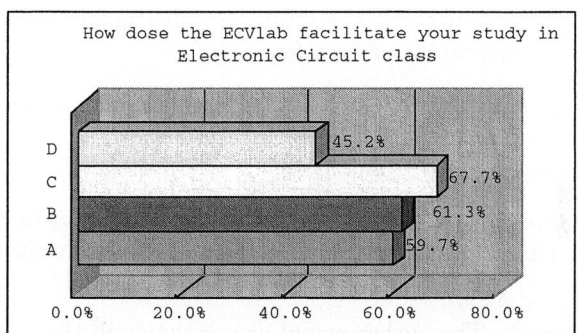

Fig. 6. Analysis of the Questionnaire

From the histogram, we can see that around 45.2% students agree that the ECVlab helps them to comprehend what they have learned from the class. Most of them (67.7%) mention that by using the ECVlab, they can improve their capability of operation and experiment. While 61.3% students consider it as a good way to improve their interests in digital circuit. Around 59.7% think the ECVlab offers them opportunity to inspire innovation and exploration. Altogether, up to 95% students hold positive attitude towards the ECVlab.

6 Conclusions and Future Work

Web-based education is becoming more and more popular. In this paper, we described the design and implementation of a web-based Vlab system for Electronic Circuit simulation. This ECVlab demonstrates not only a successful integration of web servers, simulation servers, database servers, and flash-based client, but also a successful exportation to the great potential of web-based Vlab in web-based education. It can help the students to learn electronic circuit simulation by using the following features:

- Real impression and easy operation of the experiment environment.
- Immediate execution of the simulation process and direct access to the results.

- Viewing and editing the circuit template through convenient saving and loading mechanism
- No harm to the experimental equipment compared with conducting in laboratory.

The design and implementation of the ECVlab is open to improvement. The virtual environment for digital circuit is under construction which is based on the XSPICE simulator engine. The main difference between analog circuit and digital circuit is the simulation mode. In digital circuit, the circuit state is in a incessant mode and the simulator engine needs to continuously receive circuit data and calculate the result, thus to design a arithmetic for incessant simulation is essential to our future work. In the future we will also focus on developing the performance of the system, to name a few:

- ECVlab management system including experiments management, online user management and analysis of experiment results
- New load balancing algorithms based on load average, I/O statistics.

With rapid development of Grid computing and Web services in resent years, we plan to add data grid and web services to the ECVlab in future.

Acknowledgement. The author would like to thank all the colleagues in this project. The supported of this project by a grant from the National Key Technologies R & D Program of China is also gratefully acknowledged.

References

1. John C.Waller and Natalie Foster. Training via the web: A virtual instrument. *Computers& Education*, pages 161–167, 2002.
2. Jasna Kuljis and Ray J.Paul. A review of web based simulation:whither we wander. *Proceedings of the 2000 Winter Simulation Conference*, pages 1872–1881, 2000.
3. Cheng K.W.E and Chan C.L. Virtual laboratory development for teaching power electronics. *Power Electronics Specialists Conference ,IEEE 33rd Annual*, pages 461 – 466, 2002.
4. P.Fishwick. Web based simulations: Some personal observations. *Proceedings of the 1996 Winter Simulation Conference*, pages 772–779, 1996.
5. X.Huang and J.Miller. Building a web based federated simulation system with jini and xml. *Proceedings 34th AnnualSimulation Symposium*, pages 143–150, 2001.
6. Xiaorong Xiang Yingping Huang and Gregory Madey. A self manageable infrastructure for supporting web based simulation. *Proceedings of the 37th Annual Simulation Symoposium*, pages 149–156, 2004.

MTES: Visual Programming Environment for Teaching and Research in Image Processing

JeongHeon Lee, YoungTak Cho, Hoon Heo, and OkSam Chae

Department of Computer Engineering, KyungHee University,
Seochun-ri, Kiheung-eup, Yongin-si, Kyunggi-do, South Korea
opendori@paran.com, greizen@vision.khu.ac.kr, hhoon@naver.com,
oschae@khu.ac.kr

Abstract. In this paper, we present a visual-programming environment for teaching and research referred to as "MTES". MTES(Multimedia Education System) is the system designed to support both lecture and laboratory experiment simultaneously in one environment. It provides tools to prepare on-line teaching materials for the lecture and the experiment. It also provides a suitable teaching environment where lecturers can present the online teaching materials effectively and demonstrate new image processing concepts by showing real examples. In the same teaching environment, students can carry out experiments interactively by following the online instruction prepared by lecturer. In addition to support for the image processing education, MTES is also designed to assist research and application development with visual-programming environment. By supporting both teaching and research, MTES provides an easy bridge between learning and developing applications.

1 Introduction

Image processing plays important roles not only in the areas of object recognition and scene understanding but also in other areas including virtual reality, databases, and video processing[1][2]. With the growing demand for engineers with knowledge of image computation, the number of schools with undergraduate elective courses on topics related to image computation, such as image processing and computer vision, is increasing rapidly and many educators are trying to find an effective method of teaching such courses. Some faculties even tried to integrate the image computation problems into the core curriculum as examples. However, the image computation involves the image, which is difficult to describe verbally, and the theory is defined on the strong mathematical basis. In the previous studies on image processing education, researchers point out the followings [3]:

- The image processing theory can be explained most effectively by visual means.
- It is necessary to complement the theory with computer-based experiments. Some researchers found that the laboratory exposure before the theoretical

treatment begins is most effective. It helps students understand the theory more clearly and easily.
- Hands-on experiment is essential to demonstrate the image computation concepts through examples.
- Image processing industries want to hire the students exposed to extensive laboratory work to be prepared for real applications.

The researchers also recommend to develop courses more accessible to students with computer science background who are more familiar with computer programming than applied mathematics, which is the basis of most image processing and pattern recognition theory.

Taking these into consideration, many researches have been done to develop the teaching environment that many lecturer and students in the field of image computation want to have. Those researches can be divided into two groups. The first group is the library systems with simple execution tools such as menus and script languages. The library consists of visualization functions, basic image processing functions, and interface functions defined to access image data [2][4][5]. These library systems are designed to simplify laboratory experiment by providing basic functions frequently required for the experiment. However, these library systems do not provide the programming environment with which student can implement and analyze their ideas without writing complex programs. The researches in the second group tried to solve these problems by taking advantages of the general research tools [2][6-10]. These systems are designed to support research and education. They are equipped with simple programming tools, such as a visual-programming tool and an interpreter, and various image analysis tools. With those, users can easily write a code to test their idea. However, most of these systems are designed for idea verification not for algorithm development. The structures of image processing functions in those systems are not transparent to the users [2]. And none of these systems provide either the tools to prepare online materials for lecture and experiment or the tools to present them simultaneously in the same environment.

In this paper, we propose a new visual-programming environment, referred to as "MTES", supporting both lectures and laboratory for image computation. MTES provides lecturers tools to create and register viewgraphs for the lectures and the environment to present them to the students with real examples. It is also designed as an efficient environment for research and application development by supporting the development of true object-oriented algorithm in image processing, systematic algorithm management, and easy generation of application.

To support easy creation of reusable image processing algorithms, MTES provides data classes with resources for essential image processing operations. With class libraries and online function management tools, users create an image processing software component that can be recognized as an icon on a function icon window. To promote the reuse of the user-defined functions, MTES provides an intelligent visual-programming environment which allows users to create their own image processing applications by simply dragging functions from a function icon window and dropping to a visual workspace. For the transparency of codes, the source code of the data class library and basic image processing functions pre-registered in MTES are open to users.

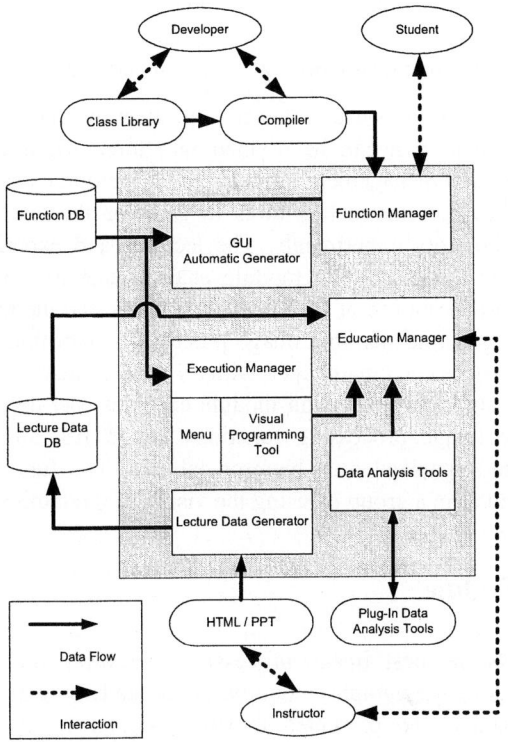

Fig. 1. Overall structure of MTES

2 Design of Visual Programming Environment for Teaching and Research of Image Processing

The requirements for an ideal visual-programming environment for teaching and research have been studied by many researchers and some of them are summarized in the previous section. In this paper, we present a new visual-programming environment designed to best fit for those requirements. In the process of designing the system, we tried to meet the following goals.

- Lecture should be given with real world examples supported by various analysis tools that can visualize the concept of the theory in the lecture.
- Lecture and hands-on experiment for the lecture should be carried out simultaneously in the same environment.
- Students should be able to easily program and test their algorithms.

- Students should be able to create new applications by making use of user-defined functions and existing functions.
- The environment should provide instructors an easy way to prepare and register lecture materials.
- The environment should be easy to learn so as to minimize "learning overhead".
- The environment should support both research and teaching.

The overall structure of the system, designed to meet the above goals is shown in Fig. 1. The proposed system can be divided into three parts: education module, algorithm generation and management module, and algorithm execution module. The education module provides instructors tools to generate the teaching materials for both lecture and laboratory experiments. The lectures and experiments are tightly integrated. The algorithm generation module enables user to create reusable user-defined functions by making use of commercial compiler and the image class library, which is defined to generate sharable image processing algorithm. The user-defined functions can be registered and maintained with all the information necessary to make reuse of them. The algorithm execution module executes user-defined functions and pre-registered basic image processing functions. In MTES, all the functions are represented as icons organized as a hierarchical tree. The functions are executed individually by menu or in a group by using the visual-programming environment.

3 Education Module

To provide students the best image processing education, the teaching material consisting of the lecture viewgraphs, examples, and hands-on experiments should be integrated to an e-book and presented to students interactively. The educational environment is divided into two parts: a lecture generation module and a learning module. The lecture generation module provides instructors the tools necessary to create the interactive teaching material. The learning module enables students to study new theory with the on-line lecture viewgraphs, interactively follow through the examples, and implement a new algorithm and apply it to solve real world problem.

To create the integrated teaching material, a lecturer prepares lecture viewgraphs and examples separately by using the commercial software of viewgraph generation like MS-PowerPoint and MTES's visual-programming tool. And then integrate them in the form of an e-book as shown in Fig. 2. MTES provides the functions necessary for the integration as shown in the pop-up menu in Fig. 2.

The teaching material may have the structure of the conventional paper-based textbook as shown in Fig. 2. However, a chapter is defined to cover one week of class work. Each chapter consists of lecture, example, and practice all of which can be covered in a week. Fig. 3 shows a lecture screen with the example menu activated. Student and instructor go through the viewgraphs to study a new topic and then students can carry out experiments interactively by either selecting an example on the example menu or a practice problem on the practice menu.

MTES: Visual Programming Environment for Teaching and Research

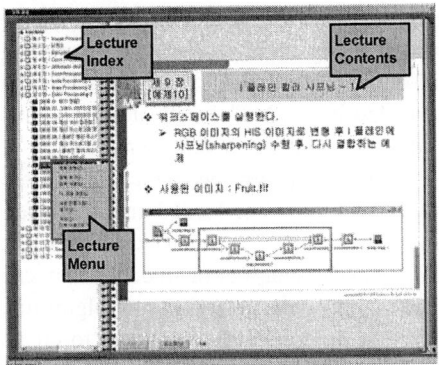
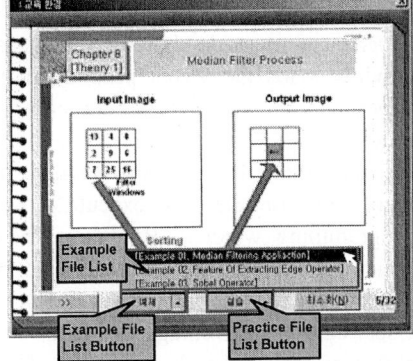

Fig. 2. Lecture screen with the command menu

Fig. 3. Lecture Screen

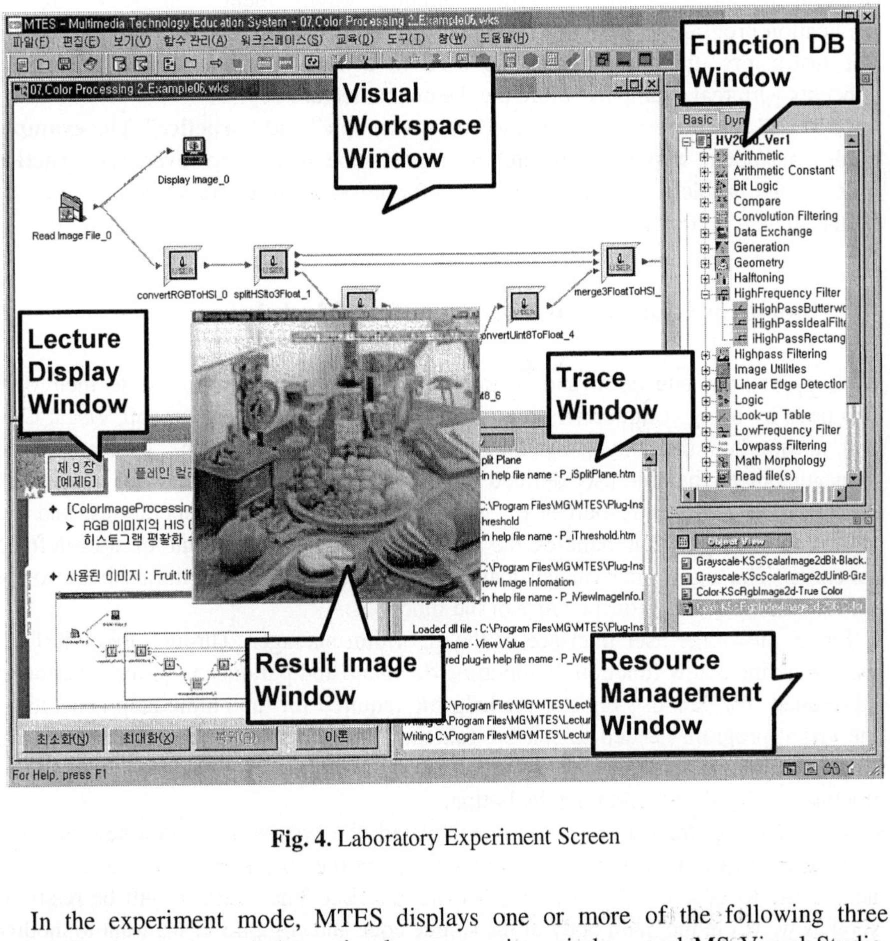

Fig. 4. Laboratory Experiment Screen

In the experiment mode, MTES displays one or more of the following three windows: instruction window, visual-programming window, and MS Visual Studio.

The instruction window displays viewgraphs showing the detail procedure of the experiment to guide the experiment as shown in Fig. 4. The visual-programming window loads the visual program that will be needed in the experiment. For the experiment involving the programming of new algorithms, MTES activates the Visual Studio loaded with the source code that includes all the codes except the main body of the algorithm. Student can complete the program by typing the main logic of the algorithm, compile, and execute it with the visual program loaded in the visual-programming window. By taking the burden of writing the house-keeping code, which dealing with the image I/O and memory management, out of students, they can concentrate all their efforts on developing the logic of image processing algorithm. Users can take advantage of MTES's algorithm-development environment to create and test user-defined functions. They can write their own image processing algorithms by using the commercial compilers such as MS visual C++. And generate a new application program by combining them with the functions registered in the function database in the visual-programming environment. The result of the application program is analyzed by using various image analysis tools in the system. For instructors, this mode could be used to demonstrate new image processing concepts with real examples before the theoretical lecture begins.

In MTES, the experiment is divided into "**example**" and "**practice**". The **example** guides a student step by step until he reaches the result. However, the **practice** presents the problem description and essential information to solve the problem and let the student do the rest.

4 Algorithm Generation and Execution Module

To support both education and research, MTES allows users to create their own functions by using commercial programming compiler such as Microsoft Visual C++ and accumulate them systematically for reuse. To simplify the generation and management of sharable user-defined functions, it provides the library of image-data class, which provides resources for essential image processing operations, and the algorithm manager that handles the registration and management of user-defined functions. The algorithm manager maintains not only the function library but also the information necessary for the reuse of the functions.

Fig. 5 shows the user interface of the algorithm manager. Through this interface, we can define a new function by inputting the name and parameters of the function to be defined. We can also register the help file required for the online help generation, the visual program needed to test the function, and the source code of the function. The visual program is displayed as a button on the online help window. User can test the function by simply clicking the button.

When the registration procedure is completed, the system registers a new function icon on the function database and creates a project file consisting of a source file and header file to generate the DLL file for the function. The function will be ready as soon as we fill in the main body of the source code and compile. If we want to modify the function later, we can reload the source code by selecting the function on the function icon window and execute a modify command on the pop-up menu. It will reload the project file for the function. All that we need to do is "modify and compile."

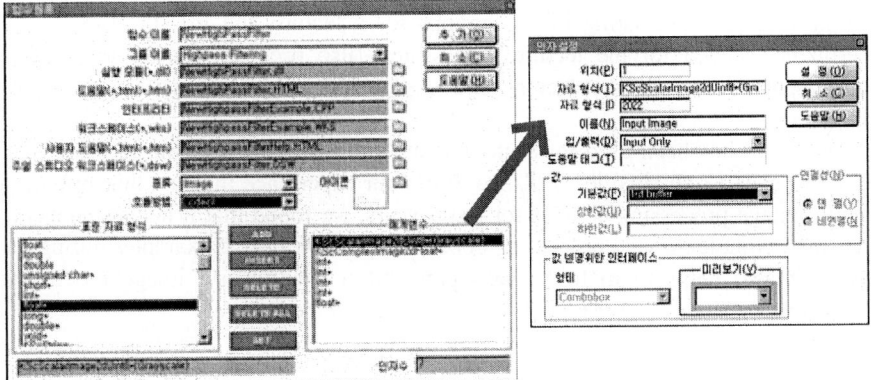

Fig. 5. Function Registration

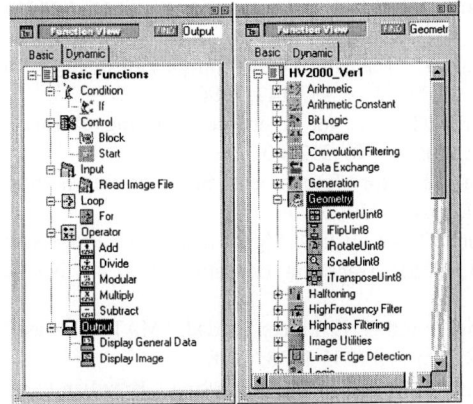

Fig. 6. Function DB Window

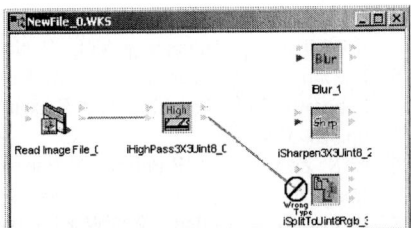

Fig. 7. Compatibility Checks

The command(function) DB window of visual program displays the list of basic commands and I/O commands for visual programming as shown in Fig. 6. The visual-programming window is the workspace that allows users to create image processing applications by connecting visual command icons and user-defined function icons in the function window. Unlike most of existing visual-programming environment, which troubleshoots parameter compatibility in run time, the proposed system troubleshoots the parameter compatibility while connecting icons for programming. This allows less chance of error in run time. Fig. 7 shows the highlighted input nodes compatible for the output node to be connected.

5 Conclusions

In this paper, we present a visual-programming environment for image processing that supports both research and education. It helps lecturers to generate online teaching materials consisting of viewgraphs for a lecture and the direction of the experiments needed for the lecture. The system also provides a suitable teaching environment to present the lecture and experiment simultaneously in the same environment.

This system has been used to teach graduate and undergraduate image processing courses in over 13 universities in Korea. From our experience in using this system in teaching image processing courses for last 4 years, we found that it helps students to understand complex image processing concepts by showing real examples during the lecture. It also helps them to be well prepared for real world image processing applications by integrating the hands-on experiment to the lecture.

References

1. G. J. Awcock and R. Thomas, "The Teaching of Applied Image Processing at thee University of Brighton," The institution of electrical engineers. Savoy Place, London WC2R 0BL, UK, 1993
2. Gregory W. Donhoe, and Patrick F, Valdez, "Teaching Digital Image Processing with Khoros," IEEE Trans. on Education, Vol. 39, No, 2, pp.137-142, 1996
3. K. Bowyer, G. Stockman, and L. Stark, "Themes for Improved Teaching of Image Computation," IEEE Trans. on Education, Vol. 43, No. 2, 2000
4. J. A. Robinson, "A Software System for Laboratory Experiments in Image Processing," IEEE Trans. on Education, Vol. 40, No, 4, pp.455-459, 2000
5. J. Campbell, F. Murtagh, and M. Kokuer, "DataLab-J: A Signal and Image Processing Laboratory for Teaching and Research," IEEE Trans. on Education, Vol. 44, No. 4, pp.329-335, 2001
6. C. S. Zuria, H. M. Ramirez, D. Baez-Lopez, and G. E. Flores-Verdad, "MATLAB based Image Processing Lab Experiments," FIE Conf., pp.1255-1258, 1998
7. R. Lotufo, R. Jordan, "WWW Khorosware on Digital Image Processing," Brazilian Conf. on Computer Graphics and Image Processing, Sibgraphi, 1995
8. H. Bassman and P. Besslich, "Ad Oculos: Digital Image Processing/Book and Software"
9. S. E. Umbaugh, So.Illinois Univ., Edwardsville, Illinois, "Computer Vision and Image Processing: A Practical Approach Using CVIPTools," Prentice Hall.
10. W. Rasband, NIH Image, available online at URL http://rsb.info.nih.gov./nih-image

Advancing Scientific Computation by Improving Scientific Code Development: Symbolic Execution and Semantic Analysis

Mark Stewart

QSS Group Inc. at NASA Glenn Research Center
Brook Park, Ohio, USA 44135
Mark.E.Stewart@grc.nasa.gov

Abstract. This paper presents an implementation of a technique for automated, rigorous scientific program comprehension and error detection. The procedure analyzes fundamental semantic concepts during a symbolic execution of a user's code. Since program execution is symbolic, the analysis is general and can replace many test cases. The prototype of this procedure is demonstrated on two test cases including a 5k line of code (LOC) program. Although this technique promises a powerful tool, several challenges remain.

1 Introduction

As scientific computing matures, a lingering problem is the manual nature of code development and testing. Arguably, the underlying issue is code semantics—the what, why, and how of computer code—and how to automate recognition of code (perhaps even synthesis) thereby reducing the time and effort of code development, testing, and maintenance. Certainly, code developers can manually translate between their code and classical mathematical and physical formulae and concepts (with reasonable reliability). However, formalization and automation of this process has not occurred.

To address these concerns, this paper reports on an effort to formalize and automatically analyze these scientific code semantics using symbolic execution and semantic analysis.

The thesis of symbolic execution is that the semantics of a programming language's construct (the variables, data structures, and operators) can be faithfully represented symbolically. Further, this symbolic representation can be propagated during an execution of the code to provide a general and rigorous analysis.

However, symbolic execution of program statements generates symbolic expressions that will grow exponentially—unless simplified. Here semantic analysis simplifies these expressions by recognizing formulae and concepts. Semantic analysis stores and recognizes the classical mathematical, logical, and physical concepts and notation that code developers and engineers are familiar with.

The concept of symbolic execution was introduced by King [1] in 1976. In a review article, Coward [2] suggests symbolic execution has languished due to the difficulty of implementation, and cites four problems:

1) evaluating array references dependent on symbolic values,
2) the evaluation of loops where the number of iterations is unknown,
3) checking the feasibility of paths: how to process branch conditions dependent on symbolic expressions,
4) how to process module calls: symbolically execute each call or execute once and abstract,

Code semantics have been the focus of some work [3,4,5] including the use of an ontology and parsers for natural language understanding [6]. Petty [7] presents an impressive procedure where—during numerical execution—the units of variables and array elements are analyzed. The procedure can be easily applied to a user's code; however the numerical execution results in high wall-time and memory requirements.

This symbolic execution / semantic analysis procedure is not only intellectually appealing as a formalism; it can be fast and very general! Human programmers analyze code at approximately 0.5 LOC per minute; symbolic execution runs quickly—approximately 1000 times faster than a human—often faster than numerically executing the code itself! The procedure is general and rigorous because it uses the abstraction of symbols—not numbers; a single symbolic analysis can replace testing with a suite of conventional test cases. In Code 1, for example, if the search fails, a memory bounds error occurs. Symbolic execution detected this error, but numerical execution would require a specific set of inputs before this error occurred.

Code 1: *Analysis detects how the search failure results in a memory access error*

```
      Dimension array(100)
      ...
      Do 10 I = 1, 100
      If ( test_value .le. array(i) ) goto 20
10    Continue
20    value = array(I)
```

Symbolic execution / semantic analysis is not without challenges, and in the following sections, the symbolic execution procedure is explained, key problems and solutions are presented, and results are demonstrated including results for a 5k LOC scientific code.

2 Symbolic Execution/Semantic Analysis Procedure

This symbolic execution / semantic analysis procedure has two preliminary steps. First, the user makes semantic declarations (1) that provide the fundamental semantic identity of primitive variables in the code including any program inputs.

$$A <= \text{acceleration, m/s2};$$
$$M <= \text{mass, kg};$$
(1)

Second, a parser converts the user's code and semantic declarations into a tree representation in a language independent form. Symbolic execution / semantic analysis start from this basis.

2.1 Symbolic Execution

Symbolic execution is similar to the ubiquitous numerical execution. In both cases statements from the user's program are executed. However, instead of loading into memory numerical values of variables, a symbolic execution emulator uses symbolic values that describe the variables. This emulator takes statements from the user's program, performs the operations on these symbols, and generates symbolic expressions. This symbolic execution emulator has a prescribed action or response to each operation encountered in a user's program, including +, -, *, /, **, array references, logical operators and loops. Table 1 contrasts numerical and symbolic execution.

Table 1. Comparison of numerical and symbolic execution for code statements. For numerical execution, the input file contains "4 5"; the semantic declarations are (1)

Code Statement	Numerical Execution	Symbolic Execution/ Semantic Analysis
READ M, A	Place 4 into M, 5 into A	Attach Semantic Declaration to Instance of M and A; Ignore input file
B = M	Transfer Number Value	Transfer Symbolic Value
M * A	Calculate 4 * 5	Form "mass * acceleration", "kg * m/s2", and attempt simplification by semantic analysis
If (A.eq.5) then B=5	A.eq.5 is True, so Transfer 5 to B	Form "A.EQ.5 $\Rightarrow$ 5	B" and attempt simplification by semantic analysis

2.2 Semantic Analysis

As statements are symbolically executed, the generated symbolic expressions become larger—unless they are simplified. The role of semantic analysis is to recognize and simplify the fundamental mathematical, logical, and physical formulae used in these expressions. Here, parsers [8,9] are used to recognize formulae in expressions and simplify them. For example, the physical formula "Force = Mass * Acceleration" is one of many formulae encoded in one expert parser. If the parser examines the expressions in Table 1, it will recognize "Mass * Acceleration" and replace it with "Force". More details of how formulae are recognized in parsers are given in [10].

Semantic analysis is not only a vital simplification tool for successful symbolic execution; symbolic representations exist for many semantic aspects of scientific and engineering code, including units, dimensions, vector analysis, and physical and mathematical equations. Table 2 provides a comprehensive list.

Table 2. Scientific semantic properties analyzed by the procedure, including sample equations and number of parsers

Property Analyzed	Sample Equation	Parsers
Physical Equation	force $\Leftarrow$ mass * accel	3
Math Equation	$\Delta\phi \Leftarrow \phi - \phi$	5
Logical Expression	$\phi \Leftarrow$ If (True) ϕ else θ	2
Value / Interval	$[1,50] \Leftarrow [0,49] + 1$	2
Grid Location	$\phi_i \Leftarrow \phi_{i+1} + \phi_{i-1}$	4
Vector Analysis	$\phi \cdot \phi \Leftarrow \phi_x^2 + \phi_y^2 + \phi_z^2$	1
Non-Dimensional	$\phi/A \Leftarrow \chi/A + \varphi/A$	1
Dimensions	$L \Leftarrow (L/T) * T$	1
Unit	$m \Leftarrow m/s * s$	1
Object	fluid $\Leftarrow$ fluid * anything	1
Data Type	Real $\Leftarrow$ Real * Integer	1

3 Symbolic Execution of Array References, Loops, and Conditional Statements

Coward [2] noted that array references, loops, and conditional expressions are a challenge in symbolic execution. The challenge is that execution of these constructs depends on the numerical value of variable(s). For example, while numerical execution of an array evaluation involves retrieving a value at a known numerical index, symbolic execution only knows what is symbolically possible for the index and must retrieve array elements within the corresponding range of index values.

Symbolic execution requires more complex analysis than for numerical execution, however advantages exist. The principle advantage is greater generality and rigorousness. The following three sections will pursue this issue for array representation, loops, and logical expressions.

3.1 Array Assignments and References

Code 2 demonstrates symbolic execution of a simple loop. After symbolic execution, the array A is represented as in Figure 1 where symbolically identical array elements—the fourth through N^{th} (and 2nd, 3rd and N-1st to 100th)— have been grouped together, while the first array element has not.

Code 2. Simple Loop shows how loops and array references are symbolically executed

```
         Integer  A(100)
         Read  N
         A(1) = 5
         Do 10 i=4,N
             A(i) = 1
10       continue
```

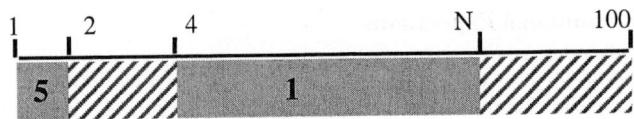

Fig. 1. Symbolic representation of the array A after execution of Code 2. Undefined values are diagonally shaded; array values are in bold; array indices are above

This grouping of symbolically identical array elements uses an ontology for array indices; the ontology entities are shown in Table 3. In Code 2, scientific code developers will easily recognize that the array variable "i" is a Counter, variable "N" is a Number.

Table 3. Entities in the Array Index Ontology. All are integer valued

Index Entity	Role of Index Entity
Integer Constant	A Known, Unchanging Value
Number	A Variable: Unspecified, Unchanging Value
Counter	A Variable: Taking on all Integer Values in a Range
Compressed Counter	Scalar Representation of Counters for Multiple Array Indices
Enumeration	A Product of Integer Constant and Number Expressions
Offset	Delineates Multiple Arrays Stored in 1-D Array
Offset Index	Offset plus Compressed Counter
Index Number	A Variable: Unspecified, Unchanging Value, in a Range
Compressed Index Number	Scalar Representation of Index Numbers for Multiple Array Indices

These index entities organize an array's representation. To evaluate an array reference, the procedure compares the array indices—symbolically—with the index entities that bound the groupings in the array representation. For example, to reference A(N+1) in Figure 1, the procedure compares N+1 with 1, 2, 4, N, and 100, and concludes A(N+1) is in the final grouping of array entries from N+1 to 100.

Not only does the grouping of symbolically identical array elements use memory efficiently, it eliminates duplicate analysis. Where loops apply an identical operation over large parts of an array—as is so common in scientific computing—the semantic analysis is reduced to one analysis of an array assignment or reference. This is the principle reason why symbolic execution / semantic analysis can be faster than numerical execution of the same code!

3.2 Loop Evaluation

Loop evaluation is a further hurdle in symbolic execution. The issue is whether dependencies exist *between* loop iterations. Often no dependencies exist between iterations, and straight line symbolic execution is possible. If dependencies do exist between loop iterations, an attempt is made to use Mathematical Induction to deduce values.

3.3 Conditional Expressions

Symbolic execution of conditional expressions is a challenging issue since symbolic values in the condition force the examination of each block of conditional code. The current procedure symbolically executes the statements from each possible code block. Then, for each variable, the procedure forms a conditional expression that is valid following the conditional expression. This expression is propagated through the following code.

3.4 Subroutine Calls

The current procedure deals with routine calls by symbolically transferring call line parameters and global variables to a child symbol table, and then symbolically executing the routine. Upon completion of the routine, call line parameters and global variables are updated in the parent symbol table.

In the test cases studied, repeat calls to routines were not excessive—since routine calls within loops are executed once or a few times. In principle, symbolically identical routine calls need not be repeated, but this feature has not been implemented yet.

3.5 Speed of Symbolic Execution/Semantic Analysis

The wall time requirements for symbolic execution of a code are fundamentally different from the wall time requirements of numerical execution. Two opposing issues influence the wall time—and decide the economics—of symbolic execution.

First, symbolic execution is considerably more expensive than numerical execution on an operation by operation basis. Numerical execution of "A*B" includes memory accesses and a floating point operation—usually within optimized software and hardware. Symbolic execution of "A*B" includes constructing a data structure representation of the expression, and its examination by several expert parsers.

Second, the computationally demanding parts of scientific codes are usually the iterations of code within loops. Yet, in symbolic execution of a loop, symbolically equivalent (and numerically different) iterations can be grouped together and analyzed once. Consequently, symbolic execution is very attractive for loop intensive code. In particular, symbolic execution can be faster than symbolic execution for loop intensive code. Conversely, codes with fewer loops can execute more slowly symbolically than numerically.

4 Demonstration of Results

This symbolic execution / semantic analysis procedure has been developed and tested with two codes.

COMDES is a 1-dimensional aerodynamic design code written in FORTRAN77 with extensive use of aerodynamic formulae, relatively less use of mathematical formulae, and minimal use of subroutines. Symbolic execution completes successfully with 100% semantic analysis of units (Table 4).

STAGE2 is a 5k LOC, 2-dimensional computational fluid dynamics (CFD) code that solves turbulent, aerodynamic flow over compressor blade sections. Written in

FORTRAN77, it is aggressively coded and makes extensive use of mathematical formulae, loops, array references and assignments (compacting multi-dimensional arrays into a 1D array, and multiple blocks of data into a 1D array), conditional expressions, and routine calls. The symbolic execution and semantic analysis of units completes almost completely. Details are shown in Table 4. For realistic grids and number of iterations, the resulting loop sizes make symbolic execution much faster than conventional numerical execution.

Table 4. Current performance results for the semantic analysis program's analysis of two test cases. Max. memory is the gross memory required to represent and retain all local and global semantic information during the semantic analysis; the executable size is 5.0 MByte. Calculations performed on a PC with a Pentium 4 2.2 GHz processor with 512 MByte of RAM. The analysis results reflect the semantic analysis code's quality and not the quality or ability of the tested codes

Code	Lines (k loc)	Semantic Declarations	Symb Exec Wall Time (s)	Unit Recognition Rate (%)	Statements Executed (k)	Max Memory (MBytes)
Comdes	0.4	42	15.1	100.	.5	5.5
STAGE2	4.9	87	199.4	93.9	8.2	65.3

5 Discussion

5.1 Semantic Complexity

Refinement of this symbolic execution / semantic analysis procedure revealed increasingly complex semantic concepts corresponding to aggressive programming techniques. For example, array indices were encountered that store a multi-dimensional array in a 1D array, and store multiple blocks of data in a 1D array.

This observation poses several questions: "Is the population of semantic concepts used in code limited or bounded?", "Does clear, well written code use only a bounded set of basic semantic concepts?", and "What are the limits on human programmers' knowledge, comprehension, and reliability?"

5.2 Inference Chains and Recognition Reliability

Reliability is a big challenge for symbolic execution. Recognizing and simplifying one expression depends on successfully recognizing and simplifying *preceding* expressions; conversely, failing to recognize one result usually prevents any further recognition. For example, in Code 2, a failure to locate and assign to A(1) compromises the remainder of the analysis.

This dependency is called the inference chain or inference tree. A code's inference chains can be exceedingly long. In COMDES, chains as long as 140 inferences have been measured, and the longest chains in STAGE2 are probably at least an order of magnitude greater. As code size and number of inferences increase, the chance of a recognition failure also increases and reliability decreases.

5.3 Problem Difficulty

The difficulty of implementing symbolic execution / semantic analysis is a further challenge. Complete symbolic execution of the STAGE2 code involved previous work [10] plus a large extension effort that was completed (part-time) over 3 years. Achieving symbolic execution of the next code is easier, but still a major effort. The expectation is that refinement efforts accumulate so that development time drops for each successive user's code until closure is reached.

6 Conclusions

This work demonstrates that rigorous, automated symbolic analysis of scientific code is possible with a formalization of code semantics. Further, this work reveals the challenges of symbolic execution / semantic analysis, in particular, semantic complexity and recognition reliability.
 Lastly, this work emphasizes the fundamental role of semantics in software, and how this role has been obscured by our contentment with manual software development.

References

1. King, J. C., "Symbolic Execution and Program Testing," Communications of ACM, 1976, **19**, (7), pp. 385-394
2. Coward, P. D., "Symbolic Execution Systems-A Review," Software Engineering Journal, Nov. 1988. P 229-239
3. L. M. Wills, "Automated Program Recognition: A Feasibility Demonstration," *Artificial Intelligence* 45 (1-2): 113-172 (1990).
4. A. Quilici, "A Memory-Based Approach to Recognizing Programming Plans," *Comm. of the ACM*, 37(5):84 (1994).
5. B. Di Martino, C. W. Keßler, "Two Program Comprehension Tools for Automatic Parallelization," *IEEE Concurrency*, Jan.-March 2000.
6. J. Allen, Natural Language Understanding (Benjamin/Cummings, Menlo Park, 1987)
7. Petty, G. W., "Automated computation and consistency checking of physical dimensions and units in scientific programs," Software—Practice and Experience, 2001; **31** pp.1067-1076
8. S. C. Johnson, "Yacc--Yet Another Compiler-Compiler," Comp. Sci. Tech. Rep. No. 32. (AT&T Bell Laboratories, Murray Hill, 1977).
9. A.V. Aho, R. Sethi, J. D. Ullman, Compilers: Principles, Techniques, and Tools (Addison-Wesley, Reading, 1986).
10. M.E.M. Stewart, S. Townsend, "An Experiment in Automated, Scientific-Code Semantic Analysis," AIAA-99-3276, June 1999

Scale-Free Networks: A Discrete Event Simulation Approach

Rex K. Kincaid[1,*] and Natalia Alexandrov[2]

[1] The College of William and Mary,
Department of Mathematics,
Williamsburg, VA 23187-8795
rrkinc@math.wm.edu

[2] NASA Langley Research Center, Hampton, VA 23681-2199
n.alexandrov@nasa.gov

Abstract. This work is motivated by the need to reconsider the methods for the analysis and design of air transportation networks in order to meet increasing demands in the face of the current hub-and-spoke network near-saturation. In the late 1990s a number of researchers noticed that networks in biology, sociology, and telecommunications exhibited similar characteristics unlike traditional random networks. Three properties—small-world, power law, and constant clustering coefficient—describe what are now most commonly referred to as scale-free networks. How do scale-free networks form? It is well documented that a network generated by adding nodes and edges preferentially will be scale-free. Are there other mechanisms? Why do networks organize themselves in this way? What causes a scale-free network to degrade? The focus of our research is to understand what drives a collection of nodes to organize as a scale-free network. Furthermore, once a network is scale-free what disrupts this apparently natural structure. To answer these questions we build a discrete-event simulation, nominally of an air transport system. The simulation is written in C.

1 Introduction

The approaching saturation of the existing U.S. hub-and-spoke air transportation system is the source of growing interest in methods for the analysis and design of transportation architectures. The ambitious goals for the development of the next generation air transportation system set out by the Joint Planning and Development Office[1] require re-thinking about the nature of dynamic networks that have evolved over time rather than appeared through a centralized, planned design activity. The hub-and-spoke air transportation system is a scale-free network of a certain kind [7, 6]; hence our interest in the nature of scale-free networks.

---

[*] The first author gratefully acknowledges the support of the NASA-NFFP program.
[1] http://www.jpdo.aero/site_content/index.html

How and why do scale-free networks form? What causes a scale-free network to degrade? Part of the difficulty in answering these questions is limited understanding of the essential properties of scale-free networks. The relatively young literature (nearly all of the references are in the last seven years) on scale-free networks has primarily focused on three properties—average shortest path lengths, cumulative degree distributions, and clustering coefficients—to define these networks. However, it is as yet unclear whether or not these are the necessary and sufficient ingredients or are merely byproducts of some other unknown mechanisms.

Given the ubiquity of scale-free networks in nature and technology, as well as their desirable and undesirable properties, one would like to know whether, in some sense, this is the best possible structure a network can attain. If so, this has profound implications for the present (nearly) scale-free structure of air transportation networks: does the small number of hops between any pair of airports and clustering that lead to high connectivity and robustness to random attacks necessarily entail vulnerability to targeted attacks? Can one preserve the desirable properties of scale-free networks without the drawbacks? How can one increase the scalability of the current system in terms of its expandability? These are just a few questions we are attempting to explore in simplified networks in anticipation of dealing eventually with more realistic transportation systems.

Several books have been dedicated to the topic of scale-free networks, including the popular books *Linked* by Barabási [4] and *Six Degrees* by Watts [14]. In addition, a number of excellent review articles exist including Newman [11], Strogatz [13], Albert and Barabási [1] and Dorogovtsev and Mendes [5]. These review articles contain hundreds of references. We note, as have many other authors, that networks whose cumulative degree distributions follow a power law are not new. One early example is Milgram's [10] work with acquaintance networks in the United States which led to his conclusion there were *six degrees of separation* among the individuals studied. However, the early references to networks following power laws are small in number when compared to the current explosion in interest.

In the next section we describe key properties of scale-free networks as well as computational mechanisms that can be used to generate two types of cumulative degree distributions observed in real networks. Section three examines the structure of simulated networks for a generic air transport systems. Section four provides results for a mechanism that degrades the power law of the degree distribution for scale-free networks.

2 Properties of Scale-Free Networks

Table 1 provides a summary of essential characteristics of four *real* networks. The interested reader is referred to Newman [11] for a more exhaustive table. The column labeled $C^{(1)}$ records one type of clustering coefficient—the number distinct paths of length two in triangles (three times the number of triangles) divided by the total number of distinct paths of length two in the network. For random networks clustering coefficients are known to approach zero as the

Table 1. Basic statistics for four published Networks

Networks	Nodes	Edges	AvgDeg	AvgSPD	$C^{(1)}$
protein	2,115	2,240	2.12	6.80	0.072
Internet	10,697	31,992	5.98	3.31	0.035
film actor	449,913	25,516,482	113.443	3.48	0.20
Altavista	203,549,046	2,130,000,000	10.46	16.18	—

number of nodes grows large. The expected $C^{(1)}$ value if the film actor network were randomly constructed is .000252. The Table 1 value of 0.20 is three orders of magnitude larger. Column 5 lists the average shortest path lengths for the networks in Table 1. The value of 16.18 for the Altavista world wide web network means that Altavista web pages are on average 16 clicks away even though there are potentially 2.1 billion edges to traverse. If the average shortest path length scales logarithmically or slower with the number of nodes in the network then it is called a *small-world* network (see Watts [14]). The natural logarithms of the number of nodes for the networks listed in Table 1 are 7.66, 9.28, 13.02 and 19.13 respectively. Consequently, all networks listed in Table 1 exhibit the small-world property. However, this property alone does not capture the essence of scale-free networks as traditional random networks are also small-world. The average degree of each network is recorded in column 4. The average degree can be used to estimate the clustering coefficient when the number of nodes is large.

The tail of the cumulative degree distributions for a scale-free network follows a power law. This is easiest to see if $P(k)$ versus k is plotted on a log-log scale. If the tail follows a power law the graph of the tail will be a line on a log-log scale and the slope of the line is the exponent of the power function. If instead the tail follows an exponential law then a log-linear plot will be nearly linear. Both types of behavior have been observed in *real* networks. Networks *grown* by adding nodes and edges via preferential attachment are known to follow a power law. Researchers [12] have noted, however, that preferential attachment does not adequately explain large clustering coefficients.

Why do the degree distributions of some networks follow an exponential law rather than a power law? A network *grown* by adding nodes one at a time with edges placed between the new node and existing nodes by selecting the nodes closest, with respect to Euclidean distance, to the new node. Networks formed is this way are shown in [9] to follow an exponential law. Consequently, one partial answer to the above question is that the actual distances between nodes (not just the number of edges) play a critical role in the formation of exponential tails. A real network with an exponential tail is the power grid of the Western United States (see [11] page 187).

3 Simulated Networks for an Air Transport System

Air transportation networks have been analyzed in [7] and [6]. These authors show that the world-wide air transport system is a scale-free network although the cumulative degree distribution follows a truncated power law. It is speculated that the truncated effects are due to capacity restrictions at the airports. The network modeled in [7] and [6] consists of 3,883 cities (airports in the same city are aggregated) and 27,051 distinct city pairs having non-stop connections. One might have expected the cumulative degree distribution to have an exponential tail as did the U.S. power grid since the distance traveled by aircraft would seem to be a critical component. In fact, prior to the deregulation of the U.S. air transport system in 1978 air fares were closely tied to the distance traveled (see [3]). After deregulation fares for short-haul flights increased while denser markets for long-haul fares dramatically decreased. Consequently, distance traveled is no longer the primary mover in the U.S. air transport network. In contrast, Amaral et al. [2] analyze a network of world airports and conclude that the cumulative degree distribution follows an exponential distribution. However, the authors of [2] did not have access to the number of distinct flights between cities but rather the number of passengers in transit through each airport. The number of airports examined in [2] is not given.

The goal of our discrete event simulation is to answer two questions. What gross topological features trigger a set of nodes to organize as a scale-free network? What causes a scale-free network to degrade to some other network structure? We are not concerned with modeling the precise operational details of aircraft routing and scheduling (see [8]) but rather the gross level network structure and how it arises.

Our simulation begins with a set of randomly generated points on a 5000 by 5000 unit square. Each point is assigned a weight generated from a statistical distribution. We think of the points as cities and the weights as relative population measures. We consider the entities that drive the simulation to be indistinguishable aircraft. The computational results reported here model 1000 cities and 2000 aircraft. Initially each aircraft selects one of the cities as an origination node (O) and one as a destination (D) node. If the Euclidean distance between O and D is larger than some user specified threshold then the flight route must go through an intermediate city (H for hub). H is selected from a set of candidate cities as the one with minimum total weighted travel distance (O to H to D) for all combinations of H with O and D. Each plane is assigned a normally distributed speed with mean 400 and a standard deviation of 20. The travel time is the travel distance divided by the speed. In addition there is a fixed time parameter that can be set to add a constant to each route to reflect time on the ground. When a simulated flight is completed a new O/D pair is selected. The simulation begins by scheduling an arrival for each of the 2000 planes.

During the simulation three pieces of data are collected. The number of times a given city is selected as an O, H, or D (called the hit distribution) and the number of times a flight leg is traversed between each possible pair of cities. If there are n cities then there $n(n-1)/2$ possible flight legs. Lastly, the number of times each city is a part of a distinct flight leg is recorded (called the degree distribution). Upon completion of the simulation a log-log plot of both the cumulative degree and hit distributions are plotted. If the tails of either (or both) are roughly linear then there is evidence that the underlying network is scale-free. If neither of the cumulative distributions follow a power law then there is no need to compute the average hop length and clustering coefficient. Statistics are not collected until after an initial warmup period to avoid the transient startup behavior.

Each simulation must specify the population distribution for the cities and the placement of the cities on the square grid. We experimented with two grid placements for the cities. The first follows a uniform distribution and the second follows a beta distribution. Sampling from $Beta(0.5, 0.5)$ for the x and y coordinates of each city pushes the cities closer to the boundaries of the square region imitating the clustering effects of cities near coastal boundaries (see Figure 1a). Several different population distributions were tested. For example, the 1000 cities were divided into three distinct population sizes–small, medium and large. Fifty-six percent of the cities were small with populations drawn from a normal distribution with a mean of 200 and a standard deviation of 5. Thirty-two percent of the cities were medium with populations drawn from a normal distribution with a mean of 600 and a standard deviation of 20. Lastly, twelve percent of the cities were large with populations drawn from a normal distribution with a mean of 1800 and a standard deviation of 80. The simulation selects origin and destination nodes preferentially based upon population size. Once the origin is selected candidate nodes for the destination must meet a user specified minimum distance threshold. Hubs were not part of these simulation experiments. Consequently, all routes are point to point. We found no examples in which the cumulative degree or hit distributions followed a power law. Further experimentation resulted in the assignment of a fitness value to each city. Initially all cities have a fitness of 1.0. Each time a city is selected as O or D its fitness is increased by 10 percent. After a certain amount of simulated time the fitness values follow a power law.

Next, we began a new series of simulation experiments by assigning fitness values following a power to each city as part of the initialization procedure. In addition the 100 most fit cities (out of 1000) were identified and set aside as potential hubs. During the simulation routes are selected point to point if the distance between origin and destination is less than a user specified threshold (recall that O/D pairs already must meet a minimum distance threshold). If the O/D distance is larger than the threshold then the route must travel through one of the 100 candidate hubs. The hub candidate list is further restricted by removing any hub cities that are too close to either the origin or the destination. Among the remaining hubs, the city that minimizes the O-H-D route distance weighted by $(1 + log \text{ (hub fitness)})^{-1}$ is selected. The origin is selected as before, preferentially based on population, but the destination selection includes the fitness information. That is, destinations are selected preferentially based on the fitness weighted population values. For all population distributions tested the resulting cumulative degree and hit distributions followed power laws. For example, if the population of every city is normally distributed with a mean of

400 and a standard deviation of 20, the city locations are uniformly distributed, and all routes are point to point (no hubs) then the resulting cumulative degree distributions follows a truncated power law and the hit distribution follows a power law.

As we add the more realistic small, medium and large population center distributions, the beta distributed city locations, and allow hubs to part of the route structure both the cumulative degree distribution and hit distribution follow power laws. Clearly the power law distribution of the fitness values is the key to observing power law behavior for the cumulative degree and hit distributions. There are a variety of interpretations of these fitness values. For example, highly fit cities may be viewed as popular final destinations, or popular jump off points for cheaper fares to other locations.

The next two figures capture several key features of the simulated scale-free network. Due to space limitations we provide only one example. Figures 1(a) and 1(b) are associated with output from the simulation model in which the power law fitness values are input values. The x and y coordinates of the cities are generated with $Beta(0.5, 0.5) * 5000.0$ and the cities population values follow the normally distributed small, medium and large values discussed earlier. Hubs are allowed to form and roughly 72% of the routes chosen make use of a hub. The remaining routes are point to point. Both hubs and destinations take into account fitness information in the selection process while the origins are selected preferentially based on population. About 68,000 flight routes are simulated in one run.

Figure 1(a) shows the geographical locations of each of the 1000 cities. Notice the clustering near the boundaries due to the beta distribution. The green triangles are placed at cities whose degree values are small. The blue circles are placed at cities whose degree values are average while the 15 red squares are placed at cities whose degree values are quite large. The red square labeled with

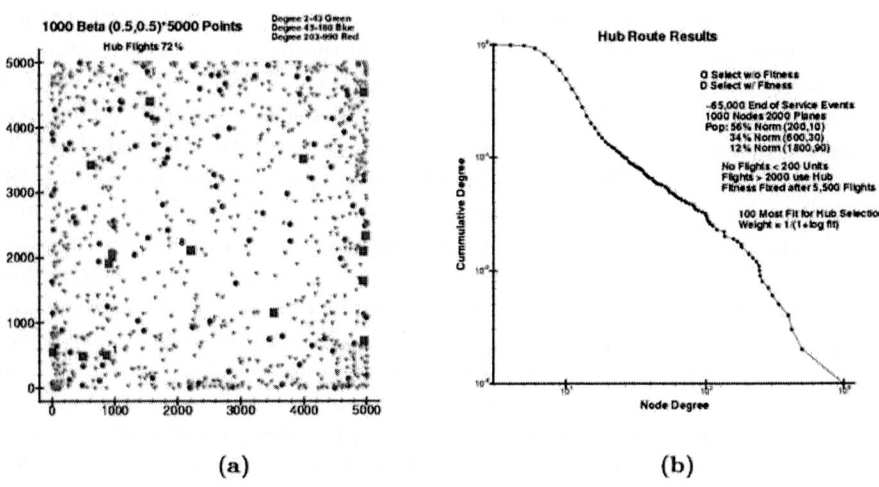

Fig. 1. 1000 cities—(a) Spatial degree dist.; (b)cummulative degree dist

a 1 in figure 1(a) is the largest degree node while the red square labeled with an 8 has the eighth largest degree. If figure 1(a) were plotted for the hit distribution the red square with an 8 would have a 2. That is, the eighth largest degree node had the second largest number of flights routed through it. This observation is in the same spirit as found in [7] where it was observed that the highest degree nodes were not necessarily the most critical for maintaining the network structure. Figure 1(b) displays the cumulative degree distribution on a log-log scale. It is easy to see that the tail of the plot is fit well by a linear function. The log-log plot of the cumulative hit distribution follows a similar pattern.

4 Degradation of Power Laws

The last question we address is what causes the power laws associated with a scale-free network to degrade. The authors of [7], [6] and [2] suggest that the scale-free structure of air transport networks is curtailed, either by a truncation of the power law or a switch to an exponential tail, when capacity restrictions ensue. The capacity restrictions might take the form of fewer flights along existing routes (changing the number of hits but not the degree) thereby forcing overflow demand to be routed to other nearby cities, or by limiting the number of routes that can be flown into the airport. One approach to capture this effect in a simulation is to model each of the cities as a finite capacity queue and develop rules for selecting hubs and destinations based on the status of the queues. In an attempt to keep things as simple as possible in our simulation we have avoided queues. Instead we allow each city to keep a short list (of length three in our

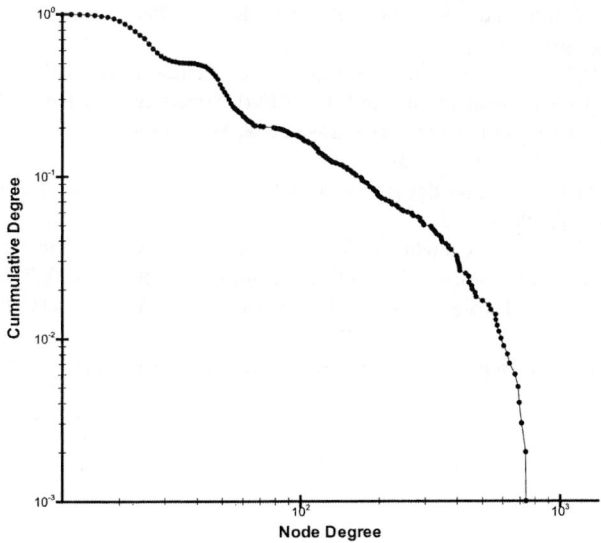

Fig. 2. Degree distribution: local capacity restrictions

example) of the most recent events (arrivals and departures). When considering selection of hubs and destination nodes if the time between the last event time on the city's list and the current simulation time is too small then the node is removed from consideration for selection. The cumulative degree distribution for this simulation run is displayed in figure 2. The truncated effect is easily seen when compared to figure 1(b).

References

1. Albert, R. and A.-L. Barabási, "Statistical Mechanics of Complex Networks," *Rev. Modern Physics*, **74**:42-47 (2002).
2. Amaral, L.A.N., A. Scala, M. Barthelemy and H.E. Stanley, "Classes of small-world networks," *Proc. Natl. Acad. Sci. USA*, **97**:11149-11152 (2000).
3. Bailey, E.E., D.R. Graham, and D.P. Kaplan, Deregulating the Airlines, MIT Press, 1985.
4. Barabási, A.-L., Linked: The New Science of Networks, Perseus, Cambridge, MA, 2002
5. Dorogovtsev, S.N. and J.F.F. Mendes, "Evolution of Networks," *Advances in Phys.*, **51**:1079-1187 (2002).
6. Guimera, R., S. Mossa, A. Turtschi and L.A.N. Amaral, "Structure and Efficiency of the World-Wide Airport Network," Preprint 0312535 (2003) available from http://arxiv.org/abs/cond-mat/.
7. Guimera, R. and L.A.N. Amaral, "Modeling the world-wide airport network," *European Physical Journal B*, **38**, 381-385 (2004).
8. Johnson, E., J. Banks, L.H. Lee, H.C. Huang, C. Lee, E.P. Chew, W. Jaruphongsa, Y.Y. Yong, Z. Liang, C.H. Leong, Y.P. Tan, and K. Namburi, "Discrete Event Simulation Model for Airline Operations: SIMAIR," *Proceedings of the 2003 Winter Simulation Conference*, S. Chick, P.J. Sanchez, D. Ferrin, and D.J. Morrice (editors), pp. 1656-1662.
9. Kincaid, R.K., "Scale-free Graphs for General Aviation Flight Schedules," NASA technical memorandum (2003) November, NASA/CR-2003-212648. (http://techreports.larc.nasa.gov/ltrs/ltrs.html)
10. Milgram, S., *Psych. Today*, **2**:60 (1967).
11. Newman, M.E.J., "The Structure and Function of Complex Networks," *SIAM Review*, **45**:167-256 (2003).
12. Solé, R.V., R. Ferrier-Cancho, J. Montoya and S. Valverde, "Selection, Tinkering and Emergence in Complex Networks," *Complexity*, **8**:1-32 (2003).
13. Strogatz, S.H., "Exploring Complex Networks," *Nature*, **410**:268-276, March (2001).
14. Watts, D.J., Six Degrees: The Science of a Connected Age, Norton, New York, 2003.

Impediments to Future Use of Petaflop Class Computers for Large-Scale Scientific/Engineering Applications in U.S. Private Industry

Myron Ginsberg

ACM Fellow and HPC Consultant
HPC Research and Education
Farmington Hills, Michigan 48335-1222 USA
m.ginsberg@ieee.org

Abstract. The environment in government/research HPC sectors is markedly different from that in private industry. Although both involve many of the same applications, the mindset and people/computer resources are significantly different. In this paper, we focus on the barriers to using future HPC machines in the private sector and some current actions and suggestions to overcome these problems. Experts generally agree that the realistic and/or real-time solution of industrial problems will require the use of computers about three orders of magnitude faster than current industrial machines. Impediments discussed include: limitations of ISV-based commercial software, inadequate benchmarking techniques for industrial size problems, limited access to the latest computer architectures and support facilities, paucity of computational science personnel, slow tech transfer of algorithms and modeling techniques from government/research facilities to private industry, and unexplored utilization of Blue Collar Computing™ in private industry.

1 Introduction

This paper focuses on the leading edge of the supercomputing frontier primarily in the U.S. industrial sector. Most of the applications involve the solution of realistic, three-dimensional simulations (modeling) of physical phenomena such as automotive/aerospace designs, climate and weather prediction, or a diverse collection of bioinformatics applications. All of these involve extremely large amounts of floating-point computations and data manipulation of massive databases. This work is typically performed in government/military research labs and/or in academic research facilities and increasingly often by a combination of such organizations. In the United States, such activities are performed in government facilities supported by the National Science Foundation, Department of Defense, NASA, Department of Energy, and other government agencies in cooperation with a variety of academic institutions. To obtain very accurate and fast and/or real-time three-dimensional solutions requires very effective utilization of computer resources several orders of magnitude faster than those typically now in use in private industry and most government labs. Experts estimate that petaflop class machines (capable of performing 10^{15} floating-point operations/second) are needed to effectively solve such problems but current technology is about three orders of magnitude less in the teraflop range, 10^{12} floating-point operations/second.

2 Government Versus Industrial HPC Environments

There is a significant difference between the HPC environment in government/academic research facilities and that encountered in most private industry facilities. For example, in the former category there is a tacit tolerance to try to utilize the latest, most advanced computer systems, typically serial – 1, i.e. often a prototype or beta category system in which the system software is often relatively primitive, somewhat unstable, constantly in a state of transition and the application software is home-grown, in-house developed rather than commercial software. Despite this somewhat chaotic state, there is much excitement and creativity as well as patience trying to make significant leaps forward in the solution process for problems previously unsolvable and/or inadequately or incompletely formulated to provide realistic solutions. There is usually a large cadre of support people available on site to cope with a plethora of systems software, applications software and/or computational physics problems and/or computer hardware/networking difficulties which may arise during the solution process.

In sharp contrast to the government/academic research environment described above, most industrial facilities in the U.S. do not have significant in-house HPC research facilities nor a large support staff. Thus most industrial organizations are much more conservative, with little tolerance to cope with prototype/unstable systems in their industrial production environments. These organizations are almost entirely dependent upon the use of commercial independent software vendor (ISV) based applications. The ISVs are usually unwilling and financially unable to port and optimize their codes for a new architecture until they are relatively certain that there will be sufficient paying customers to ensure the port will be successful. The net effect of this understandable mindset is that industrial sites might often have to wait a year or more before new HPC architectures have available ported and optimized versions of their ISV-based codes, a significant bottleneck to the use of the latest HPC technology.

3 Accurate Benchmarking Needs for Industrial Customers

The situation is further complicated by the inability to accurately benchmark a new architecture until the ISV-based software is ported and optimized. At the present time there is no means of even getting good estimates of how industrial codes might perform on the new platform. Kernels of large codes can be initially tested but there is no way of accurately accessing overall code performance when the entire application is run. The overall performance can vary substantially depending upon which path through the code is traversed and which one can best utilize both the hardware and software on the new hardware platform; thus it is indeed possible to get drastically different performance on the same hardware platform with the same commercial code applied to two very different applications. Current benchmarking strategies have to be considerably improved so that a potential industrial user of a new HPC platform can obtain a reliable indication of its value for a specific problem.

To help alleviate this situation there is a need for industry benchmark suites evaluated on diverse HPC hardware platforms using ISV-based software. Furthermore, individual companies need to devise and test their own benchmark codes and/or establish multi-industry benchmark tests. Independent benchmarking is needed rather than relying on testing conducted solely by the hardware and software vendors. A good example of such testing is the extensive efforts that DOE has expended in assessing upper end HPC machines such as the Cray X1 These benchmark tests [1] on in-house developed codes expose the good, bad, and the ugly traits of new HPC platforms with no noticeable vendor spin on the results. Such additional benchmarking is needed on an industry by industry basis. Furthermore, such benchmarks must also be able to provide some meaningful guidance to industrial users with their ISV-based codes.

4 Tech Transfer Between Government and Private Industry

In a previous paper [2] the author indicated the need in the U.S. auto industry for much faster tech transfer of algorithms from government labs to private industry. This can be accomplished in general by much closer interactions between the research community and the industrial ISVs. This could become a win win situation for both sides with U.S. industry the main beneficiary. This activity can be speeded up by increased industrial use of existing government subsidized math software libraries which could be embedded within their ISV-based codes. Increased industrial use of such libraries would also help in industrial benchmarking of new HPC hardware platforms on which those software libraries have already been ported.

5 Feedback of Recent Study of U.S. Industrial HPC Users

In addition to the obstacles to industrial HPC users mentioned in the previous sections above, a recent investigation of 33 U.S. industrial organizations [3], [4] offers its perspective on this subject. Here are some of the comments mentioned in the study: (1) 65% of respondents could not quantify direct benefit of HPC to bottom line of corporate expenses even though most survey responders admit that "high performance computing is essential to business survival"; (2) Security concerns are an important inhibiting factor to outsourcing HPC competitive sensitive problems; (3) There are current important HPC problems that are not being solved today because of one or more of the following factors: problems are too large and/or require too much computing time or memory with current in-house machines; (4) "Companies are failing to use HPC as aggressively as possible" because of corporate financial restrictions, management limited vision of HPC return on investment and/or limited trained technical personnel to deal with all aspects of HPC in-house; (5) "Most companies don't have the HPC [hardware and software] tools they want and need;"

(6) "Dramatically more powerful and easy to use computers would deliver strategic competitive benefits"; (7) "High-performance computers are desired based on actual delivered results on end-users computational problems, but most {industrial] sites [especially small companies] cannot afford to purchase the fastest computers available in the market today.

6   Help Is Coming Now

Aid for many of the problems mentioned in the preceding sections is slowly coming via renewed government initiatives to help the U.S. to be more globally competitive and also prompted by concern from the U.S. Congress as well as from prominent scientists and engineers. For example, a recent report [5] by the Committee on the Future of Supercomputing of the National Research Council has outlined a series of recommendations to rejuvenate national supercomputing efforts partly in response to the Japanese success for the past few years with the Earth Simulator overshadowing performance of U.S. based supercomputers [6]; the recommendations address hardware, software and networking concerns with particular emphasis on assuring the success of multiple domestic suppliers. DARPA continues with its efforts with IBM, Cray, and Sun with government and university partners to establish a petaflop class supercomputer by the end of this decade [7], [8], [9], [10], [11]. Additional plans for government consolidation of HPC efforts across government agencies are outlined in the Federal Plan for High-End Computing Report [12]. Industry leader Steve Wallach and others point out concerns in developing software for Petaflop class machines [13].

Several efforts are focused on improving HPC benchmarking techniques. One of those efforts, HPC Challenge Benchmarks [14], is an expansion of the Top500 [15] criterion to include such machine attributes as sustainable memory bandwidth and latency as well as bandwidth of various simultaneous communication patterns and measures of the rate of integer random updates of memory. Such attributes when combined with LINPACK TPP benchmark gives a much better perspective of machine performance. Another benchmark test is the IDC balanced rating [16] system for comparing various machines.

An interesting HPC performance metric is system balance. This can be expressed as ratios comparing resources to CPU performance. For example, ratio of bytes of memory to flops, ratio of memory bandwidth to flops, ratio of interprocessor communications bandwidth to flops, or the ratio of disk I/O bandwidth to flops. For more details including ratios for some specific machines, see [17].

Several HPC performance measurements and new tools for more accurately assessing high performance computers are being developed at Lawrence Berkeley National lab [18] by several groups including The Performance Evaluation Research Center (PERC) [19] which is trying to develop a science for better comprehending HPC performance of scientic applications. The Berkeley Benchmarking and Optimization Group (BeBOP) [20] is focusing on the interaction between application software, compilers, and hardware as well as automating the performance tuning process.

One of the concerns mentioned in Section 6 above as an impediment to vigorous use of HPC in private industry is the lack of sufficient trained HPC personnel. This is very noticeable when comparing the environments in government/research facilities vs. U.S. private industry. The former sector usually has large in-house, multidisciplinary HPC support staffs whereas in most industrial sites there are very few such people. This problem is becoming acute because as the complexity of HPC problems and computers grows, it is no longer possible for an application specialist to also be an expert in a growing variety of computer hardware, software, and networking directly impacting on the performance speed and correction of the application solution process. The long-term solution to the problem is training more people in multidisciplinary computational science programs. Too many academic departments, unfortunately, have established feudal domain mindsets which have greatly inhibited multidisciplinary projects across departments and colleges of the same or different universities. This mindset must cease if the U.S. is to be successful in the global competition arena. At present there are well over 30 computational science programs in the U.S. and Europe at both the undergraduate and graduate levels [21].

The current and future government/research HPC activities are likely to follow the pattern established by previous generations of HPC efforts (such as the transition to vector machines): innovations first accomplished by explicit tedious hand coding followed by development of math software libraries, explicit compiler directives, then automatic compiler optimizations of user source code. Thus we can expect adaptive techniques to effectively utilize the coming generation of HPC with heterogeneous multithreading processors utilizing both vector and scalar modes and incorporating automatic use of PIMS (processors in memory) and FPGA (field programmable gate arrays) technology. Even if this evolution occurs we will still need to develop new HPC programming languages [13], [22] to promote transportability and interoperability amongst HPC hardware platforms and to permit the movement of legacy codes to these new machines in a manner that will let such codes effectively utilize the new hardware and software with minimal explicit user provided modifications. This scenario will be critical to success of HPC in the private industry sector especially if this sector continues to lack a critical mass of trained computational science personnel. HPC problems have already become much too complex to expect the application specialist to correctly anticipate all factors which directly impact the correct numerical and computer optimization of the entire problem solution process.

7 Blue Collar Computing™

The HPC private industry community tacitly assumed in this paper has had to slowly emerge from observing and leaning from the government/research HPC community. Most of the companies that have immersed themselves in HPC have generally been large organizations with the financial and people resources to make the necessary commitment to HPC. These pioneers have slowly discovered that HPC definitely can have significant return on investment. For example, the U. S. automotive industry in 1980 had a 60 month (5 years) lead time between concept and production of a new vehicle; now in 2004 that lead time has shrunk to under 18 months in large part due to HPC efforts and math-based computer modeling techniques which drastically reduced the amount of physical prototyping while at the same time allowing engineers and scientists to consider more design alternatives, improve the quality and safety of the new vehicles, and contribute to improving global product competitiveness [23], [24].

Now as Ohio Supercomputer Center (OSC) Executive Director Stan Ahalt pointed out in his SC2004 speech [25], there still remains many segments of U.S. private industry untouched by the innovative potential of HPC which could improve their products and services (especially in many manufacturing areas) as well as contribute to improving U.S. global competitiveness.

Many of the comments in this paper about HPC in industry apply to both the established HPC users as well as future novices such as those depicted as part of the Blue Collar Computing™ community. A few words of encouragement here for the latter group. In some ways your HPC journey will be both easier and harder than that of your previous HPC industry pioneers. For example, the latter group had to generally have access to multimillion dollar supercomputers which were rare in U.S. industry in the 1980's and 1990's but today the hardware prices of some Linux based cluster desktop machines are less than $10,000 with deskside models under $100,000 [26], [27]. Also thanks to growing availability of access to larger HPC machines in grid environments within government and research facilities, the Blue Collar Community will have additional non-in house resources for potential use. On the negative side, the most crucial issue facing the Blue Collar Computing™ community may well be the lack of sufficient computational science personnel; see comments in Section 6. The short-term solution will require financed interactions between the HPC government/research community and the Blue Collar Computing™ community. An example of one such interaction is NCSA's Private Sector Partner Program [28] which focuses on real-world industrial challenges and currently involves Allstate, Boeing Phantom Works, Caterpillar, IBM, and Motorola Labs and will be using NCSA's newly installed 7 teraflop, 512 node Dell Cluster which will enable these industrial partners to reap the benefits of early access to breakthroughs.

8 Summary and Conclusions

This paper has spotlighted the roadblocks to U.S. industry use of future petaflop class HPC including: (1) the need for more meaningful industrial benchmarking techniques to help in the selection process for new machines; (2) Faster and more effective tech transfer of application oriented algorithms between government/research facilities and private industry via closer relationships with commercial ISVs; (3) influx of more trained computational science personnel in the private sector to cope with increasingly more complex and multidisciplinary oriented HPC applications; (4) Government support for the creation of much faster and more efficient industrial friendly supercomputers (with easy to use software and hardware tools); (5) Much improved early industrial access to new HPC hardware platforms via improved techniques for fast porting and optimization of important industrial widely used commercial ISV-based codes.

Government and industrial attention and meaningful follow up to the issues in the previous paragraph will benefit both the established industrial HPC users as well as the next generation depicted in the Blue Collar Computing™ community. It would be very helpful in the interim period if the former would help the latter via the creation of industrial mentoring activities.

References

1. Oak Ridge National Laboratory: Papers and Presentations on Cray X1 Evaluation. http://www.csm.ornl.gov/evaluation/PHOENIX/index.html
2. Ginsberg, M.: Influences on the Solution Process for Large, Numeric-Intensive Automotive Simulations. In: Alexandrov, V. N., Dongarra, J. J., Juliano, B. J., Renner, R. B., Tan, C.J. K. (eds.): Computational Science – ICCS 2001 International Conference Proceedings, Part 1, Lecture Notes in Computer Science, Vol. 2073. Springer-Verlag, Berlin Heidelberg New York (2001) 1189-1198
3. Joseph, E., Snell, A., Willard, C.G.:Council on Competitiveness Study of U.S. Industrial HPC Users. White Paper, IDC, Framingham, MA (July 2004) http://www.compete.org/pdf/HPC_Users_Survey.pdf
4. The Council on Competitiveness: First Annual HPC Users Conference: Supercharging U.S. Innovation & Competitiveness. Council on Competitiveness, Washington D.C. (2004)
5. Graham, S.L., Snir, M., Patterson, C.A. (eds.): Getting Up to Speed: The Future of Supercomputing. The National Academies Press, Washington, D.C. (2005)
6. The Earth Simulator. http://www.es.jamstec.go.jp
7. DARPA: DARPA Selects Three High Productivity Computing Systems (HPCS) Projects. http://www.darpa.mil/body/NewsItems/pdf/hpcs_phii_4.pdf (July 8, 2003)
8. Cray, Inc.: DARPA HPCS Cray Cascade Project. http://www.cray.com/cascade/
9. IBM, Inc.: DARPA HPCS IBM PERCS Project. http://www.research.ibm.com/resources/news/20030710_darpa.shtml
10. Sun Microsystems Inc.: DARPA HPCS Sun Hero Project. http://www.ncsc.org/casc/meetings/CASC2.pdf
11. Ricadela, A.: Petaflop Imperative. Information Week (June 21, 2004)http://www.informationweek.com/story/showArticle.jhtml?articleID=22100641
12. HECRTF: Federal Plan for High-End Computing: Report of the High-End Computing Revitalization Task Force (HECRTF). Office of Science and Technology Policy, Executive Office of the President, Washington, D.C. (May 10, 2004)http://www.itrd.gov/pubs/2004-hecrtf/20040510_hecrtf.pdf
13. Wallach, S.: Searching for the SOFTRON: Will We Be Able to Develop Software for PetaFlop Computing?. Keynote Talk, ISC2004, Heidelberg, Germany (June 23, 2004)
14. ICL: HPC Challenge Benchmark. The Innovative Computing Lab, U of Tennessee, Knoxville, TN, benchmarks and results available at http://icl.cs.utk.edu/hpcc/index.html and http://icl.cs.utk.edu/hpcc/hpcc_results.cgi
15. Top500 Supercomputer Sites. http://www.top500.org
16. HPC User Forum: IDC Balanced HPC Benchmark Rating Report.http://www.hpcuserforum.com/benchmark/benchmarkresults.asp
17. Cray, Inc.: Balance – the Key to Exceptional Application Performance.http://www.cray.com/products/xd1/balance.html
18. HPCWIRE: LBNL Revamps HPC Performance Measurements. Article 108895, HPCWIRE (Dec. 3, 2004)
19. The Performance Evaluation Research Center (PERC). http://perc.nersc.gov/main.htm

20. The Berkeley Benchmarking and Optimization Group (BeBOP). http://bebop.cs.berkeley.edu/
21. SIAM: List of Graduate and Undergraduate Programs in Computational Science. http://www.siam.org/cse/cse_programs.htm
22. Unified Parallel C. http://upc.lbl.gov/
23. Ginsberg, M.: An Overview of Supercomputing at General Motors Corporation. In: Ames, K.R., Brenner, A.G. (eds.): Frontiers of Supercomputing II: A National Reassessment. Volume in the Los Alamos Series in Basic and Applied Sciences, University of California Press, Berkeley (1994) 359-371
24. Ginsberg, M.: Supercomputers Help Auto Manufacturers Decrease Lead Time. In: Redelfs, A. (ed.): HPC Contributions to Society. Tabor Griffin Communications, San Diego, (November 1998) 74-79
25. Ahalt, S.C.: Towards a High Performance Computing Economy: Blue Collar Computing™. http://www.osc.edu/hpc/blue_collar/docs/stans_bc_speech_04.pdf
26. Cray, Inc.: The Cray XD1 High Performance Computer: Closing the Gap between Peak and Achievable Performance in High Performance Computing. White Paper, WP-0020404. (2004) http://www.cray.com/products/systems/xd1/whitepaper.pdf
27. HPCWIRE: Orion Brings Supercomputing to Your Desktop. Article 108305, HPCWIRE (September 3, 2004)
28. HPCWIRE: NCSA Adds Dell Cluster to Private Sector Resources. Article 108899, HPCWIRRE (December 3, 2004)

The SCore Cluster Enabled OpenMP Environment: Performance Prospects for Computational Science

H'sien. J. Wong and Alistair P. Rendell

Department of Computer Science, Australian National University,
Canberra ACT0200, Australia
alistair.rendell@anu.edu.au

Abstract. The OpenMP shared memory programming paradigm has been widely embraced by the computational science community, as has distributed memory clusters. What are the prospects for running OpenMP applications on clusters? This paper gives an overview of the SCore cluster enabled OpenMP environment, provides performance data for some of the fundamental underlying operations, and reports overall performance for a model computational science application (the finite difference solution of the 2D Laplace equation).

1 Introduction

The two main classes of parallel computers available in today's markets are clusters and hardware enabled Shared Memory Systems (SMS). Clusters are assembled from multiple disjoint computers and are generally programmed using some form of message passing. SMS on the other hand have a single common address space and can be programmed using either message passing or various threaded programming models. In general SMS are considered easier to use, but due to the need for specialized hardware they are also more expensive, and this is especially true for high processor counts.

OpenMP [1] is a threaded programming model widely used on SMSs. Essentially it provides a compiler based interface to an underlying thread library. The model is attractive since it permits the incremental parallelization of existing sequential application codes and, as it consists largely of compiler directives, a developer can easily support both parallel and sequential versions of the same code at the same time.

Given the cost advantages of clusters, but the programming advantages of OpenMP and the existence of a large body of OpenMP code, it is not surprising that various groups have been attempting to develop OpenMP programming environments for clusters [2-5]. Most of these attempts have been based on layering OpenMP on top of some existing Software Distributed Shared Memory (SDSM) environment, although we note interesting recent work by Huang *et al* to implement OpenMP over Global Arrays [6]. In either case, to obtain reasonable performance from an OpenMP code on a cluster it is likely that the application programmer will require some knowledge of the implementation. The aim of this paper is to discuss these issues in relation to the SCore cluster enabled OpenMP environment [2, 7] and to analyze its performance on a cluster comprising dual 550MHz Pentium III processors linked via a 100MBit Ethernet interconnect.

2 The SCore Distributed Shared Memory Environment

The SCore cluster enabled OpenMP is layered on top a page based SDSM environment called SCASH [2]. A basic understanding of SCASH is critical to understanding the overall performance of the SCore cluster enabled OpenMP. Essentially SCASH separates the address space of each process into global and local memory pages. Data assigned to local memory pages is private to each process, while data in the global memory pages can be shared between all processes. To facilitate sharing of the global memory pages, read and write accesses to the corresponding address ranges are protected using mprotect (on Unix). This means that when any process first accesses data in a global memory page an interrupt is triggered and this induces execution of the relevant SCASH interrupt handler. The interrupt handler determines the location of the required page and makes it available to the requesting process. There are three possible home locations for the requested page; i) it has been assigned to memory associated with the calling process; ii) it has been assigned to memory associated with another process that resides on the same physical computer as the calling process; iii) it has been assigned to memory associated with a process that resides on another physical computer. In case i the requested memory page is available immediately; in case ii the page might be available in, e.g. a shared memory segment that is linked to both processes; while in case iii a transfer of the page over the communication network is required.

The interrupt is further characterized by the type of interrupt. Global memory pages assigned to the calling process or to a process located on the same node as the calling process are initially given read access, thus interrupts involving these "home" or "local" pages only occur when writing to these pages. All "remote" global memory pages are initially marked as "unmapped", so depending on the operation transitions unmapped-read, unmapped-write or read-write can occur. Indeed if a process first reads from one of these pages and then shortly afterwards writes to the same page two interrupts will occur, one causes an unmapped-read transition and the other causes a read-write transition. In this case it would obviously be better to have a single interrupt with unmapped-write transition, and while in some cases the compiler may be able to make such optimizations this should not be taken for granted.

To permit multiple processes to simultaneously access global data, copies of the same memory page can be transferred to multiple different processes. If the requesting process only requires read access this does not present a problem. If, however, it requires write access then there are two considerations. First, when do the modifications become visible to all other processes. Second and since memory pages are typically large (e.g. 4096bytes or greater), how to support multiple simultaneous writes to disjoint regions of the same memory page.

The first issue, also known as the memory consistency model, is enforced by SCASH at synchronization points. This means that modifications made to any of the global memory pages are propagated back to the page owners at every synchronization point. Moreover if one process has modified a page, but another process has a read only copy of that page, then the read only copy must be invalidated and the protection on that page reset. This requires inter-process communication to communicate the changes, and some book keeping to keep track of which processes have copies of which pages. Whether this requires communication over the cluster interconnect will depend on the exact location of the process modifying a page compared to the owner of the page.

To handle multiple simultaneous updates to the same memory page, SCASH employs a "twinning and diffing" procedure. This means that if a write fault is encountered, as well as locating and fetching a copy of that page (if it is not already available to that process either because it is owned by that process or has been fetched via a read fault) the handler will create two copies. One is modified in subsequent write operations, while the other is left unmodified. At the next synchronization point, a "diff" is made between the modified and unmodified copy and the changes relayed to the process owning that page. The time required to communicate the differences will depend in part on the number of changes made to that page.

Table 1. The different memory page faults encountered in SCASH, details of what communications are required, and approximate times as recorded on a cluster of dual 550MHz Pentium III processor nodes linked via 100MBit/sec Ethernet. See text for further details

Access Type	Current Permission	Page Owner	Abbreviation	Send Request	Fetch Page	Create Twin	Time (usec)
Write	ReadOnly	Home	WROH				8
Write	ReadOnly	Local	WROL	Yes		Yes	44
Write	ReadOnly	Remote	WROR	Yes		Yes	51
Write	ReadWrite	Home	WRWH				7
Write	ReadWrite	Local	WRWL	Yes			16
Write	ReadWrite	Remote	WRWR	Yes			24
Write	UnMapped	Remote	WUMR	Yes	Yes	Yes	599
Read	ReadWrite	Remote	WRWR	Yes			25
Read	ReadOnly	Remote	RROR	Yes			18
Read	UnMapped	Remote	RUMR	Yes	Yes		587

Table 1 summaries the above and reports timing data for a variety of different page fault transitions obtained by running a set of specially designed OpenMP benchmarks under SCore version 5.4.0 on the Pentium III cluster. "Access type" denotes whether the interrupt was caused by a read or a write fault, "current permission" reflects the page permissions prior to the memory fault, and "page owner" denotes which process is ultimately responsible for this memory page. In this respect "home" implies that the process posting the interrupt is also the owner of the memory page, while "local" implies that the page is owned by a process running on the same physical node (noting the use of dual CPU nodes). Within SCASH not all transitions are possible, for example an RUMH transition is impossible since the default permission for a home page is ReadOnly.

By way of contrast the timing data given in Table 1 should be compared to the latency of a "normal" memory access. This was measured using LMbench [8] as roughly 0.14usec. Within SCASH this would be the cost of accessing local memory, the cost of making a read access to a global memory page owned by the calling process, or the cost of reading (writing) to a page of global memory once a local copy had been created and assigned read (readwrite) protection. Clearly, from the data given in Table 1 the cost of the first write to any page of global memory, or first read of a remote page is significantly more expensive than 0.14usec. For example the cost of the first write to a global memory page owned by the calling process (WROH) is roughly 50 times greater at 8usec. While writing to a global memory page located within the same node, but not owned by the calling process (WROL) is longer still, since it requires some book-keeping (denoted by "send request") and creation of a twin. Not surprisingly the most costly page faults involve read or write requests to an unmapped global memory page (WUMR or RUMR) as it requires that page to be transferred across the interconnect; at around 600usecs on a machine with a clock cycle of ≈2nsec this corresponds to roughly half a million clock cycles!

For the application programmer it is important to realize that the costs given in Table 1 will, in general, be encountered for the first access to global memory after every synchronization point. That is following a synchronization point any local copy of a memory page is likely to be invalidated, so subsequent reads or writes to that page will encounter a new page fault and cost penalty as detailed in Table 1.

A clear implication from the timing data is that codes which access memory by jumping from page to page with little or no reuse of data in the same page will perform very poorly (e.g. pointer chasing). Conversely to obtain reasonable performance the costs given in Table 1 need to be amortized over many subsequent data accesses to the same page, and this is especially true for remote page accesses. In short if you drag a memory page over the network you'd better make good use of it!

Finally, we note that some of the events given in Table 1 may appear a little strange. For instance, a WRWH page fault is encountered when two processes share the same computer, the owner of the page has not yet written to it, but a companion thread on the same node has. Thus when the owner thread accesses this page an interrupt is triggered, but the page has actually already been marked with readwrite access.

3 SCore Cluster Enabled OpenMP

With a basic understanding of SCASH we can now consider the performance of some key OpenMP synchronization directives. To do this we have used the OpenMP microbenchmark tests suite [9] developed at Edinburgh Parallel Computing Centre (EPCC). Before presenting the results, however, it is pertinent to outline briefly how OpenMP is mapped onto the underlying SCASH SDSM.

Not surprisingly OpenMP data quantities that are declared to be shared are stored in global SCASH memory pages. Thus if every thread in an OpenMP parallel region accesses the same global variable this will induce an interrupt on virtually all threads with requests for the relevant page to be transferred to the calling thread. The only exceptions are for read accesses to data stored in memory pages that are either owned by the calling thread or by a thread co-located on the same node. Thus as implemented the cost of transferring shared data from the master to child threads scales as *O(No_Threads)*. Of course if multiple shared data quantities are stored in the same page then accesses to these other quantities will be cheap once the initial page transfer has occurred.

The cluster enabled OpenMP compiler also uses a portion of the SCASH global shared memory space for administrative purposes. For example if the "`#pragma omp parallel`" directive is combined with a "`copyin`" clause then threadprivate data is transferred between the master and child threads by placing the relevant data items into a global memory page and having each child thread retrieve this data as required. Similarly information relating to the scheduling of parallel loops and reduction operations are communicated through global memory pages. Also and as is common to most OpenMP implementations, threads once created are kept alive but dormant between parallel regions. The parallel regions are "outlined" as functions that are then called as applicable by the child threads on entry to a parallel region. To communicate the name of the relevant function between the master and child threads the name of the outlined function is placed in an global memory page.

Synchronization operations in OpenMP map to synchronization operations in SCASH. As part of the SCASH consistency model discussed above, this is where memory pages are flushed and updated. An SCASH barrier consists of 5 phases.

1. Synchronization
2. Flushing of modified memory pages
3. Synchronization
4. Invalidation of pages
5. Synchronization

Here flushing of the modified memory pages involves transferring the differences that have resulted from write operations back to the owning page, while invalidation of pages involves communicating with remote processes so that they can update their page tables based upon which pages have been modified. Just from this basic understanding of what is involved it is clear that synchronizations will be expensive.

In Table 2 we report the timing results for the EPPC OpenMP synchronization microbenchmarks run on the Pentium III cluster. Some minor modifications to the test suite were made. In particular the iteration counter used in these benchmarks is assigned global scope. Since the first access to this variable (or more precisely the global memory page containing this variable) occurs within the timing routine this induces an interrupt and transfer of the associated memory page to the requesting page. To avoid this additional overhead we have defined this variable as threadprivate with a copyin clause to transfer it from master to child before the start of the timed loop.

The results obtained on the Pentium cluster and using between 1 and 8 OpenMP threads are compared with similar data obtained on a Sun V1280 system with 12 900MHz UltraSparc III processors and hardware shared memory. On the cluster results were obtained using both 1 and 2 threads per dual processor Pentium III node. Comparing a single thread run on the cluster (denoted 1x1) with a single thread run on the Sun (denoted 1) we see that the overheads associated with inclusion of the OpenMP directives are roughly equivalent (especially when the faster clock rate of the Sun processor is considered). As soon as we move to multiple threads, however, the situation changes dramatically with significantly larger overheads recorded on the cluster. Moreover this is even true when running 2 OpenMP threads on 1 node of the cluster (denoted 1x2) where the cost of the `parallel/for/barrier/single/ reduction` constructs are typically two orders of magnitude larger than the equivalent results obtained on the Sun. If we move to 2 threads running across 2 nodes of the cluster (denoted 2x1) the performance of these operations gets even worse. In comparison the overhead associated with the `critical/ordered/ atomic` directives is relatively good for two threads within the same nodes, but increases dramatically when the threads are located on different nodes.

Table 2. Overhead (usec) for OpenMP synchronization directives on cluster with dual processor Pentium III nodes linked via 100MBit/sec Ethernet and a 12 900MHz CPU Sun hardware shared memory V1280 system

Directive	Pentium III (nodes x threads/node)					Sun V1280 (Threads)			
	1x1	2x1	1x2	4x1	4x2	1	2	4	8
parallel	0.8	1762	474	13556	43571	0.3	5.3	6.8	10.0
for	0.5	662	221	7731	17033	0.5	2.1	2.9	4.3
parallel for	1.2	1797	471	13498	42571	0.7	6.0	7.3	12.8
barrier	0.2	661	225	7305	17631	0.1	1.1	1.8	2.9
single	0.9	4371	304	15330	47804	0.1	0.9	1.3	2.0
critical	1.3	179	6	260	976	0.2	0.3	0.5	0.5
lock	0.6	52	4	74	204	0.2	0.4	0.5	0.5
ordered	1.7	1154	8	3712	6454	0.2	0.6	0.6	0.6
atomic	1.3	3446	7	4776	5023	0.1	0.5	0.8	1.1
reduction	1.1	29994	966	47836	98553	0.5	5.5	7.8	12.0

Table 2 shows that on the cluster the cost of a `parallel/parallel for` directive is roughly twice the cost of a `barrier`, while the cost of an isolated `for` directive is roughly equal to the cost of a `barrier`. This is easily explained by the existence of an implicit barrier at both the start and end of the `parallel` and `parallel for` directives, but only one implicit barrier at the end of the `for` directive. In contrast the `single` directive, while also containing a barrier at the end of the associated region of code, also requires some additional book-keeping to ensure that just one thread executes this portion of code. This is handled by using a shared counter, with access to this counter controlled by locks. As the shared counter is stored in global memory, page faults are encountered when each thread accesses it giving rise to extra cost. Also the overall cost of the single directive appears to scale rather poorly with increasing thread count, thus depending on the context, it may be better to specifically assign the work/code associated with this directive to one thread.

The most expensive operation is `reduction`. This requires two barrier calls, and also makes use of the global administrative pages. Specifically, prior to the first barrier call, all threads place their partial results into unique locations in the global administrative page. The first barrier serves to propagate these partial contributions back to the thread owning that memory page – in this case to thread 0. When this barrier is complete thread 0 then combines the partial contributions and writes the result to another location in the same page. The second barrier is used to indicate that this operation is complete and that the child processes can now access the final result. As with the basic SCASH synchronization mechanism, the overall cost of a reduction operation will scale as $O(No\_Threads)$.

4 Case Study

To illustrate the likely performance of SCore cluster enabled OpenMP on a real computational science application code we consider heat distribution in a two dimensional conducting plate. In this problem the temperature of a conducting plate is held constant at the edges and the aim is to determine the temperature of the interior of the plate. The problem is described by the 2-D Laplace equation, and as such is similar to a number of other related problems. The equations are solved iteratively using a finite difference approach with a regular rectangular grid. During each iteration a new value of the temperature at a given grid point is computed based on the average of the temperatures of the four surrounding grid points, with iterations continuing until some agreed convergence is reached. Ignoring convergence testing and imposition of the boundary conditions the basic sequential code is as follows:

```
/*Line1*/ for (i=0; i<no_of_iterations; i++){
/*Line2*/   for (y=0; y<no_of_rows; x++)
/*Line3*/     for (x=0; x< no_of_columns; x++)
/*Line4*/       new[x,y] = (old[x+1,y]+old[x-1,y]+
                            old[x,y+1]+old[x,y-1])/4.0
/*Line5*/   tmp=new; new=old; old=tmp;
/*Line6*/ } /* next iteration */
```

Four parallel implementations were considered:

1. **Naïve:** A "`#pragma omp parallel for`" directive combining thread creation and work division is placed immediately before line2.
2. **Barrier minimization:** A "`#pragma omp parallel`" directive is placed before line 1, and a "`#pragma omp for`" before line 2. The rational for this is that it reduces the number of barriers per iteration from two to one.
3. **Page alignment and fault minimization:** the memory associated with arrays new and old is carefully allocated so that threads maximize use of "home" data.
4. **Barrier and page fault minimzation:** optimizations 2 and 3 are combined

Timing results for the four different implementations run on 1, 2 and 4 nodes of the cluster are given in Table 3. From this it is immediately apparent that a naïve inclusion of OpenMP directives into the sequential code is not a good idea. Adjusting the code with the aim of reducing the number of associated barrier calls results in a slight performance gain on 2 nodes, but worse performance on 4 nodes. Since both cases are still much slower than the sequential code there are clearly other factors affecting performance. If we now adjust memory to ensure that data quantities are optimally aligned we see a dramatic performance increase, with the code now showing some performance benefit from running on multiple nodes. Finally combining barrier minimization with page placement we obtain the best performance result – albeit only a speedup of 1.9 on 4 nodes of the cluster.

Table 3. Performance comparison of sequential heat code with three alternative OpenMP parallel algorithms run using SCore on the Pentium III cluster for a grid size of 1024x1024 and 100 iterations

Implementation	Time (sec)			Speedup	
	1x1	2x1	4x1	2x1	4x1
0 Sequential	8.2	-	-		
1 Naïve	8.2	75.7	86.9	0.11	0.09
2 Barrier Opt.	8.2	63.7	101.1	0.13	0.08
3 Page Fault Opt.	8.2	6.9	5.3	1.19	1.55
4 Barrier&Page Opt.	8.2	6.7	4.3	1.22	1.91

5 Conclusions

This paper provides a brief overview of the SCore cluster enabled OpenMP environment. The performance of some of the key underlying operations on a cluster of Pentium III processors is evaluated and compared with OpenMP running on a dedicated hardware shared memory system. Using this information and a knowledge of the SCore implementation we were able to obtain an acceptable level of performance for a computational science kernel running on the Pentium III cluster.

Acknowledgements. The authors gratefully acknowledge discussions with J. Antony and A. Over. This work was supported in part by the Australian Research Council through Linkage Grant LP0347178.

References

1. OpenMP Forum, "OpenMP: A proposed industry standard api for shared memory programming", http://www.openmp.org, Oct. 1997.
2. Y. Ojima, M. Sato, H. Harada, Y. Ishikawa, "Performance of Cluster-enabled OpenMP for the SCASH Software Distributed Shared Memory System", Proc. of the 3rd IEEE/ACM Int. Sym. on Cluster Computing and the Grid, 450-456 (2003).
3. Y.C. Hu, H. Lu, A.L. Cox, and W. Zwaenepoel, "OpenMP for Networks of SMPs", J. Parallel Dist. Computing, **60**, 1512-1530 (2000).
4. S-J. Min, A. Basumallik and R. Eigenmann, "Optimizing OpenMP Programs on Software Distributed Shared Memory Systems", Int. J. Parallel Programming, **31**, 225-249 (2003).
5. D. Margery, G. Vallée, R. Lottiaux, C. Morin, and J-Y. Berthou, "Kerrighed: a SSI Cluster OS Running OpenMP". Proc. 5th European Workshop on OpenMP (EWOMP '03), Sept. 2003.

6. L. Huang, B. Chapman, Z. Liu and R. Kendall, "Efficient Translation of OpenMP to Distributed Memory", Lecture Notes in Computer Science, **3038**, 408 (2004).
7. See http://www.pccluster.org/
8. L. McVoy, "LMBench – Tools for Performance Analysis", http://www.bitmover.com/lmbench
9. See http://www.epcc.ed.ac.uk/research/openmpbench/openmp_index.html

Author Index

Abawajy, J.H.   III-205, III-213, III-447, III-457
Abrahamyan, Lilit   I-287
Absil, P.-A.   I-33
Adamidis, Panagiotis   II-1064
Agron, Paul   I-1019
Ahn, JinHo   III-679
Ahn, Seongjin   III-1072
Ahn, Tae-Chon   I-798
Ahn, Youngjin   III-796
Akay, Bulent   I-147
Akbari, Mohammad K.   III-205
Akinlar, Cuneyt   I-396
Akyol, Derya Eren   III-562
Alam, Sadaf   I-304
Alberti, Pedro V.   I-229
Alexandrov, Natalia   I-1051
Alexandrov, Vassil N.   III-359, III-367, III-350, III-743, III-752, III-766
Alfonsi, Giancarlo   I-623
Ali, Hesham H.   II-927
Alique, J.R.   III-627
Alique, Angel   III-1056
Allen, Robert B.   II-976
Aloisio, Giovanni   II-10
Alonso, Pedro   I-220
Alpbaz, Mustafa   I-147
Altinakar, Mustafa Siddik   III-33
Al Zain, A.   II-748
Amik, St.-Cyr   III-57
Ammari, Nisreen   I-493
An, Beongku   III-1060
Anagnostopoulos, Christos-Nikolaos   I-695
Anagnostopoulos, Ioannis   I-695
Anai, Hirokazu   III-602
Andrés, Mirian   III-635
Andrews, Christopher   I-1011
Anh, Le Tuan   II-436
Anthes, Christoph   III-350, III-383
Arickx, F.   II-1080
Asirvatham, Arul   II-265
Atanassov, Emanouil   III-735, III-752
Attali, Isabelle   I-526
Aznar, Fidel   I-828

Babu, V.   III-72
Bação, Fernando   III-476
Bachmair, Leo   I-1019
Backeljauw, Franky   I-295
Bae, Hae-Young   I-405
Bae, Juhee   III-842
Bagheri, Babak   II-44
Bai, Li   II-273
Baik, Ran   I-1, III-899
Baik, Sung Wook   III-850, III-1016, III-1064
Baker, C.G.   I-33
Bala, Jerzy   III-1016, III-1064
Baldridge, Kim   II-672
Balas, Can Elmar   I-892, III-1108
Balint-Kurti, Gabriel   III-933
Bamha, M.   II-755
Bang, Young-Cheol   III-796
Bangerth, Wolfgang   II-656
Banicescu, Ioana   I-237
Bansevičius, Ramutis   III-643
Bao, Lichun   II-485
Bao, Hujun   II-248
Barros, Ligia   I-727
Barsky, Brian A.   II-224
Bartholet, Robert   II-721
Barvík, Ivan   I-860
Basak, Tanmay   II-814
Basney, Jim   I-501, II-729
Bastiaans, R.J.M.   III-64
Basu, P.K.   I-172
Batchelor, Donald B.   I-372
Bauer, Sebastian   II-1064
Baumgartner, Gerald   I-155
Bawa, Rajesh K.   III-1104
Bayhan, Gunhan Mirac   I-843, III-562
Beerli, Peter   III-775
Bein, Doina   I-560, II-535
Bein, Wolfgang W.   II-535
Bektaş, Tolga   I-188
Belkasim, Saeid O.   II-256

Bellomo, Carryn   III-1096
Benassarou, A.   II-314
Bennethum, Lynn S.   II-632
Benoit, Anne   II-764
Bereg, Sergey   II-851
Berkenbrock, C.D.M.   III-987
Bernholdt, David E.   I-155, I-372
Bernreuther, Martin   II-1
Berry, Lee A.   I-372
Berzins, M.   II-36
Beyls, Kristof   II-166
Bhana, Ismail   III-391
Bianchini, Germán   I-427
Billiet, David   II-1046
Bindel, David S.   I-50
Birnbaum, Adam   II-672
Bischof, Hans-Peter   I-703
Bittar, E.   II-314
Blais, J.A.R.   I-74
Blin, Guillaume   II-860
Bloomfield, Max O.   III-49
Boettcher, Stefan   II-386
Bogdanov, Alexander   III-933
Bolvin, Hervé   I-271, I-639
Bonizzoni, Paola   II-952
Borisov, Sergey   III-143
Bourchtein, Andrei   I-131
Bourdot, Patrick   II-290, II-339
Bourilkov, Dimitri   III-342
Box, Frieke M.A.   I-287
Branford, Simon   III-743, III-752
Breitenbach, Mark L.   III-41
Brenk, Markus   II-1
Brinza, Dumitru   II-1011
Broeckhove, Jan   II-1072, II-1080
Brogan, David   II-721
Broszkiewicz, Magdalena   III-1112
Browne, James C.   I-347
Brzeziński, Jerzy   III-423
Bu, Jiajun   III-806
Buckley, William R.   II-395
Bungartz, Hans-Joachim   II-1
Byrski, Aleksander   III-703
Byun, Daewon   II-814
Byun, Yanga   I-948

Cafaro, Massimo   II-10
Cai, Guoyin   III-496, III-883
Cai, Hongming   II-335
Cai, Keke   III-806

Cai, Liming   II-968
Cale, Timothy S.   III-41, III-49
Calleja, Mark   III-359
Campa, Sonia   II-772
Canning, Andrew   III-317
Cantillo, Karina   III-1056
Canton-Ferrer, Cristian   II-281
Cao, Chunxiang   III-464, III-472
Cao, Wuchun   III-464, III-472
Cariño, Ricolindo L.   I-237
Carmichael, Gregory R.   II-648
Carnahan, Joseph   II-721
Caromel, Denis   I-526
Casale, Giuliano   III-147
Casas, Josep R.   II-281
Castelló, Pascual   II-240
Catalyurek, Umit   II-656
Caymes, Paola   II-132
Čepulkauskas, Algimantas   III-643
Cetnarowicz, Krzysztof   III-711
Chae, OkSam   I-1035
Chai, Tianfeng   II-648
Chambarel, André   I-271, I-639, I-647
Chandra, Krishnendu   I-493
Chaturvedi, A.R.   II-695
Chen, J.   III-1076
Chen, Chongcheng   I-979
Chen, Chun   I-552, III-806
Chen, Deren   III-221
Chen, Huoping   I-615
Chen, Jian   III-578
Chen, Meng Chang   II-444
Chen, Mingshi   II-632
Chen, Pu   III-300
Chen, Cai Min Yu   III-187
Chen, Ying   III-187
Chen, Shangjian   I-568
Chen, Shudong   I-544
Chen, Yen Hung   II-845
Chen, Zhengxin   III-548
Chen, Zizhong   I-115
Cheng, Hwai-Ping   I-460
Cheng, Jing-Ru C.   I-460
Chi, Hongmei   III-775
Childs, Stephen   III-870
Chin, Francis Y.L.   II-985
Chinchalkar, Shirish   II-76
Chinnusamy, Malar   II-60
Cho, Ju   III-1016, III-1064
Cho, Haengrae   III-1012

Cho, Jung-Wan   III-407
Cho, Sunyoung   I-941
Cho, We Duke   I-576
Cho, YoungTak   I-1035
Choi, Dong Ju   II-672
Choi, Hongsik   I-419
Choi, Hyoung-Kee   II-453
Choi, Jaeyoung   III-916
Choi, Jihyun   III-196
Choi, Kee-Hyun   III-346, III-950, III-963
Choi, Ki-Young   III-866
Choi, Min-Hyung   I-735
Choi, Sung Chune   III-1000
Choo, Hyunseung   II-468, II-510, II-559, III-796, III-1120
Chover, Miguel   II-240
Chtepen, Maria   III-1116
Chung, Min Young   II-559, II-601, III-1120
Cicalese, Ferdinando   II-1029
Cięciwa, Renata   III-711
Çinar, Ahmet   III-945
Claeys, Filip   III-1116
Clark, Terry   II-44
Coen, Janice L.   II-632
Coghlan, Brian   III-870
Cohen, Jared   III-838
Cole, Martin J.   II-640
Cole, Murray   II-764
Coleman, Thomas F.   II-76
Coles, Jonathan   I-703
Collier, Rem   III-695
Collins, Lori   III-983
Collura, F.   III-267
Constantinescu, Emil M.   II-648, II-798
Contes, Arnaud   I-526
Convard, Thomas   II-290
Cooke, Daniel E.   III-891
Cooper, Rodney   I-1011
Cortés, Ana   I-427
Cortas, Maria   I-58
Costa, Rosa Maria   I-727
Cucos, Laurentiu   I-322, III-991
Cui, Yong   II-551
Cunha, Gerson   I-727
Cuyt, Annie   I-295
Cyganek, Bogusław   I-757

Dăescu, Dacian N.   II-648, II-837
Dai, Hongning   III-875

Dai, Yang   II-903
Damaschke, Peter   II-1029
Dantas, Mario A.R.   III-858, III-971, III-987
D'Apice, C.   III-594
Darema, Frederica   II-610
Das, Kamakhya   I-493
DasGupta, B.   II-1020
Dass, Rajanish   III-818
Datta, Ajoy K.   I-560
Datta, Karabi   I-1, III-899
Dauger, Dean E.   II-84
Day, Mitch D.   II-68
D'Azevedo, Eduardo F.   I-99, I-372
Deconinck, Herman   I-279
Decyk, Viktor K.   II-84
de Doncker, Elise   I-123, I-165, I-322, III-991
de Goey L.P.H.   III-64
delaRosa, J.J.   I-585
de-la-Rosa, Juan-José González   I-900
De Leenheer, Marc   III-250
Demeester, Piet   III-250, III-1116
Demmel, James W.   I-50
Deng, Ansheng   I-783
Deng, Zhiqun   III-854
Deris M. Mat   III-447
Deslongchamps, Ghislain   I-1011
Desovski, D.   I-180
De Turck, Filip   III-250
Dhoedt, Bart   III-250, III-1116
D'Hollander, Erik H.   II-166
Dikaiakos, Marios   I-534, III-870
Dimov, I.   III-752
Ding, Koubao   III-954
Ding, Yongsheng   I-517
Dobson, James E.   II-99
Dondi, Riccardo   II-952
Dong, JinXiang   I-671
Dongarra, Jack   I-115, III-317
Donnell, Barbara P.   I-66
Douglas, Craig C.   II-632, II-640
Dove, Martin T.   III-359
Droegemeier, Kelvin   II-624
Durvasula, Shravan   I-493
Dyapur, Kaviraju Ramanna   III-879

Efendiev, Yalchin   II-640
Effinger-Dean, Laura   II-107

El-Aker, Fouad   III-788
Elwasif, Wael R.   I-372
Engelmann, Christian   I-313
Epicoco, Italo   II-10
Erciyes, Kayhan   I-196, I-388
Eriksen, Jeff   I-631
Ertunc, Suna   I-147
Evans, Deidre W.   III-775
Ewing, Richard   II-640

Fabricius, Uwe   II-27
Fahey, Mark R.   I-99
Fang, Liqun   III-464, III-472
Fang, Yong   III-554
Farago, Paula   I-727
Farhat, C.   II-616
Farid, Hany   II-99
Fayyad, Dolly   I-58
Feng, Dingwu   III-887
Feng, Shengzhong   III-979
Feng, Yusheng   I-347
Fertin, Guillaume   II-860
Filatyev, S.A.   II-695
Fleming, Charles   III-760
Flores-Becerra, G.   I-17
Floudas, Christodoulos A.   II-680
Fox, Geoffrey   II-576, III-275, III-431
Frączek, Jacek   III-334
Franca, Leopoldo P.   II-632
Freedman, Jim   II-703
Freeman, T.L.   I-364
Frels, Judy   II-378
Freundl, Christoph   II-27
Friedman, Mark J.   I-50, I-263
Fu, Chong,   III-1044
Fujimoto, Junpei   I-165
Funika, Włodzimierz   II-158

Galán, Ramón   III-1056
Galiana, Isidro Lloret   I-900
Galis, Alex   III-259
Gallivan, K.A.   I-33
Galvez, Akemi   III-651
Gannon, Dennis   II-624
Gansterer, Wilfried N.   I-25
Gao, Chongnan   III-163
Gao, Lei   I-517
García, Victor M.   I-17, I-229
García, Pedro   III-246
Gargantini, Irene   II-331

Gargiulo, G.   III-594
Gashkov, Igor   III-663
Gaudiot, Jean-Luc   I-212
Gava, Frédéric   II-1046
Gaynor, Mark   II-703
Geist, Al   I-313
Gerasimova, Olesya   III-143
Gevorkyan, Ashot   III-933
Ghosh, Debi Prasad   III-1
Giesbrecht, Mark   III-619
Ginsberg, Myron   I-1059
Ginting, Victor   II-640
Glasner, Christian   II-124
Gobbert, Matthias K.   III-41
Gorbachev, Yuriy   III-933
Gore, J.P.   II-695
Gorissen, Dirk   II-1072
Goscinski, Andrzej M.   I-435
Govaerts, Willy J.F.   I-50, I-263
Grama, Ananth   II-664
Graves, Sara   II-624
Grimshaw, Andrew   II-729
Grochowski, M.   III-727
Gross, Murray   III-983
Grzymkowski, Radosław   III-895
Guan, Xiaohui   I-743
Guan, Yanning   I-908
Guleren, Kursad Melih   III-130
Gullaud, T.   II-616
Guo, S.M.   III-104
Guo, Jianping   III-464, III-472
Guo, Shan   I-908
Guo, Yuanbo   III-229
Gurd, John R.   I-364
Gyllenhaal, John   II-140
Górriz, Juan Manuel   I-585, I-900

Ha, Sang Yong   II-510
Haber, R.H.   III-627
Haber, Rodolfo E.   III-627, III-1056
Hadjarian, A.   III-1064
Haffegee, Adrian   III-350
Hains, G.   II-755
Hakobyan, Tigran   III-933
Han, Hyuck   III-179
Han, Kijun   II-585
Han, Kyungsook   I-711, I-948, III-1024, III-1028
Han, Seung Kee   I-941
Hanna, A.   II-695

Hapoglu, Hale   I-147
Hardie, Patrick   I-364
Härdtlein, Jochen   II-1055
Hariri, Salim   I-615
Hartono, Albert   I-155
Harvill, Jane L.   I-237
Hasan, S. Mehmood   III-359
Haupt, Tomasz   I-493
Hayashida, Ulisses Kendi   I-509
He, Jingwu   II-1011
He, You   I-812
He, Yuanjun   II-335
Heath, Michael T.   II-52
Heine, Felix   III-155
Heisler, Debra   II-378
Hellinckx, P.   II-1080
Hensley, Jeffrey L.   I-66
Heo, Hoon   I-1035
Herrero, Pilar   III-171
Heřman, Pavel   I-860
Hiebeler, David   II-360
Hilaire, Vincent   III-719
Hirata, So   I-155
Hoekstra, Alfons G.   I-287
Hoffmann, Christoph   II-664
Hong, Choong Seon   II-436
Hong, Feng   III-875
Hong, Helen   I-719, III-834, III-842
Hong, Jinsun   III-1024
Hong, Min   I-735
Hong, Yoopyo   I-1, III-899
Hoppe, Hugues   II-265
Horie, Ken   III-570
Horiguchi, Susumu   II-781
Horntrop, David J.   I-852
Hou, Qibin   II-273
Houlberg, Wayne A.   I-372
Houstis, E.   II-616
Hovestadt, Matthias   III-155
Howington, Stacy E.   I-66
H'sien, J. Wong   I-1067
Hu, Bao-Gang   II-322
Hu, Hualiang   III-221
Hu, Jinfeng   III-163
Hu, Xiaohua   II-976
Hu, Yincui   III-496, III-883
Hua, Wei   II-248
Huang, Changqin   III-221, III-887
Huang, He   III-578
Huang, Houkuan   I-995

Huang, Kuen-Yu   III-292
Huang, Linpeng   III-875, III-1032
Huang, Tianqiang   I-979
Hueso, E.   II-689
Humphrey, Marty   I-477, I-501, II-729
Hung, Shao-Shin   III-830
Hunter, Robert M.   I-460
Huo, Mingxu   III-954
Hwang, In-Chul   III-407
Hwang, Kai   III-187
Hwang, Sang-Jun   I-380
Hwang, Tsung-Min   III-908
Hyndman, Rob J.   III-792

Iglesias, Andrés   III-651
Ivanovska, Sofiya   III-735

Jackson, Steven Glenn   III-611
Jaeger, E.F.   I-372
Jaeger, Marc   II-322
Jamieson, Ronan   III-350
Janicki, Aleksander   III-1112
Janik, Arkadiusz   II-158
Jardin, S.C.   III-1076
Javadi, Bahman   III-205
Jeffrey, D.J.   III-586, III-667
Jenkins, Jerry   III-309
Jeon, Hoseong   II-468
Jeon, Il-Soo   III-912
Jeon, Sung-Eok   III-279
Jeong, Chang-Sung   III-862, III-866
Jeong, Hae-Duck J.   I-655
Jeong, Hong-Jong   II-477
Jeong, Kwang Cheol   II-510
Jessup, E.R.   II-91
Jho, CheungWoon   II-327
Ji, Chuanyi   III-279
Jia, Jinyuan   II-298
Jiang, Hong   II-519
Jiang, Qingshan   III-801
Jiao, Xiangmin   II-52
Jin, Xiaogang   I-599
John, N.W.   II-314
Johns, Craig J.   II-632
Johnson, Chris R.   II-36, II-640
Johnson, David   III-391
Joneja, Ajay   II-298
Jones, Greg   II-640
Ju, Byoung-Hyon   I-711

Ju, Jianwei   I-82
Jung, Hanjo   III-407
Jung, Hyungsoo   III-179
Jung, Hyunjoon   III-179
Jung, Moon-Ryul   II-216
Jung, Sunhwa   I-735
Jung, Youn Chul   II-568

Kacsuk, Peter   III-367
Kaiser, Tim   I-469
Kalyanasundaram, Anand   I-493
Kang, Hwan Il   I-593
Kang, Jung-Yup   I-212
Kang, Sanggil   I-971
Kang, Yan   I-783
Kao, Odej   III-155
Kara, İmdat   I-188
Karaata, Mehmet H.   I-560
Karaivanova, Aneta   III-735, III-766
Karl, Wolfgang   II-174, II-182
Karniadakis, G.E.   II-689
Kasperska, Elżbieta   I-837, III-1040
Kasprowski, Paweł   III-334
Katz, Paul S.   II-347
Kaugars, Karlis   I-123
Kayafas, Eleftherios   I-695
Kemmler, Dany   II-1064
Kenny, Eamonn   III-870
Kim, Byung-yeub   II-477
Kim, Chong-Kwon   II-527
Kim, Do-Hyeon   III-1060
Kim, Dongkyun   II-477
Kim, Hyun-Ki   III-1100
Kim, Hyun-Sung   III-912
Kim, Hyunjue   II-493
Kim, Hyunsook   II-585, III-1125
Kim, Intaek   I-593
Kim, Jai-Hoon   II-576, III-275
Kim, Jang-Sub   II-601
Kim, Jung Ae   I-941
Kim, Jungkee   III-431
Kim, Kyung-ah   II-527
Kim, Minjeong   II-632
Kim, Moonseong   III-796
Kim, Munchurl   I-971
Kim, Sangjin   III-958
Kim, Seungjoo   II-493
Kim, Sun Yong   II-543
Kim, Taekyun   III-958
Kim, Tongsok   II-527

Kim, Ungmo   II-568, III-813
Kim, Yongkab   I-792
Kim, Yoonhee   III-920
Kimpe, Dries   I-279
Kincaid, Rex K.   I-1051
Kirby, R.M.   II-36
Kisiel-Dorohinicki, Marek   III-703
Klie, Hector   II-656
Kluge, Michael   I-330
Knight, John C.   II-729
Knüpfer, Andreas   I-330, II-116
Knyazev, Andrew V.   II-632
Ko, Hanseok   I-139
Ko, Sunghoon   II-576, III-275
Kohl, James A.   I-372
Koker, Utku   III-562
Kolditz, Olaf   II-1064
Kolli, Vijaya Smitha   II-1003
Konwar, K.M.   II-1020
Koo, Jahwan   III-1072
Kosloff, Todd J.   II-224
Köstler, Harald   II-27
Kotulski, Leszek   III-1008
Kou, Gang   III-548
Koukam, Abder   III-719
Koutsonas, Athanassios   I-695
Kozlowski, Alex   II-224
Krause, Tara   II-428
Kremens, Robert   II-632
Krysl, Petr   II-672
Kuhara, Satoru   II-911
Kulikov, Gennady Yur'evich   I-42
Kulkarni, Vaibhav   II-632
Kulvietienė, Regina   III-643
Kulvietis, Genadijus   III-643
Kuo, Ting-Chia   III-830
Kurc, Tahsin   II-656
Kuznetsov, Yuri A.   I-50, I-263
Kwon, Ki Woon   III-850
Kwok, Yu-Kwong   III-187
Kye, Heewon   III-834, III-842

Labahn, George   III-619
Lai, Kin Keung   III-523
Laidlaw, D.H.   II-689
Lam, Chi-Chung   I-155
Lambrakos, Sam   II-738, III-80
Lambris, John D.   II-680
Landau, Luiz   I-727

Lane, Terran II-894
Langemyr, Lars III-129
Langer, Malgorzata I-607, I-876
Langou, Julien III-317
Lani, Andrea I-279
Lapenta, Giovanni I-82, III-88
Lazarov, Raytcho II-640
LeBoeuf, Eugene J. I-172
Leduc, Guy III-237
Lee, Donghoon I-711
Lee, DongWoo III-196
Lee, Eunseok III-1052
Lee, Heungkyu I-139
Lee, In-Kwon I-916
Lee, Jee-Hyong II-543
Lee, JeongHeon I-1035
Lee, Jeongjin I-719
Lee, Jong-Suk R. I-655
Lee, Sang Kun I-941
Lee, Sangho III-1012
Lee, Sangkeon III-916
Lee, Seong-Whan III-850
Lee, Seunghwa III-1052
Lee, Seungsoo II-559
Lee, Soo Myoung I-576
Lee, Tae-Dong III-862, III-866
Lee, Tae-Jin II-559, III-1120
Lees, J.M. I-751
Lefeuve-Mesgouez, Gaëlle I-647
Lei, Zhengdeng II-903
Lestrade, John Patrick I-237
Leupi, Célestin III-33
Lewis, Gareth J. III-359, III-367
Li, Degao III-783
Li, Guoqing III-484, III-492
Li, Jianping III-531
Li, Minglu III-875
Li, Peiyu I-568
Li, Shanping III-995
Li, Shuhui I-372
Li, Shujun I-123, I-165
Li, Su-Ju III-1044
Li, Xiaowen III-464, III-472
Li, Xin III-801
Li, Yifeng II-927
Li, Yiming III-292, III-300
Li, Yusong I-172
Lian, HeSong I-593
Liao, Shenghui I-671
Liao, Wenyuan II-648, II-806

Lim, Byong-In III-346, III-963
Lim, Jeongyeon I-971
Lim, Joong-Ho III-862
Lim, Jungmuk II-468
Lin, Huaizhong I-552, II-461
Lin, Lanfen I-671
Lin, Xin III-995
Linke, Alexander II-1055
Lisik, Zbigniew I-607, I-876
Liu, Chunmei II-968
Liu, Damon Shing-Min III-830
Liu, Dingsheng III-484, III-492
Liu, Haifeng II-877
Liu, Hua II-248
Liu, Hui II-1003
Liu, Jiangui I-908
Liu, Mingzhe II-420
Liu, Qi II-368
Liu, Tom III-112
Liu, Xinchun II-869
Liu, Xumin I-995
Liu, Yanfei III-1068
Liu, Zheng II-829
Liu, Zhuo II-837
Lobo, Victor III-476
Loidl, H-W. II-746
Loitiére, Yannick II-721
Lombardo, S. III-267
Lopez-Parra, Fernando III-120
Loulergue, Frédéric II-1046
Loumos, Vassily I-695
Lu, Hsueh-I II-845
Lucas, L. II-314
Luján, Mikel I-364
Lukac, Rastislav I-679, I-687, II-886
Luke, Edward A. II-790
Luo, Jiancheng I-963
Lou, Xiasong III-187
Luo, Ying III-496
Luo, Yingwei III-511, III-515
Luque, Emilio I-427, II-132
lv, Yong III-937

Ma, Bin II-960
Ma, Fanyuan I-544
Ma, Jianfeng III-229
Ma, Lizhuang III-846
Ma, Yongquan III-163
Machì, A. III-267
Maeng, Seung-Ryoul III-407

Mahanti, Ambuj   III-818
Mahapatra, Debiprosad Roy   III-1, III-25
Mahawar, Hemant   I-107
Mahmood, Nasim   I-347
Majumdar, Amit   II-672
Malkowski, Konrad   I-245
Malladi, Srilaxmi   II-535
Malm, Nils   III-129
Malmberg, Russell L.   II-968
Malony, Allen   I-631
Mamat, R.   III-447
Mandel, Jan   II-632
Măndoiu, Ion I.   II-994, II-1020
Mansfield, Peter   II-76
Manzo, R.   III-594
Mao, Weidong   II-1011
Mao, Zhihong   III-846
Marchesini, John C.   II-99
Margalef, Tomàs   I-427, II-132
Mariani, Lorenzo   II-952
Marín, Mauricio   I-411, I-1003
Markidis, Stefano   III-88
Marsh, David   III-687
Marshall, Geoffrey   I-388
Martin, S.M.   III-64
Martin, Jonathan   I-501
Martin, Sylvain   III-237
Maruyama, Osamu   II-911
Mascagni, Michael   III-760, III-775
Masoumi, Beeta   II-936
Mateja-Losa, Elwira   III-1040
Mathee, Kalai   II-944
Matossian, Vincent   II-656
Matsuda, Akiko   II-911
Matsuhisa, Takashi   III-570
May, John   II-140
Małysiak, Bożena   III-334
Means, J.   II-695
Mellema, A.K.   II-695
Melnik, Roderick V.N.   I-884, III-25, III-134
Merkle, Daniel   II-412
Mesgouez, Arnaud   I-647
Metaxas, Dimitris   II-712
Miaoliang, Zhu   I-1027
Michaelson, G.J.   II-746
Michaelson, Greg   II-781
Michel, Olivier   I-820

Michopoulos, John   II-616, II-738, III-80
Middendorf, Martin   II-412
Min, Kyungha   I-916, II-216
Min, Sung-Gi   III-679
Min, Yong   I-599
Ming, Dongping   I-963
Mingarelli, Angelo B.   II-351
Mondéjar, Rubén   III-246
Moon, Sanghoon   I-948, III-1064
Morajko, Anna   II-132
Morales, J.D.   I-585
Moreno-Hagelsieb, Gabriel   III-134
Morikis, Dimitrios   II-680
Morisse, Karsten   III-375
Morley, C.T.   I-751
Moulton, Steve   I-703
Mrozek, Dariusz   III-334
Mukherjee, Amar   II-395
Mukherjee, Arup   III-1084
Mukherjee, Sarit   I-396
Muldoon, Conor   III-695
Mundani, Ralf-Peter   II-1
Muntean, Ioan Lucian   II-1

Nagel, Wolfgang E.   I-330
Nakhleh, Luay   II-919
Nam, Hyunwoo   I-941
Nam, Junghyun   II-493
Nam, Young Jin   III-439
Narasimhan, Giri   II-944
Nassif, Nabil R.   I-58
Natesan, S.   III-1104
Naumann, Uwe   I-338
Navon, I. Michael   II-837
Neveux, Philippe   I-271
Newman, Harvey   III-196
Newman, Timothy S.   I-9
Nguyen-Tuong, Anh   II-729
Ni, Jun   III-326
Nie, Changhui   III-1088
Nielsen, Frank   I-1019
Nikishkov, G.P.   II-232, II-306
Ning, Ning   III-163
Nishidate, Y.   II-232
No, Jaechun   I-380, I-485
Noël, Alfred G.   III-611
Nooijen, Marcel   I-155
Nyman, Gunnar   III-933
Nystrom, J.F.   III-1096

Obeyesekere, Mandri   III-96
O'Callaghan, David   III-870
Ocampo, Roel   III-259
Oh, Eunseuk   I-419
Oh, Heekuck   III-958
Oh, Sangyoon   II-576, III-275
Oh, Sung-Kwun   I-792, I-798, III-1080, III-1100
O'Hare, Gregory M.P.   III-687, III-695
Ohn, Kyungoh   III-1012
O'Kane, Donal   III-687
Okuda, Kunio   I-509
Oliveira, Suely   I-204
Owen, G. Scott   I-451, II-256
Oysal, Yusuf   I-775

Paarhuis, B.D.   I-868
Pachowicz, P.   III-1064
Paik, Juryon   III-813
Painho, Marco   III-476
Pairot, Carles   III-246
Pakin, Scott   II-149
Pallickara, Sangmi Lee   II-576, III-275
Pamplin, Jason A.   II-347
Pan, Gang   I-743
Pan, Michelle Hong   II-1003
Pan, Xuezeng   III-937
Pan, Yi   II-1003
Pan, Yunhe   III-1020
Pan, Zhijian   II-404
Panetta, Jairo   I-509
Parashar, Manish   I-615, II-656
Pardàs, Montse   II-281
Park, Byoung-Jun   I-798, III-1100
Park, Byungkyu   III-1028
Park, Chanik   III-439
Park, Gyung-Leen   II-468
Park, Ho-Sung   I-792
Park, Hyoungwoo   I-485
Park, Jeong-Su   II-477
Park, Jong-An   I-934
Park, Sang-Min   I-477
Park, Seung-Jin   I-934
Park, Soon-Young   I-405
Park, Sung Soon   I-380
Pascual, Vico   III-635
Pasztor, Egon   II-224
Patnaik, Kiran Kumar   III-879
Patrick, Charles, Jr.   III-96
Payne, Bryson R.   I-451, II-256

Pedrycz, Witold   I-792, I-798, III-1100
Peng, Bo   I-599
Peng, Gang   III-995
Peng, Yi   III-548
Percell, Peter   II-814
Perelman, Alex   II-224
Pérez, Jesús Fabián López   III-924
Pérez, María S.   III-171
Perminov, Valeriy   III-139
Pernas, Ana Marilza   III-858
Pflaum, Christoph   II-1055
Phelan, Donnacha   III-695
Pi, Daoying   III-1036
Pieczynska-Kuchtiak, Agnieszka   III-671
Ping, Lingdi   III-937
Ping, Xiaohui   III-975
Pinto, A.R.   III-971
Pitsch, H.   III-64
Pitzer, Russell M.   I-155
Pivkin, I.V.   II-689
Plale, Beth   II-624
Plataniotis, Konstantinos N.   I-679, I-687, II-886
Pletzer, Alexander   III-838
Poedts, Stefaan   I-279
Prăjescu, Claudia   II-994
Pratibha   III-667
Praun, Emil   II-265
Primavera, Leonardo   I-623
Primeaux, David   I-419
Provins, D.A.   I-74
Pu, Jiantao   II-343
Pujol, Mar   I-828
Puntonet, Carlos García   I-585, I-900

Qiao, Ying   I-90
Qin, Guan   II-632
Qin, Xiaolin   I-979
Qiu, Shibin   II-894
Quarta, Gianvito   II-10
Quigley, Geoff   III-870
Quintino, Tiago   I-279

Raghavan, Padma   I-245
Raj, Ewa   I-876
Ramakrishna, R.S.   III-196
Ramamurthy, Mohan   II-624
Ramani, Karthik   II-343
Ramanujam, J.   I-155
Ramos, José Francisco   II-240

Reed, Dan II-624
Reggia, James II-378, II-404
Reiber, Johan H.C. I-287
Reid, G.J. III-586
Rendell, Alistair P. I-1067, II-18
Reyes, Nora I-1003
Reynolds, Paul II-721
Ribbens, Calvin J. II-60
Richards, David F. III-49
Richards, David R. I-460
Ridgway, Scott, L. II-44
Rizo, Ramón I-828
Rizzi, Romeo II-860
Robles, Víctor III-171
Rodosek, Robert I-804
Rodriguez, Sebastian III-719
Roh, Seok-Beom III-1080
Rojek, Gabriel III-711
Roman, Eric II-224
Romero, Ana III-635
Roper, James II-18
Ros, S. III-627
Rowanhill, Jonathan II-729
Rubio, Julio III-635
Rüde, Ulrich II-27
Rushton, J. Nelson III-891
Rutt, Benjamin II-656
Ryoo, SeungTaek II-327
Ryu, Jungpil II-585

Sadayappan, P. I-155
Salman, Adnan I-631
Saltz, Joel II-656
Salvadores, Manuel III-171
Sameh, Ahmed I-33, II-664
Sandu, Adrian II-648, II-798, II-806, II-829
Santini, Cindy I-469
Sarin, Vivek I-107
Sautois, B. I-263
Sazhin, Oleg III-143
Scaife, Norman II-781
Schaap, Jorrit A. I-287
Schaefer, R. III-727
Scheidler, Alexander II-412
Schreppers, Walter I-295
Schuetze, Hans-Joachim II-378
Schulz, Martin II-140
Schwarz, Phil III-112
Schwarz, Susan A. II-99

Scott, Stephen L. I-443
Seber, Dogan I-469
Seinfeld, John H. II-648
Seltzer, Margo II-703
Sempf, Thomas III-375
Sengupta, Debasis III-309
Seo, Kyung-Sik I-934, III-822
Seok, Sang-Cheol I-204
Seyfarth, Benjamin Ray I-664
Sfarti, Adrian II-224
Shamonin, Denis I-287
Shang, Hui II-373
Sharma, Arjun II-107
Shen, Hao III-826
Shen, Hong II-985
Shen, Huifeng III-1020
Shi, Liang III-1088
Shi, Wei III-995
Shi, Xiaofeng III-854
Shi, Yong III-531, III-548
Shimizu, Yoshimitsu I-165
Shin, Dong-Ryeol II-601, III-346, III-813, III-950, III-963
Shin, Ho-Jin III-950
Shin, Jitae II-453
Shin, Yeong Gil I-719, III-834, III-842
Shin, Young-Suk III-941
Shindin, Sergey Konstantinovich I-42
Shoshmina, Irina III-933
Shu, Jiwu III-399, III-415
Shvartsman, A.A. II-1020
Sibiryakov, Alexander I-155
Sim, Terence II-207
Simonov, Nikolai III-760
Sipos, Gergely III-367
Skarmeta, Antonio F. Gómez III-246
Sławińska, Magdalena I-355
Sloot, Peter M.A. I-287, I-534
Słota, Damian I-837, III-659, III-895
Smith, Kate A. III-792
Smith, Sean W. II-99
Smolka, Bogdan II-886
Smołka, M. III-727
Sobaniec, Cezary III-423
Soboleva, Olga III-9
Song, Shanshan III-187
Song, Il-Yeol II-976
Song, Jeomki II-477
Song, Mao I-1027
Song, Min II-976

Song, Siand Wun   I-509
Song, Yinglei   II-968
Soofi, M.A.   I-74
Soysert, Zehra   I-196
Spet, Olivier   I-1011
Spicher, Antoine   I-820
Spiegl, Edith   II-124
Srinivasan, Gopalakrishnan   III-1
Srinivasan, Kasthuri   I-107
St.-Cyr, Amik   II-822
Stankova, Elena   III-933
Stevens, John G.   III-1084
Stewart, Mark   I-1043
Strahan, Robin   III-695
Strauss, H.R.   III-1076
Streit, Achim   III-155
Stuer, Gunther   II-1072, II-1080
Su, Fanjun   III-937
Su, Xianchuang   I-599
Subramani, K.   I-180
Sun, Jing   III-163
Sun, Ninghui   II-869, III-979
Sun, Shuyu   III-96
Sun, Yeali S.   II-444
Sun, Yi   III-492
Sun, Youxian   III-1036
Sundaram, Shankar   III-309
Sunder, C. Shyam   III-72
Sunderraman, Rajshekhar   II-347
Swain, W. Thomas   I-443
Swaminathan, Gautam   II-60
Swartz, S.   II-689
Święcicki, Mariusz   III-810
Székely, Gábor   III-17
Szczerba, Dominik   III-17

Takeuchi, Fumihiko   I-956
Talay, A. Cagatay   III-1004
Tan, Chew Lim   II-207
Tan, Guangming   II-869, III-979
Tan, Guozhen   III-975
Tan, Haixia   II-485
Tan, Shaohua   I-90
Tang, Chuan Yi   II-845
Tang, Jiakui   III-496, III-883
Tang, Kai   II-298
Tao, Jie   II-174, II-182
Tarault, Antoine   II-339
Teng, Jun   II-322
Teow, Loo-Nin   II-877

Teresco, James D.   II-107
Thandavan, A.   III-752
Thomas, Michael A.   II-68, III-196
Thomas, Stephen J.   I-256, II-822, III-57
Thysebaert, Pieter   III-250
Tirado-Ramos, Alfredo   I-534
Todd, Chris   III-259
Tomov, Stanimire   III-317
Tong, RuoFeng   I-671
Tosik, Grzegorz   I-607, I-876
Tracy, Fred T.   I-66
Tran, Nick   III-80
Trincă, Dragoş   II-994
Trinder, P.W.   II-746
Trivedi, Abhishek   II-672
Trinitis, Carsten   II-191
Tsechpenakis, Gabriel   II-712
Tsompanopoulou, P.   II-616
Tsouloupas, George   I-534, III-870
Tucker, Don   I-631
Tufo, H.M.   II-91
Tuncel, Gonca   I-843, III-562
Turan, Ali   III-120, III-130
Turcotte, Marcel   II-936
Turovets, Sergei   I-631
Twerda, A.   I-868
Tynan, Richard   III-687

Uhruski, P.   III-727
Urmetzer, Florian   III-367
Usman, Anila   I-364
Utke, Jean   I-338

Vaccaro, Ugo   II-1029
Vandeputte, Frederik   II-166
van der Geest, Rob J.   I-287
Vandewalle, Stefan   I-279
Vanmechelen, Kurt   II-1072, II-1080
van Oijen J.A.   III-64
Vanrolleghem, Peter   III-1116
Varadarjan, Srinidhi   II-60
Varotsos, Costas   III-504
Vedova, Gianluca Della   II-952
Venetsanopoulos, Anastasios N.   II-886
Vetter, Jeffrey   I-304, I-868
Vézien, Jean-Marc   II-290, II-339
Vialette, Stéphane   II-860
Vidal, Antonio Manuel   I-17, I-220, I-229
Vodacek, Anthony   II-632
Volckaert, Bruno   III-250

Volkert, Jens II-124, III-383
Vuik, C. I-868

Wainer, Gabriel II-368, II-373
Wais, Piotr III-810
Wajs, Wiesław III-810
Walkowiak, Krzysztof III-1092
Walsh, John III-870
Wang, Dongsheng III-163
Wang, Guangming I-987
Wang, Hao II-851
Wang, Jianqin III-472, III-496
Wang, Jinlong III-1020
Wang, Lei I-568
Wang, Li-San II-919
Wang, Lin-Wang III-317
Wang, Linxiang I-884
Wang, Min I-963
Wang, Qing II-248
Wang, Qinmin I-979
Wang, Ruili II-420
Wang, Shaowen III-326
Wang, Shou-Yang III-523, III-539, III-554
Wang, Wei I-987
Wang, Weichung III-908
Wang, Wenqing II-1064
Wang, Xianqing III-887
Wang, Xiao-jing I-812
Wang, Xiaolin III-511, III-515
Wang, Xiaozhe III-792
Wang, Yadi III-229
Wang, Yangsheng II-273
Wang, Yanguang III-496
Wang, Yi III-875
Wang, Yong II-944
Warfield, Simon K. II-672
Wasson, Glenn II-729
Watson, James V.S. I-477
Wawrzyniak, Dariusz III-423
Webber, Robert E. II-331
Weber, Irene I-451
Weeks, Michael C. II-256
Wei, Guiyi I-987
Wei, Xilin III-134
Weidendorfer, Josef II-191
Weihrauch, Christian III-743, III-752
Weinstein, R. II-689
Wheeler, Mary F. II-656, III-96
Whitlock, P.A. III-983

Wilhelmson, Bob II-624
Wiszniewski, Bogdan I-355
Witułam, Roman III-659
Woitaszek, M.S. II-91
Wojtowicz, Hubert III-810
Won, Dongho II-493
Wong, Adam K.L. I-435
Woodward, Jeffrey B. II-99
Wozniak, Michal III-929
Wozny, Janusz I-607, I-876
Wu, Chaolin III-496
Wu, Jianjia II-632
Wu, Jianping II-551
Wu, Jun III-539
Wu, Lieyu II-960
Wu, Yong II-335
Wu, Zhaohui I-568, I-743, II-461, III-1068

Xian, Jun III-783
Xiao, Da III-399
Xiao, Shaoping III-284, III-326
Xie, Jinkui III-1032
Xie, Kai I-925
Xin, Jin II-502
Xiong, Guomin III-515
Xu, Baowen III-1088
Xu, Congfu III-1020
Xu, Hongtao I-9
Xu, Weixiang I-995
Xu, Weixuan III-531
Xu, Xian II-1038
Xu, Zhuoqun III-511, III-515
Xue, Huifeng III-854
Xue, Wei III-399
Xue, Yong III-464, III-472, III-496, III-883

Yabo, Dong I-1027
Yamamoto, Kenji I-956
Yan, Chang-Ching II-444
Yanami, Hitoshi III-602
Yang, Chengyong II-944
Yang, Hyungkyu II-493
Yang, Jie I-925
Yang, Jingmei I-615
Yang, Luobin II-68
Yang, Weixuan III-284
Yang, X.S. I-751, II-199
Yang, Xiaolong II-519

Yang, Xin-She III-1048
Yao, Nianmin III-415
Yao-Xue, Zhang II-502
Yaya, Wei II-502
Ye, Juntao II-331
Yeom, Heon Y. III-179
Yi, Myung-Kyu III-967
Yi, Ping II-593
Yin, Weiwei II-322
Yoon, Seokho II-543
Yoon, Yeo-Ran III-1120
Youn, Choonhan I-469
Youn, Hee Yong I-576, II-568, III-1000, III-1125
Yu, Hai III-1044
Yu, Lean III-523
Yu, Lishan III-511
Yu, Shui I-544
Yuasa, Fukuko I-165
Yue, Wuyi III-539
Yue-Zhi, Zhou II-502
Yuewei, Huang I-1027
Yunjie, Mao I-1027

Zaman, Safaa I-560
Zanero, Stefano III-147
Zarina, M. III-447
Zelikovsky, Alexander II-1011
Zeng, Qinhuai III-887
Zeng, Weilin II-485
Zhang, Aidong II-1038
Zhang, Changyong I-804
Zhang, Defu I-783, III-801
Zhang, Hua II-616, III-826
Zhang, Jian J. II-199

Zhang, Kaizhong II-960
Zhang, Liang I-544
Zhang, Min II-519
Zhang, Qiangfeng II-985
Zhang, Shiyong II-593
Zhang, Wenju I-544
Zhang, Xia I-908
Zhang, Yang II-790, III-619
Zhang, Yu II-207
Zhao, Mingxi III-846
Zhao, Qiang III-826
Zhao, Wei II-632
Zheng, Kougen II-461
Zheng, Weimin III-399, III-415
Zheng, Weiming III-163
Zheng, Yao I-987
Zheng, Zengwei I-552, II-461
Zhong, Shaobo III-464, III-496, III-883
Zhong, Weimin III-1036
Zhong, Yiping II-593
Zhou, Chenghu I-963
Zhou, Dong II-248
Zhou, Hong I-664
Zhou, Li I-812
Zhou, Runfang III-187
Zhou, W. III-586
Zhu, Changqian III-826
Zhu, Qi I-90
Zhu, Wei-Yong III-1044
Zhu, Ying II-256, II-347
Zhu, Yue Min I-925
Ziemlinski, Remik III-838
Znamirowski, Lech I-766
Zornes, Adam II-632

Lecture Notes in Computer Science

For information about Vols. 1–3397

please contact your bookseller or Springer

Vol. 3525: A.E. Abdallah, C.B. Jones, J.W. Sanders (Eds.), Communicating Sequential Processes. XIV, 321 pages. 2005.

Vol. 3517: H.S. Baird, D.P. Lopresti (Eds.), Human Interactive Proofs. IX, 143 pages. 2005.

Vol. 3516: V.S. Sunderam, G.D.v. Albada, P.M.A. Sloot, J.J. Dongarra (Eds.), Computational Science – ICCS 2005, Part III. LXIII, 1143 pages. 2005.

Vol. 3515: V.S. Sunderam, G.D.v. Albada, P.M.A. Sloot, J.J. Dongarra (Eds.), Computational Science – ICCS 2005, Part II. LXIII, 1101 pages. 2005.

Vol. 3514: V.S. Sunderam, G.D.v. Albada, P.M.A. Sloot, J.J. Dongarra (Eds.), Computational Science – ICCS 2005, Part I. LXIII, 1089 pages. 2005.

Vol. 3510: T. Braun, G. Carle, Y. Koucheryavy, V. Tsaousidis (Eds.), Wired/Wireless Internet Communications. XIV, 366 pages. 2005.

Vol. 3508: P. Bresciani, P. Giorgini, B. Henderson-Sellers, G. Low, M. Winikoff (Eds.), Agent-Oriented Information Systems II. X, 227 pages. 2005. (Subseries LNAI).

Vol. 3503: S.E. Nikoletseas (Ed.), Experimental and Efficient Algorithms. XV, 624 pages. 2005.

Vol. 3501: B. Kégl, G. Lapalme (Eds.), Advances in Artificial Intelligence. XV, 458 pages. 2005. (Subseries LNAI).

Vol. 3500: S. Miyano, J. Mesirov, S. Kasif, S. Istrail, P. Pevzner, M. Waterman (Eds.), Research in Computational Molecular Biology. XVII, 632 pages. 2005. (Subseries LNBI).

Vol. 3498: J. Wang, X. Liao, Z. Yi (Eds.), Advances in Neural Networks – ISNN 2005, Part III. L, 1077 pages. 2005.

Vol. 3497: J. Wang, X. Liao, Z. Yi (Eds.), Advances in Neural Networks – ISNN 2005, Part II. L, 947 pages. 2005.

Vol. 3496: J. Wang, X. Liao, Z. Yi (Eds.), Advances in Neural Networks – ISNN 2005, Part II. L, 1055 pages. 2005.

Vol. 3495: P. Kantor, G. Muresan, F. Roberts, D.D. Zeng, F.-Y. Wang, H. Chen, R.C. Merkle (Eds.), Intelligence and Security Informatics. XVIII, 674 pages. 2005.

Vol. 3494: R. Cramer (Ed.), Advances in Cryptology – EUROCRYPT 2005. XIV, 576 pages. 2005.

Vol. 3492: P. Blache, E. Stabler, J. Busquets, R. Moot (Eds.), Logical Aspects of Computational Linguistics. X, 363 pages. 2005. (Subseries LNAI).

Vol. 3489: G.T. Heineman, J.A. Stafford, H.W. Schmidt, K. Wallnau, C. Szyperski, I. Crnkovic (Eds.), Component-Based Software Engineering. XI, 358 pages. 2005.

Vol. 3488: M.-S. Hacid, N.V. Murray, Z.W. Raś, S. Tsumoto (Eds.), Foundations of Intelligent Systems. XIII, 700 pages. 2005. (Subseries LNAI).

Vol. 3483: O. Gervasi, M.L. Gavrilova, V. Kumar, A. Laganà, H.P. Lee, Y. Mun, D. Taniar, C.J.K. Tan (Eds.), Computational Science and Its Applications – ICCSA 2005, Part IV. XXVII, 1362 pages. 2005.

Vol. 3482: O. Gervasi, M.L. Gavrilova, V. Kumar, A. Laganà, H.P. Lee, Y. Mun, D. Taniar, C.J.K. Tan (Eds.), Computational Science and Its Applications – ICCSA 2005, Part III. LXVI, 1340 pages. 2005.

Vol. 3481: O. Gervasi, M.L. Gavrilova, V. Kumar, A. Laganà, H.P. Lee, Y. Mun, D. Taniar, C.J.K. Tan (Eds.), Computational Science and Its Applications – ICCSA 2005, Part II. LXIV, 1316 pages. 2005.

Vol. 3480: O. Gervasi, M.L. Gavrilova, V. Kumar, A. Laganà, H.P. Lee, Y. Mun, D. Taniar, C.J.K. Tan (Eds.), Computational Science and Its Applications – ICCSA 2005, Part I. LXV, 1234 pages. 2005.

Vol. 3479: T. Strang, C. Linnhoff-Popien (Eds.), Location- and Context-Awareness. XII, 378 pages. 2005.

Vol. 3477: P. Herrmann, V. Issarny (Eds.), Trust Management. XII, 426 pages. 2005.

Vol. 3475: N. Guelfi (Ed.), Rapid Integration of Software Engineering Techniques. X, 145 pages. 2005.

Vol. 3468: H.W. Gellersen, R. Want, A. Schmidt (Eds.), Pervasive Computing. XIII, 347 pages. 2005.

Vol. 3467: J. Giesl (Ed.), Term Rewriting and Applications. XIII, 517 pages. 2005.

Vol. 3465: M. Bernardo, A. Bogliolo (Eds.), Formal Methods for Mobile Computing. VII, 271 pages. 2005.

Vol. 3463: M. Dal Cin, M. Kaâniche, A. Pataricza (Eds.), Dependable Computing - EDCC 2005. XVI, 472 pages. 2005.

Vol. 3462: R. Boutaba, K. Almeroth, R. Puigjaner, S. Shen, J.P. Black (Eds.), NETWORKING 2005. XXX, 1483 pages. 2005.

Vol. 3461: P. Urzyczyn (Ed.), Typed Lambda Calculi and Applications. XI, 433 pages. 2005.

Vol. 3460: Ö. Babaoglu, M. Jelasity, A. Montresor, C. Fetzer, S. Leonardi, A. van Moorsel, M. van Steen (Eds.), Self-star Properties in Complex Information Systems. IX, 447 pages. 2005.

Vol. 3459: R. Kimmel, N.A. Sochen, J. Weickert (Eds.), Scale Space and PDE Methods in Computer Vision. XI, 634 pages. 2005.

Vol. 3456: H. Rust, Operational Semantics for Timed Systems. XII, 223 pages. 2005.

Vol. 3455: H. Treharne, S. King, M. Henson, S. Schneider (Eds.), ZB 2005: Formal Specification and Development in Z and B. XV, 493 pages. 2005.

Vol. 3454: J.-M. Jacquet, G.P. Picco (Eds.), Coordination Models and Languages. X, 299 pages. 2005.

Vol. 3453: L. Zhou, B.C. Ooi, X. Meng (Eds.), Database Systems for Advanced Applications. XXVII, 929 pages. 2005.

Vol. 3452: F. Baader, A. Voronkov (Eds.), Logic for Programming, Artificial Intelligence, and Reasoning. XI, 562 pages. 2005. (Subseries LNAI).

Vol. 3450: D. Hutter, M. Ullmann (Eds.), Security in Pervasive Computing. XI, 239 pages. 2005.

Vol. 3449: F. Rothlauf, J. Branke, S. Cagnoni, D.W. Corne, R. Drechsler, Y. Jin, P. Machado, E. Marchiori, J. Romero, G.D. Smith, G. Squillero (Eds.), Applications of Evolutionary Computing. XX, 631 pages. 2005.

Vol. 3448: G.R. Raidl, J. Gottlieb (Eds.), Evolutionary Computation in Combinatorial Optimization. XI, 271 pages. 2005.

Vol. 3447: M. Keijzer, A. Tettamanzi, P. Collet, J.v. Hemert, M. Tomassini (Eds.), Genetic Programming. XIII, 382 pages. 2005.

Vol. 3444: M. Sagiv (Ed.), Programming Languages and Systems. XIII, 439 pages. 2005.

Vol. 3443: R. Bodik (Ed.), Compiler Construction. XI, 305 pages. 2005.

Vol. 3442: M. Cerioli (Ed.), Fundamental Approaches to Software Engineering. XIII, 373 pages. 2005.

Vol. 3441: V. Sassone (Ed.), Foundations of Software Science and Computational Structures. XVIII, 521 pages. 2005.

Vol. 3440: N. Halbwachs, L.D. Zuck (Eds.), Tools and Algorithms for the Construction and Analysis of Systems. XVII, 588 pages. 2005.

Vol. 3439: R.H. Deng, F. Bao, H. Pang, J. Zhou (Eds.), Information Security Practice and Experience. XII, 424 pages. 2005.

Vol. 3437: T. Gschwind, C. Mascolo (Eds.), Software Engineering and Middleware. X, 245 pages. 2005.

Vol. 3436: B. Bouyssounouse, J. Sifakis (Eds.), Embedded Systems Design. XV, 492 pages. 2005.

Vol. 3434: L. Brun, M. Vento (Eds.), Graph-Based Representations in Pattern Recognition. XII, 384 pages. 2005.

Vol. 3433: S. Bhalla (Ed.), Databases in Networked Information Systems. VII, 319 pages. 2005.

Vol. 3432: M. Beigl, P. Lukowicz (Eds.), Systems Aspects in Organic and Pervasive Computing - ARCS 2005. X, 265 pages. 2005.

Vol. 3431: C. Dovrolis (Ed.), Passive and Active Network Measurement. XII, 374 pages. 2005.

Vol. 3429: E. Andres, G. Damiand, P. Lienhardt (Eds.), Discrete Geometry for Computer Imagery. X, 428 pages. 2005.

Vol. 3427: G. Kotsis, O. Spaniol (Eds.), Wireless Systems and Mobility in Next Generation Internet. VIII, 249 pages. 2005.

Vol. 3423: J.L. Fiadeiro, P.D. Mosses, F. Orejas (Eds.), Recent Trends in Algebraic Development Techniques. VIII, 271 pages. 2005.

Vol. 3422: R.T. Mittermeir (Ed.), From Computer Literacy to Informatics Fundamentals. X, 203 pages. 2005.

Vol. 3421: P. Lorenz, P. Dini (Eds.), Networking - ICN 2005, Part II. XXXV, 1153 pages. 2005.

Vol. 3420: P. Lorenz, P. Dini (Eds.), Networking - ICN 2005, Part I. XXXV, 933 pages. 2005.

Vol. 3419: B. Faltings, A. Petcu, F. Fages, F. Rossi (Eds.), Constraint Satisfaction and Constraint Logic Programming. X, 217 pages. 2005. (Subseries LNAI).

Vol. 3418: U. Brandes, T. Erlebach (Eds.), Network Analysis. XII, 471 pages. 2005.

Vol. 3416: M. Böhlen, J. Gamper, W. Polasek, M.A. Wimmer (Eds.), E-Government: Towards Electronic Democracy. XIII, 311 pages. 2005. (Subseries LNAI).

Vol. 3415: P. Davidsson, B. Logan, K. Takadama (Eds.), Multi-Agent and Multi-Agent-Based Simulation. X, 265 pages. 2005. (Subseries LNAI).

Vol. 3414: M. Morari, L. Thiele (Eds.), Hybrid Systems: Computation and Control. XII, 684 pages. 2005.

Vol. 3412: X. Franch, D. Port (Eds.), COTS-Based Software Systems. XVI, 312 pages. 2005.

Vol. 3411: S.H. Myaeng, M. Zhou, K.-F. Wong, H.-J. Zhang (Eds.), Information Retrieval Technology. XIII, 337 pages. 2005.

Vol. 3410: C.A. Coello Coello, A. Hernández Aguirre, E. Zitzler (Eds.), Evolutionary Multi-Criterion Optimization. XVI, 912 pages. 2005.

Vol. 3409: N. Guelfi, G. Reggio, A. Romanovsky (Eds.), Scientific Engineering of Distributed Java Applications. X, 127 pages. 2005.

Vol. 3408: D.E. Losada, J.M. Fernández-Luna (Eds.), Advances in Information Retrieval. XVII, 572 pages. 2005.

Vol. 3407: Z. Liu, K. Araki (Eds.), Theoretical Aspects of Computing - ICTAC 2004. XIV, 562 pages. 2005.

Vol. 3406: A. Gelbukh (Ed.), Computational Linguistics and Intelligent Text Processing. XVII, 829 pages. 2005.

Vol. 3404: V. Diekert, B. Durand (Eds.), STACS 2005. XVI, 706 pages. 2005.

Vol. 3403: B. Ganter, R. Godin (Eds.), Formal Concept Analysis. XI, 419 pages. 2005. (Subseries LNAI).

Vol. 3402: M. Daydé, J.J. Dongarra, V. Hernández, J.M.L.M. Palma (Eds.), High Performance Computing for Computational Science - VECPAR 2004. XI, 732 pages. 2005.

Vol. 3401: Z. Li, L.G. Vulkov, J. Waśniewski (Eds.), Numerical Analysis and Its Applications. XIII, 630 pages. 2005.

Vol. 3400: J.F. Peters, A. Skowron (Eds.), Transactions on Rough Sets III. IX, 461 pages. 2005.

Vol. 3399: Y. Zhang, K. Tanaka, J.X. Yu, S. Wang, M. Li (Eds.), Web Technologies Research and Development - APWeb 2005. XXII, 1082 pages. 2005.

Vol. 3398: D.-K. Baik (Ed.), Systems Modeling and Simulation: Theory and Applications. XIV, 733 pages. 2005. (Subseries LNAI).